FEATURES AND BENEFITS

Houghton Mifflin Unified Mathematics, Book 2

Comprehensive content coverage shows the interrelationship of logic, algebra, geometry, probability, and statistics with special attention to problem solving skills and concepts. (Table of Contents, pp. iii – vii)

Special end-of-chapter features include an enrichment page, Vocabulary Review, and Mixed Review Exercises. These pages accompany Computer Exercises, Chapter Review, Chapter Test, and periodic Cumulative Review pages. (pp. 91– 97, 296–305)

The **Preparing for Regents Examinations** appendix helps students prepare for the exams and includes topic checklists and sample exam questions. (pp. 629–640)

Algebra Review Exercises and **Extra Practice** appendices contain additional exercises on algebra and other important topics. (pp. 614–619, 600–613)

Applications of the computer to lesson concepts can be found at the end of each chapter in **Computer Exercises** and in the new **Programming in BASIC** appendix. (pp. 39–40, 168, 272, 620–628)

Supplementary materials include a **Teacher's Manual with Solutions** and a **Resource Book** that contains permission-to-copy blackline master pages of tests (quizzes, chapter tests, and cumulative tests), practice exercises, enrichment activities, and diagrams.

Table of Contents

7 QUADRATIC EQUATIONS

8 SIMILARITY

9 RIGHT TRIANGLES

10 CONSTRUCTIONS AND LOCI

11 COORDINATE GEOMETRY

12 QUADRATIC EQUATIONS IN TWO VARIABLES

13 PROBABILITY, PERMUTATIONS, AND COMBINATIONS

14 ALGEBRAIC SYSTEMS

SYMBOLS

Symbol	Meaning	Page
$\in$	is an element of	p. 1
$\notin$	is not an element of	p. 1
$\{1, 2, 3\}$	the set containing the elements 1, 2, and 3	p. 2
$\ldots$	and so on	p. 2
$\emptyset$ or $\{\ \}$	the empty set	p. 2
$\subset$	is a subset of	p. 2
$A \cap B$	intersection of A and B	p. 3
$A \cup B$	union of A and B	p. 3
$>$	is greater than	p. 21
$<$	is less than	p. 21
$\neq$	does not equal	p. 21
$\leqslant$	is less than or equal to	p. 22
$\geqslant$	is greater than or equal to	p. 22
$\sim$	the negation of	p. 51
$\wedge$	and	p. 53
$\vee$	or	p. 53
$\rightarrow$	implies	p. 58
$\leftrightarrow$	if and only if	p. 59
$\forall$	for all	p. 62
$\exists$	there exists	p. 62
$\overleftrightarrow{AB}$	line AB	p. 101
$\overline{AB}$	line segment AB	p. 101
AB	the length of $\overline{AB}$	p. 101
$\overrightarrow{AB}$	ray AB	p. 102
$\angle$	angle	p. 107
$m\angle A$	the measure of $\angle A$	p. 108
28°	28 degrees	p. 108
$\perp$	is perpendicular to	p. 114
$\cong$	is congruent to	p. 119
$\triangle$	triangle	p. 129
$\parallel$	is parallel to	p. 154
$\square$	parallelogram	p. 188
$\sqrt{\ }$	the principal square root of	p. 267
$\pm$	plus or minus	p. 269
$a : b$	the ratio of a to b	p. 307
$\sim$	is similar to	p. 314
$\lvert x \rvert$	the absolute value of x	p. 429
$P(E)$	the probability of an event E	p. 523
$!$	factorial	p. 537
${}_nP_r$	the number of permutations of n things taken r at a time	p. 538
${}_nC_r$ or $\binom{n}{r}$	the number of combinations of n things taken r at a time	p. 542 p. 543
$(W, +)$	the set of whole numbers under addition	p. 557
Z_N	$0, 1, 2, 3, \ldots, N-1$	p. 565
$Z_N \backslash \{0\}$	$1, 2, 3, \ldots, N-1$	p. 571
$\begin{pmatrix} 1 & 3 & 2 \\ 2 & 3 & 1 \end{pmatrix}$	an element of the permutation group of the numbers 1, 2, and 3	p. 576

METRIC UNITS OF MEASURE

Prefixes	kilo	hecto	deka	deci	centi	milli
	1000	100	10	0.1	0.01	0.001

Length
1 centimeter (cm) = 10 millimeters (mm)
1 meter (m) = 100 cm
1 kilometer (km) = 1000 m

Area
1 square centimeter (cm^2) = 100 square millimeters (mm^2)
1 square meter (m^2) = 10,000 cm^2

Volume
1 cubic centimeter (cm^3) = 1000 cubic millimeters (mm^3)
1 cubic meter (m^3) = 1,000,000 cm^3
1 liter (L) = 1000 cm^3
1 liter = 1000 milliliters (mL)

Mass
1 gram (g) = 1000 milligrams (mg)
1 kilogram (kg) = 1000 g
1 metric ton (t) = 1000 kg

ACKNOWLEDGMENTS

Cover concept by Richard Hannus
Biographical portraits by Gary Torrisi (pages 43, 183, 381, 471, and 551)

Photographs
page x: Frank Siteman / THE PICTURE CUBE
page 46: © Sandra Johnson 1981 / THE PICTURE CUBE
page 98: Barbara Marshall
page 149: © Bruce Roberts 1978 / PHOTO RESEARCHERS
page 152: Barbara Marshall
page 186: © Fritz Menle 1976 / PHOTO RESEARCHERS
page 213: John McGrail / WHEELER PICTURES
page 216: Barbara Marshall
page 255: courtesy of Bausch and Lomb Company
page 258: Colonial Williamsburg Photograph
page 303: Gene Alexander / USDA Soil Conservation Service
page 306: W.B. Finch / STOCK BOSTON
page 349: set design for *Les Comtes d'Hoffman*, by Guenther Schneider-Siemssen, The Metropolitan Opera of New York.
page 352: © Phaneuf-Gurdziel / THE PICTURE CUBE
page 384: Peter Southwick / STOCK BOSTON
page 423: Woods Hole Oceanographic Institution
page 426: © E. Williamson / THE PICTURE CUBE
page 474: NASA
page 515: courtesy of Bethlehem Steel Company
page 518: Mike Mazzaschi / STOCK BOSTON
page 554: Paul S. Conklin / Monkmeyer Press Photo Service
page 597: © Bruce Roberts, Rapho / PHOTO RESEARCHERS

The ability to solve linear equations is an essential mathematical skill that has a variety of applications. For example, linear equations are used in solving uniform motion problems involving airplane travel.

Review of Essentials 1

It was not until the 1700's that mathematicians began to use the same notation the world over. Before the 1700's every mathematician devised and defined his or her own symbols. If you had been studying algebra with the mathematician Ghaligai in 1521, for example, you would have learned to solve an equation like this one:

$$6\text{C}^\circ \text{ e } 9 \text{ numeri} - 3\text{C}^\circ \text{ e } 24 \text{ numeri}$$

In 1559, the same equation would have been written by Buteo as

$$6\rho \text{ P } 9 \text{ [} 3\rho \text{ P } 24.$$

Viete, in 1591, would have written this as

$$6\text{N} + 9 \text{ aequatur } 3\text{N} + 24.$$

Today we write

$$6x + 9 = 3x + 24.$$

In this chapter you will review solving equations like the one above.

Section 1-1 SETS AND GRAPHING

In this section you will learn about:
- sets
- subsets
- union of sets
- intersection of sets
- graphing sets of numbers on the number line

Set is an undefined term, but we will use it to mean a well-defined collection of objects. Each member of a set is called an *element* of the set. In general we will denote sets by capital letters, such as A or B. To designate the elements of a given set we will use lower-case letters.

$x \in A$ is read x is an element of the set A.
$y \notin B$ is read y is not an element of the set B.

Some sets of numbers with which you are familiar are:

$\mathcal{R} = \{\text{the real numbers}\}$

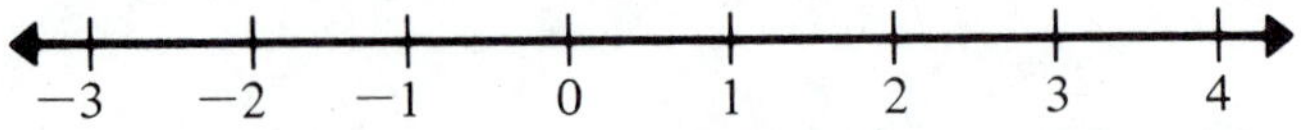

$Z = \{\text{the integers}\} = \{\ldots, -2, -1, 0, 1, 2, \ldots\}$

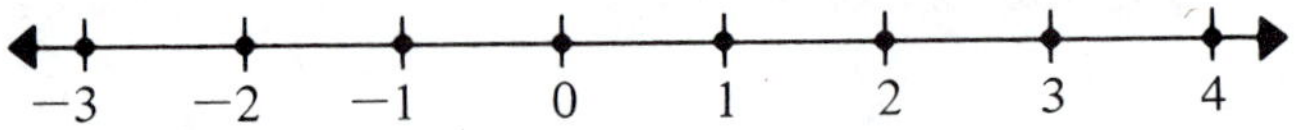

$W = \{\text{the whole numbers}\} = \{0, 1, 2, 3, \ldots\}$

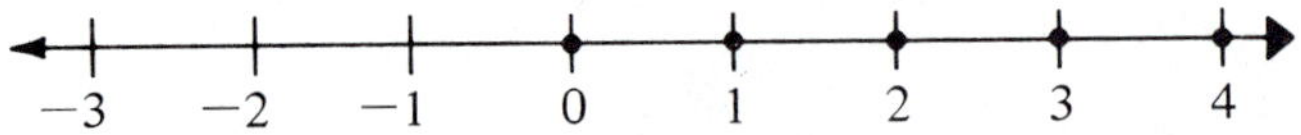

Another set of numbers whose name you will recognize is

$$Q = \{\text{the rational numbers}\}.$$

A rational number is a number that can be expressed in the form $\frac{a}{b}$, where a and b are integers (with $b \neq 0$, of course). Any terminating or repeating decimal is a rational number. Some elements of the set of rational numbers are:

$$\frac{2}{3}, \quad -4 = \frac{-4}{1}, \quad 0.27 = \frac{27}{100}, \quad \text{and} \quad 0.111\ldots = \frac{1}{9}$$

Any real number that is not rational is *irrational.* Some examples of irrational numbers are:

$$\sqrt{2}, \frac{\sqrt{7}}{3}, \text{ and } \pi$$

Any nonrepeating, nonterminating decimal, such as 0.121121112. . . , is irrational.

The set containing no elements is called the *null set,* or the *empty set,* denoted by $\emptyset$ or $\{\ \}$.

It is sometimes convenient to use *set-builder notation* to define a set. For example, if $A = \{\text{the integers less than } -1\}$, you can use set-builder notation to write

$$A = \{x\colon x < -1, x \in Z\}.$$

This is read as "the set of all x such that x is less than -1 and x is an integer."

A set A is a *subset* of a set B, symbolized by $A \subset B$, if every element of A is an element of B. Thus, for every set A, $A \subset A$.

EXAMPLES

$$\{1, 2, 3\} \subset \{1, 2, 3, \ldots\}$$
$$\{1, 2, 3\} \subset \{1, 2, 3\}$$
$$W \subset Z, \quad Z \subset Q, \quad Q \subset \mathcal{R}$$

The null set, $\emptyset$, is considered to be a subset of every set.

EXAMPLES

$\emptyset \subset \mathcal{R}$
$\emptyset \subset \emptyset$

Two sets A and B are *equal,* symbolized by $A = B$, if they contain exactly the same elements.

EXAMPLES

$\{1, 2\} = \{2, 1\}$
$\{3, 4, 5\} = \{5, 3, 4\}$
$\emptyset \neq \{0\}$ because $0 \in \{0\}$, but $0 \notin \emptyset$.

Two or more sets can be combined to form new sets. You may already be familiar with two common operations, *intersection* and *union.*

The *intersection* of set A and set B is symbolized by $A \cap B$.

$$A \cap B = \{x: x \in A \text{ and } x \in B\}$$

In other words, $A \cap B$ contains the elements common to both A *and* B. You can represent $A \cap B$ pictorially as shown below.

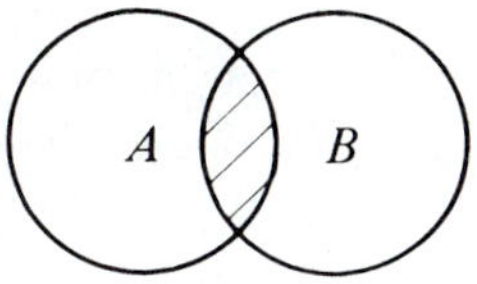

$A \cap B$

EXAMPLES

$A = \{0, 2, 4, 6\}, \quad B = \{1, 3, 5\}$
$A \cap B = \emptyset$

$C = \{2, 3, 5, 7\}, \quad D = \{-2, 0, 2, 3\}$
$C \cap D = \{2, 3\}$

$X = \{\text{the even integers}\}, \quad Y = \mathcal{R}$
$X \cap Y = X$

The *union* of set A and set B is symbolized by $A \cup B$.

$$A \cup B = \{x: x \in A \text{ or } x \in B\}$$

Here the word "or" is used in the inclusive sense, meaning $A \cup B$ contains all the elements in either A *or* B *or* both. You can represent $A \cup B$ pictorially as shown below.

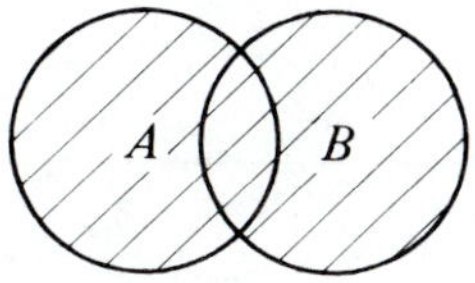

$A \cup B$

EXAMPLES

$A = \{2, 3, 5, 7\}, \quad B = \{-2, 0, 2, 3\}$
$A \cup B = \{-2, 0, 2, 3, 5, 7\}$

$C = \{1, 2, 3\}, \quad D = \{1, 2, 3, 4, 5\}$
$C \cup D = \{1, 2, 3, 4, 5\} = D$

$Q = \{\text{the rational numbers}\}, \quad I = \{\text{the irrational numbers}\}$
$Q \cup I = \mathcal{R}$

Graphing sets of numbers can help you see the relationships that may exist between two or more sets. In graphing sets of numbers you use the following assumptions:

> To every real number there corresponds exactly one point on the number line.
>
> To every point on the number line there corresponds exactly one real number.

On the number line the point that corresponds to a given number is called the *graph* of the number. The number that corresponds to a given point is called the *coordinate* of the point.

EXAMPLE If $A = \{\ldots, -3, -1, 1, 3, \ldots\}$ and $B = \{1, 2, 3, \ldots\}$, graph $A \cap B$.

SOLUTION $A \cap B = \{1, 3, 5, \ldots\}$

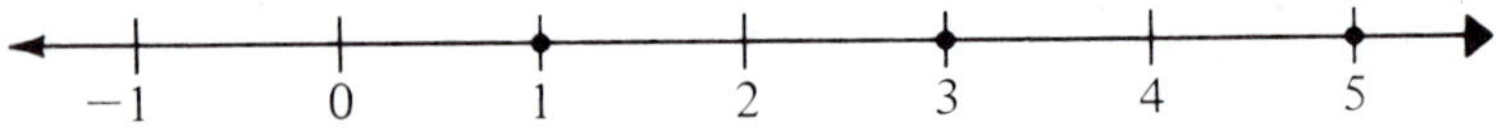

The heavy arrow shows that the graph continues to the right in the same pattern.

EXAMPLE If $C = \left\{-4, -1.4, 0, \frac{1}{3}, 2.7, 6\right\}$, graph the subset of C that contains all the even integers in C.

SOLUTION

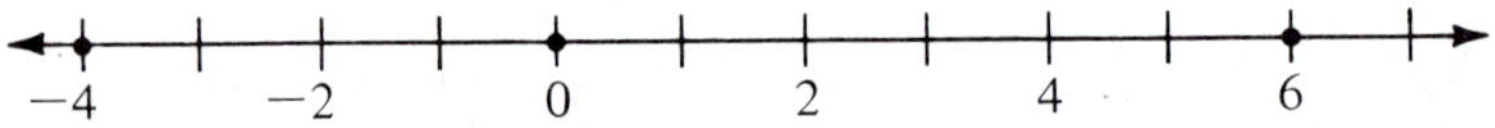

EXAMPLE Name the set represented by this graph:

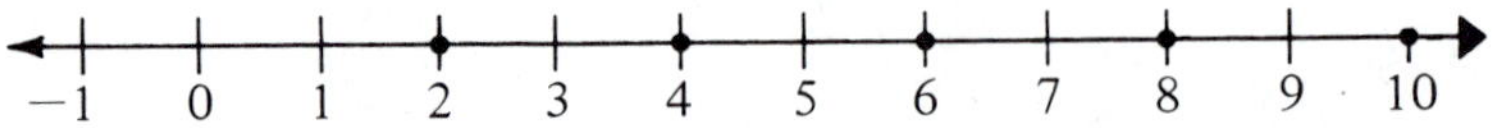

SOLUTION $\{\text{the positive even integers}\}$

ORAL EXERCISES

Tell whether the statement is true or false.

1. $\{-3, -2, -1\} \subset \{\text{the odd integers}\}$
2. $\emptyset \subset \{1, 2, 3, 4\}$
3. $\{0\} \subset \emptyset$
4. $\{\ldots, -2, -1, 0\} \subset \{\ldots, -2, -1, 0\}$
5. $0 \in \{\text{the positive integers}\}$
6. $\emptyset = \{0\}$
7. $\{2, 3, 7\} = \{7, 2, 3\}$
8. $\{1\} \in \{1, 3, 5\}$

Match each graph with the set it represents.

9. 

10. 

11. 

12. 

13. 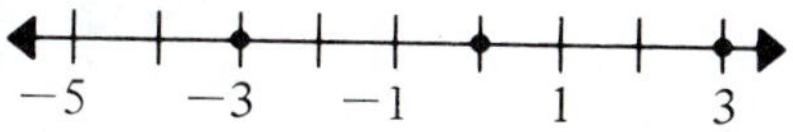

14. 

a. {the nonpositive integers}

b. {1, 2, 3}

c. {the positive integers}

d. {the multiples of 3}

e. {the negative integers}

f. {the nonnegative real numbers}

Complete.

15. If $A \subset B$, then $A \cap B =$ __?__.
16. If $A \subset B$, then $A \cup B =$ __?__.

WRITTEN EXERCISES

For each pair of sets A and B, find **(a)** $A \cup B$ and **(b)** $A \cap B$.

A

1. $A = \{-2, -1, 0, 1\}$
   $B = \{-1, 1, 3\}$
2. $A = \{3, -5, 0, 7, -2\}$
   $B = \{-5, -2, 0, 3, 7\}$
3. $A = \{-5, -4, -3\}$
   $B = \{2, 3\}$
4. $A = \{0\}$
   $B = \{-1, -2\}$
5. $A = \{\text{the integers}\}$
   $B = \{\text{the rational numbers}\}$
6. $A = \{1, 3, 5, 7\}$
   $B = \{7, 3, 1, 5, 9\}$

Graph each set.

7. $\{-2, 0.5, 3, 4\}$

8. $\left\{\frac{0}{3}, \frac{1}{3}, \frac{2}{3}, \frac{3}{3}\right\}$

9. $\{\ldots, -4, 0, 4, 8, \ldots\}$

10. $\{10, 11, 12, \ldots\}$

11. $\{\ldots, -7, -5, -3\}$

12. {the integers less than 2}

For each pair of sets C and D, graph **(a)** $C \cup D$ and **(b)** $C \cap D$.

13. $C = \{-3.5, 1.25, 4\}$
$D = \left\{\frac{1}{2}, -5, -\frac{7}{2}, 2\right\}$

14. $C =$ {the positive even integers}
$D =$ {the positive multiples of 4}

State if each given number is an element of **(a)** $\mathcal{R}$, **(b)** Z, **(c)** W, **(d)** Q, **(e)** {the irrational numbers}. (You may want to refer to page 2 for the definitons of $\mathcal{R}$, Z, W, and Q.)

15. $\frac{1}{3}$

16. -2

17. $\sqrt{3}$

18. 6

19. $\frac{8}{2}$

20. $-\frac{2}{3}$

21. 0

22. $\frac{\sqrt{2}}{3}$

List all the subsets of each given set.

23. $\{-2, -1\}$

24. $\{0, 1, 2\}$

Describe each set.

EXAMPLE $\{2, 4, 8, 16, 32, \ldots\}$

SOLUTION {the positive powers of 2}

B 25. $\{1, 4, 9, 16, \ldots\}$

26. $\{2, 3, 5, 7, 11, 13, 17, \ldots\}$

27. $\{3, 9, 27, 81, \ldots\}$

28. $\{0, 5, 10, 15, \ldots\}$

Give an example to show that each of the following statements is false.

EXAMPLE If $A \subset B$ and $B \not\subset C$, then $A \not\subset C$.
($\not\subset$ is read "is not a subset of.")

SOLUTION One possible example is
$A = \{1, 2\}$
$B = \{1, 2, 3\}$
$C = \{1, 2, 4\}$
$A \subset B$ and $B \not\subset C$, but $A \subset C$.

29. If $A \not\subset B$ and $B \not\subset C$, then $A \not\subset C$.

30. If $A \neq B$ and $B \neq C$, then $A \neq C$.

31. If $A \subset B$ and $B \neq C$, then $A \neq C$.

32. If $x \notin A$ and $A \subset B$, then $x \notin B$.

33. If $x \in A$ and $B \subset A$, then $x \in B$.

34. If $A \subset B$, then $B \subset A$.

Let D, E, and F represent the sets given below.

$$D = \{1, 2, 3, 4, 5, 6\}$$
$$E = \{2, 4, 5, 7, 8\}$$
$$F = \{1, 3, 9\}$$

List the elements in each of the following sets.

EXAMPLE $(D \cap E) \cup F$

SOLUTION $D \cap E = \{2, 4, 5\}$
$(D \cap E) \cup F = \{2, 4, 5\} \cup \{1, 3, 9\}$
$(D \cap E) \cup F = \{1, 2, 3, 4, 5, 9\}$

35. $D \cap (E \cap F)$ **36.** $(D \cap E) \cap F$ **37.** $(D \cup E) \cap F$

38. $D \cup (E \cap F)$ **39.** $D \cap (E \cup F)$ **40.** $(D \cup E) \cup F$

Graph each set.

C 41. $\{x: x > 1, x \in Z\}$ **42.** $\{x: x < 5, x \in W\}$

43. $\{x: x > -4, x \in W\}$

44. $\{x: x > 2, x \in Z\} \cap \{x: x < 4, x \in Z\}$

45. $\{x: x > 5, x \in Z\} \cap \{x: x < 5, x \in Z\}$

46. $\{x: x > 5, x \in Z\} \cup \{x: x < 5, x \in Z\}$

Section 1-2 PROPERTIES OF REAL NUMBERS

In this section you will study:

- closure
- commutativity
- associativity
- identities
- inverses
- other properties of real numbers

In this chapter *properties* may be either postulates (statements that we accept as true without proof) or theorems (statements that can be proved). Because much of this material is review, we shall simply present the major points without differentiating between those that can and cannot be proved.

You may already be familiar with many of the following properties. In each case a, b, and c represent real numbers.

Properties of Addition

Closure Property	$a + b$ is a unique (exactly one) real number.
Commutative Property	$a + b = b + a$
Associative Property	$a + (b + c) = (a + b) + c$
Identity Property	$\mathcal{R}$ contains a unique element 0, called the *identity element for addition,* such that for any a, $a + 0 = a$ and $0 + a = a$.
Inverse Property	For every $a \in \mathcal{R}$, there is a unique real number, $-a$, called the *additive inverse* of a, such that $a + (-a) = 0$ and $(-a) + a = 0$.

The additive inverse of a is sometimes called the *opposite* of a.

Properties of Multiplication

Closure Property	ab is a unique real number.
Commutative Property	$ab = ba$
Associative Property	$a(bc) = (ab)c$
Identity Property	$\mathcal{R}$ contains a unique element 1, called the *identity element for multiplication,* such that for any a, $a \cdot 1 = a$ and $1 \cdot a = a$.
Inverse Property	For every nonzero $a \in \mathcal{R}$, there is a unique real number, $\frac{1}{a}$, called the *multiplicative inverse* of a, such that $a \cdot \frac{1}{a} = 1$ and $\frac{1}{a} \cdot a = 1$.

The multiplicative inverse of a is sometimes called the *reciprocal* of a.

Two other numbers that have special properties related to multiplication are 0 and -1.

Multiplicative Property of 0

$a \cdot 0 = 0$ and $0 \cdot a = 0$

Multiplicative Property of -1

$a(-1) = -a$ and $(-1)a = -a$

Here is one property that involves both multiplication and addition.

Distributive Property

$a(b + c) = ab + ac$ and $(b + c)a = ba + ca$

There is also a distributive property for multiplication and subtraction.

$$a(b - c) = ab - ac \quad \text{and} \quad (b - c)a = ba - ca.$$

Another property that you will use frequently is given below.

Substitution Property

If $a = b$, you can replace a by b in any expression without changing the value of the expression.

Algebraic expressions are expressions containing only algebraic symbols and operations.

EXAMPLES OF ALGEBRAIC EXPRESSIONS

$x\sqrt{5}$ $\quad$ $2x + 3$ $\quad$ $3(y - 4)$ $\quad$ $\frac{2}{5}r + 7s$ $\quad$ $3ab$

You can use the distributive property to simplify algebraic expressions containing parentheses. For example:

$$\begin{aligned} 3(y-4) &= 3(y) - 3(4) && \text{Distributive Property} \\ &= 3y - 12 && \text{Substitution Property} \end{aligned}$$

Many algebraic expressions contain *powers.* Here are the first three powers of 4.

First power: $4^1 = 4$
Second power: $4^2 = 4 \cdot 4 = 16$
Third power: $4^3 = 4 \cdot 4 \cdot 4 = 64$

An algebraic expression using numbers or variables or both to indicate a product or quotient is called a *term.* For example, $4rs$, $7x^3$, 5, and $\frac{m}{3}$ are all terms. The numerical part of a term is the coefficient of the term. For example, 4 is the coefficient of $4rs$ and 7 is the coefficient of $7x^3$. When we refer to *like* terms, we mean terms that are equal or that differ only in their coefficients. For example:

$7x$ and $5x$ are like terms.
$4rs$ and $-2sr$ are like terms.
$-3m$ and $6mn$ are *not* like terms.
$2a^2$ and $2a$ are *not* like terms.

You can also use the distributive property to combine like terms.

$$7x + 5x = (7 + 5)x \qquad \text{Distributive Property}$$
$$= 12x \qquad \text{Substitution Property}$$

$$4rs - 2rs = (4 - 2)rs \qquad \text{Distributive Property}$$
$$= 2rs \qquad \text{Substitution Property}$$

ORAL EXERCISES

Name the property illustrated.

1. $5x + 2x = (5 + 2)x$

2. $(5 + 2)x = 7x$

3. $7x + 7y = 7y + 7x$

4. $0(8x) = 0$

5. $(8x + 7) + (-7) = 8x + (7 + (-7))$

6. $8x + (7 + (-7)) = 8x + 0$

7. $a(b + c) = (b + c)a$

8. $\left(\frac{1}{3} \cdot 3\right)w = 1 \cdot w$

9. $1 \cdot w = w$

10. $2(3y) = (2 \cdot 3)y$

11. $-1(x + y) = (-1)x + (-1)y$

12. $(-1)x + (-1)y = -x + (-y)$

13. $8x + 0 = 8x$

14. If $y \in \mathcal{R}$, then $y + 2 \in \mathcal{R}$.

WRITTEN EXERCISES

Use the indicated property to make a true statement.

A **1.** $7m + 6m = \underline{\ ?\ }$ (Commutative Property of Addition)

2. $7m + 6m = \underline{\ ?\ }$ (Distributive Property)

3. $8(17x) = \underline{\ ?\ }$ (Associative Property of Multiplication)

4. $8(17x) = \underline{\ ?\ }$ (Commutative Property of Multiplication)

5. $((-w) + w) + 3 = \underline{\ ?\ }$ (Inverse Property of Addition)

6. $1 \cdot y + 0 = \underline{\ ?\ }$ (Identity Property of Multiplication)

7. $1 \cdot y + 0 = \underline{\ ?\ }$ (Identity Property of Addition)

8. $-(q + r) = \underline{\ ?\ }$ (Multiplicative Property of -1)

9. $(-x) + (x + 27) = \underline{\ ?\ }$ (Associative Property of Addition)

10. $\left(5 \cdot \frac{1}{5}\right)r = \underline{\ ?\ }$ (Inverse Property of Multiplication)

11. $(7 + 6)m = \underline{\ ?\ }$ (Substitution Property)

12. If $x \in \mathcal{R}$, then $3x$ $\underline{\ ?\ }$ (Closure Property for Multiplication)

Write in simplest form without parentheses.

13. $4(y - x)$

14. $5(q - 2)$

15. $(x - 7)3$

16. $(a + 4)10$

17. $\frac{1}{2}(b + 2)$

18. $(4c + 4)\frac{1}{4}$

19. $0.3(m - 5)$

20. $-2.5(6 + p)$

21. $a(b+7)$ 22. $p(3+q)$ 23. $2y(2+y)$ 24. $c(c-d)$

Combine like terms.

EXAMPLE $4x + 7y - 3x + 2y$

SOLUTION
$$
\begin{aligned}
4x + 7y - 3x + 2y &= 4x - 3x + 7y + 2y \\
&= (4-3)x + (7+2)y \\
&= (1)x + (9)y \\
&= x + 9y
\end{aligned}
$$

B 25. $4b + 3b$ 26. $2r + 8s - 3s$ 27. $12m + 23n + 3m - 8n$

28. $7xy + 2x - 3xy$ 29. $5q + 11r - 4q - 2r$ 30. $16w - 3aw + 4a - 9w + a$

Write an equivalent expression without parentheses and combine like terms if possible.

EXAMPLE $(x+2)(y+3)$

SOLUTION Apply the distributive property twice.
$$
\begin{aligned}
(x+2)(y+3) &= x(y+3) + 2(y+3) \\
&= xy + x \cdot 3 + 2y + 6 \\
&= xy + 3x + 2y + 6
\end{aligned}
$$

C 31. $(a+5)(b+7)$ 32. $(m-1)(n+2)$ 33. $(x+4)(x+3)$

34. $(4+c)(g+2)$ 35. $(9+x)(11+w)$ 36. $(y+7)(y-2)$

Find a value for x that will make the statement true.

37. $x + (-3) = -3$ 38. $x + 7 = 0$ 39. $(-3) + 8 + 3 = x$

40. $10x = 0$ 41. $5x = 5$ 42. $6x = 1$

Section 1-3 BASIC OPERATIONS

In this section you will review:

- adding and subtracting real numbers
- multiplying and dividing real numbers
- grouping symbols
- order of operations

Recalling what you have learned about computing sums of positive numbers, you can apply the following property to find the sum of any two real numbers.

Property of the Negative of a Sum

For all real numbers a and b, $-(a+b) = (-a) + (-b)$

EXAMPLES

1. $(-27) + (-15) = -(27 + 15)$
$= -42$

2. $38 + (-42) = 38 + \big(-(38 + 4)\big)$
$= 38 + \big((-38) + (-4)\big)$
$= \big(38 + (-38)\big) + (-4)$
$= 0 + (-4)$
$= -4$

Subtracting a number is the same as adding its opposite.

Rule for Subtraction

For all real numbers a and b, $\quad a - b = a + (-b)$

EXAMPLES

1. $5 - 8 = 5 + (-8)$
$= 5 + \big(-(5 + 3)\big)$
$= 5 + (-5) + (-3)$
$= -3$

2. $-5 - (-8) = -5 + 8$
$= 8 + (-5)$
$= 8 - 5$
$= 3$

3. $8 - (-5) = 8 + 5$
$= 13$

To multiply positive and negative numbers the following properties are essential.

Properties of Negatives in Products

For all real numbers a and b,

$a(-b) = -ab \qquad (-a)b = -ab \qquad (-a)(-b) = ab$

EXAMPLES

1. $3(-6) = -(3 \cdot 6)$
$= -18$

2. $-4(7) = -(4 \cdot 7)$
$= -28$

3. $(-9)(-11) = 9 \cdot 11$
$= 99$

Dividing by a number is the same as multiplying by its reciprocal.

Rule for Division

For all real numbers a and all real nonzero numbers b,

$$a \div b = a\left(\frac{1}{b}\right) \quad \text{or} \quad \frac{a}{b} = a\left(\frac{1}{b}\right)$$

Remember that $\frac{a}{b}$ is called the *quotient* of a and b. Notice that the rule for division states that you cannot divide by zero. Why? Suppose that $\frac{a}{0} = x$ for some real number x. Then $a = 0 \cdot x$. If $a \neq 0$, then no value of x would make this equation true. If $a = 0$, then every real value of x would make the equation true. Thus, $\frac{a}{0}$ would have either no value or many values. Therefore, dividing by zero would produce a meaningless expression.

EXAMPLES

1. $$\begin{aligned}\frac{-12}{27} &= -12\left(\frac{1}{27}\right)\\ &= -3 \cdot 4\left(\frac{1}{3 \cdot 9}\right)\\ &= -3 \cdot 4 \cdot \frac{1}{3} \cdot \frac{1}{9}\\ &= -3 \cdot \frac{1}{3} \cdot 4 \cdot \frac{1}{9}\\ &= -1 \cdot \frac{4}{9} = -\frac{4}{9}\end{aligned}$$

2. $$\begin{aligned}\frac{-105}{-35} &= -105\left(-\frac{1}{35}\right)\\ &= 105\left(\frac{1}{35}\right)\\ &= 3 \cdot 35\left(\frac{1}{35}\right)\\ &= 3\left(35 \cdot \frac{1}{35}\right)\\ &= 3 \cdot 1\\ &= 3\end{aligned}$$

You can use the properties of real numbers to simplify algebraic expressions.

EXAMPLE Combine like terms: $5m - 6y - 12m - (-2y)$

SOLUTION
$$\begin{aligned}5m - 6y - 12m - (-2y) &= 5m - 6y - 12m + 2y\\ &= 5m - 12m - 6y + 2y\\ &= (5 - 12)m + \big((-6) + 2\big)y\\ &= -7m - 4y\end{aligned}$$

EXAMPLE Simplify: $-3(x - y)$

SOLUTION
$$\begin{aligned}-3(x - y) &= (-3)x - (-3)y\\ &= -3x + 3y\end{aligned}$$

To simplify more complicated numerical expressions, you must first understand the use of *grouping symbols.*

EXAMPLE Simplify: $27 - (7 - 8)$

SOLUTION Here parentheses, ( ), are used as grouping symbols showing you which operation to perform first.

$$27 - (7 - 8) = 27 - (-1) = 27 + 1 = 28$$

Brackets, [ ], are also used like parentheses. Always simplify expressions in the innermost parentheses or brackets first.

EXAMPLE Simplify: $3[(12-17)+(28-9)]$

SOLUTION
$$\begin{aligned} 3[(12-17)+(28-9)] &= 3[(-5)+19] \\ &= 3(14) \\ &= 42 \end{aligned}$$

The fraction bar is another familiar grouping symbol.

EXAMPLE Simplify: $\dfrac{29-18}{3}$

SOLUTION $\dfrac{29-18}{3} = \dfrac{(29-18)}{3} = \dfrac{11}{3}$

How would you evaluate this expression?

$$16 - 10 \div 2 + 3(5^2 - 4)$$

Your answer would depend on the order in which you performed the indicated operations. To avoid confusion, mathematicians have agreed to follow these rules in this order:

1. Perform operations within grouping symbols, including powers.
2. Compute powers.
3. Compute products and quotients in order from left to right.
4. Compute sums and differences in order from left to right.

Thus,
$$\begin{aligned} 16 - 10 \div 2 + 3(5^2 - 4) &= 16 - 10 \div 2 + 3(25 - 4) \\ &= 16 - 10 \div 2 + 3(21) \\ &= 16 - 5 + 63 \\ &= 11 + 63 \\ &= 74 \end{aligned}$$

ORAL EXERCISES

Calculate.

1. $3+(-5)$ **2.** $-3+(-5)$ **3.** $3-(-5)$

4. $-3-(-5)$ **5.** $7(-8)$ **6.** $-3(-9)$

7. $0(-2)$ **8.** $-12 \div 4$ **9.** $12 \div (-4)$

10. $48 \div (-6)$ **11.** $-48 \div (-6)$ **12.** $(-48) \div 6$

13. $5-(3-2)$ **14.** $5-3-2$ **15.** $4+3 \cdot 0$

State the first step you would use to simplify the expression.

16. $-9 \div 3 + 2^3$ **17.** $4(12-6) \div 8$ **18.** $\dfrac{2+1}{3} - 5$

WRITTEN EXERCISES

Calculate.

A **1.** $105 + (-14)$ **2.** $-17 + 130$ **3.** $-17 - 130$

4. $105 - (-14)$ **5.** $-91 + (-12)$ **6.** $-12 + 11 - 33$

7. $-33 - 29 + 62$ **8.** $-91 - (-12)$ **9.** $25 + [-(-32 + 46)]$

10. $-7(2)(-2)$ **11.** $3(-1)\,8$ **12.** $-9 + 10 + (-12) + 8$

13. $6(-2)(-5)$ **14.** $-3(2 - 8)(-1)$ **15.** $-18 \div 3 - 4$

16. $36 \div (3 - 7)$ **17.** $-162 \div 6 \div (-3)$ **18.** $-54(-8) \div (-4)$

Simplify.

19. $6a - 8a$ **20.** $4x - 8x$ **21.** $-7w - 11w$

22. $-3t - 4t$ **23.** $-2x - (-3x)$ **24.** $-4w - (-3w)$

25. $8t - (-2t)$ **26.** $7y - (-10y)$ **27.** $-3(-4x)(-1)$

28. $7(-9x)^2$ **29.** $-4(2m - n)$ **30.** $-6(4z - 7z)$

Calculate.

B **31.** $\big(-3(7 + 5)\big) \div 2 + 1$ **32.** $(3 - 5)^2 \div \big(4(-7 - 1)\big)$

33. $10^2 \div (-5) \cdot (-2) + 10$ **34.** $-18 + 4 \cdot 2 - 14 - (-6)^2$

35. $\frac{34}{-7}[-20 - (-6)] - \frac{7 + 2}{3}$ **36.** $(18 - 3^2 \cdot 2)(5^2 - (-3) + 4)$

Simplify and combine like terms.

37. $-8y - (-12) + 3 - (-15y)$ **38.** $13x - 28y - 49x - (-19y)$

39. $-6y - (-10w) - 11w - (-12y)$ **40.** $7a - 25b - 3a + 14b$

41. $31m - (-7n) + (-6m) + (-15n)$ **42.** $-7x + 3x - (-6x) + 15y$

43. $(-3x + 4) - (2 - 5x)$ **44.** $(2 - 8y) - (6 + 5y)$

45. $(4a + b) - (a + b) - (a - 2b)$ **46.** $(3x - 4y) - (2x - 3y) + (4x - 5y)$

47. $-8 \cdot 5(-9y) + 13(x + y) - 12x$ **48.** $4\left(\frac{x + y}{2}\right) - 5x$

49. $\frac{3}{5}(10a - 15b) - \frac{2}{3}(6a - 3b)$ **50.** $-\frac{4}{9}\left(18a - \frac{9b}{4}\right) + b + 8a$

Use the distributive property to write an equivalent expression without parentheses.

C **51.** $(2x - 3y)(4x - y)$ **52.** $(a - 7b)(3a + 4b)$

53. $(5w + v)(-6w + v)$ **54.** $(-1)(6q - 5r)(7q - 9r)$

Combine like terms.

55. $3\big(4x - 3(x + 2)\big) - 5\big(-3x - 4(2x - 1)\big)$

56. $-7\big(2y - 5(4 - y)\big) + 4\big(8y - 7(3 + y)\big)$

SELF-TEST 1

Fill in the blank with one of the symbols $\in$, $\subset$, $\cup$, or $\cap$, to make a true statement.

1. $\emptyset$ _?_ $\mathcal{R}$ **2.** 3 _?_ $\{-3, 0, 3\}$ **3.** $\{1, 2, 3\}$ _?_ $\{0, 2, 4\} = \{2\}$ Section 1-1

For each pair of sets A and B, graph **(a)** $A \cup B$ and **(b)** $A \cap B$.

4. $A = \{-5, -3, -1, 1, 3\}$
$B = \{-5, -4, -3, -2, -1\}$

5. $A = \{\text{the positive integers}\}$
$B = \{\text{the nonnegative real numbers}\}$

Use the indicated property to make a true statement. Section 1-2

6. $2(r + s) =$ _?_ (Commutative Property of Multiplication)

7. $2(r + s) =$ _?_ (Commutative Property of Addition)

8. $2(r + s) =$ _?_ (Distributive Property)

9. $12t(-1) =$ _?_ (Multiplicative Property of -1)

10. Simplify: $(j - 3)4$

11. Combine like terms: $9y + 5z - 4z - 8yz + y$

Simplify. Section 1-3

12. $57 + (-75) \div (-3)$

13. $(-23 - 17)(-2 + 12)$

14. $-9v - (-v) + (-3v)$

15. $13x + 25 - 25x$

16. $-(-8)^2 \div 4 \cdot (1 - 4)$

17. $\frac{10 - 4}{-2} - 3 \cdot 2^2$

Section 1-4 LINEAR EQUATIONS

In this section you will review:

- equations
- domain of a variable
- solution sets
- solving linear equations
- least common multiple
- least common denominator

A sentence that states two expressions are equal is called an *equation*. Here are some examples:

$$2(14 + 7) = 6 \cdot 7 \qquad 5 - 2 = 7 \qquad 3x = x + 12$$

The first equation is true and the second is false. Note that the third equation contains a *variable*, in this case x. The truth or falsity of the last sentence depends on what replacement you make for the variable x. If $x = 6$, the sentence is true. If $x \neq 6$, the sentence is false.

A sentence containing a variable is called an *open sentence.* The variable may represent any element of a specified set, called the *domain* of the variable. The set of values in the domain of the variable that make an open sentence true is called the *solution set* of the sentence. Each element of the solution set is called a *root* or *solution* of the equation. If there is no replacement that makes the sentence true, the solution set is the *null set,* $\emptyset$. An equation, such as $x = x$, that is true for every element of the domain, is called an *identity.*

EXAMPLE Find the solution set of $3x - 4 = 11$ if the domain of x is $\{3, 4, 5, 6\}$.

SOLUTION $3(3) - 4 = 5$
$3(4) - 4 = 8$
$3(5) - 4 = 11$
$3(6) - 4 = 14$

The only replacement for x that makes the sentence $3x - 4 = 11$ true is 5. Therefore, the solution set is $\{5\}$.

EXAMPLE Find the solution set of $3x = 7$ if the domain of x is $Z = \{\text{the integers}\}$.

SOLUTION There is no integral value of x that makes the sentence $3x = 7$ true. Therefore, the solution set is $\emptyset$.

In this text you may assume the domain of x is $\mathcal{R}$ unless otherwise specified.

If two equations have the same domain and the same solution set, they are said to be *equivalent.* The following equations are all equivalent since the solution set of each is $\{7\}$.

$$4x + 3 = 5x - 4$$
$$2(x - 3) = x + 1$$
$$\frac{x}{3} - \frac{5}{6} = \frac{3}{2}$$
$$x = 7$$

Note that the solution of the last equation is obvious. The process of solving equations involves transforming equations into equivalent equations until finally you arrive at one like "$x = 7$" whose solution set can be determined by inspection. The following properties can be used in solving equations.

Properties of Equality

Let a, b, c, and d represent real numbers.

Addition	If $a = c$ and $b = d$, then $a + b = c + d$.
Subtraction	If $a = c$ and $b = d$, then $a - b = c - d$.
Multiplication	If $a = c$ and $b = d$, then $ab = cd$.
Division	If $a = c$ and $b = d$ (b and $d \neq 0$), then $\frac{a}{b} = \frac{c}{d}$.

The following example illustrates how some of these properties can be applied in solving equations.

EXAMPLE Solve $7x - 2 = 4x + 10$.

SOLUTION

$7x - 2 + 2 = 4x + 10 + 2$	Addition Prop. of Equality
$7x = 4x + 12$	Substitution Property
$7x - 4x = 4x - 4x + 12$	Subtraction Prop. of Equality
$3x = 12$	Substitution Property
$\frac{3x}{3} = \frac{12}{3}$	Division Prop. of Equality
$x = 4$	Substitution Property

Check Substitute 4 for x in the *original* equation.

$7(4) - 2 \ ? \ 4(4) + 10$

$28 - 2 \ ? \ 16 + 10$

$26 = 26$

Therefore, 4 is the solution of $7x - 2 = 4x + 10$.

An equation like $7x - 2 = 4x + 10$ is called a *linear equation in one variable* because the variable appears raised to the first power only. Some equations that are *not* linear are:

$$x^2 - 3 = 2x \qquad x^3 = 8 \qquad \frac{1}{x} = x$$

If an equation contains *parentheses,* it is best to simplify the equation by applying the distributive property as a first step.

EXAMPLE Find the solution set of $3(x - 4) + 9 = 8 - 2(x - 2)$.

SOLUTION

$3(x - 4) + 9 = 8 - 2(x - 2)$	
$3x - 12 + 9 = 8 - 2x + 4$	Distributive Property
$3x - 3 = 12 - 2x$	Commutative Prop. of Add. and Substitution
$5x - 3 = 12$	Addition Prop. of Equality
$5x = 15$	Addition Prop. of Equality
$x = 3$	Division Prop. of Equality

Check Substitute 3 for x in the original equation.

$3(3 - 4) + 9 \ ? \ 8 - 2(3 - 2)$

$3(-1) + 9 \ ? \ 8 - 2(1)$

$-3 + 9 \ ? \ 8 - 2$

$6 = 6$

Therefore, the solution set is $\{3\}$.

When solving linear equations, you generally will not give a reason to justify each step in the process. Be sure to show enough steps, however, so that if you should have to go back to find where you made an error, you can follow your own reasoning clearly.

The *least common multiple* (LCM) of a set of numbers is the least positive integer that can be divided evenly by each member of the set. The *least common denominator* (LCD) of a set of fractions is the LCM of their denominators. If a linear equation contains fractions, multiplying both sides of the equation by the LCD will transform the equation into one without fractions.

EXAMPLE Solve $\frac{2}{3} - \frac{x}{2} = -\frac{5}{6}$.

SOLUTION Multiply both sides of the equation by the least common denominator. The least common multiple of 2, 3, and 6 is 6. Therefore, the LCD = 6.

$$\begin{aligned} 6\left(\frac{2}{3} - \frac{x}{2}\right) &= 6\left(-\frac{5}{6}\right) \\ 6\left(\frac{2}{3}\right) - 6\left(\frac{x}{2}\right) &= 6\left(-\frac{5}{6}\right) \\ 4 - 3x &= -5 \\ -3x &= -9 \\ x &= 3 \end{aligned}$$

Check

$$\begin{aligned} \frac{2}{3} - \frac{3}{2} &\overset{?}{=} -\frac{5}{6} \\ \left(\frac{2}{3}\right)\left(\frac{2}{2}\right) - \left(\frac{3}{2}\right)\left(\frac{3}{3}\right) &\overset{?}{=} -\frac{5}{6} \\ \frac{4}{6} - \frac{9}{6} &= -\frac{5}{6} \end{aligned}$$

Therefore, the solution of $\frac{2}{3} - \frac{x}{2} = -\frac{5}{6}$ is 3.

ORAL EXERCISES

In each equation the domain of x is $\mathcal{R}$. State whether the equation is always, sometimes, or never true.

1. $x + 3 = 3 + x$
2. $7x - 2 = 12$
3. $2x - 3 = x - 3$
4. $x + 2 = x + 3$
5. $6x - 2 = 8x$
6. $7(x + 8) = 7x + 56$
7. $x - 7 = x + 1$
8. $6x = 0$
9. $12x = 5$
10. $x - 4 = -4$
11. $2(x + 4) = 3x$
12. $8 - (x + 2) = 10 - x$

State whether or not each equation is linear.

13. $x^2 - 16 = 0$
14. $3y + 9 = 2y + 5$
15. $\frac{1}{w} + 18 = 2$
16. $m + m^3 = 12m^2$
17. $3(x + 5) = 21$
18. $\frac{y + 3}{5} = \frac{y}{2} - \frac{y}{8}$

WRITTEN EXERCISES

Solve and check.

A

1. $9 - 7x = -68$
2. $9y - 20 = 34$
3. $12a + 5 = -1$
4. $-13 = 2 + 5t$
5. $8x - 15 = 25$
6. $7 - 6y = 49$
7. $15 + 3m - 2 = 4$
8. $18 - 5p - 2 = -4$
9. $13r - 16 = 5r$
10. $15x = 9x + 24$
11. $35 - 2z = 5z$
12. $25a - 58 = 8a + 44$
13. $7p + 23 = p + 65$
14. $19x - 41 = 12x + 15$
15. $22d + 42 = 17d + 27$
16. $-56 - 25y = 16y + 108$
17. $8 - 5u = 8u - 31$
18. $11 - 14w = 2w + 27$
19. $3x - 14 + 2x = -14 + 5x$
20. $12 + 8k - 4 = 4k - 12$
21. $18 - 6y + 7 = 4y - 25$
22. $8t + 14 + 2t = 7t + 38$
23. $39 + 12q + 65 = 8 + 4q$
24. $24y + 29 - 2y = 22y + 11$
25. $2(x - 4) = 22$
26. $19z = 4(11 + 2z)$
27. $3w - (6 + w) = 12$
28. $5m - (m - 3) = -21$
29. $15 - (a - 2) = 2(a + 1)$
30. $2(20 - 6x) = 8 - 4(3x - 8)$

B

31. $6s - 4(1 - 3s) = 5(s + 3) + 13s$
32. $3(k + 1) - 2(3k - 2) = -4(k - 5)$
33. $19y - 3(4 - 3y) = 2(9 + 7y) + 19$
34. $4(z - 7) = 9z - (20 - z)$
35. $28 - 6(w + 3) = 16 + 3(2w - 12)$
36. $37 - 4x = 2(3x + 10) - (x - 1)$
37. $\frac{m}{5} - \frac{m}{3} = 10$
38. $\frac{y}{4} + \frac{y}{14} = -\frac{9}{2}$
39. $\frac{2w - 3}{5} = \frac{w}{2}$
40. $\frac{4b}{3} - \frac{2b}{7} = -\frac{2}{3}$
41. $\frac{3a + 7}{4} + \frac{a - 3}{2} = -9$
42. $\frac{2x + 5}{3} - \frac{4 - x}{4} = \frac{3}{4}$
43. $\frac{2}{3}(r + 4) + \frac{1}{2}(r - 3) = \frac{7}{6}$
44. $\frac{2}{5}(3 - m) - \frac{1}{2}(4 - 3m) = \frac{1}{10}(6m + 17)$

EXAMPLE Solve $0.82x + 17.7 = 317$

SOLUTION Some people prefer to work with an equation that does not contain decimal fractions. To find an equivalent equation involving only integers, multiply each side of the equation by the appropriate power of 10. In this case, multiply by 100, then solve.

$$\begin{aligned} 100(0.82x) + 100(17.7) &= 100(317) \\ 82x + 1770 &= 31{,}700 \\ 82x &= 29{,}930 \\ x &= 365 \end{aligned}$$

Solve and check.

45. $0.6x + 2.8 = 1.3x$

46. $58 - 0.7y = 1.1y - 104$

47. $0.7w + 20 = 0.25w + 11$

48. $0.75z + 3 = 8 - 0.5z$

Solve for the indicated variable.

C 49. $F = ma$; a

50. $D = rt$; r

51. $C = 2\pi r$; r

52. $V = lwh$; h

53. $P = 2l + 2w$; l

54. $C = \frac{5}{9}(F - 32)$; F

55. $A = \frac{1}{2}(b + c)h$; b

56. $mx + t = rx + s$; x

Section 1-5 LINEAR INEQUALITIES

In this section you will review:

- properties of inequalities
- solving and graphing linear inequalities

A *simple linear inequality* contains one of the following inequality signs:

$>$ is greater than
$<$ is less than
$\neq$ does not equal

As is true of a linear equation in one variable, a *linear inequality in one variable* contains a variable raised to the first power only.

In solving inequalities you will use properties much like those for equations.

Properties of Inequality

Let a, b, and c represent real numbers.

Addition	If $a > b$, then $a + c > b + c$. If $a < b$, then $a + c < b + c$.
Subtraction	If $a > b$, then $a - c > b - c$. If $a < b$, then $a - c < b - c$.

EXAMPLE Graph the solution set of $x + 5 > 7$.

SOLUTION

$$x + 5 > 7$$
$$x + 5 + (-5) > 7 + (-5)$$
$$x > 2$$

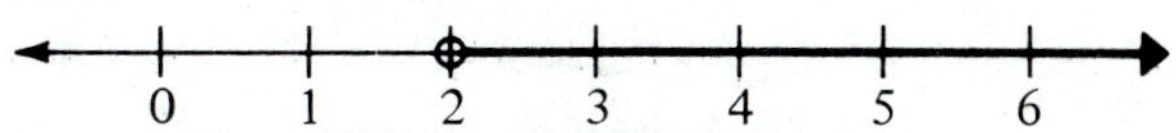

The open dot means that 2 is not included in the solution set.

A linear inequality containing either the symbol $\leqslant$ or the symbol $\geqslant$ is solved in the same manner as a simple linear inequality. $x \leqslant 3$ is read as "$x < 3$ or $x = 3$," or more commonly "x is less than or equal to 3." The solution set $\{x: x \leqslant 3\}$ is equal to $\{x: x < 3\} \cup \{x: x = 3\}$. (Note that the symbol $\leqslant$ means the same as $\not>$ and the symbol $\geqslant$ means the same as $\not<$.)

EXAMPLE Graph the solution set of $2x + 6 \leqslant x + 3$.

SOLUTION $2x + 6 \leqslant x + 3$

$x + 6 \leqslant 3$

$x \leqslant -3$

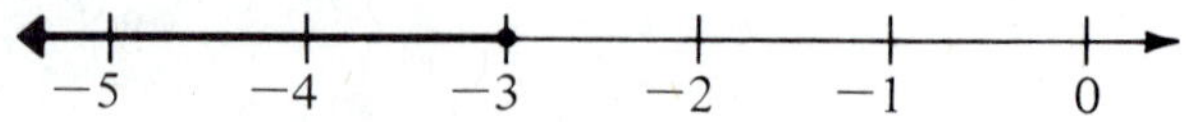

The closed dot means that -3 is included in the solution set.

To multiply or divide both sides of an inequality by a nonzero number you must use a set of rules different from those for equations.

Properties of Inequality

Let a, b, and c represent real numbers.

Multiplication
- If $a > b$ and $c > 0$, then $ac > bc$.
- If $a < b$ and $c > 0$, then $ac < bc$.
- If $a > b$ and $c < 0$, then $ac < bc$.
- If $a < b$ and $c < 0$, then $ac > bc$.

Division
- If $a > b$ and $c > 0$, then $\frac{a}{c} > \frac{b}{c}$.
- If $a < b$ and $c > 0$, then $\frac{a}{c} < \frac{b}{c}$.
- If $a > b$ and $c < 0$, then $\frac{a}{c} < \frac{b}{c}$.
- If $a < b$ and $c < 0$, then $\frac{a}{c} > \frac{b}{c}$.

Notice that if you multiply or divide both sides of an inequality by a positive number you keep the inequality sign unchanged; if you multiply or divide both sides by a negative number you reverse the inequality sign.

EXAMPLE Find and graph the solution set of $2 - 3x > 17$.

SOLUTION $2 - 3x > 17$

$-3x > 15$

$\frac{-3x}{-3} < \frac{15}{-3}$ Reverse the inequality sign. (Why?)

$x < -5$

The solution set is $\{x: x < -5\}$, whose graph is shown.

EXAMPLE Find and graph the solution set of $4(5-x) \leqslant 3(x+2)$.

SOLUTION

$$\begin{aligned} 4(5-x) &\leqslant 3(x+2) \\ 4 \cdot 5 - 4x &\leqslant 3x + 3 \cdot 2 \\ 20 - 4x &\leqslant 3x + 6 \\ -4x &\leqslant 3x - 14 \\ -7x &\leqslant -14 \\ x &\geqslant 2 \end{aligned}$$

The solution set is $\{x: x \geqslant 2\}$, whose graph is shown.

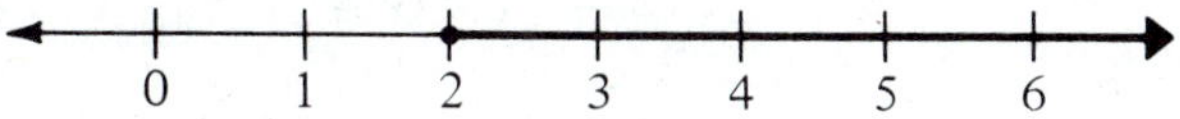

ORAL EXERCISES

Complete by replacing the blank with $<$, $>$, or $=$ to make a true statement.

1. -4 __?__ -3
2. 2 __?__ -1
3. 0 __?__ -10
4. $10 - 18$ __?__ $5 - 9$
5. $-1 - (-2)$ __?__ $-1 + (-2)$
6. $3(-4)$ __?__ $(-6)2$
7. $(-1)(-9)$ __?__ $\left(-\frac{1}{3}\right)33$
8. $0(-6)$ __?__ $(-2)10$
9. $(-1)(-5)(-3)$ __?__ $20 - 32$
10. $0(-100)$ __?__ $0(1000)$

State the solution set.

11. $-x < -5$
12. $y + 3 > -6$
13. $\frac{1}{2}w > -4$
14. $-3z < 27$
15. $a - 5 < 10$
16. $b + 12 > -20$
17. $-81 < 3y$
18. $7 > -\frac{1}{3}x$
19. $w > w + 5$
20. $m < 2m$

WRITTEN EXERCISES

Find and graph the solution set of each inequality.

A

1. $2x - 14 < 16$
2. $21 + 3x > 15$
3. $28 \leqslant 7y + 7$
4. $11 \geqslant 3y + 5$
5. $5 - 4y \leqslant 29$
6. $3 - 8x > -21$
7. $47 - 17t \geqslant 13$
8. $38 - 11w \leqslant -17$
9. $13y - 5 > 9y + 11$
10. $6w + 14 > 2w - 26$
11. $9x - 23 \leqslant 7x - 18$
12. $29 - 8y < 2y - 21$

Solve.

13. $44 - 3z \geqslant 4z - 19$

14. $6x - 3 < 7x + 12$

15. $-2z - 4 > -5z + 2$

16. $13m + 4 \geqslant 16m - 8$

17. $12 \leqslant 6(2n - 3)$

18. $7(8 - z) \geqslant 21$

19. $2(y - 3) < y - 15$

20. $4(s + 6) - 3 > 2s - 7$

B **21.** $12w + 8 > 5(3w - 5)$

22. $3(x - 4) - 8x \leqslant 4 + 6(x + 1)$

23. $\frac{y}{5} \leqslant -1 + y$

24. $4 - \frac{2}{3}x < x + 2$

25. $\frac{m}{3} - \frac{1}{5} > \frac{2m}{5}$

26. $\frac{s}{5} + \frac{3s}{4} \leqslant \frac{s}{2}$

27. $16\left(\frac{3}{4} + \frac{x}{2}\right) \geqslant -x$

28. $20\left(\frac{3}{5} - \frac{y}{4}\right) > -2y$

29. $-0.3x + 2.5x < 1.1$

30. $3(1.5 - 0.4x) \geqslant 0.6x$

31. $1 - 2(j + 3) < j + 16$

32. $4 - 2(3 - 2w) > 3(w + 1) - 10$

C **33.** $\frac{2}{3}(5x - 12) < -\frac{1}{2}(5x + 1)$

34. $\frac{3(4 - a)}{2} + 5 \geqslant \frac{2(3a + 1)}{3} - \frac{1}{6}$

35. $\frac{7(5 - 2y)}{5} + \frac{8y - 3}{4} \leqslant \frac{73}{20}$

36. $\frac{12w - 3}{4} < 6 - \frac{3(3 - 4w)}{2}$

37. $\frac{5}{3}(x - 3) - \frac{2}{5}(2x - 5) \ngtr \frac{1}{3}(2x + 9) - (x - 12)$

38. $\frac{2}{3}(2b + 3) + \frac{3}{4}(2 - 3b) \nless \frac{5}{6}(b + 6) + 9$

Section 1-6 COMBINING INEQUALITIES

In this section you will learn how to solve combined inequalities such as:

- $a < x$ and $x < b$
- $x < a$ or $x > b$

A sentence formed by joining two clauses with the word *and* is called a *conjunction.* For a conjunction to be true, *both* clauses must be true.

EXAMPLES

$3 < 0$ and $4 > 2$	False
$-1 > -2$ and $-3 < -2$	True

The solution set of a conjunction is the set of values that makes both clauses true. That is, the solution set of an inequality like $a < x$ and $x < b$, abbreviated $a < x < b$, is the *intersection* of the solution sets of the two inequalities.

$$\{x\colon a < x < b\} = \{x\colon a < x\} \cap \{x\colon x < b\}$$

EXAMPLE Graph the solution set of $-1 \leqslant x < 3$.

SOLUTION $-1 \leqslant x < 3$ can be written as $-1 \leqslant x$ and $x < 3$. The graph of the solution set is the graph of $\{x: -1 \leqslant x\} \cap \{x: x < 3\}$.

$\{x: -1 \leqslant x\}$

$\{x: x < 3\}$

$\{x: -1 \leqslant x < 3\}$

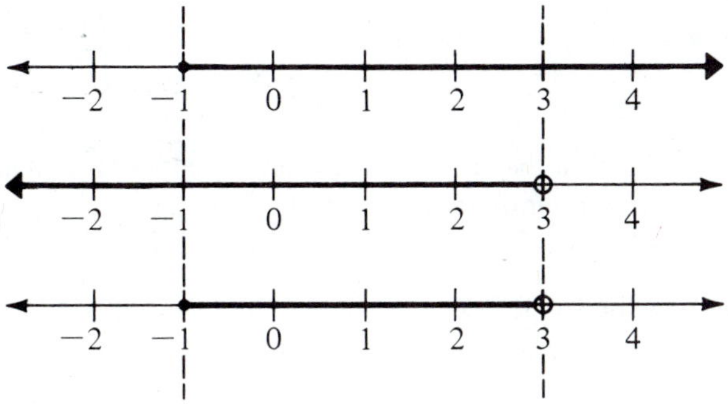

$\{x: -1 \leqslant x\} \cap \{x: x < 3\}$

EXAMPLE Solve and graph the solution set: $7 > 3x - 5 > -2$

SOLUTION

Method 1

$7 > 3x - 5 > -2$ can be written as

$$7 > 3x - 5 \qquad \textit{and} \qquad 3x - 5 > -2$$

To solve $7 > 3x - 5 > -2$, you can solve the separate statements and find the intersection of their solution sets.

$7 > 3x - 5$	and	$3x - 5 > -2$
$7 + 5 > 3x - 5 + 5$		$3x - 5 + 5 > -2 + 5$
$12 > 3x$		$3x > 3$
$\frac{12}{3} > \frac{3x}{3}$		$\frac{3x}{3} > \frac{3}{3}$
$4 > x$	and	$x > 1$

The intersection of $\{x: 4 > x\}$ and $\{x: x > 1\}$ is $\{x: 4 > x > 1\}$, which can be written $\{x: 1 < x < 4\}$. Thus, the solution set is $\{x: 1 < x < 4\}$.

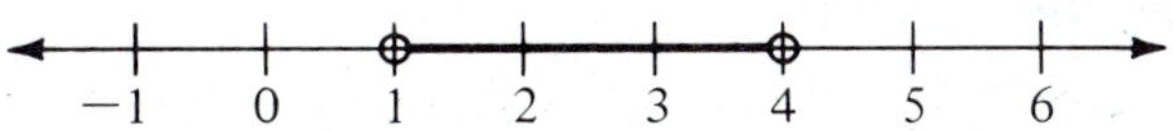

Method 2

Another method of solving this type of inequality is to consider the inequality as having three "parts" rather than two "sides." You then apply the properties of inequality to each "part" to solve.

$$7 > 3x - 5 > -2$$

Add 5 to each part. $\quad 7 + 5 > 3x - 5 + 5 > -2 + 5$

$$12 > 3x > 3$$

Divide each part by 3. $\quad 4 > x > 1$

The solution set is $\{x: 1 < x < 4\}$.

A sentence formed by joining two clauses with the word *or* is called a *disjunction.* For a disjunction to be true, *at least one* of the clauses must be true.

EXAMPLES

$3 < 0$ or $4 > 2$ True

$-5 > -4$ or $1 < -3$ False

The solution set of a disjunction is the set of values that makes either clause true. That is, the solution set of an inequality like

$$x < a \quad \text{or} \quad x > b$$

is the *union* of the solution sets of the two inequalities.

$$\{x: x < a \text{ or } x > b\} = \{x: x < a\} \cup \{x: x > b\}$$

EXAMPLE Graph the solution set of $x < -2$ or $x \geqslant 1$.

SOLUTION Graph $\{x: x < -2\} \cup \{x \geqslant 1\}$.

$\{x: x < -2\}$

$\{x: x \geqslant 1\}$

$\{x: x < -2 \text{ or } x \geqslant 1\}$

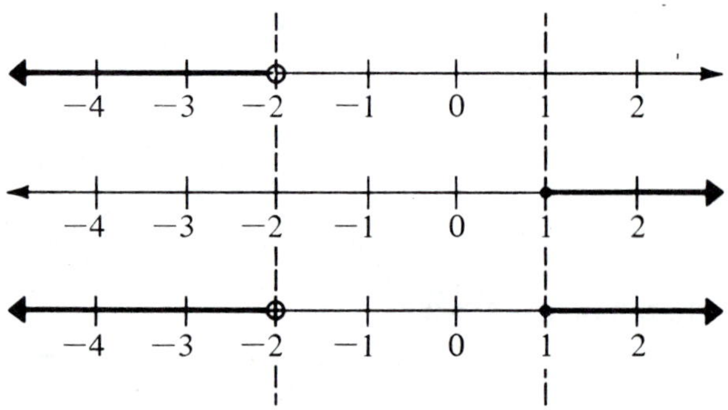

$\{x: x < -2\} \cup \{x: x \geqslant 1\}$

EXAMPLE Solve and graph the solution set: $3x + 4 < 16$ or $2x - 7 \geqslant 3$

SOLUTION

$$\begin{aligned} 3x + 4 &< 16 \quad \text{or} \quad & 2x - 7 &\geqslant 3 \\ 3x &< 12 \quad \Big| & 2x &\geqslant 10 \\ x &< 4 \quad \text{or} & x &\geqslant 5 \end{aligned}$$

The solution set is $\{x: x < 4 \text{ or } x \geqslant 5\}$, whose graph is shown.

ORAL EXERCISES

Tell whether the statement is true or false.

1. $-3 < 0$ and $4 > -1$

2. $3 < -6$ or $5 \geqslant 2$

3. $-1.5 \leqslant -2$ or $6 > 8$

4. $\frac{1}{8} < 8$ and $-3 \leqslant -5$

5. $1 < 2$ or $5 = -5$

6. $-0.5 \leqslant -\frac{1}{2}$ and $3.2 \geqslant \frac{5}{3}$

Determine if the given value satisfies the inequality.

7. $-1 < 3y \leqslant 6; 2$

8. $-12 < w + 8 < 16; -18$

9. $-4 \leqslant -2x < 8; -4$

10. $27 \geqslant -3m \geqslant -18; 9$

11. $2x + 1 \geqslant 3$ or $x - 5 \leqslant -5; 0$

12. $-2y < -16$ or $-3 > y + 2; 10$

13. $3y + 8 > 22$ or $2y - 1 < 6; 5$

14. $5 < 8 - x$ or $8 > 12 - x; 5$

WRITTEN EXERCISES

Match each inequality with the graph of its solution set.

A 1. $2 < x + 3 \leqslant 5$

2. $0 \leqslant -x + 1 < 3$

3. $x + 2 > 3$ or $-x \geqslant 2$

4. $1 \geqslant -(x + 1) > -2$

5. $2x < -4$ or $x + 1 \geqslant 2$

6. $-3x \geqslant 3$ or $x - 2 > 0$

a.

b.
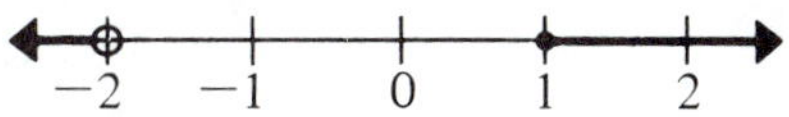

c.
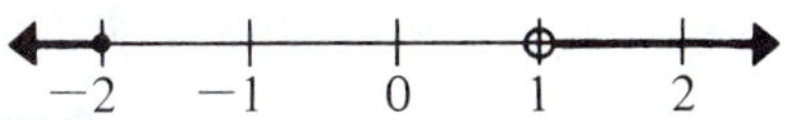

d.

e.

f.
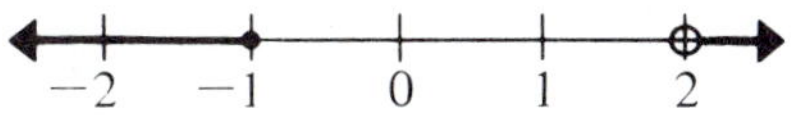

Solve each inequality and graph any solution set that is not the null set.

7. $-1 \leqslant x + 4 < 2$

8. $0 > x - 3 > -5$

9. $-7 < -4 + x \leqslant -2$

10. $6 \leqslant 5 + x \leqslant 10$

11. $1 > 2x + 1 > -7$

12. $5 \leqslant 3x - 1 < 11$

13. $y + 3 > 9$ or $y - 2 < 4$

14. $-5 + y \leqslant -4$ or $2y + 1 > 7$

15. $6y - 3 \geqslant 12$ or $5y - 2 \leqslant -7$

16. $-7 + 3y < 20$ or $4y + 5 > 53$

17. $-7 + 2w > 3$ and $-28 > 3w + 1$

18. $-28 - 5w \leqslant 12$ and $4w + 3 < -13$

19. $17 - 2y \geqslant 5$ or $\frac{1}{2}y > 5$

20. $2y - 8 \leqslant -12$ or $8 - y \leqslant 2$

21. $3x - 5 > 10$ or $3 - 2x < -13$

22. $-5w + 7 < 17$ and $17 < 4w + 33$

Solve.

B **23.** $3a \leqslant -2 - 3a$ or $3a > 2 - a$

24. $11 - 2z < 7$ or $2z - 8 > 5z + 1$

25. $-8 < 2(x + 3)$ and $2(x + 3) < 8$

26. $-6 \leqslant 3(r - 2)$ and $3(r - 2) < 6$

27. $9 > -3(x - 1) \geqslant -12$

28. $16 \geqslant 4(x - 2) > -24$

29. $2 + 7y \leqslant 17 + 2y$ or $4y \geqslant 49 - 3y$

30. $24 + 7b < -24 + b$ or $9b > -35 + 4b$

31. $d + 6 \leqslant 4d - 9$ or $3d - 1 < 2d + 4$

32. $-x + 4 > -4$ or $3x - 5 \leqslant x + 7$

33. $3(t - 3) + 1 \geqslant 7$ and $2(t + 1) + 3 \leqslant 1$

34. $2(m + 3) - 3 \geqslant 19$ or $5 + 4(1 - m) < -7$

C **35.** $-4 < -\frac{2}{3}(y + 5) - 1 \leqslant 6$

36. $-10 \leqslant \frac{2}{5}(3 - x) + 2 \leqslant -6$

37. $2x - 7 < x + 8 \leqslant 2x + 5$

38. $-x + 1 \geqslant 3x + 5 > x + 2$

39. $-8 + x < 3(x + 2) \leqslant -2(2 + x)$

40. $12 - x \leqslant 4(x - 2) < 3(25 - x)$

SELF-TEST 2

Solve and check.

Section 1-4

1. $14 - 5x = 7x + 110$

2. $8a + 9 - a = 74 + 2a$

3. $6 - (-y + 4) = 3(y - 2)$

4. $\frac{d}{8} + \frac{5}{2} = \frac{d}{4}$

Find and graph the solution set of each inequality.

Section 1-5

5. $-4z + 18 \geqslant -2$

6. $4(x + 6) < 20$

7. $5 - 3t > 4t - 37$

8. $21 - 7n \leqslant 3(3 - n)$

Section 1-6

9. $8 > 1 - 7b \geqslant -13$

10. $y + 10 \leqslant 9$ or $-33 \geqslant -11y$

11. $-6 < 2(x - 3) \leqslant 4$

12. $9 < 2x + 5$ and $2x + 5 < 11$

Section 1-7 TRANSLATING WORDS INTO SYMBOLS

An important skill in solving word problems is the ability to translate English phrases into algebraic expressions. Certain key words indicate which operation symbol to use in your translations.

English Phrase	*Algebraic Expression*
Addition	
x *increased* by 4	$x + 4$
4 *more than* x	$x + 4$
the *sum* of x and 4	$x + 4$
x *exceeded* by 4	$x + 4$

English Phrase	*Algebraic Expression*
Subtraction	
x decreased by 4	$x - 4$
4 *less than x*	$x - 4$
4 *subtracted from x*	$x - 4$
the *difference of x minus* 4	$x - 4$
x diminished by 4	$x - 4$
Multiplication	
the *product* of x and 4	$4x$
4 *times x*	$4x$
$\frac{1}{4}$ *of x*	$\frac{1}{4}x$
Division	
the *quotient* of x divided by 4	$\frac{x}{4}$

Study the following examples illustrating how to translate more complicated phrases.

English Phrase	*Algebraic Expression*
twice x	$2x$
3 less than twice x	$2x - 3$
-4 increased by the product of 3 and x	$-4 + 3x$
one half the difference of x minus 7	$\frac{1}{2}(x - 7)$
the quotient of 3 divided by twice x	$\frac{3}{2x}$
the product of -2 and 10 more than x	$(-2)(x + 10)$

To solve word problems you must be able to write an equation or inequality that accurately describes the known facts of the problem. How can you tell when to use an equation as opposed to an inequality? Certain expressions in English indicate which to use. Study the following list:

English	*Algebra*
is	$=$
is more than	$>$
is less than	$<$
is no more than	$\leqslant$
is at most	$\leqslant$
is no less than	$\geqslant$
is at least	$\geqslant$

Now let's do a few examples.

The *difference* of 7 *minus* three *times x is* 43. $7 - 3x = 43$

The *product* of 2 and 3 *more than x is at least* 12. $2(x + 3) \geq 12$

One third *of* two *less than x is less than* 5. $\frac{1}{3}(x - 2) < 5$

Note the distinction between *less than* and *is less than.* Generally, when *less than* stands alone, it implies subtraction. When *less than* is preceded by the verb *is,* it implies the use of an inequality. The same is true of *more than* and *is more than.*

Three *more than* twice *x is more than x.* $3 + 2x > x$

Now we'll show you some examples of how to translate the information presented in a problem into an equation or inequality.

EXAMPLE One number is 3 times as great as another number and their sum is 112.

SOLUTION Let x represent the lesser number.
Then $3x$ represents the greater number.
The problem states their sum *is* 112.
Therefore, $x + 3x = 112$.

EXAMPLE The sum of two consecutive even integers is at most 78.

SOLUTION Let y represent the first integer.
Then $y + 2$ represents the next consecutive even integer.
$y + (y + 2) \leq 78$

EXAMPLE If 4 times a number is decreased by 3, the result is less than twice the difference of the number minus 1.

SOLUTION Let a represent the number. $4a - 3 < 2(a - 1)$

EXAMPLE One number is 3 more than another. Four times the greater number exceeds 3 times the lesser number by 10.

SOLUTION Let x represent the lesser number. Then $x + 3$ represents the greater number.
$4(x + 3) = 3x + 10$

ORAL EXERCISES

Translate each word phrase into an algebraic expression.

1. twice x
2. 2 more than x
3. 10 more than x
4. one half x
5. 8 less than x
6. x diminished by 7
7. the difference of 3 minus x
8. the quotient of x divided by 3
9. number of cents in n nickels
10. number of cents in d dimes
11. number of cents in q quarters
12. number of days in w weeks

13. If Carlos is c years old now, how old will he be in 4 years?

14. If Leona is $x + 2$ years old now, how old was she 3 years ago?

15. If Kim is k years old now, then 1 year ago what was twice Kim's age?

16. If Dom is d years old now, then 5 years from now what will be three times Dom's age?

17. If x is an even integer, what is the next even integer?

18. If x is an even integer, what is the next odd integer?

19. If x is the greatest of three consecutive integers, what are the other two integers?

20. The sum of two integers is 23. If one integer is x, what is the other integer?

21. The difference of a number minus y is 18. What is the number?

22. The product of two numbers is 8. If one number is z, what is the other number?

23. The quotient of a number divided by w is 10. What is the number?

WRITTEN EXERCISES

Translate each sentence into an equation. Identify the variable you are using and what it represents.

A 1. A number increased by 7 is 21.

2. A number diminished by 12 is 18.

3. The product of 6 and a number is 72.

4. The quotient of a number divided by 5 is 15.

5. Two consecutive odd integers have a sum of 0.

6. The sum of two consecutive integers is 45.

7. In 4 years Miguel will be 19.

8. Seven years ago Tina was 4.

9. A piggy bank contains only quarters worth a total of \$20.75.

10. The dimes in Barbara's pocket are worth 90¢.

Translate each sentence into an inequality. Identify the variable you are using and what it represents.

11. The product of 5 more than a number and 3 is more than 12.

12. The quotient of 2 less than a number divided by 4 is less than 7.

13. The sum of two consecutive integers diminished by 1 is at least twice the smaller.

14. The lengths of the sides of a triangle are represented by consecutive integers and the perimeter of the triangle is at most 12.

15. Five less than 8 times a number is no more than 32.

Translate each sentence into an inequality. Identify the variable you are using and what it represents.

16. Sixteen more than twice a number is no less than 18.

17. The cost of 45 L of gasoline is at least $15.60.

18. It will cost no more than $59.40 to give a luncheon for 12 people.

Each problem contains two unknowns. Represent each unknown in terms of the same variable and translate each problem into an equation or inequality. Identify the variable you are using and what it represents.

EXAMPLE The length of a rectangle is twice its width. The perimeter of the rectangle is at most 54 cm.

SOLUTION Let w = width of the rectangle in centimeters.
Then $2w$ = length of the rectangle.
$2(2w) + 2(w) \leqslant 54$

B 19. A certain number of pennies and 3 times as many nickels are worth at least $1.92.

20. A parking meter that takes only dimes and nickels contains 8 more nickels than dimes for a total of $3.85.

21. Nancy is twice as old as Bill and the difference in their ages is at least 32.

22. Fred is 5 years younger than Toni and the sum of their ages is at most 29.

23. A board 421 cm long is cut into two pieces so that one piece is 4 cm more than twice as long as the other.

24. The sum of two numbers is 27. Their difference is 19.

25. Sidney walks to and from school in under 45 min. It takes him 10 min longer to walk home than it does to walk to school.

26. The length of a rectangle is 2 cm less than 3 times the width. The perimeter is at most 46 cm.

27. A square and a rectangle have the same width, but the length of the rectangle is 6 more than the width. The sum of their perimeters is 76.

28. The difference of two numbers is 11. The larger is 6 less than twice the smaller.

29. Tim's age is one third his father's. The difference in their ages is 24 years.

30. The cost of a ballpoint pen is four fifths that of a felt-tipped pen. Two felt-tipped pens and one ballpoint pen cost less than $1.30.

C 31. The sum of three consecutive even integers is at least 156.

32. The sum of five consecutive odd integers is less than 208.

33. The sum of three consecutive even integers is 12 more than twice the fourth consecutive even integer.

34. Of two consecutive integers, four times the lesser exceeds three times the greater by 30.

35. Kay has one half as many nickels as dimes and 15 fewer quarters than nickels. Altogether she has $11.25.

36. Dave is 5 years older than Angie. In 8 years 3 times Angie's age will exceed twice Dave's age by 10.

Section 1-8 PROBLEM SOLVING

It is difficult to write an exact procedure for problem solving because no two problems are exactly the same, but the following steps are often helpful in solving problems.

Step 1 Choose a variable and state what the variable represents.

Step 2 Write an equation or inequality that describes the facts presented in the problem.

Step 3 Solve the equation or inequality.

Step 4 Check your solution with the wording of the problem to see if it satisfies the stated conditions.

Step 5 Write your answer.

EXAMPLE The length of a rectangle is 28 cm and its perimeter is 106 cm. What is its area?

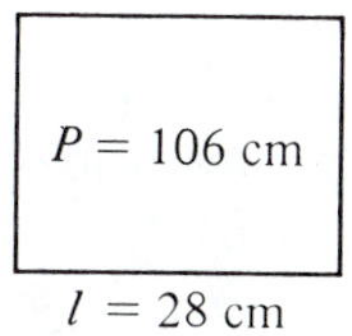

SOLUTION Follow the steps outlined above.

Step 1 Let w represent the width, in centimeters, of the rectangle.

Step 2 We know the perimeter is twice the length plus twice the width, so
$2(28) + 2w = 106$

Step 3
$$\begin{aligned} 56 + 2w &= 106 \\ 2w &= 106 - 56 \\ 2w &= 50 \\ w &= 25 \end{aligned}$$

Step 4 If the length is 28 cm and the width is 25 cm, then the perimeter is $2(28\text{ cm}) + 2(25\text{ cm}) = 56\text{ cm} + 50\text{ cm} = 106\text{ cm}$.

Step 5 The problem asks for area. For a rectangle, Area $= l \cdot w$. In this case, Area $= 28\text{ cm} \cdot 25\text{ cm} = 700\text{ cm}^2$. The answer is 700 cm^2.

In subsequent examples we will not specifically identify each step of this method.

EXAMPLE The sum of three consecutive integers is 72. Find the integers.

SOLUTION Let x = the first integer.
Then $x + 1$ = second consecutive integer,
and $x + 2$ = third consecutive integer.

$$\begin{aligned} x + (x + 1) + (x + 2) &= 72 \\ x + x + x + 1 + 2 &= 72 \\ 3x + 3 &= 72 \\ 3x &= 69 \\ x &= 23 \\ x + 1 &= 24 \\ x + 2 &= 25 \end{aligned}$$

Check $23 + 24 + 25 = 72$ The three consecutive integers are 23, 24, and 25.

EXAMPLE Anne received scores of 93, 79, and 58 on her first three math tests. What score does she need on the next test in order to have at least an 80 average?

SOLUTION To find the average of 4 scores you add them and divide by 4.
Let x = the score needed on the next test.

$$\begin{aligned} \frac{93 + 79 + 58 + x}{4} &\geqslant 80 \\ \frac{230 + x}{4} &\geqslant 80 \\ 230 + x &\geqslant 320 \\ x &\geqslant 90 \end{aligned}$$

Anne needs at least a 90 on the next test.

A problem like the following one is called a *uniform motion* problem because the object moves at a constant (or uniform) speed. To solve, you apply the formula $D = rt$, where D = distance (in kilometers), r = rate (in km/h), and t = time (in hours).

EXAMPLE Bob traveled a certain distance in 4 h and returned the same distance in 3.5 h. If his speed on the return trip was 5 km/h more than his speed on the first half of the trip, find the rate at which he returned home.

SOLUTION Let r = Bob's rate outbound.
Then $r + 5$ = Bob's rate on the return trip.

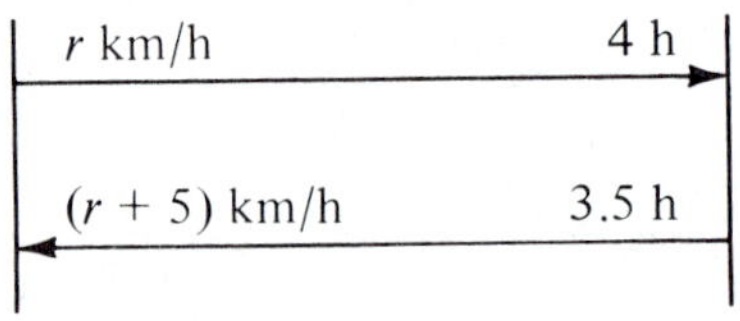

You may wish to use a table to help you organize the data.

	Rate (r)	*Time (t)*	*Distance (D = rt)*
Outbound	r	4	$4r$
Return	$r + 5$	3.5	$3.5(r + 5)$

$$
\begin{aligned}
4r &= 3.5(r+5) \\
40r &= 35(r+5) \\
40r &= 35r + 175 \\
5r &= 175 \\
r &= 35 \quad \text{and} \quad r + 5 = 40
\end{aligned}
$$

Check 35 km/h for 4 h is 140 km traveled.
40 km/h for 3.5 h is 140 km traveled.
Therefore, Bob's rate on the return trip was 40 km/h.

Problems like the following are often referred to as *mixture problems.*

EXAMPLE One solution is 10% acid and another is 25% acid. How many liters of each should be mixed to form 20 L of a solution which is 16% acid?

SOLUTION Write an equation stating that the sum of the number of liters of acid in the solutions that are mixed together equals the number of liters of acid in the resulting solution.

Let x be the number of liters of the 10% solution.
Then $20 - x =$ the number of liters of the 25% solution.

How many liters of acid are in the 10% solution? If the solution is 10% acid and there are x liters of the solution, then there are $0.10x$ liters of acid.

How many liters of acid are in the 25% solution? If the solution is 25% acid and there are $(20 - x)$ liters of the solution, then there are $0.25(20 - x)$ liters of acid.

How many liters of acid are in the resulting solution? The solution is 16% acid and there are 20 L of the solution. Therefore, there are 0.16(20) L of acid in the resulting solution.

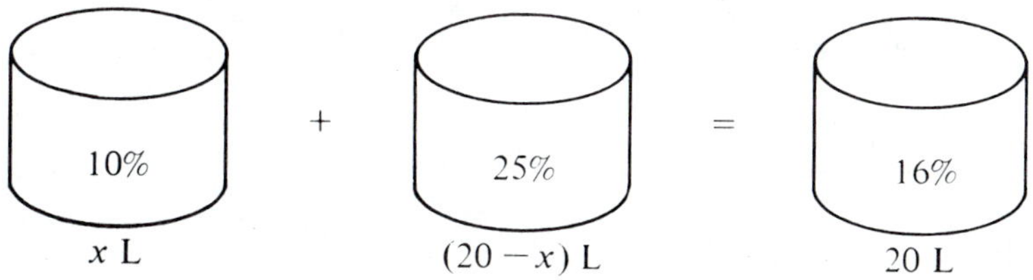

Thus, your equation is

$$
\begin{aligned}
0.10x + 0.25(20 - x) &= 0.16(20) \\
10x + 25(20 - x) &= 16(20) \\
10x + 500 - 25x &= 320 \\
-15x &= -180 \\
x &= 12 \\
20 - x &= 8
\end{aligned}
$$

Check $0.10(12) + 0.25(8) \ ? \ 0.16(20)$
$1.2 + 2 = 3.2$

You should mix 12 L of the 10% acid solution and 8 L of the 25% acid solution to form 20 L of a 16% acid solution.

ORAL EXERCISES

1. How many hours does it take to travel 300 km at 75 km/h?
2. How many kilometers do you travel in 4 h at 46 km/h?
3. At what rate do you travel if you drive 260 km in 5 h?
4. How many hours does it take to travel 540 km at 60 km/h?
5. At what rate do you travel if you walk 30 km in 4 h?
6. How many liters of acid are in 20 L of a 15% acid solution?
7. How many grams of salt are in 150 g of a 25% salt solution?
8. How many liters of *water* are in 400 L of a 10% acid solution?
9. How many liters of a 10% acid solution do you have if the solution contains 3 L of acid?
10. How many liters of a 20% acid solution do you have if the solution contains 8 L of water?
11. Let x be a multiple of 3. What are the next two consecutive multiples of 3?
12. Let y be a multiple of 5. What are the next three consecutive multiples of 5?

Complete the table for a rectangle.

	l	w	$P = 2l + 2w$	$A = lw$
13.	12 cm	4 cm	?	?
14.	?	15 cm	?	300 cm^2
15.	15 m	?	50 m	?
16.	$40x$	?	$80x + 20$	?
17.	?	25	?	$50x$

WRITTEN EXERCISES

Complete the table.

	rate	*time*	*Distance*
A 1.	?	7.5	420
2.	37	x	?
3.	?	5	x
4.	200	?	x
5.	60	?	$300x$

6. The sum of two consecutive even integers is 242. Find the integers.
7. The sum of three consecutive odd integers is -105. Find the integers.

8. The width of a rectangle is 15 cm and the perimeter is 78 cm. Find the length.

9. The width of a rectangle is 35 cm and the area is 1400 cm^2. Find the perimeter.

10. The high school baseball team plays 20 games in one season. If they win 9 of the first 15 games, how many more must they win to have a winning percentage of at least 60% for the season?

11. A company spends \$8 to manufacture a calculator that it sells for \$10.50. How many calculators must the company sell to make a profit of more than \$10,000?

12. Three consecutive even integers have a sum of 0. What are the integers?

13. Terry has 3 more quarters than dimes for a total of \$3.55. How many quarters does he have?

14. Chris has 12 fewer nickels than dimes for a total of \$6.00. How many dimes does she have?

15. Laverne is 3 years older than Shirley and the sum of their ages is 75 years. How old is Shirley?

16. Ozzie is three times as old as Benji. The difference in their ages is 18 years. How old is Ozzie?

17. The length of a rectangle is 12 cm and its area at least 54 cm^2. What do you know about the width of the rectangle?

18. A student says that he has a passing average after taking 5 exams in a certain class. If the first four test scores were 69, 52, 40, and 65, what do you know about his fifth test score? (Assume that "passing" means having an average of 60 or more.)

B 19. Four consecutive multiples of 10 have a sum of 20. What are the numbers?

20. One alloy contains 20% silver and another contains 8% silver. How many grams of each should be mixed to give 120 g of an alloy which is 15% silver?

21. One solution contains 10% alcohol while another contains 30% alcohol. How many liters of each should be mixed to give 10 L of a solution which is 25% alcohol?

22. A nut mixture contains 10% cashews. How many kilograms of the mixture and how many kilograms of cashews should be mixed to give 6 kg of a nut mixture which is 25% cashews?

23. Sixty grams of an alloy that contains less than 15% gold is melted together with 45 g of an alloy that is 20% gold. What do you know about the amount of gold in the resulting alloy?

24. A train leaves New York City for Buffalo, 630 km away, at an average rate of 68 km/h. At the same time a train leaves Buffalo for New York City at an average rate of 58 km/h. In how many hours will they pass each other?

25. Two planes head in opposite directions at rates of 700 km/h and 850 km/h. In how many hours will they be 3875 km apart?

26. Moose leaves Philadelphia at 11:30 A.M. An hour later Jacky leaves Philadelphia at a rate 22 km/h faster than Moose and catches up with Moose at 3:30 P.M. How fast was Jacky driving?

27. It took Andy 1 h longer to drive to his cousins' chicken farm than it did to return home. If it took Andy 6 h to make the round trip and he traveled 20 km/h faster on the return trip, how fast did he drive on the way to the farm?

28. Find three consecutive integers such that the sum of the first and the third diminished by 12 is 14 more than three times the second.

C 29. Donald is 3 years older than Mickey. In 2 years Donald will be twice as old as Mickey was 5 years ago. How old is Donald?

30. Wendy is twice as old as Paul, and Nancy is 5 years younger than Wendy. The sum of Paul's and Wendy's ages is 2 less than twice Nancy's age. How old is Wendy?

31. What is the least and the most water that must be added to 25 L of a 30% acid solution to form a solution that is at least 5% and at most 8% acid?

32. Harriet has $\frac{1}{3}$ as many dimes as nickels and 3 more quarters than dimes, for a total of $2.75. How many of each coin does she have?

SELF-TEST 3

Translate each problem into an equation or an inequality. Begin by identifying the variable you are using and what it represents. (In Exercise 4, express both unknowns in terms of the same variable.)

1. The sum of two consecutive odd integers is at least 64. Section 1-7
2. In 5 years Meena will be 7.
3. Roberto has a bank filled with only half dollars worth a total of $12.50.
4. The length of one side of a rectangle is 7 cm less than twice the length of an adjacent side. The perimeter is greater than 40 cm.

Solve.

5. The sum of two consecutive integers is 10 more than the third. Find the integers. Section 1-8
6. The perimeter of a square is less than 6 m. What can you say about the area of the square?
7. Pat drove to the dentist in 0.75 h. The return trip took 1.25 h due to icy roads. If her average speed on the return trip was 20 km/h less than that on the first part of the trip, how far did she travel each way?
8. One solution is 40% iodine and another is 25% iodine. How many milliliters of each should be mixed to produce 100 mL of a 28% iodine solution?

Computer Exercises

Here are some of the kinds of statements that are used in the BASIC language. The letters A and B are variables.

PRINT "INPUT A";	This prints the material between the quotation marks. The semicolon holds the printer at the end of the line.
INPUT A	This prints a question mark. The computer waits for the value of A to be typed.
INPUT A, B	This prints *one* question mark, but the computer waits to be given *two* values.
FOR I = M TO N STEP S [Body of loop] NEXT I	This arrangement causes the body of the loop to be performed for the values of I from M to N in increments of S. "FOR I = 3 TO 11 STEP 2" gives the values 3, 5, 7, 9, 11. If S = 1, then "STEP S" may be omitted.
PRINT A	This prints the value of A.
Subscripted variables A(1), A(2), . . . , A(I)	These are used to handle lists of values.
IF (a) THEN (x)	If the statement (a) is true, then the computer goes to line (x). Otherwise, it goes to the next line in sequence.
END	Many computers require this as the final statement of a BASIC program.

EXAMPLE Find the union of the sets $A = \{2, 5, -1\}$ and $B = \{-1, -2, 3, 5\}$.

SOLUTION Write the elements of A, and follow them with the elements of B that are not elements of A:

$$A \cup B = \{2, 5, -1, -2, 3\}$$

Using the method of the preceding example, write the union of sets A and B.

1. $A = \{1, 2, 3\}$, $B = \{4, 5, 6\}$ **2.** $A = \{1, 2, 3\}$, $B = \{2, 3, 1\}$

The following computer program uses the method of the preceding example. Lines 10 through 100 input the data. Lines 130 through 150 print the elements of set A. Lines 160 through 210 print the elements of B that are not elements of A.

```
10  PRINT "INPUT THE NUMBER OF ELEMENTS   ";
15  PRINT "IN EACH OF SETS A AND B";
20  INPUT M,N
30  PRINT "INPUT THE ELEMENTS OF A:"
40  FOR I=1 TO M
50  INPUT A[I]
60  NEXT I
```

(Program continued on next page.)

```
70    PRINT "INPUT THE ELEMENTS OF B:"
80    FOR I=1 TO N
90    INPUT B[I]
100   NEXT I
110   PRINT
120   PRINT "A ∪ B =    ";
130   FOR I=1 TO M
140   PRINT A[I];
150   NEXT I
160   FOR I=1 TO N
170   FOR J=1 TO M
180   IF B[I] =A[J] THEN 210
190   NEXT J
200   PRINT B[I];
210   NEXT I
220   END
```

3. Copy and RUN the preceding program for the data of the example.

4. RUN the program for the data in Exercises 1 and 2.

5. Modify the program to print out the intersection of any two sets A and B.

CHAPTER REVIEW

Tell whether the statement is true or false. If it is false, explain why.

1. $Z \cup W = \{1, 2, 3, \ldots\}$ **2.** $\pi \in \mathcal{R}$ **3.** $\mathcal{R} \subset Q$ Section 1-1

For each pair of sets A and B, graph **(a)** $A \cup B$ and **(b)** $A \cap B$.

4. $A = \{-1, 0, 1, 2\}$
$B = \{-4, -1, 1, 4\}$

5. $A = \{$the integers$\}$
$B = \{$the whole numbers$\}$

Use the indicated property to make a true statement.

6. If $x \neq 0$, then $t\left(\frac{1}{x} \cdot x\right) = \underline{\ ?\ }$ (Inverse Property of Multiplication) Section 1-2

7. If $x \neq 0$, then $t\left(\frac{1}{x} \cdot x\right) = \underline{\ ?\ }$ (Associative Property of Multiplication)

8. $-z + 0 = \underline{\ ?\ }$ (Identity Property of Addition)

9. Simplify: $2(r - 1)$

10. Combine like terms: $3br - 8bw - 2b + 10bw + 3b$

Simplify.

11. $\frac{-9-3}{2(-2)}$ **12.** $-4 + 9 \div (-3)^2$ **13.** $[15 - (-5)] \cdot (-5)\left(-\frac{1}{4}\right)$ Section 1-3

14. $2x(-7) - 2x(-5)$ **15.** $4k - (5k - 3j) + 9j$

Solve and check.

16. $14 + 6r = -4$ | 17. $14x + 35 = 5x + 71$ | Section 1-4

18. $8(2 - w) - 3w = 5$ | 19. $7.85 - 0.5y = 0.35 + y$

Solve.

20. $18 + 5a \geqslant 3a - 6$ | 21. $29 - 3t < -t + 1$ | Section 1-5

22. $7 - (2m - 4) \geqslant m + 17$ | 23. $5n < 53 - 3(3n - 1)$

Solve the inequality and graph its solution set.

24. $6 < 1 - 5t < 21$ | 25. $-\frac{1}{2}b \geqslant -1$ or $3b - 8 > 7$ | Section 1-6

26. $9 \leqslant 4x + 3$ and $4x + 3 \leqslant 19$

Translate the problem into an equation or an inequality. Identify the variable you are using and what it represents. (In Exercise 29 represent each unknown in terms of the same variable.)

27. Four less than the quotient of a number and 3 is 1. Section 1-7

28. In 14 years Donald will be no less than three times as old as he is now.

29. A dryer in a self-service laundry accepts only dimes and quarters. The machine contains 26 coins worth at most \$3.95.

Solve.

30. The sum of three consecutive even integers is 102. Find the integers. Section 1-8

31. One alloy contains 18% copper and another contains 8% copper. How many grams of each should be mixed to produce 150 g of an alloy which is 11% copper?

32. The difference between Toby's running rate and walking rate is 210 m/min. If Toby ran a race in 1.5 min and walked back to her starting point in 5 min, find her running rate.

CHAPTER TEST

For each pair of sets A and B, graph **(a)** $A \cup B$ and **(b)** $A \cap B$.

1. $A = \{-2, -1, 0, 1\}$
   $B = \{-3, -2, -1, \ldots, 3\}$

2. $A = \{\text{nonnegative integers}\}$
   $B = \{\text{even integers}\}$

State which of the sets, $\mathcal{R}$, Z, W, and Q, contain the given number.

3. $\frac{-27}{-9}$ 4. 2π 5. $-\frac{3}{6}$

Use the indicated property to make a true statement.

6. $\left(\frac{1}{8} \cdot 8\right) + n = \underline{\ ?\ }$ (Commutative Property of Multiplication)

7. $0(-z + z) = \underline{\ ?\ }$ (Inverse Property of Addition)

8. $(9 \cdot 7)x = \underline{\ ?\ }$ (Substitution Property)

Simplify.

9. $\dfrac{9 - 8^2}{-6 - (-1)}$

10. $7a - (9b + a) + 9b$

11. $(3k + 6)\frac{1}{3}$

12. $(4 - 12)(-3) \div (-1 + 7)$

13. $4x - y - 2x + x$

14. $(5r - 3) - (-2r + 4)$

Solve.

15. $7x \leqslant 5(x - 4)$

16. $-38 + 2y = 5y - 53$

17. $-1 \geqslant 3 - 4z > -21$

18. $3(2a + 5) + a = -13$

19. $-\frac{3}{4}b > 12$ or $2b + 9 > 1$

20. $0.27c \leqslant 0.35c - 1.6$

Find and graph each solution set.

21. $\dfrac{3d}{4} + \dfrac{d}{6} = -11$

22. $18 - 3f > 2f + 8$

23. $-5(q + 3 - 2q) \geqslant 25$

24. $17r - 2(3r - 1) = r + 27$

In each exercise, express every unknown in terms of the same variable. Then write and solve an equation or an inequality to answer the question.

25. Twice the sum of two consecutive odd integers is 9 more than the third consecutive odd integer. What are the integers?

26. Tim is 3 years older than his sister. Five years ago, the sum of their ages was at least 19. At least how old is Tim now?

27. The length of a rectangle is 7 cm more than its width. The perimeter of the rectangle is 58 cm. What are the dimensions of the rectangle?

28. Marjory bought some 18¢ stamps and some 28¢ stamps. If she bought 13 stamps in all and paid $2.74, how many of each kind did she buy?

Biographical Note

Srinivasa Ramanujan

The British mathematician G. H. Hardy described the work of Srinivasa Ramanujan (1887-1920) as having a "profound and invincible originality." Although his formal mathematical training was limited, Ramanujan, who delighted in the properties of numbers, made significant contributions to the branch of mathematics known as number theory.

Born in southern India, Ramanujan had a remarkable memory and an aptitude for numerical calculation. When he was fifteen, he acquired Carr's *Synopsis of Pure Mathematics*, which fascinated him. This book contained approximately 6000 formulas that Ramanujan proceeded to verify.

The following year he won a scholarship to Government College in India, but because he was so preoccupied with mathematics, he neglected his other subjects and eventually had to leave.

In 1913 Ramanujan wrote to G. H. Hardy, a leading authority on number theory, to ask him if any of the approximately 120 theorems that he had enclosed were good enough to be published. Hardy, a professor at Trinity College, was amazed by Ramanujan's untaught brilliance and was anxious for Ramanujan to study with him.

The next year Ramanujan entered Trinity College. Although Ramanujan possessed profound mathematical insight and remarkable intuition, there were surprising gaps in his background. He was often embarrassed to learn that a theorem he had discovered on his own had already been proved.

During his stay in England, Ramanujan published twenty-one papers, some in collaboration with Hardy. In recognition of his mathematical achievements, in 1918 he was elected a fellow of the Royal Society and of Trinity College. The following year Ramanujan returned to India, where he died in 1920.

Although his life was brief, his work has contributed to a number of fields. Some of the formulas he developed made advances in statistics and physics possible. Other work, only recently discovered, may prove useful in the study of many physical phenomena.

VOCABULARY REVIEW

Be sure that you understand the meaning of these terms:

set, p. 1
element, p. 1
rational number, p. 2
irrational number, p. 2
null (empty) set, p. 2
equal sets, p. 3
intersection, p. 3
union, p. 3
graph of a number, p. 4
coordinate of a point, p. 4
opposite (additive inverse), p. 8
reciprocal (multiplicative inverse), p. 8
algebraic expression, p. 9
term, p. 9
like terms, p. 9
equation, p. 16
variable, p. 16
open sentence, p. 17
domain of a variable, p. 17
solution, p. 17
identity, p. 17
equivalent equations, p. 17
least common multiple, p. 19
least common denominator, p. 19
simple linear inequality, p. 21
conjunction, p. 25
disjunction, p. 26

MIXED REVIEW

Solve.

1. $3(1-x)=4(x-2)-10$

2. $\frac{2a-1}{5}=\frac{a+1}{2}$

3. $-(2x-5)=-6(x+1)$

4. $5y-8>y+12$ and $3-3y<9$

5. $\frac{b-4}{2}+\frac{2b+1}{3}=\frac{b-2}{4}+\frac{5b+1}{3}$

6. $2(6y-5)-2\leqslant 36$

7. $-18\leqslant 4(x+3)<20$

8. $2(5y-1)+2>3(y-7)$

For each pair of sets C and D, graph **(a)** $C\cup D$ and **(b)** $C\cap D$.

9. $C=\{2,4,6,8,10\}$
$D=\{0,1,2,3,6,8\}$

10. $C=\{$the whole numbers$\}$
$D=\{$the positive odd integers$\}$

Simplify.

11. $16\div 2^2+4(-2)$

12. $3b-c+2(c-b)$

13. $2(x-y)-(y-2x)$

14. $3-4(3-4)$

Name the property illustrated.

15. $3x+(7x-2y)=(3x+7x)-2y$

16. $3w+3y=3(w+y)$

Give an example to show that each statement is false.

17. If $x\notin A$ and $A\subset B$, then $x\notin A\cup B$.

18. If $A\subset B$ and $B\subset C$, then $A\cap B=B\cap C$.

Represent each unknown in terms of the same variable. Then write and solve an equation to answer the question.

19. When twice the larger of two numbers is added to three times the smaller of the numbers, the sum is 106. Find the two numbers if their sum is 47.

20. A dairy farmer wants to make an 8% butterfat mixture. How many pounds of milk containing 3% butterfat must she add to 100 pounds of cream containing 20% butterfat?

21. Of three numbers, the first is three times the second and the third is 4 less than the first. If their sum is 108, what are the numbers?

Let A, B, and C represent the following sets:

$$A = \{\text{the positive multiples of } 2\}$$
$$B = \{\text{the positive multiples of } 3\}$$
$$C = \{\text{the counting numbers less than } 11\}$$

List the elements in each of the following sets.

22. $(A \cup B) \cap C$

23. $A \cap C$

24. $(A \cap C) \cap (B \cap C)$

Understanding principles of logical reasoning can help you analyze the structure of arguments presented in a political debate, such as the one taking place in the state legislature shown above.

Logic

2

A popular sport in the cities and towns of ancient Greece was getting into and winning arguments. It didn't much matter what two people chose to argue about; winning was everything. The sport even had its professionals, a class of verbal athletes called Sophists who traveled from place to place and, for a fee, taught students how to argue persuasively on any subject at all. The rules of the game were fuzzy, and the Sophists often resorted to trickery and deceit.

Around 450 B.C., Zeno of Elea gave the Sophists a hard time with a group of four paradoxes that still give us trouble today. In one of them, Achilles, the swiftest of runners, races with a tortoise, the slowest of creatures. If Achilles gives the tortoise a head start, then, Zeno stated, Achilles can never overtake the tortoise!

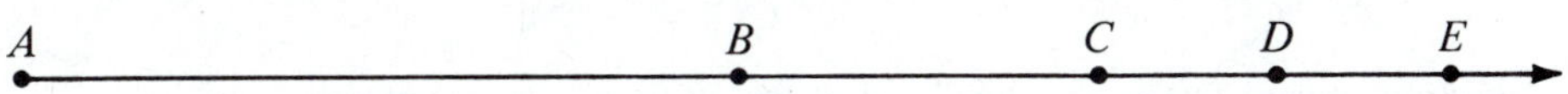

Suppose Achilles starts at point A and the tortoise at point B. By the time Achilles gets to point B, the tortoise will have moved to some farther point, say C. By the time Achilles gets to C, the tortoise will be at D. By the time Achilles gets to D, the tortoise will be at E, and so on. Thus the tortoise always stays ahead of Achilles!

Zeno wasn't fooling himself. He knew as well as you do that Achilles could overtake the tortoise, but in the game of verbal tennis that the Sophists played, his argument was a service ace. His opponent was left speechless. Even today it's not easy to see where his argument goes wrong.

Section 2-1 STATEMENTS AND NEGATION

The earliest recorded rules for logical arguments were first worked out by Aristotle about one hundred years after Zeno devised his paradoxes. The building blocks for logical arguments are sentences called statements. In this section you will cover:

- statements
- equivalent sentences
- contradictory sentences
- quantifiers
- negation

A *statement* is a sentence that can be determined to be true or false. The truth or falsity of a statement is called its *truth value.* Here are some examples of statements:

It never snows in New York. (F)

$3 + 8 = 11$ (T)

Water freezes at 0°C. (T)

The sentences below are *not* statements. Can you tell why?

1. Is it time for lunch yet?
2. Way to go!
3. Your stereo is too loud.
4. "Mathematics" has ten letters.

Sentence 3 is not a statement because its truth value is a matter of opinion, not fact. Sentence 4 is not a statement because its meaning is unclear. Does it mean "mathematics" has *exactly* ten letters? If so, the sentence is false. Or does it mean "mathematics" has *at least* ten letters? If so, it is true. In order for a sentence to be a statement, it must be unambiguous.

Any *True-False* test should consist only of statements. The test taker is supposed to assign the correct truth value to each statement.

EXAMPLE Answer True (T) or False (F).

"Mathematics" has exactly ten letters. F

"Mathematics" has at least ten letters. T

The *always-sometimes-never* test is another kind of test with which you may be familiar. In such tests, *sometimes* has the meaning "sometimes, but not always."

EXAMPLE Complete each sentence with *always, sometimes,* or *never.*

A rectangle is always a parallelogram.

A parallelogram is sometimes a rectangle.

A parallelogram sometimes has congruent diagonals.

A trapezoid never has diagonals that bisect each other.

Closely associated with always-sometimes-never sentences are sentences beginning with the *quantifying* words *All, Some,* and *No.* In ordinary English, people often use the word *some* to mean "at least two but not all." In logic, however, the word *some* means "at least one and perhaps all."

EXAMPLES *All* rectangles are parallelograms.

Some parallelograms are rectangles.

Some numbers satisfy $x = 2x$.

No trapezoid has diagonals that bisect each other.

Logically equivalent statements are statements that always have the same truth value. All of the statements above are true and therefore are equivalent. The statements "Today is Monday" and "Tomorrow is Tuesday" are equivalent because no matter on what day these statements are made, they will either be both true or both false.

The quantifiers *All, Some,* and *No* are used often in logic and mathematics. It is important that you understand the rules governing their use. Here are four ways of translating a sentence containing the quantifier *All.*

All squares *are* rectangles.
Each square *is* a rectangle.
Every square *is* a rectangle.
No square *is not* a rectangle.

Here are three ways of translating a sentence containing the quantifier *Some.*

Some rectangles *are* squares.
There exists a rectangle that *is* a square.
At least one rectangle *is* a square.

The statement *No* trapezoid *is* a rectangle
is equivalent to *All* trapezoids *are not* rectangles.

You can rewrite any sentence beginning with *No* as an equivalent sentence beginning with *All* by changing *No* to *All* and negating the verb.

EXAMPLE Write a sentence beginning with *All* that is equivalent to "No gleep is gloppy."

SOLUTION All gleeps are not gloppy.

Logically contradictory statements are statements that always have opposite truth values. For example, "Today is Monday" and "Today is not Monday" are contradictory. Whenever the first is true (on Mondays), the second is false. Whenever the first is false (on any day of the week but Monday), the second is true. Note, however, the statements "Today is Monday" and "Today is Tuesday" are not contradictory because both can be false simultaneously (on Wednesdays, for example).

A sentence and its *negation* must be logically contradictory. You can always negate a sentence by preceding the sentence with *It is not true that* or *It is false that,* but there are easier ways. Sometimes (but not always) you need only include the word *not* with the verb.

EXAMPLES Sentence: This rectangle is a square.
Negation: *It is false that* this rectangle is a square
or This rectangle is *not* a square.

Sentence: These two lines intersect.
Negation: These two lines do *not* intersect.

Negating *quantified* sentences is a little more complex. If you try to negate the true quantified statement

Some rectangles are squares (T)

by including *not* with the verb, you will get a second true statement:

Some rectangles are not squares. (T)

If you try to negate the false quantified statement

All rectangles are squares (F)

by changing *All* to *No*, you will get a second false statement!

No rectangle is a square. (F)

The following examples illustrate the correct method of negating quantified sentences.

EXAMPLE 1 Negate the statement: *All* rectangles *are* squares. (F)

SOLUTION To negate this sentence, you must indicate that there is *at least one* rectangle that is not a square. To do so, substitute *Some* for *All* and negate the verb.

Some rectangles *are not* squares. (T)

EXAMPLE 2 Negate the statement: *Some* triangles *are* isosceles. (T)

SOLUTION To negate the idea that at least one triangle possesses a certain property, you must indicate that all triangles do not possess this property. Thus you substitute the word *All* for *Some* and negate the verb.

All triangles *are not* isosceles. (F)

or

No triangle *is* isosceles. (F)

EXAMPLE 3 Negate the statement: *No* trapezoids *are* rectangles. (T)

SOLUTION First rewrite the statement as an equivalent statement beginning with *All.*

All trapezoids *are not* rectangles. (T)

Then use the method described in Example 1. The negation is

Some trapezoids *are* rectangles. (F)

Sometimes you can negate a sentence by replacing a single word with its opposite. But be careful! This method does not always work.

EXAMPLE Sentence: The integer is even. Negation: The integer is odd.

EXAMPLE Sentence: The number is positive. Negation: The number is not positive.

Note that "The number is negative" is not a negation of "The number is positive" because the two sentences are not contradictory. If the number were 0, both sentences would be false.

As you may recall, it is customary in logic to use the letters p, q, and r to represent sentences, just as in algebra we use x and y to represent real numbers. If p is a sentence, we denote its negation by $\sim p$, which we read as "not p." But $\sim p$ is itself a sentence, so it too has a negation, which we denote by $\sim(\sim p)$ and read as "not not p." The *Law of Double Negation* says that $\sim(\sim p)$ is equivalent to p. This can all be compactly summarized in the following *truth table*:

p	$\sim p$	$\sim(\sim p)$
T	F	T
F	T	F

ORAL EXERCISES

Tell whether the sentence is a statement. If it is not, explain why.

1. It rained in Albany today.

2. It is clear.

3. Wait 'til the sun shines, Nellie.

4. August has 30 days.

5. "Logic" is a five-letter word.

6. Logic doesn't make sense.

Negate each statement. Give the truth values of the statement and its negation.

7. Seven is even.

8. The number 7 is positive.

9. The numbers 13 and 91 do not have a common factor.

10. A hexagon has ten sides.

11. $\pi = \frac{22}{7}$

12. π is no greater than $\frac{22}{7}$.

13. $\pi > 3$

14. It never rains in California.

WRITTEN EXERCISES

Rewrite each statement as a related equivalent statement beginning with *All, Some,* or *No.*

A **1.** A square is always a rhombus.

2. A rectangle is sometimes a rhombus.

3. A regular pyramid never has a square base.

4. A triangle never has two obtuse angles.

5. A trapezoid must have two opposite sides that are not parallel.

6. An integer may have a reciprocal that is an integer.

Rewrite each sentence as a related true statement beginning with *All, Some,* or *No.*

7. A prime number is odd.

8. Three angles of a triangle are acute.

9. The reciprocal of a negative number is negative.

10. The square root of at least one counting number is irrational.

11. The square of any number is positive.

12. An even number divisible by three is divisible by six.

Rewrite each *All* statement as an equivalent *No* statement and each *No* statement as an equivalent *All* statement.

13. No politician is dishonest.

14. All logicians are rational.

15. All students will pass this course.

16. No student will not pass.

17. No rhombus fails to have perpendicular diagonals.

18. No polygon can be a rectangle without having congruent diagonals.

Negate each sentence.

19. 5 is not a multiple of 3.

20. $\triangle ABC$ is isosceles.

21. No tall men are fat.

22. Some people live in glass houses.

23. All my cousins are cheerful.

24. Some eggs are not scrambled.

25. No physicists are philosophers.

26. All rabbits are not violent.

Write the negation of each statement and say which of the pair is true and which is false.

B **27.** All integers are rational.

28. Some irrational numbers have terminating decimals.

29. No irrational numbers have repeating decimals.

30. All prime numbers have exactly two positive integral factors.

31. Some positive integers have an odd number of divisors.

32. No prime number is the sum of two prime numbers.

For each pair of sentences, state whether the sentences are contradictory. If the sentences are not contradictory, explain how they can both have the same truth value.

33. (a) The counting number n is even.
(b) The counting number n is divisible by 3.

34. (a) The integer x is positive.
(b) The integer x is negative.

35. (a) The counting number p is prime.
(b) The counting number p is a multiple of 5.

36. (a) The number x is rational.
(b) The number x is irrational.

37. (a) $\angle A$ is acute.
(b) $\angle A$ is obtuse.

38. (a) Planes P and Q are parallel.
(b) Planes P and Q intersect.

C **39.** Complete the following to get the Law of Extended Negation:
a. A statement p preceded by an __?__ number of $\sim$'s is equivalent to p.
b. A statement p preceded by an __?__ number of $\sim$'s is equivalent to $\sim p$.

Section 2-2 CONJUNCTION AND DISJUNCTION

As you may recall from your previous studies, two or more sentences may be combined to form a compound sentence whose truth value depends upon the truth or falsity of its components. In this section you will consider:

- the conjunction of sentences
- the disjunction of sentences
- DeMorgan's Laws

The *conjunction* of two sentences p and q is the compound sentence p *and* q. Here are four examples of conjunctions with varying truth values of the two components. What truth value would you assign to each?

1. Seven is prime *and* seven is odd.
2. Two is prime *and* two is odd.
3. Nine is prime *and* nine is odd.
4. Eight is prime *and* eight is odd.

As you may remember, the conjunction of two statements is true only when both components are true. Otherwise it is false. Using the logical symbol $\wedge$ for *and*, you have the following:

TRUTH TABLE FOR CONJUNCTION

p	q	$p \wedge q$
T	T	T
T	F	F
F	T	F
F	F	F

Do you see that statements 1-4, above, correspond to the four rows in the truth table? Of these statements, only statement 1 is true because it is the only statement whose clauses are *both* true.

The *disjunction* of two sentences p and q is the compound sentence p *or* q. Here are the same four examples with *and* replaced by *or*. What truth value would you assign to each?

5. Seven is prime *or* seven is odd.
6. Two is prime *or* two is odd.
7. Nine is prime *or* nine is odd.
8. Eight is prime *or* eight is odd.

The disjunction of two statements is false only when both components are false. Otherwise it is true. Statements 5, 6, and 7 are true because each contains at least one true clause. Using the logical symbol $\vee$ for *or*, you have the following:

TRUTH TABLE FOR DISJUNCTION

p	q	$p \vee q$
T	T	T
T	F	T
F	T	T
F	F	F

It is easy to confuse the logical symbols for *and* and *or* unless you happen to know that the Latin word for *or* is *vel.* One trick you may find helpful is Winston Churchill's famous slogan, "**V** f**OR** vict**OR**y." Perhaps "**V** f**OR**" will be enough to remind you what $\vee$ means. Another trick is to remember that $\wedge$ looks like the **A** in "**A**nd." Note also that two symbols that are very closely related to $\wedge$ and $\vee$ are $\cap$ and $\cup$. Recall the definitions of intersection and union.

$$x \in A \cap B \text{ is equivalent to } x \in A \wedge x \in B.$$
$$x \in A \cup B \text{ is equivalent to } x \in A \vee x \in B.$$

How do you negate conjunctions or disjunctions? Examine their truth tables closely.

$(p \wedge q)$ is false when p is false *or* q is false.

$(p \vee q)$ is false when p is false *and* q is false.

Thus, the negation of $(p \wedge q)$ is $(\sim p \vee \sim q)$ and the negation of $(p \vee q)$ is $(\sim p \wedge \sim q)$. Informally speaking, to negate a symbolic conjunction or disjunction, simply turn the symbol over and negate both parts.

These rules for the negation of a conjunction and of a disjunction are known as *DeMorgan's Laws.*

DeMorgan's Laws

$\sim(p \wedge q)$ is equivalent to $(\sim p \vee \sim q)$.

$\sim(p \vee q)$ is equivalent to $(\sim p \wedge \sim q)$.

The following example illustrates how to apply DeMorgan's Laws to a sentence written in symbolic form.

EXAMPLE Use symbols to write a statement equivalent to

$$\sim(p \vee \sim q).$$

SOLUTION By DeMorgan's Laws,

$$\sim(p \vee \sim q) \text{ is equivalent to } \sim p \wedge \sim(\sim q).$$

By the Law of Double Negation,

$$\sim p \wedge \sim(\sim q) \text{ is equivalent to } \sim p \wedge q.$$

Therefore, $\sim(p \vee \sim q)$ is equivalent to $\sim p \wedge q$.

The examples below show how to use DeMorgan's Laws to negate conjunctions and disjunctions written in words.

EXAMPLE Negate: Ducks moo and cows quack. (F)

SOLUTION Ducks don't moo or cows don't quack. (T)

EXAMPLE Negate: Tomatoes are not blue or coal is black. (T)

SOLUTION Tomatoes are blue and coal is not black. (F)

Sometimes a sentence that appears to say only one thing actually says two things and must be treated as a conjunction.

EXAMPLE Negate: Five is an even prime.

SOLUTION Rewrite the sentence as a conjunction and then write its negation.

Five is an even prime

is equivalent to

Five is even and five is prime.

Therefore, the negation is

Five is not even or five is not prime.

ORAL EXERCISES

Given statements p and q as follows, convert each compound statement to words and state its truth value.

p: Some rectangles are squares. q: No prime number is even.

1. $p \vee q$
2. $p \wedge q$
3. $p \vee \sim q$
4. $\sim p \wedge q$
5. $\sim p \vee q$
6. $p \wedge \sim q$
7. $\sim p \vee \sim q$
8. $\sim p \wedge \sim q$

Given statements p and q as follows, state the truth value of the compound sentence for the given replacement value of x.

p: $x < 5$ q: $x > 2$

9. $p \wedge q; x = 3$
10. $p \vee q; x = 6$
11. $p \wedge q; x = 5$
12. $\sim p \wedge q; x = 5$

WRITTEN EXERCISES

In Exercises 1-6, match each symbolic sentence with its negation.

A

1. $\sim r \wedge s$
2. $r \vee s$
3. $\sim r \wedge \sim s$
4. $\sim(r \wedge s)$
5. $r \wedge \sim s$
6. $\sim r \vee s$

a. $r \wedge s$
b. $r \wedge \sim s$
c. $\sim r \wedge \sim s$
d. $r \vee \sim s$
e. $r \vee s$
f. $\sim r \vee s$
g. $\sim r \vee \sim s$

Negate each sentence.

7. Harry is hairy and Leroy is not short.
8. Mary is not merry or Flora is happy.
9. Leo or Mimi plays football.
10. Donald speaks French and Beth speaks German.
11. Snow is white or the Empire State Building is tall.
12. $\triangle ABC$ is an isosceles right triangle.

Given statements r, s, and t as follows, **(a)** translate each statement into symbolic notation and **(b)** give its truth value.

r: $\frac{2}{3}$ is rational.

s: $20 - 7 > 27$

t: A triangle has 4 sides.

13. A triangle has 4 sides and $20 - 7 \leqslant 27$.

14. $20 - 7 > 27$ or $\frac{2}{3}$ is irrational.

15. A triangle does not have 4 sides and $\frac{2}{3}$ is rational.

16. $20 - 7 \leqslant 27$ and a triangle does not have 4 sides.

17. $\frac{2}{3}$ is irrational or $20 - 7 \leqslant 27$.

18. A triangle has 4 sides or $20 - 7 > 27$.

Given statements p, q, and r as follows, translate each symbolic sentence into good English.

p: It is March.

q: The wind is blowing.

r: The sky is clear.

19. $p \wedge q$ **20.** $\sim r \wedge p$ **21.** $q \vee r$ **22.** $\sim p \vee q$

B **23.** $\sim(q \wedge r)$ **24.** $\sim q \wedge \sim r$ **25.** $\sim(p \vee \sim q)$ **26.** $(p \wedge \sim q) \vee r$

Given statements p and q as follows, **(a)** translate each statement into symbolic notation and **(b)** give its truth value.

p: Six is prime. q: Six is even.

27. Six is prime and six is even.

28. It is not true that six is prime and even.

29. Six is prime or six is even.

30. Six is not prime or not even.

31. It is false that six is prime or even.

32. Six is not prime and six is even.

33. Six is prime or six is not even.

34. Six is an even prime.

35-42. Write in words the negation of each statement in Exercises 27-34.

C **43.** Construct a truth table to verify DeMorgan's Laws. (Show the equivalency of $\sim(p \wedge q)$ and $\sim p \vee \sim q$; $\sim(p \vee q)$ and $\sim p \wedge \sim q$.)

Suppose that the truth values of p, q, r, and s are T, F, T, and F, respectively. State the truth value of each compound sentence.

44. $(p \vee q) \wedge r$ **45.** $(p \wedge q) \vee (r \wedge s)$ **46.** $(p \vee \sim s) \wedge (q \vee r)$

47. $\sim(q \wedge r) \wedge (p \vee \sim s)$ **48.** $\sim((p \wedge \sim q) \wedge s)$ **49.** $\sim(\sim(r \vee \sim s) \wedge p)$

Section 2-3 CONDITIONALS

Zeno's paradox, which we discussed in the introduction to this chapter, asserts that given the *condition* that Achilles gives the tortoise a head start, Achilles will never be able to overtake the tortoise and so must lose the race. In your study of mathematics you will often encounter conditional relationships between statements. In this section you will consider the truth and falsity of statements joined by:

- If . . . then . . .
- only if
- if and only if

You will also learn the difference between *necessary* and *sufficient* conditions.

We don't believe Zeno, of course, or at least we don't want to. We believe that if Achilles gives the tortoise a head start, Achilles can still go on to win the race. We believe this sentence is true even if the first part is false. That is, we believe that even if Achilles doesn't give the tortoise a head start, Achilles will still win the race. On the other hand, if both parts of the sentence are false (suppose the organizing committee calls off the race), then Achilles will not win. However, in none of the preceding cases does the truth of the following compound statement change:

If Achilles gives the tortoise a head start, then Achilles wins the race.

The only way the statement above can be false is for the first part to be true and the second part false. That is, Achilles gives the tortoise a head start and Achilles does not win.

Any statement that can be written in the form

If p, then q

is called a *conditional.* The logical abbreviation for this is $p \rightarrow q$. Thus,

If Achilles gives the tortoise a head start, then Achilles wins the race

is a *conditional,* or *implication,* where

p represents: Achilles gives the tortoise a head start

and

q represents: Achilles wins the race.

It should be clear from the preceding discussion that the only time the conditional $p \rightarrow q$ is *false* is when p is true and q is false. Thus we have the following:

TRUTH TABLE FOR THE CONDITIONAL

p	q	$p \rightarrow q$
T	T	T
T	F	F
F	T	T
F	F	T

In the conditional statement $p \rightarrow q$, the first component, p, is called the *hypothesis* (or *antecedent*), and the second component, q, is called the *conclusion* (or *consequent*).

Here is a list of the most common ways in which $p \rightarrow q$ is translated:

If p, then q.
p implies q.
p only if q.
q if p.
p is sufficient for q.
q is necessary for p.

There are many ways in English to express the relationship symbolized in logic by $p \rightarrow q$.

EXAMPLES All of the following mean the same thing as $\underbrace{\text{If it rains}}_{p}$, *then* $\underbrace{\text{I stay home}}_{q}$.

$\underbrace{\text{It rains}}_{p}$ *only if* $\underbrace{\text{I stay home}}_{q}$.

$\underbrace{\text{I stay home}}_{q}$ *if* $\underbrace{\text{it rains}}_{p}$.

$\underbrace{\text{Rain}}_{p}$ *is a sufficient condition for* $\underbrace{\text{my staying home}}_{q}$.

$\underbrace{\text{My staying home}}_{q}$ *is a necessary condition for* $\underbrace{\text{rain}}_{p}$.

You may have noticed that in the truth table for the conditional and the truth table for disjunction the right-hand columns have three T's and only one F.

CONDITIONAL

p	q	$p \rightarrow q$
T	T	T
T	F	F
F	T	T
F	F	T

DISJUNCTION

p	q	$p \vee q$
T	T	T
T	F	T
F	T	T
F	F	F

This suggests that $p \rightarrow q$ is equivalent to some disjunction. Let's examine the truth table for the disjunction $\sim p \vee q$.

p	$\sim p$	q	$\sim p \vee q$	$p \rightarrow q$
T	F	T	T	T
T	F	F	F	F
F	T	T	T	T
F	T	F	T	T

Since $\sim p \vee q$ and $p \rightarrow q$ have the same truth values, they are logically equivalent.

EXAMPLE Rewrite the following conditional as an equivalent disjunction:

If this quadrilateral is a square, then it is a rectangle.

SOLUTION This quadrilateral is not a square or it is a rectangle.

The equivalence of $p \to q$ and $\sim p \vee q$ provides a means other than "It is not true that p implies q" of expressing the negation of a conditional statement. Since $p \to q$ is equivalent to $\sim p \vee q$, their negations must be equivalent. This means that

$$\sim(p \to q) \text{ is equivalent to } \sim(\sim p \vee q).$$

Using DeMorgan's Laws and the Law of Double Negation, you have

$$\sim(p \to q) \text{ is equivalent to } p \wedge \sim q.$$

EXAMPLE Negate the following sentence: If I save my money, then I can buy a car.

SOLUTION Let p represent "I save my money" and let q represent "I can buy a car." Recall that the negation of $p \to q$ is $p \wedge \sim q$ or

I save my money and I cannot buy a car.

Any statement that is written in the form p if and only if q (sometimes abbreviated p iff q) is called a *biconditional* and has the same meaning as

(p implies q) and (q implies p)

or p is a necessary and sufficient condition for q.

We abbreviate the conjunction $(p \to q) \wedge (q \to p)$ by $p \leftrightarrow q$.

From what you have learned about the truth tables for the implication and for conjunction, you can easily derive the following:

TRUTH TABLE FOR THE BICONDITIONAL

p	q	$p \leftrightarrow q$
T	T	T
T	F	F
F	T	F
F	F	T

EXAMPLES OF BICONDITIONALS

That $\angle A$ is a right angle (p) *is a necessary and sufficient condition for* it not to be acute (q). (F)

$x = 2$ (p) *iff* $2x + 3 = 7$ (q). (T)

$\triangle ABC$ is an isosceles triangle (p) *if and only if* it has two congruent sides (q). (T)

The last biconditional is the *definition* of an isosceles triangle. All definitions are biconditionals.

ORAL EXERCISES

Identify the hypothesis and the conclusion in each sentence.

1. If an integer is divisible by 6, then it is divisible by 2.
2. If m and n are both integers, then $m + n$ is an integer.
3. x^2 is irrational implies that x is irrational.
4. $x = y$ implies that $ax = ay$.

5-8. Read each of the conditionals given in Exercises 1-4 in four ways using
 a. only if **b.** if **c.** is sufficient for **d.** is necessary for

WRITTEN EXERCISES

Rewrite each of the following conditionals in *If. . . then . . .* form.

A 1. Two triangles have the same area if they are congruent.

2. For a quadrilateral to be a trapezoid, it is necessary that it have at least one pair of parallel sides.
3. x is rational if x is an integer.
4. $x^2 - 25 = 0$ if $x = 5$
5. It snows only if it is cold.
6. "Two lines are parallel" implies that they are coplanar.
7. Wagging his tail implies that Fido is happy.
8. For an animal to be a horse it is necessary that it have four legs.
9. For two lines to be coplanar it is sufficient that they intersect.
10. A quadrilateral has congruent diagonals only if it is a rectangle.

Write the negation of each conditional.

11. If you turn down the thermostat, you use less heating oil.
12. If it snows, then the game is cancelled.
13. I can go to the movies if I do my homework.
14. The phone rings only if I take a bath.
15. That fish can fly is a sufficient condition for birds to swim.
16. That mares eat oats is a necessary condition for little lambs to eat ivy.

Let p represent "It is snowing today," q represent "It is cold today," and r represent "It is going to rain tomorrow." Translate each of the following into English.

17. $p \rightarrow q$
18. $q \rightarrow r$
19. $p \leftrightarrow r$
20. $(p \wedge q) \rightarrow r$
21. $\sim q \rightarrow \sim p$
22. $\sim r \leftrightarrow \sim q$
23. $(q \wedge \sim p) \rightarrow r$
24. $(\sim r \vee q) \rightarrow p$

In Exercises 25-32: **(a)** Use each pair of sentences to write a biconditional sentence containing the words "if and only if." **(b)** State the truth value of the biconditional.

B 25. p: $y > 3$
 q: $3 < y$

26. p: $1 = 2$
 q: $3 = 6$

27. p: Two rectangles have equal areas.
 q: Two rectangles have equal perimeters.

28. p: a is even.
 q: $a + 1$ is odd.

29. p: x is even.
 q: $2x$ is even.

30. p: y^2 is an odd integer.
 q: y is an odd integer.

31. p: Line l is parallel to line m.
q: Line m is parallel to line l.

32. p: $\triangle ABC$ is a right triangle.
q: $\angle A$ is a right angle.

Rewrite each (false) conditional as an equivalent disjunction.

33. If 4 is divisible by 2, then 4 is divisible by 6.

34. If the sum of 3.5 and 2.5 is an integer, then 3.5 and 2.5 are both integers.

35. If $\sqrt{2}$ is irrational, then $(\sqrt{2})^2$ is irrational.

36. If $0 \cdot 3 = 0 \cdot 4$, then $3 = 4$.

37–40. Write the negation of each conditional in Exercises 33–36 as a (true) conjunction.

Suppose that the truth values of $p, q, r,$ and s are T, T, F, and F, respectively. State the truth value of each compound sentence.

C 41. $(p \vee \sim q) \rightarrow r$

42. $(\sim r \wedge q) \leftrightarrow s$

43. $(p \rightarrow q) \leftrightarrow (r \rightarrow s)$

44. $(p \rightarrow r) \vee (q \leftrightarrow s)$

45. $(s \wedge \sim p) \leftrightarrow (q \rightarrow \sim r)$

46. $(q \vee r) \rightarrow (p \leftrightarrow \sim s)$

SELF-TEST 1

Rewrite each statement as an equivalent statement beginning with *All, Some,* or *No.*

Section 2-1

1. An integer is never irrational.
2. A rhombus is sometimes a square.
3. A square is always a rhombus.

Negate each statement.

4. All prime numbers are even.
5. Some integers are positive.
6. No rectangles are triangles.

Given statements p and q as follows, **(a)** translate each statement into symbolic notation and **(b)** give its truth value.

p: Seven is prime. q: Seven is even.

Section 2-2

7. Seven is prime or not even.
8. Seven is an odd prime.
9. Seven is not prime and not even.

Negate each symbolic sentence.

10. $r \wedge s$

11. $r \vee \sim s$

12. $\sim s \wedge \sim r$

Rewrite each conditional in *If . . . then . . .* form.

Section 2-3

13. A cat is a canary only if it has wings.

14. $x > 5$ is a sufficient condition for x to be positive.

15. Negate the conditional: If $x = 3$, then $3x - 2 = 5$.

16. Use the given statements to write a biconditional.
The number is divisible by three.
The sum of the digits of the number is divisible by three.

Section 2-4 OPEN SENTENCES AND QUANTIFIERS

You should already be familiar with open sentences from algebra and from your study of logic in *Book 1.* Following a review of the concepts associated with open sentences, you will learn how they are related to the logic of *All, Some,* and *No* statements by means of:

- the universal quantifier
- the existential quantifier

An *open sentence* is a sentence that contains a *variable.* The variable may be a symbol such as x or y, or it may be a pronoun such as "it," "she," or "they."

EXAMPLES OF OPEN SENTENCES

1. *She* got 100 on the test.
2. x is a prime number.
3. The square root of y is rational.
4. The reciprocal of z is positive.

The *domain,* or *replacement set,* of a variable is a set whose members can be used as replacements for the variable. The only replacements allowed are those that make the sentence a statement, that is, that make the sentence true or false.

EXAMPLES The following are possible domains for the variables in the open sentences of the examples above:

1. All females who took the test
2. All counting numbers
3. All nonnegative real numbers (Why is the restriction "nonnegative" necessary?)
4. All real numbers except zero (Why is the restriction "except zero" necessary?)

Finally, recall that the *solution set* of an open sentence consists of all replacements from the domain that make the sentence true, and that two open sentences whose solution sets are the same are said to be *equivalent.* (You may assume that the domain of a variable is the set of real numbers unless otherwise specified.)

Now consider the following three open sentences:

1. $x^2 - 9 = (x + 3)(x - 3)$
2. $x^2 - 9 = 0$
3. $x^2 + 9 = 0$

If all three have the same domain, the set of real numbers, the first is *always* true, the second is *sometimes* true, and the third is *never* true. We can use quantifiers in these open sentences to obtain the following true statements:

1. For *all* x, $x^2 - 9 = (x + 3)(x - 3)$.
2. For *some* x, $x^2 - 9 = 0$.
3. For *no* x does $x^2 + 9 = 0$.

Statement 3, however, can be rewritten as 3. For *all* x, $x^2 + 9 \neq 0$.

Thus, you can change an open sentence into a statement by asserting either that it (or its negation) is true for *all* values of the variable or that it is true for *some* value of the variable—that is, there *exists* at least one replacement for the variable that makes the open sentence true. The key words here are *All* and *Exist,* and their initial letters are what logicians have adapted as symbols for

the *Universal Quantifier:* $\forall$ (Read as "For all," "For every," "For each.")
the *Existential Quantifier:* $\exists$ (Read as "There exists" or "For some.")

Using these symbols, you can rewrite the three statements as follows:

1. $\forall_x\ x^2 - 9 = (x + 3)(x - 3)$ 2. $\exists_x\ x^2 - 9 = 0$ 3. $\forall_x\ x^2 + 9 \neq 0$

Just as we let p and q represent sentences, we will let $P(x)$ and $Q(x)$ represent open sentences with the variable x.

EXAMPLE Let $P(x)$ represent "$x^2 \geqslant 0$" and $Q(x)$ represent "$2x = x + 1$." Translate each symbolic sentence into English and state its truth value.

1. $P(2)$ 2. $Q(3)$ 3. $\forall_x\ P(x)$ 4. $\exists_x\ P(x)$ 5. $\forall_x\ Q(x)$ 6. $\exists_x\ Q(x)$

SOLUTION 1. $2^2 \geqslant 0$. True 2. $2 \cdot 3 = 3 + 1$. False

3. The square of every real number is nonnegative. True
4. The square of some real number is nonnegative. True
5. For all real numbers x, $2x = x + 1$. False
6. There exists a real number x such that $2x = x + 1$. True

EXAMPLE Let $Z(x)$ represent "x is an integer" and $Q(x)$ represent "x is rational." Translate each English sentence into a symbolic sentence.

1. All integers are rational numbers.
2. Some rational numbers are integers.

SOLUTION 1. You can translate the sentence "All integers are rational numbers" as "For every x, if x is an integer, then x is a rational number." This can be represented symbolically as

$$\forall_x\ \big(Z(x) \rightarrow Q(x)\big).$$

2. The sentence "Some rational numbers are integers" may be translated as "There exists an x such that x is rational and x is an integer." This can be represented symbolically as

$$\exists_x\ \big(Q(x) \wedge Z(x)\big).$$

Perhaps you have already noticed that for the sets of numbers Z and Q,

$$\forall_x\ \big(Z(x) \rightarrow Q(x)\big) \text{ is equivalent to } Z \subset Q.$$

Similarly,

$$\exists_x\ \big(Q(x) \wedge Z(x)\big) \text{ means that } Q \cap Z \neq \emptyset.$$

A common error is to translate a sentence like

"Some rational numbers are integers" (1)

as $\exists_x\ \big(Q(x) \rightarrow Z(x)\big)$, which is equivalent to $\exists_x\ \big(\sim Q(x) \vee Z(x)\big)$. The preceding statement is translated as

"There exists a number that is not a rational number or that is an integer." (2)

Clearly sentences (1) and (2) do not mean the same thing. To show that the first sentence is true, you *must* be able to name a number that is rational and an integer. To show that the second sentence is true, you could simply name a number that is not rational.

How do you negate quantified sentences in symbolic form? Follow the same principles you learned about negations in Section 2-1. Study the examples below where $P(x)$ represents $x^2 \geqslant 0$ and $Q(x)$ represents $2x = x + 1$.

	English	*Symbols*
Sentence:	For all x, $x^2 \geqslant 0$.	$\forall_x\ P(x)$
Negation:	For some x, $x^2 < 0$.	$\exists_x \sim P(x)$
Sentence:	There exists an x such that $2x = x + 1$.	$\exists_x\ Q(x)$
Negation:	For all x, $2x \neq x + 1$.	$\forall_x \sim Q(x)$

In general, the negation of a universally quantified sentence is an existentially quantified sentence, and the negation of an existentially quantified sentence is a universally quantified sentence.

EXAMPLE Write the negation of $\forall_x\ x < 3$.

SOLUTION $\exists_x\ x \not< 3$ or $\exists_x\ x \geqslant 3$

EXAMPLE Write the negation of $\exists_x\ x \cdot 0 \geqslant 1$.

SOLUTION $\forall_x\ x \cdot 0 < 1$

Study the following examples carefully. Notice how the placement of the negation symbol, $\sim$, affects the meanings of the sentences.

EXAMPLE Let $P(x)$ represent "x is an integer." Translate each of the following into English.

1. $\sim\forall_x\ P(x)$ 2. $\forall_x \sim P(x)$ 3. $\sim\exists_x\ P(x)$ 4. $\exists_x \sim P(x)$

SOLUTION
1. Not all x are integers.
2. All x are not integers.
3. There does not exist an x that is an integer.
4. Some x are not integers.

Notice that sentences (1) and (4) are equivalent, as are (2) and (3).

ORAL EXERCISES

State a possible domain for each open sentence.

1. $x^2 > 0$ **2.** n^2 is even. **3.** It barks.

4. $2k + 1$ is odd. **5.** $\frac{1}{z} = z$ **6.** It is a Great Lake.

State whether you would use the universal quantifier, $\forall$, or the existential quantifier, $\exists$, in rewriting each sentence in symbolic form.

7. Some rivers flow into the Atlantic Ocean.

8. All residents of Long Island are residents of New York State.

9. Some residents of New York State live in Lake Placid.

10. No residents of New York City live in Houston.

11. Every right angle has measure 90.

12. No right triangle is obtuse.

13. There is at least one person who lives in Connecticut and works in New York City.

14. There exists a state in New England whose name begins with M.

15. Even numbers are divisible by two.

Read each statement aloud and state its truth value.

16. $\forall_x\ x + x = 2x$

17. $\forall_y\ 3y > y$

18. $\exists_t\ \sqrt{t} = 2$

19. $\exists_x\ (x \in Z \wedge 2x = 1)$

20. $\forall_x\ (x \in Z \rightarrow 2x \in Z)$

21. $\exists_x\ (x \in Z \rightarrow 2x = 1)$

WRITTEN EXERCISES

Given $P(x)$, $Q(x)$, and $R(x)$ as follows, write each statement in words and state its truth value.

$P(x)$: $\frac{1}{x} > 0 \quad (x \neq 0)$ $\qquad$ $Q(x)$: $x \leqslant 0$ $\qquad$ $R(x)$: $x > 0$

A 1. $\forall_x\ P(x)$

2. $\exists_x\ P(x)$

3. $\forall_x\ Q(x)$

4. $\exists_x\ \sim P(x)$

5. $\exists_x\ \sim R(x)$

6. $\forall_x\ \sim R(x)$

7. $\forall_x (Q(x) \vee R(x))$

8. $\exists_x (P(x) \wedge R(x))$

9. $\exists_x (Q(x) \wedge P(x))$

Write the negation of each sentence and give the truth value of the negation.

10. $\forall_x\ x \leqslant -1$

11. $\exists_x\ 2x = x + 1$

12. $\exists_x\ x \cdot 0 = 1$

13. $\forall_x\ x + 0 = x$

14. $\exists_y\ y \cdot 0 = y$

15. $\forall_y\ y^2 \neq y$

16. $\exists_y\ y^2 < y$

17. $\forall_y\ 3y + 2 = 7$

18. $\forall_y \left(y \neq 0 \rightarrow \frac{1}{y} < y\right)$

In Exercises 19-24: **(a)** Use a variable and either the universal quantifier or the existential quantifier to rewrite each statement in symbolic form; **(b)** give the truth value of the statement.

B 19. There exists a number whose square is less than itself.

20. No real number has a negative square.

21. Every number is greater than three times itself.

22. Some number is two less than twice itself.

23. All numbers are greater than their opposites.

24. At least one number is less than -2.

25-30. Use a variable and a quantifier to write the negation of each sentence in Exercises 19-24.

In Exercises 31-33, the domain of the open sentences given below is the set of all angles.

$P(t)$: t is acute $\qquad$ $Q(t)$: t is obtuse

(a) Write each statement entirely in symbols; **(b)** state its truth value.

C **31.** Every angle is either acute or obtuse.

32. No angle is both acute and obtuse.

33. Some angle is not acute and not obtuse.

Let $P(x)$ represent "x is a multiple of 4" and let $Q(x)$ represent "x is a multiple of 2." Translate each symbolic sentence into an English sentence and state its truth value.

34. $\forall_x(P(x) \rightarrow Q(x))$

35. $\exists_x(P(x) \wedge \sim Q(x))$

36. $\exists_x(Q(x) \wedge \sim P(x))$

37. $\forall_x(Q(x) \rightarrow P(x))$

38. $\forall_x(P(x) \vee \sim Q(x))$

39. $\exists_x(P(x) \rightarrow Q(x))$

Negate each sentence and state which is true, the original sentence or its negation.

40. All cowboys wear spurs or are bowlegged.

41. Some cats have long hair and some do not.

42. All elephants are green or some sheep are black.

43. Some integers are prime and no triangles are acute.

Section 2-5 RELATED CONDITIONALS

This section is concerned with three conditionals related to $p \rightarrow q$:

- the converse
- the inverse
- the contrapositive

You will see how these four conditionals are related, in pairs, by the *Law of the Contrapositive.*

EXAMPLE 1

Statement:	If $x = 3$, then $x^2 = 9$.	(T)
Converse:	If $x^2 = 9$, then $x = 3$.	(F)
Inverse:	If $x \neq 3$, then $x^2 \neq 9$.	(F)
Contrapositive:	If $x^2 \neq 9$, then $x \neq 3$.	(T)

From the example above, you can see that even when $p \rightarrow q$ is true, its *converse,* $q \rightarrow p$, may not be true. What can be said of the *inverse,* $\sim p \rightarrow \sim q$, and the *contrapositive* (the converse of the inverse), $\sim q \rightarrow \sim p$? Try to draw your own conclusions from the following examples.

EXAMPLE 2

Statement:	If the weather is cloudy, then it is raining.	(F)
Converse:	If it is raining, then the weather is cloudy.	(T)
Inverse:	If the weather is not cloudy, then it is not raining.	(T)
Contrapositive:	If it is not raining, then the weather is not cloudy.	(F)

EXAMPLE 3

Statement:	Fish love onions.	(F)
Converse:	Those who love onions are fish.	(F)
Inverse:	Those who are not fish do not love onions.	(F)
Contrapositive:	Those who do not love onions are not fish.	(F)

EXAMPLE 4

Statement: A triangle with two congruent sides has two congruent angles. (T)

Converse: A triangle with two congruent angles has two congruent sides. (T)

Inverse: If a triangle does not have two congruent sides, then it does not have two congruent angles. (T)

Contrapositive: If a triangle does not have two congruent angles, then it does not have two congruent sides. (T)

The conclusion that you should have reached from the examples is that a conditional and its contrapositive always have the same truth value. You probably noticed that the inverse is the contrapositive of the converse, so these statements are also covered by the following law:

The Law of the Contrapositive

A conditional and its contrapositive are logically equivalent.

To verify this law, you can construct a truth table.

p	q	$\sim p$	$\sim q$	$p \to q$	$\sim q \to \sim p$	$q \to p$	$\sim p \to \sim q$
T	T	F	F	T	T	T	T
T	F	F	T	F	F	T	T
F	T	T	F	T	T	F	F
F	F	T	T	T	T	T	T

This shows that $p \to q$ is equivalent to $\sim q \to \sim p$ and that $q \to p$ is equivalent to $\sim p \to \sim q$.

ORAL EXERCISES

Match each conditional from the first column with an equivalent conditional from the second column.

1. $p \to q$

2. $\sim q \to p$

3. $p \to \sim q$

4. $\sim p \to \sim q$

a. $q \to p$

b. $q \to \sim p$

c. $\sim q \to \sim p$

d. $\sim p \to q$

State the **(a)** converse, **(b)** inverse, and **(c)** contrapositive of each conditional and compare their truth values.

5. If a number is divisible by 6, then it is divisible by 5.

6. If $1 = 2$, then $3 = 6$.

7. If x is a prime number greater than 2, then x is odd.

8. If two triangles have the same area, then they are congruent.

WRITTEN EXERCISES

Write the **(a)** converse, **(b)** inverse, and **(c)** contrapositive of each of the following conditionals in symbolic form.

A 1. $p \rightarrow q$ 2. $q \rightarrow p$ 3. $q \rightarrow \sim p$ 4. $\sim p \rightarrow q$ 5. $\sim q \rightarrow \sim p$ 6. $\sim p \rightarrow \sim q$

Write the **(a)** converse, **(b)** inverse, and **(c)** contrapositive of each statement and compare their truth values.

7. If $a = 0$, then $ab = 0$.
8. If $x < 0$, then $x^2 < 0$.
9. y is an integer if y is rational.
10. $\triangle ABC$ is isosceles if $AB = BC$.
11. If $x \neq 2$, then $x + 3 \neq 5$.
12. If an integer is not divisible by 2, then it is not divisible by 4.
13. $x = y$ implies $a + x = a + y$.
14. $x = y$ implies $ax = ay$.

B 15. $x \neq 3$ only if $x > 3$.

16. Two lines are parallel only if they do not intersect.
17. For n^2 to be even, it is sufficient for n to be even.
18. For $m + 2$ to be odd, it is necessary that m be odd.
19. Even numbers are divisible by 2.
20. That $x + 1$ is even implies that x is even.
21. Being born in New York is a sufficient condition for being a citizen of the United States.
22. A triangle is not isosceles only if it is not equilateral.
23. If $a = 0$ or $b = 0$, then $ab = 0$.
24. $1 = 2$ or $3 \neq 6$.
25. If the product of two numbers is divisible by 6, then at least one of the numbers must be divisible by 6.
26. $ab = a$ only if $a = 0$ or $b = 1$.
27. No irrational numbers have terminating decimals.
28. If either of two numbers is divisible by 3, then their product is divisible by 3.
29. Paris is the capital of England or London is the capital of France.

In a *compound conditional,* either the hypothesis or the conclusion (or both) is a conjunction. A *partial converse* of a compound conditional is a statement obtained by interchanging one part of the hypothesis with one part of the conclusion. The most commonly seen compound conditionals have the form:

$$(p \wedge q) \rightarrow r$$

Such a conditional has two partial converses:

$$(p \wedge r) \rightarrow q \qquad (r \wedge q) \rightarrow p$$

Write both partial converses for each of the following compound conditionals and indicate their truth values.

C 30. If p is odd and prime, then p is greater than 2.

31. If $ab = 0$ and $a \neq 0$, then $b = 0$.
32. If a is positive and b is negative, then ab is negative.

Section 2-6 LAWS OF INFERENCE

In a valid argument you use statements called *premises* to deduce other statements called *conclusions.* In this section you will study several well-known forms of valid argument, as well as some related fallacies that Sophists, both ancient and modern, use to dupe the unwary.

One valid form of argument is the *Law of Detachment,* which states that if you can assert both $p \rightarrow q$ and p, then you can infer q. The form of the argument is this:

Premise 1:	$p \rightarrow q$
Premise 2:	p
Conclusion C:	q

EXAMPLE P_1: If this quadrilateral is a square, then it is a rectangle.
P_2: This quadrilateral is a square.
C: This quadrilateral is a rectangle.

The Law of Detachment applied to the contrapositive, $\sim q \rightarrow \sim p$, of the conditional $p \rightarrow q$ yields:

Premise 1:	$\sim q \rightarrow \sim p$
Premise 2:	$\sim q$
Conclusion C:	$\sim p$

Since $\sim q \rightarrow \sim p$ is equivalent to $p \rightarrow q$, this leads us to a rule of inference called the *Law of Contrapositive Inference.* This is the form of the argument:

Premise 1:	$p \rightarrow q$
Premise 2:	$\sim q$
Conclusion C:	$\sim p$

EXAMPLE P_1: If this quadrilateral is a square, then it is a rectangle.
P_2: This quadrilateral is not a rectangle.
C: This quadrilateral is not a square.

If you apply the Law of Detachment to the conditional $\sim p \rightarrow q$, you will get the following:

Premise 1:	$\sim p \rightarrow q$
Premise 2:	$\sim p$
Conclusion C:	q

Since $\sim p \rightarrow q$ is equivalent to $p \vee q$, this leads to the *Law of Disjunctive Inference.* Here is the form of the argument:

Premise 1:	$p \vee q$
Premise 2:	$\sim p$
Conclusion C:	q

In other words, the Law of Disjunctive Inference states that you can deduce one of two alternatives if you are able to deny the other.

EXAMPLE P_1: 327 is divisible by 3 or by 7.
P_2: 327 is not divisible by 7.

C: 327 is divisible by 3.

These three laws can also be expressed as logical "formulas."

Law of Detachment: $[(p \rightarrow q) \wedge p] \rightarrow q$
Law of Contrapositive Inference: $[(p \rightarrow q) \wedge \sim q] \rightarrow \sim p$
Law of Disjunctive Inference: $[(p \vee q) \wedge \sim p] \rightarrow q$

If you construct truth tables for these laws, you will find that the final column in each case will contain only T's. Hence, these laws are examples of what logicians call *tautologies.* The laws of Double Negation, DeMorgan, and Contraposition that you studied earlier are also examples of tautologies. A *tautology* is any symbolic statement that is true regardless of the truth or falsity of its component statements.

EXAMPLE Show that $(p \wedge q) \rightarrow p$, called the *Law of Conjunctive Simplification,* is a tautology.

SOLUTION Construct a truth table.

p	q	$p \wedge q$	$(p \wedge q) \rightarrow p$
T	T	T	T
T	F	F	T
F	T	F	T
F	F	F	T

Since $(p \wedge q) \rightarrow p$ is always true, regardless of the truth values of p and q, it is a tautology.

Each of the first three laws of inference considered above can be tampered with to produce what might sound like a tautology to those who haven't studied logic, but which you will no doubt spot as fallacies. Here they are displayed in argument form:

The Fallacy of the Converse
Premise 1: $p \rightarrow q$
Premise 2: q

Conclusion C: p

The Fallacy of the Inverse
Premise 1: $p \rightarrow q$
Premise 2: $\sim p$

Conclusion C: $\sim q$

The Fallacy of Disjunction

Premise 1:	$p \vee q$
Premise 2:	p
Conclusion C:	$\sim q$

Now you are getting to the heart of logic, the determination of whether or not an argument is valid.

EXAMPLE Is the following argument valid?

P_1: If Fred does not clean his room, then he can't go to the game.
P_2: Fred can go to the game.

C: Fred cleaned his room.

SOLUTION Write the argument in symbolic form. For this problem, let
c represent "clean his room" and
g represent "can go to the game."
Then the argument in symbolic form becomes

P_1: $\sim c \rightarrow \sim g$
P_2: g

C: c

This is the Law of Contrapositive Inference. Therefore, the argument is valid.

EXAMPLE Is the following argument valid?

P_1: If Mrs. C doesn't use Wax-It, her floors will have yellow waxy build-up.
P_2: Mrs. C's floors have yellow waxy build-up.

C: Mrs. C doesn't use Wax-It.

SOLUTION Let w represent "use Wax-It."
Let y represent "have yellow waxy build-up."
The argument in symbolic form is

P_1: $\sim w \rightarrow y$
P_2: y

C: $\sim w$

This is the Fallacy of the Converse. Therefore, the argument is invalid.

Note that the validity of an argument does not depend upon the truth of the statements it contains.

EXAMPLE P_1: If $3 = 2 - 1$, then $4 = 7 + 3$.
P_2: $3 = 2 - 1$

C: $4 = 7 + 3$

This is an example of the Law of Detachment. Even though the conclusion is a false statement, the argument is valid.

Now we will present one more rule of inference. Consider the following two examples.

P_1: If 156 is divisible by 2, then 156 is even.
P_2: 156 is divisible by 2.

C: 156 is even.

P_1: For all x, if x is divisible by 2, then x is even.
P_2: 156 is divisible by 2.

C: 156 is even.

The first example is an illustration of the Law of Detachment. The second example involves quantification. The statement "For all x, if x is divisible by 2, then x is even" is true for all substitutions of x. Thus we can substitute 156 for x and infer that 156 is even. This is an example of the law of inference called the *Principle of Substitution.*

Principle of Substitution

If an open sentence is true for all values of the variable in a given domain, then substituting a value from the given domain for the variable will produce a true statement.

EXAMPLE P_1: All men are mortal.
P_2: Socrates is a man.

C: Socrates is mortal.

ORAL EXERCISES

For each pair of premises state what conclusion, if any, can be drawn using the Law of Detachment.

1. P_1: $(p \wedge q) \rightarrow r$
P_2: $p \wedge q$

2. P_1: $\sim p$
P_2: $\sim p \rightarrow \sim s$

3. P_1: $a \rightarrow \sim b$
P_2: $\sim a$

For each pair of premises, state what conclusion, if any, can be drawn using the Law of Contrapositive Inference.

4. P_1: $\sim w$
P_2: $r \rightarrow w$

5. P_1: $\sim q \rightarrow \sim s$
P_2: $\sim s$

6. P_1: $\sim m \rightarrow n$
P_2: $\sim n$

For each pair of premises, state what conclusion, if any, can be drawn using the Law of Disjunctive Inference.

7. P_1: m
P_2: $\sim m \vee w$

8. P_1: $(a \vee b) \vee c$
P_2: $\sim c$

9. P_1: $\sim p$
P_2: $\sim p \vee \sim q$

For each invalid argument illustrated below, state the name of the fallacy involved.

10. $\triangle ABC$ is isosceles or equilateral. $\triangle ABC$ is isosceles. Therefore, $\triangle ABC$ is not equilateral.

11. If Tom doesn't use Bleacho, then he will have grime around the collar. Tom uses Bleacho. Therefore, Tom will not have grime around the collar.

12. If Darlene uses Bright 'n' Shiny shampoo, she will have bouncy hair. Darlene has bouncy hair. Therefore, Darlene uses Bright 'n' Shiny shampoo.

WRITTEN EXERCISES

If an argument is valid, identify which rule of inference is being used. If an argument is invalid, identify which fallacy is being committed.

A 1. P_1: All sophomores love logic.
P_2: Mimi is a sophomore.
C: Mimi loves logic.

2. P_1: I should use Odorguard or my friends will avoid me.
P_2: I don't want my friends to avoid me.
C: I should use Odorguard.

3. P_1: If I use Odorguard, then I won't offend my friends.
P_2: I don't want to offend my friends.
C: I had better use Odorguard.

4. P_1: 7 is prime or 6 is prime.
P_2: 7 is prime.
C: 6 is not prime.

5. P_1: If $\frac{2}{5}$ is rational, then $\frac{2}{5}$ has a nonterminating decimal.
P_2: $\frac{2}{5}$ has a terminating decimal.
C: $\frac{2}{5}$ is irrational.

6. P_1: If my cat, Leonardo, meows, then he is hungry.
P_2: Leonardo is not meowing.
C: Leonardo is not hungry.

7. a. Which of the arguments above show that a valid argument can reach a false conclusion?
   b. Which show that an invalid argument can reach a true conclusion?
   c. Does this mean logic can't be trusted? Explain.

Write the conclusion that follows logically from the premises and state which rule of inference is being used. If no conclusion follows, explain why.

8. P_1: All dees are mees.
   P_2: Booboo is a dee.

9. P_1: If $\sqrt{2}$ is a terminating or a repeating decimal, then $\sqrt{2}$ is rational.
   P_2: $\sqrt{2}$ is irrational.

10. P_1: Mark passes the exam.
    P_2: If Mark doesn't pass the exam, he gets grounded for a week.

11. P_1: Hamburgers are delicious if hot dogs are tasty.
    P_2: Hot dogs are tasty.

12. P_1: Mohammed will not go to the mountain.
    P_2: The mountain must go to Mohammed or Mohammed must go to the mountain.

13. P_1: If you don't want to suffer the heartache of terminal acne, lay off the sweets.
    P_2: I never touch sweets.

14. P_1: Flour is good for food.
    P_2: Oatmeal is a kind of flour.

15. P_1: Cindy is a member of the basketball team or the field hockey team.
    P_2: Cindy is a member of the field hockey team.

16. P_1: All that shines is silver.
    P_2: This bracelet does not shine.

17. P_1: If x^2 is not even, then x is not even.
    P_2: x is even.

18. P_1: No crocodile is both hungry and friendly.
    P_2: Ollie is a friendly crocodile.

B 19. Construct a truth table to show that $p \rightarrow (p \vee q)$, the *Law of Addition,* is a tautology.

20. Construct a truth table to show that the Law of Detachment is a tautology.

21. Construct a truth table to show that the Law of Contrapositive Inference is a tautology.

22. Construct a truth table to show that the Law of Disjunctive Inference is a tautology.

Complete the following arguments.

23. P_1: $(p \wedge q) \rightarrow (r \vee s)$
    P_2: $\sim(r \vee s)$

24. P_1: $(p \rightarrow q) \vee (r \rightarrow s)$
    P_2: $\sim(p \rightarrow q)$

25. P_1: $[\sim(p \vee q)] \rightarrow (r \wedge s)$
    P_2: $\sim p \wedge \sim q$

26. P_1: $p \rightarrow (q \vee r)$
    P_2: $\sim q \wedge \sim r$

27. P_1: $(p \vee q) \vee r$
P_2: $\sim p \wedge \sim q$

28. P_1: $p \rightarrow q$
P_2: $(\sim p \vee q) \rightarrow \sim r$

What conclusion(s) follow(s) logically from the given premises? Give your reasons.

EXAMPLE P_1: $p \rightarrow q$
P_2: $p \vee r$
P_3: $\sim r$

SOLUTION
P_2: $p \vee r$
P_3: $\sim r$
C_1: p — Law of Disjunctive Inference
P_1: $p \rightarrow q$
C_2: q — Law of Detachment

C 29. P_1: $(p \vee q) \vee r$
P_2: $\sim r$

30. P_1: $p \wedge (q \vee r)$
P_2: $\sim r$

31. P_1: $p \vee (q \wedge r)$
P_2: $\sim r$

32. P_1: $p \rightarrow q$
P_2: $p \vee r$
P_3: $\sim q$

33. P_1: $p \vee q$
P_2: $\sim p \vee r$
P_3: $\sim r$

34. P_1: $p \rightarrow q$
P_2: $p \rightarrow r$
P_3: $p \wedge s$

SELF-TEST 2

Write the negation of each statement and give the truth value of the negation.

Section 2-4

1. $\forall_x \; x < x^2$

2. $\exists_x \; \frac{1}{x} = -x$

Use a variable and either the universal quantifier or the existential quantifier to rewrite the sentence in symbolic form.

3. There exists a number that is two less than three times itself.
4. Every number is less than three times itself.

Section 2-5

5. Write the converse, inverse, and contrapositive of the following statement and give the truth value of each:

If the ground is wet, then it is raining.

Write the conclusion that follows logically from the premises.

Section 2-6

6. P_1: If Rob eats properly, then Rob will stay healthy.
P_2: Rob is ill.

7. P_1: It rains in Spain or it snows in Florida.
P_2: It doesn't rain in Spain.

Section 2-7 CHAINS OF REASONING

In this section you will learn about the *Law of the Syllogism* and how to use it along with the other rules of inference to construct valid arguments. You will also see where Zeno went wrong.

When the conclusion of one conditional ($p \longrightarrow q$) is the hypothesis of a second conditional ($q \longrightarrow r$), the *Law of the Syllogism* allows you to infer the conditional $p \longrightarrow r$. In other words, $[(p \longrightarrow q) \wedge (q \longrightarrow r)] \longrightarrow (p \longrightarrow r)$. In argument form it looks like this:

Premise 1: $p \longrightarrow q$
Premise 2: $q \longrightarrow r$

Conclusion C: $p \longrightarrow r$

EXAMPLE P_1: If magpies like pie, then coyotes drink tea.
P_2: If coyotes drink tea, then poodles recite poetry.

C: If magpies like pie, then poodles recite poetry.

Notice that even though an argument is valid, its conclusion is not necessarily true.

The following examples illustrate how chains of reasoning may be used to draw conclusions.

EXAMPLE Show that the indicated conclusion follows from the given premises.

P_1: $p \longrightarrow q$
P_2: $\sim p \longrightarrow r$

C: $\sim q \longrightarrow r$

SOLUTION

1. $p \longrightarrow q$	1. Given
2. $\sim q \longrightarrow \sim p$	2. Law of the Contrapositive
3. $\sim p \longrightarrow r$	3. Given
4. $\sim q \longrightarrow r$	4. Law of the Syllogism

EXAMPLE Show that the indicated conclusion follows from the given premises.

P_1: $\sim p \longrightarrow r$
P_2: $\sim(p \vee s)$

C: r

SOLUTION

1. $\sim(p \vee s)$	1. Given
2. $\sim p \wedge \sim s$	2. DeMorgan's Laws
3. $\sim p$	3. Law of Simplification
4. $\sim p \longrightarrow r$	4. Given
5. r	5. Law of Detachment

EXAMPLE What can you conclude about Chaucer from the following premises?

P_1: Chaucer is ferocious. P_2: If an animal is ferocious, it does not purr.
P_3: If an animal is a cat, it purrs. P_4: Chaucer is either a dog or a cat.

SOLUTION One way to solve a puzzle like this is to assign letters to the attributes discussed and then convert each sentence into symbolic form. For the purposes of this text, we can be fairly informal. For this problem, let

f stand for "is ferocious" p stand for "purrs"
c stand for "is a cat" d stand for "is a dog"

Then the four premises, symbolically, become:

P_1: f P_2: $f \rightarrow \sim p$
P_3: $c \rightarrow p$ P_4: $d \vee c$

1. f	1. Given
2. $f \rightarrow \sim p$	2. Given
3. $\sim p$	3. Law of Detachment
4. $c \rightarrow p$	4. Given
5. $\sim c$	5. Law of Contrapositive Inference
6. $d \vee c$	6. Given
7. d	7. Law of Disjunctive Inference

Therefore, you can conclude that Chaucer is a dog.

Note that the premises need not be in *If . . . then . . .* form, nor must they be in the right order to construct a chain. Lewis Carroll, in his book *Symbolic Logic,* devised a series of 60 puzzles ranging in difficulty from 3 premises to 10 premises. He helped solvers get started by assigning each feature a letter. Here is one puzzle.

EXAMPLE What conclusion can be drawn from the following premises?

P_1: Everyone who is sane can do logic.
P_2: No lunatic is fit to serve on a jury.
P_3: None of your brothers can do logic.

SOLUTION Let l stand for "is able to do logic"
j stand for "is fit to serve on a jury"
s stand for "is sane"
b stand for "is your brother"

To solve the puzzle,
(1) express each premise in symbols,
(2) rewrite with contrapositives as necessary, and
(3) rearrange the order to form a chain.

(1)		(2)		(3)	
P_1:	$s \rightarrow l$	P_1:	$\sim l \rightarrow \sim s$	P_3:	$b \rightarrow \sim l$
P_2:	$\sim s \rightarrow \sim j$			P_1:	$\sim l \rightarrow \sim s$
P_3:	$b \rightarrow \sim l$			P_2:	$\sim s \rightarrow \sim j$

Applying the Law of the Syllogism twice, you have $b \rightarrow \sim j$.
One way to state the conclusion in words is
None of your brothers is fit to serve on a jury.

The Sophists, of course, were adept at constructing arguments that sounded plausible but were in fact fallacious.

EXAMPLE What is wrong with this argument?

P_1: I never neglect important business.

P_2: Your business is unimportant.

C: Therefore, I neglect your business.

SOLUTION Let n = neglect

i = important business

y = your business

Then, in symbols, the argument is

$$\begin{array}{ll} P_1: & i \rightarrow \sim n \\ P_2: & y \rightarrow \sim i \\ \hline C: & y \rightarrow n \end{array}$$

and the conclusion does not follow from the premises. (Why not?)

Before reading further, go back to the beginning of the chapter and look at how Zeno constructs his argument as a chain of reasoning. Can you now spot the flaw?

There are really two problems with Zeno's argument, one logical and the other mathematical. The logical flaw is hidden in those three little words "and so on." What Zeno tries to do is use the Law of the Syllogism to take a short cut to infinity. It is not so much that Achilles can never catch the tortoise as that Zeno can never get through arguing about it.

The mathematical part of the paradox raises the question of what happens if you try to add up infinitely many numbers that keep getting smaller and smaller. The answer is that, in general, you can't say. Sometimes the sum is finite, and sometimes it just keeps growing bigger without limit. In the case of Achilles and the tortoise, the sum is finite, but Zeno argues that it is not.

ORAL EXERCISES

Examine Zeno's argument mathematically by running the race on a number line. Let Achilles start at point A with coordinate -1 and have the tortoise start at point B with coordinate 0. Suppose that Achilles runs only twice as fast as the tortoise.

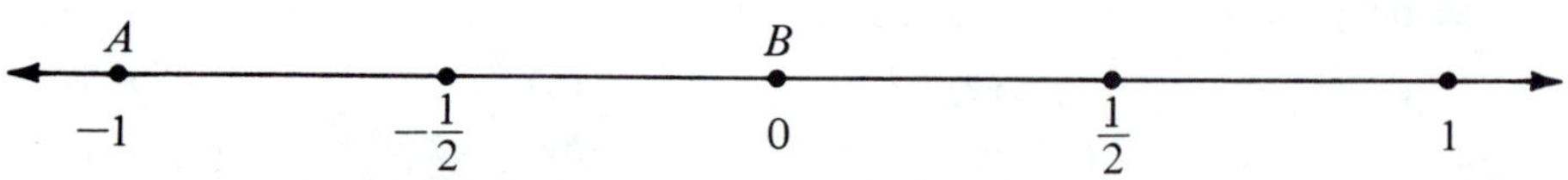

1. Using coordinate numbers instead of letters of the alphabet, reconstruct the first five steps of Zeno's argument as stated at the start of the chapter. The first step is:

   By the time Achilles gets to 0, the tortoise will be at $\frac{1}{2}$.

2. What conclusion do you come to if you add "and so on" after the fifth step of the previous exercise?

3. Zeno neglects to mention anything about where the finish line of the race is, implying that no matter where it is, the tortoise will get there first. Who would win, and where would the loser be at that time, if the finish line is at the point with coordinate:

   a. 2? b. 1? c. $\frac{63}{64}$?

4. Answer Questions 1 and 2 if Achilles and the tortoise start at points A and B as before, but Achilles runs ten times as fast as the tortoise.

WRITTEN EXERCISES

Write a conclusion that follows from each pair of premises.

A 1. $p \rightarrow q$
$\sim r \rightarrow \sim q$

2. $\sim q \rightarrow r$
$s \rightarrow \sim r$

3. $r \rightarrow s$
$t \rightarrow \sim s$

4. If you live in Moxee City, then you live in the Rattlesnake Hills.
If you don't live in Washington, then you don't live in the Rattlesnake Hills.

5. If I pass this course, I will eat my hat.
If I eat my hat, I will get sick.

6. If Tom doesn't get an A in French, then he can't try out for the basketball team.
If Tom doesn't study, then he won't get an A in French.

7. Everyone who likes logic likes Latin.
If you don't admire Julius Caesar, you won't like Latin.

8. s is a perfect square only if s is odd.
If s is odd, then s is not negative.

9. You must let me carry your books or I won't let you help me with my homework.
If you don't help me with my homework, I'm quite sure to fail the quiz.

10. All A's are B's.
No C's are B's.

11. If it rains in Spain, then it doesn't snow in Buffalo.
If it's windy in Walla Walla, then it snows in Buffalo.

(Exercises 12–16 are from Lewis Carroll's *The Game of Logic.*)

12. All ducks waddle.
Nothing that waddles is graceful.

13. No lobsters are unreasonable.
No reasonable creatures expect impossibilities.

14. No kind deed is unlawful.
What is lawful may be done without fear.

15. A prudent man shuns hyenas.
No banker is imprudent.

16. No country that has been explored is infested by dragons.
Unexplored countries are fascinating.

Show that the indicated conclusion follows from the given premises.

B **17.** P_1: $p \rightarrow q$
P_2: $q \rightarrow r$
P_3: p
C: r

18. P_1: $(\sim p \wedge s) \rightarrow r$
P_2: $\sim r$
C: $p \vee \sim s$

19. P_1: $p \vee \sim q$
P_2: $r \rightarrow \sim p$
C: $q \rightarrow \sim r$

20. P_1: $p \rightarrow \sim q$
P_2: $r \vee p$
P_3: q
P_4: $r \rightarrow t$
C: t

In Exercises 21-24, use the letters given to represent each premise and **(a)** express each sentence in symbolic form, then **(b)** use the Laws of Inference to show that the indicated conclusion follows from the premises.

21. No football play that has not been tried in practice is ever used in a game.
The rules do not allow any play that involves two forward passes.
No plays not allowed by the rules are ever tried in practice.

Let p = tried in practice
g = used in a game
f = involves two forward passes
r = allowed by the rules

Show that no play involving two forward passes is ever used in a game.

22. Only contemptible people are unkind to animals.
No one who treats animals kindly is ridiculous.
All pompous people are ridiculous.

Let c = contemptible
k = kind to animals
r = ridiculous
p = pompous

Show that all pompous people are contemptible.

23. All rich foods are fattening.
This food is rich.
No fattening food is wholesome.
This food is either wholesome or delicious.

Let r = rich
f = fattening
w = wholesome
d = delicious

Show that this food is delicious.

C **24.** I need to borrow some money from my sister or I won't be able to go to the game.
My sister will lend me the money only if I let her come with me.
Unless I fix my car, I'll have to ride my bike.
If I ride my bike, I can't take my sister.
There's no way I won't make it to the game.

Let m = borrow some money
g = go to the game f = fix my car
s = bring my sister b = ride my bike

Show that I fix my car.

25. From the infinite sequence of square numbers $1, 4, 9, 16, \ldots, m^2, \ldots, n^2, \ldots,$ select two so that the larger is double the smaller.

If $n^2 = 2m^2$, then $\frac{n^2}{m^2} = 2$.

If $\frac{n^2}{m^2} = 2$, then $\frac{n}{m} = \sqrt{2}$.

Therefore, $\sqrt{2}$ is a rational number (ratio of two integers).

The bit of trickery above is an example of the *Fallacy of Circular Reasoning,* otherwise known as assuming what you are trying to prove. A circular argument, when analyzed, reduces to the form $p' \rightarrow p$, where p' is just a disguised form of p. What are the p' and p of this example?

26. Here are the first four of an unending sequence of statements:

1. The chord joining two points of a circle divides the circle into two regions.
2. All chords joining three points of a circle divide it into four regions.
3. All chords joining four points of a circle divide it into eight regions.
4. All chords joining five points of a circle divide it into 16 regions.

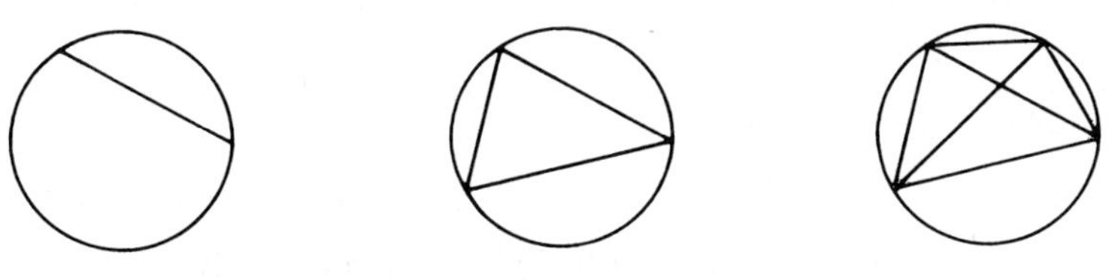

a. Verify statement (4) on a circle about 10 cm in diameter.
b. Are you willing to conclude "and so on *ad infinitum*"?
c. What do you think the next statement in the sequence would be?
d. Add one more point to the circle in your diagram for part (a) and complete the diagram for six points. How many regions do you get?
e. Do you want to change your answer to part (b)?

Section 2-8 DIRECT PROOF

You are now ready to use the rules of logic to write proofs. In this and subsequent chapters, proofs will be used to show the validity of theorems, all of which can be written in *If . . . then . . .* form. In a *direct proof* of a theorem having the structure $p \rightarrow q$ you take the hypothesis, p, as a premise and then use a chain of inferences to establish the conclusion, q. You can then infer that $p \rightarrow q$.

Before starting the proof itself you list the GIVEN information (the hypothesis) and what you are trying to PROVE (the conclusion). Most direct proofs in this book will be written in a *two-column format* with statements on the left and a reason for each statement on the right. The reason "Given" corresponds to the hypothesis. Every other step in a proof must be justified by a previous definition or principle that you have accepted as (or proved to be) true.

Before you begin writing a direct proof it is a good idea to work out an outline of the proof on scratch paper. Frequently even those who are experienced at writing proofs make many false starts before finding a successful path to follow.

Whereas you must supply a reason for every step in a proof, it is not always necessary (or desirable) to list every possible step. In the final version of your proof you should show enough steps so that the reader will be able to follow your argument and see why the theorem is true (or so that your teacher will see that you know what you are talking about). The following detailed proof might be appropriate for someone just learning how to solve equations.

EXAMPLE Prove that if $2x + 3 = 7$, then $x = 2$.

SOLUTION GIVEN: $2x + 3 = 7$ PROVE: $x = 2$

Statements	Reasons
1. $2x + 3 = 7$	1. Given
2. $2x + 3 + (-3) = 7 + (-3)$	2. Addition Prop. of Equality
3. $2x + 0 = 7 + (-3)$	3. Inverse Prop. of Addition
4. $2x = 7 + (-3)$	4. Identity Prop. of Addition
5. $2x = 4$	5. Substitution Property
6. $\frac{1}{2} \cdot 2x = \frac{1}{2} \cdot 4$	6. Multiplication Prop. of Equality
7. $1x = \frac{1}{2} \cdot 4$	7. Inverse Prop. of Multiplication
8. $x = \frac{1}{2} \cdot 4$	8. Identity Prop. of Multiplication
9. $x = 2$	9. Substitution Property

Since you have already learned how to solve linear equations, you probably will be able to follow easily a less detailed version of the preceding proof.

Statements	Reasons
1. $2x + 3 = 7$	1. Given
2. $2x = 4$	2. Addition Prop. of Equality
3. $x = 2$	3. Multiplication Prop. of Equality

The proofs in the examples that follow will depend upon these true statements:

Definition of Even	x is even iff $x = 2k$ for some $k \in Z$.
Closure for Z under Addition	If $x \in Z$ and $y \in Z$, then $x + y \in Z$.
Definition of complementary ∠s	$\angle 1$ and $\angle 2$ are complementary iff $m\angle 1 + m\angle 2 = 90$.

We will also use the properties listed on pages 8, 9, and 17.

EXAMPLE Prove the following theorem: The sum of any two even numbers is even.

SOLUTION Rewrite the theorem in *If . . . then . . .* form.
If x and y are even, then $x + y$ is even.

GIVEN: x and y are even.
PROVE: $x + y$ is even.

Statements	Reasons
1. x and y are even.	1. Given
2. $x = 2k$ for some $k \in Z$; $y = 2m$ for some $m \in Z$.	2. Definition of Even
3. $x + y = 2k + 2m$	3. Addition Prop. of Equality
4. $x + y = 2(k + m)$	4. Distributive Property
5. $k + m \in Z$	5. Closure for Z under Addition
6. $x + y$ is even.	6. Definition of Even

A mathematician might write a proof of this theorem in *paragraph form:*

If x and y are even, then $x + y = 2k + 2m$ where $k, m \in Z$. Thus, $x + y = 2(k + m) = 2u$ where $u \in Z$. Therefore, $x + y$ is even.

You will study proofs more thoroughly in the next few chapters on geometry. In geometric proofs, a picture describing the given information may help you to organize your thoughts.

EXAMPLE Prove the following theorem:
Two angles complementary to the same angle have equal measures.

SOLUTION Draw a diagram like the one at the right.

GIVEN: $\angle 1$ and $\angle 2$ are complementary.
$\angle 3$ and $\angle 2$ are complementary.
PROVE: $m\angle 1 = m\angle 3$

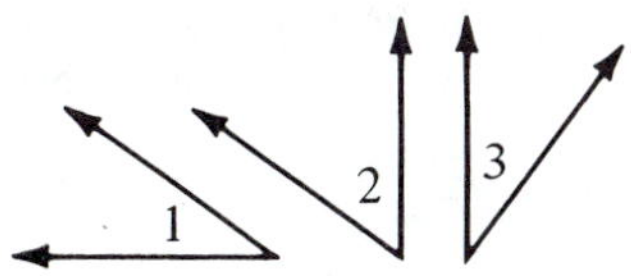

Statements	Reasons
1. $\angle 1$ and $\angle 2$ are complementary. $\angle 3$ and $\angle 2$ are complementary.	1. Given
2. $m\angle 1 + m\angle 2 = 90$ $m\angle 3 + m\angle 2 = 90$	2. Definition of complementary ∠s
3. $m\angle 1 + m\angle 2 = m\angle 3 + m\angle 2$	3. Substitution Property
4. $m\angle 1 = m\angle 3$	4. Subtraction Prop. of Equality

As you read the proofs in this book and as you do the exercises, keep in mind that two people seldom write a proof in exactly the same manner. If your ideas of how a certain proof should be written differ from what is printed here, your proof is not necessarily incorrect. If you have any questions about the validity of the proof you have written, be sure to ask your teacher.

Another form of direct proof is a *Proof by Contrapositive.* As its name implies, it uses the Law of the Contrapositive, $p \rightarrow q$ is equivalent to $\sim q \rightarrow \sim p$. To write such a proof, you begin by assuming the negation of the conclusion and then argue directly to the negation of the hypothesis. Sometimes it is much easier to write a proof of the contrapositive of a theorem than it is to write a proof of the original theorem.

EXAMPLE Prove the following theorem:
If $a^2 \neq b^2$, then $a \neq b$.

SOLUTION Prove the contrapositive of the theorem.
If $a = b$, then $a^2 = b^2$.

GIVEN: $a = b$
PROVE: $a^2 = b^2$

Statements	Reasons
1. $a = b$	1. Given
2. $ab = b^2$	2. Multiplication Prop. of Equality
3. $a^2 = b^2$	3. Substitution Property

ORAL EXERCISES

Supply the reason for each step in the following proof.

Theorem: Two angles complementary to angles of equal measures have equal measures.

GIVEN: $m\angle 1 = m\angle 3$
$\angle 1$ and $\angle 2$ are complementary.
$\angle 3$ and $\angle 4$ are complementary.
PROVE: $m\angle 2 = m\angle 4$

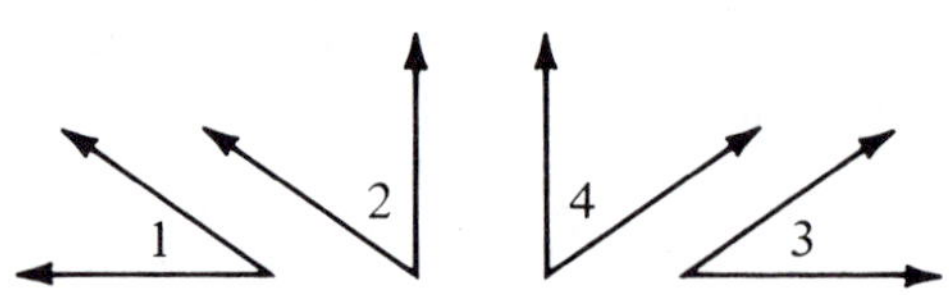

Statements	Reasons
1. $\angle 1$ and $\angle 2$ are complementary. $\angle 3$ and $\angle 4$ are complementary.	1. (a)
2. $m\angle 1 + m\angle 2 = 90$ $m\angle 3 + m\angle 4 = 90$	2. (b)
3. $m\angle 1 + m\angle 2 = m\angle 3 + m\angle 4$	3. (c)
4. $m\angle 1 = m\angle 3$	4. (d)
5. $m\angle 2 = m\angle 4$	5. (e)

WRITTEN EXERCISES

Fill in the lettered spaces to complete each proof. Choose a reason from the list below or from pages 8, 9, and 17.

Given	
Definition of supplementary ∠s	∠1 and ∠2 are supplementary iff $m\angle 1 + m\angle 2 = 180$.
Closure for Z under Addition	If $x \in Z$ and $y \in Z$, then $x + y \in Z$.
Closure for Z under Multiplication	If $x \in Z$ and $y \in Z$, then $xy \in Z$.
Definition of Even	x is even iff $x = 2k$ for some $k \in Z$.
Definition of Odd	x is odd iff $x = 2k + 1$ for some $k \in Z$.

A **1.** The product of two even numbers is even.

GIVEN: x and y are even. PROVE: xy is even.

Statements	Reasons
1. x and y are even.	1. (a)
2. $x = 2k$ for some $k \in Z$; $y = 2m$ for some $m \in Z$	2. (b)
3. $xy = 2k \cdot 2m$	3. (c)
4. $xy = 2 \cdot 2km$	4. Commutative Prop. of Mult. and Associative Prop. of Mult.
5. $2km$ is an integer.	5. (d)
6. (e)	6. (f)

2. The sum of two odd numbers is even.

GIVEN: x and y are odd. PROVE: $x + y$ is even.

Statements	Reasons
1. x and y are odd.	1. (a)
2. (b)	2. (c)
3. $x + y = 2k + 1 + 2m + 1$	3. (d)
4. $x + y = 2k + 2m + 1 + 1$	4. (e)
5. $x + y = 2k + 2m + 2$	5. (f)
6. $x + y = 2(k + m + 1)$	6. (g)
7. $k + m + 1$ is an integer.	7. (h)
8. (i)	8. (j)

3. The product of two odd numbers is odd.

GIVEN: x and y are odd. PROVE: ____(a)____

Statements	Reasons
1. x and y are odd.	1. (b)
2. (c)	2. (d)
3. $xy = (2k + 1)(2m + 1)$	3. (e)
4. $xy = 4km + 2k + 2m + 1$	4. (f)
5. $xy = 2(2km + k + m) + 1$	5. (g)
6. (h)	6. (i)
7. (j)	7. (k)

4. The difference of an odd and an even number is odd.

GIVEN: x is odd and y is even.
PROVE: ____(a)____

Statements	Reasons
1. (b)	1. (c)
2. $x = 2k + 1$ for some $k \in Z$.	2. (d)
3. $y = 2m$ for some $m \in Z$.	3. (e)
4. (f)	4. (g)
5. $x - y = 2(k - m) + 1$	5. (h)
6. (i)	6. (j)
7. (k)	7. (l)

5. The product of an odd and an even number is even.

GIVEN: ____(a)____
PROVE: ____(b)____

Statements	Reasons
1. (c)	1. (d)
2. (e)	2. (f)
3. (g)	3. (h)
4. $xy = (2k + 1)2m$	4. (i)
5. $xy = 2m(2k + 1)$	5. (j)
6. (k)	6. Closure for Z under Addition and Multiplication
7. (l)	7. (m)

6. Two angles supplementary to the same angle have equal measures.

GIVEN: $\angle 1$ and $\angle 2$ are supplementary.
$\angle 3$ and $\angle 2$ are supplementary.
PROVE: ____(a)____

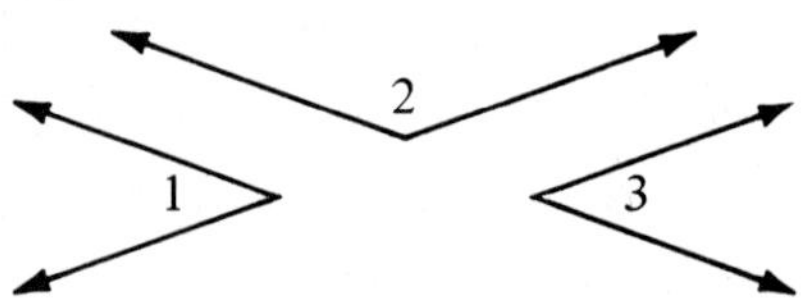

Statements	Reasons
1. (b)	1. (c)
2. (d)	2. (e)
3. (f)	3. (g)
4. (h)	4. (i)

7. Use the fact that the product of two rational numbers is rational to prove that if x^2 is irrational, then x is irrational.

8. Prove by contrapositive: If $b \neq d$, then $a \neq c$ or $a + b \neq c + d$.

Prove each theorem.

B 9. Two angles supplementary to angles of equal measures have equal measures. (Draw a picture and label the angles.)

10. The square of an even number is even.

11. The square of an odd number is odd.

12–13. For the theorems in Exercises 1 and 2, **(a)** state the converse of each theorem, and **(b)** give an example that shows the converse of each theorem is false.

14. Use the theorem you proved in Exercise 4 to prove that if the difference of two integers is even, then both integers are even or both integers are odd.

15. Prove by contrapositive: If the square of an integer is even, then the integer, n, is even.

C 16. Write the conditional that the following argument is intended to prove. Is there something wrong with this proof? Explain.

Statements	Reasons
1. $3x = 6x$	1. Given
2. $3 = 6$	2. Division Prop. of Equality

17. Write the conditional that the following argument is intended to prove. Is there something wrong with this proof? Explain.

Statements	Reasons
1. $3x^2 = 6x$	1. Given
2. $3x = 6$	2. Division Prop. of Equality
3. $x = 2$	3. Division Prop. of Equality

18. Prove by contrapositive: If $x^2 - 1$ is even, then the integer x is odd. (You may use the theorems you proved in Exercises 4 and 10.)

19. A Sophist offers these three arguments. How do you refute each argument?

a. $\frac{1}{2}$ empty $= \frac{1}{2}$ full
Multiplying each side by 2, you are forced to conclude that
Empty = Full.

b. 2 lb = 32 oz
$\frac{1}{2}$ lb = 8 oz
Multiplying the left sides and right sides of those two equations, you are forced to conclude that
1 lb = 256 oz.

c. $\frac{1}{4}$ dollar = 25 cents
Taking the square root of both sides, you are forced to conclude that
$\frac{1}{2}$ dollar = 5 cents.

Section 2-9 INDIRECT PROOF

In a direct proof, the argument proceeds directly from hypothesis to conclusion. In an *indirect proof* (also called a *proof by contradiction* or *reductio ad absurdum proof*) you begin by assuming as premises the hypothesis and the negation of the conclusion, and then show that this assumption leads to some contradiction of the form $s \wedge \sim s$, although, at the outset, you may have no idea of just what that contradiction will be.

Why is this type of proof valid? An indirect proof of

$$p \rightarrow q$$

asserts that for some statement s,

$$(p \wedge \sim q) \rightarrow (s \wedge \sim s),$$

which is equivalent to

$$\sim(p \rightarrow q) \rightarrow (s \wedge \sim s).$$

By the Law of the Contrapositive, this is equivalent to

$$(\sim s \vee s) \rightarrow (p \rightarrow q).$$

$(\sim s \vee s)$ is a tautology (always true), so by the Law of Detachment you can infer $p \rightarrow q$.

The first step in an indirect proof is to assume the hypothesis of the conditional and the negation of the conclusion. Here is a simple example of an indirect proof.

EXAMPLE GIVEN: $x + 3 = 7$;
$y - 2 = 7$
PROVE: $x \neq y$

We shall use indirect proof, therefore we temporarily assume that $x = y$.

SOLUTION

Statements	Reasons
1. $x = y$; $x + 3 = 7$; $y - 2 = 7$	1. Hypothesis and assumption
2. $x + 3 = y - 2$	2. Substitution Property
3. $y + 3 = y - 2$	3. Substitution Property
4. $0 = -5$	4. Subtraction Prop. of Equality
5. Contradiction	5. $0 \neq -5$

Therefore, the temporary assumption that $x = y$ must be false, and it follows that $x \neq y$.

An indirect proof is frequently written in paragraph form. Here are some well-known examples.

EXAMPLE Prove that $\sqrt{2}$ is irrational.

SOLUTION Temporarily assume that $\sqrt{2}$ is rational. You know that any rational number can be written as the quotient of two integers in lowest terms. Therefore, you can say that $\sqrt{2} = \frac{m}{n}$ $(n \neq 0)$, where m and n are integers with no common factor other than 1.

Squaring both sides, you have $2 = \frac{m^2}{n^2}$, or $2n^2 = m^2$. This implies that m^2 is even. By Exercise 15 on page 87, you can conclude that m is even, or that $m = 2k$ for some $k \in Z$. Thus,

$$m^2 = (2k)^2 = 4k^2$$

and by the Substitution Property,

$$2n^2 = 4k^2$$

or

$$n^2 = 2k^2.$$

Thus, you have that n^2 and as a result, n, are even. But this contradicts the assumption that $\frac{m}{n}$ is in lowest terms. Therefore, the temporary assumption must be false. Hence, $\sqrt{2}$ is irrational.

One of the earliest indirect proofs on record is Euclid's proof that the number of primes is infinite. Here is a summary of his proof.

Assume there is a finite number of primes. Call the largest prime P. Let N be the product of all the prime numbers.

$$N = 2 \cdot 3 \cdot 5 \cdot 7 \cdot 11 \cdots P$$

Then N is divisible by every prime number. So $N + 1$ is not divisible by any prime, since dividing by each prime would leave a remainder of 1. Since every number greater than 1, even a prime number, is divisible by at least one prime, $N + 1$ must be divisible by some prime Q. Since Q is a prime that cannot be less than or equal to P, Q must be a prime greater than P. This contradicts the assumption that P is the largest prime. Therefore, the number of primes must be infinite.

ORAL EXERCISES

State what you would assume to begin an indirect proof of each of the following.

1. If a is rational and x is irrational, then $a + x$ is irrational.
2. If a or b is even, then ab is even.
3. Through a point not on a line, there is only one line perpendicular to the given line.
4. The measure of an exterior angle of a triangle is greater than the measure of either remote interior angle.
5. If a triangle is not isosceles, then no two angles are congruent.
6. If two lines are not parallel, then corresponding angles are not congruent.
7. Andy wants to prove that $x > y$. He has shown that it is false that $x \leqslant y$. Can he conclude that $x > y$?
8. Norma wants to prove that l is perpendicular to m. She has shown that it is false that l is parallel to m. Can she conclude that l is perpendicular to m?

WRITTEN EXERCISES

Justify each step in the indirect proofs of the following theorems.

A 1. If $i^2 = -1$, then $i \neq 0$.

1. $i = 0$ and $i^2 = -1$	1. Hypothesis and assumption
2. $i \cdot i = 0 \cdot i$	2. (a)
3. $-1 = 0$	3. (b)
4. Contradiction	4. (c)

Therefore, the temporary assumption that $i = 0$ is false. It follows that $i \neq 0$.

2. If $i^2 = -1$, then $i \not> 0$.

1. $i > 0$ and $i^2 = -1$.	1. (a)
2. $i \cdot i > 0 \cdot i$	2. (b)
3. $-1 > 0$	3. (c)
4. Contradiction	4. (d)

Therefore, the temporary assumption that $i > 0$ is false. It follows that $i \not> 0$.

3. If $i^2 = -1$, then $i \not< 0$.

1. $i < 0$ and $i^2 = -1$.	1. (a)
2. $i \cdot i > 0 \cdot i$	2. (b)
3. $-1 > 0$	3. (c)
4. Contradiction	4. (d)

Therefore, the temporary assumption that $i < 0$ is false. It follows that $i \not< 0$.

4. In the proof of the prime number theorem on page 89, what value do you get for $N + 1$ when $P = 7$? What corresponding value do you get for Q?

5. Write an indirect proof of the following theorem: If $x^2 > 0$, then $x \neq 0$.

6. Write an indirect proof of the following theorem: If $x < 0$, then $\frac{1}{x} < 0$.

(*Hint:* Consider Case I: Assume that $\frac{1}{x} = 0$; Case II: Assume that $\frac{1}{x} > 0$.)

B 7. Prove that there is no largest counting number.

8. Prove that zero is not a divisor of any counting number.

C 9. Prove that the compound conditional $(p \land q) \to r$ is equivalent to its *partial contrapositive* $(p \land \sim r) \to \sim q$ as follows: Recall that the conditional $p \to q$ is equivalent to the disjunction $\sim p \lor q$ and use this to write each compound conditional above as a disjunction.

10. Prove by partial contraposition: If n is even and $m + n$ is even, then m is even.

11. As a final problem for this chapter, consider the following sentence: "The statement I am now making is a lie." Explain how this sentence seems to contradict itself.

SELF-TEST 3

Write a conclusion that follows from the given premises.

1. $p \rightarrow q$ Section 2-7
$r \rightarrow \sim q$

2. If $3 = 6$, all triangles are isosceles. All triangles are isosceles only if the moon is made of green cheese.

3. Fill in the lettered spaces to complete the proof. Section 2-8

GIVEN: $2x + 5 = 11$
PROVE: $x = 3$

Statements	Reasons
1. $2x + 5 = 11$	1. (a)
2. $2x = 6$	2. (b)
3. (c)	3. (d)

4. Prove by contrapositive: If $x^3 \leqslant 0$, then $x \leqslant 0$.

5. Without using the Multiplicative Property of 0, write an indirect proof of the following theorem: Section 2-9
If $a \neq 0$, then $0 \cdot a \neq a$.

CHAPTER REVIEW

1. Rewrite as a true statement beginning with *All, Some,* or *No*: Section 2-1

A perfect square is a prime number.

2. Change to a related equivalent statement beginning with *All*:

No hexagon has five sides.

Negate each sentence.

3. Every cloud has a silver lining.

4. Some people shouldn't throw stones.

5. Fair is foul and foul is fair. Section 2-2

6. You help pay expenses or you can't go.

Given statements p and q as follows, **(a)** translate each statement into symbolic notation and **(b)** give its truth value.

p: Oxygen is a mineral. q: $-8 \div 4 = -2$

7. Oxygen is not a mineral or $-8 \div 4 \neq -2$.

8. Oxygen is a mineral and $-8 \div 4 \neq -2$.

Rewrite each conditional in *If . . . then . . .* form. Then rewrite the conditional as an equivalent disjunction.

9. $3x = 12$ if $x = 4$.

10. $x < 0$ implies that $x^2 > 0$. Section 2-3

11. Negate the conditional: If 350 is divisible by ten, then the units' digit is zero.

12. Use the given sentences to write a biconditional and give its truth value:
Quadrilateral $ABCD$ is a parallelogram.
Quadrilateral $ABCD$ has congruent diagonals.

Write the negation of each statement and give the truth value of the negation.

13. $\forall_x \; x \neq x + 2$ 14. $\exists_y \; \frac{1}{y} = 0$ Section 2-4

Use a variable and either the universal quantifier or the existential quantifier to rewrite the sentence in symbolic form.

15. Some numbers have negative cubes.

16. For every real number, the product of −1 and that number is the opposite of the number.

Write the converse, inverse, and contrapositive of each sentence.

17. $p \rightarrow \sim q$ 18. $\sim q \rightarrow p$ Section 2-5

19. The cow jumps over the moon only if the little dog laughs.

Write the conclusion that follows logically from the premises. If no conclusion follows, state why.

20. P_1: If I don't clean my room, then I can't go out. Section 2-6
P_2: I clean my room.

21. P_1: That dog has fleas or poison ivy.
P_2: That dog does not have poison ivy.

Write a conclusion that follows from each pair of premises.

22. P_1: No cat has six legs. Section 2-7
P_2: Charlotte has six legs.

23. P_1: If the swallows are in Capistrano, then the lilies are blooming.
P_2: If there is snow on the ground, the lilies are not blooming.

24. Fill in the lettered spaces to complete the proof that if two numbers are equal, then their difference is 0. Section 2-8

GIVEN: $x = y$
PROVE: $x - y = 0$

Statements	Reasons
1. (a)	1. (b)
2. $x + (-y) = y + (-y)$	2. (c)
3. $x - y = y + (-y)$	3. Definition of subtraction
4. (d)	4. (e)

25. Write an indirect proof of the following theorem: Section 2-9
If $x = y + 1$, then $x \neq y$.

CHAPTER TEST

Write each conditional in *If . . . then . . .* form. Then rewrite the conditional as an equivalent disjunction.

1. The fair will be held if it does not rain.
2. $4 > 0$ is necessary for 4 to be a whole number.

Write the negation of each statement.

3. Every good boy deserves fudge.
4. Elephants can't fly or centipedes can't dance.
5. Students are industrious and teachers aren't dull.
6. No real number is a solution of $x^2 = -1$.
7. If $m + n = 0$, then $m = -n$.
8. Some story has a happy ending.

In Exercises 9-12, given statements p and q as follows, **(a)** translate each sentence into symbolic notation and **(b)** give its truth value.

p: $\frac{1}{2}$ is an integer. q: $\frac{1}{2}$ is rational.

9. If $\frac{1}{2}$ is an integer, then $\frac{1}{2}$ is rational.
10. The inverse of the statement in Exercise 9.
11. $\frac{1}{2}$ is an integer if and only if $\frac{1}{2}$ is rational.
12. $\frac{1}{2}$ is not an integer or $\frac{1}{2}$ is rational.

Use a variable and either the universal quantifier or the existential quantifier to write a true statement in symbolic form.

13. For all real numbers, the product of a real number and 1 is the number itself.
14. The difference of a number and twice itself is less than 0.
15. Write the (a) converse, (b) inverse, and (c) contrapositive of the given statement and compare their truth values.

$$\text{If } A \cap B \neq \emptyset, \text{ then } A \cup B \neq \emptyset.$$

If an argument is valid, identify which rule of inference is being used. If an argument is invalid, identify which fallacy is being committed.

16. If $\pi = \frac{22}{7}$, then π is rational. $\pi \neq \frac{22}{7}$, so π is irrational.
17. If Ralph is a lawyer, then he sometimes wins an argument. Ralph never wins an argument. Therefore, Ralph is not a lawyer.

Write a conclusion that follows from the given premises.

18. P_1: $\sim p \rightarrow q$
P_2: $\sim q$

19. P_1: $\sim q \rightarrow r$
P_2: $r \rightarrow s$

20. Fill in the lettered spaces to complete the proof.

GIVEN: $a + 1 = 1$
PROVE: $a = 0$

Statements	Reasons
1. $a + 1 = 1$	1. (a)
2. $(a + 1) + (-1) = 1 + (-1)$	2. (b)
3. $a + [1 + (-1)] = 1 + (-1)$	3. (c)
4. $a + 0 = 0$	4. (d)
5. (e)	5. (f)

21. Write an indirect proof of the following: If $x \neq 0$, then $-x \neq 0$.

Application — Logic and Electrical Circuits

Switches in electrical circuits determine the path that electrical current takes. In the diagram at left below, which represents part of an electrical circuit, an open switch p prevents electricity from flowing from A to B. When the switch is closed, the current reaches B.

Open Switch 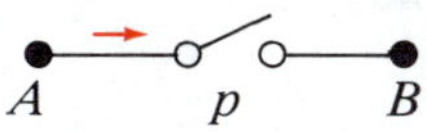

Closed Switch 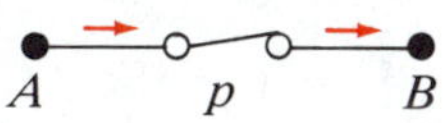

Often more than one switch controls the flow of electricity. In the first diagram below, the switches p and q are connected in series, which means that current will flow if and only if both switches are closed. The switches connected in parallel allow the current to flow if either switch is closed or if both switches are closed.

Series Circuit 

Parallel Circuit 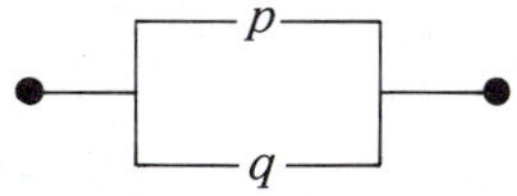

Determining whether electricity will flow through a circuit is similar to determining the truth value of compound statements in logic. To see this, let us say that closed switches will be labeled T and that open switches will be labeled F. If current will flow in the circuit, label the circuit T. If current will not flow in the circuit, label it F. Using these conventions and the diagrams of the series and parallel circuits shown above, fill in the truth tables below.

Series Circuit

p	q	circuit
T	T	
T	F	
F	T	
F	F	

Parallel Circuit

p	q	circuit
T	T	
T	F	
F	T	
F	F	

Notice that the truth table for the series circuit is the same as the truth table for the conjunction of two statements, $p \wedge q$, and that the truth table for the parallel circuit is the same as the truth table for the disjunction of two statements, $p \vee q$. (See page 53.)

In your study of logic, you have seen that truth tables can sometimes help you discover how to express a complex statement in simpler form. For example, by constructing a truth table, you can see that $(p \wedge q) \vee (p \wedge \sim q)$ is logically equivalent to p. Since electrical circuits can be represented by logical statements, truth tables can also be used to simplify electrical circuits!

VOCABULARY REVIEW

Be sure that you understand the meaning of these terms:

statement, p. 48
truth value, p. 48
logically equivalent statements, p. 49
logically contradictory statements, p. 49
conditional (implication), p. 57
biconditional, p. 59
universal quantifier, p. 62
existential quantifier, p. 62
converse, p. 66
inverse, p. 66
contrapositive, p. 66
tautology, p. 70

MIXED REVIEW

Given statements p, q, and r as follows, translate each symbolic sentence into good English.

p: It is Thursday.
q: The book is interesting.
r: A circle is round.

1. $p \vee q$

2. $p \wedge q$

3. $\sim r$

4. $\sim(q \wedge r)$

Solve.

5. $(2y - 3) - 4 \geqslant 3(y - 6)$

6. $\frac{3}{4}t + 2 = \frac{2}{3}t - 7$

7. $13 \geqslant -(3 - 2x) > 7$

8. $\frac{2n - 1}{4} + \frac{2n + 1}{3} = \frac{2}{3}$

For each pair of sets A and B, find **(a)** $A \cup B$ and **(b)** $A \cap B$.

9. $A = \{-4, -2, 0, 2, 4\}$
$B = \{0, 1, 2, 3\}$

10. $A = \{$the whole numbers$\}$
$B = \{$the negative integers$\}$

Use the indicated property to make a true statement.

11. $(-8 + 7)t = \underline{\ ?\ }$ (Substitution Property)

12. $5x + (2x + y) = \underline{\ ?\ }$ (Associative Property of Addition)

Rewrite each statement as a related equivalent statement beginning with *All*, *Some*, or *No*.

13. A trapezoid is never a parallelogram.

14. A rectangle has 4 right angles.

15. Graph: $\{$the integers greater than $-5\}$

Negate each statement.

16. Some months have 30 days.

17. No figs are big.

Each problem contains two unknowns. Represent each unknown in terms of the same variable and translate each problem into an equation or inequality. Identify the variable you are using and what it represents.

18. The length of a rectangle is 4 cm more than twice the width. The perimeter is at least 56 cm.

19. The sum of two numbers is 44. One is 5 less than twice the other.

Rewrite each conditional in *If . . . then . . .* form.

20. People become successful only if they work hard.

21. I will become an A student if I work hard.

22. Prove that the cube of an even number is even.

Write the **(a)** converse, **(b)** inverse, and **(c)** contrapositive of each statement and compare their truth values.

23. If x is even, then x^2 is even.

24. A quadrilateral is a rectangle if it has a right angle.

25. Graph the solution set: $5 - \frac{1}{2}x \leqslant 13$

Determine the conclusion that follows logically from the premises.

26. P_1: If x is an irrational number, then x is real.
P_2: x is not real.

27. P_1: Rich people are happy.
P_2: If money can't buy happiness, then rich people aren't happy.

Solve.

28. Steve is 6 times as old as his son Aaron. In 4 years he will be only 4 times as old as Aaron. How old is each today?

29. Donna and Vera leave from the same point at the same time and travel in opposite directions. After one hour they are 165 km apart. If Donna travels 5 km/h slower than Vera, find the rate of speed of each.

Simplify completely.

30. $2(x + 2y) - 3(x + 2y) - (x + 2y)$

31. $6 - 12a - 10b + 2a - (-b)$

32. Write the negation of "Every positive rational number has a rational square root." Which is true, the statement or its negation?

Acute, right, and obtuse angles can all be seen in this photograph showing a stairway at the Olympic Stadium in Tokyo, Japan. Types of angles and relationships between pairs of angles are among the topics covered in this chapter.

Geometry and Proof 3

Geometry is the branch of mathematics that deals with the properties and relationships of figures that are made up of points. Unlike the natural sciences, which are derived experimentally, geometry is built up logically. This fact ensures that no discovery in the universe can prove geometry wrong. On the other hand, physicists sometimes discover that we have not been applying geometry accurately to the real world.

Over 2000 years ago a Greek mathematician named Euclid first set down principles of geometry in a systematic way. This system, now called *Euclidean geometry,* is the one we will study in this book.

Until recently Euclidean geometry was believed to be the mathematical "model" of the universe. About seventy-five years ago, however, the greatest physicist of the twentieth century, Albert Einstein, discovered that when very large distances are involved, the universe does not obey Euclidean geometry exactly. This was an important discovery, but it in no way diminished the value of Euclidean geometry, which remains valid for all purposes we are likely to encounter on Earth. Indeed, anyone who wishes to understand Einstein's discoveries must understand Euclidean geometry first.

If at each stage of our progress we insist on *proving* each new fact, it is because we want to be certain that each step, however obvious, will remain true mathematically no matter what scientific discoveries future ages may bring.

Section 3-1 POINTS AND LINES

We will begin our development of Euclidean geometry by reviewing some basic concepts. In this section you will learn what is meant by:

- undefined terms
- betweenness
- distance
- line segments
- rays

If you did not understand English, it would do you no good to look up an English word in an ordinary dictionary, where the definition as well as the word being defined would be in English. If you knew the meaning of a certain number of English words, however, and if somehow all the definitions in the dictionary used only those words, the dictionary might be useful to you.

In geometry the situation is similar. If you assume no knowledge of any words, you will simply go around in circles trying to define one unknown word with another. Fortunately, most of us already know, for example, that a *point* is simply a location, with no width or length. A dot drawn on paper is only a physical picture of a point. The point, like all the concepts of geometry, can be accurately drawn only in the mind.

Another familiar notion is a *line,* a set of points extending without end in opposite directions. The arrowheads in the diagram of the line shown below suggest this idea.

You already know what it means to say that a point is on a line, or that a line passes through a point. If points are on the same line, we say that they are *collinear.*

The third term we will take as undefined is *plane.* We will think of a *plane* as a set of points that form a flat surface extending without end in all directions. Points that lie in the same plane are said to be *coplanar.*

Any three points are coplanar—that is, there exists a plane that contains them. The three points may or may not be collinear. If they are noncollinear, then there is only one plane that contains them. (Try, for example, to balance a piece of cardboard on the tips of three pencils.) We say that there is *one and only one* plane (or that there is *a unique* plane or that there is *exactly one* plane) through three noncollinear points. In this book, we will assume that all our work is in one fixed plane, unless otherwise specified.

Any two points that lie on a line can be used to name the line. For example, $\overleftrightarrow{PQ}$, $\overleftrightarrow{QR}$, $\overleftrightarrow{PR}$, $\overleftrightarrow{QP}$, $\overleftrightarrow{RQ}$, $\overleftrightarrow{RP}$, and line l are all correct names for the line shown below.

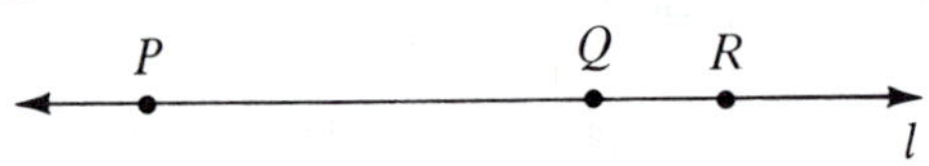

We will also assume that for any two points there is a unique positive real number called the *distance* between the points. The distance between points P and Q is denoted by PQ or QP.

We will use the word *between* in the sense of "on the same line and between."

Property of Betweenness

Point B is between points A and C if and only if $AB + BC = AC$.

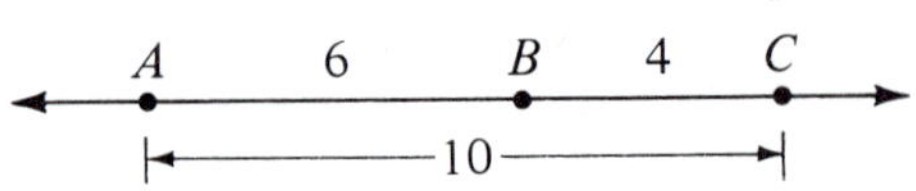

$$AB + BC = AC$$
$$6 + 4 = 10$$

In your study of geometry, there are certain things you may conclude from a diagram and other things you may not. We shall make the following agreements concerning diagrams.

You may conclude that:

1. Points drawn on one line in a diagram do lie on one line. (*Collinearity*)
2. A point drawn on the same line as two other points and "between" them in the usual English sense does lie between them in the mathematical sense. (*Betweenness*)
3. Two lines drawn as intersecting at a point do intersect at that point. (*Incidence*)
4. Unless otherwise stated, all points and lines in any diagram lie in one plane.

You may not conclude that:

Two distances that appear to be equal actually are equal.

In the following example of a proof, for instance, you may conclude from the diagram that:

1. $\overleftrightarrow{AC}$ and $\overleftrightarrow{BC}$ are lines.
2. X is between A and C; Y is between B and C.
3. $\overleftrightarrow{AX}$ and $\overleftrightarrow{BY}$ intersect at C.
4. $\overleftrightarrow{AC}$ and $\overleftrightarrow{BC}$ are in the same plane.

You may not conclude from the diagram alone that $AX = BY$ or that $XC = YC$, even though these relationships appear to be true.

EXAMPLE

GIVEN: $AX = BY$; $XC = YC$

PROVE: $AC = BC$

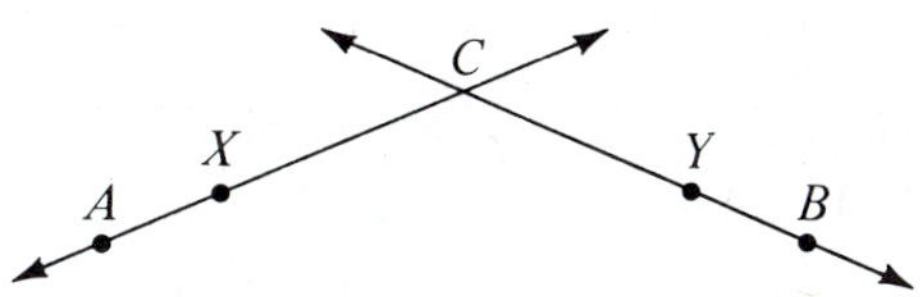

Statements	Reasons
1. $AX = BY$; $XC = YC$	1. Given
2. $AX + XC = BY + YC$	2. Addition Property of Equality
3. $AX + XC = AC$; $BY + YC = BC$	3. Property of Betweenness
4. $AC = BC$	4. Substitution Property

Often in geometry you will use two important types of subsets of lines: line segments and rays. A *line segment* (or simply a *segment*) consists of two distinct points, called the *endpoints*, and all the points between these two points. The line segment with endpoints P and Q is denoted by $\overline{PQ}$ or $\overline{QP}$. The *length* of a line segment is the distance between its endpoints.

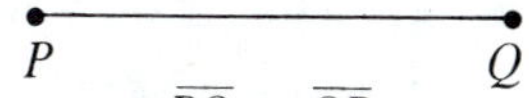

$\overline{PQ}$ or $\overline{QP}$

A *ray* is a part of a line that starts at one point (called the *endpoint* of the ray) and extends endlessly in one direction. We denote a ray by giving the name of the endpoint *first* and the name of any other point on the ray next, with an arrow above pointing from left to right. A more formal definition of a ray would be: $\overrightarrow{BA}$ is the union of $\overline{BA}$ and all the points X such that A is between B and X.

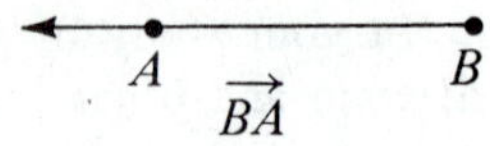

If $\overrightarrow{PQ}$ and $\overrightarrow{PR}$ are two rays such that P is between Q and R, then $\overrightarrow{PQ}$ and $\overrightarrow{PR}$ are called *opposite rays.*

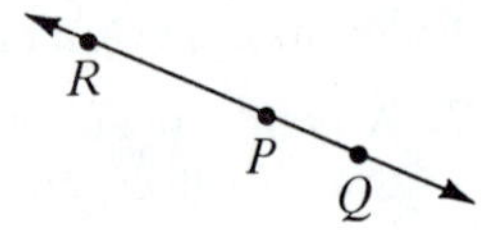

ORAL EXERCISES

Explain each symbol in words and illustrate it with a diagram if possible.

1. $\overrightarrow{AB}$ **2.** $\overrightarrow{BA}$ **3.** $\overline{AB}$ **4.** $\overleftrightarrow{AB}$ **5.** AB **6.** $\overline{BA}$

Classify each statement as true or false.

7. *Line* is an undefined term.

8. If $\overrightarrow{AB}$ and $\overrightarrow{AC}$ are opposite rays, then A, B, and C are collinear.

9. If B is between A and C, then $\overrightarrow{AB}$ and $\overrightarrow{AC}$ are opposite rays.

10. If P is between Q and R and $PQ = PR$, then $QR = 2PQ$.

11. $\overline{AB} \subset \overrightarrow{AB}$ **12.** $\overrightarrow{AB} \subset \overleftrightarrow{AB}$ **13.** $\overrightarrow{AB} \cup \overrightarrow{BA} = \overleftrightarrow{AB}$ **14.** $\overrightarrow{AB} \cap \overrightarrow{BA} = \overline{AB}$

15. If $\overrightarrow{AB}$ and $\overrightarrow{AC}$ are opposite rays, no point of $\overrightarrow{AB}$ is a point of $\overrightarrow{AC}$.

WRITTEN EXERCISES

Use the figure and the notation of this section to name each set as a line, a segment, or a ray.

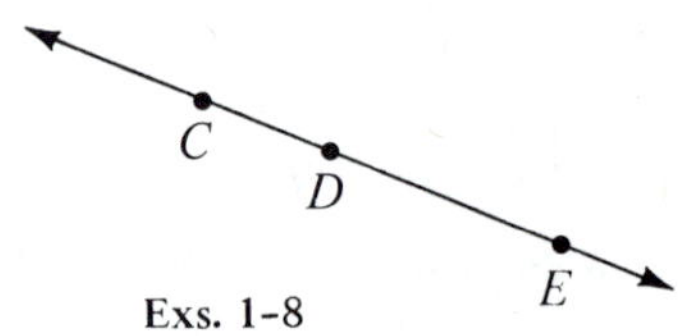

Exs. 1–8

EXAMPLE All points X such that X is between C and E, together with the points C and E

SOLUTION $\overline{CE}$

A **1.** Points C and D and all points between C and D

2. All points on $\overline{CD}$, together with all points on $\overrightarrow{DE}$

3. All points on $\overrightarrow{DC}$, together with all points on $\overrightarrow{DE}$

4. The points on $\overrightarrow{DC}$, together with those on $\overline{DE}$

5. All points X such that X is between C and D or X is between D and E, together with the points C, D, and E

6. All points X such that D is between X and E, together with point D

7. The points on $\overline{EC}$, together with all points X such that E is between D and X

8. All points X such that X is between D and E or D is between X and E, together with the points D and E

Find the required length.

9. $PQ = 5$, $QR = 13$, $PR =$ __?__

10. $PQ = 7$, $PR = 19$, $QR =$ __?__

11. $QR = 6$, $PR = 10$, $PQ =$ __?__

12. $PQ = 2$, $PR = 7\frac{1}{2}$, $QR =$ __?__

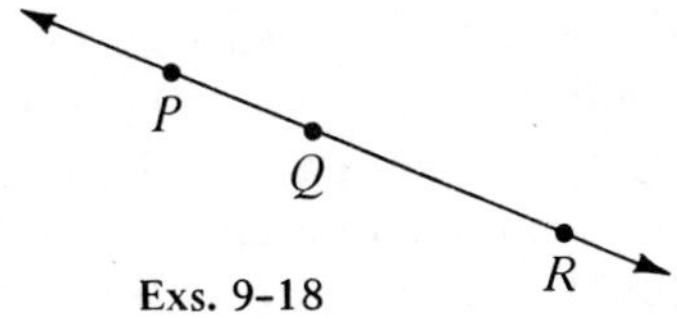

Exs. 9–18

Find the value of x.

B 13. $PQ = x$, $PR = 11$, $QR = x + 1$

14. $QR = x - 1$, $PQ = 2x$, $PR = 17$

15. $PR = x + 23$, $QR = 4x - 3$, $PQ = 5$

16. $PQ = 5x - 1$, $PR = 25 - x$, $QR = 2$

17. $QR = 3x - 5$, $PQ = x + 6$, $PR = 36 - x$

18. $PQ = 4x$, $QR = x - 7$, $PR = 2x + 35$

Choose a reason for each step of the following proofs from the list below.

Reasons: Given
Addition Property of Equality
Subtraction Property of Equality
Substitution Property
Property of Betweenness

19. GIVEN: $AB = CD$;
$AX = CX$
PROVE: $XB = XD$

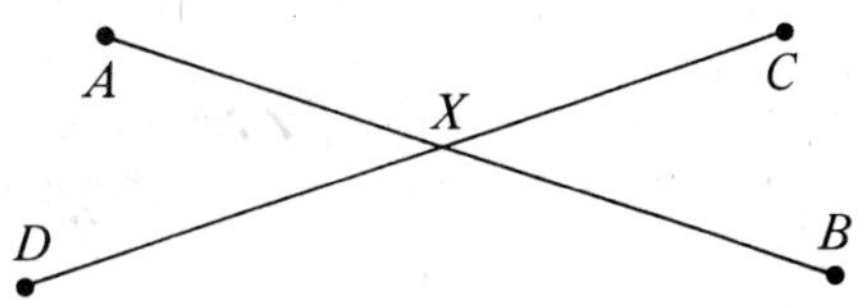

Statements	Reasons
1. $AX + XB = AB$; $CX + XD = CD$	1. __?__
2. $AB = CD$	2. __?__
3. $AX + XB = CX + XD$	3. __?__
4. $AX = CX$	4. __?__
5. $XB = XD$	5. __?__

20. GIVEN: $\overline{EH}$
PROVE: $EF + FG + GH = EH$

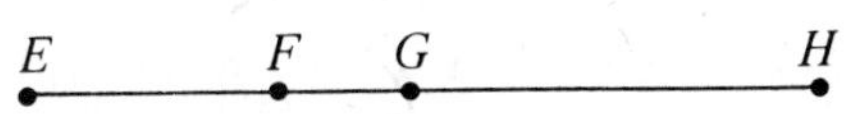

Statements	Reasons
1. $EF + FH = EH$	1. __?__
2. $FH = FG + GH$	2. __?__
3. $EF + FG + GH = EH$	3. __?__

State which of the three points P, Q, and R is between the other two under the given conditions.

C 21. $\overline{PQ} \cup \overline{QR} = \overline{PQ}$

22. $\overrightarrow{QP} \cup \overrightarrow{RP} = \overleftrightarrow{QP}$

23. $\overrightarrow{QR} \cup \overrightarrow{PR} = \overrightarrow{QR}$

24. $\overrightarrow{PR} \cap \overrightarrow{QR} = \overrightarrow{QR}$

25. $\overrightarrow{PR} \cap \overrightarrow{QR} = \overline{PQ}$

26. $\overrightarrow{RQ} \cap \overline{PR} = \{R\}$

Section 3-2 POSTULATES AND THEOREMS

A *postulate,* or *axiom,* is a statement that is accepted without proof. The chief requirement of a postulate is that it should not contradict another postulate. Of course, we could state every apparent truth as a postulate, but in doing so we would learn nothing. Instead, we try to assert as postulates only those facts about which there can be no doubt, and we prove everything else. In this way we minimize the part of our knowledge for which we do not have definite proof.

We begin our development of a system of postulates with the following:

Postulate 1 Through any two points there is exactly one line. (Two points determine a line.)

Postulate 2 There is exactly one point on a ray at a given distance from the endpoint of the ray.

Postulate 2 says that given $\overrightarrow{PQ}$ and $d > 0$, there is exactly one point X on $\overrightarrow{PQ}$ such that $PX = d$.

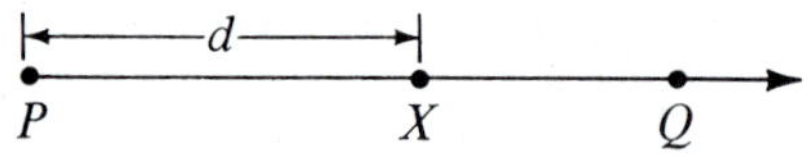

A *theorem* is a statement that can be proved using postulates, properties, definitions, and previously proved theorems. We begin with three simple theorems.

Theorem 3-1 If two lines intersect, they intersect in exactly one point.

Proof: We will give an indirect proof. We temporarily assume that the two lines intersect in two points, P and Q. Then there are two lines passing through P and Q. This contradicts Postulate 1. Therefore, the temporary assumption that the lines intersect in two points is false, and the two lines intersect in exactly one point.

Before we discuss the next two theorems, we need a definition. Point P is a *midpoint* of $\overline{AB}$ if P is between A and B and $PA = PB$.

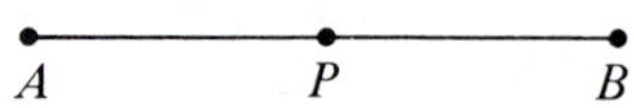

Theorem 3-2 A midpoint divides a segment into two segments, each half the length of the given segment.

GIVEN: X is a midpoint of $\overline{AB}$.

PROVE: $AX = \frac{1}{2}AB$; $BX = \frac{1}{2}AB$

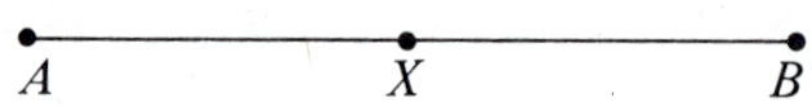

Statements	Reasons
1. X is a midpoint of $\overline{AB}$.	1. Given
2. $AX = XB$	2. Def. of a midpoint
3. $AX + XB = AB$	3. Prop. of Betweenness
4. $AX + AX = AB$ or $2AX = AB$	4. Substitution Property
5. $AX = \frac{1}{2}AB$	5. Multiplication Prop. of Equality

Similarly, you can prove that $BX = \frac{1}{2}AB$ by substituting XB for AX in the equation $AX + XB = AB$.

The following theorem, whose proof will be omitted, permits us to talk about *the* midpoint of a line segment.

Theorem 3-3 Every segment has exactly one midpoint.

A line, ray, or segment is a *bisector* of a segment if it passes through the midpoint of the segment. Suppose that, in the figure, E is the midpoint of $\overline{AB}$ but not of $\overline{CD}$. Then $\overline{CD}$ *bisects* $\overline{AB}$, but $\overline{AB}$ does not bisect $\overline{CD}$.

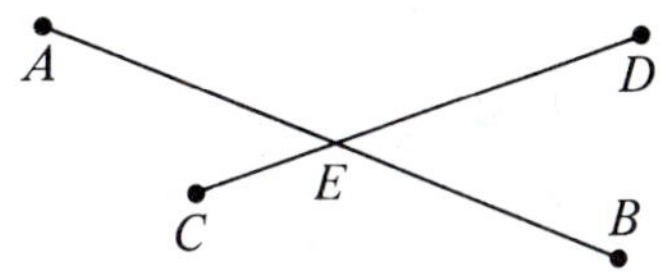

The following example demonstrates how definitions, theorems, and properties may be used in a proof.

EXAMPLE

GIVEN: $DE = AF$;
$\overline{EF}$ bisects $\overline{AB}$.

PROVE: $DE = \frac{1}{2}AB$

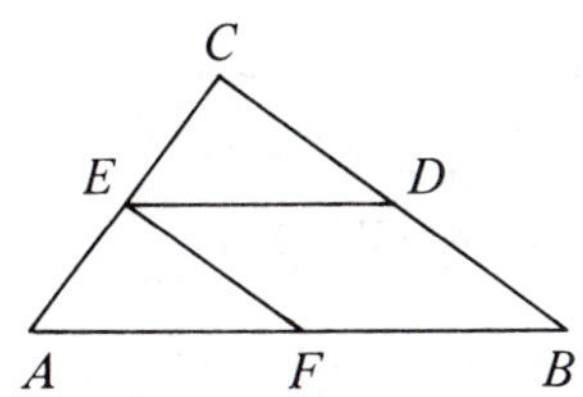

Statements	Reasons
1. $\overline{EF}$ bisects $\overline{AB}$.	1. Given
2. F is the midpoint of $\overline{AB}$.	2. Def. of a bisector of a segment
3. $AF = \frac{1}{2}AB$	3. A midpoint divides a segment into 2 segments, each half the length of the given segment. (Theorem 3-2)
4. $DE = AF$	4. Given
5. $DE = \frac{1}{2}AB$	5. Substitution Prop. (DE for AF in Statement 3)

In general, each reason justifying a statement in a proof must fall into one of the following categories:

1. Given information
2. Definitions
3. Postulates
4. Properties
5. Theorems that have already been proved

ORAL EXERCISES

Classify each statement as true or false.

1. A line segment has exactly one bisector.
2. A postulate cannot be used in a proof of a theorem.
3. A theorem is a statement that can be proved.
4. There is exactly one point X on $\overleftrightarrow{AB}$ such that $AX = 3$.
5. A postulate is a statement that is accepted without proof.
6. If B is the midpoint of $\overline{AC}$, then $BC = \frac{1}{2}AC$.
7. There is exactly one point X on $\overleftrightarrow{AB}$ such that $AX = \frac{1}{2}AB$.
8. A line may bisect a segment, but a segment may not bisect a line.

WRITTEN EXERCISES

Find the required length, given that B is the midpoint of $\overline{AC}$ and C is the midpoint of $\overline{BD}$.

A

1. $AB = 15$, $AC = \underline{\ ?\ }$
2. $BD = 18$, $CD = \underline{\ ?\ }$
3. $AC = 12$, $AD = \underline{\ ?\ }$
4. $CD = 8$, $AB = \underline{\ ?\ }$
5. $AC = 2k + 6$, $AB = \underline{\ ?\ }$
6. $BC = 8 - 2k$, $AC = \underline{\ ?\ }$

If points P, Q, R, S, and T lie on a line in the order shown, state a midpoint relationship that is implied by the given information.

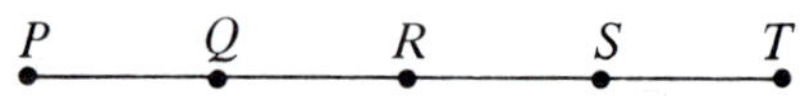

EXAMPLE $PQ = 5$, $QR = 7$, $RT = 12$

SOLUTION Since $PR = 12$, R is the midpoint of $\overline{PT}$.

7. $QR = 6$, $RS = 2$, $ST = 4$
8. $PR = 15$, $PQ = 5$, $RS = 10$
9. $RS = 7$, $QS = 19$, $PQ = 12$
10. $PT = 24$, $PR = 18$, $PS = 21$

If F is the midpoint of $\overline{EG}$, find the value of x.

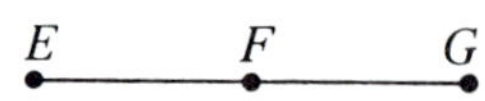

11. $EF = 2x + 11$, $FG = 53$
12. $EG = 3x - 5$, $FG = 17$
13. $EG = 5x + 4$, $EF = 2x + 37$
14. $FG = x - 5$, $EF = 3x - 19$

For Exercises 15-19, choose reasons from the following list.

Given
Definition of
  a midpoint
  a bisector of a segment
Property of Betweenness

Theorem 3-1
Theorem 3-2
Theorem 3-3
Postulate 1
Postulate 2

Properties of Equality
  Addition
  Subtraction
  Multiplication
  Division
Substitution Property

B **15.** GIVEN: $\overline{XB}$ bisects $\overline{AC}$;
$\overline{XC}$ bisects $\overline{BD}$.
PROVE: $AC = BD$

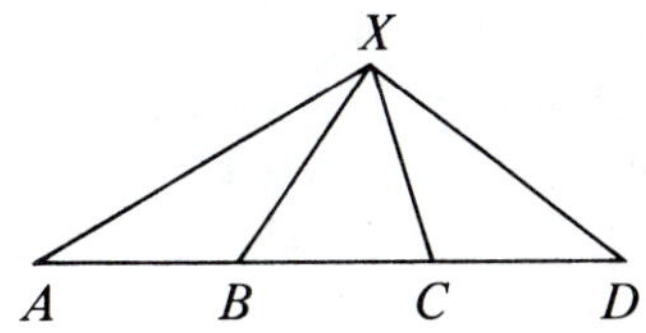

Statements	Reasons
1. $\overline{XB}$ bisects $\overline{AC}$.	1. ?
2. B is the midpoint of $\overline{AC}$.	2. ?
3. $AB = BC$	3. ?
4. $AB + BC = AC$	4. ?
5. $BC + BC = AC$	5. ?
6. $\overline{XC}$ bisects $\overline{BD}$.	6. ?
7. C is the midpoint of $\overline{BD}$.	7. ?
8. $BC = CD$	8. ?
9. $BC + CD = BD$	9. ?
10. $BC + BC = BD$	10. ?
11. $AC = BD$	11. ?

Write two-column proofs, choosing reasons from the list at the bottom of page 106.

16. Given: $XK = YL$; $KZ = LZ$
Prove: $XZ = YZ$

17. Given: $XZ = YZ$; $KZ = LZ$
Prove: $XK = YL$

18. Given: $YZ = XZ$;
K is the midpoint of $\overline{XZ}$.
Prove: $YZ = 2KZ$

19. Given: $XZ = YZ$;
K is the midpoint of $\overline{XZ}$;
L is the midpoint of $\overline{YZ}$.
Prove: $KZ = LZ$

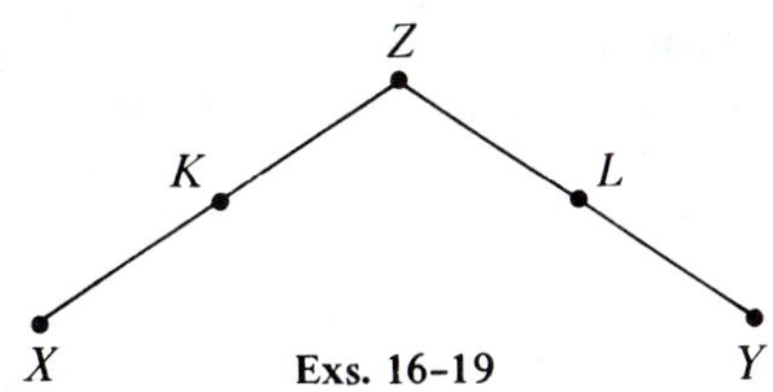

Exs. 16–19

Section 3-3 ANGLES AND ANGLE MEASURE

In this section you will review the definition of an angle, how to measure angles, and some basic properties of angles.

As you may recall from *Book 1*, an *angle* is the union of two rays with a common endpoint. The common endpoint is called the *vertex* of the angle, and the two rays are called its *sides*.

The angle at the right is called $\angle AXB$ or $\angle BXA$ or $\angle 1$. Since there is only one angle shown with the vertex X, it could also be called $\angle X$.

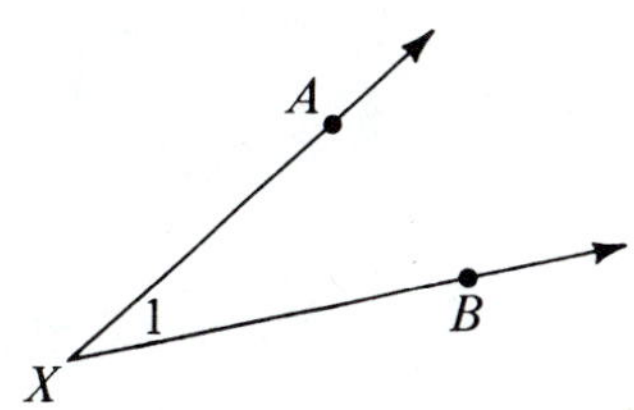

We need a way to measure angles. The most common way to do this is to use a system of measurement in *degrees* (°), in which each angle has a *degree measure* that is a real number greater than 0 and less than or equal to 180. We denote the degree measure of $\angle A$ by $m\angle A$.

EXAMPLES

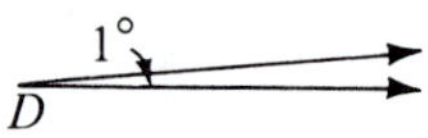

$m\angle D = 1$

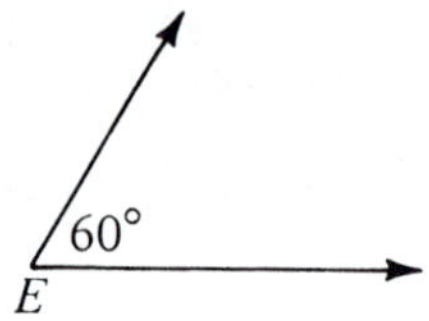

$m\angle E = 60$

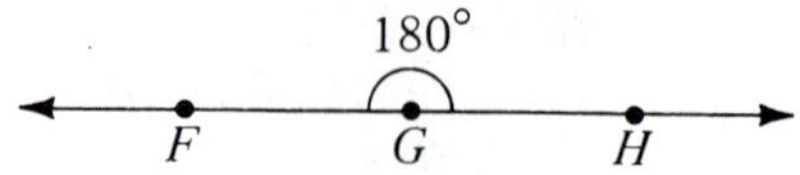

$m\angle FGH = 180$

An angle whose measure is less than 180 divides a plane into three sets of points—the angle itself, the points in the *interior* of the angle, and the points in the *exterior* of the angle. In the figure, point X is in the interior of $\angle ABC$. Points Y and Z are in its exterior.

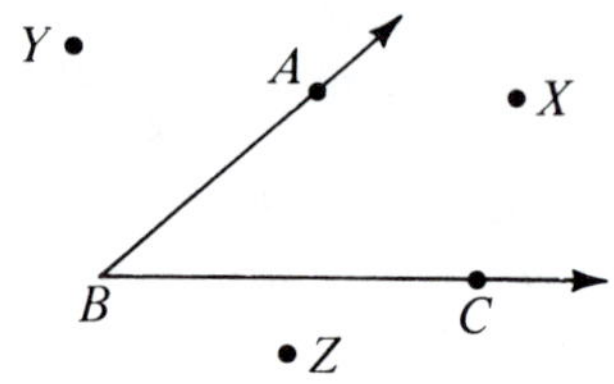

Postulate 3 If D is in the interior of $\angle ABC$, then $m\angle ABD + m\angle DBC = m\angle ABC$. (Angle Addition Postulate)

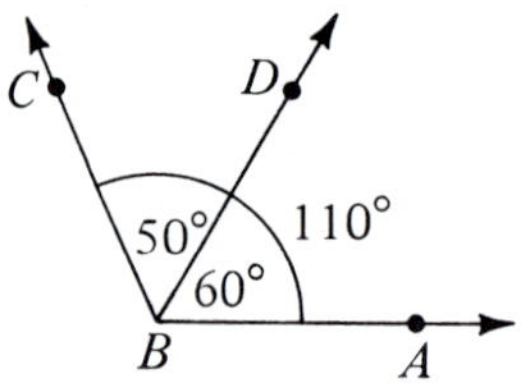

$$m\angle ABD + m\angle DBC = m\angle ABC$$
$$60 \quad + \quad 50 \quad = \quad 110$$

A line divides a plane into three sets of points—the line itself, the points on one side of the line, and the points on the other side of the line. Each set of points on one side of the line is called a *half-plane.*

Postulate 4 Given a real number k such that $0 < k < 180$ and a half-plane determined by the line containing $\overrightarrow{XA}$, there is exactly one ray $\overrightarrow{XB}$ such that B is in the given half-plane and $m\angle AXB = k$. (Angle Construction Postulate)

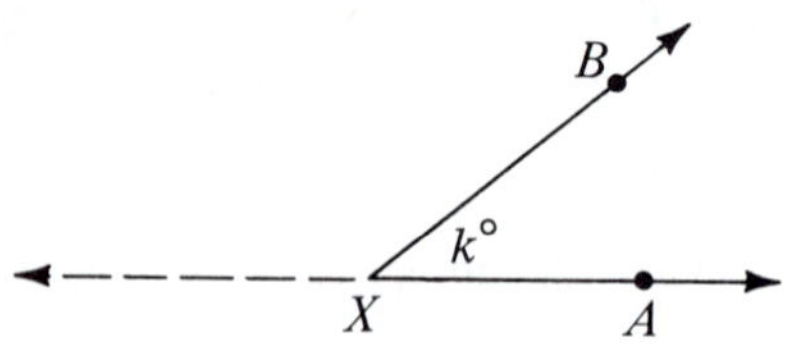

We say that $\overrightarrow{KM}$ *bisects* $\angle NKL$ if M is in the interior of $\angle NKL$ and $m\angle NKM = m\angle MKL$.

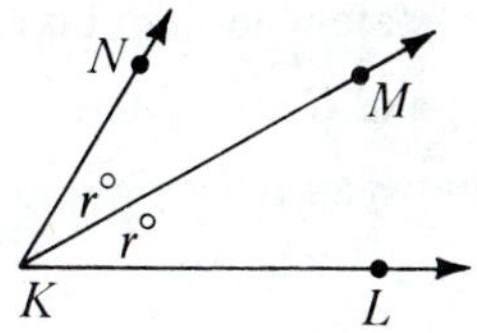

The two theorems that follow are analogous to Theorems 3-2 and 3-3, respectively. A proof of the first theorem is suggested in Exercise 23. The proof of the second is more difficult and will be omitted.

Theorem 3-4 A bisector of an angle divides the angle into two angles, each of which has measure half that of the given angle.

Theorem 3-5 Any angle of measure less than 180 has exactly one bisector.

We can now supplement the list in Section 3-1 of what you may and may not conclude from a diagram.

You may conclude that:

A point drawn in the interior of an angle does lie in the interior of the angle.

You may not conclude that:

1. An angle has a particular measure (unless the measure is specified in the diagram).
2. Two angles that appear to have the same measure actually do so (or that a ray bisects a given angle).

ORAL EXERCISES

For Exercises 1-7, refer to the diagram and complete each statement.

1. If $m\angle QXR = 40$ and $m\angle RXS = 25$, then $m\angle QXS =$ __?__.
2. If $m\angle QXS = 75$ and $m\angle RXS = 20$, then $m\angle QXR =$ __?__.
3. If $m\angle PXS = 103$ and $m\angle QXS = 60$, then $m\angle PXQ =$ __?__.
4. If $m\angle PXQ = 38$, $m\angle QXR = 42$, and $m\angle RXS = 21$, then $m\angle PXS =$ __?__.
5. If $\overrightarrow{XQ}$ bisects $\angle PXR$ and $m\angle PXR = 72$, then $m\angle PXQ =$ __?__.
6. If $\overrightarrow{XQ}$ bisects $\angle PXR$ and $m\angle PXQ = 25$, then $m\angle PXR =$ __?__.
7. If $m\angle QXR = m\angle RXS$, then __?__ bisects $\angle$ __?__.

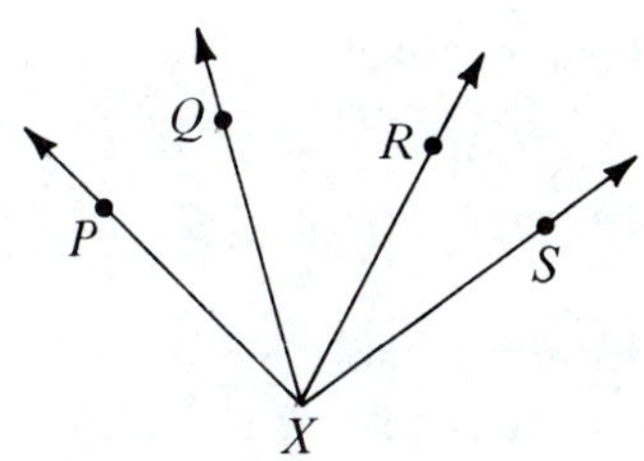

Exs. 1-7

8. In order to know that $m\angle WXY + m\angle YXZ = m\angle WXZ$, you need to know that point __?__ is in the interior of $\angle$ __?__.
9. Draw an example of three angles, $\angle AXB$, $\angle BXC$, and $\angle AXC$, such that it is *not* true that $m\angle AXB + m\angle BXC = m\angle AXC$.

Given the figure, state whether you may conclude the information listed.

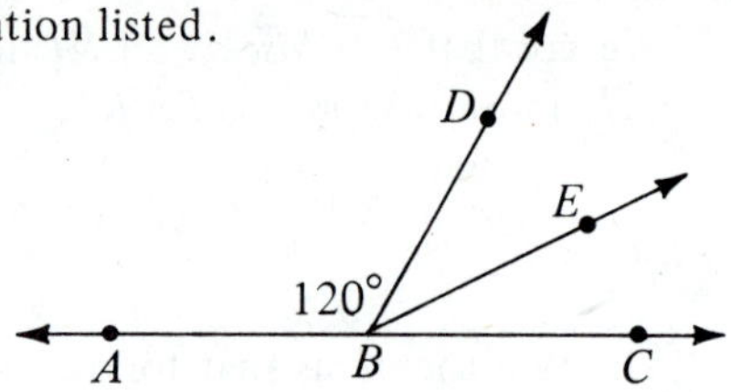

10. Points A, B, and C are collinear.

11. Point B is between points A and C.

12. $AB = BC$

13. Point D is in the interior of $\angle ABE$.

14. $m\angle DBE = m\angle EBC$ 15. $m\angle ABD = 120$

16. $\overrightarrow{BA}$ and $\overrightarrow{BC}$ are opposite rays.

WRITTEN EXERCISES

Complete.

A

	1.	2.	3.	4.
$m\angle AOB$	32	27	42	?
$m\angle BOC$	28	15	?	21
$m\angle AOC$	?	?	59	75

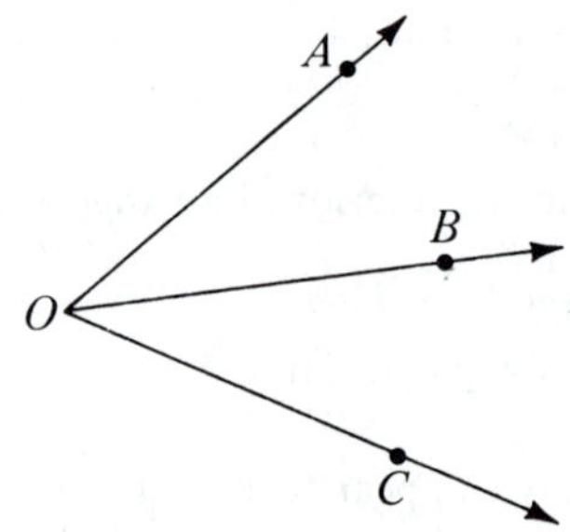

Exs. 1–8

If $\overrightarrow{OB}$ bisects $\angle AOC$, complete.

	5.	6.	7.	8.
$m\angle AOB$	36	?	?	?
$m\angle BOC$	?	42	?	?
$m\angle AOC$	?	?	54	62

Find the value of x.

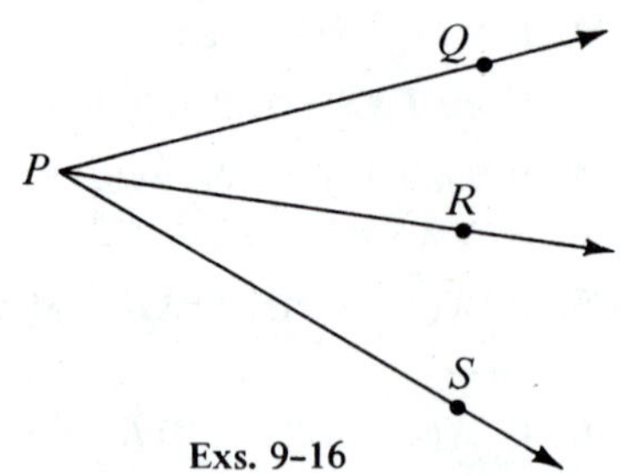

Exs. 9–16

9. $m\angle QPR = 2x - 3$; $m\angle RPS = x - 1$; $m\angle QPS = 50$

10. $m\angle QPR = 4x - 2$; $m\angle RPS = x - 1$; $m\angle QPS = 62$

11. $m\angle QPR = \frac{1}{2}x + 3$; $m\angle RPS = \frac{1}{2}x - 1$; $m\angle QPS = 3x - 54$

12. $m\angle QPR = \frac{1}{4}x - 5$; $m\angle RPS = \frac{1}{4}x + 4$; $m\angle QPS = x - 31$

If $\overrightarrow{PR}$ bisects $\angle QPS$, find the value of y and the measure of $\angle QPS$.

13. $m\angle QPR = y - 8$; $m\angle QPS = 5y - 100$

14. $m\angle QPR = 3y - 1$; $m\angle QPS = 3y + 70$

15. $m\angle QPR = \frac{1}{2}y + 1$; $m\angle QPS = 6y - 28$

16. $m\angle QPR = \frac{1}{2}y - 3$; $m\angle QPS = 16 - y$

17. Supply reasons for the following proof.

GIVEN: $\overrightarrow{XB}$ bisects $\angle AXC$;
$\overrightarrow{XC}$ bisects $\angle BXD$.

PROVE: $m\angle AXC = m\angle BXD$

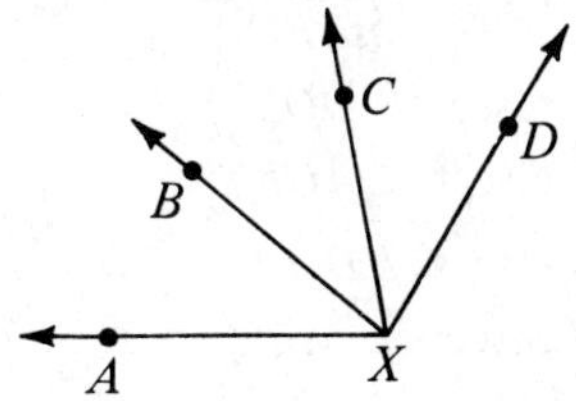

Statements	Reasons
1. $\overrightarrow{XB}$ bisects $\angle AXC$.	1. ?
2. $m\angle AXB = m\angle BXC$	2. ?
3. $\overrightarrow{XC}$ bisects $\angle BXD$.	3. ?
4. $m\angle BXC = m\angle CXD$	4. ?
5. $m\angle AXB = m\angle CXD$	5. ?
6. $m\angle AXB + m\angle BXC = m\angle CXD + m\angle BXC$	6. ?
7. $m\angle AXB + m\angle BXC = m\angle AXC$	7. ?
8. $m\angle CXD + m\angle BXC = m\angle BXD$	8. ?
9. $m\angle AXC = m\angle BXD$	9. ?

18. What theorem allows us to draw the bisector of a given angle?

B **19.** $\overrightarrow{JL}$ bisects $\angle KJN$ and $\overrightarrow{JM}$ bisects $\angle LJN$. If $m\angle KJM = 72$, find $m\angle MJN$.

20. $\overrightarrow{JM}$ bisects $\angle LJN$ and the measure of $\angle KJM$ is four times that of $\angle MJN$. If the measure of $\angle KJL$ is 60, find the measure of $\angle KJN$.

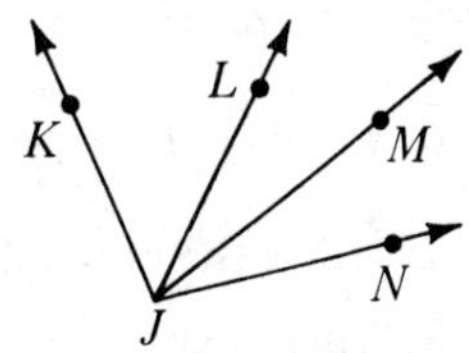

Exs. 19, 20

Write two-column proofs.

21. Given: $m\angle PXQ = m\angle RXS$
Prove: $m\angle PXR = m\angle QXS$

22. Given: $m\angle PXR = m\angle QXS$
Prove: $m\angle PXQ = m\angle RXS$

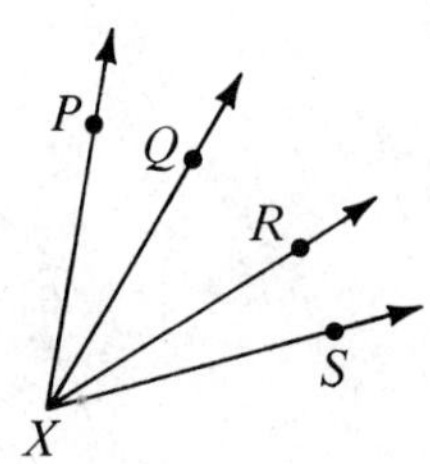

Exs. 21, 22

23. Given: $\overrightarrow{OB}$ bisects $\angle AOC$.

Prove: $m\angle AOB = \frac{1}{2}m\angle AOC$;

$m\angle BOC = \frac{1}{2}m\angle AOC$

24. Given: $m\angle AOB = \frac{1}{2}m\angle AOC$

Prove: $\overrightarrow{OB}$ bisects $\angle AOC$.

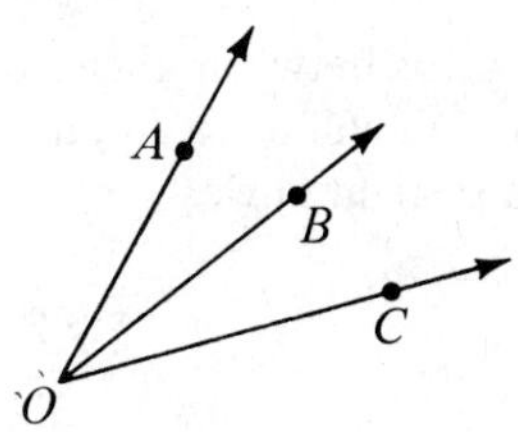

Exs. 23, 24

C **25.** Given: $m\angle EOD = m\angle BOA$;
$\overrightarrow{OC}$ bisects $\angle DOB$.
Prove: $\overrightarrow{OC}$ bisects $\angle AOE$.

26. Given: $\overrightarrow{OC}$ bisects $\angle DOB$;
$\overrightarrow{OC}$ bisects $\angle AOE$.
Prove: $m\angle EOD = m\angle BOA$

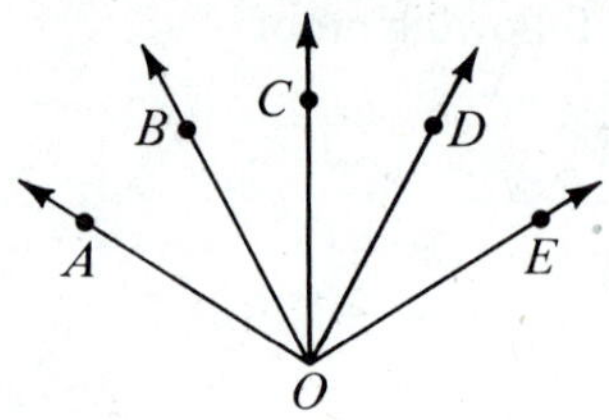

Exs. 25, 26

SELF-TEST 1

Use the diagram at the right for Exercises 1-8.
Does the set of points form a line, a segment, or a ray?

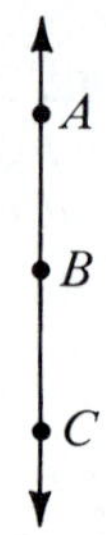

Exs. 1-8

1. The points on $\overline{AB}$, together with those on $\overrightarrow{BC}$ Section 3-1
2. The points on $\overrightarrow{BC}$, together with those on $\overrightarrow{CA}$
3. If $AB = 8.5$ and $AC = 12$, find BC.
4. If $AB = x + 4$, $BC = 2x - 3$, and $AC = 16$, find the value of x.
5. If B is the midpoint of $\overline{AC}$, $AB = 3y - 1$, and $AC = 5y + 7$, find the value of y. Section 3-2

State the postulate or theorem that justifies each statement.

6. If B is the midpoint of $\overline{AC}$, then $BC = \frac{1}{2}AC$.
7. $\overleftrightarrow{AB}$ is the only line that contains points A and B.
8. There exists a point X on $\overrightarrow{BA}$ such that $BX = AC$.
9. If $\overrightarrow{OE}$ bisects $\angle DOF$, $m\angle DOE = 2z + 48$, and $m\angle EOF = 7z - 22$, find the value of z and the measure of $\angle DOE$. Section 3-3
10. Write a two-column proof.
Given: $\overrightarrow{BX}$ bisects $\angle ABC$.
Prove: $m\angle ABC = 2m\angle 1$

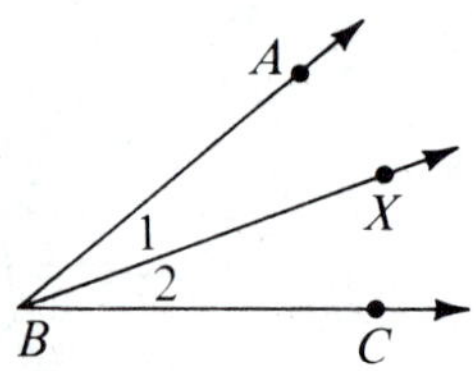

Section 3-4 ANGLE RELATIONSHIPS

Certain angles and relationships between angles occur so often that they have been given names. In this section you will learn about:

- acute, right, obtuse, and straight angles
- adjacent angles
- complementary angles
- supplementary angles
- vertical angles

We begin with some types of angles that you will probably recall from *Book 1:*

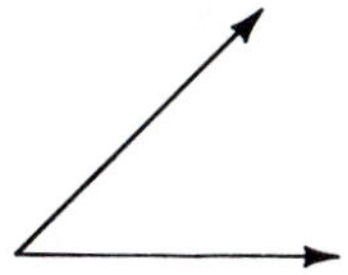

An *acute* angle has measure less than 90.

A *right* angle has measure exactly 90.

An *obtuse* angle has measure greater than 90 and less than 180.

A *straight* angle has measure exactly 180.

Two coplanar angles are *adjacent angles* (adj. $\angle$s) if they have a common vertex and a common side but no common interior points. The sides of two adjacent angles that are *not* common to both angles are called *exterior sides.*

In the figure, $\angle AXB$ and $\angle BXC$ are adjacent angles with exterior sides $\overrightarrow{XA}$ and $\overrightarrow{XC}$.

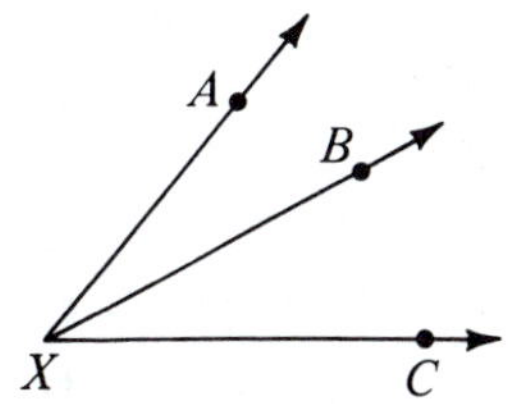

Two angles are *complementary* if the sum of their measures is 90. Each angle is called a *complement* of the other.

EXAMPLES

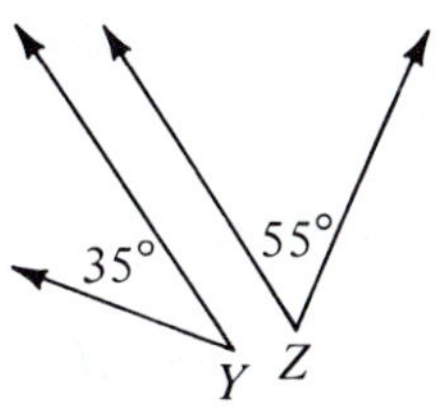

$\angle Y$ and $\angle Z$ are comp. $\angle$s.

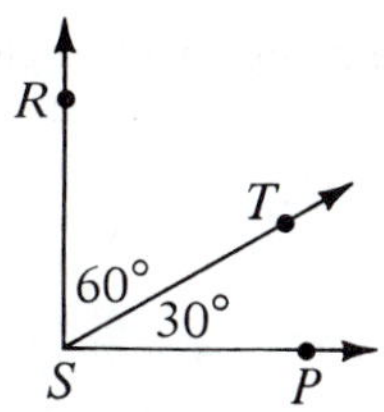

$\angle RST$ and $\angle TSP$ are comp. $\angle$s.

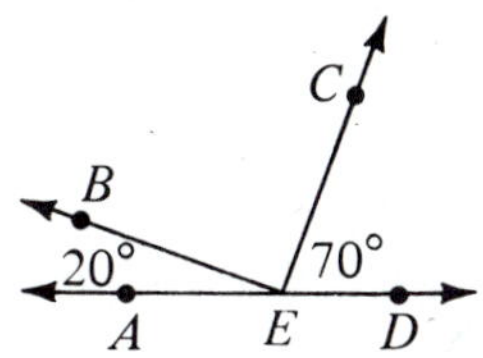

$\angle AEB$ and $\angle CED$ are comp. $\angle$s.

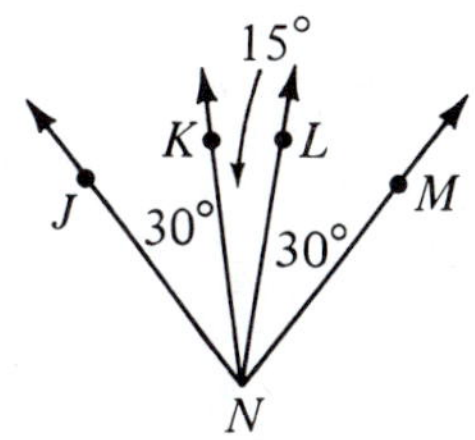

$\angle JNL$ and $\angle KNM$ are comp. $\angle$s.

Two angles are *supplementary* if the sum of their measures is 180. Each angle is called a *supplement* of the other.

EXAMPLES

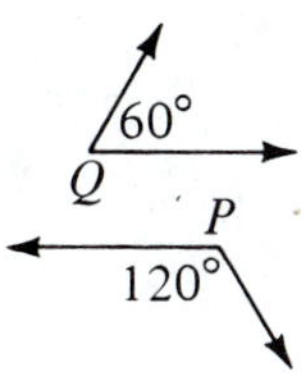

$\angle P$ and $\angle Q$ are supp. $\angle$s.

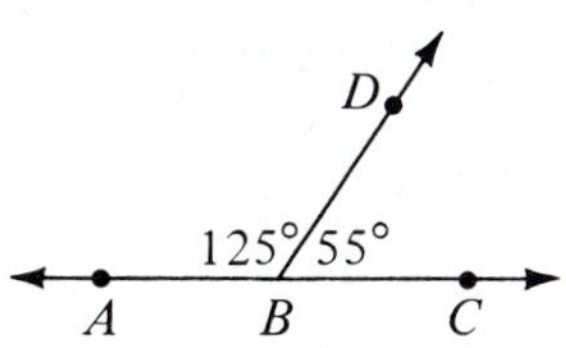

$\angle ABD$ and $\angle DBC$ are supp. $\angle$s.

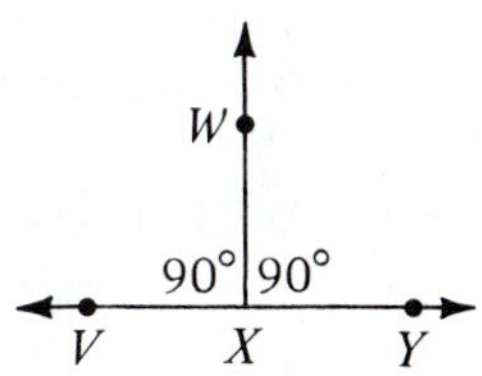

$\angle VXW$ and $\angle WXY$ are supp. $\angle$s.

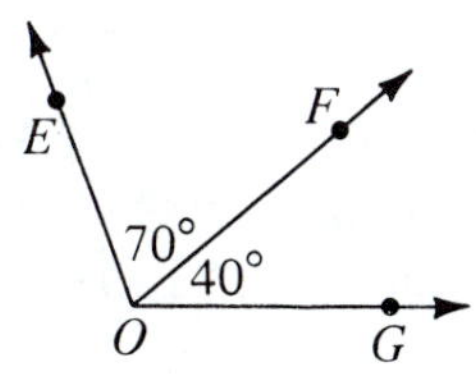

$\angle EOF$ and $\angle EOG$ are supp. $\angle$s.

The second and third examples directly above suggest the postulate that is stated at the top of the following page.

Notice that Postulate 5 is similar to the Angle Addition Postulate.

Postulate 5 If the exterior sides of two adjacent angles are opposite rays, then the angles are supplementary.

The next two theorems were proved in the text and exercise sets of Section 2-8.

Theorem 3-6 Two angles complementary to the same angle or to angles of equal measure have equal measures.

EXAMPLES

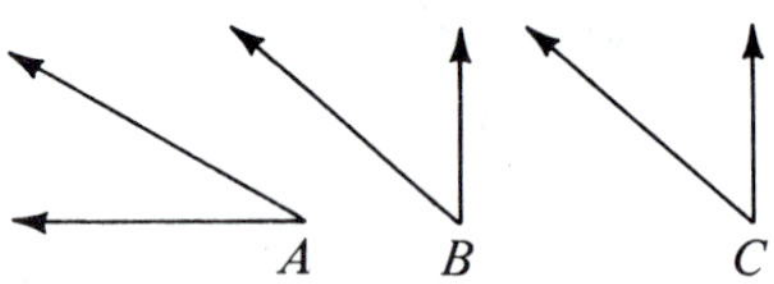

If $\angle B$ and $\angle C$ are each comp. to $\angle A$, then $m\angle B = m\angle C$.

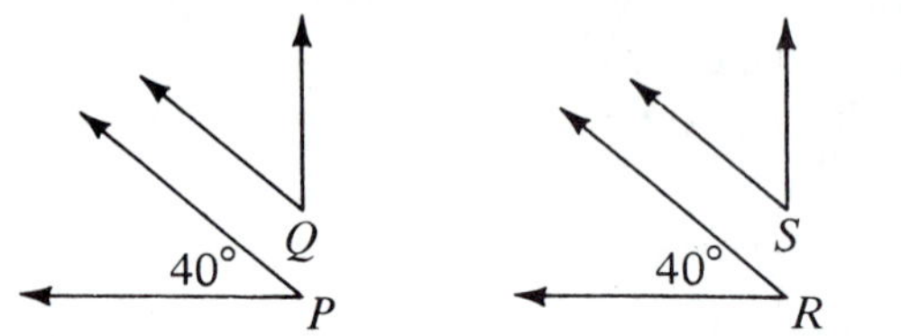

If $\angle P$ and $\angle Q$ are comp., $\angle R$ and $\angle S$ are comp., and $m\angle P = m\angle R$, then $m\angle Q = m\angle S$.

Theorem 3-7 Two angles supplementary to the same angle or to angles of equal measure have equal measures.

EXAMPLES

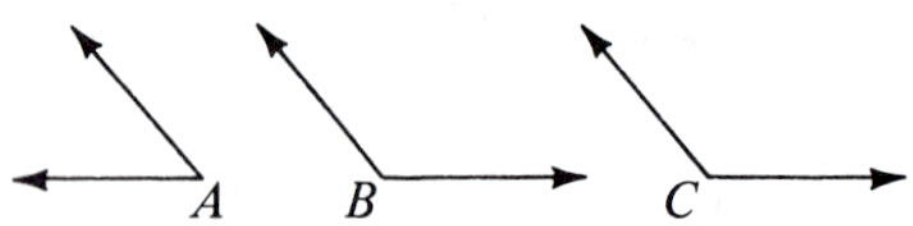

If $\angle B$ and $\angle C$ are each supp. to $\angle A$, then $m\angle B = m\angle C$.

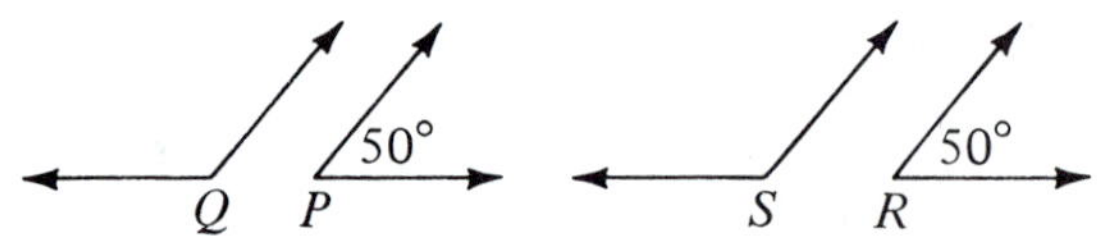

If $\angle P$ and $\angle Q$ are supp., $\angle R$ and $\angle S$ are supp., and $m\angle P = m\angle R$, then $m\angle Q = m\angle S$.

Since a supplement of a right angle is a right angle, we know that if two lines intersect to form a right angle they actually form four right angles. We say that two lines (or rays or segments) are *perpendicular* ($\perp$) if they intersect to form right angles.

In the figure, we use a small square to indicate that $\angle COB$ is a right angle and $\overleftrightarrow{AB} \perp \overleftrightarrow{CD}$.

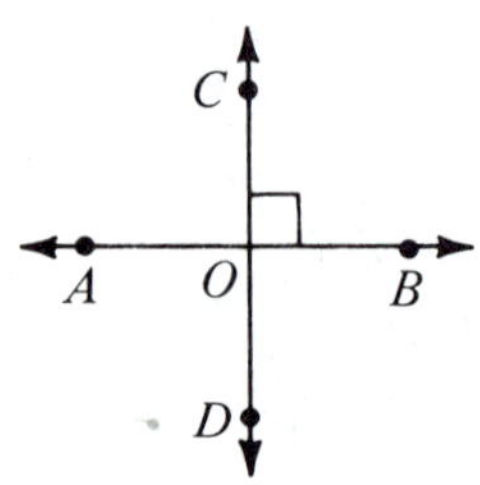

Recall that we cannot conclude fiom the *appearance* of two angles in a diagram that the angles are complementary, since this would be a conclusion about the *measures* of the angles. The next theorem, however, provides a way of drawing such a conclusion, given a situation that occurs often.

Theorem 3-8 If the exterior sides of two adjacent acute angles are perpendicular, then the angles are complementary.

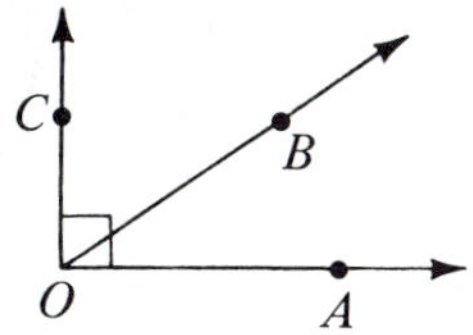

GIVEN: $\overrightarrow{OA} \perp \overrightarrow{OC}$
PROVE: $\angle AOB$ and $\angle BOC$ are complementary $\angle$s.

Statements	Reasons
1. $m\angle AOB + m\angle BOC = m\angle AOC$	1. Angle Addition Postulate
2. $\overrightarrow{OA} \perp \overrightarrow{OC}$	2. Given
3. $m\angle AOC = 90$	3. Def. of $\perp$ rays and def. of a right $\angle$
4. $m\angle AOB + m\angle BOC = 90$	4. Substitution Property
5. $\angle AOB$ and $\angle BOC$ are comp. $\angle$s.	5. Def. of comp. $\angle$s

The proofs of the following theorems are left as Oral Exercise 1 and Written Exercises 15 and 16.

Theorem 3-9 All right angles have equal measures.

Theorem 3-10 Two adjacent angles formed by perpendicular lines have equal measures.

Theorem 3-11 If two lines form adjacent angles with equal measures, then the lines are perpendicular.

Two angles whose sides form two pairs of opposite rays are called *vertical angles.*

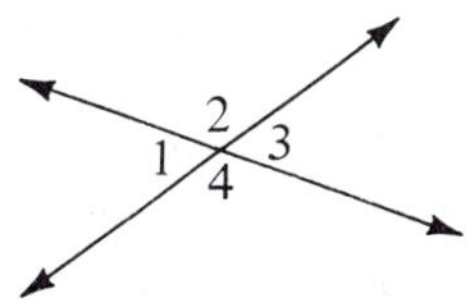

$\angle 1$ and $\angle 3$ are vertical angles.
$\angle 2$ and $\angle 4$ are vertical angles.

Note that the *definition* of vertical angles says nothing about their having equal measures, but as you probably can guess, this is the case. For a proof of the following theorem, see Oral Exercise 2.

Theorem 3-12 Vertical angles have equal measures.

ORAL EXERCISES

1. Supply reasons for the following proof of Theorem 3-9.

GIVEN: $\angle A$ and $\angle B$ are right $\angle$s.
PROVE: $m\angle A = m\angle B$

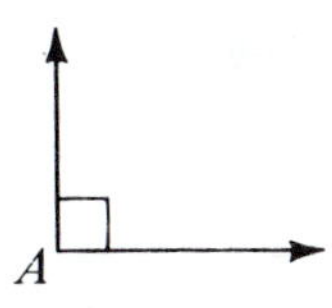

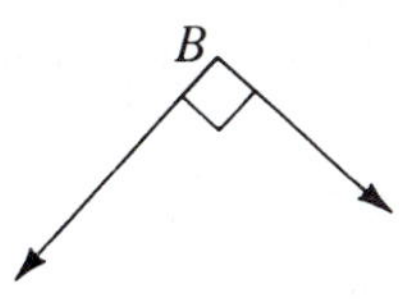

Statements	Reasons
1. $\angle A$ and $\angle B$ are right $\angle$s.	1. ___?___
2. $m\angle A = 90$; $m\angle B = 90$	2. ___?___
3. $m\angle A = m\angle B$	3. ___?___

2. Supply reasons for the following proof of Theorem 3-12.

GIVEN: $\overleftrightarrow{AB}$ and $\overleftrightarrow{CD}$ intersect at X.

PROVE: $m\angle 1 = m\angle 2$

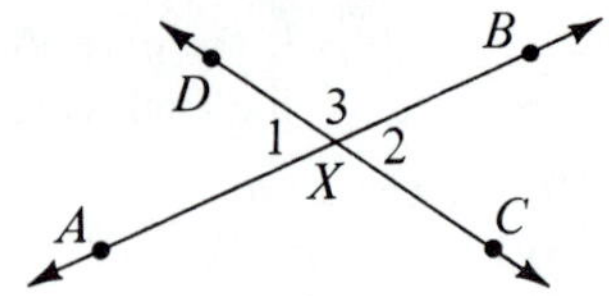

Statements	Reasons
1. $\overleftrightarrow{AB}$ and $\overleftrightarrow{CD}$ intersect at X.	1. __?__
2. $\angle 1$ and $\angle 3$ are supplementary; $\angle 2$ and $\angle 3$ are supplementary.	2. __?__
3. $m\angle 1 = m\angle 2$	3. __?__

3. If $m\angle A = m\angle B$, $\angle A$ is complementary to $\angle P$, and $\angle B$ is complementary to $\angle Q$, give a reason why $m\angle P = m\angle Q$.

4. Suppose points A, B, and C are collinear, with B between A and C. If D is any point not on $\overleftrightarrow{AB}$, give a reason why $\angle ABD$ and $\angle DBC$ are supplementary.

Name the angle that forms a pair of vertical angles with the given angle.

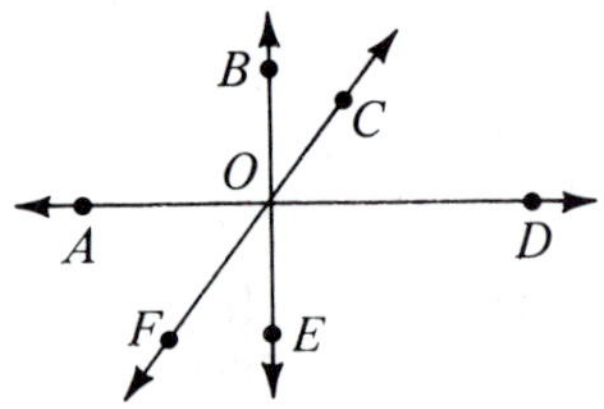

Exs. 5-10

5. $\angle AOB$ 6. $\angle AOE$

7. $\angle FOE$ 8. $\angle AOC$

9. If $\overleftrightarrow{AD} \perp \overleftrightarrow{BE}$ and $m\angle EOF = 40$, then $m\angle AOF =$ __?__.

10. If $\overleftrightarrow{AD} \perp \overleftrightarrow{BE}$ and $m\angle BOC = 45$, then $m\angle AOC =$ __?__.

Classify each statement as true or false.

11. Any two right angles are supplementary.
12. Each of two complementary angles is acute.
13. Two obtuse angles cannot be supplementary.
14. Two complementary angles cannot have equal measures.

WRITTEN EXERCISES

If $\angle 1$ and $\angle 2$ are complementary, find the value of x and the measures of the angles.

A 1. $m\angle 1 = x + 4;\ m\angle 2 = x - 4$ 2. $m\angle 1 = x + 3;\ m\angle 2 = 4x - 8$

3. $m\angle 1 = 3x - 23;\ m\angle 2 = 4x + 1$ 4. $m\angle 1 = 26 - x;\ m\angle 2 = 4x + 13$

If $\angle 3$ and $\angle 4$ are supplementary, find the value of x and the measures of the angles.

5. $m\angle 3 = 90 - x;\ m\angle 4 = 6x$ 6. $m\angle 3 = 3x - 17;\ m\angle 4 = 5x + 21$

7. $m\angle 3 = 75 - x;\ m\angle 4 = 10x - 3$ 8. $m\angle 3 = 79 - x;\ m\angle 4 = 50 - 2x$

Find the measures of angles A and B under the given conditions.

9. $\angle A$ and $\angle B$ are complementary and the measure of $\angle A$ is 5 times the measure of $\angle B$.

10. $\angle A$ and $\angle B$ are supplementary and the measure of $\angle B$ is 11 times the measure of $\angle A$.

11. $\angle A$ and $\angle B$ are supplementary and the measure of $\angle A$ is 12 less than twice the measure of $\angle B$.

12. $\angle A$ and $\angle B$ are complementary and the measure of $\angle B$ is 2 greater than three times the measure of $\angle A$.

Supply reasons for each proof.

13. GIVEN: $\overline{PX} \perp \overline{XR}$; $\overline{QX} \perp \overline{XS}$

PROVE: $m\angle 1 = m\angle 3$

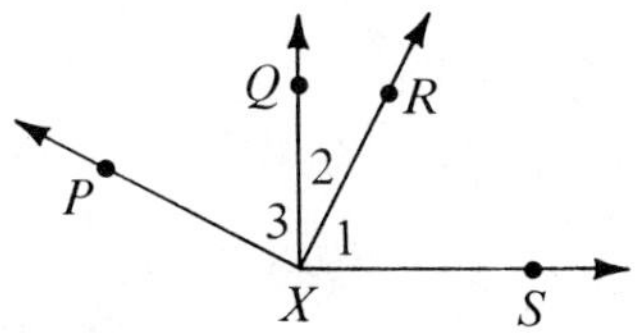

Statements	Reasons
1. $\overline{PX} \perp \overline{XR}$; $\overline{QX} \perp \overline{XS}$	1. ?
2. $\angle 1$ and $\angle 2$ are complementary; $\angle 2$ and $\angle 3$ are complementary.	2. ?
3. $m\angle 1 = m\angle 3$	3. ?

14. GIVEN: $m\angle 1 = m\angle 3$; $\overleftrightarrow{EA} \perp \overleftrightarrow{CO}$

PROVE: $\overleftrightarrow{FB} \perp \overleftrightarrow{DO}$

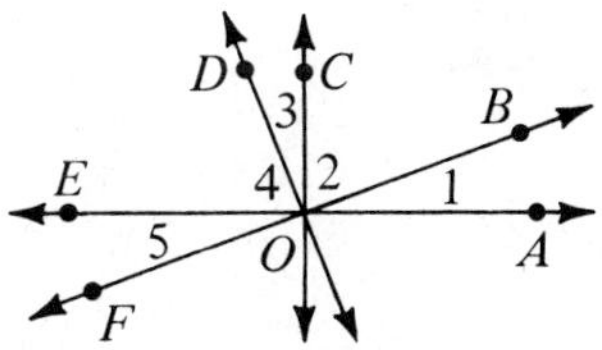

Statements	Reasons
1. $\overleftrightarrow{EA} \perp \overleftrightarrow{CO}$	1. ?
2. $\angle 1$ and $\angle 2$ are complementary; $\angle 3$ and $\angle 4$ are complementary.	2. ?
3. $m\angle 1 = m\angle 3$	3. ?
4. $m\angle 2 = m\angle 4$	4. ?
5. $m\angle 1 = m\angle 5$	5. ?
6. $m\angle 3 = m\angle 5$	6. ?
7. $m\angle 2 + m\angle 3 = m\angle 4 + m\angle 5$	7. ?
8. $m\angle 2 + m\angle 3 = m\angle BOD$; $m\angle 4 + m\angle 5 = m\angle DOF$	8. ?
9. $m\angle BOD = m\angle DOF$	9. ?
10. $\overleftrightarrow{FB} \perp \overleftrightarrow{DO}$	10. ?

B 15. Write a two-column proof of Theorem 3-10.
Given: $a \perp b$
Prove: $m\angle 1 = m\angle 2$

16. Write a two-column proof of Theorem 3-11.
Given: $m\angle 1 = m\angle 2$
Prove: $a \perp b$

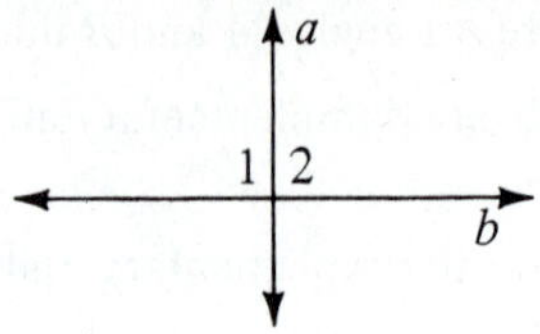

Exs. 15, 16

Write two-column proofs.

17. Given: $\angle 2$ and $\angle 3$ are supplementary.
Prove: $m\angle 1 = m\angle 3$

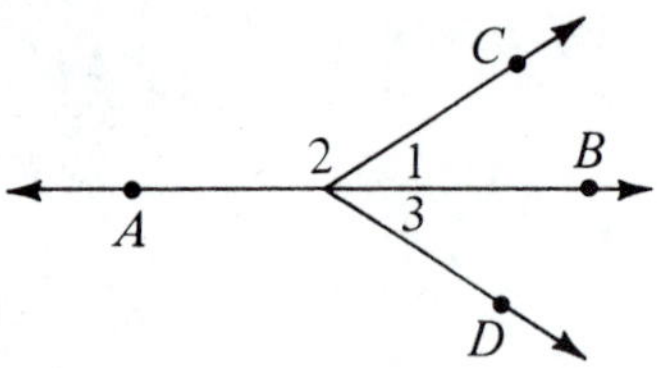

18. Given: $m\angle 5 = m\angle 6$
Prove: $m\angle 4 = m\angle 7$

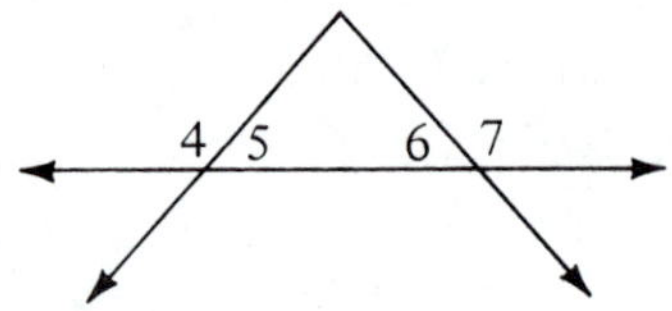

19. Given: $\overline{PB} \perp \overline{AD}$; $m\angle 1 = m\angle 3$;
$\overline{QC} \perp \overline{AD}$
Prove: $m\angle 2 = m\angle 4$

20. Given: $\overline{PB} \perp \overline{AD}$;
$\angle 2$ and $\angle 3$ are complementary.
Prove: $m\angle 1 = m\angle 3$

Exs. 19, 20

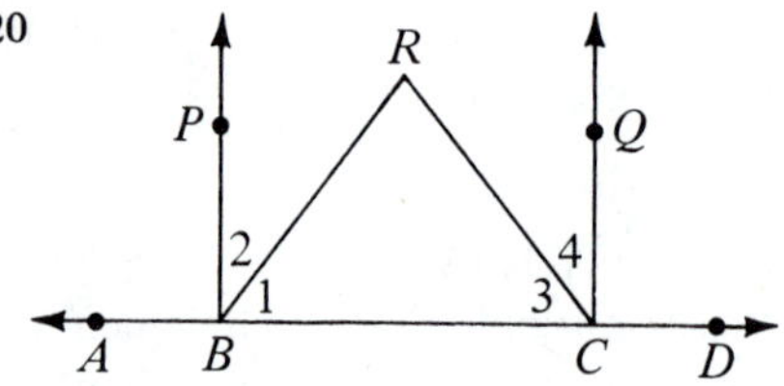

21. Given: $m\angle 1 = m\angle 4$;
$\angle 3$ and $\angle 4$ are supplementary.
Prove: $\angle 2$ and $\angle 3$ are supplementary.

22. Given: $m\angle 2 = m\angle 4$;
$\angle 3$ and $\angle 4$ are supplementary.
Prove: $\angle 1$ and $\angle 3$ are supplementary.

Exs. 21, 22

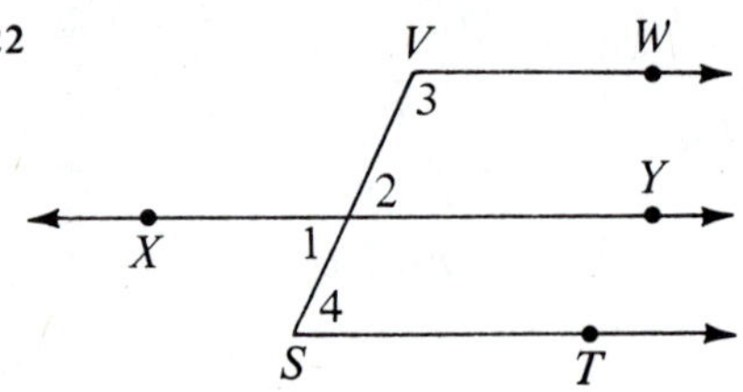

23. Given: $\overline{EJ} \perp \overline{EK}$
Prove: $\angle 1$ and $\angle 2$ are complementary.

24. Given: $\angle 1$ and $\angle 2$ are complementary.
Prove: $\overline{EJ} \perp \overline{EK}$

Exs. 23, 24

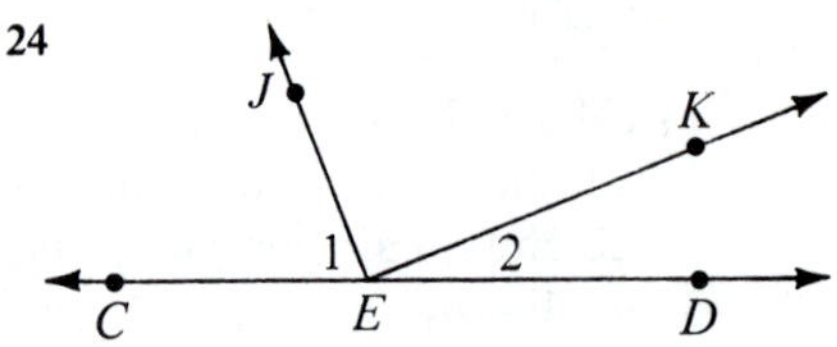

C 25. Given: $\overrightarrow{OX}$ bisects $\angle AOB$; $\overrightarrow{OY}$ bisects $\angle BOC$.
Prove: $\overline{OX} \perp \overline{OY}$

26. Given: $\overrightarrow{OX}$ bisects $\angle AOB$; $\overline{OX} \perp \overline{OY}$
Prove: $\overrightarrow{OY}$ bisects $\angle BOC$.

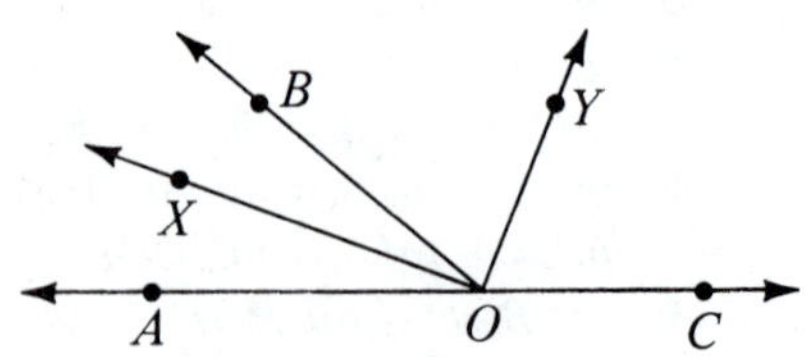

Exs. 25, 26

Section 3-5 CONGRUENCE

In this section you will learn what it means for two segments, angles, or triangles to be congruent. You will also study some properties of congruence and use the following postulates to prove that one triangle is congruent to another triangle:

- SSS Postulate
- ASA Postulate

In general, two geometric figures are *congruent* (denoted $\cong$) if they have the same size and shape. Early geometers defined two figures to be congruent if either one could be placed on top of the other so that the figures exactly coincided. This is a useful way to think about congruence, but we will need the more precise definitions that follow.

Two segments are *congruent* if they have equal lengths. That is, the statements

$$AB = CD \quad \text{and} \quad \overline{AB} \cong \overline{CD}$$

are equivalent.

Two angles are *congruent* if they have equal measures. That is, the statements

$$m\angle 1 = m\angle 2 \quad \text{and} \quad \angle 1 \cong \angle 2$$

are equivalent.

Using these definitions, we can simplify the statements of Theorems 3-6, 3-7, and 3-9 through 3-12. For example, Theorem 3-6 could be written:

Theorem 3-6 Two angles complementary to the same angle or to congruent angles are congruent.

As you probably recall, a *triangle* consists of three noncollinear points and the three line segments each of which has a pair of these points as its endpoints. The three noncollinear points are called the *vertices* of the triangle, and the segments are called the *sides* of the triangle. We denote the triangle having the points A, B, and C as its vertices by $\triangle ABC$.

Two triangles are *congruent* if the vertices of one triangle can be matched with the vertices of the other triangle in such a way that corresponding angles are congruent and corresponding sides are congruent.

In the diagram, corresponding congruent line segments and angles have been marked alike.

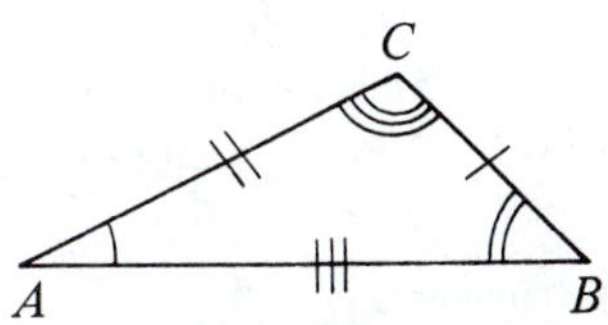

Corresponding Angles	*Corresponding Sides*
$\angle A \cong \angle D$	$\overline{AB} \cong \overline{DE}$
$\angle B \cong \angle E$	$\overline{BC} \cong \overline{EF}$
$\angle C \cong \angle F$	$\overline{AC} \cong \overline{DF}$

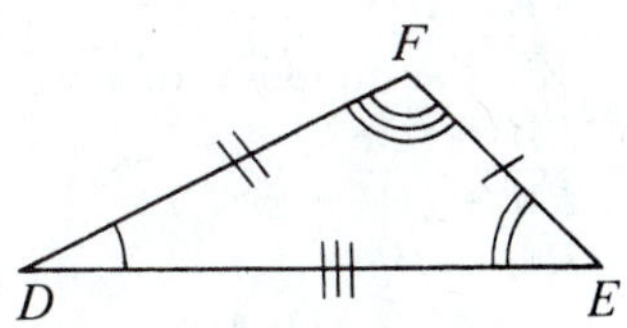

Thus $\triangle ABC \cong \triangle DEF$.

You can write a congruence in more than one way, as long as you write corresponding vertices in the same order. Thus, $\triangle BCA \cong \triangle EFD$ is also a correct statement, but $\triangle CAB \cong \triangle FED$ is incorrect.

Notice that in congruent triangles the corresponding sides are *included by* corresponding angles, and corresponding angles are included by corresponding sides. For example, $\overline{AB}$, which is included by $\angle A$ and $\angle B$, corresponds to $\overline{DE}$, which is included by $\angle D$ and $\angle E$. Also, corresponding angles are *opposite* corresponding sides.

In our proofs we will use the following properties of congruence of segments, angles, and triangles.

Reflexive Property	A figure is always congruent to itself.
Symmetric Property	If Figure I is congruent to Figure II, then Figure II is congruent to Figure I.
Transitive Property	If Figure I is congruent to Figure II and Figure II is congruent to Figure III, then Figure I is congruent to Figure III.

In order to prove two triangles congruent *by definition,* you would have to prove all six parts (sides and angles) of one triangle congruent respectively to all six parts of the other triangle. However, you can use less information to conclude that two triangles are congruent. Two such methods are given in this section, and the next section covers two more.

It seems reasonable that three rods of given lengths will form at most one triangle. We begin by adopting this as a postulate.

Postulate 6 If the three sides of one triangle are congruent to the three sides of another triangle, then the two triangles are congruent. (SSS Postulate)

EXAMPLE

GIVEN: $\overline{RP} \cong \overline{RQ}$;
M is the midpoint of $\overline{PQ}$.
PROVE: $\triangle PRM \cong \triangle QRM$

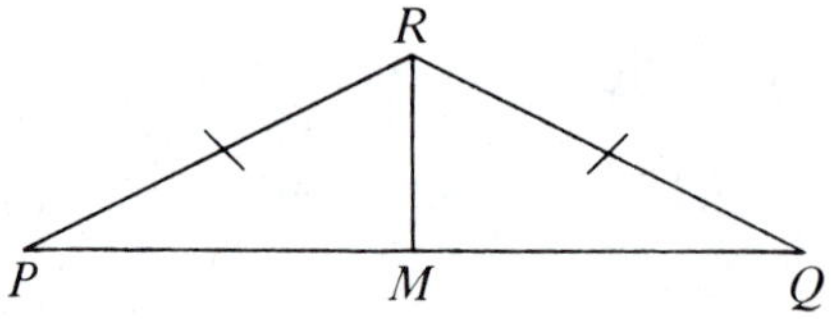

Statements	Reasons
1. $\overline{RP} \cong \overline{RQ}$	1. Given
2. M is the midpoint of $\overline{PQ}$.	2. Given
3. $PM = QM$	3. Def. of a midpoint
4. $\overline{PM} \cong \overline{QM}$	4. Def. of $\cong$ segments
5. $\overline{RM} \cong \overline{RM}$	5. Reflexive Prop. of Congruence
6. $\triangle PRM \cong \triangle QRM$	6. SSS Postulate

Note: In future proofs we may omit a step like Step 3 above and then give as a reason for Step 4: "Definition of a midpoint," thus using the statements "$PM = QM$"

and "$\overline{PM} \cong \overline{QM}$" interchangeably when no confusion results. We will also use statements such as "$m\angle A = m\angle B$" and "$\angle A \cong \angle B$" interchangeably.

If two rods are placed at given angles at the endpoints of a rod of given length, then there exists at most one triangle whose third vertex is the intersection point of the two rods. We adopt this as our next postulate.

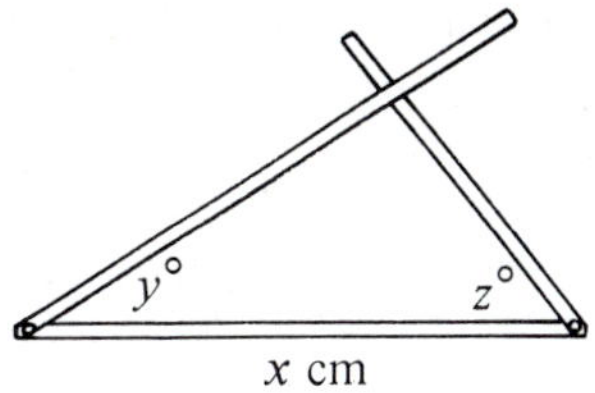

Postulate 7 If two angles and the included side of one triangle are congruent to two angles and the included side of another triangle, then the two triangles are congruent. (ASA Postulate)

EXAMPLE

GIVEN: $\angle A \cong \angle C$;
X is the midpoint of $\overline{AC}$.
PROVE: $\triangle ABX \cong \triangle CDX$

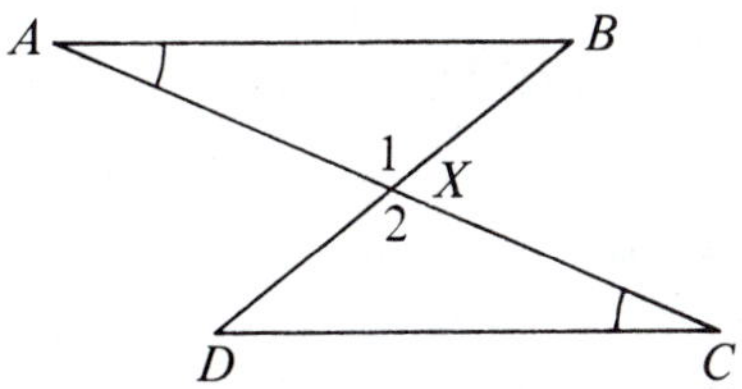

Statements	Reasons
1. $\angle A \cong \angle C$	1. Given
2. X is the midpoint of $\overline{AC}$.	2. Given
3. $\overline{AX} \cong \overline{CX}$	3. Def. of a midpoint
4. $\angle 1 \cong \angle 2$	4. Vertical angles are congruent.
5. $\triangle ABX \cong \triangle CDX$	5. ASA Postulate

ORAL EXERCISES

1. If $\triangle JKL \cong \triangle PQR$, name:
   a. three pairs of corresponding vertices
   b. three pairs of corresponding congruent sides
   c. three pairs of corresponding congruent angles

2. If $\triangle JKL \cong \triangle PQR$, then:
   a. $\triangle KLJ \cong \triangle$ __?__
   b. $\triangle LJK \cong \triangle$ __?__
   c. $\triangle LKJ \cong \triangle$ __?__

3. In $\triangle DEF$, the side included by $\angle D$ and $\angle E$ is __?__. The side included by $\angle D$ and $\angle F$ is __?__.

4. In $\triangle ABC$, the angle included by $\overline{AB}$ and $\overline{BC}$ is __?__. The angle included by $\overline{AC}$ and $\overline{BC}$ is __?__.

Complete.

5. $\overline{RS} \cong$ __?__
6. $\overline{ST} \cong$ __?__
7. $\overline{TR} \cong$ __?__
8. $\triangle RST \cong$ __?__
9. $m\angle Y =$ __?__ (numerical value)
10. $m\angle Z =$ __?__ (numerical value)

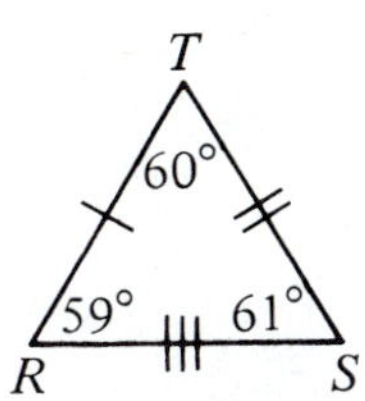

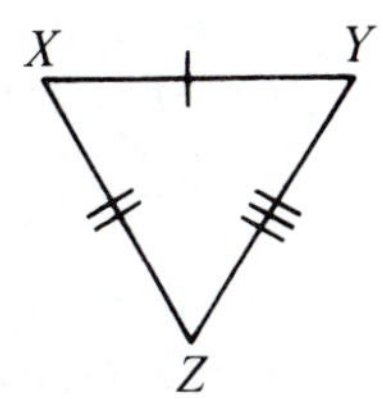

Given congruent sides and congruent angles as marked, can the two triangles be proved congruent using the SSS Postulate or the ASA Postulate? If so, state which postulate you would use.

11.

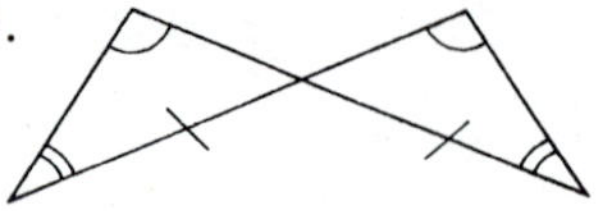

12.

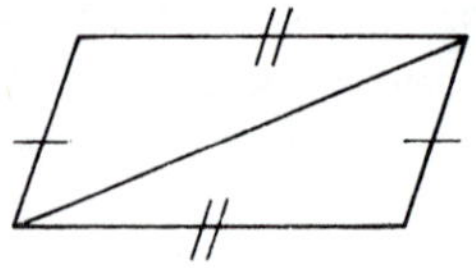

13.

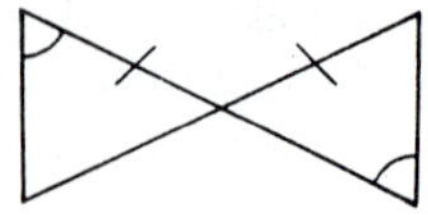

WRITTEN EXERCISES

Name a pair of triangles you could prove congruent using the given information. Also state whether you would use the SSS Postulate or the ASA Postulate.

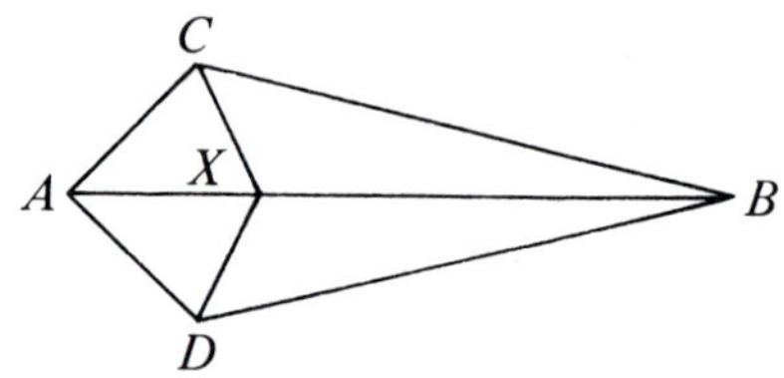

A 1. Given: $\overline{AC} \cong \overline{AD}$; $\overline{CX} \cong \overline{DX}$

2. Given: $\angle CXB \cong \angle DXB$; $\overrightarrow{BX}$ bisects $\angle CBD$.

3. Given: $\overrightarrow{AB}$ bisects $\angle CAD$; $\overrightarrow{BA}$ bisects $\angle CBD$.

4. Given: $\overline{AC} \cong \overline{AD}$; $\overline{BC} \cong \overline{BD}$

5. Given: $\angle AXC \cong \angle AXD$; $\overrightarrow{AB}$ bisects $\angle CAD$.

6. Given: $\overline{AC} \cong \overline{AD}$; $\overrightarrow{AB}$ bisects $\angle CAD$; $\angle ACB \cong \angle ADB$

Write two-column proofs.

7. Given: $\overline{AB}$ bisects $\overline{CD}$;
$\angle C \cong \angle D$
Prove: $\triangle ACP \cong \triangle BDP$

8. Given: $\overline{AB}$ bisects $\overline{CD}$;
$\overline{CD}$ bisects $\overline{AB}$;
$AC = BD$
Prove: $\triangle ACP \cong \triangle BDP$

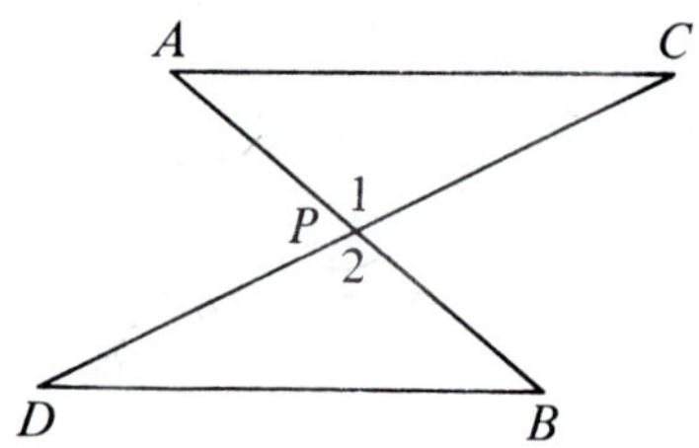

Exs. 7, 8

9. Given: $\overline{XW} \cong \overline{XZ}$;
$\overline{WY} \cong \overline{ZY}$
Prove: $\triangle XWY \cong \triangle XZY$

10. Given: $\overrightarrow{XY}$ bisects $\angle WXZ$;
$\overrightarrow{YX}$ bisects $\angle WYZ$.
Prove: $\triangle XWY \cong \triangle XZY$

11. Given: $\angle WYV \cong \angle ZYV$;
$\overrightarrow{XY}$ bisects $\angle WXZ$.
Prove: $\triangle XWY \cong \triangle XZY$

12. Given: $\overline{WY} \cong \overline{ZY}$;
$\angle W \cong \angle Z$;
$\angle VYW \cong \angle VYZ$
Prove: $\triangle XWY \cong \triangle XZY$

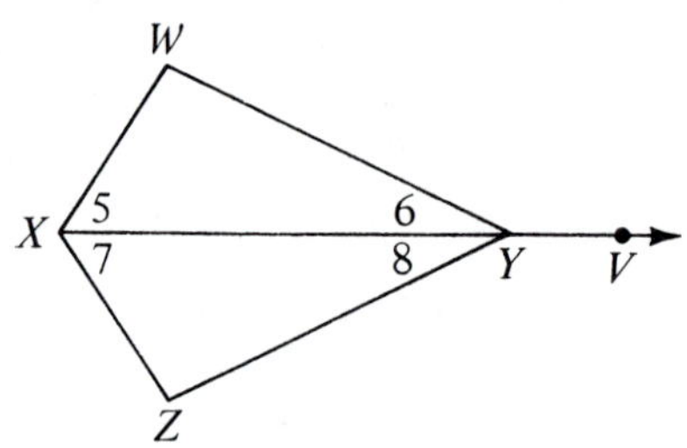

Exs. 9–12

13. Given: $FA = FB$;
$AD = BE$;
G is the midpoint of $\overline{DE}$.
Prove: $\triangle DFG \cong \triangle EFG$

14. Given: $FA = FB$;
$AD = BE$;
$m\angle 1 = m\angle 2$;
$m\angle D = m\angle E$
Prove: $\triangle DFG \cong \triangle EFG$

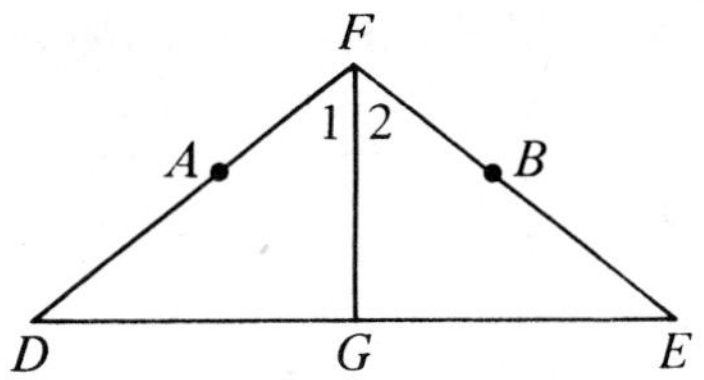

Exs. 13, 14

15. Given: $\overline{PQ} \perp \overline{AB}$;
$\overrightarrow{PQ}$ bisects $\angle APB$.
Prove: $\triangle APQ \cong \triangle BPQ$

16. Given: $\angle A \cong \angle B$;
$\overline{PQ} \perp \overline{AB}$;
Q is the midpoint of $\overline{AB}$.
Prove: $\triangle APQ \cong \triangle BPQ$

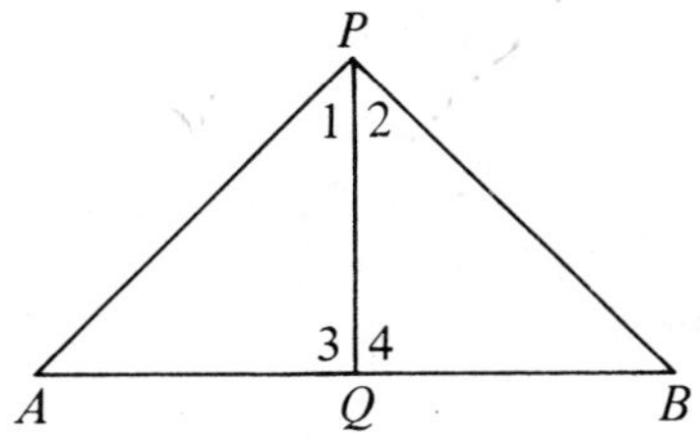

Exs. 15–18

B 17. Given: $\overline{AQ} \cong \overline{PQ}$;
$\overline{BQ} \cong \overline{PQ}$;
$\overline{AP} \cong \overline{BP}$
Prove: $\triangle APQ \cong \triangle BPQ$

18. Given: $\angle A \cong \angle 1$; $\angle B \cong \angle 2$;
$\overrightarrow{PQ}$ bisects $\angle APB$;
$\overline{AP} \cong \overline{BP}$
Prove: $\triangle APQ \cong \triangle BPQ$

19. Given: $\overline{RS} \perp \overline{XY}$;
$\overline{RS} \perp \overline{PQ}$;
$\angle 1 \cong \angle 4$
Prove: $\triangle PRS \cong \triangle QRS$

20. Given: $\overline{RS} \perp \overline{XY}$;
$\angle 1 \cong \angle 4$;
$\angle P \cong \angle 1$;
$\angle Q \cong \angle 4$;
$\overline{PS} \cong \overline{QS}$
Prove: $\triangle PRS \cong \triangle QRS$

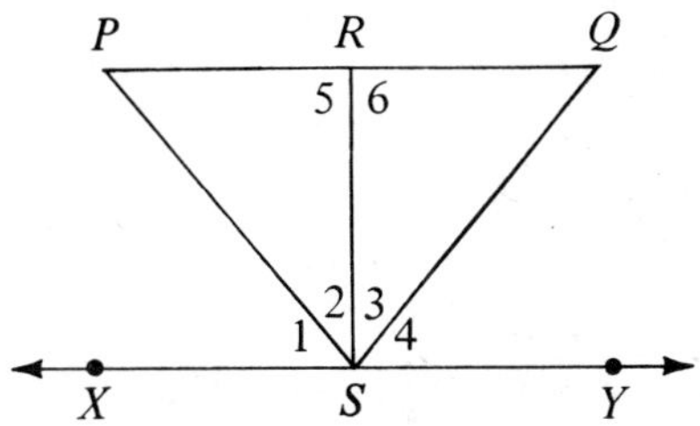

Exs. 19, 20

Determine which two triangles you can prove congruent using the given information, and then do so.

21. Given: $\overline{AC} \cong \overline{BC}$;
$\overline{AX} \cong \overline{BX}$

22. Given: $\overrightarrow{CD}$ bisects $\angle ACB$;
$\angle ADC \cong \angle BDC$

23. Given: $\angle 1 \cong \angle 2$;
$\angle 3 \cong \angle 4$

24. Given: $\overline{CD} \perp \overline{AB}$;
$\angle AXC \cong \angle BXC$

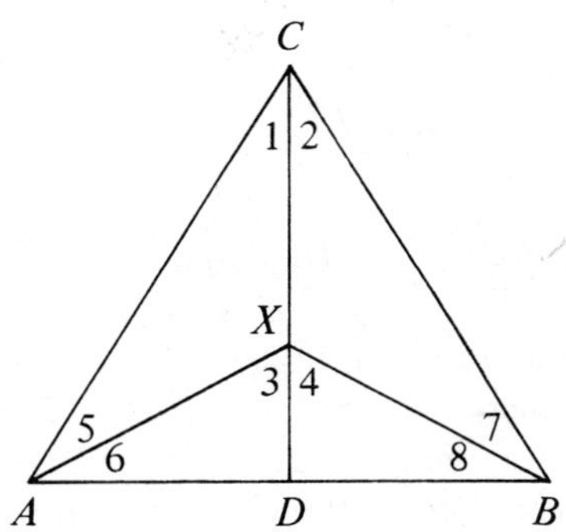

Exs. 21–24

Section 3-6 MORE WAYS OF PROVING CONGRUENCE

In this section we will expand the list of ways to prove two triangles congruent. The two new postulates are the:

- SAS Postulate
- HL Postulate

If two rods of given lengths are rigidly joined so that the angle between them is fixed, then there is only one way to finish forming a triangle by joining the other endpoints of the rods. This is stated formally as the next postulate.

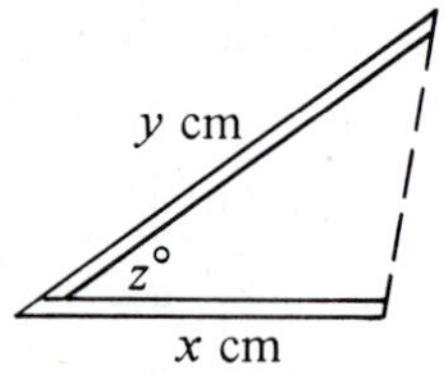

> **Postulate 8** If two sides and the included angle of one triangle are congruent to two sides and the included angle of another triangle, then the two triangles are congruent. (SAS Postulate)

EXAMPLE

GIVEN: $\overrightarrow{CD}$ bisects $\angle ACB$;
$\overline{AC} \cong \overline{BC}$

PROVE: $\triangle ACD \cong \triangle BCD$

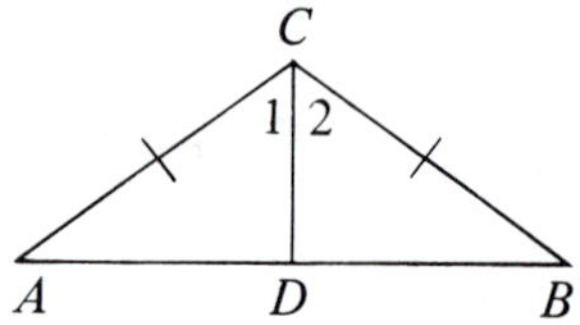

Statements	Reasons
1. $\overline{AC} \cong \overline{BC}$	1. Given
2. $\overrightarrow{CD}$ bisects $\angle ACB$.	2. Given
3. $\angle 1 \cong \angle 2$	3. Def. of an angle bisector
4. $\overline{CD} \cong \overline{CD}$	4. Reflexive Prop. of $\cong$
5. $\triangle ACD \cong \triangle BCD$	5. SAS Postulate

The diagram below illustrates why it is important that the congruent angles be *included* between the congruent pairs of sides in order to apply the SAS Postulate.

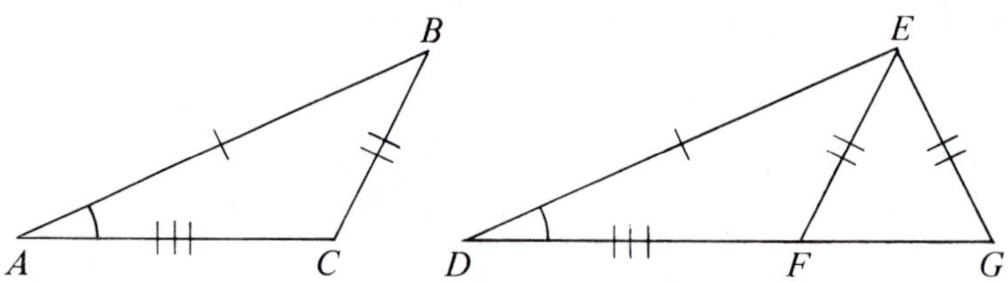

In the diagram, $\triangle ABC \cong \triangle DEF$ and $\overline{EF} \cong \overline{EG}$. Obviously, $\triangle ABC$ and $\triangle DEG$ are *not* congruent, although two sides and a *non-included* angle of $\triangle ABC$ are congruent to two sides and a non-included angle of $\triangle DEG$.

However, this situation cannot arise if the two triangles are *right* triangles. Recall that any triangle with one right angle is called a *right* triangle. The two sides of the triangle that include the right angle are called *legs*. The third side is called the *hypotenuse*.

Postulate 9 If the hypotenuse and a leg of one right triangle are congruent to the hypotenuse and a leg of another right triangle, then the two triangles are congruent. (HL Postulate)

It is important to note that Postulate 9 can be used *only* if the two triangles are *right* triangles. In order to use this postulate in a proof, we must show that each triangle contains a right angle.

EXAMPLE

GIVEN: $\overline{RS} \perp \overline{PQ}$;
$\overline{PR} \cong \overline{QR}$

PROVE: $\triangle PRS \cong \triangle QRS$

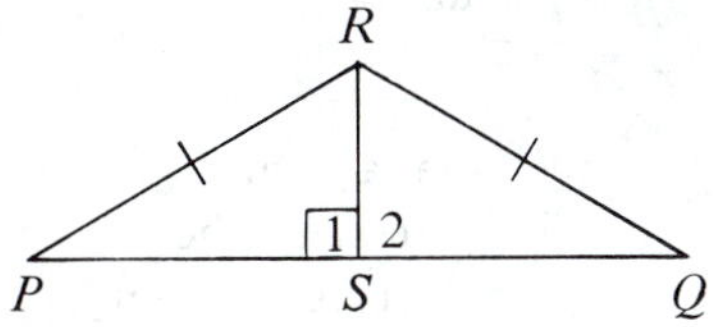

Statements	Reasons
1. $\overline{RS} \perp \overline{PQ}$	1. Given
2. $\angle 1$ is a right angle; $\angle 2$ is a right angle.	2. Def. of $\perp$ lines
3. $\overline{PR} \cong \overline{QR}$	3. Given
4. $\overline{RS} \cong \overline{RS}$	4. Reflexive Prop. of $\cong$
5. $\triangle PRS \cong \triangle QRS$	5. HL Postulate

It may happen that two triangles to be proved congruent overlap. In such a case, look for a common side or angle and use the Reflexive Property of Congruence. You may find it helpful at first to redraw the overlapping triangles as separate figures.

EXAMPLE

GIVEN: $\overline{MK} \cong \overline{LJ}$;
$\angle 1 \cong \angle 2$

PROVE: $\triangle MJK \cong \triangle LKJ$

Redraw $\triangle MJK$ and $\triangle LKJ$ as separate figures.

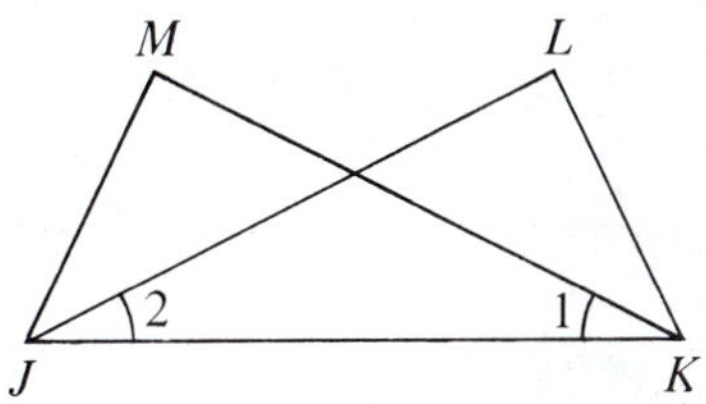

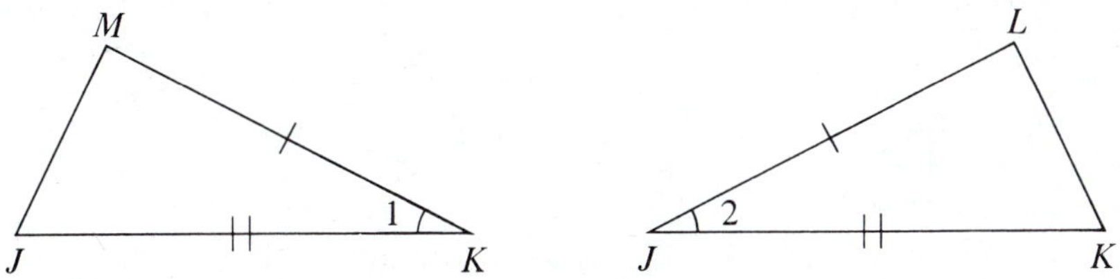

Statements	Reasons
1. $\overline{MK} \cong \overline{LJ}$	1. Given
2. $\angle 1 \cong \angle 2$	2. Given
3. $\overline{JK} \cong \overline{KJ}$	3. Reflexive Prop. of $\cong$
4. $\triangle MJK \cong \triangle LKJ$	4. SAS Postulate

EXAMPLE

GIVEN: $\overline{ZA} \cong \overline{ZB}$;
$\overline{AX} \cong \overline{BY}$

PROVE: $\triangle XZB \cong \triangle YZA$

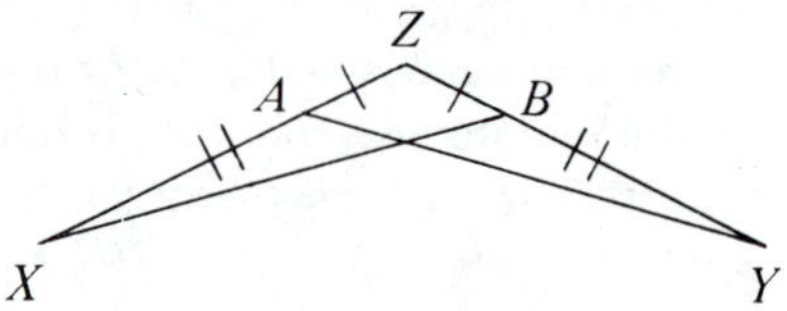

Statements	Reasons
1. $\overline{ZA} \cong \overline{ZB}$, or $ZA = ZB$	1. Given
2. $\overline{AX} \cong \overline{BY}$, or $AX = BY$	2. Given
3. $ZA + AX = ZB + BY$	3. Addition Prop. of Equality
4. $ZA + AX = ZX$; $ZB + BY = ZY$	4. Prop. of Betweenness
5. $ZX = ZY$, or $ZX \cong ZY$	5. Substitution Property
6. $\angle Z \cong \angle Z$	6. Reflexive Prop. of $\cong$
7. $\triangle XZB \cong \triangle YZA$	7. SAS Postulate

Although for the sake of simplicity we have stated as postulates the facts that two triangles are congruent by SAS, ASA, SSS, and HL, some of these postulates can be proved as theorems. This idea is explored in the appendix, The Triangle Congruence Postulates Revisited.

ORAL EXERCISES

Given the information shown in the diagram, state which of the four postulates SSS, ASA, SAS, or HL you would use to show that the triangles are congruent.

1.

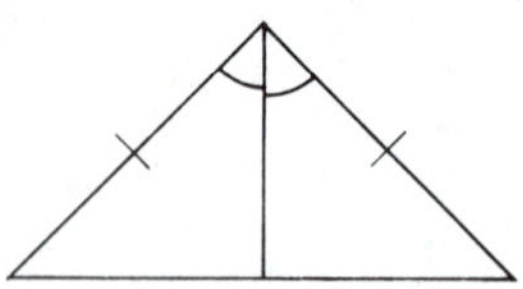

2.

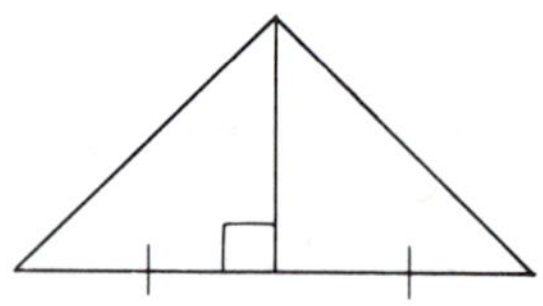

3.

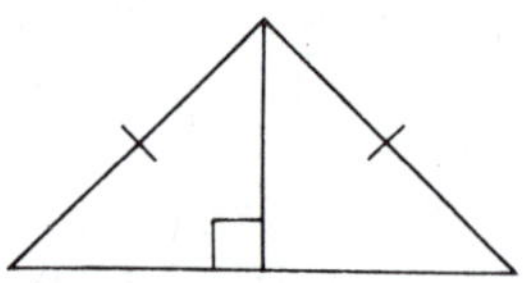

4.

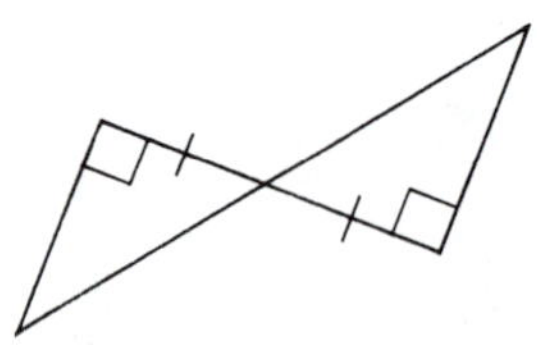

5.

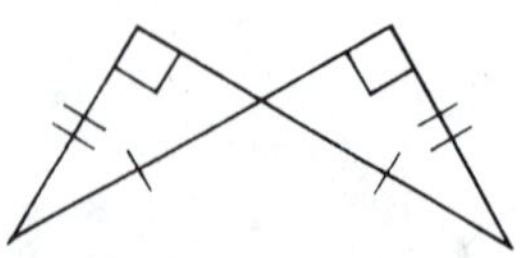

6. $\triangle RST \cong \triangle$ ___?___

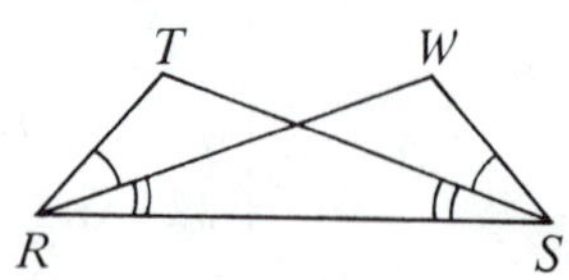

WRITTEN EXERCISES

Name a pair of triangles you could prove congruent using the given information. Also state which of the four postulates SSS, ASA, SAS, or HL you would use.

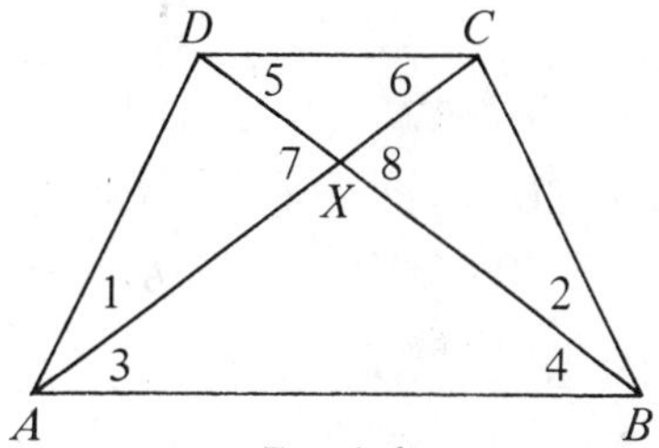

Exs. 1-6

A 1. Given: $\angle 7$ and $\angle 8$ are right angles; $\overline{DX} \cong \overline{CX}$; $\overline{AD} \cong \overline{BC}$

2. Given: $\overline{AX} \cong \overline{BX}$; $\angle 1 \cong \angle 2$

3. Given: $\overline{AD} \cong \overline{BC}$; $\angle ADC \cong \angle BCD$

4. Given: $\overline{AC} \cong \overline{BD}$; $\angle 3 \cong \angle 4$

5. Given: $\angle 5 \cong \angle 6$; $\overline{AC} \cong \overline{BD}$

6. Name *two* pairs of triangles that could be proved congruent, given $\overline{AD} \cong \overline{BC}$ and $\overline{AC} \cong \overline{BD}$.

Use the given information to write a two-column proof that $\triangle ABD \cong \triangle CBD$.

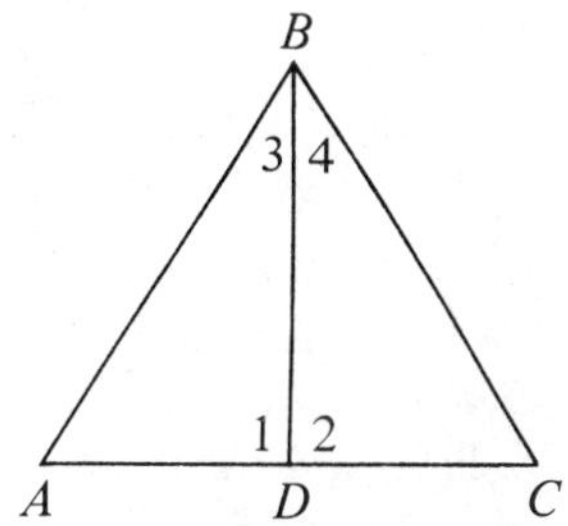

Exs. 7-12

7. Given: $\overline{AB} \cong \overline{CB}$; $\angle 3 \cong \angle 4$

8. Given: $\overline{AD} \cong \overline{CD}$; $\overline{AB} \cong \overline{CB}$

9. Given: $\overline{BD} \perp \overline{AC}$; $\overline{AB} \cong \overline{CB}$

10. Given: $\overline{BD} \perp \overline{AC}$; $\overline{AD} \cong \overline{CD}$

11. Given: $\angle 1 \cong \angle 2$; $\angle 3 \cong \angle 4$

12. Given: $\overline{AB} \cong \overline{CB}$; $\overrightarrow{BD}$ bisects $\angle ABC$.

Write two-column proofs.

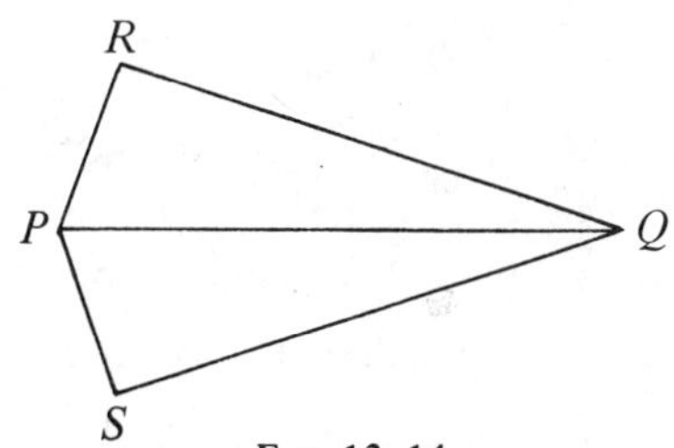

Exs. 13, 14

13. Given: $\overline{PR} \perp \overline{RQ}$; $\overline{PS} \perp \overline{SQ}$; $\overline{PR} \cong \overline{PS}$
Prove: $\triangle PRQ \cong \triangle PSQ$

14. Given: $\overline{PR} \perp \overline{RQ}$; $\overline{PS} \perp \overline{SQ}$; $\overline{RQ} \cong \overline{SQ}$
Prove: $\triangle PRQ \cong \triangle PSQ$

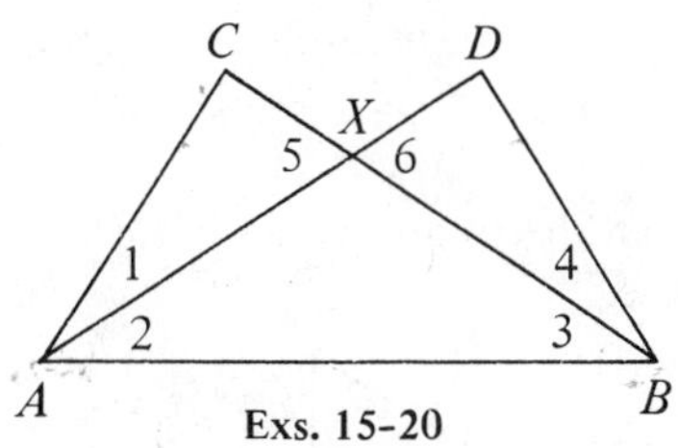

Exs. 15-20

15. Given: $\overline{BC} \cong \overline{AD}$; $\overline{AC} \cong \overline{BD}$
Prove: $\triangle ABC \cong \triangle BAD$

16. Given: $\overline{AX} \cong \overline{BX}$; $\overline{CX} \cong \overline{DX}$
Prove: $\triangle ACX \cong \triangle BDX$

17. Given: $\overline{AC} \cong \overline{BD}$; $\angle CAB \cong \angle DBA$
Prove: $\triangle BAC \cong \triangle ABD$

18. Given: $\overline{AC} \perp \overline{BC}$; $\overline{BD} \perp \overline{AD}$; $\overline{AC} \cong \overline{BD}$
Prove: $\triangle ABC \cong \triangle BAD$

B 19. Given: $\angle 1 \cong \angle 4$; $\angle CAB \cong \angle DBA$
Prove: $\triangle ABD \cong \triangle BAC$

20. Given: $\angle 1 \cong \angle 4$; $\angle 2 \cong \angle 3$
Prove: $\triangle ABD \cong \triangle BAC$

21. Given: $\angle P \cong \angle Q$; $\overline{PR} \cong \overline{QR}$
Prove: $\triangle PRY \cong \triangle QRX$

22. Given: $\overline{RY} \cong \overline{RX}$; $\overline{PR} \cong \overline{QR}$
Prove: $\triangle PRY \cong \triangle QRX$

23. Given: $\overline{PY} \perp \overline{RQ}$; $\overline{QX} \perp \overline{PR}$; $\overline{RY} \cong \overline{RX}$
Prove: $\triangle PRY \cong \triangle QRX$

24. Given: $\overline{PR} \cong \overline{QR}$; $\overline{PX} \cong \overline{QY}$
Prove: $\triangle PRY \cong \triangle QRX$

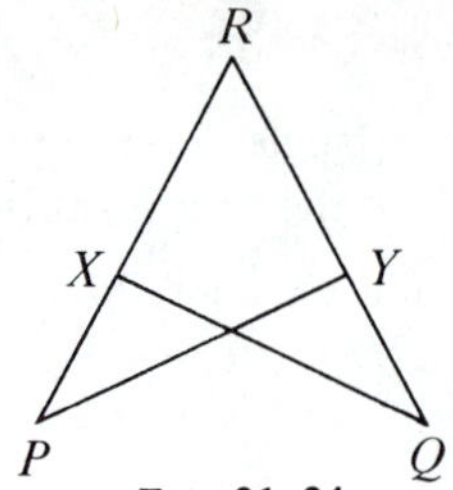

Exs. 21–24

25. Given: $\overline{AB} \cong \overline{CB}$; $\angle 1 \cong \angle 2$
Prove: $\triangle ABD \cong \triangle CBD$

26. Given: $\overline{AB} \cong \overline{CB}$;
$m\angle 1 = 90$
Prove: $\triangle ABD \cong \triangle CBD$

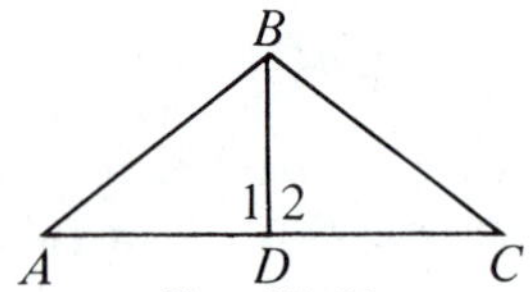

Exs. 25, 26

27. Given: $\overline{JK} \perp \overline{KM}$; $\overline{LM} \perp \overline{KM}$; $\angle 4 \cong \angle 1$
Prove: $\triangle JKM \cong \triangle LMK$

28. Given: $\angle 1$ and $\angle 2$ are complementary;
$\angle 3$ and $\angle 4$ are complementary;
$\overline{JM} \cong \overline{LK}$
Prove: $\triangle JKM \cong \triangle LMK$

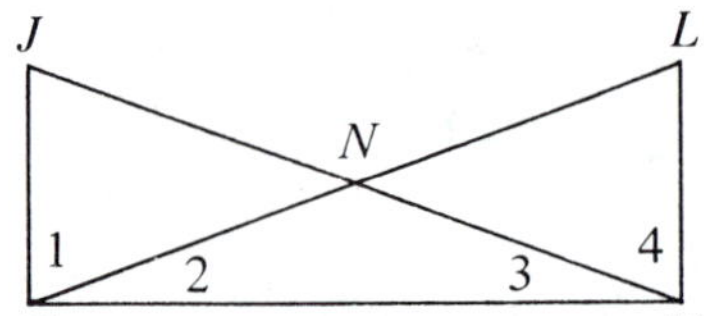

Exs. 27, 28

SELF-TEST 2

If $m\angle 1 = 6x - 23$ and $m\angle 2 = 2x - 5$, find the numerical measure of each angle when:

1. $\angle 1$ and $\angle 2$ are supplementary. Section 3-4
2. $\angle 1$ and $\angle 2$ are vertical angles.
3. Write a two-column proof.
Given: $\overrightarrow{OC} \perp \overleftrightarrow{AE}$;
$m\angle AOB = m\angle DOE$
Prove: $\overrightarrow{OC}$ bisects $\angle BOD$.

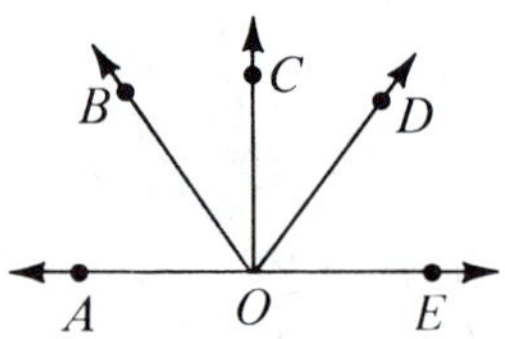

Can the two triangles be proved congruent? If so, state which postulate you would use.

4. 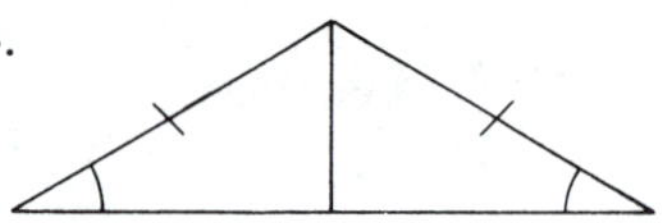

5. 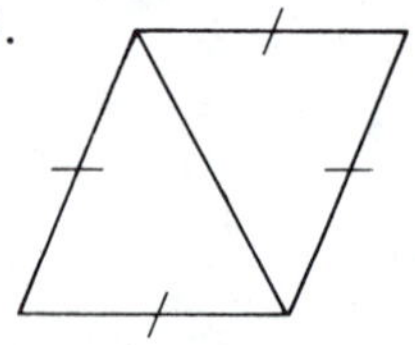

Section 3-5

6. Write a two-column proof.

Given: $\overline{RL}$ bisects $\overline{ST}$;

$\angle S \cong \angle T$

Prove: $\triangle RMS \cong \triangle LMT$

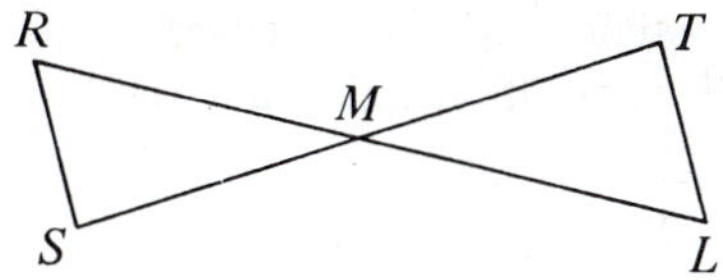

State which postulate you could use to prove that $\triangle EFG \cong \triangle GHE$.

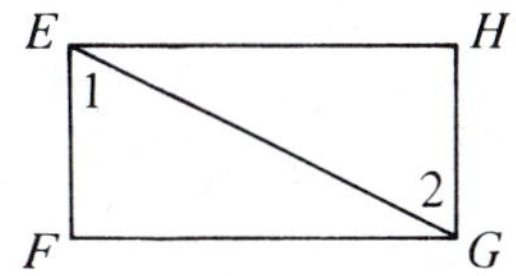

7. $\overline{EF} \perp \overline{FG}$; $\overline{EH} \perp \overline{HG}$; $\overline{EF} \cong \overline{GH}$ Section 3-6

8. $\angle 1 \cong \angle 2$; $\overline{EF} \cong \overline{GH}$

9. Write a two-column proof.

Given: $\angle 3 \cong \angle 4$;

$\overline{WX} \cong \overline{YX}$

Prove: $\triangle XWZ \cong \triangle XYZ$

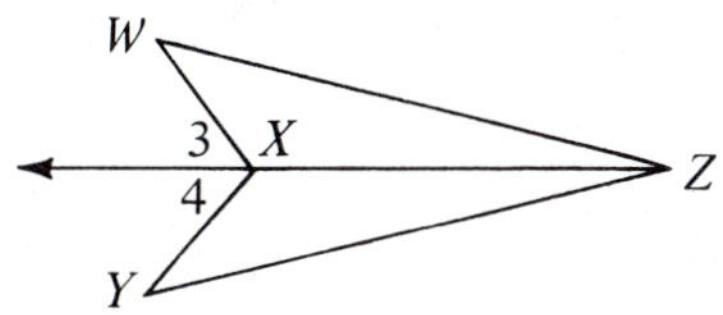

Section 3-7 CORRESPONDING PARTS OF CONGRUENT TRIANGLES

If you have proved two triangles congruent by one of the four postulates, SSS, ASA, SAS, or HL, then *all the other* corresponding parts are also congruent, by the definition of congruent triangles. You can therefore give as a reason for stating such a congruence in a proof:

Corresponding parts of congruent triangles are congruent.

(You can abbreviate this: "Corr. parts of $\cong$ ▲s are $\cong$.")

EXAMPLE

GIVEN: $\overline{AC} \cong \overline{BC}$;

$\overline{AD} \cong \overline{BD}$

PROVE: $\angle 1 \cong \angle 2$

Method: Find two congruent triangles of which $\angle 1$ and $\angle 2$ are corresponding parts, then prove that these triangles are congruent.

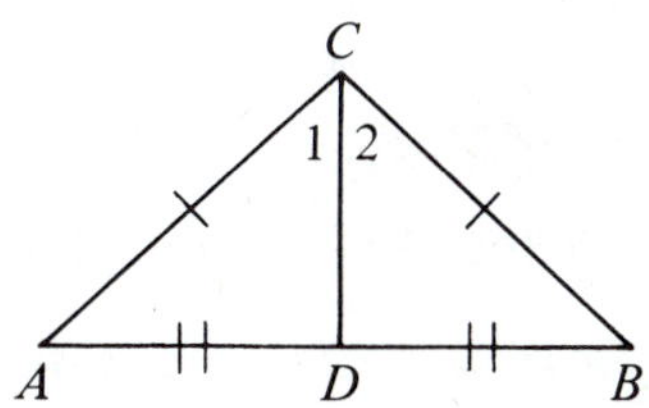

Statements	Reasons
1. $\overline{AC} \cong \overline{BC}$; $\overline{AD} \cong \overline{BD}$	1. Given
2. $\overline{CD} \cong \overline{CD}$	2. Reflexive Prop. of $\cong$
3. $\triangle ACD \cong \triangle BCD$	3. SSS Postulate
4. $\angle 1 \cong \angle 2$	4. Corr. parts of $\cong$ ▲s are $\cong$.

It is often possible to prove that two triangles are congruent if they have parts in common with a pair of triangles that are known to be congruent.

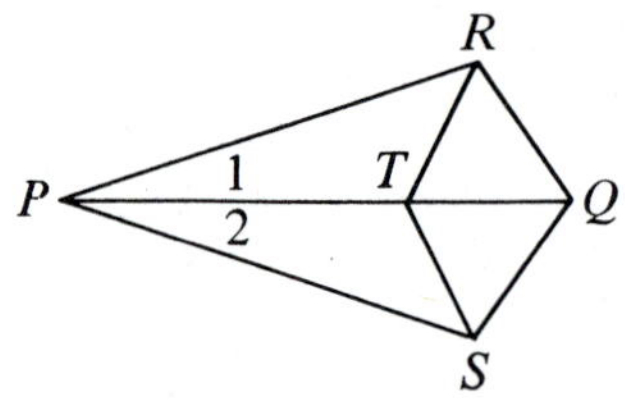

EXAMPLE

GIVEN: $\triangle PRQ \cong \triangle PSQ$
PROVE: $\triangle PRT \cong \triangle PST$

Method: First use the given congruence to get $\overline{PR} \cong \overline{PS}$ and $\angle 1 \cong \angle 2$.

Statements	Reasons
1. $\triangle PRQ \cong \triangle PSQ$	1. Given
2. $\overline{PR} \cong \overline{PS}$; $\angle 1 \cong \angle 2$	2. Corr. parts of $\cong$ ▲ are $\cong$.
3. $\overline{PT} \cong \overline{PT}$	3. Reflexive Prop. of $\cong$
4. $\triangle PRT \cong \triangle PST$	4. SAS Postulate

ORAL EXERCISES

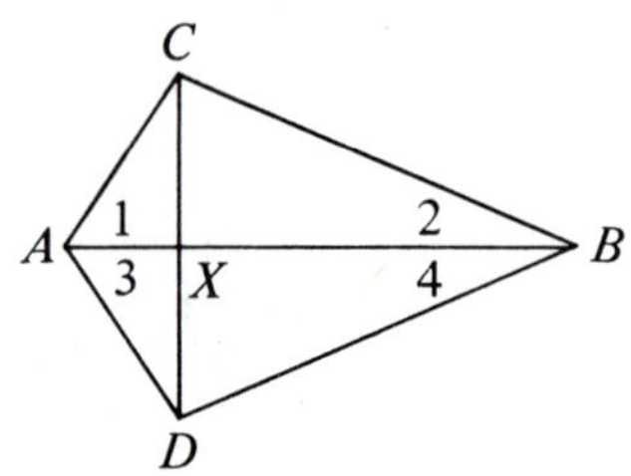

State all parts common to the two given triangles.

1. $\triangle ACB$ and $\triangle ACX$ **2.** $\triangle ADX$ and $\triangle BDX$

3. $\triangle ACX$ and $\triangle BCX$ **4.** $\triangle ACB$ and $\triangle BCX$

5. $\triangle ADX$ and $\triangle ADB$ **6.** $\triangle BDX$ and $\triangle ADB$

Which corresponding congruent parts of the triangles that are known to be congruent are also corresponding parts of the triangles you would like to prove congruent?

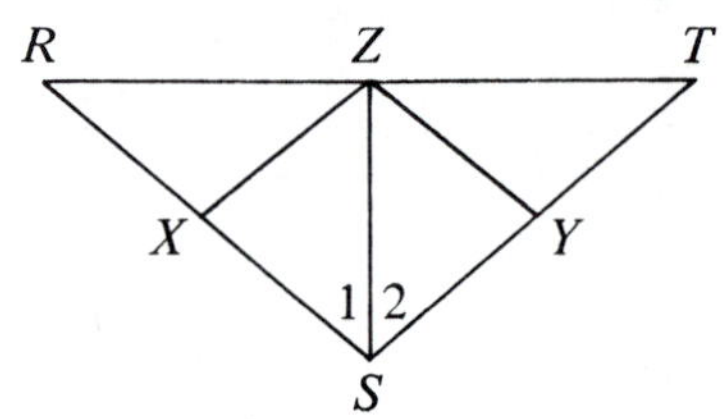

7. Given: $\triangle RZX \cong \triangle TZY$
Prove: $\triangle SZX \cong \triangle SZY$

8. Given: $\triangle SXZ \cong \triangle SYZ$
Prove: $\triangle SRZ \cong \triangle STZ$

9. Given: $\triangle SRZ \cong \triangle STZ$
Prove: $\triangle RZX \cong \triangle TZY$

WRITTEN EXERCISES

Name all pairs of triangles that have the given angles or line segments as corresponding parts.

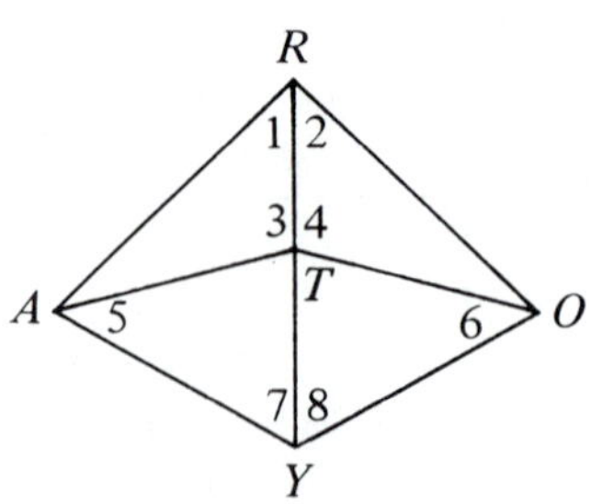

A **1.** $\angle 3$ and $\angle 4$

2. $\angle 5$ and $\angle 6$

3. $\overline{AY}$ and $\overline{OY}$

4. $\overline{AT}$ and $\overline{OT}$

5. $\angle 1$ and $\angle 2$

6. $\angle 7$ and $\angle 8$

Write two-column proofs.

7. Given: $\overline{AP} \cong \overline{DP}$;
$\angle A \cong \angle D$
Prove: $\overline{BP} \cong \overline{CP}$

8. Given: $\overline{AP} \cong \overline{DP}$;
$\overline{BP} \cong \overline{CP}$
Prove: $\overline{AB} \cong \overline{DC}$

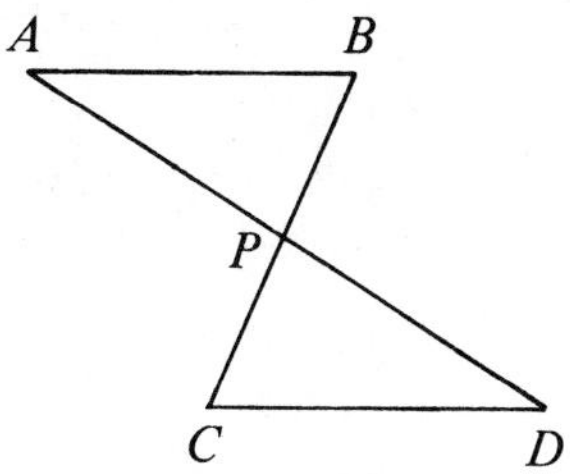

Exs. 7, 8

9. Given: $\overline{AX} \cong \overline{BX}$; $\overline{AY} \cong \overline{BY}$
Prove: $\angle 1 \cong \angle 2$

10. Given: $\overline{AX} \cong \overline{BX}$; $\angle 1 \cong \angle 2$
Prove: $\angle 3 \cong \angle 4$

11. Given: $m\angle A = 90$; $m\angle B = 90$;
$\overline{AY} \cong \overline{BY}$
Prove: $\overline{AX} \cong \overline{BX}$

12. Given: $\angle 1 \cong \angle 2$; $\angle 3 \cong \angle 4$
Prove: $\overline{AX} \cong \overline{BX}$

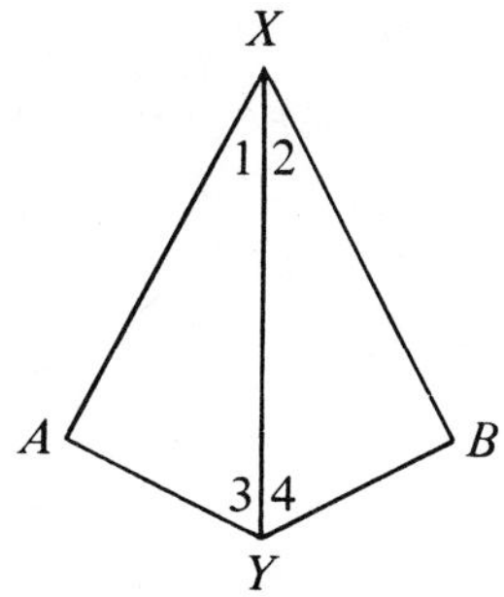

Exs. 9–12

13. Given: $\overline{GH} \perp \overline{JK}$; $\overline{JH} \cong \overline{KH}$
Prove: $\overline{JG} \cong \overline{KG}$

14. Given: $\overline{GH} \perp \overline{JK}$; $\overline{JG} \cong \overline{KG}$
Prove: $\angle J \cong \angle K$

B 15. Given: $\overline{JG} \cong \overline{KG}$;
$\overrightarrow{GH}$ bisects $\angle JGK$.
Prove: $\overline{GH} \perp \overline{JK}$

16. Given: $\overline{JG} \cong \overline{KG}$;
H is the midpoint of $\overline{JK}$.
Prove: $\overrightarrow{GH}$ bisects $\angle JGK$.

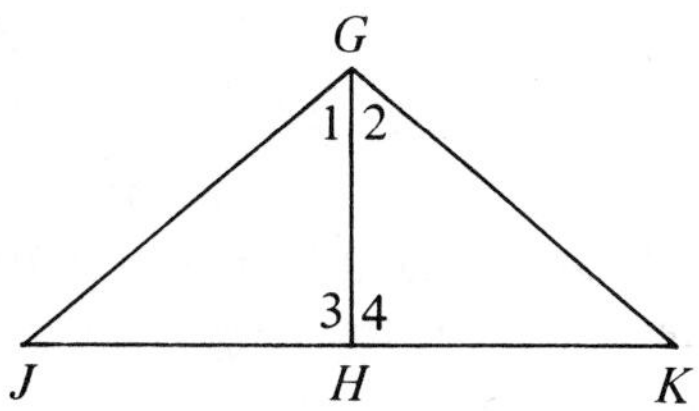

Exs. 13–16

17. Given: $\triangle PRQ \cong \triangle PSQ$
Prove: $\triangle QRX \cong \triangle QSX$

18. Given: $\triangle PRX \cong \triangle PSX$
Prove: $\triangle PRQ \cong \triangle PSQ$

19. Given: $\triangle PRX \cong \triangle PSX$
Prove: $\triangle QRX \cong \triangle QSX$
(*Hint:* $\angle 3$ and $\angle RXQ$ are supp.;
$\angle 4$ and $\angle SXQ$ are supp.)

20. Given: $\triangle QRX \cong \triangle QSX$
Prove: $\triangle PRX \cong \triangle PSX$
(*Hint:* See the hint for Exercise 19.)

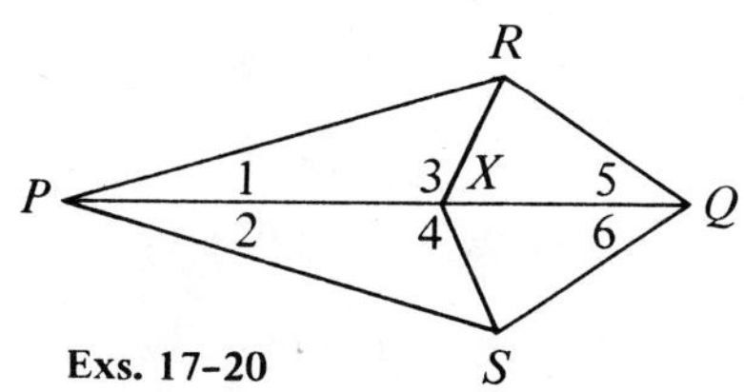

Exs. 17–20

21. Given: $\triangle DAF \cong \triangle DBF$; $\overline{DE} \cong \overline{DC}$
Prove: $\triangle DEF \cong \triangle DCF$

22. Given: $\triangle DEF \cong \triangle DCF$; $\overline{DA} \cong \overline{DB}$
Prove: $\triangle DAF \cong \triangle DBF$

23. Given: $\triangle DEF \cong \triangle DCF$; $\angle 1 \cong \angle 4$
Prove: $\triangle EFA \cong \triangle CFB$

24. Given: $\triangle EFA \cong \triangle CFB$; $\overline{FA} \perp \overline{DE}$; $\overline{FB} \perp \overline{DC}$
Prove: $\triangle DAF \cong \triangle DBF$

25. Given: $\triangle EFA \cong \triangle CFB$; $\overline{DF} \perp \overline{EC}$
Prove: $\triangle DAF \cong \triangle DBF$

26. Given: $\triangle DEF \cong \triangle DCF$; $\angle 2 \cong \angle 3$
Prove: $\triangle EFA \cong \triangle CFB$

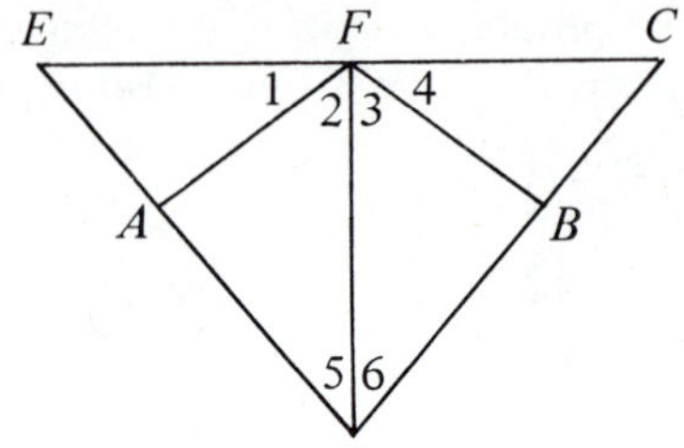

Exs. 21-26

C **27.** Given: $\overline{GH} \perp \overline{YZ}$; $\angle 1 \cong \angle 2$
Prove: $\triangle GKZ \cong \triangle HKZ$

28. Given: $\overline{YG} \cong \overline{YH}$; $\overline{ZG} \cong \overline{ZH}$
Prove: $\overline{GH} \perp \overline{YZ}$

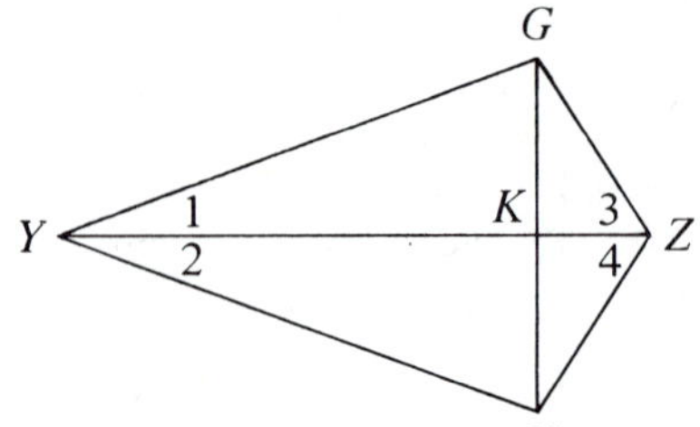

Exs. 27, 28

29. Given: $\overline{MR} \cong \overline{PR}$; $\overline{MS} \cong \overline{PS}$
Prove: $\overline{MX} \cong \overline{PX}$

30. Given: $\overline{MR} \cong \overline{PR}$; $\overline{MX} \cong \overline{PX}$
Prove: $\angle 3 \cong \angle 4$

31. Given: $\overline{RS} \perp \overline{MP}$; $\overline{MX} \cong \overline{PX}$
Prove: $\angle 1 \cong \angle 2$

32. Given: $\overline{RS} \perp \overline{MP}$; $\overline{MR} \cong \overline{PR}$
Prove: $\overline{MX} \cong \overline{PX}$

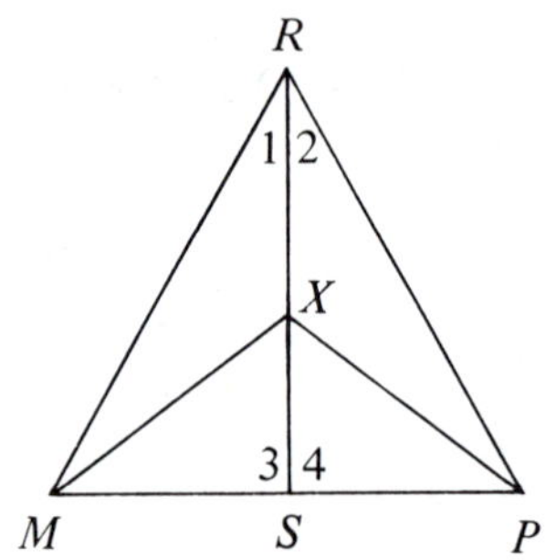

Exs. 29-32

Section 3-8 ISOSCELES TRIANGLES

Certain types of triangles play an important role in geometry. In this section you will learn about the following triangles and special segments associated with every triangle:

- isosceles triangles
- equilateral triangles
- medians of a triangle

An *isosceles triangle* is a triangle that has at least two congruent sides. The congruent sides are called the *legs* of the isosceles triangle, and the angle included between them is called the *vertex angle*. The side opposite this angle is called the *base*. The two angles of the triangle that include the base are called *base angles*.

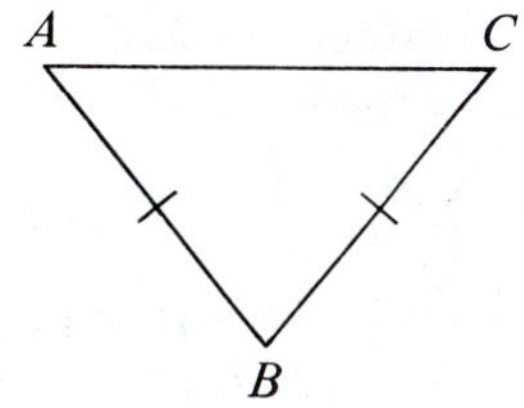

$\triangle ABC$ is an isosceles triangle since $\overline{AB} \cong \overline{CB}$.

legs:	$\overline{AB}$, $\overline{CB}$
vertex angle:	$\angle B$
base:	$\overline{AC}$
base angles:	$\angle A$, $\angle C$

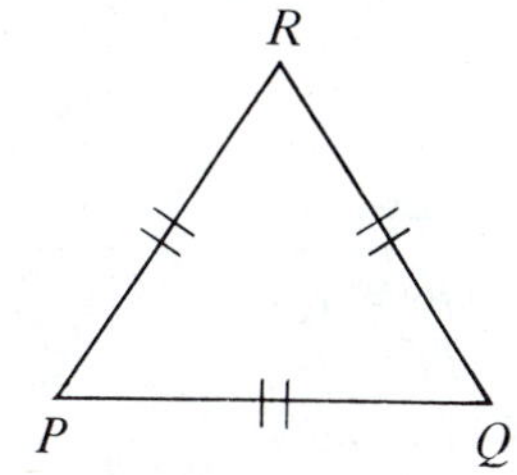

An *equilateral triangle* is a triangle that has three congruent sides. $\triangle PRQ$ is equilateral since $\overline{PR} \cong \overline{RQ} \cong \overline{PQ}$.

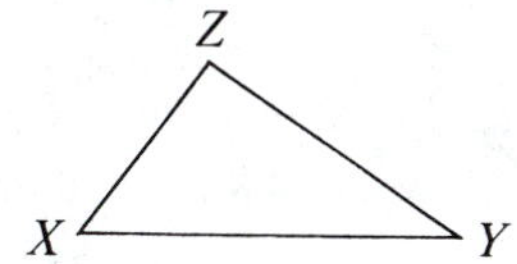

A triangle with no two sides congruent is called a *scalene triangle*. $\triangle XYZ$ is a scalene triangle.

An isosceles triangle is *defined* as a triangle with at least two congruent sides, but it also has at least two congruent angles. Our proof of this theorem requires the use of a segment that does not appear in the given geometric situation. Such segments (or lines) are called *auxiliary lines*. A justification must be given for the existence of any auxiliary line used.

Theorem 3–13 If two sides of a triangle are congruent, then the angles opposite these sides are congruent. (Base angles of an isosceles triangle are congruent.)

GIVEN: $\overline{AB} \cong \overline{CB}$
PROVE: $\angle A \cong \angle C$

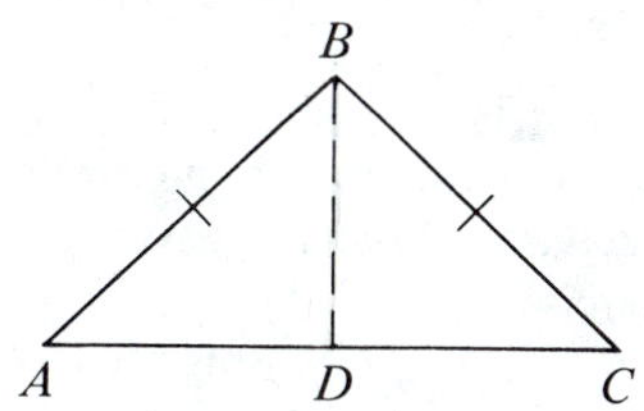

Statements	Reasons
1. Draw the bisector of $\angle ABC$ and let D be the point where it intersects $\overline{AC}$.	1. Any angle of measure less than 180 has exactly one bisector.
2. $\angle ABD \cong \angle CBD$	2. Def. of angle bisector
3. $\overline{AB} \cong \overline{CB}$	3. Given
4. $\overline{BD} \cong \overline{BD}$	4. Reflexive Prop. of $\cong$
5. $\triangle ABD \cong \triangle CBD$	5. SAS Postulate
6. $\angle A \cong \angle C$	6. Corr. parts of $\cong$ $\triangle$s are $\cong$.

In order to use Theorem 3-13, the two congruent sides must be sides of the *same* triangle. In the following example, the theorem is applied to $\triangle PQR$.

EXAMPLE

GIVEN: $\overline{PS} \cong \overline{QT}$;
$\overline{PR} \cong \overline{QR}$

PROVE: $\triangle PRS \cong \triangle QRT$

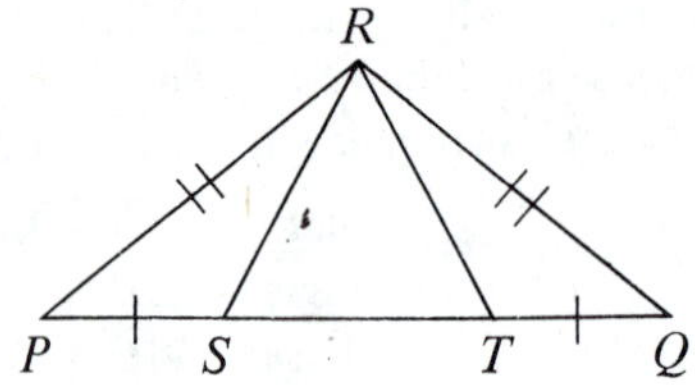

Statements	Reasons
1. $\overline{PS} \cong \overline{QT}$; $\overline{PR} \cong \overline{QR}$	1. Given
2. $\angle P \cong \angle Q$	2. Base $\angle$s of an isos. $\triangle$ are $\cong$.
3. $\triangle PRS \cong \triangle QRT$	3. SAS Postulate

The converse of Theorem 3-13 is also true.

Theorem 3-14 If two angles of a triangle are congruent, then the sides opposite these angles are congruent.

GIVEN: $\angle A \cong \angle C$

PROVE: $\overline{AB} \cong \overline{CB}$

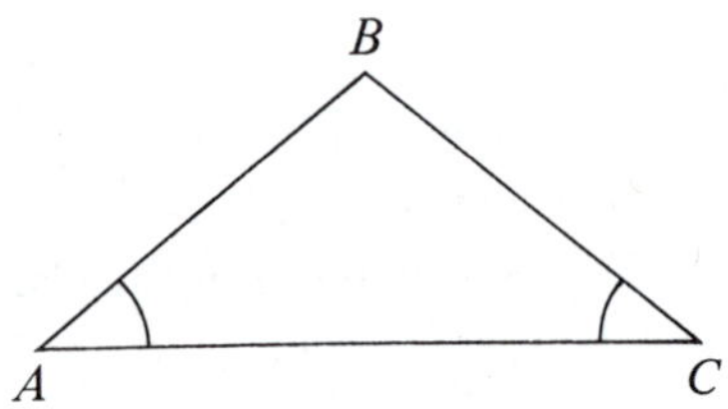

Theorem 3-14 will be easier to prove after some other theorems have been proved, and a simple proof of it will be found in Section 4-4.

As you will show in the exercise set, the bisector of the vertex angle of an isosceles triangle has some interesting properties. It is perpendicular to the base of the triangle (See Exercise 22.) and it passes through the midpoint of the base, as you will show in Exercise 19.

A segment that joins *any* vertex of a triangle to the midpoint of the opposite side is called a *median* of the triangle. Thus every triangle has exactly three medians. The medians of $\triangle DEF$ are $\overline{DP}$, $\overline{EQ}$, and $\overline{FR}$.

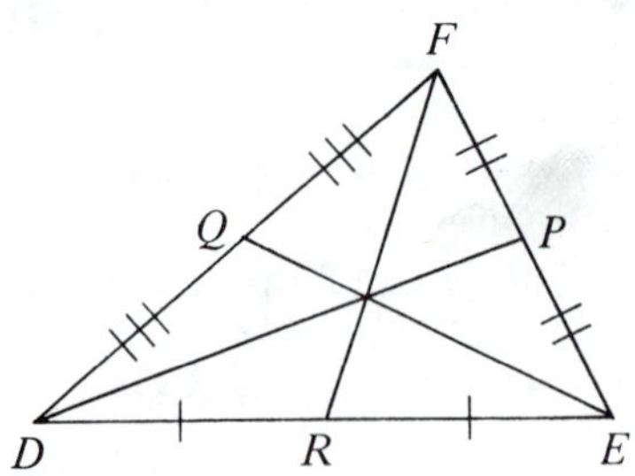

ORAL EXERCISES

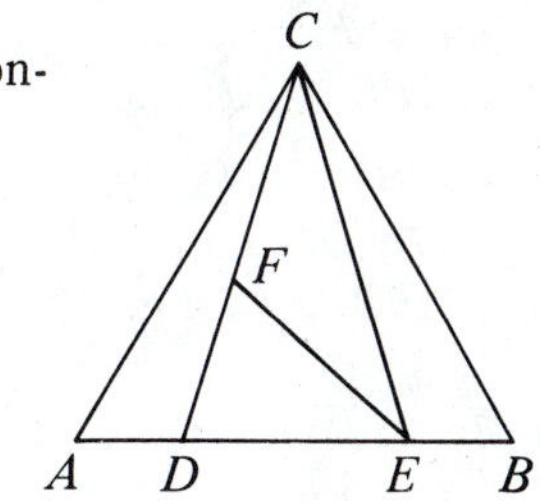

If the given line segments are congruent, which angles are congruent?

1. $\overline{AC} \cong \overline{BC}$
2. $\overline{EF} \cong \overline{CF}$
3. $\overline{EF} \cong \overline{ED}$
4. $\overline{DC} \cong \overline{EC}$ (Name two pairs of congruent angles.)

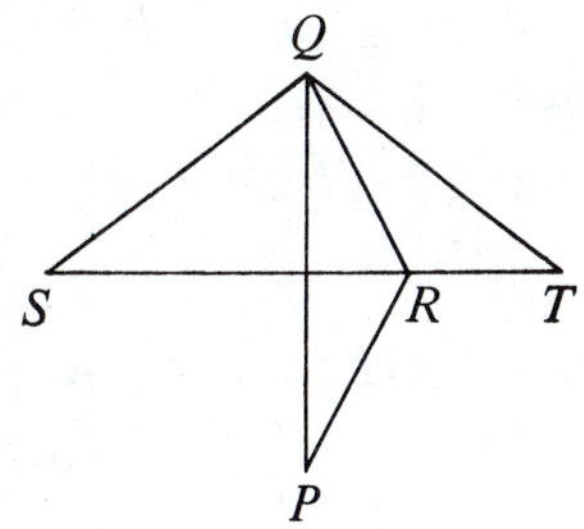

If the given angles are congruent, which segments are congruent?

5. $\angle S \cong \angle T$
6. $\angle P \cong \angle PQR$
7. $\angle SRQ \cong \angle SQR$
8. $\angle T \cong \angle RQT$

If the sides of $\triangle DEF$ have the given lengths, state whether $\triangle DEF$ must be: (a) equilateral; (b) isosceles, but not equilateral; (c) scalene.

9. $DE = 3,\ EF = 4,\ DF = 5$
10. $DE = 5,\ EF = 7,\ DF = 5$
11. $DE = 2x,\ EF = 2x,\ DF = 2x$
12. $DE = x,\ EF = x - 4,\ DF = x + 4$

Draw a picture to illustrate that each of the following statements is *false*.

13. A median of a triangle must be perpendicular to the side to which it is drawn.
14. A median of a triangle must bisect the angle at the vertex from which it is drawn.

WRITTEN EXERCISES

In $\triangle ABC$, $\overline{AC} \cong \overline{BC}$. Find the value of x.

A

1. $m\angle A = 2x - 25;\ m\angle B = x + 15$
2. $m\angle A = 2x + 19;\ m\angle B = 3x - 12$
3. $m\angle A = 5x - 28;\ m\angle B = 3x$
4. $m\angle A = 128 - x;\ m\angle B = x + 50$

In $\triangle ABC$, $\angle A \cong \angle B$. Find the value of x.

5. $AC = 2x + 17;\ BC = 5x - 25$
6. $AC = 7 - x;\ BC = 3x - 9$
7. $AC = 14 + 2x;\ BC = 59 - 3x$
8. $AC = 12 - 3x;\ BC = 12 + x$

Write two-column proofs.

9. Given: $\overline{RS} \cong \overline{RT}$
   Prove: $\angle 5 \cong \angle 6$
10. Given: $\angle 5 \cong \angle 6$
    Prove: $\overline{RS} \cong \overline{RT}$
11. Given: $\angle 1 \cong \angle 4$
    Prove: $\overline{RS} \cong \overline{RT}$
12. Given: $\overline{RS} \cong \overline{RT}$
    Prove: $\angle 1 \cong \angle 4$

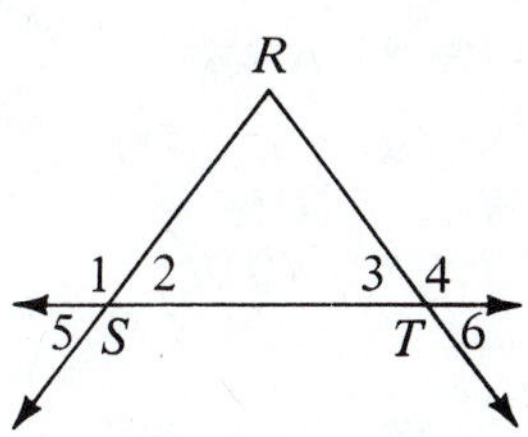

13. Given: $\overline{AE} \cong \overline{DE}$;
$\angle 1 \cong \angle 2$
Prove: $\triangle AEB \cong \triangle DEC$

14. Given: $\angle A \cong \angle D$;
$\angle 1 \cong \angle 2$
Prove: $\triangle AEB \cong \triangle DEC$

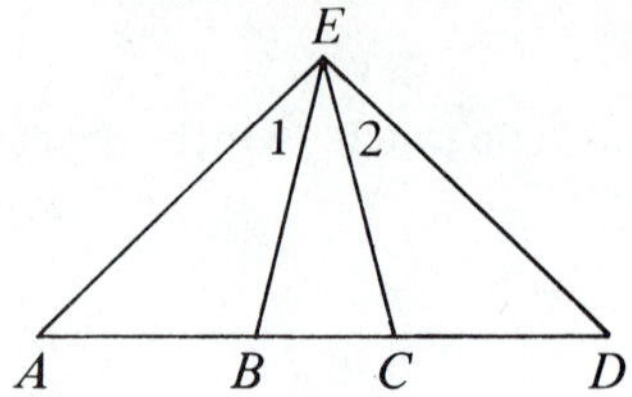

Exs. 13, 14

15. Given: $\overline{PS} \cong \overline{RS}$;
$\overline{PT} \cong \overline{VR}$;
$\angle 3 \cong \angle 4$
Prove: $\angle 1 \cong \angle 2$

16. Given: $\overline{PS} \cong \overline{RS}$;
$\angle 1 \cong \angle 2$;
$\overline{PQ} \cong \overline{RQ}$
Prove: $\angle 3 \cong \angle 4$

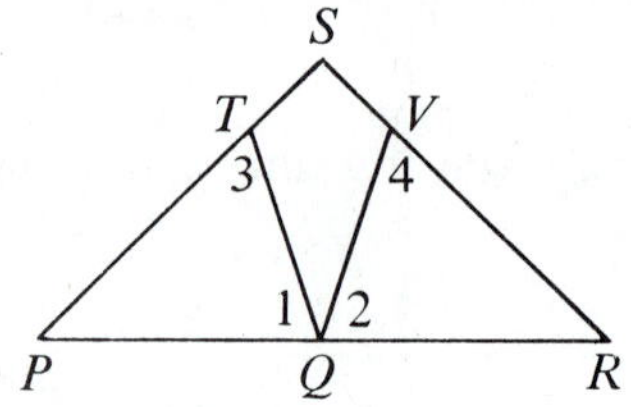

Exs. 15, 16

17. Given: $\angle 1 \cong \angle 2$;
$\angle 3 \cong \angle 4$;
$\angle 5 \cong \angle 6$
Prove: $\overline{LM} \cong \overline{LN}$

18. Given: $\overline{LM} \cong \overline{LN}$
$\angle 5 \cong \angle 6$;
$\angle 1 \cong \angle 2$
Prove: $\overline{JM} \cong \overline{KN}$

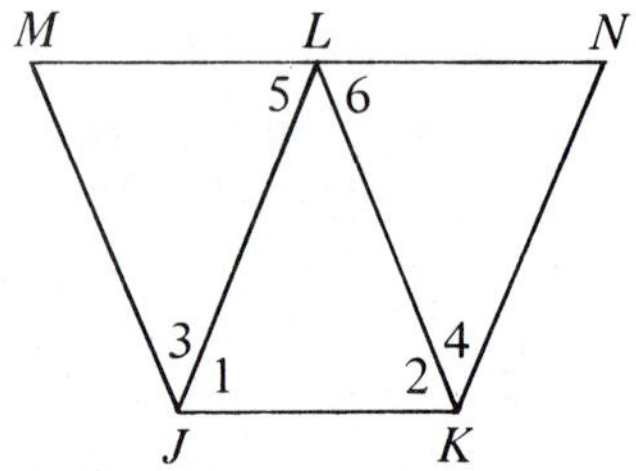

Exs. 17, 18

B 19. Given: $\overline{AB} \cong \overline{CB}$; $\overrightarrow{BD}$ bisects $\angle ABC$.
Prove: $\overline{BD}$ is a median of $\triangle ABC$.

20. Given: $\overline{AB} \cong \overline{CB}$; $\overline{BD}$ is a median of $\triangle ABC$.
Prove: $\overrightarrow{BD}$ bisects $\angle ABC$.

21. Given: $\overline{AB} \cong \overline{CB}$; $\overline{BD} \perp \overline{AC}$
Prove: $\overrightarrow{BD}$ bisects $\angle ABC$.

22. Given: $\overline{AB} \cong \overline{CB}$; $\overrightarrow{BD}$ bisects $\angle ABC$.
Prove: $\overline{BD} \perp \overline{AC}$

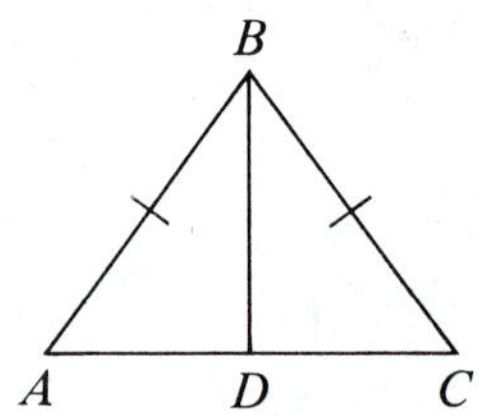

Exs. 19–22

23. Given: $\overline{ML} \cong \overline{NL}$;
$\overline{PM} \cong \overline{QN}$
Prove: $\triangle MPN \cong \triangle NQM$

24. Given: $\overline{ML} \cong \overline{NL}$;
$\angle PNM \cong \angle QMN$
Prove: $\triangle MPN \cong \triangle NQM$

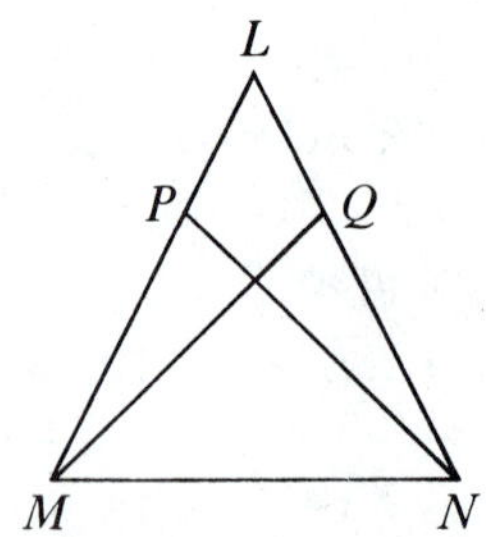

Exs. 23, 24

25. Given: $\angle 1 \cong \angle 2$;
$\overline{AX} \cong \overline{BX}$
Prove: $\overline{AC} \cong \overline{BD}$

26. Given: $\overline{CX} \cong \overline{DX}$;
$\overline{AD} \cong \overline{BC}$
Prove: $\overline{AC} \cong \overline{BD}$

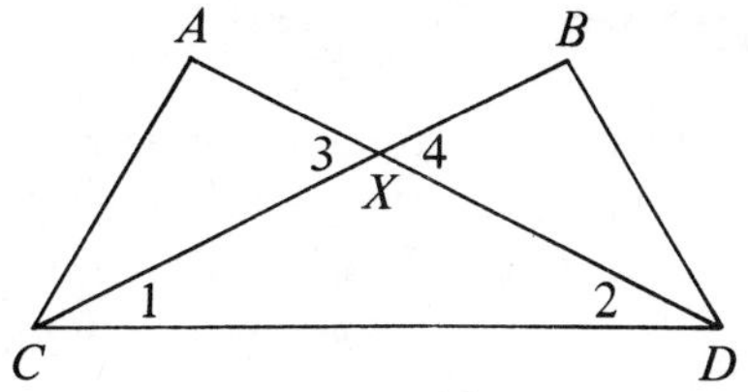

Exs. 25, 26

27. Given: $\overline{SZ} \cong \overline{TZ}$;
$\overline{RS} \cong \overline{WT}$
Prove: $\triangle RZW$ is isosceles.

28. Given: $\overline{RZ} \cong \overline{WZ}$;
$\overline{RT} \cong \overline{WS}$
Prove: $\triangle STZ$ is isosceles.

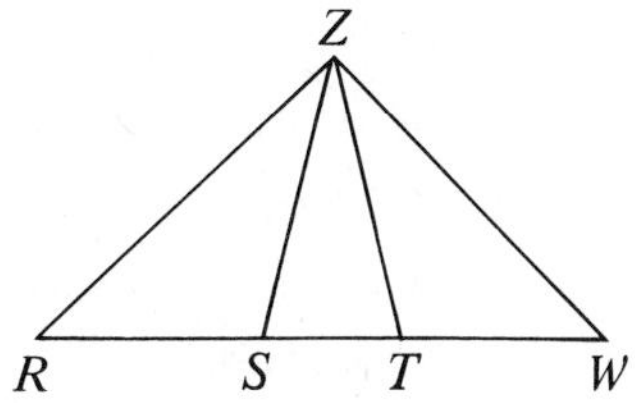

Exs. 27, 28

29. Given: $\overline{DG} \cong \overline{DH}$;
$\overline{AG} \cong \overline{BH}$;
$\overline{AC} \cong \overline{BC}$
Prove: $\overline{AE} \cong \overline{BF}$

30. Given: $\overline{AE} \cong \overline{BF}$;
$\overline{AC} \cong \overline{BC}$;
$\overline{AG} \cong \overline{BH}$
Prove: $\overline{DG} \cong \overline{DH}$

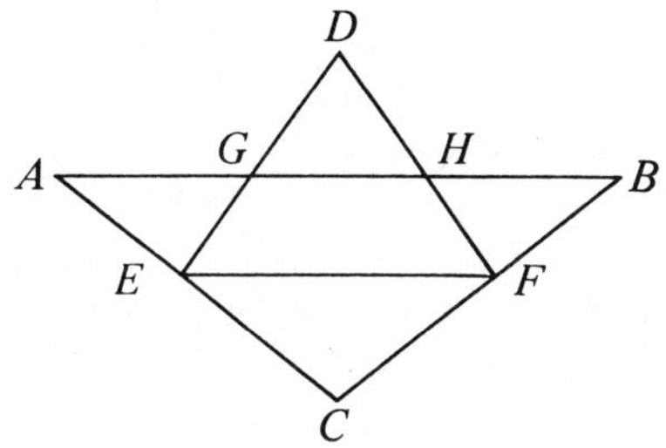

Exs. 29, 30

31. Given: $\angle 1 \cong \angle 2$;
$\angle 5 \cong \angle 6$
Prove: $\angle 3 \cong \angle 4$

C 32. Given: $\overline{RS} \perp \overline{PQ}$;
$\angle 3 \cong \angle 4$
Prove: $\overline{RQ} \cong \overline{SQ}$
(*Hint:* Prove $\triangle PRX \cong \triangle PSX$ to get $\angle 1 \cong \angle 2$. Then prove $\triangle PRQ \cong \triangle PSQ$.)

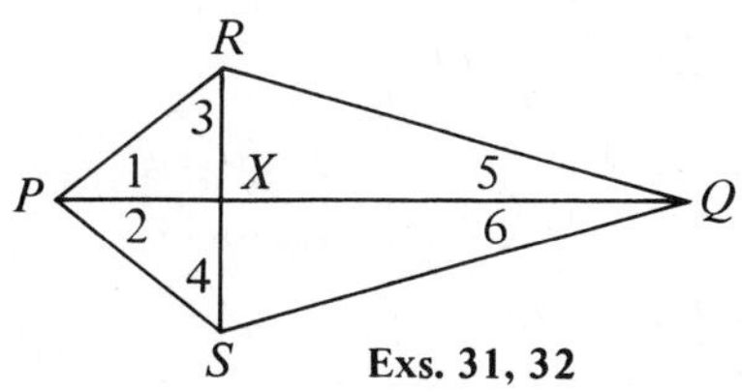

Exs. 31, 32

33. Given: $\overline{LM} \cong \overline{LN}$;
$\angle ZMN \cong \angle ZNM$
Prove: $\overline{ZJ} \cong \overline{ZK}$

34. Given: $\overline{NJ} \perp \overline{ML}$;
$\overline{MK} \perp \overline{NL}$;
$\overline{MK} \cong \overline{NJ}$
Prove: $\triangle LMN$ is isosceles.

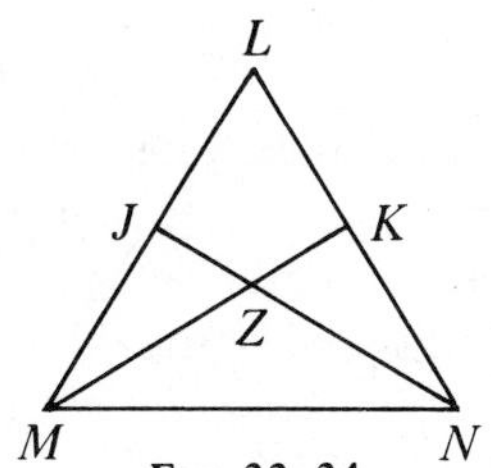

Exs. 33, 34

Section 3-9 GEOMETRIC INEQUALITIES

Many of the facts stated thus far in our study of geometry have expressed equality of two numbers—equality of lengths or of angle measures. In this section you will learn about geometric inequalities. In particular, you will learn:

- how the measure of an exterior angle of a triangle compares with the measure of either remote interior angle
- what can be concluded from the presence of two angles of unequal measure or two sides of unequal length in a triangle
- how the lengths of the sides of any triangle are related

In addition to the properties of inequality listed in Section 1-5, we have the following properties for real numbers a, b, and c.

Properties of Inequality

If $a = b + c$ and c is a positive number, then $a > b$.

If $a > b$ and $b > c$, then $a > c$. (Transitive Property of Inequality)

Exactly one of the following is true: $a > b$, $a = b$, $a < b$ (Comparison Property)

An *exterior angle* of a triangle is any angle that is both adjacent to and supplementary to an angle of the triangle. The two angles of the triangle that are not adjacent to an exterior angle are called *remote interior angles* with regard to that exterior angle.

In the diagram, the six numbered angles are the exterior angles of $\triangle ABC$. The remote interior angles with regard to $\angle 4$, for example, are $\angle CAB$ and $\angle ACB$.

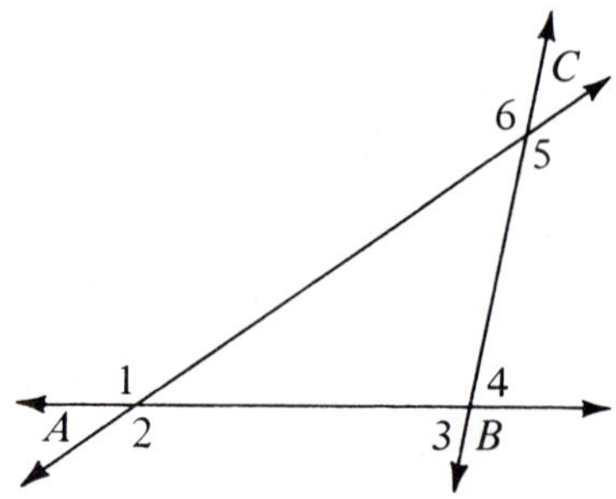

Theorem 3-15 The measure of an exterior angle of a triangle is greater than the measure of either remote interior angle.

GIVEN: $\triangle ABC$ with exterior angle BCD
PROVE: $m\angle BCD > m\angle B$;
$m\angle BCD > m\angle A$

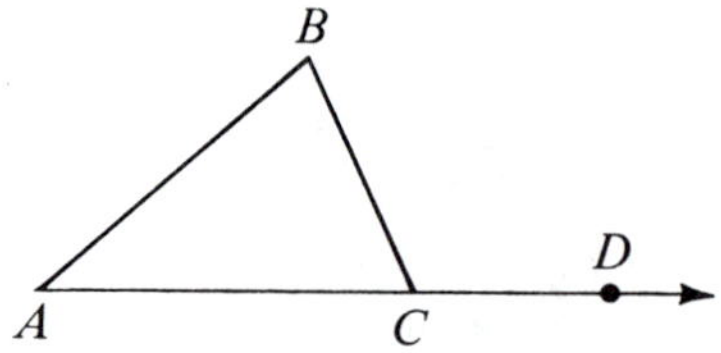

The proof of Theorem 3-15 is left as Exercise 29, and the proof of the following theorem is left as Exercise 30.

Theorem 3-16 If the length of one side of a triangle is greater than the length of another side, then the measure of the angle opposite the longer side is greater than that of the angle opposite the shorter side. (In a $\triangle$ the greater angle is opposite the longer side.)

GIVEN: $\triangle ABC$ with $AB > BC$
PROVE: $m\angle C > m\angle A$

The converse of Theorem 3-16 is also true.

Theorem 3-17 If the measure of one angle of a triangle is greater than the measure of another angle, then the side opposite the greater angle is longer than the side opposite the smaller angle. (In a $\triangle$ the longer side is opposite the greater angle.)

GIVEN: $\triangle ABC$ with $m\angle C > m\angle A$
PROVE: $AB > BC$

Proof: We will use an indirect proof, and therefore we temporarily assume that AB is *not* greater than BC. Then $BC \geqslant AB$. If $BC > AB$, then by Theorem 3-16, $m\angle A > m\angle C$. This contradicts the given information. If $BC = AB$, then by Theorem 3-13, $m\angle A = m\angle C$. This also contradicts the given information. Therefore the temporary assumption is false, and it must be true that $AB > BC$.

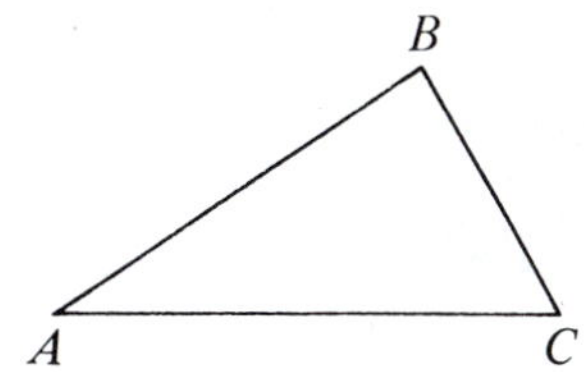

EXAMPLE

GIVEN: $m\angle 1 > m\angle 2$
PROVE: $AC > BC$

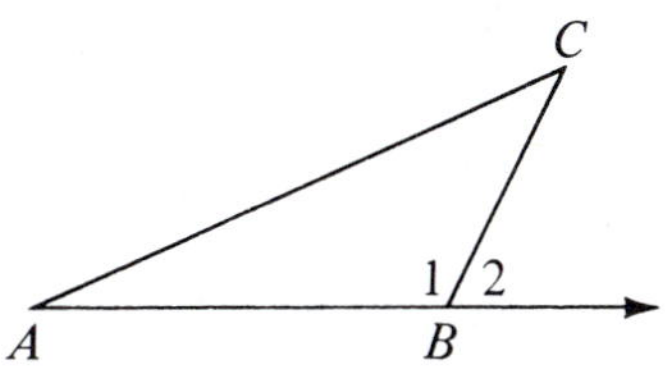

Statements	Reasons
1. $m\angle 1 > m\angle 2$	1. Given
2. $m\angle 2 > m\angle A$	2. The measure of an exterior $\angle$ is greater than that of either remote interior $\angle$. (Theorem 3-15)
3. $m\angle 1 > m\angle A$	3. Trans. Prop. of Inequality
4. $AC > BC$	4. In a $\triangle$ the longer side is opposite the greater $\angle$. (Theorem 3-17)

Our next theorem, called the Triangle Inequality, states a fundamental relationship between the lengths of the sides of any triangle.

Theorem 3-18 The sum of the lengths of any two sides of a triangle is greater than the length of the third side. (Triangle Inequality)

GIVEN: $\triangle PQR$
PROVE: $PR + RQ > PQ$

Outline of Proof: On the ray opposite to $\overrightarrow{RP}$, choose point S so that $RS = RQ$ (Postulate 2). Then $m\angle RQS = m\angle S$, by Theorem 3-13. Since $m\angle PQS = m\angle PQR + m\angle RQS$, you have $m\angle PQS > m\angle RQS$. By substitution, $m\angle PQS > m\angle S$. By Theorem 3-17, $PS > PQ$. But $PS = PR + RS$, or $PS = PR + RQ$. Hence $PR + RQ > PQ$.

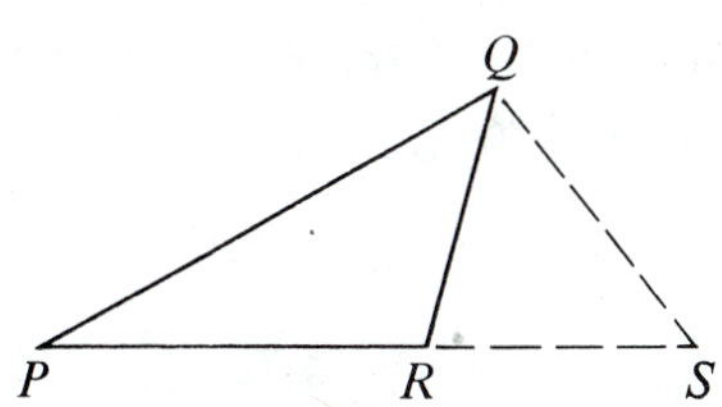

EXAMPLE In $\triangle ABC$, $AB = 12$ and $BC = 7$. Find all possible lengths of $\overline{AC}$.

SOLUTION

$AB + BC > AC$
$12 + 7 > AC$
$19 > AC$

$AC + BC > AB$
$AC + 7 > 12$
$AC > 5$

$AB + AC > BC$
$12 + AC > 7$
$AC > -5$

Therefore $19 > AC > 5$.

ORAL EXERCISES

Draw a $\triangle ABC$ that satisfies the given condition. Use the given condition to state an inequality involving other parts of the triangle.

1. $AB > BC$
2. $BC < AC$
3. $AB > AC$
4. $m\angle A > m\angle B$
5. $m\angle A < m\angle C$
6. $m\angle B > m\angle A$

Use the diagram to name all the triangles of which the given angle is an exterior angle.

7. $\angle 1$
8. $\angle 2$
9. $\angle 3$
10. $\angle 4$

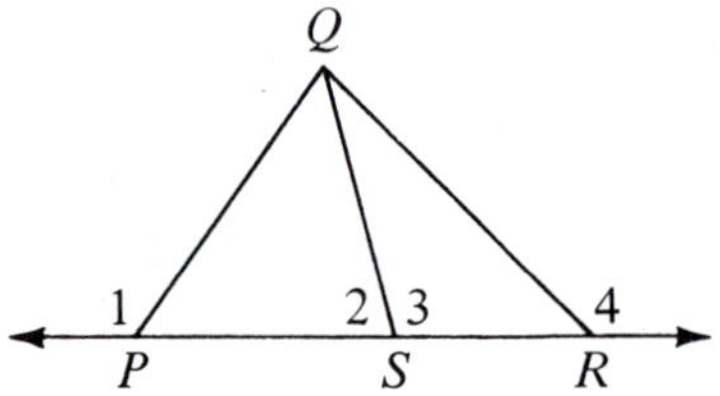

For each exercise, state whether the three numbers can represent the lengths of the sides of a triangle.

11. 3, 7, 1
12. 5, 6, 8
13. 12, 7, 9
14. 4, 11, 15

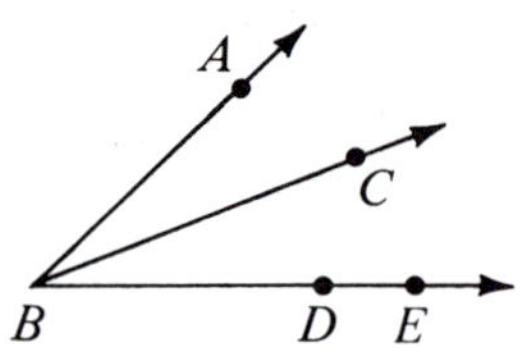

15. Explain why $BE > BD$ in the figure.
16. Explain why $m\angle ABD > m\angle ABC$ in the figure.

WRITTEN EXERCISES

In $\triangle EFG$, name the longest side of the triangle under the given conditions.

A 1. $m\angle E = 70$, $m\angle F = 80$, $m\angle G = 30$
2. $m\angle E = 25$, $m\angle F = 45$, $m\angle G = 110$

In $\triangle KLM$, name the largest angle of the triangle under the given conditions.

3. $KL = 12$, $LM = 15$, $MK = 13$
4. $KL = 6.05$, $LM = 6.005$, $MK = 6.5$

In $\triangle ABC$ with the given values of AC and BC, find all possible values of AB.

5. $AC = 9$, $BC = 15$
6. $AC = 11$, $BC = 4$
7. $AC = \frac{15}{4}$, $BC = \frac{1}{2}$
8. $AC = \frac{5}{6}$, $BC = \frac{7}{3}$
9. $AC = m + 2$, $BC = m - 1$
10. $AC = 5 - n$, $BC = 5 + n$ $(n > 0)$

Write two-column proofs.

11. Given: $m\angle 1 > m\angle A$
 Prove: $AC > BC$

12. Given: $AC > BC$
 Prove: $m\angle 1 > m\angle A$

13. Given: $AC > AB$; $m\angle C > m\angle A$
 Prove: $m\angle 2 > m\angle A$

14. Given: $m\angle 2 > m\angle 3$
 Prove: $m\angle 2 > m\angle A$

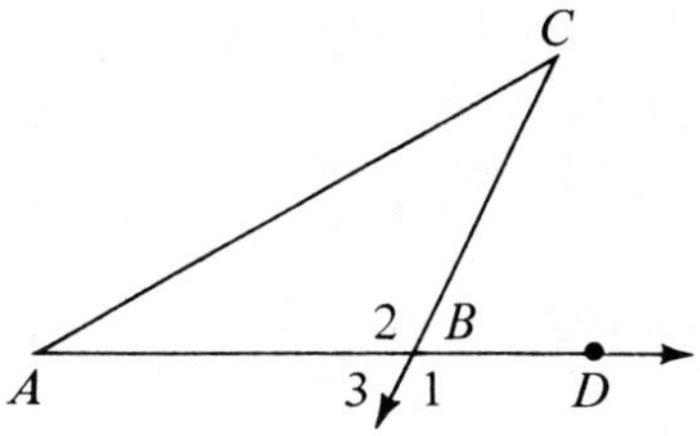

Exs. 11–14

15. Given: $KL > ML$
 Prove: $m\angle 4 > m\angle 5$

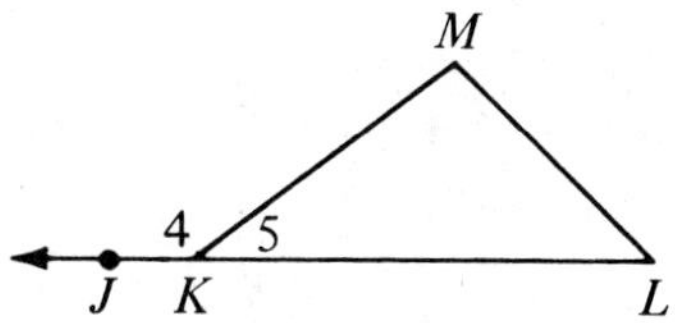

16. Given: $m\angle 2 > m\angle 1$
 Prove: $EF > FG$

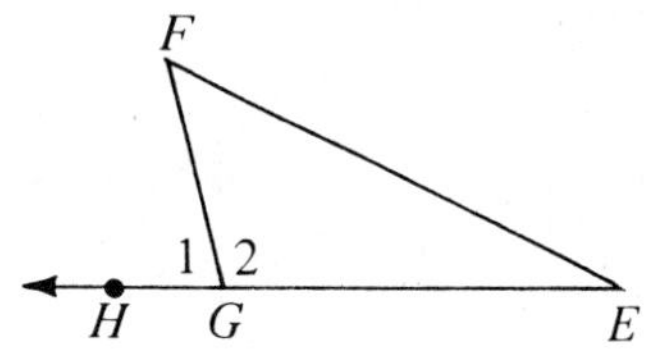

B 17. Complete, using the lengths of the line segments shown.
 $\underline{\ ?\ } > \underline{\ ?\ } > \underline{\ ?\ } > \underline{\ ?\ } > \underline{\ ?\ }$

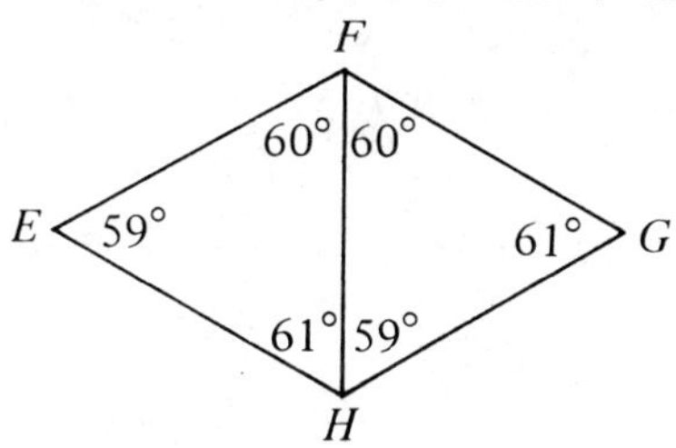

18. Complete, using the measures of the following angles: $\angle A$, $\angle C$, $\angle ADC$, $\angle BDC$, $\angle DBC$
 $\underline{\ ?\ } > \underline{\ ?\ } > \underline{\ ?\ } > \underline{\ ?\ } > \underline{\ ?\ }$

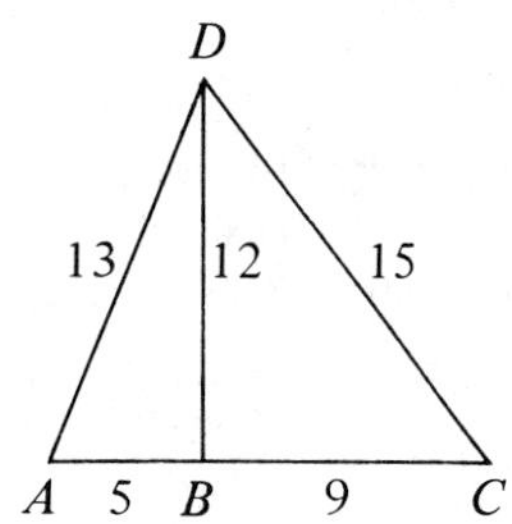

Write two-column proofs.

19. Given: $ZW = ZY$
 Prove: $ZX > ZY$

20. Given: $m\angle Y > m\angle 1$
 Prove: $XZ > YZ$

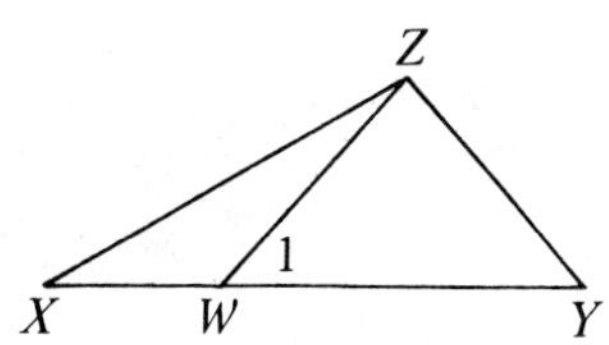

Exs. 19, 20

21. Given: $CD = CB$
 Prove: $m\angle 2 > m\angle 1$

22. Given: $CD = CB$; $m\angle A > m\angle C$
 Prove: $AC > AB$

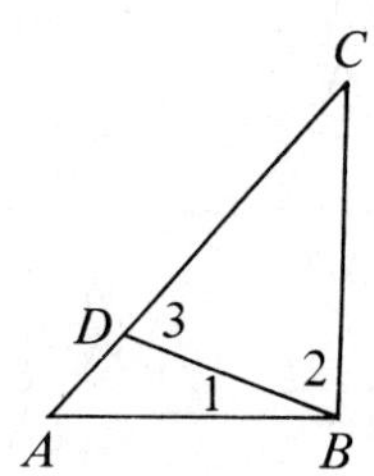

Exs. 21, 22

23. Given: $m\angle Q > m\angle 2$
Prove: $PR > RQ$

24. Given: $PR > RQ$; $RS > PS$
Prove: $m\angle Q > m\angle 3$

25. Given: $PR > RQ$
Prove: $PR > RS$

26. Given: $RS > RQ$
Prove: $PR > RQ$

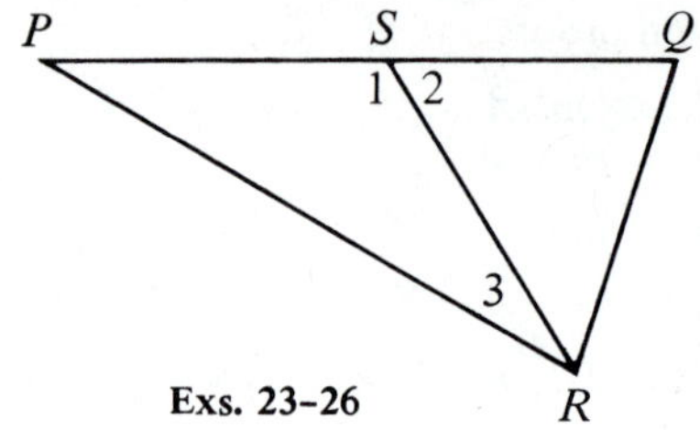

Exs. 23–26

C 27. Given: $m\angle Y > m\angle 2$
Prove: $XZ > WZ$

28. Given: $XZ = XY$;
$ZW = ZY$
Prove: $ZY > WY$

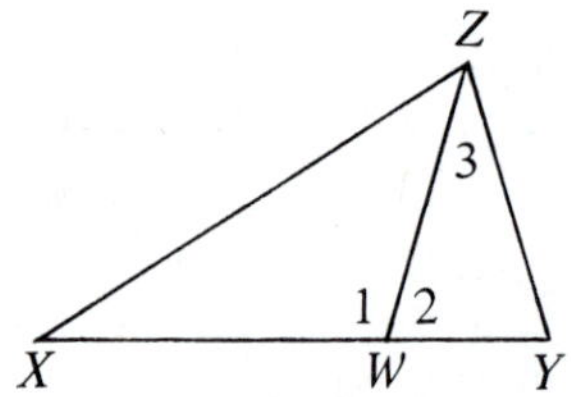

Exs. 27, 28

29. a. To prove the first half of Theorem 3-15, use the following auxiliary lines:
Let X be the midpoint of $\overline{BC}$ and draw $\overrightarrow{AX}$. On the ray opposite to $\overrightarrow{XA}$, choose point E so that $XE = XA$. Draw $\overline{EC}$. Now prove that $m\angle BCD > m\angle B$. (*Hint:* Show that $m\angle B = m\angle 1$.)

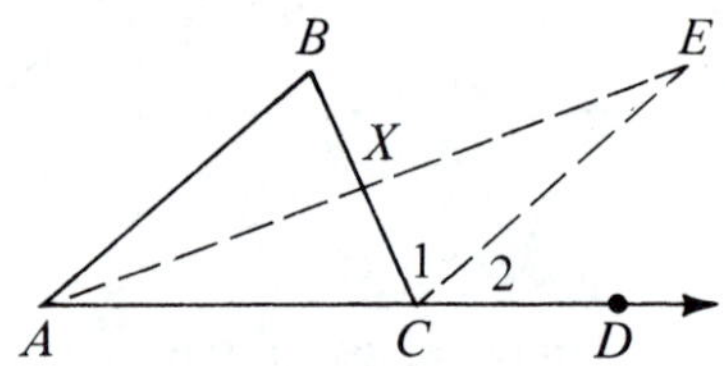

b. Explain how you would show that $m\angle BCD > m\angle A$. (*Hint:* Let $\overrightarrow{CF}$ be the ray opposite to $\overrightarrow{CB}$. Use a procedure similar to the one in part (a) to show that $m\angle ACF > m\angle A$.)

30. Write a two-column proof of Theorem 3-16. Begin by choosing point D on $\overrightarrow{BA}$ so that $DB = BC$.

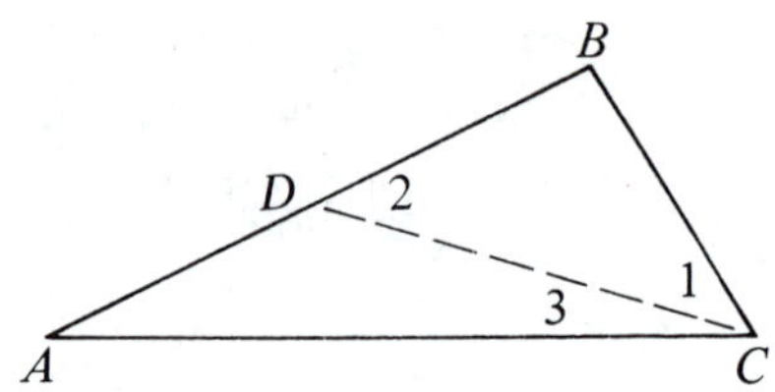

SELF-TEST 3

Section 3-7

1. Write a two-column proof.
Given: $\angle X$ and $\angle Z$ are right angles;
$\overline{WX} \cong \overline{WZ}$
Prove: $\angle 1 \cong \angle 2$

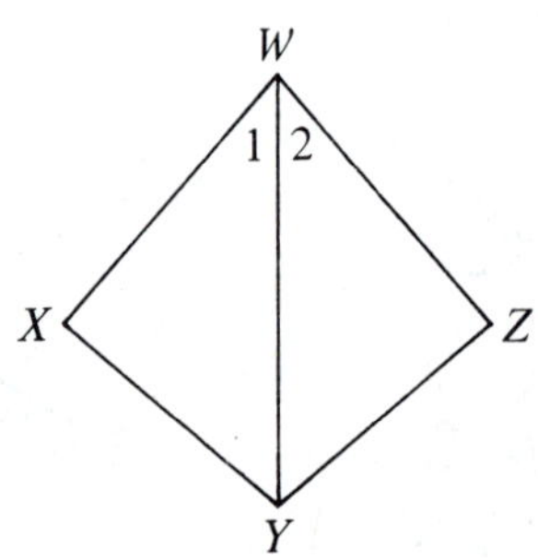

Find the value of x.

Section 3-8

2.

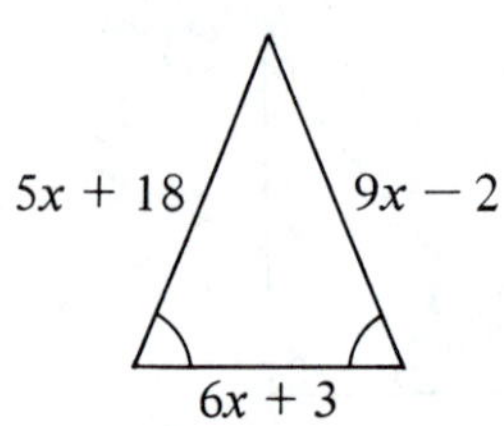

3.

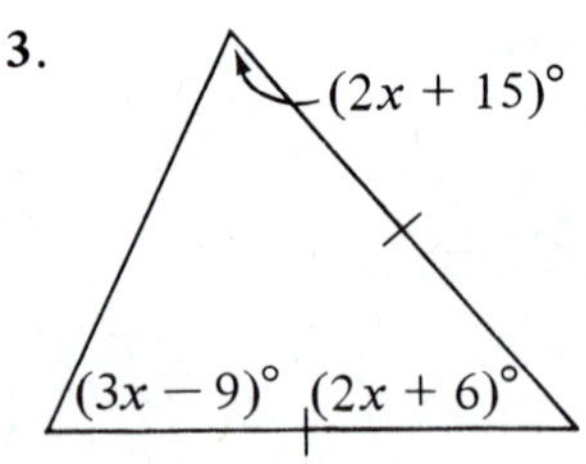

4. Prove: The median to the base of an isosceles triangle is perpendicular to the base.
Given: $\overline{AB} \cong \overline{CB}$; $\overline{AD} \cong \overline{CD}$
Prove: $\overline{BD} \perp \overline{AC}$

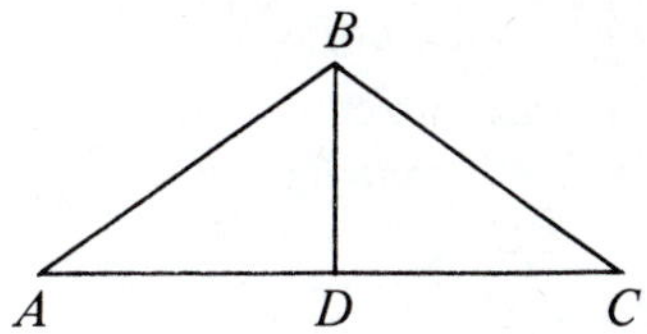

Exercises 5-7 refer to the diagram shown.

5. If $EF = 9$ and $FG = 8$, state an inequality relating the measures of two angles. Section 3-9

6. If $EF = 9$ and $FG = 8$, find all possible values of EG.

7. Explain why $m\angle 3 > m\angle 2$.

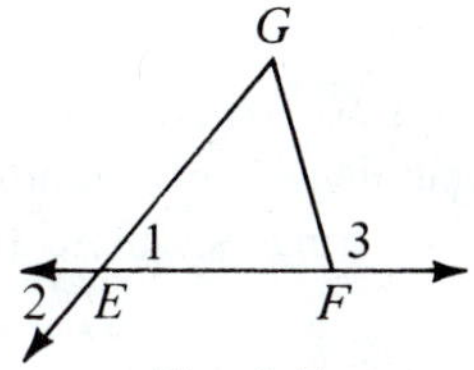

Exs. 5-7

CHAPTER REVIEW

1. How many endpoints does each of the following have? Section 3-1
   a. $\overline{AB}$ b. $\overleftrightarrow{AB}$ c. $\overrightarrow{AB}$

2. What is the length of $\overline{RT}$ if $RS = 5$, $ST = 8$, and
   a. S is between R and T?   b. R is between S and T?

3. If point B is between points A and C, $AB = y - 1$, and $BC = 2y + 1$, express AC in terms of y.

4. Complete: There is __?__ point on a ray at a given distance from the endpoint of the ray. Section 3-2

5. Write a two-column proof.
   Given: S is the midpoint of $\overline{RT}$; $ST = SQ$
   Prove: $RT = SQ + ST$

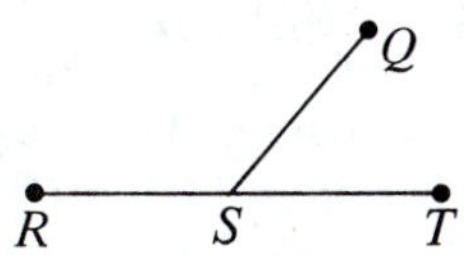

In the diagram, $\overrightarrow{OY}$ bisects $\angle XOZ$.

6. State the postulate or theorem that allows you to conclude that: **(a)** $m\angle 1 = \frac{1}{2}m\angle XOZ$; **(b)** $m\angle 1 + m\angle 2 = m\angle XOZ$. Section 3-3

7. If $m\angle 1 = 5x - 3$ and $m\angle 2 = x + 33$, find the value of $m\angle XOZ$.

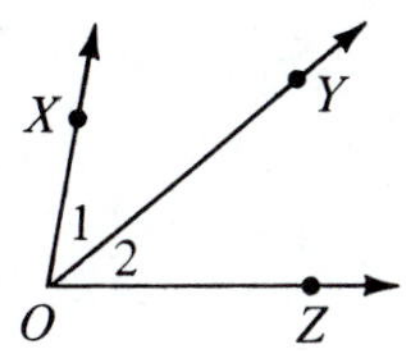

State the postulate or theorem that justifies each statement.

8. $m\angle 1 = m\angle 4$ Section 3-4

9. If $m\angle 1 = m\angle 5$, then $j \perp k$.

10. $\angle 1$ and $\angle 5$ are supplementary angles.

11. If $j \perp k$, $m\angle 2 = 7x + 4$, and $m\angle 3 = 2x + 5$, find the value of x.

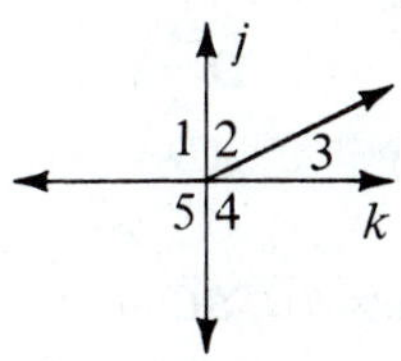

Exs. 8-11

12. Write a two-column proof. Section 3-5

Given: $\angle 1$ and $\angle 3$ are comp. $\angle$s;
$\angle 4$ and $\angle 2$ are comp. $\angle$s;
$\overline{AC} \cong \overline{DC}$

Prove: $\triangle ABC \cong \triangle DEC$

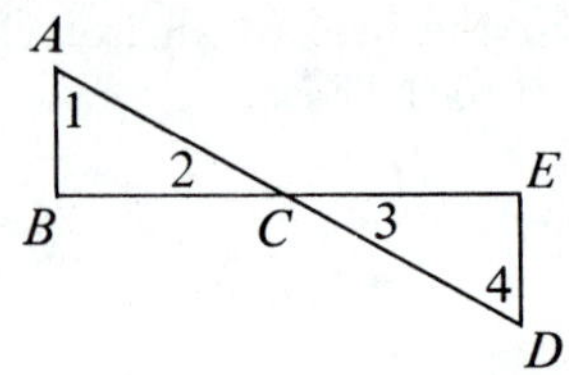

In the figure, $\overline{GI}$ bisects $\overline{HJ}$. What additional information concerning congruent corresponding parts is needed to use the following postulate to show that $\triangle GHI \cong \triangle GJI$:

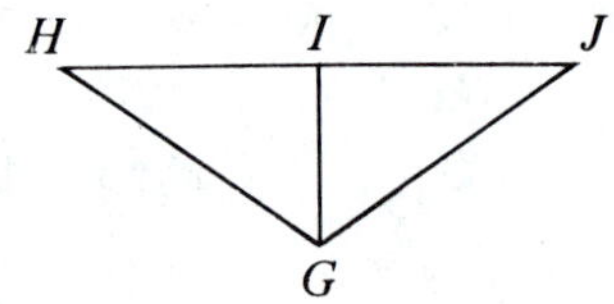

13. the SSS Postulate? Section 3-6

14. the SAS Postulate?

15. Write a two-column proof.

Given: $\angle KLM$ and $\angle NML$ are rt. $\angle$s;
$\overline{KM} \cong \overline{NL}$

Prove: $\triangle KLM \cong \triangle NML$

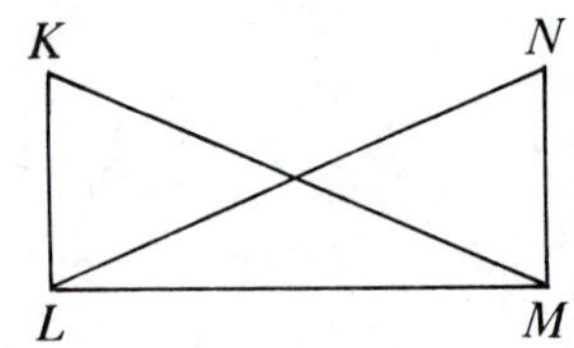

16. If $\triangle RST \cong \triangle XYZ$, what statement justifies the assertion that $\angle R \cong \angle X$? Section 3-7

17. If $\overline{XB} \cong \overline{XD}$, name two *pairs* of congruent angles. Section 3-8

18. If $\angle A \cong \angle E$, name two congruent segments.

19. Write a two-column proof:

Given: $\overline{AB} \cong \overline{ED}$; $\triangle XCB \cong \triangle XCD$

Prove: $\triangle XAB \cong \triangle XED$

20. Refer to the figure. What reason justifies the statement that $m\angle XBA > m\angle BXC$? Section 3-9

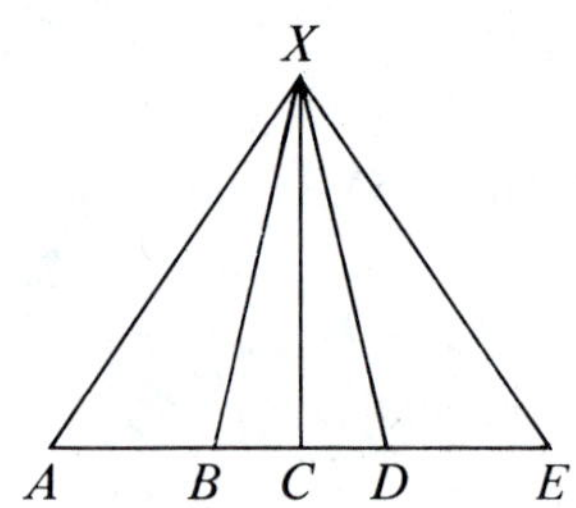

Exs. 17–21

21. Complete: If $XA > XD$, then $m\angle$ _?_ $> m\angle$ _?_.

22. In $\triangle RST$, if $RT = 5$ and $ST = 7$, find all possible lengths of the third side, $\overline{RS}$.

CHAPTER TEST

In the figure, $\angle 1 \cong \angle 2$.

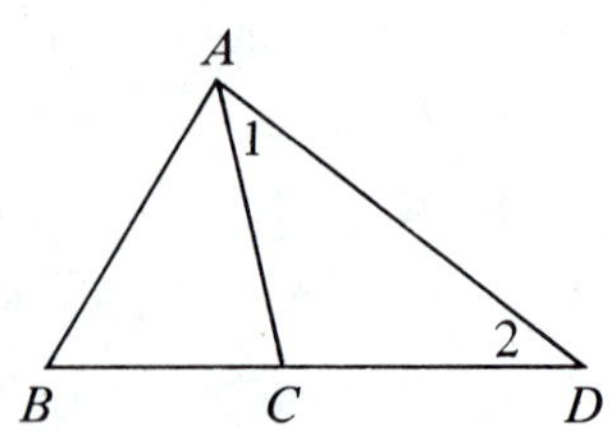

Exs. 1–9

1. If $BC = 8x - 7$, $CD = 6x$, and $BD = 21$, find the value of x.
2. If $AC = 4y + 2$, $AD = 5y - 1$, and $CD = y + 14$, find the value of y.
3. If $\overline{AC}$ bisects $\overline{BD}$, $BC = 2z - 15$, and $BD = z + 9$, find BD.
4. If $AB = 14$ and $BD = 21$, what can you conclude about AD?

State the postulate, property, or theorem that justifies the statement.

5. $m\angle ACB > m\angle 1$

6. $\angle ACB$ and $\angle ACD$ are supplementary.

7. Point C is the only point common to $\overleftrightarrow{AC}$ and $\overleftrightarrow{BD}$.

8. $m\angle BAD > m\angle 1$

9. If $\overrightarrow{AC}$ bisects $\angle BAD$, then $m\angle 1 = \frac{1}{2}m\angle BAD$.

10. Two angles are complementary. The measure of one angle is 3 more than twice the measure of the other. Find the measure of each angle.

11. Write six equations that follow from the fact that $\triangle ABC \cong \triangle XYZ$.

12. In $\triangle JKL$, $m\angle J = 50$, $m\angle K = 28$, and $m\angle L = 102$. Write an inequality relating the lengths of the sides of $\triangle JKL$.

Write two-column proofs.

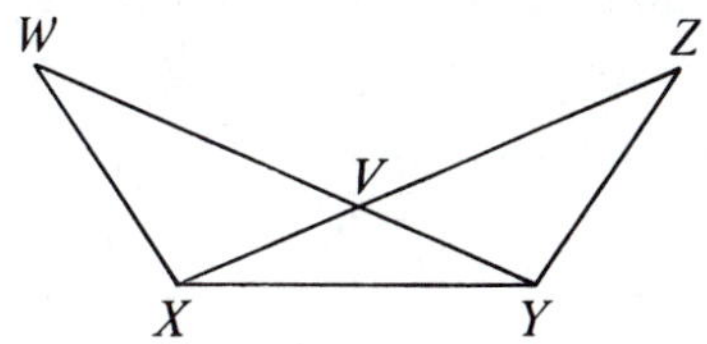

Exs. 13–15

13. Given: $\overline{WX} \cong \overline{ZY}$; $\overline{WY} \cong \overline{ZX}$
Prove: $\triangle WXY \cong \triangle ZYX$

14. Given: $\overline{WV} \cong \overline{ZV}$; $\overline{XV} \cong \overline{YV}$
Prove: $\overline{WX} \cong \overline{ZY}$

15. Given: $\angle W \cong \angle Z$; $\overline{WV} \cong \overline{ZV}$
Prove: $\triangle XVY$ is isosceles.

16. Given: $\overline{AB} \perp \overline{BC}$;
$\overline{AD} \perp \overline{DC}$;
$\angle 1 \cong \angle 2$
Prove: $\triangle ABC \cong \triangle ADC$

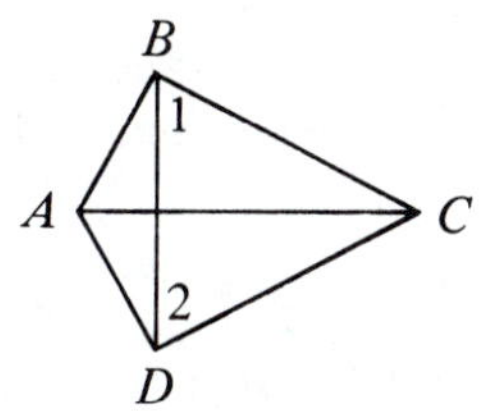

CUMULATIVE REVIEW: CHAPTERS 1-3

Indicate the best answer by writing the appropriate letter.

1. The Commutative Property of Addition guarantees that $(-1)(-r + r) = \underline{\ ?\ }$.
a. $(-1)[r + (-r)]$ b. $(-1)(0)$ c. $(-r + r)(-1)$ d. $-(-r + r)$

2. Simplify: $-(8x - 3y) + (4y + x) - (-5x + 7y)$
a. $-2x$ b. $4x + 8y$ c. $-12x + 14y$ d. $14x - 6y$

3. If S is the midpoint of $\overline{RT}$, $RT = 13y - 50$, and $ST = 20 - y$, find RT.
a. 28 b. 15 c. 6 d. 5

4. Translate into an equation: Gretchen has two more dimes than quarters. These coins are worth 90¢.
a. $25z + 10(2z) = 90$ b. $10(z + 2) + 25z = 90$
c. $10z + 25(z + 2) = 90$ d. $10z + 25(2z) = 90$

5. Which inequality is equivalent to $-\frac{2}{3}b + 5 < 11$?

a. $b < 9$ b. $b < -9$ c. $b > 9$ d. $b > -9$

6. Simplify: $7 + (-3)^3 - 2(-5) \div \frac{1}{3}$

a. 64 b. -50 c. 10 d. $-16\frac{2}{3}$

7. Which statement must have the same truth value as $p \rightarrow q$?

a. $q \rightarrow p$ b. $\sim p \rightarrow \sim q$ c. $\sim q \rightarrow \sim p$ d. $\sim q \rightarrow p$

8. Which of the following can be assumed from the appearance of a diagram?

a. Two angles are congruent. b. Two lines are perpendicular.

c. Two lines intersect at a particular point.

d. A particular point is the midpoint of a segment.

9. Which statement is logically equivalent to $p \rightarrow q$?

a. $\sim p \vee q$ b. $\sim q \vee p$ c. $p \wedge \sim q$ d. $\sim p \wedge q$

10. Which is *not* an undefined term?

a. ray b. line c. plane d. point

11. In $\triangle XYZ$, $\overline{XY} \cong \overline{XZ}$, $m\angle Y = 3a + 14$, and $m\angle Z = 5a$. Find $m\angle Y$.

a. 10 b. 35 c. 20 d. $\frac{80}{13}$

12. Which of the following cannot be used as a reason in the proof of a theorem?

a. a corollary of the theorem being proved b. a property

c. given information d. a postulate

13. Given: $\overline{AB} \cong \overline{DE}$ and $\overline{AC} \cong \overline{DF}$. Which additional information would *not* guarantee that $\triangle ABC \cong \triangle DEF$?

a. $\overline{BC} \cong \overline{EF}$ b. $\angle A \cong \angle D$

c. $\angle C \cong \angle F$ d. $\overline{AB} \perp \overline{BC}$; $\overline{DE} \perp \overline{EF}$

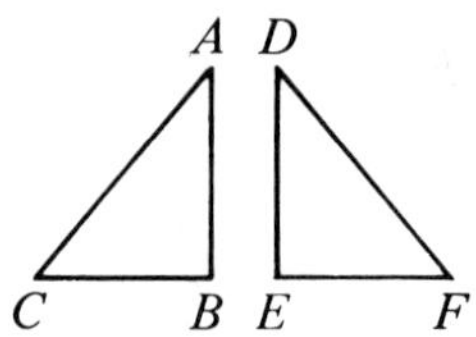

14. $\{\text{nonpositive even integers}\} \cap \{\text{nonnegative even integers}\} = \underline{\quad ? \quad}$

a. $\emptyset$ b. $\{0\}$ c. $\{\text{even integers}\}$ d. $\{\text{integers}\}$

15. Which statement is logically equivalent to $\sim(q \wedge \sim p)$?

a. $\sim q \rightarrow \sim p$ b. $q \rightarrow p$ c. $p \rightarrow q$ d. $\sim p \wedge q$

16. Solve: $3 - 4(x - 2) = -17$

a. $x = 3$ b. $x = -3$ c. $x = -7$ d. $x = 7$

17. Which is the graph of the solution set of $(2 - k > -6) \wedge \left(\frac{1}{3}k > 2\right)$?

a. 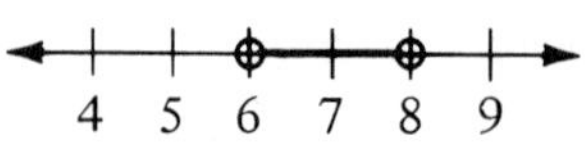

b. 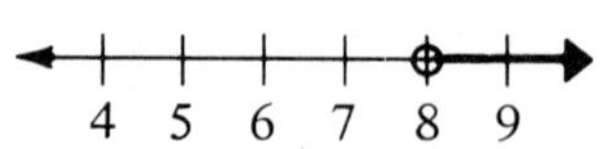

c. 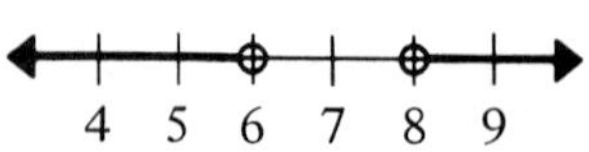

d. 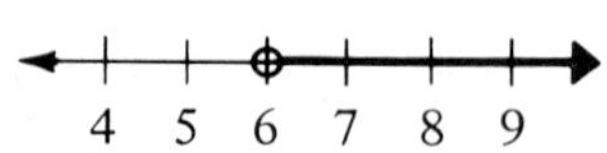

18. If the lengths of two sides of a triangle are 7 and 10, the length of the third side may *not* be:
a. 15 b. 10 c. 7 d. 3

19. Which statement is always true?
a. $p \vee \sim p$ b. $(p \vee q) \rightarrow p$ c. $p \rightarrow (p \wedge q)$ d. $p \wedge \sim p$

20. Which statement is true?
a. $\forall_x\ x + 3 = 2x$ b. $\exists_x\ x = x + 2$ c. $\forall_x\ x + 1 \neq 2x$ d. $\exists_x\ x + 3 = 2x$

21. Negate: $\exists_y\ y \geqslant 2$
a. $\exists_y\ y < 2$ b. $\forall_y\ y \geqslant 2$ c. $\forall_y\ y < 2$ d. $\sim\forall_y\ y \geqslant 2$

22. Find the solution set of $3x + 1 < -11$ or $3x + 1 \geqslant 4$.
a. $\{x: -4 < x \leqslant 1\}$ b. $\{x: x \geqslant 1 \text{ or } x < -4\}$
c. $\{x: -4 \leqslant x < 1\}$ d. $\{x: x > 1 \text{ or } x \leqslant -4\}$

23. What is the negation of the statement "All cows are brown"?
a. Some cows are brown. b. No cows are brown.
c. All cows are not brown. d. Some cows are not brown.

24. Translate into an inequality: The sum of two consecutive odd integers is at least 40.
a. $n + (n + 1) \leqslant 40$ b. $n + (n + 1) \geqslant 40$
c. $n + (n + 2) \leqslant 40$ d. $n + (n + 2) \geqslant 40$

25. Given the true statements: "If $a > b$, then $c > d$" and "$c \leqslant d$." A valid conclusion is:
a. $a > b$ b. $a < b$ c. $a \leqslant b$ d. $a < d$

26. In $\triangle RST$, $RS < ST < RT$. Name the smallest angle of $\triangle RST$.
a. $\angle R$ b. $\angle S$ c. $\angle T$ d. cannot be determined

27. Negate the sentence "No idea is unacceptable."
a. All ideas are acceptable. b. Some ideas are unacceptable.
c. Some ideas are acceptable. d. All ideas are not unacceptable.

28. Given: $\angle 1 \cong \angle 2$, X and Y are midpoints of $\overline{KL}$ and $\overline{ST}$, respectively, and $\overline{KX} \cong \overline{SY}$. What other information is needed to show that $\triangle JKL \cong \triangle RST$ by the ASA Postulate?
a. $\angle L \cong \angle T$ b. $\angle J \cong \angle R$
c. $\overline{JK} \cong \overline{RS}$ d. $\overline{JL} \cong \overline{RT}$

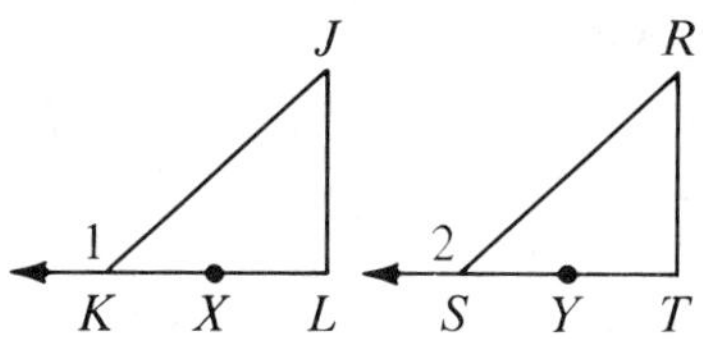

29. Negate: I will study hard and I will get an A.
a. I will not study hard or I will not get an A.
b. I will not study hard or I will get an A.
c. I will not study hard and I will not get an A.
d. I will study hard and I will not get an A.

30. If $\overline{QR} \cong \overline{UT}$ and $\overline{RS} \cong \overline{TS}$, which postulate could be used to prove two triangles congruent?
a. ASA b. HL c. SSS d. SAS

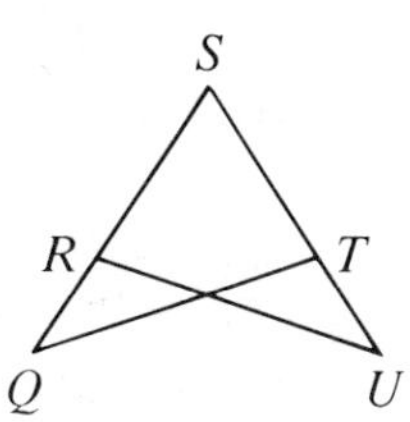

31. Which statement must have the same truth value as $\sim(p \wedge q)$?

a. $\sim p \vee \sim q$ b. $p \leftrightarrow q$ c. $q \rightarrow p$ d. $\sim p \wedge \sim q$

32. Given premises P_1: $\sim(p \wedge q)$ and P_2: q. What can you conclude?

a. $p \wedge q$ b. $p \rightarrow q$ c. p d. $\sim p$

33. Given premises P_1: $q \rightarrow \sim p$ and P_2: p. What can you conclude?

a. q b. $\sim q$ c. $p \rightarrow q$ d. $\sim p \vee q$

34. Sam is twice as old as Pam. In three years the sum of their ages will be 30. Find the difference in their ages.

a. 6 b. 7 c. 8 d. 10

35. How many liters of a 30% acid solution should be mixed with a 60% acid solution to form 12 L of a 40% acid solution?

a. 4 b. 5 c. 7 d. 8

36. If the lengths of two sides of an isosceles triangle are 4 cm and 11 cm, find the length of the third side.

a. 4 cm b. 11 cm c. 7 cm d. 15 cm

37. Given the following statements:

If Steve joins the volleyball team, then he won't have enough time to practice playing the trumpet.
If Steve joins the band, then he will have enough time to practice playing the trumpet.
Steve will join the volleyball team.

Let v represent "join the volleyball team."
Let b represent "join the band."
Let t represent "have enough time to practice playing the trumpet."

a. Using v, b, and t, and proper connectives, express each statement in symbolic form.
b. Using laws of inference, show that Steve will not join the band.

Choose one of Exercises 38–40 and write a two-column proof.

38. Given: $\overline{AC}$ bisects $\overline{BD}$;
$\angle ADB \cong \angle CBD$
Prove: $\overline{AD} \cong \overline{CB}$

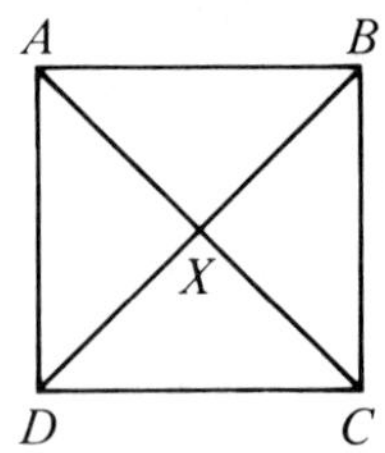

39. Given: $\overline{AD} \cong \overline{BC}$; $\overline{AC} \cong \overline{BD}$
Prove: $\angle ADC \cong \angle BCD$

40. Show that the medians from corresponding vertices of congruent triangles are congruent.

Given: $\triangle JKL \cong \triangle MNO$;
$\overline{JY}$ and $\overline{MZ}$ are medians.
Prove: $\overline{JY} \cong \overline{MZ}$

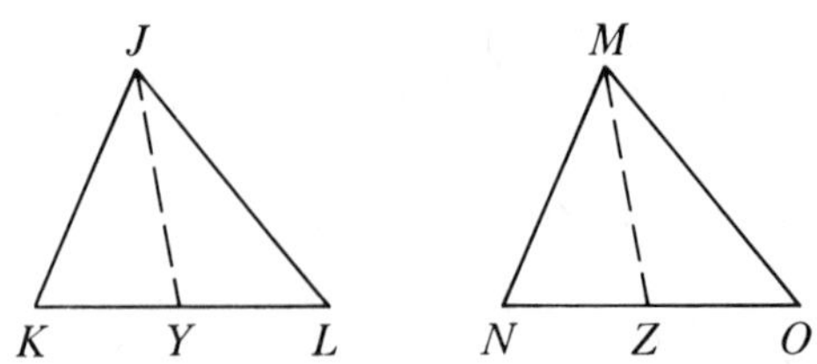

Careers — Carpenter

Carpentry is applied geometry. A cabinet, for instance, has sides that lie in parallel planes, doors in the shape of rectangles or squares, and circular nail holes. Usually the geometric principles of congruence and similarity also apply. Drawers in a chest, for instance, are very often the same size and shape, or else the same shape but a smaller size.

Carpenters make up the largest group of building trade workers. The skills required depend on the type of construction involved. "Rough" carpentry is what it sounds like – the scaffolding, subflooring, rafters, and tunnel supports that have to be functional but not beautiful. "Finish" carpentry includes hardwood floors, stairwork, and windows.

The layout, or detailed measuring or marking, is the first step in a carpentry job. The instructions for size, shape, building material, and purpose come from a blueprint or from a supervisor. The materials, which include wood, plastic, fiberglass, and drywall, must then be cut, shaped, checked, and assembled. Carpenters check their accuracy with such instruments as the right-angled T square, the ruler, and the level.

You can sharpen your eye for geometric shapes by taking high school courses in carpentry, shop, mechanical drawing, and mathematics. An apprenticeship is the recommended training for the occupation, but many carpenters learn the trade informally. Examples of apprenticeship programs are those run by Associated Builders and Contractors, Inc., and the United Brotherhood of Carpenters and Joiners.

Building contractors employ most carpenters. Others work for schools, government agencies, utility companies, and manufacturing firms. Still others are self-employed, offering services from rough carpentry to expensive reproductions of antique cabinets and chairs.

VOCABULARY REVIEW

Be sure that you know the meaning of these words:

point, p. 100
line, p. 100
collinear points, p. 100
plane, p. 100
coplanar points, p. 100
line segment, p. 101
ray, p. 102
opposite rays, p. 102
postulate (axiom), p. 104
theorem, p. 104
midpoint of a segment, p. 104
bisector of a segment, p. 105
angle, p. 108
half-plane, p. 108
complementary angles, p. 113
supplementary angles, p. 113
perpendicular, p. 114
vertical angles, p. 115
congruent figures, p. 119
isosceles triangle, p. 133
equilateral triangle, p. 133
median of a triangle, p. 134

MIXED REVIEW

In the diagram at the right, $\overrightarrow{OP}$ bisects $\angle BOC$. Find the value of x and the measure of $\angle BOC$.

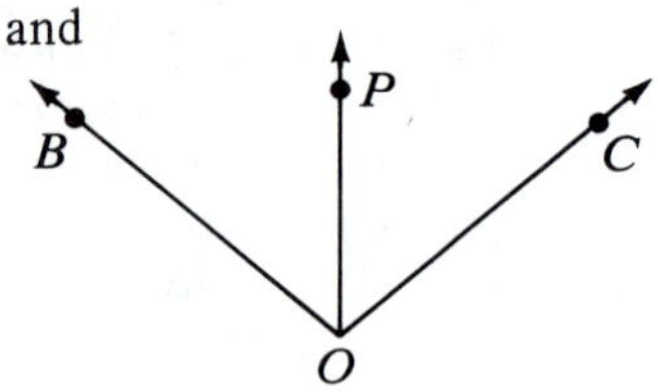

1. $m\angle BOP = 70 - 3x$; $m\angle POC = 2x + 50$

2. $m\angle BOP = \frac{2}{3}x$; $m\angle BOC = 2x - 46$

Solve.

3. $5(x - 2) - 4(2x + 3) + 50x = 260$

4. $-4 \leqslant \frac{2}{3}(x + 5) < 2$

5. $6(3 + y) < 2(y - 5)$

6. $\frac{1}{2}(c + 6) - \frac{1}{3}(c - 8) = 0$

Exercises 7–9 refer to the diagram at the right.

7. If $BC = 9$ and $AC = 13.5$, find AB.

8. If $AB = x - 3$ and $BC = 2x + 5$, find x if $AC = 17$.

9. If C is the midpoint of $\overline{BD}$, $BC = 4t - 1$, and $CD = 7t - 11$, find the value of t.

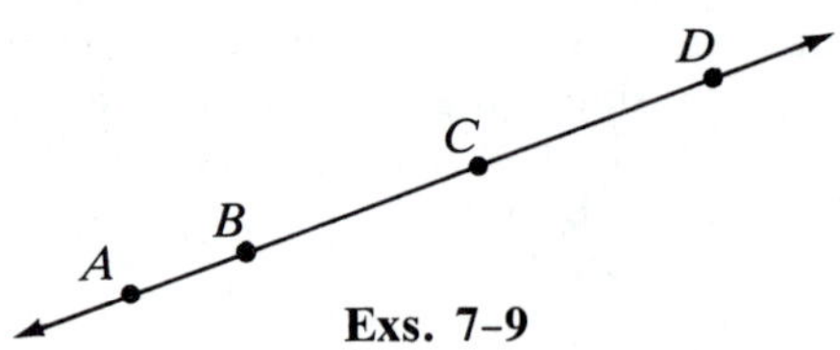

Exs. 7–9

Rewrite each conditional as an equivalent disjunction.

10. If a number is divisible by 6, then it is divisible by 2.

11. If a home has aluminum siding, then it doesn't need painting.

Write two-column proofs.

12. Given: lines j, k, and l; $m\angle 2 = m\angle 3$
Prove: $m\angle 1 = m\angle 4$

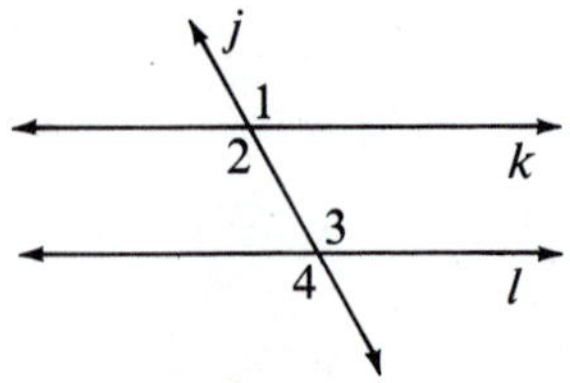

13. Given: $\overline{AC} \cong \overline{BD}$; $\angle A \cong \angle B$; $\angle EDA \cong \angle FCB$
Prove: $\triangle ADE \cong \triangle BCF$

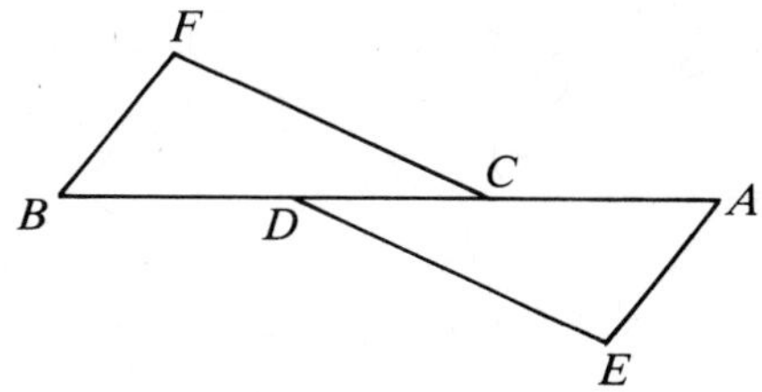

14. Given: $\overline{QS} \perp \overline{PR}$; $\overline{QR} \cong \overline{QT}$; $\angle PRS \cong \angle QTP$
Prove: $\triangle PQT \cong \triangle SQR$

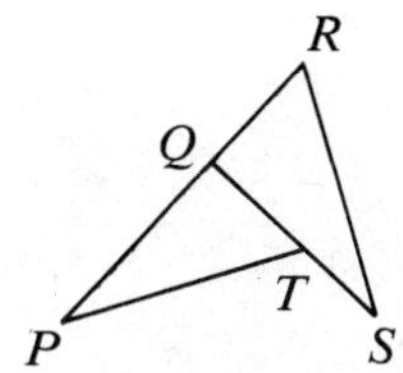

15. Given: $m\angle 1 = m\angle 2$
Prove: $\angle 2$ and $\angle 3$ are supplementary.

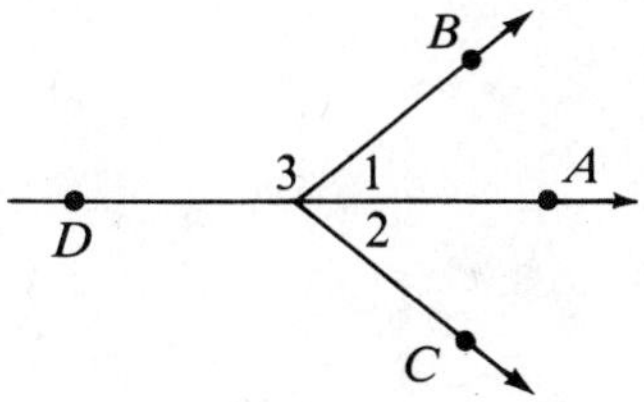

Write the negation of each statement and give the truth value of the negation.

16. $\forall_x\ x + (-x) = 0$

17. $\exists_y\ y = -y$

Solve.

18. A plot of land is 7 times as long as it is wide. The difference between its length and its width is 300 m. Find the dimensions.

19. Two trains are 1120 km apart and are traveling toward each other on side-by-side tracks. If one train is traveling at 120 km/h and the other at 100 km/h, in how many hours will they meet?

Hexagons are one of the types of polygons you will study in this chapter. Hexagons can be seen in a variety of natural and manufactured objects – for example, honeycombs, crystals, tiles, soccer balls, and chicken wire.

Parallel Lines 4

A subject of much controversy in the history of mathematics has been the *Parallel Postulate* of Euclid. This states that in the plane containing a given line and a point not on the line there is *at most one* line that can be drawn through the given point and not intersecting the given line—that is, *parallel* to the given line. For more than two thousand years after Euclid stated the Parallel Postulate (in about 300 B.C.), mathematicians believed that it could be *proved* from the other postulates—but no one succeeded in doing so.

Finally, in the first half of the nineteenth century, two mathematicians—Nikolai Lobachevsky, a Russian, and Janos Bolyai, a Hungarian—each working without knowledge of the other, took a different approach. They began to develop a new geometry in which the Euclidean Parallel Postulate was replaced by a postulate that states there are *at least two* lines parallel to a given line through a given point not on the line. In doing so they suggested that the Parallel Postulate is independent of the other postulates. They were later vindicated when it was shown that no contradiction could result from their new geometry.

Earlier, Karl Friedrich Gauss, the great German mathematician, had reached many of the same conclusions, although he had not published them for fear of ridicule. Gauss is said to have gone so far as to measure large geometric figures with surveyors' instruments in an attempt to find out whether the Parallel Postulate is true in the physical world. Modern physicists have discovered that the Euclidean notion of parallelism does *not* hold throughout the universe. However, in relatively small regions (like Earth!) *all* the postulates of Euclidean geometry are true for all practical purposes.

Section 4-1 PROVING LINES PARALLEL

Whenever two lines intersect a third line in different points, eight angles are formed. If these two lines are *parallel*, certain of these angles are related in important ways. In this section you will learn what is meant by:

- a transversal of two or more lines
- parallel lines
- alternate interior angles
- corresponding angles
- a corollary of a theorem

Any line that intersects two or more coplanar lines in different points is called a *transversal* of the lines. In the diagram at the right, $\overleftrightarrow{AB}$ is a transversal of lines l and n. When two lines are cut by a transversal, the pairs of angles formed that have the same relative positions as the following pairs are called *alternate interior angles* and *corresponding angles*:

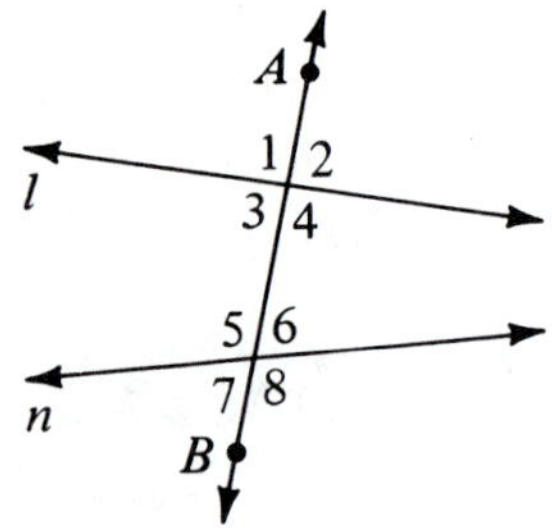

Alternate Interior Angles	*Corresponding Angles*
$\angle 3$ and $\angle 6$	$\angle 1$ and $\angle 5$
$\angle 4$ and $\angle 5$	$\angle 2$ and $\angle 6$
	$\angle 3$ and $\angle 7$
	$\angle 4$ and $\angle 8$

Two coplanar lines that do not intersect are called *parallel lines.* We write "$l \parallel n$" to mean "line l is parallel to line n." Two line segments or rays are said to be parallel if they are contained in parallel lines.

The next two theorems enable you to conclude that two lines are parallel given that certain angles are congruent.

Theorem 4-1 If two lines are cut by a transversal so that two alternate interior angles are congruent, then the lines are parallel. (If 2 lines are cut by a trans. so that alt. int. $\angle$s are $\cong$, the lines are $\parallel$.)

GIVEN: $\angle 1 \cong \angle 2$
PROVE: $l \parallel n$

Proof: We will use an indirect proof, and thus we make the temporary assumption that l and n are not parallel. Then they must intersect at some point X. Therefore $\angle 1$ is an exterior angle of $\triangle ABX$, and $\angle 2$ is a remote interior angle. Hence $m\angle 1 > m\angle 2$ by Theorem 3-15. But this contradicts the hypothesis that $\angle 1 \cong \angle 2$. Therefore our temporary assumption must be false, and $l \parallel n$.

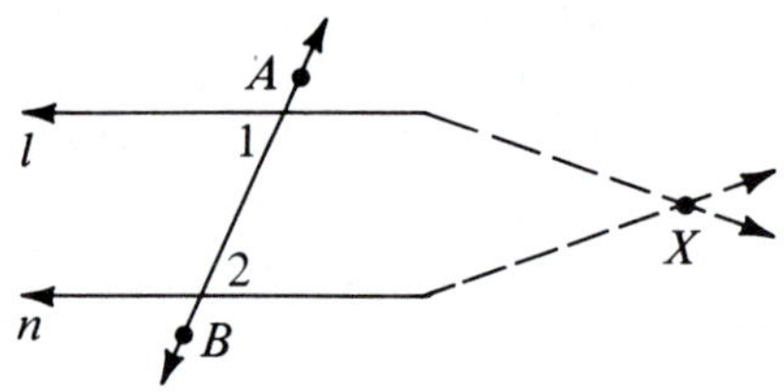

A *corollary* to a theorem is a statement that can be proved directly using the theorem. The proof of the first of the following two important corollaries of Theorem 4-1 is left to you as Exercise 19.

Corollary 1 In a plane, two lines perpendicular to the same line are parallel.

GIVEN: $l \perp t$; $n \perp t$
PROVE: $l \parallel n$

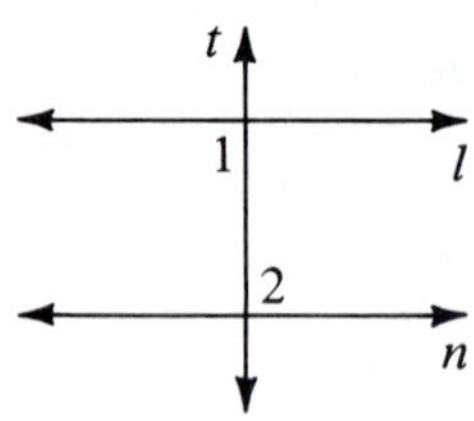

Corollary 2 Through a given point not on a given line, there is *at least one* line parallel to the given line.

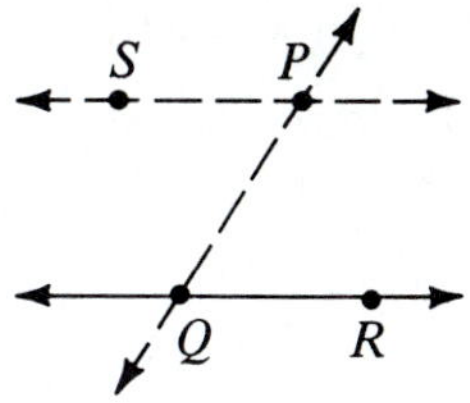

GIVEN: Point P not on $\overleftrightarrow{QR}$

PROVE: There is a line through P parallel to $\overleftrightarrow{QR}$.

Proof: By Postulate 1, you can draw $\overleftrightarrow{PQ}$. By the Angle Construction Postulate, there exists a ray, $\overrightarrow{PS}$, such that $m\angle QPS = m\angle PQR$. Draw $\overleftrightarrow{PS}$. By Theorem 4-1, $\overleftrightarrow{PS} \parallel \overleftrightarrow{QR}$.

You may have already guessed the fact stated in the next theorem.

Theorem 4-2 If two lines are cut by a transversal so that corresponding angles are congruent, then the lines are parallel. (If 2 lines are cut by a trans. so that corr. ∠s are ≅, the lines are ∥.)

GIVEN: $\angle 1 \cong \angle 2$

PROVE: $l \parallel n$

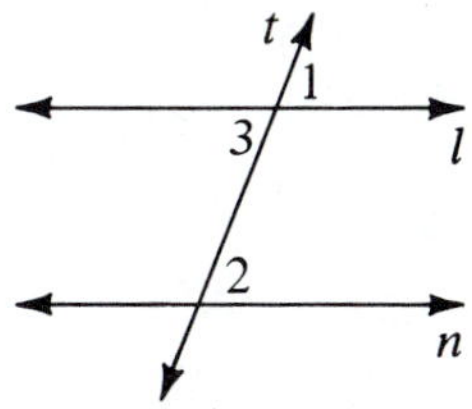

Statements	Reasons
1. $\angle 1 \cong \angle 2$	1. Given
2. $\angle 1 \cong \angle 3$	2. Vertical ∠s are ≅.
3. $\angle 2 \cong \angle 3$	3. Transitive Prop. of Congruence (or Substitution Property)
4. $l \parallel n$	4. If 2 lines are cut by a trans. so that alt. int. ∠s are ≅, the lines are ∥.

Corollary If two lines are cut by a transversal so that interior angles on the same side of the transversal are supplementary, then the lines are parallel.

GIVEN: $\angle 1$ and $\angle 2$ are supplementary.

PROVE: $l \parallel n$

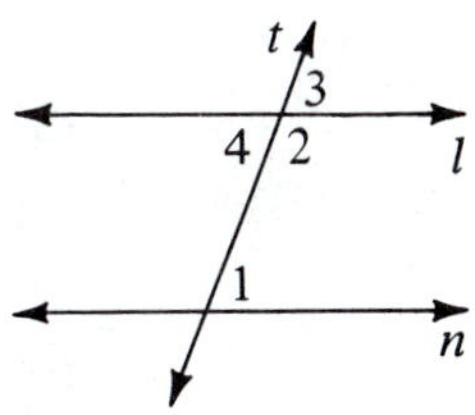

Statements	Reasons
1. $\angle 1$ and $\angle 2$ are supplementary.	1. Given
2. $\angle 3$ and $\angle 2$ are supplementary.	2. If the ext. sides of 2 adj. ∠s are opp. rays, the ∠s are supp.
3. $\angle 1 \cong \angle 3$	3. Two ∠s supp. to the same ∠ are ≅.
4. $l \parallel n$	4. If 2 lines are cut by a trans. so that corr. ∠s are ≅, the lines are ∥.

ORAL EXERCISES

1. Tim proved Theorem 4-1 as follows. Was his proof a good one? Explain.

If two lines are cut by a transversal so that two alternate interior angles are congruent, then the lines are parallel.

GIVEN: $\angle 1 \cong \angle 2$
PROVE: $l \parallel n$

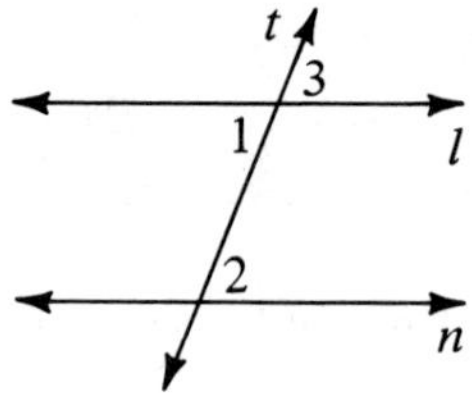

Statements	Reasons
1. $\angle 1 \cong \angle 2$	1. Given
2. $\angle 3 \cong \angle 1$	2. Vertical ∡ are ≅.
3. $\angle 3 \cong \angle 2$	3. Transitive Prop. of Congruence
4. $l \parallel n$	4. If 2 lines are cut by a trans. so that corr. ∡ are ≅, the lines are ∥. (Theorem 4-2.)

Classify each pair of angles as alternate interior angles, corresponding angles, interior angles on the same side of the transversal, or none of these.

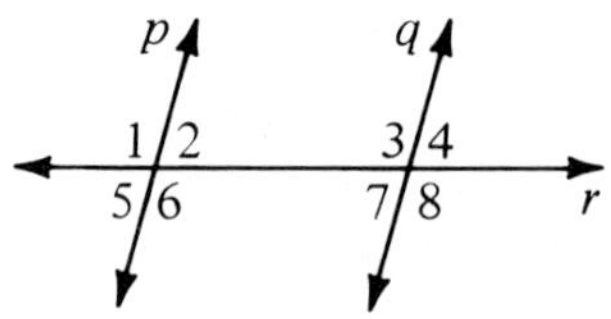

2. $\angle 5$ and $\angle 7$
3. $\angle 1$ and $\angle 5$
4. $\angle 2$ and $\angle 4$
5. $\angle 6$ and $\angle 3$
6. $\angle 6$ and $\angle 7$
7. $\angle 2$ and $\angle 7$
8. $\angle 1$ and $\angle 3$
9. $\angle 6$ and $\angle 8$
10. $\angle 2$ and $\angle 3$
11. $\angle 2$ and $\angle 5$

Which pair of lines, if any, must be parallel, given the information below?

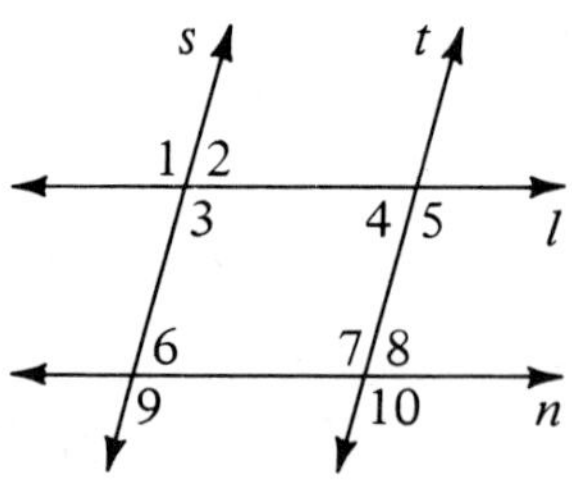

12. $\angle 2 \cong \angle 6$
13. $\angle 6 \cong \angle 8$
14. $\angle 3 \cong \angle 7$
15. $\angle 7 \cong \angle 9$
16. $\angle 4 \cong \angle 8$
17. $\angle 1 \cong \angle 10$
18. $\angle 5 \cong \angle 7$
19. $\angle 3 \cong \angle 5$
20. $\angle 3 \cong \angle 4$
21. $\angle 3$ and $\angle 4$ are supplementary.
22. $\angle 5$ and $\angle 8$ are supplementary.
23. $\angle 2$ and $\angle 6$ are supplementary.
24. $\angle 4$ and $\angle 10$ are supplementary.

WRITTEN EXERCISES

Can the given information be used to show that two lines are parallel? If so, name the lines and state the theorem or corollary you would use.

A **1.** $m\angle 2 = 70$, $m\angle 7 = 70$

2. $\angle 5 \cong \angle BAC$

3. $\angle 5 \cong \angle 9$

4. $\angle 2 \cong \angle 3$

5. $m\angle 3 = 60$, $m\angle 6 = 60$

6. $\angle 6 \cong \angle 2$

7. $\angle 1 \cong \angle 10$

8. $\angle 5 \cong \angle 8$

9. $\angle 5$ and $\angle 8$ are supplementary.

10. $\angle ABD$ and $\angle 10$ are supplementary.

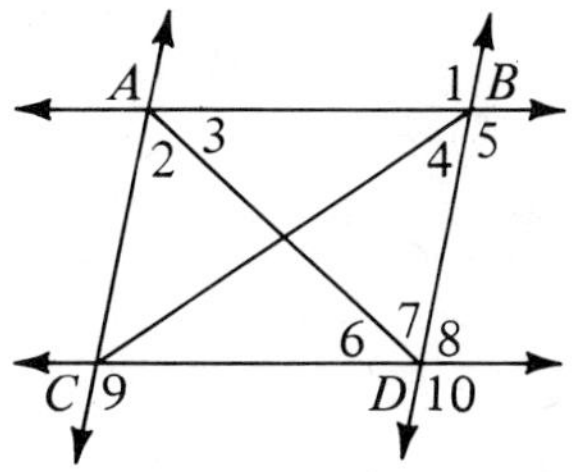

Write two-column proofs.

11. Given: $\angle 1 \cong \angle 4$
Prove: $l \parallel m$

12. Given: $\angle 1$ and $\angle 3$ are supplementary.
Prove: $l \parallel m$

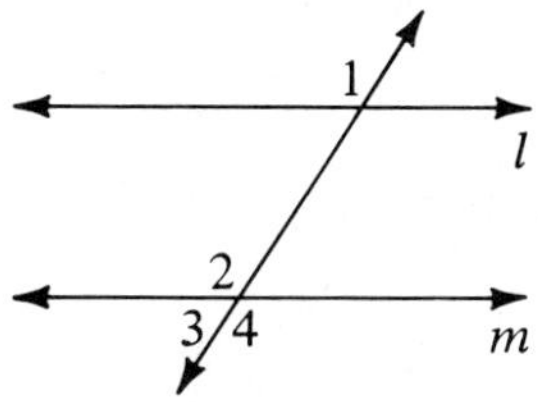

Exs. 11, 12

13. Given: $\overrightarrow{QT}$ bisects $\angle PQS$; $\angle 1 \cong \angle 4$
Prove: $\overleftrightarrow{QT} \parallel \overleftrightarrow{RS}$

14. Given: $\angle 2 \cong \angle 3$; $\overline{QR} \cong \overline{QS}$
Prove: $\overleftrightarrow{QT} \parallel \overleftrightarrow{RS}$

15. Given: $\angle 1 \cong \angle 4$; $\overline{QR} \cong \overline{QS}$
Prove: $\overleftrightarrow{QT} \parallel \overleftrightarrow{RS}$

16. Given: $\angle 2 \cong \angle 3$; $\overrightarrow{QT}$ bisects $\angle PQS$.
Prove: $\overleftrightarrow{QT} \parallel \overleftrightarrow{RS}$

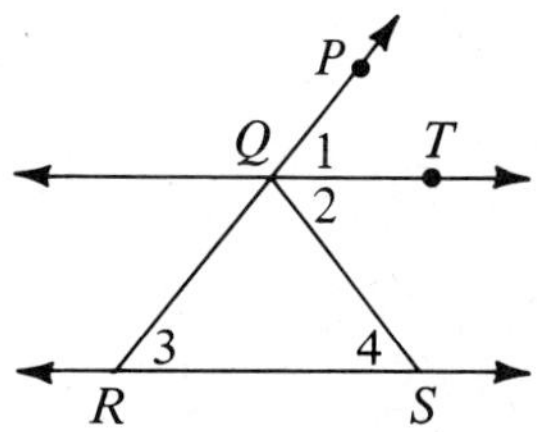

Exs. 13–16

17. Given: $\angle 5 \cong \angle 8$;
$\angle 7 \cong \angle 8$
Prove: $a \parallel b$

18. Given: $\angle 5 \cong \angle 7$;
$\angle 6 \cong \angle 8$
Prove: $a \parallel c$

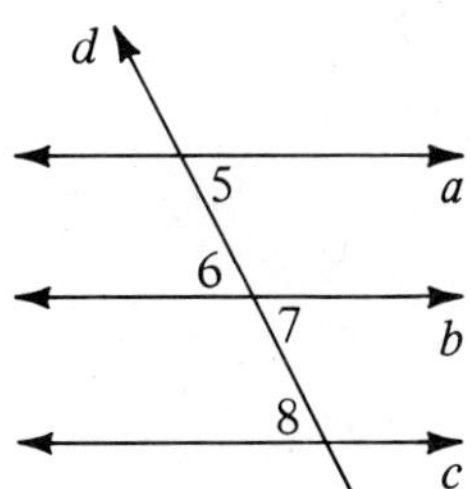

Exs. 17, 18

19. Prove Corollary 1 to Theorem 4-1, using the figure that is shown on page 154.

20. Use Theorem 4-1 to prove the corollary to Theorem 4-2. Use the figure on page 155.

21. Given: $\angle 2 \cong \angle 7$; $\angle 3 \cong \angle 7$
Prove: $c \parallel d$

22. Given: $\angle 1 \cong \angle 8$; $\angle 1 \cong \angle 5$
Prove: $a \parallel b$

B 23. Given: $\angle 1 \cong \angle 8$; $\angle 5 \cong \angle 9$
Prove: $c \parallel d$

24. Given: $\angle 2 \cong \angle 7$; $\angle 1 \cong \angle 4$
Prove: $a \parallel b$

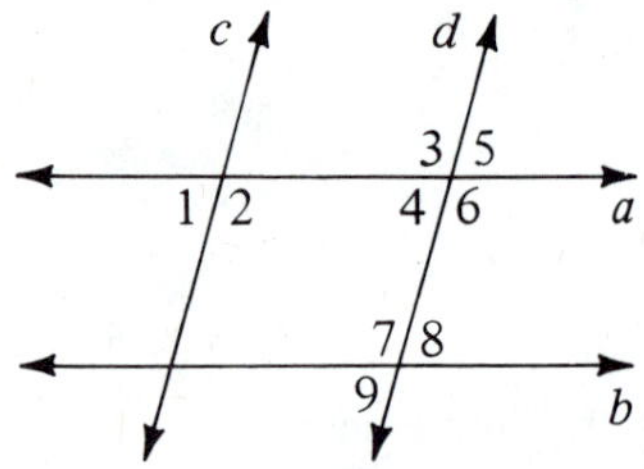

Exs 21–24

25. Given: $\overline{AD}$ and $\overline{BC}$ bisect each other.
Prove: $\overline{AB} \parallel \overline{CD}$

26. Given: $\overline{AD} \perp \overline{BC}$;
$\overline{AE} \cong \overline{DE}$;
$\overline{AB} \cong \overline{CD}$
Prove: $\overline{AB} \parallel \overline{CD}$

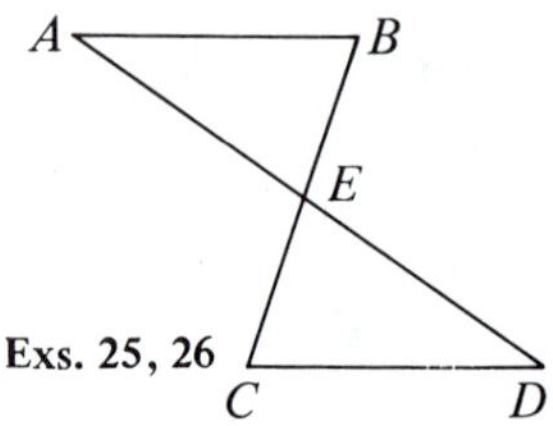

Exs. 25, 26

27. Given: $\angle 1$ and $\angle 2$ are complementary;
$\angle 3$ and $\angle 4$ are complementary.
Prove: $\overrightarrow{KR} \parallel \overrightarrow{LX}$

28. Given: $\overrightarrow{KR} \perp \overrightarrow{PZ}$;
$m\angle 1 = m\angle 3$;
$m\angle 2 = m\angle 4$
Prove: $\overrightarrow{KR} \parallel \overrightarrow{LX}$

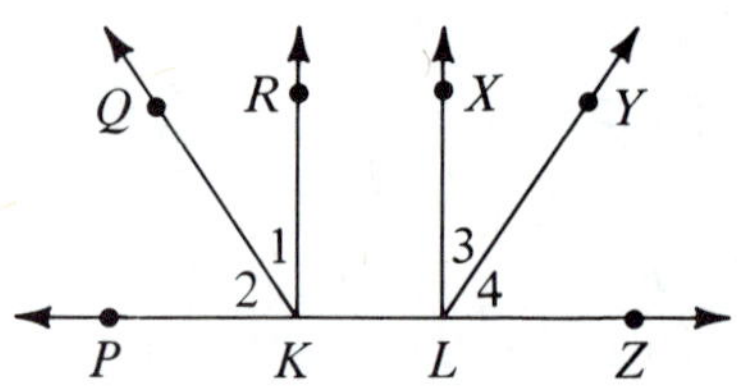

Exs. 27, 28

29. Given: $\overrightarrow{CB}$ bisects $\angle ACD$;
$\overline{AC} \cong \overline{AB}$
Prove: $\overline{AB} \parallel \overline{CD}$

30. Given: $\overline{AC} \cong \overline{BD}$; $\overline{AB} \cong \overline{CD}$
Prove: $\overline{AB} \parallel \overline{CD}$

31. Given: $\overline{AB} \cong \overline{CD}$; $\overline{AC} \cong \overline{BD}$
Prove: $\overline{AC} \parallel \overline{BD}$

32. Given: $\angle 1 \cong \angle 4$; $\overline{AC} \cong \overline{BD}$
Prove: $\overline{AB} \parallel \overline{CD}$

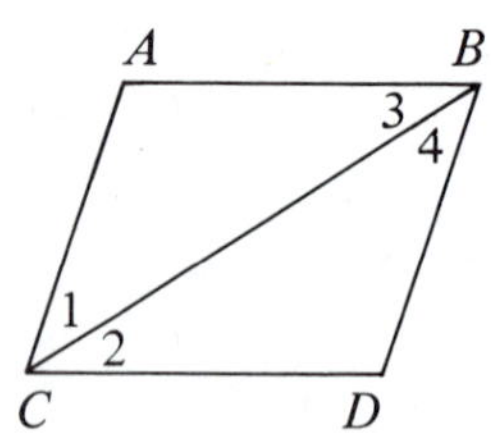

Exs. 29–32

Section 4-2 THE PARALLEL POSTULATE

In the last section you learned some ways to prove that two lines are parallel. In this section you will learn what conclusions can be drawn from *knowing* that two lines are parallel.

We have already proved that through a given point not on a given line, there is *at least* one line parallel to the given line (Corollary 2 to Theorem 4-1). Although mathematicians tried for centuries to prove that there is *at most* one such line, it can be shown that this cannot be proved from the other postulates. We therefore state as a postulate the *Parallel Postulate* of Euclid:

Postulate 10 Through a given point not on a given line, there is at most one line parallel to the given line.

Combining this postulate with Corollary 2 to Theorem 4-1, we have the following theorem:

Theorem 4-3 Through a given point not on a given line, there is exactly one line parallel to the given line.

Postulate 10 enables us to prove the following:

Theorem 4-4 If two parallel lines are cut by a transversal, then the alternate interior angles formed are congruent. (If 2 lines are $\parallel$, alt. int. $\angle$s are $\cong$.)

GIVEN: $k \parallel n$
PROVE: $\angle 1 \cong \angle 2$ (or $m\angle 1 = m\angle 2$)

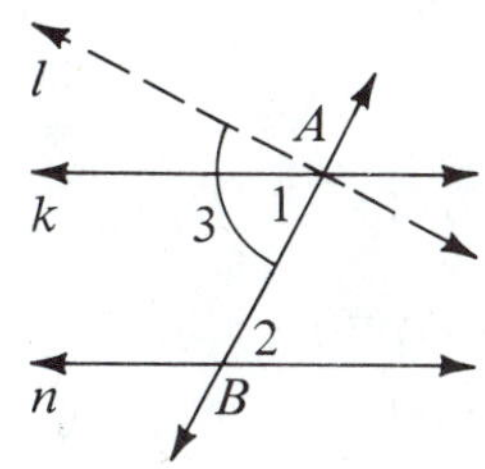

Proof: We will use an indirect proof. Thus we temporarily assume that $m\angle 1 \neq m\angle 2$. Then, by the Angle Construction Postulate, there is another line l through A such that $m\angle 3 = m\angle 2$. By Theorem 4-1, line l is parallel to line n. This, however, contradicts Postulate 10 since there would then be two lines, l and k, through A parallel to n. Therefore our temporary assumption is false, and $m\angle 1 = m\angle 2$.

EXAMPLE If $l \parallel n$, $m\angle 1 = 3x - 17$, and $m\angle 2 = x + 45$, find the value of x.

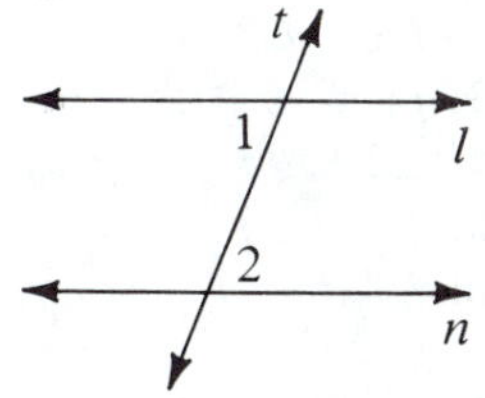

SOLUTION Since $l \parallel n$, $m\angle 1 = m\angle 2$.

$$3x - 17 = x + 45$$
$$2x = 62$$
$$x = 31$$

Theorem 4-4 has the corollary stated on the following page.

Corollary In a plane, a line perpendicular to one of two parallel lines is also perpendicular to the other line.

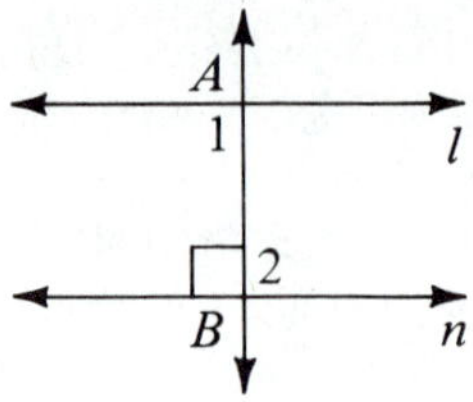

GIVEN: $\overleftrightarrow{AB} \perp n$;
$l \parallel n$
PROVE: $\overleftrightarrow{AB} \perp l$

Statements	Reasons
1. $l \parallel n$	1. Given
2. $\angle 1 \cong \angle 2$, or $m\angle 1 = m\angle 2$	2. If 2 lines are $\parallel$, alt. int. $\angle$s are $\cong$.
3. $\overleftrightarrow{AB} \perp n$	3. Given
4. $\angle 2$ is a right $\angle$.	4. Def. of $\perp$ lines
5. $m\angle 2 = 90$	5. Def. of right $\angle$
6. $m\angle 1 = 90$	6. Substitution Property
7. $\overleftrightarrow{AB} \perp l$	7. Def. of $\perp$ lines

As you may have noticed, Theorem 4-4 is the converse of Theorem 4-1. The converses of Theorem 4-2 and its corollary are also true. The proofs of the following theorem and Corollary 1 are left as Exercises 27 and 28.

Theorem 4-5 If two parallel lines are cut by a transversal, then the corresponding angles formed are congruent. (If 2 lines are $\parallel$, corr. $\angle$s are $\cong$.)

Corollary 1 If two parallel lines are cut by a transversal, then interior angles on the same side of the transversal are supplementary.

Corollary 2 Two lines parallel to a third line are parallel to each other.

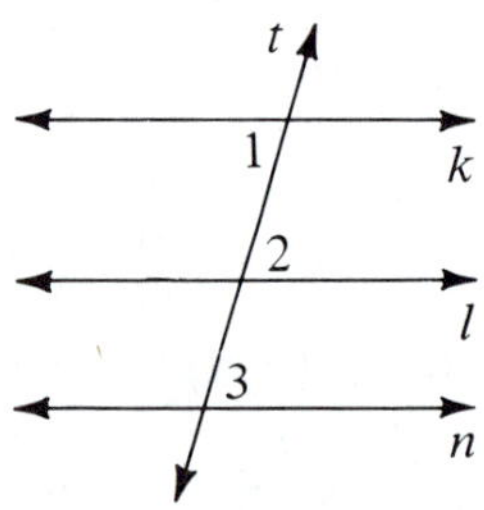

GIVEN: $k \parallel l$;
$l \parallel n$
PROVE: $k \parallel n$

Statements	Reasons
1. $k \parallel l$	1. Given
2. $\angle 1 \cong \angle 2$	2. If 2 lines are $\parallel$, alt. int. $\angle$s are $\cong$.
3. $l \parallel n$	3. Given
4. $\angle 2 \cong \angle 3$	4. If 2 lines are $\parallel$, corr. $\angle$s are $\cong$.
5. $\angle 1 \cong \angle 3$	5. Trans. Prop. of $\cong$
6. $k \parallel n$	6. If 2 lines are cut by a trans. so that alt. int. $\angle$s are $\cong$, the lines are $\parallel$.

Compare the statements of the following theorems to the statement of Theorem 4-3 and to each other. Theorem 4-6 follows from the Angle Construction Postulate. The proof of Theorem 4-7 is left as Exercises 41 and 42.

Theorem 4-6 In a plane containing a given line and a given point on the line, there is exactly one line perpendicular to the line at that point.

GIVEN: A plane containing line l;
point P on l

PROVE: In that plane there exists exactly one line perpendicular to l at point P.

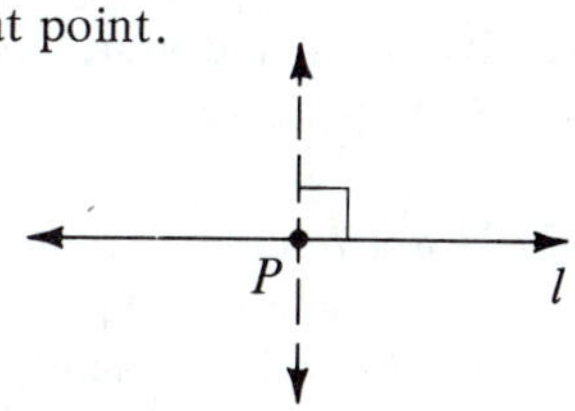

Theorem 4-7 Through a given point not on a given line, there is exactly one line perpendicular to the given line.

GIVEN: Point P not on line l

PROVE: Through P there exists exactly one line perpendicular to l.

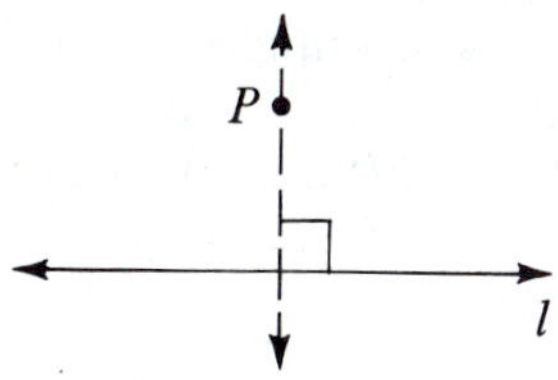

Theorem 4-7 enables us to define an important new term. An *altitude* of a triangle is a segment from a vertex that is perpendicular to the line containing the opposite side. Thus, every triangle has three altitudes. The second endpoint of an altitude does not always lie on a side of the triangle, as the diagram shown below on the right illustrates.

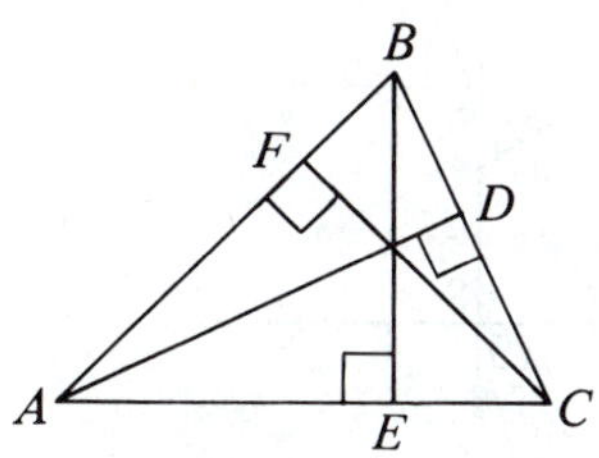

Altitudes of $\triangle ABC$: $\overline{AD}$, $\overline{BE}$, $\overline{CF}$

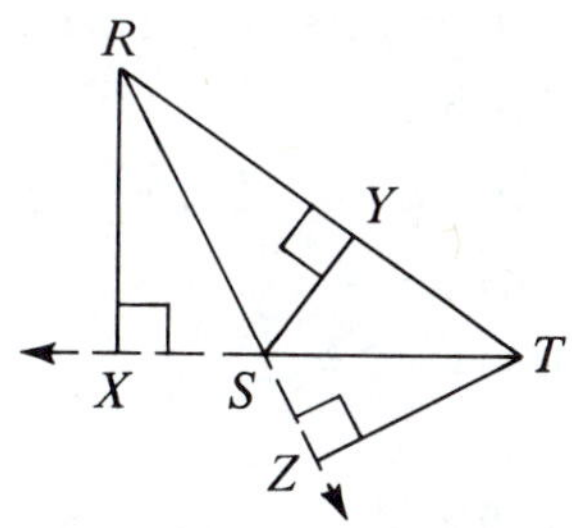

Altitudes of $\triangle RST$: $\overline{RX}$, $\overline{SY}$, $\overline{TZ}$

ORAL EXERCISES

Given $a \parallel b$, name two angles that are congruent to the given angle but do not have the same vertex as the given angle.

1. $\angle 3$ **2.** $\angle 4$ **3.** $\angle 7$ **4.** $\angle 8$

Given $c \parallel d$, name two angles that are congruent to the given angle but do not have the same vertex as the given angle.

5. $\angle 3$ **6.** $\angle 4$ **7.** $\angle 7$ **8.** $\angle 8$

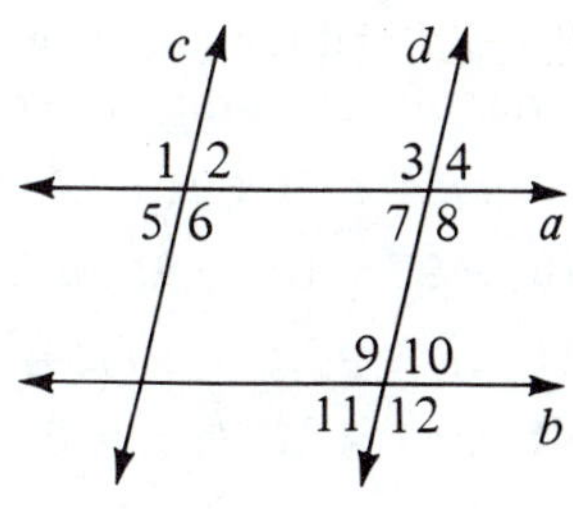

Exs. 1-10

9. Given $a \parallel b$, name two angles that are supplementary to $\angle 7$ but do not have the same vertex as $\angle 7$.

10. Given $c \parallel d$, name two angles that are supplementary to $\angle 7$ but do not have the same vertex as $\angle 7$.

11. Complete: If $k \perp l$ and $l \parallel n$, with k, l, and n coplanar lines, then k __?__ n.

12. If lines a and b are both parallel to line c, can a and b intersect at a point P? State a theorem or a postulate that justifies your answer.

13. If $a \parallel b$ and $b \parallel c$, use the numbered angles in the diagram to explain why $a \parallel c$. (The notation $a \parallel b \parallel c$ is frequently used to indicate that the three lines are parallel.)

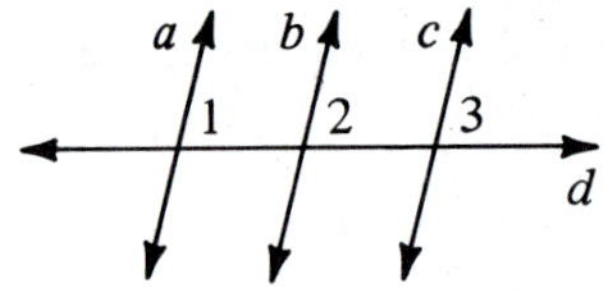

14. If $\triangle XYZ$ is a right triangle with right angle $\angle X$, which two sides of $\triangle XYZ$ are also altitudes of $\triangle XYZ$? Why?

15. If $\overline{BD}$ is both an altitude and a median of $\triangle ABC$, what kind of triangle must $\triangle ABC$ be? Explain.

WRITTEN EXERCISES

In the diagram, $r \parallel s$. Complete each statement.

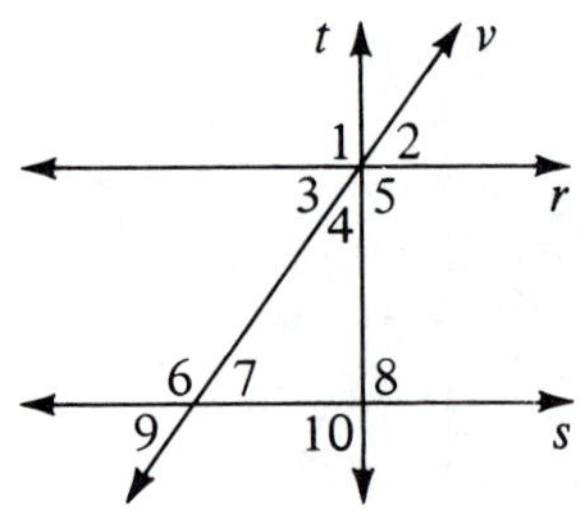

Exs. 1-8

A 1. If $m\angle 2 = 57$, then $m\angle 7 =$ __?__.

2. If $m\angle 3 = 62$, then $m\angle 7 =$ __?__.

3. If $m\angle 2 = x$, then $m\angle 9 =$ __?__.

4. If $m\angle 1 = y$, then $m\angle 10 =$ __?__.

5. If $m\angle 6 = 118$, then $m\angle 2 =$ __?__.

6. If $m\angle 4 = 30$ and $\angle 4 \cong \angle 9$, then $m\angle 2 =$ __?__.

7. If $r \perp t$, what, if anything, can you conclude about line s? Explain.

8. If $\angle 5 \cong \angle 8$, what, if anything, can you conclude about lines r and t? Explain.

If $c \parallel d$ and the angles in the diagram have the given measures in terms of x, find the value of x.

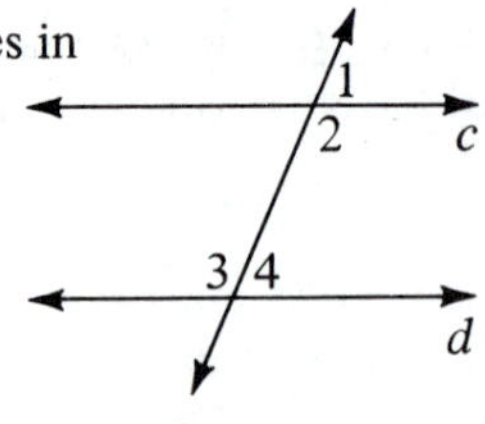

Exs. 9-14

9. $m\angle 1 = 7x + 1$; $m\angle 4 = 5x + 19$

10. $m\angle 2 = 10x - 9$; $m\angle 3 = 8x + 15$

11. $m\angle 2 = 3x + 5$; $m\angle 4 = x - 9$

12. $m\angle 1 = 2x + 6$; $m\angle 3 = 3x - 11$

13. If $c \parallel d$ in the diagram and the measure of $\angle 2$ is twice the measure of $\angle 4$, find the measure of $\angle 2$.

14. If $c \parallel d$ in the diagram and the measure of $\angle 3$ is 30 less than four times the measure of $\angle 1$, find the measure of $\angle 1$.

Write two-column proofs.

15. Given: $l \parallel m$; $\angle 1 \cong \angle 2$
    Prove: $\angle 1 \cong \angle 3$

16. Given: $l \parallel m$; $\angle 1 \cong \angle 2$
    Prove: $\angle 1 \cong \angle 4$

17. Given: $l \parallel m$; $\angle 4 \cong \angle 5$
    Prove: $\angle 1 \cong \angle 3$

18. Given: $l \parallel m$; $\angle 3 \cong \angle 5$
    Prove: $\angle 1 \cong \angle 2$

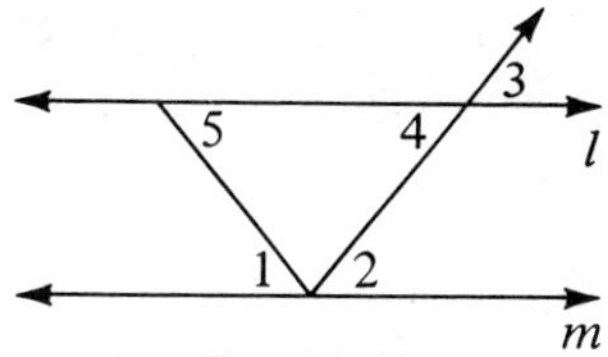

Exs. 15–18

19. Given: $\angle 1 \cong \angle 2$
    Prove: $\angle 3 \cong \angle 4$

20. Given: $\angle 3 \cong \angle 4$
    Prove: $\angle 1 \cong \angle 2$

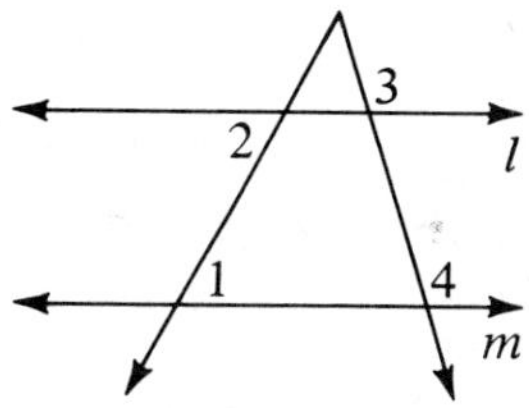

Exs. 19, 20

21. Given: $\angle 1$ and $\angle 3$ are right angles.
    Prove: $\angle 2 \cong \angle 4$

22. Given: $\angle 1 \cong \angle 3$; $\overline{XY} \perp \overline{RS}$
    Prove: $\overline{ST} \perp \overline{RS}$

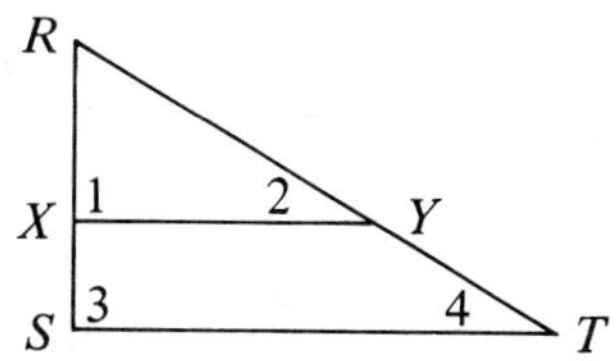

Exs. 21, 22

23. Given: $\overrightarrow{KJ} \perp \overleftrightarrow{KL}$;
    $\overrightarrow{KJ} \parallel \overrightarrow{LM}$
    Prove: $\angle 3$ and $\angle 4$ are complementary.

24. Given: $\angle 3$ and $\angle 4$ are complementary;
    $\overrightarrow{KJ} \parallel \overrightarrow{LM}$
    Prove: $\angle 1$ and $\angle 2$ are complementary.

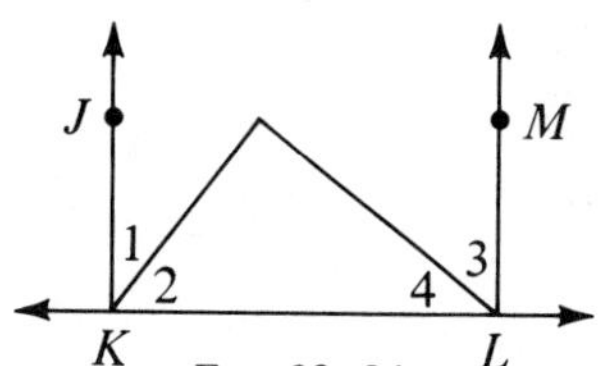

Exs. 23, 24

B 25. Given: $\overline{AE} \parallel \overline{BC}$;
    $\angle 1 \cong \angle 2$
    Prove: $\triangle ABC$ is isosceles.

26. Given: $\overline{AE} \parallel \overline{BC}$;
    $\overline{AB} \cong \overline{AC}$
    Prove: $\overrightarrow{AE}$ bisects $\angle DAC$.

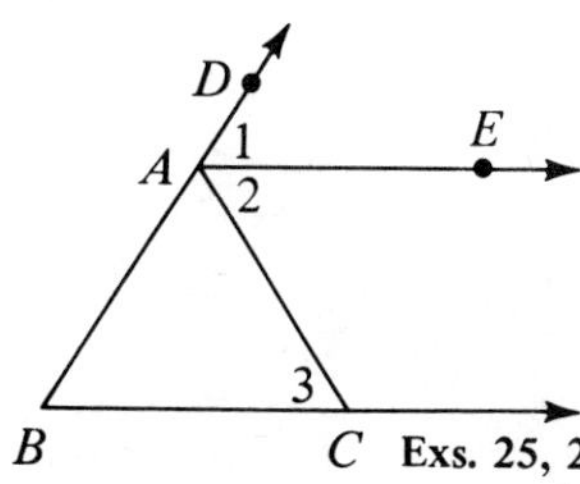

Exs. 25, 26

27. Prove Theorem 4-5, using the figure at the right. Begin by listing what is given and what is to be proved.

28. Prove Corollary 1 to Theorem 4-5, using the figure at the right. Begin by listing what is given and what is to be proved.

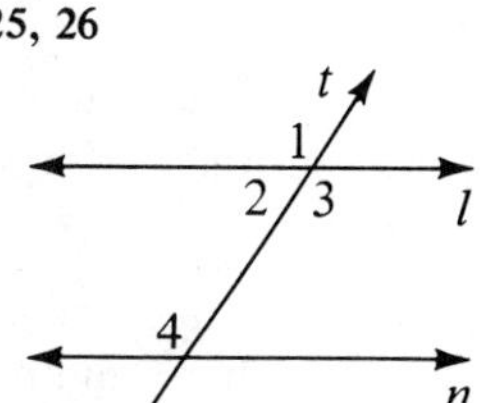

Exs. 27, 28

29. Given: $\angle W$, $\angle X$, and $\angle Y$ are right angles.
Prove: $\angle Z$ is a right angle.

30. Given: $\overline{WX} \perp \overline{XY}$; $\overline{WZ} \perp \overline{ZY}$;
$\overline{WZ} \parallel \overline{XY}$
Prove: $\overline{WX} \parallel \overline{ZY}$

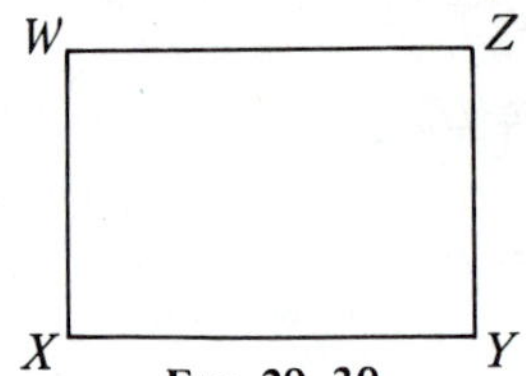

Exs. 29, 30

31. Given: $\overline{AB} \parallel \overline{CD}$; $\overline{AD} \parallel \overline{BC}$
Prove: $\overline{AD} \cong \overline{CB}$

32. Given: $\overline{AB} \parallel \overline{CD}$; $\overline{AB} \cong \overline{CD}$
Prove: $\overline{AD} \cong \overline{CB}$

33. Given: $\overline{AB} \cong \overline{CD}$; $\overline{AB} \parallel \overline{CD}$
Prove: $\overline{AD} \parallel \overline{BC}$

34. Given: $\overline{AD} \parallel \overline{BC}$; $\overline{AD} \cong \overline{CB}$
Prove: $\overline{AB} \parallel \overline{CD}$

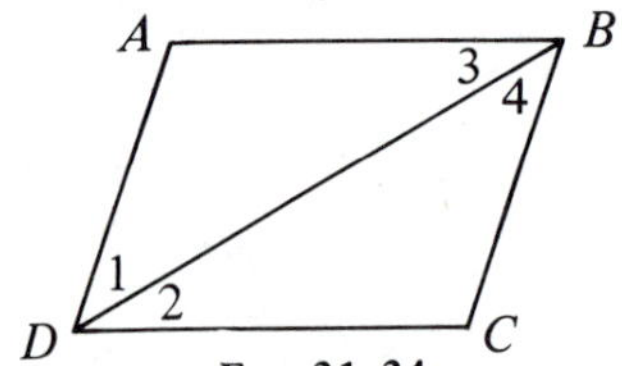

Exs. 31–34

35. Given: $\overline{XO} \cong \overline{YO}$;
$\overline{QO} \cong \overline{RO}$
Prove: $\angle P \cong \angle S$

36. Given: $\overline{PO} \cong \overline{SO}$;
$\overline{QO} \cong \overline{RO}$
Prove: $\angle 1 \cong \angle 2$

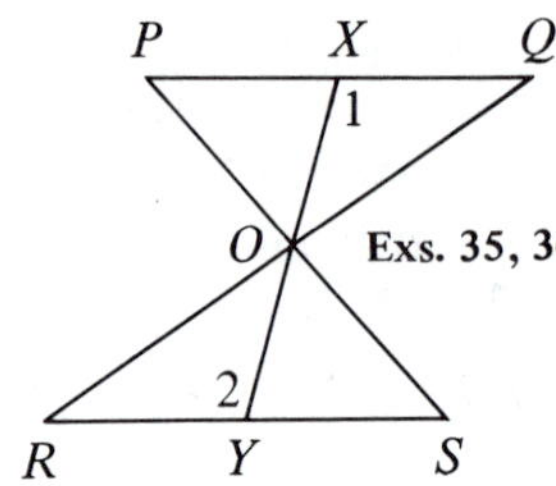

Exs. 35, 36

C 37. Given: $\overline{GH} \parallel \overline{EF}$;
$\angle 1 \cong \angle 2$
Prove: $\overline{GF} \cong \overline{HE}$

38. Given: $\overline{GP} \cong \overline{HP}$;
$\overline{GH} \parallel \overline{EF}$
Prove: $\angle GFE \cong \angle HEF$

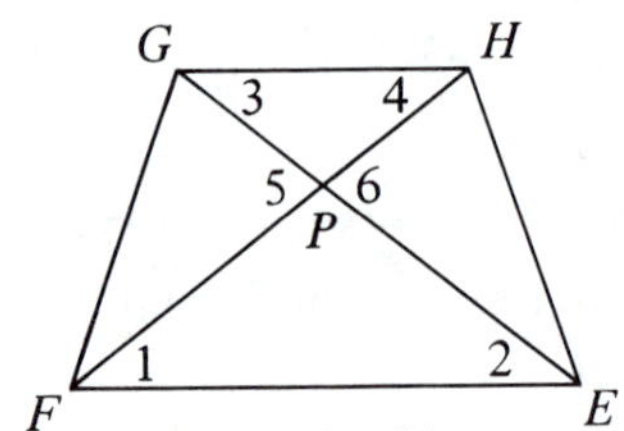

Exs. 37, 38

39. Given: $\overline{AB} \parallel \overline{CD}$;
$\overline{XY} \perp \overline{CD}$;
$\overline{AX} \cong \overline{BX}$
Prove: $\angle AYC \cong \angle BYD$

40. Given: $\overline{XY} \perp \overline{AB}$; $\overline{XY} \perp \overline{CD}$;
$\overline{AX} \cong \overline{BX}$; $\overline{CY} \cong \overline{DY}$
Prove: $\angle C \cong \angle D$

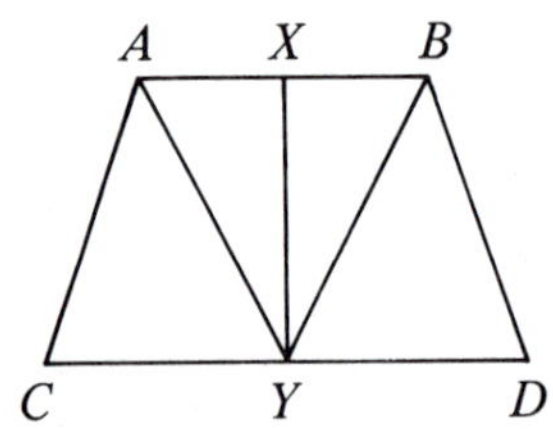

Exs. 39, 40

41. Use a two-column proof to prove the *existence* part of Theorem 4-7; that is, given line l and point P not on l, prove that *there exists* a line through P that is perpendicular to l.

42. Use an indirect proof to prove the *uniqueness* part of Theorem 4-7; that is, given line l and point P not on l, prove that *there is at most one* line through P that is perpendicular to l.

SELF-TEST 1

State the theorem or corollary that could be used with the given information to prove that $a \parallel b$.

1. $\angle 3 \cong \angle 6$ 2. $a \perp t$; $b \perp t$ 3. $\angle 6 \cong \angle 8$ Section 4-1

4. Write a two-column proof.
   Given: $\angle 1$ and $\angle 7$ are supplementary.
   Prove: $a \parallel b$

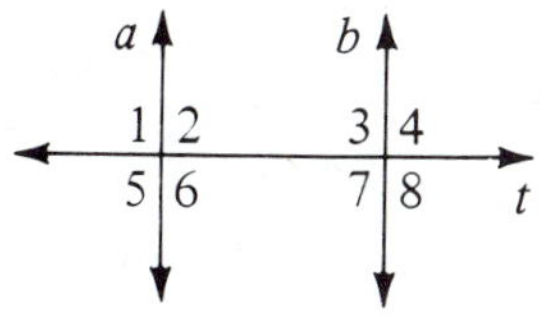

Exs. 1–7

Section 4-2

5. If $a \parallel b$ and $m\angle 1 = x$, express the measures of the other numbered angles in terms of x.
6. If $a \parallel b$, $m\angle 1 = 3y - 5$, and $m\angle 4 = 2y + 15$, find the value of y.
7. Write a two-column proof.
   Given: $a \parallel b$
   Prove: $\angle 1 \cong \angle 8$
8. Given a line k and a point X not on k, how many lines contain X and are perpendicular to k? Explain.

Section 4-3 THE TRIANGLE ANGLE-SUM THEOREM

In this section you will learn one of the most basic and most useful theorems in Euclidean geometry: the Triangle Angle-Sum Theorem.

Theorem 4-8 The sum of the measures of the interior angles of any triangle is 180. (Triangle Angle-Sum Theorem)

GIVEN: $\triangle ABC$
PROVE: $m\angle A + m\angle B + m\angle ACB = 180$

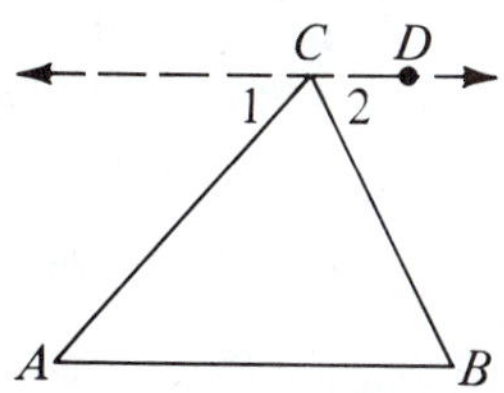

Statements	Reasons
1. Let $\overleftrightarrow{CD}$ be the line through C and parallel to $\overleftrightarrow{AB}$.	1. Through a given point not on a given line, there is exactly one line parallel to the given line.
2. $\angle 1 \cong \angle A$; $\angle 2 \cong \angle B$ (or $m\angle 1 = m\angle A$; $m\angle 2 = m\angle B$)	2. If 2 lines are $\parallel$, alt. int. $\angle$s are $\cong$.
3. $m\angle 1 + m\angle ACD = 180$	3. If the ext. sides of 2 adj. $\angle$s are opp. rays, the $\angle$s are supp.
4. $m\angle ACD = m\angle 2 + m\angle ACB$	4. Angle Addition Postulate
5. $m\angle 1 + m\angle 2 + m\angle ACB = 180$	5. Substitution Property
6. $m\angle A + m\angle B + m\angle ACB = 180$	6. Substitution Property

EXAMPLE In $\triangle ABC$, $m\angle A = x - 5$, $m\angle B = 2x + 3$, and $m\angle C = 4x$. Find the value of x and the measures of the three angles.

SOLUTION

$$\begin{aligned} m\angle A + m\angle B + m\angle C &= 180 \\ (x-5) + (2x+3) + 4x &= 180 \\ 7x - 2 &= 180 \\ 7x &= 182 \\ x &= 26 \end{aligned}$$

$$\begin{aligned} m\angle A &= x - 5 \\ &= 26 - 5 \\ &= 21 \end{aligned} \qquad \begin{aligned} m\angle B &= 2x + 3 \\ &= 2(26) + 3 \\ &= 55 \end{aligned} \qquad \begin{aligned} m\angle C &= 4x \\ &= 4(26) \\ &= 104 \end{aligned}$$

EXAMPLE The measure of the vertex angle of an isosceles triangle is 40 less than twice the measure of a base angle. Is the base longer or shorter than a leg of the triangle?

SOLUTION Let x = the measure of each base angle.
Then $2x - 40$ = the measure of the vertex angle.

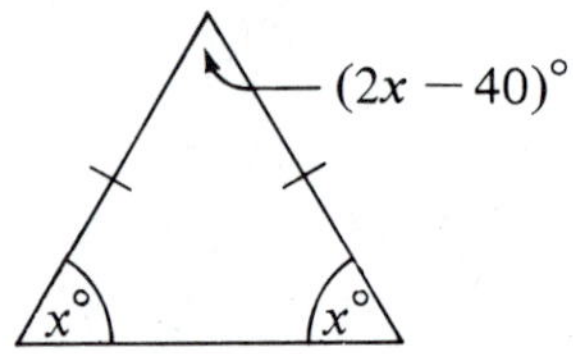

$$\begin{aligned} x + x + (2x - 40) &= 180 \\ 4x &= 220 \\ x &= 55 \end{aligned}$$

$$\begin{aligned} 2x - 40 &= 2(55) - 40 \\ &= 70 \end{aligned}$$

Since the measure of the vertex angle is greater than the measure of a base angle, by Theorem 3-17 the base (the side opposite the vertex angle) is longer than each leg (a side opposite a base angle).

The Triangle Angle-Sum Theorem has several important corollaries:

Corollary 1 If two angles of one triangle are congruent to two angles of another triangle, then the third angles of the triangles are congruent.

Corollary 2 The two acute angles of a right triangle are complementary.

Corollary 3 In a triangle, there can be at most one right or obtuse angle.

The proofs of Corollaries 1 and 2 are left as Exercises 21 and 22. For a justification of Corollary 3, see the Oral Exercises.

Corollary 4 Each angle of an equilateral triangle has measure 60.

GIVEN: Equilateral $\triangle ABC$
PROVE: $m\angle A = 60$; $m\angle B = 60$; $m\angle C = 60$

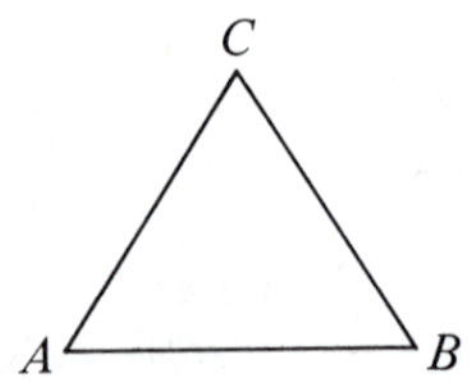

Statements	Reasons
1. $\triangle ABC$ is equilateral.	1. Given
2. $\overline{AC} \cong \overline{BC}$; $\overline{AB} \cong \overline{BC}$	2. Def. of equilateral $\triangle$
3. $m\angle A = m\angle B$; $m\angle A = m\angle C$	3. Base $\angle$s of an isos. $\triangle$ are $\cong$.
4. $m\angle A + m\angle B + m\angle C = 180$	4. The sum of the measures of the $\angle$s of a $\triangle$ is 180.
5. $m\angle A + m\angle A + m\angle A = 3m\angle A = 180$	5. Substitution Property
6. $m\angle A = 60$	6. Division Prop. of Equality
7. $m\angle B = 60$; $m\angle C = 60$	7. Substitution Property

You know that a triangle that contains a right angle is called a *right triangle.* Similarly, a triangle that contains an obtuse angle is called an *obtuse triangle.* A triangle that contains only acute angles is called an *acute triangle.*

ORAL EXERCISES

The measures of two angles of $\triangle ABC$ are given.

a. Find the measure of the third angle.
b. Name the longest side of $\triangle ABC$.
c. Classify $\triangle ABC$ as acute, obtuse, or right.

1. $m\angle A = 40$, $m\angle B = 60$
2. $m\angle B = 30$, $m\angle C = 60$
3. $m\angle A = 45$, $m\angle C = 90$
4. $m\angle A = 120$, $m\angle B = 30$

5. Justify Corollary 3 of Theorem 4-8: Explain why a triangle cannot have more than one right or obtuse angle.

6. Explain why the longest side of a right triangle must be opposite the right angle.

WRITTEN EXERCISES

Find the measure of $\angle C$ of $\triangle ABC$.

A **1.** $m\angle A = 33$, $m\angle B = 98$
2. $m\angle A = 57$, $m\angle B = 105$
3. $m\angle A = 62$ and $\overline{AC} \cong \overline{BC}$
4. $m\angle B = 37$ and $\overline{AC} \cong \overline{BC}$
5. $m\angle A = 70$ and $\overline{AC} \cong \overline{AB}$
6. $m\angle B = 40$ and $\overline{BC} \cong \overline{AB}$

Name **(a)** the longest side and **(b)** the shortest side of $\triangle PQR$ if the given angles have the given measures.

7. $m\angle P = 30$, $m\angle Q = 80$
8. $m\angle Q = 50$, $m\angle R = 65$
9. $m\angle R = 55$, $m\angle P = 65$
10. $m\angle P = 62$, $m\angle R = 60$

Find the value of x and the measures of $\angle A$, $\angle B$, and $\angle C$ if the angles of $\triangle ABC$ have the given measures.

11. $m\angle A = x + 3$, $m\angle B = 2x - 8$, $m\angle C = 2x - 5$

12. $m\angle A = 3x$, $m\angle B = 3x$, $m\angle C = x - 2$

13. $m\angle A = 40 - x$, $m\angle B = 20 - x$, $m\angle C = 10x$

14. $m\angle A = 7x$, $m\angle B = 35 - x$, $m\angle C = 30 - x$

15. $m\angle A = \frac{1}{2}x + 17$, $m\angle B = \frac{1}{2}x + 11$, $m\angle C = 3x$

16. $m\angle A = \frac{3}{2}x - 8$, $m\angle B = \frac{1}{2}x + 2$, $m\angle C = x$

Find the measures of the two acute angles of $\triangle ABC$ if $m\angle C = 90$.

17. $m\angle A = 3x - 12$, $m\angle B = x - 6$

18. $m\angle A = 85 - 2x$, $m\angle B = 50 - x$

Write two-column proofs.

19. Given: $\angle 2 \cong \angle A$
Prove: $\angle 1 \cong \angle C$

20. Given: $\overline{ED} \perp \overline{AB}$;
$\angle C$ is a right angle.
Prove: $\angle 2 \cong \angle A$

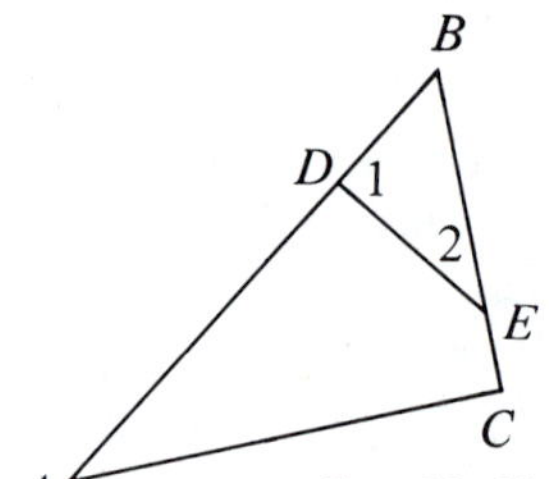

Exs. 19, 20

21. Prove Corollary 1 to Theorem 4-8.
Given: $m\angle R = m\angle X$;
$m\angle S = m\angle Y$;
Prove: $m\angle T = m\angle Z$

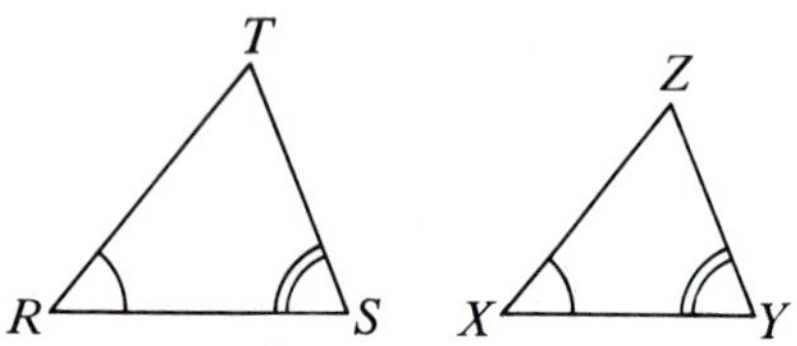

22. Prove Corollary 2 to Theorem 4-8.
Given: $m\angle C = 90$
Prove: $\angle A$ and $\angle B$ are complementary.

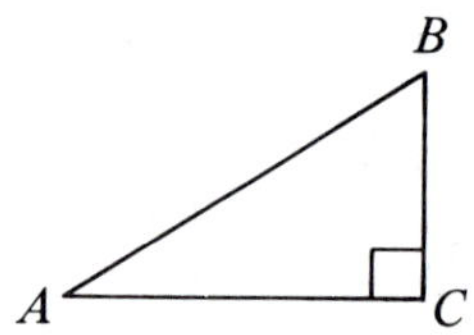

B **23.** Given: $\overleftrightarrow{XY} \parallel \overleftrightarrow{VW}$;
$\angle 3 \cong \angle 5$
Prove: $\overline{XV} \parallel \overline{YZ}$

24. Given: $\overline{XV} \parallel \overline{YZ}$;
$\angle 3 \cong \angle 5$
Prove: $\overleftrightarrow{XY} \parallel \overleftrightarrow{VW}$

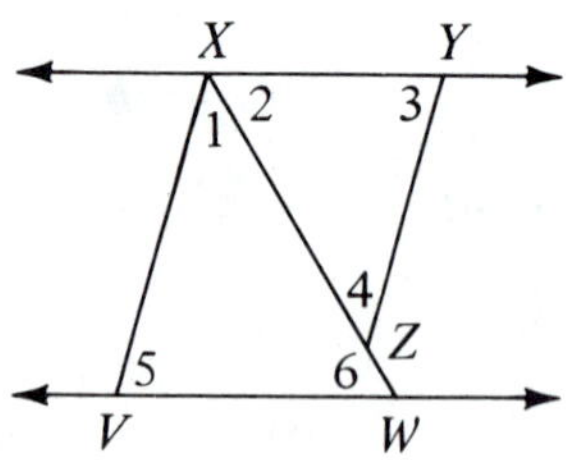

Exs. 23, 24

25. Given: $\overline{ST} \perp \overline{SY}$;
$\overline{PQ} \perp \overline{SY}$;
$\angle 1 \cong \angle 2$
Prove: $\angle T \cong \angle Y$

26. Given: $\angle T \cong \angle Y$;
$\angle 1 \cong \angle 2$;
$\overline{PQ} \perp \overline{SY}$
Prove: $\overline{ST} \perp \overline{SY}$

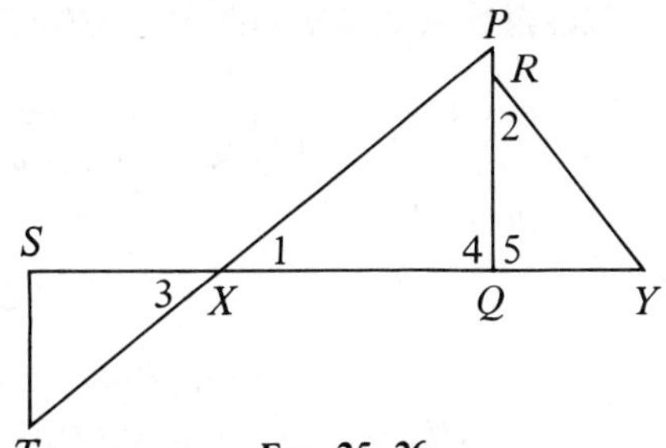

Exs. 25, 26

27. The measure of the vertex angle of an isosceles triangle is 12 less than the measure of each base angle. Find the measures of all three angles.

28. The measure of each base angle of an isosceles triangle is 15 less than twice the measure of the vertex angle. Find the measures of all three angles.

29. The measure of one of the acute angles of a right triangle is 12 greater than the other. Find the measures of the two acute angles.

30. In $\triangle RST$, the measure of $\angle S$ is twice that of $\angle T$ and the measure of $\angle R$ is 45 less than that of $\angle S$. What special kind of triangle is $\triangle RST$? Explain.

31. Use the diagram to explain why $m\angle A + m\angle B + m\angle C + m\angle D = 360$.

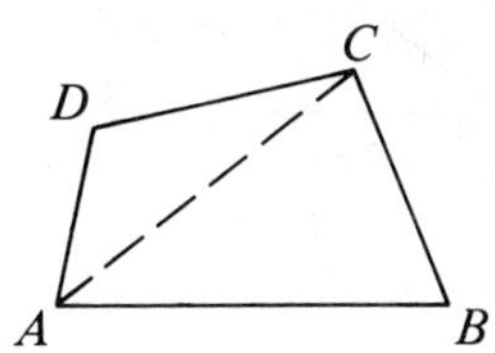

32. Use the diagram to explain why $m\angle P + m\angle Q + m\angle R + m\angle S = 360$.

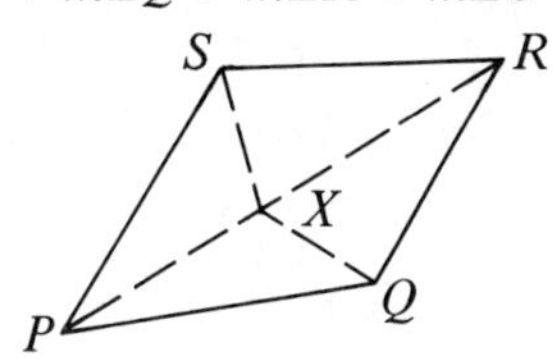

Find the values of x and y.

C 33.

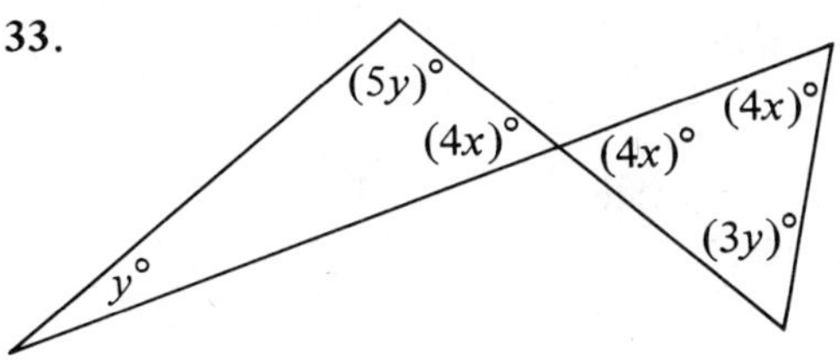

34.

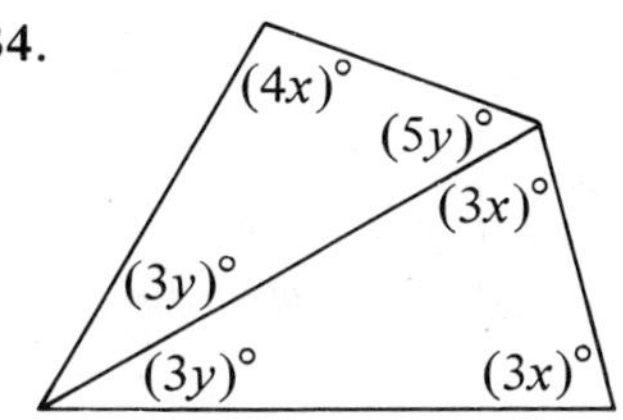

Write two-column proofs.

35. Given: $\overleftrightarrow{XA} \parallel \overleftrightarrow{YB}$;
$\overrightarrow{XZ}$ bisects $\angle AXY$;
$\overrightarrow{YZ}$ bisects $\angle BYX$.
Prove: $\angle Z$ is a right angle.

36. Given: $m\angle 1 = m\angle 2$; $m\angle 3 = m\angle 4$;
$m\angle Z = 90$
Prove: $\overleftrightarrow{XA} \parallel \overleftrightarrow{YB}$

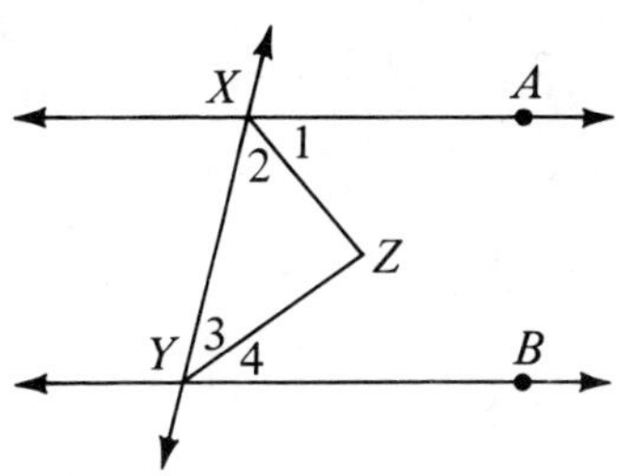

Section 4-4 ANOTHER WAY OF PROVING CONGRUENCE

The Triangle Angle-Sum Theorem has several other consequences, at least two of which are too important to be called corollaries. These two theorems, which you will learn in this section, are the:

- AAS Theorem
- Exterior Angle Theorem

So far we have four methods of establishing congruence without showing that two triangles satisfy the *definition* of congruent triangles:

The SSS Postulate
The ASA Postulate
The SAS Postulate
The HL Postulate for right triangles

We now add to this list a fifth method.

Theorem 4-9 If two angles and a non-included side of one triangle are congruent to two angles and the corresponding non-included side of another triangle, the two triangles are congruent. (AAS Theorem)

GIVEN: $\triangle ABC$ and $\triangle DEF$;
$\angle A \cong \angle D$;
$\angle B \cong \angle E$;
$\overline{BC} \cong \overline{EF}$
PROVE: $\triangle ABC \cong \triangle DEF$

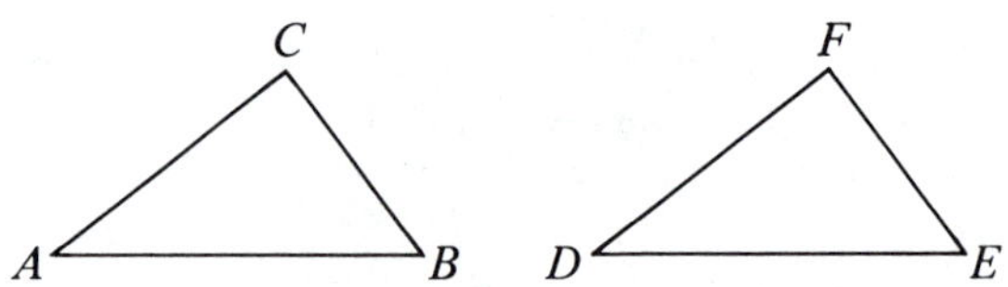

Statements	Reasons
1. $\angle A \cong \angle D$; $\angle B \cong \angle E$	1. Given
2. $\angle C \cong \angle F$	2. If 2 ∠s of one △ are ≅ to 2 ∠s of another △, the third ∠s are ≅.
3. $\overline{BC} \cong \overline{EF}$	3. Given
4. $\triangle ABC \cong \triangle DEF$	4. ASA Postulate

One immediate benefit of the AAS Theorem is that it provides an easy way of proving Theorem 3-14, which we now show as an example.

EXAMPLE

Theorem 3-14 If two angles of a triangle are congruent, then the sides opposite these angles are congruent.

GIVEN: $\triangle ABC$ with $\angle A \cong \angle C$
PROVE: $\overline{AB} \cong \overline{CB}$

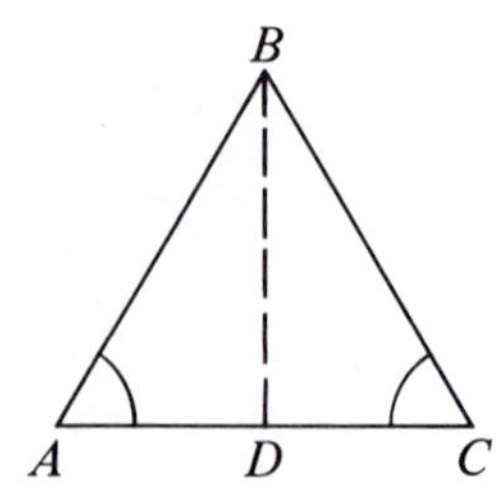

Statements	Reasons
1. Draw the bisector of $\angle ABC$ and let D be the point where it intersects $\overline{AC}$.	1. Any angle of measure less than 180 has exactly one bisector.
2. $\angle ABD \cong \angle CBD$	2. Def. of an angle bisector
3. $\angle A \cong \angle C$	3. Given
4. $\overline{BD} \cong \overline{BD}$	4. Reflexive Prop. of $\cong$
5. $\triangle ABD \cong \triangle CBD$	5. AAS Theorem
6. $\overline{AB} \cong \overline{CB}$	6. Corr. parts of $\cong$ $\triangle$s are $\cong$.

The second theorem that follows from the Triangle Angle-Sum Theorem is the following theorem.

Theorem 4-10 The measure of an exterior angle of a triangle is equal to the sum of the measures of the remote interior angles. (Exterior Angle Theorem)

GIVEN: $\triangle ABC$

PROVE: $m\angle 1 = m\angle A + m\angle B$

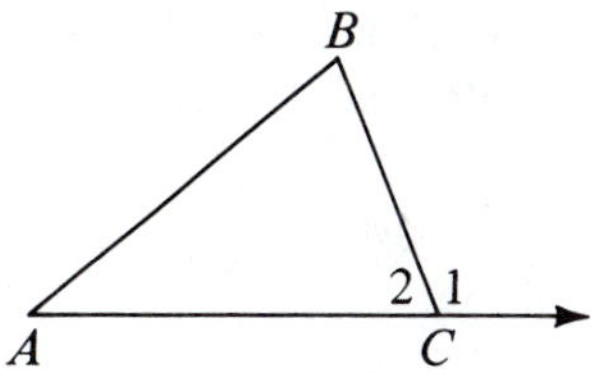

Statements	Reasons
1. $\triangle ABC$	1. Given
2. $m\angle A + m\angle B + m\angle 2 = 180$	2. Triangle Angle-Sum Theorem
3. $m\angle 1 + m\angle 2 = 180$	3. If the ext. sides of 2 adj. $\angle$s are opp. rays, the $\angle$s are supp.
4. $m\angle 1 + m\angle 2 = m\angle A + m\angle B + m\angle 2$	4. Substitution Property
5. $m\angle 1 = m\angle A + m\angle B$	5. Subtraction Prop. of Equality

EXAMPLE Find $m\angle B$ if $m\angle BCD = 150$ and $m\angle A = 35$.

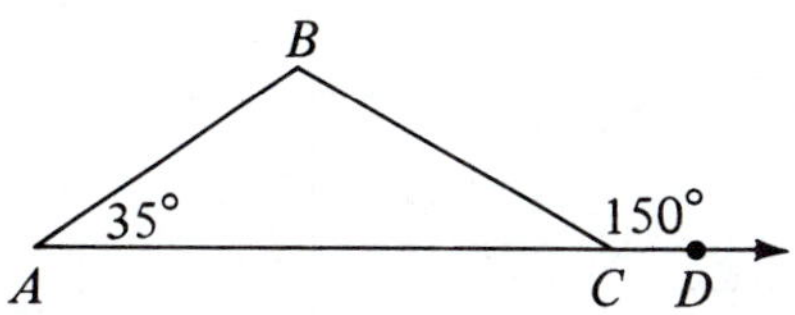

SOLUTION

$$\begin{aligned} m\angle BCD &= m\angle A + m\angle B \\ 150 &= 35 + m\angle B \\ 115 &= m\angle B \end{aligned}$$

EXAMPLE In the diagram, $m\angle 1 = 5x - 4$, $m\angle 2 = x + 30$, and $m\angle 3 = x + 20$. Find the value of x and the measures of the three angles.

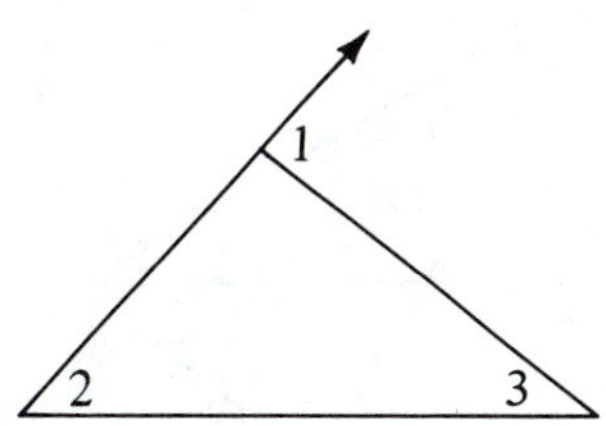

SOLUTION Since $m\angle 1 = m\angle 2 + m\angle 3$,

we have

$$\begin{aligned} 5x - 4 &= (x + 30) + (x + 20) \\ 5x &= 2x + 54 \\ 3x &= 54 \\ x &= 18 \end{aligned}$$

$$\begin{aligned} m\angle 1 &= 5(18) - 4 \\ &= 86 \end{aligned} \qquad \begin{aligned} m\angle 2 &= 18 + 30 \\ &= 48 \end{aligned} \qquad \begin{aligned} m\angle 3 &= 18 + 20 \\ &= 38 \end{aligned}$$

ORAL EXERCISES

Complete.

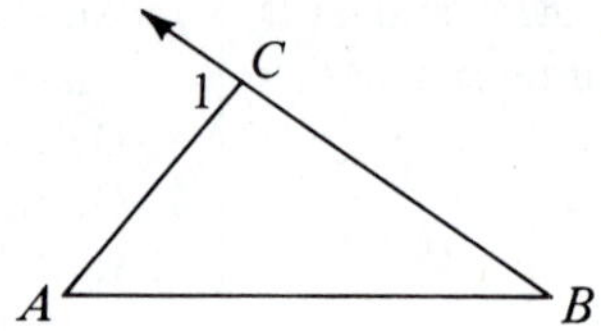

1. $m\angle A = 40$, $m\angle B = 30$, $m\angle 1 =$ __?__
2. $m\angle A = 45$, $m\angle B = 35$, $m\angle 1 =$ __?__
3. $m\angle A = 50$, $m\angle 1 = 85$, $m\angle B =$ __?__
4. $m\angle B = 50$, $m\angle 1 = 75$, $m\angle A =$ __?__

State two angles whose measures have a sum equal to the measure of the given angle.

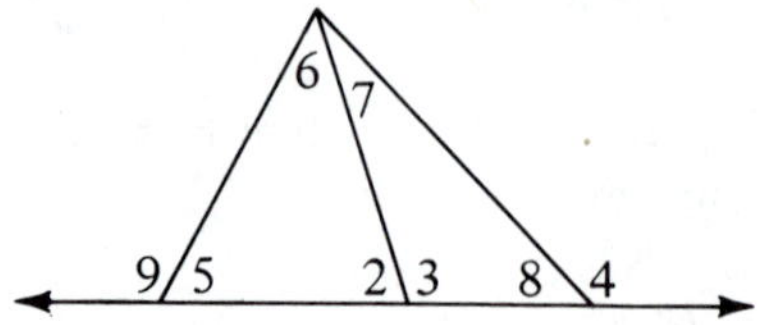

5. $\angle 9$ 6. $\angle 2$ 7. $\angle 3$ 8. $\angle 4$

9. Explain how the following statement follows directly from the AAS Theorem: If the hypotenuse and an acute angle of one right triangle are congruent to the hypotenuse and an acute angle of another right triangle, then the triangles are congruent.

10. Explain how the AAS Theorem can be used in the following proof:

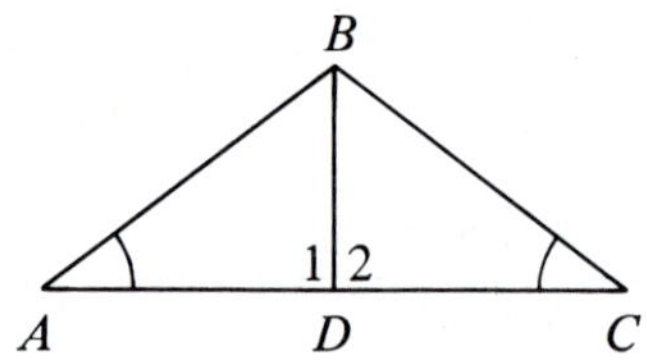

Given: $\angle A \cong \angle C$;
$\overline{BD} \perp \overline{AC}$

Prove: $\triangle ABD \cong \triangle CBD$

WRITTEN EXERCISES

State the postulate or theorem, if any, that could be used to prove two triangles congruent.

A 1.

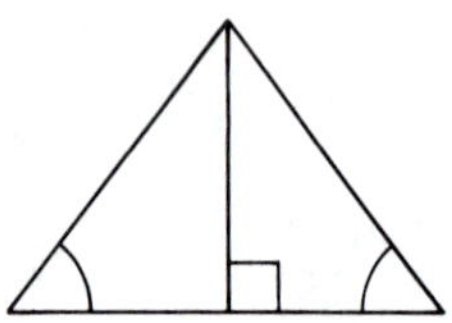

2.

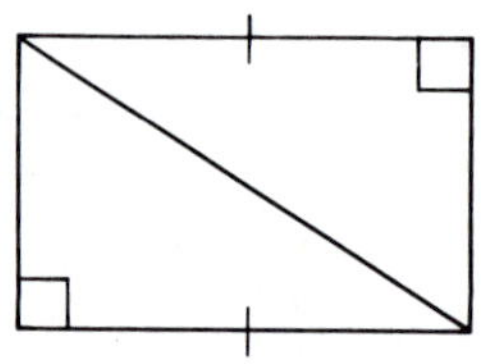

3.

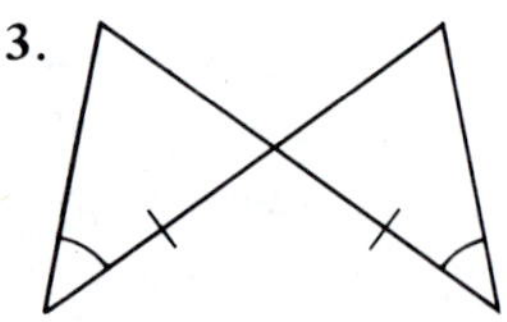

4.

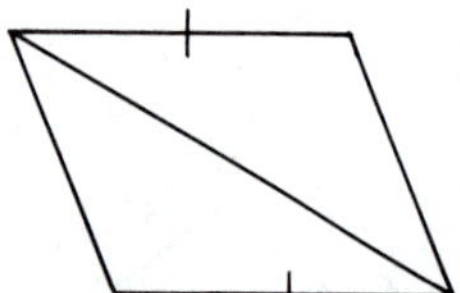

5.

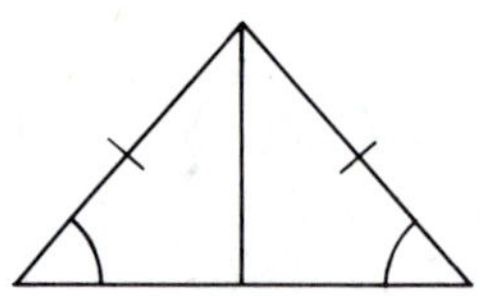

6.

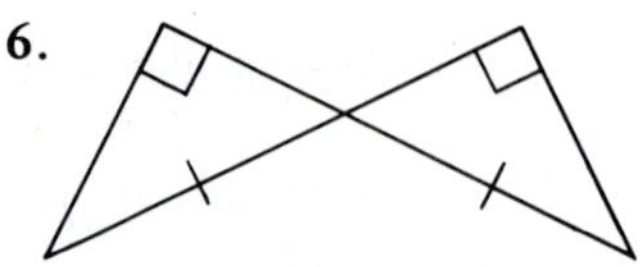

7.

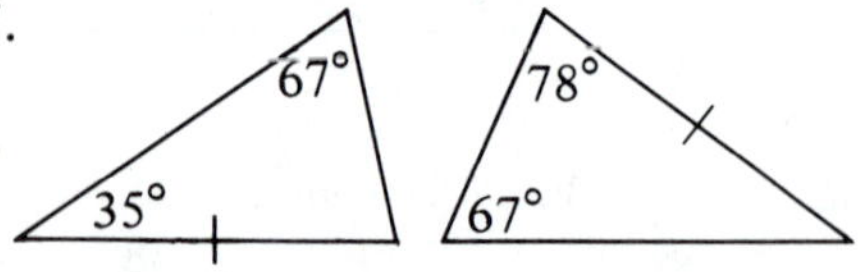

8.

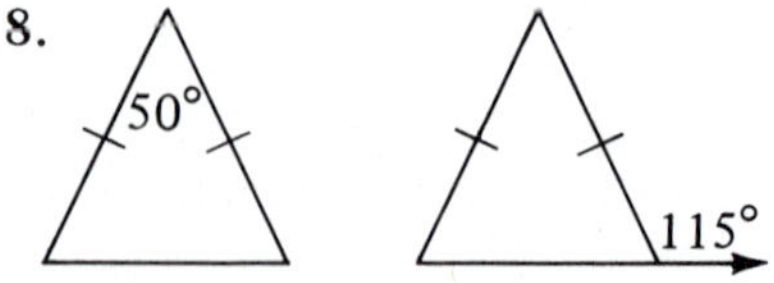

Complete.

9. $m\angle 2 = 45$, $m\angle 3 = 60$, $m\angle 5 =$ __?__

10. $m\angle 3 = 70$, $m\angle 4 = 80$, $m\angle 1 =$ __?__

11. $m\angle 1 = 140$, $m\angle 3 = 55$, $m\angle 4 =$ __?__

12. $m\angle 3 = 84$, $m\angle 5 = 110$, $m\angle 2 =$ __?__

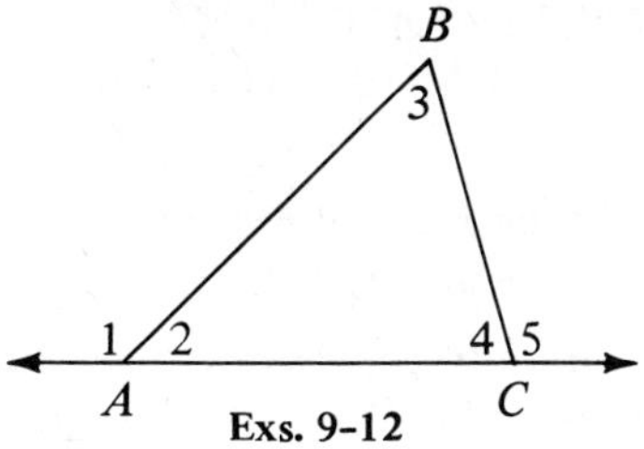

Exs. 9–12

Find the measures of $\angle 6$, $\angle 7$, and $\angle 8$.

13. $m\angle 6 = 5x + 17$, $m\angle 7 = 2x$, $m\angle 8 = 35$

14. $m\angle 6 = 4x - 80$, $m\angle 7 = x$, $m\angle 8 = x - 10$

15. $m\angle 6 = 3x + 7$, $m\angle 7 = 40 - x$, $m\angle 8 = 32 - x$

16. $m\angle 6 = \frac{1}{2}x + 15$, $m\angle 7 = 21 - \frac{1}{2}x$, $m\angle 8 = 50 - x$

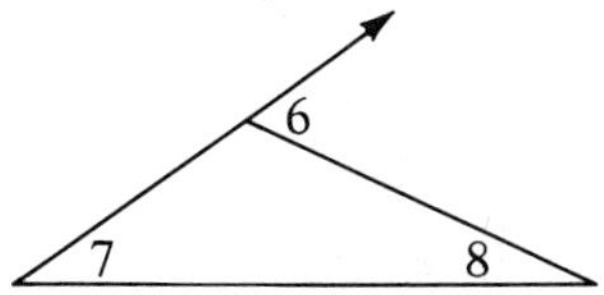

Exs. 13–16

Write two-column proofs.

17. Given: $\angle P \cong \angle R$; $\overline{PQ} \cong \overline{RS}$
Prove: $\overline{PX} \cong \overline{RX}$

18. Given: $\overline{PX} \cong \overline{RX}$; $\angle Q \cong \angle S$
Prove: $\overline{PQ} \cong \overline{RS}$

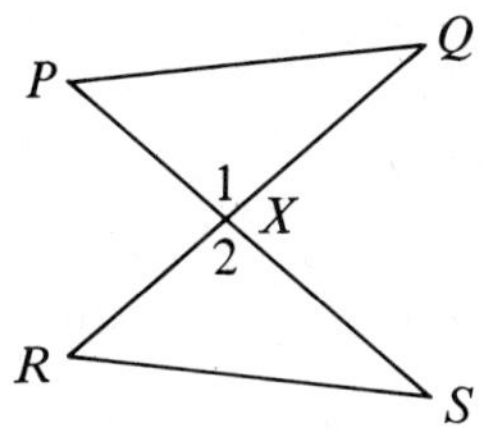

Exs. 17, 18

19. Given: $\overline{AB} \cong \overline{CB}$; $\angle 2 \cong \angle 1$
Prove: $\overline{EB} \cong \overline{DB}$

20. Given: $\angle A \cong \angle C$; $\overline{BE} \cong \overline{BD}$
Prove: $\overline{AB} \cong \overline{CB}$

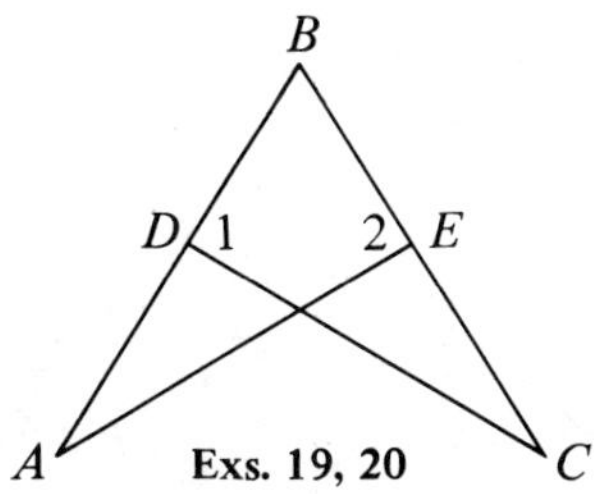

Exs. 19, 20

21. Given: Diagram as shown
Prove: $m\angle 1 + m\angle 2 = m\angle ACB + m\angle ADB$

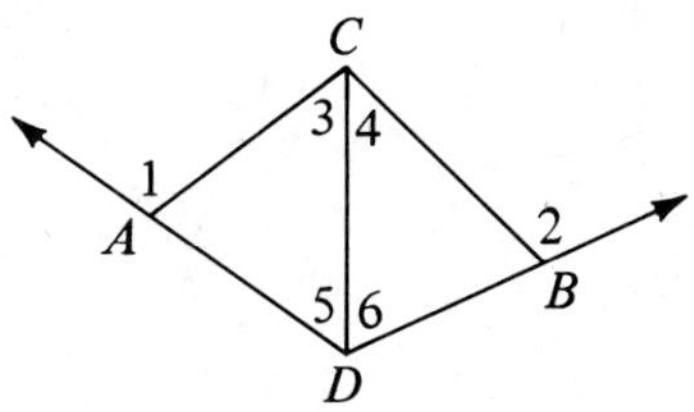

22. Given: Diagram as shown
Prove: $m\angle 2 + m\angle 3 = m\angle PQR + m\angle 4$

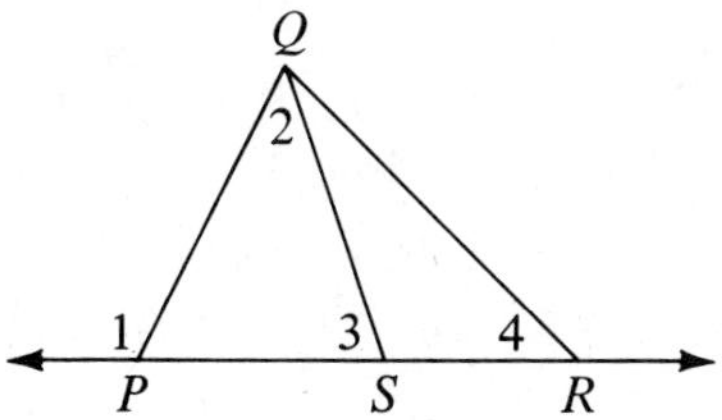

B 23. In $\triangle KLM$, the measure of $\angle K$ is 55 and the measure of an exterior angle adjacent to $\angle KLM$ is 105. What is the shortest side of $\triangle KLM$ and what is the longest side?

24. In $\triangle RST$, the measure of an exterior angle adjacent to $\angle RST$ is 4 less than twice the measure of $\angle T$. Is $\overline{RS}$ longer or shorter than $\overline{ST}$? Explain.

25. Find the measure of the vertex angle of an isosceles triangle if the measure of an exterior angle adjacent to a base angle is 145.

26. In isosceles $\triangle PQR$, the measure of an exterior angle adjacent to $\angle PRQ$ is three times the measure of the vertex angle, $\angle P$. Find the measures of the angles of the triangle.

Write two-column proofs.

27. Given: $\angle A \cong \angle B$;
$\overline{DC} \parallel \overline{AB}$;
$\angle 1 \cong \angle 2$
Prove: $\overline{AD} \cong \overline{BC}$

28. Given: $\angle A \cong \angle B$;
$\angle ADT \cong \angle BCT$;
T is the midpoint of $\overline{AB}$.
Prove: $\angle 1 \cong \angle 2$

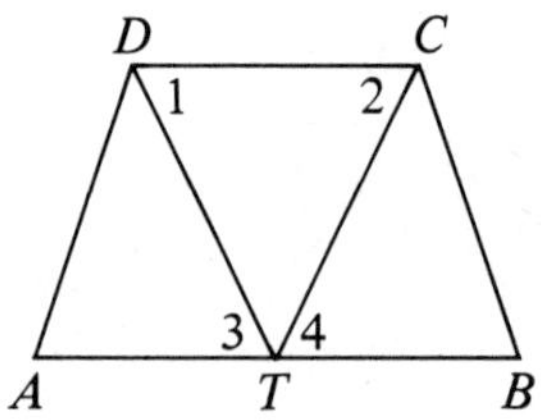

Exs. 27, 28

29. Given: $\overline{AJ} \cong \overline{AK}$;
$\angle 1 \cong \angle 2$;
$\angle 3 \cong \angle 4$
Prove: $\triangle JLB \cong \triangle KMB$

30. Given: $\angle J \cong \angle K$;
$\angle 3 \cong \angle 4$;
$\angle 1 \cong \angle 2$
Prove: $\overline{JL} \cong \overline{KM}$

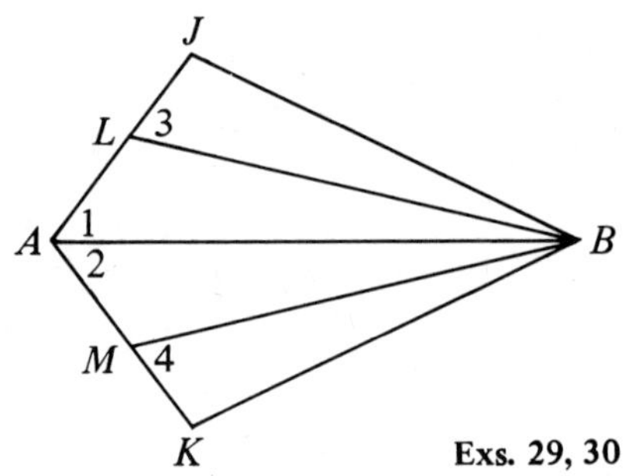

Exs. 29, 30

31. Given: $\overline{LN} \cong \overline{MN}$;
$\angle 3 \cong \angle 4$;
$\overline{OP} \parallel \overline{LM}$
Prove: Q is the midpoint of $\overline{LM}$.

32. Given: $\overline{LN} \cong \overline{MN}$;
$\overline{LO} \cong \overline{MP}$;
$\angle 1 \cong \angle 2$
Prove: $\angle 3 \cong \angle 4$

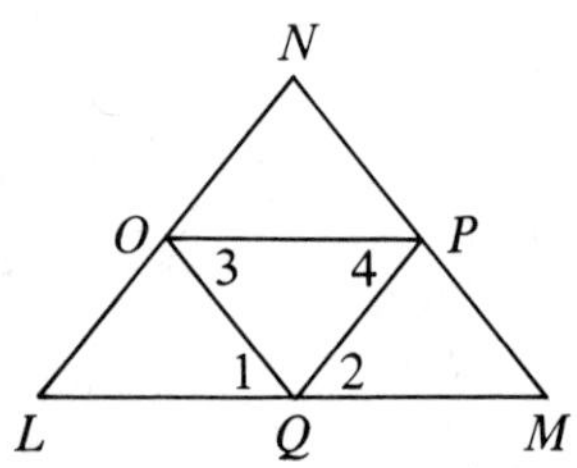

Exs. 31, 32

33. Given $m\angle 6 = 125$, $m\angle 5 = 85$, $\overline{FD} \perp \overleftrightarrow{AE}$, and $m\angle CFE = 115$, find $m\angle 3$, $m\angle 2$, and $m\angle 1$.

34. Given $m\angle 1 = 150$, $m\angle CFE = 110$, $\overline{FD} \perp \overleftrightarrow{AE}$, and $m\angle 5 = 80$, find $m\angle 3$, $m\angle 4$, and $m\angle 6$.

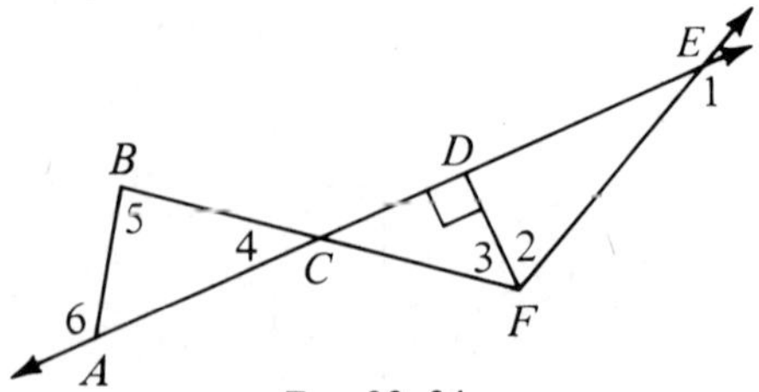

Exs. 33, 34

C 35. Given $RS = SW$, $SW = WT$, and $RT = ST$, find the values of x, y, and z in the diagram at the right.

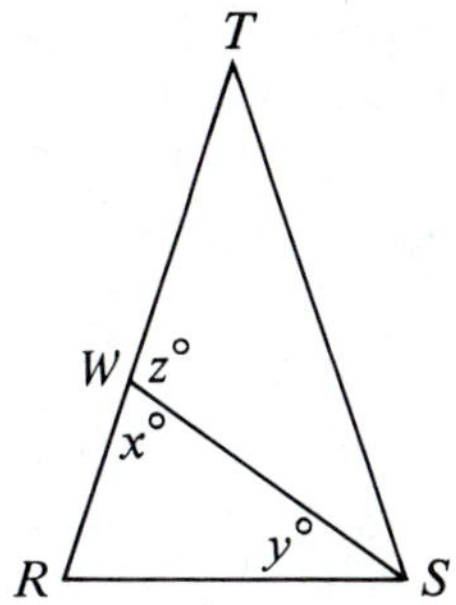

Section 4-5 POLYGONS

A triangle is the simplest example of a more general figure called a *polygon.* In this section you will learn what a polygon and a regular polygon are and will study theorems about:

- the sum of the measures of the interior angles of a polygon
- the sum of the measures of the exterior angles of a polygon
- the measure of each interior angle and each exterior angle of a regular polygon

If $P_1, P_2, \ldots, P_n$ (with $n \geqslant 3$) are coplanar points such that the segments $\overline{P_1P_2}$, $\overline{P_2P_3}, \ldots, \overline{P_nP_1}$ satisfy the following conditions, then the union of these segments is called a *polygon*:

1. Each segment intersects exactly two other segments, one at each endpoint.
2. No two segments with a common endpoint are collinear.

Each segment is called a *side* of the polygon, and its endpoints are *vertices* of the polygon. A segment joining two nonconsecutive vertices is a *diagonal* of the polygon.

EXAMPLES All the figures below are polygons.

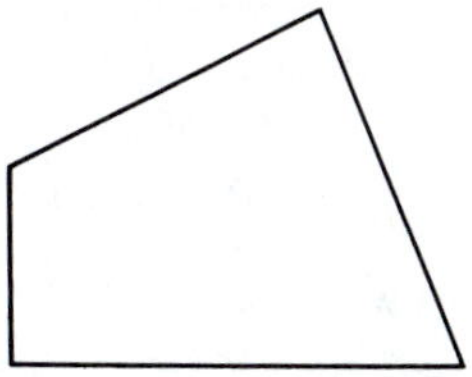

4 sides and 4 vertices

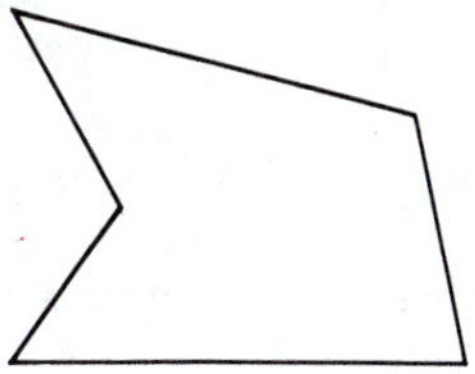

5 sides and 5 vertices

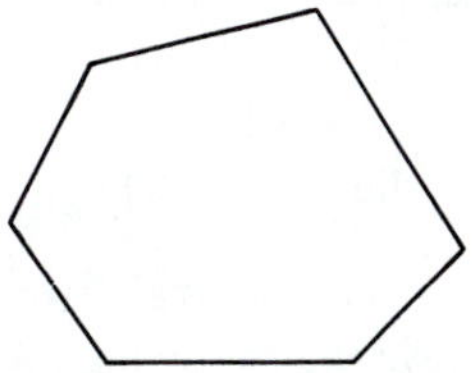

6 sides and 6 vertices

EXAMPLES The figures below are *not* polygons.

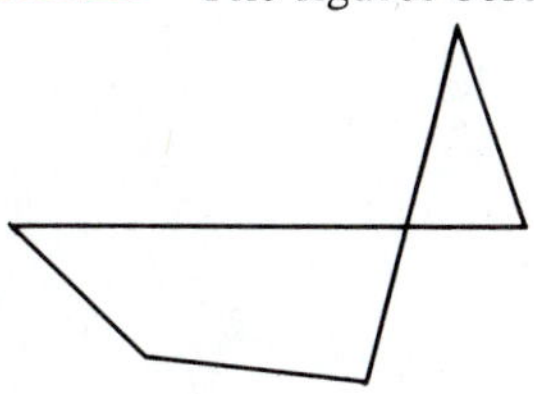

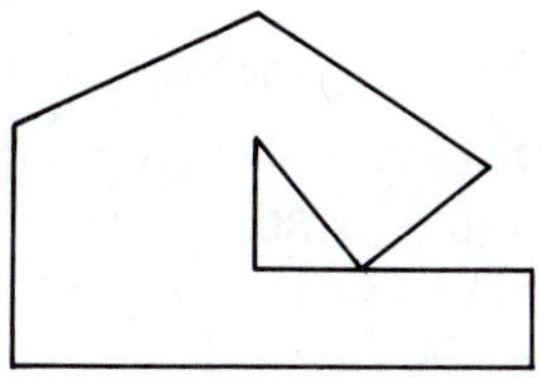

Some polygons have special names according to the number of sides they have. The most common names are listed below.

Number of Sides	*Name of Polygon*
3	triangle
4	quadrilateral
5	pentagon
6	hexagon
8	octagon

A polygon with n sides is called an n-gon.

A polygon is *convex* if any line segment whose endpoints are on different sides of the polygon intersects the polygon only at those two points. The polygon shown is not convex since $\overline{RS}$ intersects the polygon at T. When we refer to a polygon in this book, we will mean a convex polygon.

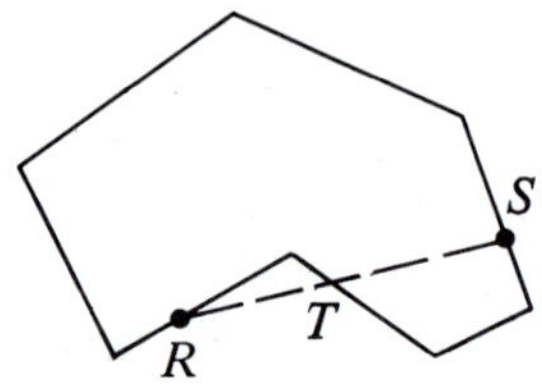

The angles determined by the sides of a polygon are called the *angles* or the *interior angles* of the polygon. The interior angles of polygon $ABCDE$ are $\angle A$, $\angle B$, $\angle C$, $\angle D$, and $\angle E$.

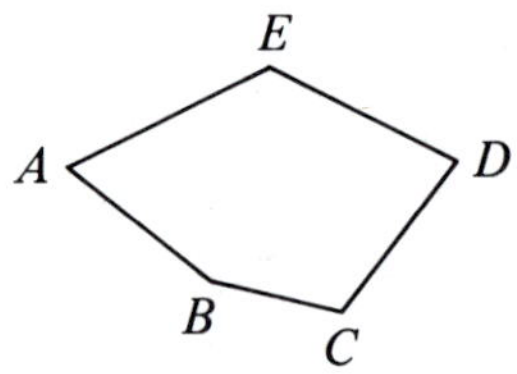

If you draw all the diagonals from one vertex of a polygon, as shown below, you can use the Triangle Angle-Sum Theorem to find the sum of the measures of the interior angles of the polygon.

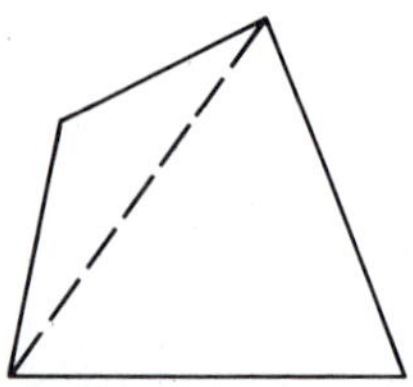

Sides: 4
$\triangle$s: 2
Sum: 2(180)

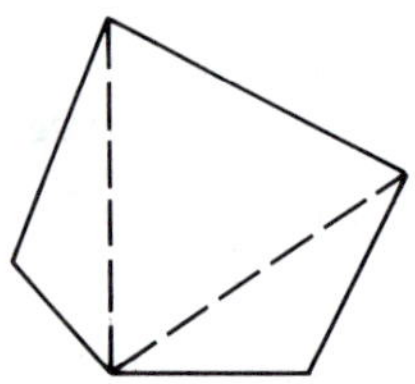

Sides: 5
$\triangle$s: 3
Sum: 3(180)

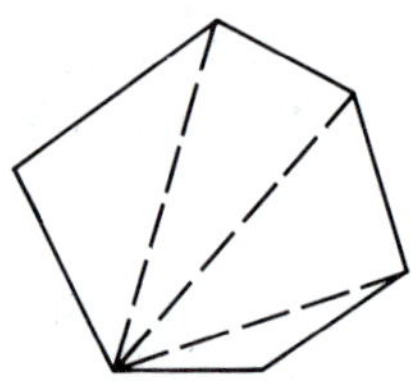

Sides: 6
$\triangle$s: 4
Sum: 4(180)

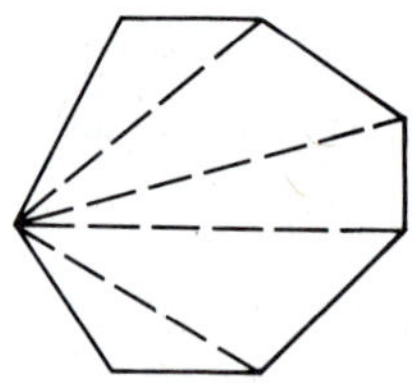

Sides: 7
$\triangle$s: 5
Sum: 5(180)

The diagonals from one vertex will divide an n-gon into $n - 2$ triangles. Thus we have the following theorem.

Theorem 4–11 The sum of the measures of the interior angles of a polygon with n sides is $(n - 2)180$.

EXAMPLE Find the sum of the measures of the interior angles of a hexagon.

SOLUTION A hexagon has 6 sides, so $n = 6$.

$$\begin{aligned} \text{Sum} &= (6 - 2)180 \\ &= 4 \cdot 180 \\ &= 720 \end{aligned}$$

EXAMPLE If the sum of the measures of the interior angles of a polygon is 1440, how many sides does the polygon have?

SOLUTION 1 Let n be the number of sides.

$$(n-2)180 = 1440$$
$$n-2 = \frac{1440}{180}$$
$$n-2 = 8$$
$$n = 10$$

SOLUTION 2 Let n be the number of sides.

$$(n-2)180 = 1440$$
$$180n - 360 = 1440$$
$$180n = 1800$$
$$n = 10$$

A *regular polygon* is a polygon in which all sides are congruent and all interior angles are congruent. Therefore Theorem 4-11 has the following corollary.

Corollary The measure of each interior angle of a regular polygon with n sides is $\frac{(n-2)180}{n}$.

An angle that is both adjacent to and supplementary to an interior angle of a polygon is called an *exterior angle* of the polygon. Surprisingly, the sum of the measures of the exterior angles of a polygon does not depend on the number of sides of the polygon!

Theorem 4-12 For any polygon, the sum of the measures of the exterior angles, one angle at each vertex, is 360.

Explanation:

At each vertex of a polygon, the sum of the measures of the interior angle and one exterior angle is 180.

Therefore, in an n-gon the sum of the measures of the interior angles and the exterior angles, one angle at each vertex, is $n(180)$. Since $n(180)$ = (sum of measures of interior angles) + (sum of measures of exterior angles), we have:

$$\text{sum of measures of exterior angles} = n(180) - [(n-2)180]$$
$$= 180n - [180n - 360]$$
$$= 360$$

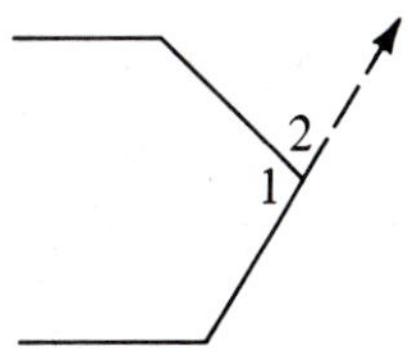

$m\angle 1 + m\angle 2 = 180$

Corollary The measure of each exterior angle of a regular polygon with n sides is $\frac{360}{n}$.

EXAMPLE Find the measure of **(a)** one interior angle and **(b)** one exterior angle of a regular hexagon.

SOLUTION A hexagon has 6 sides, so $n = 6$.

(a) $m(\text{int. } \angle) = \frac{(6-2)180}{6} = \frac{4(180)}{6} = 120$

(b) $m(\text{ext. } \angle) = \frac{360}{6} = 60$

ORAL EXERCISES

Find the sum of the measures of the interior angles of each polygon.

1. quadrilateral
2. octagon
3. dodecagon (a polygon with 12 sides)
4. a polygon with 102 sides
5. What is the sum of the measures of the exterior angles, one angle at each vertex, of each polygon in Exercises 1-4?
6. How many exterior angles does an n-gon have?
7. What is the sum of the measures of *all* the exterior angles of an n-gon?
8. Explain how the diagram below can be used to give an intuitive "proof" of Theorem 4-12.

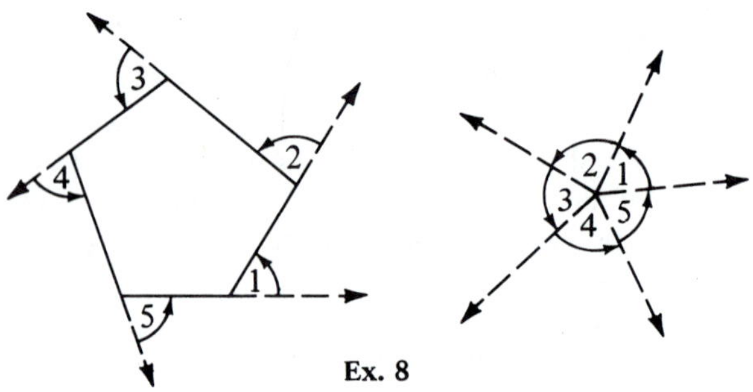

Ex. 8

9. Verify the formula given in the corollary to Theorem 4-11 for the case of $n = 3$ (an equilateral triangle) and $n = 4$ (a square).

WRITTEN EXERCISES

Find the sum of the measures of the interior angles of a polygon with the given number of sides.

A 1. 9 2. 7 3. 10 4. 11 5. 14 6. 18

For a polygon with the given number of sides, find the sum of the measures of the exterior angles, one angle at each vertex.

7. 15 8. 22 9. 40 10. 100

Find the measure of one interior angle of a regular polygon with the given number of sides.

11. 5 12. 8 13. 10 14. 12 15. 20 16. 24

Find the measure of one exterior angle of a regular polygon with the given number of sides.

17. 12 **18.** 18 **19.** 20 **20.** 24

Find the number of *sides* of a polygon whose interior angles have the given number as the sum of their measures.

B **21.** 1080 **22.** 1800 **23.** 1980 **24.** 2340

25. Find the measure of the sixth interior angle of a hexagon if the measures of the other five interior angles are 70, 90, 140, 150, and 160.

26. Find the measure of the fifth interior angle of a pentagon if the measure of each of the other four angles is 115.

Find the measure of one exterior angle of a regular polygon, if the *sum* of the measures of the interior angles is the given number.

C **27.** 1260 **28.** 2880 **29.** 3240 **30.** 3960

SELF-TEST 2

If the given angles of $\triangle ABC$ have the given measures, name **(a)** the numerical measure of $\angle C$ and **(b)** the shortest side of $\triangle ABC$.

1. $m\angle A = 53$, $m\angle B = 61$ Section 4-3

2. $m\angle A = 90$, $m\angle B = 2x + 13$, $m\angle C = 4x - 13$

3. $m\angle A = 3y - 26$, $m\angle B = 2y + 5$, $m\angle C = y + 27$

Section 4-4

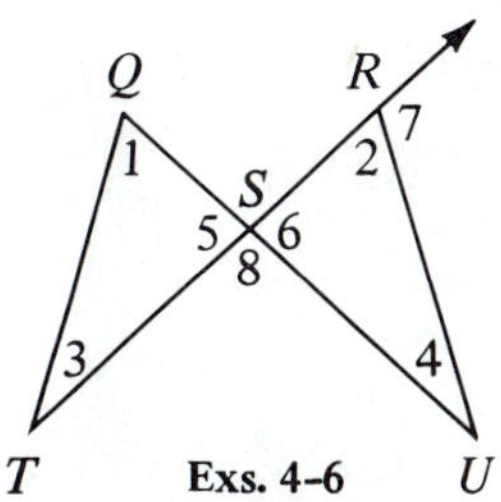

Exs. 4-6

4. Express the measure of $\angle 7$ in terms of the measures of two other angles.

5. If $m\angle 8 = 7x - 4$, $m\angle 2 = 2x + 17$, and $m\angle 4 = 3x + 11$, find the value of x.

6. Write a two-column proof.
Given: $\angle 1 \cong \angle 2$; $\overline{ST} \cong \overline{SU}$
Prove: $\overline{QT} \cong \overline{RU}$

7. Find the sum of the measures of the interior angles of a pentagon. Section 4-5

8. Find the sum of the measures of the exterior angles of a triangle, one angle at each vertex.

9. Find the measures of each interior angle and each exterior angle of a regular octagon.

Computer Exercises

Here is another kind of statement that is used in BASIC.

Let A = (expression) This tells the computer to give A the value of the expression.

1. Copy and RUN the following program for the values of N given.

```
10   PRINT "TO FIND THE MEASURES OF THE"
20   PRINT "INTERIOR AND EXTERIOR ANGLES OF A"
30   PRINT "REGULAR POLYGON HAVING N SIDES:"
40   PRINT
50   PRINT "INPUT N";
60   INPUT N
70   PRINT
80   PRINT "EACH INTERIOR ANGLE HAS MEASURE";
90   LET X=(N-2)*180/N  (* means "multiplied by"; / means "divided by.")
100  PRINT X
110  PRINT "EACH EXTERIOR ANGLE HAS MEASURE";
120  LET E=360/N
130  PRINT E
140  END
```

a. N=3   b. N=4

c. N=5   d. N=6

e. N=7   f. N=8

g. N=9   h. N=10

2. What happens if you input 2, 1, 0, or −1 for N? To avoid this, insert this statement:

```
65   IF N<3 THEN 50
```

3. In the preceding program, delete lines 65, 80, and 110, and make these changes:

```
50   PRINT "   N","   INT. ANGLE","   EXT. ANGLE","   SUM"
```

(Notice the spaces.)

```
60   FOR N=3 TO 15
100  PRINT N,X,
130  PRINT E,X+E
135  NEXT N
```

RUN the program, and complete these statements.

a. As the number of sides increases, the measures of the interior angles __?__.

b. As the number of sides increases, the measures of the exterior angles __?__.

c. As the number of sides increases, the sum of the measures of the interior and exterior angles, one at each vertex, __?__.

CHAPTER REVIEW

Use the given information to name two parallel lines. State the theorem or corollary that justifies your answer.

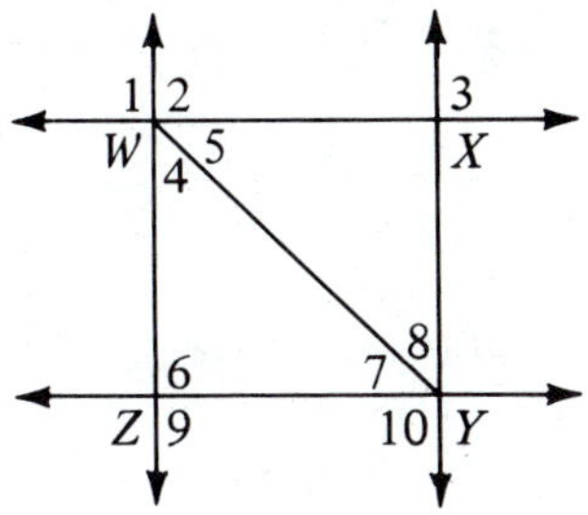

Exs. 1-7

1. $\angle 1$ and $\angle 3$ are right angles. Section 4-1
2. $\angle 2 \cong \angle 6$
3. $\angle 9$ and $\angle 10$ are supplementary.
4. Write a two-column proof.
   Given: $\triangle WXY \cong \triangle YZW$
   Prove: $\overleftrightarrow{WZ} \parallel \overleftrightarrow{XY}$

If $\overleftrightarrow{WX} \parallel \overleftrightarrow{ZY}$, find the value of x.

5. $m\angle 5 = 8x - 5$, $m\angle 7 = 5x + 28$ Section 4-2
6. $m\angle 2 = 9x - 23$, $m\angle 9 = 4x + 21$
7. Write a two-column proof.
   Given: $\overleftrightarrow{WZ} \parallel \overleftrightarrow{XY}$; $\overline{ZW} \cong \overline{ZY}$
   Prove: $\overrightarrow{YW}$ bisects $\angle XYZ$.

Find the numerical measure of $\angle C$ in $\triangle ABC$. Then classify $\triangle ABC$ as acute, right, or obtuse.

8. $m\angle A = 28$, $m\angle B = 62$ Section 4-3
9. $m\angle A = 4x - 4$, $m\angle B = 3x + 2$, $m\angle C = 2x + 11$
10. The measure of one of the acute angles of a right triangle is 3 more than twice the measure of the other. Find the measures of the two angles.
11. Describe how you could show that $\overline{QS} \cong \overline{TR}$, given that $\overline{QR} \perp \overline{RS}$, $\overline{TS} \perp \overline{RS}$, and $\angle Q \cong \angle T$. Section 4-4

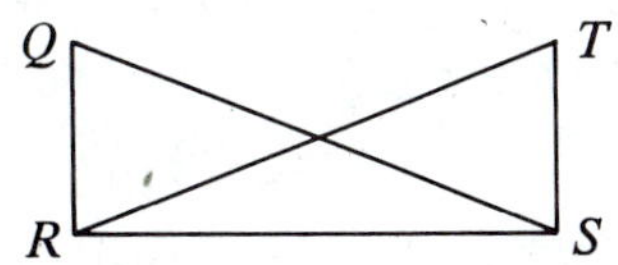

12. If $m\angle 6 = 118$ and $m\angle 4 = 61$, name the longest side of $\triangle ACD$.
13. Write a two-column proof.
    Given: $m\angle 3 = m\angle 1 + m\angle 2$
    Prove: $\overline{BE} \parallel \overleftrightarrow{CD}$

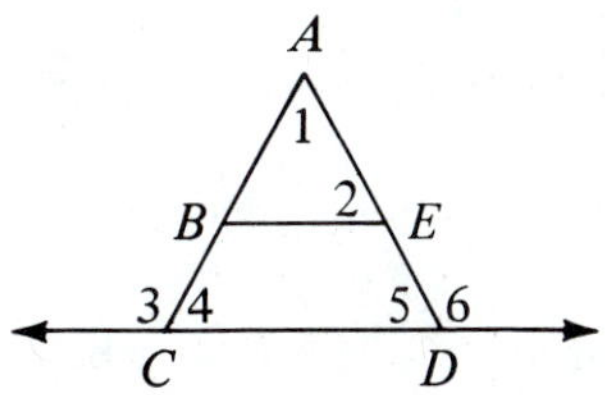

Exs. 12, 13

14. A polygon has 16 sides. Find the sum of the measures of **(a)** the interior angles and **(b)** the exterior angles, one angle at each vertex. Section 4-5
15. What is the measure of each exterior angle of a regular polygon with 18 sides?
16. The sum of the measures of the interior angles of a polygon is 2700. How many sides does the polygon have?

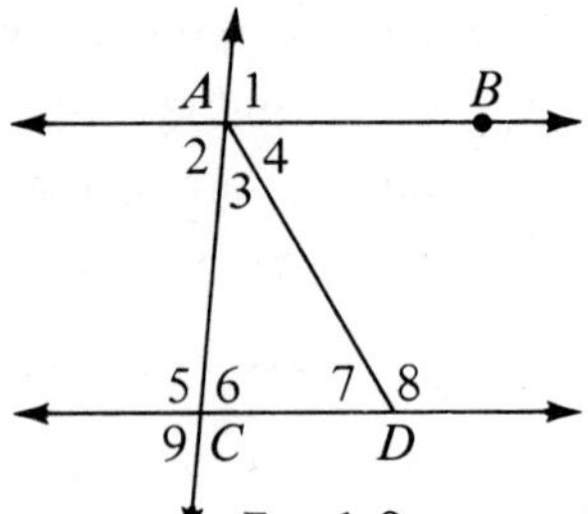

Exs. 1–9

1. Name two alternate interior angles with respect to transversal $\overleftrightarrow{AD}$.
2. If $m\angle 3 = x + 18$, $m\angle 6 = 2x - 1$, and $m\angle 7 = 2x + 3$, find the value of x.
3. If $\overleftrightarrow{AB} \parallel \overleftrightarrow{CD}$, $m\angle 5 = 103$, and $m\angle 3 = 28$, find the measures of $\angle 4$ and $\angle 8$.

If $\overleftrightarrow{AB} \parallel \overleftrightarrow{CD}$, find the value of y.

4. $m\angle 5 = 5y - 8$, $m\angle 2 = 2y + 20$
5. $m\angle 1 = \frac{1}{2}y + 50$, $m\angle 9 = \frac{3}{2}y - 10$

State the theorem or corollary that justifies the given statement.

6. If $\angle 1 \cong \angle 6$, then $\overleftrightarrow{AB} \parallel \overleftrightarrow{CD}$.
7. If $\overleftrightarrow{AC} \perp \overleftrightarrow{CD}$, then $m\angle 7 \neq 90$.
8. If $m\angle 4 + m\angle 8 = 180$, then $\overleftrightarrow{AB} \parallel \overleftrightarrow{CD}$.
9. If $\overleftrightarrow{AB} \parallel \overleftrightarrow{CD}$ and $\overleftrightarrow{AB} \perp \overleftrightarrow{AC}$, then $\overleftrightarrow{CD} \perp \overleftrightarrow{AC}$.

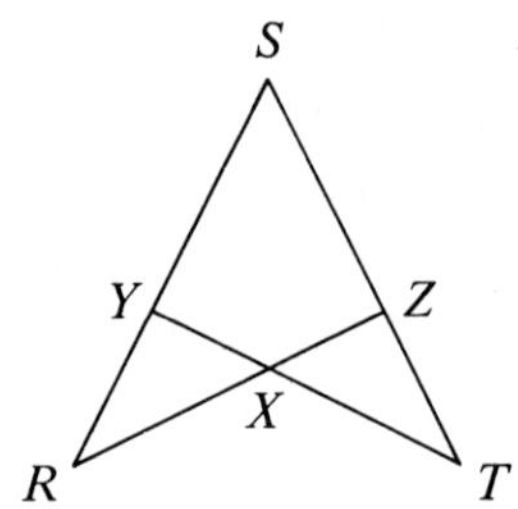

Exs. 10, 11

10. Given: $\overline{TY} \perp \overline{RS}$; $\overline{RZ} \perp \overline{TS}$
    a. Name two pairs of triangles that have $\angle R$ and $\angle T$ as corresponding parts.
    b. Explain why $\angle R \cong \angle T$.

Write two-column proofs.

11. Given: $\overline{RZ} \cong \overline{TY}$;
    $\angle SYT \cong \angle SZR$
    Prove: $\overline{RS} \cong \overline{TS}$

12. Given: $\overleftrightarrow{AB} \parallel \overleftrightarrow{CD}$;
    $\overline{AC} \cong \overline{AD}$
    Prove: $m\angle 1 + m\angle 2 + m\angle 3 = 180$

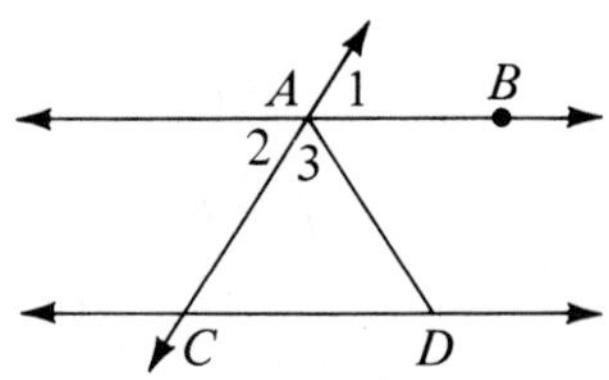

13. An exterior angle adjacent to a base angle of an isosceles triangle has measure 102. Find the measure of the vertex angle.
14. A polygon has 52 sides. Find the sum of the measures of **(a)** the interior angles and **(b)** the exterior angles, one angle at each vertex.
15. Find the measure of each interior angle of a regular polygon with 30 sides.
16. If each exterior angle of a regular polygon has measure 20, how many sides does the polygon have?

Biographical Note

Sophie Germain

Stories about Sophie Germain (1776–1831) leave us with a memorable image of a talented young girl whose determination to study mathematics paid off. Her papers on number theory and elasticity helped launch the study of mathematical physics.

Germain grew up in Paris during the French Revolution. Because it was dangerous to venture outside, she spent a great deal of time in the family library. She is thought to have read that the famous Greek mathematician Archimedes was killed by a soldier because he was too absorbed in solving a geometry problem to answer the soldier's question. Germain was intrigued by a field of study that could command such a high level of concentration. Despite her parents' initially strong objections, she set out to learn as much as she could about mathematics.

In 1794 the École Polytechnique opened in Paris. Although female students were not admitted, Germain was able to get lecture notes from friends. She was particularly interested in the lectures of the French mathematician Lagrange. Using the name M. Le Blanc, she submitted a paper which so impressed Lagrange that he insisted on meeting Monsieur Le Blanc. After learning her identity, he introduced her to many of the leading French scientists of the day. From then on Sophie Germain corresponded with a number of prominent scholars, one of whom was the eminent German mathematician Gauss.

Around 1808 Parisian mathematicians became interested in the work of Ernst Chladni, the German physicist who founded the science of acoustics. In 1811 the French Academy of Sciences offered a prize for the best essay that would mathematically explain and predict the unusual patterns Chladni was demonstrating. Five years later Germain won the grand prize, receiving world-wide recognition. She continued her studies, publishing several papers on the theory of elasticity.

Gauss recommended that Sophie Germain be granted an honorary doctorate degree from the University of Göttingen, but unfortunately, she died before the degree could be awarded.

VOCABULARY REVIEW

Be sure that you understand the meaning of these terms:

transversal, p. 154
parallel lines, p. 154
corollary, p. 154
alternate interior angles, p. 154
corresponding angles, p. 154
altitude of a triangle, p. 161
right triangle, p. 161
obtuse triangle, p. 167
acute triangle, p. 167
polygon, p. 175
diagonal of a polygon, p. 175
convex polygon, p. 175
regular polygon, p. 177

MIXED REVIEW

Find the numerical measure of $\angle C$ in $\triangle ABC$. Then classify $\triangle ABC$ as acute, right, or obtuse.

1. $m\angle A = 46$, $m\angle B = 74$
2. $m\angle A = 3x - 4$, $m\angle B = x + 5$, $m\angle C = 6x - 1$

Construct a truth table to show that each of the following is a tautology.

3. $r \rightarrow (s \rightarrow r)$
4. $r \rightarrow (\sim r \rightarrow r)$

Exercises 5–8 refer to the diagram at the right, in which $l \parallel m$, and $m\angle 1 = 65$.

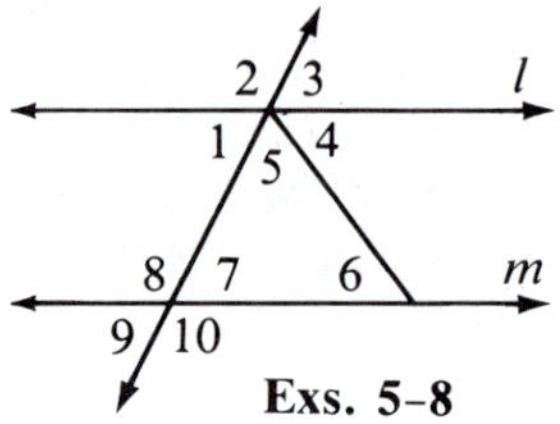

Exs. 5–8

5. Find $m\angle 2$.
6. Name all the pairs of corresponding angles.
7. Find $m\angle 9$ and $m\angle 8$.
8. If $\angle 5 \cong \angle 6$, find $m\angle 4$ and $m\angle 6$.
9. Find the number of sides of a regular polygon if the sum of the measures of the interior angles is 3960.

Write two-column proofs.

10. Given: $\overline{AC} \cong \overline{CE}$; $\overline{BC} \cong \overline{CD}$
    Prove: $\overline{AB} \parallel \overline{DE}$
11. Given: $\overline{AB} \parallel \overline{DE}$; $\overline{AB} \cong \overline{DE}$
    Prove: $\overline{AC} \cong \overline{EC}$; $\overline{BC} \cong \overline{DC}$

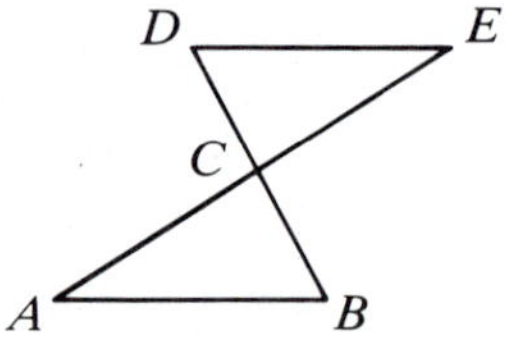

Exs. 10, 11

12. Given: Isosceles $\triangle AEC$; $\overline{AE} \cong \overline{AC}$; $\overline{BD} \parallel \overline{AE}$
    Prove: $\overline{BD} \cong \overline{BC}$

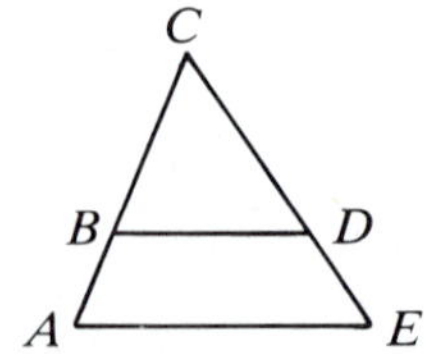

Simplify. In Exercise 14, combine like terms.

13. $(5^2 - 3 \cdot 4)(-4 - (-2^2) + 3)$

14. $3(5a - 3b) - (2b - 3a) + 4b$

15. Name the largest angle of $\triangle ABC$ if $AB = 13$, $BC = 8$, and $AC = 11$.

16. Negate the statement "If I don't clean up the spilled paint, it will harden on the floor."

Solve.

17. $-9 \leqslant 3(1 + y) \leqslant -3$

18. $5t - (2 - 3t) < 22$

19. $4 - 2x > 3x + 14$

20. $\frac{2}{3}z + 4 = \frac{1}{4}z + 5$

21. Find the sum of the measures of the interior angles of a polygon with 52 sides.

22. In $\triangle ABC$, $\overline{AB} \cong \overline{AC}$. Find the value of x if $m\angle B = 3x + 62$ and $m\angle C = 7x - 22$.

Write the conclusion that follows logically from the premises and state which rule of inference is being used. If no conclusion follows, explain why.

23. P_1: Ed will visit us on Saturday or Sunday.
    P_2: Ed didn't visit us on Saturday.

24. P_1: I don't understand the subject.
    P_2: If I read the book, I will understand the subject.

25. Supply reasons for the proof.

GIVEN: $\angle B \cong \angle E$; $\overline{BD} \cong \overline{CE}$
PROVE: $\angle C \cong \angle D$

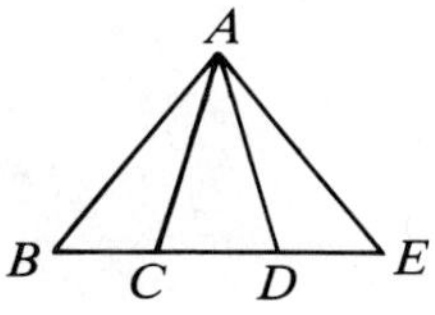

Statements	Reasons
1. $\overline{BD} \cong \overline{CE}$ or $BD = CE$	1. ?
2. $BC + CD = BD$; $CD + DE = CE$	2. ?
3. $BC + CD = CD + DE$	3. ?
4. $BC = DE$ or $\overline{BC} \cong \overline{DE}$	4. ?
5. $\angle B \cong \angle E$	5. ?
6. $\overline{AB} \cong \overline{AE}$	6. ?
7. $\triangle ABC \cong \triangle AED$	7. ?
8. $\overline{AC} \cong \overline{AD}$	8. ?
9. $\angle C \cong \angle D$	9. ?

Navigators of ships sailing the high seas made use of the properties of a parallelogram when they determined their ship's heading using the parallel rule described on the facing page.

Quadrilaterals

5

For centuries mariners have used a simple device called a *parallel rule* to determine the *heading*, or direction in relation to north, that a ship should take in order to reach its destination. Two positions of a parallel rule are shown below.

Open **Shut**

In *any* position of the parallel rule, the arms and indicators form a quadrilateral such as $RSTW$ shown in the diagram at the right, with $RS = WT$ and $RW = ST$. As you will learn in this chapter, this means that $RSTW$ is a figure called a *parallelogram*, and $\overleftrightarrow{RS}$ must be parallel to $\overleftrightarrow{WT}$.

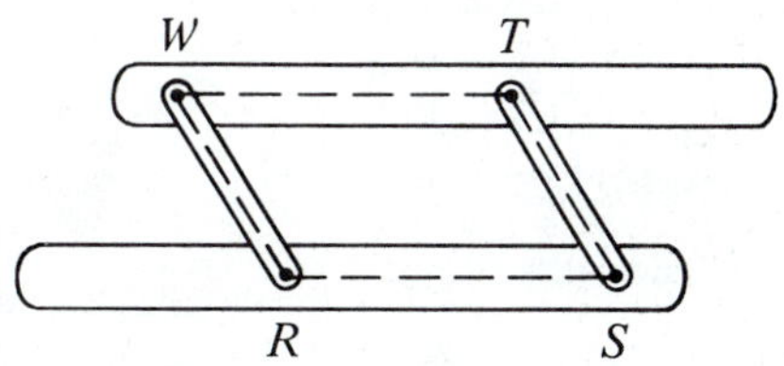

On a *chart*, or nautical map, the navigator plots the course $\overline{SD}$ from the present position of the ship, S, to the destination, D. The heading the ship should take is determined by a line parallel to $\overline{SD}$ through the center of the *compass rose*, a diagram of the points of the compass that is printed on the chart. The navigator places one indicator of the parallel rule along $\overline{SD}$. By successively opening and closing the parallel rule, keeping one indicator *fixed* at each step, the desired course is transferred across the chart to the compass rose.

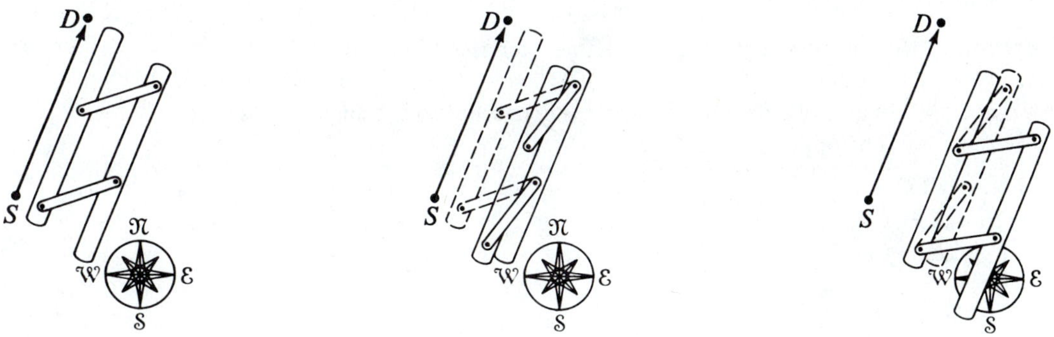

Section 5-1 PARALLELOGRAMS AND THEIR PROPERTIES

Recall that any four-sided polygon is called a quadrilateral. A *parallelogram* (▱) is a quadrilateral in which each side is parallel to the side opposite it. In this section you will learn some properties of parallelograms.

EXAMPLES

If $\overline{AB} \parallel \overline{DC}$ and $\overline{AD} \parallel \overline{BC}$, then $ABCD$ is a ▱.

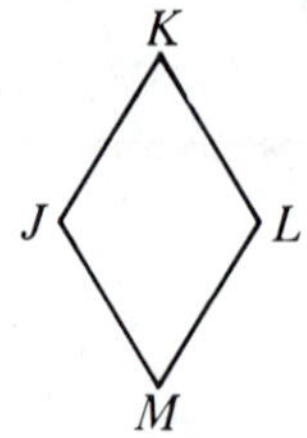

If $\overline{JK} \parallel \overline{ML}$ and $\overline{JM} \parallel \overline{KL}$, then $JKLM$ is a ▱.

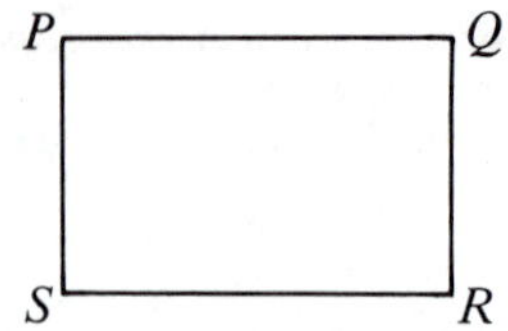

If $PQRS$ is a ▱, then $\overline{PQ} \parallel \overline{SR}$ and $\overline{PS} \parallel \overline{QR}$.

The next three theorems state important properties of parallelograms.

Theorem 5-1 In a parallelogram each side is congruent to the side opposite it. (The opp. sides of a ▱ are ≅.)

GIVEN: $ABCD$ is a ▱.
PROVE: $\overline{AB} \cong \overline{DC}$;
$\overline{AD} \cong \overline{BC}$

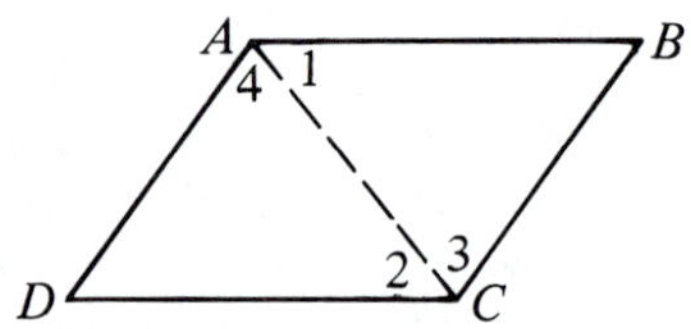

Statements	Reasons
1. $ABCD$ is a ▱.	1. Given
2. Draw $\overline{AC}$.	2. Two points determine a line.
3. $\overline{AB} \parallel \overline{DC}$; $\overline{AD} \parallel \overline{BC}$	3. Def. of a ▱
4. $\angle 1 \cong \angle 2$; $\angle 3 \cong \angle 4$	4. If 2 lines are ∥, alt. int. ∠s are ≅.
5. $\overline{AC} \cong \overline{AC}$	5. Reflexive Prop. of Congruence
6. $\triangle ABC \cong \triangle CDA$	6. ASA Postulate
7. $\overline{AB} \cong \overline{DC}$; $\overline{AD} \cong \overline{BC}$	7. Corr. parts of ≅ △s are ≅.

The proofs of the following theorems are left as Exercises 17 and 18.

Theorem 5-2 In a parallelogram each angle is congruent to the angle opposite it. (The opp. ∠s of a ▱ are ≅.)

GIVEN: $ABCD$ is a ▱.
PROVE: $\angle A \cong \angle C$;
$\angle B \cong \angle D$

Theorem 5-3 The diagonals of a parallelogram bisect each other.

GIVEN: $ABCD$ is a ▱.
PROVE: $\overline{AC}$ bisects $\overline{BD}$;
$\overline{BD}$ bisects $\overline{AC}$.

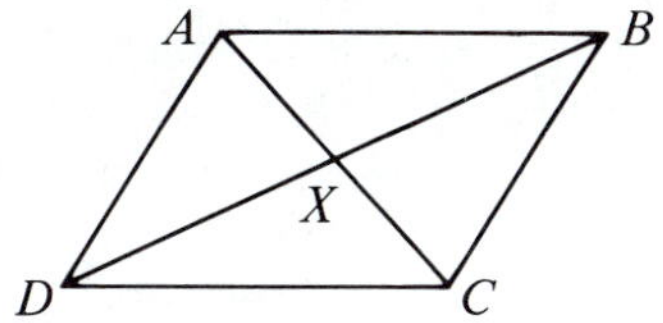

The following example demonstrates how one of these theorems can be used in a proof.

EXAMPLE

GIVEN: $\overline{PQ} \parallel \overline{SR}$;
$\overline{PS} \parallel \overline{QR}$
PROVE: $\overline{PS} \cong \overline{QR}$

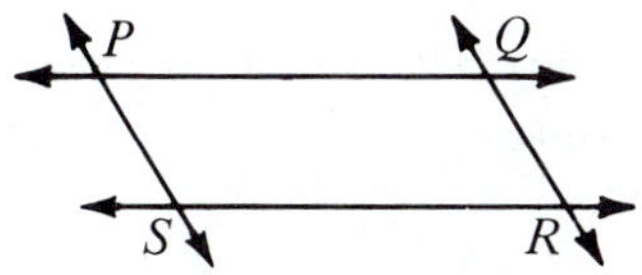

Statements	Reasons
1. $\overline{PQ} \parallel \overline{SR}$; $\overline{PS} \parallel \overline{QR}$	1. Given
2. $PQRS$ is a ▱.	2. Def. of a ▱
3. $\overline{PS} \cong \overline{QR}$	3. The opp. sides of a ▱ are ≅.

ORAL EXERCISES

In the diagram shown, $\overline{AB} \parallel \overline{DC}$, $\overline{AD} \parallel \overline{BC}$, and $\overline{AF} \parallel \overline{EC}$.

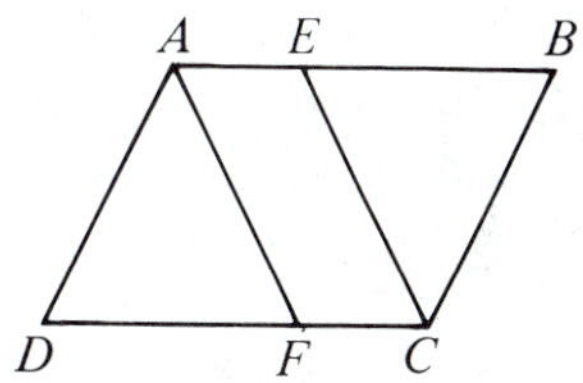

1. Name two parallelograms in the diagram.

2. Name five pairs of congruent segments.

3. Name six pairs of congruent angles.

4. Name the pairs of diagonals (not drawn) of the two parallelograms.

Exercises 5-10 refer to ▱$PQRS$.

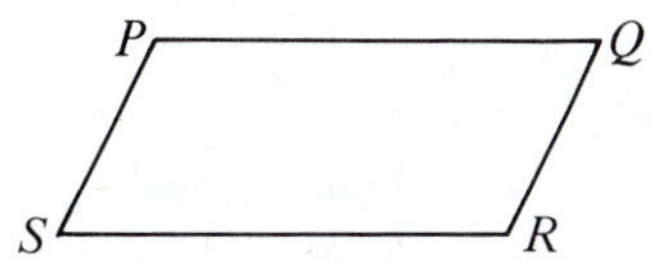

5. Name two pairs of parallel segments.

6. Explain why $\angle P$ and $\angle Q$ are supplementary.

7. Complete: $m\angle Q + m\angle R =$ __?__

8. Complete: Two consecutive angles of a parallelogram are __?__.

9. If $m\angle S = 60$, state the measures of $\angle P$, $\angle Q$, and $\angle R$.

10. If $m\angle P = x$, state the measures of $\angle Q$, $\angle R$, and $\angle S$ in terms of x.

11. Referring to the diagram, give reasons for the following proof.

GIVEN: $\overleftrightarrow{AB} \parallel \overleftrightarrow{DC}$; $\overline{AD} \perp \overleftrightarrow{DC}$; $\overline{BC} \perp \overleftrightarrow{DC}$
PROVE: $\overline{AD} \cong \overline{BC}$

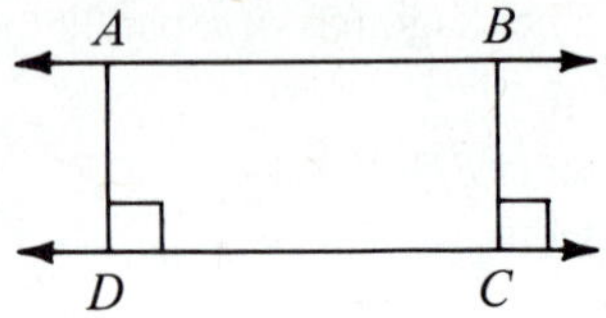

Statements	Reasons
1. $\overline{AD} \perp \overleftrightarrow{DC}$; $\overline{BC} \perp \overleftrightarrow{DC}$	1. _?_
2. $\overline{AD} \parallel \overline{BC}$	2. _?_
3. $\overleftrightarrow{AB} \parallel \overleftrightarrow{DC}$	3. _?_
4. $ABCD$ is a ▱.	4. _?_
5. $\overline{AD} \cong \overline{BC}$	5. _?_

12. The *distance* from a point to a line is defined as the length of the perpendicular segment from the point to the line.
    a. Use the results of Exercise 11 to complete: If $l \parallel m$, then the distances to line m from any two points on line l are _?_.
    b. How would you define the distance between two parallel lines?

WRITTEN EXERCISES

Exercises 1-12 refer to ▱$JKLM$.

A 1. Name four pairs of congruent line segments.

2. If $m\angle MJK = 65$, state the measures of $\angle JKL$, $\angle KLM$, and $\angle LMJ$.

3. If $NJ = 7$, then $JL =$ _?_.

4. If $MK = 10$, then $NK =$ _?_.

5. If $m\angle MJL = 37$ and $m\angle LJK = 27$, then $m\angle JKL =$ _?_.

6. If $m\angle JMK = 71$ and $m\angle KML = 42$, then $m\angle JKL =$ _?_ and $m\angle MKL =$ _?_.

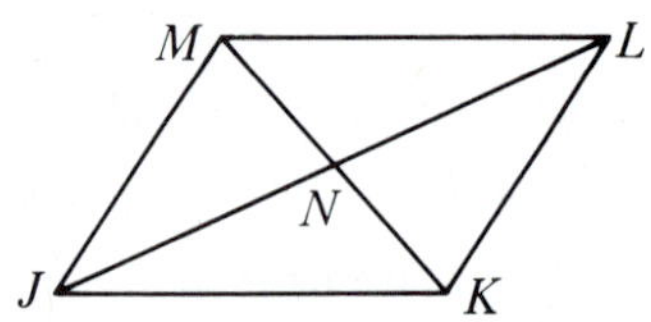

Exs. 1-12

Find the value of x or y.

7. $MJ = 2y + 5$; $LK = 14 - y$

8. $m\angle MJK = 4x + 4$; $m\angle KLM = 74 - x$

9. $m\angle MLK = 2x + 9$; $m\angle JKL = 5x + 3$

10. $JL = 4y + 6$; $NL = 3y - 1$

11. $MN = y + 4$; $MK = 5y - 10$

12. $m\angle MJL = 4x + 7$; $m\angle LJK = 5x - 8$; $m\angle MLK = 7x + 13$

Write two-column proofs.

13. Given: $\overleftrightarrow{AB} \parallel \overleftrightarrow{DC}$; $\overleftrightarrow{AD} \parallel \overleftrightarrow{BC}$
    Prove: $\angle 1 \cong \angle 2$

14. Given: $\overleftrightarrow{AB} \parallel \overleftrightarrow{DC}$; $\overleftrightarrow{AD} \parallel \overleftrightarrow{BC}$
    Prove: $\angle 3 \cong \angle 5$

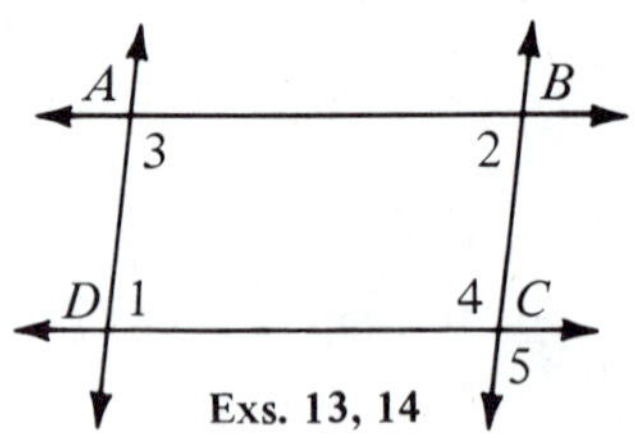

Exs. 13, 14

15. Given: $PQRS$ is a ▱;
$\overline{PS} \cong \overline{QT}$
Prove: $\triangle QRT$ is isosceles.

16. Given: $PQRS$ is a ▱;
$\overline{QR} \cong \overline{QT}$
Prove: $\angle S \cong \angle T$

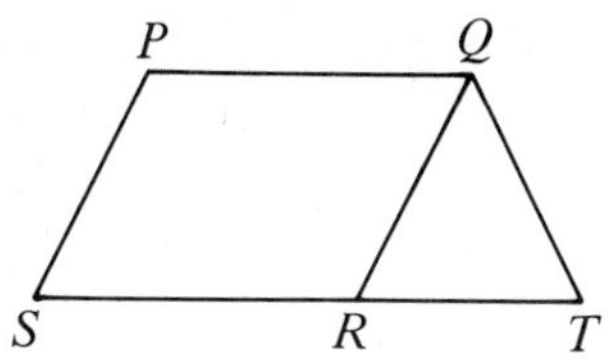

Exs. 15, 16

B 17. Prove Theorem 5-2. (*Hint*: Draw a diagonal of the parallelogram.)

18. Prove Theorem 5-3. (*Hint*: Show that $\triangle AXB \cong \triangle CXD$ in the diagram accompanying the theorem.)

19. Given: $JKLM$ is a ▱;
$\overline{JO} \cong \overline{OL}$
Prove: $\overline{OP} \cong \overline{OQ}$

20. Given: $JKLM$ is a ▱;
$\overline{JP} \cong \overline{QL}$
Prove: $\overline{JL}$ and $\overline{QP}$ bisect each other.

Exs. 19, 20
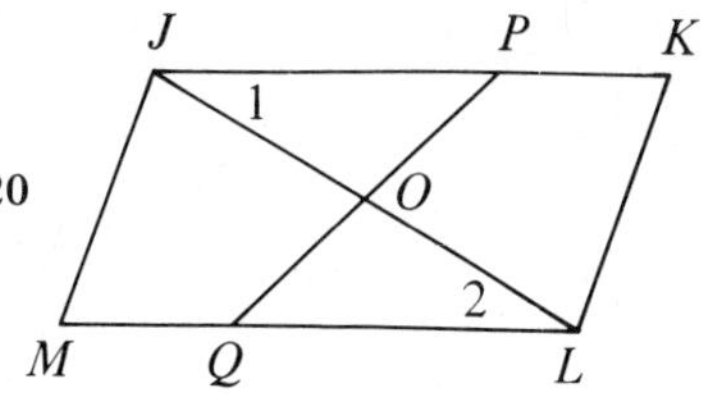

21. Given: $AECF$ is a ▱;
$\overline{FD} \cong \overline{BE}$
Prove: $\overline{AD} \cong \overline{BC}$

22. Given: $ABCD$ is a ▱;
$\overline{FD} \cong \overline{BE}$
Prove: $\overline{AF} \cong \overline{EC}$

Exs. 21, 22
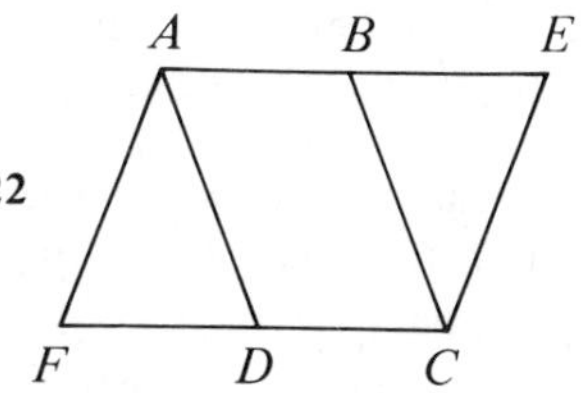

23. Given: $PQRS$ is a ▱.
Prove: $\overline{JX} \cong \overline{KX}$

24. Given: $PQRS$ is a ▱.
Prove: $\overline{JQ} \cong \overline{KS}$

Exs. 23, 24
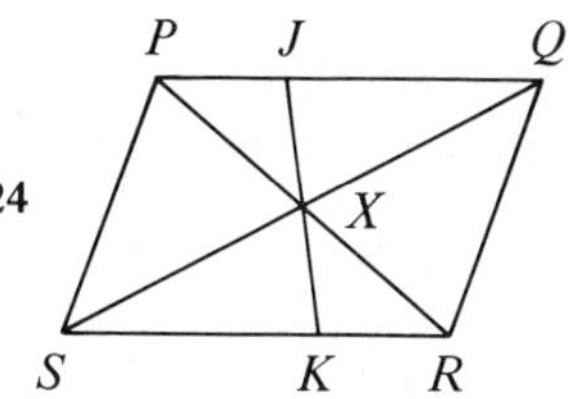

25. Given: $DEFG$ is a ▱;
$\overline{DR} \perp \overline{GE}$;
$\overline{FS} \perp \overline{GE}$
Prove: $\overline{DR} \cong \overline{FS}$

26. Given: $DEFG$ is a ▱;
$\overline{DR} \parallel \overline{FS}$
Prove: $\angle EDR \cong \angle GFS$

Exs. 25, 26
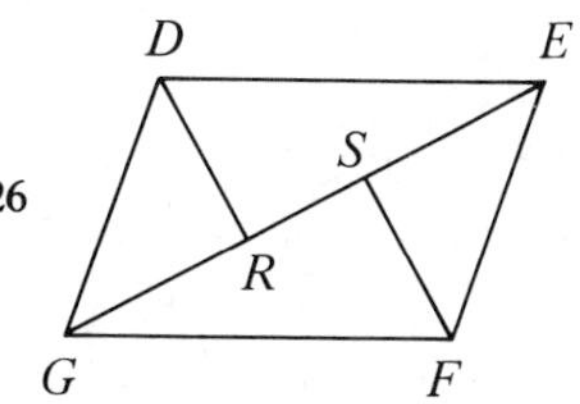

27. Given: $XYZA$ is a ▱;
$AX = XC$
Prove: $\overline{CB} \parallel \overline{XZ}$

28. Given: $XYZA$ and $XYBZ$ are parallelograms.
Prove: $\angle 3 \cong \angle 5$

Exs. 27, 28
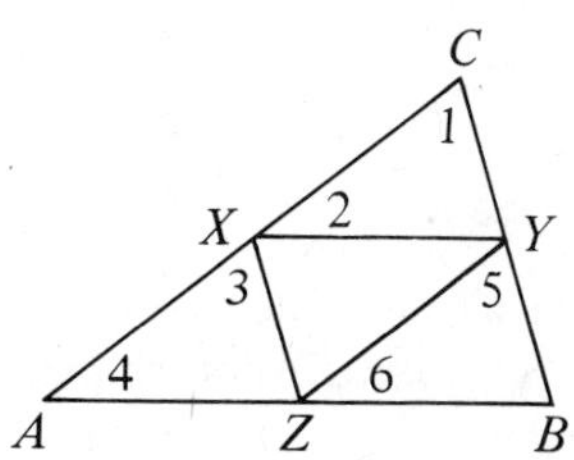

C 29. Given: *ABCD, CPDX,* and *AQBX* are parallelograms.
Prove: $\overline{DP} \cong \overline{BQ}$
(Can be done in very few steps!)

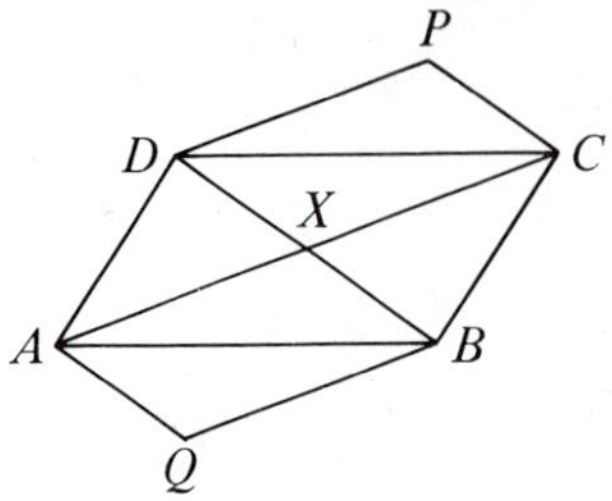

30. Given: *ABCD* and *DEBF* are parallelograms.
Prove: $\overline{AX} \cong \overline{CY}$

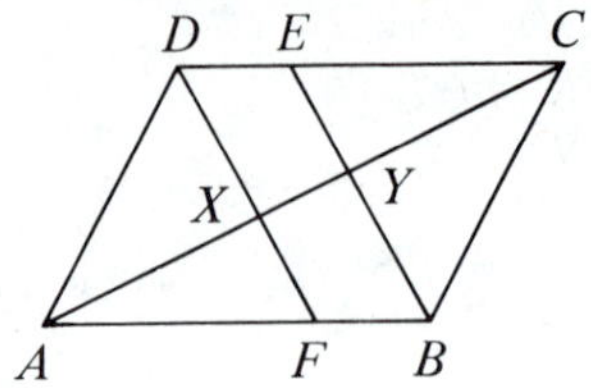

Section 5-2 PROVING THAT A QUADRILATERAL IS A PARALLELOGRAM

Do you need to know that each side of a quadrilateral is parallel to its opposite side in order to conclude that the quadrilateral is a parallelogram? As you will see in this section, the answer to the question is "No." In the text and exercises, you will learn other ways to show that a quadrilateral is a parallelogram.

Theorem 5-4 If each side of a quadrilateral is congruent to the side opposite it, then the quadrilateral is a parallelogram.

GIVEN: $\overline{AB} \cong \overline{CD}$;
$\overline{CB} \cong \overline{AD}$
PROVE: *ABCD* is a ▱.

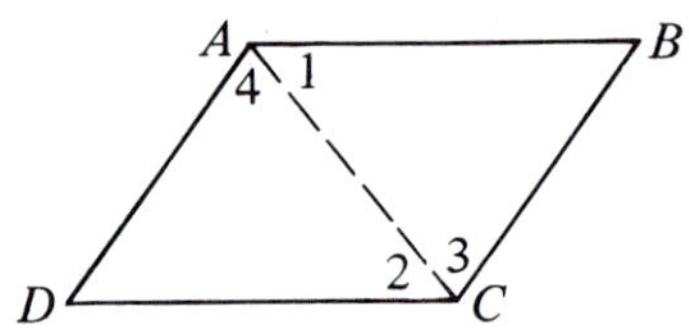

Statements	Reasons
1. $\overline{AB} \cong \overline{CD}$; $\overline{CB} \cong \overline{AD}$	1. Given
2. Draw $\overline{AC}$.	2. Two points determine a line.
3. $\overline{AC} \cong \overline{AC}$	3. Reflexive Prop. of Congruence
4. $\triangle ABC \cong \triangle CDA$	4. SSS Postulate
5. $\angle 1 \cong \angle 2$; $\angle 3 \cong \angle 4$	5. Corr. parts of ≅ ▵ are ≅.
6. $\overline{AB} \parallel \overline{DC}$; $\overline{AD} \parallel \overline{BC}$	6. If 2 lines are cut by a trans. so that alt. int. ∠s are ≅, the lines are ∥.
7. *ABCD* is a ▱.	7. Def. of a ▱

Note that in order to use the theorem above, you must know that *both* pairs of *opposite* sides are congruent. Another property that guarantees that a quadrilateral is a parallelogram is given in the following theorem. Its proof is left as Exercise 15.

Theorem 5-5 If two sides of a quadrilateral are both congruent and parallel, then the quadrilateral is a parallelogram.

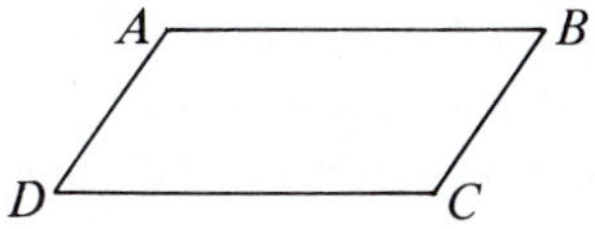

GIVEN: $\overline{AB} \cong \overline{CD}$;
$\overline{AB} \parallel \overline{DC}$

PROVE: $ABCD$ is a ▱.

Note that in the preceding theorem the two sides of the quadrilateral that are given to be congruent are the same as the two that are given to be parallel. The diagram at the right illustrates that when two opposite sides of a quadrilateral are congruent and the *other* two opposite sides are parallel, the quadrilateral is *not* necessarily a parallelogram.

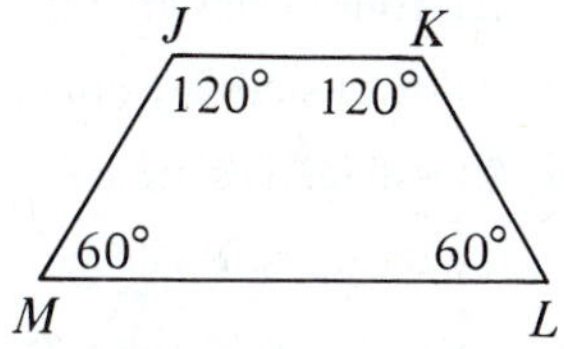

Here $\overline{JM} \cong \overline{KL}$ and $\overline{JK} \parallel \overline{ML}$, but $JKLM$ is not a ▱.

Theorem 5-6 If each angle of a quadrilateral is congruent to the angle opposite it, then the quadrilateral is a parallelogram. (If the opp. ∠s of a quad. are ≅, the quad. is a ▱.)

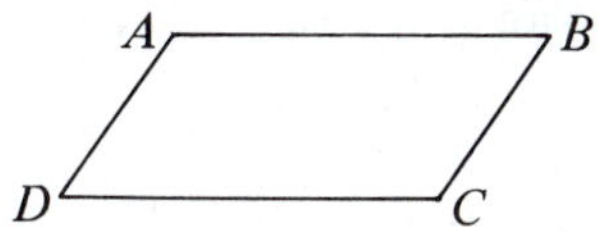

GIVEN: $\angle A \cong \angle C$;
$\angle B \cong \angle D$

PROVE: $ABCD$ is a ▱.

The proof is left as Exercise 19.

Note that Theorems 5-4 and 5-6 are the converses of Theorems 5-1 and 5-2, respectively. The next theorem is the converse of Theorem 5-3.

Theorem 5-7 If the diagonals of a quadrilateral bisect each other, then the quadrilateral is a parallelogram.

GIVEN: $\overline{AC}$ and $\overline{BD}$ bisect each other.

PROVE: $ABCD$ is a ▱.

The proof is left as Exercise 16.

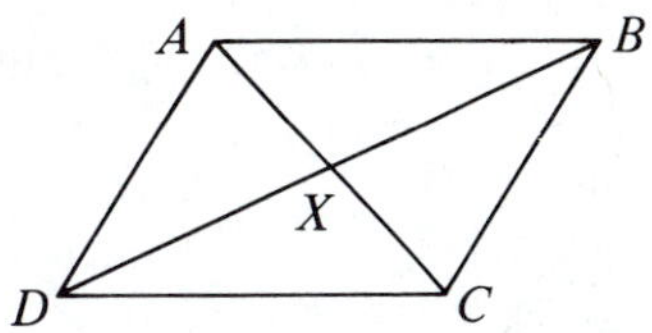

Thus, to prove that a quadrilateral is a parallelogram you may use any of the methods listed below. In Exercises 17, 18, 20, and 21 you will learn some additional ways.

A quadrilateral is a parallelogram if and only if:

1. Both pairs of opposite sides are parallel.
2. Both pairs of opposite sides are congruent.
3. Two sides are both parallel and congruent.
4. Both pairs of opposite angles are congruent.
5. The diagonals bisect each other.

ORAL EXERCISES

Tell whether each condition is sufficient to guarantee that a quadrilateral satisfying that condition is a parallelogram. If it is not sufficient, draw a quadrilateral that satisfies the condition but is not a parallelogram.

1. Two opposite angles are congruent.
2. All four angles are right angles.
3. All four sides are congruent.
4. The diagonals are perpendicular.
5. Two opposite sides are parallel and two adjacent sides are congruent.
6. Two opposite sides are congruent to the same diagonal and those two sides are also parallel.
7. Two pairs of sides are congruent.
8. Two pairs of angles are congruent.
9. Name some examples of parallelograms in everyday life. Can you find parallelograms, for example, in a folding metal gate, aluminum lawn furniture, a basketball backboard support?

WRITTEN EXERCISES

For what values of x and y is quadrilateral $RSTW$ a parallelogram?

A 1. $RS = 2x + 7$, $ST = 3y - 5$, $TW = 25$, $WR = 16$

2. $WZ = 4x - 3$, $ZS = 13$, $RZ = 17$, $ZT = 7y + 3$
3. $m\angle WRS = 24x$, $m\angle RST = 14y - 4$, $m\angle STW = 15x + 27$, $m\angle TWR = 11y + 20$
4. $m\angle WRS = 75$, $m\angle RST = 7x$, $m\angle STW = 11y + 9$

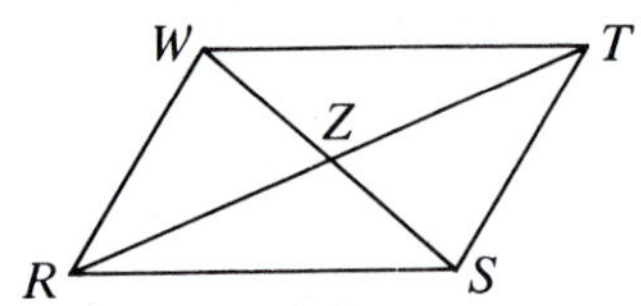

Exs. 1–4

Write two-column proofs.

5. Given: $\overline{PQ} \parallel \overline{SR}$; $\overline{PQ} \cong \overline{SR}$
   Prove: $\overline{PS} \cong \overline{QR}$
6. Given: $\angle P \cong \angle R$; $\angle Q \cong \angle S$
   Prove: $\overline{PQ} \parallel \overline{RS}$

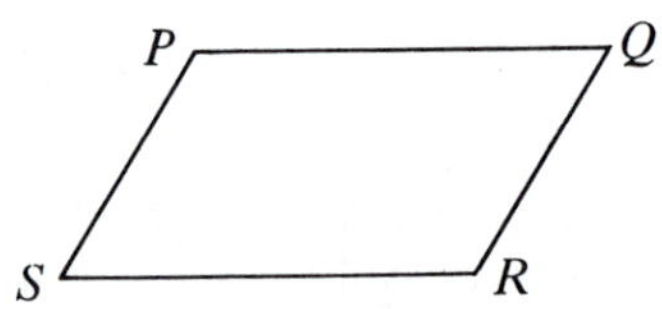

Exs. 5, 6

7. Given: $\angle 1 \cong \angle 2$; $\angle 2 \cong \angle 3$; $\overline{AK} \cong \overline{BL}$
   Prove: $ABJK$ is a ▱.
8. Given: $\overline{AK} \cong \overline{BJ}$; $\overline{BJ} \cong \overline{BL}$; $\angle 1 \cong \angle 3$
   Prove: $ABJK$ is a ▱.

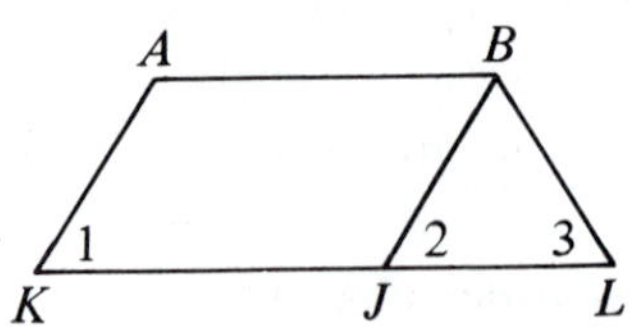

Exs. 7, 8

9. Given: $ABCD$ is a ▱;
$\overline{AD} \cong \overline{DE}$
Prove: $DBCE$ is a ▱.

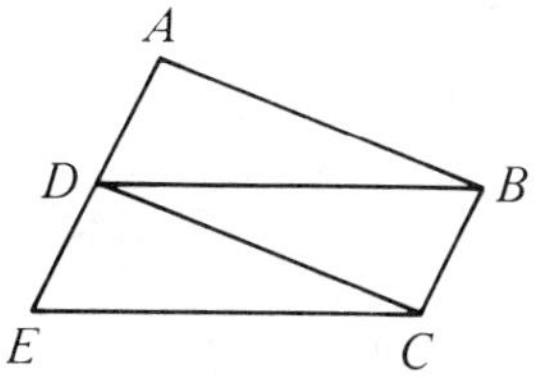

10. Given: $JKLM$ is a ▱;
$\overline{PX} \cong \overline{QX}$
Prove: $JPLQ$ is a ▱.

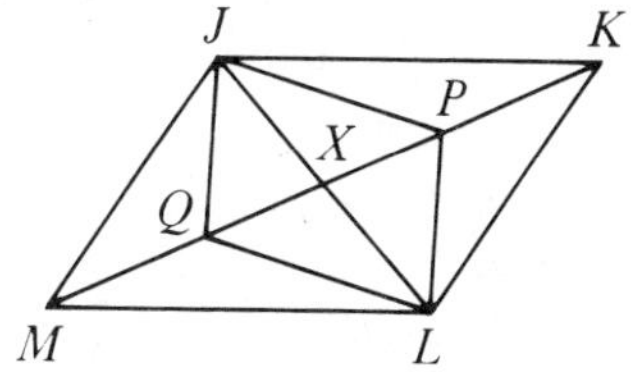

B 11. Given: $\overline{AF} \cong \overline{CF}$;
$\overline{BF} \cong \overline{DF}$;
$\overline{AB} \cong \overline{BE}$
Prove: $\overline{CD} \cong \overline{BE}$

12. Given: $\overline{CD} \cong \overline{BE}$;
$\overline{CD} \parallel \overleftrightarrow{BE}$;
$\overline{AB} \cong \overline{BE}$
Prove: $\overline{AF} \cong \overline{CF}$

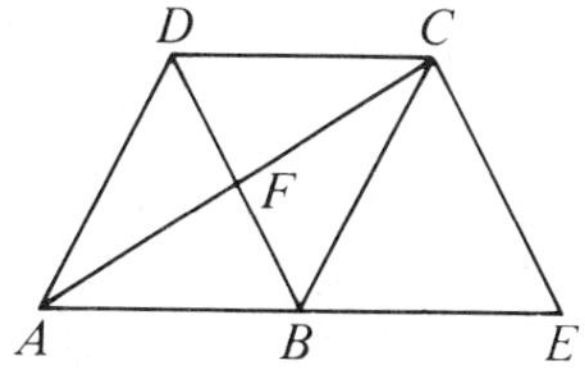

Exs. 11, 12

13. Given: $\overline{RX} \cong \overline{RY}$;
$\triangle RYS \cong \triangle TZS$
Prove: $RTZX$ is a ▱.

14. Given: $\overline{RX} \cong \overline{RY}$;
$\overline{RT} \cong \overline{XZ}$;
$\overline{RT} \parallel \overline{XZ}$
Prove: $\triangle RYS \cong \triangle TZS$

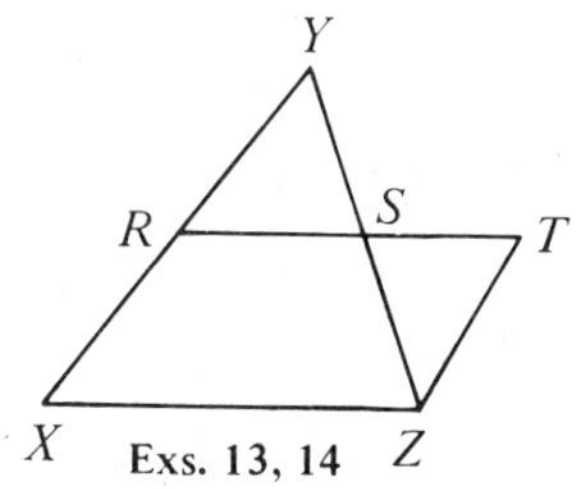

Exs. 13, 14

15. a. Prove Theorem 5-5, using the definition of a parallelogram as the reason for the last step. (*Hint*: Draw a diagonal of the quadrilateral.)
b. Describe how you would revise your proof in part **(a)** so that Theorem 5-4 is the reason for the last step.

16. Prove Theorem 5-7.

17. Prove: If two sides of a quadrilateral are parallel and *one* diagonal bisects the other, then the quadrilateral is a parallelogram.

GIVEN: $\overline{AB} \parallel \overline{DC}$;
$\overline{AC}$ bisects $\overline{BD}$.
PROVE: $ABCD$ is a ▱.

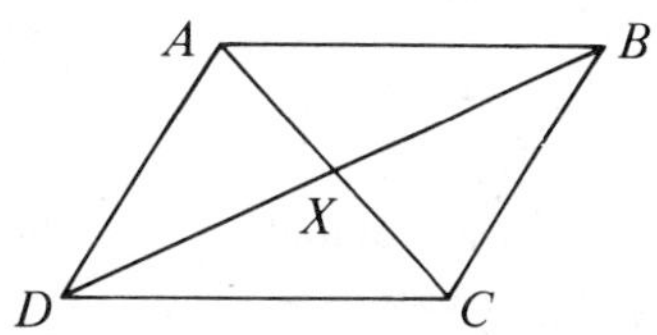

18. Prove: If two opposite angles of a quadrilateral are congruent and two opposite sides are parallel, then the quadrilateral is a parallelogram.

GIVEN: $\angle B \cong \angle D$;
$\overline{AB} \parallel \overline{DC}$
PROVE: $ABCD$ is a ▱.

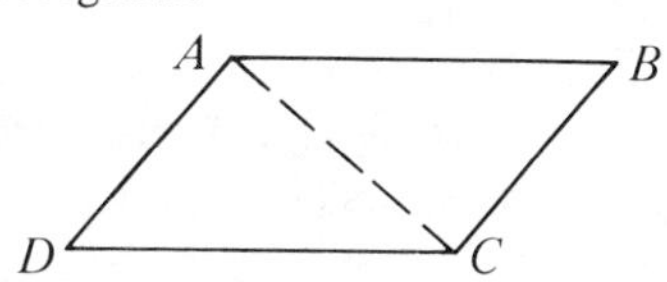

19. Prove Theorem 5-6. (*Hint*: Use the fact that the sum of the measures of the angles of a quadrilateral is 360.)

Exercises 20 and 21 prove: If two opposite sides of a quadrilateral are congruent and two opposite angles are congruent, then the quadrilateral is a parallelogram.

C 20. Given: $\overline{AD} \cong \overline{BC}$; $\angle A \cong \angle C$
Prove: $ABCD$ is a ▱.
(*Hint*: Let E and F be the points where the perpendiculars from D and B intersect $\overleftrightarrow{AB}$ and $\overleftrightarrow{DC}$, respectively. First prove that $\triangle AED \cong \triangle CFB$, then that $\triangle DEB \cong \triangle BFD$.)

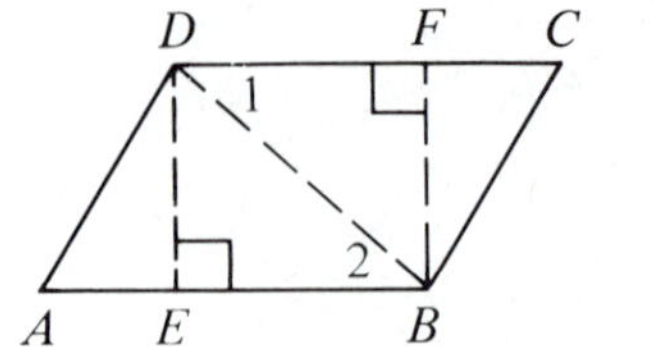

21. Given: $\overline{AD} \cong \overline{BC}$; $\angle B \cong \angle D$
Prove: $ABCD$ is a ▱.
(*Hint*: Let G and H be the points where the perpendiculars from A and C intersect $\overleftrightarrow{DC}$ and $\overleftrightarrow{AB}$, respectively. First prove that $\triangle AGD \cong \triangle CHB$, then that $\triangle AGC \cong \triangle CHA$.)

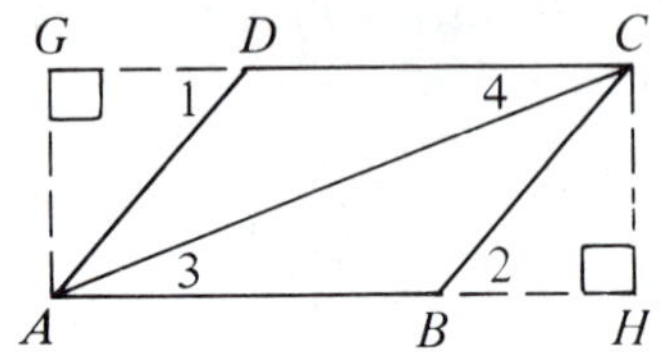

Section 5-3 SPECIAL PARALLELOGRAMS

Certain special kinds of parallelograms are of particular importance in geometry. These parallelograms are:

- rectangles
- rhombuses
- squares

A *rectangle* is a quadrilateral with four right angles.

EXAMPLES OF RECTANGLES

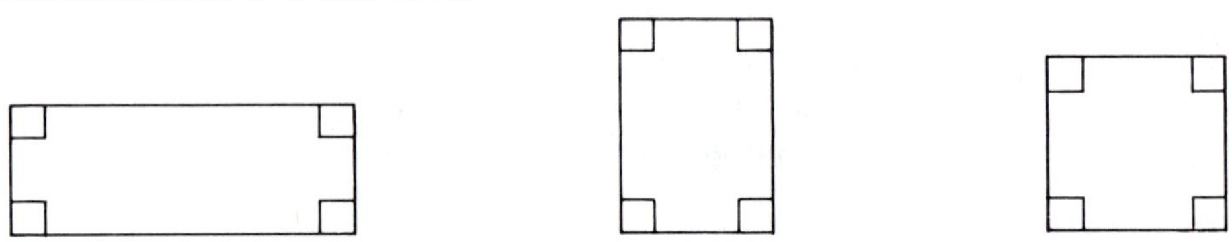

Since each angle of a rectangle is congruent to the angle opposite it, every rectangle is also a parallelogram by Theorem 5-6. In Oral Exercise 7 you will justify the following theorem.

Theorem 5-8 If one angle of a parallelogram is a right angle, then the parallelogram is a rectangle.

A *rhombus* is a quadrilateral with four congruent sides. Since each side of a rhombus is congruent to the side opposite it, every rhombus is a parallelogram by Theorem 5-4.

EXAMPLES OF RHOMBUSES

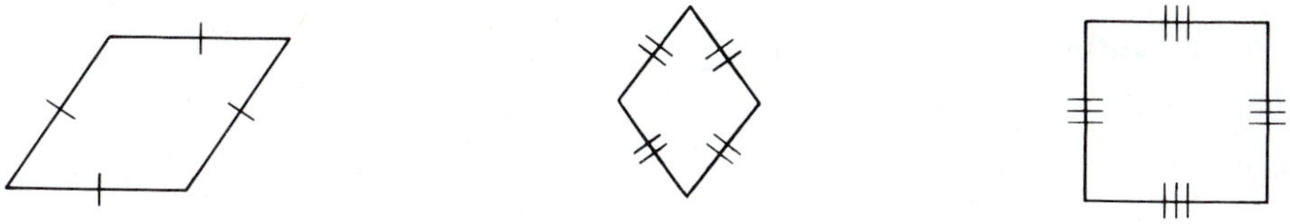

Theorem 5-9 If two consecutive sides of a parallelogram are congruent, then the parallelogram is a rhombus.

You will justify Theorem 5-9 in Oral Exercise 8.

A *square* is a quadrilateral with four right angles and four congruent sides. Thus a square is a parallelogram that is both a rectangle and a rhombus.

The following theorems state some of the properties possessed by special parallelograms. The proofs of the theorems are left as Exercises 25, 21, and 23.

Theorem 5-10 The diagonals of a rectangle are congruent.

GIVEN: $ABCD$ is a rectangle.
PROVE: $\overline{AC} \cong \overline{BD}$

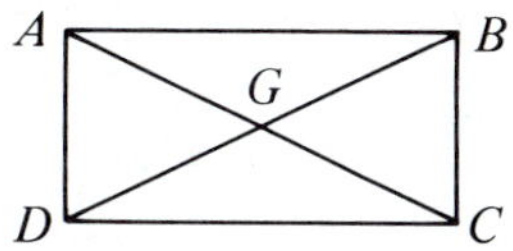

Theorem 5-11 Each diagonal of a rhombus bisects two angles of the rhombus.

GIVEN: $PQRS$ is a rhombus.
PROVE: $\overline{PR}$ bisects $\angle SPQ$;
$\overline{PR}$ bisects $\angle SRQ$.

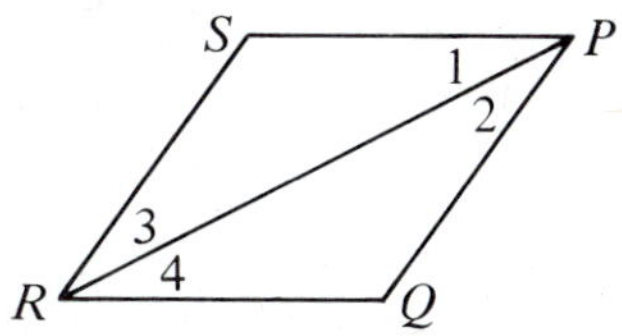

Theorem 5-12 The diagonals of a rhombus are perpendicular.

GIVEN: $EFGH$ is a rhombus.
PROVE: $\overline{EG} \perp \overline{HF}$

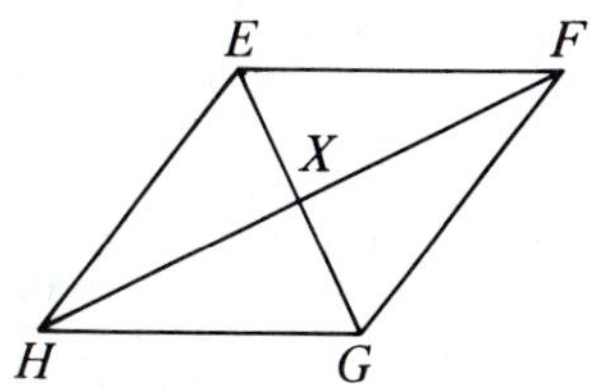

Since all rectangles, rhombuses, and squares are parallelograms, you may use any of the properties of parallelograms if you know that a given figure is a rectangle, rhombus, or square. For example, the proof of Theorem 5-12 might begin with the following steps:

Statements	Reasons
1. $EFGH$ is a rhombus.	1. Given
2. $\overline{EG}$ bisects $\overline{HF}$.	2. The diagonals of a ▱ bisect each other.

ORAL EXERCISES

Classify each statement as true or false.

1. Every rhombus is a parallelogram.
2. Every parallelogram is a rectangle.
3. The diagonals of a rhombus are congruent.
4. A quadrilateral with one right angle is a rectangle.
5. The diagonals of a square are congruent and perpendicular.
6. A quadrilateral with four congruent angles is a rectangle.
7. Explain why a parallelogram with one right angle must have *four* right angles (and is therefore a rectangle).
8. Explain why if two consecutive sides of a parallelogram are congruent, then all *four* sides must be congruent (and hence the parallelogram is a rhombus).
9. What's wrong with the following "proof" that every parallelogram is a rhombus?

   GIVEN: *ABHK* is a ▱.
   PROVE: *ABHK* is a rhombus.

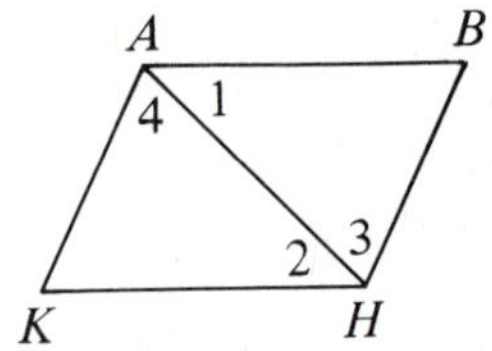

Statements	Reasons
1. *ABHK* is a ▱.	1. Given
2. $\overline{AB} \parallel \overline{KH}$; $\overline{AK} \parallel \overline{BH}$	2. Def. of a ▱
3. $\angle 1 \cong \angle 2$; $\angle 3 \cong \angle 4$	3. If 2 lines are ‖, alt. int. ∠s are ≅.
4. $\overline{AH} \cong \overline{AH}$	4. Reflexive Prop. of Congruence
5. $\triangle ABH \cong \triangle AKH$	5. ASA Postulate
6. $\overline{AB} \cong \overline{AK}$	6. Corr. parts of ≅ △s are ≅.
7. *ABHK* is a rhombus.	7. If 2 consecutive sides of a ▱ are ≅, the ▱ is a rhombus.

WRITTEN EXERCISES

Exercises 1-6 refer to rectangle *RSTW* shown at the right.

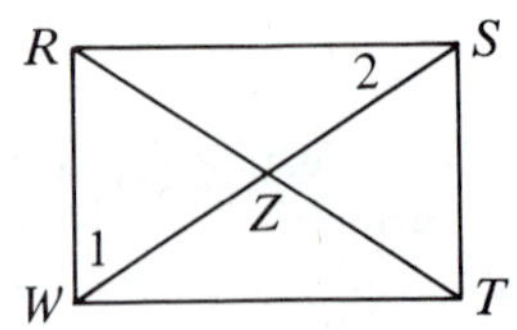

Exs. 1-6

A 1. What is $m\angle RWT$?

2. Find $m\angle 1 + m\angle 2$.
3. If $m\angle 1 = 3x + 12$ and $m\angle 2 = 2x - 7$, find **(a)** the value of x and **(b)** $m\angle 1$.
4. If $RT = 5y + 2$ and $WS = 11y - 10$, find **(a)** the value of y and **(b)** WS.
5. If $RT = 7a - 2$ and $WZ = 4a - 3$, find **(a)** the value of a and **(b)** RZ.
6. If $m\angle 1 = 61$, what is $m\angle RZW$?

Exercises 7-12 refer to rhombus $ABCD$ shown at the right.

7. What is $m\angle AEB$?

8. If $m\angle 1 = 50$, find $m\angle 2$ and $m\angle 3$.

9. If $m\angle DAB = 7x + 14$ and $m\angle 2 = 5x - 5$, find **(a)** the value of x and **(b)** $m\angle 1$.

10. If $AD = 3w + 7$ and $AB = 2(w + 8)$, find **(a)** the value of w and **(b)** $AD + DC$.

11. If $AC = 2y + 8$ and $EC = 2y - 1$, find **(a)** the value of y and **(b)** AE.

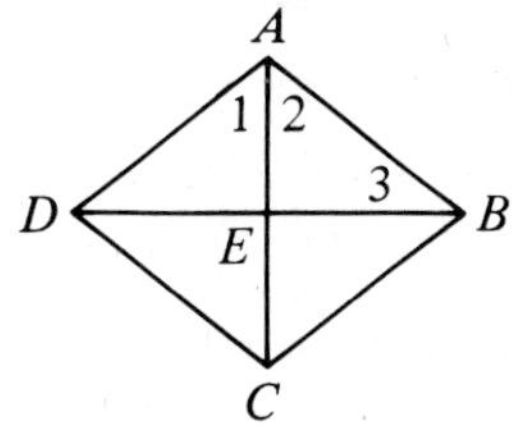

Exs. 7 - 12

12. If $AB = 3k + 1$ and the perimeter of rhombus $ABCD$ is $13k - 1$, find AB.

13. In $\square TUVW$, $TU = 3z - 14$ and $WV = 2z - 6$.
    a. Find the value of z.
    b. If $UV = z + 2$, what kind of parallelogram is $TUVW$? Why?

14. In $\square TUVW$, $TX = 2y + 11$, $VX = y + 9$, and $WU = y + 18$.
    a. Find TV.
    b. Can $\square TUVW$ be a rectangle? Why?

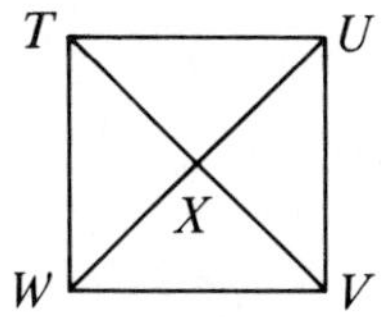

Exs. 13, 14

15. In $\square ABCD$, $m\angle A = 2z + 20$ and $m\angle D = 3z - 15$. Show that $\square ABCD$ is a rectangle.

16. In $\square ABCD$, $AB = 2x + 3$, $BC = 4x - 5$, and $CD = 5x - 9$. Show that $\square ABCD$ is a rhombus.

Write two-column proofs.

17. Given: $ACEF$ is a rhombus;
    $\overline{AC} \cong \overline{BC}$
    Prove: $\angle 1 \cong \angle 2$

18. Given: $ACEF$ is a $\square$;
    $\overline{AC} \cong \overline{BC}$;
    $\angle 1 \cong \angle 2$
    Prove: $ACEF$ is a rhombus.

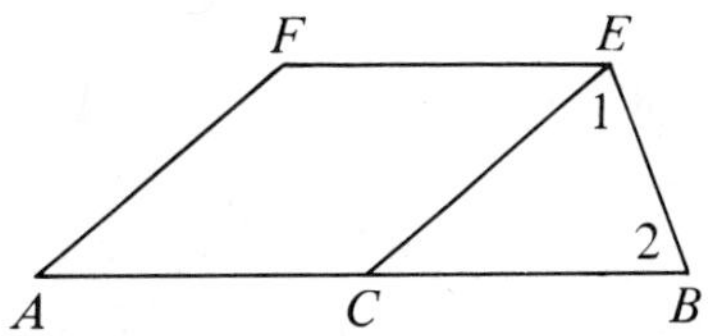

Exs. 17, 18

19. Given: $XYRS$ is a rectangle;
    M is the midpoint of $\overline{YR}$.
    Prove: $\overline{XM} \cong \overline{SM}$

20. Given: $XYRS$ is a rectangle;
    $\overline{XM} \cong \overline{SM}$
    Prove: M is the midpoint of $\overline{YR}$.

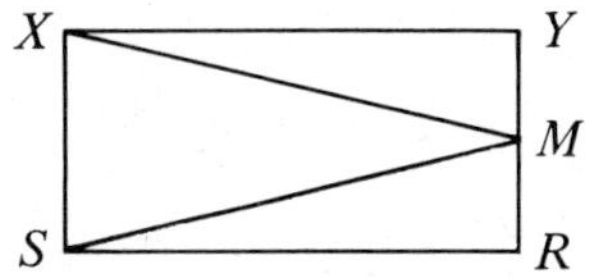

Exs. 19, 20

B 21. Prove Theorem 5-11.

22. Prove that if a diagonal of a parallelogram bisects one angle of the parallelogram, then the parallelogram is a rhombus.

23. Prove Theorem 5-12.

24. Prove that if the diagonals of a parallelogram are perpendicular, then the parallelogram is a rhombus.

25. Prove Theorem 5-10.

26. Prove that if the diagonals of a parallelogram are congruent, then the parallelogram is a rectangle.

In the remainder of the exercise set you may use the results of Exercises 22, 24, and 26.

27. Given: $\angle 1 \cong \angle 2$;
$\angle 2 \cong \angle 3$;
$\angle 3 \cong \angle 4$
Prove: *PQRS* is a rhombus.

28. Given: $\angle 1 \cong \angle 2$;
$\angle 2 \cong \angle 3$;
$\overline{PQ} \cong \overline{RS}$
Prove: *PQRS* is a rhombus.

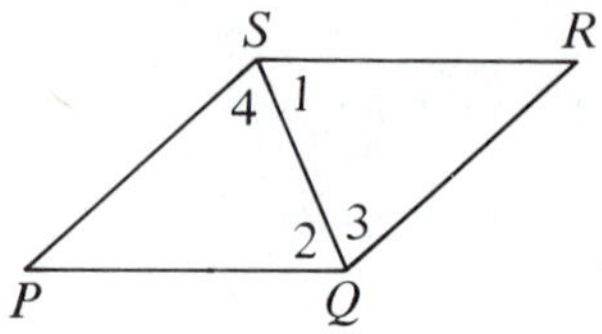

Exs. 27, 28

29. Given: *ABCD* is a rectangle;
$\angle ABX \cong \angle XBC$
Prove: *ABCD* is a square.

30. Given: *ABCD* is a rhombus;
$\angle ABX \cong \angle BAX$
Prove: *ABCD* is a square.

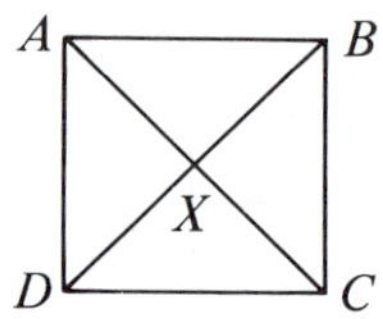

Exs. 29, 30

31. *RSTW* is a quadrilateral such that
$RS = 7x - 3,\ ST = 6(x + 1),\ TW = 10(x - 3)$, and $RW = 4(x + 6)$.
Is there a value of x such that *RSTW* is a rhombus? Explain.

32. *ABCD* is a quadrilateral such that
$AB = y + 2,\ BC = 2y - 1,\ CD = 3y - 2,\ AD = y + 1,\ AC = y + 3$, and $BD = 3y - 1$.
Is there a value of y such that *ABCD* is a rectangle? Explain.

33. Show that the area of a rhombus is half the product of the length of its diagonals.
Given: *PQRS* is a rhombus.
Prove: area of $PQRS = \frac{1}{2}(PR)(SQ)$
(Recall that the area of a right triangle is half the product of the lengths of the legs.)

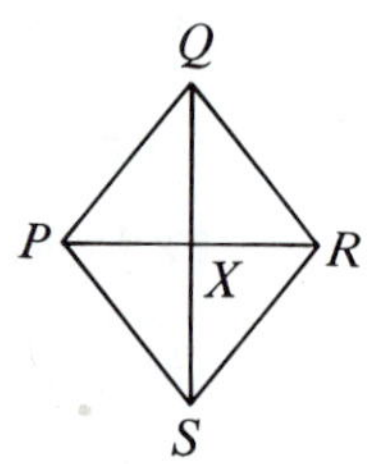

Write two-column proofs.

C 34. Given: $\overline{AC}$ bisects $\angle DAB$;
$\overline{DB}$ bisects $\angle ADC$;
$\overline{AB} \parallel \overline{DC}$
Prove: *ABCD* is a rhombus.
(*Hint*: Do *not* use congruent triangles.)

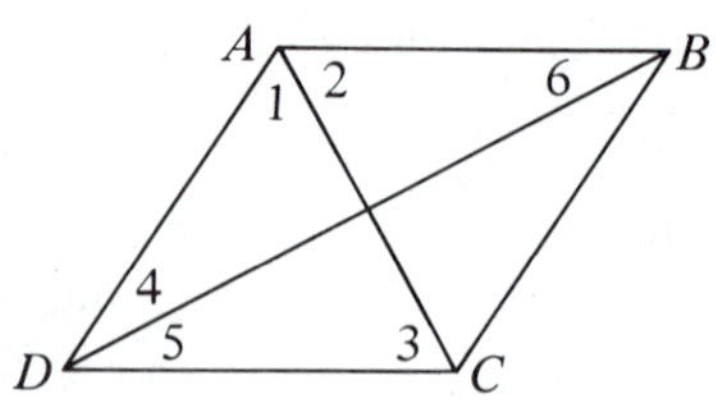

35. Given: $VPQY$ is a rectangle; $\overline{VX} \cong \overline{WY}$
Prove: $VWXY$ is a rectangle.

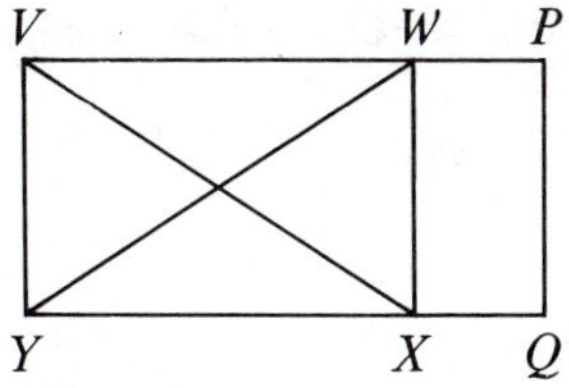

SELF-TEST 1

Exercises 1-3 refer to ▱$ABCD$.

Section 5-1

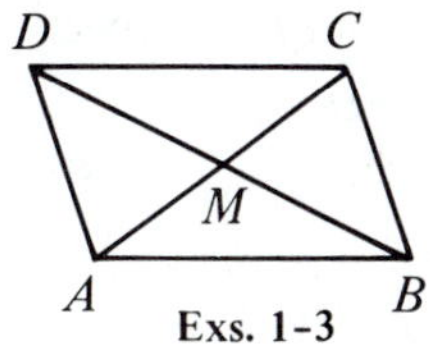

Exs. 1-3

1. If $m\angle DAB = 3(3x + 1)$ and $m\angle ABC = 5x + 9$, find the value of x.
2. If $AB = 17 + 2y$ and $CD = 7 - 3y$, find AB.
3. If $DM = 4z + 3$ and $MB = 8z - 9$, find the numerical value of DB.
4. Write a two-column proof.
   Given: $PRSU$ is a ▱; $\angle SRT \cong \angle PUQ$
   Prove: $\overline{RT} \cong \overline{UQ}$

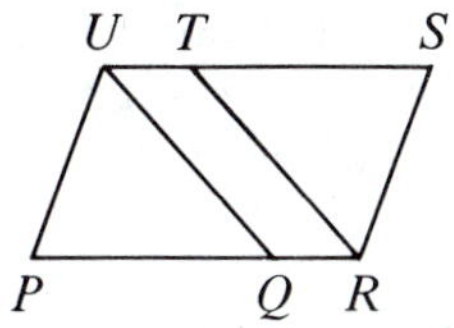

True or false?

Section 5-2

5. If $\overline{AC}$ bisects $\overline{BD}$, then quadrilateral $ABCD$ must be a parallelogram.
6. If a pair of opposite sides of quadrilateral $PQRS$ are congruent and a pair of opposite sides are parallel, then $PQRS$ must be a parallelogram.
7. If quadrilateral $MNOP$ has four congruent sides, then $MNOP$ must be a parallelogram.
8. If $\angle A \cong \angle B$ and $\angle C \cong \angle D$, then quadrilateral $ABCD$ must be a parallelogram.
9. If M is the midpoint of $\overline{PR}$ and $\overline{QS}$, then quadrilateral $PQRS$ must be a parallelogram.

Write two-column proofs.

Section 5-3

10. Given: $MNOP$ is a rectangle.
    Prove: $\triangle MXN$ is isosceles.

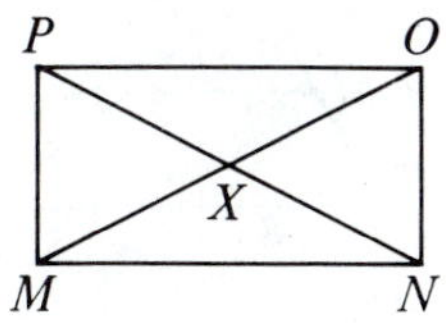

11. Given: $ABCD$ is a rhombus; $\overline{MB} \cong \overline{NB}$
    Prove: $\overline{MD} \cong \overline{ND}$

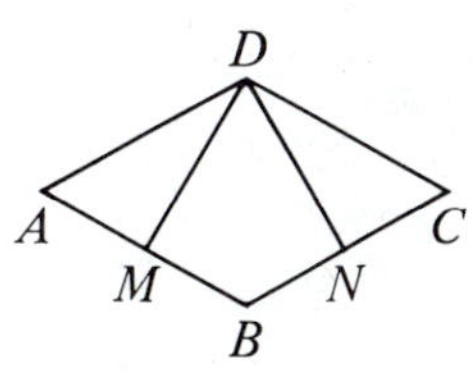

True or false?

12. Some rectangles are squares.
13. Some rhombuses are rectangles.
14. No square is a rhombus.
15. All squares are rectangles and rhombuses.

Section 5-4 TRIANGLES AND PARALLEL LINES

In this section you will learn two important theorems about triangles. Here is the first.

Theorem 5-13 The segment joining the midpoints of two sides of a triangle is parallel to the third side. Its length is half the length of the third side.

GIVEN: D and E are the midpoints of $\overline{AC}$ and $\overline{BC}$, respectively.
PROVE: $\overline{DE} \parallel \overline{AB}$;
$DE = \frac{1}{2}AB$

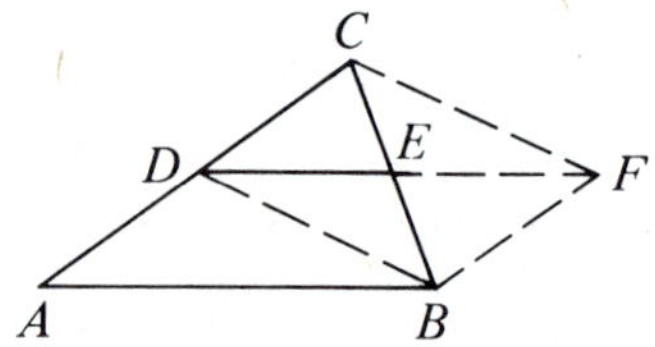

Outline of proof: On the ray opposite to $\overrightarrow{ED}$ let F be the point such that $EF = DE$. Draw $\overline{CF}$, $\overline{BF}$, and $\overline{BD}$. Then $BFCD$ is a parallelogram because the diagonals bisect each other, and hence $BF = DC$ and $\overline{BF} \parallel \overleftrightarrow{DC}$. But since D is the midpoint of $\overline{AC}$, you know that $BF = AD$ and $\overline{BF} \parallel \overline{AD}$. Then, by Theorem 5-5, $ABFD$ is a parallelogram, and hence $\overline{DE} \parallel \overline{AB}$, and $DF = AB$. Because $EF = DE$, you can conclude that $DE = \frac{1}{2}AB$.

EXAMPLE

GIVEN: X, Y, and Z are the midpoints of $\overline{PQ}$, $\overline{QR}$, and $\overline{PR}$, respectively.
PROVE: $PXYZ$ is a ▱.

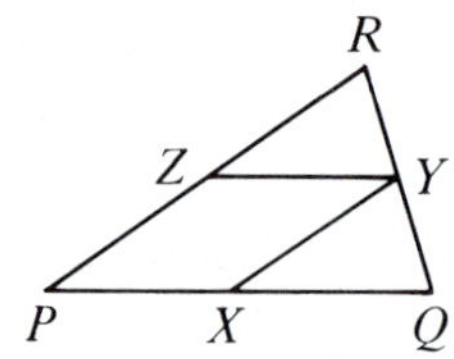

Statements	Reasons
1. X, Y, and Z are midpoints of $\overline{PQ}$, $\overline{QR}$, and $\overline{PR}$.	1. Given
2. $\overline{XY} \parallel \overline{PZ}$; $\overline{XP} \parallel \overline{YZ}$	2. The segment joining the midpoints of 2 sides of a $\triangle$ is $\parallel$ to the third side.
3. $PXYZ$ is a ▱.	3. Def. of a ▱

EXAMPLE Find the value of x if M and N are midpoints of $\overline{JL}$ and $\overline{KL}$, respectively, $MN = 4x$, and $JK = 3x + 10$.

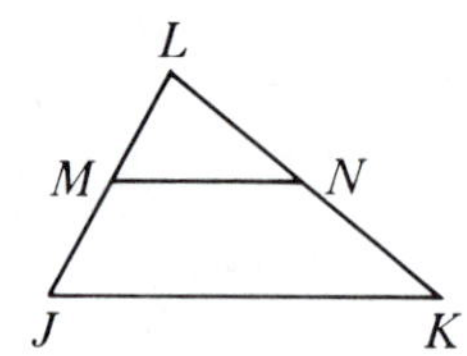

SOLUTION By Theorem 5-13, you have $MN = \frac{1}{2}JK$, or

$$4x = \frac{1}{2}(3x + 10)$$
$$8x = 3x + 10$$
$$5x = 10$$
$$x = 2$$

The next theorem is a partial converse of Theorem 5-13.

Theorem 5-14 A line that contains the midpoint of one side of a triangle and is parallel to another side bisects the third side.

GIVEN: R is the midpoint of $\overline{AC}$;
$\overleftrightarrow{RS} \parallel \overline{AB}$

PROVE: $\overleftrightarrow{RS}$ bisects $\overline{BC}$.

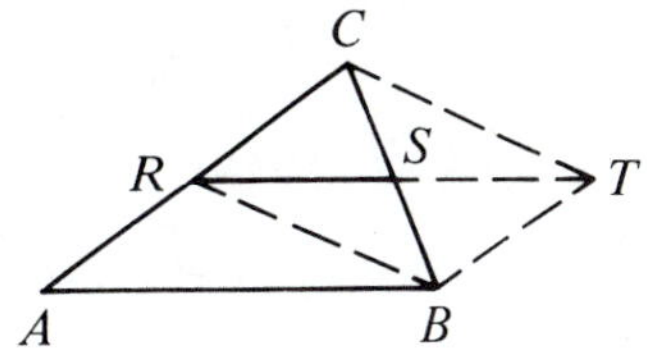

Outline of proof: On $\overrightarrow{RS}$ let T be the point such that $RT = AB$. Draw $\overline{RB}$, $\overline{BT}$, and $\overline{CT}$. Then $ABTR$ is a parallelogram by Theorem 5-5, and so $\overline{BT} \cong \overline{AR}$. Hence $\overline{BT} \cong \overline{RC}$, and since $\overline{BT} \parallel \overline{RC}$, you know that $BTCR$ is a parallelogram by Theorem 5-5. Therefore, $\overleftrightarrow{RS}$ bisects $\overline{BC}$ since $\overline{CS} \cong \overline{SB}$ by Theorem 5-3.

Corollary If three parallel lines cut off congruent segments on one transversal, they cut off congruent segments on any transversal.

For a proof of the corollary, see Oral Exercise 1.

ORAL EXERCISES

1. State the reasons for each step in the following proof of the corollary to Theorem 5-14.

GIVEN: $\overleftrightarrow{AX} \parallel \overleftrightarrow{BY} \parallel \overleftrightarrow{CZ}$;
$\overline{AB} \cong \overline{BC}$

PROVE: $\overline{XY} \cong \overline{YZ}$

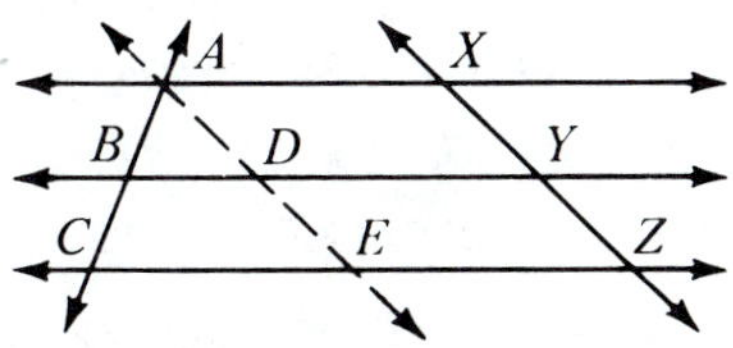

Statements	Reasons
1. Draw $\overleftrightarrow{AE}$ through A parallel to $\overleftrightarrow{XZ}$.	1. ?
2. $\overline{AB} \cong \overline{BC}$	2. ?
3. B is the midpoint of $\overline{AC}$.	3. ?
4. $\overleftrightarrow{BY} \parallel \overleftrightarrow{CZ}$	4. ?
5. $\overleftrightarrow{BY}$ bisects $\overline{AE}$.	5. ?
6. $\overline{AD} \cong \overline{DE}$	6. ?
7. $\overleftrightarrow{AX} \parallel \overleftrightarrow{BY}$	7. ?
8. $DYXA$ and $EZYD$ are parallelograms.	8. ?
9. $\overline{XY} \cong \overline{AD}$ and $\overline{DE} \cong \overline{YZ}$	9. ?
10. $\overline{XY} \cong \overline{YZ}$	10. ?

Exercises 2-5 refer to the diagram, in which A and B are the midpoints of $\overline{XY}$ and $\overline{XZ}$, respectively.

2. If $YZ = 15$, then $AB =$ __?__.
3. If $AB = 23$, then $YZ =$ __?__.
4. If $YZ = 4m + 10$, then $AB =$ __?__.
5. If $AB = 6 - 3n$, then $YZ =$ __?__.

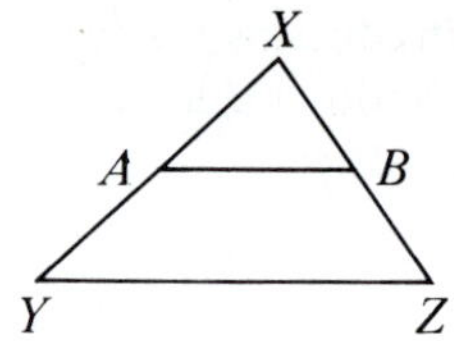

WRITTEN EXERCISES

In the figure, $\overleftrightarrow{AJ}$, $\overleftrightarrow{BK}$, $\overleftrightarrow{CL}$, $\overleftrightarrow{DM}$, and $\overleftrightarrow{EN}$ are parallel lines, with $\overline{AB} \cong \overline{BC} \cong \overline{CD} \cong \overline{DE}$. Complete.

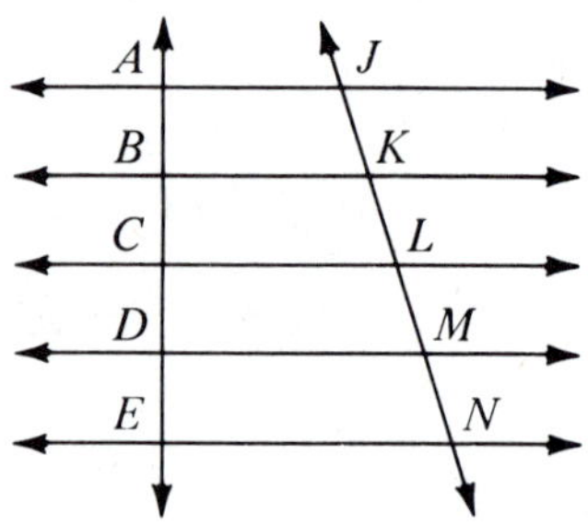

A 1. If $JK = 3.5$, then $JL =$ __?__.

2. If $LN = 11$, then $MN =$ __?__.
3. If $KM = 2x - 4$, then $KL =$ __?__.
4. If $LM = 6y + 2$, then $KM =$ __?__.
5. If $JK = 13$, then $JL =$ __?__ and $JM =$ __?__.
6. If $LM = 7$, then $JN =$ __?__.
7. If $MN = 2x + 1$ and $JN = 5x + 19$, then $x =$ __?__.
8. If $JK = 5y + 7$ and $KL = 6y - 12$, then $y =$ __?__.

Exercises 9-16 refer to the diagram, in which S and T are the midpoints of $\overline{PR}$ and $\overline{QR}$, respectively. Find the value of x.

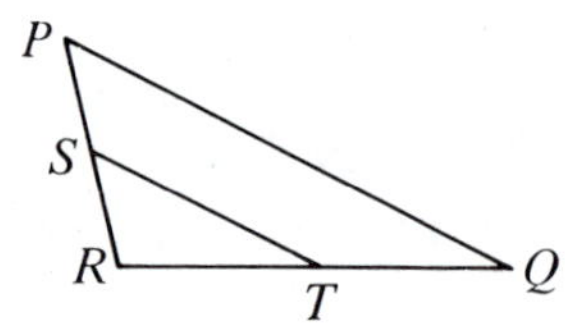

9. $ST = 3x$; $PQ = 24$
10. $ST = 40$; $PQ = 5x$
11. $ST = 12$; $PQ = 2x + 4$
12. $ST = 3x - 1$; $PQ = 34$
13. $ST = x$; $PQ = 3x - 5$
14. $ST = x - 8$; $PQ = x + 3$
15. $ST = 5x$; $PQ = x + 36$
16. $ST = 2x - 7$; $PQ = x + 1$

In Exercises 17-19 the midpoints of $\overline{XY}$, $\overline{XZ}$, and $\overline{YZ}$ are A, B, and C, respectively.

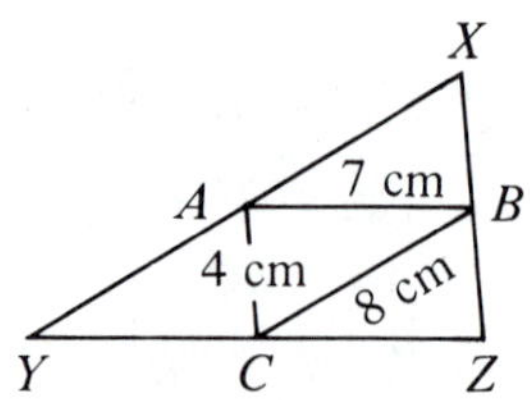

17. Find the lengths of the sides of $\triangle XYZ$.
18. Find the perimeter of $\triangle XYZ$.
19. Find the perimeter of ▱$ABCY$.
20. A triangle has sides of length 8, 9, and 11. Find the perimeter of the triangle formed by joining the midpoints of its sides.
21. $\triangle ABC$ has perimeter 24 cm. Find the perimeter of the triangle formed by joining the midpoints of the sides of $\triangle ABC$.
22. When the midpoints of the sides of $\triangle PQR$ are joined, a triangle of perimeter 18 cm is formed. Find the perimeter of $\triangle PQR$.

23. Given: P is the midpoint of $\overline{AO}$;
R is the midpoint of $\overline{OC}$;
$\overline{PQ} \parallel \overline{AB}$
Prove: $\overline{QR} \parallel \overline{BC}$

24. Given: P is the midpoint of $\overline{AO}$;
$\overline{PQ} \parallel \overline{AB}$;
$\overline{QR} \parallel \overline{BC}$
Prove: R is the midpoint of $\overline{OC}$.

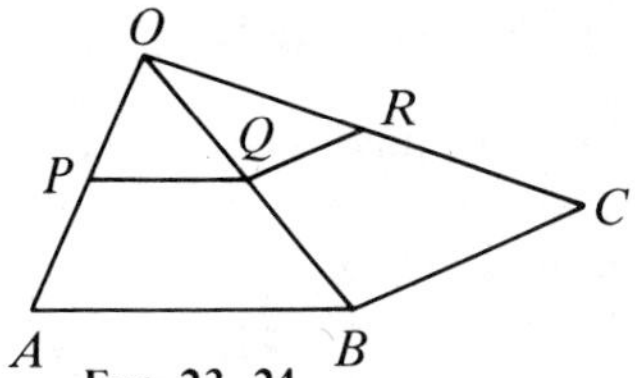

Exs. 23, 24

B 25. a. Given: W, X, Y, and Z are midpoints of $\overline{AB}$, $\overline{BC}$, $\overline{CD}$, and $\overline{AD}$, respectively.
Prove: $WXYZ$ is a ▱.
(*Hint*: Draw $\overline{AC}$. Prove that $\overline{ZY} \parallel \overline{WX}$ and $\overline{ZY} \cong \overline{WX}$.)

b. State as a theorem the result you proved in part **(a)**.

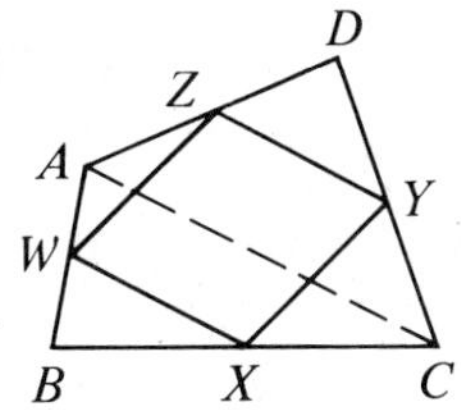

26. Write a two-column proof of Theorem 5-13.

27. Write a two-column proof of Theorem 5-14.

28. Given: $UWXZ$ is a ▱; A, B, C, and D are midpoints of $\overline{VW}$, $\overline{XY}$, $\overline{YZ}$, and $\overline{UV}$, respectively.
Prove: $\overline{AB} \cong \overline{DC}$
(*Hint*: Draw $\overline{AD}$ and $\overline{BC}$.)

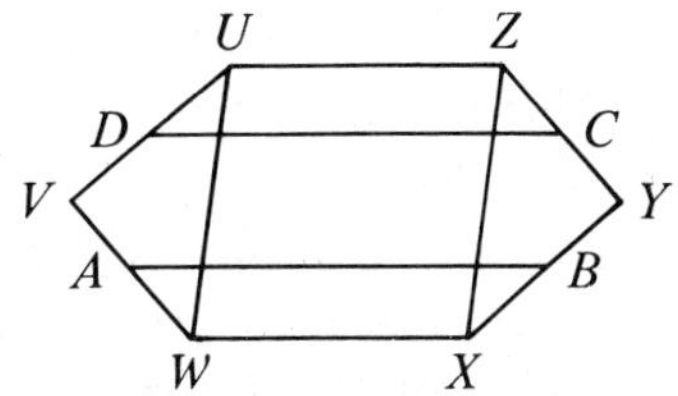

C 29. Given: $\overline{AB} \cong \overline{BC}$;
$\overline{CD} \cong \overline{DE}$;
$\overline{BH} \parallel \overline{CF} \parallel \overline{DG}$
Prove: $\overline{AE} \parallel \overline{HG}$

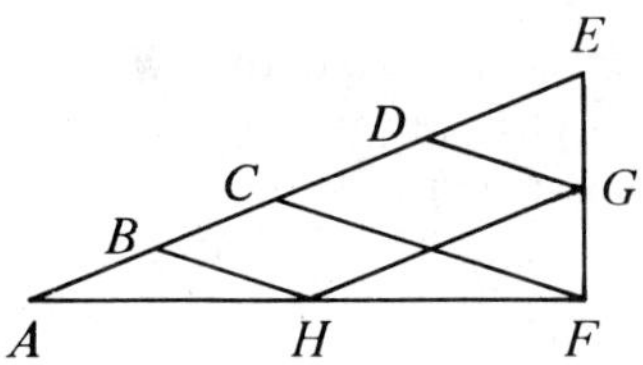

30. This exercise proves that the three medians of any triangle intersect in a point.
Given: $\overline{AM}$ and $\overline{BN}$ are medians of $\triangle ABC$, and O is the point where they intersect; $\overrightarrow{CO}$ intersects $\overline{AB}$ at P.
Prove: $\overline{CP}$ is a median of $\triangle ABC$.
On the ray opposite to $\overrightarrow{NB}$ let D be the point such that $\overline{DN} \cong \overline{ON}$. Draw $\overline{AD}$ and $\overline{DC}$.

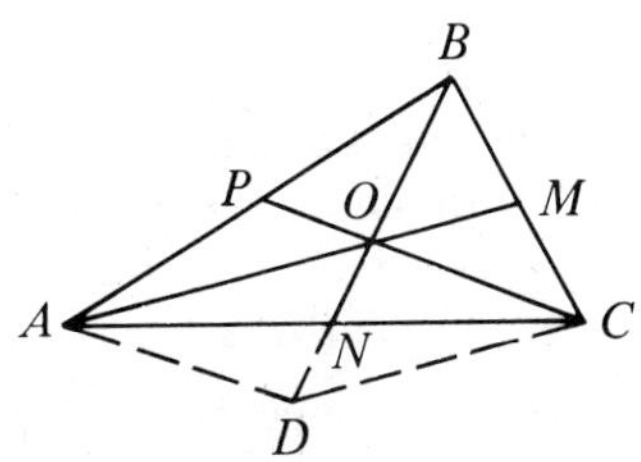

a. With the given information and the given extra line segments, prove that $ADCO$ is a parallelogram.
b. Then prove that O is the midpoint of $\overline{BD}$. (*Hint*: Remember that $\overline{AM}$ is a median of $\triangle ABC$.)
c. Then prove that P is the midpoint of $\overline{AB}$, and hence $\overline{CP}$ is a median of $\triangle ABC$.

Section 5-5 TRAPEZOIDS

An important kind of quadrilateral that is not a parallelogram is called a *trapezoid.* In this section you will learn:

- the definition of a trapezoid
- the definition of an isosceles trapezoid
- some properties of trapezoids and isosceles trapezoids

A *trapezoid* is a quadrilateral with two parallel sides and two nonparallel sides. The parallel sides are called the *bases* of the trapezoid.

EXAMPLES OF TRAPEZOIDS

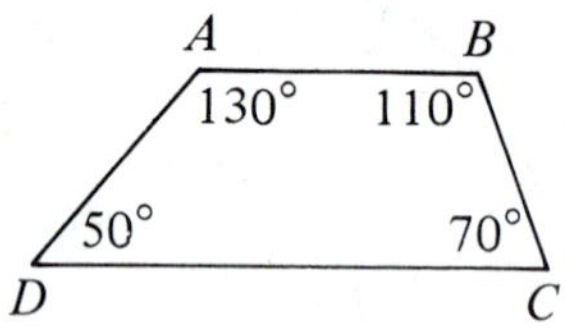

Bases: $\overline{AB}$ and $\overline{DC}$

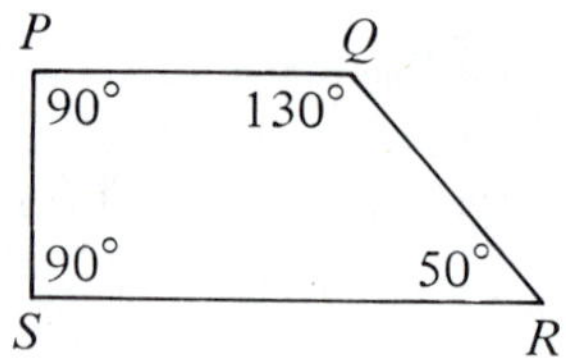

Bases: $\overline{SR}$ and $\overline{PQ}$

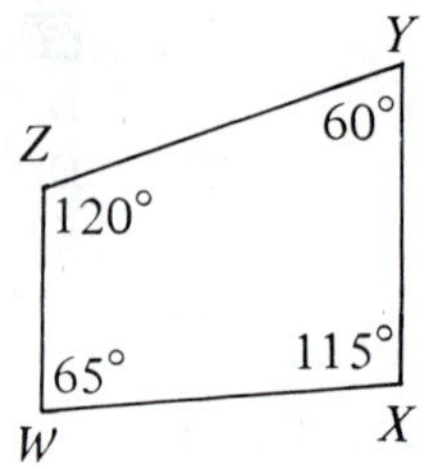

Bases: $\overline{WZ}$ and $\overline{XY}$

A trapezoid whose nonparallel sides are congruent is called an *isosceles trapezoid.* The next two theorems state properties of isosceles trapezoids. The first theorem should remind you of a theorem about isosceles triangles.

Theorem 5-15 In an isosceles trapezoid the two angles whose vertices are the endpoints of either base are congruent. (Base $\angle$s of an isos. trapezoid are congruent.)

GIVEN: Isosceles trapezoid $EFGH$ with $\overline{EF} \parallel \overline{HG}$ and $\overline{HE} \cong \overline{GF}$

PROVE: $\angle E \cong \angle F$; $\angle G \cong \angle H$

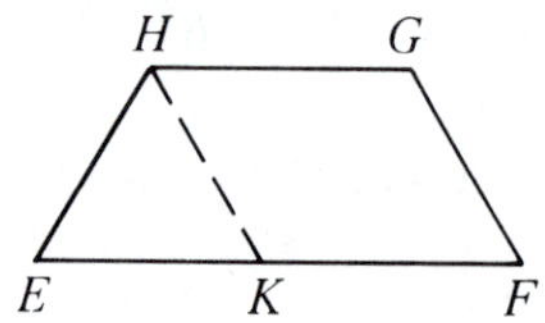

Proof: Draw $\overline{HK}$ through H parallel to $\overline{GF}$ and intersecting $\overleftrightarrow{EF}$ at K. The remainder of the proof is left as Exercise 17.

Theorem 5-16 The diagonals of an isosceles trapezoid are congruent.

GIVEN: Isosceles trapezoid $MNOP$ with $\overline{MN} \parallel \overline{PO}$ and $\overline{MP} \cong \overline{NO}$

PROVE: $\overline{MO} \cong \overline{NP}$

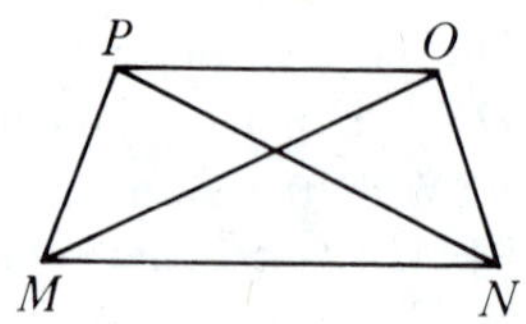

The proof is left as Exercise 18.

The *median* of a trapezoid is the line segment connecting the midpoints of the nonparallel sides.

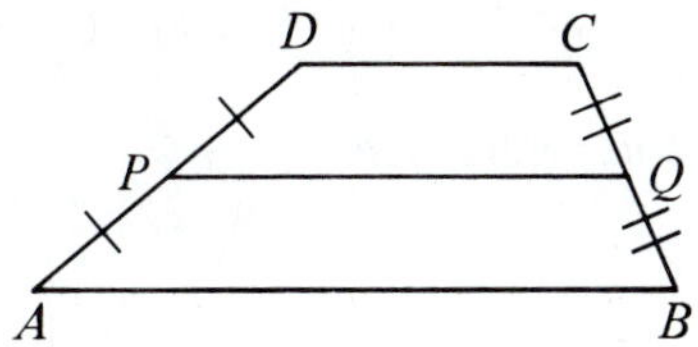

$\overline{PQ}$ is the median of trapezoid *ABCD*.

The properties of the median of a trapezoid are stated in the next theorem.

Theorem 5-17 The median of a trapezoid is parallel to the bases, and its length is half the sum of the lengths of the bases.

The proof of Theorem 5-17 is left as Exercise 20.

EXAMPLE One base of a trapezoid is 6 units longer than the other. The median has length 17. Find the lengths of the bases.

SOLUTION Let x be the length of the shorter base.
Then $x + 6$ is the length of the longer base.

$$\text{length of median} = \frac{\text{sum of lengths of bases}}{2}$$

$$17 = \frac{x + (x + 6)}{2}$$

$$34 = 2x + 6$$

$$28 = 2x$$

$$14 = x$$

The shorter base has length 14. The longer base has length 20.

ORAL EXERCISES

Answer each question *yes* or *no*. If the answer is *no*, explain.

1. Can a trapezoid also be a parallelogram?
2. Can the diagonals of a trapezoid bisect each other?
3. Can the bases of a trapezoid be congruent?
4. Can two angles of a trapezoid be right angles?
5. State a theorem that justifies the assertion that a line perpendicular to one base of a trapezoid is also perpendicular to the other base.
6. State a theorem that justifies the assertion that if a line is perpendicular to two sides of a quadrilateral whose other two sides are not parallel, then the quadrilateral is a trapezoid.
7. Justify the assertion that any two line segments joining the bases of a given trapezoid and perpendicular to the bases are congruent.

WRITTEN EXERCISES

Exercises 1–10 refer to trapezoid $ABCD$, with $\overline{AB} \parallel \overline{DC}$ and median $\overline{EF}$. Complete.

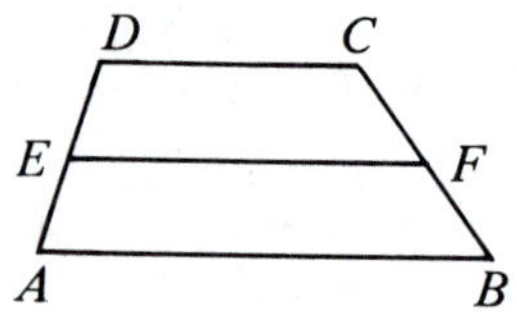

A 1. $AB = 25$, $DC = 13$, $EF =$ __?__

2. $AB = 12$, $DC = 8$, $EF =$ __?__

3. $AB = 26$, $EF = 18$, $DC =$ __?__

4. $DC = 17$, $EF = 21$, $AB =$ __?__

5. $AB = 5 + n$, $DC = 17 - n$, $EF =$ __?__

6. $AB = 2m - 19$, $EF = m - 2$, $DC =$ __?__

Find the value of x.

7. $AB = x$, $CD = x + 8$, $EF = 13$

8. $AB = 21$, $CD = 4x - 11$, $EF = 3x$

9. AB is 10 greater than CD, and $EF = 17$. Find AB and CD.

10. AB is 6 greater than twice CD, and $EF = 15$. Find AB and CD.

Write two-column proofs.

11. Given: Isosceles trapezoid $ABCD$, with $\overline{AB} \parallel \overline{DC}$ and $\overline{AD} \cong \overline{BC}$
    Prove: $\overline{AE} \cong \overline{BE}$

12. Given: Isosceles trapezoid $ABCD$, with $\overline{AB} \parallel \overline{DC}$ and $\overline{AD} \cong \overline{BC}$
    Prove: $\angle 1 \cong \angle 2$

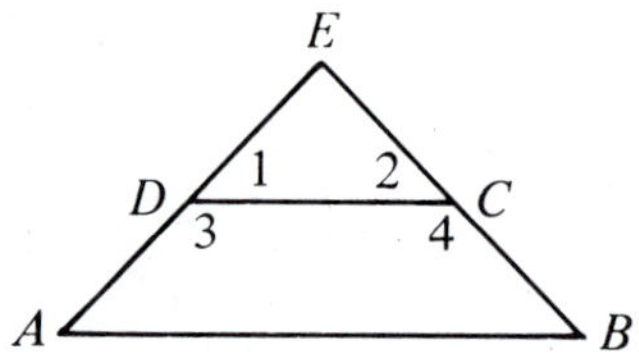

Exs. 11, 12

13. Given: J is the midpoint of $\overline{PR}$; $PQKJ$ is a trapezoid.
    Prove: $\overline{RK} \cong \overline{KQ}$

14. Given: J and K are midpoints of $\overline{PR}$ and $\overline{QR}$, respectively.
    Prove: $PQKJ$ is a trapezoid.

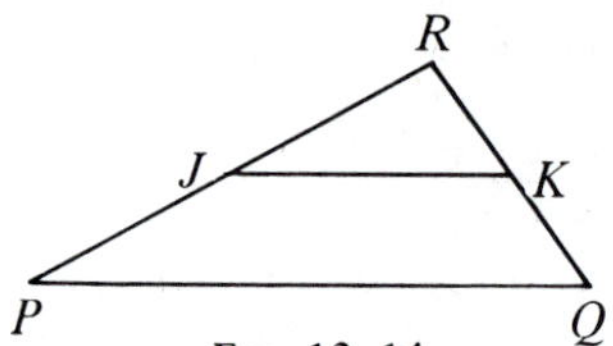

Exs. 13, 14

B 15. Given: Isosceles trapezoid $ABCD$, with $\overline{AB} \parallel \overline{DC}$ and $\overline{AD} \cong \overline{BC}$; M is the midpoint of $\overline{AB}$.
    Prove: $\angle 1 \cong \angle 2$

16. Given: Trapezoid $ABCD$, with $\overline{AB} \parallel \overline{DC}$; $\angle 1 \cong \angle 2$; M is the midpoint of $\overline{AB}$.
    Prove: Trapezoid $ABCD$ is isosceles.

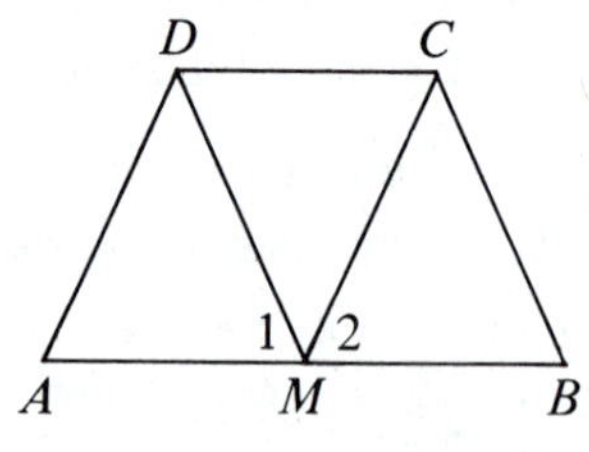

Exs. 15, 16

17. Complete the proof of Theorem 5-15.

18. Prove Theorem 5-16.

19. This exercise proves that if two base angles of a trapezoid are congruent, then the trapezoid is isosceles.

a. Given: Trapezoid $ABCD$, with $\overline{AB} \parallel \overline{DC}$; $\angle A \cong \angle B$
Prove: Trapezoid $ABCD$ is isosceles.
(*Hint:* Draw $\overline{CE} \parallel \overline{DA}$.)

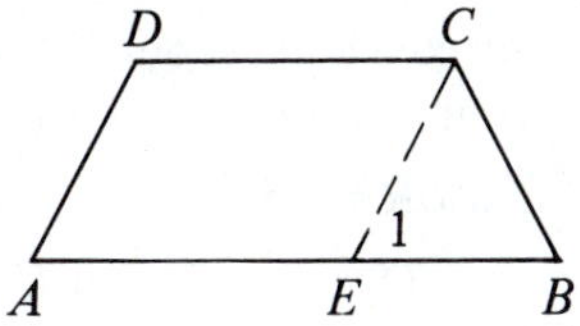

b. Given: Trapezoid $ABCD$, with $\overline{AB} \parallel \overline{DC}$; $\angle D \cong \angle DCB$
Prove: Trapezoid $ABCD$ is isosceles.
(*Hint:* Prove $\angle A \cong \angle B$ and then use part **(a)** above.)

C **20.** Prove Theorem 5-17.
Given: Trapezoid $ABCD$, with $\overline{AB} \parallel \overline{DC}$ and median $\overline{EF}$
Prove: $\overline{EF} \parallel \overline{AB}$; $\overline{EF} \parallel \overline{DC}$; $EF = \frac{1}{2}(AB + DC)$
(*Hint:* Let G be the point where $\overleftrightarrow{DF}$ and $\overleftrightarrow{AB}$ intersect and prove that $\triangle DCF \cong \triangle GBF$. Then $CD = BG$ and $DF = GF$. Apply Theorem 5-13 to $\triangle DAG$.)

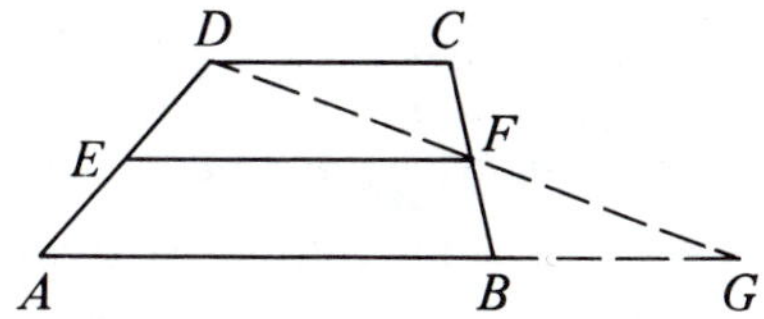

21. Prove: If the diagonals of a trapezoid are congruent, then the trapezoid is isosceles.
Given: Trapezoid $DEFG$, with $\overline{DE} \parallel \overline{GF}$ and $\overline{DF} \cong \overline{EG}$
Prove: Trapezoid $DEFG$ is isosceles.
(*Hint:* Draw $\overline{GX}$ and $\overline{FY}$ perpendicular to $\overleftrightarrow{DE}$. Prove $\triangle DFY \cong \triangle EGX$ to get $\angle FDE \cong \angle GED$. Then show that $\triangle DFE \cong \triangle EGD$.)

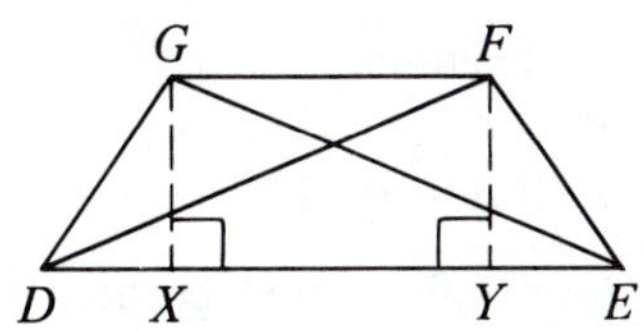

SELF-TEST 2

Section 5-4

In $\triangle ABC$, $\overline{AX} \cong \overline{BX}$ and $\overline{AY} \cong \overline{CY}$.

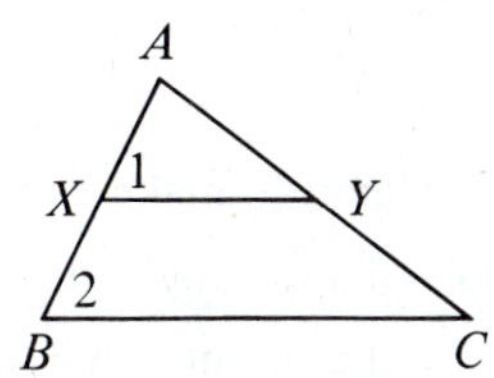

1. Explain why $\angle 1 \cong \angle 2$.

2. If $XY = \frac{1}{2}k + 9$ and $BC = 5k - 2$, find the value of k and BC.

3. If $\overleftrightarrow{DK} \parallel \overleftrightarrow{EJ} \parallel \overleftrightarrow{FH}$ and $DE = EF$, justify the assertion that $KJ = JH$.

4. Write a two-column proof.
Given: $\overline{EF} \cong \overline{FG}$; $\overline{FH} \parallel \overline{EJ}$
Prove: $\overline{EH}$ is a median of $\triangle EGJ$.

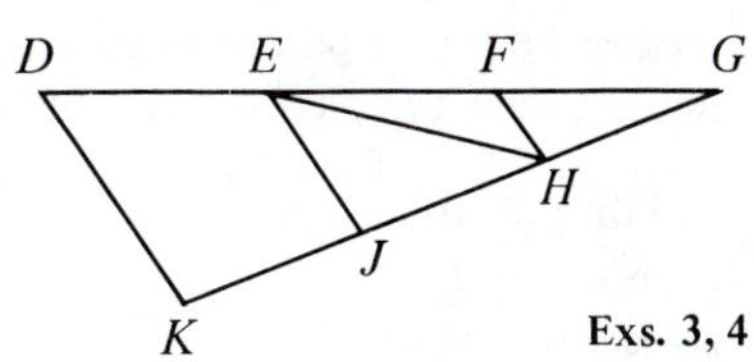

Exs. 3, 4

(*Self-Test continued on next page.*)

Exercises 5-7 refer to trapezoid $PQST$ with median $\overline{UR}$.

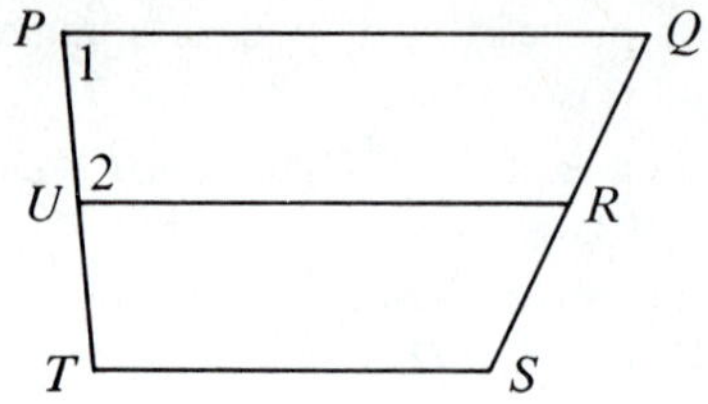

5. What can you conclude about $\angle 1$ and $\angle 2$? Justify your answer. Section 5-5
6. If $PQ = 18$ and $UR = 15$, find TS.
7. If $PQ = 3x - 7$, $TS = x + 3$, and $UR = x + 4$, find the value of x.
8. Write a two-column proof.
   Given: $WXYZ$ is an isosceles trapezoid, with $\overline{WX} \parallel \overline{ZY}$ and $\overline{WZ} \cong \overline{XY}$.
   Prove: $\triangle WZY \cong \triangle XYZ$

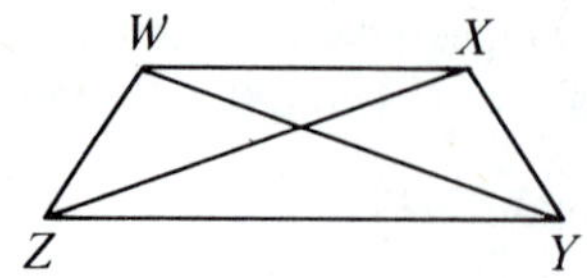

CHAPTER REVIEW

If $ABCD$ is a parallelogram, find the value of x.

1. $m\angle A = 18x - 11$; $m\angle C = 3x + 49$ Section 5-1
2. $AB = 13 - 2x$; $BC = x + 4$; $CD = 5x - 1$
3. $m\angle B = 6x + 26$; $m\angle C = 16x - 44$
4. Write a two-column proof.
   Given: $WXYZ$ is a ▱.
   Prove: $\triangle WVX \cong \triangle YVZ$

Could the given information be used to prove that quadrilateral $WXYZ$ is a parallelogram? Explain.

5. $\overline{ZV} \cong \overline{XV}$; $\overline{YV} \cong \overline{WV}$ Section 5-2
6. $\overline{ZW} \cong \overline{YX}$; $\overline{ZY} \parallel \overline{WX}$
7. $\angle ZWX \cong \angle ZYX$; $\overline{ZY} \parallel \overline{WX}$
8. Write a two-column proof.
   Given: $\overline{WX} \cong \overline{ZY}$; $\overline{XY} \cong \overline{WZ}$
   Prove: $\overline{ZY} \parallel \overline{WX}$

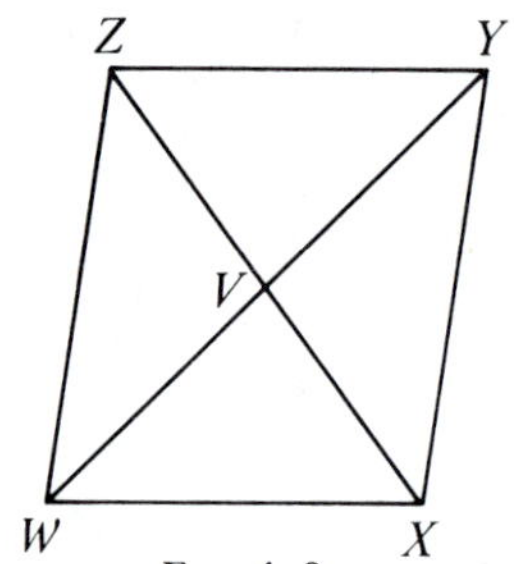

Exs. 4-8

Complete each sentence to make a true statement.

9. The diagonals of a rectangle are __?__. Section 5-3
10. The diagonals of a rhombus are __?__.
11. If two consecutive sides of a parallelogram are congruent and perpendicular, then the parallelogram is a __?__.
12. Write a two-column proof.
    Given: $EFGH$ is a ▱; $\angle 1 \cong \angle 2$
    Prove: $EFGH$ is a rhombus.

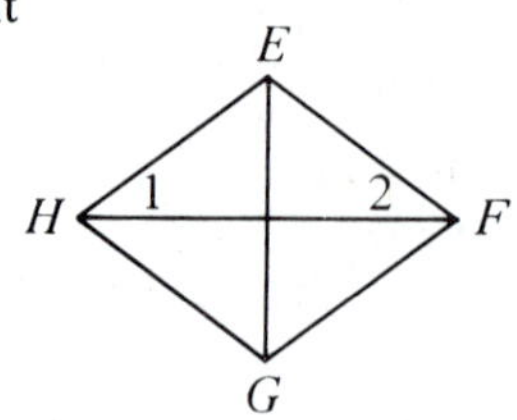

In Exercises 13 and 14, you may assume that $\overline{RS} \cong \overline{ST}$ and $\overline{RV} \cong \overline{VU}$.

Section 5-4

13. If $SV = 3y + 5$ and $TU = 9y - 11$, find the value of y.

14. State the theorems that justify the statement that $\angle 3 \cong \angle 4$.

15. Write a two-column proof.
Given: $\overline{RS} \cong \overline{ST}$; $\overline{SV} \perp \overline{RT}$; $\overline{TU} \perp \overline{RT}$
Prove: $TU = 2 \cdot SV$

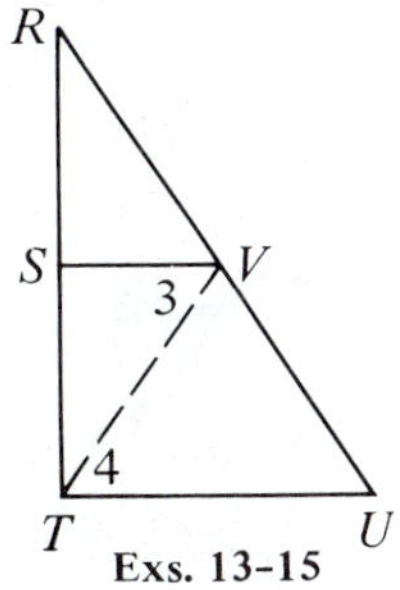

Exs. 13-15

Exercises 16-18 refer to trapezoid *KLNO* with median $\overline{PM}$.

Section 5-5

16. If $m\angle MPK = 4x$ and $m\angle K = 6x$, find the measure of $\angle K$.

17. If $KN = 3y - 5$ and $LO = 3y + 5$, can *KLNO* be an isosceles trapezoid? Explain.

18. The length of $\overline{ON}$ is 2 less than twice that of $\overline{KL}$. If the length of $\overline{PM}$ is 17, how long are $\overline{KL}$ and $\overline{ON}$?

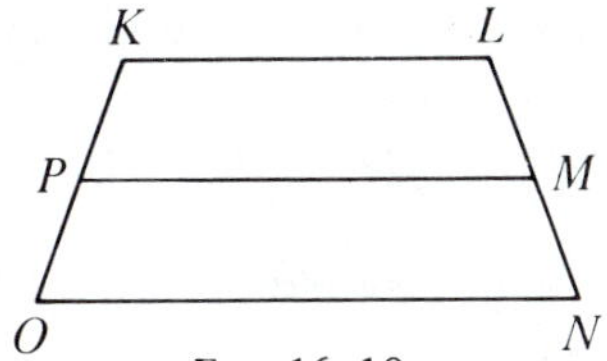

Exs. 16-18

19. Write a two-column proof.
Given: Isosceles trapezoid *ABCD*, with $\overline{AB} \parallel \overline{DC}$ and $\overline{AD} \cong \overline{BC}$; *E* is the midpoint of $\overline{AB}$.
Prove: $\angle 1 \cong \angle 2$

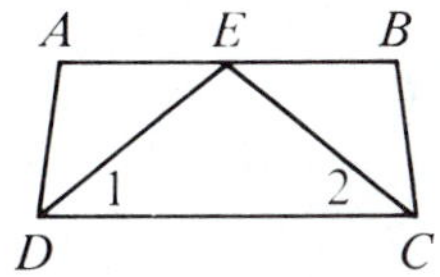

CHAPTER TEST

Exercises 1-7 refer to the diagram, in which quadrilateral *ABDE* is a rhombus and $\angle 1 \cong \angle 2$.

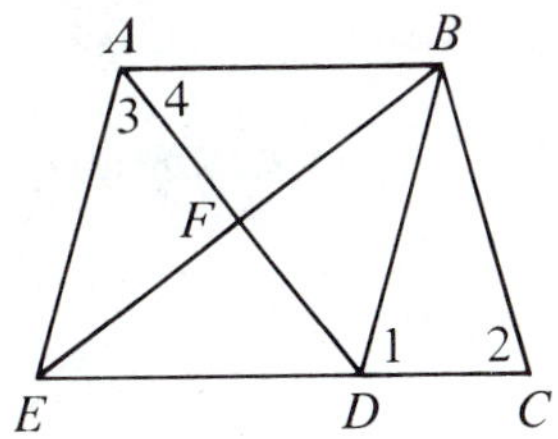

Exs. 1-7

1. State two relationships between $\overline{AE}$ and $\overline{BD}$.
2. State two relationships between $\overline{AD}$ and $\overline{BE}$.
3. State the theorem that justifies the assertion that $\angle 3 \cong \angle 4$.
4. If $m\angle DBC = 32$, find the measures of $\angle DBA$, $\angle BAE$, $\angle AED$, and $\angle ADB$.
5. If $AF = 2x + 7$ and $AD = 6x - 5$, find the value of x.
6. What kind of quadrilateral is *ABCE*? Be as specific as possible and justify your answer.
7. State the theorem that justifies the assertion that $\overline{AC} \cong \overline{BE}$.

The diagonals of quadrilateral *RSTU* intersect at point *Z*. What additional information, together with that given in each exercise, would enable you to prove that *RSTU* is a parallelogram?

8. $\overline{RS} \cong \overline{TU}$

9. $\angle R \cong \angle T$

10. $\overline{ST} \parallel \overline{RU}$

11. *Z* is the midpoint of $\overline{RT}$.

In $\triangle JMN$, $\overline{JK} \cong \overline{KL} \cong \overline{LM}$. Explain why each statement is true.

12. If $\angle 1 \cong \angle 2 \cong \angle 3$, then $\overline{PO} \cong \overline{ON}$.

13. If $\overline{JP} \cong \overline{PO}$, then $\overline{KP} \parallel \overline{LO}$.

14. If $\overline{PO} \cong \overline{ON}$ and $\overline{KP} \parallel \overline{MN}$, then $\overline{LO} \parallel \overline{MN}$.

15. If $\overline{JK} \cong \overline{KL}$, $\overline{JP} \cong \overline{PO}$, and $KP = 2r + 2$, express LO in terms of r.

16. If $\overline{LO}$ is the median of trapezoid $KMNP$, with $KP = 13$ and $LO = 20$, find MN.

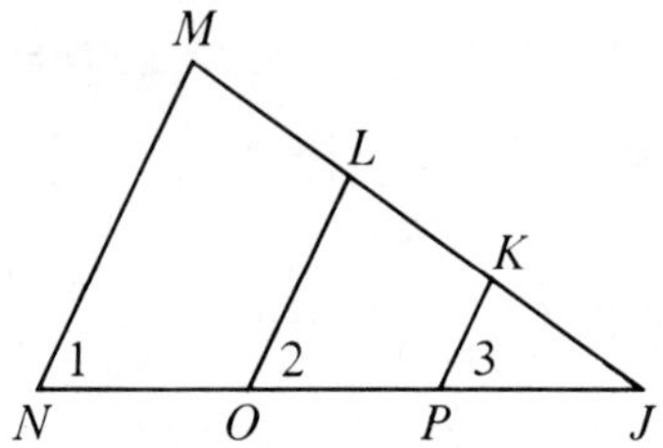

Exs. 12-16

17. Refer to right triangle ABC. Explain why the midpoint, D, of hypotenuse $\overline{AC}$ is equally distant from the vertices of $\triangle ABC$.
(*Hint*: Consider rectangle $ABCE$.)

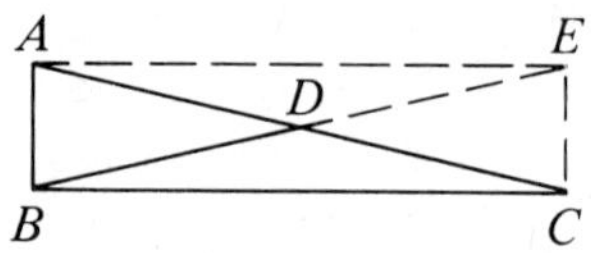

Write two-column proofs for any *two* of Exercises 18-20.

18. Given: $RSTW$ and $SXTW$ are parallelograms.
Prove: $\triangle RSW \cong \triangle SXT$

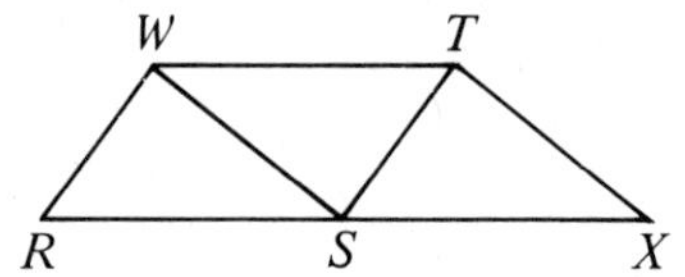

19. Given: $FGJK$ is a rectangle; $\angle 1 \cong \angle 2$
Prove: $\overline{KJ} \parallel \overline{EH}$; $\overline{KE} \cong \overline{JH}$

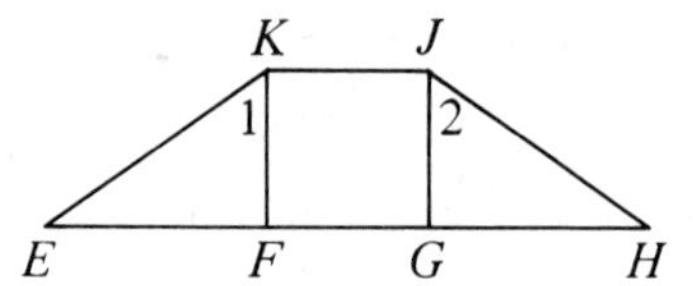

20. Given: $\triangle LNO$ is equilateral;
$\overline{MP}$ bisects $\overline{LN}$ and $\overline{LO}$.
Prove: $\triangle LMP$ is equilateral.

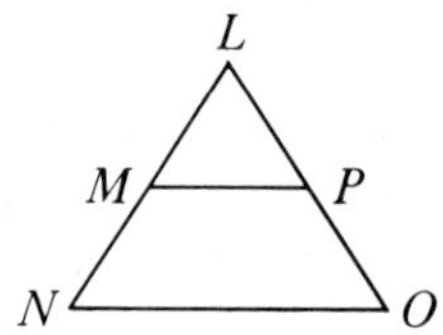

Careers — Computer Software Designer

In the sixties and seventies the computer software field was fluid and informal. Companies were scrambling for qualified software personnel, and many jobs were neither structured nor well defined. As the industry has matured, it has become more structured, and a number of different software-related jobs have evolved.

A computer software designer has responsibilities that fall in between those of a systems analyst and a programmer. As an illustration of the difference between these three jobs, imagine that the Olympic volleyball team asks a software firm for help. Sports teams are increasingly getting involved with computer analysis, so this is not a far-fetched example.

A systems analyst's job would be to decide what kind of computer and graphics terminal the team needs, details about the camera that would take pictures of the players in motion, and what kinds of data the team wants to gather and analyze (for instance, how successful the different plays are or which players have wasted motion in their jumps). The software designer would then break down these decisions into blocks, or what are called subroutines. He or she would identify the logic behind analyzing a player's motion and how each logical step relates to the ones before and after it. Programmers would actually write and debug the programs.

Software designers can be found in most large companies and in virtually all fields, ranging from private industry to government to consulting. Growth areas for designers include computer-aided design and manufacture (CAD/CAM), medical diagnosis by computer, computer games, and simulation programs. A designer can use a graphics display terminal and CAD/CAM to analyze a new manufacturing design – for instance, of an automobile bumper.

Software designers need to take a number of courses in engineering, computer science, and mathematics. A major in an applied field, such as biology, is a good choice if you know what kind of software design you want to do. Designers often start out as programmers. Programming courses are available at most colleges, as well as in night and adult education programs.

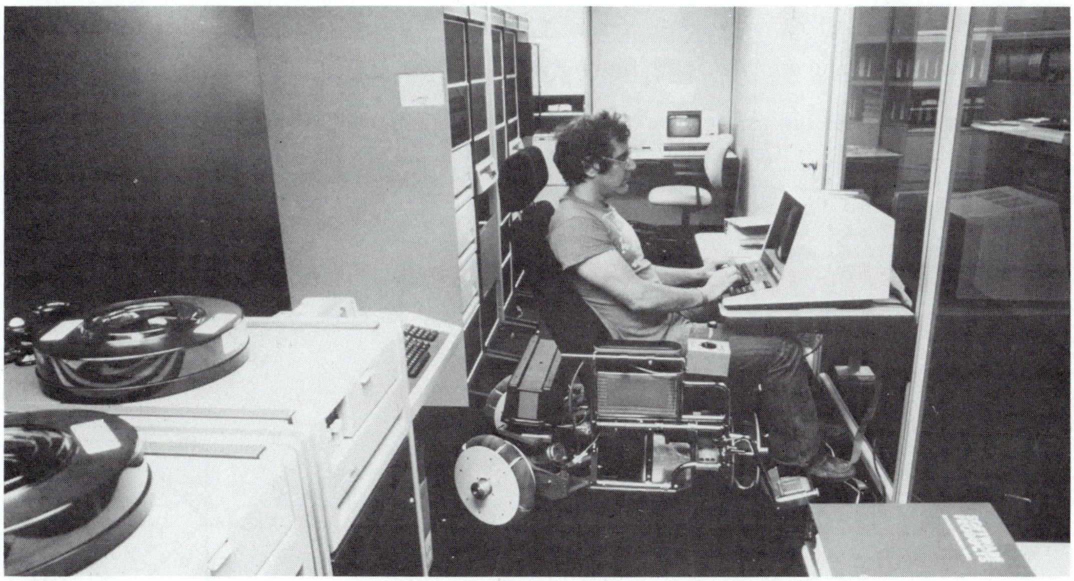

VOCABULARY REVIEW

Be sure that you understand the meaning of these terms:

parallelogram, p. 188
rectangle, p. 196
rhombus, p. 196
square, p. 197
trapezoid, p. 206
isosceles trapezoid, p. 206
median of a trapezoid, p. 206

MIXED REVIEW

Write two-column proofs.

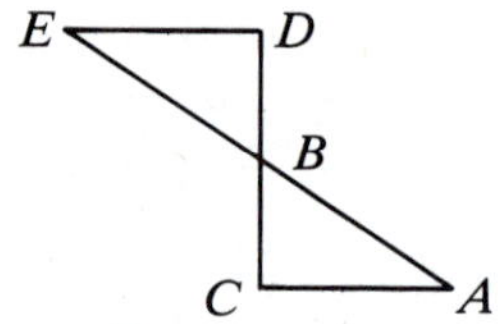

Exs. 1, 2

1. Given: $\overline{ED} \perp \overline{DC}$; $\overline{AC} \perp \overline{DC}$; B is the midpoint of $\overline{DC}$.
   Prove: $\triangle EBD \cong \triangle ABC$
2. Given: B is the midpoint of $\overline{DC}$; B is the midpoint of $\overline{EA}$.
   Prove: $\triangle DEB \cong \triangle CAB$

Exercises 3–7 refer to $\square ABCD$.

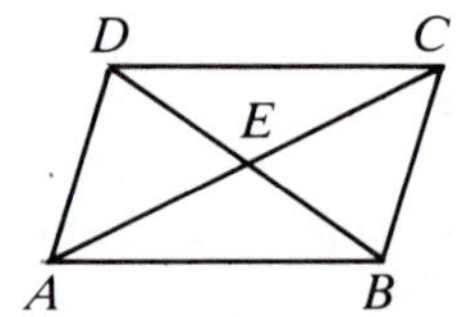

3. If $m\angle ADC = 125$, find $m\angle DAB$, $m\angle ABC$, and $m\angle BCD$.
4. Name an angle congruent to $\angle BAC$.
5. If $AB = 3x - 2$ and $DC = 7x - 6$, find x.
6. If $AE = 6x + 1$ and $AC = 8x + 10$, find x.
7. If $m\angle BAC = 25$ and $m\angle CDB = 20$, find $m\angle AEB$.
8. Annette is twice as old as Fran. Next year the sum of their ages will be 5 times Fran's age 4 years ago. How old are Fran and Annette now?

Write two-column proofs.

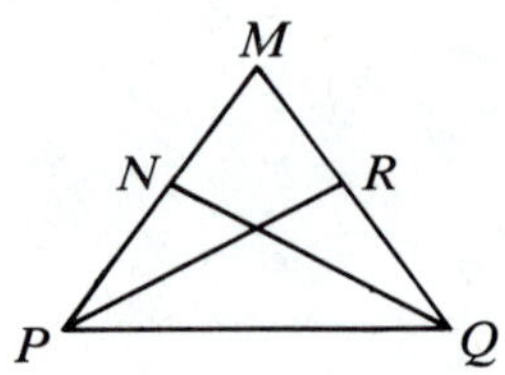

9. Given: $\overline{MN} \cong \overline{MR}$; $\angle P \cong \angle Q$
   Prove: $\overline{NQ} \cong \overline{RP}$

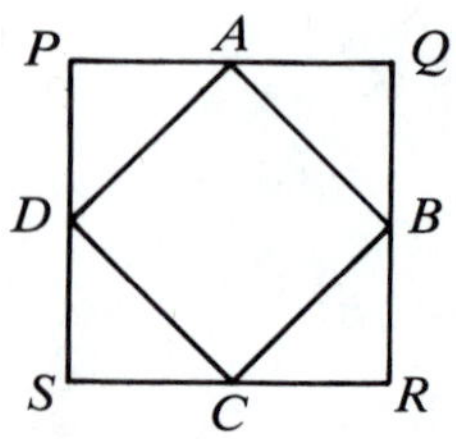

10. Given: $PQRS$ is a square; A, B, C, and D are the midpoints of $\overline{PQ}$, $\overline{QR}$, $\overline{RS}$, and $\overline{SP}$, respectively.
    Prove: $ABCD$ is a rhombus.

11. Given: $ABCD$ is a trapezoid with median $\overline{EF}$.
Prove: $\overline{EF}$ bisects both diagonals.

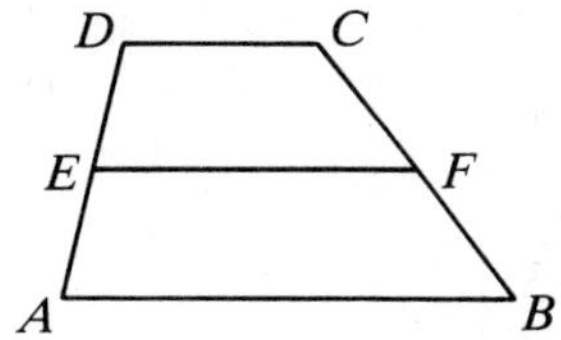

12. Write a conclusion that follows logically from the given premises.

P_1: The early bird gets the worm.
P_2: The robin does not have a worm.

13. Simplify: $\dfrac{4 - 12 + 6}{-(-3 + 1)}$

Find the numerical measure of $\angle C$ in $\triangle ABC$. Then classify $\triangle ABC$ as acute, right, or obtuse.

14. $m\angle A = m\angle C$, $m\angle B = 86$

15. $m\angle A = 10x - 5$, $m\angle B = 4x + 1$, $m\angle C = x + 4$

Exercises 16-18 refer to the diagram. Could the given information be used to prove $QRST$ is a parallelogram? Explain.

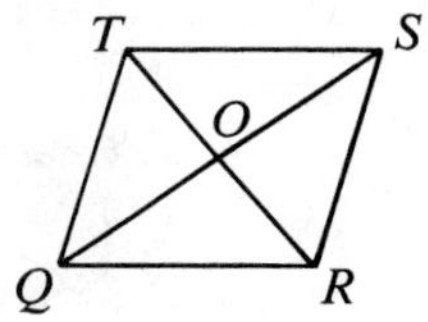

16. $\triangle QRS \cong \triangle STQ$

17. $\angle SQR \cong \angle TSQ$; $\angle TQS \cong \angle QSR$

18. $\overline{TO} \cong \overline{RO}$; $\overline{QO} \cong \overline{SO}$

19. Negate the conjunction: $\sim r \vee s$

Solve.

20. $2(2a + 3) - (a + 2) = a + 1$

21. $-3 \leqslant \frac{1}{2}t + 4 < 6$

22. Write the **(a)** converse, **(b)** inverse, and **(c)** contrapositive of "If x is even, then $2x + 1$ is odd," and **(d)** compare their truth values.

23. **a.** Use a variable and either the universal quantifier or the existential quantifier to rewrite the following statement in symbolic form: At least one number has a positive square.

    b. Give the truth value of the statement.

    c. Write the negation of the given statement and repeat part (a) for the negation.

24. In quadrilateral $ABCD$, $AB = 2x - 6$, $BC = 3x - 12$, $CD = 9 - x$, $AD = x - 2$, $AC = 2x - 5$, and $BD = 5x - 20$. Find the value of x that will make $ABCD$ a rectangle.

A number of the techniques for multiplying and factoring polynomials that you will learn in this chapter can be illustrated using diagrams involving rectangles and squares. For example, the formula for squaring a binomial is illustrated on the facing page.

Polynomials and Rational Expressions 6

In contrast to the algebra you see today, the algebra of the early Greeks (500-200 B.C.) was geometric. Because of the clumsiness of their system of numeration (Greek numerals were similar to Roman numerals), the lack of any adequate algebraic notation, and their philosophy about the nature of numbers (the Pythagoreans believed that everything mathematical could be explained using *whole* numbers only), the Greeks represented a number by a length. For example, the early Greeks thought of the product ab as a rectangle of base b and height a and referred to it as "the rectangle contained by CD and DE."

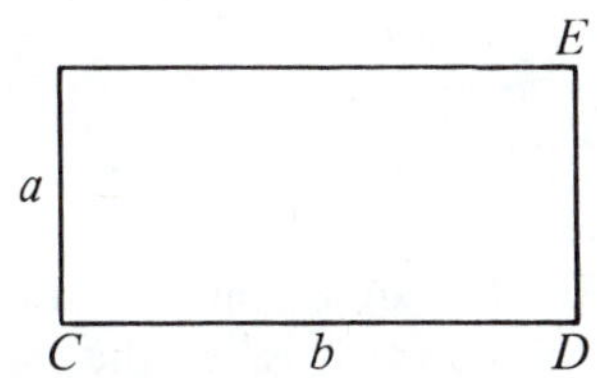

The Greek mathematician Euclid stated the following proposition:

"If a straight line (segment) is divided into any two parts, the square on the whole line (segment) is equal to the sum of the squares on the two parts together with twice the rectangle contained by the two parts."

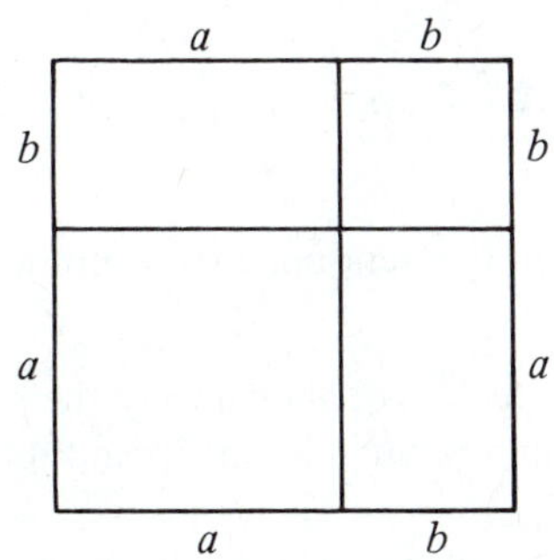

How would the algebra student of today state the same proposition? Simply,

$$(a + b)^2 = a^2 + 2ab + b^2.$$

You will see this identity again in Section 6-4.

Section 6-1 SUMS AND DIFFERENCES OF POLYNOMIALS

In Section 1-2 we briefly discussed powers. Now we will review that concept and extend it to:

- monomials
- binomials
- trinomials
- polynomials
- sums and differences of polynomials

Consider the following expressions:

$$x \cdot x = \underset{\uparrow \text{ base}}{x}^{2 \,\swarrow \text{ exponent}}$$ read "x squared" or "x to the second power"

$$y \cdot y \cdot y = \underset{\uparrow \text{ base}}{y}^{3 \,\swarrow \text{ exponent}}$$ read "y cubed" or "y to the third power"

To generalize, if x is a real number and n is a positive integer, then for x^n, x is the *base*, n is the *exponent*, and

$$x^n = \underbrace{x \cdot x \cdot x \cdots x}_{n \text{ factors}}$$

A *monomial* is a number, a variable, or a product of a number and powers of one or more variables. The numerical part of a monomial is called the *coefficient*. The *degree* of a nonzero monomial is the sum of the powers of the variables. A nonzero constant monomial, like -5, has degree 0. The monomial 0 has no degree.

EXAMPLES

Monomial	*Coefficient*	*Degree*
$17x^2$	17	2
xy	1	$1 + 1 = 2$
$-8x^3yz^2$	-8	$3 + 1 + 2 = 6$

An expression like $\frac{2}{x}$, containing a variable in the denominator, is not a monomial.

Two monomials are *like*, or *similar*, if they are equal or if only their coefficients differ. Thus, $9xy^2$, $-3y^2x$, and xy^2 are similar, whereas $5a^2b^3$ and $4a^3b^2$ are not similar.

A *polynomial* is a monomial or a sum of monomials. Each monomial forming the polynomial is called a *term* of the polynomial. A polynomial is in *simplified form* when it contains no like terms. The *degree* of a polynomial is the greatest of the degrees of its terms when it is written in simplified form.

EXAMPLES

Polynomial	Number of Terms	Degree of the Polynomial
$a^3 - 2$	2	3
$4a^2b$	1	3
-6	1	0
$x^2 + 3xy + 4y^2$	3	2
$12x^5 - 7x^4 - 8x^3 + 9$	4	5
$x^3 + x^2yz - 2xy^3 + 4yz^4 + 7xz - 8$	6	5

A polynomial that has two terms is called a *binomial*; one with three terms is a *trinomial*.

In Chapter 7 you will study some of the applications of polynomials. The principal objective of this chapter is to teach you how to manipulate polynomials.

To find the sum of two polynomials, remove parentheses and combine like terms:

EXAMPLE

$$\begin{aligned}(3x + 6y - 8z) + (2x - 9y + 3z) &= 3x + 6y - 8z + 2x - 9y + 3z \\ &= 3x + 2x + 6y - 9y - 8z + 3z \\ &= 5x - 3y - 5z\end{aligned}$$

For convenience, we sometimes arrange the work vertically as shown below.

$$\begin{array}{r} 3x + 6y - 8z \\ +\ 2x - 9y + 3z \\ \hline 5x - 3y - 5z \end{array}$$

EXAMPLE Simplify: $(3a - 10b - 9c) + (2a - 2c) + (2a + 4b)$

SOLUTION Write vertically, lining up like terms.

$$\begin{array}{rrr} 3a & -\ 10b & -\ 9c \\ 2a & & -\ 2c \\ +\ 2a & +\ 4b & \\ \hline 7a & -\ 6b & -\ 11c \end{array}$$

EXAMPLE Simplify: $(4x^3 - 2x^2y + 7y^2) + (8x^3 + 3xy^2 - 3y^2)$

SOLUTION In problems such as this, examine the terms very carefully to see which ones are like terms.

$$\begin{aligned}(4x^3 - 2x^2y + 7y^2) + (8x^3 + 3xy^2 - 3y^2) &= 4x^3 - 2x^2y + 7y^2 + 8x^3 + 3xy^2 - 3y^2 \\ &= 4x^3 + 8x^3 - 2x^2y + 3xy^2 + 7y^2 - 3y^2 \\ &= 12x^3 - 2x^2y + 3xy^2 + 4y^2\end{aligned}$$

Or you can arrange the work vertically. In this example the middle terms of the polynomials are not like terms so they should not be written in the same column.

$$\begin{array}{rrrr} 4x^3 & -\ 2x^2y & & +\ 7y^2 \\ +\ \ 8x^3 & & +\ 3xy^2 & -\ 3y^2 \\ \hline 12x^3 & -\ 2x^2y & +\ 3xy^2 & +\ 4y^2 \end{array}$$

When subtracting polynomials, remember that subtracting a number is the same as adding its opposite. In practice, a minus sign between two polynomials says "change the signs and add."

EXAMPLE Simplify: $(12a - 3b + 4c) - (7a + b - 2c)$

SOLUTION Write the problem vertically.

$$\begin{array}{r} 12a - 3b + 4c \\ \underline{-\ (7a + \ \ b - 2c)} \end{array}$$

Change the signs and add.

$$\begin{array}{rr} & 12a - 3b + 4c \\ + & -7a - \ \ b + 2c \\ \hline & 5a - 4b + 6c \end{array}$$

To do the same problem horizontally, change the signs as you remove the parentheses, then combine like terms.

$$\begin{aligned} (12a - 3b + 4c) - (7a + b - 2c) &= 12a - 3b + 4c - 7a - b + 2c \\ &= 12a - 7a - 3b - b + 4c + 2c \\ &= 5a - 4b + 6c \end{aligned}$$

EXAMPLE Simplify: $(2x - 3y) - (4y + 7x) - (3x - 8y)$

SOLUTION

$$\begin{aligned} (2x - 3y) - (4y + 7x) - (3x - 8y) &= 2x - 3y - 4y - 7x - 3x + 8y \\ &= 2x - 7x - 3x - 3y - 4y + 8y \\ &= -8x + y \end{aligned}$$

ORAL EXERCISES

For each of the following polynomials state

a. the coefficient of the x^2 term,
b. the coefficient of the x term,
c. the exponent in the first term,
d. the exponent in the last term,
e. the degree of the polynomial.

1. $12x^7 - 3x^4 + 7x^2 - 3x$
2. $x^3 + 8x^2 - x + x^4$
3. $5x^8 - 6 + 4x + x^2$
4. $-x^2 + x^3 - x^3 + x$
5. $3x - 4x^2 + 6x^3$
6. $y^3 - 2y + 3 - x + x^2$

Identify each of the following as a monomial, binomial, trinomial or none of these.

7. $x^4 - 3x^2 + x^3 - 2x$
8. $x^4 + 6$
9. $149x^2y^3z^{10}$
10. $x^2y - xy^2 + y^3$
11. $\frac{3}{x^2} + \frac{2}{x}$
12. $x^2 - 3x^2yz$

Simplify.

13. $3a + 4b - 5a - 2b$
14. $7x + 8y - 6y - 2x$
15. $-x + z - 3x + y$
16. $2x^2 - 1 + x - 3 + 4x$
17. $9 - 2t - 4 + 7t + 10$
18. $x - y + xy + x$

WRITTEN EXERCISES

Simplify.

A

1. $2x - 3y + y - x$
2. $4a - 2b - 3a + 5b$
3. $7m + 8n + 3m - 8n$
4. $5p^2 + 2p - 3p - 4p^2$
5. $2 - 4y^2 + 6 + 5y$
6. $7t - 4v - 10t + 6v$
7. $7x^2 - 8x + 2x^2 - 5x$
8. $4z - 2y - 4z - 9y$
9. $12 - 4b - 6 + 5b$
10. $9xy^2 + 12x - 3x - 4x^2y$
11. $(17x - 21) + (3x + 19)$
12. $(21x + 15) + (13x - 24)$
13. $(2a + 3b) + (4a - 7b)$
14. $(7x - 3y) + (12x + 11y)$
15. $(a + b^2 - c) + (3a - 4b^2 + c) + (7a + 3b^2 - 3c)$
16. $(a^2 - 2b) + (3b - 2c) + (4a + 8c)$
17. $(2x^2 - 3y) + (3x^2 + 4z) + (7y^2 - 9z)$
18. $(2x^2y - 3x^2 + 2y) + (5yz^2 + 9x^2 + 11y)$
19. $(7x^2 - 3xy + 4y^2) + (6x^2 - 12xy + 17y^2)$
20. $(19a^2 - 13ab + 10b^2) + (3a^2 - 18ab - 9b^2)$
21. $(2x^2 - 4y^3) - (3x^2 + 7y^3)$
22. $(a + 2b) - (7a - 3b)$
23. $(7ab + 2c) - (6a - 2ba - 4c)$
24. $(2x - 4y + 6z) - (8x - 11y + 13z)$
25. $(14a - 16c) - (19b + 12d)$
26. $(a^2 - 2ab + b^2) - (2a^2 - 3b^2)$
27. $28z - (17x + 19y - 13z)$
28. $8 - (2x^2 - 3x + 3)$
29. $(3x^2 - 2xy + 7y^2) - (7x^2 + 11xy - 11y^2)$
30. $(6 + 12x^2 - 13x^3) - (2x^3 + 19x - 15)$

Supply the missing exponent(s) or variable(s) to make a pair of similar terms.

31. $3x^2y^3$; $12x^2y^?$
32. $-5ab^3c^5$; $b^? \underline{\ ?\ } c^5$
33. $2r^4s^6$; $3\underline{\ ?\ }^4s^?$
34. $-8m^2n$; $\underline{\ ?\ }m^2$
35. $3q^5s^2$; $-2q^?s^?$
36. p^6t^3; $2\underline{\ ?\ }^3\underline{\ ?\ }^6$

Simplify.

B

37. $(42c^3 - 26c^2 + 38c - 18) + (35c^2 + 16c^3 - 49c - 19)$
38. $(43x^2y + 16xy^2 - 93y^3) - (12x^3 + 14xy^2 - 100y^3)$
39. $(4x^3 - 12x^2 + 17x - 23) + (3x^2 - 11x^3 - 12 - 24x)$
40. $(12a + 16b - 3c) + (21a + 32c - 19b) + (24b - 39c)$
41. $(75a^3 - 49a^2 + 61a - 101) - (80a + 24a^2 + 210)$
42. $(21x^2 - 32xy - 64y^2) + (19x^2 + 18xy - 50)$
43. $(53p^3 + 27p^2q - 39pq^2 + 18) - (-16p^3 - 39pq^2 - 27p^2q - 31)$
44. $(193x^2y + 232y^2 + 64xy) - (72xy - 123xy^2 + 300y^2)$

Simplify.

45. $4 - (3x + 1) + (2x - 6)$

46. $7y - (3x + 2y) - (8x - 3y)$

47. $12a + b - (3a - 4b) - (5a + b)$

48. $13 - [6x - (x + 8)]$

49. $(17x - 2y) - [x - (y - x)]$

50. $9k - [6 - (3k - 1)]$

C 51. $[12 - (3x + 4)] - [7 + (8 - 9x)]$

52. $(9 - x) - [12 - (3 - 4x) + (8 - 2x)]$

53. $7 - 3x^2 - (12 - [4 - (3x^2 + 1) + 2] - 3x^2)$

54. $12 - 3y - (6 - 9y - [8 - (4y + 3)])$

Section 6-2 PRODUCTS OF POLYNOMIALS

In this section you will review:

- three laws of exponents for multiplication
- multiplying two polynomials
- the FOIL method for multiplying two binomials

As you know, $x^2 = x \cdot x$ and $x^4 = x \cdot x \cdot x \cdot x$. Thus,

$$x^2 \cdot x^4 = (x \cdot x) \cdot (x \cdot x \cdot x \cdot x) = x^6 = x^{2+4}$$

Similarly,

$$y^3 \cdot y^2 = (y \cdot y \cdot y) \cdot (y \cdot y) = y^5 = y^{3+2}$$

and

$$m \cdot m^4 = m \cdot (m \cdot m \cdot m \cdot m) = m^5 = m^{1+4}$$

The results of these examples are stated in the first law of exponents.

> If a and b are positive integers, then
> $$x^a \cdot x^b = x^{a+b}.$$

EXAMPLE Simplify: $(2x^4)(3x^2)$

SOLUTION $(2x^4)(3x^2) = 2 \cdot 3 \cdot x^4 \cdot x^2$
$= 6x^6$

EXAMPLE Simplify: $(-8a)(3b^3)(5a^2b)$

SOLUTION $(-8a)(3b^3)(5a^2b) = -8 \cdot 3 \cdot 5 \cdot a \cdot a^2 \cdot b^3 \cdot b$
$= -120a^3b^4$

You could apply the first law of exponents to simplify products such as $(z^3)^4$.

$$(z^3)^4 = z^3 \cdot z^3 \cdot z^3 \cdot z^3 = z^{3+3+3+3} = z^{12}$$

It is easier, however, to use the following law of exponents to simplify this type of product.

> If a and b are positive integers, then
> $$(x^a)^b = x^{ab}.$$

Applying this law to the previous example, you get

$$(z^3)^4 = z^{3 \cdot 4} = z^{12}.$$

You can apply the definition of an exponent to simplify products such as $(xy)^4$.

$$(xy)^4 = xy \cdot xy \cdot xy \cdot xy$$
$$= (x \cdot x \cdot x \cdot x) \cdot (y \cdot y \cdot y \cdot y) = x^4y^4$$

Similarly, $(2y)^3 = (2y)(2y)(2y) = 2^3y^3 = 8y^3$.

The results of these examples suggest the third law of exponents.

> If a is a positive integer, then
> $$(xy)^a = x^a y^a.$$

EXAMPLE Simplify: $(4y^3)^2$

SOLUTION $(4y^3)^2 = 4^2 \cdot (y^3)^2 = 16y^6$

To multiply a polynomial by a monomial, use the distributive property to multiply each term of the polynomial by the monomial.

EXAMPLE Simplify: $2x(3x^3 - 4x^2 + x - 8)$

SOLUTION
$$2x(3x^3 - 4x^2 + x - 8) = 2x(3x^3) + 2x(-4x^2) + 2x(x) + 2x(-8)$$
$$= 6x^4 - 8x^3 + 2x^2 - 16x$$

The distributive property is used to multiply two polynomials.

EXAMPLE Simplify: $(x + 4)(x^2 - 2x + 3)$

SOLUTION By the distributive property,
$$(x + 4)(x^2 - 2x + 3) = x(x^2 - 2x + 3) + 4(x^2 - 2x + 3)$$
$$= x(x^2) + x(-2x) + x(3) + 4(x^2) + 4(-2x) + 4(3)$$
$$= x^3 - 2x^2 + 3x + 4x^2 - 8x + 12$$
$$= x^3 + 2x^2 - 5x + 12$$

Multiplication of two polynomials can be organized in vertical form as illustrated below.

EXAMPLE Simplify: $(x^2 - 3x + 4)(x - 2)$

SOLUTION

$$\begin{array}{rcr} & & x^2 - 3x + 4 \\ & & x - 2 \\ \hline x(x^2 - 3x + 4) & \longrightarrow & x^3 - 3x^2 + 4x \\ -2(x^2 - 3x + 4) & \longrightarrow & -2x^2 + 6x - 8 \\ \hline (x - 2)(x^2 - 3x + 4) & \longrightarrow & x^3 - 5x^2 + 10x - 8 \end{array}$$

EXAMPLE Simplify: $(2x^3 - x + 5)(x^2 - 3)$

SOLUTION As you multiply, pay careful attention to the exponents and remember that you can combine *like terms only.*

$$\begin{array}{r} 2x^3 - x + 5 \\ x^2 - 3 \\ \hline 2x^5 - x^3 + 5x^2 \qquad\qquad \\ -6x^3 \qquad\quad + 3x - 15 \\ \hline 2x^5 - 7x^3 + 5x^2 + 3x - 15 \end{array}$$

Because you will need to multiply binomials so often, it will help to be able to write such products at sight. The **FOIL** method, illustrated below, may help shorten your work.

By the distributive property,

$$\begin{aligned} (x - 2)(x + 6) &= x(x + 6) - 2(x + 6) \\ &= x^2 + 6x - 2x - 12 \\ &= x^2 + 4x - 12 \end{aligned}$$

$(x - 2)(x + 6)$

F x is the **F**irst term of each binomial. $x \cdot x = x^2$

$(x - 2)(x + 6)$

O x and 6 are the **O**uter terms. $x \cdot 6 = 6x$

$(x - 2)(x + 6)$

I -2 and x are the **I**nner terms. $-2 \cdot x = -2x$

$(x - 2)(x + 6)$

L -2 and 6 are the **L**ast terms. $-2 \cdot 6 = -12$

$(x - 2)(x + 6) \quad = \quad x^2 + 6x - 2x - 12$

As a final step, combine the outer-term product and the inner-term product, $6x$ and $-2x$, to obtain $4x$. Thus, the simplified product is $x^2 + 4x - 12$.

EXAMPLE Simplify: $(3x + 4)(2x + 5)$

SOLUTION First terms: $3x \cdot 2x = 6x^2$ **F**

Outer plus Inner products: $15x + 8x = 23x$ **O + I**

Last terms: $4 \cdot 5 = 20$ **L**

$(3x + 4)(2x + 5) = 6x^2 + 23x + 20$

ORAL EXERCISES

Simplify.

1. $x^4 \cdot x^9$
2. $a \cdot a^8$
3. $y \cdot y^2$
4. $x \cdot x^2 \cdot y \cdot y^2$
5. $(x^3)^5$
6. $(y^2)^2$
7. $(xy)^5$
8. $(ab)^3$

Complete.

9. $(t + 3)(t + 10) = t^2 + \underline{\ ?\ }\,t + 30$
10. $(x - 2)(x + 7) = x^2 + 5x - \underline{\ ?\ }$
11. $(m + 5)(m - 9) = m^2 - \underline{\ ?\ }\,m - \underline{\ ?\ }$
12. $(3z - 2)(4z - 3) = 12z^2 - \underline{\ ?\ }\,z + \underline{\ ?\ }$
13. $(2r - 3)(2r + 3) = 4r^2 - \underline{\ ?\ }$
14. $(7x - 5)(7x - 5) = 49x^2 - \underline{\ ?\ }\,x + \underline{\ ?\ }$
15. $(s + 2t)(2s - t) = 2s^2 + \underline{\ ?\ }\,st - 2t^2$
16. $(2a + 5b)(3a + b) = 6a^2 + \underline{\ ?\ }\,ab + \underline{\ ?\ }\,b^2$

WRITTEN EXERCISES

Simplify.

A

1. $3x^2 \cdot x^3$
2. $2y^2 \cdot y^8$
3. $(2z^2)^4$
4. $(-x^3)^4$
5. $(-x^2)^3$
6. $x \cdot x^2 \cdot x^2$
7. $(y^2)^3 y$
8. $z(z^3)^4$

Use the FOIL method to simplify each product.

9. $(y + 1)(y - 8)$
10. $(m + 5)(m + 8)$
11. $(q + 2)(q + 9)$
12. $(y - 4)^2$
13. $(r - 6)(r + 5)$
14. $(6 - t)(6 + t)$
15. $(v + 9)(v - 9)$
16. $(3x + 2)(3x - 1)$
17. $(x - 5)(5x - 2)$
18. $(3y - 4)(3y + 4)$
19. $(7p + 2)^2$
20. $(2q - 3r)^2$
21. $(y - 4)(y + 4)$
22. $(2u + 3)(3u - 1)$
23. $(4z + 3)(z - 2)$
24. $(x + 8y)(x + 9y)$
25. $(y + 2z)^2$
26. $(y - 6)(6y - 1)$
27. $(5a - 3b)(a - b)$
28. $(2x - 5)(3x - 4)$
29. $(3h + 1)(4h + 5)$
30. $(4n - 3)(2n + 1)$
31. $(y - 7z)(y + z)$
32. $(6c - 5f)(2c - f)$

Simplify.

33. $(y^2 - 2)(y + 1)$
34. $(3t + 2w)(2t - 5w)$
35. $2x(x^2 - 3x + 4)$
36. $3x^2(x^3 - 2x^2 + 7)$
37. $4a^3(7a^2 + 9a - 1)$
38. $xy^2(2x^2 + 3xy - 2y^2)$
39. $6a^2bc(a^2 - 3ab + 4b^2)$
40. $xyz(x^2yz^3 - 3xy^2z^2 + 6y^4)$
41. $4(a^2 - 3a + 4) - 3a(2a + 7)$
42. $6(b^2 - 3b + 2) + 3b(2b - 5)$

Represent the area of each large rectangle as the product of two polynomials.

43. 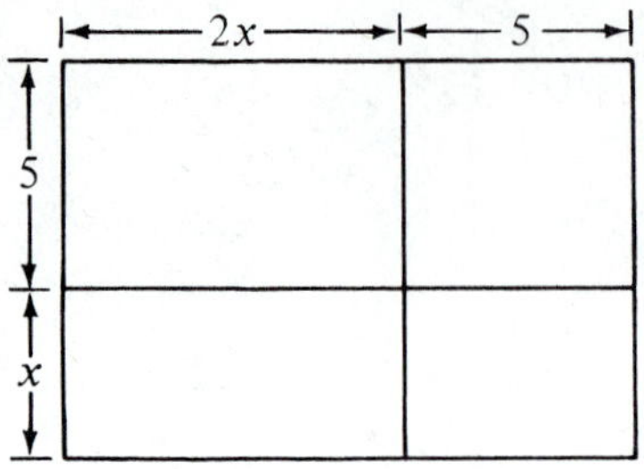

44. 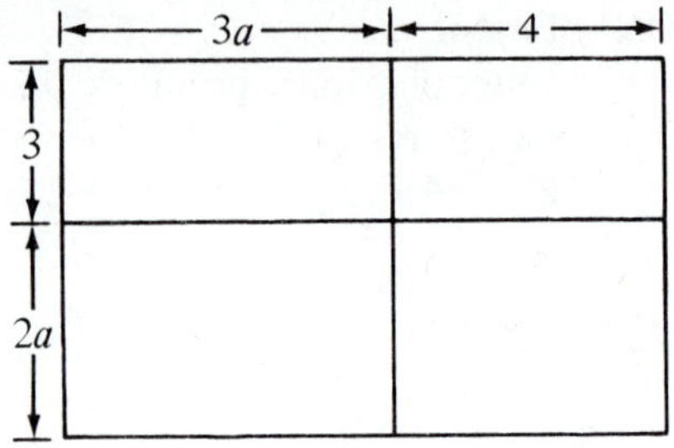

Multiply.

45. $\begin{array}{r} 3x - 5 \\ \underline{x + 2} \end{array}$

46. $\begin{array}{r} 2y^2 - 5 \\ \underline{y \ + 2} \end{array}$

47. $\begin{array}{r} 3m^2 - 7m + 1 \\ \underline{m \ + 3} \end{array}$

48. $\begin{array}{r} x^2 + 2x - 1 \\ \underline{x \ - 4} \end{array}$

49. $\begin{array}{r} a^2 - 2a + 1 \\ \underline{a^2 + 2} \end{array}$

50. $\begin{array}{r} 3y^2 - 4y + 5 \\ \underline{2y \ - 3} \end{array}$

B 51. $\begin{array}{r} x^2 + 3xy + y^2 \\ \underline{2x \ - y} \end{array}$

52. $\begin{array}{r} 2b^2 + 7b - 3 \\ \underline{4b \ - 5} \end{array}$

53. $\begin{array}{r} 5z^2 - 3z - 2 \\ \underline{2z^2 + 1} \end{array}$

54. $\begin{array}{r} -3x^2 + 4x + 1 \\ \underline{2x^2 + \ x} \end{array}$

55. $\begin{array}{r} 5y^2 - 2y + 4 \\ \underline{y^2 + 3y + 1} \end{array}$

56. $\begin{array}{r} 4t^2 + 5t - 2 \\ \underline{2t^2 - \ t + 3} \end{array}$

Simplify.

57. $(2xy^2)^3$

58. $2(xy^2)^3$

59. $(-4ab^2)^2$

60. $-4(ab^2)^2$

61. $2y^2(3y)^3$

62. $(5x)^2(-2y)^3$

63. $(2m)^3(6n)^2$

64. $(-4r^2s)^2 3s^4$

65. $2a^2(3a^2b)^3(-ab)$

66. $(s + 7)^2 - 10s$

67. $(x + 4)(x - 5) - 2(x - 7)$

68. $(a - 6)(a + 3) + 2a^2$

69. $(x + y)(x - y) + 2(x^2 + y^2)$

70. $4(x - 6)(x + 4)$

71. $-2(y - 3)(y - 5)$

72. $-5(2b + 1)^2$

73. $-3t^2(4t - 1)(4t + 1)$

Find the area of each rectangle in terms of x.

74. 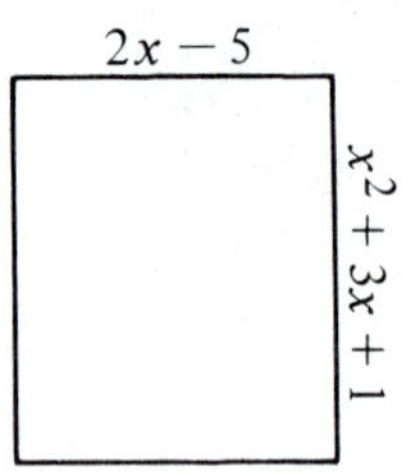

75. 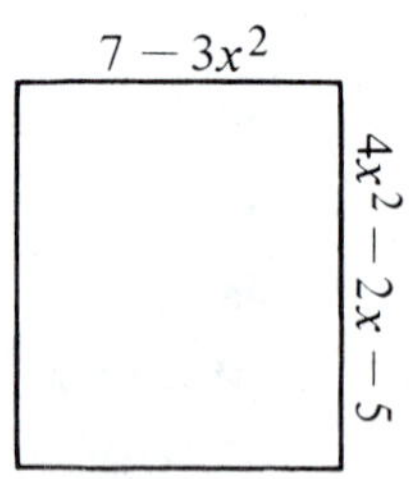

76. 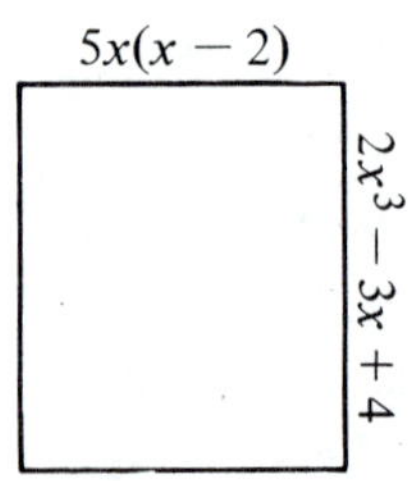

Find the area of the shaded region in terms of x.

C 77.

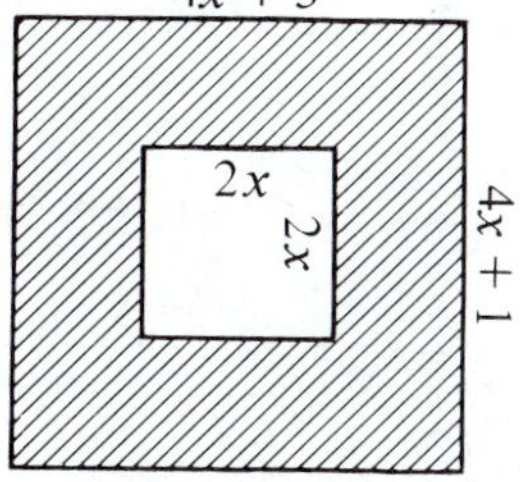

78.

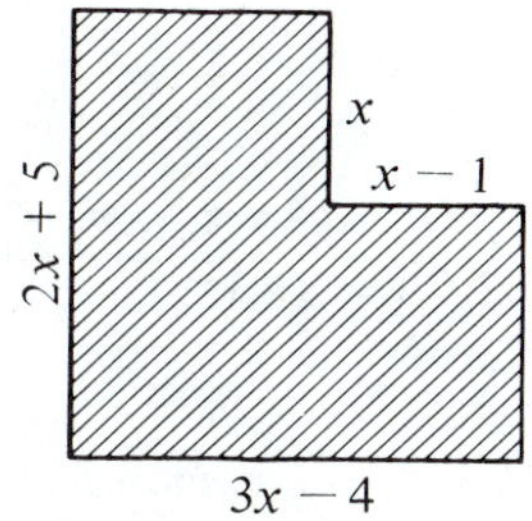

79.

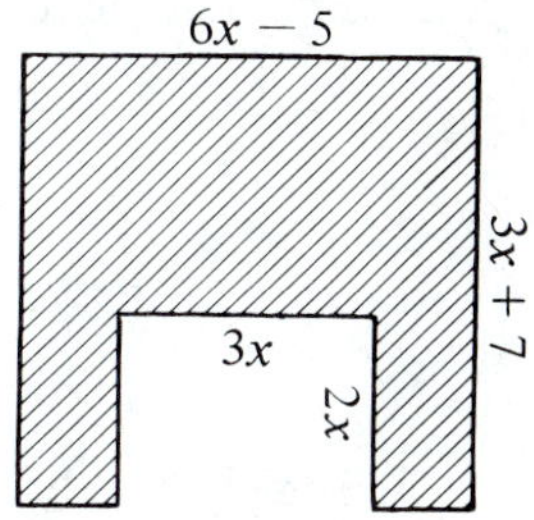

Simplify.

80. $3(x + 3)(x - 2) - (x + 4)(x - 1)$

81. $4(n - 3)(n + 1) - 2(n - 4)(n - 5)$

82. $(y + 1)^2 - 2(y - 2)^2$

83. $(4a + 1)(3a - 2) + (3 - 2a)(4a + 1)$

84. $-(2m + 3)(3m - 4) + (7m - 6)(m + 2)$

85. $5(r - 3)(r + 3) - 2(3r + 4)(r + 5)$

86. $6(2a - 1)^2 + 2(3a - 2)^2$

87. $x^2(x - 4)^2 + x(2 - x)^2$

SELF-TEST 1

Simplify and state the degree of the simplified polynomial.

Section 6-1

1. $(5a - 3b + 7ab) + (5ab - 7a - 2b^2)$
2. $(9x^2 - xy - y^2) - (-2y^2 + xy - x^2)$
3. $(4d^3 + 7d - 8) + (-d^2 - 7d + 6)$
4. $(4d^3 + 7d - 8) - (-d^2 - 7d + 6)$

Simplify.

Section 6-2

5. $(-3x^2y)^3$
6. $-4k^2(3k^2 - 2k + 1)$
7. $(2g + 7h)(3g - 5h)$
8. $(3m - 2)(m^2 - 2m - 3)$

Section 6-3 FACTORING POLYNOMIALS

In this section you will review:

- factors
- prime polynomials
- how to factor polynomials with integral coefficients

When you express an integer as a product of integers, you are *factoring* the number. For example, here are several ways to factor 12.

$$12 = 12 \cdot 1 \qquad 12 = 2 \cdot 6 \qquad 12 = (-3)(-4)$$

An integer greater than 1 which has only two positive factors, itself and 1, is called a *prime number.* The first five prime numbers are 2, 3, 5, 7, and 11.

When you express an integer as a product of prime numbers, you are writing the *prime factorization* of the number.

EXAMPLES

$$12 = 2 \cdot 2 \cdot 3 = 2^2 \cdot 3$$
$$18 = 2 \cdot 3 \cdot 3 = 2 \cdot 3^2$$
$$30 = 2 \cdot 3 \cdot 5$$

Notice in the examples above that 2 and 3 are *common factors* of 12, 18, and 30. The *greatest common factor* (GCF) of a set of numbers is the greatest positive integer that will divide each member of the set evenly. The greatest common factor of 12, 18, and 30 is $2 \cdot 3$, or 6.

When you write a polynomial as the product of other polynomials, you are factoring the polynomial. The principal uses of factoring polynomials are to simplify algebraic expressions and to solve certain types of equations. A polynomial with integral coefficients is *prime* if it has no factors except itself, its opposite, 1, and -1. (In this section we will work with polynomials with integral coefficients only.) The factorization of a polynomial is *complete* when the polynomial is expressed as the product of a monomial and one or more prime polynomials.

In general, it is not easy to factor polynomials of high degree, but the discussion of factoring in this and the next section should help you. The first step in factoring a polynomial is to look for a possible *common monomial factor* of the terms.

EXAMPLE Factor: $x^3y^2 - 3xy$

SOLUTION Examine each term and look for the greatest common monomial factor. In this case, it is xy. Applying the distributive property, you have:

$$x^3y^2 - 3xy = xy(x^2y - 3)$$

EXAMPLE Factor: $4a^3b^2 - 8a^3bc + 12a^2b^2c$

SOLUTION Examining each term of the expression you see that the greatest common monomial factor of $4a^3b^2$, $8a^3bc$, and $12a^2b^2c$ is $4a^2b$. The expression in factored form is:

$$4a^2b(ab - 2ac + 3bc)$$

A *quadratic polynomial in x* is one of the form $ax^2 + bx + c$, where $a \neq 0$. The term of degree 2 is the *quadratic term*. The term of first degree is the *linear term*, and the numerical term is the *constant term*.

EXAMPLES

Quadratic Polynomial	*Quadratic Term*	*Linear Term*	*Constant Term*
$x^2 - 3x - 18$	x^2	$-3x$	-18
$3y^2 - 2y$	$3y^2$	$-2y$	0
$4w^2$	$4w^2$	none	0

Sometimes trinomials of the form $x^2 + bx + c$ can be factored as the product of two binomials. You use the fact that

$$(x + m)(x + n) = x^2 + (m + n)x + mn.$$

So, to factor $x^2 + bx + c$, you must find two integers, m and n, such that

$$m + n = b \quad \text{and} \quad mn = c.$$

EXAMPLE Factor: $x^2 + 5x + 6$

SOLUTION Find two integers m and n such that

$$m + n = 5 \quad \text{and} \quad mn = 6.$$

If you cannot immediately think of two such numbers, write down all the possible pairs of factors of 6, check their sums and products, and then pick the two whose sum is 5. The pairs of positive factors of 6 are 1 and 6 and 2 and 3.

$$2 + 3 = 5 \quad \text{and} \quad 2(3) = 6$$

Therefore, $m = 3$ and $n = 2$ (or $m = 2$ and $n = 3$).

$$x^2 + 5x + 6 = (x + 3)(x + 2)$$

Check your answer by multiplying out the right-hand side of the equation.

EXAMPLE Factor: $x^2 - 5x - 24$

SOLUTION Find two integers m and n such that $m + n = -5$ and $mn = -24$.

$$m = -8 \quad \text{and} \quad n = 3$$

$$x^2 - 5x - 24 = (x - 8)(x + 3)$$

How do you factor a quadratic polynomial like $6x^2 + x - 12$ in which the coefficient of the quadratic term is not 1? Let's begin the analysis of this problem by recalling how to multiply two binomials.

$$\begin{aligned}(3x + 2)(5x + 3) &= 3x(5x + 3) + 2(5x + 3) && (1)\\ &= (15x^2 + 9x) + (10x + 6) && (2)\\ &= 15x^2 + 19x + 6 && (3)\end{aligned}$$

Reversing the process,

$$\begin{aligned}15x^2 + 19x + 6 &= 15x^2 + 9x + 10x + 6 && (4)\\ &= 3x(5x + 3) + 2(5x + 3) && (5)\\ &= (3x + 2)(5x + 3) && (6)\end{aligned}$$

In (4), we knew that we should express the linear term, $19x$, as the sum of $9x$ and $10x$ because we wanted to duplicate line (2) above in order to reverse the process. But is there some other way of knowing how to split up the linear term? Note that the product of 9 and 10 equals the product of the coefficient of the quadratic term and the constant term: $9 \cdot 10 = 15 \cdot 6$.

Let's look at another example.

$$\begin{aligned} 20x^2 + 31x + 12 &= 20x^2 + (15x + 16x) + 12 \\ &= (20x^2 + 15x) + (16x + 12) \\ &= 5x(4x + 3) + 4(4x + 3) \\ &= (5x + 4)(4x + 3) \end{aligned}$$

Here we split the linear term $31x$ into the sum of $15x$ and $16x$. How did we choose those particular numbers? We found the product of the coefficient of the quadratic term and the constant term, $20 \cdot 12 = 240$, and looked for the factors of 240 whose sum was 31, the coefficient of the linear term. Since $15 \cdot 16 = 240$ and $15 + 16 = 31$, 15 and 16 are the desired numbers.

EXAMPLE Factor: $6x^2 + x - 12$

SOLUTION

1. The product of the coefficients of the quadratic term and the constant term is $6(-12) = -72$.
2. Find the factors of -72 whose sum is 1, the coefficient of the linear term.
$$9(-8) = -72 \quad \text{and} \quad 9 + (-8) = 1$$
3. Express the linear term as the sum of $9x$ and $(-8x)$.
$$\begin{aligned} 6x^2 + x - 12 &= 6x^2 + 9x + (-8x) - 12 \\ &= (6x^2 + 9x) - (8x + 12) \\ &= 3x(2x + 3) - 4(2x + 3) \\ &= (3x - 4)(2x + 3) \end{aligned}$$

The following example illustrates another popular method of factoring polynomials.

EXAMPLE Factor: $6x^2 - 17x + 5$

SOLUTION Possible ways to factor $6x^2$ are $6x \cdot x$ and $2x \cdot 3x$. The constant term, 5, may be factored as $1 \cdot 5$ or $(-1)(-5)$. In this method you must consider all arrangements of the factors of the constant term:

$$\begin{array}{cc} 1 \cdot 5 & (-1)(-5) \\ 5 \cdot 1 & (-5)(-1) \end{array}$$

Then the possible factorizations of $6x^2 - 17x + 5$ are:

$$\begin{array}{ll} (6x + 1)(x + 5) & (2x + 1)(3x + 5) \\ (6x + 5)(x + 1) & (2x + 5)(3x + 1) \\ (6x - 1)(x - 5) & (2x - 1)(3x - 5) \\ (6x - 5)(x - 1) & (2x - 5)(3x - 1) \end{array}$$

Choose the factorization that has $-17x$ as the sum of the inner and outer products:

$$6x^2 - 17x + 5 = (2x - 5)(3x - 1)$$

Note: If you observe that the linear term has a *negative* coefficient, it is unnecessary to write down the first two rows of possible factors because, in each case, you can see that the sum of the inner and outer products will always be *positive.*

EXAMPLE Factor: $3x^2 + 7x - 2$

SOLUTION The factors of $3x^2$ are $3x$ and x. Next consider all possible arrangements of the factors of -2.

$$(-2)(1) \qquad (2)(-1)$$
$$(1)(-2) \qquad (-1)(2)$$

Then the possible factorizations of $3x^2 + 7x - 2$ are:

$$(3x - 2)(x + 1) \qquad (3x + 2)(x - 1)$$
$$(3x + 1)(x - 2) \qquad (3x - 1)(x + 2)$$

In none of these factorizations is $7x$ the sum of the inner and outer products. We therefore say that $3x^2 + 7x - 2$ is *not* factorable.

Determining the factors of a quadratic trinomial will become easier with practice.

No matter which method you use to factor quadratic polynomials, it is always wise to check your answer by multiplying the factors.

The following example incorporates both of the techniques you have studied in this section: factoring out the greatest common monomial and factoring a quadratic polynomial.

EXAMPLE Factor completely: $6y^3 - 27y^2 - 15y$

SOLUTION The GCF of $6y^3$, $27y^2$, and $15y$ is $3y$.

$$\begin{aligned} 6y^3 - 27y^2 - 15y &= 3y(2y^2 - 9y - 5) \\ &= 3y(2y^2 - 10y + y - 5) \\ &= 3y[2y(y - 5) + 1(y - 5)] \\ &= 3y(2y + 1)(y - 5) \end{aligned}$$

ORAL EXERCISES

Find the prime factorization of each number.

1. 18 **2.** 28 **3.** 15 **4.** 42 **5.** 36 **6.** 19

Find the GCF.

7. $2xy$ and 4

8. x^2 and xy

9. y^3 and y^2

10. $3m$ and $-6m^2$

11. $-x^2y$ and xy

12. $15a^2b^3$ and $6ab^2$

13. $8m^2$ and $16m$

14. $28xy^3z^2$ and $12x^2yz^3$

15. Find two numbers whose product is 24 and whose sum is 11.

16. Find two numbers whose product is 12 and whose sum is -7.

17. Find two numbers whose product is -56 and whose sum is 1.

18. Find two numbers whose product is -120 and whose sum is -2.

Complete.

19. $x^2 + 7x + 12 = (x + 3)(x + \underline{\ ?\ })$

20. $y^2 - y - 30 = (y - \underline{\ ?\ })(y + 5)$

21. $a^2 - 12a + 20 = (a - 2)(a - \underline{\ ?\ })$

22. $6x^2 + 7x - 3 = (2x + 3)(\underline{\ ?\ } - 1)$

23. $36b^2 - 25 = (6b - 5)(6b + \underline{\ ?\ })$

24. $8y^2 + 13y + \underline{\ ?\ } = (8y + 5)(y + \underline{\ ?\ })$

WRITTEN EXERCISES

Factor completely. If the polynomial is not factorable, so state.

A

1. $22x^2y^3 + 14xy$
2. $30a^4b^2 + 45ab^3$
3. $14r^2s - 21rst$
4. $54x^2y^3z - 18y^2z^2$
5. $m^2n - mn^3 + m^3n^4$
6. $6w^3x^2y - 9w^2x^3 - 3x^4y^2$
7. $12a^3b^2 - 6ab + 9a^3$
8. $8p^4q^2r^3 + 12p^4q^4r^2 - 10p^8q$
9. $-36x^2y^3 + 12y^6z^4 - 18x^3y^3$
10. $x^2 + 15x + 56$
11. $y^2 + 20y + 36$
12. $a^2 + 14a + 48$
13. $w^2 + 11w + 24$
14. $x^2 - 18x + 45$
15. $x^2 - 10x + 27$
16. $m^2 - 5m + 12$
17. $x^2 - 12x - 28$
18. $a^2 + 8a - 20$
19. $x^2 - x - 6$
20. $2k^2 - 9k + 9$
21. $3s^2 + 2s - 16$
22. $2a^2 + 11a + 12$
23. $2q^2 - 13q - 15$
24. $5m^2 - 7m - 6$

EXAMPLE Factor completely: $q^2 - 8qr + 7r^2$

SOLUTION Follow basically the same procedure you use to factor a quadratic polynomial in one variable. Find two integers m and n such that

$$m + n = -8 \quad \text{and} \quad mn = 7$$

In this case, $m = -7$ and $n = -1$.

$$q^2 - 8qr + 7r^2 = (q - 7r)(q - r)$$

25. $x^2 + xy - y^2$
26. $a^2 - 6ab - 16b^2$
27. $m^2 - 16mn + 48n^2$
28. $z^2 + 17wz + 42w^2$
29. $q^2 - 6qr - 72r^2$
30. $12c^2 + cd - d^2$
31. $21x^2 + 4xy - y^2$
32. $24v^2 + 5wv - w^2$
33. $g^2 - 11gh + 18h^2$
34. $p^2 + 10pq + 24q^2$
35. $2x^2 - 4xy - 3y^2$
36. $3m^2 - 14mn - 5n^2$

B

37. $7x^2 + 6x - 16$
38. $5x^2 - 6xy - 6y^2$
39. $6r^2 + 11r - 7$
40. $6w^2 - 11w + 3$
41. $12b^2 + 28b + 15$
42. $10w^2 + 23w + 12$
43. $16y^2 - 18y + 3$
44. $6z^2 - 7z - 10$
45. $4x^2 + 25x + 6$
46. $16m^2 - 14m + 3$
47. $5a^2 - 19ab + 12b^2$
48. $24x^2 - 5x - 36$
49. $8q^2 - 29q - 12$
50. $10y^2 + 29y + 12$
51. $15r^2 - 49rs + 24s^2$
52. $x(2 + y) - 3(2 + y)$
53. $a(b - 1) + 5(b - 1)$
54. $(2r + 3)s - (2r + 3)2$

Factor completely. (*Hint:* First find the greatest common monomial factor.)

55. $6x^2 + 18x - 24$ **56.** $x^3 - 3x^2 + 2x$ **57.** $y^3 + 11y^2 + 18y$

58. $2q^2 + qr - r^2$ **59.** $2q^2r^2 + qr^2 - 6r^2$ **60.** $2w^2 - 2w - 12$

61. $24a^2 + 20ab + 4b^2$ **62.** $x^2y^2 - 6x^2y + 5x^2$ **63.** $30x^3 + 2x^2 - 4x$

64. $w^2z^3 - 6w^2z^2 + 8w^2z$ **65.** $4z^2 - 4$ **66.** $y^3 - 4y$

Factor by grouping.

EXAMPLE Factor $xy - 3x + 4y - 12$.

SOLUTION This is not like the problems seen previously since there are four terms in the polynomial. To factor this polynomial you use a technique called *factoring by grouping* which is similar to the technique you learned for factoring quadratic polynomials. First group those terms having a common factor. Consider this possible grouping of the given terms:

$$(xy - 3x) + (4y - 12)$$

In the first group you have a common factor, x,

$$xy - 3x = x(y - 3)$$

and in the second group there is a common factor, 4,

$$4y - 12 = 4(y - 3)$$

so,

$$\begin{aligned} xy - 3x + 4y - 12 &= x(y - 3) + 4(y - 3) \\ &= (x + 4)(y - 3) \end{aligned}$$

C **67.** $xy + 2x + 3y + 6$ **68.** $xy - x + 2y - 2$ **69.** $uv - 4u - v + 4$

70. $2ab + 4a + b + 2$ **71.** $3xy - x + 3y - 1$ **72.** $2pq + 3p - 2q - 3$

73. $x^3 + x^2 + 5x + 5$ **74.** $y^3 + 3y^2 - 2y - 6$ **75.** $v^2w - 6 + 3w - 2v^2$

Section 6-4 FACTORING SPECIAL POLYNOMIALS

There are two types of quadratic polynomials and two types of third-degree polynomials that can be identified easily on sight. They are:

- trinomial squares
- the difference of two squares
- the sum of two cubes
- the difference of two cubes

A polynomial is a *trinomial square* if it can be written in either of the following forms:

$$a^2 + 2ab + b^2 = (a + b)^2 \qquad a^2 - 2ab + b^2 = (a - b)^2$$

EXAMPLES

$$x^2 + 6x + 9 = (x + 3)^2 \qquad y^2 - 2ay + a^2 = (y - a)^2$$

$$4w^2 + 28w + 49 = (2w + 7)^2 \qquad 16z^2 - 40z + 25 = (4z - 5)^2$$

A polynomial is called a *difference of two squares* if it is in this form:

$$a^2 - b^2 = (a + b)(a - b)$$

EXAMPLES

$$x^2 - 9 = (x + 3)(x - 3) \qquad 4y^2 - 25 = (2y + 5)(2y - 5)$$
$$36w^2 - b^2 = (6w + b)(6w - b) \qquad 49 - 100w^2 = (7 + 10w)(7 - 10w)$$

The patterns for factoring the *sum of two cubes* and the *difference of two cubes* are shown below.

$a^3 + b^3 = (a + b)(a^2 - ab + b^2)$ Sum of two cubes

$a^3 - b^3 = (a - b)(a^2 + ab + b^2)$ Difference of two cubes

EXAMPLES

$$\begin{aligned} x^3 + 27 &= (x + 3)(x^2 - 3x + 9) \\ 8y^3 + 1 &= (2y + 1)\big((2y)^2 - 2y \cdot 1 + 1^2\big) \\ &= (2y + 1)(4y^2 - 2y + 1) \\ w^3 - 64 &= (w - 4)(w^2 + 4w + 16) \\ 27z^3 - 8 &= (3z - 2)\big((3z)^2 + 3z \cdot 2 + 2^2\big) \\ &= (3z - 2)(9z^2 + 6z + 4) \end{aligned}$$

Remember that the first step in factoring any polynomial is to look for a common monomial factor of its terms. You can then express the polynomial as a product of the greatest common monomial factor of its terms and another polynomial. Check the remaining polynomial factor to see if it is factorable.

EXAMPLE Factor: $2x^2 + 28x + 98$

SOLUTION
$$\begin{aligned} 2x^2 + 28x + 98 &= 2(x^2 + 14x + 49) \\ &= 2(x + 7)^2 \end{aligned}$$

EXAMPLE Factor: $75y^4 - 48y^2$

SOLUTION
$$\begin{aligned} 75y^4 - 48y^2 &= 3y^2(25y^2 - 16) \\ &= 3y^2(5y + 4)(5y - 4) \end{aligned}$$

EXAMPLE Factor: $128r^4 - 250r$

SOLUTION
$$\begin{aligned} 128r^4 - 250r &= 2r(64r^3 - 125) \\ &= 2r(4r - 5)\big((4r)^2 + (4r)5 + 5^2\big) \\ &= 2r(4r - 5)(16r^2 + 20r + 25) \end{aligned}$$

In the next example the parentheses indicate an easily factorable group of terms, a trinomial square. By factoring the polynomial in parentheses you can see that the original polynomial can be expressed as a difference of two squares.

EXAMPLE Factor: $x^2 + 6x + 9 - y^2$

SOLUTION
$$\begin{aligned} x^2 + 6x + 9 - y^2 &= (x^2 + 6x + 9) - y^2 \\ &= (x + 3)^2 - y^2 \\ &= \big((x + 3) + y\big)\big((x + 3) - y\big) \\ &= (x + y + 3)(x - y + 3) \end{aligned}$$

ORAL EXERCISES

Factor completely. If the polynomial is not factorable, so state.

1. $s^2 - t^2$ **2.** $x^2 + 2xy + y^2$ **3.** $c^3 + d^3$ **4.** $m^2 + n^2$

5. $8x^3 - z^3$ **6.** $b^2 - 8b + 16$ **7.** $m^2 + 6m - 9$ **8.** $a^2b^2 - 1$

WRITTEN EXERCISES

Factor completely. If the polynomial is not factorable, so state. (In Exercises 44 and 45, assume that n is a positive integer.)

A **1.** $25x^2 - 1$ **2.** $144 + y^2$ **3.** $16b^2 - 121$

4. $36 - 9w^2$ **5.** $r^2 - 10r + 25$ **6.** $81 - 18b + b^2$

7. $64 + 16x + x^2$ **8.** $9m^2 + 12m + 4$ **9.** $s^3 - 1$

10. $y^3 + 8$ **11.** $125k^3 + 27$ **12.** $64n^3 - 27$

13. $27c^3 + 100$ **14.** $8f^3 - 125$ **15.** $x^2y^2 - a^2$

16. $z^3 + c^3d^3$ **17.** $98h^2 - 28h + 2$ **18.** $2s^3 - 16$

B **19.** $100y^2 - 25z^2$ **20.** $16p^3 - 54q^3$ **21.** $169x^4 + 78x^2 + 9$

22. $y^3 - 11y^2m + 121ym^2$ **23.** $192a + 375az^3$ **24.** $32a^2 - 108a^2b^3$

25. $4rs^2 - 104rs + 676r$ **26.** $27ax^2 + 12a^3$ **27.** $c^2 - 2c + 1 - d^2$

28. $m^2 - 10m + 25 - n^2$ **29.** $p^2 - 4q^2 - 12q - 9$ **30.** $16g^2 - 9h^2 + 30h - 25$

C **31.** $m^4 - 1$ **32.** $16 - x^4$ **33.** $2y^4 - 162$

34. $y^4 - 8y^2 + 16$ **35.** $81m^4 + 18m^2 + 1$ **36.** $128a^2 - 54a^2b^6$

37. $x^3 + 2x^2 - 16x - 32$ **38.** $4y^3 + 20y^2 - y - 5$ **39.** $m^4 + 2m^3 - m - 2$

40. $x^2y - 4y + 2x^2 - 8$ **41.** $2 - 2b^3 + a - ab^3$ **42.** $(y^2 + 4y + 3)^2 - (y^2 - y - 12)^2$

43. $(a^2 - 8a + 12)^2 - (a^2 - a - 30)^2$ **44.** $x^{2n} - y^{4n}$ **45.** $t^{3n} + 216v^{9n}$

Section 6-5 DIVIDING POLYNOMIALS

In this section you will learn:

- laws of exponents for division
- how to divide a polynomial by a monomial
- how to divide a polynomial by a polynomial

The following examples illustrate one method of simplifying quotients of monomials. (Note: In this section we will assume that no divisor equals zero.)

$$\frac{x^3}{x^3} = \frac{x \cdot x \cdot x}{x \cdot x \cdot x} = \frac{x}{x} \cdot \frac{x}{x} \cdot \frac{x}{x} = 1$$

$$\frac{x^5}{x^2} = \frac{x^2 \cdot x^3}{x^2} = \frac{x^2}{x^2} \cdot x^3 = x^3$$

$$\frac{x^2}{x^4} = \frac{x^2}{x^2 \cdot x^2} = \frac{x^2}{x^2} \cdot \frac{1}{x^2} = \frac{1}{x^2}$$

These examples would seem to suggest a pattern for simplifying quotients of the form $\frac{x^a}{x^b}$, where a and b are positive integers.

> If $a = b$, then $\dfrac{x^a}{x^b} = 1$.
>
> If $a > b$, then $\dfrac{x^a}{x^b} = x^{a-b}$.
>
> If $a < b$, then $\dfrac{x^a}{x^b} = \dfrac{1}{x^{b-a}}$.

EXAMPLES Simplify.

1. $\dfrac{2y^2}{6y^4}$ 2. $\dfrac{x^2y^7}{x^6y^3}$ 3. $\dfrac{x^2yz^6}{x^2y^5z^5}$

SOLUTIONS

1. $\dfrac{2y^2}{6y^4} = \dfrac{2}{6} \cdot \dfrac{y^2}{y^4} = \dfrac{1}{3} \cdot \dfrac{1}{y^{4-2}} = \dfrac{1}{3y^2}$

2. $\dfrac{x^2y^7}{x^6y^3} = \dfrac{x^2}{x^6} \cdot \dfrac{y^7}{y^3} = \dfrac{1}{x^{6-2}} \cdot y^{7-3} = \dfrac{y^4}{x^4}$

3. $\dfrac{x^2yz^6}{x^2y^5z^5} = \dfrac{x^2}{x^2} \cdot \dfrac{y}{y^5} \cdot \dfrac{z^6}{z^5} = 1 \cdot \dfrac{1}{y^{5-1}} \cdot z^{6-5} = \dfrac{z}{y^4}$

Consider the following example.

$$\left(\frac{z}{x}\right)^3 = \frac{z}{x} \cdot \frac{z}{x} \cdot \frac{z}{x} = \frac{z \cdot z \cdot z}{x \cdot x \cdot x} = \frac{z^3}{x^3}$$

This example is an illustration of another law of exponents.

If a is a positive integer, then

$$\left(\frac{x}{y}\right)^a = \frac{x^a}{y^a}.$$

For example,

$$\left(\frac{2}{3}\right)^4 = \frac{2^4}{3^4} = \frac{16}{81}$$

$$\left(\frac{4x}{5y}\right)^3 = \frac{(4x)^3}{(5y)^3} = \frac{4^3x^3}{5^3y^3} = \frac{64x^3}{125y^3}$$

To divide a polynomial by a monomial, apply the distributive property.

EXAMPLE Express as a sum in simplest form: $\dfrac{x^2y - 4x^5y^2 + 7x^2y^5}{x^2y}$

SOLUTION
$$\frac{x^2y - 4x^5y^2 + 7x^2y^5}{x^2y} = \frac{x^2y}{x^2y} - \frac{4x^5y^2}{x^2y} + \frac{7x^2y^5}{x^2y}$$
$$= 1 - 4x^3y + 7y^4$$

EXAMPLE Express as a sum in simplest form: $\dfrac{12x^5 - 6x^3 + 1}{4x^2}$

SOLUTION
$$\frac{12x^5 - 6x^3 + 1}{4x^2} = \frac{12x^5}{4x^2} - \frac{6x^3}{4x^2} + \frac{1}{4x^2} = 3x^3 - \frac{3x}{2} + \frac{1}{4x^2}$$

Study the following equations.

$$\frac{86}{7} = 12 + \frac{2}{7} \quad \text{and} \quad 86 = 12 \cdot 7 + 2$$

$$\frac{29}{3} = 9 + \frac{2}{3} \quad \text{and} \quad 29 = 9 \cdot 3 + 2$$

These equations illustrate the *Division Algorithm:*

$$\frac{\text{Dividend}}{\text{Divisor}} = \text{Quotient} + \frac{\text{Remainder}}{\text{Divisor}}$$

or

$$\text{Dividend} = \text{Quotient} \cdot \text{Divisor} + \text{Remainder}$$

To divide one polynomial by another you use the Division Algorithm and a procedure very similar to division with whole numbers. Before beginning the division process, be sure that the dividend and divisor are written in *standard form,* that is, with terms in decreasing degree. Write your answer in the form

$$\frac{\text{Dividend}}{\text{Divisor}} = \text{Quotient} + \frac{\text{Remainder}}{\text{Divisor}}.$$

If the remainder is zero you express the result as

$$\frac{\text{Dividend}}{\text{Divisor}} = \text{Quotient}.$$

EXAMPLE Divide $15x^2 + 7x - 4$ by $3x - 1$.

SOLUTION

1. Divide the first term of the dividend by the first term of the divisor. ⟶
   $15x^2 \div 3x = 5x$

$$\begin{array}{r} 5x \\ 3x-1\overline{)\,15x^2+7x-4} \end{array}$$

2. Multiply the divisor by $5x$. ⟶
   $5x(3x - 1) = 15x^2 - 5x$

$$\begin{array}{r} 5x \\ 3x-1\overline{)\,15x^2+7x-4} \\ 15x^2-5x \end{array}$$

3. Subtract. ⟶
   $15x^2 + 7x - 4 - (15x^2 - 5x) = 12x - 4$

$$\begin{array}{r} 5x \\ 3x-1\overline{)\,15x^2+7x-4} \\ \underline{15x^2-5x} \\ 12x-4 \end{array}$$

4. Divide the first term of the divisor, $3x$, into the first term of the difference, $12x$. Repeat the process until the remainder is zero *or* until the degree of the remainder is less than the degree of the divisor.

$$\begin{array}{r} 5x+4 \\ 3x-1\overline{)\,15x^2+7x-4} \\ \underline{15x^2-5x} \\ 12x-4 \\ \underline{12x-4} \\ 0 \end{array} \quad \textbf{Remainder}$$

The result is written $\dfrac{15x^2 + 7x - 4}{3x - 1} = 5x + 4$.

Check Divisor · Quotient + Remainder = Dividend $\quad (3x - 1)(5x + 4) + 0 = 15x^2 + 7x - 4$

In the example above, we say that $3x - 1$ is a *factor* of $15x^2 + 7x - 4$ because when $15x^2 + 7x - 4$ is divided by $3x - 1$ the remainder is 0.

EXAMPLE Divide $4x^3 - 7x + 3$ by $2x + 3$.

SOLUTION Insert $0x^2$ as a placeholder and rewrite the dividend as $4x^3 + 0x^2 - 7x + 3$.

$$\begin{array}{r} 2x^2-3x+1 \\ 2x+3\overline{)\,4x^3+0x^2-7x+3} \\ \underline{4x^3+6x^2} \\ -6x^2-7x+3 \\ \underline{-6x^2-9x} \\ 2x+3 \\ \underline{2x+3} \\ 0 \end{array} \quad \textbf{Remainder}$$

Therefore, $\dfrac{4x^3 - 7x + 3}{2x + 3} = 2x^2 - 3x + 1$.

Check

$$\begin{aligned}(2x + 3)(2x^2 - 3x + 1) + 0 &= 2x(2x^2 - 3x + 1) + 3(2x^2 - 3x + 1) \\ &= 4x^3 - 6x^2 + 2x + 6x^2 - 9x + 3 \\ &= 4x^3 - 7x + 3\end{aligned}$$

EXAMPLE Divide $10x^2 + x^4 - 5x^3 - 10 - 4x$ by $x^2 - 2x$.

SOLUTION Rewrite the dividend with terms in descending degree.

$$\begin{array}{r} x^2 - 3x + 4 \\ x^2 - 2x \overline{)x^4 - 5x^3 + 10x^2 - 4x - 10} \\ \underline{x^4 - 2x^3 \qquad\qquad\qquad\quad} \\ -3x^3 + 10x^2 - 4x - 10 \\ \underline{-3x^3 + \ \ 6x^2 \qquad\qquad\ } \\ 4x^2 - 4x - 10 \\ \underline{4x^2 - 8x \qquad} \\ 4x - 10 \end{array} \quad \textbf{Remainder}$$

Therefore, $\dfrac{x^4 - 5x^3 + 10x^2 - 4x - 10}{x^2 - 2x} = x^2 - 3x + 4 + \dfrac{4x - 10}{x^2 - 2x}$.

Check

$$\begin{aligned} (x^2 - 2x)(x^2 - 3x + 4) + 4x - 10 &= x^2(x^2 - 3x + 4) - 2x(x^2 - 3x + 4) + 4x - 10 \\ &= x^4 - 3x^3 + 4x^2 - 2x^3 + 6x^2 - 8x + 4x - 10 \\ &= x^4 - 5x^3 + 10x^2 - 4x - 10 \end{aligned}$$

ORAL EXERCISES

Simplify.

1. $\dfrac{36ab}{9a}$
2. $\dfrac{-18x^2}{6x}$
3. $\dfrac{p^3q^2}{2pq^4}$
4. $\dfrac{-12x^3y^5}{-48x^3}$
5. $\dfrac{9w + 12z}{3}$
6. $\dfrac{24f^2 - 18f}{6f}$
7. $\dfrac{x^3 + 2x^2 + x}{x}$
8. $\dfrac{a^5 - a^4 + 3a^2}{a^2}$
9. $\left(-\dfrac{1}{4}\right)^2$
10. $\left(\dfrac{k}{2}\right)^4$
11. $\left(-\dfrac{2}{3}\right)^3$
12. $\dfrac{1}{4}\left(\dfrac{2}{5}\right)^2$

Express each dividend and divisor in standard form.

13. $(-19x - 21x^2 + 6 + 4x^3) \div (x - 6)$
14. $(-4 - 2x + 3x^2 + x^3) \div (x + 1)$
15. $(x^4 - 4x^2 + 10 + 4x^3 - 11x) \div (x + x^2 - 2)$

WRITTEN EXERCISES

Simplify.

A

1. $\dfrac{12x^2}{6x^3}$
2. $\dfrac{-14xy^2}{21y^3}$
3. $(-3ab^2)^3 \div 6(ab^2)^2$
4. $\left(\dfrac{b}{3}\right)^3 \div b^2$
5. $\dfrac{25x^5y^2z}{-15x^2z^2}$
6. $\dfrac{(2c^2d^3)^2}{8b^2c^4}$
7. $\dfrac{28(x + y)^3}{-12(x + y)^4}$
8. $\dfrac{8(2k - 1)^8}{32(2k - 1)^2}$
9. $\dfrac{-9r^2s^3t^4}{-12r^4s^3t}$
10. $\dfrac{16x^5y^3z}{10y^2z^4}$
11. $\dfrac{2x + 6y}{2}$
12. $\dfrac{18x - 12y}{-6}$
13. $(x^2y + 4x^3) \div x$
14. $(2z^4 - 3z^2) \div z^2$
15. $(4a^2b - 6ab^3) \div 12a^3$
16. $\dfrac{6x^2y^2 - 2xy^3 + 8y^4}{-2y^2}$
17. $\dfrac{3x^4y - 5x^3y^2 + 6x^2y^3}{xy}$
18. $\dfrac{27x^3y^4 - 15x^5y^2 + 3x^6y}{3x^3y}$

Divide the first polynomial by the second. Write your answer in the form of an equation.

19. $x^2 + 12x - 28;\ x - 2$

20. $a^2 + 6a - 72;\ a + 12$

21. $x^3 - 6x^2 + 13x - 12;\ x - 3$

22. $y^3 + 4y^2 - 5y + 42;\ y + 6$

23. $2m^3 - 3m^2 + 5m - 2;\ 2m - 1$

24. $3z^3 + 11z^2 + 3z - 2;\ 3z + 2$

25. $5t^3 + t^2 + t - 4;\ 5t - 4$

26. $2r^3 - r^2 - 12r - 9;\ 2r + 3$

27. $x^3 - 9x^2 + 10x + 1;\ x - 1$

28. $a^3 + 4a^2 - 16a + 4;\ a - 2$

29. $2y^3 - 11y^2 + 14y - 8;\ y - 4$

30. $3x^3 - 14x^2 + x - 4;\ 3x + 1$

31. $4z^3 + 4z^2 - z + 1;\ 2z + 1$

32. $2w^3 - 23w^2 + 21w - 5;\ 2w - 1$

33. $2a^3 + a^2 - 14a + 13;\ a + 3$

34. $12b^4 + 7b^3 - b^2 - 6b;\ 3b^2 - 2b$

B 35. $y^3 - 8y^2 + 15y - 18;\ y^2 - 2y + 3$

36. $2z^3 - 5z^2 - 11z - 4;\ z^2 - 3z - 4$

37. $w^3 + w^2 + w + 21;\ w^2 - 2w + 7$

38. $3x^3 + 11x^2 + 2x - 2;\ 2 + x^2 + 4x$

39. $5m^3 + 2m^2 - 15m + 6;\ m^2 - 3$

40. $3w^3 - 4w^2 + 3w - 4;\ w^2 + 1$

41. $6r^3 - 3r + 3;\ 3 + 3r$

42. $7x - 15x^2 + 4x^3 + 6;\ x - 3$

43. $6y^3 + 8y - 17y^2 + 8;\ 2y - 3$

44. $4q^2 + 3q^3 - 1;\ q + 1$

45. $2 - z - 4z^2 + 3z^3;\ z^2 - 2z + 1$

46. $x^4 - 13x^2 - x^3 + 11x - 2;\ x^2 - 4x + 1$

47. $-6y + 3y^3 - 6y^2 + y^4 + 8;\ y^2 + 3y - 4$

48. $6w^4 - w^2 + w^3 + 1;\ 2w^2 - w + 1$

C 49. $x^4 + x^3y - 2x^2y^2 + 3xy^3 + y^4;\ x - y$

50. $a^5 + a^4b + a^3b^2 + a^2b^3 - ab^4 - b^5;\ a + b$

51. $x^4 - y^4;\ x + y$

52. $m^4 - 3m^2;\ 2 + m^2$

53. $2x^4 - 3x^3;\ x^2 - 3$

54. $m + 5$ is a factor of $2m^3 + 16m^2 + 29m - k$. Find the value of k.

55. $x + 2$ is a factor of $3x^3 - 2x^2 + kx - 6$. Find the value of k.

56. $b - 3$ is a factor of $4b^3 - kb^2 + 23b - 15$. Find the value of k.

SELF-TEST 2

Factor completely.

Section 6-3

1. $21a^3b^2c^5 - 56a^5bc^3 - 35a^4b^3c^2$

2. $t^2 - 2t - 63$

3. $7b^2 - 9b + 2$

4. $6s^3 + 7s^2 - 5s$

Section 6-4

5. $1 + 125n^3$

6. $54w^2 - 6x^2$

7. $16d^2 - 88d + 121$

8. $j^3 - 8k^3$

Simplify.

Section 6-5

9. $\left(\dfrac{-3m}{5n^2}\right)^3$

10. $\dfrac{36x^3y(x-y)^2}{27x^4y(x-y)}$

11. $\dfrac{12z^5 - z^3 + 6z}{4z^3}$

Divide the first polynomial by the second. Write your answer in the form of an equation.

12. $2y^3 - 3y^2 - 8y - 3;\ y - 3$

13. $10x^3 + 21x^2 - 4;\ 5x - 2$

Section 6-6 SIMPLIFYING RATIONAL EXPRESSIONS

Now you will learn how to apply the laws of exponents and the methods of factoring polynomials to simplify some fairly complicated expressions.

A *rational expression* is any quotient of two polynomials. Some examples of rational expressions are:

$$\frac{3x}{2y} \qquad \frac{a^2 - 3a}{a^3 - 2a} \qquad \frac{q^2 - 2qr + r^2}{q + r}$$

A rational expression is *simplified* (or is in *lowest terms*) when the greatest common factor of the numerator and denominator is 1. You simplify a rational expression the same way you simplify a fraction. That is, you factor the numerator and denominator and then divide them by their greatest common factor.

In each example we will state restrictions on the values of the variable(s). Why? Consider a simple example.

Are the following expressions equal for all $x \in \mathcal{R}$?

$$\frac{(x+1)(x+2)}{x+1} \quad \text{and} \quad x+2$$

No. $\frac{(x+1)(x+2)}{x+1}$ is not defined for $x = -1$, whereas $x + 2$ is defined for all $x \in \mathcal{R}$. Thus we can state that

$$\frac{(x+1)(x+2)}{x+1} = x + 2 \text{ if and only if } x \neq -1.$$

In other words, we must restrict the domain of each variable to values that give us a nonzero denominator.

EXAMPLE Simplify: $\frac{4a^2 - 2a}{4a^2 - 1}$

SOLUTION Factor the numerator and denominator.

$$\frac{4a^2 - 2a}{4a^2 - 1} = \frac{2a(2a-1)}{(2a+1)(2a-1)}$$

Dividing the numerator and denominator by their greatest common factor, (GCF), $2a - 1$, you obtain

$$\frac{2a}{2a+1}, \quad a \neq \frac{1}{2} \quad \text{and} \quad a \neq -\frac{1}{2}.$$

You can use slant bars to indicate which factors you are cancelling. For example,

$$\frac{2x^2 + 6x}{x^3 - 9x} = \frac{2\cancel{x}\cancel{(x+3)}}{\cancel{x}\cancel{(x+3)}(x-3)} = \frac{2}{x-3}, \quad x \neq 3,\ x \neq -3,\ x \neq 0$$

It is most important to understand that you can cancel equal *factors* only. Here is a common error.

This is **INCORRECT.** $\dfrac{4a^2 - 2a}{4a^2 - 1} = \dfrac{\cancel{4a^2} - 2a}{\cancel{4a^2} - 1} = 2a$ This is **INCORRECT.**

The *term* $4a^2$ is *not* a *factor* of the numerator and the denominator. Therefore, you *cannot* cancel $4a^2$ from the numerator and denominator.

If you can recognize that $\dfrac{a-b}{b-a} = \dfrac{(-1)(b-a)}{b-a} = -1$ for $a \neq b$, it will save you time and difficulty when simplifying rational expressions.

EXAMPLE Simplify: $\dfrac{2-x}{2x^2 + 3x - 14}$

SOLUTION
$$\begin{aligned}\frac{2-x}{2x^2+3x-14} &= \frac{2-x}{(x-2)(2x+7)}\\ &= \frac{(-1)\cancel{(x-2)}}{\cancel{(x-2)}(2x+7)}\\ &= \frac{-1}{2x+7}, \quad x \neq 2, \quad x \neq -\frac{7}{2}\end{aligned}$$

EXAMPLE Simplify: $\dfrac{2xy^2 - 3x^2y}{6x^2 - 4xy}$

SOLUTION
$$\begin{aligned}\frac{2xy^2 - 3x^2y}{6x^2 - 4xy} &= \frac{xy(2y-3x)}{2x(3x-2y)}\\ &= \frac{\cancel{x}y(-1)\cancel{(3x-2y)}}{2\cancel{x}\cancel{(3x-2y)}}\\ &= \frac{(-1)y}{2}\\ &= -\frac{y}{2}, \quad x \neq 0 \text{ and } 3x \neq 2y\end{aligned}$$

ORAL EXERCISES

Find the GCF of the numerator and denominator and state all restrictions on the values of the variables.

1. $\dfrac{6xy}{12y}$
2. $\dfrac{15m^2}{5mn}$
3. $\dfrac{(x+1)(x-2)}{(x+1)}$
4. $\dfrac{4a-1}{4a}$
5. $\dfrac{30(c-4)}{12(c-4)}$
6. $\dfrac{m-5}{(m+5)(m-5)}$

7. $\dfrac{b-6}{(b+2)(6-b)}$

8. $\dfrac{2x(x+1)}{3x(2x+1)}$

9. $\dfrac{(p+2)^2}{p(p+2)^3}$

WRITTEN EXERCISES

Simplify. (Assume that no denominator is zero.)

A

1. $\dfrac{3x}{9x^2}$
2. $\dfrac{16a^2}{24ab}$
3. $\dfrac{28abc^2}{49a^3b}$
4. $\dfrac{2a(a+b)}{a^2(a+b)}$
5. $\dfrac{-7m(m+4)}{14m(m+4)}$
6. $\dfrac{2p-6q}{4p-12q}$
7. $\dfrac{y^2+y}{y}$
8. $\dfrac{k^2+7k}{10(k+7)}$
9. $\dfrac{4m^2+8m}{6m^3+12m^2}$
10. $\dfrac{3w^2+15w}{w^2}$
11. $\dfrac{4x-16x^2}{4x}$
12. $\dfrac{x-y}{2(y-x)}$
13. $\dfrac{2-n}{n^2-4}$
14. $\dfrac{x^2-25}{3x-15}$
15. $\dfrac{p^2+8p+16}{8(p+4)}$
16. $\dfrac{2y-6}{y^2-y-6}$
17. $\dfrac{q^2-8q+15}{q^2-3q}$
18. $\dfrac{4a+12}{a^2+6a+9}$
19. $\dfrac{h^2+4h+3}{h^2+2h+1}$
20. $\dfrac{n^2+8n+16}{n^2-16}$
21. $\dfrac{m^3+8}{m^2-4}$
22. $\dfrac{25-w^2}{125-w^3}$
23. $\dfrac{x^3-1}{2x^2+2x+2}$
24. $\dfrac{9+3b}{b^3+27}$

B

25. $\dfrac{c^3d+c^2d^2}{c^3d-c^2d^2}$
26. $\dfrac{4x^2y-4xy^2}{12x^3+6x^2y}$
27. $\dfrac{y^2+y-2}{y^2-4y+3}$
28. $\dfrac{x^2-1}{x^2+3x+2}$
29. $\dfrac{2b^2-9b-5}{4b^2-1}$
30. $\dfrac{12a+10}{6a^2-13a-15}$
31. $\dfrac{3z^2-48}{z^2-2z-8}$
32. $\dfrac{2w^2-72x^2}{w^2-12wx+36x^2}$
33. $\dfrac{2z^2-3z-5}{25-4z^2}$
34. $\dfrac{m^3+1}{m^2-m+1}$
35. $\dfrac{27k^3-8}{15k^2+8k-12}$
36. $\dfrac{16m^2+8m+1}{64m^3+1}$

C

37. $\dfrac{12x^3-2x^2-30x}{12x^3+4x^2-40x}$
38. $\dfrac{(3y^2-108)(y^3+2y^2-24y)}{y(y+6)^2(3y^2-30y+72)}$
39. $\dfrac{m^4-1}{m^3-m^2+m-1}$
40. $\dfrac{a^4-5a^2+4}{a^2-a-2}$

Simplify. (Assume that n is a positive integer and that no denominator is zero.)

41. $\dfrac{x^{2n}-y^{2n}}{x^{2n}-2x^ny^n+y^{2n}}$
42. $\dfrac{x^{6n}-y^{6n}}{x^{2n}-y^{2n}}$

Section 6-7 PRODUCTS AND QUOTIENTS OF RATIONAL EXPRESSIONS

The product of two rational numbers is defined as

$$\frac{a}{b} \cdot \frac{c}{d} = \frac{ac}{bd}, \quad b, d \neq 0$$

Finding the product of two or more rational expressions is based on the same property: multiply the numerators, multiply the denominators, and take the quotient of these products. For example,

$$\frac{x-3}{x} \cdot \frac{x+5}{x+2} = \frac{(x-3)(x+5)}{x(x+2)} = \frac{x^2+2x-15}{x^2+2x}, \quad x \neq 0, x \neq -2$$

To find the simplest form of the product of two or more rational expressions, factor each numerator and denominator and then divide them by their common factors. You may leave your answer in factored form.

EXAMPLE Simplify: $\frac{3xy}{7wy} \cdot \frac{14w^2}{18x}$

SOLUTION $$\frac{3xy}{7wy} \cdot \frac{14w^2}{18x} = \frac{\cancel{3} \cdot \cancel{x} \cdot \cancel{y}}{\cancel{7} \cdot \cancel{w} \cdot \cancel{y}} \cdot \frac{\cancel{2} \cdot \cancel{7} \cdot \cancel{w} \cdot w}{\cancel{2} \cdot \cancel{3} \cdot 3 \cdot \cancel{x}}$$

$$= \frac{w}{3}, \quad w \neq 0, x \neq 0, y \neq 0$$

From this point on we will assume that, for any rational expression in which the denominator does not simplify to zero, the domains of the variables exclude all values of the variables for which the denominator equals zero. Thus, it will not be necessary to indicate the restrictions on the variables.

EXAMPLE Simplify: $\frac{3x^2+3x}{6x^2-6x} \cdot \frac{7xy-7y}{5xy-10y}$

SOLUTION $$\frac{3x^2+3x}{6x^2-6x} \cdot \frac{7xy-7y}{5xy-10y} = \frac{\cancel{3x}(x+1)}{2 \cdot \cancel{3} \cdot \cancel{x}\cancel{(x-1)}} \cdot \frac{7\cancel{y}\cancel{(x-1)}}{5\cancel{y}(x-2)}$$

$$= \frac{7(x+1)}{2 \cdot 5(x-2)}$$

$$= \frac{7(x+1)}{10(x-2)}$$

EXAMPLE Simplify: $\frac{2n-m}{2m^2+7mn+3n^2} \cdot \frac{2m^3+m^2n}{m^3-2m^2n}$

SOLUTION $$\frac{2n-m}{2m^2+7mn+3n^2} \cdot \frac{2m^3+m^2n}{m^3-2m^2n} = \frac{-1\cancel{(m-2n)}}{\cancel{(2m+n)}(m+3n)} \cdot \frac{\cancel{m} \cdot \cancel{m}\cancel{(2m+n)}}{\cancel{m} \cdot \cancel{m}\cancel{(m-2n)}}$$

$$= \frac{-1}{m+3n}$$

The quotient of two rational numbers is defined as

$$\frac{a}{b} \div \frac{c}{d} = \frac{a}{b} \cdot \frac{d}{c} = \frac{ad}{bc}$$

To divide two rational expressions, you use basically the same property: invert the divisor and find the product of the resulting expressions.

EXAMPLE Simplify: $\frac{2x^2 + xy}{x^3 + x^2y} \div \frac{xy - y^2}{x^2 - y^2}$

SOLUTION
$$\begin{aligned}\frac{2x^2 + xy}{x^3 + x^2y} \div \frac{xy - y^2}{x^2 - y^2} &= \frac{2x^2 + xy}{x^3 + x^2y} \cdot \frac{x^2 - y^2}{xy - y^2} \\ &= \frac{\cancel{x}(2x + y)}{\cancel{x} \cdot x\cancel{(x + y)}} \cdot \frac{\cancel{(x + y)}\cancel{(x - y)}}{y\cancel{(x - y)}} \\ &= \frac{2x + y}{xy}\end{aligned}$$

EXAMPLE Simplify: $\frac{10xy}{2x - 4y} \cdot \frac{x + y}{x^2 - xy} \div \frac{xy + y^2}{y - x}$

SOLUTION
$$\begin{aligned}\frac{10xy}{2x - 4y} \cdot \frac{x + y}{x^2 - xy} \div \frac{xy + y^2}{y - x} &= \frac{10xy}{2x - 4y} \cdot \frac{x + y}{x^2 - xy} \cdot \frac{y - x}{xy + y^2} \\ &= \frac{\cancel{2} \cdot 5\cancel{x}\cancel{y}}{\cancel{2}(x - 2y)} \cdot \frac{\cancel{x + y}}{\cancel{x}\cancel{(x - y)}} \cdot \frac{-1\cancel{(x - y)}}{\cancel{y}\cancel{(x + y)}} \\ &= \frac{-5}{x - 2y}\end{aligned}$$

(Note that $\frac{-5}{x - 2y}$ can also be written as $\frac{5}{2y - x}$ since $\frac{-a}{b} = \frac{a}{-b}$.)

ORAL EXERCISES

Simplify.

1. $\frac{m}{n} \div \frac{n}{m}$
2. $3 \div \frac{12}{x}$
3. $\frac{r^3}{s^2} \cdot \frac{s^2}{r}$
4. $\frac{3m}{2} \cdot \frac{4}{3m^2}$
5. $\frac{a}{2b} \cdot \frac{2}{a}$
6. $\frac{3}{4x} \cdot \frac{x^2}{3}$
7. $\frac{2x}{3} \div \frac{2x^2}{3}$
8. $\frac{c - 1}{2} \div \frac{1 - c}{3}$
9. $\frac{3(x - 5)}{4(x + 1)} \cdot \frac{(x + 1)}{3(x - 5)}$
10. $\frac{a - 3}{5} \cdot \frac{a + 3}{3} \cdot \frac{5}{a + 3}$
11. $\frac{x - 1}{4y} \div \frac{y}{x - 1}$
12. $\frac{b^2}{b + 1} \div \frac{b}{b + 1}$

WRITTEN EXERCISES

Simplify.

A

1. $\frac{30}{7} \cdot \frac{14}{27}$
2. $\frac{25}{18} \cdot \frac{27}{10}$
3. $\left(-\frac{3}{5}x\right)^2 \left(\frac{20}{81}\right)$
4. $\frac{7r}{4} \div \frac{21r^2}{2}$
5. $\frac{9c^2}{2} \div \frac{15c}{4}$
6. $\frac{15}{8m} \cdot \frac{m^2}{2}$
7. $-12f^5 \div \frac{f^3}{3}$
8. $\frac{m^3}{n} \cdot \frac{n^5}{m^2} \div \frac{n^3}{m^4}$
9. $\frac{26h^2}{15k^2} \cdot \frac{20k^2}{39h} \div \frac{2k}{3h}$
10. $\frac{ab^2c}{18} \div \frac{abc^2}{12} \cdot \frac{15bc}{2}$
11. $\frac{2x-10}{x-1} \cdot \frac{x-1}{x^2-5x}$
12. $\frac{6m-18n}{9m+9n} \cdot \frac{4m-4n}{8m-24n}$
13. $\frac{4x^2-4x}{3y^2+6y} \div \frac{x-1}{y+2}$
14. $\frac{3a^2b-3ab^2}{4c^2} \div \frac{3ab}{2c}$
15. $\frac{2t-8}{3t+6} \cdot \frac{t^2-4}{t-4}$
16. $\frac{5-5x^2}{6x+12} \cdot \frac{x+2}{x^2-1}$
17. $\frac{2x^2-4x}{3xy-9y} \cdot \frac{3xy^2-9y^2}{4x^3+4x^2}$
18. $\frac{3a^2-3ab}{4a^2+8ab} \cdot \frac{ab+2b^2}{2ab-b^2}$
19. $\frac{12x+12y}{7x-7y} \div \frac{18x+18y}{7x+14}$
20. $\frac{xy-x}{2y+4} \div \frac{x^2y+3x^2}{2y+4}$
21. $\frac{a^2-1}{b^2+3b} \div \frac{6a^2+6a}{2b^2+6b}$
22. $\frac{9m^2-16}{m^2-25} \cdot \frac{3m+15}{6m-8}$
23. $(3p-12) \div \frac{8-2p}{3}$
24. $\frac{2x^2}{x^2-25} \cdot \frac{25-5x}{x-2}$
25. $\frac{r^2+s^2}{r^2-s^2} \cdot \frac{r-s}{r+s}$
26. $\frac{(a+b)^2}{a-b} \div \frac{a^2-b^2}{(a-b)^2}$

B

27. $\frac{x^2-x-30}{2x^2-11x-6} \cdot \frac{4x^2-4x-3}{2x^2-11x+12}$
28. $\frac{a^2-36}{2a^2-15a+18} \cdot \frac{4a^2-9}{a^2-4a-12}$
29. $\frac{x^2+3x-4}{x^2-4x+3} \div \frac{x^2+2x-8}{x^2-9}$
30. $\frac{a^2+a-2}{a^2+2a-3} \div \frac{a^2-4a-12}{a^2-7a+6}$
31. $\frac{6u^2-7u-3}{2u^2-17u+21} \div \frac{9u^2+9u+2}{2u^2-11u-21}$
32. $\frac{6a^2-11a+3}{8a^2-10a-3} \div \frac{6a^2+7a-3}{8a^2+14a+3}$
33. $\frac{t^2-9}{2t^2-11t+15} \cdot \frac{6t^2-17t+5}{9t^2-1}$
34. $\frac{10k^2-k-2}{15k^2+26k+8} \cdot \frac{6k^2-7k-20}{4k^2-12k+5}$
35. $\frac{k^3+8}{k-2} \cdot \frac{k+2}{k^2+4k+4}$
36. $\frac{b^2-25}{b^3-125} \cdot \frac{b^2-8b+15}{b^2+2b-15}$
37. $\frac{6x+6}{x-2} \cdot \frac{x^2+2x-3}{x^2-1} \div \frac{9x+18}{2x-4}$
38. $\frac{r^2-25}{r-3} \cdot \frac{6r+12}{r^2+7r+10} \div \frac{r+3}{4r-20}$

C

39. $\frac{2x^2y-2xy}{3x^3+6x^2} \cdot \frac{x^2-9}{4xy+16y} \div \frac{x^2+2x-3}{x^2+6x+8}$
40. $\frac{3m^2n-6mn^2}{m^5+3m^4n} \cdot \frac{m^3n+3m^2n^2}{6mn^2-3n^3} \div \frac{m^2-8mn+12n^2}{2m^2+3mn-2n^2}$
41. $\frac{x^2y-xy^2}{x^2-xy-6y^2} \div \frac{x^3y^2+x^2y^3}{x^2-3xy} \cdot \frac{x^2+3xy+2y^2}{x^2-2xy+y^2}$
42. $\frac{a^2b-2ab^2}{a^2+2ab-3b^2} \div \frac{a^2+6ab}{a^2+11ab-12b^2} \div \frac{ab-2b^2}{a^2+9ab+18b^2}$

Section 6-8 SUMS AND DIFFERENCES OF RATIONAL EXPRESSIONS

In this section you will learn about

- the least common multiple
- the least common denominator
- adding and subtracting rational expressions

As you may recall, the *least common multiple* (LCM) of a set of numbers is the least number that can be divided evenly by each member of the set.

EXAMPLE Find the LCM of 8, 9, and 12.

SOLUTION Write the prime factorization of each number.

$$8 = 2^3 \qquad 9 = 3^2 \qquad 12 = 2^2 \cdot 3$$

Now take each factor the maximum number of times it appears in any one set of factors.

$$\text{LCM} = 2^3 \cdot 3^2 = 72$$

You can use the same technique to find the LCM of algebraic expressions.

EXAMPLE Find the LCM of $2x + 2$, $3x + 3$, and $6x$.

SOLUTION

$$\begin{aligned} 2x + 2 &= 2(x + 1) \\ 3x + 3 &= 3(x + 1) \\ 6x &= 2 \cdot 3 \cdot x \\ \text{LCM} &= 2 \cdot 3 \cdot x(x + 1) \\ &= 6x(x + 1) \end{aligned}$$

To express the sum or difference of two fractions as a single fraction you use the following rule, which is an application of the distributive property: For all real numbers a, b, and c, $c \neq 0$,

$$\frac{a}{c} + \frac{b}{c} = \frac{a+b}{c} \quad \text{and} \quad \frac{a}{c} - \frac{b}{c} = \frac{a-b}{c}$$

For example,

$$\frac{3}{5} + \frac{1}{5} = \frac{3+1}{5} = \frac{4}{5} \quad \text{and} \quad \frac{9}{11} - \frac{3}{11} = \frac{9-3}{11} = \frac{6}{11}$$

If the fractions have unequal denominators, you use the *least common denominator* (LCD), which is the LCM of the denominators.

EXAMPLE Simplify: $\frac{5}{6} + \frac{1}{4}$

SOLUTION

$$\left.\begin{aligned} 6 &= 2 \cdot 3 \\ 4 &= 2 \cdot 2 \end{aligned}\right\} \quad \text{LCD} = 2 \cdot 2 \cdot 3 = 12$$

$$\frac{5}{6} + \frac{1}{4} = \frac{5 \cdot 2}{6 \cdot 2} + \frac{1 \cdot 3}{4 \cdot 3} = \frac{10}{12} + \frac{3}{12} = \frac{10+3}{12} = \frac{13}{12}$$

To add or subtract rational expressions, use the same procedure.

EXAMPLE Simplify: $\frac{2}{3x} - \frac{3}{4x^2}$

SOLUTION $\text{LCD} = 2 \cdot 2 \cdot 3 \cdot x^2 = 12x^2$

$$\frac{2}{3x} - \frac{3}{4x^2} = \frac{2 \cdot 4x}{3x \cdot 4x} - \frac{3 \cdot 3}{4x^2 \cdot 3}$$
$$= \frac{8x}{12x^2} - \frac{9}{12x^2} = \frac{8x - 9}{12x^2}$$

EXAMPLE Simplify: $\frac{2x - 3}{2x^2 + 2x} + \frac{4}{3x + 3}$

SOLUTION $\frac{2x - 3}{2x^2 + 2x} + \frac{4}{3x + 3} = \frac{2x - 3}{2x(x + 1)} + \frac{4}{3(x + 1)}$

$\text{LCD} = 6x(x + 1)$

$$\frac{(2x - 3)3}{2x(x + 1)3} + \frac{4(2x)}{3(x + 1)2x} = \frac{(2x - 3)3 + 4(2x)}{6x(x + 1)}$$
$$= \frac{6x - 9 + 8x}{6x(x + 1)} = \frac{14x - 9}{6x(x + 1)}$$

EXAMPLE Simplify: $\frac{7}{x^2 + 8x + 16} - \frac{1}{x + 4}$

SOLUTION
$$\frac{7}{x^2 + 8x + 16} - \frac{1}{x + 4} = \frac{7}{(x + 4)^2} - \frac{1}{x + 4}$$
$$= \frac{7}{(x + 4)^2} - \frac{1(x + 4)}{(x + 4)^2}$$
$$= \frac{7 - 1(x + 4)}{(x + 4)^2}$$
$$= \frac{7 - x - 4}{(x + 4)^2} = \frac{3 - x}{(x + 4)^2}$$

EXAMPLE Simplify: $x^2 + \frac{2x^3}{x + 3}$

SOLUTION
$$x^2 + \frac{2x^3}{x + 3} = \frac{x^2}{1} + \frac{2x^3}{x + 3}$$
$$= \frac{x^2(x + 3)}{x + 3} + \frac{2x^3}{x + 3}$$
$$= \frac{x^2(x + 3) + 2x^3}{x + 3}$$
$$= \frac{x^3 + 3x^2 + 2x^3}{x + 3}$$
$$= \frac{3x^3 + 3x^2}{x + 3} = \frac{3x^2(x + 1)}{x + 3}$$

ORAL EXERCISES

In each exercise, find the LCM.

1. 3, 8, 6
2. 4, 8, 16
3. $2x$, $3x$
4. $7y$, $14y$
5. 3, x, x^2
6. 4, y^2, $5y$
7. $3a$, $4a$, $6a^2$
8. $b + 1$, $b - 1$, $b^2 - 1$

Simplify.

9. $\frac{5x}{4} - \frac{2x}{4}$
10. $\frac{9}{x} + \frac{4}{x}$
11. $\frac{3}{2x} - \frac{5}{2x}$
12. $\frac{3a}{5} + \frac{2a}{5} - \frac{a}{5}$
13. $\frac{7}{a} - \frac{1}{a} - \frac{2}{a}$
14. $\frac{2}{3b} + \frac{4}{3b} + \frac{b}{3b}$

State the LCD of each of the following.

15. $\frac{2}{3x} + \frac{x}{6} - \frac{-1}{x}$
16. $\frac{x + y}{2} + \frac{3x}{4} - \frac{x^3y}{8}$
17. $\frac{5}{x + 3} + \frac{7}{x - 3}$

WRITTEN EXERCISES

Simplify. Write each answer in lowest terms.

A

1. $\frac{1}{4x} - \frac{1}{3x^2}$
2. $\frac{y}{2} + \frac{1}{y^2}$
3. $\frac{3}{8z} - \frac{5}{xy}$
4. $\frac{2}{5m} + \frac{3}{m^2}$
5. $\frac{3y}{x^2} + \frac{1}{3x} + \frac{5}{2x}$
6. $\frac{x}{8} + \frac{5}{6} - \frac{x^2}{4}$
7. $\frac{a}{5} + \frac{a^2}{10} + \frac{1}{20}$
8. $\frac{2}{mn} - \frac{6}{m^2} - \frac{m}{3n}$
9. $\frac{1}{2x} - \frac{2}{3x^2} + \frac{1}{6x^3}$
10. $\frac{x - 1}{x^2} + \frac{3}{2x}$
11. $\frac{a - 6}{3b} + \frac{7}{9b}$
12. $\frac{2y + 1}{3y^2} - \frac{5}{4}$
13. $\frac{x}{x - y} - \frac{y}{x - y}$
14. $\frac{a^2}{a + b} - \frac{b^2}{a + b}$
15. $\frac{p^2}{p - 5} - \frac{5}{p - 5}$
16. $\frac{4}{7} - \frac{2x + 1}{14}$
17. $\frac{3}{5} - \frac{4 - 5b}{15}$
18. $\frac{2}{3} + m$
19. $x + \frac{3}{4}$
20. $z^2 - \frac{7}{10}$
21. $\frac{2}{3} + \frac{x}{y}$
22. $\frac{5}{6} - \frac{a}{b}$
23. $\frac{a + b}{2} + \frac{3}{4}$
24. $\frac{5}{6} - \frac{x + y}{3}$
25. $\frac{x + y}{5} - \frac{3x - 2y}{15}$
26. $\frac{2m + n}{3} + \frac{m - 3n}{18}$
27. $\frac{3z - 1}{12} + \frac{2z + 5}{6}$

Simplify. Write each answer in lowest terms.

28. $\frac{a-3b}{20}+\frac{3a-2b}{5}$

29. $\frac{2}{a}+\frac{3}{a^2}+\frac{4}{a^3}$

30. $\frac{5}{x^3}-\frac{2a}{x}+\frac{4}{x^2}$

B 31. $\frac{5}{6x}+\frac{3}{4y}$

32. $\frac{7y}{18x^2}-\frac{7}{12xy}$

33. $\frac{3}{ab}-\frac{2}{bc}+\frac{1}{ac}$

34. $\frac{2}{xy}-\frac{1}{x+y}$

35. $\frac{x}{x+2}-\frac{1}{x^2-4}$

36. $\frac{2}{y+2}-\frac{3}{4y}$

37. $\frac{2x}{3x-3}+\frac{1}{6x-6}$

38. $\frac{4}{9x-45}-\frac{1}{6x}$

39. $\frac{a+2}{a-2}+\frac{a-2}{a+2}$

40. $\frac{5}{x^2+6x+9}+\frac{2}{x+3}$

41. $\frac{7}{a^2-7a+12}-\frac{5}{a-3}$

42. $\frac{3}{z+6}-\frac{7}{z^2-36}$

43. $\frac{7}{m^2-100}+\frac{4}{m^2+11m+10}$

44. $\frac{5x}{x^2+5x+6}-\frac{3x}{x^2-4}$

45. $\frac{2b}{b^2+4b-21}-\frac{b}{b^2-9}$

C 46. $\frac{1}{2x^2}+\frac{1}{8x}-\frac{1}{x-1}$

47. $\frac{1}{4x^2-12x}+\frac{x}{8x-24}-\frac{1}{6x^2}$

48. $\frac{3}{y^2-2y}-\frac{y}{y^2-4}+\frac{1}{y^2+2y}$

49. $\frac{1}{x^2-2x}+\frac{1}{x^2-4}+\frac{2}{x^2+2x}$

50. $\frac{5}{m^2-16}+\frac{3}{4-m}+\frac{1}{m+4}$

51. $\frac{7}{4x^2-1}-\frac{2}{1-2x}-\frac{3}{2x+1}$

52. $\frac{2}{x^2-2x-8}+\frac{3}{x^2-5x+4}+\frac{3}{x^2+x-2}$

53. $\frac{11}{a^2-a-30}-\frac{6}{a^2-3a-18}-\frac{2}{a^2+8a+15}$

54. $\frac{6}{y^2-10y+16}-\frac{8}{y^2-64}+\frac{10}{y^2+6y-16}$

55. $\frac{2}{2m^2+3m+1}+\frac{18}{2m^2-7m-4}-\frac{3}{m^2-3m-4}$

56. $\left(\frac{1}{x+2}-\frac{2}{x-2}\right)\left(\frac{x^2+2x}{x+6}\right)$

57. $\left(\frac{2r+1}{r+5}-\frac{r}{r-7}\right)\left(\frac{r^2-25}{r-18}\right)$

SELF-TEST 3

Simplify. (Assume that no denominator is zero.)

Section 6-6

1. $\dfrac{5b^2}{25b^2 - 10b}$
2. $\dfrac{1 - 2r}{4r^2 - 1}$
3. $\dfrac{s^2 + 4s - 32}{s^2 + 8s}$
4. $\dfrac{x^3y - xy^3}{x^4 + 2x^3y + x^2y^2}$

Section 6-7

5. $\dfrac{8b^3c}{6b^2cd} \div \dfrac{4c^2}{abd^2}$
6. $\dfrac{9t^2 + 12t + 4}{-9t - 6} \cdot \dfrac{9t - 6}{9t^2 - 4}$
7. $\dfrac{8r^2 + 2r - 3}{6r^2 - 3r} \cdot \dfrac{r^3}{4r^2 + 3r}$
8. $\dfrac{2k^2 + 4k + 8}{k^3 - 8} \div \dfrac{1}{2 - k}$

Section 6-8

9. $\dfrac{xy}{2} + \dfrac{3xy}{5} - \dfrac{xy}{10}$
10. $\dfrac{3}{a^2b} - \dfrac{2}{4ab} + \dfrac{1}{6b^2}$
11. $\dfrac{2}{m^2 + 6m + 5} + \dfrac{1}{3m + 3}$
12. $\dfrac{q - 9}{2q - 3} - \dfrac{5q - 15}{2q - 3}$

Computer Exercises

1. Copy and RUN the following program to print the positive factors of a positive integer. Try 576.

```
10   PRINT "INPUT POSITIVE INTEGER A";
20   INPUT A
30   FOR F=1 TO SQR(A)                (SQR(A) means √A.)
40   LET Q=A/F
50   IF Q<> INT(Q) THEN 70
60   PRINT F;Q
70   NEXT F
80   END
```

2. Copy and RUN the following program to factor the polynomials given below. (Compare the program of Exercise 1 with lines 70–140 of this program. H is a counter.)

```
10   PRINT "TO FACTOR   AX ↑ 2   +   BX   +   C:"
20   PRINT "INPUT A(>0), B, C(<>0):";
30   INPUT A,B,C
40   PRINT A; "X ↑ 2   +   (";B;")X   +   (";C;")   ";
50   DIM L[50],M[50],R[100],S[100]      (Dimension as needed.)
60   REM***FIND POS. FACTORS OF A       (REM statements are printed when the program
70   LET H=0                             is listed, but do not affect the run.)
80   FOR F=1 TO SQR(A)
90   LET Q=A/F
```

(Program continued on the next page.)

```
100  IF Q<> INT(Q) THEN 140          (INT(Q) means the greatest integer less than or
110  LET H=H+1                        equal to Q.)
120  LET L[H]=F
130  LET M[H]=Q
140  NEXT F
150  REM***FIND POS. AND NEG. FACTORS OF C
160  LET K=-1
170  FOR F=1 TO ABS(C)               (ABS(C) = C if C ≥ 0;
180  LET Q=C/F                        ABS(C) = -C if C < 0.)
190  IF Q<> INT(Q) THEN 250
200  LET K=K+2
210  LET R[K]=F
220  LET R[K+1]=-F
230  LET S[K]=Q
240  LET S[K+1]=-Q
250  NEXT F
260  REM***PRINT POLYNOMIAL FACTORS
270  FOR I=1 TO H
280  FOR J=1 TO K+1
290  LET B1=L[I]*S[J]+M[I]*R[J]
300  IF B1=B THEN 350
310  NEXT J
320  NEXT I
330  PRINT "IS NOT FACTORABLE."
340  STOP
350  PRINT "=  (";L[I];"X  +  (";R[J];"))";
360  PRINT "(";M[I];"X  +  (";S[J];"))"
370  END
```

a. $x^2 + 5x + 6$ b. $x^2 - 5x - 24$ c. $20x^2 + 31x + 12$

d. $6x^2 + x - 12$ e. $6x^2 - 17x + 5$ f. $3x^2 + 7x - 2$

CHAPTER REVIEW

State the degree of the polynomial.

1. $5 - 2t^3 - 7t + 9t^2$ 2. $-y + 5x^3y - x^5y^2 + y^3$ Section 6-1

Simplify.

3. $(-a^2 + ab - 3b^2) + (-2a^2 - ab + b^2)$
4. $(4x^2 - 5 + 3x) - (2 - x^3 + 8x^2)$

5. $c(-2cd^3)^2$ 6. $(3p - 2)(4p - 3)$ Section 6-2
7. $5wy^2(3w - y) + 2wy^3$ 8. $(7a - 1)(a^2 + 3a - 2)$

9. Find the GCF of $52ab^3c^4$ and $26a^5c^2$. Section 6-3

Factor completely.

10. $60d^3e - 25de^2$ 11. $f^2 + 12f - 28$ 12. $4g^2 - 25gh + 6h^2$

13. $j^2k^2 + 6jk + 9$ 14. $1000 - m^3$ 15. $12r^2 - 75s^2$ Section 6-4

Simplify.

16. $\frac{-30cd(c + d)^5}{18c^2d^3(c + d)}$ 17. $\frac{8y^7 - 2y^4 + 40y}{4y^3}$ Section 6-5

18. Divide $9a^3 - 3a^2 - 8a - 1$ by $3a - 2$. Express your answer in the form of an equation.

Simplify. (Assume that no denominator is zero.)

19. $\frac{4x + 32}{64 - x^2}$ 20. $\frac{2q^3 - 5q^2}{2q^2 - 7q + 5}$ 21. $\frac{k^2 + 8k + 7}{k^2 + 2k - 35}$ Section 6-6

22. $\frac{4}{a^2b} \div \frac{2}{ab^2}$ 23. $\frac{72r}{45r^2s^3} \cdot \frac{5s^4}{16r^3}$ Section 6-7

24. $\frac{x^2 - 25y^2}{x^2y + 5xy^2} \cdot \frac{10xy^2}{10y - 2x}$ 25. $\frac{z^2 + 4z + 4}{3z + 6} \div \frac{z^3 + 8}{12}$

26. $\frac{2w}{3} - \frac{w}{4} + \frac{5w}{6}$ 27. $3d + \frac{2}{cd}$ Section 6-8

28. $\frac{t}{4 - t^2} - \frac{2}{4 - t^2}$ 29. $\frac{1}{b - 3} - \frac{3b - 7}{2b - 6}$

CHAPTER TEST

Find **(a)** the GCF and **(b)** the LCM.

1. $24xz^2$, $60y^3z^2$ 2. $3m^2 - 3$, $7m + 7$

Factor completely.

3. $3p^2 - 4p - 4$ 4. $9a^2 - 64c^2$ 5. $s^3 - 14s^2 + 40s$

6. $z(w + 1) - 2(w + 1)$ 7. $t^2 + 20ty + 100y^2$ 8. $6m^2 + m - 15$

Simplify. (Assume that no denominator is zero.)

9. $2k^3(k + 3)^2$ 10. $(-7x^3y)^2$ 11. $\frac{-30m^2t(3m - t)^5}{-36mt^3(3m - t)^3}$

12. $\frac{36c^2 - 4d^2}{2d + 6c}$ 13. $\frac{5b - 10}{6 - 3b}$ 14. $(9s - 1)(5s + 4)$

15. $\frac{2r^2}{-3s} \cdot \frac{9s^5}{8r^4}$ 16. $(5h^2 - 3h - 7) + (2h + h^4 - 9h^2)$

17. $\frac{a^2 + 8a + 12}{a^2 + 2a - 24}$ 18. $\frac{2}{e - f} + \frac{3}{e + f} + \frac{4f}{f^2 - e^2}$

Simplify. (Assume that no denominator is zero.)

19. $\dfrac{g^3 - 8}{g - 2} \div \dfrac{g^2 + 2g + 4}{g + 2}$

20. $\dfrac{8w^4 - 6w^3 + 4w^2}{2w^2}$

21. $\dfrac{3}{y - 1} - \dfrac{1}{2}$

22. $\dfrac{j^2 - 10j + 25}{5j - j^2} \cdot \dfrac{j^2 + 5j}{j^2 - 25}$

23. $(t^2 + 2t - 1)(t - 3)$

24. $(7x^2y^2 + 4 - 8xy) - (x^2 - xy - 3y^2)$

25. $\dfrac{d^2}{9} - \dfrac{7d}{12} + \dfrac{5d}{6}$

26. $(3b + 4a)(2b - 7a)$

Divide the first polynomial by the second. Write your answer in the form of an equation.

27. $2x^3 + 5x^2 - 13x - 30$; $\quad 2x - 5$

28. $21a^3 - 43a^2 + 4$; $\quad 7a + 2$

Application—Optics

Prisms spread or disperse beams of light. If you put one in your window, it will break the sunlight passing through it into a rainbow of color on your wall. Beams of light can also be bent by lenses, which are curved pieces of glass or plastic that can be found in cameras and eyeglasses. A common type of lens, the converging lens, brings parallel rays of light together as shown in the photo below.

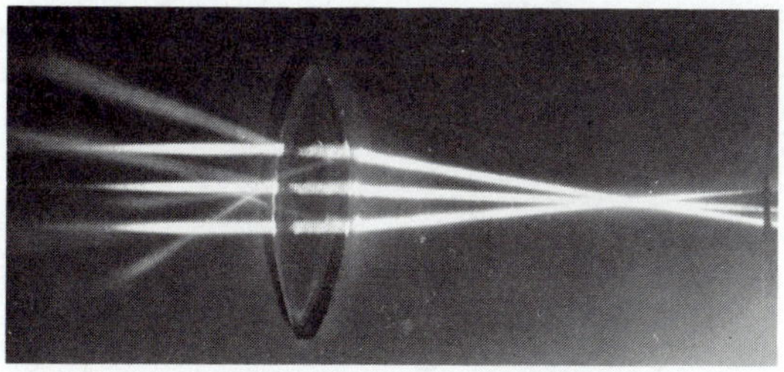

You can see the point where the rays converge in the photo and in the diagram below. A lens has two of these principal focus points (F) because light can pass through in either direction. In the diagram the light is shown passing from left to right.

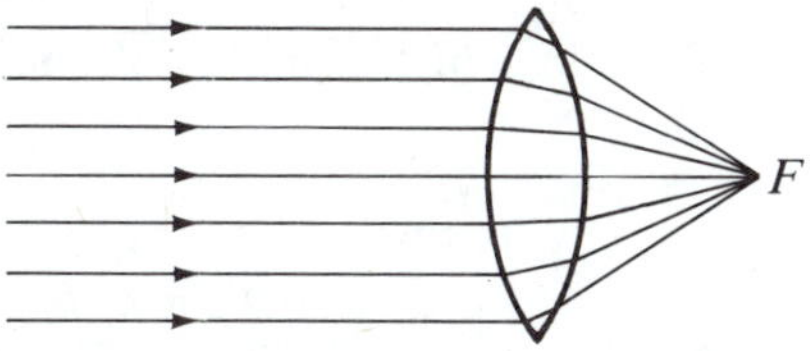

The distance from either principal focus to the lens center is called the focal length (f). Suppose that light is passing through a converging lens. The size of any image created is determined by the distance of the lens to the object (d_o), the distance from the image to the lens (d_i), and the focal length (f).

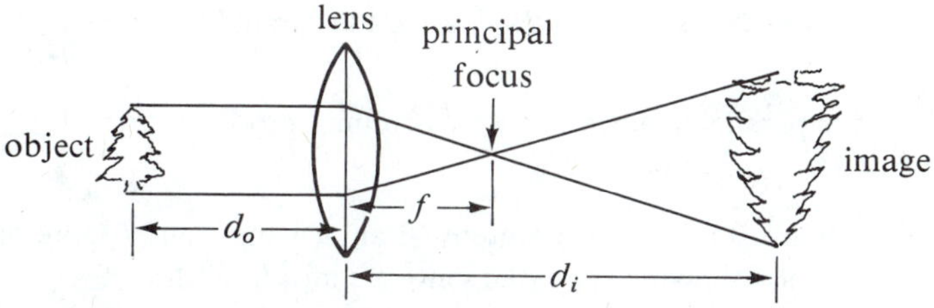

The relationship among d_o, d_i, and f is given by the following formula:

$$\frac{1}{d_o} + \frac{1}{d_i} = \frac{1}{f}$$

Objects viewed through a lens often appear to be bigger than they really are. You can find the amount of magnification (M) by the following formula:

$$M = \frac{d_i}{d_o}$$

VOCABULARY REVIEW

Be sure that you understand the meaning of these terms:

monomial, p. 218
coefficient, p. 218
polynomial, p. 218
degree of a polynomial, p. 218
binomial, p. 220
trinomial, p. 220
prime number, p. 227
prime factorization, p. 228
greatest common factor, p. 228
quadratic polynomial, p. 228
trinomial square, p. 233
rational expression, p. 241
least common multiple, p. 247
least common denominator, p. 247

MIXED REVIEW

Solve.

1. $-3 \leqslant 4t + 1 < 5$
2. $6(j + 8) - 5(2j - 5) = 15$

Factor completely.

3. $a - a^3$
4. $(a + b)^2 - 1$
5. $16a^2 + 24a - 7$

Exercises 6–8 refer to the diagram, in which $\overline{QW} \cong \overline{TW}$ and $\angle 3 \cong \angle 8$. Write two-column proofs.

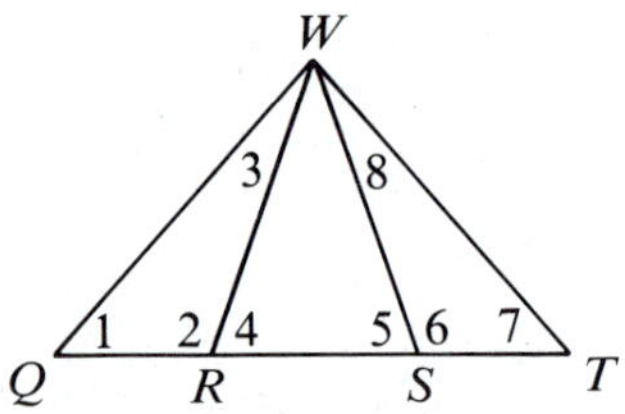

6. Prove: $\angle 4 \cong \angle 5$
7. Prove: $\triangle WQS \cong \triangle WTR$
8. Prove: $m\angle 5 = m\angle 1 + m\angle 3$

Exercises 9–11 refer to quadrilateral $QRST$.

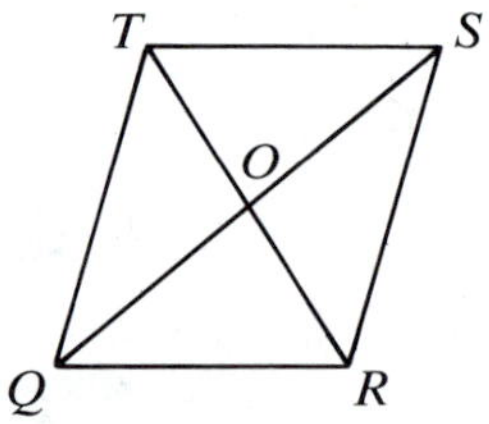

9. If $\triangle QOR \cong \triangle SOT$, is $QRST$ a parallelogram? Explain.
10. If $\overline{TQ} \cong \overline{SR}$ and $\angle TQS \cong \angle RSQ$, is $QRST$ a parallelogram? Explain.
11. Given that $QRST$ is a parallelogram, write a two-column proof that $\triangle QOT \cong \triangle SOR$.
12. Let p represent "you are staying," q represent "I am leaving," and r represent "it is late." Translate the following symbolic sentence into English: $(r \wedge \sim p) \rightarrow q$
13. Divide $b^4 + 3b^2 - 55$ by $b^2 - 6$.

Find the number of sides of a regular polygon if each exterior angle has the given measure.

14. 10
15. 45
16. 12
17. In $\triangle RST$, $\angle R \cong \angle S$. If $TR = 12x - 21$ and $TS = 3x + 15$, find the value of x.

Simplify. Write answers in lowest terms.

18. $(6a^2 - 13a + 6) - (4a^2 - 12a + 9)$

19. $-(x^2 - 1)(x^2 + x - 1)$

20. $\dfrac{3a}{a^2 - 4} + \dfrac{2}{2 - a}$

21. $j + 2 - \dfrac{5}{2j + 1}$

22. The measure of each base angle of an isosceles triangle is 6 less than half the measure of the vertex angle. Find the measure of each angle.

Negate.

23. $\sim r \rightarrow \sim s$

24. $\exists_x \; \dfrac{1}{x} > x$

Reduce to lowest terms.

25. $\dfrac{b - 3}{b^2 + 3b - 18}$

26. $\dfrac{5ab}{-10a^2b^2}$

27. In $\triangle ABC$, if $AC = 7$ and $BC = 5$, find all possible values of AB.

28. If quadrilateral $ABCD$ is a parallelogram, which postulate could you use to prove that $\triangle ABC \cong \triangle CDA$?

Exercises 29-31 refer to trapezoid $EFGH$ with bases $\overline{EF}$ and $\overline{GH}$ and median $\overline{AB}$.

29. If $HG = 14$ and $EF = 26$, find AB.

30. If $HG = 4 - 3x$, $EF = 6 + 5x$, and $AB = 11x$, find x.

31. If $AB = 13$ and $EF = 22$, find HG.

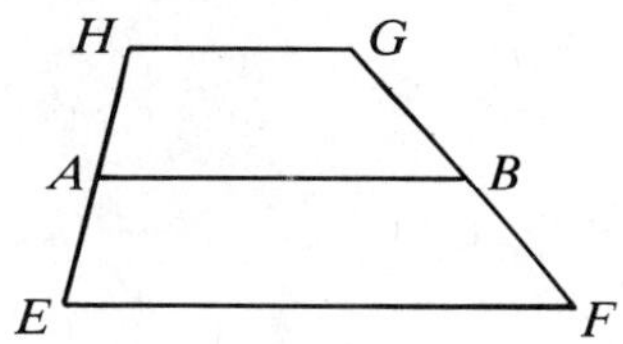

32. Write the conclusion that follows logically from the given premises and state which rule of inference is being used. If no conclusion follows, explain why.

P_1: If a fish can't fly, then a fly can't fish.
P_2: A fly can fish.

33. The sum of two consecutive integers is 5 more than the third. Find the integers.

Find the measures of $\angle A$ and $\angle B$ under the given conditions.

34. $\angle A$ and $\angle B$ are complementary; $m\angle A = 4x - 8$; $m\angle B = x + 8$.

35. $\angle A$ and $\angle B$ are supplementary and the measure of $\angle B$ is 18 more than twice the measure of $\angle A$.

Quadratic equations are often used in solving problems involving areas of rectangular regions – for instance, problems involving a rectangular garden bordered by a hedge or a walkway. The gardens shown are in Colonial Williamsburg.

Quadratic Equations 7

When you use algebra to solve a word problem, you usually write one or more equations to describe the known facts and then solve the equations. For example, suppose you know that the length of a rectangle is 2 more than its width and that the area is 14. Can you find the dimensions of the rectangle? If x is the width, then $x + 2$ is the length. The expression $x(x + 2)$ represents the area, so

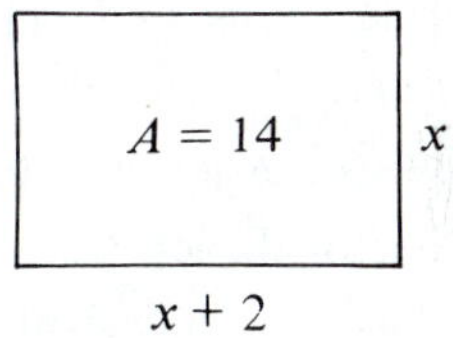

$$x(x + 2) = 14.$$

Simplifying, you have

$$x^2 + 2x - 14 = 0.$$

This is an example of a quadratic equation, which, unlike the linear equations you have learned to solve, contains a term with a variable to the second power. In this chapter you will learn several methods of solving quadratic equations. These methods will enable you to find the dimensions of the rectangle described above.

Section 7-1 SOLVING QUADRATIC EQUATIONS BY FACTORING

A *quadratic equation in x* is any equation that can be written in the *standard form*

$$ax^2 + bx + c = 0 \qquad a, b, c \in \mathcal{R}, a \neq 0$$

Here are some examples of quadratic equations:

$$2x - 5 + x^2 = 0$$
$$4y^2 = 6y - 1$$
$$8 = t^2$$
$$\frac{x^2}{4} - \frac{1}{2} = \frac{x}{3}$$
$$r(r - 1) = 6$$
$$s(s + 2) = 2s(s + 6)$$

The last two equations don't seem to be quadratic equations, but if you use the distributive property to eliminate the parentheses, it should be clear that each equation can be transformed to a quadratic equation in standard form.

The following are examples of equations that are *not* quadratic:

$$2x - 3 = 0 \qquad \frac{1}{x} - x^2 = 2$$

$$4y - 6 = 7 - y \qquad \sqrt{z} - z = -6$$

$$\frac{1}{m} - \frac{3}{m} = \frac{1}{2}$$

Later in this chapter you will learn how to apply the *techniques* of solving quadratic equations to solve the last equation.

Suppose you know that the product of two numbers is 0. What can you say about the numbers? It is true that *at least one* of the numbers must equal 0.

Zero-Product Property

For all real numbers a and b, $ab = 0$ if and only if $a = 0$ or $b = 0$.

For example, the Zero-Product Property implies that the equation $(x-2)(x-5) = 0$ is equivalent to the disjunction $x - 2 = 0$ or $x - 5 = 0$.

This is the basic idea behind the method of solving quadratic equations by *factoring*. To use this method:

1. Write the equation in standard form, $ax^2 + bx + c = 0$.
2. Factor the left side of the equation.
3. Solve the equivalent disjunction.
4. Check the roots in the *original* equation.

EXAMPLE Solve: $x^2 - 3x = 18$

SOLUTION Following the previously outlined steps:

1. $x^2 - 3x - 18 = 0$
2. $(x-6)(x+3) = 0$
3. $x - 6 = 0$ or $x + 3 = 0$
   $x = 6$ or $x = -3$
4. Check the roots by substituting 6 and -3 for x in the original equation.

$$x^2 - 3x = 18 \qquad x^2 - 3x = 18$$

$$6^2 - 3(6) = 36 - 18 = 18 \qquad (-3)^2 - 3(-3) = 9 + 9 = 18$$

Therefore, the solution set of $x^2 - 3x = 18$ is $\{6, -3\}$.

EXAMPLE Solve: $y^2 + 9 = 6y$

SOLUTION

$$y^2 - 6y + 9 = 0$$

$$(y-3)(y-3) = 0$$

$$y - 3 = 0 \quad \text{or} \quad y - 3 = 0$$

$$y = 3 \quad \text{or} \quad y = 3$$

The solution set is $\{3\}$.

A member of the solution set of a quadratic equation is called a *root* of the equation. In the example above we say that 3 is a *double root* of the equation.

EXAMPLE Solve: $6x^2 + 5x = 4$

SOLUTION
$$6x^2 + 5x - 4 = 0$$
$$(3x + 4)(2x - 1) = 0$$
$$3x + 4 = 0 \quad \text{or} \quad 2x - 1 = 0$$
$$x = -\frac{4}{3} \quad \text{or} \quad x = \frac{1}{2}$$

The check is left to you.

The solution set is $\left\{-\frac{4}{3}, \frac{1}{2}\right\}$

You can also use the Zero-Product Property to solve some equations that are not quadratic.

EXAMPLE Solve: $m^3 - 14m = 5m^2$

SOLUTION Rewrite as an equivalent equation in which the right side is 0.
$$m^3 - 5m^2 - 14m = 0$$
$$m(m^2 - 5m - 14) = 0$$
$$m(m - 7)(m + 2) = 0$$
$$m = 0 \quad \text{or} \quad m - 7 = 0 \quad \text{or} \quad m + 2 = 0$$
$$m = 0 \quad \text{or} \quad m = 7 \quad \text{or} \quad m = -2$$

The solution set of $m^3 - 14m = 5m^2$ is $\{0, 7, -2\}$.

If you know one root of a quadratic equation, you can find the other root of the equation.

> m is a root of $ax^2 + bx + c = 0$ $(a, b, c \in \Re, a \neq 0)$
> if and only if $x - m$ is a factor of $ax^2 + bx + c$.

This biconditional is known as the *Factor Theorem,* the proof of which is beyond the scope of this book. The following example shows how you may use this theorem to find a second root of a quadratic equation given one root of the quadratic equation.

EXAMPLE Use the fact that $r_1 = -2$ is a root of $4x^2 + 3x - 10 = 0$ to find the other root.

SOLUTION The Factor Theorem states that if -2 is a root of $4x^2 + 3x - 10 = 0$, then $x + 2$ is a factor of $4x^2 + 3x - 10$. Use factoring to find the second factor of $4x^2 + 3x - 10$.
$$4x^2 + 3x - 10 = (x + 2)(\underline{\ ?\ }\, x - \underline{\ ?\ })$$
$$4x^2 + 3x - 10 = (x + 2)(4x - 5)$$

To find the second root, let $4x - 5 = 0$.
$$4x - 5 = 0$$
$$x = \frac{5}{4}$$

The second root is $\frac{5}{4}$.

ORAL EXERCISES

Is the equation a quadratic equation?

1. $x - 2 = x^2$
2. $x^3 - 4x = 0$
3. $y^2 = 0$
4. $\sqrt{t} - t = 1$
5. $2y + 4 = 7y$
6. $x^2 = 1$
7. $3z^2 - \frac{2}{z} = 0$
8. $\frac{1}{x} = 3$
9. $n - 4 = n^2 - 6$

Express each of the following quadratic equations in standard form.

10. $2x^2 = 3x + 1$
11. $x - 2 = x^2$
12. $4 = m^2$
13. $x(x + 2) = 3$
14. $7t = t^2$
15. $-y^2 = 3y - 10$

State the roots of each equation.

16. $(x - 3)(x - 15) = 0$
17. $(x + 5)(x + 17) = 0$
18. $(x + 2)(x - 2) = 0$
19. $(x - 10)x = 0$
20. $(x - 4)(x + 6) = 0$
21. $(x + 4)(x + 4) = 0$

State **(a)** the factors of the quadratic polynomial and **(b)** the roots of the equation.

22. $x^2 - 10x - 24 = 0$
23. $t^2 - 19t + 18 = 0$
24. $y^2 + 9y - 36 = 0$
25. $z^2 + 18z + 32 = 0$
26. $h^2 - 3h + 2 = 0$
27. $r^2 + 2r - 24 = 0$

28. Elwood solved the following equation as shown: Is his solution correct? Explain.

$$\begin{aligned} x^2 - 8x &= 0 \\ \frac{x^2}{x} - \frac{8x}{x} &= \frac{0}{x} \\ x - 8 &= 0 \\ x &= 8 \end{aligned}$$

29. Chloe solved the following equation as shown: Is her solution correct? Explain.

$$\begin{aligned} x^2 - 4x &= 5 \\ x(x - 4) &= 1 \cdot 5 \\ x = 1 \quad &\text{or} \quad x - 4 = 5 \\ x = 1 \quad &\text{or} \quad x = 9 \end{aligned}$$

WRITTEN EXERCISES

Solve.

A

1. $x^2 + x - 2 = 0$
2. $x^2 - 9x + 18 = 0$
3. $q^2 - 4q - 5 = 0$
4. $x^2 - 2x - 8 = 0$
5. $t^2 + 5t = 0$
6. $w^2 + 7w - 8 = 0$
7. $h^2 + h - 6 = 0$
8. $c^2 - 3c = 0$
9. $u^2 + 7u + 12 = 0$
10. $m^2 - 14m + 49 = 0$
11. $z^2 - 8z + 16 = 0$
12. $x^2 - 25 = 0$
13. $n^2 - n - 90 = 0$
14. $y^2 + y - 20 = 0$
15. $r^2 + 4r = 12$
16. $y^2 - y = 12$
17. $h^2 = 3h$
18. $x^2 - 2x = 24$

19. $z^2 = 12z - 35$

20. $-6q = q^2$

21. $4 = p^2$

22. $8x - 15 = x^2$

23. $10b = b^2 + 25$

24. $x(x + 2) = 63$

25. $-16 = r^2 + 8r$

26. $y^2 = 14y - 24$

27. $3x^2 + 2x - 8 = 0$

28. $m(m - 4) = 77$

29. $(w - 5)w = 6$

30. $(b + 12)b = -32$

Use the root r_1 to find the other root of the equation.

31. $d^2 + 2d - 8 = 0;\ r_1 = -4$

32. $2y^2 - 9y - 18 = 0;\ r_1 = 6$

33. $5n^2 - 7n = 6;\ r_1 = 2$

34. $8n^2 + 19n = 15;\ r_1 = -3$

Solve.

B 35. $5k^2 = -13k + 6$

36. $2x^2 - 5x + 2 = 0$

37. $6t^2 + t = 1$

38. $2p^2 + 9p = 5$

39. $9a^2 - 2 = -3a$

40. $2x^2 = 11x$

41. $12m^2 + 6 = 17m$

42. $\frac{x^2}{9} = x + 4$

43. $3y^2 + 17y = 0$

44. $3q^2 + \frac{1}{4} = 2q$

45. $4h^2 + 6 = 11h$

46. $11z^2 + z = 10$

47. $m^3 + 15m^2 + 56m = 0$

48. $a^3 - 48a = 2a^2$

49. $2y^3 + y^2 = 6y$

50. $3b^3 + 4b = 7b^2$

51. $14x^3 + 31x^2 = 10x$

52. $12n^3 + 7n^2 = 12n$

Write a quadratic equation in standard form with the given solution set. Use x as the variable.

53. $\{-5, 4\}$

54. $\{-7, -11\}$

55. $\left\{\frac{1}{3}, -2\right\}$

56. $\left\{\frac{2}{5}\right\}$

EXAMPLE Solve: $x^3 + x^2 - 4x - 4 = 0$

SOLUTION Factoring by grouping you have:

$$x^2(x + 1) - 4(x + 1) = 0$$
$$(x + 1)(x^2 - 4) = 0$$
$$(x + 1)(x + 2)(x - 2) = 0$$

Thus,

$x + 1 = 0$ or $x + 2 = 0$ or $x - 2 = 0$

$x = -1$ or $x = -2$ or $x = 2$

The solution set of $x^3 + x^2 - 4x - 4 = 0$ is $\{-2, -1, 2\}$.

Solve.

C 57. $x^3 + 3x^2 - x - 3 = 0$

58. $t^3 - 4t^2 - 4t + 16 = 0$

59. $y^3 - 8y^2 + y - 8 = 0$

60. $x^3 + 3x^2 - 25x - 75 = 0$

61. $2a^3 - 3a^2 - 18a + 27 = 0$

62. $2f^3 + f^2 - 8f - 4 = 0$

Given one root, r_1, of the equation, find the value of c.

63. $x^2 - x + c = 0;\ r_1 = -2$

64. $m^2 - 2m + c = 0;\ r_1 = 7$

65. $p^2 - 9p + c = 0;\ r_1 = 3$

66. $w^2 - 5w + c = 0;\ r_1 = 9$

Section 7-2 SOLVING FRACTIONAL EQUATIONS

An equation that has a variable in the denominator of one or more terms is called a *fractional equation.* To solve a fractional equation, first multiply each side of the equation by the LCD to clear the equation of fractions.

EXAMPLE Solve: $\frac{1}{2y} - \frac{1}{y+1} = \frac{3}{2y^2 + 2y}$

SOLUTION Express each denominator in factored form.

$$\frac{1}{2y} - \frac{1}{y+1} = \frac{3}{2y(y+1)}$$

LCD $= 2y(y + 1)$

$$2y(y+1)\left[\frac{1}{2y} - \frac{1}{y+1}\right] = 2y(y+1) \cdot \frac{3}{2y(y+1)}$$

$$2y(y+1) \cdot \frac{1}{2y} - 2y(y+1) \cdot \frac{1}{y+1} = 2y(y+1) \cdot \frac{3}{2y(y+1)}$$

$$y + 1 - 2y = 3$$

$$-y = 2$$

$$y = -2$$

Check

$$\frac{1}{2(-2)} - \frac{1}{(-2)+1} \ ? \ \frac{3}{2(-2)^2 + 2(-2)}$$

$$-\frac{1}{4} - \frac{1}{-1} \ ? \ \frac{3}{8-4}$$

$$-\frac{1}{4} + 1 \ ? \ \frac{3}{4}$$

$$\frac{3}{4} = \frac{3}{4}$$

The solution set is $\{-2\}$.

In the preceding example, multiplying both sides of the fractional equation by the LCD produced a *linear equation.* This is not always the case, as the next example illustrates.

EXAMPLE Solve: $\frac{5}{m+3} - \frac{m+1}{m-1} = \frac{3m-16}{m^2 + 2m - 3}$

SOLUTION LCD $= (m + 3)(m - 1)$

$$(m+3)(m-1)\left[\frac{5}{m+3} - \frac{m+1}{m-1}\right] = (m+3)(m-1)\frac{3m-16}{(m+3)(m-1)}$$

$$5(m-1) - (m+1)(m+3) = 3m - 16$$

$$5m - 5 - m^2 - 4m - 3 = 3m - 16$$

$$-m^2 - 2m + 8 = 0$$

$$m^2 + 2m - 8 = 0$$

$$(m+4)(m-2) = 0$$

$$m = -4 \text{ or } m = 2$$

The check is left to you. The solution set is $\{-4, 2\}$.

When you multiply both sides of a fractional equation by the LCD, be careful! As the next example shows, the solution set of the transformed equation is not always the same as the solution set of the original equation.

EXAMPLE Solve: $\frac{x}{x-2} + \frac{x+3}{x+1} = \frac{2}{x-2}$

SOLUTION LCD $= (x-2)(x+1)$

$$(x-2)(x+1)\left[\frac{x}{x-2} + \frac{x+3}{x+1}\right] = (x-2)(x+1)\frac{2}{x-2}$$

$$x(x+1) + (x+3)(x-2) = 2(x+1)$$

$$x^2 + x + x^2 + x - 6 = 2x + 2$$

$$2x^2 - 8 = 0$$

$$2(x^2 - 4) = 0$$

Divide both sides by 2. $(x+2)(x-2) = 0$

$$x = -2 \text{ or } x = 2$$

But x cannot equal 2 or we would have a zero denominator in the original equation. Therefore, 2 is *not* a root of the original equation. Check the other root, -2.

Check

$$\frac{-2}{-2-2} + \frac{-2+3}{-2+1} \overset{?}{=} \frac{2}{-2-2}$$

$$\frac{-2}{-4} + \frac{1}{-1} \overset{?}{=} \frac{2}{-4}$$

$$\frac{1}{2} - 1 \overset{?}{=} \frac{2}{-4}$$

$$-\frac{1}{2} = -\frac{1}{2}$$

The solution set is $\{-2\}$.

To summarize, when you multiply both sides of an equation by a polynomial, the solution set of the resulting equation will always contain the roots of the original equation. But it may also contain some numbers that are *not* roots of the original equation. So, *always check your solutions in the original equation.*

ORAL EXERCISES

State all restrictions on the values of the variables.

1. $\frac{6}{x} + \frac{2}{x} = \frac{11}{x^2}$
2. $\frac{7y}{2} + \frac{3}{y} = \frac{7}{2y}$
3. $\frac{z+1}{z} = \frac{2}{2-z}$
4. $\frac{12}{a+3} - \frac{2}{a-2} = 0$
5. $\frac{7}{m-1} + \frac{3}{m} = \frac{18}{m+1}$
6. $\frac{b+2}{b-2} + \frac{b-3}{b} = \frac{5}{b^2-2b}$

State the LCD of each of the following.

7. $\frac{7}{c} + \frac{3}{c^2} = \frac{5}{2}$

8. $\frac{6}{x} + \frac{2}{3x^2} = \frac{1}{2}$

9. $\frac{5}{3} - \frac{2}{a^2} = \frac{7}{4}$

10. $\frac{d}{d+1} - \frac{d}{2} = \frac{4}{3}$

11. $\frac{z+1}{z-1} + \frac{z+2}{z} = 3$

12. $\frac{b-2}{b+1} - \frac{5}{b-1} = \frac{b}{4}$

13. $\frac{7}{a-2} - \frac{a}{a+1} = \frac{a-1}{5}$

14. $\frac{x+1}{x^2-4} + \frac{5}{x+2} = \frac{3}{x-2}$

WRITTEN EXERCISES

Solve.

A 1. $\frac{4}{s} - \frac{1}{3} = 1$

2. $\frac{5}{h} + \frac{1}{2} = -2$

3. $\frac{7}{k} + 2 = \frac{1}{4}$

4. $\frac{7}{x} + \frac{3}{x} = \frac{10}{x}$

5. $\frac{1}{y} + \frac{2}{3} = \frac{1}{3y}$

6. $\frac{5}{4z} + \frac{5}{2z} = \frac{5}{4}$

7. $\frac{3}{x} - \frac{x-3}{2x} = \frac{10}{x^2}$

8. $\frac{1}{m+2} - \frac{2}{m} = \frac{1}{m^2+2m}$

9. $\frac{6y}{y-1} = \frac{5y+1}{y-1}$

10. $\frac{3}{c} - \frac{2}{c-1} = \frac{1}{c^2-c}$

11. $\frac{b-3}{4b} - \frac{2}{b+3} = 0$

12. $\frac{6-a}{6a} = \frac{1}{a+1}$

13. $\frac{3d-1}{d+3} = \frac{d+3}{d}$

14. $\frac{1}{z^2-z} = \frac{3-z}{z}$

15. $\frac{8}{y+2} = 3 - \frac{y+6}{y}$

16. $\frac{90}{x} = \frac{90}{x+5} + 3$

17. $t = \frac{25}{12} - \frac{1}{t}$

18. $v - \frac{9}{v} = -\frac{9}{2}$

19. $n - \frac{1}{n} = \frac{5}{6}$

20. $x + \frac{4}{x} = \frac{25}{6}$

21. $\frac{14}{n-6} - \frac{1}{2} = \frac{n-19}{2}$

22. $\frac{r+2}{r-2} + \frac{7}{3} = \frac{5-r}{3}$

23. $\frac{4}{c-1} = \frac{4}{c^2-c} - 2$

24. $\frac{10}{x^2-1} = \frac{5}{x-1} - 3$

B 25. $\frac{-2}{m^2-4m+4} = \frac{2m}{m-2}$

26. $\frac{2a+18}{a^2+10a+25} = \frac{a}{a+5}$

27. $\frac{-2x-8}{x^2-8x+16}=\frac{3x}{x-4}$

28. $\frac{5g}{g+3}=\frac{-8g+10}{g^2+6g+9}$

29. $\frac{6}{s-1}+2=\frac{12}{s^2-1}$

30. $\frac{5}{r-3}-\frac{30}{r^2-9}=1$

31. $\frac{5}{2t+6}-\frac{1-2t}{4t}=2$

32. $\frac{2}{n+1}+\frac{1-n}{n}=\frac{1}{n^2+n}$

33. $\frac{12}{y^2-4}=\frac{3}{y-2}-1$

34. $\frac{2}{x^2-x}=1+\frac{2}{x-1}$

35. $\frac{6x}{x+3}-\frac{x-6}{x-3}=\frac{5x^2-12}{x^2-9}$

36. $\frac{z^2+1}{z^2-1}=\frac{z}{z-1}+\frac{2}{z+1}$

C 37. $\frac{r+3}{r^2-1}+\frac{r-3}{r^2-r}=\frac{2r}{r^2+r}$

38. $\frac{2}{a-3}-\frac{2}{a+3}=\frac{3}{a-1}$

39. $\frac{2x}{x^2-4}-\frac{x-3}{x^2+2x-8}=\frac{7}{x^2+2x-8}$

40. $\frac{4}{c+1}-\frac{c^2-2c+2}{c^2-2c-3}=\frac{c}{3-c}$

41. $\frac{b}{b+1}-\frac{b+1}{b-4}=\frac{5}{b^2-3b-4}$

42. $\frac{a}{a-3}+\frac{a^2}{a^2-7a+12}=\frac{2a+1}{a-4}$

Section 7-3 SOLVING QUADRATIC EQUATIONS BY TAKING SQUARE ROOTS

In Section 7-1 you learned to solve quadratic equations by factoring. In this section you will learn a second method of solving quadratic equations as well as the meaning of the following terms:

- radical
- radicand
- square root
- perfect square
- simplified radical

An expression such as $\sqrt{y}$ is called a *radical.* The symbol $\sqrt{}$ is a *radical sign* and the expression under the radical sign is the *radicand.*

A number x is defined to be a *square root* of a nonnegative number y if

$$x^2=y.$$

Since the square of every real number is nonnegative, only nonnegative numbers have square roots in the set of real numbers. Since $y=x^2=(-x)^2$, every positive number has two square roots that are additive inverses. The positive square root is called the *principal square root* and is denoted by

$$\sqrt{y}.$$

The *negative square root* is denoted by

$$-\sqrt{y}.$$

EXAMPLES

1. $\sqrt{49} = 7$
2. $-\sqrt{49} = -7$
3. $-\sqrt{16} = -4$
4. $\sqrt{0} = -\sqrt{0} = 0$
5. $\sqrt{-16}$ does not represent a real number.

When working with radicals in this book we will assume that a radicand containing variables is never negative and that all variables in the radicand represent nonnegative integers.

Some numbers have square roots that are rational. Such numbers are called *perfect squares.* A square root radical with a monomial radicand is *simplified* if the radicand contains no factor that is a perfect square. For example, $\sqrt{22}$ is simplified, but $\sqrt{12}$ is not because $\sqrt{12} = \sqrt{4 \cdot 3}$.

EXAMPLE Simplify: $\sqrt{441}$

SOLUTION $441 = 9 \cdot 49 = 3^2 \cdot 7^2$

$$\sqrt{441} = \sqrt{3^2 \cdot 7^2} = \sqrt{3^2} \cdot \sqrt{7^2} = 3 \cdot 7 = 21$$

Check $21 \cdot 21 = 441$

EXAMPLE Simplify: $-\sqrt{2704}$

SOLUTION $2704 = 4 \cdot 676 = 4 \cdot 4 \cdot 169 = 4^2 \cdot 13^2$

$$-\sqrt{2704} = -\sqrt{4^2 \cdot 13^2} = -\sqrt{4^2} \cdot \sqrt{13^2} = -(4 \cdot 13) = -52$$

Check $(-52)(-52) = 2704$

The method of simplification just illustrated is based on the following property:

Product Property of Square Roots

If $a \geqslant 0$ and $b \geqslant 0$, then

$$\sqrt{ab} = \sqrt{a} \cdot \sqrt{b}$$

EXAMPLE Simplify: $\sqrt{72y^2}$

SOLUTION $72y^2 = 36y^2 \cdot 2$

$$\sqrt{72y^2} = \sqrt{36y^2 \cdot 2} = \sqrt{36y^2} \cdot \sqrt{2} = 6y\sqrt{2}$$

To simplify a square-root radical with a fractional radicand, use this property:

Quotient Property of Square Roots

If $a \geqslant 0$ and $b > 0$, then

$$\sqrt{\frac{a}{b}} = \frac{\sqrt{a}}{\sqrt{b}}$$

EXAMPLE Simplify: $\sqrt{\frac{3}{4}}$

SOLUTION $\sqrt{\frac{3}{4}} = \frac{\sqrt{3}}{\sqrt{4}} = \frac{\sqrt{3}}{2}$

EXAMPLE Simplify: $\sqrt{\frac{16a^4}{81x^2}}$

SOLUTION $\frac{16a^4}{81x^2} = \frac{(4a^2)^2}{(9x)^2}$

$$\sqrt{\frac{16a^4}{81x^2}} = \sqrt{\frac{(4a^2)^2}{(9x)^2}} = \frac{\sqrt{(4a^2)^2}}{\sqrt{(9x)^2}} = \frac{4a^2}{9x}$$

A square-root radical expression is in *simplified form* (or *simplest radical form*) when:

1. the radicand contains no factor, other than 1, that is a perfect square;
2. the radicand contains no fractions;
3. no radical appears in a denominator.

EXAMPLES Simplify.

1. $\sqrt{\frac{3}{5}}$ **2.** $\sqrt{\frac{7}{12}}$ **3.** $\frac{5}{\sqrt{18}}$

SOLUTION

1. $\sqrt{\frac{3}{5}} = \frac{\sqrt{3}}{\sqrt{5}} = \frac{\sqrt{3}}{\sqrt{5}} \cdot \frac{\sqrt{5}}{\sqrt{5}} = \frac{\sqrt{15}}{5}$

2. $\sqrt{\frac{7}{12}} = \frac{\sqrt{7}}{\sqrt{12}} = \frac{\sqrt{7}}{2\sqrt{3}} = \frac{\sqrt{7}}{2\sqrt{3}} \cdot \frac{\sqrt{3}}{\sqrt{3}} = \frac{\sqrt{21}}{2 \cdot 3} = \frac{\sqrt{21}}{6}$

3. $\frac{5}{\sqrt{18}} = \frac{5}{3\sqrt{2}} = \frac{5}{3\sqrt{2}} \cdot \frac{\sqrt{2}}{\sqrt{2}} = \frac{5\sqrt{2}}{6}$

Certain types of quadratic equations can be solved by taking the square root of each side. This method may be simpler and faster than the factoring method described in Section 7-1.

For example, consider the equation

$$x^2 = 25.$$

This equation states that x is a square root of 25. Since 25 has two square roots, 5 and -5, we know that

$$x = 5 \quad \text{or} \quad x = -5,$$

which is often abbreviated

$$x = \pm 5.$$

EXAMPLE Solve: $9x^2 = 49$

SOLUTION

$$9x^2 = 49$$

Divide both sides by 9. $x^2 = \frac{49}{9}$

Take the square root of both sides. $x = \pm\sqrt{\frac{49}{9}}$

Simplify. $x = \pm\frac{7}{3}$

The solution set of $9x^2 = 49$ is $\left\{\pm\frac{7}{3}\right\}$.

EXAMPLE Solve: $w^2 - 40 = 0$

SOLUTION

$$w^2 - 40 = 0$$
$$w^2 = 40$$
$$w = \pm\sqrt{40} = \pm\sqrt{4 \cdot 10}$$
$$w = \pm 2\sqrt{10}$$

The solution set of $w^2 - 40 = 0$ is $\{\pm 2\sqrt{10}\}$.

EXAMPLE Solve: $(x - 3)^2 = 16$

SOLUTION

$$(x - 3)^2 = 16$$

Take the square root of both sides. $x - 3 = \pm 4$

Therefore, $x - 3 = 4$ or $x - 3 = -4$

$x = 7$ or $x = -1$

Check the roots in the original equation.

$(7 - 3)^2 = 16$ $\quad (-1 - 3)^2 = 16$

The solution set of $(x - 3)^2 = 16$ is $\{7, -1\}$.

ORAL EXERCISES

Simplify each expression if possible. If the radical expression does not represent a real number, so state.

1. $\sqrt{64}$ **2.** $-\sqrt{144}$ **3.** $\sqrt{-4}$ **4.** $\pm\sqrt{4}$ **5.** $\frac{2}{\sqrt{3}}$

6. $\pm\sqrt{0}$ **7.** $-\sqrt{400}$ **8.** $\sqrt{121}$ **9.** $\sqrt{-25}$ **10.** $\sqrt{\frac{1}{2}}$

11. State the root(s) of each equation.

a. $x = \sqrt{16}$ **b.** $y^2 = 16$ **c.** $x = \sqrt{0}$ **d.** $y^2 = 0$

Solve.

12. $x^2 = 36$ **13.** $r^2 - 49 = 0$ **14.** $3n^2 = 27$ **15.** $\frac{1}{5}w^2 = 80$

16. Does the equation $x^2 + 4 = 0$ have any real roots? Explain.

WRITTEN EXERCISES

Simplify.

A 1. $\sqrt{484}$ 2. $-\sqrt{225}$ 3. $-\sqrt{256}$

4. $\sqrt{\frac{100}{49}}$ 5. $\sqrt{\frac{81}{400}}$ 6. $-\sqrt{\frac{7}{16}}$

7. $\sqrt{36x^2}$ 8. $\sqrt{16y^2}$ 9. $-\sqrt{50x^4}$

10. $-\sqrt{98y^4}$ 11. $-\sqrt{80x^2y^8}$ 12. $-\sqrt{120a^4b^4}$

13. $-\sqrt{\frac{13}{25}}$ 14. $\sqrt{\frac{27}{100t^2}}$ 15. $\frac{\sqrt{60a^4}}{\sqrt{441}}$

16. $\frac{\sqrt{11}}{\sqrt{24}}$ 17. $\frac{\sqrt{12}}{\sqrt{75}}$ 18. $\frac{9}{\sqrt{48}}$

Solve by taking square roots.

19. $a^2 = 121$ 20. $x^2 - 0.01 = 0$ 21. $y^2 = 2.89$

22. $4c^2 = 25$ 23. $16b^2 = 196$ 24. $9z^2 = 0$

25. $(v - 5)^2 = 49$ 26. $(t + 7)^2 = 625$ 27. $(w + 11)^2 = 900$

28. $(m - 4)^2 = 121$ 29. $(a + 6)^2 = 144$ 30. $r^2 + 4r = 4r + 81$

31. $5x + x^2 = 144 + 5x$ 32. $80 - b = 2b^2 - b$ 33. $f - 240 = f + 4f^2$

34. $r^2 - 75 = 0$ 35. $\frac{c^2}{2} = 54$ 36. $\frac{f^2}{3} = 32$

37. $9x^2 = 80$ 38. $25z^2 = 243$ 39. $81y^2 - 75 = 0$

40. $121c^2 - 140 = 0$ 41. $36z^2 - 1 = 0$ 42. $25c^2 = 1$

Simplify.

B 43. $\sqrt{98x^3y^4}$ 44. $\sqrt{54a^2b^5}$ 45. $-\sqrt{36x^7}$ 46. $-\sqrt{64y^5}$

47. $(-4 + \sqrt{6})^2$ 48. $(3 - \sqrt{2})^2$ 49. $(5 + \sqrt{3})(5 - \sqrt{3})$ 50. $(4 + \sqrt{7})(4 - \sqrt{7})$

Solve by taking square roots.

51. $z^2 - 50 = 6$ 52. $r^2 + 60 = 12$ 53. $b^2 - 5 = 81$

54. $d^2 - 6 = 42$ 55. $2x^2 - 9 = 89$ 56. $5z^2 - 8 = 92$

57. $8c^2 - 5 = 59$ 58. $7s^2 - 11 = 87$ 59. $4r^2 - 5 = 18$

60. $9t^2 - 10 = 64$ 61. $25e^2 - 6 = 54$ 62. $16h^2 - 24 = 84$

63. $15t^2 = 3$ 64. $7m^2 - 4 = 0$ 65. $5r^2 - 7 = 9$

66. $(2w + 3)^2 = 81$ 67. $(7b - 6)^2 = -1$ 68. $(9 + 4u)^2 = 25$

69. $(4 - x)^2 = 96$ 70. $(6 - x)^2 = 120$ 71. $(r + 3)^2 = 32$

72. $(s + 7)^2 = 147$ 73. $(11 + z)^2 = 500$ 74. $(13 + c)^2 = 88$

75. $\left(x - \frac{1}{2}\right)^2 = \frac{1}{4}$ 76. $\left(y - \frac{1}{3}\right)^2 = \frac{1}{9}$ 77. $\left(v + \frac{2}{3}\right)^2 = \frac{4}{9}$

EXAMPLE $(x\sqrt{5} - 1)^2 = 4$

SOLUTION

$$x\sqrt{5} - 1 = 2 \quad \text{or} \quad x\sqrt{5} - 1 = -2$$
$$x\sqrt{5} = 3 \quad\quad x\sqrt{5} = -1$$
$$x = \frac{3}{\sqrt{5}} \quad\quad x = \frac{-1}{\sqrt{5}}$$
$$x = \frac{3}{\sqrt{5}} \cdot \frac{\sqrt{5}}{\sqrt{5}} \quad\quad x = \frac{-1}{\sqrt{5}} \cdot \frac{\sqrt{5}}{\sqrt{5}}$$
$$x = \frac{3\sqrt{5}}{5} \quad \text{or} \quad x = \frac{-\sqrt{5}}{5}$$

C **78.** $(y\sqrt{2} + 3)^2 = 9$ **79.** $(m\sqrt{7} + 6)^2 = 49$ **80.** $(n\sqrt{3} - 2)^2 = 16$

81. $(b\sqrt{13} - 4)^2 = 25$ **82.** $(c\sqrt{5} - 6)^2 = 100$ **83.** $(r\sqrt{2} + 10)^2 = 8$

Section 7-4 SOLVING PROBLEMS WITH QUADRATIC EQUATIONS

In order to solve some kinds of problems, it may be necessary to write and solve quadratic equations. You should follow the same general procedure outlined in Section 1-8.

Step 1 Choose a variable and state what the variable represents.

Step 2 Write an equation that describes the facts presented in the problem.

Step 3 Solve the equation.

Step 4 Check your solution(s). This step is particularly important when working with a quadratic equation since the solution set may contain two different roots. It is possible that only one, both, or neither of these numbers satisfies the conditions of the problem.

Step 5 Write your answer.

EXAMPLE The base of a parallelogram is twice as long as the altitude to that base. If the area is 72 cm², what is the length of the altitude?

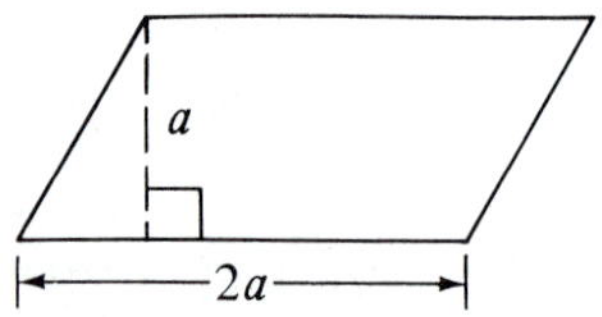

SOLUTION Choose a variable, say a, to represent the length of the altitude of the parallelogram. Let a = the length, in centimeters, of the altitude. Then $2a$ = the length of the base. For a parallelogram,

$$\text{(base)(altitude)} = \text{area}.$$

An equation describing the known facts is

$$2a \cdot a = 72.$$

Simplify.

$$2a^2 = 72$$
$$a^2 = 36$$
$$a = \pm 6$$

Since the length of a line segment cannot be negative, reject the root -6. Therefore, $a = 6$.

Check If the length of the altitude is 6 cm, the length of the base is 12 cm and the area is

$$(12 \text{ cm})(6 \text{ cm}) = 72 \text{ cm}^2.$$

Thus the length of the altitude is 6 cm.

EXAMPLE The difference in the lengths of two legs of a right triangle is 5 cm. If the area of the triangle is 150 cm², what is the length of each leg?

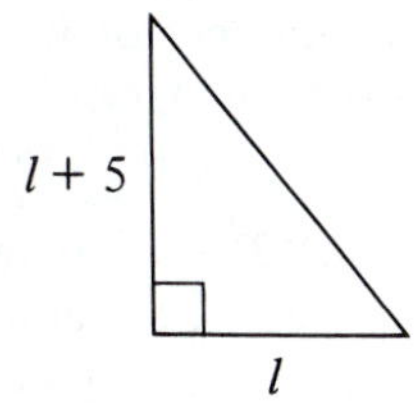

SOLUTION Let l = the length in centimeters of one leg.
Then $l + 5$ = the length of the other leg.
The formula for the area of a triangle is

$$\frac{1}{2}(\text{base})(\text{height}) = \text{area}.$$

Since the lengths of the legs of a right triangle may be used as its base and height, you have

$$\frac{1}{2}l(l + 5) = 150.$$

Multiply both sides by 2. $l(l + 5) = 300$

Simplify. $l^2 + 5l = 300$

$$l^2 + 5l - 300 = 0$$

$$(l + 20)(l - 15) = 0$$

$$l + 20 = 0 \quad \text{or} \quad l - 15 = 0$$

$$l = -20 \quad \text{or} \quad l = 15$$

As in the previous example, you reject the negative root. Thus $l = 15$.

Check If $l = 15$, then $l + 5 = 20$.

$$\frac{1}{2}(15 \text{ cm})(20 \text{ cm}) = 150 \text{ cm}^2$$

Therefore the lengths of the legs are 15 cm and 20 cm.

EXAMPLE The sum of the squares of two consecutive even integers is 52. What are the integers?

SOLUTION Let x = the first even integer.
Then $x + 2$ = the next consecutive even integer.

The problem states that: $x^2 + (x + 2)^2 = 52$

Simplify. $x^2 + x^2 + 4x + 4 = 52$

Express in standard form. $2x^2 + 4x - 48 = 0$

Divide each side by 2 to simplify. $x^2 + 2x - 24 = 0$

$$(x + 6)(x - 4) = 0$$

$$x + 6 = 0 \quad \text{or} \quad x - 4 = 0$$

$$x = -6 \quad \text{or} \quad x = 4$$

Check If $x = -6$, then $x + 2 = -4$. $(-6)^2 + (-4)^2 = 36 + 16 = 52$
If $x = 4$, then $x + 2 = 6$. $4^2 + 6^2 = 16 + 36 = 52$

The two integers are -6 and -4 or 4 and 6.

One application of quadratic equations is solving a type of problem known as a *work problem.* One method of solving a work problem involves first writing an equation of the form

$$\frac{1}{t_1} + \frac{1}{t_2} = \frac{1}{T}$$

in which t_1 is the number of hours required by one worker to do the job alone, t_2 is the number of hours required by the second worker to do the job alone, and T is the number of hours required to do the job if they work together.

EXAMPLE Staley and Beth can install 35 m^2 of carpet in 6 h working together. If Staley works alone, he requires 5 h more than Beth doing the same job alone. How long would it take Beth to install 35 m^2 of carpet working alone?

SOLUTION Let x = the number of hours Beth needs to do the job alone.
Then $x + 5$ = the number of hours Staley needs to do the job alone.

If Beth does the whole job in x hours, then the part of the job she does in 1 h is $\frac{1}{x}$.

If Staley does the whole job in $(x + 5)$ hours, then the part of the job he does in 1 h is $\frac{1}{x+5}$.

If together they do the whole job in 6 hours, then the part of the job they do together in 1 h is $\frac{1}{6}$.

You may wish to use a table to help you organize the information.

	Beth	Staley	Together
Time needed (h)	x	$x + 5$	6
Part done in 1 h	$\frac{1}{x}$	$\frac{1}{x+5}$	$\frac{1}{6}$

$$\underbrace{\text{Part Beth does in 1 h}} + \underbrace{\text{Part Staley does in 1 h}} = \underbrace{\text{Part both do in 1 h}}$$

$$\frac{1}{x} + \frac{1}{x+5} = \frac{1}{6}$$

Now that you have found an equation that describes the problem, the next step is to solve the equation.

LCD $= 6x(x + 5)$ $\qquad 6x(x+5)\left(\frac{1}{x} + \frac{1}{x+5}\right) = 6x(x+5)\frac{1}{6}$

$$6(x + 5) + 6x = x(x + 5)$$
$$6x + 30 + 6x = x^2 + 5x$$
$$12x + 30 = x^2 + 5x$$
$$x^2 - 7x - 30 = 0$$
$$(x - 10)(x + 3) = 0$$
$$x = 10 \quad \text{or} \quad x = -3$$

Reject the root -3.

Check $x = 10$

Beth does the job in 10 h working alone.
Staley does the job in 15 h working alone.

$$\frac{1}{10} + \frac{1}{15} \stackrel{?}{=} \frac{1}{6}$$

$$\frac{1}{10} + \frac{1}{15} = \frac{3}{30} + \frac{2}{30}$$

$$= \frac{5}{30} = \frac{1}{6}$$

Therefore, it would take Beth 10 h to install 35 m^2 of carpet when working alone.

ORAL EXERCISES

State an equation needed to solve the problem.

1. Let a represent the altitude, in meters, of a parallelogram whose base is three times as long as the altitude to that base. If the area of the parallelogram is 108 m^2, what is the length of the altitude?
2. Let x represent the length in centimeters of a leg of an isosceles right triangle. If the area of the triangle is 32 cm^2, what is the length of a leg?

Given the area of the shaded portion, state a quadratic equation you could use to find the value of x.

3. 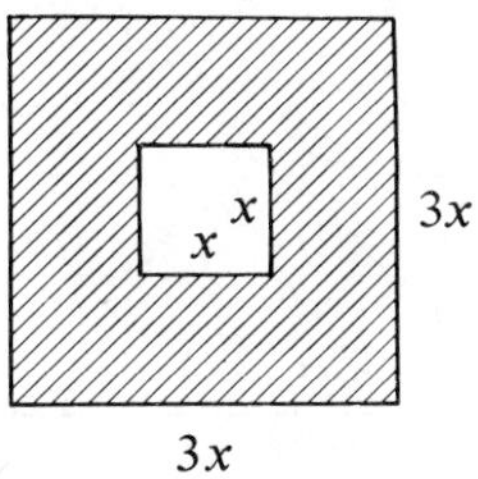

Area = 72 cm^2

4. 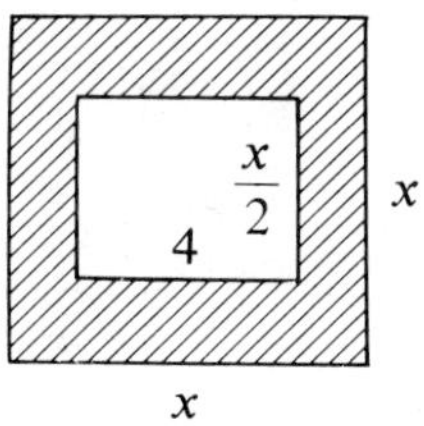

Area = 63 cm^2

5. 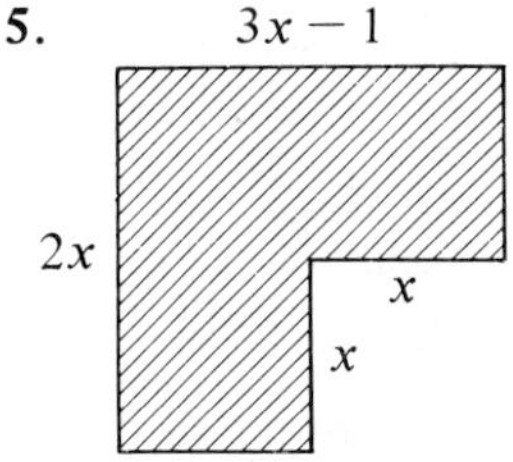

Area = 3 m^2

6. The area of a square is 121 m^2.
   a. What is the length of each side?
   b. What is the perimeter of the square?
7. Emanuel and Jorge can load a truck in 2 h working together. If Emanuel works alone he requires 3 h less than Jorge to do the same job alone. Complete the table.

	Jorge	Emanuel	Together
Time needed (h)	x	?	2
Part done in 1 h	?	?	?

WRITTEN EXERCISES

A **1–3.** Solve Oral Exercises 1–3.

4. One leg of a right triangle is twice as long as the other leg. The area of the triangle is 16 m^2. What is the length of the shorter leg?

5. The length of a rectangle is 3 cm longer than the width. If the area of the rectangle is 40 cm^2, find the dimensions of the rectangle.

6. The sum of the squares of two consecutive integers is 113. What are the integers?

7. The sum of the squares of two consecutive odd integers is 202. Find the lesser integer.

8. If 5 is subtracted from twice the square of a negative number, the difference is 27. What is the number?

9. If a positive number is subtracted from the square of the number, the difference is 20. What is the number?

10. The square of a number decreased by twice the number is 35. What is the number?

11. The square of a number increased by three times the number is 40. What is the number?

12, 13. Solve Oral Exercises 4 and 5.

14. Find an expression in simplest form for the area of each shaded region below.

a.

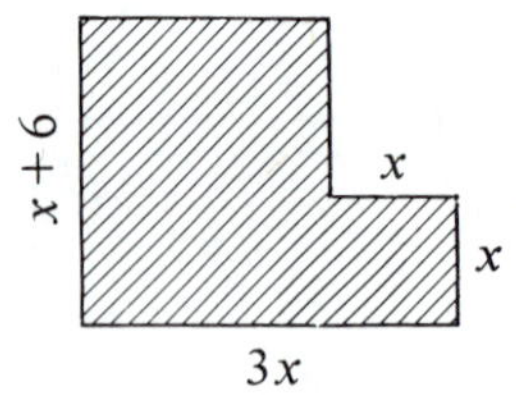

b.

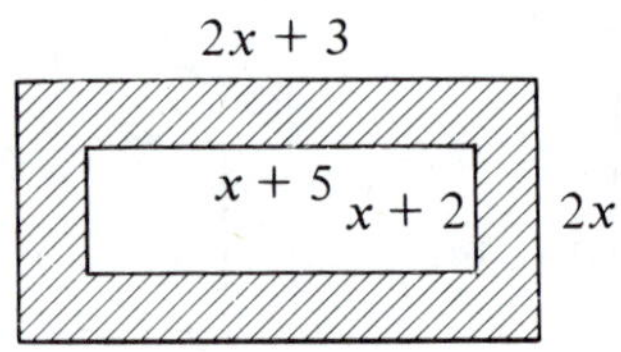

15. If the area of the shaded region in Exercise 14a is 96 m^2, find x.

16. If the area of the shaded region in Exercise 14b is 92 cm^2, find x.

17. A swimming pool 3 m by 6 m is surrounded by a square lawn. The area of the lawn is seven times the area of the pool. Find the dimensions of the lawn.

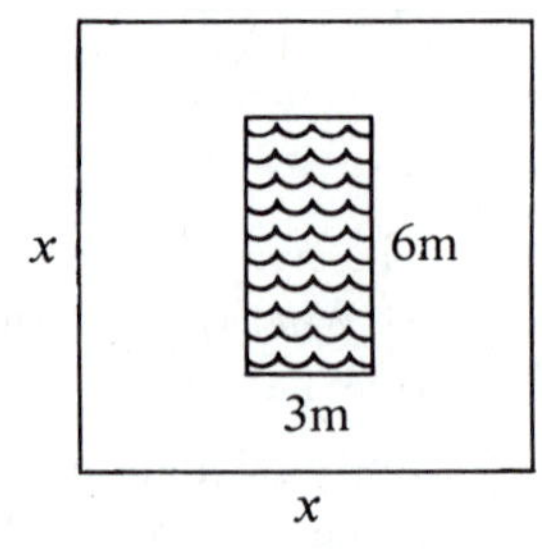

18. Refer to Oral Exercise 7. How long would each take to do the job alone?

B **19.** If a positive number is added to the square of the number, the sum is eight times the number. What is the number?

20. If the square of a number is subtracted from seven times the number, the difference is 12. What is the number?

21. If the square of a negative number is subtracted from twenty times the number, the difference is -44. What is the number?

22. If twice the square of a positive number is added to the number, the sum is 21. What is the number?

23. Wanda and her younger brother, Jim, can rake the yard in 1.6 h working together. Working alone, it takes Jim 6 h more than Wanda to do the job. How long does it take each of them to do the raking alone?

24. A house painter and the house painter's apprentice can paint one side of a house in 3 h. The apprentice alone takes 8 h more than the master painter to paint one side of a house. How long does it take the apprentice?

25. Two presses, working together, can stamp 1000 pieces of metal in 24 min. The newer machine can do the job alone in 20 min less than the older machine. How long does it take the older machine working alone to stamp 1000 pieces of metal?

26. One pipe can fill an empty tank in $\frac{2}{3}$ the time it takes a second pipe to do the same job. Working together, the two pipes can fill the empty tank in 12 min. How long does it take each pipe to fill the tank?

Use a quadratic equation to solve each problem. Some problems may contain more information than is necessary to solve the problem.

27. One side of a rectangle is $\frac{2}{3}$ the length of the other side. The perimeter is 5 times the width. If the area is 108, what are the dimensions of the rectangle?

28. The width of a rectangle is $\frac{3}{4}$ the length. The perimeter is $3\frac{1}{2}$ times the length. If the area is 108, what are the dimensions of the rectangle?

EXAMPLE The sum of two numbers is 3 and the sum of their squares is 269. Find the numbers.

SOLUTION Let x = the first number.
Then $3 - x$ = the second number.

$$x^2 + (3 - x)^2 = 269$$
$$x^2 + 9 - 6x + x^2 = 269$$
$$2x^2 - 6x - 260 = 0$$
$$x^2 - 3x - 130 = 0$$
$$(x - 13)(x + 10) = 0$$
$$x = 13 \quad \text{or} \quad x = -10$$

The numbers are 13 and -10.

Solve.

29. The sum of two numbers is 15. The sum of their squares is 153. Find the numbers.

30. The sum of two numbers is -4. The sum of their squares is 170. Find the numbers.

31. The difference of two numbers is 4. The sum of their squares is 250. Find the numbers.

32. The difference of two numbers is -11. The sum of their squares is 61. Find the numbers.

33. Each side of one square is 2 cm longer than each side of a smaller square. If the sum of their areas is 130 cm^2, find the length of a side of the smaller square.

34. If the area of the square equals the area of the rectangle, find the value of x.

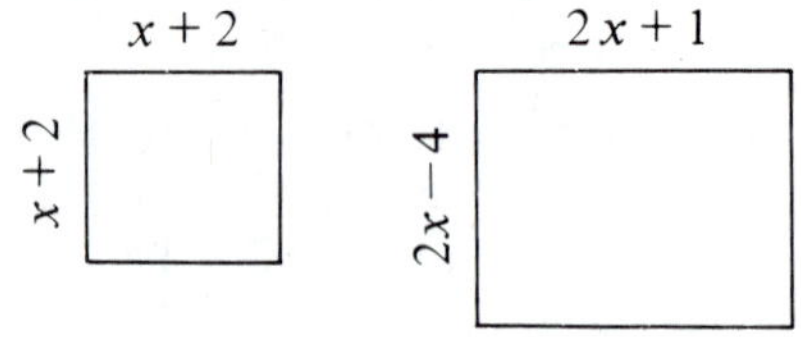

C 35. Find three consecutive odd integers such that the sum of the squares of the first and third is 89 more than the square of the second.

36. Find three consecutive even integers such that the square of the sum of the first and second is 420 more than the square of the third.

You can solve work problems by using a formula similar to the one you use for motion problems.

Rate	· Time	= Work completed
(part of job/h)	(h)	(part of job done)

EXAMPLE A farmer and her son plow a field together for 3 h, then she leaves and he finishes the job 1.5 h later. If it takes him 3 h more than it takes his mother to plow the field alone, how many hours would it take him to plow the field by himself?

SOLUTION Let $x =$ the number of hours it takes the mother to do the job.
Then $x + 3 =$ the number of hours it takes her son to do the job.
(If the job is completed, then the "Work completed" must equal 1.)

$$\frac{1}{x} \cdot 3 + \frac{1}{x+3} \cdot 3 + \frac{1}{x+3} \cdot \frac{3}{2} = 1$$

$$\frac{3}{x} + \frac{3}{x+3} + \frac{3}{2(x+3)} = 1$$

$$3 \cdot 2(x+3) + 3(2x) + 3x = 2x(x+3)$$

$$6x + 18 + 6x + 3x = 2x^2 + 3x$$

$$2x^2 - 9x - 18 = 0$$

$$(2x+3)(x-6) = 0$$

$$x = 6$$

$$x + 3 = 9$$

The son would take 9 h to plow the field.

37. Elinor needs 2 h more than Luke does to trim the hedges. Luke works alone trimming the hedges for 2 h, then Elinor joins him, and they finish the job 1.75 h later. How long would it take Luke to do the job by himself?

38. Leona is studying to be a mechanic. It takes her 12 h more than her boss to reassemble an engine. Leona started helping her boss reassemble an engine 2 h after her boss started the job. Together they completed the job 3 h later. How long does it take Leona to do the job alone?

SELF-TEST 1

Solve by factoring.

1. $a^2 + 10a - 24 = 0$ 2. $b^2 + 45 = 14b$ Section 7-1

3. $c^2 + 7c = 30$ 4. $4d^2 - 1 = 0$

Solve.

5. $\frac{9}{5} - \frac{2}{r} = \frac{17}{5r}$ 6. $\frac{8}{3s} + \frac{1}{6} = \frac{2}{s}$ Section 7-2

7. $\frac{4}{t} + \frac{3}{t+1} = 3$ 8. $\frac{8}{u-1} - 7 = \frac{3}{u^2 - 1}$

Simplify.

9. $-\sqrt{48x^4y^2}$ 10. $\sqrt{\frac{16}{5}}$ Section 7-3

Solve by taking square roots.

11. $9k^2 = 100$ 12. $(p-1)^2 = 49$

13. $g^2 - 63 = 0$ 14. $2 = \frac{1}{3}z^2 - 2$

Solve.

15. The sum of the squares of two consecutive integers is 61. Find the integers. Section 7-4

16. The length of a rectangle is 4 cm less than twice its width. If its area is 70 cm^2, find its dimensions.

17. Two air conditioners can cool a house in 1.2 h. The smaller air conditioner, working alone, takes 1 h more than the larger one to do the job. How many hours does each need to cool the house?

Section 7-5 COMPLETING THE SQUARE

From the last example of Section 7-3 you know that a quadratic equation can be solved if it can first be written in the form

$$(x + m)^2 = d \quad d \geqslant 0$$

The process of adding a constant to a quadratic polynomial in the form $x^2 + 2mx$ to make a trinomial square, $x^2 + 2mx + m^2$, is called *completing the square.*

Study these squares of binomials and look for patterns.

	Coefficient of x	*Constant term*
$(x-3)^2 = x^2 - 6x + 9$	-6	9
$(x+7)^2 = x^2 + 14x + 49$	14	49
$\left(x+\frac{1}{2}\right)^2 = x^2 + x + \frac{1}{4}$	1	$\frac{1}{4}$

Notice that in each case the constant term is the square of half the coefficient of x. That is, $9 = \left(\frac{-6}{2}\right)^2$, $49 = \left(\frac{14}{2}\right)^2$, and $\frac{1}{4} = \left(\frac{1}{2}\right)^2$. This is because

$$(x \pm m)^2 = x^2 \pm 2mx + m^2.$$

EXAMPLE Find a value for k so that $x^2 - \frac{1}{5}x + k$ is a trinomial square.

SOLUTION The coefficient of x is $-\frac{1}{5}$, so $k = \left[\frac{1}{2}\left(-\frac{1}{5}\right)\right]^2 = \left(-\frac{1}{10}\right)^2 = \frac{1}{100}$.

Check $x^2 - \frac{1}{5}x + \frac{1}{100} = \left(x - \frac{1}{10}\right)^2$

You can use the process of completing the square to solve quadratic equations.

EXAMPLE Solve by completing the square: $y^2 - 12y + 27 = 0$

SOLUTION

1. Write the equation with the constant on the right side. $\quad y^2 - 12y = -27$
2. Add to each side the square of half the coefficient of y. $\quad y^2 - 12y + 36 = -27 + 36$
3. Factor the left side. $\quad (y-6)^2 = 9$
4. Take the square root of both sides. $\quad y - 6 = 3 \quad$ or $\quad y - 6 = -3$
5. Solve. $\quad y = 9 \quad$ or $\quad y = 3$
6. Check the solutions.

$9^2 - 12 \cdot 9 + 27 = 81 - 108 + 27 = -27 + 27 = 0$

$3^2 - 12 \cdot 3 + 27 = 9 - 36 + 27 = -27 + 27 = 0$

The solution set is $\{9, 3\}$.

Solving by completing the square also allows you to solve quadratic equations whose roots are irrational.

EXAMPLE Solve by completing the square: $x^2 + 8x + 10 = 0$

SOLUTION

1. $x^2 + 8x = -10$
2. $x^2 + 8x + 16 = -10 + 16$
3. $(x+4)^2 = 6$
4. $x + 4 = \sqrt{6} \quad$ or $\quad x + 4 = -\sqrt{6}$
5. $x = -4 + \sqrt{6} \quad$ or $\quad x = -4 - \sqrt{6}$

Check $(-4+\sqrt{6})^2 + 8(-4+\sqrt{6}) + 10 = (16 - 8\sqrt{6} + 6) + (-32 + 8\sqrt{6}) + 10$
$= 22 - 8\sqrt{6} - 32 + 8\sqrt{6} + 10$
$= 0$

You should check the other root, $-4 - \sqrt{6}$, yourself.
The solution set is $\{-4 \pm \sqrt{6}\}$.

If the coefficient of x^2 does not equal 1, divide every term of the equation by the coefficient of x^2. For example, before completing the square to solve

$$3x^2 - 2x - 5 = 0$$

you would divide each term by 3 to obtain

$$x^2 - \frac{2}{3}x - \frac{5}{3} = 0.$$

Now you can solve this equation by completing the square.

$$x^2 - \frac{2}{3}x = \frac{5}{3}$$

Add to each side $\left[\frac{1}{2}\left(-\frac{2}{3}\right)\right]^2 = \frac{1}{9}$.

$$x^2 - \frac{2}{3}x + \frac{1}{9} = \frac{5}{3} + \frac{1}{9}$$

$$x^2 - \frac{2}{3}x + \frac{1}{9} = \frac{15}{9} + \frac{1}{9}$$

$$\left(x - \frac{1}{3}\right)^2 = \frac{16}{9}$$

$$x - \frac{1}{3} = \frac{4}{3} \quad \text{or} \quad x - \frac{1}{3} = -\frac{4}{3}$$

$$x = \frac{5}{3} \quad \text{or} \quad x = -1$$

ORAL EXERCISES

In each case state the term necessary to complete the square.

1. $y^2 - 18y + \underline{\ ?\ }$
2. $p^2 - 14p + \underline{\ ?\ }$
3. $x^2 + x + \underline{\ ?\ }$
4. $z^2 + 5z + \underline{\ ?\ }$
5. $r^2 - \frac{1}{2}r + \underline{\ ?\ }$
6. $t^2 - \frac{3}{4}t + \underline{\ ?\ }$
7. $y^2 + 2cy + \underline{\ ?\ }$
8. $z^2 + az + \underline{\ ?\ }$

WRITTEN EXERCISES

Find a value for k so that the polynomial is a trinomial square.

A 1. $x^2 + 16x + k$
2. $b^2 + 22b + k$
3. $r^2 - \frac{2}{3}r + k$

Solve by completing the square.

4. $x^2 - 4x = 0$
5. $a^2 - 2a = 0$
6. $z^2 + 5z = 0$
7. $v^2 + 7v = 0$
8. $r^2 - 2r = 15$
9. $z^2 + 18z = -80$
10. $t^2 + 14t + 49 = 0$
11. $c^2 + 26c = -120$
12. $b^2 + 10b = -21$

13. $x^2 + x - 6 = 0$

14. $y^2 + 9y + 20 = 0$

15. $v^2 - 3v - 28 = 0$

16. $u^2 + 11u + 18 = 0$

17. $a^2 + 4a - 24 = 0$

18. $x^2 - 10x + 10 = 0$

19. $z^2 - 8z = -4$

20. $c^2 + 6c = 31$

21. $3u^2 + 4u + 1 = 0$

22. $2e^2 - 3e = 5$

23. $2c^2 + 5c = 3$

24. $4r^2 - 11r + 6 = 0$

B 25. $2r^2 + 8r + 3 = 0$

26. $3y^2 + 12y = 2$

27. $5 - 6t - 3t^2 = 0$

28. $2a^2 - 7 - 10a = 0$

29. $5d^2 + 4d - 1 = 0$

30. $7x + 3x^2 - 10 = 0$

31. $4r^2 - 2 - r = 0$

32. $6b^2 + 2b - 3 = 0$

33. $\dfrac{1}{m-4} - \dfrac{1}{m+2} = \dfrac{m}{m+2}$

34. $\dfrac{3x+5}{x+1} = \dfrac{2x-5}{x+2}$

35. $\dfrac{2n}{n+3} - \dfrac{n-3}{n+6} = 0$

36. $\dfrac{r+1}{r-3} + \dfrac{3r+1}{r+1} = 0$

C 37. $\dfrac{1}{5}c^2 + \dfrac{3}{4}c - \dfrac{1}{2} = 0$

38. $x^2 + 2x\sqrt{2} - 4 = 0$

39. $a^2 + 4a\sqrt{3} - 6 = 0$

40. $4h^2 + 2h + 1 = h^2 - 4h + 2$

41. $4q + 3 = 2 - 2q^2 - 2q$

42. $2 + z - z^2 = 2z^2 - 3z$

Solve for x.

43. $x^2 + bx + c = 0$

44. $x^2 - mx + 2n = 0$

45. $2x^2 + ax - c = 0$

Section 7-6 THE QUADRATIC FORMULA

Suppose we have an equation of the form

$$ax^2 + bx + c = 0 \quad (a \neq 0)$$

As shown in Section 7-5, the method of completing the square can be used to solve quadratic equations with either rational or irrational roots. If we apply the method of completing the square to the general quadratic equation, shown above, we can derive a formula that expresses the roots in terms of the constants a, b, and c.

$$ax^2 + bx + c = 0$$

Subtract c from both sides.

$$ax^2 + bx = -c$$

Divide each term by a.

$$x^2 + \frac{b}{a}x = -\frac{c}{a}$$

Complete the square. $\left(\frac{1}{2} \cdot \frac{b}{a}\right)^2 = \frac{b^2}{4a^2}$

$$x^2 + \frac{b}{a}x + \frac{b^2}{4a^2} = -\frac{c}{a} + \frac{b^2}{4a^2}$$

Factor and simplify.

$$\left(x + \frac{b}{2a}\right)^2 = \frac{b^2 - 4ac}{4a^2}$$

Take the square root of each side.

$$x + \frac{b}{2a} = \pm\sqrt{\frac{b^2 - 4ac}{4a^2}}$$

Simplify.

$$x + \frac{b}{2a} = \pm\frac{\sqrt{b^2 - 4ac}}{2a}$$

Subtract $\frac{b}{2a}$ from both sides.

$$x = -\frac{b}{2a} \pm \frac{\sqrt{b^2 - 4ac}}{2a}$$

Simplify.

$$x = \frac{-b \pm \sqrt{b^2 - 4ac}}{2a}$$

The solution of $ax^2 + bx + c = 0$ $(a \neq 0)$ is given by the *quadratic formula*:

$$x = \frac{-b \pm \sqrt{b^2 - 4ac}}{2a}$$

Application of the quadratic formula is one of the most widely used methods of solving quadratic equations.

EXAMPLE Use the quadratic formula to solve $x^2 - 3x - 10 = 0$.

SOLUTION In this case $a = 1$, $b = -3$, and $c = -10$.
The quadratic formula gives

$$x = \frac{-(-3) \pm \sqrt{(-3)^2 - 4(1)(-10)}}{2(1)}$$

Simplify. $x = \dfrac{3 \pm \sqrt{9 + 40}}{2}$

$$x = \frac{3 \pm 7}{2}$$

So, $x = \dfrac{3 + 7}{2}$ or $x = \dfrac{3 - 7}{2}$

$x = 5$ or $x = -2$

The solution set of $x^2 - 3x - 10 = 0$ is $\{5, -2\}$.

Notice that you could have solved $x^2 - 3x - 10 = 0$ by factoring or by completing the square. The answer, of course, would be the same.

EXAMPLE Use the quadratic formula to solve $y^2 - 7y + 9 = 0$.

SOLUTION $a = 1$, $b = -7$, $c = 9$

$$y = \frac{-(-7) \pm \sqrt{(-7)^2 - 4(1)(9)}}{2(1)}$$

$$y = \frac{7 \pm \sqrt{49 - 4(9)}}{2} = \frac{7 \pm \sqrt{49 - 36}}{2}$$

$$y = \frac{7 \pm \sqrt{13}}{2}$$

So $y = \dfrac{7 + \sqrt{13}}{2}$ or $y = \dfrac{7 - \sqrt{13}}{2}$.

The solution set of $y^2 - 7y + 9 = 0$ is $\left\{\dfrac{7 + \sqrt{13}}{2}, \dfrac{7 - \sqrt{13}}{2}\right\}$.

The check is left to you in Exercise 57.

If an equation is not in standard form, then you must rewrite it in standard form before applying the quadratic formula.

EXAMPLE Use the quadratic formula to solve $4q^2 + 1 = 2q^2 - 6q$.

SOLUTION Rewriting $4q^2 + 1 = 2q^2 - 6q$ in standard form, you obtain $2q^2 + 6q + 1 = 0$.

$a = 2,\ b = 6,\ c = 1$

$$q = \frac{-6 \pm \sqrt{36 - 8}}{4}$$

$$q = \frac{-6 \pm \sqrt{28}}{4}$$

Notice that $\sqrt{28}$ is not in simplified form. Rewrite it as $\sqrt{28} = \sqrt{4 \cdot 7} = 2\sqrt{7}$.

Thus, $$q = \frac{-6 \pm 2\sqrt{7}}{4}$$

Simplify. $$q = \frac{2(-3 \pm \sqrt{7})}{2(2)} = \frac{-3 \pm \sqrt{7}}{2}$$

So, the solution set is $\left\{\frac{-3 + \sqrt{7}}{2}, \frac{-3 - \sqrt{7}}{2}\right\}$.

The check is left to you in Exercise 58.

The quadratic formula is one of the most important tools of mathematics because you can use it to solve any quadratic equation. It would be wise to memorize the quadratic formula. But if you should forget this formula, you can still solve a given quadratic equation by completing the square.

ORAL EXERCISES

Put each equation not in standard form into the form $ax^2 + bx + c = 0$ and state the values of a, b, and c.

1. $2x^2 + 4x + 7 = 0$
2. $5x^2 - 2x - 6 = 0$
3. $2x - 3 = 5x^2$
4. $7 - 6x^2 - 4x = 0$
5. $9x^2 + 5 = -3x$
6. $-8x - 3x^2 = 6$
7. Find the dimensions of the rectangle described on page 259.

WRITTEN EXERCISES

Use the quadratic formula to solve each equation.

A

1. $x^2 + 7x + 12 = 0$
2. $c^2 + 5c + 6 = 0$
3. $r^2 - 3r - 10 = 0$
4. $d^2 - 3d - 4 = 0$
5. $y^2 - 8y + 7 = 0$
6. $z^2 - 10z + 9 = 0$
7. $0 = a^2 - 6a + 9$
8. $0 = v^2 - 10v + 25$
9. $t^2 + 18t + 81 = 0$
10. $n^2 + 22n + 121 = 0$
11. $x^2 - 3x - 1 = 0$
12. $r^2 - 7r - 3 = 0$
13. $a^2 + 5a - 4 = 0$
14. $u^2 + 6u - 2 = 0$
15. $c^2 + 8c + 5 = 0$
16. $v^2 + 10v + 6 = 0$
17. $2t^2 + t = 6$
18. $2y^2 + 3y = 5$
19. $3d^2 + 10d + 3 = 0$
20. $3a^2 + 14a + 8 = 0$
21. $5z^2 + 2z = 3$
22. $0 = 9 - 10y + y^2$
23. $3f = 2f^2 - 1$
24. $0 = 7 - 6z + z^2$

B 25. $4c + 2c^2 = 3$ 26. $6x + 3x^2 = 2$ 27. $3r^2 - 4r = 6$

28. $4h^2 = 7 + 2h$ 29. $3a^2 - 6 + 5a = 2 + 3a$ 30. $5z^2 - 4 + 3z = 2 - z^2 + 2z$

31. $15 + 3u + 2u^2 = 11u + 21 - 3u^2$ 32. $2r - 1 + 3r^2 = 2 - r^2 + 6r$

33. $-2v + 7v^2 - 6 = 3v^2 - 9 + 8v$ 34. $-5 + b^2 + 2b = 4b - 2b^2 + 2$

35. $\frac{2t}{2t + 1} = \frac{2t - 1}{3t + 2}$ 36. $\frac{7 + 2x}{2} = \frac{x - 1}{x}$ 37. $\frac{3}{y - 1} - \frac{1}{y + 1} = 4$

38. $\frac{3}{2 - m} + \frac{2}{m - 1} = -2$ 39. $\frac{3a - 2}{a} - \frac{a + 3}{a^2} = 0$ 40. $\frac{2b + 3}{b + 2} = \frac{b + 4}{b - 1}$

41. $\frac{e}{e - 3} + \frac{2e}{e + 3} = \frac{1}{e^2 - 9}$ 42. $\frac{6}{x - 3} + \frac{2}{x + 4} = \frac{25}{x}$

C 43. $b^2 + 1 = b\sqrt{6}$ 44. $2c^2 - c\sqrt{10} + 1 = 0$ 45. $d^2 + d\sqrt{19} = 4$

EXAMPLE Find the roots of $\sqrt{z} - z = -6$.

SOLUTION

Isolate the radical. $\sqrt{z} = z - 6$

Square both sides. $z = z^2 - 12z + 36$

Write in standard form. $z^2 - 13z + 36 = 0$

Solve. $z = \frac{13 \pm \sqrt{169 - 144}}{2} = \frac{13 \pm \sqrt{25}}{2}$

$z = \frac{13 + 5}{2}$ or $z = \frac{13 - 5}{2}$

$z = 9$ or $z = 4$

Check the solutions in the *original* equation.

$\sqrt{9} - 9 \ ? \ -6$ $\sqrt{4} - 4 \ ? \ -6$

$3 - 9 = -6$ $2 - 4 = -2$

Therefore, the solution set of $\sqrt{z} - z = -6$ is $\{9\}$.

46. $a + 15 = 8\sqrt{a}$ 47. $b - 12 = \sqrt{b}$ 48. $x + 2\sqrt{x} - 8 = 0$

49. $y - 5\sqrt{y} + 6 = 0$ 50. $3\sqrt{z} + z = 4$ 51. $-7\sqrt{c} + c = 18$

Given the equation $ax^2 + bx + c = 0$ with $a \neq 0$ and $b^2 - 4ac > 0$, show that:

52. The sum of the roots of the equation is $-\frac{b}{a}$.

53. The product of the roots of the equation is $\frac{c}{a}$.

Find **(a)** the sum of the roots and **(b)** the product of the roots of each equation.

54. $3x^2 + 12x - 15 = 0$ 55. $5x^2 - 7x - 1 = 0$ 56. $2x^2 - x = 6$

57. Check the solutions to the second example on page 283.

58. Check the solutions to the example on page 284.

Section 7-7 CHOOSING A METHOD OF SOLUTION

In this section you will learn:

- which method to choose to solve a particular quadratic equation
- to identify the discriminant of a quadratic equation
- to use the discriminant to determine the nature of the roots of a quadratic equation

You have learned several methods of solving quadratic equations: factoring, completing the square and taking square roots, and applying the quadratic formula. Remember that you can always use the quadratic formula to solve a given quadratic equation, but if the equation can be solved by factoring or by taking square roots it may be simpler and faster to do so. You may wish to go through the reasoning process outlined below when trying to decide which method to use.

Write the equation in standard form: $ax^2 + bx + c = 0$.

Case 1 The equation simplifies to one of the form $ax^2 + c = 0 \quad (c \leqslant 0)$.
Solve by taking square roots.

Case 2 The equation simplifies to one of the form $ax^2 + bx = 0$.
Solve the equation by factoring.

Case 3 The simplest form of the equation is $ax^2 + bx + c = 0$.
Try factoring. If the factorization is not immediately obvious (especially if $a \neq 1$), use the quadratic formula.

EXAMPLE 1 Solve: $3x^2 - 12 = 0$

SOLUTION The equation is in the form $ax^2 + c = 0$. *Case 1* indicates you can solve by taking square roots.

$$\begin{aligned} 3x^2 - 12 &= 0 \\ 3x^2 &= 12 \\ x^2 &= 4 \\ x &= \pm 2 \end{aligned}$$

The solution set of $3x^2 - 12 = 0$ is $\{2, -2\}$.

EXAMPLE 2 Solve: $x^2 = 8x$

SOLUTION Rewrite in standard form as $x^2 - 8x = 0$. The equation is in the form $ax^2 + bx = 0$. *Case 2* indicates you can solve by factoring.

$$\begin{aligned} x^2 - 8x &= 0 \\ x(x - 8) &= 0 \\ x = 0 \quad \text{or} \quad x - 8 &= 0 \\ x = 0 \quad \text{or} \quad x &= 8 \end{aligned}$$

The solution set of $x^2 = 8x$ is $\{0, 8\}$.

EXAMPLE 3 Solve: $x^2 + 20 = 5 - 8x$

SOLUTION

$$\begin{aligned} x^2 + 20 &= 5 - 8x \\ x^2 + 8x + 15 &= 0 \end{aligned}$$

You may solve the equation by factoring or by applying the quadratic formula.

Since it is not difficult to find the factors of $x^2 + 8x + 15$, factorization may be the most efficient method to use.

$(x + 3)(x + 5) = 0$

$x = -3$ or $x = -5$

The solution set is $\{-3, -5\}$.

EXAMPLE Solve: $1 - 4x = x^2$

SOLUTION $x^2 + 4x - 1 = 0$

Since the factorization is not immediately obvious, apply the quadratic formula.

$$x = \frac{-4 \pm \sqrt{4^2 - 4(1)(-1)}}{2 \cdot 1}$$

$$x = \frac{-4 \pm \sqrt{20}}{2}$$

$$x = \frac{-4 \pm 2\sqrt{5}}{2}$$

$$x = -2 \pm \sqrt{5}$$

The solution set is $\{-2 + \sqrt{5}, -2 - \sqrt{5}\}$.

The expression, $b^2 - 4ac$, under the radical sign in the quadratic formula is called the *discriminant* of the quadratic equation. The discriminant allows us to determine the nature of the roots of a quadratic equation without solving it.

The value of the discriminant must be zero, greater than zero, or less than zero. Let's assume that a, b, and c represent rational numbers and consider the three cases separately.

Case 1 $b^2 - 4ac > 0$

If $b^2 - 4ac > 0$, then the quadratic formula yields two real roots of the equation.

How do we know whether the roots will be rational or irrational?

If $b^2 - 4ac$ is a perfect square, then the roots will be rational. Otherwise, the roots will be irrational.

EXAMPLES

1. $2x^2 + 3x + 1 = 0$

$b^2 - 4ac = 3^2 - 4(2)(1)$

$= 9 - 8 = 1 = 1^2$

The roots are rational.

2. $2x^2 - 5x + 1 = 0$

$b^2 - 4ac = (-5)^2 - 4(2)(1)$

$= 25 - 8 = 13$

The roots are irrational.

Check

1. $x = \dfrac{-3 \pm \sqrt{9 - 8}}{4}$

$x = \dfrac{-3 \pm 1}{4}$

$x = -1$ or $x = -\dfrac{1}{2}$

These roots are rational.

2. $x = \dfrac{5 \pm \sqrt{25 - 8}}{4}$

$x = \dfrac{5 \pm \sqrt{17}}{4}$

$x = \dfrac{5 + \sqrt{17}}{4}$ or $x = \dfrac{5 - \sqrt{17}}{4}$

Since $\sqrt{17}$ is irrational, these roots are irrational.

Case 2 $b^2 - 4ac = 0$

If $b^2 - 4ac = 0$, then the expression $\sqrt{b^2 - 4ac}$ equals zero. Thus, the quadratic formula yields $x = \frac{-b}{2a}$.

The quadratic equation has one rational root (the root is a double root).

EXAMPLE $9x^2 - 12x + 4 = 0$

$$\begin{aligned} b^2 - 4ac &= (-12)^2 - 4 \cdot 9 \cdot 4 \\ &= 144 - 144 \\ &= 0 \end{aligned}$$

The root is a rational double root.

Check You may solve the equation by factoring or by applying the quadratic formula.

$$x = \frac{12 \pm \sqrt{144 - 4 \cdot 9 \cdot 4}}{2 \cdot 9}$$

$$x = \frac{12 \pm 0}{18}$$

$x = \frac{12}{18} = \frac{2}{3}$. The solution, $\frac{2}{3}$, is a double root.

Case 3 $b^2 - 4ac < 0$

If $b^2 - 4ac < 0$, then we cannot evaluate $\sqrt{b^2 - 4ac}$ because negative numbers do not have real square roots. If $b^2 - 4ac < 0$, the equation has *no real roots.* You will study this case in more advanced courses.

EXAMPLES Without solving the equation, determine the nature of its roots.

1. $3y^2 + y - 3 = 0$
2. $2m^2 - 9m = -4$
3. $4g^2 + g + 2 = 0$
4. $x^2 + 49 = 14x$

SOLUTION

1. $3y^2 + y - 3 = 0$

$$\begin{aligned} b^2 - 4ac &= 1^2 - 4 \cdot 3(-3) \\ &= 1 + 36 \\ &= 37 \end{aligned}$$

The roots are irrational.

2. $2m^2 - 9m = -4$

$2m^2 - 9m + 4 = 0$

$$\begin{aligned} b^2 - 4ac &= (-9)^2 - 4 \cdot 2 \cdot 4 \\ &= 81 - 32 \\ &= 49 = 7^2 \end{aligned}$$

The roots are rational.

3. $4g^2 + g + 2 = 0$

$$\begin{aligned} b^2 - 4ac &= 1^2 - 4 \cdot 4 \cdot 2 \\ &= 1 - 32 \\ &= -31 \end{aligned}$$

There are no real roots.

4. $x^2 + 49 = 14x$

$x^2 - 14x + 49 = 0$

$$\begin{aligned} b^2 - 4ac &= (-14)^2 - 4 \cdot 1 \cdot 49 \\ &= 196 - 196 \\ &= 0 \end{aligned}$$

The root is a rational double root.

ORAL EXERCISES

State the value of the discriminant of each of the following. Is the value of the discriminant a perfect square?

1. $x^2 - 4x + 4 = 0$
2. $r^2 + 2r - 4 = 0$
3. $2d^2 + d - 1 = 0$
4. $g^2 + 2g + 5 = 0$
5. $c^2 + 6c + 9 = 0$
6. $7f^2 + 6f - 1 = 0$

What method would you use to solve the given equation? Explain.

7. $y^2 = 16$
8. $x^2 + 2x = 0$
9. $(b + 3)^2 = 36$
10. $x^2 + 2x + 1 = 0$
11. $3c^2 - 6c - 2 = 0$
12. $m^2 + 5m - 2 = 0$

WRITTEN EXERCISES

Solve.

A

1. $x^2 - 12x + 36 = 0$
2. $a^2 + 5a = 0$
3. $r^2 = 6r$
4. $x^2 - 100 = 0$
5. $c^2 - 10c + 16 = 0$
6. $a^2 - 4a + 1 = 0$
7. $r^2 = 20$
8. $3y^2 + 5y - 2 = 0$
9. $4e^2 - 25 = 0$
10. $9m^2 = 4$
11. $9h^2 = 4h$
12. $v^2 + 6v + 9 = 0$
13. $0 = 4r^2 + 12r + 9$
14. $0 = 3m^2 - 5m + 1$
15. $(x - 4)^2 = 32$
16. $27 = (y + 2)^2$
17. $25x^2 + 4 = 20x$
18. $16b^2 + 1 = 10b$
19. $m^2 + 2m + 1 = 4$
20. $9 = g^2 + 10g + 25$
21. $16 = (2a + 1)^2$
22. $81 = (3b - 4)^2$
23. $2c^2 + 3c - 10 = 0$
24. $5d^2 + 6d = 2$

Use the discriminant to determine the nature of the roots of each quadratic equation. You do not need to solve the equation.

25. $x^2 + 8x - 20 = 0$
26. $z^2 + 6z - 1 = 0$
27. $2p^2 - 2p - 1 = 0$
28. $9y^2 + 6y + 1 = 0$
29. $w^2 - 14w + 40 = 0$
30. $3x^2 + x = 0$
31. $u^2 - 7 = 0$
32. $6m^2 - 32 = 0$
33. $-2r^2 + 10r - 11 = 0$
34. $-9n^2 + 12n - 4 = 0$
35. $t - t^2 = 8t + 8$
36. $8x^2 - 8x = 6x - 3$

Solve.

B

37. $\frac{w^2}{2} = 8 - 3w$
38. $6 = 3x - \frac{x^2}{4}$
39. $\frac{4}{3}a^2 = 4a - 3$
40. $24(6 - b^2) = 9$
41. $m^2 - 8 = \frac{m}{2}$
42. $5(2y^2 + y) = 3y^2$
43. $\frac{z^2}{6} + 3z + 12 = \frac{z}{6} + \frac{1}{3}$
44. $n^2 + 2n = -3(n + 1)$
45. $48(2d^2 - d) = 32d^2 - 9$

Find the positive root of each equation.

46. $2x^2 + 3x = 18$
47. $2(b^2 - 3) = -11b$
48. $3e^2 - 10e = 7$

Find the negative root of each equation.

49. $7x(x+13) = 16x^2 + 55x - 189$ **50.** $3t^2 + 11t - 4 = t(t+2)$ **51.** $3e(e-4) = 2(4-e)$

Use the discriminant to determine the nature of the roots of each quadratic equation.

52. $2x(5-x) = 7$ **53.** $5(5-2y) = 3y^2$ **54.** $\frac{4-m^2}{13} = m+2$

55. $\frac{t}{t+13} + \frac{2}{t-13} = 0$ **56.** $u(u-1) = 14(u-4)$ **57.** $25(r+1) = r(5-4r)$

Write an equation of the form $ax^2 + bx + c = 0$ with the given roots.

58. $\frac{2}{5}, \frac{1}{2}$ **59.** $-\sqrt{3}$ (double root) **60.** $\frac{\sqrt{2}}{2}, \frac{-\sqrt{2}}{2}$

EXAMPLE Write an equation of the form $ax^2 + bx + c = 0$ that has $\left\{\frac{7+\sqrt{19}}{10}, \frac{7-\sqrt{19}}{10}\right\}$ as its solution set.

SOLUTION In Exercises 52 and 53 on page 285, you were asked to show that if $ax^2 + bx + c = 0$ with $a \neq 0$ and $b^2 - 4ac > 0$, then the sum of the roots is $-\frac{b}{a}$ and the product of the roots is $\frac{c}{a}$. Applying these facts, we have

$$-\frac{b}{a} = \frac{7+\sqrt{19}}{10} + \frac{7-\sqrt{19}}{10} = \frac{14}{10}$$

and

$$\frac{c}{a} = \frac{7+\sqrt{19}}{10} \cdot \frac{7-\sqrt{19}}{10} = \frac{49-19}{100} = \frac{30}{100} = \frac{3}{10}$$

Thus, $-\frac{b}{a} = \frac{14}{10}$ and $\frac{c}{a} = \frac{3}{10}$.

If $a = 10$, then $b = -14$ and $c = 3$. (This is only one of many possible solutions for a, b, and c.)

Therefore, $10x^2 - 14x + 3 = 0$ has $\left\{\frac{7+\sqrt{19}}{10}, \frac{7-\sqrt{19}}{10}\right\}$ as its solution set.

C 61. $\{3+\sqrt{2}, 3-\sqrt{2}\}$ **62.** $\{5-\sqrt{3}, 5+\sqrt{3}\}$

63. $\left\{\frac{2+\sqrt{6}}{2}, \frac{2-\sqrt{6}}{2}\right\}$ **64.** $\left\{\frac{3-3\sqrt{5}}{4}, \frac{3+3\sqrt{5}}{4}\right\}$

Section 7-8 SOLVING PROBLEMS WITH QUADRATIC EQUATIONS

In this section you will study more problems that can be solved with quadratic equations. Again remember to follow the basic procedure for problem solving.

The first type of problem uses the idea of a *reciprocal* of a number. Recall that two numbers are reciprocals of each other if their product is 1. For example,

$\frac{3}{8}$ and $\frac{8}{3}$ are reciprocals since $\frac{3}{8} \cdot \frac{8}{3} = 1$. The reciprocal of 2 is $\frac{1}{2}$; the reciprocal of $-\frac{3}{4}$ is $-\frac{4}{3}$ or $-1\frac{1}{3}$. (Every number except 0 has a reciprocal.)

EXAMPLE The sum of a number and its reciprocal is $2\frac{4}{15}$. What are the numbers?

SOLUTION Let x be one of the numbers. Then $\frac{1}{x}$ is its reciprocal.

The problem states that the sum of x and $\frac{1}{x}$ is $2\frac{4}{15}$.

So you have
$$x + \frac{1}{x} = 2\frac{4}{15}$$
$$x + \frac{1}{x} = \frac{34}{15}$$

LCD = $15x$
$$15x\left(x + \frac{1}{x}\right) = 15x\left(\frac{34}{15}\right)$$
$$15x^2 + 15 = 34x$$
$$15x^2 - 34x + 15 = 0$$
$$(5x - 3)(3x - 5) = 0$$
$$5x - 3 = 0 \quad \text{or} \quad 3x - 5 = 0$$
$$x = \frac{3}{5} \quad \text{or} \quad x = \frac{5}{3}$$

Check
$$\frac{3}{5} + \frac{5}{3} = \frac{9}{15} + \frac{25}{15}$$
$$= \frac{34}{15}$$
$$= 2\frac{4}{15}$$

The numbers are $\frac{5}{3}$ and $\frac{3}{5}$.

EXAMPLE A rectangular parking lot 40 m × 60 m is to be doubled in area by adding a strip of uniform width to all sides of the rectangle. How wide will the strip be?

SOLUTION Let x = width in meters of the strip.

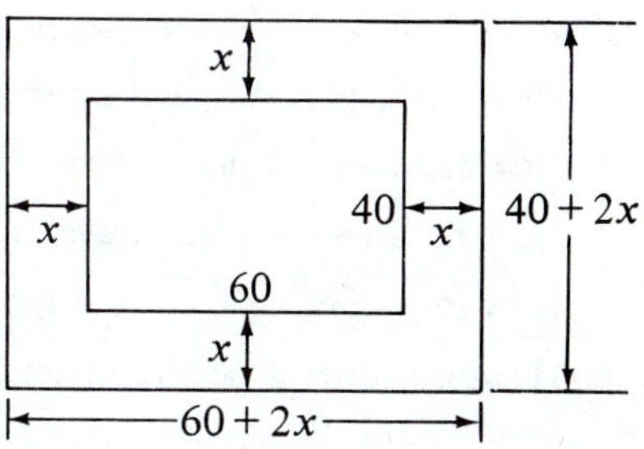

$$(60 + 2x)(40 + 2x) = 2(60)(40)$$
$$2400 + 200x + 4x^2 = 4800$$
$$4x^2 + 200x - 2400 = 0$$
$$x^2 + 50x - 600 = 0$$
$$(x - 10)(x + 60) = 0$$

Since length cannot be negative, we reject the solution $x = -60$. The strip will be 10 m wide.

EXAMPLE A small motorboat makes a trip 48 km upstream and back in 2.8 h. If the rate of the boat in still water is 35 km/h, find the rate of the current.

SOLUTION Let r = the rate of the current. Recall the formula

$$\text{rate} \cdot \text{time} = \text{distance}$$

or

$$\text{time} = \frac{\text{distance}}{\text{rate}}$$

	rate (km/h)	distance (km)	time (h)
upstream (against the current)	$35 - r$	48	$\frac{48}{35 - r}$
downstream (with the current)	$35 + r$	48	$\frac{48}{35 + r}$

$$\underbrace{\text{Time going}}_{\frac{48}{35-r}} + \underbrace{\text{Time returning}}_{\frac{48}{35+r}} = \underbrace{\text{Total traveling time}}_{2.8}$$

$$\text{LCD} = (35 - r)(35 + r)$$

$$(35 - r)(35 + r)\left[\frac{48}{35 - r} + \frac{48}{35 + r}\right] = (35 - r)(35 + r)(2.8)$$

$$\begin{aligned}(35 + r)(48) + (35 - r)(48) &= (1225 - r^2)(2.8)\\ 1680 + 48r + 1680 - 48r &= 3430 - 2.8r^2\\ 3360 &= 3430 - 2.8r^2\\ 2.8r^2 &= 70\\ r^2 &= 25\\ r &= \pm 5\end{aligned}$$

Reject the negative root (since the current cannot have a negative rate). The rate of the current is 5 km/h.

ORAL EXERCISES

A plane travels 500 km/h in still air and the wind speed is 40 km/h.

1. What is the plane's speed when it is flying against the wind?
2. What is the plane's speed when it is flying with the wind?
3. How far does the plane travel in 2 h with the wind?
4. How far does the plane travel in 3 h against the wind?
5. How long does it take the plane to travel 2160 km with the wind?
6. How long does it take the plane to travel 1150 km against the wind?

Given the area of the shaded region, state an equation you could use to find the value of x.

7.

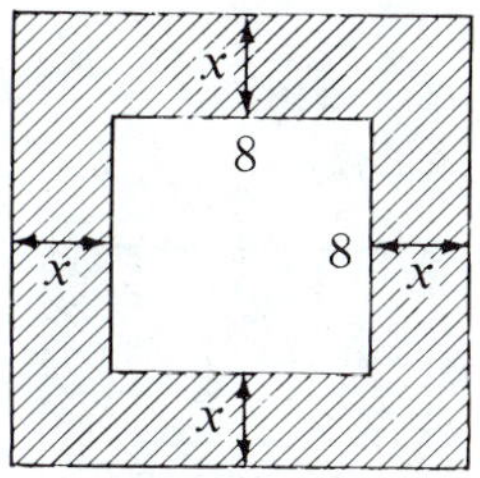

Area = 105

8.

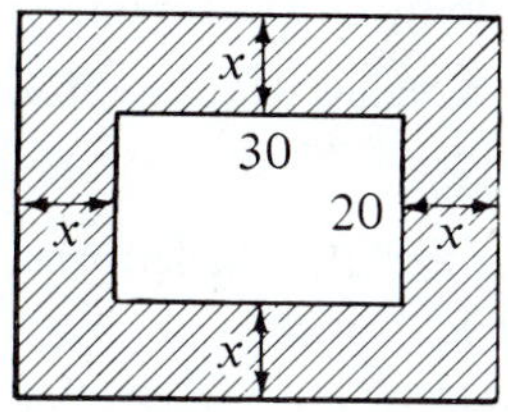

Area = 104

9.

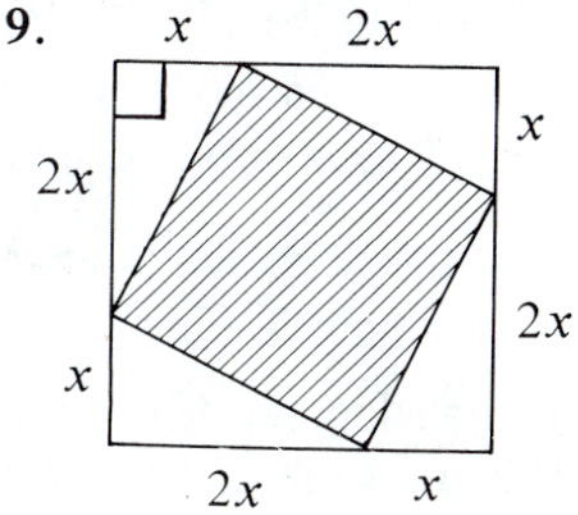

Area = 125

State an equation you could use to solve the problem. You do not need to solve the equation.

10. The length of a rectangle is 4 more than the width. If the area of the rectangle is 10, find its width.

11. The sum of a number and its reciprocal is $\frac{29}{10}$. Find the number.

12. In 6 h Donald paddles his canoe 28 km upstream and 34 km downstream. He can paddle at a rate of 12 km/h in still water. Find the speed of the current.

WRITTEN EXERCISES

Solve, rejecting inappropriate roots. Express irrational numbers in simplified form.

A 1–6. Solve Oral Exercises 7–12.

7. When a certain number is added to the square of that number, the sum is $1\frac{1}{9}$. Find the number.

8. When a certain number is subtracted from the square of the number, the difference is $-\frac{6}{25}$. Find the number.

9. The sum of a number and its reciprocal is $-\frac{5}{2}$. Find the number.

10. The sum of a number and its reciprocal is $\frac{25}{12}$. Find the number.

11. When the reciprocal of a positive number is subtracted from the number, the difference is $2\frac{1}{10}$. Find the number.

12. A photograph is 24 cm long and 20 cm wide. When it is placed in a frame of constant width, the area of the framed photo is 300 cm^2 more than the area of the unframed photo. Find the width of the frame.

13. A square lawn has area 100 m^2. A sidewalk of uniform width surrounds the lawn. If the area of the sidewalk alone is 69 m^2, how wide is the sidewalk?

14. When the park shown at the right lost the shaded region, its area was halved. Find the value of x.

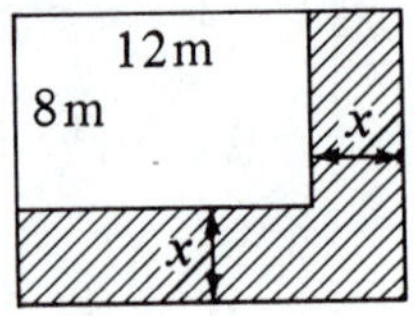

15. In parallelogram $ABCD$, the measure of $\angle A$ is half the square of the measure of $\angle B$. Find the measures of $\angle A$ and $\angle B$.

16. In the diagram below, $l \parallel m$, $m\angle 1 = 3x^2 + 2x - 23$, and $m\angle 2 = 2x^2 + 40$. Find the value of x.

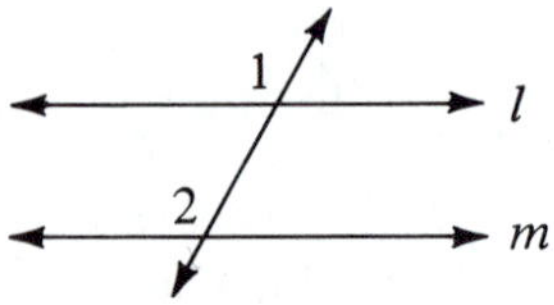

17. Two numbers have a sum of 6 and a product of 1. Find the numbers.

18. Two numbers have a sum of -5 and a product of -5. Find the numbers.

19. The sum S of the first n positive integers can be computed using the formula $S = \frac{n(n+1)}{2}$. For what value of n is $S = 91$?

20. Refer to Exercise 19. Explain why there is no value of n for which $S = 19$.

B 21. The sum of two numbers is 8 and the sum of their reciprocals is $\frac{2}{3}$. Find the numbers.

22. The product of two numbers is 5 and the sum of their reciprocals is $\frac{9}{10}$. Find the numbers.

23. Antonia can row 8 km/h in still water. If it takes her 6.4 h to row 24 km up the river and back, find the rate of the current.

24. In 4 h Lauren travels 64 km by car and 222 km by bus. If the bus travels at a rate 10 km/h faster than the car, find the rate of the bus.

25. Ray walked 1 km to a bus stop and rode in a bus for 8 km. If his total traveling time was 18 min, and the bus's average speed was 54 km/h faster than Ray's walking speed, find the bus's average speed.

26. A plane travels 2100 km with the wind. The return trip against the wind is 1 h longer. If the rate of the plane in still air is 325 km/h, find the rate of the wind.

C 27. The area of a rhombus is 18 mm^2. One diagonal is 4 mm longer than the other. Find the length of each diagonal.

In Exercises 28–30, use the following information:
If an object is thrown upward with an initial velocity of v m/s from a height of d m, then its height h after t s is given by the formula

$$h = d + vt - 4.9t^2.$$

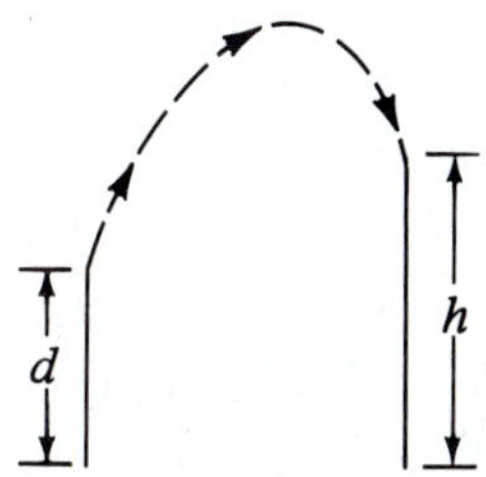

28. A rocket is fired from a height of 200 m above ground level with an initial velocity of 147 m/s. After how many seconds will the rocket be 1302.5 m above the ground?

29. A ball is thrown upward from a height of 1 m with an initial velocity of 24.5 m/s. After how many seconds will it reach a height of 30.4 m?

30. A rocket is fired with an initial velocity of 59.5 m/s from a platform 250 m above ground level. For how many seconds will the altitude of the rocket be above 258.4 m?

SELF-TEST 2

Solve by completing the square.

1. $x^2 + 10x = -16$	2. $s^2 - 2s - 2 = 0$	Section 7-5
3. $4y^2 + 4y = -1$	4. $9z^2 - 12z + 2 = 0$	

Solve by using the quadratic formula.

5. $t^2 - 12t + 36 = 0$	6. $v^2 + v = 7$	Section 7-6
7. $2n^2 + 4n + 1 = 0$	8. $4b^2 = 5b + 6$	

State the value of the discriminant and the nature of the roots. Then solve the equation.

9. $h^2 = 25h$	10. $4q^2 = 1 - 4q$	Section 7-7
11. $5j^2 - 9 = 0$	12. $9(m - 2) = 2m - m^2$	

Solve.

13. The sum of a number and twice its reciprocal is $3\frac{3}{10}$. What is the number? Section 7-8

14. The sum of two numbers is 6 and their product is 7. Find the numbers.

15. A plane that flies 580 km/h in still air traveled 4200 km against the wind and returned with the wind. The return trip took 0.5 h less than the trip going. Find the speed of the wind.

Computer Exercises

It is easy to write a program that will solve quadratic equations by means of the quadratic formula. By computing and analyzing the discriminant, it is possible to classify the roots as well as compute them.

1. Copy and RUN the following program to solve the equations given below.

```
10  PRINT "TO SOLVE A QUADRATIC EQUATION:"
20  PRINT "INPUT A(<>0),B,C:"
30  INPUT A,B,C
40  LET D=B*B-4*A*C
50  IF D<0 THEN 230
60  LET A1=2*A
70  IF D=0 THEN 210
80  REM***D IS POSITIVE.
90  LET S=SQR(D)
100  IF S=INT(S) THEN 150
110  PRINT "THE ROOTS ARE IRRATIONAL."
120  PRINT "   R1  =  (";-B;"  +  SQR(";D;"))/";A1
130  PRINT "   R2  =  (";-B;"  -  SQR(";D;"))/";A1
140  STOP
150  PRINT "THE ROOTS ARE RATIONAL."
160  PRINT "   R1  =";-B/A1;"  +";S/A1;
170  PRINT "   =";(-B+S)/A1
180  PRINT "   R2  =";-B/A1;"  -";S/A1;
190  PRINT "   =";(-B-S)/A1
200  STOP
210  PRINT -B/A1;"  IS A DOUBLE ROOT."
220  STOP
230  PRINT "THERE ARE NO REAL ROOTS."
240  END
```

a. $3x^2 - 12 = 0$
b. $x^2 - 8x = 0$
c. $x^2 + 8x + 15 = 0$
d. $x^2 + 4x - 1 = 0$
e. $x^2 + 4 = 0$
f. $4x^2 - 20x + 25 = 0$

2. RUN the preceding program for the equations in Written Exercises 25-36, Section 7-7.

CHAPTER REVIEW

Solve by factoring.

1. $m^2 + 169 = 26m$
2. $5q^2 = -35q$
3. $2 + 7c + 3c^2 = 0$ Section 7-1

Solve.

4. $\frac{n+1}{4n} + \frac{1}{n} = \frac{2}{3}$
5. $\frac{r}{3} - \frac{2}{r} = \frac{5}{3}$
6. $\frac{y}{y+2} - \frac{8}{y^2-4} = 1$ Section 7-2

7. Simplify: **(a)** $\sqrt{150x^2y^6}$; **(b)** $-\sqrt{\frac{49}{3}}$ Section 7-3

Solve by taking square roots.

8. $\frac{z^2}{8} = 20$

9. $81 = (a + 4)^2$

10. $3s^2 = \frac{200}{3}$

Solve.

11. The product of 5 and the square of a number is $\frac{5}{4}$. Find the number. Section 7-4

12. One pipe can fill a large tank in 25 h. If a second pipe is used with the first, the tank can be filled in 15 h. How long would it take the second pipe alone to fill the tank?

13. The dimensions of a rectangle are consecutive even integers. When each dimension is increased by 3 cm, the area is 3 cm^2 more than twice the area of the original rectangle. Find the dimensions of the original rectangle.

Solve by completing the square.

14. $x^2 + 16x = -28$

15. $w^2 - 4w = 7$

16. $t^2 + 7t = 30$ Section 7-5

Solve by using the quadratic formula.

17. $b^2 = 36 + 9b$

18. $4f^2 + 5f = 11$ Section 7-6

19. $3u^2 - 8u + 4 = 0$

20. $2d = 5d^2 - 6$

In Exercises 21-23, **(a)** use the discriminant to determine the nature of the roots of each equation and **(b)** solve each equation.

21. $20x^2 - 4x = 0$

22. $4y^2 + 9 = 12y$

23. $z^2 + z = 1$ Section 7-7

Solve.

24. One integer is 4 more than another integer. The sum of their reciprocals is $\frac{2}{3}$. Find the integers. Section 7-8

25. A rectangular pool is 12 m long and 6 m wide. A walkway of uniform width surrounds the pool. If the area of the walkway is 9 m^2 less than that of the pool, how wide is the walkway?

CHAPTER TEST

1. Simplify: **(a)** $\frac{\sqrt{10}}{\sqrt{32}}$; **(b)** $-\sqrt{\frac{x^2}{2}}$; **(c)** $\sqrt{48x^4y^2}$

2. Solve by factoring: $6k^2 - k - 15 = 0$

3. Solve by taking square roots: $(7 - q)^2 = 100$

4. Solve by completing the square: $2n^2 + 8n + 5 = 0$

5. Solve by using the quadratic formula: $3s^2 + s - 1 = 0$

Solve.

6. $\frac{3}{b-2} + \frac{5}{2b} = \frac{7}{b}$
7. $\frac{3}{p} + \frac{p}{p+2} = 2$
8. $\frac{9}{z} = \frac{7}{z^2 + 3z} - \frac{z}{z+3}$

Use the discriminant to determine the nature of the roots of the equation.

9. $2 - x^2 = 3x - 5$
10. $25g^2 + 1 = 10g$
11. $(c - 4)^2 = 2c$

Solve.

12. $25a^2 - 8 = 0$
13. $40 + 14m + m^2 = 0$
14. $d^2 - 1 = 5d$
15. $3r^2 - 10r + 2 = 0$
16. $3y - 2y^2 = 3y - 90$
17. $9j^2 - 12j = -4$
18. The altitude of a triangle is twice as long as the corresponding base. If the area of the triangle is 363, how long is the altitude?
19. The square of a negative number increased by four times the number is 60. Find the number.
20. Brad needs 3 h more than Jeff to load a truck. Together they can do the job in 3.6 h. How long would each take to load the truck alone?
21. The sum of a number and 2 is equal to twice the reciprocal of the number. Find the number.
22. In 4.75 h a small plane flew 900 km against the wind and returned with the wind. If the plane's speed in still air was 380 km/h, find the wind speed.

CUMULATIVE REVIEW: CHAPTERS 1–7

Indicate the best answer by writing the appropriate letter.

1. The degree of $6x^5 + 3x^4y^2 - 4y^3$ is:
   a. 14 b. 6 c. 5 d. 8
2. If two parallel lines are cut by a transversal, a pair of corresponding angles are:
   a. congruent b. complementary c. supplementary d. acute
3. One base of a trapezoid has length 27. If the length of the median is 35, find the length of the other base.
   a. 31 b. 39 c. 4 d. 43
4. Which of the following is a factor of $7n^2 - 8n - 12$?
   a. $7n + 3$ b. $7n + 4$ c. $7n + 6$ d. $7n - 2$
5. Negate the sentence "Some cars are unreliable."
   a. Some cars are reliable. b. All cars are unreliable.
   c. All cars are reliable. d. No car is reliable.
6. Find the value of the discriminant of $3g^2 + 2g - 5 = 0$.
   a. 8 b. −56 c. 62 d. 64

7. In isosceles trapezoid $ABCD$, $\overline{AB} \parallel \overline{CD}$. Which statement is false?
a. The median is parallel to $\overline{AB}$.
b. $\overline{AC} \cong \overline{BD}$
c. $\overline{AC}$ and $\overline{BD}$ bisect each other.
d. $\angle C \cong \angle D$

8. Given that $\overline{AB} \cong \overline{ST}$ and $\angle A \cong \angle S$. What additional information do you need to prove $\triangle ABC \cong \triangle STR$ by the SAS postulate?
a. $\overline{AC} \cong \overline{RT}$
b. $\overline{AC} \cong \overline{SR}$
c. $\overline{BC} \cong \overline{ST}$
d. $\overline{BC} \cong \overline{TR}$

9. Simplify: $(9 - 5b - b^2) - (7b + 5 - 6b^2)$
a. $5b^2 - 12b + 4$
b. $-7b^2 + 2b + 14$
c. $-7b^2 - 12b + 4$
d. $5b^2 + 2b + 14$

10. Two angles of a triangle have measures 45 and 70. Which of the following is *not* the measure of an exterior angle of the triangle?
a. 110
b. 115
c. 125
d. 135

11. If $ABCD$ is a parallelogram, which statement is *not* always true?
a. $\overline{AC} \perp \overline{BD}$
b. $\overline{BC} \cong \overline{AD}$
c. $\overline{AC}$ bisects $\overline{BD}$.
d. $\angle B$ and $\angle C$ are supplementary.

12. Given premises P_1: $\sim q \rightarrow p$ and P_2: $\sim p$. What can you conclude?
a. p
b. q
c. $p \wedge q$
d. $\sim q$

13. Simplify: $16 - 24\left(-\frac{1}{2}\right)^3 - (-2)$
a. 3
b. 15
c. 21
d. 17

14. The lengths of the sides of a triangle can *not* be:
a. 6, 11, 18
b. 13, 19, 31
c. 5, 8, 4
d. $\frac{1}{4}, \frac{1}{3}, \frac{1}{2}$

15. Solve: $\frac{x}{3} - 15 = 2(x - 5)$
a. $x = -15$
b. $x = -35$
c. $x = -9$
d. $x = -3$

16. Simplify: $\frac{6 - 3m}{m^2 - 4m + 4}$
a. $\frac{3}{2 - m}$
b. $\frac{3}{m - 2}$
c. $\frac{3}{m + 2}$
d. $-\frac{3}{m + 2}$

17. Which equation has 6 and -8 as roots?
a. $x^2 + 14x - 48 = 0$
b. $x^2 - 2x - 48 = 0$
c. $x^2 + 2x - 48 = 0$
d. $-x^2 + 2x + 48 = 0$

18. $\angle 1$ and $\angle 2$ are adjacent angles whose exterior sides form opposite rays. The measure of $\angle 1$ is twice the measure of $\angle 2$. Find the measure of $\angle 1$.
a. 30
b. 60
c. 90
d. 120

19. Solve: $(x + 3)^2 = 13$
a. $\{-3 \pm \sqrt{13}\}$
b. $\{3 \pm \sqrt{13}\}$
c. $\{-16, 10\}$
d. $\{-10, 16\}$

20. Simplify: $(3r - 2)(2r^2 - r + 1)$
a. $2r^2 + 2r - 1$
b. $6r^3 + r^2 + r - 2$
c. $2r^3 - r^2 + r$
d. $6r^3 - 7r^2 + 5r - 2$

21. Which information is not sufficient to guarantee that $l \parallel n$?

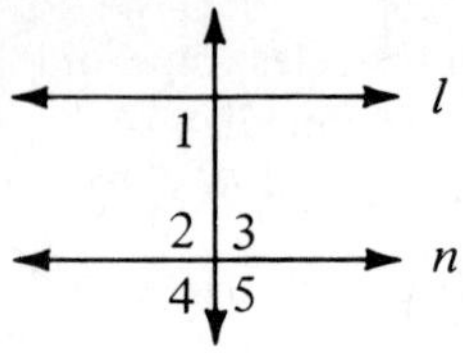

a. $\angle 1 \cong \angle 4$
b. $\angle 1$ and $\angle 3$ are supplementary.
c. $\angle 1$ and $\angle 2$ are supplementary.
d. $\angle 1$ and $\angle 5$ are right angles.

22. In rhombus $ABCD$, $m\angle ABC = 60$. Find $m\angle ACD$.

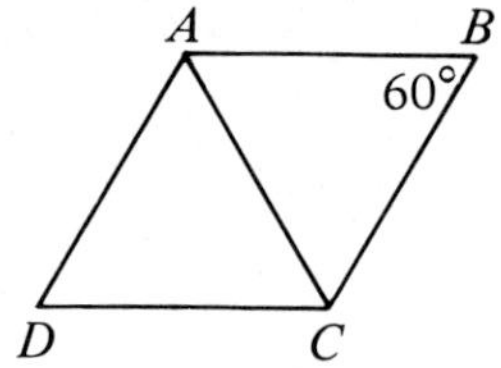

a. 30 b. 60 c. 90 d. 120

23. Simplify: $(3x^2y)(-2xy^3)^2$
a. $-6x^3y^4$ b. $-6x^4y^7$ c. $12x^4y^7$ d. $-12x^4y^6$

24. Find the sum of the measures of the interior angles of an octagon.
a. 360 b. 1080 c. 1800 d. 1440

25. The graph shows the solution set of which inequality?

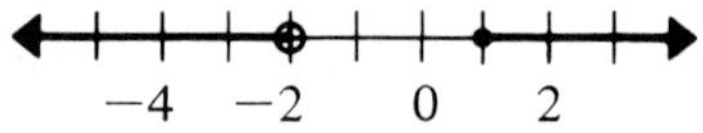

a. $-5 < 3y + 1 \leqslant 4$
b. $-1 - y < -2$ or $1 \leqslant -1 - y$
c. $-\frac{y}{2} > 1$ or $4y - 3 \geqslant 1$ d. $2 \geqslant -y$ and $-y > -1$

26. Find the greatest common factor of $28s^3(s + 1)$ and $14s(s + 1)(s - 1)$.
a. $14s(s + 1)^2$ b. $14s(s + 1)$
c. $28s^3(s + 1)(s - 1)$ d. $28s^3(s + 1)^2(s - 1)$

27. $\triangle RST$ is equilateral. M and N are the midpoints of $\overline{RS}$ and $\overline{ST}$, respectively. If $MN = 6$, find the perimeter of $\triangle RST$.
a. 36 b. 18 c. 27 d. 9

28. If p is true and q is false, which statement must be true?
a. $\sim(p \wedge \sim q)$ b. $\sim(\sim p \vee q)$ c. $\sim q \rightarrow \sim p$ d. $\sim(q \rightarrow p)$

29. Name the property of the real numbers illustrated below.

$$(a + b)(cd) = [(a + b)c]d$$

a. Commutative Property of Addition
b. Commutative Property of Multiplication
c. Associative Property of Multiplication
d. Distributive Property

30. Given that $68 - 3y$ and $5y + 12$ represent the measures of a pair of alternate interior angles formed by two parallel lines and a transversal, find the measure of one of the angles.
a. 50 b. 7 c. 47 d. 137

31. Write the negation of the following sentence: $\exists_x\, x \geqslant 3$
a. $\exists_x\, x < 3$ b. $\forall_x\, x < 3$ c. $\forall_x\, x \leqslant 3$ d. $\forall_x\, x \geqslant 3$

32. In the accompanying diagram, $ABCD$ is a rectangle. Which statement must be true?

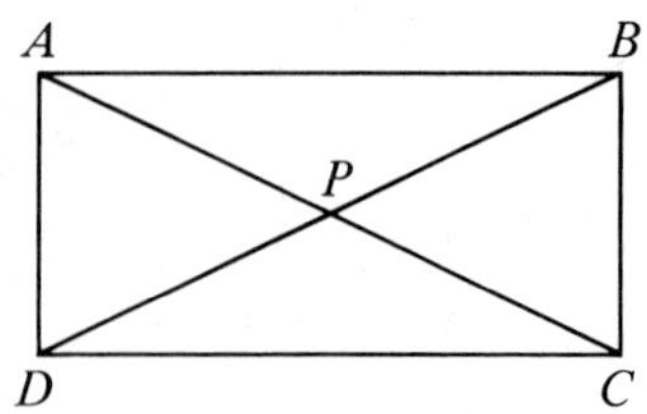

a. $\angle PAB \cong \angle PBA$ b. $\overrightarrow{DP}$ bisects $\angle ADC$.
c. $\triangle APD \cong \triangle APB$ d. $\angle BAP$ and $\angle CDP$ are complementary.

33. Solve: $x^2 - 5x - 1 = 0$

a. $\left\{\frac{5 \pm \sqrt{29}}{2}\right\}$ b. $\left\{\frac{-5 \pm \sqrt{29}}{2}\right\}$ c. $\left\{\frac{5 \pm \sqrt{21}}{2}\right\}$ d. $\left\{\frac{-5 \pm \sqrt{21}}{\cdot 2}\right\}$

34. In quadrilateral $WXYZ$, $\overline{WX} \parallel \overline{YZ}$, and $\overline{WY}$ and $\overline{XZ}$ intersect at point P. Which additional information would *not* permit you to conclude that $WXYZ$ must be a parallelogram?

a. $\angle XYW \cong \angle YWZ$ b. $\overline{XY} \cong \overline{WZ}$ c. $\overline{WZ} \parallel \overline{XY}$ d. $WX = YZ$

35. Which of the following is the inverse of $\sim r \rightarrow s$?

a. $\sim s \rightarrow \sim r$ b. $\sim s \rightarrow r$ c. $r \rightarrow s$ d. $r \rightarrow \sim s$

36. Find the measure of each exterior angle of a regular polygon with 30 sides.

a. 6 b. 12 c. 15 d. 18

37. Simplify: $\frac{3}{y+1} - \frac{2}{y} + \frac{2}{y^2 + y}$

a. $\frac{y-4}{y^2+y}$ b. $\frac{5y+4}{y^2+y}$ c. $\frac{y+4}{y^2+y}$ d. $\frac{1}{y+1}$

38. Which statement is equivalent to $\sim(p \vee \sim q)$?

a. $p \rightarrow q$ b. $\sim p \wedge q$ c. $\sim p \vee q$ d. $p \wedge \sim q$

39. How many roots does the following equation have?

$$\frac{3}{x^3 - 1} - \frac{3x}{x^2 + x + 1} = \frac{1}{x - 1}$$

a. 0 b. 1 c. 2 d. 3

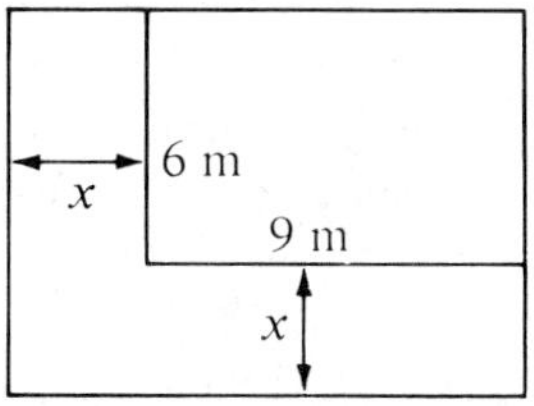

40. The area of a rectangular playground, 9 m by 6 m, is doubled when a strip of uniform width is added to two consecutive sides. How wide is the strip?

a. 2 m b. 2.5 m c. 3 m d. 3.5 m

41. Which statement about $15x^2 - x - 2 = 0$ is true?

a. The sum of the roots is $\frac{1}{15}$. b. The product of the roots is -2.

c. The roots are irrational. d. The discriminant is negative.

42. In $\triangle DEF$, $m\angle D = 40$, $m\angle E = 40$, and $m\angle F = 100$. Which statement is false?

a. $DF > \frac{1}{2}DE$ b. $DE > EF$

c. $DE + EF > DF$ d. Each exterior angle is obtuse.

43. Simplify: $\frac{5p^3(p+2)^2}{10p^4} \div \frac{4 + 2p}{p}$

a. $\frac{p+2}{4}$ b. $\frac{(p+2)^3}{p^2}$

c. $p + 2$ d. $\frac{(p+2)^2 \div 4(p+2)}{2p}$

44. When $6k^3 - 9k^2 - 4$ is divided by $2k + 1$, what is the remainder?

a. $3k^2 - 6k + 3$ b. -7 c. -1 d. 0

45. In $\triangle ABC$, $\overline{AB} \cong \overline{BC}$. If $m\angle ABD = x$, find $m\angle A$ in terms of x.

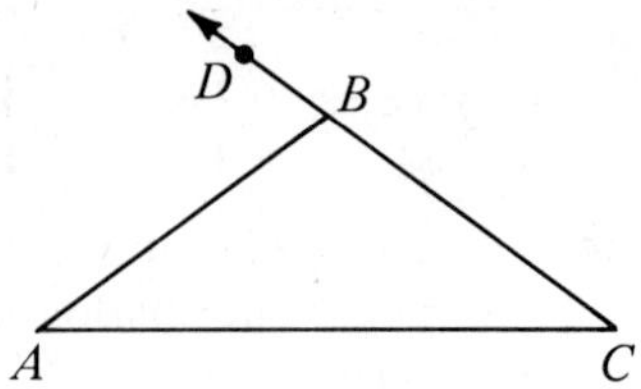

a. $180 - x$ b. x c. $\frac{x}{2}$ d. $180 - 2x$

46. Which figure shows two triangles that *must* be congruent?

a.
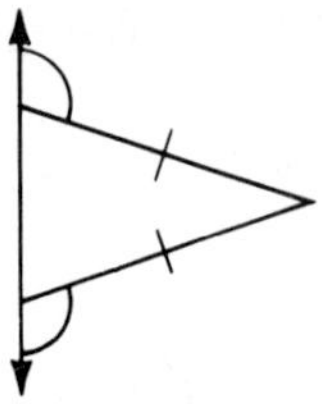

b.
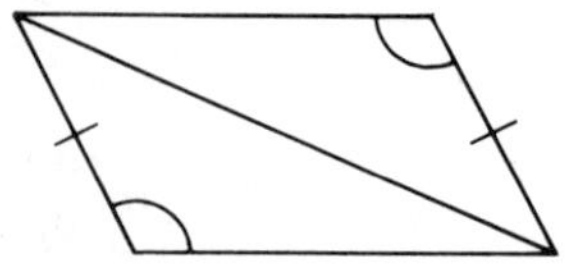

c.
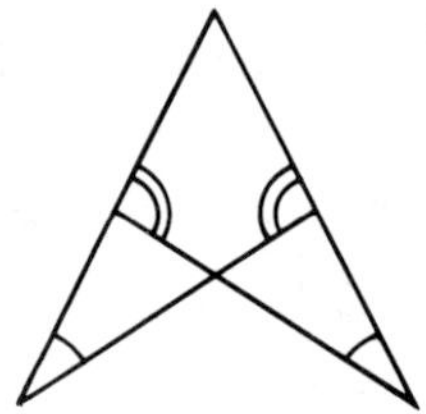

d.
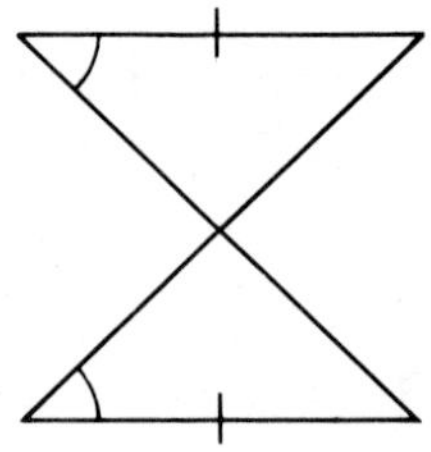

47. Given the following statements:

If sea cucumbers don't eat lily pads, then they don't get indigestion.
Sea cucumbers are befuddled or rebellious.
Sea cucumbers eat lily pads only if they are not befuddled.
Sea cucumbers are not rebellious.

Let p represent "sea cucumbers eat lily pads."
Let i represent "sea cucumbers get indigestion."
Let b represent "sea cucumbers are befuddled."
Let r represent "sea cucumbers are rebellious."

a. Using p, i, b, and r and the proper connectives, express each statement in symbolic form.
b. Use the laws of inference to show that sea cucumbers don't get indigestion.

Choose *one* of Exercises 48 through 50 and write a two-column proof.

48. Given: X is the midpoint of $\overline{RS}$; $\overline{XY} \parallel \overline{ST}$; $\overline{YZ} \parallel \overline{RS}$

Prove: $SZ = \frac{1}{2}ST$

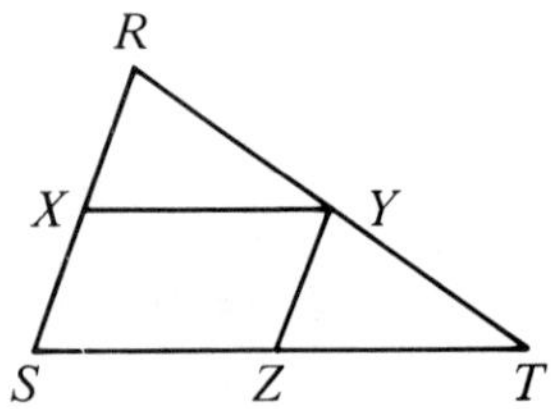

49. Given: $\triangle ADC \cong \triangle ABC$

Prove: $\overline{AC} \perp \overline{BD}$

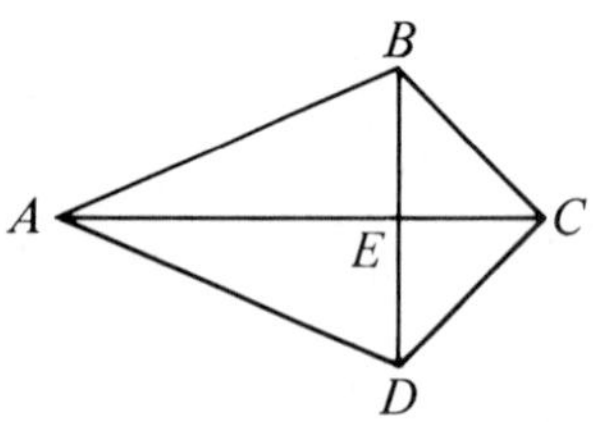

50. Given: $RSWY$ is a ▱; $\angle 1 \cong \angle 2$

Prove: $\overline{XY} \cong \overline{SZ}$

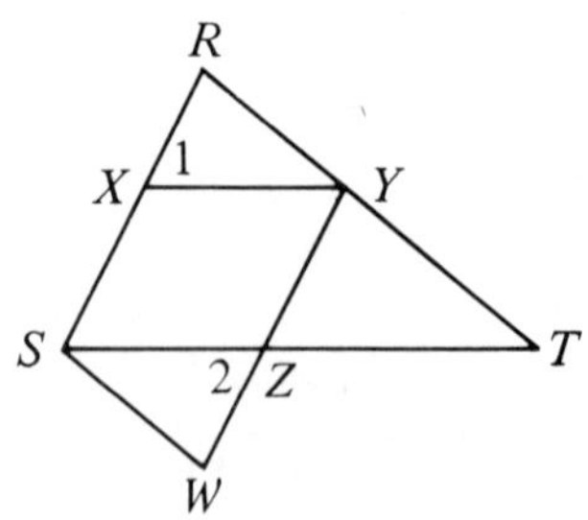

Careers — Agricultural Scientist

As farming has moved away from being a family operation to being big business, a number of support services have grown up around it. Agricultural scientists analyze problems of food, fiber, and horticulture, with the purpose of making agribusiness as profitable an enterprise as possible. You can specialize in the field by becoming a soil or an animal scientist, or a horticulturist.

Soil scientists confer with farmers about the physical, chemical, and biological characteristics of their soil. Studying the form, origin, and distribution of soil helps them advise farmers about the most productive use of land. A common specialty in soil science is agronomy. Agronomists try to improve crop yields and control diseases and pests.

Animal scientists do a similar job for the farmer but with animals instead of plants. Their research into the breeding, feeding, and health of farm animals helps increase farm productivity.

Horticulturists manage orchard and garden plants such as fruit trees, vegetables, and flowers. Their job is to improve plant culture methods and increase crop quality and yields. As a horticulturist you would be concerned with increasing the beauty of public areas, such as parks and gardens, as well as with productivity.

Many occupations, such as forestry, range management, and soil conservation, are related to agricultural science. A broad background in biology, chemistry, physics, and mathematics is helpful for all of them. Many colleges and universities offer agricultural programs, including courses in plant breeding, husbandry, and soil science.

Many agricultural scientists teach or do research at the college level. Other options include working for the United States Department of Agriculture and the National Institutes of Health, or in private industry or zoo management.

VOCABULARY REVIEW

Be sure that you understand the meaning of these terms:

quadratic equation, p. 259
root, p. 260
fractional equation, p. 264
principal square root, p. 267
perfect square, p. 268
completing the square, p. 279
discriminant, p. 287

MIXED REVIEW

Simplify:

1. $\sqrt{72a^3b^2c}$
2. $(3x - y)(y - 2x)$
3. $-24 \div 12 - 8$
4. $\angle A$ and $\angle B$ are the acute angles of a right triangle. If $m\angle A = 12x + 7$ and $m\angle B = 9x + 4$, find the measure of each angle.

Solve.

5. The sum of two positive numbers is 15 and the difference of their squares is 15. What are the numbers?
6. A vegetable garden 12 m × 16 m is surrounded by a flower border of uniform width. If the area of the flower border is $\frac{2}{3}$ the area of the vegetable garden, how wide is the flower border?
7. Let p represent "the sky is blue," q represent "it is raining," and r represent "it is cloudy." Translate the following sentence into symbolic notation: If the sky is not blue and it is raining, then it is cloudy.

Reduce to lowest terms.

8. $\dfrac{4 - a}{a^2 - 16}$
9. $\dfrac{x^2 - xy}{xy(x - y)^2}$

Write two-column proofs.

10. Given: $\square ABCD$; E, F, G, and H are the midpoints of $\overline{AO}$, $\overline{BO}$, $\overline{CO}$, and $\overline{DO}$, respectively.
Prove: $EFGH$ is a $\square$.

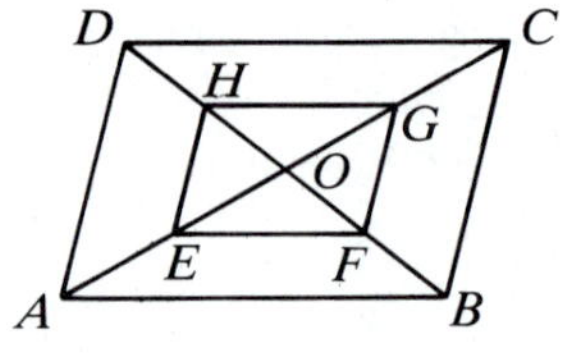

11. Given: $\triangle ACD$ with $\overline{AC} \cong \overline{AD}$; B and E are the midpoints of $\overline{AC}$ and $\overline{AD}$, respectively.
Prove: $BCDE$ is an isosceles trapezoid.

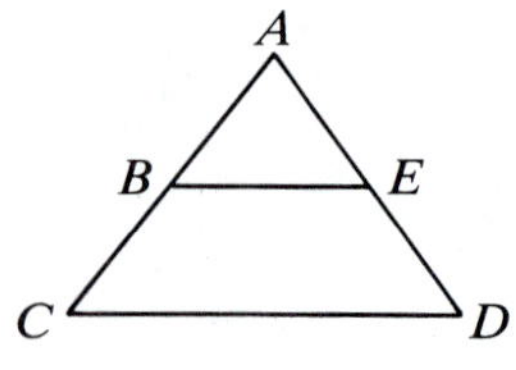

12. Find the measure of one interior angle of a polygon with 40 sides.

Solve.

13. $3(1 - x) > -2x$

14. $\frac{4}{x - 3} = \frac{3}{x - 2} + \frac{6}{x^2 - 5x + 6}$

15. $\frac{5}{y} - \frac{4}{2y} - \frac{1}{5} = 0$

16. $2x^2 - 10x = 0$

17. If $C =$ {the positive odd integers} and $D =$ {the positive multiples of 3}, graph **(a)** $C \cup D$ and **(b)** $C \cap D$.

18. Solve by completing the square: $y^2 + 6y = 16$

19. Write the **(a)** converse, **(b)** inverse, and **(c)** contrapositive of "If $0 < x < 1$, then $\frac{1}{x} > x$" and **(d)** compare their truth values.

Factor completely.

20. $4e^2 - 4ef + f^2$

21. $24a^2b - 6b$

Write a quadratic equation in standard form with the given solution set.

22. $\{-1, -3\}$

23. $\{4, -3\}$

24. If $\angle 1$ and $\angle 2$ are supplementary angles, $m\angle 1 = 2x + 15$, and $m\angle 2 = 4x - 21$, find the value of x and the measures of the angles.

Without solving the equation, determine the nature of the roots.

25. $2a^2 - a - 5 = 0$

26. $3 - 4b^2 = b$

27. In the figure at the right, $AB = 8$ and $AC = 13\frac{1}{2}$. Find BC.

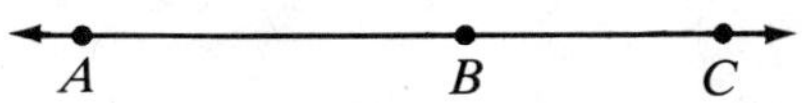

Simplify.

28. $\frac{\sqrt{25b^5}}{\sqrt{5b}}$

29. $\frac{a + b}{b - a} \div \frac{b - a}{a + b}$

30. $\frac{12}{\sqrt{288}}$

31. Find the number of sides of a polygon if the sum of the measures of the interior angles is 7740.

32. Solve by the quadratic formula: $x^2 + 7x + 11 = 0$

33. Negate: $\forall_x\ 2(x - 1) = 2x - 2$

Similar hexagons are a striking architectural feature of the roof of this exhibition pavilion. The hexagons have been positioned so that each vertex of a hexagon lies on the midpoint of a side of the next larger hexagon. Notice the pattern of similar triangles created by this arrangement.

Similarity 8

Roughly speaking, two figures are similar if they have the same shape but not necessarily the same size. A pantograph, pictured below, is a mechanical drawing instrument that is used to enlarge or reduce a design. It is made of four rods held together with adjustable pins so that $WXYZ$ is a parallelogram. One end of the pantograph is fixed. One tracing point is moved over the design to be copied while the other point describes the enlarged (or reduced) outline. By adjusting the pins, the pantograph can be set to copy in any size.

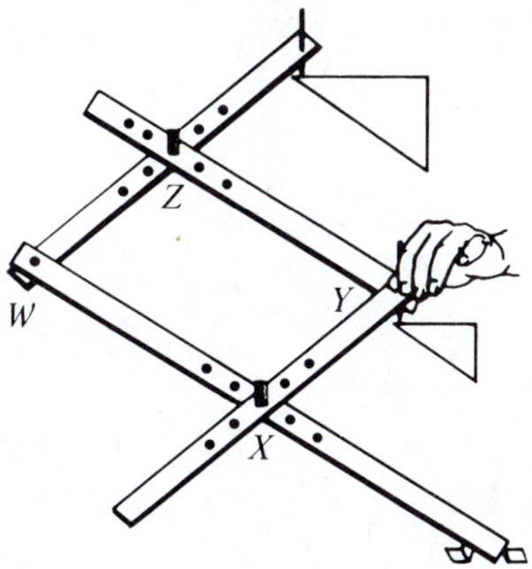

Section 8-1 RATIO AND PROPORTION

In this section you will learn the meaning, uses, and properties of:

- ratios
- proportions
- geometric means

The *ratio* of one number to another is the quotient of the first number divided by the second. The ratio of a to b ($a, b \in R$, $b \neq 0$) may be expressed as $\frac{a}{b}$ or $a:b$.

For example, the ratio of 2 to 3 is 2:3 or $\frac{2}{3}$. In this chapter you may assume that the denominator of any ratio is not zero. A ratio should be expressed in *simplest form* or *lowest terms.* (Recall that a fraction is in lowest terms when the greatest common factor of the numerator and denominator is 1.) For example, in $\triangle ABC$:

The ratio of AB to BC is $\frac{16}{6}$, or $\frac{8}{3}$.

The ratio of BC to AC is $\frac{6}{12}$, or $\frac{1}{2}$.

The ratio of AC to AB is $\frac{12}{16}$, or $\frac{3}{4}$.

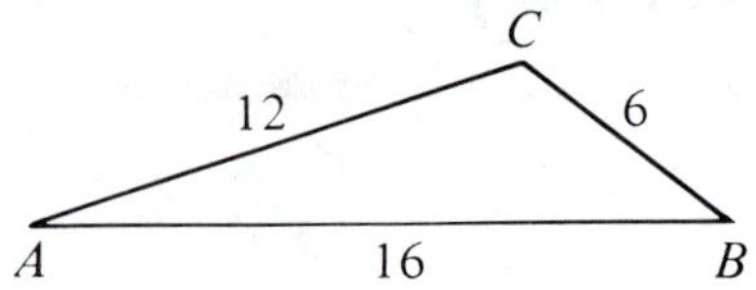

An equation that states that two ratios are equal is called a *proportion.*

$$\frac{a}{b} = \frac{c}{d} \text{ and } a:b = c:d \quad (b,\ d \neq 0) \quad \text{are equivalent proportions,}$$

because in each case there exists a constant k, called the *constant of proportionality*, such that $a = kc$ and $b = kd$. In each proportion above, a, b, c, and d are the first, second, third, and fourth *terms*, respectively, of the proportion. The first and fourth terms are called the *extremes* of the proportion; the second and third are called the *means* of the proportion.

The following useful properties of proportions will be proved in Exercises 37-42 and 50.

Properties of Proportions

1. $\frac{a}{b} = \frac{c}{d}$ if and only if $ad = bc$.
2. $\frac{a}{b} = \frac{c}{d}$ if and only if $\frac{b}{a} = \frac{d}{c}$.
3. $\frac{a}{b} = \frac{c}{d}$ if and only if $\frac{a}{c} = \frac{b}{d}$.
4. $\frac{a}{b} = \frac{c}{d}$ if and only if $\frac{a+b}{b} = \frac{c+d}{d}$.
5. $\frac{a}{b} = \frac{c}{d}$ if and only if $\frac{a-b}{b} = \frac{c-d}{d}$.
6. If $\frac{a}{b} = \frac{x}{y}$ and $\frac{c}{d} = \frac{x}{y}$, then $\frac{a+c}{b+d} = \frac{x}{y}$.

Property 1 is sometimes phrased as *the product of the extremes equals the product of the means.* This property enables you to solve a proportion by "cross-multiplying."

EXAMPLE Solve the proportion $\frac{6}{5} = \frac{4}{x-2}$ for x.

SOLUTION

$$\frac{6}{5} = \frac{4}{x-2}$$

Apply Property 1.
$$6(x-2) = 5 \cdot 4$$
$$6x - 12 = 20$$
$$6x = 32$$
$$x = \frac{32}{6} = \frac{16}{3}$$

If you know that $\frac{a}{b} = \frac{2}{3}$, then, since $\frac{a}{b}$ and $\frac{2}{3}$ are equivalent fractions, you can represent a and b as $2x$ and $3x$, respectively, for some nonzero real number x. We will use this idea in the first solution of the next example.

EXAMPLE Angles 1 and 2 are supplementary and $m\angle 1 : m\angle 2 = 2:7$. Find the measures of the two angles.

SOLUTION

Method 1

Let $m\angle 1 = 2x$ and $m\angle 2 = 7x$.
Since $\angle 1$ and $\angle 2$ are supplementary,

$$m\angle 1 + m\angle 2 = 180$$
$$2x + 7x = 180$$
$$9x = 180$$
$$x = 20$$
$$2x = 40$$
$$7x = 140$$

The two angles have measures 40 and 140.

Method 2

Let $m\angle 1 = y$. Then $m\angle 2 = 180 - y$.
Thus, you can write:

$$\frac{y}{180-y} = \frac{2}{7}$$
$$7y = 2(180-y)$$
$$7y = 360 - 2y$$
$$9y = 360$$
$$y = 40$$
$$180 - y = 140$$

The result, of course, is the same.

A relationship among three or more quantities can be stated using an *extended ratio*, that is, an expression of the form

$$a:b:c \quad (b \neq 0, c \neq 0)$$

The *extended proportion* $a:b:c = d:e:f$ means that there exists a constant k such that $a = kd$, $b = ke$, and $c = kf$. If none of the terms equals 0, then $a:b:c = d:e:f$ is equivalent to $\frac{a}{d} = \frac{b}{e} = \frac{c}{f}$.

EXAMPLE The measures of the angles of a triangle are in the ratio 1:3:5. Find the measures of the angles.

SOLUTION Let x = the measure of the smallest angle. Then $3x$ and $5x$ are the measures of the other two angles. Thus, you have:

$$x + 3x + 5x = 180$$
$$9x = 180$$
$$x = 20$$
$$3x = 60; \ 5x = 100$$

The three angle measures are 20, 60, and 100.

Suppose a, x, and d are positive real numbers. If $\frac{a}{x} = \frac{x}{d}$, then x is called the *geometric mean*, or *mean proportional*, between a and d. Since $x^2 = ad$, you have that $x = \sqrt{ad}$.

EXAMPLE Find the geometric mean between 6 and 24.

SOLUTION Let x = the geometric mean between 6 and 24.

Then, $x = \sqrt{6 \cdot 24}$

$x = \sqrt{144}$

$x = 12$

The geometric mean between 6 and 24 is 12.

You will study the geometric mean again in Chapters 9 and 10.

One ratio that is historically important is called the *Golden Ratio* or the *Golden Mean.* The Golden Mean can be defined as the (unique) positive real number x that satisfies the proportion

$$\frac{1-x}{x} = \frac{x}{1}.$$

This proportion can be interpreted geometrically as follows:

Let $AC = 1$ and let B be between A and C such that $BC:AB = AB:AC$. In other words, AB is the geometric mean between BC and AC. Then the length AB is the Golden Mean. In Exercise 47 you will show that the value of the Golden Mean is $\frac{\sqrt{5}-1}{2}$.

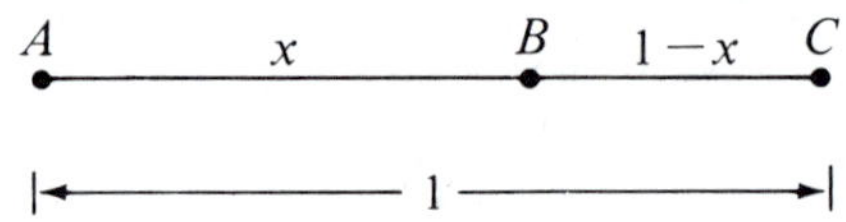

A rectangle whose sides have lengths in the ratio of $\frac{\sqrt{5}-1}{2}:1$ is called a *golden rectangle.* Many consider a golden rectangle to have the most pleasing proportions of all rectangles. A 3-by-5 or a 5-by-8 file card is a close approximation of a golden rectangle.

ORAL EXERCISES

1. In the proportion $\frac{1}{4} = \frac{x}{8}$, **(a)** name the means; **(b)** name the extremes; **(c)** state how the product of the means is related to the product of the extremes.

State each ratio in lowest terms.

2. $\frac{16}{24}$ **3.** $\frac{45}{25}$ **4.** $30:75$ **5.** $14:49$

6. $\frac{3x}{15x}$ **7.** $m^3:4m^2$ **8.** $12:18:6$ **9.** $2a:4a^2:2a$

True or false?

10. If $\frac{x}{y} = \frac{3}{4}$, then $x = 3$ and $y = 4$.

11. $\frac{x}{5} = \frac{2}{3}$ if and only if $3x = 10$.

12. $\frac{5}{2} = \frac{1}{x}$ if and only if $\frac{2}{5} = x$.

13. If $x:y = 2:3$, then $x = 2$ and $y = 3$.

14. If $x = 2$ and $y = 3$, then $x:y = 2:3$.

15. If $a:b:c = 1:3:5$, then $a = 1$, $b = 3$, and $c = 5$.

16. If $\frac{x}{3} = \frac{1}{2}$, then $\frac{x+3}{3} = \frac{3}{2}$.

17. If $\frac{a}{b} = \frac{4}{5}$, then $\frac{a}{5} = \frac{b}{4}$.

18. If $\frac{a}{b} = \frac{2}{3}$, then $\frac{a+2}{b+3} = \frac{2}{3}$.

19. If $\frac{x}{4} = 5$, then $5x = 20$.

20. The geometric mean between 3 and 12 is 6.

21. The geometric mean between 2 and 8 is 5.

WRITTEN EXERCISES

In the figure at the right $AC = BC$, and D is the midpoint of $\overline{AB}$. If $AC = 15$ and $AB = 12$, write the value of each ratio in simplest form.

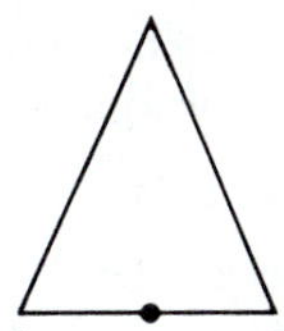

A **1.** $\frac{AC}{AB}$ **2.** $\frac{AD}{AC}$ **3.** $AC:BC$

4. $AD:BD$ **5.** $\frac{AB}{DB}$ **6.** $\frac{BC}{DB}$

Complete.

7. If $\frac{x}{y} = \frac{5}{2}$, then $\frac{y}{x} = \underline{\ ?\ }$.

8. If $\frac{a}{3} = \frac{b}{4}$, then $\frac{a}{b} = \underline{\ ?\ }$.

9. If $\frac{m}{5} = \frac{n}{6}$, then $5n = \underline{\ ?\ }$.

10. If $\frac{x+5}{5} = \frac{y+2}{2}$, then $\frac{x}{5} = \underline{\ ?\ }$.

11. If $\frac{4}{a} = \frac{b}{12}$, then $\frac{a}{12} = \underline{\ ?\ }$.

12. If r is the geometric mean between 5 and 9, then $r^2 = \underline{\ ?\ }$.

Find the value of $\frac{x}{y}$ in each of the following.

13. $2x = 3y$ **14.** $4x - 7y = 0$ **15.** $\frac{1}{2}x = \frac{3}{4}y$

16. In the figure at the right, $AB:BC = 4:3$. Find $AB:AC$ and $BC:AC$.

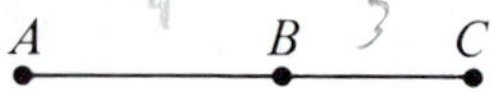

Find the value of x.

17. $\frac{x}{15} = \frac{2}{3}$ **18.** $21:10 = 14:x$ **19.** $(x-3):35 = 8:5$ **20.** $\frac{5}{3} = \frac{2x-8}{6}$

Find the values of x and y.

EXAMPLE $4:10:12 = 2:x:y$

SOLUTION From the definition of extended proportion, you know that there exists a constant k such that

$$4 = k \cdot 2 \qquad 10 = k \cdot x \qquad 12 = k \cdot y.$$

Since $4 = k \cdot 2$, you know that $k = 2$. Therefore,

$$10 = 2 \cdot x \quad \text{and} \quad 12 = 2 \cdot y, \quad \text{or}$$
$$5 = x \quad \text{and} \quad 6 = y.$$

21. $3:4:5 = 6:x:y$ **22.** $8:x:20 = 2:3:y$

23. $(x+1):12:6 = 12:y:2$ **24.** $2:3:4 = (x-1):18:(y+1)$

Write each ratio in simplest form.

25. $\dfrac{2\frac{1}{2}}{3\frac{1}{4}}$ **26.** $\dfrac{1\frac{1}{3}}{2\frac{1}{6}}$ **27.** $\dfrac{5\frac{1}{4}}{2\frac{5}{12}}$ **28.** $3\frac{1}{5}:1\frac{3}{4}$

29. Two complementary angles have measures that are in the ratio 5:13. Find the measures of the two angles.

30. Two supplementary angles have measures that are in the ratio 7:8. Find the measures of the two angles.

31. Two numbers are in the ratio 3:5. Their sum is 16. Find the numbers.

32. Two positive numbers are in the ratio 5:12. The difference of the smaller minus the larger is −56. Find the numbers.

33. $\overline{AB}$ contains a point C such that $AC:BC = 3:10$. If $AB = 65$, find AC and BC.

34. $\overline{PQ}$ has length 48, and a point R is between P and Q so that $PR:QR = 3:13$. Find PR.

35. The measures of the angles of a triangle are in the ratio 2:3:5. Find the three angle measures.

36. The sides of a triangle have lengths in the ratio 3:4:5. Find the three lengths if the perimeter of the triangle is 48.

In Exercises 37–42, assume that a, b, c, and d are non-zero real numbers.

Exercises 37 and 38 prove Property 1.

B **37.** Prove that if $\frac{a}{b} = \frac{c}{d}$, then $ad = bc$. (*Hint*: Use the Multiplication Property of Equality.)

38. Prove that if $ad = bc$, then $\frac{a}{b} = \frac{c}{d}$. (*Hint*: Use the Division Property of Equality.)

39. This exercise proves Property 2.

a. Prove that if $\frac{a}{b} = \frac{c}{d}$, then $\frac{b}{a} = \frac{d}{c}$. (*Hint*: Use Property 1 first, then divide both sides by the same quantity.)

b. Prove the converse of part **a**.

40. This exercise proves Property 3.

a. Prove that if $\frac{a}{b} = \frac{c}{d}$, then $\frac{a}{c} = \frac{b}{d}$.

b. Prove the converse of part **a**.

Exercises 41 and 42 prove Property 4.

41. Prove that if $\frac{a}{b} = \frac{c}{d}$, then $\frac{a+b}{b} = \frac{c+d}{d}$. (The proof of Property 5 is similar.)

42. Prove that if $\frac{a+b}{b} = \frac{c+d}{d}$, then $\frac{a}{b} = \frac{c}{d}$. (The proof of Property 5 is similar.)

Solve for x.

43. $\frac{x-1}{4} = \frac{5}{x+7}$ **44.** $\frac{2}{x-3} = \frac{x+2}{12}$ **45.** $\frac{x}{x-2} = \frac{10}{x+1}$ **46.** $\frac{x-4}{x} = \frac{6}{x-3}$

47. If x is the geometric mean between $1 - x$ and 1, show that the Golden Mean is $\frac{\sqrt{5}-1}{2}$.

48. Show that the Golden Mean is the unique positive real number whose reciprocal is 1 greater than the number itself. (*Hint*: Solve the equation $\frac{1}{x} = x + 1$.)

49. If you have a calculator with a reciprocal key, get the Golden Mean on the display. Press the reciprocal key several times. How does the display change each time you press it? Can you explain why?

C **50.** Prove Property 6: If $\frac{a}{b} = \frac{x}{y}$ and $\frac{c}{d} = \frac{x}{y}$, then $\frac{a+c}{b+d} = \frac{x}{y}$. (*Hint*: Use Property 3 to rewrite each of the given proportions with a, b, c, and d in the numerators. Add the respective sides of these two new proportions.)

51. Use a technique similar to the one suggested in the hint for Exercise 50 above to show that if $\frac{a}{b} = \frac{x}{y}$ and $\frac{c}{d} = \frac{x}{y}$, then $\frac{a-c}{b-d} = \frac{x}{y}$.

Section 8-2 SIMILAR POLYGONS

Two figures that have the same shape but not necessarily the same size are called similar figures. Two (convex) polygons are *similar* ($\sim$) if their consecutive vertices can be paired so that:

1. Corresponding angles are congruent.
2. The lengths of corresponding sides are proportional (have the same ratio).

EXAMPLE

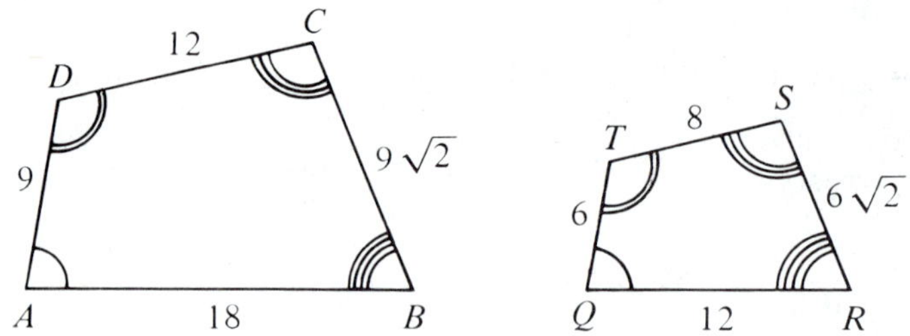

1. $\angle A \cong \angle Q$ $\quad\angle B \cong \angle R$ $\quad\angle C \cong \angle S$ $\quad\angle D \cong \angle T$
2. $\frac{AB}{QR} = \frac{BC}{RS} = \frac{CD}{ST} = \frac{DA}{TQ} = \frac{3}{2}$

Therefore $ABCD \sim QRST$.

In any statement that two polygons are similar the corresponding vertices should always be listed *in order*. The statement "$ABCDE \sim FGHJK$" means that polygons $ABCDE$ and $FGHJK$ are similar *and* that A corresponds to F, B corresponds to G, and so on. Another way to write this statement is "$BCDEA \sim GHJKF$," since this establishes the same correspondence of vertices.

If two polygons are similar, the ratio of the lengths of any two corresponding sides is called the *constant of proportionality*. In the preceding example the constant of proportionality is $\frac{3}{2}$ since $\frac{AB}{QR} = \frac{3}{2}$.

EXAMPLE The sides of a quadrilateral have lengths 8, 10, 12, and 15. Find the lengths of the sides of a larger similar quadrilateral if the constant of proportionality is $\frac{2}{5}$.

SOLUTION If x = the length of the shortest side of the larger quadrilateral, you have:

$$\frac{8}{x} = \frac{2}{5}$$

Using Property 1 (cross-multiplying), you obtain:

$$2x = 40$$
$$x = 20$$

Using proportions involving each of the other three sides, you will find that the remaining three sides of the larger quadrilateral have lengths 25, 30, and 37.5.

EXAMPLE In $\triangle ABC$, $AB = 9$, $BC = 15$, and $AC = 18$. If $\triangle ABC \sim \triangle DEF$ and $DE = 12$, find EF and DF.

SOLUTION If $\triangle ABC \sim \triangle DEF$, then $\overline{AB}$ corresponds to $\overline{DE}$, and $\overline{BC}$ corresponds to $\overline{EF}$. Therefore, you have:

$$\frac{AB}{DE} = \frac{BC}{EF}$$

Substituting the given values, you obtain:

$$\frac{9}{12} = \frac{15}{EF}, \quad \text{or} \quad \frac{3}{4} = \frac{15}{EF}$$

Therefore, $3(EF) = 60$ and $EF = 20$.

From the proportion $\frac{AB}{DE} = \frac{AC}{DF}$, you find that $DF = 24$.

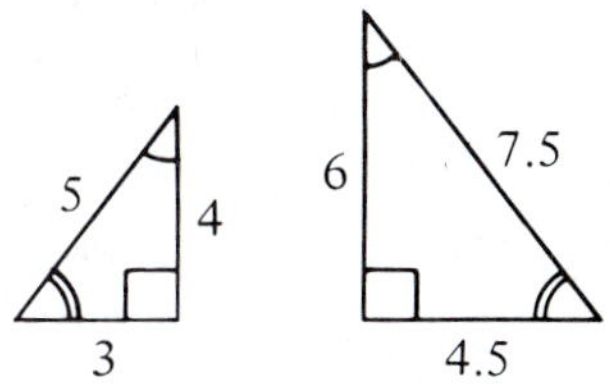

The triangles shown at the right are similar. The ratio of the lengths of any two corresponding sides is $\frac{2}{3}$. What is the ratio of their perimeters?

$$\frac{3 + 4 + 5}{4.5 + 6 + 7.5} = \frac{12}{18} = \frac{2}{3}$$

As you can see, the ratio of their perimeters equals the ratio of the lengths of any two corresponding sides. This is true for any two similar polygons. We will prove the specific case in which the polygons are triangles.

Theorem 8-1 If two polygons are similar, the ratio of their perimeters equals the ratio of the lengths of any two corresponding sides.

GIVEN: $\triangle ABC \sim \triangle DEF$

PROVE: $\frac{AB + BC + AC}{DE + EF + DF} = \frac{AB}{DE}$

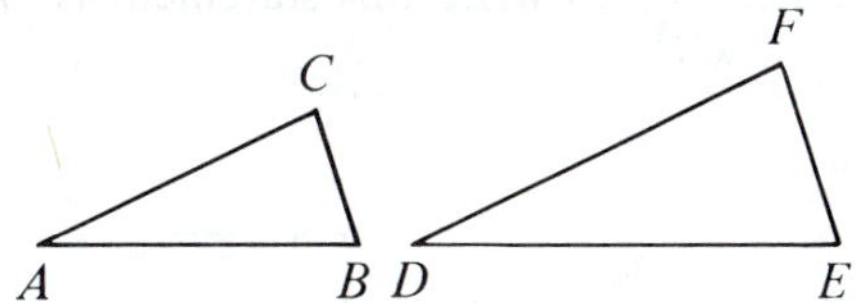

Statements	Reasons
1. $\triangle ABC \sim \triangle DEF$	1. Given
2. $\frac{BC}{EF} = \frac{AB}{DE}$	2. The lengths of corr. sides of ~ polygons are proportional. (Def. of ~ polygons)
3. $\frac{AB + BC}{DE + EF} = \frac{AB}{DE}$	3. Property 6 of Proportions, page 308
4. $\frac{AC}{DF} = \frac{AB}{DE}$	4. Def. of ~ polygons
5. $\frac{AB + BC + AC}{DE + EF + DF} = \frac{AB}{DE}$	5. Property 6 of Proportions, page 308

EXAMPLE The sides of a pentagon have lengths 4, 5, 6, 8, and 9. The perimeter of a similar pentagon is 24. Find the length of the shortest side of the second pentagon.

SOLUTION Let x = the length of the shortest side of the second pentagon. Then by Theorem 8-1 you have:

$$\frac{\text{Perimeter of given pentagon}}{\text{Perimeter of similar pentagon}} = \frac{4}{x}$$

$$\frac{4+5+6+8+9}{24} = \frac{4}{x}$$

$$\frac{32}{24} = \frac{4}{x}$$

$$x = 3$$

The length of the shortest side of the second pentagon is 3.

Like equality and congruence, similarity is reflexive, symmetric, and transitive.

Reflexive Property	A polygon is always similar to itself.
Symmetric Property	If Polygon I is similar to Polygon II, then Polygon II is similar to Polygon I.
Transitive Property	If Polygon I is similar to Polygon II and Polygon II is similar to Polygon III, then Polygon I is similar to Polygon III.

You will prove transitivity for similar *triangles* in Exercise 24.

ORAL EXERCISES

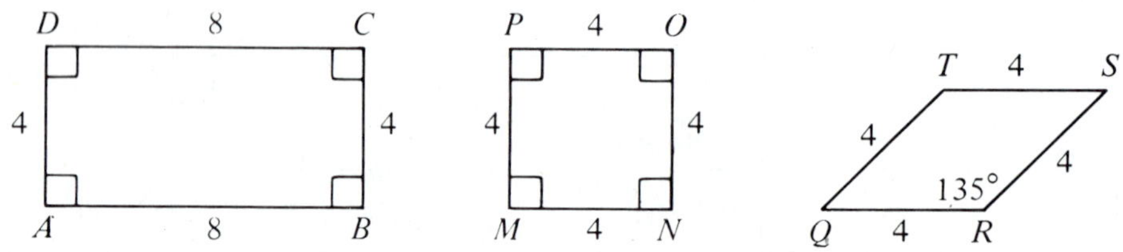

1. Compare quadrilateral *ABCD* and quadrilateral *MNOP*.
   a. Are corresponding angles congruent?
   b. Are the lengths of corresponding sides proportional?
   c. Is $ABCD \sim MNOP$?
2. Compare quadrilateral *MNOP* and quadrilateral *QRST*.
   a. Are corresponding angles congruent?
   b. Are the lengths of corresponding sides proportional?
   c. Is $MNOP \sim QRST$?

Tell whether the polygons are *always, sometimes,* or *never* similar.

3. Two squares

4. Two rhombuses

5. Two isosceles triangles

6. A trapezoid and a rhombus

7. Two equilateral triangles

8. Two regular hexagons

9. Two isosceles trapezoids

10. A square and a hexagon

11. If two corresponding sides of similar pentagons have lengths 9 and 15, respectively, what is the ratio of their perimeters?

12. If the perimeters of two similar hexagons are 21 and 24, respectively, and one side of the smaller one has length 5, state a proportion you could use to find the length x of the corresponding side of the larger hexagon.

13. Draw two parallelograms with congruent acute angles. Explain why corresponding angles of the two parallelograms must be congruent. Must any two such parallelograms be similar? Explain.

True or false?

14. Every polygon is similar to itself.

15. If two polygons are congruent, then they are similar.

16. If two polygons are similar, then they are congruent.

17. Two squares have sides of lengths 9 and 12, respectively.

a. What is the constant of proportionality?

b. What is the ratio of their perimeters?

c. What is the ratio of their areas?

WRITTEN EXERCISES

In Exercises 1–4, $\triangle ABC \sim \triangle DEF$. Find **(a)** the constant of proportionality for each pair of triangles and **(b)** the values of x and y.

A **1.**

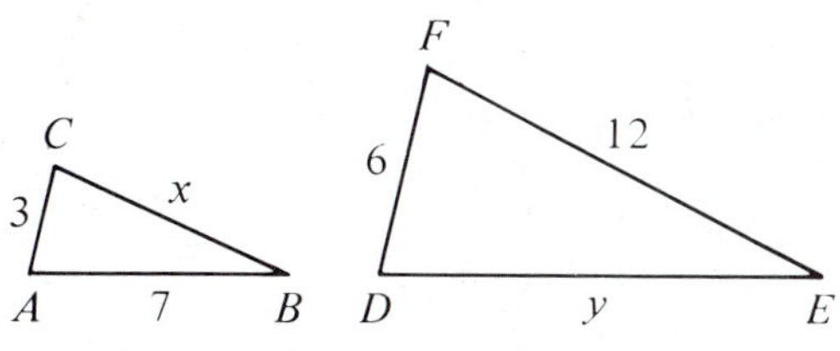

2.

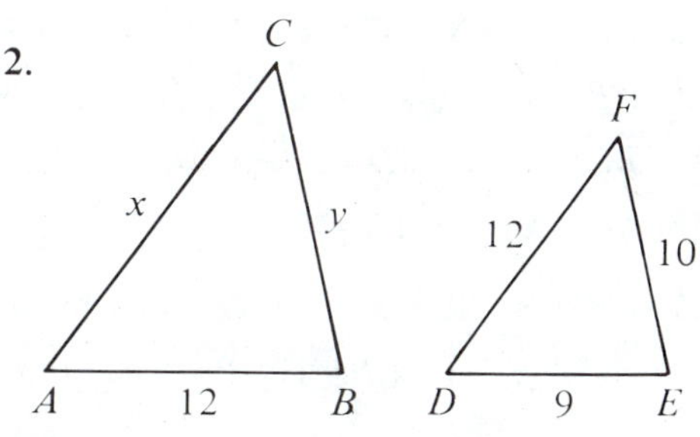

3.

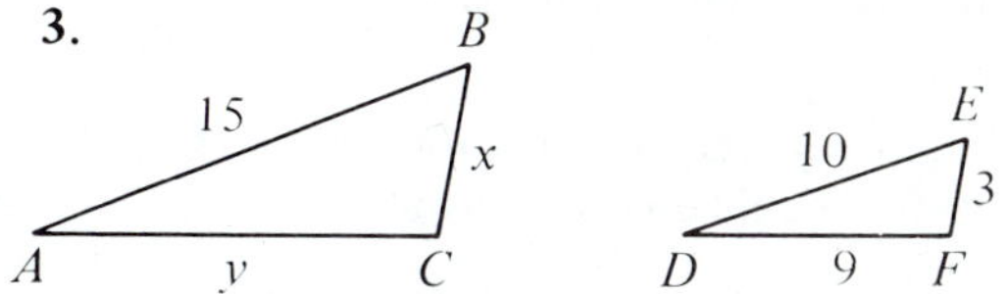

4.

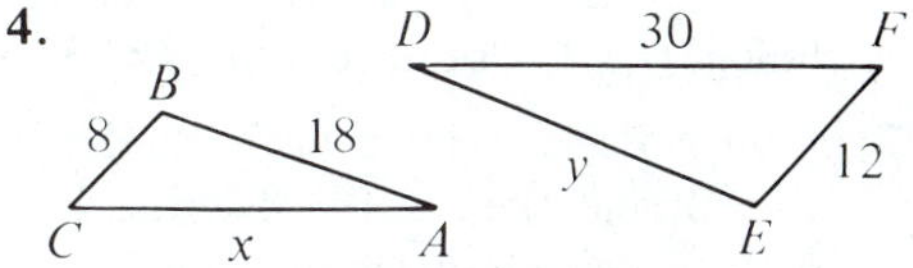

Given that $\triangle DEF \sim \triangle RST$, complete the following table.

	DE	EF	DF	Perimeter of $\triangle DEF$	RS	ST	RT	Perimeter of $\triangle RST$
5.	12	15	?	?	8	?	6	?
6.	24	?	10	?	30	35	?	?
7.	?	7.5	9	?	8	20	?	?
8.	17.5	?	12.5	?	?	18	10	?
9.	?	?	12	36	?	20	?	48
10.	3	?	?	15	?	?	24	60

11. The sides of a triangle have lengths 6, 8, and 10. What is the length of the shortest side of a similar triangle whose perimeter is 18?

12. The sides of a pentagon have lengths 5, 8, 9, 11, and 17. What is the length of the shortest side of a similar pentagon whose perimeter is 40?

13. If $\triangle ABC \sim \triangle DEF$, $AB = 5$, $BC = 7$, $CA = 10$, and the perimeter of $\triangle DEF$ is 33, find the lengths of the sides of $\triangle DEF$.

14. In the figure at the right, $\triangle ADE \sim \triangle ABC$.

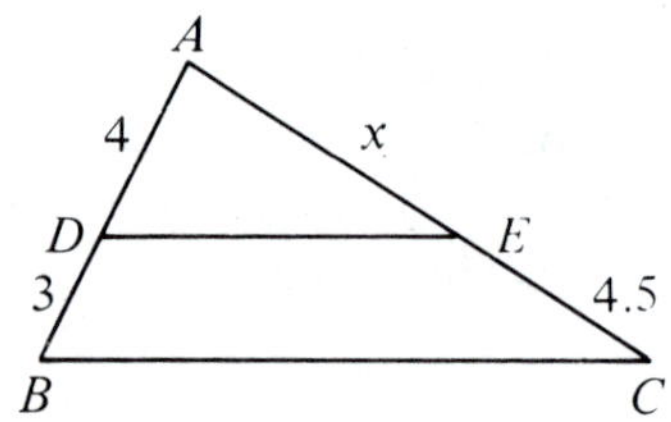

a. Express the length of $\overline{AC}$ in terms of x.

b. Complete: $\dfrac{AD}{AB} = \dfrac{AE}{?}$

c. Complete: $\dfrac{4}{7} = \dfrac{x}{?}$

d. Find the value of x.

e. Is $\overline{DE} \parallel \overline{BC}$? Explain.

Exercises 15–20 refer to the diagram below, in which $ABCD \sim PQRS$. In each exercise find all possible values of x.

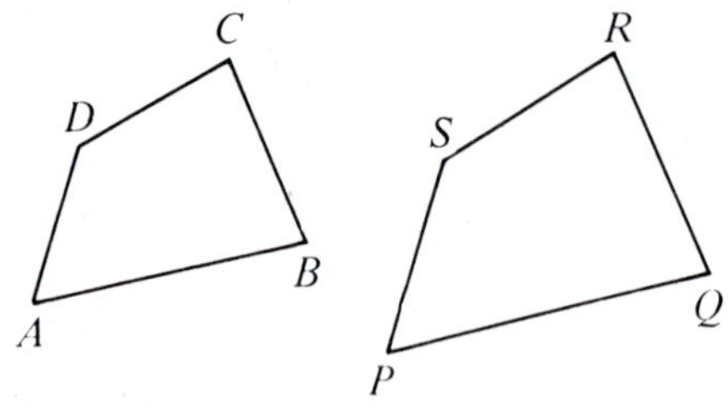

15. $AB = x + 3$, $BC = x - 1$, $PQ = 5$, $QR = 4$

16. $AB = x - 2$, $AD = x - 3$, $PQ = 7$, $PS = 5$

17. $BC = x - 2$, $CD = x - 1$, $RS = x + 4$, $QR = x$

18. $BC = x + 3$, $AD = 2x$, $PS = 2x - 3$, $QR = x + 1$

B 19. $AB = 8$, $BC = x + 1$, $PQ = x - 1$, $QR = x - 2$

20. $AD = x + 1$, $AB = 10$, $PS = x - 1$, $PQ = x + 2$

21. Explain why any two *regular* polygons with the same number of sides are similar. (*Hint*: Use the corollary to Theorem 4-11.)

22. In an isosceles right triangle the length of the hypotenuse equals the product of $\sqrt{2}$ and the length of either leg. Use this fact to explain why any two isosceles right triangles are similar.

23. Prove that if $\triangle ABC \cong \triangle DEF$ and $\triangle DEF \sim \triangle JKL$, then $\triangle ABC \sim \triangle JKL$.

C 24. Prove the Transitive Property of Similarity for triangles.
Given: $\triangle ABC \sim \triangle DEF$; $\triangle DEF \sim \triangle JKL$
Prove: $\triangle ABC \sim \triangle JKL$
(*Hint*: If $\frac{AB}{DE} = \frac{BC}{EF}$ and $\frac{DE}{JK} = \frac{EF}{KL}$, then $\frac{AB}{DE} \cdot \frac{DE}{JK} = \frac{BC}{EF} \cdot \frac{EF}{KL}$ by the Multiplication Property of Equality.)

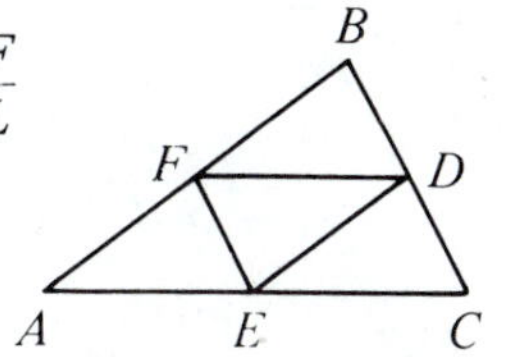

25. In $\triangle ABC$, D, E, and F are the midpoints of $\overline{BC}$, $\overline{AC}$, and $\overline{AB}$. Prove that $\triangle DEF \sim \triangle ABC$.

Section 8-3 THE TRIANGLE PROPORTIONALITY THEOREM

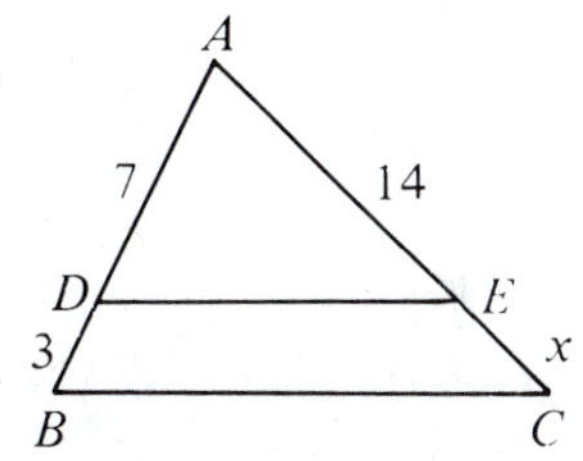

Suppose you know that in the diagram at the right, $\overline{DE} \parallel \overline{BC}$. Is it possible to find the value of x? The following theorem will enable you to do so.

Theorem 8-2 If a line parallel to one side of a triangle intersects the other two sides, then it divides them proportionally. (Triangle Proportionality Theorem)

GIVEN: $\overline{DE} \parallel \overline{BC}$

PROVE: $\frac{AD}{DB} = \frac{AE}{EC}$

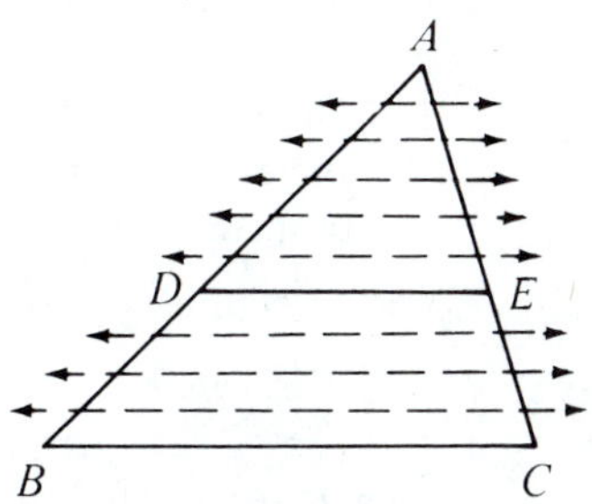

Outline of proof: Assume that $\frac{AD}{DB}$ is a positive rational number. (If it is not, the theorem is still true but its proof is beyond the scope of this book.) Then there exist positive integers x and y such that $\frac{AD}{DB} = \frac{x}{y}$. Therefore, $\frac{1}{x} \cdot AD = \frac{1}{y} \cdot DB$. Thus if $\overline{AD}$ is subdivided into x segments of length $\frac{1}{x} \cdot AD$ and $\overline{DB}$ is subdivided into y segments of length $\frac{1}{y} \cdot DB$, then all these segments will be congruent. The corollary to Theorem 5-14 can be generalized to prove that if lines are drawn parallel to $\overline{BC}$ through the endpoints of these segments to intersect $\overline{AC}$, then they will cut off x segments on $\overline{AE}$ and y segments on $\overline{EC}$, all of the same length. Therefore,

$$\frac{AE}{EC} = \frac{x}{y} \text{ and thus, } \frac{AD}{DB} = \frac{AE}{EC}.$$

In this book, if $\overline{DE} \parallel \overline{BC}$, you may use the Triangle Proportionality Theorem to support any proportion that is equivalent to $\frac{AD}{DB} = \frac{AE}{EC}$. Here are some proportions that are equivalent to $\frac{AD}{DB} = \frac{AE}{EC}$.

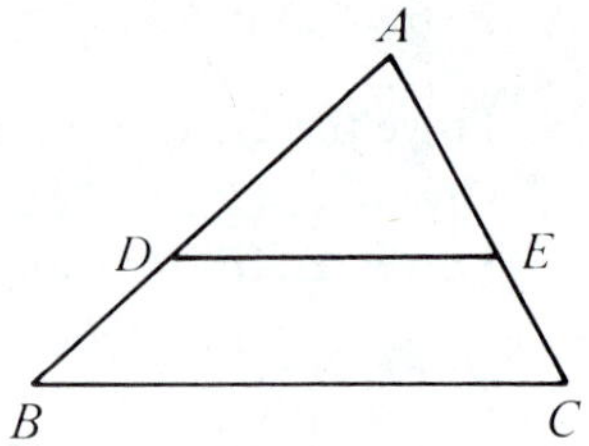

$$\frac{DB}{AB} = \frac{EC}{AC}, \quad \frac{AD}{AE} = \frac{DB}{EC}, \quad \frac{DB}{EC} = \frac{AB}{AC}$$

You will be asked to prove that some of these proportions are equivalent in Exercise 25.

EXAMPLE Given $\overline{DE} \parallel \overline{AB}$, $DC = 8$, $EC = 6$, and $BE = 5$. Find AD.

SOLUTION Let $AD = x$. Then you have

$$\frac{8}{x} = \frac{6}{5}$$
$$6x = 40$$
$$x = \frac{40}{6} = \frac{20}{3}$$

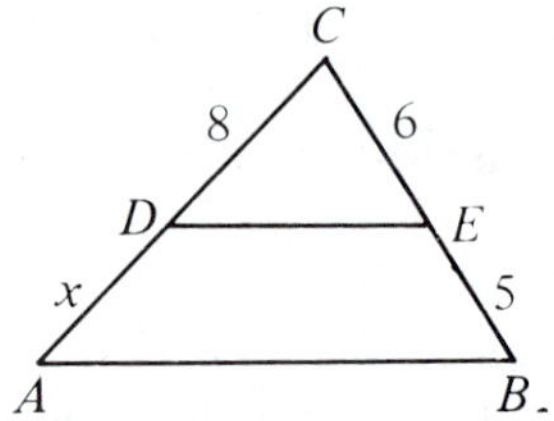

Corollary Three (or more) parallel lines cut off proportional segments on any two transversals.

GIVEN: $\overleftrightarrow{AC} \parallel \overleftrightarrow{EF} \parallel \overleftrightarrow{BD}$

PROVE: $\frac{BE}{EA} = \frac{DF}{FC}$

Proof: See Exercise 39.

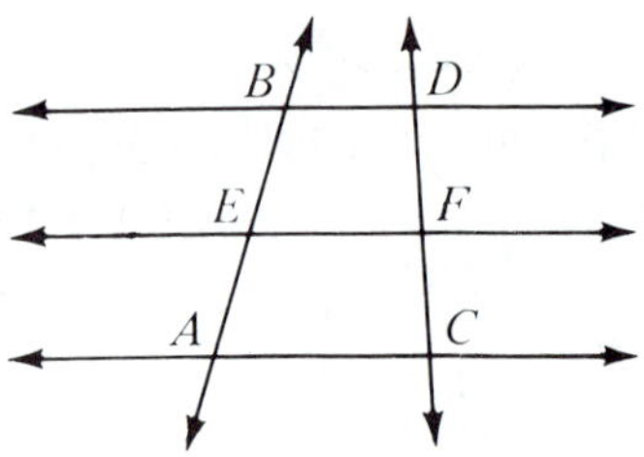

Theorem 8-3 The bisector of an angle of a triangle divides the side opposite the angle in the ratio of the lengths of the other two sides of the triangle.

GIVEN: $\triangle ABC$; $\overline{CD}$ bisects $\angle ACB$.

PROVE: $\frac{AD}{DB} = \frac{AC}{BC}$

Proof: Through B draw a line parallel to $\overline{CD}$. Let E be the point where $\overrightarrow{AC}$ intersects the line. By the Triangle Proportionality Theorem,

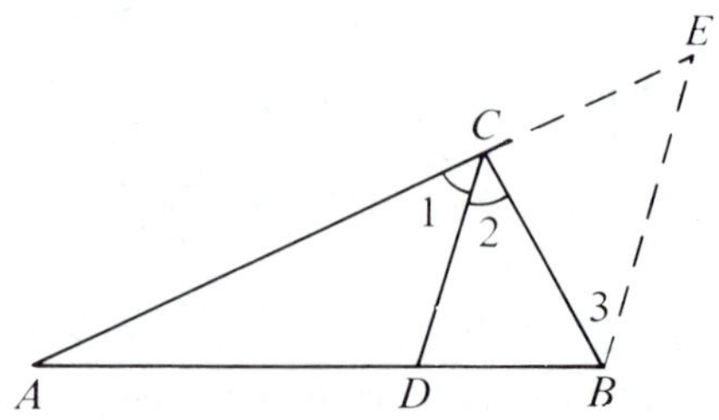

$$\frac{AD}{DB} = \frac{AC}{CE}$$

$\angle 1 \cong \angle 2$	(Def. of $\angle$ bisector)
$\angle 2 \cong \angle 3$	(If 2 lines are $\parallel$, alt. int. $\angle$s are $\cong$.)
$\angle 1 \cong \angle E$	(If 2 lines are $\parallel$, corr. $\angle$s are $\cong$.)
$\angle 3 \cong \angle E$	(Subst. prop.)

Therefore, $BC = CE$, and by substitution,

$$\frac{AD}{DB} = \frac{AC}{BC}$$

The converse of the Triangle Proportionality Theorem is also true.

Theorem 8-4 If a line divides two sides of a triangle proportionally, then it is parallel to the third side.

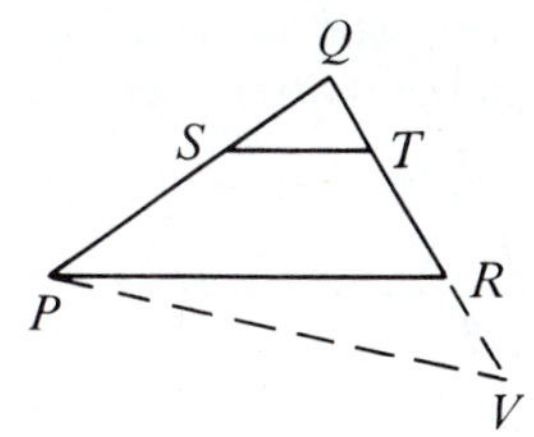

GIVEN: $\frac{QS}{SP} = \frac{QT}{TR}$

PROVE: $\overline{ST} \parallel \overline{PR}$

Indirect proof: Temporarily assume that $\overline{ST}$ is not parallel to $\overline{PR}$. Then by Theorem 4-3, you can draw a line through P and parallel to $\overline{ST}$. Let V be the point where this line intersects $\overrightarrow{QR}$. Then,

$$\frac{QS}{SP} = \frac{QT}{TV} \quad \text{(Triangle Proportionality Theorem)}$$

But, $$\frac{QS}{SP} = \frac{QT}{TR} \quad \text{(Given)}$$

Therefore, $TV = TR$. But this contradicts Postulate 2: There is exactly one point on a ray at a given distance from the endpoint of the ray. Therefore, our temporary assumption that $\overline{ST}$ is not parallel to $\overline{PR}$ is false. Hence, $\overline{ST} \parallel \overline{PR}$.

ORAL EXERCISES

Exercises 1-6 refer to the diagram at the right, in which $\overline{KL} \parallel \overline{EF}$. Complete each proportion.

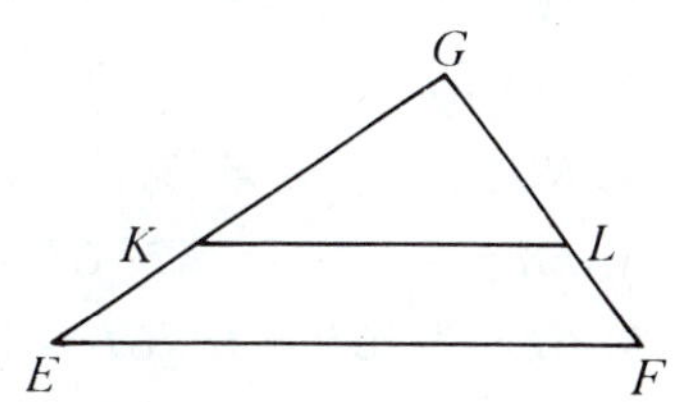

1. $\frac{EK}{KG} = \underline{\ ?\ }$
2. $\frac{GL}{LF} = \underline{\ ?\ }$
3. $\frac{GK}{GL} = \underline{\ ?\ }$
4. $\frac{FL}{EK} = \underline{\ ?\ }$
5. $\frac{GK}{GE} = \underline{\ ?\ }$
6. $\frac{GF}{GL} = \underline{\ ?\ }$

State whether $\overline{ST}$ is parallel to $\overline{PQ}$.

7.

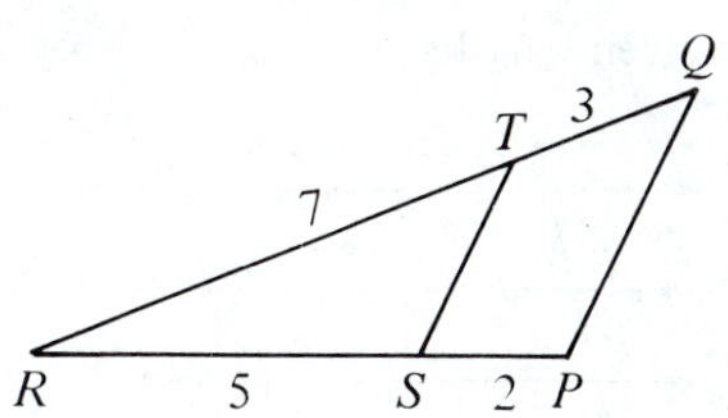

8.

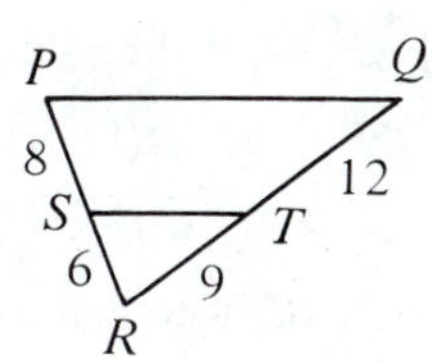

State whether $\overline{ST}$ is parallel to $\overline{PQ}$.

9

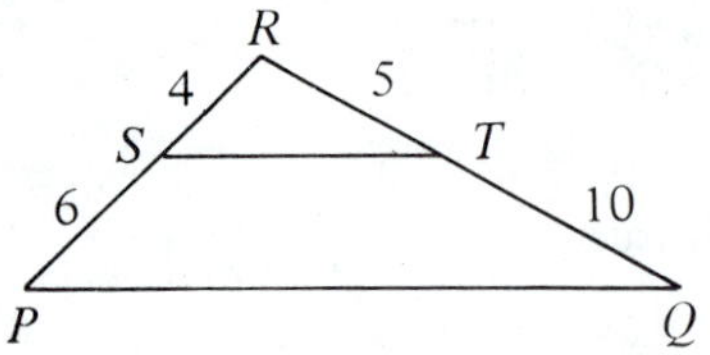

10.

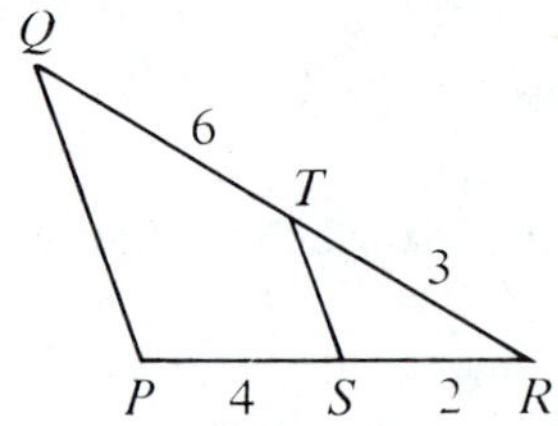

State a proportion you could use to find the value of x. You will solve the proportions in the Written Exercises.

11.

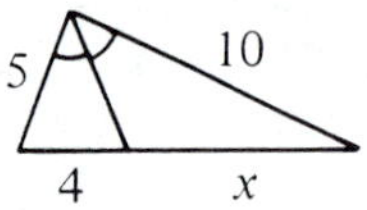

12.

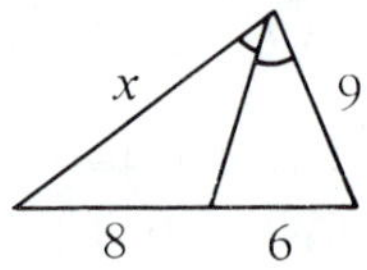

13.

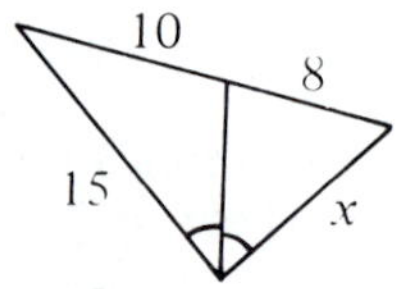

14.

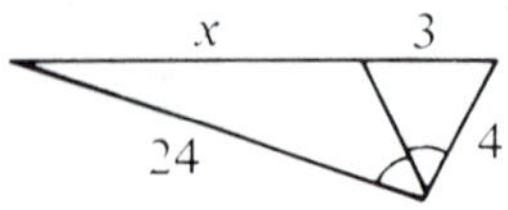

15.

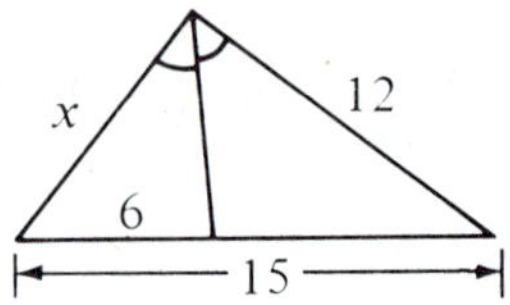

16.

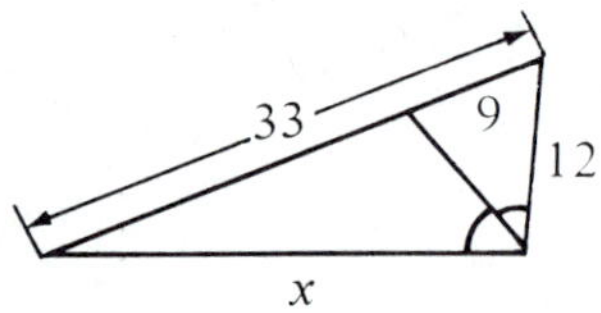

17. True or false? If a line is parallel to one base of a trapezoid and intersects the legs of the trapezoid, then it divides them proportionally. Explain.

WRITTEN EXERCISES

In $\triangle GEF$, $\overline{KL} \parallel \overline{EF}$. Find the desired length.

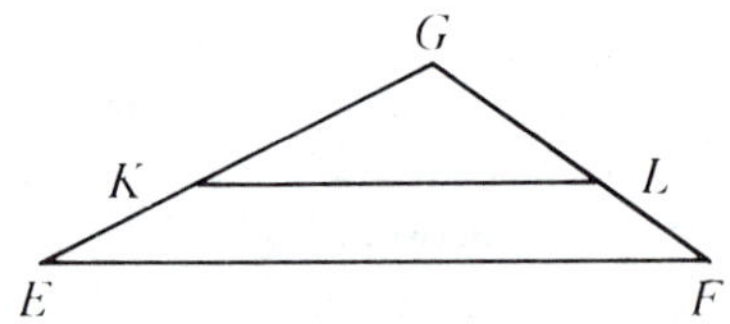

A **1.** $EK = 6$, $KG = 9$, $FL = 8$, $LG =$ __?__

2. $EK = 12$, $FL = 21$, $LG = 35$, $KG =$ __?__

3. $KG = 10$, $GL = 15$, $LF = 8$, $EK =$ __?__

4. $EK = 7.5$, $KG = 12$, $GL = 4$, $LF =$ __?__

5. $GK = 8$, $GE = 10$, $GL = 12$, $GF =$ __?__

6. $GL = 9$, $GF = 12$, $GK = 6$, $GE =$ __?__

7. $GK = 12$, $GE = 15$, $GF = 25$, $GL =$ __?__

8. $GE = 21$, $GF = 24$, $GL = 12$, $GK =$ __?__

9. $KE = 12$, $GL = 15$, $GK = 10$, $GF =$ __?__

10. $GK = 6$, $KE = 4$, $GL = 8$, $LF =$ __?__

In the diagram, $\overleftrightarrow{AB} \parallel \overleftrightarrow{CD} \parallel \overleftrightarrow{EF}$. Find the value of x, using the given lengths.

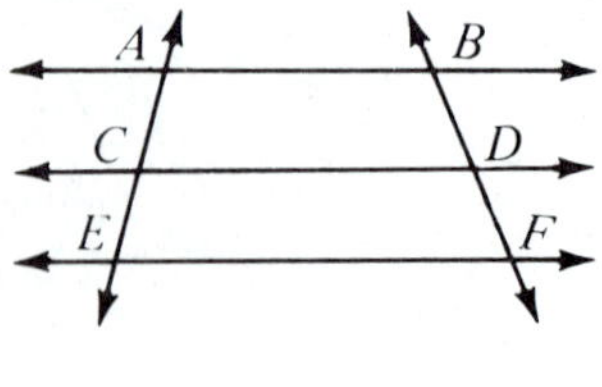

11. $AC = x$, $CE = x + 2$, $DF = 15$, $BD = 10$

12. $CE = x$, $FD = x + 3$, $AC = 12$, $BD = 16$

13. $AC = x - 4$, $CE = x - 2$, $BD = x - 1$, $DF = x + 3$

14. $CE = x$, $AC = x - 5$, $DF = x - 3$, $BD = x - 6$

Use the given information to determine if $\overline{DE} \parallel \overline{AC}$.

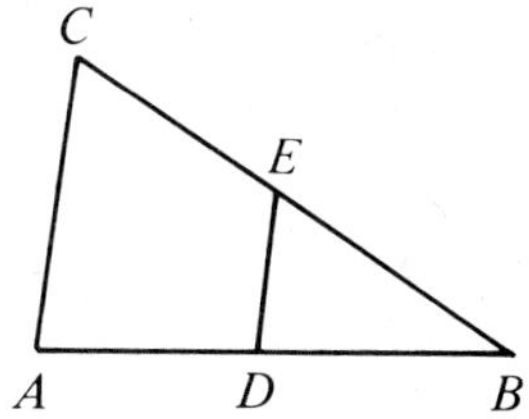

15. $AD = 4$, $BD = 5$, $CE = 6$, $EB = 7.5$

16. $BD = 6$, $CE = 8$, $AD = 4$, $BE = 12$

17. $BE = 2$, $BC = 10$, $BD = 3$, $AD = 12$

18. $AB = 15$, $AD = 5$, $BC = 21$, $BE = 7$

19–24. Solve the proportions in Oral Exercises 11–16.

25. Given: $\overline{DE} \parallel \overline{BC}$

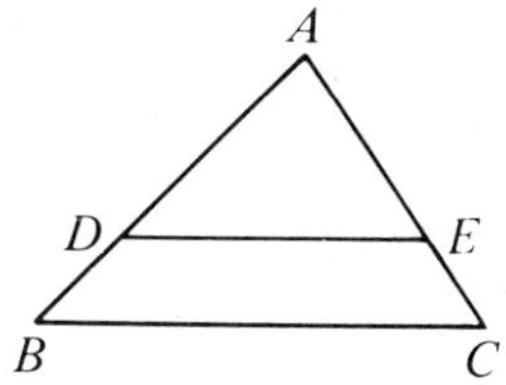

a. Show that $\dfrac{DB}{AB} = \dfrac{EC}{AC}$.

b. Show that $\dfrac{AD}{AB} = \dfrac{AE}{AC}$.

Write two-column proofs.

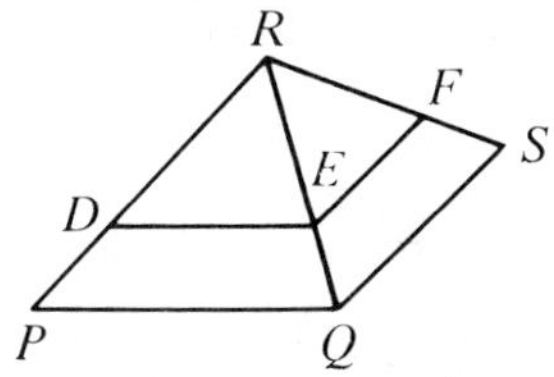

Exs. 26, 27

26. Given: $\overline{DE} \parallel \overline{PQ}$; $\overline{EF} \parallel \overline{QS}$

Prove: $\dfrac{RD}{DP} = \dfrac{RF}{FS}$

27. Given: $\overline{DE} \parallel \overline{PQ}$; $\dfrac{RD}{DP} = \dfrac{RF}{FS}$

Prove: $\dfrac{RE}{EQ} = \dfrac{RF}{FS}$

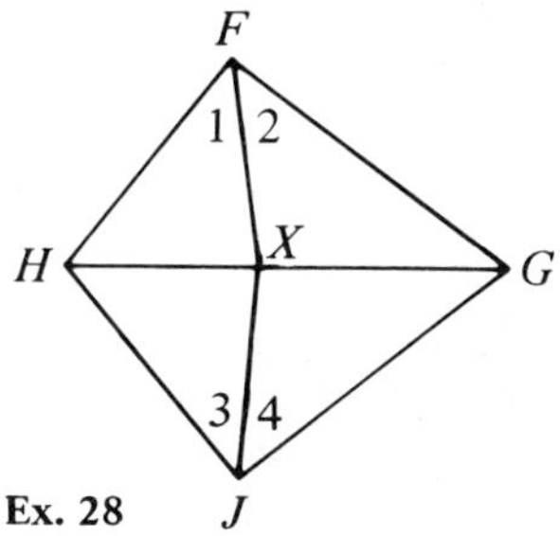

Ex. 28

28. Given: $\angle 1 \cong \angle 2$; $\angle 3 \cong \angle 4$

Prove: $\dfrac{FH}{FG} = \dfrac{JH}{JG}$

Exercises 29–36 refer to the diagram at the right below, in which $\angle 1 \cong \angle 2$. Complete.

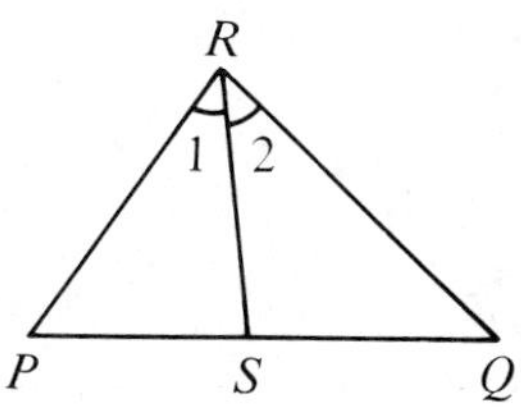

29. $PR = 6$, $PQ = 10$, $QR = 9$, $PS =$ ___?___

30. $PQ = 14$, $PR = 5$, $QR = 15$, $QS =$ ___?___

B 31. $PS = 4$, $RQ = 9$, $PR = SQ$, $PR =$ ___?___

32. $PS = RQ$, $SQ = 18$, $PR = 8$, $PS =$ ___?___

33. $PR = PS + 3$, $SQ = 6$, $RQ = 7$, $PS =$ ___?___ (*Hint:* Let $PS = x$.)

34. $SQ = RQ - 6$, $PR = 12$, $PS = 10$, $RQ =$ ___?___ (*Hint:* Let $RQ = x$.)

35. $PS = 5$, $RQ = 6$, $SQ = PR + 7$, $PR =$ ___?___

36. $SQ = 4$, $PR = 10$, $RQ = PS - 3$, $RQ =$ ___?___

Write two-column proofs.

37. Given: $\angle 1 \cong \angle 2$; $\overline{EF} \parallel \overline{CD}$

Prove: $\frac{AB}{AC} = \frac{BF}{FD}$

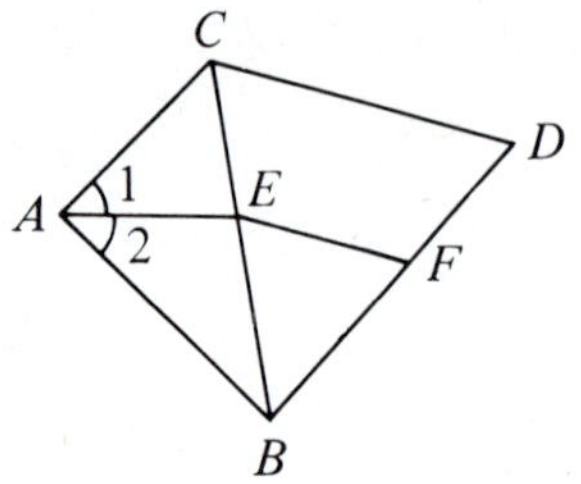

38. Given: $\frac{CJ}{JA} = \frac{CK}{KB} = \frac{AL}{LB}$

Prove: $AL = JK$

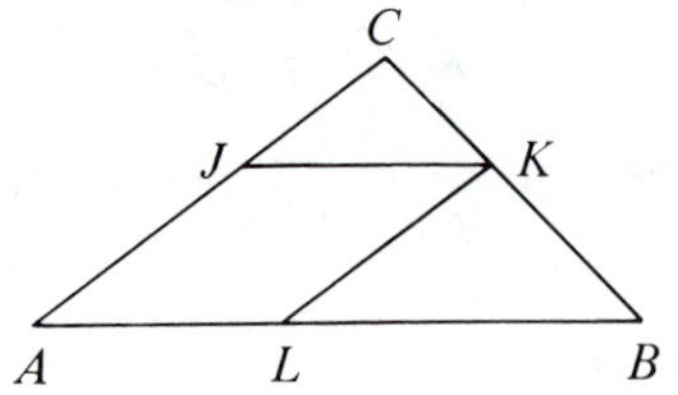

39. Write a two-column proof of the corollary to Theorem 8-2. (*Hint*: Draw $\overline{BC}$ and let G be the point where $\overline{BC}$ intersects $\overline{EF}$.)

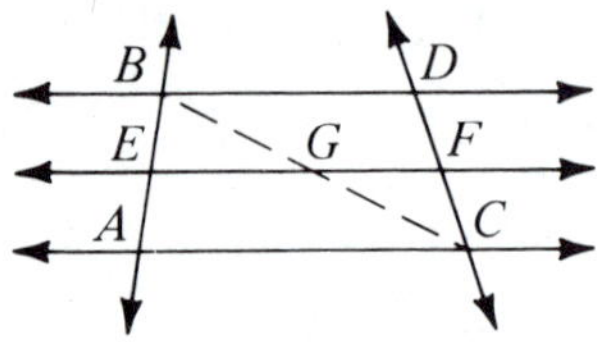

In Exercises 40 and 41, find x given that $\overline{DE} \parallel \overline{AB}$.

40.

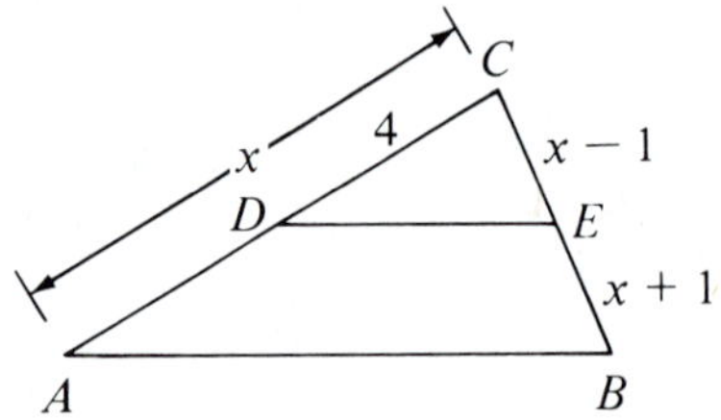

41.

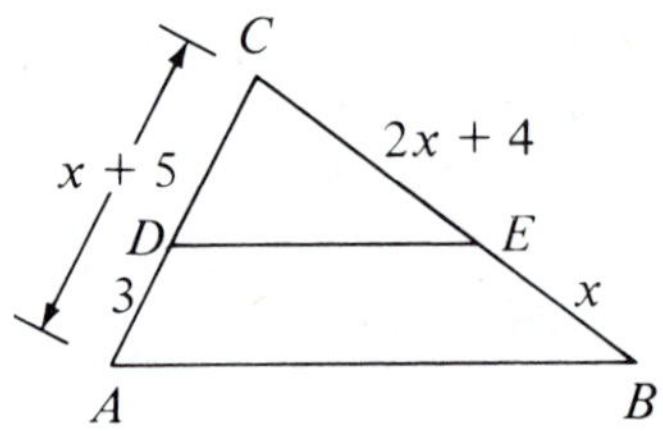

Write two-column proofs.

C **42.** Given: $\overline{GF} \parallel \overline{AB}$; $\overline{GE} \parallel \overline{BC}$; $\overline{EF} \parallel \overline{AC}$

Prove: E is the midpoint of $\overline{AB}$ (and hence F and G are also midpoints).

(*Hint*: Prove that $\frac{AE}{EB} = \frac{EB}{AE}$ and conclude that $AE = EB$.)

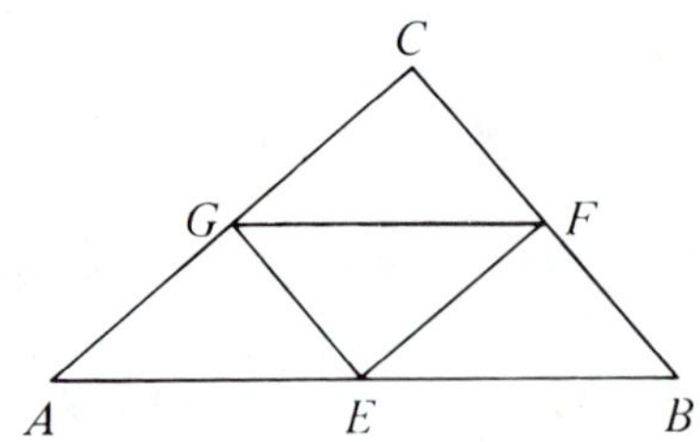

43. Given: $\triangle DEF$ with exterior angle $\angle FEG$; $\overrightarrow{EH}$ bisects $\angle FEG$.

Prove: $\frac{DH}{FH} = \frac{DE}{FE}$

(*Hint*: Draw $\overline{FJ}$ through F parallel to $\overline{HE}$.)

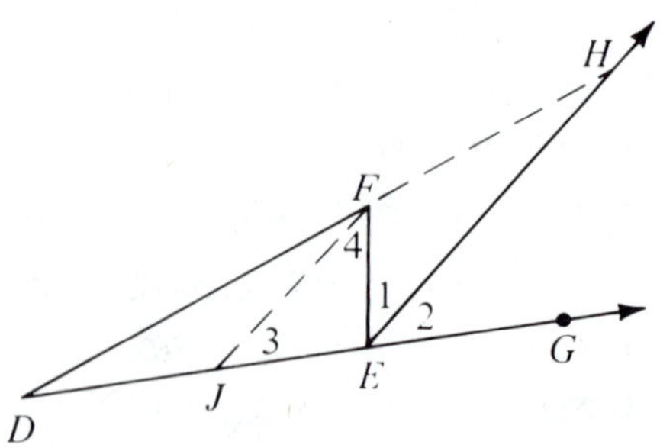

SELF-TEST 1

1. In an isosceles right triangle, what is the ratio in lowest terms of the measure of a base angle to the measure of the vertex angle? Section 8-1

2. Complete: If $\frac{r}{s} = \frac{7}{9}$, then $\frac{7}{r} = \underline{\ ?\ }$.

3. Find the value of x if $9:(x+2) = 5:2$.

In Exercises 4–6, $\triangle ABC \sim \triangle FDE$. Section 8-2

4. Find the constant of proportionality.

5. Find the value of y.

6. Find the perimeter of $\triangle FDE$.

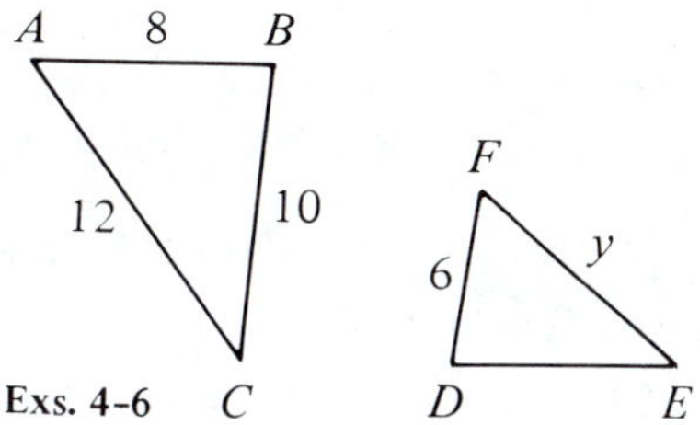

Exs. 4-6

7. $\overline{SV} \parallel \overline{TU}$, $RT = 9$, $RU = 15$, and $RS = 3$. Find RV and VU. Section 8-3

8. Write a two-column proof.

   Given: $\frac{RT}{ST} = \frac{RU}{VU}$

   Prove: $\angle RVS \cong \angle U$

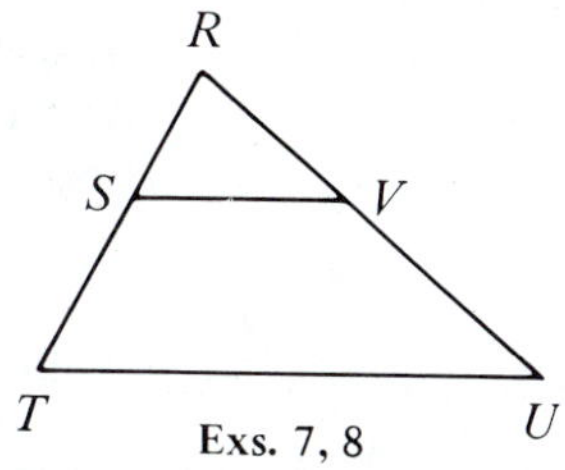

Exs. 7, 8

9. In $\triangle XYZ$, the bisector of $\angle X$ intersects $\overline{YZ}$ at point P. If $YP = 4$, $XY = 6$, and $PZ = 9$, find XZ.

Section 8-4 PROVING TRIANGLES SIMILAR

In order to prove that two triangles are similar you do not need to prove all corresponding angles congruent and the lengths of all corresponding sides proportional. In this section you will learn a simpler method called the AA Similarity Theorem.

Theorem 8-5 If two angles of one triangle are congruent to two angles of another triangle, then the triangles are similar. (AA Similarity Theorem)

GIVEN: $\triangle ABC$ and $\triangle DEF$;
$\angle A \cong \angle D$;
$\angle B \cong \angle E$

PROVE: $\triangle ABC \sim \triangle DEF$

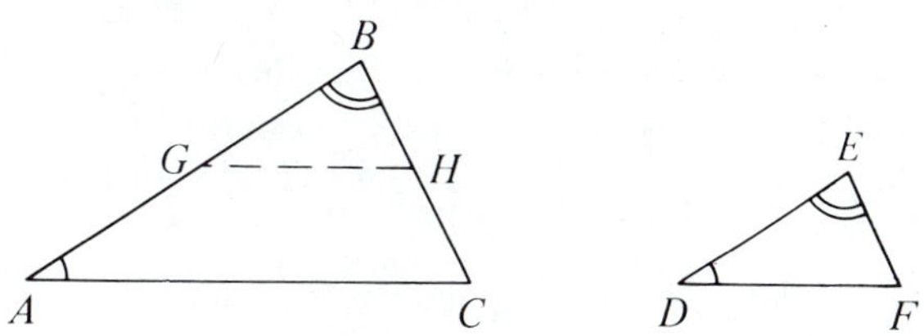

(The proof is on the next page.)

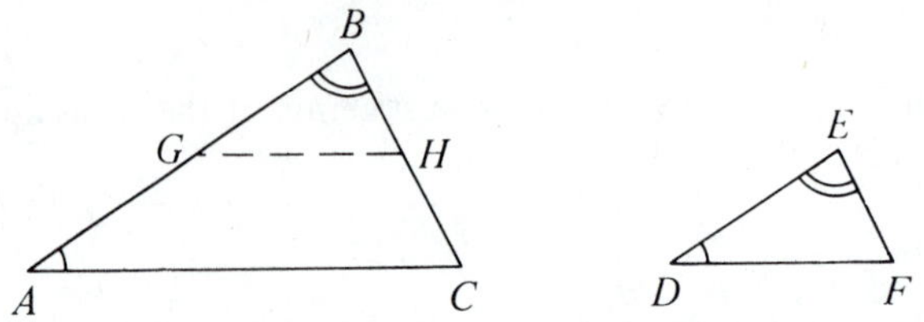

Proof: Assume $AB > DE$. Then locate G on $\overline{BA}$ so that $BG = ED$. Draw the line through G parallel to $\overline{AC}$. Let H be the point where this line intersects $\overline{BC}$. Then,

$\frac{BA}{BG} = \frac{BC}{BH}$	(Triangle Proportionality Theorem)
$\angle A \cong \angle D$	(Given)
$\angle BGH \cong \angle A$	(If 2 lines are $\parallel$, corr. ∠s are $\cong$.)
$\angle BGH \cong \angle D$	(Subst. Prop.)
$\angle B \cong \angle E$	(Given)
$\triangle GBH \cong \triangle DEF$	(ASA Post.)
$BH = EF$	(Corr. parts of $\cong$ △s are $\cong$.)

Since $BG = ED$,

$$\frac{BA}{BG} = \frac{BC}{BH} \text{ is equivalent to } \frac{BA}{ED} = \frac{BC}{EF}.$$

Similarly, by cutting off a segment congruent to $\overline{FE}$ along $\overline{BC}$, you can show that

$$\frac{BC}{EF} = \frac{AC}{DF}.$$

Therefore, $\frac{BA}{ED} = \frac{BC}{EF} = \frac{AC}{DF}$.

Since $\angle A \cong \angle D$ and $\angle B \cong \angle E$, you know that $\angle C \cong \angle F$ by Corollary 1 to Theorem 4-8. Thus, $\triangle ABC \sim \triangle DEF$ by definition of similar triangles.

EXAMPLE

GIVEN: $\angle A \cong \angle C$
PROVE: $\triangle ABX \sim \triangle CDX$

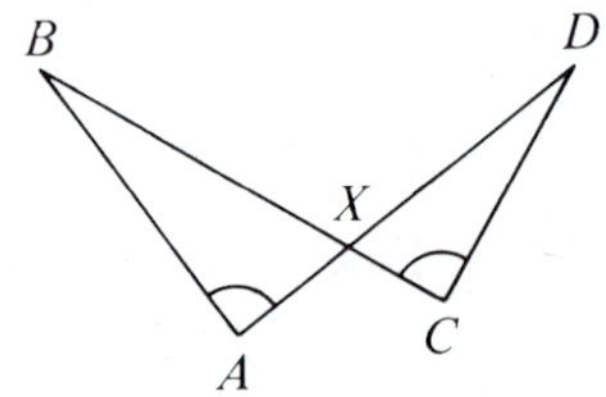

Statements	Reasons
1. $\angle A \cong \angle C$	1. Given
2. $\angle BXA \cong \angle DXC$	2. Vertical ∠s are $\cong$.
3. $\triangle ABX \sim \triangle CDX$	3. AA Similarity Theorem

Once you know that two triangles are similar, you can use the definition of similarity, just as you did the definition of congruence, to draw conclusions about the sides and angles of the two triangles. In particular, you can use the following facts:

1. The lengths of corresponding sides of similar triangles are proportional.
2. Corresponding angles of similar triangles are congruent.

EXAMPLE

GIVEN: $\angle PST \cong \angle PRQ$

PROVE: $\frac{PS}{PR} = \frac{PT}{PQ}$

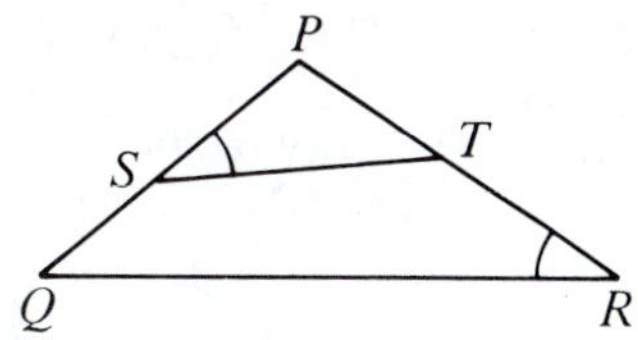

Statements	Reasons
1. $\angle PST \cong \angle PRQ$	1. Given
2. $\angle P \cong \angle P$	2. Reflexive prop. of $\cong$
3. $\triangle PST \sim \triangle PRQ$	3. AA Similarity Theorem
4. $\frac{PS}{PR} = \frac{PT}{PQ}$	4. The lengths of corr. sides of $\sim$ $\triangle$s are prop.

If you want to prove a statement like $AB \cdot CX = CD \cdot AX$, then rewrite the conclusion as a proportion and look for a pair of similar triangles whose sides correspond to the terms of the proportion.

EXAMPLE

GIVEN: $\angle G \cong \angle J$

PROVE: $GK \cdot KI = JK \cdot KH$

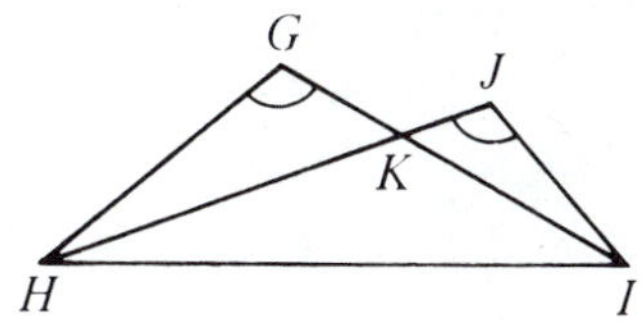

Analysis: Since $GK \cdot KI = JK \cdot KH$ is equivalent to $\frac{GK}{JK} = \frac{KH}{KI}$, we will try to prove that $\triangle GKH \sim \triangle JKI$.

Statements	Reasons
1. $\angle G \cong \angle J$	1. Given
2. $\angle GKH \cong \angle JKI$	2. Vert. $\angle$s are $\cong$.
3. $\triangle GKH \sim \triangle JKI$	3. AA Similarity Theorem
4. $\frac{GK}{JK} = \frac{KH}{KI}$	4. The lengths of corr. sides of $\sim$ $\triangle$s are prop.
5. $GK \cdot KI = JK \cdot KH$	5. Property 1 of Proportions

The next theorem states a relationship that you might expect between the altitudes of similar triangles.

Theorem 8-6 If two triangles are similar, the ratio of their corresponding altitudes equals the ratio of the lengths of any two corresponding sides.

GIVEN: $\triangle ABC \sim \triangle DEF$;
$\overline{CG}$ is the altitude from C;
$\overline{FH}$ is the altitude from F.

PROVE: $\frac{CG}{FH} = \frac{AC}{DF} = \frac{CB}{FE} = \frac{AB}{DE}$

Proof: See Exercise 24.

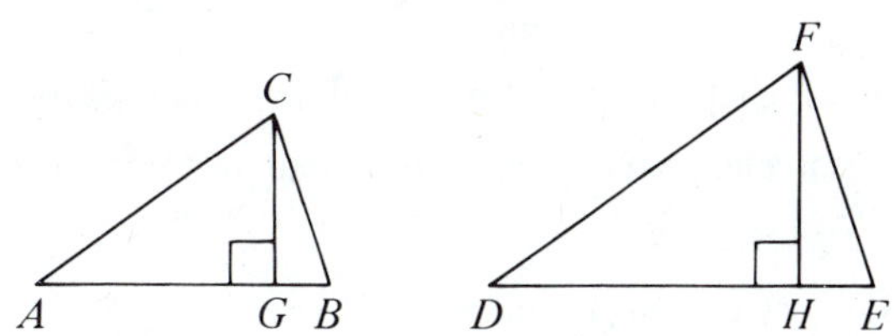

You will study some important applications of this theorem in Section 8-6, which concerns areas of similar polygons.

ORAL EXERCISES

State a similarity relationship between a pair of triangles in each diagram.

1.

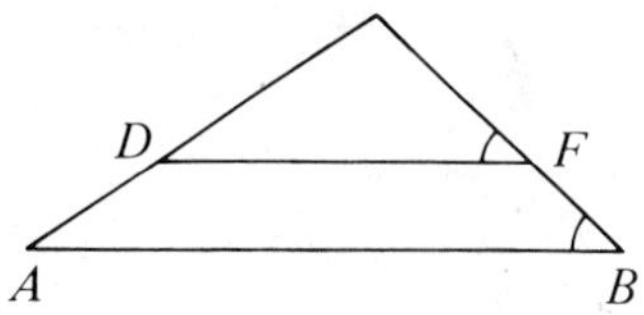

2.

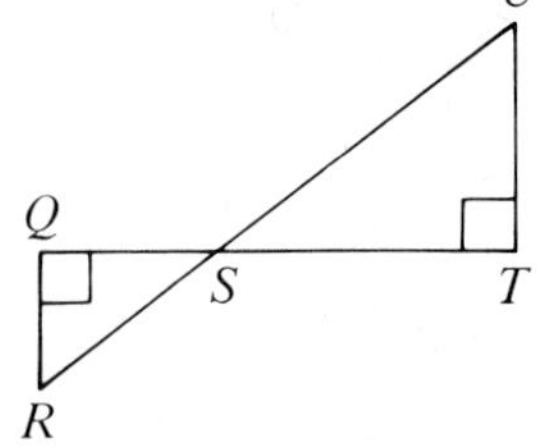

3.

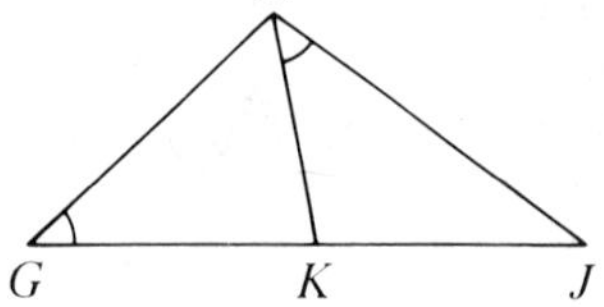

4.

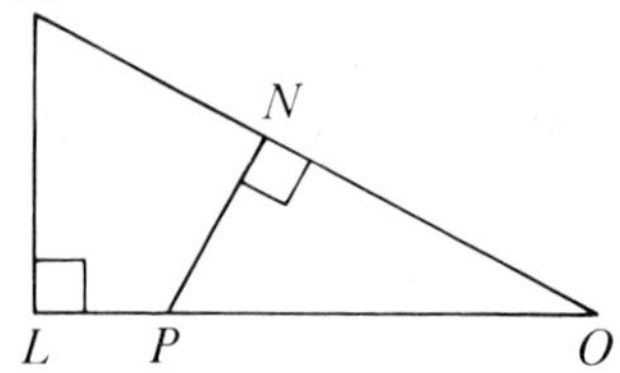

$\triangle ABC \sim \triangle DEF$, and $\overline{CG}$ and $\overline{FH}$ are altitudes. State a proportion you could use to find the value of x. You will solve the proportions in the Written Exercises.

5. $DF = 6$, $CG = 10$, $FH = 4$, $AC = x$
6. $AC = 24$, $CG = 16$, $DF = 21$, $FH = x$
7. $AB = 45$, $CG = 25$, $FH = 10$, $DE = x$
8. $FH = 9$, $DE = 36$, $CG = 11$, $AB = x$

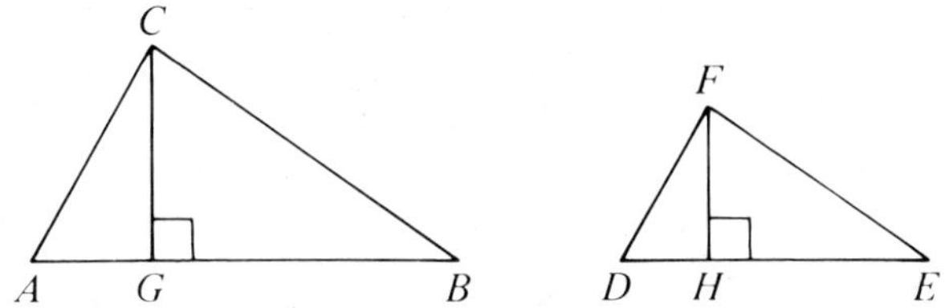

Name two triangles which, if shown to be similar, would yield the given proportion.

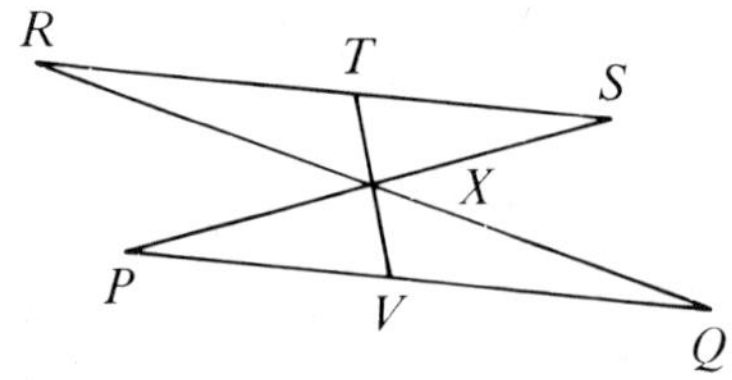

9. $\dfrac{RS}{PQ} = \dfrac{RX}{QX}$
10. $\dfrac{RT}{VQ} = \dfrac{TX}{VX}$
11. $\dfrac{RX}{QX} = \dfrac{SX}{PX}$
12. $\dfrac{TS}{PV} = \dfrac{TX}{VX}$

Answer *yes* if each of the following pairs of triangles *must be* similar, *no* otherwise.

13. Two isosceles triangles with congruent vertex angles
14. Two right triangles
15. Two isosceles right triangles
16. Two equilateral triangles
17. Two acute isosceles triangles

WRITTEN EXERCISES

Complete.

A **1.** Given: $\triangle STY \sim \triangle QRY$
$\frac{ST}{?} = \frac{TY}{RY}$

2. Given: $\triangle TYS \sim \triangle QYR$
$\frac{TY}{QY} = \frac{SY}{?}$

3. Given: $\triangle SXY \sim \triangle SPQ$
$\frac{SY}{SQ} = \frac{XY}{?}$

4. Given: $\triangle SXY \sim \triangle SPQ$
$\frac{PQ}{XY} = \frac{?}{SX}$

5. Given: $\triangle STY \sim \triangle SQP$
$\frac{ST}{?} = \frac{?}{PQ}$

6. Given: $\triangle STY \sim \triangle SQP$
$\frac{?}{TY} = \frac{?}{SY}$

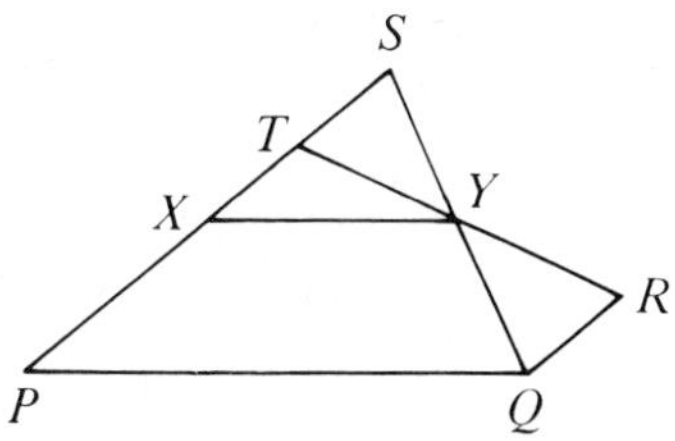

Exercises 7–10 refer to $\triangle ABC$, in which $\overline{DE} \parallel \overline{AB}$.

7. If $CD = 5$, $AD = 10$, and $AB = 18$, find DE.

8. If $AD = DC$ and $DE = 8$, find AB.

9. If $DE = 15$, $AB = 20$, and $DC = 12$, find AC.

10. If $CB = 10$, $AB = 16$, and $DE = 12$, find CE.

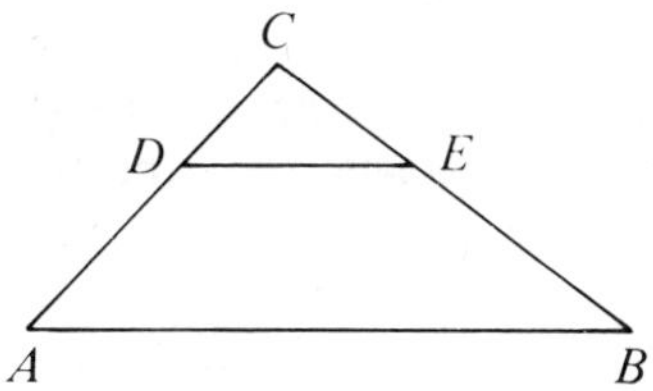

11–14. Solve the proportions in Oral Exercises 5–8.

15. Corresponding altitudes of two similar triangles have lengths 12 and 15. If the perimeter of the smaller triangle is 56, find the perimeter of the larger.

16. The perimeters of two similar triangles are 32 and 80. If an altitude of the smaller triangle has length 6, what is the length of the corresponding altitude of the larger triangle?

Write two-column proofs.

17. Given: $\angle 1 \cong \angle 2$
Prove: $\triangle MLN \sim \triangle JLK$

18. Given: $\overline{MN} \parallel \overline{JK}$
Prove: $\triangle MLN \sim \triangle JLK$

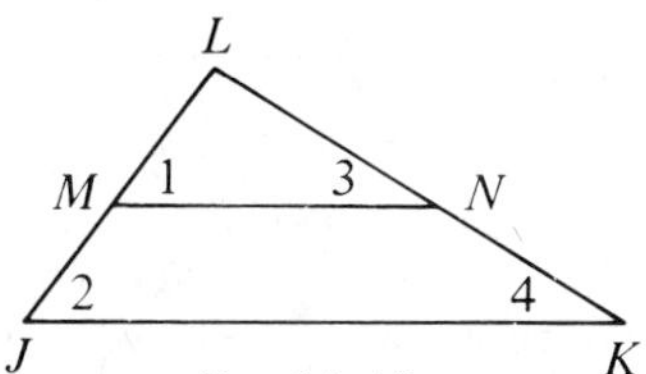

Exs. 17, 18

19. Given: $\angle 1 \cong \angle A$
Prove: $\triangle ABE \sim \triangle CDE$

20. Given: $\overline{AB} \parallel \overline{CD}$
Prove: $\triangle ABE \sim \triangle CDE$

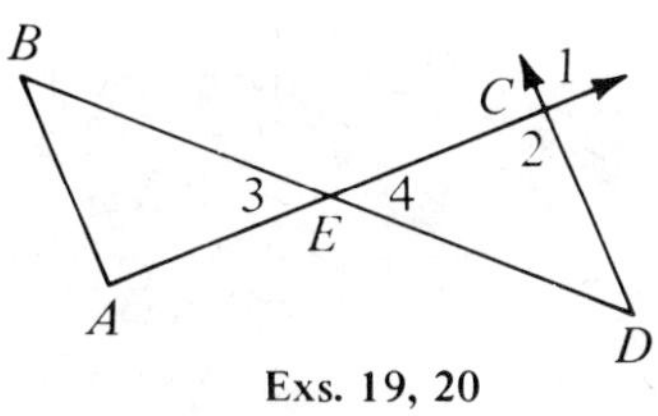

Exs. 19, 20

B **21.** Given: $\angle 1 \cong \angle C$
Prove: $AC \cdot AD = (AB)^2$

22. Given: $\overline{BD} \perp \overline{AC}$;
$\angle 1 \cong \angle C$
Prove: $BC \cdot BD = AB \cdot CD$

Exs. 21, 22
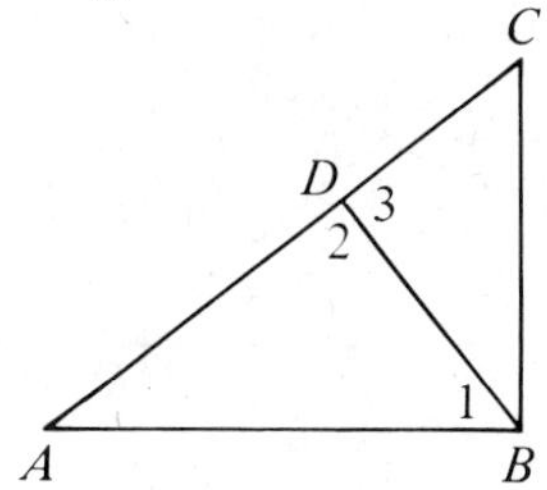

23. Given: $\overline{XZ} \parallel \overline{DE}$

Prove: $\dfrac{XY}{DG} = \dfrac{YZ}{GE}$

(*Hint*: Consider $\dfrac{FY}{FG}$.)

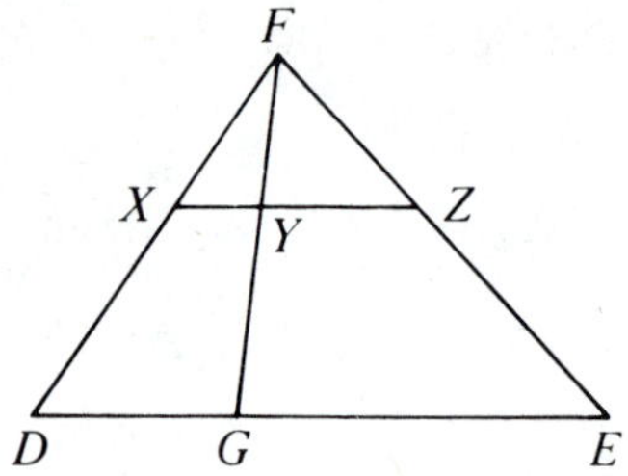

24. Write a two-column proof of Theorem 8-6.
(*Hint*: First prove that $\triangle ACG \sim \triangle DFH$.)

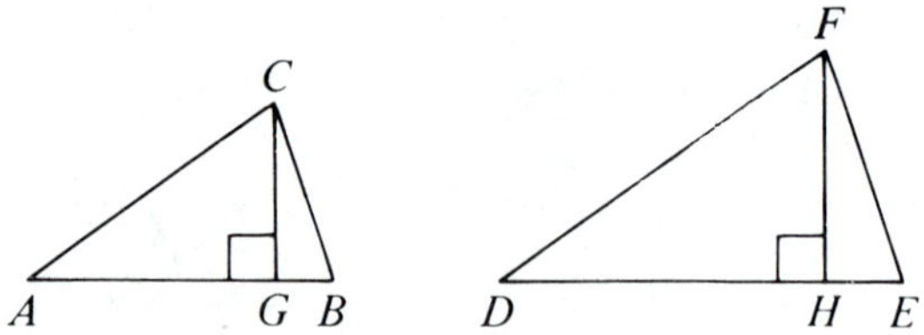

25. Prove that if an acute angle of one right triangle is congruent to an acute angle of another right triangle, then the triangles are similar. Refer to the diagram below.

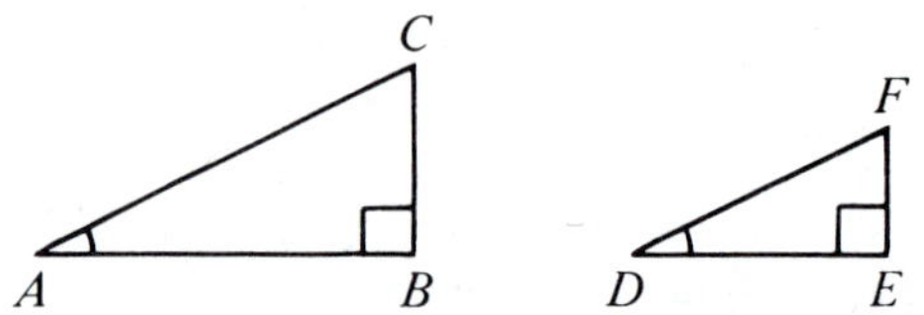

C 26. Given: $\overline{SP} \parallel \overline{RT}$;
$\overline{SR} \parallel \overline{PQ}$

Prove: $RT \cdot PQ = PS \cdot RS$

27. Given: $\overline{SP} \parallel \overline{RT}$;
$\overline{SR} \parallel \overline{PQ}$

Prove: SV is the geometric mean between QV and TV.

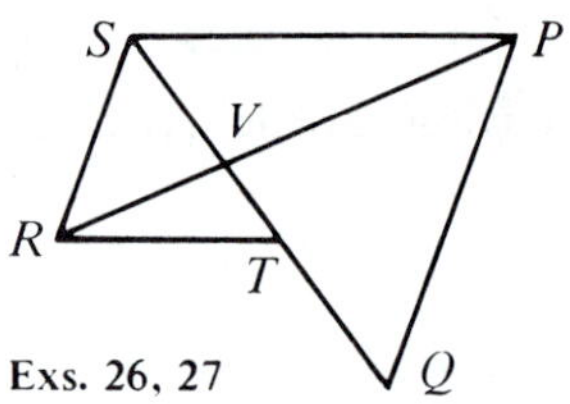

Exs. 26, 27

28. Given: $\overline{MN} \parallel \overline{JK}$

Prove: $\dfrac{LM}{LJ} = \dfrac{MO}{KO}$

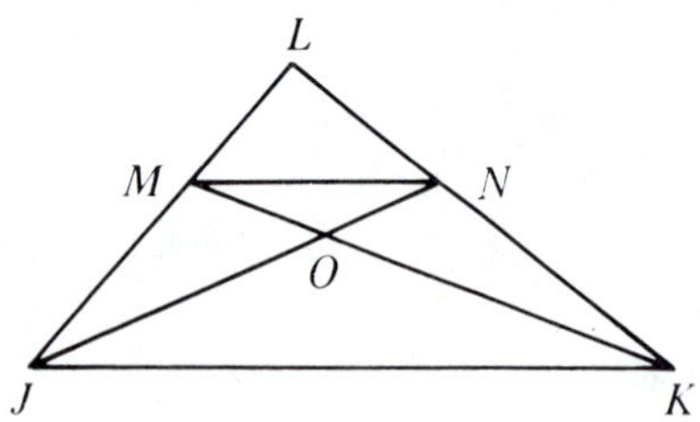

29. Given: $\overline{PQ} \parallel \overline{AD}$;
$\overline{AB} \cong \overline{BD}$

Prove: $\dfrac{CP}{CA} = \dfrac{PR}{RD}$

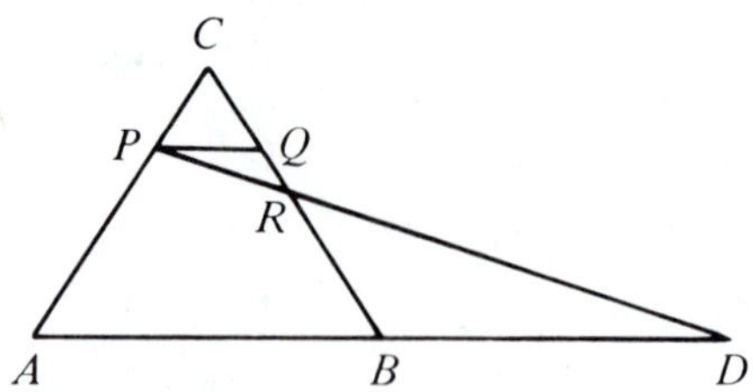

Section 8-5 OTHER WAYS OF PROVING TRIANGLES SIMILAR

So far you know that you can prove two triangles similar by the AA Similarity Theorem. In this section you will learn two more ways of proving triangles similar:

- the SAS Similarity Theorem
- the SSS Similarity Theorem

Theorem 8-7 If one angle of one triangle is congruent to one angle of another triangle and the lengths of the sides including these angles are proportional, then the triangles are similar. (SAS Similarity Theorem)

GIVEN: $\triangle ABC$ and $\triangle DEF$;
$\angle C \cong \angle F$;
$\frac{CA}{FD} = \frac{CB}{FE}$

PROVE: $\triangle ABC \sim \triangle DEF$

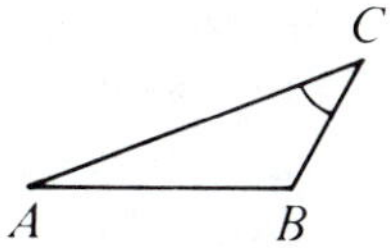

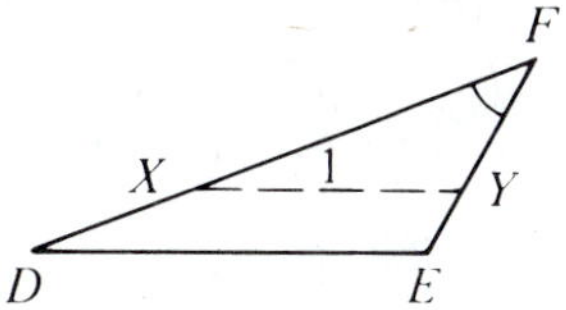

Outline of proof: Assume $FD > CA$. (Then the given proportion implies that $FE > CB$.) Choose X on $\overline{FD}$ and Y on $\overline{FE}$ so that $FX = CA$ and $FY = CB$.

$\triangle CAB \cong \triangle FXY$	(SAS Post.)
$\angle A \cong \angle 1$	(Corr. parts of $\cong$ ⚠ are $\cong$.)
$CA = FX,\ CB = FY$	(By construction)

But it is given that $\frac{CA}{FD} = \frac{CB}{FE}$, so $\frac{FX}{FD} = \frac{FY}{FE}$.

Therefore, $\frac{FD}{FX} = \frac{FE}{FY}$ and $\frac{FD - FX}{FX} = \frac{FE - FY}{FY}$, or $\frac{XD}{FX} = \frac{YE}{FY}$.

By the converse of the Triangle Proportionality Theorem, $\overline{XY} \parallel \overline{DE}$, and hence $\angle 1 \cong \angle D$. Thus, $\angle A \cong \angle D$, and $\triangle ABC \sim \triangle DEF$ by the AA Similarity Theorem.

EXAMPLE

GIVEN: $\frac{VZ}{XZ} = \frac{WZ}{YZ}$

PROVE: $\angle 1 \cong \angle 2$

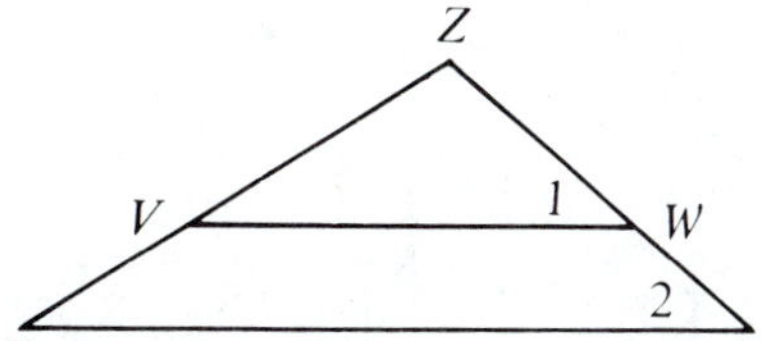

Statements	Reasons
1. $\frac{VZ}{XZ} = \frac{WZ}{YZ}$	1. Given
2. $\angle Z \cong \angle Z$	2. Reflexive Prop. of $\cong$
3. $\triangle VWZ \sim \triangle XYZ$	3. SAS Similarity Theorem
4. $\angle 1 \cong \angle 2$	4. Corr. ⦣ of $\sim$ ⚠ are $\cong$.

Theorem 8-8 If the lengths of the corresponding sides of two triangles are proportional, then the triangles are similar. (SSS Similarity Theorem)

GIVEN: $\frac{AB}{DE} = \frac{BC}{EF} = \frac{AC}{DF}$

PROVE: $\triangle ABC \sim \triangle DEF$

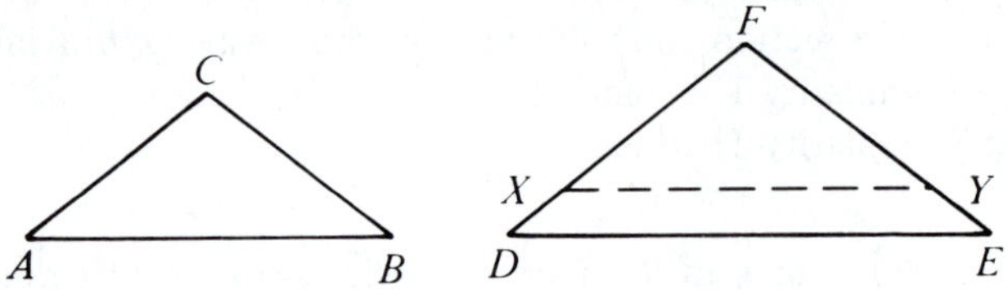

Outline of proof: Assume $DF > AC$. (Then the given proportion implies that $EF > BC$.) Choose X on $\overline{DF}$ and Y on $\overline{EF}$ so that $AC = XF$ and $\overline{XY} \parallel \overline{DE}$.

$$\frac{YF}{EF} = \frac{XF}{DF} \qquad \text{(Triangle Proportionality Theorem)}$$

$$\angle XFY \cong \angle DFE \qquad \text{(Reflexive Prop. of } \cong\text{)}$$

Therefore, $\triangle XYF \sim \triangle DEF$ by the SAS Similarity Theorem.

Hence, $\frac{XY}{DE} = \frac{XF}{DF}$. But $XF = AC$. Therefore, $\frac{XY}{DE} = \frac{AC}{DF}$.

But it is given that $\frac{AC}{DF} = \frac{AB}{DE}$. Therefore, $\frac{XY}{DE} = \frac{AB}{DE}$ and $XY = AB$.

Thus $\triangle XYF \cong \triangle ABC$ (SSS). Therefore, $\triangle ABC \sim \triangle DEF$. (See Exercise 23, Section 8-2.)

ORAL EXERCISES

Name a pair of similar triangles and state a theorem that supports your answer.

1.

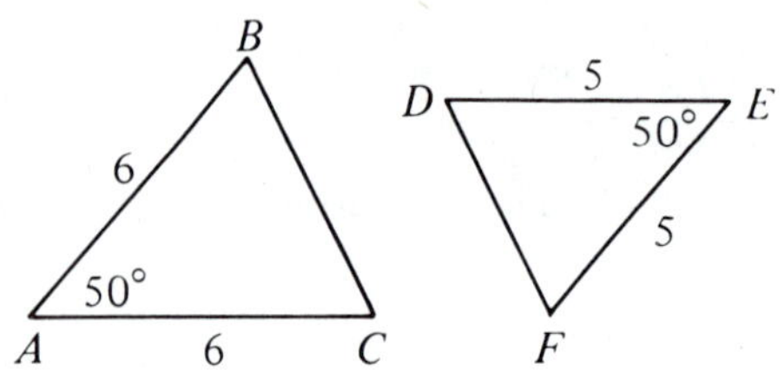

2.

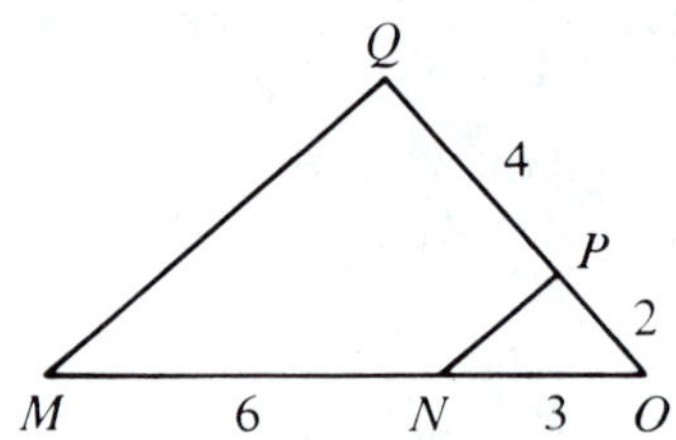

3.

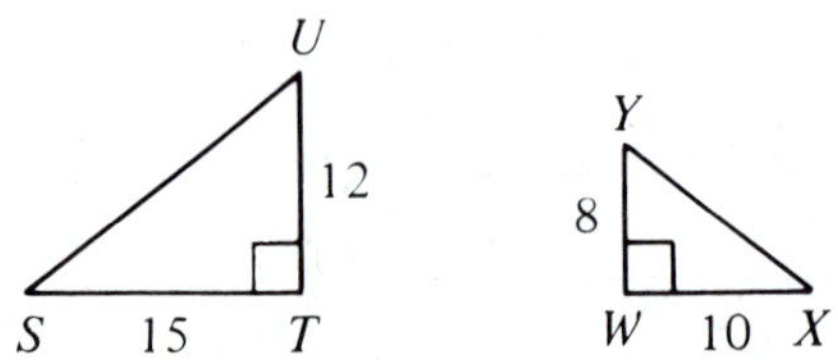

4.

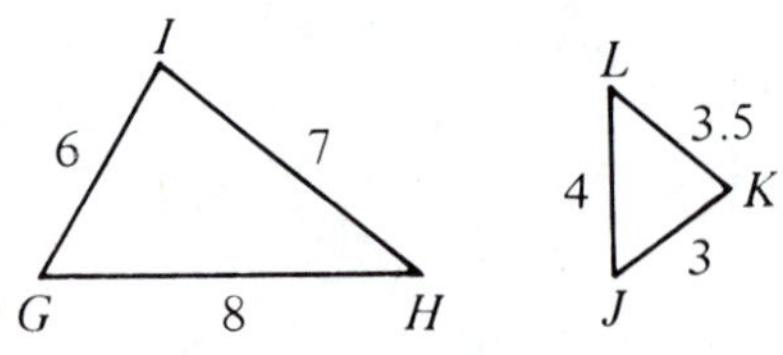

5.

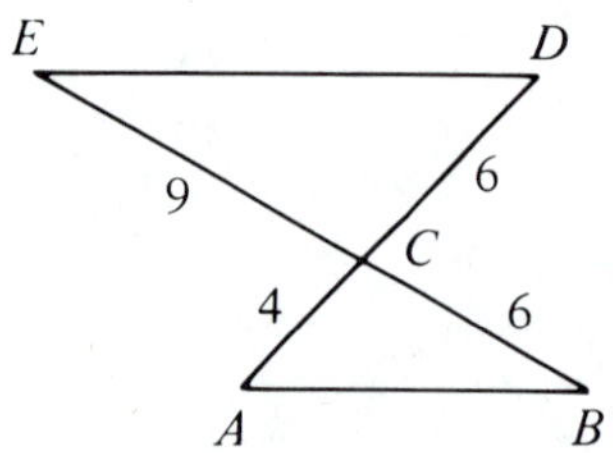

6.

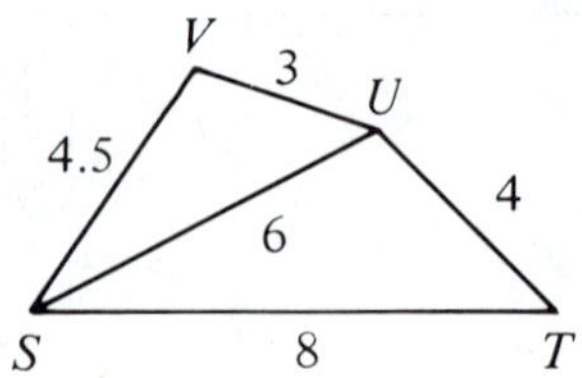

The lengths of corresponding sides of two triangles are given. Are the triangles similar?

7. 7, 9, 4; 21, 27, 12

8. 12, 5, 8; 6, 2.5, 4

9. 3, 4, 5; 6, 7, 8

10. 12, 15, 18; 9, 10, 12

WRITTEN EXERCISES

In Exercises 1–4, for each pair of triangles, name two angles that are congruent. State a proportion that, along with the congruent angles, could be used to prove that the two triangles are similar by the SAS Similarity Theorem.

A **1.** $\triangle CDF$ and $\triangle BEF$

2. $\triangle ABC$ and $\triangle DFC$

3. $\triangle ABC$ and $\triangle GDC$

4. $\triangle DFC$ and $\triangle GDC$

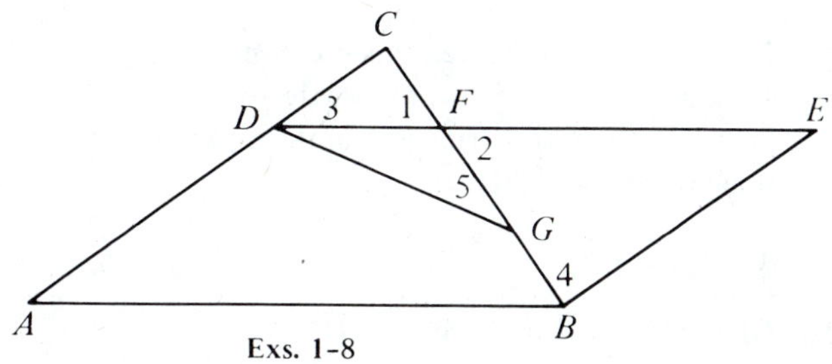

Exs. 1–8

For each pair of triangles, write an extended proportion that could be used to prove the triangles similar by the SSS Similarity Theorem.

5. $\triangle ABC \sim \triangle DFC$

6. $\triangle CDF \sim \triangle BEF$

7. $\triangle ABC \sim \triangle GDC$

8. $\triangle GDC \sim \triangle DFC$

In each diagram state whether or not two triangles can be proved similar. If so, state all the proportion(s) and/or congruent angles you would use to prove this, and the method (AA, SAS, or SSS Similarity).

9.

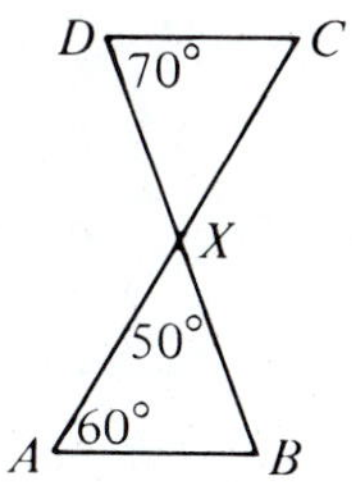

10.

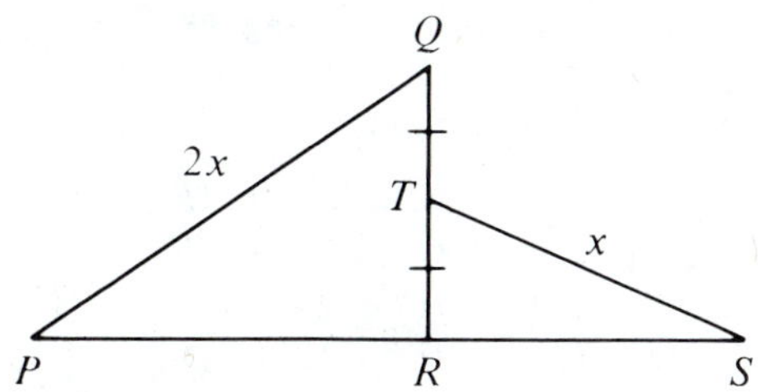

11. $\overline{ZW} \parallel \overline{XY}$

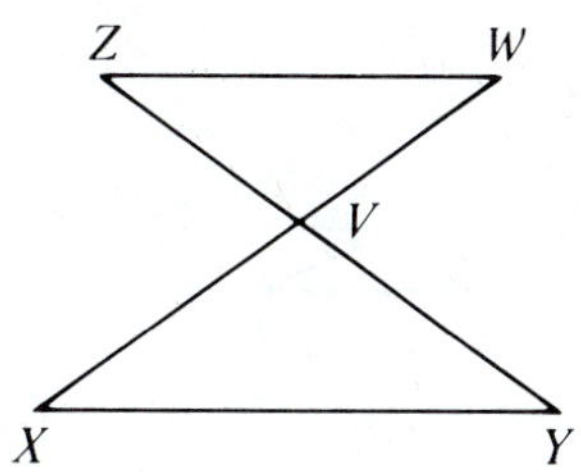

12. $\dfrac{AB}{DE} = \dfrac{BC}{EF}$

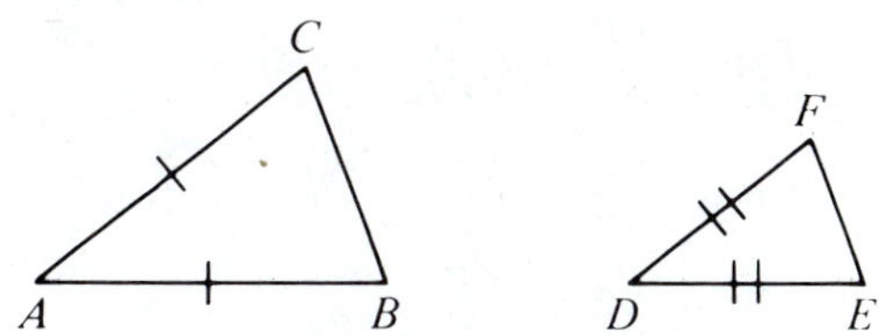

Write two-column proofs.

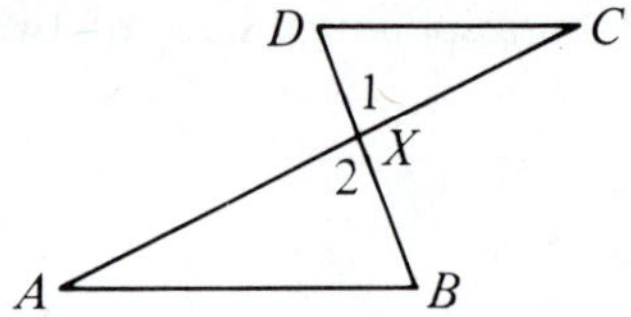

13. Given: $\frac{AX}{CX} = \frac{BX}{DX}$

Prove: $\angle B \cong \angle D$

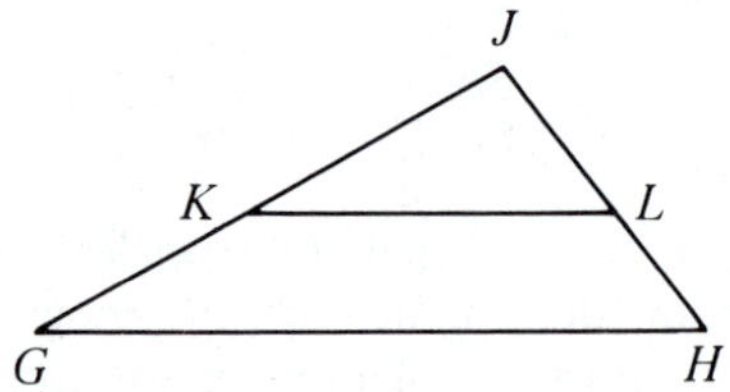

14. Given: $\frac{GJ}{KJ} = \frac{HJ}{LJ}$

Prove: $\frac{GH}{KL} = \frac{GJ}{KJ}$

15. Prove that if the vertex angle of one isosceles triangle is congruent to the vertex angle of another isosceles triangle, then the triangles are similar. Refer to the diagram at the right.

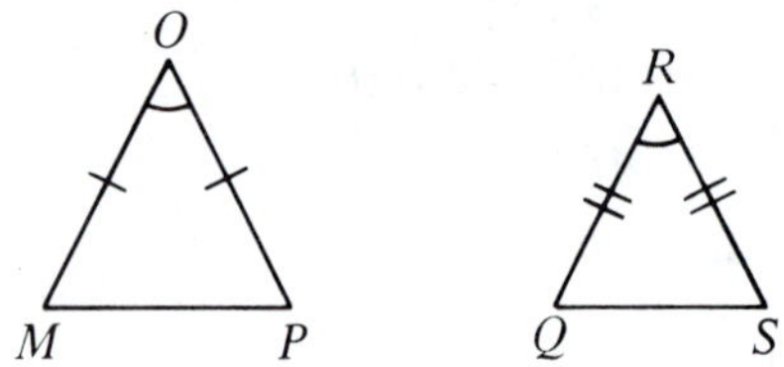

16. Given: E is the midpoint of $\overline{AC}$;
$AB = 2 \cdot DE$;
$BC = 2 \cdot AD$

Prove: $\triangle ABC \sim \triangle EDA$

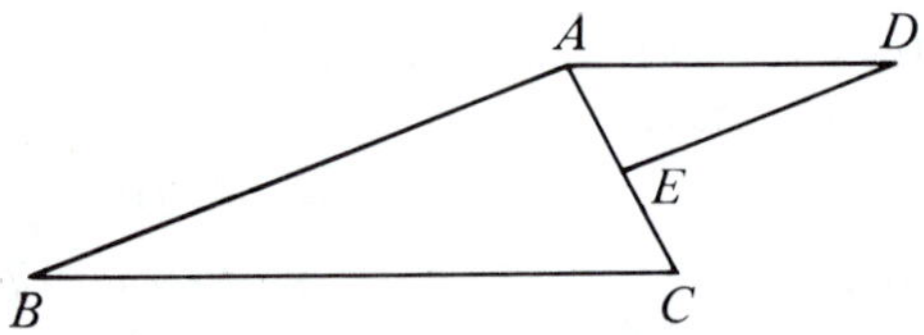

B **17.** This exercise proves that the lengths of corresponding medians of similar triangles have the same ratio as the lengths of corresponding sides.

Given: $\triangle ABC \sim \triangle DEF$;
$\overline{CM}$ is the median to $\overline{AB}$;
$\overline{FN}$ is the median to $\overline{DE}$.

Prove: $\frac{CM}{FN} = \frac{AC}{DF} = \frac{CB}{FE} = \frac{AB}{DE}$

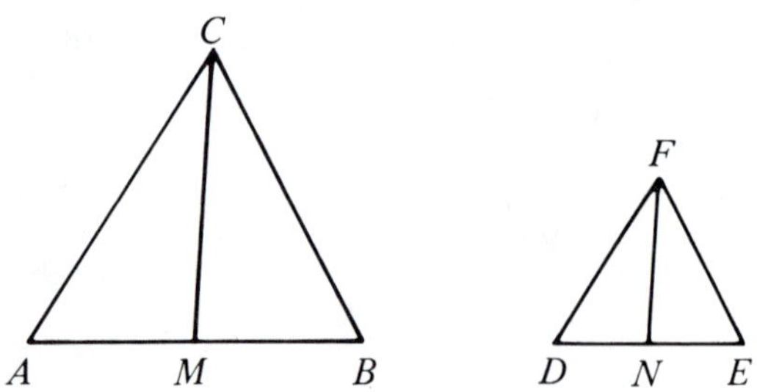

18. Given: $\frac{CX}{BY} = \frac{CD}{BA} = \frac{DX}{AY}$

Prove: $\frac{CP}{BP} = \frac{DP}{AP}$

19. Given: $\frac{CP}{BP} = \frac{DP}{AP}$;

$\frac{CX}{BY} = \frac{CP}{BP}$

Prove: $\triangle CDX \sim \triangle BAY$

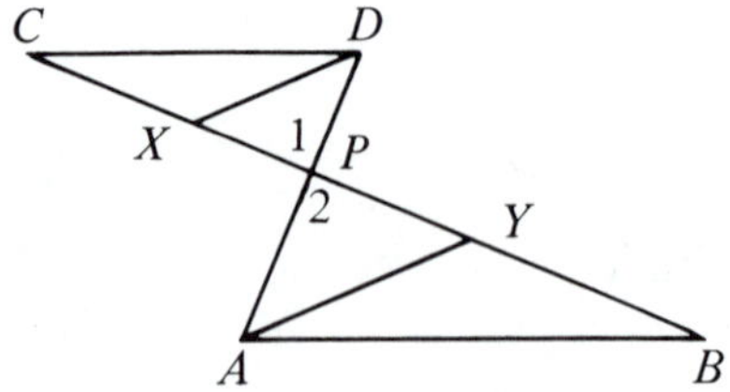

Exs. 18, 19

20. Prove that the lengths of the corresponding diagonals of similar pentagons have the same ratio as the lengths of a pair of corresponding sides.

Given: $ABCDE \sim PQRST$

Prove: $\frac{AC}{PR} = \frac{AB}{PQ}$

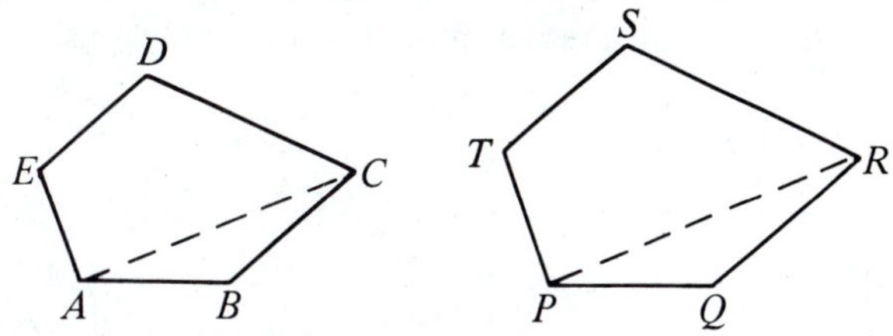

C 21. Use the diagram of Exercise 20 to prove that if the lengths of the sides of two pentagons are proportional and the lengths of all pairs of corresponding diagonals have the same ratio as the lengths of a pair of corresponding sides, then the pentagons are similar.

Given: $\frac{AB}{PQ} = \frac{BC}{QR} = \frac{CD}{RS} = \frac{DE}{ST} = \frac{EA}{TP} = \frac{AC}{PR}$

Prove: $ABCDE \sim PQRST$

SELF-TEST 2

Section 8-4

1. In $\triangle ABC$, $\overline{EF} \parallel \overline{BC}$, $EF = 6$, and $BC = 9$. Find the ratio of **(a)** AE to AB; **(b)** AE to EB.

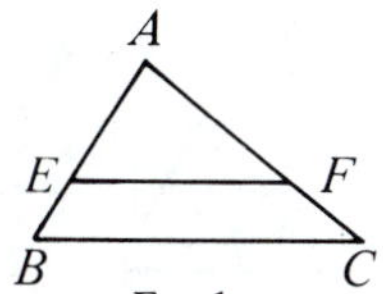

Ex. 1

2. An altitude of $\triangle ABC$ is 8 cm longer than the corresponding altitude of $\triangle DEF$. If $\triangle ABC \sim \triangle DEF$, $BC = 30$, and $EF = 18$, find the lengths of the altitudes.

3. Write a two-column proof.

Given: $\angle 1 \cong \angle 2$

Prove: $\triangle JKN \sim \triangle JLM$

Ex. 3

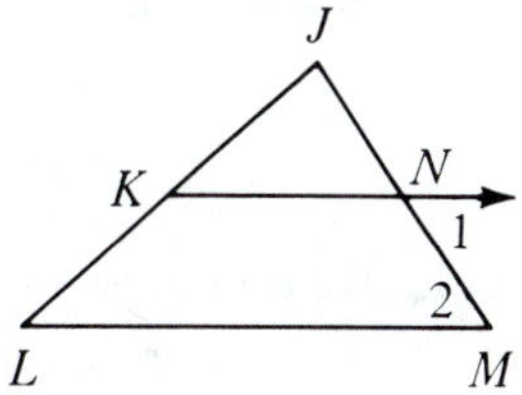

Could two triangles be proved similar? If so, which theorem could be used?

Section 8-5

4.

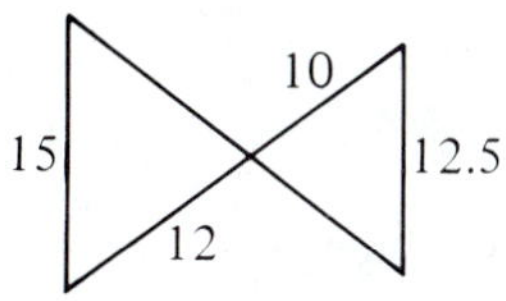

5.

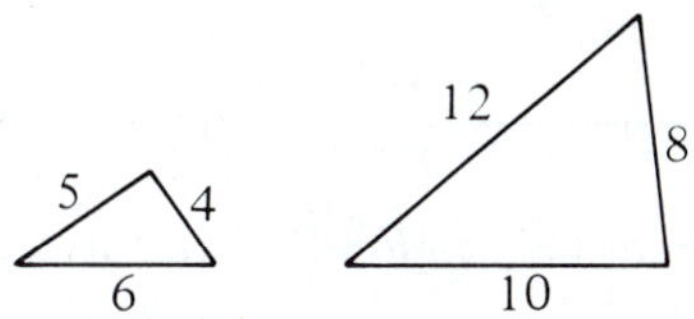

6.

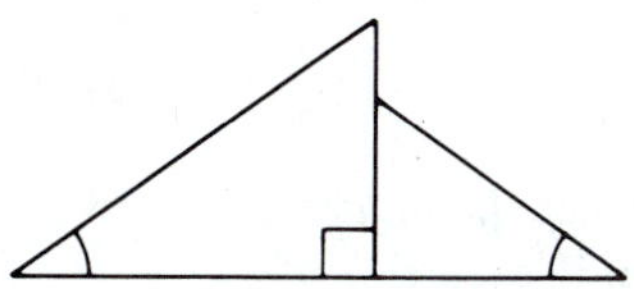

7.

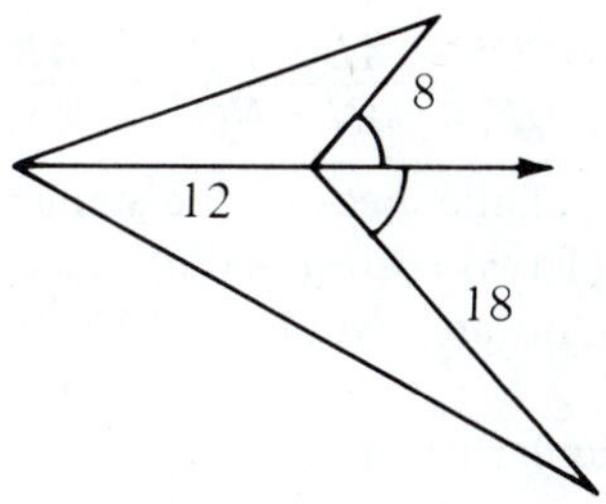

Section 8-6 AREAS OF SIMILAR POLYGONS

Before we begin to study the relationship between the areas of similar polygons we will review the area formulas for some polygons.

Rectangle

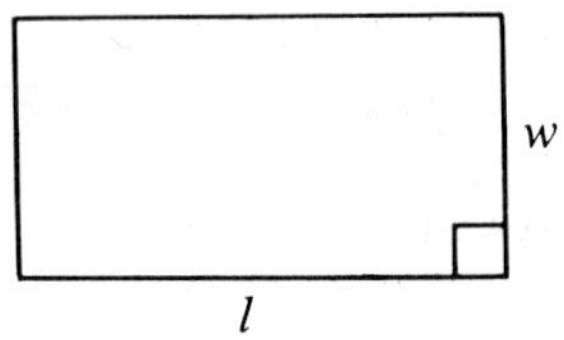

Area = length · width

$A = lw$

Parallelogram

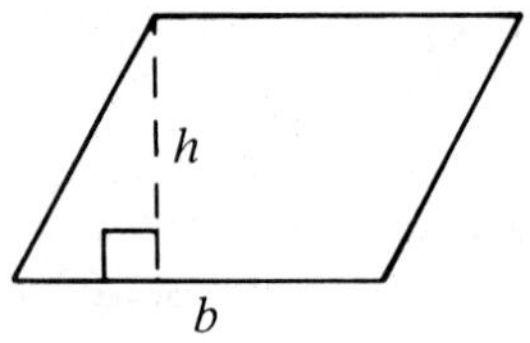

Area = base · height

$A = bh$

Triangle

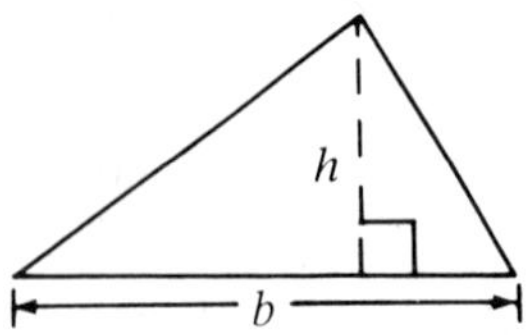

Area $= \frac{1}{2}$ · base · height

$A = \frac{1}{2}bh$

Trapezoid

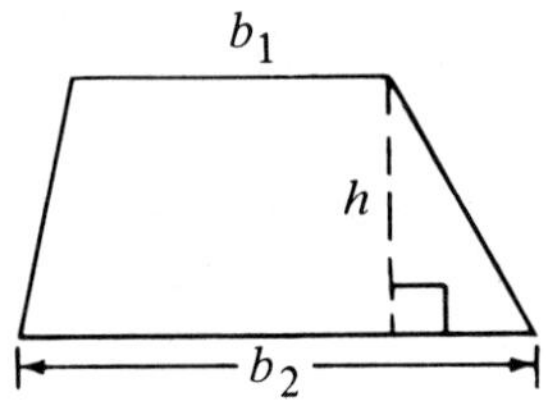

Area $= \frac{1}{2}$ · height(base$_1$ + base$_2$)

$A = \frac{1}{2}h(b_1 + b_2)$

Now let's consider the case of two similar rectangles.

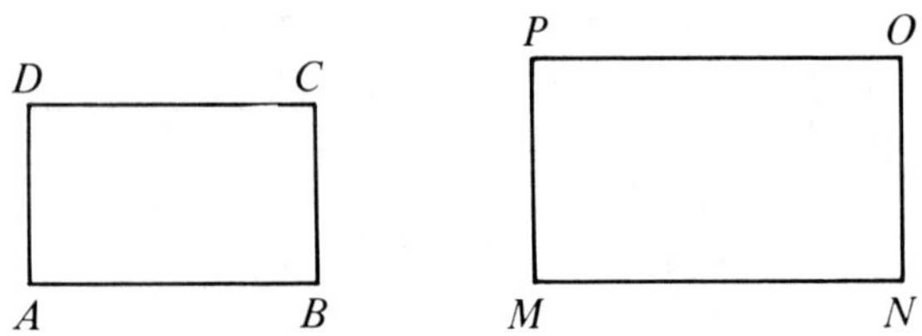

If rectangle $ABCD \sim$ rectangle $MNOP$, then there exists some positive number k, the constant of proportionality, such that $\frac{AB}{MN} = k$ and $\frac{BC}{NO} = k$.

$$\frac{\text{area } ABCD}{\text{area } MNOP} = \frac{AB \cdot BC}{MN \cdot NO} = \frac{AB}{MN} \cdot \frac{BC}{NO} = k^2$$

In other words, the ratio of the areas of the similar rectangles equals the *square* of the ratio of the lengths of two corresponding sides. For example, if each side of a rectangle were doubled, the area would *not* be doubled. It would be multiplied by 4.

The same is true for similar triangles.

Theorem 8-9 If two triangles are similar, then the ratio of their areas equals the square of the ratio of the lengths of any two corresponding sides.

GIVEN: $\triangle ABC \sim \triangle DEF$

PROVE: $\dfrac{\text{area } \triangle ABC}{\text{area } \triangle DEF} = \left(\dfrac{AB}{DE}\right)^2 = \left(\dfrac{BC}{EF}\right)^2 = \left(\dfrac{AC}{DF}\right)^2$

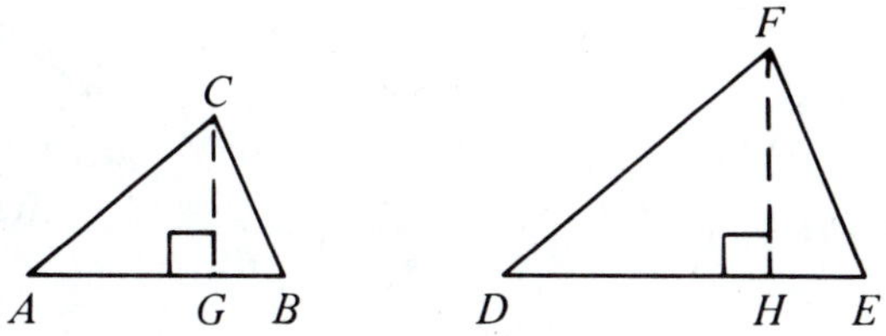

Proof: Let $\overline{CG}$ and $\overline{FH}$ be the altitudes from C and F, respectively.

$$\frac{\text{area } \triangle ABC}{\text{area } \triangle DEF} = \frac{\frac{1}{2}AB \cdot CG}{\frac{1}{2}DE \cdot FH} = \frac{AB}{DE} \cdot \frac{CG}{FH}$$

$$\frac{CG}{FH} = \frac{AB}{DE} \qquad \text{(Theorem 8-6)}$$

$$\frac{\text{area } \triangle ABC}{\text{area } \triangle DEF} = \left(\frac{AB}{DE}\right)^2 \qquad \text{(Substitution)}$$

$$\frac{AB}{DE} = \frac{BC}{EF} = \frac{AC}{DF} \qquad \text{(Def. of} \sim \triangle\text{s)}$$

$$\frac{\text{area } \triangle ABC}{\text{area } \triangle DEF} = \left(\frac{AB}{DE}\right)^2 = \left(\frac{BC}{EF}\right)^2 = \left(\frac{AC}{DF}\right)^2 \qquad \text{(Substitution)}$$

EXAMPLE Two similar triangles have areas 12 and 27. One side of the smaller triangle has length 6. Find the length of the corresponding side of the larger triangle.

SOLUTION Let $x =$ the length of the corresponding side.
Then, by Theorem 8-9,

$$\frac{12}{27} = \left(\frac{6}{x}\right)^2$$

or

$$\frac{4}{9} = \frac{36}{x^2}.$$

Cross-multiply. $4x^2 = 9 \cdot 36$

$x^2 = 9 \cdot 9$

$x = \pm 9$

Since x represents length, it must be positive, so $x = 9$. The length of the corresponding side of the larger triangle is 9.

A more general result for any two similar polygons can be proved using Theorem 8-9. An outline of a proof for two similar pentagons is given on the next page.

Theorem 8–10 If two polygons are similar, the ratio of their areas equals the square of the ratio of the lengths of any two corresponding sides.

GIVEN: $ABCDE \sim FGHJK$

PROVE: $\dfrac{\text{area } ABCDE}{\text{area } FGHJK} = \left(\dfrac{AB}{FG}\right)^2 = \left(\dfrac{BC}{GH}\right)^2 = \cdots = \left(\dfrac{EA}{KF}\right)^2$

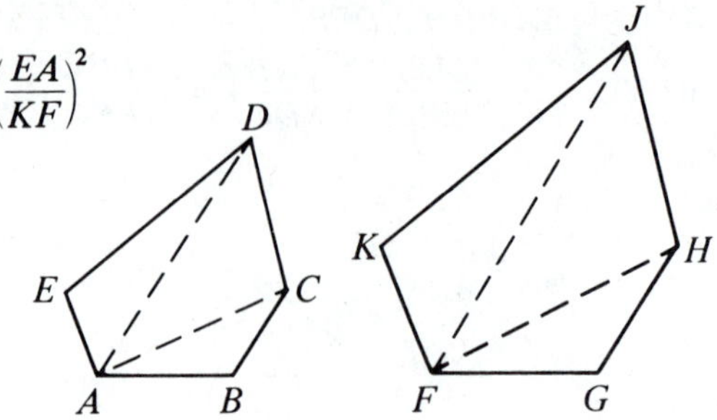

Outline of proof: Draw $\overline{AC}$, $\overline{AD}$, $\overline{FH}$, and $\overline{FJ}$.

$\triangle ABC \sim \triangle FGH$
$\triangle ACD \sim \triangle FHJ$
$\triangle ADE \sim \triangle FJK$
} You will prove these in Exercises 17 and 18.

Therefore, there exists some positive number k such that

$$\frac{AB}{FG} = k, \ \frac{AC}{FH} = k, \text{ and } \frac{AD}{FJ} = k.$$

$$\frac{\text{area } \triangle ABC}{\text{area } \triangle FGH} = \left(\frac{AB}{FG}\right)^2 = k^2$$

$$\frac{\text{area } \triangle ACD}{\text{area } \triangle FHJ} = \left(\frac{AC}{FH}\right)^2 = k^2$$

$$\frac{\text{area } \triangle ADE}{\text{area } \triangle FJK} = \left(\frac{AD}{FJ}\right)^2 = k^2$$

Therefore, by Property 6 of Proportions,

$$\frac{\text{area } \triangle ABC + \text{area } \triangle ACD + \text{area } \triangle ADE}{\text{area } \triangle FGH + \text{area } \triangle FHJ + \text{area } \triangle FJK} = k^2$$

$$\frac{\text{area } ABCDE}{\text{area } FGHJK} = k^2 = \left(\frac{AB}{FG}\right)^2 = \left(\frac{BC}{GH}\right)^2 = \cdots = \left(\frac{EA}{KF}\right)^2$$

ORAL EXERCISES

The following are the ratios of the lengths of a pair of corresponding sides of two similar polygons. State the ratio of the areas of the polygons.

1. $\frac{1}{2}$ **2.** $\frac{3}{5}$ **3.** $3:2$ **4.** $a:3b$

The following are the ratios of the areas of two similar polygons. State the ratio of the lengths of a pair of corresponding sides.

5. $\frac{25}{16}$ **6.** $\frac{1}{9}$ **7.** $4:1$ **8.** $4a^2:b^2$ $(a, b > 0)$

WRITTEN EXERCISES

In $\triangle ABC$, $\overline{DE} \parallel \overline{BC}$. Use the given information to find the ratio in lowest terms of the area of $\triangle ADE$ to the area of $\triangle ABC$.

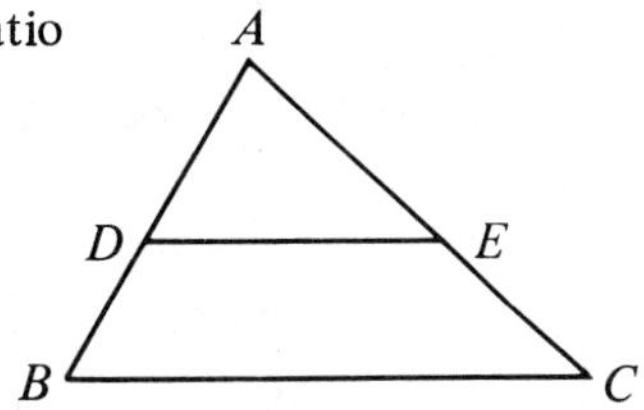

A 1. $DE = 5,\ BC = 6$ 2. $AD = 4,\ AB = 6$

3. $AE = 5,\ EC = 3$ 4. $BD = 1,\ AD = 2$

5. $DE = \sqrt{2},\ BC = 4$ 6. $AE : EC = 3:2$

7. The perimeters of two similar polygons are 12 and 16.
   a. What is the ratio of the lengths of two corresponding sides?
   b. What is the ratio of the areas?
   c. If the area of the smaller polygon is 54, what is the area of the larger?

8. The areas of two similar triangles are 50 and 18.
   a. What is the ratio of the lengths of two corresponding sides?
   b. If the perimeter of the larger triangle is 40, what is the perimeter of the smaller triangle?

9. The ratio of the lengths of corresponding sides of two similar pentagons is 4 : 7. If the area of the larger pentagon is 441, find the area of the smaller pentagon.

10. The ratio of the areas of two similar parallelograms is 9 : 25. The length of one side of the larger parallelogram is 20. Find the length of the corresponding side of the smaller parallelogram.

11. Two equilateral triangles have sides of lengths 5 cm and 6 cm. Find the ratio of their areas.

12. Two similar right triangles have hypotenuses of lengths 6 and 8. If the area of the smaller triangle is 9, find the area of the larger triangle.

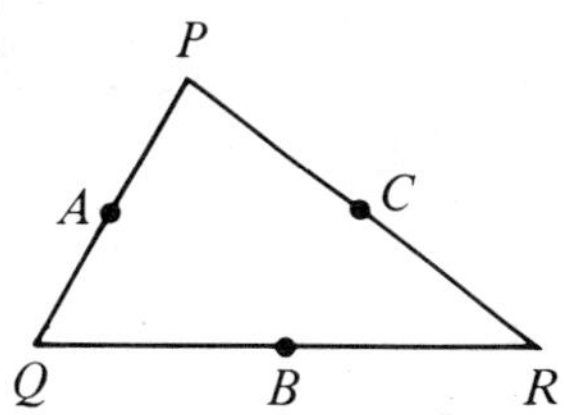

13. A, B, and C are the midpoints of the sides of $\triangle PQR$. If the area of $\triangle ABC$ is 16, find the area of $\triangle PQR$.

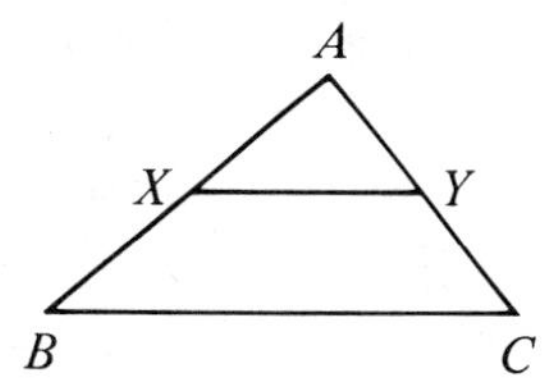

14. X and Y are the midpoints of sides $\overline{AB}$ and $\overline{AC}$, respectively, of $\triangle ABC$. If the area of $\triangle AXY$ is 24, find the area of $\triangle ABC$.

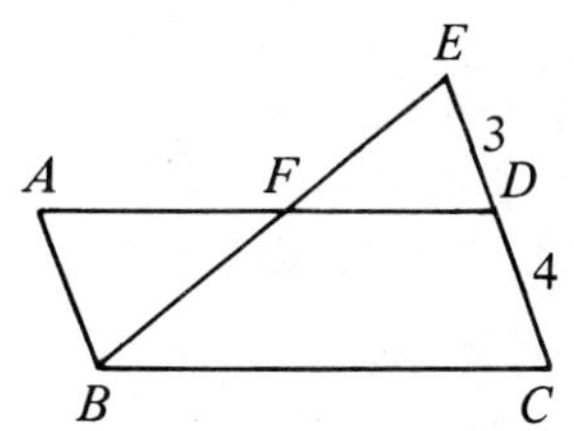

15. $ABCD$ is a parallelogram. Find the following ratios.
    a. area $\triangle DEF$: area $\triangle CEB$
    b. area $\triangle ABF$: area $\triangle DEF$

16. In the figure at the right, $\overline{JH} \parallel \overline{KG}$, $JI = 5$, $KJ = 4$, and area $\triangle KGI = 108$.

a. Find the area of $\triangle HIJ$.

b. Find the area of trapezoid $GHJK$.

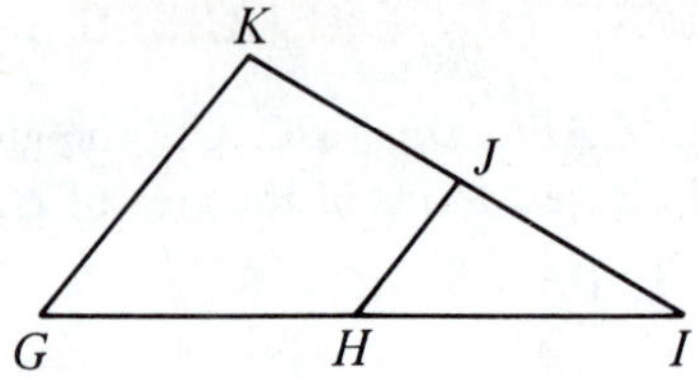

Exercises 17 and 18 prove that the diagonals from each of two corresponding vertices subdivide similar polygons into similar triangles for the case of two pentagons.

B 17. Given: $ABCDE \sim FGHJK$

Prove: $\triangle ABC \sim \triangle FGH$ and $\triangle ACD \sim \triangle FHJ$

(*Hint*: You may use the results of Exercise 20, Section 8-5.)

18. Refer to the diagram for Exercise 17. Give the outline of a proof of the fact that $\triangle ADE \sim \triangle FJK$, using the same given information.

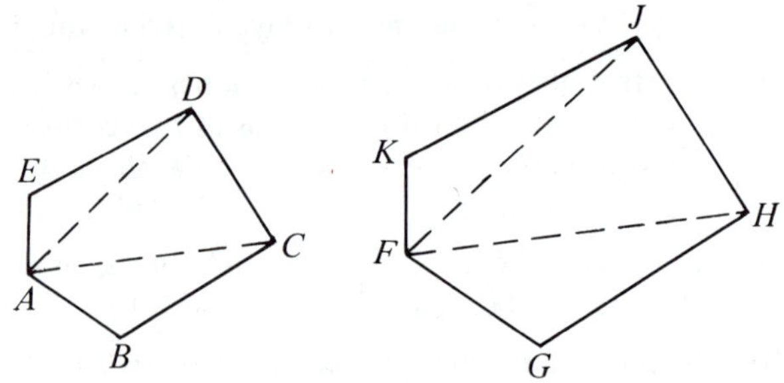

Exs. 17, 18

19. One side of a triangle is 1 cm longer than the corresponding side of a similar triangle. The areas of the two triangles are 9 cm^2 and 25 cm^2, respectively. Find the lengths of the two corresponding sides.

20. One side of a pentagon is 1 cm shorter than the corresponding side of a similar pentagon. The areas of the pentagons are 2 cm^2 and 18 cm^2, respectively. Find the lengths of the two corresponding sides.

C 21. Given that $\overline{DE} \parallel \overline{BC}$, show that

$$\frac{\text{Area of } BCED}{\text{Area of } \triangle ADE} = \frac{2xy + y^2}{x^2}$$

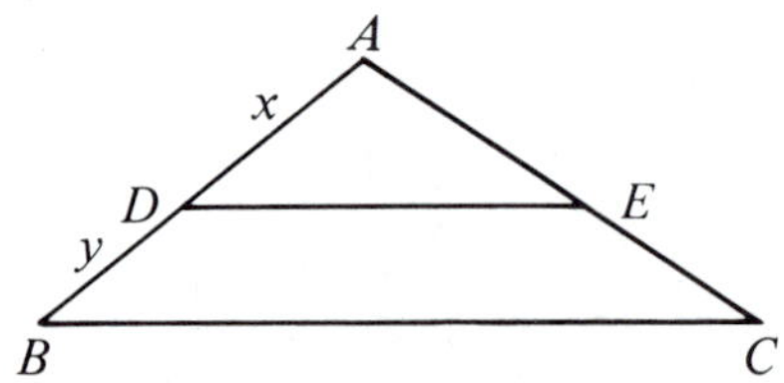

22. a. Show that if rectangle $ABCD$ is similar to rectangle $BCFE$, then x equals the Golden Mean, $\dfrac{\sqrt{5} - 1}{2}$.

b. Show that the ratio of the areas of the two rectangles, $ABCD$ and $BCFE$, is $3 + \sqrt{5} : 2$.

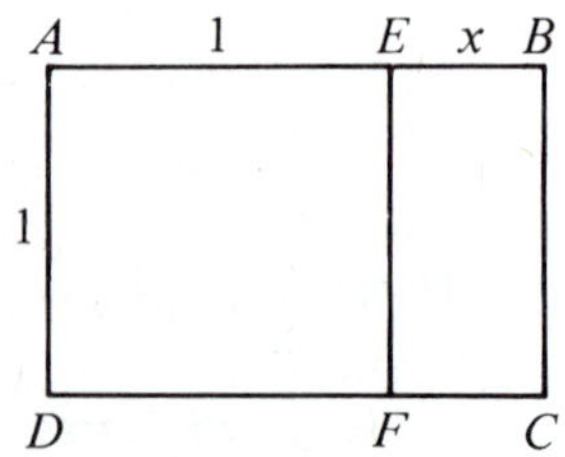

Section 8-7 APPLYING SIMILAR TRIANGLES

You can use similar triangles to find distances that cannot be measured directly. The next two examples will show you two such applications.

EXAMPLE To find the height of a telephone pole, a girl 1.6 m tall stands so that her head just touches the guy wire. She is standing 6 m from the spot where the guy wire meets the ground and 24 m from the base of the pole. How high is the telephone pole?

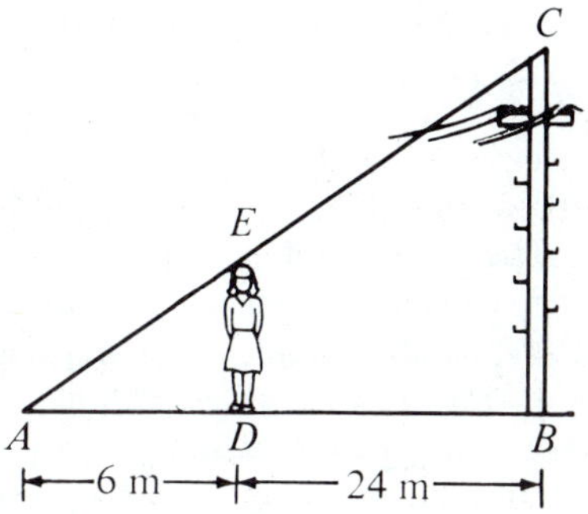

SOLUTION Since $\triangle ADE \sim \triangle ABC$, you have:

$$\frac{AD}{AB} = \frac{DE}{BC}$$

$$\frac{6}{30} = \frac{1.6}{BC}$$

Thus, $6(BC) = 48$, or $BC = 8$.

The height of the telephone pole is 8 m.

The similar triangles used to find distances do not have to be right triangles.

EXAMPLE Use the information in the diagram to find the distance from P to Q across part of a lake.

SOLUTION $\triangle PQR \sim \triangle TSR$ by the AA Similarity Theorem.

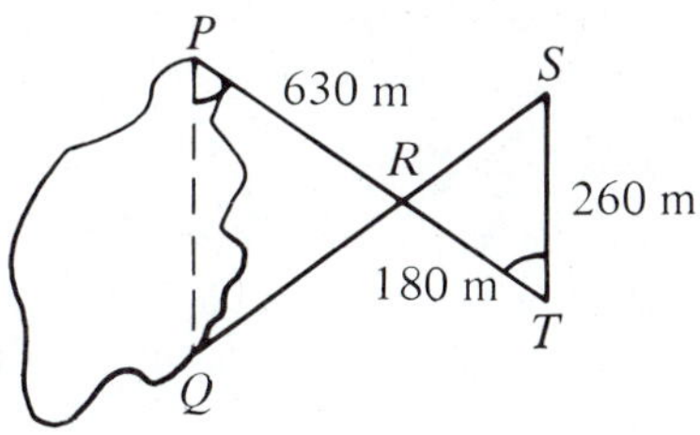

Therefore, $\dfrac{TR}{PR} = \dfrac{TS}{PQ}$.

$$\frac{180}{630} = \frac{260}{PQ}, \text{ so } PQ = 910.$$

The distance from P to Q is 910 m.

ORAL EXERCISES

1. a. Copy the diagram and choose a point P outside of $\triangle ABC$. Locate point A' on $\overline{PA}$ such that $PA' = \frac{1}{2}PA$. Locate point B' on $\overline{PB}$ and point C' on $\overline{PC}$ such that $PB' = \frac{1}{2}PB$ and $PC' = \frac{1}{2}PC$.

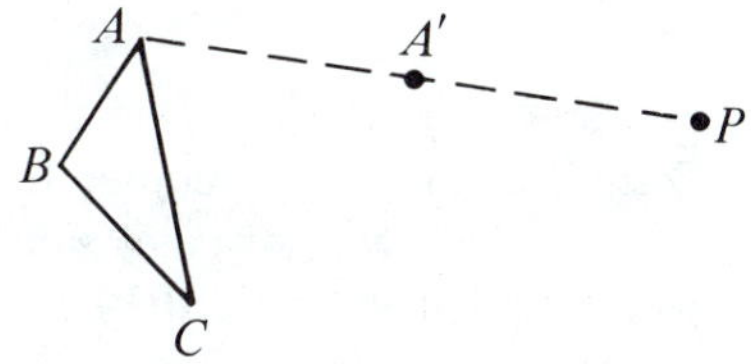

   b. Explain why $\triangle A'B'C' \sim \triangle ABC$.
   c. What is the constant of proportionality?

2. a. Copy the diagram and choose a point O inside polygon $PQRS$. Locate point R' such that $OR' = 3(OR)$. Locate points S', P', and Q' in the same manner.
   b. Explain why $P'Q'R'S' \sim PQRS$.
   c. What is the constant of proportionality?

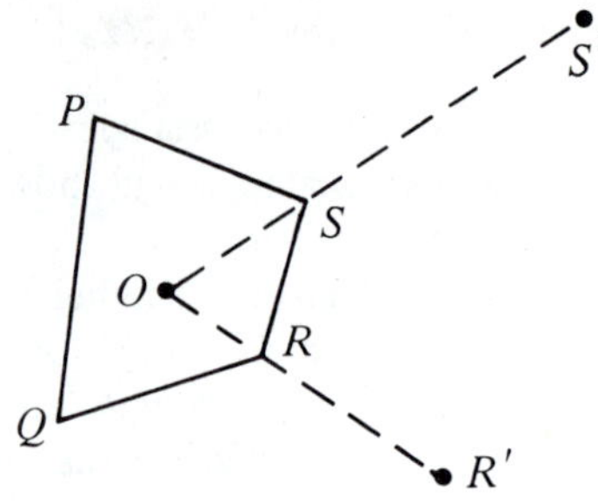

3. The pantograph is being used to draw figure II similar to figure I. Point A is fixed. Pins are placed at E and G so that $\overline{AB} \parallel \overline{FH}$ and $\overline{DF} \parallel \overline{BC}$. Thus $EFGB$ is a parallelogram. A, F, and C are collinear.
   a. Name three pairs of similar triangles formed by the dashed line segment and the segments forming the pantograph.
   b. Which pair of similar triangles has the same constant of proportionality as similar figures I and II?
   c. If $AB = 25$ and $AE = 15$, find the ratio of the lengths of a pair of corresponding sides of figures I and II.
   d. If $CG = 8$ and $FE = 20$, find the ratio of the perimeters of figures I and II.
   e. If $AE = 16$ and $EB = 10$, find the ratio of the areas of figures I and II.

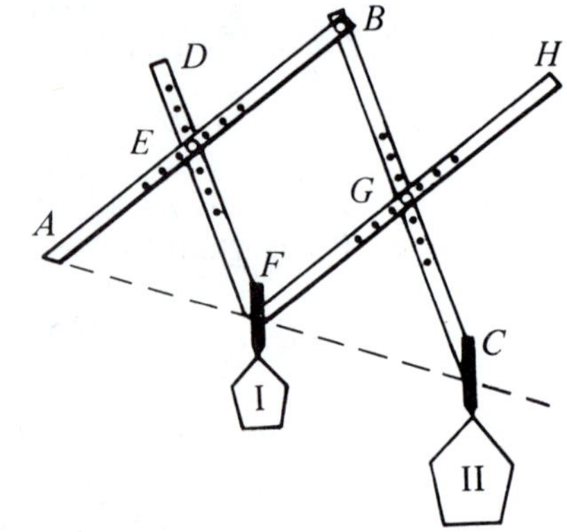

WRITTEN EXERCISES

A 1. At the same moment that a person 1.8 m tall casts a shadow 2.4 m long, a rocket at a nearby launch site casts a shadow 48 m long. Find the height of the rocket.

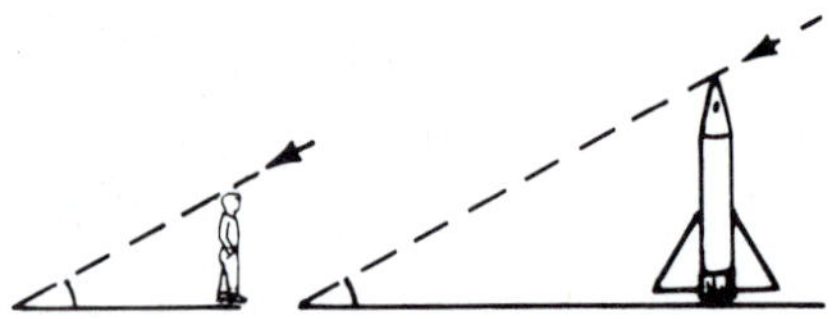

2. Find the width, AB, of the river given that $BE = 72$ m, $CE = 32$ m, and $CD = 92$ m.

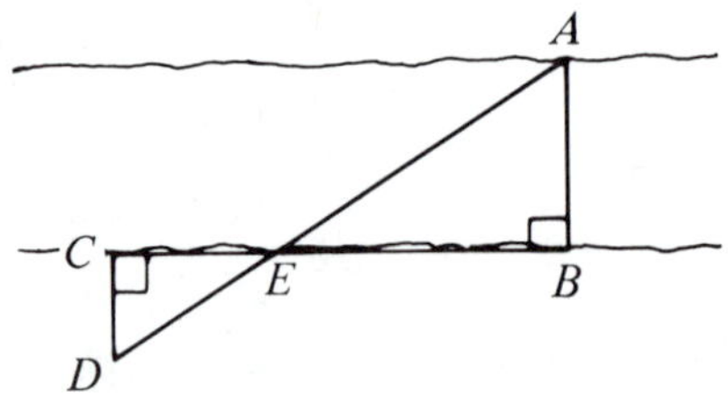

3. During a full solar eclipse, the moon blocks the rays of the sun headed toward point P on Earth. The approximate distances from Earth to the moon and to the sun are 380,000 km and 140,000,000 km, respectively. If the diameter of the moon is approximately 3500 km, find the approximate diameter of the sun.

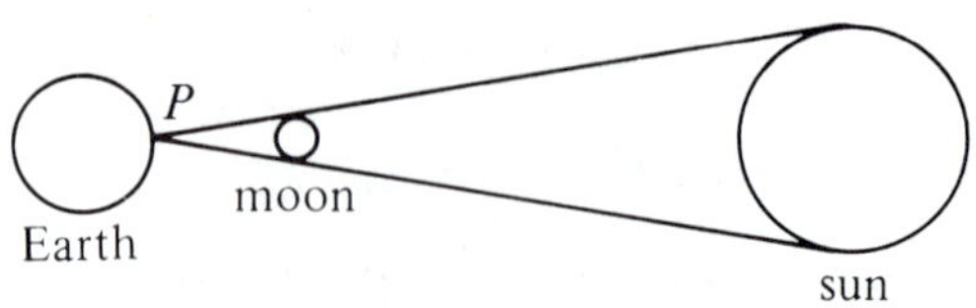

4. Light rays are reflected from a mirror at an angle congruent to the angle at which they strike the mirror. In the diagram at the right a man whose eye level is 1.8 m above the ground finds that if he places a mirror flat on the ground and backs up 4.5 m, he can just see the reflection of the top of a building located 63 m from the mirror. How tall is the building?

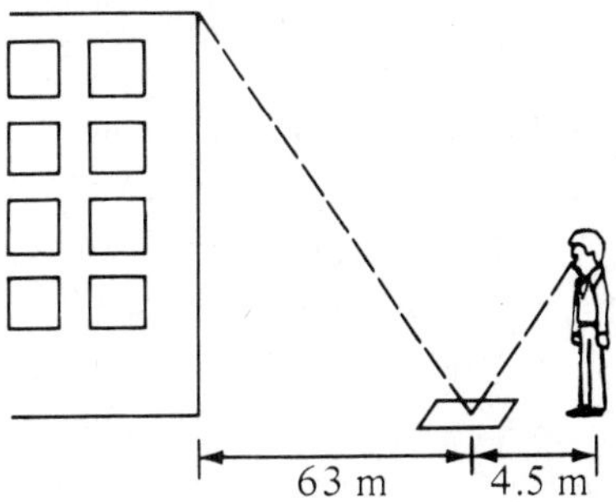

5. In a pinhole camera, light reflected from the object enters the small hole in the box and forms an inverted image on the opposite side. If the penguin shown in the diagram is 1.2 m tall and the camera is 12 cm long and 10 cm high, how far from the penguin should the photographer place the camera so that the image will be 10 cm high?

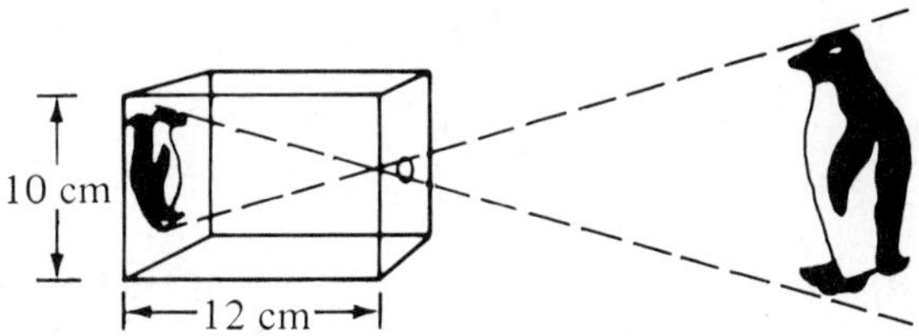

6. A map maker wants to triple the area of a rectangular map while keeping all dimensions in proportion. By what number should the map maker multiply the length and width of the rectangular map?

B 7. A man whose eye level is 2 m above ground notices that he can just see the top of a tree above a nearby flagpole. If the flagpole is 11 m high and the man is 18 m from the flagpole and 30 m from the tree, how tall is the tree?

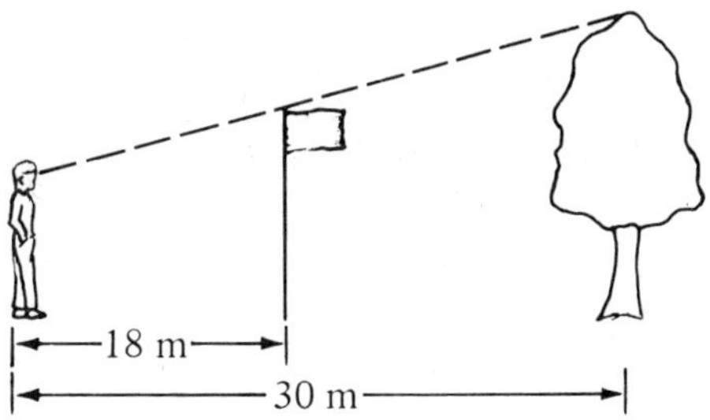

8. The towns of Truckstop (T) and Cowpoke (C) are located on the same side of a railroad track. The distance from Truckstop to the track, TT', is 7.5 km and the distance from Cowpoke to the track, CC', is 5 km. $T'C' =$ 30 km. The towns want to build a station along the track such that the sum of the distances from the station to the two towns will be as small as possible. (See Exercise 9.) This will occur when the station is located at point S such that $\angle TST' \cong \angle CSC'$. How far from T' and C' should the station be located?

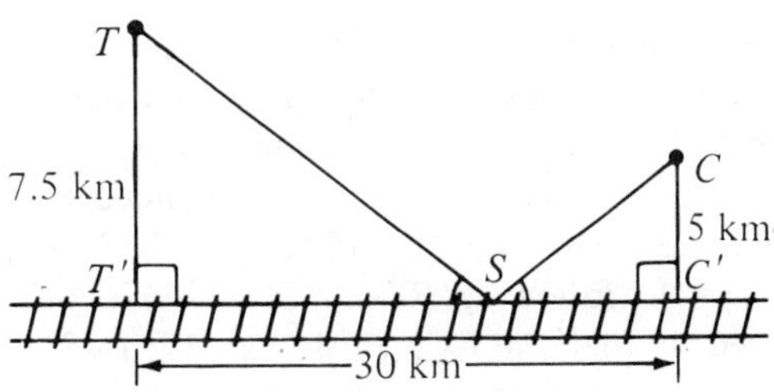

C 9. Write a two-column proof.

Given: $\angle APQ \cong \angle BPR$

Prove: $AP + PB < AQ + QB$ for any point Q on $\overleftrightarrow{PR}$ other than P.

(*Hint*: Extend $\overline{AP}$ to B' so that $PB = PB'$. Draw $\overline{BB'}$ and $\overline{QB'}$. Label the point of intersection of $\overline{BB'}$ and $\overleftrightarrow{PR}$ as X. Prove that $QB = QB'$.)

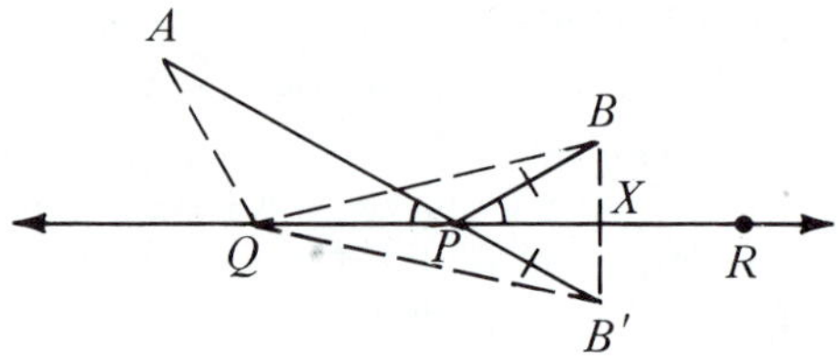

SELF-TEST 3

1. Two regular hexagons have sides of length 2 m and 1.5 m. Find the ratio of their areas in simplest form. Section 8-6

The areas of two similar parallelograms are 72 cm^2 and 98 cm^2.

2. Find the ratio of the lengths of two corresponding sides.
3. If an altitude of the smaller parallelogram is 3 cm, find the corresponding altitude of the larger parallelogram.
4. A tree casts a shadow 18 m long at the same time that a nearby lamppost casts a shadow 1.4 m long. If the lamppost is 2.1 m high, find the height of the tree. Section 8-7
5. A 4 m ramp is used to load appliances on a truck. A point 50 cm along the ramp is 15 cm above the ground. How far above the ground is the upper end of the ramp?

Computer Exercises

The following program finds the ratios in lowest terms of given lengths of the sides of two triangles. The program uses a procedure called the *Euclidean algorithm* to find the greatest common factor (GCF) of the numerator and the denominator of a given fraction. The ratio is reduced to lowest terms by dividing the numerator and denominator of the fraction by the GCF.

In the following display, Steps (1)-(3) illustrate how the Euclidean algorithm is used to find the GCF of 42 and 15. In Step (1), 42 is divided by 15, which produces a quotient of 2 and a remainder of 12. In Step (2), the divisor in Step (1), 15, becomes the dividend, and the remainder in Step (1), 12, becomes the divisor.

As you can see, the remainder in Step (1) becomes the divisor in Step (2) and then the dividend in Step (3). This shifting process,

Remainder ⟶ Divisor ⟶ Dividend,

continues until the remainder equals 0, as in Step (3). The divisor in the final step is the GCF of the two original integers. Thus, 3 is the GCF of 42 and 15.

	Dividend	=	Divisor	·	Quotient	+	Remainder
Step (1)	42	=	[15]	·	2	+	(12)
Step (2)	[15]	=	(12)	·	1	+	3
Step (3)	(12)	=	3	·	4	+	0
			↑ GCF of 42 and 15				

Copy and RUN the program for the given lengths of the sides of the two triangles. (A[1], A[2], and A[3] are the sides of one triangle. B[1], B[2], and B[3] are the sides of the other triangle.) Examine the output and state whether the numbers represent the lengths of the sides of similar triangles.

```
10   PRINT "INPUT A[1],A[2],A[3]"
20   INPUT A[1],A[2],A[3]
30   PRINT "INPUT B[1],B[2],B[3]"
40   INPUT B[1],B[2],B[3]
50   PRINT
60   PRINT "THE RATIOS OF THE SIDES ARE:"
70   PRINT
80   FOR I=1 TO 3
90   LET N=A[I]                        (N = numerator of fraction)
100   LET D=B[I]                       (D = denominator)
110   LET Q=INT (N/D)    }
120   LET R=N−Q*D        }
130   IF R=0 THEN 170    }             (Euclidean algorithm is
140   LET N=D            }             performed in lines 110-160.)
150   LET D=R            }
160   GOTO 110           }
170   PRINT A[I]/D;":";B[I]/D
180   NEXT I
190   END
```

1. A[1] = 26, A[2] = 18, A[3] = 14; B[1] = 39, B[2] = 27, B[3] = 21
2. A[1] = 21, A[2] = 19, A[3] = 8; B[1] = 42, B[2] = 38, B[3] = 24
3. A[1] = 30, A[2] = 45, A[3] = 25; B[1] = 21, B[2] = 27, B[3] = 15
4. A[1] = 230, A[2] = 724, A[3] = 822; B[1] = 805, B[2] = 2534, B[3] = 2877

1. If $\frac{x}{3} = \frac{y}{4}$, complete each statement. Section 8-1

 a. $4x = \underline{\ ?\ }$ b. $\frac{3}{4} = \underline{\ ?\ }$ c. $\frac{x-3}{3} = \underline{\ ?\ }$

2. Find the geometric mean between 4 and 8.

3. Find the value of x: $\frac{x-3}{x} = \frac{x}{x+1}$

4. A string 63 cm long is divided into three pieces whose lengths are in the ratio 1:2:4. How long are the pieces?

In Exercises 5-7, $\triangle ABC \sim \triangle RST$, $AB = 6$, $BC = 8$, $AC = 10$, and $ST = 12$.

5. How are $\angle C$ and $\angle T$ related? Why? Section 8-2

6. Find the ratio of the perimeters of the triangles.

7. Find RT and the perimeter of $\triangle RST$.

Find the value of t.

8. $j \parallel k \parallel l$

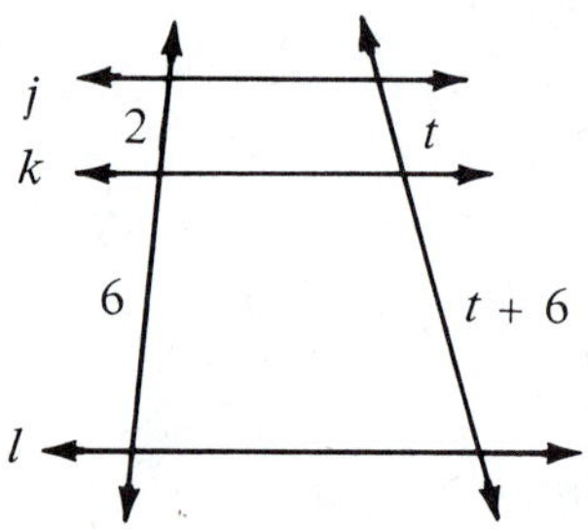

9. Section 8-3

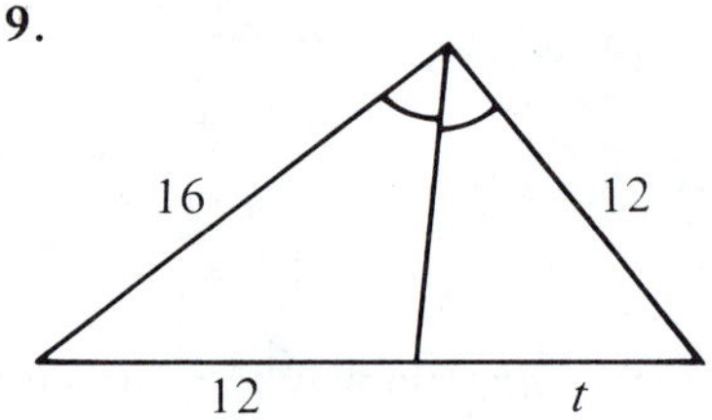

10. In the diagram below, if $\overline{WZ} \parallel \overline{XY}$, $VW = 9$, $WX = 6$, and $ZY = 4$, find VZ.

Write two-column proofs.

11. Given: $\overline{WZ} \parallel \overline{XY}$ Section 8-4

 Prove: $\frac{WZ}{XY} = \frac{VW}{VX}$

12. Given: $\frac{VW}{WX} = \frac{VZ}{ZY}$; $\angle 1 \cong \angle 2$

 Prove: $WX = WZ$

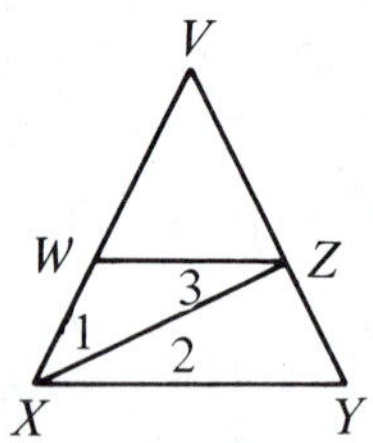

Exs. 10-12

13. The altitudes to the bases of two similar isosceles triangles are 20 cm long and 35 cm long, respectively. If the length of the base of the smaller triangle is 96 cm, what is the length of the base of the larger triangle?

In Exercises 14 and 15, are the two given triangles similar? If so, what similarity theorem would you use to prove that fact?

Section 8-5

14. $\triangle AEB$ and $\triangle CED$

15. $\triangle AED$ and $\triangle CEB$

16. What additional information would you need to prove that $\triangle AEB \sim \triangle CED$ by the SSS Similarity Theorem?

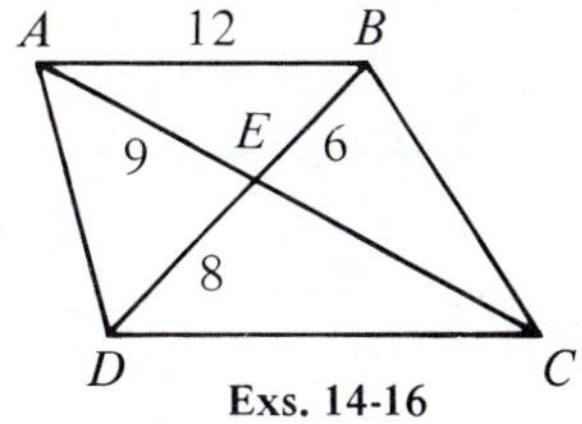

Exs. 14-16

17. Find the ratio of the areas of the triangles described in Exercise 13. Section 8-6

18. The ratio of the areas of two similar rhombuses is 9 : 1. If the perimeter of the smaller rhombus is 12, find the perimeter of the larger rhombus.

19. To estimate the height of her house, Riva measured her shadow and the shadow cast by her house. If the shadows were 2.6 m long and 11.7 m long, respectively, and Riva is 1.6 m tall, how tall is her house? Section 8-7

CHAPTER TEST

In the diagram, $\overline{KM} \parallel \overline{JN}$, $MN = 4$, and $ML = 6$.

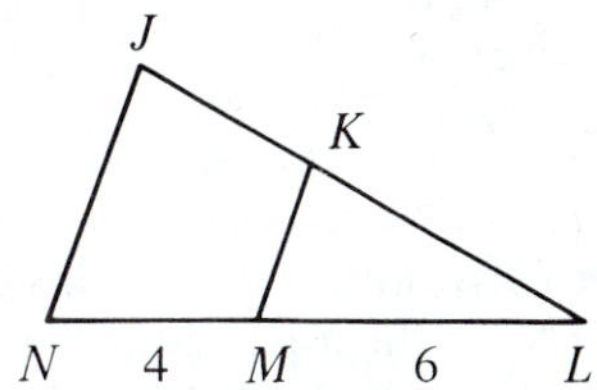

Exs. 1-3

1. Name two similar triangles. Explain why they are similar.
2. Find each ratio in lowest terms.

 a. $JK : KL$ b. $\dfrac{KM}{JN}$

 c. $\dfrac{\text{perimeter of } \triangle JLN}{\text{perimeter of } \triangle KLM}$

3. If $JL = 30$, find JK and KL.
4. The ratio of the lengths of a pair of corresponding sides of two similar triangles is 4 : 11. State the ratio of their areas.
5. Write an extended proportion that could be used to prove that $\triangle WXY \sim \triangle WYZ$ by the SSS Similarity Theorem.
6. a. State the similarity theorem you would use to prove the triangles similar.
 b. Find the value of x.

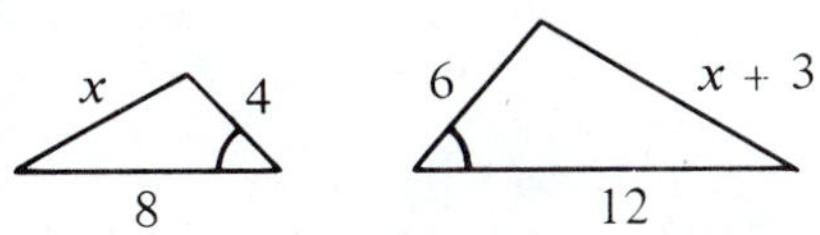

Find the value of x.

7. $a \parallel b \parallel c$

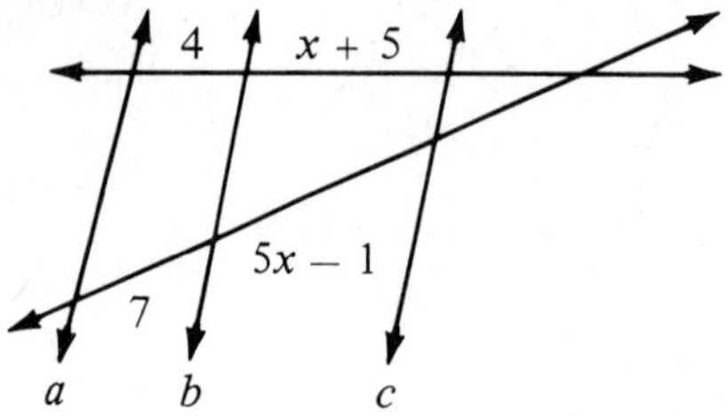

8.

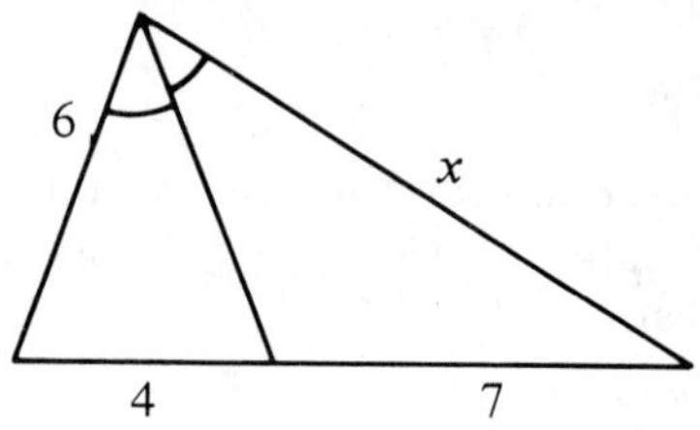

9. A quadrilateral has sides of lengths 21, 24, 26, and 27. Find the length of the shortest side of a similar quadrilateral with perimeter 126.

10. Two supplementary angles have measures that are in the ratio 2 : 13. Find the measures of the two angles.

11. The lengths of the hypotenuses of two similar right triangles are 35 and 42. If the altitude to the shorter hypotenuse has length 12, find the length of the corresponding altitude of the larger triangle.

Write two-column proofs.

12. Given: $\frac{SU}{UR} = \frac{SV}{VW}$

Prove: $\angle 1 \cong \angle 2$

13. Given: $RS = RT$; $\angle SUV \cong \angle TRW$

Prove: $SU \cdot TW = TR \cdot SV$

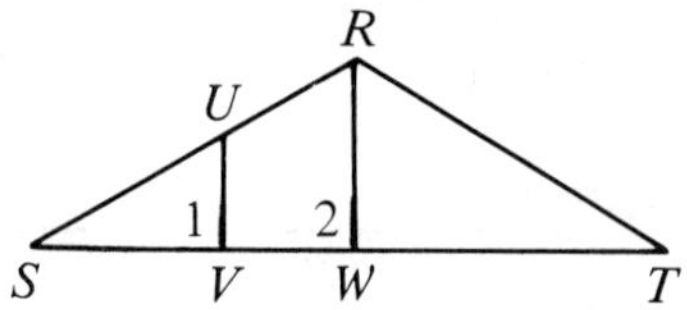

Exs. 12,13

14. When sitting 3 m from his bedroom window, Dave can just see the top and the base of a lamppost through the window. If the height of the window is 2 m and the window is 9 m from the lamppost, what is the height of the lamppost?

Application — Scale Models and Scale Drawings

The concepts and properties of similarity and proportion are used in making scale models and drawings. People in a number of quite different professions use scale models or drawings in their work.

Architects always make scale drawings for their clients to study. They make more detailed working drawings for the people who will do the actual construction. For large projects they often build scale models. These scale drawings and models enable the client to visualize the structure better and make any suggestions for changes before construction begins.

Stage-set models help directors of plays plan actors' movements on stage. If, for example, a script calls for an actor to bang a door when leaving a room, the director will check to be sure that there actually is a door in the stage model.

Models are also used in movies to create special effects. Using models to create special effects is much easier and much less expensive than constructing full-sized counterparts. Models are often filmed in fantasy or science fiction movies, because characters and props often exist only in the imagination of a writer. On film it is impossible to tell that a model has been used.

Sculptors frequently make small clay models of figures in order to decide such questions as the size of hands, how clothing should be arranged, or if a figure should be leaning or standing. Designers also study proportions and make sure that the various parts of a project are harmonious. Urban planners use scale drawings and models to determine, for example, where parks and buildings could most suitably be located and how transportation could be made more efficient. Aircraft designers study scale drawings and use models to determine the number of seats a plane will hold, the width of the aisles, the location of the exits, and so on.

Today computer graphics are often used to generate scale drawings and models. By using computer-aided design and manufacture (CAD/CAM), businesses no longer have to lose valuable time constructing costly models as a product is being developed. For example, designers can enter the specifications of a new automobile design into a computer, and the computer will generate a simulated three-dimensional image of an automobile with those dimensions. The designers can then enter information on the materials to be used in production. Using aerodynamic principles and stress analysis, the computer can run tests and point out possible flaws in the design.

VOCABULARY REVIEW

Be sure that you understand the meaning of these terms:

ratio, p. 307
proportion, p. 308
constant of proportionality, p. 308
extremes, p. 308
means, p. 308
extended ratio, p. 309
extended proportion, p. 309
geometric mean, p. 310
similar figures, p. 314

MIXED REVIEW

Solve for x.

1. $\frac{x-5}{2} = \frac{x-6}{4}$
2. $3x - 7 \leqslant 7x + 21$
3. $\frac{3}{2x+1} = \frac{2}{3x-2}$

Negate.

4. $p \vee (\sim q)$
5. $\exists_x\ 3x + 1 > -5$

Simplify.

6. $-3\sqrt{80}$
7. $\frac{2m}{5n^2} \cdot \frac{10mn}{16m}$

8. Translate the following sentence into an equation: The product of two consecutive even integers is 14 less than seven times their sum. Identify the variable you are using and what it represents.
9. Divide $t^3 + 6t^4 - 10t^2 + 5t - 7$ by $2t + 3$.

Write two-column proofs.

10. Given: $\overline{XW} \perp \overline{YZ}$;
$XW = 6$; $WY = 4$; $ZW = 9$
Prove: $\triangle XWY \sim \triangle ZWX$

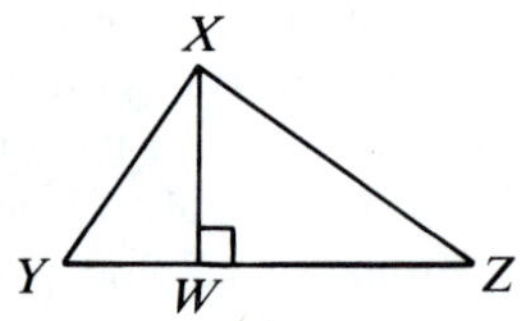

11. Given: $\triangle QRS$; X, Y, and Z are the midpoints of $\overline{RS}$, $\overline{QS}$, and $\overline{QR}$, respectively.
Prove: $\triangle XYZ \sim \triangle QRS$

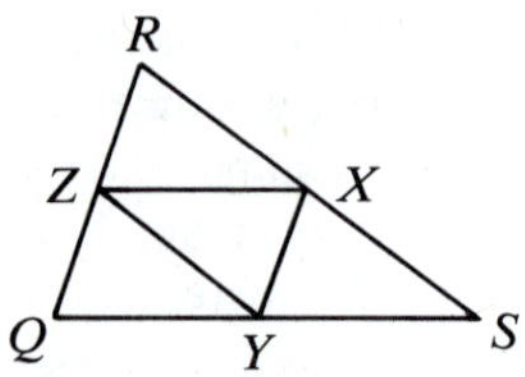

12. Given: Trapezoid $PQRS$; $\overline{SR} \perp \overline{RQ}$; $\overline{PQ} \perp \overline{RQ}$; T is the midpoint of $\overline{PS}$.
Prove: $\overline{RT} \cong \overline{QT}$

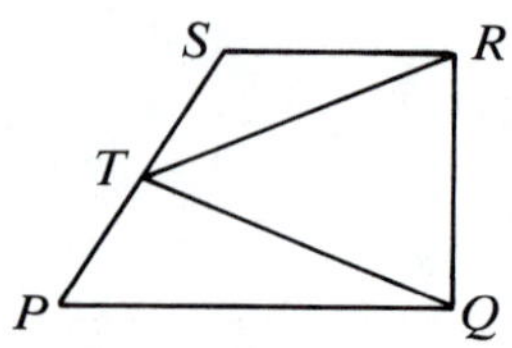

13. Given: $\overline{AC}$ bisects $\angle BAD$; $\angle B \cong \angle D$
Prove: C is the midpoint of $\overline{BD}$.

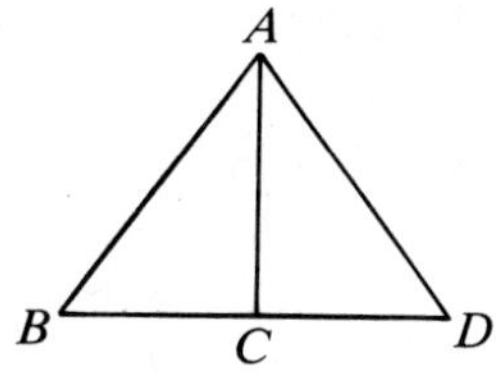

14. Solve by completing the square: $x^2 - 3x = 18$

15. Write the conclusion that follows logically from the premises and state which rule of inference is being used. If no conclusion follows, explain why.

Only fools rush in.
You are not a fool.

16. The sum of a number and its reciprocal is $\frac{17}{4}$. What is the number?

Exercises 17-20 refer to the diagram, in which $\triangle ABC \sim \triangle DEF$.

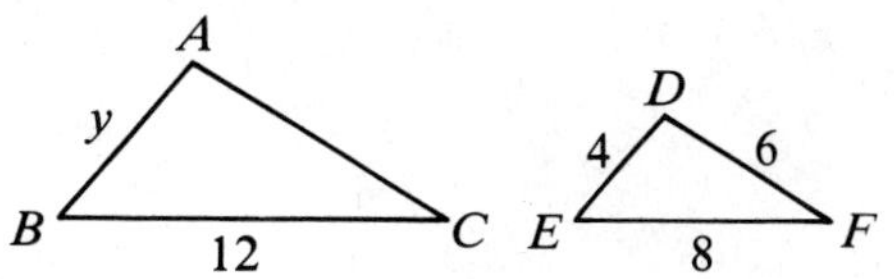

17. Find the constant of proportionality.

18. Find the value of y.

19. Find the perimeter of $\triangle ABC$.

20. Find the ratio of the perimeter of $\triangle ABC$ to the perimeter of $\triangle DEF$.

21. Solve by the quadratic formula: $9y^2 = 6y + 7$

22. One solution contains 23% salt and another contains 15% salt. How many liters of each should be mixed to give 8 L of a solution that is 20% salt?

23. The length of a rectangle is 4 cm greater than its width. The area is 45 cm^2. Find the length and width.

24. In $\triangle ABC$, $m\angle A = 3x$, $m\angle B = 6x - 11$, and $m\angle C = 2x + 4$.

 a. Find the numerical measure of each angle.

 b. Name the longest side of $\triangle ABC$.

25. Factor: $36 + 107x - 3x^2$

26. In the figure at the right, $PQ = 3x + 7$, $QR = 8x - 15$, and $PR = 9x$. Find **(a)** the value of x and **(b)** $PQ:PR$.

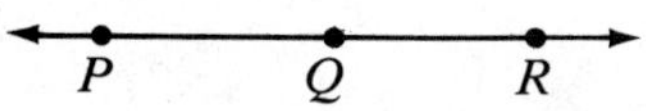

27. Two similar polygons have sides in the ratio of 5:9. The sum of their areas is 212 m^2. Find the area of the smaller figure.

Exercises 28 and 29 refer to the figure.

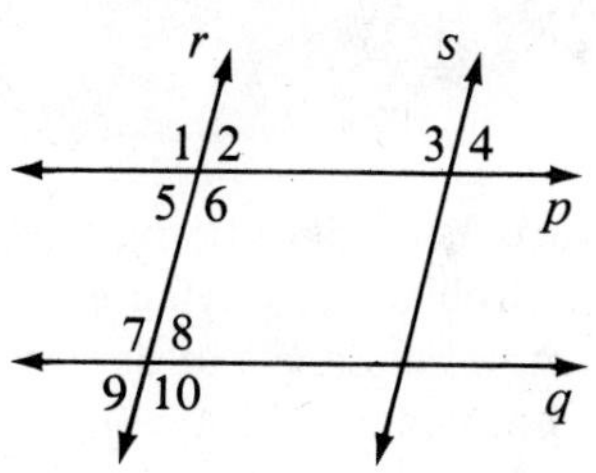

28. If $\angle 1 \cong \angle 3 \cong \angle 7$, which lines, if any, are parallel?

29. Write a two-column proof.
Given: $\angle 1 \cong \angle 10$
Prove: $p \parallel q$

The concepts of right-triangle trigonometry introduced in this chapter can be used to solve problems involving a line of sight and an angle of elevation or depression – for example, the angle of depression of a sailboat from a lighthouse. The lighthouse in the photograph is in Acadia, Maine.

Right Triangles

9

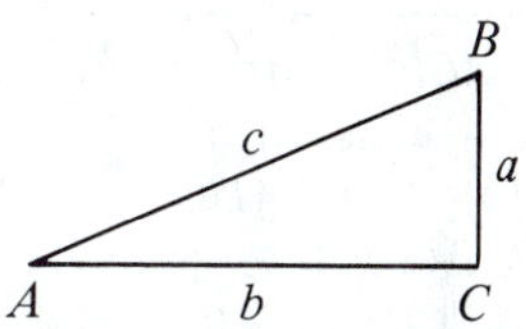

One of the most well-known and most important theorems in geometry is the Pythagorean Theorem, which states that if a and b are the lengths of the legs of any right triangle and c is the length of the hypotenuse, then

$$a^2 + b^2 = c^2.$$

Although the Babylonians may have known of this relationship as early as 1700 B.C., the first proof of the theorem is credited to Pythagoras, a Greek mathematician and philosopher of the sixth century B.C. Since then, many other proofs have been given. A particularly simple one is based on the diagrams below, in which the two large squares have equal areas of $(a + b)^2$. The area of each large square equals the sum of the areas of the right triangles, $4\left(\frac{1}{2}ab\right)$, and the areas of the shaded squares. Therefore, the sum of the areas of the shaded squares in the first diagram must equal the area of the shaded square in the second; that is, $a^2 + b^2 = c^2$.

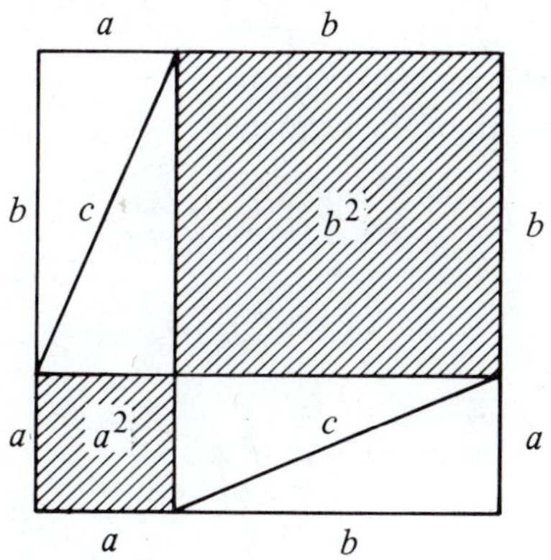

Area $= a^2 + b^2 + 4\left(\frac{1}{2}ab\right)$

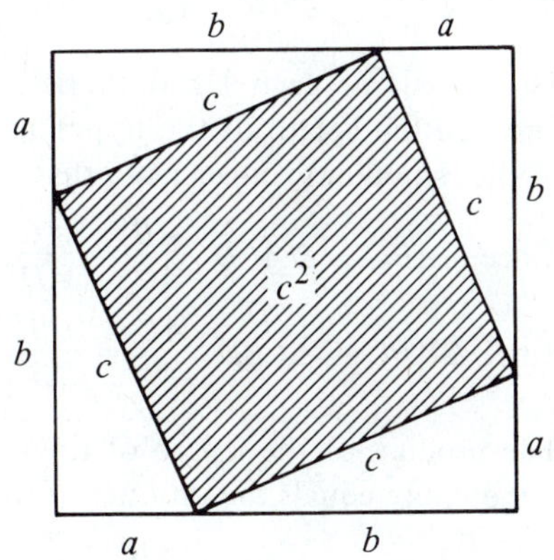

Area $= c^2 + 4\left(\frac{1}{2}ab\right)$

Section 9-1 SIMILAR RIGHT TRIANGLES

In this section you will apply to right triangles some of the facts you learned about similar triangles in the preceding chapter.

Theorem 9-1 If the altitude is drawn to the hypotenuse of a right triangle, then the two triangles formed are similar to the original triangle and to each other.

GIVEN: Right $\triangle ABC$ with right $\angle ACB$;
$\overline{CD} \perp \overline{AB}$

PROVE: (1) $\triangle ACD \sim \triangle ABC$
(2) $\triangle CBD \sim \triangle ABC$
(3) $\triangle ACD \sim \triangle CBD$

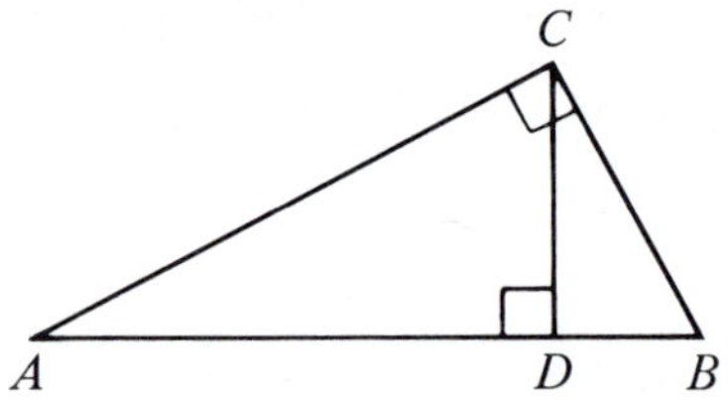

Statements	Reasons
1. $\angle ACB$ is a right $\angle$; $\overline{CD} \perp \overline{AB}$	1. Given
2. $\angle ADC$ is a right $\angle$.	2. Def. of $\perp$ lines
3. $\angle ACB \cong \angle ADC$	3. All right $\angle$s are $\cong$.
4. $\angle A \cong \angle A$	4. Reflexive Property of Congruence
5. $\triangle ACD \sim \triangle ABC$	5. AA Similarity Theorem

The proof that $\triangle CBD \sim \triangle ABC$ is left as Exercise 27. Then $\triangle ACD \sim \triangle CBD$ by the symmetric and transitive properties of similarity.

Theorem 9-1 leads directly to some important relationships between the lengths AB, BC, CA, CD, AD, and DB in the diagram above.

Corollary 1 The length of the altitude to the hypotenuse of a right triangle is the geometric mean between the lengths of the two segments of the hypotenuse.

Proof: Since $\triangle ACD \sim \triangle CBD$, $\dfrac{AD}{CD} = \dfrac{CD}{BD}$.

Corollary 2 The length of each leg of a right triangle is the geometric mean between the length of the hypotenuse and the length of the segment of the hypotenuse adjacent to that leg.

Proof: Since $\triangle ABC \sim \triangle ACD$, $\dfrac{AB}{AC} = \dfrac{AC}{AD}$.

Since $\triangle ABC \sim \triangle CBD$, $\dfrac{AB}{CB} = \dfrac{CB}{DB}$.

Corollary 3 The product of the length of the hypotenuse and the altitude to the hypotenuse equals the product of the lengths of the legs.

Proof: Since $\triangle ABC \sim \triangle ACD$, $\dfrac{AB}{AC} = \dfrac{BC}{CD}$. Thus $AB \cdot CD = AC \cdot BC$.

EXAMPLE Find QS.

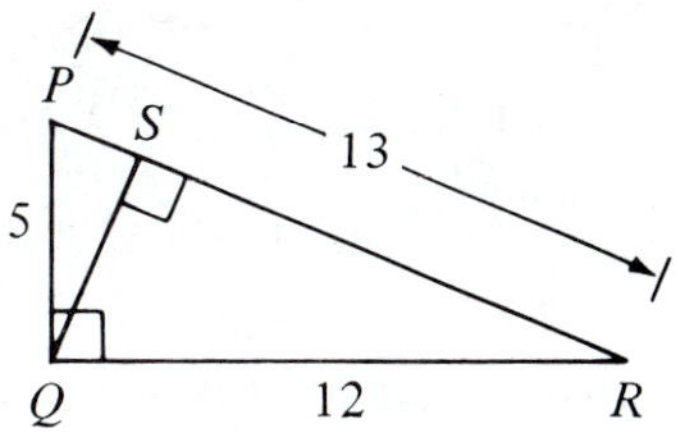

SOLUTION According to Corollary 3 of Theorem 9-1:

$$PR \cdot QS = PQ \cdot QR$$
$$13 \cdot QS = 5 \cdot 12$$
$$13(QS) = 60$$
$$QS = \frac{60}{13}, \text{ or } 4\frac{8}{13}$$

EXAMPLE Find the value of x.

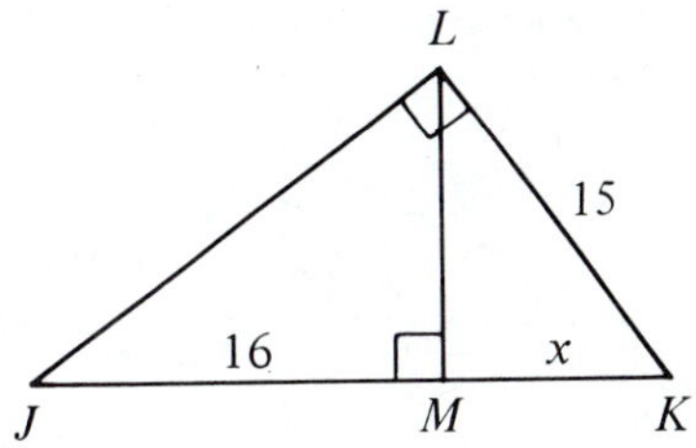

SOLUTION According to Corollary 2 of Theorem 9-1:

$$\frac{JK}{LK} = \frac{LK}{MK}$$

Substituting the given values, you obtain:

$$\frac{16 + x}{15} = \frac{15}{x}$$

Cross-multiplying, you have:

$$(16 + x)x = 15^2$$
$$16x + x^2 = 225$$
$$x^2 + 16x - 225 = 0$$
$$(x + 25)(x - 9) = 0$$
$$x + 25 = 0 \quad \text{or} \quad x - 9 = 0$$
$$x = -25 \quad \text{or} \quad x = 9$$

Since a length cannot be negative, reject $x = -25$.
Therefore: $x = 9$

Check your solution by substituting 9 for x in the proportion $\frac{16 + x}{15} = \frac{15}{x}$.

ORAL EXERCISES

State the geometric mean between the given numbers.

1. 2 and 50
2. 4 and 9
3. 1 and 2
4. $\sqrt{2}$ and $\sqrt{8}$

Name the triangles that are similar to the given one.

5. $\triangle MNP$

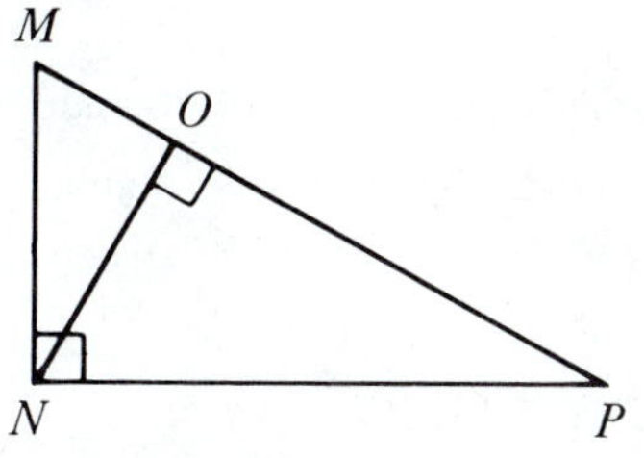

6. $\triangle TYZ$

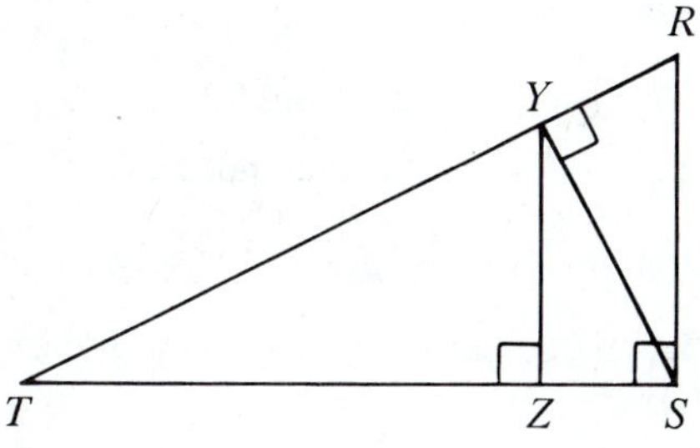

Complete.

7. a. In Exercise 5, if $m\angle P = 30$, then $m\angle M =$ _?_, $m\angle MNO =$ _?_, and $m\angle ONP =$ _?_.
   b. In Exercise 5, if $m\angle P = k$, then $m\angle M =$ _?_, $m\angle MNO =$ _?_, and $m\angle ONP =$ _?_.

8. a. In Exercise 5, ON is the geometric mean between _?_ and _?_.
   b. In Exercise 5, NP is the geometric mean between _?_ and _?_.

State an equation you could use to find the value of x. You need not solve the equation.

9.

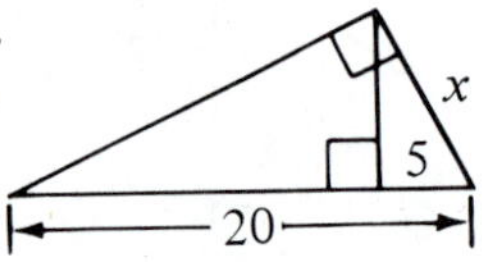

10.

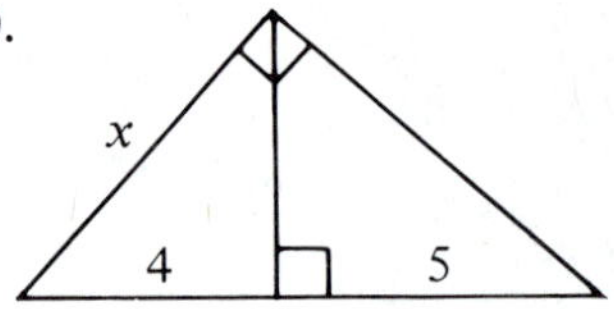

11.

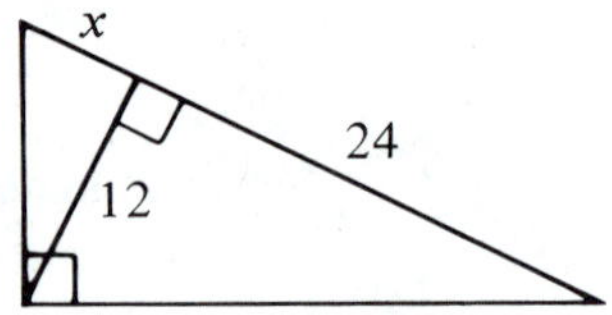

12.

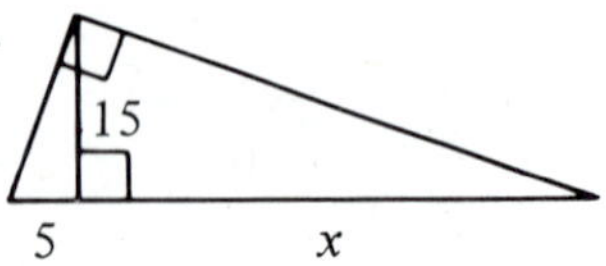

13.

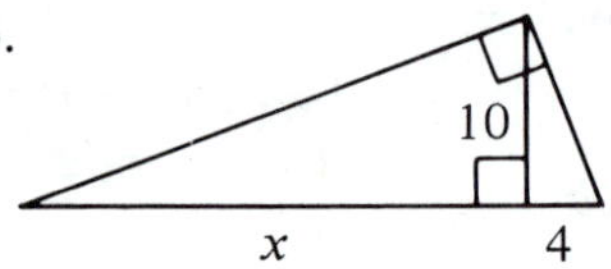

14.

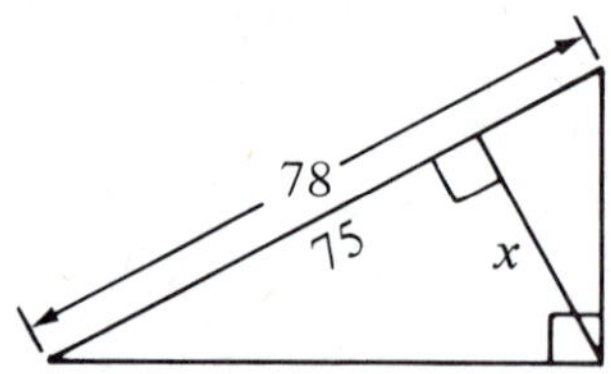

15.

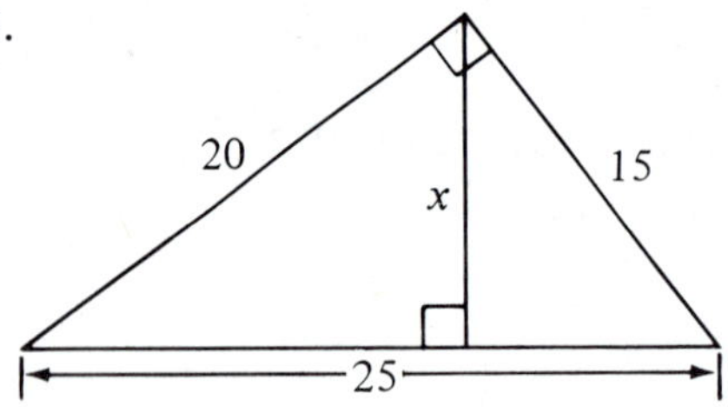

16.

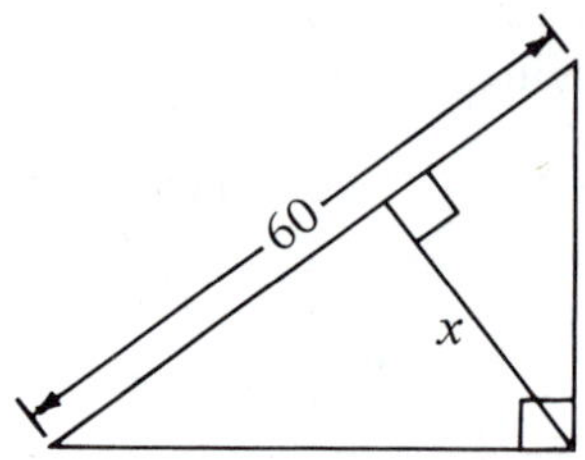

WRITTEN EXERCISES

State the geometric mean between the given numbers. (Assume that $a > 0$ and $b > 0$.)

A 1. 3 and 27   2. 2 and 72   3. 18 and 50   4. 6 and 150

5. 5 and 15   6. $\sqrt{2}$ and $2\sqrt{2}$ 7. a and ab^2 8. a and a^7

9–16. In Oral Exercises 9–16, find the value of x.

17. In right $\triangle ABC$, $\overline{CD}$ is the altitude to hypotenuse $\overline{AB}$. If $AC = 6$ and $AB = 12$, find AD.

18. In right $\triangle RST$, $\overline{TW}$ is the altitude to hypotenuse $\overline{RS}$. If $RW = 18$ and $WS = 6$, find ST.

19. The altitude to the hypotenuse of a right triangle divides the hypotenuse into segments whose lengths are 5 and 45. What is the length of the altitude?

20. In a right triangle, an altitude of length 8 divides the hypotenuse into two segments. If the length of one of these segments is 16, find the length of the other segment.

Find the value of x.

B 21.

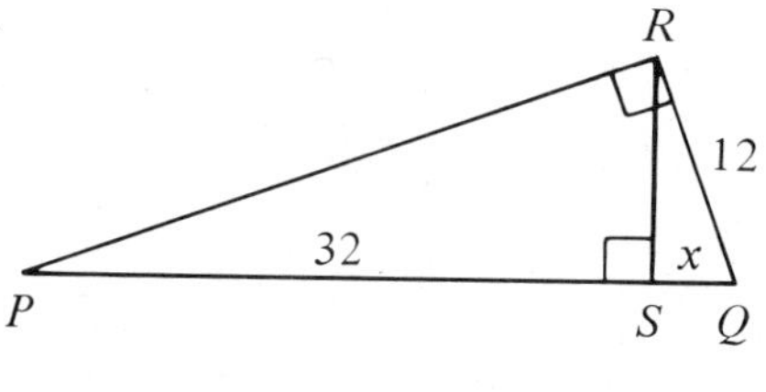

22.

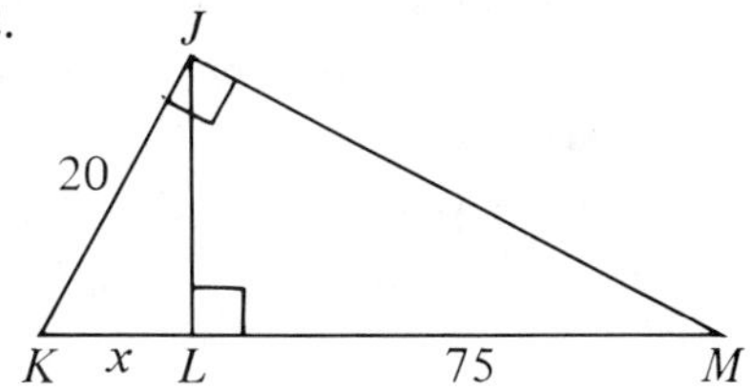

23. In right $\triangle DEF$, the length of the altitude $\overline{DG}$ to hypotenuse $\overline{EF}$ is 20. If the length of the hypotenuse is 58, find all possible lengths of $\overline{EG}$.

24. In right $\triangle PQR$, the length of the altitude $\overline{RS}$ to hypotenuse $\overline{PQ}$ is 12. The length of $\overline{PQ}$ is 26.
   a. Find all possible lengths of $\overline{PS}$.
   b. If $\overline{PS}$ is longer than $\overline{SQ}$, what are the lengths of $\overline{PS}$ and $\overline{SQ}$?

25. The altitude to the hypotenuse of a right triangle divides the hypotenuse into two segments whose lengths are in the ratio 1:9. If the length of the altitude is 9, find the lengths of the two segments and the length of the hypotenuse.

26. The altitude to the hypotenuse of a right triangle divides the hypotenuse into two segments. The length of one segment exceeds the length of the other segment by 15. If the length of the altitude is 10, find the lengths of the shorter segment and the length of the hypotenuse.

27. Prove part (2) of Theorem 9-1: $\triangle CBD \sim \triangle ABC$

C 28. Without using the transitive property of similarity, prove part (3) of Theorem 9-1: $\triangle ACD \sim \triangle CBD$

Write two-column proofs.

29. Given: $\overline{DF} \perp \overline{EF}$;
$\overline{FG} \perp \overline{DE}$;
$\overline{GH} \perp \overline{DF}$
Prove: $DG \cdot GE = DF \cdot HF$

30. Given: $\overline{DF} \perp \overline{EF}$;
$\overline{FG} \perp \overline{DE}$;
$\overline{GH} \perp \overline{DF}$
Prove: $EF \cdot HG = DF \cdot HF$

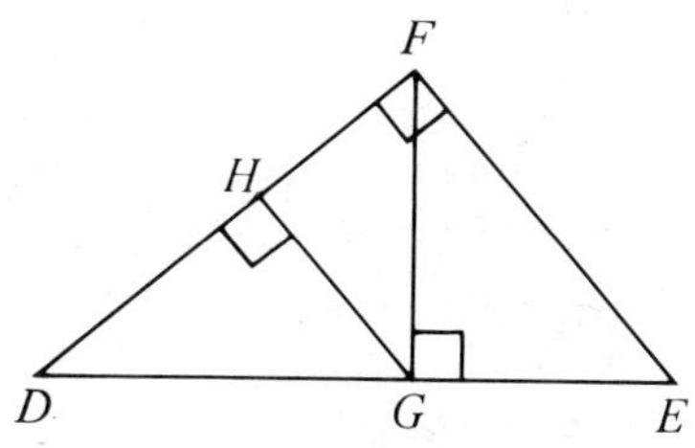

Exs. 29, 30

Section 9-2 THE PYTHAGOREAN THEOREM

In this section you will learn one of the most important theorems in geometry, the Pythagorean Theorem.

Theorem 9-2 The sum of the squares of the lengths of the legs of a right triangle equals the square of the length of the hypotenuse. (Pythagorean Theorem)

GIVEN: Right $\triangle ABC$ with right $\angle ACB$
PROVE: $a^2 + b^2 = c^2$

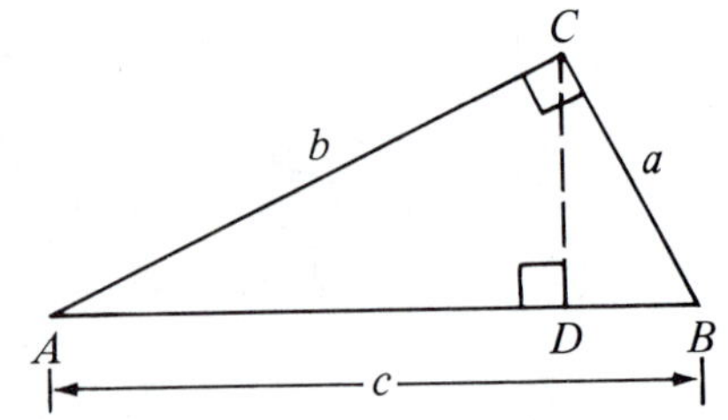

Outline of proof: Draw $\overline{CD}$, the altitude to the hypotenuse. Then, by Theorem 9-1:

$$\triangle ACD \sim \triangle CBD \sim \triangle ABC$$

By Theorem 8-9, the ratio of the areas equals the square of the ratio of the lengths of any two corresponding sides, which is the same as the ratio of the squares of the lengths of those sides. Thus:

$$\frac{\text{area } \triangle CBD}{\text{area } \triangle ABC} = \frac{a^2}{c^2}$$

and

$$\frac{\text{area } \triangle ACD}{\text{area } \triangle ABC} = \frac{b^2}{c^2}$$

Adding the respective sides of these two equations:

$$\frac{\text{area } \triangle CBD + \text{area } \triangle ACD}{\text{area } \triangle ABC} = \frac{a^2 + b^2}{c^2}$$

Since the sum of the areas of $\triangle CBD$ and $\triangle ACD$ equals the area of $\triangle ABC$, you have:

$$\frac{\text{area } \triangle ABC}{\text{area } \triangle ABC} = \frac{a^2 + b^2}{c^2}, \quad \text{or} \quad 1 = \frac{a^2 + b^2}{c^2}$$

Multiplying both sides by c^2, you have:

$$c^2 = a^2 + b^2$$

EXAMPLE The diagonal of a rectangle has length 15. If the width of the rectangle is 9, find its length.

SOLUTION Let x = the length of the rectangle. Since $\overline{PR}$ is the hypotenuse of right $\triangle PQR$, by the Pythagorean Theorem you have:

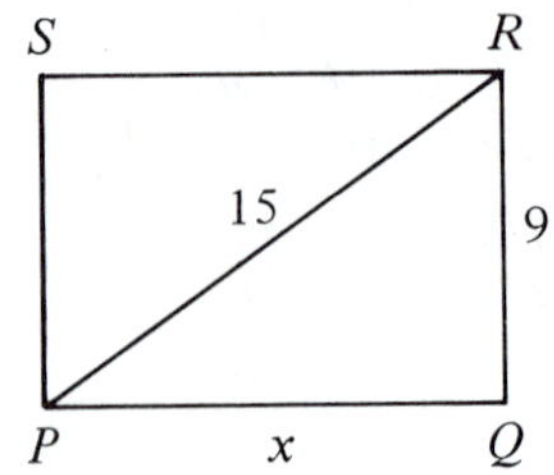

$$\begin{aligned} (PQ)^2 + (QR)^2 &= (PR)^2 \\ x^2 + 9^2 &= 15^2 \\ x^2 + 81 &= 225 \\ x^2 &= 144 \\ x &= 12 \text{ (since } x \text{ must be positive)} \end{aligned}$$

EXAMPLE Find the length of one side of a rhombus whose diagonals have lengths 8 and 12.

SOLUTION Since the diagonals of a rhombus are perpendicular and bisect each other, $\triangle ADE$ (for example) is a right triangle with legs of length 4 and 6. Therefore by the Pythagorean Theorem:

$$(AD)^2 = (AE)^2 + (DE)^2$$
$$(AD)^2 = 4^2 + 6^2$$
$$(AD)^2 = 52$$
$$AD = \sqrt{52}, \text{ or } 2\sqrt{13}$$

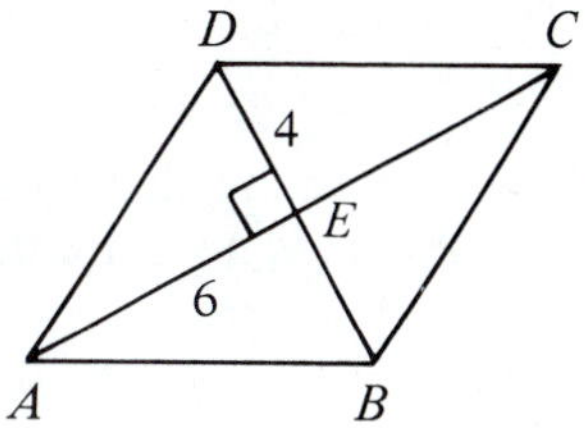

It is important to note that in order to apply the Pythagorean Theorem, the given triangle must be a *right* triangle. The fact that the theorem holds *only* for right triangles is stated in the next theorem, which is the converse of the Pythagorean Theorem.

Theorem 9-3 If the sum of the squares of the lengths of two sides of a triangle equals the square of the length of the third side, then the triangle is a right triangle (whose hypotenuse is the third side).

GIVEN: $a^2 + b^2 = c^2$
PROVE: $\triangle ABC$ is a right triangle.

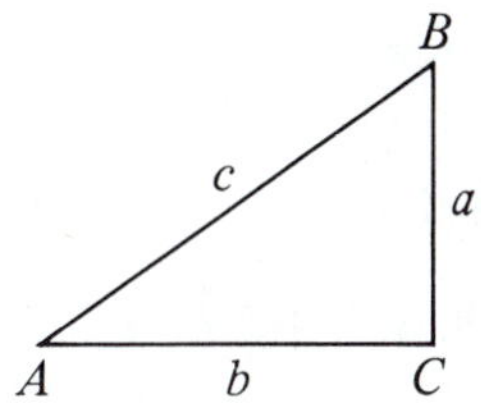

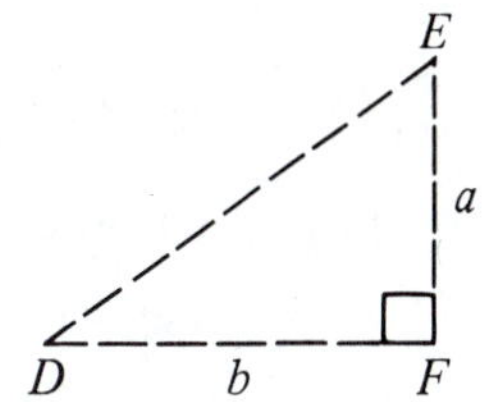

Outline of proof: Draw $\overline{DF} \perp \overline{EF}$ so that $EF = a$ and $DF = b$. Draw $\overline{DE}$. Then $\triangle DEF$ is a right triangle, and hence, by the Pythagorean Theorem,

$$a^2 + b^2 = (DE)^2.$$

Substituting in the given equation $a^2 + b^2 = c^2$, you have

$$(DE)^2 = c^2,$$

and thus

$$DE = c.$$

Hence $\triangle ABC \cong \triangle DEF$ by the SSS Postulate, and thus $\angle C \cong \angle F$. But since $\angle F$ is a right angle, $\angle C$ is also, and $\triangle ABC$ is a right triangle with hypotenuse $\overline{AB}$.

EXAMPLE Determine whether a triangle with sides of length 30, 40, and 50 is a right triangle.

SOLUTION Since

$$30^2 + 40^2 = 900 + 1600 = 25{,}000$$

and

$$50^2 = 25{,}000,$$

the triangle is a right triangle by the converse of the Pythagorean Theorem.

ORAL EXERCISES

Is the triangle whose sides have the given lengths a right triangle?

1. 2, 3, 4 **2.** 7, 24 25 **3.** $\sqrt{2}$, $\sqrt{2}$, 2 **4.** 1, 2, $\sqrt{3}$

Find the length of the hypotenuse of a right triangle whose legs have the given lengths.

5. 6, 8

6. 2, $\sqrt{5}$

Write an equation that would enable you to find the value of x in each right triangle. You need not solve the equation.

7.

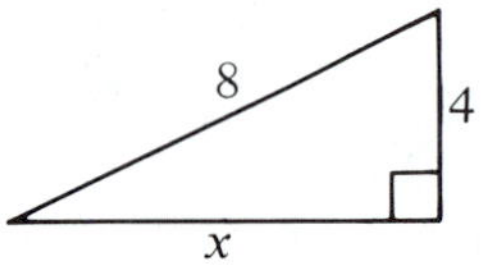

8.

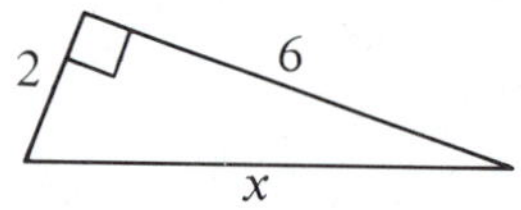

9.

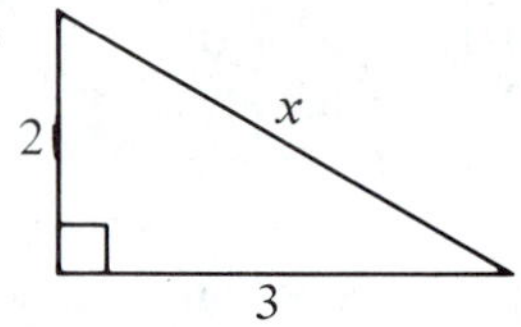

10.

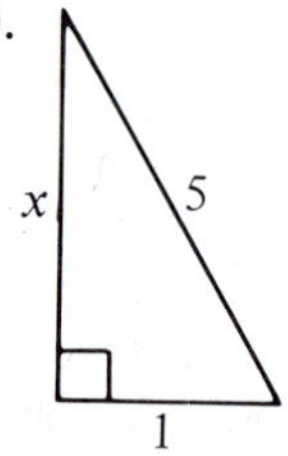

11. The size of a television screen is usually expressed as the length of its diagonal. What is the size of a rectangular screen whose dimensions are 12 in. by 16 in.?

12. What is the length of the hypotenuse of an isosceles right triangle each of whose legs has length 1?

WRITTEN EXERCISES

Give all irrational answers in simplest radical form.

Right $\triangle ABC$ has legs of length a and b and hypotenuse of length c. Find the missing length.

A 1. $a = 15$, $c = 25$, $b =$ _?_

2. $b = 21$, $c = 29$, $a =$ _?_

3. $b = 10$, $c = 26$, $a =$ _?_

4. $a = 40$, $c = 41$, $b =$ _?_

5. $a = 15$, $b = 8$, $c =$ _?_

6. $a = 14$, $b = 48$, $c =$ _?_

7–10. Find the value of x in the triangles in Oral Exercises 7–10.

Is the triangle whose sides have the given lengths a right triangle?

11. a. 3, 4, 5 b. 6, 8, 10 c. 9, 12, 15 d. $3x$, $4x$, $5x$ $(x > 0)$

12. a. 5, 12, 13 b. 0.5, 1.2, 1.3 c. 50, 120, 130 d. $5y$, $12y$, $13y$ $(y > 0)$

13. Find the width of a rectangle of length 84 whose diagonals each have length 85.

14. Find the length of a diagonal of a rectangle of length 24 and width 18.

15. The diagonals of a rhombus have lengths 16 and 12. Find the perimeter of the rhombus.

16. The perimeter of a rhombus is 68 and one diagonal has length 30. Find the length of the other diagonal.

Find the length of a side of a square whose diagonals each have the given length.

17. 6

18. $8\sqrt{6}$

If $\triangle RST$ is an isosceles triangle with base $\overline{RT}$, find the length of the altitude to $\overline{RT}$ and the area of $\triangle RST$.

19. $RT = 10$, $RS = 13$

20. $ST = 2\sqrt{13}$, $RT = 12$

B 21. If you travel 15 km north, then 7 km east, then 9 km north, how far are you from your starting point?

22. Find the height of an isosceles trapezoid each of whose legs has length 12 and whose bases have lengths 20 and 36.

23. In $\square ABCD$, $AB = 10$, and diagonals $\overline{AC}$ and $\overline{BD}$ have lengths 12 and 16, respectively. Show that $ABCD$ is a rhombus.

24. In $\triangle ABC$, $AC = 13$, $AB = 10$, and $CD = 12$, where $\overline{CD}$ is the median to side $\overline{AB}$. Show that $\triangle ABC$ is isosceles.

In $\triangle JKL$, $\angle K$ is a right angle. Find all possible values of x.

25. $JK = x$, $KL = x + 1$, $JL = x + 2$

26. $JK = x + 2$, $KL = 2x$, $JL = x + 6$

27. $JK = x - 7$, $KL = x$, $JL = x + 1$

28. $JK = x + 1$, $KL = x + 4$, $JL = 2x - 1$

29. The length of a rectangle is 5 more than 5 times the width. The length of each of its diagonals is 5 less than 6 times the width of the rectangle. Find the length and width of the rectangle.

30. The length of one leg of a right triangle is 8 more than that of the other leg. If the perimeter of the triangle is 4 times the length of the shorter leg, find the lengths of all three sides.

31. If $AB = 13$ and $BC = 12$, find:
   a. the area of $\triangle ABC$
   b. CD
   c. DB
   d. the area of $\triangle CDB$

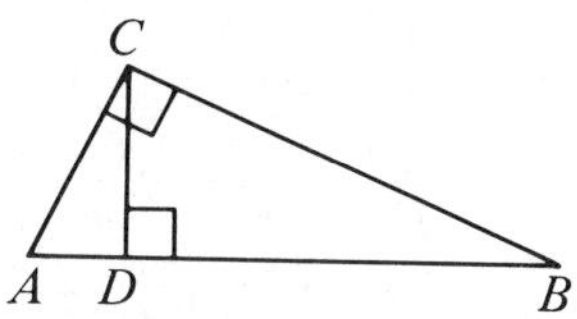

Exs. 31-34

32. If $AC = \sqrt{3}$ and $BC = \sqrt{6}$, find AB and CD.

33. If $CD = 24$ and $CB = 40$, find DB, AD, AB, and AC.

34. If $AC = 34$ and $CD = 30$, find the area of $\triangle ABC$.

35. The lengths of the sides of right $\triangle ABC$ are a, b, and c. The lengths of the sides of $\triangle DEF$ are ka, kb, and kc.
   a. Explain why $\triangle DEF$ is similar to $\triangle ABC$.
   b. Explain why $\triangle DEF$ is a right triangle.

36. Three integers form a *Pythagorean triple* if they are lengths of the sides of a right triangle.
   a. Show that if m and n are any positive integers with $m > n$, then $m^2 - n^2$, $2mn$, and $m^2 + n^2$ form a Pythagorean triple.
   b. Use part **(a)** to find all Pythagorean triples generated by positive integers m and n with $5 \geq m > n$.

C 37. In the diagram, $\overline{AC} \perp \overline{BC}$ and $\overline{CG} \perp \overline{AB}$.

a. Explain why:
area of rectangle $BDGE$ = area of square $BHJC$
area of rectangle $ADGF$ = area of square $ACKL$
(*Hint:* Use Corollary 2 of Theorem 9-1.)

b. Explain how part **(a)** can be used to give an alternative proof of the Pythagorean Theorem.

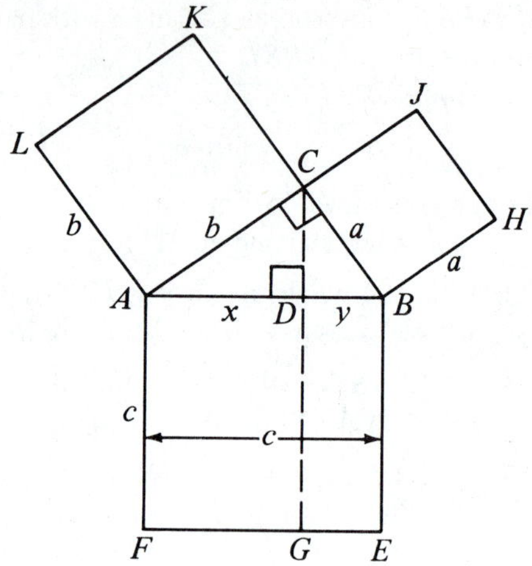

38. The following proof of the Pythagorean Theorem is essentially the one devised by James Garfield several years before he became President of the United States.

Given: Right $\triangle ABC$ with right $\angle C$
Prove: $a^2 + b^2 = c^2$

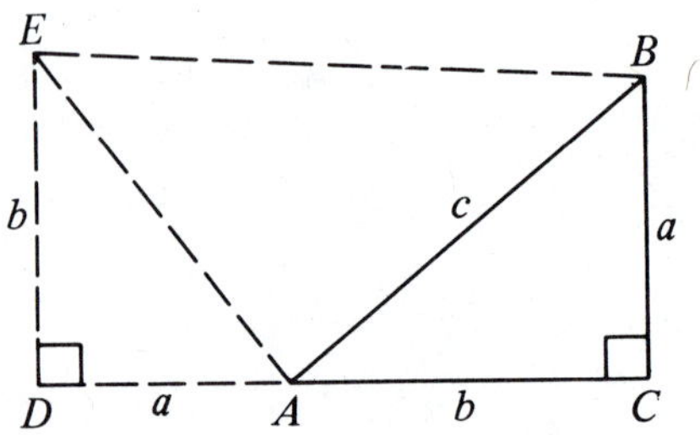

Method: On the ray opposite to $\overrightarrow{AC}$ let D be the point such that $AD = a$. On the ray perpendicular to $\overline{DC}$ at D let E be the point such that $ED = b$. Draw $\overline{AE}$ and $\overline{BE}$. Justify each of the following:

a. $AE = c$

b. $\angle EAB$ is a right angle.

c. Area of trapezoid $BCDE = \frac{1}{2}(a + b)^2$

d. Area of trapezoid $BCDE = ab + \frac{1}{2}c^2$

e. $a^2 + b^2 = c^2$

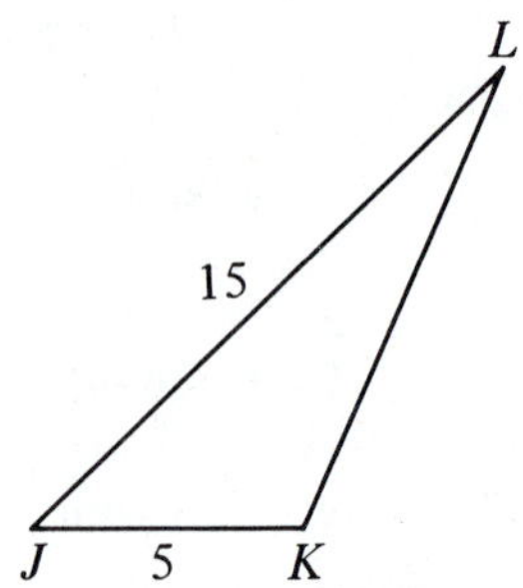

39. In $\triangle JKL$, shown at the right, if the length of the altitude from L is $5\sqrt{5}$, find the length of $\overline{KL}$.

Section 9-3 SPECIAL RIGHT TRIANGLES

Two special kinds of right triangles occur so frequently that it is important to know the relationships between the lengths of their sides. Then you won't have to use the Pythagorean Theorem to calculate them again and again. In this section you will learn about:

- 45-45-90 triangles
- 30-60-90 triangles

Each of the two base angles of an isosceles right triangle has measure 45. Therefore, such a triangle is sometimes called a "45-45-90 triangle." The following theorem, which could be regarded as a corollary of the Pythagorean Theorem, states the relationship between the lengths of the sides of any 45-45-90 triangle.

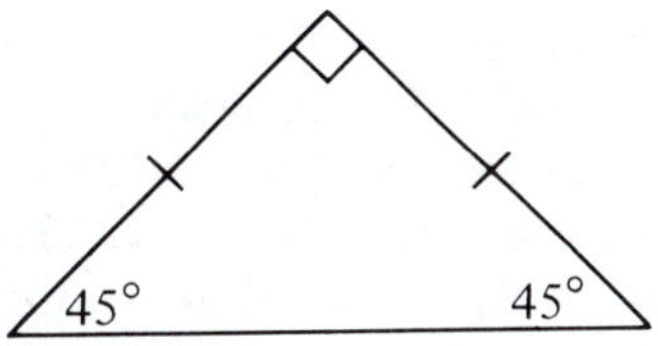

Theorem 9-4 In a 45-45-90 triangle, the length of the hypotenuse is $\sqrt{2}$ times the length of either leg.

GIVEN: Right $\triangle ABC$ with right $\angle C$;
$BC = AC = a$

PROVE: $AB = a\sqrt{2}$

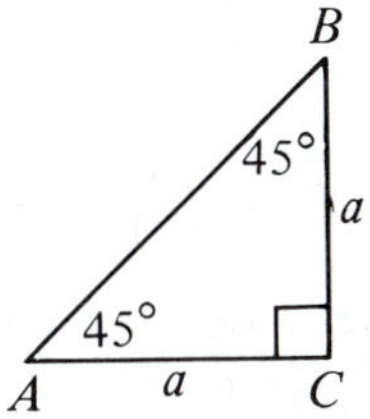

Proof: By the Pythagorean Theorem you have:

$$(AB)^2 = a^2 + a^2$$
$$(AB)^2 = 2a^2$$
$$AB = \sqrt{2a^2}$$
$$AB = a\sqrt{2}$$

EXAMPLE Find the length of the hypotenuse of a right triangle each of whose legs has length 8.

SOLUTION Let a = the length of a leg. Then the length of the hypotenuse is $a\sqrt{2}$. Since $a = 8$, the length of the hypotenuse is $8\sqrt{2}$.

EXAMPLE Find the length of a leg of an isosceles right triangle whose hypotenuse has length 8.

SOLUTION Let a = the length of a leg. Then the length of the hypotenuse is $a\sqrt{2}$. Therefore,

$$a\sqrt{2} = 8$$

Dividing both sides by $\sqrt{2}$, you have:

$$a = \frac{8}{\sqrt{2}}$$

Thus:

$$a = \frac{8}{\sqrt{2}} \cdot \frac{\sqrt{2}}{\sqrt{2}} = \frac{8\sqrt{2}}{2} = 4\sqrt{2}$$

If one of the acute angles of a right triangle has measure 30, then the measure of the other acute angle must be 60. Such a triangle is called a "30-60-90 triangle." The next theorem gives the relationship between the lengths of its sides.

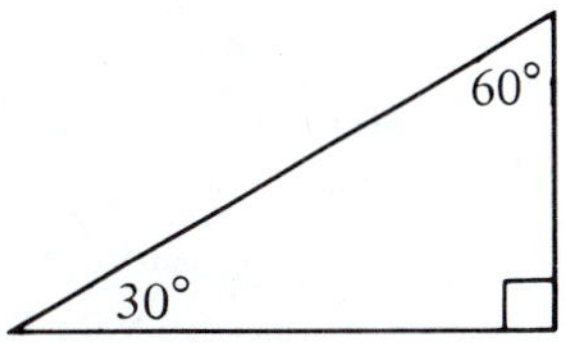

Theorem 9-5 In a 30-60-90 triangle the length of the hypotenuse is twice that of the shorter leg, and the length of the longer leg is $\sqrt{3}$ times the length of the shorter leg.

GIVEN: Right $\triangle ABC$ with right $\angle C$;
$m\angle A = 30$;
$BC = a$

PROVE: $AB = 2a$;
$AC = a\sqrt{3}$

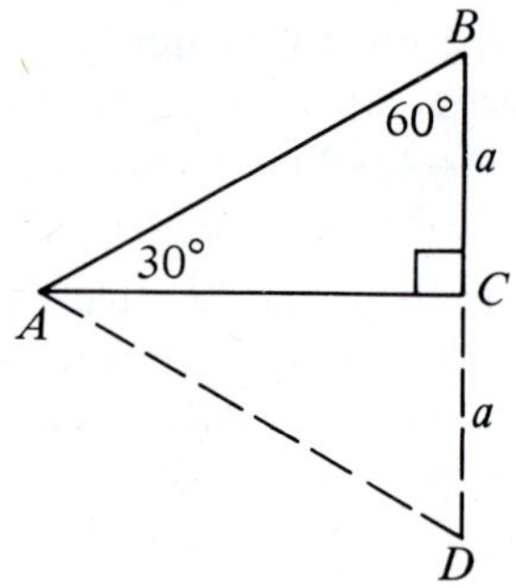

Outline of Proof: On the ray opposite to $\overrightarrow{CB}$ let D be the point such that $CD = a$. Draw $\overline{AD}$. Then $\triangle ABC \cong \triangle ADC$ by the SAS Postulate. Therefore $m\angle D = 60$. By the Triangle Angle-Sum Theorem, $m\angle BAD = 60$. Thus $\triangle ABD$ is equilateral, and hence $AB = BD = 2a$. Applying the Pythagorean Theorem to $\triangle ABC$, you have:

$$(AC)^2 + (BC)^2 = (AB)^2$$
$$(AC)^2 + a^2 = (2a)^2$$
$$(AC)^2 + a^2 = 4a^2$$
$$(AC)^2 = 3a^2$$
$$AC = \sqrt{3a^2} = a\sqrt{3}$$

The two diagrams below summarize the results of Theorems 9-4 and 9-5.

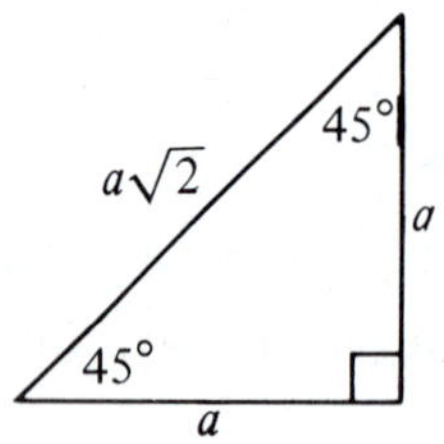

45-45-90 Triangle

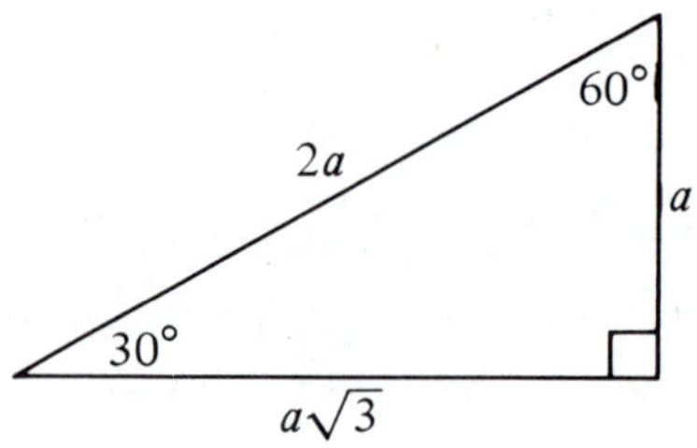

30-60-90 Triangle

EXAMPLE The longer leg of a 30-60-90 triangle has length 6. Find the length of the shorter leg and the length of the hypotenuse.

SOLUTION If $BC = a$, then, by Theorem 9-5, you know that

$$AC = a\sqrt{3}$$
$$6 = a\sqrt{3}$$
$$a = \frac{6}{\sqrt{3}} = \frac{6}{\sqrt{3}} \cdot \frac{\sqrt{3}}{\sqrt{3}}$$
$$= \frac{6\sqrt{3}}{3} = 2\sqrt{3}$$

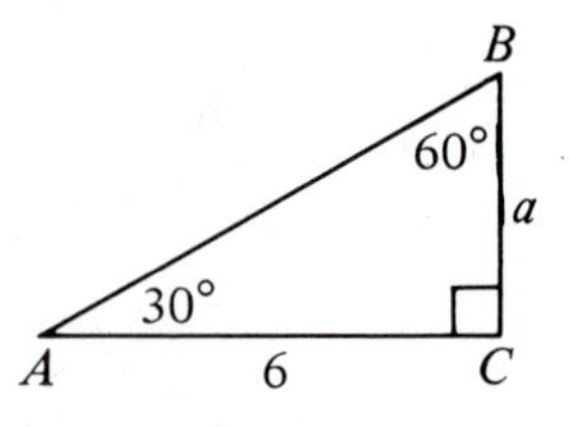

Therefore the shorter leg has length $2\sqrt{3}$, and the hypotenuse has length $2(2\sqrt{3})$, or $4\sqrt{3}$. You can check these answers by applying the Pythagorean Theorem.

ORAL EXERCISES

1. a. What is the ratio of the lengths of the sides of a 45-45-90 triangle?
   b. If the lengths of the sides of a triangle have this ratio, must the triangle be a 45-45-90 triangle? Explain.
2. a. What is the ratio of the lengths of the sides of a 30-60-90 triangle?
   b. Explain why a triangle the lengths of whose sides have this ratio must be a 30-60-90 triangle.

If the three sides of a triangle have the given lengths, state whether the triangle is a 45-45-90 triangle, a 30-60-90 triangle, or neither.

3. $3, 3, 3\sqrt{2}$ 4. $3, 4, 5$ 5. $\sqrt{2}, \sqrt{2}, 2$ 6. $2, 2\sqrt{3}, 4$
7. $5\sqrt{2}, 5\sqrt{2}, 10$ 8. $\sqrt{3}, 2\sqrt{3}, 2\sqrt{3}$ 9. $\sqrt{3}, 3, 2\sqrt{3}$ 10. $\sqrt{2}, 2\sqrt{2}, \sqrt{6}$

11. What is the length of a diagonal of a square each of whose sides has length 2?
12. What is the length of an altitude of an equilateral triangle each of whose sides has length 2?
13. If the lengths of two sides of a rectangle are 4 and $4\sqrt{3}$, what is the measure of an angle formed by a diagonal and the longer side?

WRITTEN EXERCISES

Give all answers in simplest form.

Find the length of the hypotenuse of an isosceles right triangle each of whose legs has the given length.

A 1. 6 2. 4 3. $5\sqrt{2}$ 4. $\sqrt{6}$

Find the length of one leg of an isosceles right triangle whose hypotenuse has the given length.

5. $10\sqrt{2}$ 6. $\frac{\sqrt{2}}{3}$ 7. 12 8. 16

Complete.

	9.	10.	11.	12.	13.	14.	15.	16.
RS	5	?	?	?	?	$15\sqrt{3}$	$\sqrt{2}$	?
ST	?	$8\sqrt{3}$	?	?	12	?	?	$\sqrt{2}$
RT	?	?	14	$2\sqrt{3}$	?	?	?	?

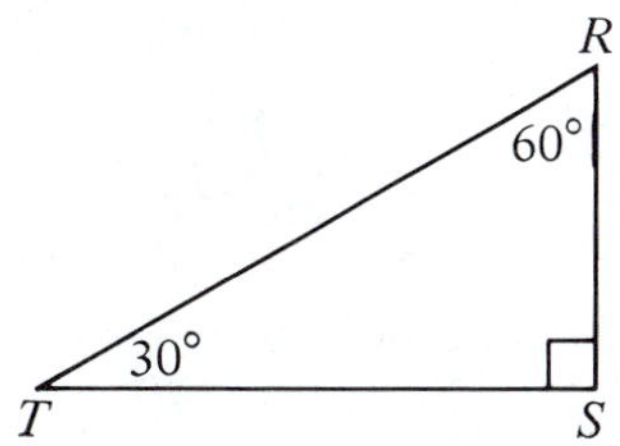

17. The length of one side of a square is $\sqrt{2}$. Find the length of a diagonal of the square.
18. The length of a diagonal of a square is $\sqrt{6}$. Find the length of each side of the square.
19. The length of each side of an equilateral triangle is 6. What is the length of each altitude of the triangle?

20. The length of each altitude of an equilateral triangle is 6. What is the length of each side of the triangle?

21. The length of each altitude of an equilateral triangle is 4.5. Find the perimeter of the triangle.

22. The perimeter of an equilateral triangle is $15\sqrt{3}$. Find the length of each altitude of the triangle.

B 23. A rhombus has a 60° angle and a side whose length is 2. How long are its diagonals?

24. One angle of a rhombus has measure 120. If the length of the longer diagonal is 12, find the length of the shorter diagonal and the perimeter of the rhombus.

25. Each base angle of an isosceles triangle has measure 30. The length of the base is 2. Find the height, area, and perimeter of the triangle.

26. A base angle of an isosceles triangle has measure 30, and the side opposite it has length 10. Find the perimeter and the area of the triangle.

27. Two adjacent sides of a parallelogram have lengths 8 and 10, respectively. If the angle determined by these sides has measure 60, find the area of the parallelogram.

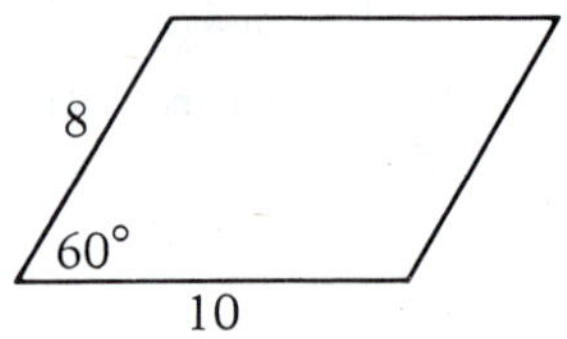

28. Two adjacent sides of a parallelogram meet at a 45° angle. The lengths of the two sides are 12 and $16\sqrt{2}$. Find the area of the parallelogram.

29. An equilateral triangle has a side of length s.
  a. Find the height of the triangle in terms of s.
  b. State a formula for the area of the triangle in terms of s.
  c. If the area of an equilateral triangle is $4\sqrt{3}$, find the length of each side of the triangle.

30. The length of a diagonal of a square is x.
  a. State a formula for the area of the square in terms of x.
  b. If the area of a square is 20, find the length of a diagonal of the square.

31. Each side of a regular hexagon has length 10. Find the area of the hexagon. (*Hint:* Divide the hexagon into triangles.)

32. If $\overline{AB}$, $\overline{BC}$, $\overline{CD}$, $\overline{DE}$, and $\overline{EA}$ each have length 4, find the area of pentagon $ABCDE$.

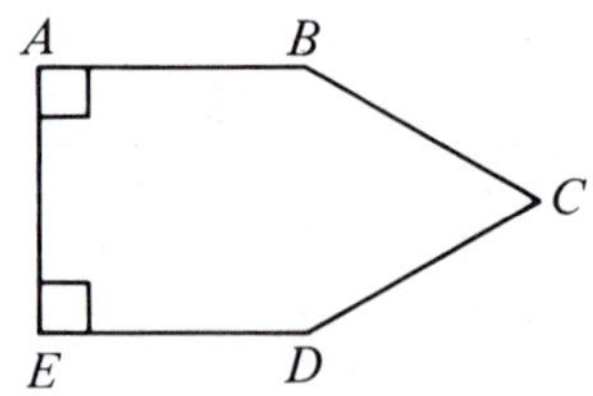

33. In $\triangle JKL$, $JL = 8\sqrt{3}$.
  a. Find JK and LK.
  b. Find the perimeter of $\triangle JKL$.
  c. Find the area of $\triangle JKL$.

34. In $\triangle JKL$, $LK = 12$.
  a. Find JL and JK.
  b. Find the perimeter of $\triangle JKL$.
  c. Find the area of $\triangle JKL$.

Exs. 33, 34

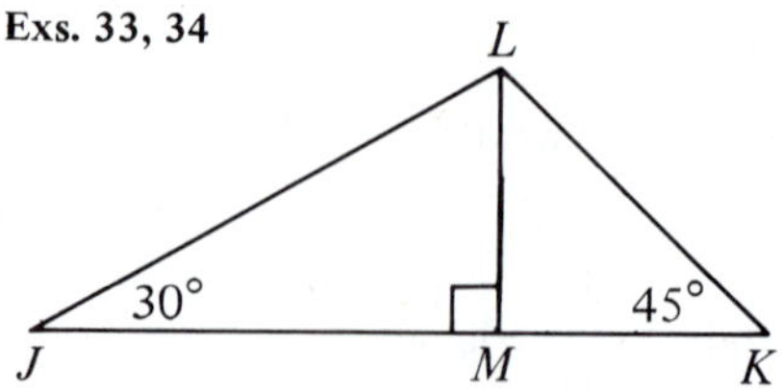

C **35.** In trapezoid $PQRS$, $\overline{PQ} \parallel \overline{RS}$, $m\angle P = 30$, $m\angle Q = 45$, and $RS = PS = 12$. Find RQ and PQ.

36. In trapezoid $ABCD$, $\overline{AB} \parallel \overline{DC}$, $m\angle A = 60$, $m\angle B = 135$, and $AB = AD = 6$. Find DC and BC.

37. If $RQ = 6$, find RS. (*Hint:* Draw the altitude to $\overline{PQ}$.)

38. If $RS = 18$, find:

a. the perimeter of $\triangle PRQ$

b. the area of $\triangle PRQ$

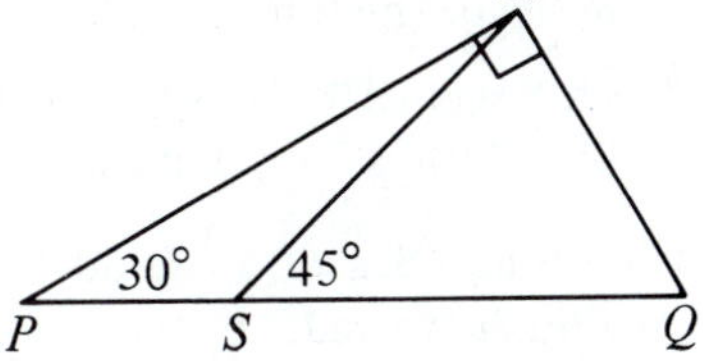

Exs. 37, 38

Find the value of x.

39.

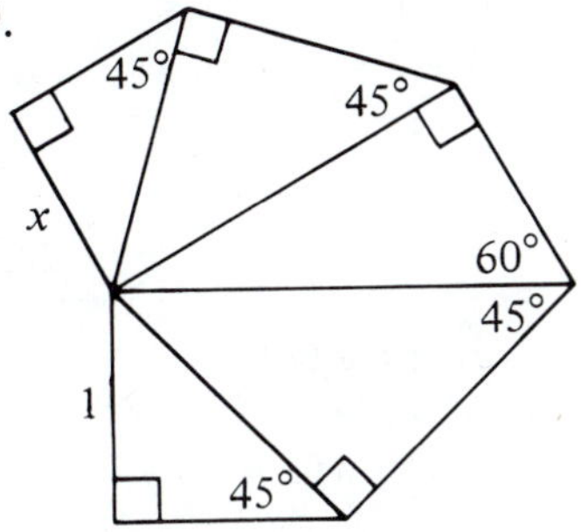

40.

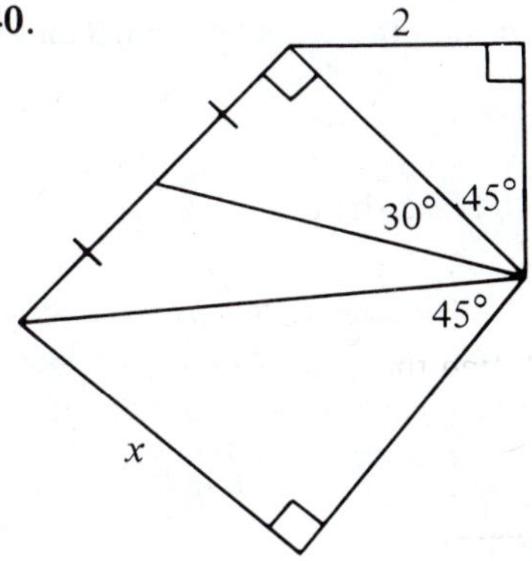

41. a. Show that the quadrilateral formed by joining the midpoints of the sides of a square with area s^2 is a square with area $\frac{s^2}{2}$.

b. Each vertex of a square divides a side of a larger square into segments whose lengths are in the ratio 1:2. If the area of the larger square is s^2, find the area of the smaller square.

SELF-TEST 1

Section 9-1

1. Use the diagram to complete:
$\triangle WXY \sim \triangle$ ___?___ $\sim \triangle$ ___?___

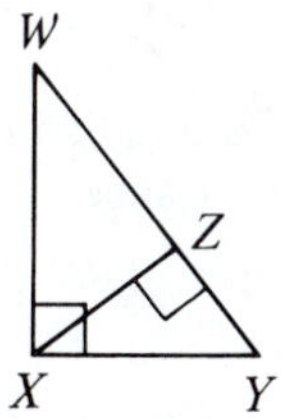

Find the values, in simplest form, of the variables.

2.

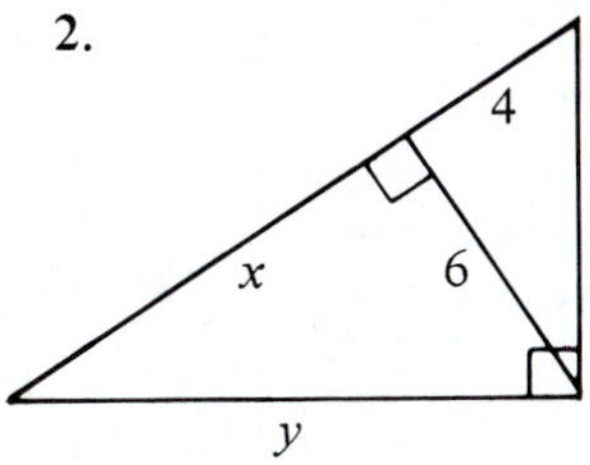

3.

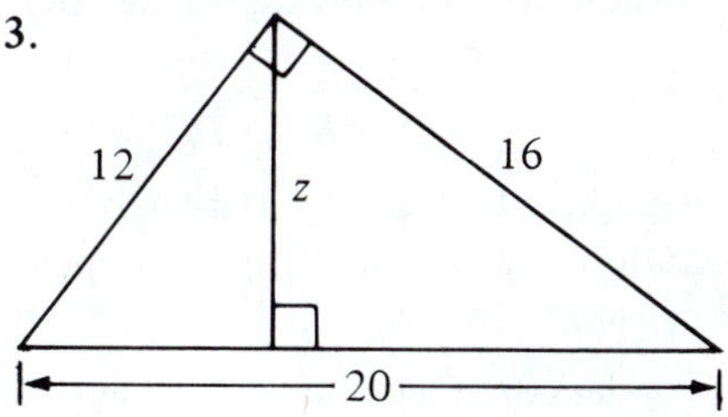

(Self-Test continued on next page.)

4. A triangle has sides of length 20, 21, and 28. Is the triangle a right triangle? Explain. Section 9-2

5. Two sides of a right triangle are 6 cm long and 8 cm long. Find two possible lengths for the third side.

6. Each diagonal of a rectangle is 1 m longer than the length of the rectangle. If the width is 5 m, find the perimeter of the rectangle.

7. The longer leg of a 30–60–90 triangle has length $4\sqrt{6}$. Find the lengths of the other two sides of the triangle. Section 9-3

8. The diagonals of square $ABCD$ intersect at point M.
   a. What kind of triangle is $\triangle AMB$? Explain.
   b. If $AC = 16$, find the areas of $\triangle AMB$ and square $ABCD$.

Section 9-4 TRIGONOMETRIC RATIOS

The branch of mathematics called *trigonometry* is based on the fact that the ratio of the lengths of any two sides of a right triangle is equal to the ratio of the lengths of the corresponding sides of any similar right triangle. In this section you will be introduced to three *trigonometric ratios:*

- the sine ratio
- the cosine ratio
- the tangent ratio

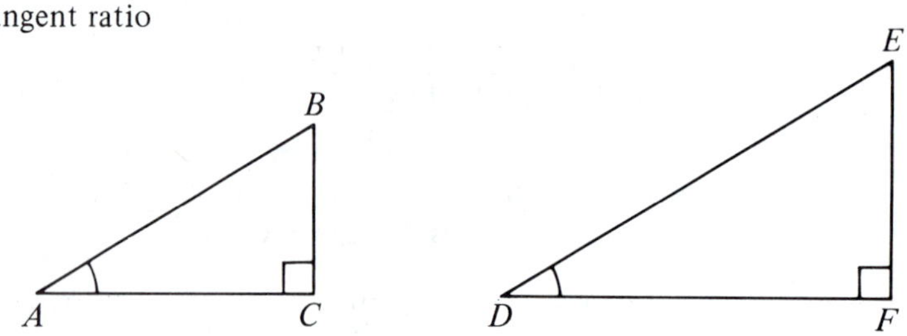

In the diagram, $\angle C$ and $\angle F$ are right angles and $\angle A \cong \angle D$. Thus $\triangle ABC \sim \triangle DEF$ by the AA Similarity Theorem. Hence

$$\frac{BC}{EF} = \frac{AB}{DE},$$

since the lengths of corresponding sides of similar triangles are proportional. By Property 3 of Proportions, this proportion can be rewritten as:

$$\frac{BC}{AB} = \frac{EF}{DE}.$$

This proportion implies that in *any* right triangle that contains an angle congruent to $\angle A$ the ratio of the length of the leg opposite the angle and the length of the hypotenuse is *constant.* That is:

$$\frac{\text{length of leg opposite the angle}}{\text{length of hypotenuse}} = \frac{BC}{AB}$$

The ratio $\frac{BC}{AB}$ is called the *sine of angle A*, and it is denoted *sin A*. Therefore, in $\triangle ABC$ with right $\angle C$, we have:

$$\sin A = \frac{\text{length of the leg opposite } \angle A}{\text{length of the hypotenuse}}$$
$$= \frac{BC}{AB}$$

Whenever $m\angle A = m\angle D$, $\sin A = \sin D$. In other words, all angles with the same measure have the same sine. The notation "sin 50°" is used to mean "the sine of any angle whose measure is 50."

The sine of 50° has been calculated to be 0.7660 to the nearest ten-thousandth. We could use the symbol ≈ (read "is approximately equal to") to write

$$\sin 50° \approx 0.7660.$$

However, it is customary to write instead $\sin 50° = 0.7660$.

EXAMPLE Find **(a)** sin 30° and **(b)** sin 60°.

SOLUTION The ratio of the lengths of the sides of any 30-60-90 triangle is $1 : \sqrt{3} : 2$. Sketch any 30-60-90 triangle.

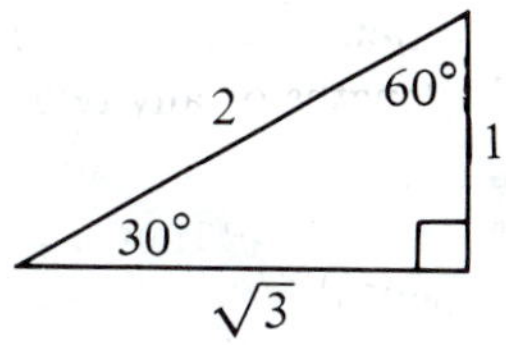

(a) $$\sin 30° = \frac{\text{length of the side opposite the } 30° \text{ angle}}{\text{length of the hypotenuse}}$$
$$= \frac{1}{2}, \text{ or } 0.5$$

(b) $$\sin 60° = \frac{\text{length of the side opposite the } 60° \text{ angle}}{\text{length of the hypotenuse}}$$
$$= \frac{\sqrt{3}}{2}$$

To find a decimal approximation, you can use the approximation 1.732 for $\sqrt{3}$.

$$\sin 60° = \frac{1.732}{2}, \text{ or } 0.866$$

EXAMPLE Find sin *P*.

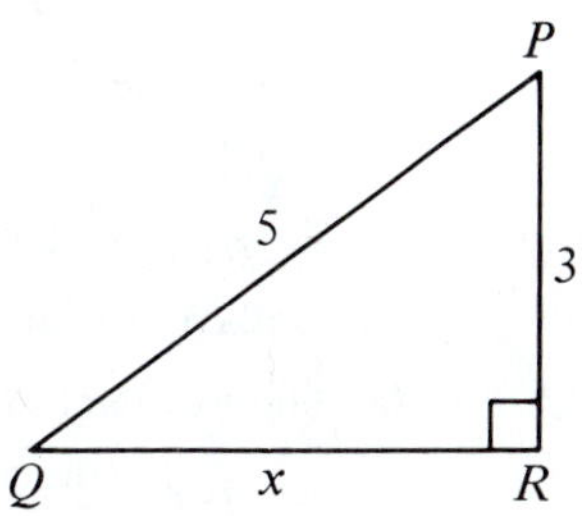

SOLUTION $$\sin P = \frac{\text{length of the side opposite } \angle P}{\text{length of the hypotenuse}}$$
$$= \frac{QR}{PQ} = \frac{x}{5}$$

Use the Pythagorean Theorem to find the value of x:

$$(PR)^2 + (QR)^2 = (PQ)^2$$
$$3^2 + x^2 = 5^2$$
$$x^2 = 16$$
$$x = 4$$

$$\sin P = \frac{4}{5}, \text{ or } 0.8$$

EXAMPLE Given that $\sin 40^\circ = 0.6428$, find the value of y to the nearest integer.

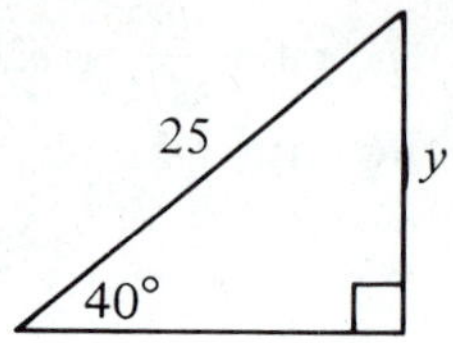

SOLUTION

$$\sin 40^\circ = \frac{\text{length of the leg opposite the } 40^\circ \text{ angle}}{\text{length of the hypotenuse}}$$

$$0.6428 = \frac{y}{25}$$

$$25(0.6428) = y$$

$$16.07 = y$$

To the nearest integer, $y = 16$.

The proportionality of the lengths of the sides of similar right triangles enables us to define other trigonometric ratios. Two of the most common ratios are the *cosine* ratio, abbreviated *cos*, and the *tangent* ratio, abbreviated *tan*. In $\triangle ABC$ with right $\angle C$, these are defined as follows:

$$\cos A = \frac{\text{length of the leg adjacent to } \angle A}{\text{length of the hypotenuse}} = \frac{AC}{AB}$$

$$\tan A = \frac{\text{length of the leg opposite } \angle A}{\text{length of the leg adjacent to } \angle A} = \frac{BC}{AC}$$

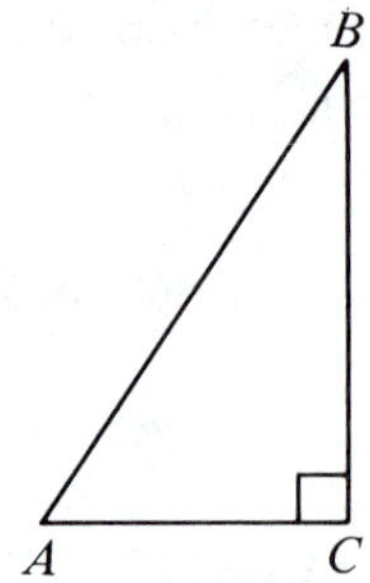

EXAMPLE Find $\cos J$ to the nearest ten-thousandth.

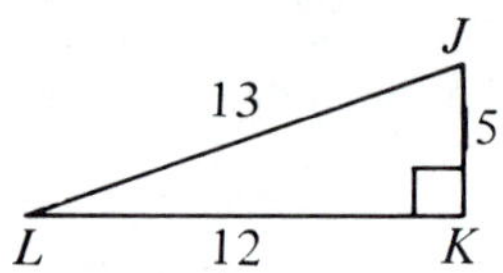

SOLUTION

$$\cos J = \frac{\text{length of the leg adjacent to } \angle J}{\text{length of the hypotenuse}} = \frac{5}{13} = 0.3846$$

EXAMPLE In $\triangle DEF$, $m\angle F = 90$, $m\angle D = 35$, and $DF = 20$. Given that $\tan 35^\circ = 0.7002$, find EF to the nearest integer.

SOLUTION Draw a diagram and let $EF = x$. Then you have:

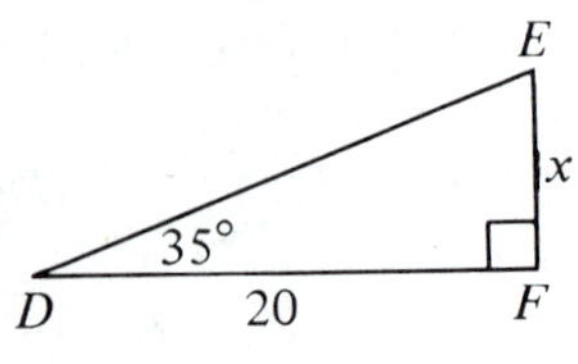

$$\tan D = \frac{\text{length of the leg opposite } \angle D}{\text{length of the leg adjacent to } \angle D}$$

$$0.7002 = \frac{x}{20}$$

$$x = 20(0.7002) = 14.004 \approx 14$$

ORAL EXERCISES

Referring to the diagram, match each trigonometric ratio with its value. Two ratios may have the same value.

1. sin Q
2. tan P
3. cos Q
4. cos P
5. tan Q
6. sin P

a. $\frac{3}{4}$

b. $\frac{3}{5}$

c. $\frac{4}{5}$

d. $\frac{4}{3}$

e. $\frac{5}{3}$

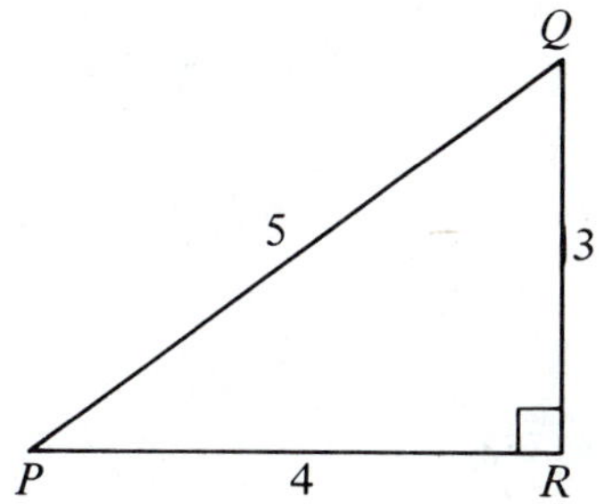

Exs. 1-8

7. a. What is the relationship between sin P and cos Q?
   b. What is the relationship between sin Q and cos P?
   c. What relationship do you think will always hold for the sine and cosine of complementary angles?
8. a. What is the relationship between tan P and tan Q?
   b. What relationship do you think will always hold for the tangents of complementary angles?
9. Explain why sin $A < 1$ and cos $A < 1$ for every acute angle A.

WRITTEN EXERCISES

Give the value of each trigonometric ratio in simplest radical form.

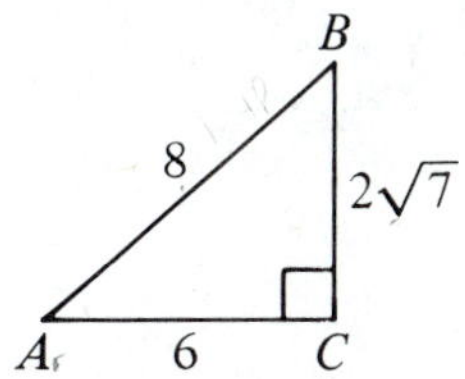

A 1. sin A 2. sin B 3. cos B

4. cos A 5. tan A 6. tan B

Find each value in simplest radical form.

7. sin 45° 8. cos 30° 9. cos 60° 10. tan 45°

Using the given lengths, find the sine, cosine, and tangent of $\angle P$ and of $\angle Q$. Leave answers as fractions in simplest form.

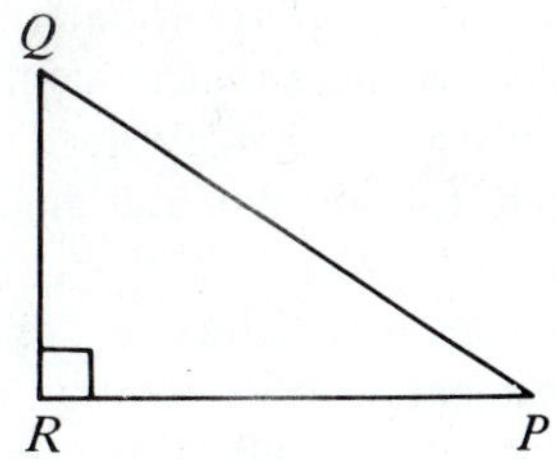

11. $QR = 6,\ PQ = 10$
12. $QR = 16,\ PQ = 20$
13. $QR = 5,\ PR = 12$
14. $PR = 8,\ PQ = 17$
15. $PR = 24,\ QR = 7$
16. $PQ = 26,\ PR = 24$

For Exercises 17-22, assume that $\sin 20° = 0.3420$, $\cos 20° = 0.9397$, and $\tan 20° = 0.3640$. Find each length to the nearest integer.

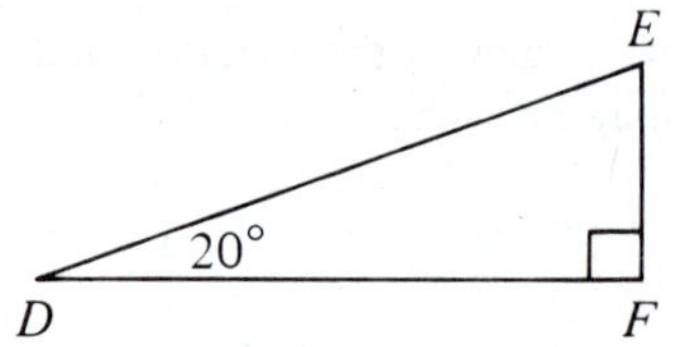

B 17. $DE = 25$, $EF =$ _?_ 18. $DF = 50$, $EF =$ _?_

19. $DE = 63$, $DF =$ _?_ 20. $EF = 171$, $DE =$ _?_

21. $DF = 125$, $EF =$ _?_ 22. $DF = 250$, $DE =$ _?_

For $\triangle JKL$ with $m\angle L = 90$ and with sides of the given lengths, find the value of the quantity $(\sin J)^2 + (\cos J)^2$.

23. $JL = 3$, $KL = 4$ 24. $JL = 12$, $JK = 13$

25. Show that $(\sin 30°)^2 + (\cos 30°)^2 = 1$.

26. For $\triangle ABC$ with $m\angle C = 90$, $BC = a$, $AC = b$, and $AB = c$, show that $(\sin A)^2 + (\cos A)^2 = 1$.

For $\triangle XYZ$ with $m\angle Z = 90$ and with sides of the given lengths, find the value of the quantity $\left(\frac{1}{\cos x}\right)^2 - (\tan x)^2$.

27. $XZ = 6$, $XY = 10$ 28. $XZ = 7$, $YZ = 24$

29. Show that $\left(\frac{1}{\cos 60°}\right)^2 - (\tan 60°)^2 = 1$.

30. For $\triangle ABC$ with $m\angle C = 90$, $BC = a$, $AC = b$, and $AB = c$, state and prove a generalization of the relationship stated in Exercise 29.

C 31. In $\triangle RST$, $m\angle T = 90$.
   a. Express $(\tan R)(\cos R)$ in terms of the lengths of sides of $\triangle RST$.
   b. Express $(\tan R)(\cos R)$ in terms of $\sin R$.
   c. Express $\tan R$ in terms of $\sin R$ and $\cos R$.

32. In $\triangle ABC$, $AB = 5$, $AC = 5$, and $BC = 6$. Find the value of:
   a. $\sin B$ b. $\cos C$ c. $\tan B$

Section 9-5 USING TRIGONOMETRY

About 150 B.C. a Greek astronomer named Hipparchus created the first known table of values of trigonometric ratios. Using these ratios, he was able to estimate celestial distances that he could not measure directly. In this section you will learn how to use a table of values of trigonometric ratios like the one Hipparchus constructed to calculate the angle measures and the lengths of the sides of a right triangle.

On page 373 there is a table of values of the trigonometric ratios, sine, cosine, and tangent, correct to four decimal places. In order to find the cosine of 35°, for example, you look in the column labeled "Cosine" for the value to the right of 35°. Thus you find that $\cos 35° = 0.8192$.

TABLE OF TRIGONOMETRIC RATIOS

Angle	Sine	Cosine	Tangent	Angle	Sine	Cosine	Tangent
1°	.0175	.9998	.0175	**46°**	.7193	.6947	1.0355
2°	.0349	.9994	.0349	**47°**	.7314	.6820	1.0724
3°	.0523	.9986	.0524	**48°**	.7431	.6691	1.1106
4°	.0698	.9976	.0699	**49°**	.7547	.6561	1.1504
5°	.0872	.9962	.0875	**50°**	.7660	.6428	1.1918
6°	.1045	.9945	.1051	**51°**	.7771	.6293	1.2349
7°	.1219	.9925	.1228	**52°**	.7880	.6157	1.2799
8°	.1392	.9903	.1405	**53°**	.7986	.6018	1.3270
9°	.1564	.9877	.1584	**54°**	.8090	.5878	1.3764
10°	.1736	.9848	.1763	**55°**	.8192	.5736	1.4281
11°	.1908	.9816	.1944	**56°**	.8290	.5592	1.4826
12°	.2079	.9781	.2126	**57°**	.8387	.5446	1.5399
13°	.2250	.9744	.2309	**58°**	.8480	.5299	1.6003
14°	.2419	.9703	.2493	**59°**	.8572	.5150	1.6643
15°	.2588	.9659	.2679	**60°**	.8660	.5000	1.7321
16°	.2756	.9613	.2867	**61°**	.8746	.4848	1.8040
17°	.2924	.9563	.3057	**62°**	.8829	.4695	1.8807
18°	.3090	.9511	.3249	**63°**	.8910	.4540	1.9626
19°	.3256	.9455	.3443	**64°**	.8988	.4384	2.0503
20°	.3420	.9397	.3640	**65°**	.9063	.4226	2.1445
21°	.3584	.9336	.3839	**66°**	.9135	.4067	2.2460
22°	.3746	.9272	.4040	**67°**	.9205	.3907	2.3559
23°	.3907	.9205	.4245	**68°**	.9272	.3746	2.4751
24°	.4067	.9135	.4452	**69°**	.9336	.3584	2.6051
25°	.4226	.9063	.4663	**70°**	.9397	.3420	2.7475
26°	.4384	.8988	.4877	**71°**	.9455	.3256	2.9042
27°	.4540	.8910	.5095	**72°**	.9511	.3090	3.0777
28°	.4695	.8829	.5317	**73°**	.9563	.2924	3.2709
29°	.4848	.8746	.5543	**74°**	.9613	.2756	3.4874
30°	.5000	.8660	.5774	**75°**	.9659	.2588	3.7321
31°	.5150	.8572	.6009	**76°**	.9703	.2419	4.0108
32°	.5299	.8480	.6249	**77°**	.9744	.2250	4.3315
33°	.5446	.8387	.6494	**78°**	.9781	.2079	4.7046
34°	.5592	.8290	.6745	**79°**	.9816	.1908	5.1446
35°	.5736	.8192	.7002	**80°**	.9848	.1736	5.6713
36°	.5878	.8090	.7265	**81°**	.9877	.1564	6.3138
37°	.6018	.7986	.7536	**82°**	.9903	.1392	7.1154
38°	.6157	.7880	.7813	**83°**	.9925	.1219	8.1443
39°	.6293	.7771	.8098	**84°**	.9945	.1045	9.5144
40°	.6428	.7660	.8391	**85°**	.9962	.0872	11.4301
41°	.6561	.7547	.8693	**86°**	.9976	.0698	14.3007
42°	.6691	.7431	.9004	**87°**	.9986	.0523	19.0811
43°	.6820	.7314	.9325	**88°**	.9994	.0349	28.6363
44°	.6947	.7193	.9657	**89°**	.9998	.0175	57.2900
45°	.7071	.7071	1.0000				

The same table can also be used to find the approximate degree measure of an angle if you *know* the value of one of its trigonometric ratios. For example, suppose you know that sin $A = 0.6$. Since 0.6 does not appear in the sine column, you look for the value in the sine column that is *closest* to 0.6. You find that 0.6 is between 0.5878, corresponding to 36°, and 0.6018, corresponding to 37°.

$$\begin{array}{r} 0.6000 \\ -0.5878 \\ \hline 0.0122 \end{array} \qquad \begin{array}{r} 0.6018 \\ -0.6000 \\ \hline 0.0018 \end{array}$$

Since 0.6 is closer to 0.6018 than to 0.5878, $m\angle A = 37$ (to the nearest degree).

EXAMPLE Find the value of x to the nearest integer.

SOLUTION The trigonometric ratio that involves the lengths of the hypotenuse of a right triangle and the leg adjacent to an acute angle is the cosine ratio.

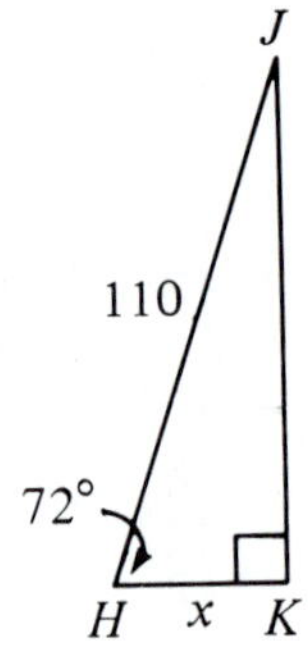

$$\cos 72° = \frac{x}{110}$$

In the table, the value of cos 72° is 0.3090.

$$0.3090 = \frac{x}{110}$$
$$x = (0.3090)110$$
$$= 33.99 \approx 34$$

EXAMPLE In $\triangle ABC$, $m\angle C = 90$, $m\angle A = 53$, and $AC = 30$. Find BC to the nearest integer.

SOLUTION Draw a diagram and let $BC = x$.

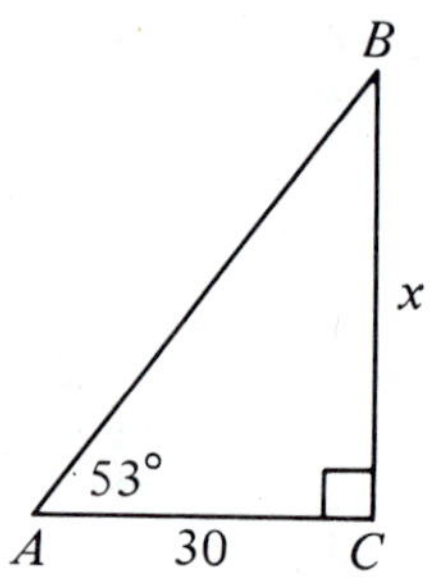

$$\tan A = \frac{BC}{AC}$$

$$\tan 53° = \frac{x}{30}$$

$$1.3270 = \frac{x}{30} \quad \text{(Find tan 53° = 1.3270 in the table.)}$$

$$x = (1.3270)(30) = 39.81 \approx 40$$

EXAMPLE Find $m\angle P$ to the nearest degree.

SOLUTION Since PQ is the length of the hypotenuse and QR is the length of the leg opposite $\angle P$, use the sine ratio.

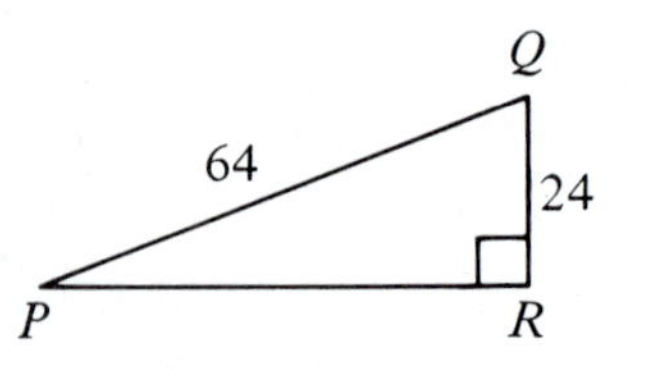

$$\sin P = \frac{QR}{PQ}$$
$$= \frac{24}{64}$$
$$= 0.375$$

The entry in the sine column of the table on page 373 closest to this value is 0.3746. Since 0.3746 is the sine of 22°, $m\angle P = 22$ (to the nearest degree).

EXAMPLE In $\triangle ABC$, $m\angle C = 90$, $AC = 5$, and $BC = 8$. Find $m\angle A$ to the nearest degree.

SOLUTION Draw a diagram.

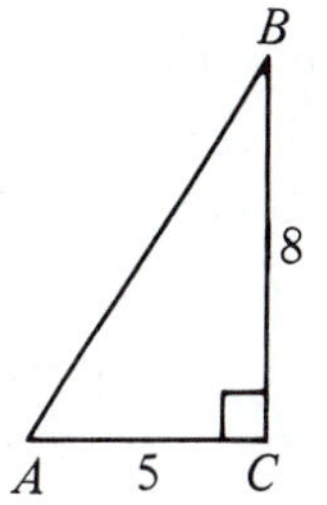

$$\tan A = \frac{BC}{AC}$$
$$= \frac{8}{5}$$
$$= 1.6$$

From the table you find that $\tan 58^\circ = 1.6003$, and so to the nearest degree, $m\angle A = 58$.

In describing physical situations involving angles, the terms "angle of elevation" and "angle of depression" are often used. An *angle of elevation* is the angle between a horizontal line and a line of sight *above* the horizontal. In the diagram, if a person at point C looks up at point A, then $\angle ACX$ is the angle of elevation of point A. An *angle of depression* is the angle between a horizontal line and a line of sight *below* the horizontal. If a person at point C looks down at point B, then $\angle BCX$ is the angle of depression of point B.

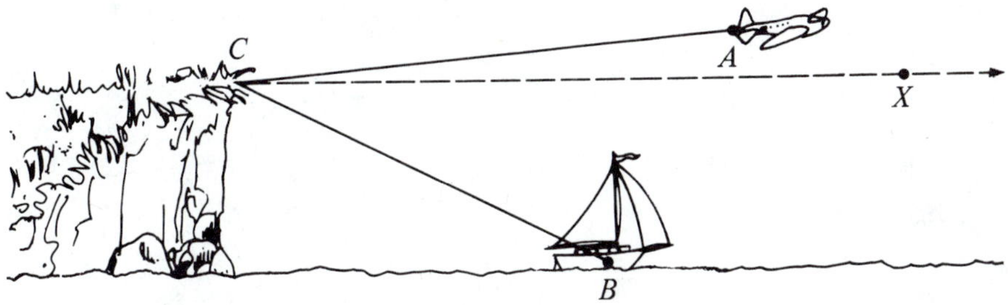

EXAMPLE The angle of elevation of the top of a building from a point on the ground 200 m from the base of the building is 36°. To the nearest meter, how tall is the building?

SOLUTION Draw a diagram.

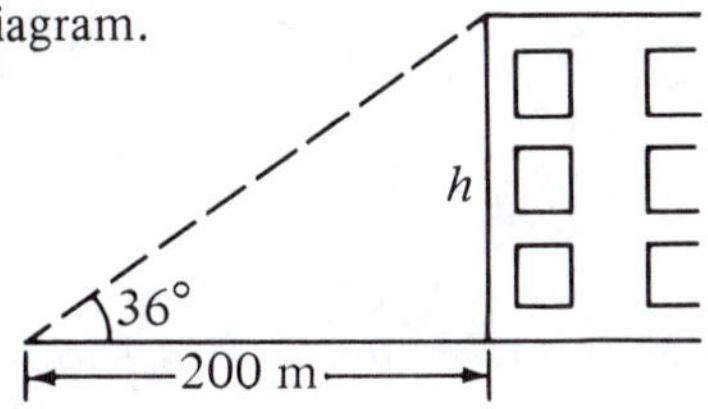

Let h = the height in meters of the building.
Then you have:

$$\tan 36^\circ = \frac{h}{200}$$
$$0.7265 = \frac{h}{200}$$
$$h = (200)(0.7265) = 145.3$$

To the nearest meter, the height of the building is 145 m.

ORAL EXERCISES

For each right triangle, state an equation that you could use to find the value of x. You need not solve the equation.

1.
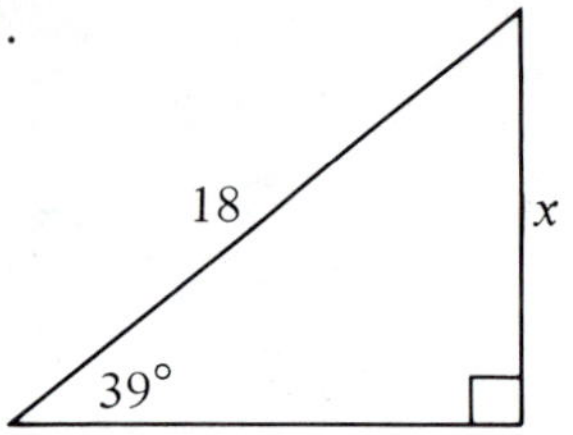

2.
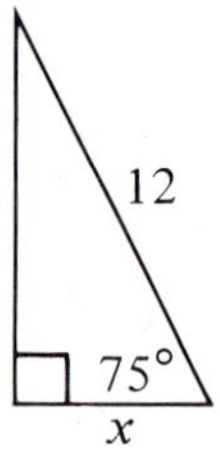

3.
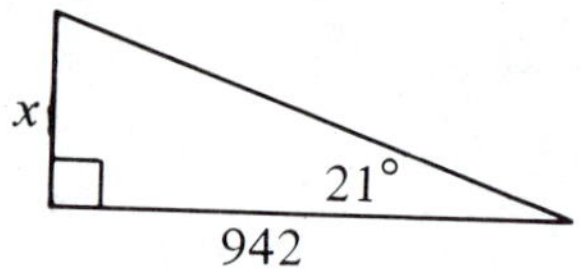

4.
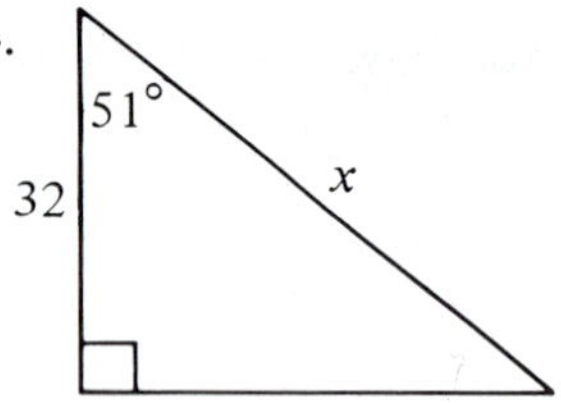

5.
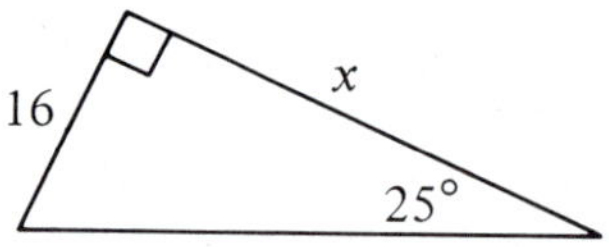

6.
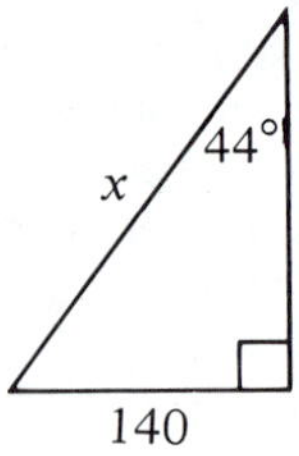

7.
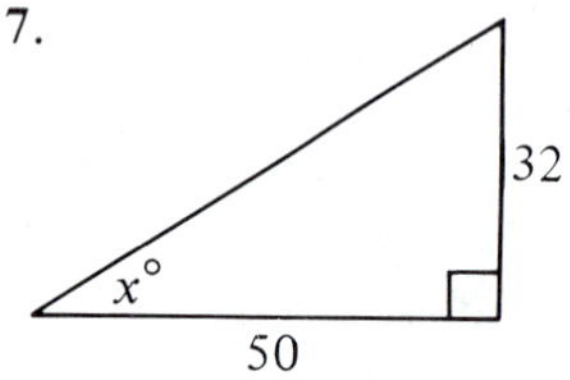

8.
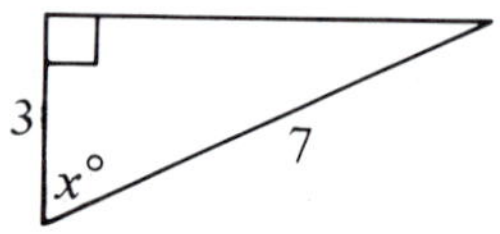

9.
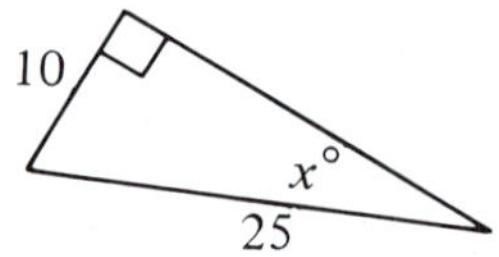

Complete.

10. $\sin 76° = \underline{\ ?\ }$ **11.** $\tan 78° = \underline{\ ?\ }$ **12.** $\cos 23° = \underline{\ ?\ }$

13. If $\cos A = 0.1045$, then $m\angle A = \underline{\ ?\ }$. **14.** If $\sin B = 0.4848$, then $m\angle B = \underline{\ ?\ }$.

15. If $\tan C = 0.8098$, then $m\angle C = \underline{\ ?\ }$. **16.** If $\sin R = 0.75$, then $m\angle R = \underline{\ ?\ }$.

WRITTEN EXERCISES

In these exercises find all lengths to the nearest integer and all angles to the nearest degree (unless otherwise specified).

Find the length of the indicated side of $\triangle PQR$. Draw a diagram for each exercise, and use the table on page 373.

A 1. $m\angle Q = 90$, $m\angle P = 38$, $PR = 25$, $PQ = \underline{\ ?\ }$

2. $m\angle Q = 90$, $m\angle P = 13$, $PR = 40$, $QR = \underline{\ ?\ }$

3. $m\angle P = 90$, $m\angle Q = 20$, $PQ = 75$, $PR = \underline{\ ?\ }$

4. $m\angle P = 90$, $m\angle R = 32$, $QR = 125$, $PR = \underline{\ ?\ }$

5. $m\angle R = 90$, $m\angle P = 31$, $PQ = 64$, $QR = \underline{\ ?\ }$

6. $m\angle R = 90$, $m\angle Q = 68$, $QR = 70$, $PR =$ __?__

7. $m\angle Q = 90$, $m\angle P = 53$, $QR = 197$, $PR =$ __?__

8. $m\angle Q = 90$, $m\angle P = 40$, $PQ = 383$, $PR =$ __?__

9. $m\angle P = 90$, $m\angle R = 23$, $PQ = 101$, $PR =$ __?__

10. $m\angle P = 90$, $m\angle Q = 4$, $PR = 12$, $PQ =$ __?__

Find the measure of the indicated angle of $\triangle DEF$ with $m\angle F = 90$.

11. $DE = 120$, $DF = 45$, $m\angle D =$ __?__

12. $DE = 80$, $DF = 47$, $m\angle E =$ __?__

13. $DF = 40$, $EF = 53$, $m\angle D =$ __?__

14. $DF = 25$, $EF = 40$, $m\angle E =$ __?__

15. $EF = 360$, $DE = 400$, $m\angle E =$ __?__

16. $EF = 82$, $DE = 125$, $m\angle D =$ __?__

B 17. The angle of elevation of the top of a building from a point on the ground 51 m from the base of one wall is 63°. How tall is the building?

18. The sloping walls of a tent each make an angle of 55° with the ground. Each wall measures 2.2 m from the ground to the ridgepole (top) of the tent. What is the height of the tent, to the nearest *tenth* of a meter?

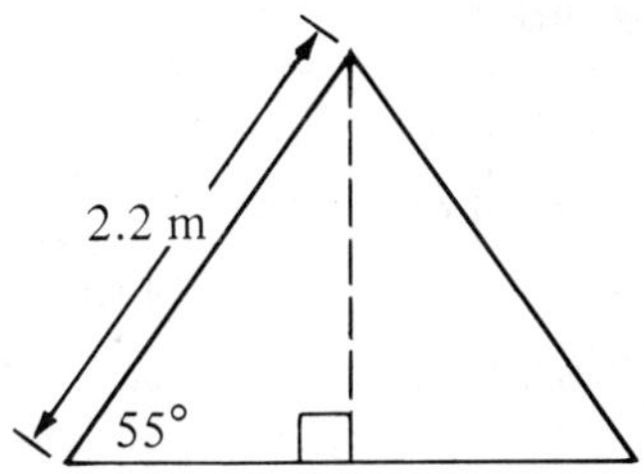

19. A sunken pirate ship on the ocean floor is sighted at an angle of depression of 42° from a sailboat. After sailing 50 m, the sailboat is directly above the sunken ship. How deep is the ocean at this point?

20. A guy wire for a radio tower makes an angle of 73° with the ground. The vertex of this angle is 24 m from the base of the tower. How long is the wire?

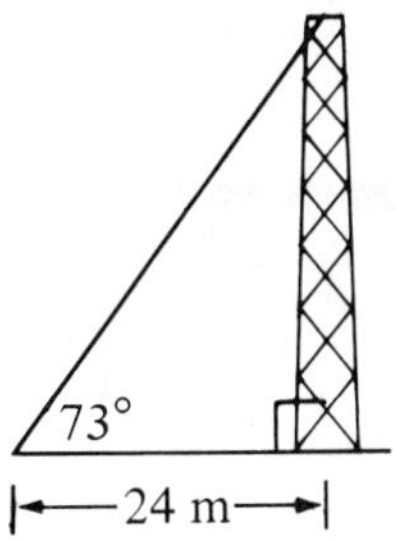

Ex. 20

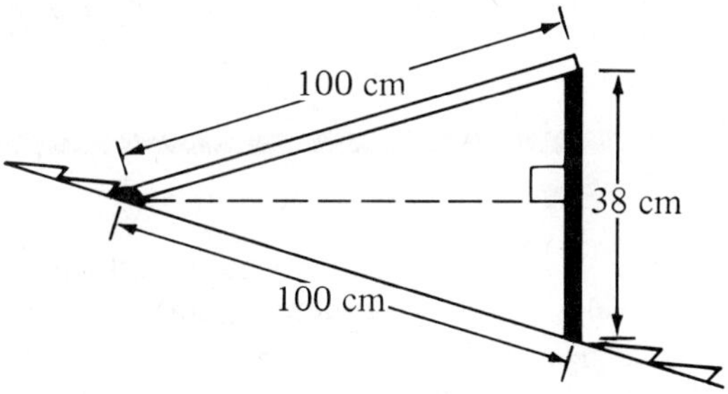

Ex. 21

21. A hinged roof window is 100 cm long and is to be held open by a rod 38 cm long. What angle does the open window make with the roof? (*Hint:* Find half the measure of this angle first.)

22. The sides of a rhombus are each 75 cm long, and the measure of one of its angles is 76. Find the length of the longer diagonal of the rhombus.

23. Each diagonal of a rectangle has length 12, and one of the acute angles formed by the two diagonals has measure 40. Find the length and width of the rectangle.

C 24. The sides of a rectangle have lengths 14.0 and 4.8. Find the measure of one of the acute angles formed by the diagonals of the rectangle.

25. The lengths of the sides of a triangle are in the ratio 4 : 4 : 7. Find the measures of its angles.

26. To the nearest integer, find the area of $\triangle JKL$, shown at the right.

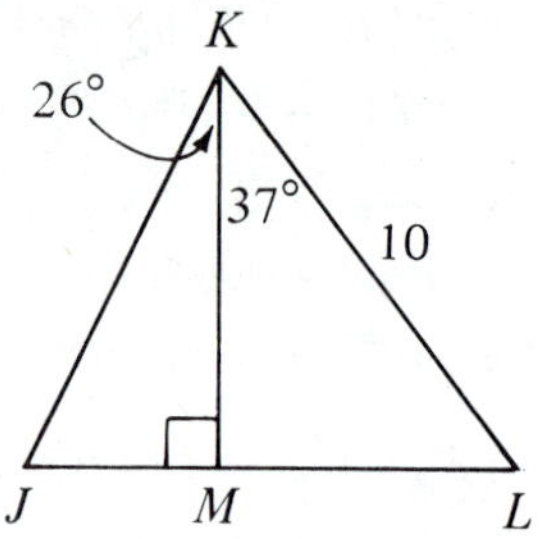

SELF-TEST 2

Find the value of each trigonometric ratio in simplest radical form.

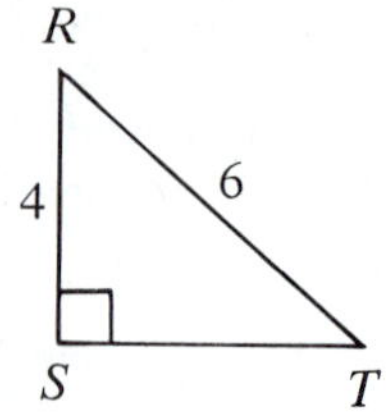

Exs. 1-5

Section 9-4

1. sin R 2. sin T

3. cos R 4. tan T

In Exercises 5-7, use the table on page 373.

Section 9-5

5. Find the measure of $\angle R$ to the nearest degree.

6. Find (a) sin 37° and (b) cos 53°.

7. Anne, standing at point A on one side of a river, sights a tree on the opposite side of the river. Her line of sight makes a 28° angle with the riverbank. Bob, at point B, is standing directly across from the tree. If AB = 500 m, find the width of the river to the nearest meter.

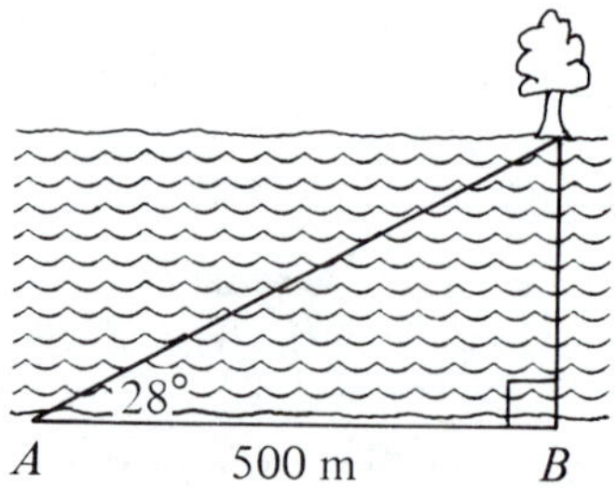

Computer Exercises

The sine, cosine, and tangent of any angle are given in BASIC by SIN(A), COS(A), and TAN(A), respectively, where A is the measure of the angle. In BASIC, the measure of the angle is given in units called *radians*. In more advanced courses you will use radians, as well as degrees, to express angle measure. It is useful to know that 1 degree $\approx$ 0.017453 radians

1 radian $\approx$ 57.2958 degrees

Thus if the degree measure of an angle is 82, then the radian measure is approximately 82(0.017453), or 1.431146.

1. The following program will cause the computer to print a table of approximate values for sine, cosine, and tangent for angles from 5° to 45° in increments of 5°. Note that some computers will print sin 5° as 8.71543E-02. This is an example of *scientific notation* in BASIC. The "-02" following the "E" means that the decimal point must be moved 2 places to the left in order to express the number in its usual decimal form. Thus, 8.71543E-02 = 0.0871543.

Copy and RUN the program.

```
10   PRINT "DEG"; TAB(10); "SINE"; TAB(20); "COS"; TAB(30); "TAN"
20   FOR D=5 TO 45 STEP 5
30   LET R=D*.017453                (Conversion from degrees to radians)
40   PRINT D; TAB(6); SIN(R); TAB(17); COS(R); TAB(29); TAN(R)
50   NEXT D
60   END
```

2. Modify line 20 of the program in Exercise 1 so that the trigonometric values are printed for angles of measure 5° to 85° in increments of 5°. RUN the program.

   Compare the values in this new table to the corresponding values in the Table of Trigonometric Ratios on page 373. Are corresponding values approximately the same?

3. Modify line 20 of the program in Exercise 1 so that the trigonometric values are printed for angles of measure 1° to 43° in increments of 3°. RUN the program.

CHAPTER REVIEW

Complete.

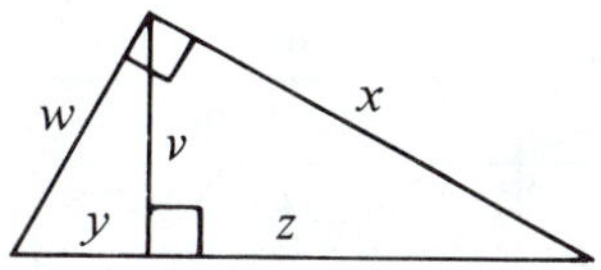

Exs. 1–4

1. v is the geometric mean between __?__ and __?__. Section 9-1
2. x is the geometric mean between __?__ and __?__.
3. $wx = (\underline{\ ?\ }) \cdot (\underline{\ ?\ })$
4. If $w = 65$ and $y = 25$, find the values of z, v, and x.

5. Is a triangle whose sides have length 1, 2, and $\sqrt{3}$ a right triangle? Explain. Section 9-2
6. A right triangle has a hypotenuse of length 53 and a leg of length 28. Find its perimeter.
7. The diagonals of a rhombus are 16 cm long and 30 cm long. How long is each side?

8. $PQRS$ is a square and $PT = ST = 5$. Find the area of pentagon $PQRST$. Section 9-3

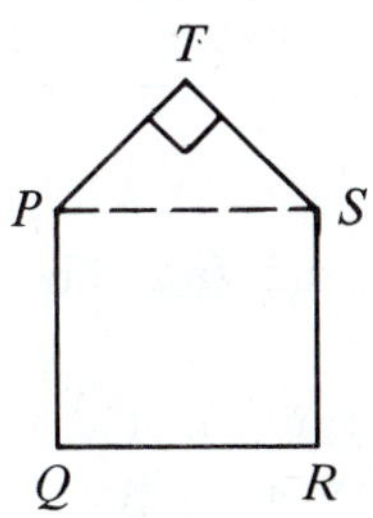

Ex. 8

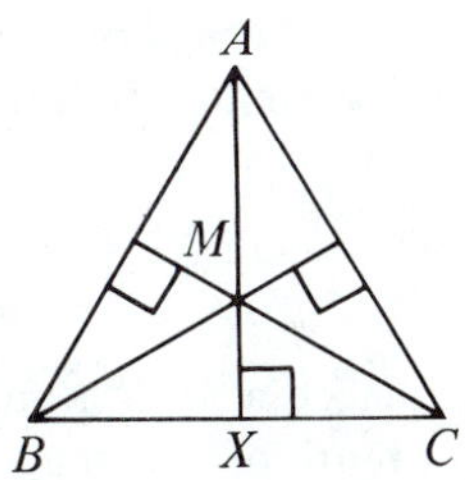

Ex. 9

9. $\triangle ABC$ is equilateral and its altitudes intersect at point M. If $AB = 12$, find XA, XM, and MA. (*Hint:* What is the measure of $\angle MBX$?)

Find each ratio in simplest radical form.

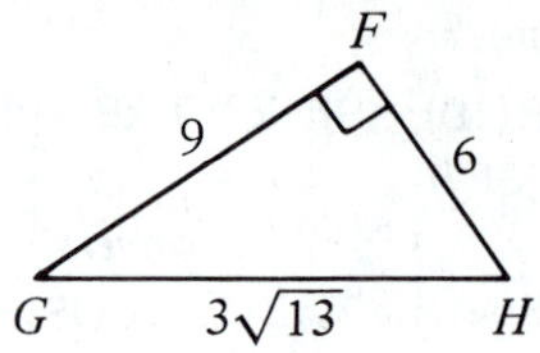

10. $\cos G$ **11.** $\sin H$ Section 9-4

12. $\tan H$ **13.** $\tan G$

14. Evaluate $(\sin 45°)^2 - (\cos 45°)^2$.

In Exercises 15-17, use the table on page 373.

15. Find sin 84° and cos 84°. Section 9-5

16. If $\tan A = \frac{7}{25}$, find the measure of $\angle A$ to the nearest degree.

17. Tom is sitting at the top of Mount Rudolf, which is 1.6 km high. The angle of depression of his house from where he sits is 80°. Find the distance, to the nearest kilometer, from Tom to his house.

CHAPTER TEST

Use the table on page 373 as needed.
Complete, giving each answer in simplest form.

	AB	*BC*	*AC*	*AD*	*DC*	*BD*
1.	15	20	?	?	?	?
2.	?	?	?	32	112.5	?
3.	?	$4\sqrt{3}$	?	2	?	?

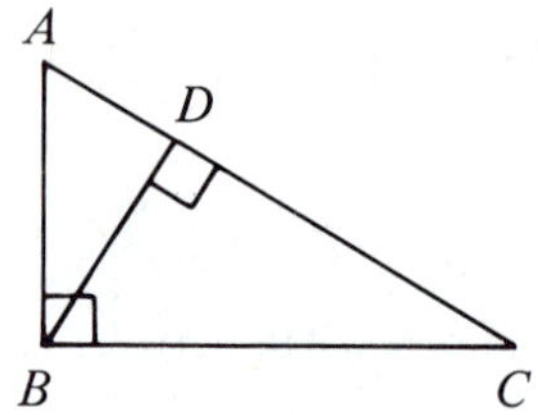

Exs. 1-5

4. If $AB = 43$ and $AC = 75$, find the measure of $\angle A$ to the nearest degree.

5. If $m\angle DBC = 62$ and $BC = 20$, find DC and BD to the nearest integer.

6. A parallelogram has sides of length 10 and 24 and a diagonal of length 26. Is the parallelogram a rectangle? Justify your answer.

7. Find the area of isosceles trapezoid *PQRS*. (*Hint:* Draw altitudes from *R* and *S*.)

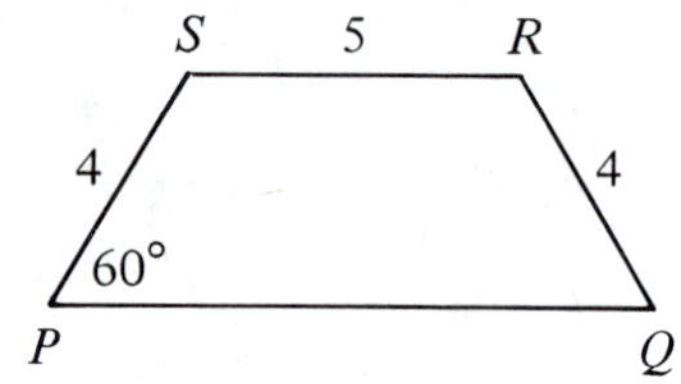

8. Find the value of tan 30° in simplest radical form.

9. A square has sides 6 cm long. Find the perimeter and the area of the quadrilateral formed by joining the midpoints of the sides of the square.

10. In $\triangle XYZ$, $m\angle Y = 90$, $XY = 6$, and $XZ = 8$. Find $\cos X$, $\tan X$, and $\tan Z$, in simplest form.

11. When a ladder 6 m long is placed against a building, it meets the ground at an angle of 70°. To the nearest tenth of a meter, find the distance from the top of the ladder to the ground.

12. In a certain right triangle, the length of the altitude to the hypotenuse is 10 cm. The hypotenuse is 25 cm long. Find the lengths of the segments into which the altitude divides the hypotenuse.

Biographical Note

Pythagoras

The famous theorem that states the relationship among the sides of a right triangle has long been attributed to Pythagoras, a Greek philosopher and mathematician who lived during the sixth century B.C.

Pythagoras was born on the Greek island of Samos, off the western coast of Turkey about 580 B.C. As a young man he seems to have traveled extensively, eventually settling in Crotona, a Greek seaport in southern Italy. Widely respected for his knowledge of mathematics, philosophy, music, and astronomy, he soon had many followers. Because Pythagoras was so venerated by his students, they often gave him credit for their discoveries, making it impossible for us to separate his achievements from those of his followers.

The Pythagoreans, as Pythagoras and the members of the Pythagorean school came to be known, thought that everything in nature could be explained using numbers. They made many contributions to the field of mathematics. Their influence changed how mathematics was understood and studied. They made geometry a part of a well-rounded education and continued the trend toward giving mathematics independence as a field of study separate from such fields as calculation and geodesy.

The Pythagoreans systematized, stated, discovered, and proved many mathematical concepts. They are generally considered to have provided the foundation for the development of number theory. In representing numbers by patterns of dots, they linked arithmetic and geometry. They are credited with discovering many relationships between musical intervals and ratios. They are also credited with important work regarding parallel lines, real roots of quadratic equations, and properties of polygons and regular solids.

Although Pythagoras died around 500, the Pythagoreans continued their studies until the middle of the fourth century B.C. As you continue to study mathematics, you will use many of the discoveries they made.

VOCABULARY REVIEW

Be sure that you understand the meaning of these terms:

sine ratio, p. 369
cosine ratio, p. 370
tangent ratio, p. 370
angle of elevation, p. 375
angle of depression, p. 375

MIXED REVIEW

Is the triangle whose sides have the given lengths a right triangle?

1. 6, 8, 11

2. 12, 15, 9

3. Find the length of a diagonal of a square with sides of length $5\sqrt{2}$.

4. Write the **(a)** converse, **(b)** inverse, and **(c)** contrapositive of the following in symbolic form: $p \rightarrow \sim q$

Exercises 5-7 refer to $\triangle ABC$ in the figure. Give the value of each trigonometric ratio in simplest radical form.

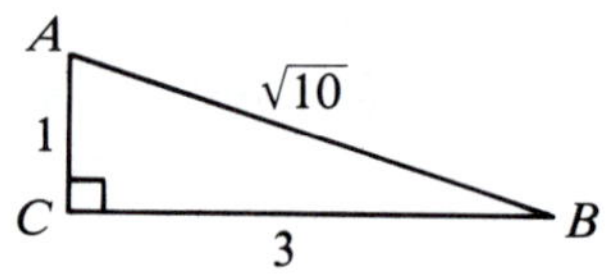

5. $\sin B$

6. $\cos A$

7. $\tan B$

8. Find the value of x in $\triangle DEF$.

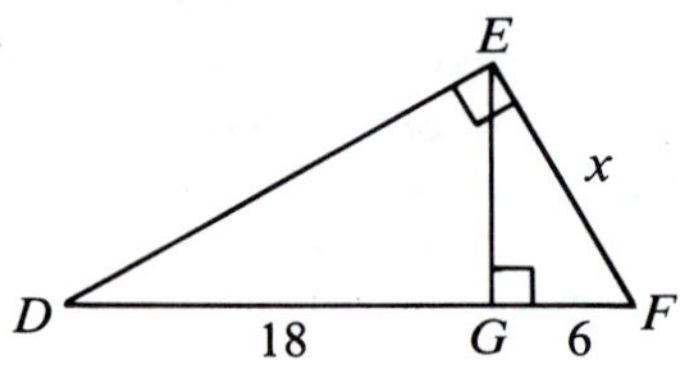

9. A tank can be filled in 2 hours when two pipes are used simultaneously. If the smaller pipe is used alone, it takes 3 hours more than the larger pipe. How long does it take the larger pipe working alone to fill the tank?

Solve.

10. $b + 1 = \dfrac{4}{b + 1}$

11. $6 \leqslant 2(x + 5) - 6 < 10$

12. $\dfrac{y}{8} = \dfrac{3}{6}$

13. $(2y - 3)^2 = 25$

14. Express in simplest form: $2\frac{1}{3} : 1\frac{1}{2}$

15. The ratio of the areas of two similar hexagons is 16:25. The length of one side of the larger hexagon is 40 cm. Find the length of the corresponding side of the smaller hexagon.

16. Use a truth table to show that $(p \wedge q) \rightarrow p$ is a tautology.

Simplify.

17. $\frac{n-3}{n-1} - \frac{n+2}{n-2}$

18. $\frac{-2}{\sqrt{10}}$

19. $(10-\sqrt{3})(10+\sqrt{3})$

20. $\frac{a^2-16}{a^2+13a+36}$

21. A monument is 48 m high. The angle of depression from the top of the monument to a road marker is 26°. Find the distance from the base of the monument to the marker to the nearest meter.

22. Steve travels 30 km/h faster than Ed and takes one hour less to travel 180 km. What are their rates of speed?

23. Let p represent "a rose is a rose" and q represent "$5 + 2 > 8$." Translate the following statement into symbolic notation and give its truth value: $5 + 2 \leqslant 8$ or a rose is not a rose.

24. Factor completely: $x^2y^3 - x^5$

25. A helicopter 35 m above sea level sites two boats in the water at points A and B. The angles of depression to the boats are 55° and 40°. If the boats and the helicopter are in the same vertical plane, and the helicopter is between the boats, find the distance between the boats to the nearest meter.

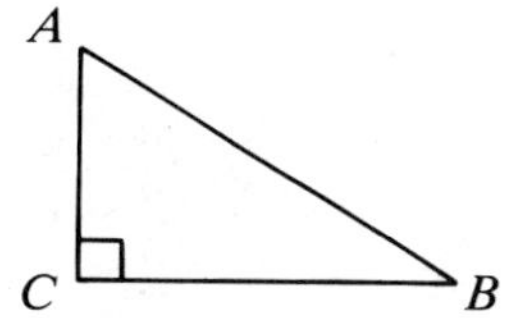

26. In right $\triangle ABC$, leg $\overline{BC}$ is 35 cm longer than leg $\overline{AC}$. If the hypotenuse is 85 cm, find the lengths of the legs.

27. In right $\triangle ABC$, $AC = 10$ cm and $\overline{AB}$ is 2 cm longer than $\overline{BC}$. Find the length of the hypotenuse.

Write two-column proofs.

28. Given: Isosceles trapezoid $ACEG$ with $\overline{AG} \parallel \overline{CE}$; $\overline{BF}$ is the median of $ACEG$; D is the midpoint of $\overline{CE}$.
Prove: $\triangle DBF$ is isosceles.

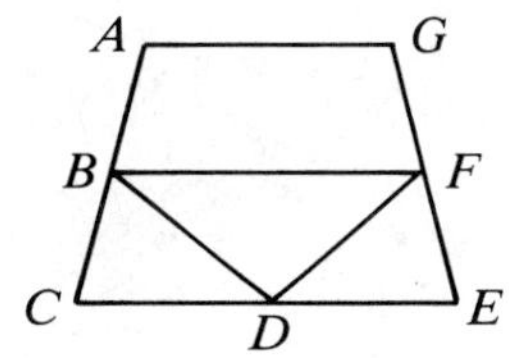

29. Given: $\angle 1 \cong \angle 4$; $\angle 2 \cong \angle 3$
Prove: $ABCD$ is a parallelogram.

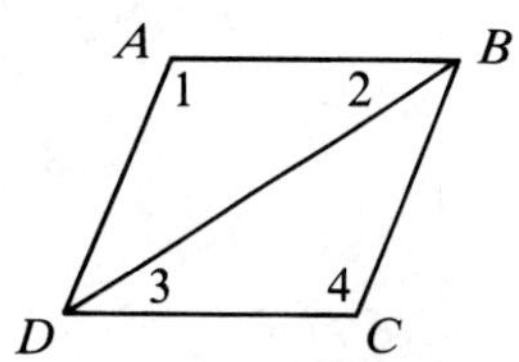

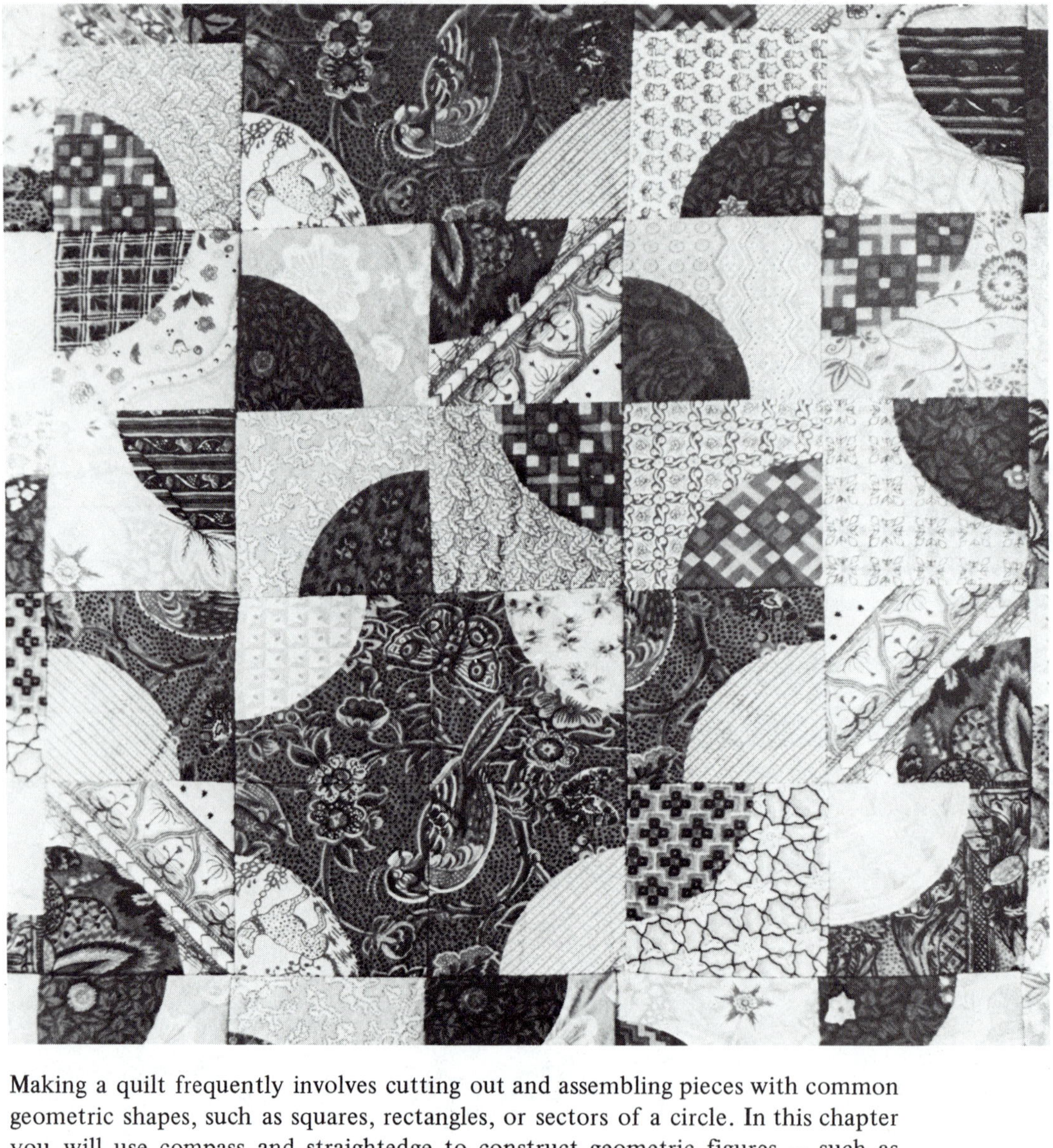

Making a quilt frequently involves cutting out and assembling pieces with common geometric shapes, such as squares, rectangles, or sectors of a circle. In this chapter you will use compass and straightedge to construct geometric figures – such as angles, triangles, and circles – having specified properties.

Constructions and Loci 10

Ancient Greek mathematicians used a straightedge and a compass to do the same geometric constructions you will study in this chapter. In fact, three of history's most famous construction problems were first proposed over 2000 years ago. These constructions are:

(1) Squaring a circle (constructing a square equal in area to that of a given circle)

(2) Doubling a cube (constructing a cube whose volume is twice that of a given cube)

(3) Trisecting an angle (dividing a given angle into three congruent angles)

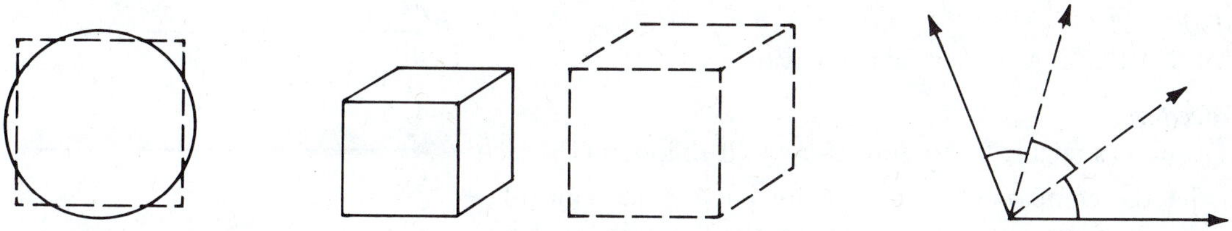

(1) (2) (3)

It was not until the nineteenth century that these three constructions were proved to be impossible using straightedge and compass alone. Even though mathematicians were unsuccessful in performing these constructions, their efforts led to discoveries in areas of mathematics other than Euclidean geometry. In fact, the methods finally used to prove that these constructions were impossible without additional tools involved algebra, *not* geometry.

Section 10-1 STRAIGHTEDGE-AND-COMPASS CONSTRUCTION

In this section you will review the uses of the two basic tools involved in constructing geometric figures: the straightedge and the compass. Using these tools, you will review how to construct:

- a line segment congruent to a given line segment
- an angle congruent to a given angle

A ruler is the most common straightedge now in use. The graduations marked on the ruler were not used in the classical Greek constructions that you will learn in this chapter, and using them does not make the constructions easier to perform. Therefore, do not use the markings on the ruler in your constructions.

A compass is commonly used to draw a circle. You may recall that a *circle* is the set of points (in a plane) a given distance from a given point, called the *center* of the circle. (Note that the center of a circle is not a point of the circle.) A *diameter* of a circle is any segment that passes through the center of the circle and whose endpoints are points of the circle. The word *diameter* may also refer to the length of any such segment. A *radius* of a circle is any segment with the center and a point on the circle as endpoints. The word *radius* may also refer to the length of any such segment. The *radius of a compass* (OP in the diagram) is the distance between the metal point of the compass and the tip of the pencil.

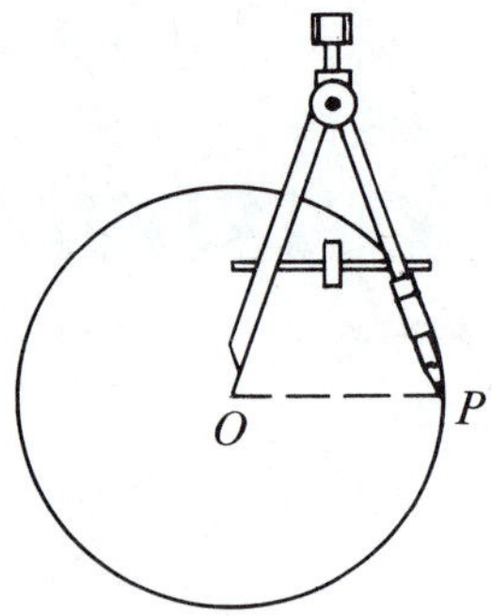

In constructions you often need to draw only part of a circle, called an *arc* of the circle. In the diagram, points A, B, and C lie on an arc with center O. All the points on an arc, as on a circle, are the same distance from (or *equidistant* from) the center. Therefore $OA = OB = OC$.

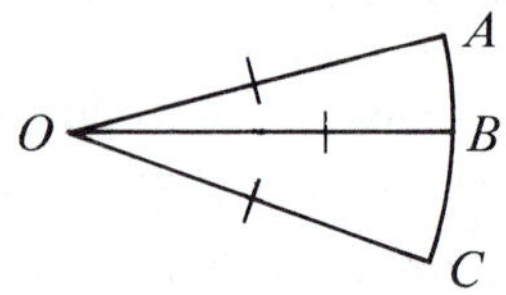

The first construction illustrates how a straightedge and a compass may be used.

Construction 1 Construct a line segment congruent to a given segment.

GIVEN: $\overline{AB}$
CONSTRUCT: A segment congruent to $\overline{AB}$

Procedure:

1. Use the straightedge to draw a ray with endpoint C.

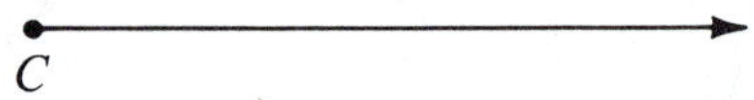

2. Set the compass to radius AB by putting the point of the compass on A and adjusting the compass so that the tip of the pencil lies on B.
3. With the point of the compass on C, draw an arc of radius AB that intersects the ray. Label the point of intersection as D.

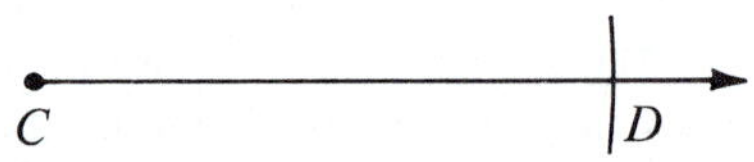

$\overline{CD} \cong \overline{AB}$

Justification: Since $CD = AB$, $\overline{CD} \cong \overline{AB}$.

Although you will not always be asked to justify your constructions, you should be able to do so. In this book, a justification will be an informal proof establishing that a particular construction does in fact satisfy the conditions of the problem.

EXAMPLE

GIVEN: Segments of lengths a, b, and c
CONSTRUCT: A triangle with sides of these lengths

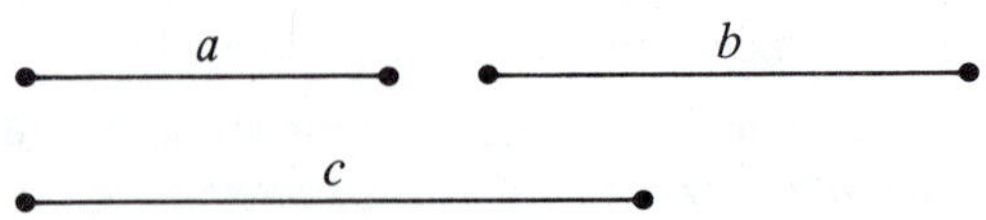

Procedure:

1. Use Construction 1 to construct $\overline{PQ}$ such that $PQ = c$.

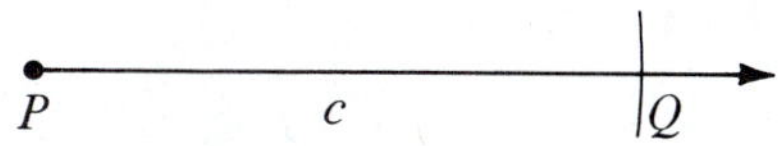

2. With P as center draw an arc of radius a.
3. With Q as center, draw an arc of radius b. Label the point of intersection of the arcs as R.
4. Use the straightedge to draw $\overline{PR}$ and $\overline{QR}$.

$\triangle PQR$ is the desired triangle.

Justification: $PR = a$, $QR = b$, and $PQ = c$ by construction.

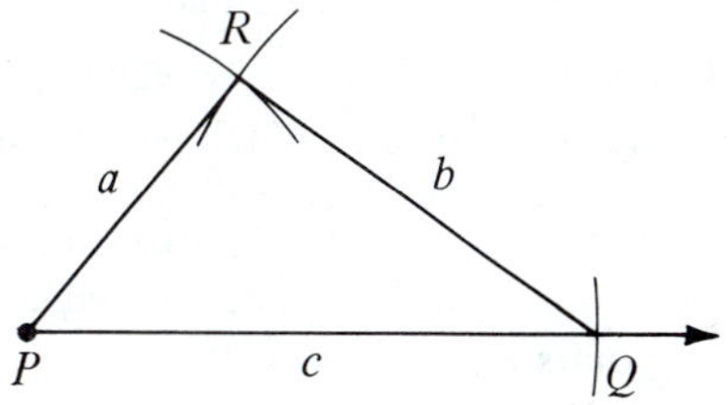

It is important that your completed construction contain all the arcs that you drew at each step. The arcs indicate whether you have done the construction correctly. Therefore, do not erase your construction arcs after you have completed your construction.

Construction 2 Construct an angle congruent to a given angle.

GIVEN: $\angle A$

CONSTRUCT: An angle congruent to $\angle A$

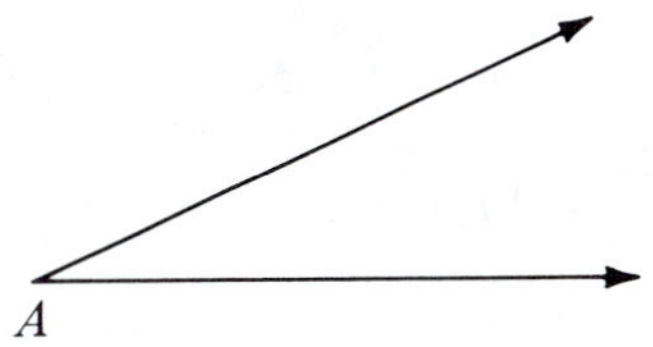

Procedure:

1. Use the straightedge to draw a ray with endpoint P.

2. With A as center and any convenient radius, draw an arc that intersects both sides of $\angle A$. Label the points of intersection as B and C.

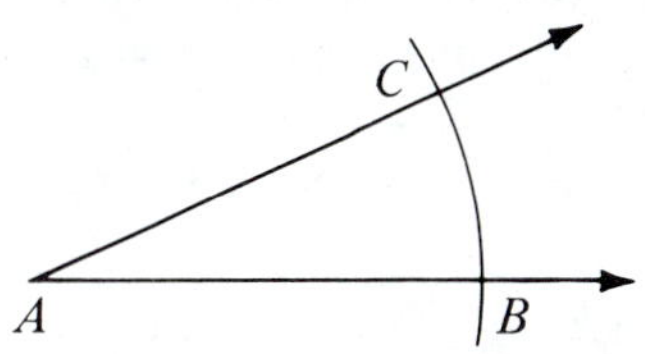

3. Using the *same radius* as in Step 2, draw an arc with center P that intersects the ray at a point Q.

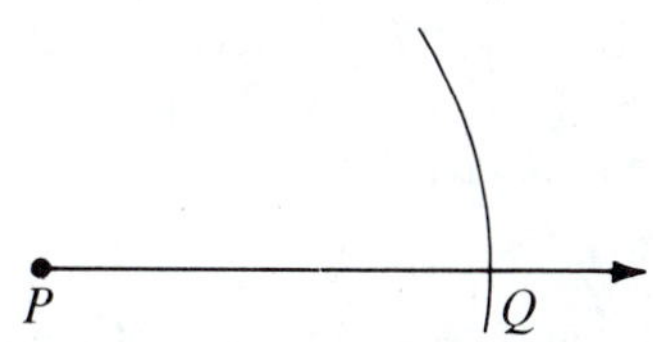

4. With Q as center and radius BC, draw an arc that intersects the first arc at R.
5. Draw $\overrightarrow{PR}$.

$\angle P \cong \angle A$

Justification: See Oral Exercise 5.

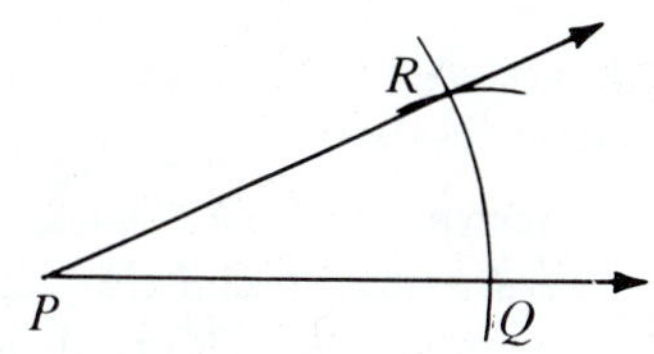

ORAL EXERCISES

1. In the diagram, $DE > DF$. Describe a procedure for constructing a point G on $\overline{DE}$ such that $\overline{DG} = \overline{DF}$.

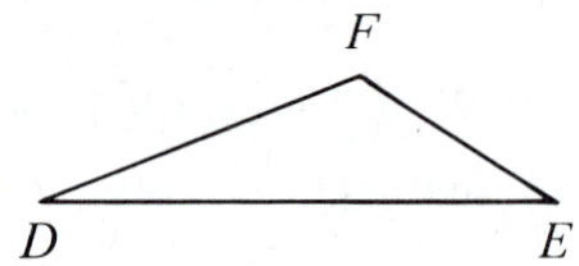

2. Given two acute angles, $\angle A$ and $\angle B$, describe a method of constructing an $\angle C$ whose measure is $(m\angle A + m\angle B)$.

3. Point O is the center of the circle shown in the diagram at the right. Two lines through O intersect the circle at points A, B, C, and D.
   a. Name four congruent segments. Why are they congruent?
   b. If $\overline{AB}$, $\overline{BC}$, $\overline{CD}$, and $\overline{DA}$ are drawn, what kind of quadrilateral is $ABCD$? Explain.

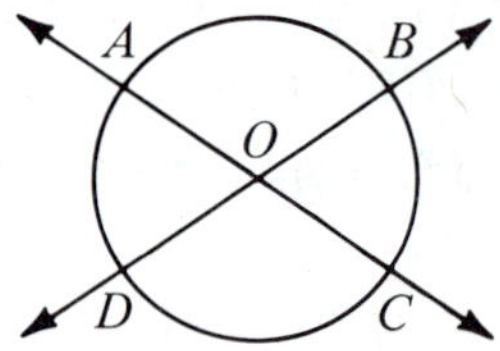

4. Is it possible to construct a triangle with sides of lengths a, b, and c, as shown at the right? Explain.

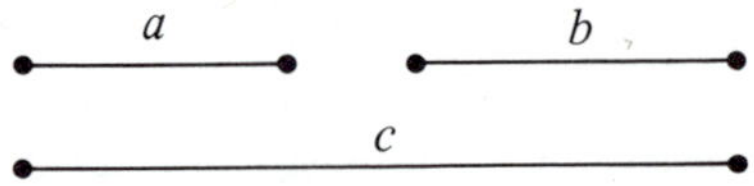

5. $\angle P$ has been constructed by the procedure in Construction 2 so that $\angle P \cong \angle A$. Supply the reason for each statement in the following justification of the construction.
   a. Draw $\overline{BC}$ and $\overline{QR}$.
   b. $PQ = AB$; $PR = AC$; $QR = BC$
   c. $\triangle PQR \cong \triangle ABC$
   d. $\angle P \cong \angle A$

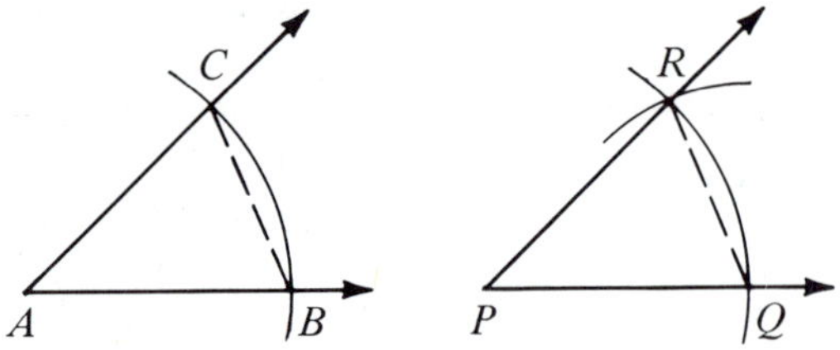

WRITTEN EXERCISES

In Exercises 1-6, begin by drawing line segments like those shown. Then construct segments of the given lengths.

A 1. a 2. b 3. $a + b$ 4. $b - a$ 5. $2a$ 6. $2b - c$

In Exercises 7-10, begin by drawing angles like those shown. Then construct angles with the given measures.

7. $m\angle P$ 8. $180 - m\angle Q$

9. $m\angle P + m\angle Q$ 10. $m\angle P - m\angle Q$

In Exercises 11-14, begin by drawing line segments and angles like those used in Exercises 1-10.

11. Construct a $\triangle XYZ$ such that $XY = a$, $YZ = b$, and $XZ = c$. Suppose you were to do this construction ten times. Which congruence postulate or theorem guarantees that all ten triangles would be congruent?

12. Repeat Exercise 11, using $XY = a$, $YZ = c$, and $m\angle Y = m\angle Q$ to construct $\triangle XYZ$.

13. Repeat Exercise 11, using $XY = c$, $m\angle X = m\angle P$, and $m\angle Y = m\angle Q$ to construct $\triangle XYZ$.

14. Construct an equilateral triangle with sides of length b.

15. Begin by drawing an acute $\triangle DEF$ and $\overline{PQ}$ so that $PQ > DE$. Construct $\triangle PQR$ such that $\triangle PQR \sim \triangle DEF$. Which similarity postulate or theorem would you use to prove that $\triangle PQR \sim \triangle DEF$?

16. Construct a rectangle using the method suggested in Oral Exercise 3.

B 17. Draw an acute $\triangle GHI$. Construct $\triangle LMN$ so that $LM = 2GH$, $LN = 2GI$, and $MN = 2HI$. Why is $\triangle LMN \sim \triangle GHI$? What is the ratio of the areas of the two triangles?

18. Draw an acute $\triangle ABC$. Label a point on $\overline{AB}$ as D. Construct $\triangle ADE$ with E on $\overline{AC}$ so that $\triangle ADE \sim \triangle ABC$.

19. Construct an angle of measure 120.

20. Draw segments like those shown. Then construct a parallelogram with sides of lengths x and y and a diagonal of length z. Justify your construction.

21. Construct a regular hexagon. A simple method is suggested by the diagram at the right. Justify your construction.

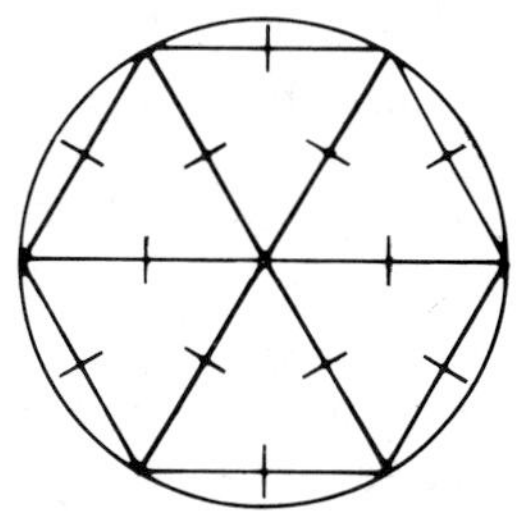

In Exercises 22 and 23, draw segments and angles like those shown.

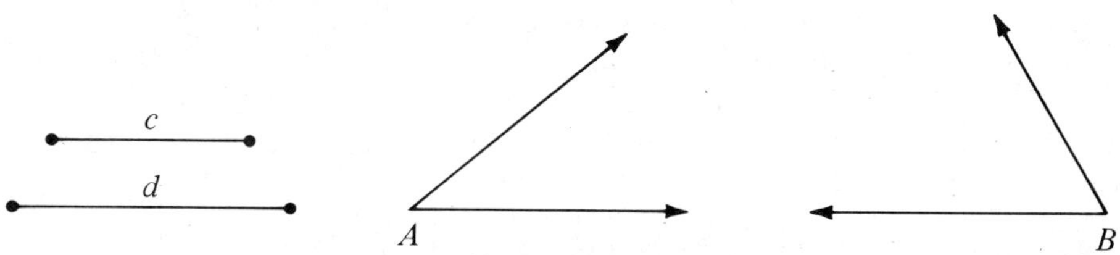

C 22. Construct a $\triangle XYZ$ so that $XY = d$, $m\angle Y = m\angle A$, and $m\angle Z = m\angle B$. Suppose you were to do this construction ten times. Which congruence postulate or theorem guarantees that all ten triangles would be congruent?

23. Construct a rectangle with a side of length d and a diagonal of length $2c$. (*Hint:* See Oral Exercise 3.)

Section 10-2 BISECTING ANGLES AND SEGMENTS

In the preceding section you learned how to construct a segment congruent to a given segment and an angle congruent to a given angle. In this section you will learn how to construct the following:

- the bisector of a given angle
- the perpendicular bisector of a given segment

In order to bisect a given angle, you construct two triangles that are congruent by the SSS Postulate.

Construction 3 Construct the bisector of a given angle.

GIVEN: $\angle A$
CONSTRUCT: A ray that bisects $\angle A$

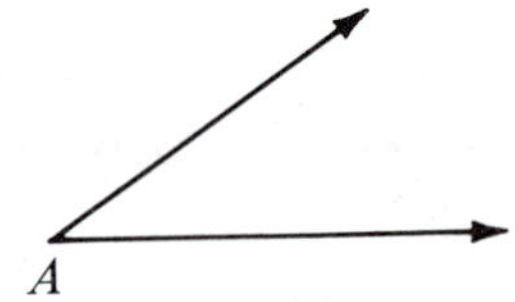

Procedure:

1. With A as center and any convenient radius, draw an arc that intersects the two sides of $\angle A$ in points X and Y, respectively.

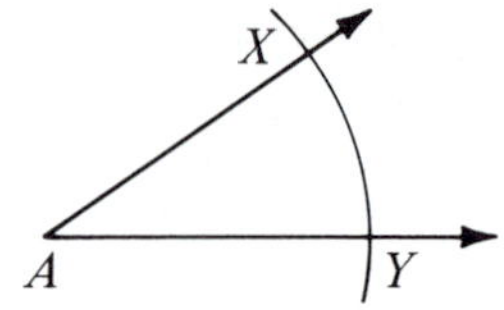

2. Using any convenient radius, draw an arc with center X. Using the same radius, draw an arc with center Y that intersects the first arc at point B.
3. Draw $\overrightarrow{AB}$.

$\overrightarrow{AB}$ bisects $\angle A$.

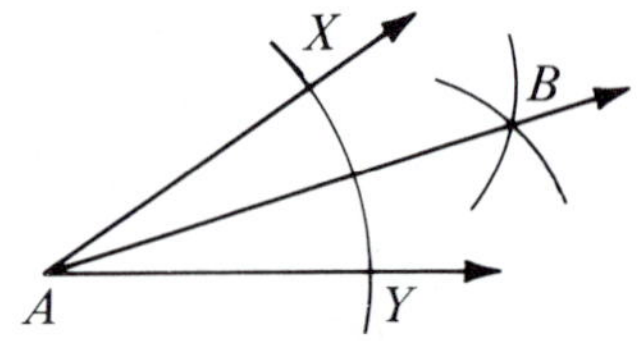

Justification: $\overline{AX} \cong \overline{AY}$ and $\overline{XB} \cong \overline{YB}$ by construction. Since $\overline{AB} \cong \overline{AB}$, you have $\triangle ABX \cong \triangle ABY$ (SSS Postulate). Therefore, $\angle XAB \cong \angle YAB$ (Corr. parts of $\cong$ ⚠ are $\cong$).

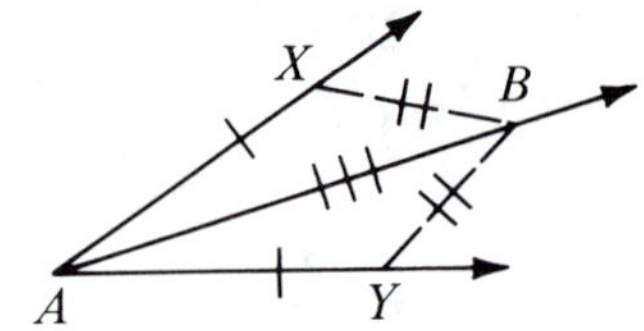

Recall that any line through the midpoint of a segment is a bisector of the segment. A *perpendicular bisector* of a segment is a line that bisects the segment *and* is perpendicular to it. By Theorem 4-6, in a plane there is exactly one line perpendicular to a line segment at its midpoint. Since our constructions are restricted to a single plane, we can refer to *the* perpendicular bisector of a segment. In the diagram at the right, both of lines m and n bisect $\overline{AB}$. Line m is the perpendicular bisector of $\overline{AB}$.

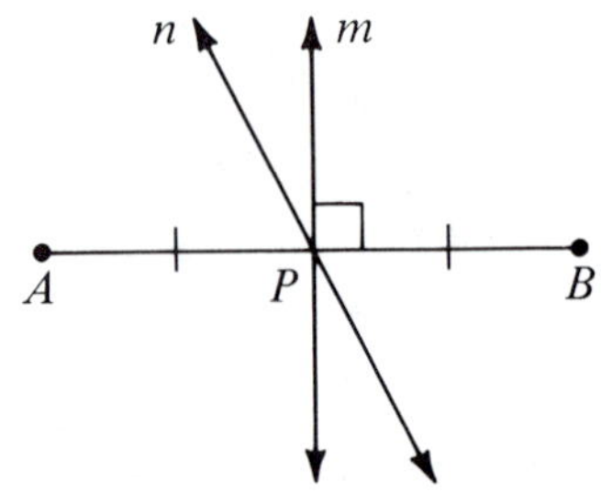

Construction 4 Construct the perpendicular bisector of a given segment.

GIVEN: $\overline{AB}$
CONSTRUCT: The perpendicular bisector of $\overline{AB}$

Procedure:

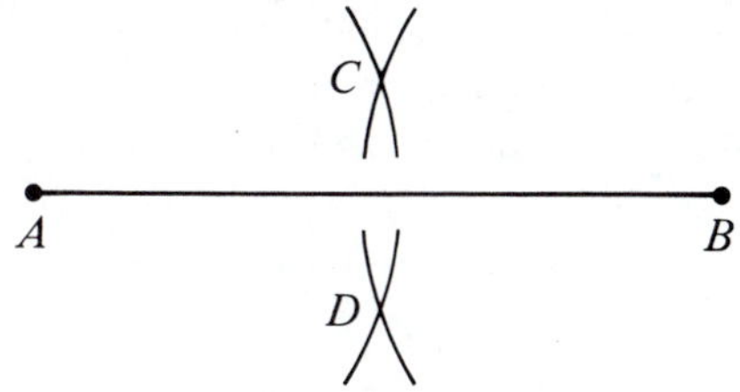

1. Using any radius greater than $\frac{1}{2}AB$, draw four arcs of equal radii, two with center A and two with center B. Label the points of intersection of these arcs as C and D.

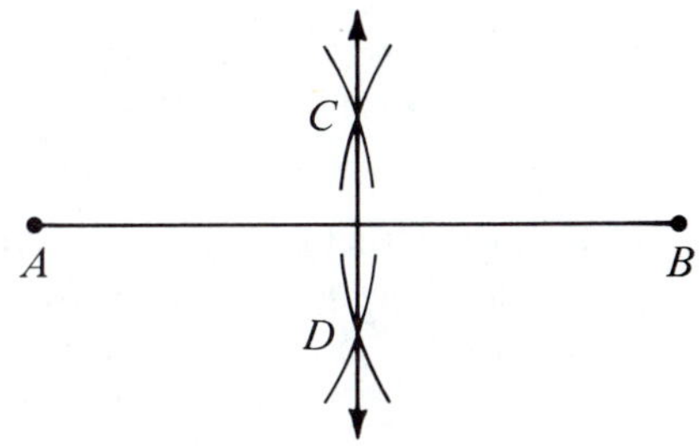

2. Draw $\overleftrightarrow{CD}$.

$\overleftrightarrow{CD}$ is the perpendicular bisector of $\overline{AB}$.

Justification: See Oral Exercise 8.

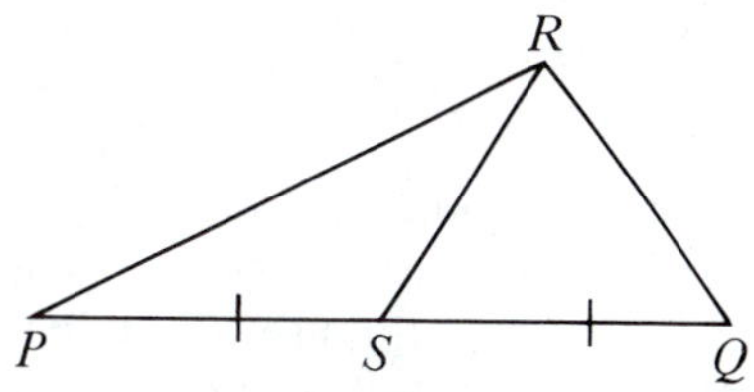

Recall that a median of a triangle is a segment whose endpoints are a vertex of the triangle and the midpoint of the opposite side. In the diagram at the right, $\overline{RS}$ is the median to side $\overline{PQ}$. You can use Construction 4 to construct a median of a triangle.

EXAMPLE

GIVEN: $\triangle ABC$
CONSTRUCT: The median to side $\overline{AB}$

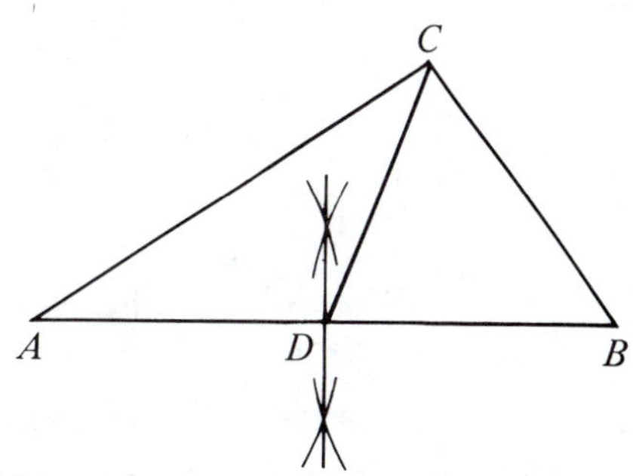

Procedure:
1. Use Construction 4 to find the midpoint, D, of $\overline{AB}$.
2. Draw $\overline{CD}$.

$\overline{CD}$ is the median to side $\overline{AB}$ by definition.

Theorem 10-1 A point lies on the perpendicular bisector of a segment if and only if it is equidistant from the endpoints of the segment.

Theorem 10-2 The perpendicular bisectors of the sides of a triangle intersect in a point that is equidistant from the vertices of the triangle.

Theorem 10-1 will be proved in Exercises 21 and 22 at the end of this section. Theorem 10-2 will be proved in Exercise 25.

You can use Theorem 10-2 to construct a circle that passes through all three vertices of a given triangle. Theorem 10-1 implies that for any given triangle there is *exactly one* point where the perpendicular bisectors intersect and therefore just one point that is equidistant from the vertices. Thus, there is *exactly one* circle that passes through all three vertices. Such a circle is said to be *circumscribed* about the triangle and is called the circumscribed circle of the triangle.

Construction 5 Given a triangle, construct its circumscribed circle.

GIVEN: $\triangle ABC$

CONSTRUCT: A circle circumscribed about $\triangle ABC$

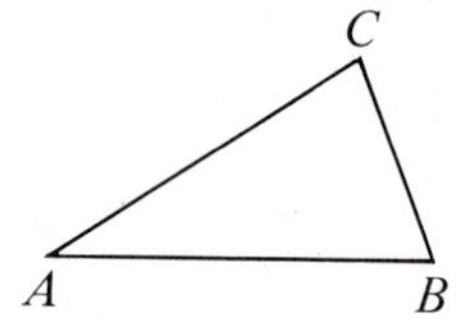

Procedure:

1. Use Construction 4 to construct the perpendicular bisectors of two sides of $\triangle ABC$. Label the point of intersection of the perpendicular bisectors as O.
2. Draw a circle with center O and radius OB (or OA or OC).

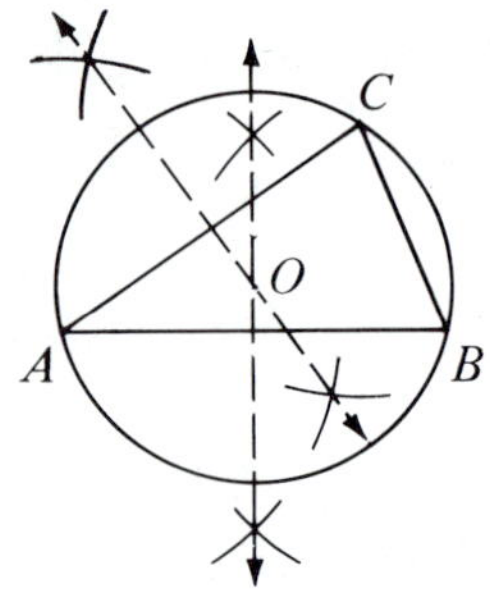

The circle with center O and radius OB is the circumscribed circle of $\triangle ABC$.

Justification: The perpendicular bisectors of $\overline{AB}$ and $\overline{BC}$ meet at O. By Theorem 10-2, O is equidistant from A, B, and C. Therefore A, B, and C lie on a circle with center O and radius OA. The circle is the circumscribed circle of $\triangle ABC$ by definition.

ORAL EXERCISES

In Exercises 1-4, classify each statement as true or false. If a statement is false, draw a diagram that illustrates a counterexample.

1. A median of a triangle is always perpendicular to the side to which it is drawn.
2. The perpendicular bisector of a side of a triangle must pass through the vertex of the angle opposite that side.
3. The perpendicular bisector of one side of a triangle may contain the median to that side.
4. The perpendicular bisector of one side of a triangle must pass through the point of intersection of the perpendicular bisectors of the other two sides.
5. Describe a procedure for constructing a 45° angle.
6. Describe a procedure for constructing a segment whose length is one fourth the length of a given segment.
7. Given a 36° angle, describe a procedure for constructing a 9° angle.

8. $\overleftrightarrow{CD}$ has been constructed as in Construction 4. Supply the reason for each statement in the following justification of the construction.

a. $AC = AD = BC = BD$

b. $ADBC$ is a rhombus.

c. $\overleftrightarrow{CD} \perp \overline{AB}$

d. $\overleftrightarrow{CD}$ bisects $\overline{AB}$.

e. $\overleftrightarrow{CD}$ is the perpendicular bisector of $\overline{AB}$.

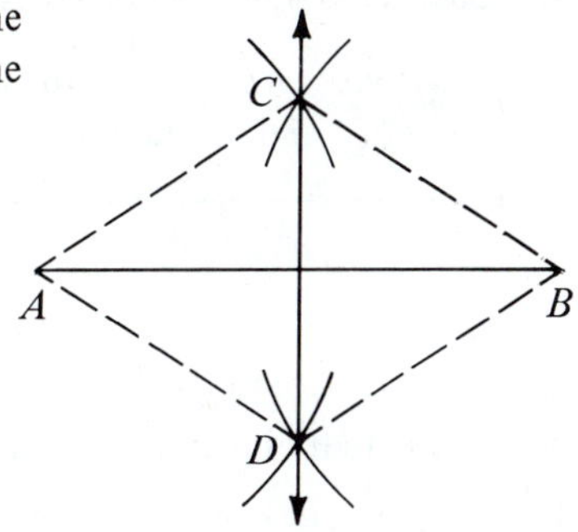

WRITTEN EXERCISES

A 1. Draw any acute angle and bisect it.

2. Draw any obtuse angle and bisect it.

3. Draw any segment and construct its perpendicular bisector.

4. Draw any acute triangle and construct the median to one side.

5. Draw any obtuse triangle and construct the median to the longest side.

6. Draw any segment and divide it into four congruent segments.

7. Draw any obtuse angle and divide it into four congruent angles.

8. Draw any acute triangle and construct the three angle bisectors.

9. Draw any acute triangle and construct the three medians.

10. Draw any acute angle and construct an angle whose measure is one half the measure of the *supplement* of the angle you drew.

11. Draw any obtuse angle and construct an angle whose measure is three eighths of the measure of the angle you drew.

Construct an angle with the given measure.

12. 45   13. $22\frac{1}{2}$ 14. $67\frac{1}{2}$

B 15. 30   16. 15   17. 75

18. Use a method suggested by the diagram to construct a 30-60-90 triangle.

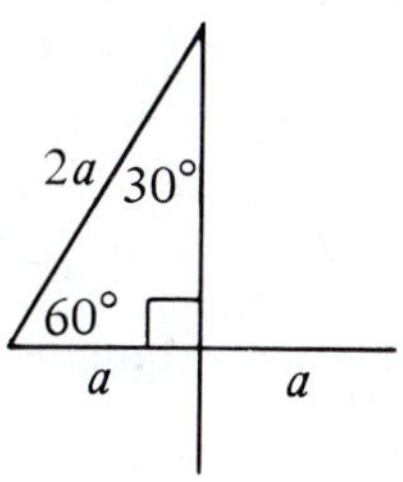

19. Construct a 45–45–90 triangle.

20. Draw two segments like those shown. Construct an isosceles triangle with base of length a and altitude of length h.

Exercises 21 and 22 prove Theorem 10-1 except for the case in which the given point is the midpoint of the given segment. (The proof of that case is quite easy.)

21. If a point lies on the perpendicular bisector of a segment, then it is equidistant from the endpoints of the segment.
    Given: $\overline{PC} \perp \overline{AB}$;
    $\overline{PC}$ bisects $\overline{AB}$.
    Prove: $PA = PB$

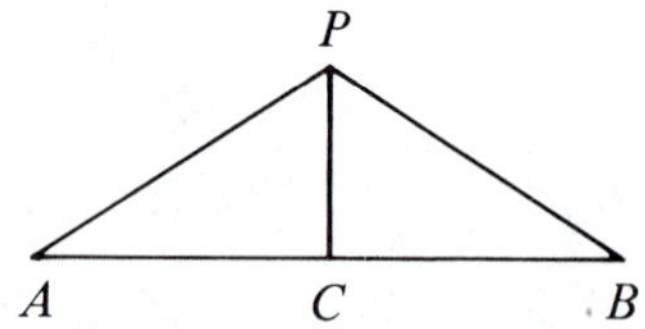

22. Prove the converse of the statement in Exercise 21: If a point is equidistant from the endpoints of a segment, then it lies on the perpendicular bisector of the segment.
    Given: $XA = XB$
    Prove: X lies on the perpendicular bisector of $\overline{AB}$.
    (*Hint:* Let C be the midpoint of $\overline{AB}$ and draw $\overline{XC}$. Prove that $\overline{XC} \perp \overline{AB}$.)

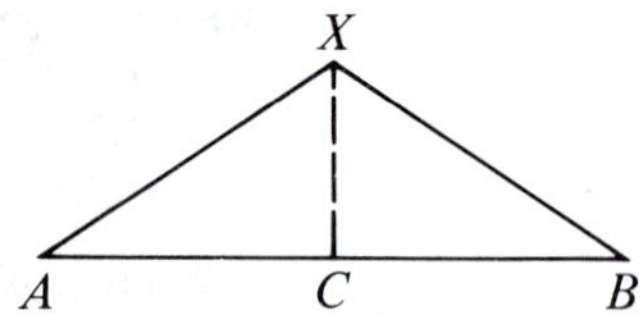

23. a. Draw an acute triangle and construct its circumscribed circle.
    b. Repeat part (a) for a right triangle.
    c. Repeat part (a) for an obtuse triangle.
    d. For each of the three triangles you constructed, state whether the center of the circumscribed circle lies inside, outside, or on the triangle.

24. Explain why the point of intersection of the perpendicular bisectors of the legs of a right triangle is the midpoint of the hypotenuse of the triangle.

C 25. Prove Theorem 10-2.
    Given: $\triangle ABC$;
    $\overleftrightarrow{OD}$ and $\overleftrightarrow{OE}$ are the perpendicular bisectors of $\overline{AB}$ and $\overline{BC}$ respectively; $\overleftrightarrow{OF} \perp \overline{AC}$
    Prove: a. $\overleftrightarrow{OF}$ bisects $\overline{AC}$.
    b. $OA = OB = OC$
    (*Hints:* 1. Draw $\overline{OA}$, $\overline{OB}$, and $\overline{OC}$.
    2. Use Theorem 10-1 to show that $OA = OC$.)

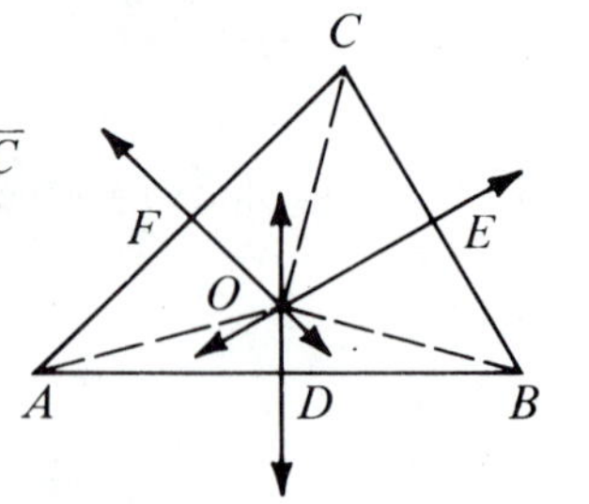

26. Draw two segments like those below and construct a rhombus whose diagonals are congruent to the segments.

27. Draw an acute angle and a line segment like those shown. In each of parts (a) and (b), construct a rhombus such that one of its angles is congruent to $\angle C$ and the indicated diagonal has length d.

a. The longer diagonal has length d.

b. The shorter diagonal has length d.

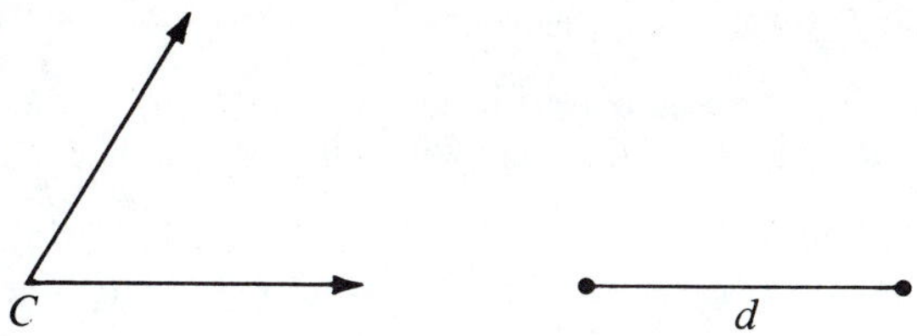

Section 10-3 CONSTRUCTING PERPENDICULAR LINES

In Section 10-2 you learned how to construct the perpendicular bisector of a segment. In this section you will learn how to apply that construction to construct a line perpendicular to a given line:

- through a given point on the line
- through a given point not on the line

Suppose you are given a line l and a point P on l. How could you construct a line through P perpendicular to l? If you first constructed on l a segment whose midpoint is P, you could then use Construction 4 to construct a line perpendicular to that segment at P.

Construction 6 Given a line and a point on the line, construct the line perpendicular to the given line through the given point.

GIVEN: Line l; point P on l

CONSTRUCT: A line perpendicular to l through P

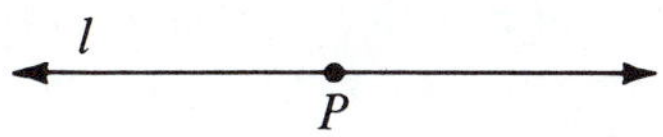

Procedure:

1. With P as center and any convenient radius, draw two arcs of equal radii that intersect l at points X and Y.

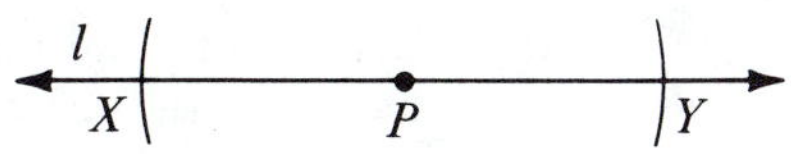

2. Using any radius greater than $\frac{1}{2}XY$, draw two arcs of equal radii with centers at X and Y that intersect at a point Q not on l.
3. Draw $\overleftrightarrow{PQ}$.

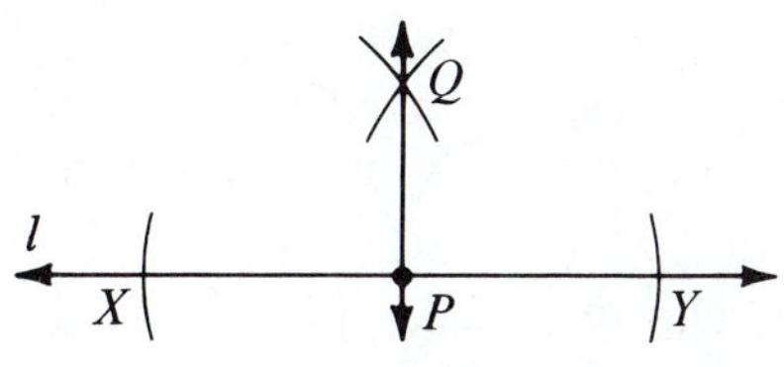

$\overleftrightarrow{PQ}$ is perpendicular to l at P.

Justification: By construction, $\overline{XP} \cong \overline{YP}$ and $\overline{XQ} \cong \overline{YQ}$. Thus, by Theorem 10-1, points P and Q lie on the perpendicular bisector of $\overline{XY}$. Since two points (P and Q) determine exactly one line, $\overleftrightarrow{PQ}$ is the perpendicular bisector of $\overline{XY}$, and $\overleftrightarrow{PQ} \perp \overleftrightarrow{XY}$ by the definition of a perpendicular bisector.

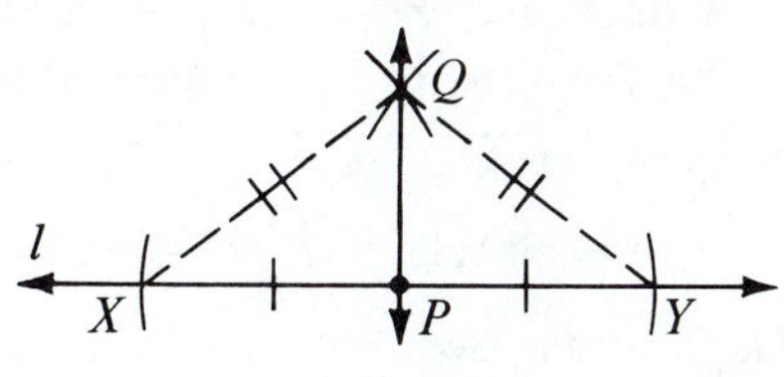

EXAMPLE

GIVEN: A segment of length x
CONSTRUCT: A square with sides of length x

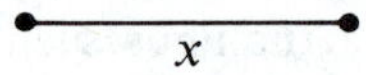

Procedure:

1. Use Construction 1 to construct $\overline{PQ}$ of length x.

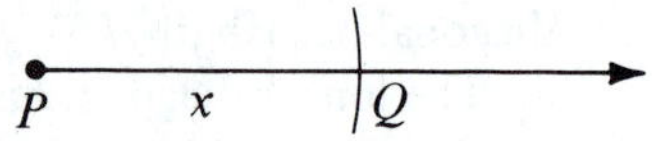

2. Use Construction 6 to construct a line k perpendicular to $\overrightarrow{PQ}$ at Q.

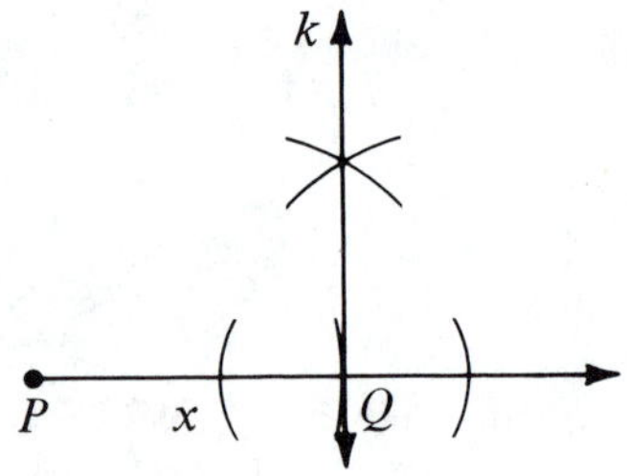

3. Construct $\overline{QR}$ of length x on line k.
4. With P and R as centers, draw two arcs of radius x that intersect at point S.
5. Draw $\overline{RS}$ and $\overline{PS}$.

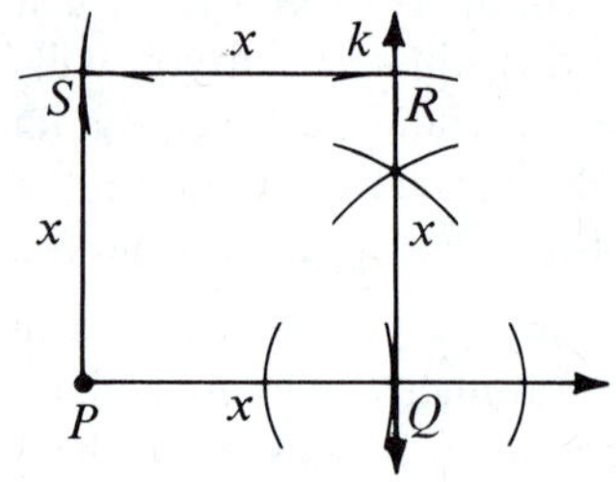

PQRS is a square with sides of length x.

Justification: $PQ = QR = RS = PS = x$ by construction. Therefore, *PQRS* is a rhombus by definition. Since $\angle PQR$ is a right angle, *PQRS* is a square.

In Construction 6, you constructed a line perpendicular to a given line through a point *on* the given line. In the next construction the given point does *not* lie on the given line.

Construction 7 Given a line and a point not on the line, construct a line perpendicular to the given line through the given point.

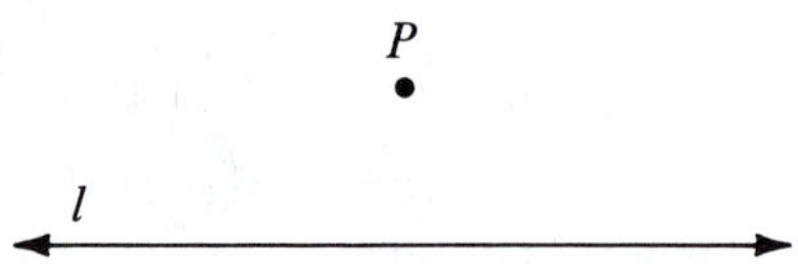

GIVEN: Line l; point P not on l
CONSTRUCT: A line through P that is perpendicular to l

Procedure:

1. With P as center, draw two arcs of equal radii that intersect l at points X and Y.
2. With X and Y as centers and any convenient radius, draw two arcs of equal radii that intersect at a point Q.
3. Draw $\overleftrightarrow{PQ}$.

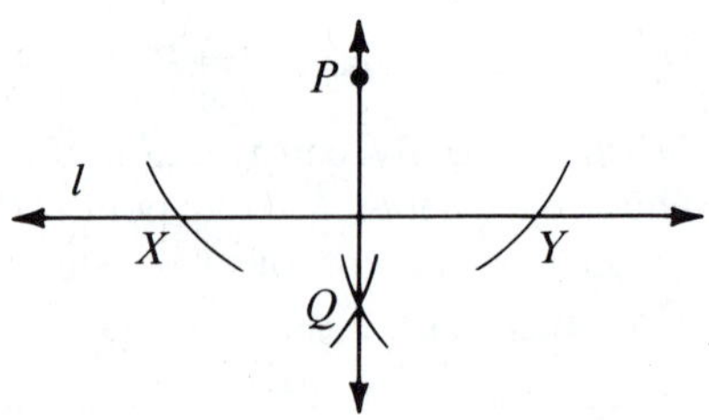

$\overleftrightarrow{PQ}$ is perpendicular to l.

Justification: See Oral Exercise 1.

EXAMPLE

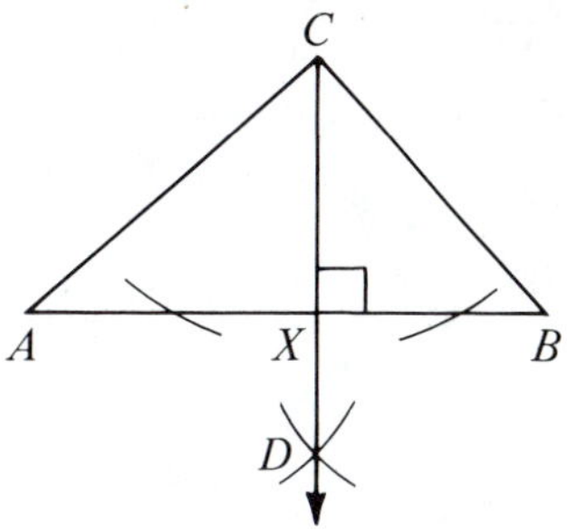

GIVEN: $\triangle ABC$
CONSTRUCT: The altitude to $\overline{AB}$

Procedure: Use Construction 7 to construct $\overleftrightarrow{CD}$ perpendicular to $\overline{AB}$. Label the point of intersection of $\overline{AB}$ and $\overleftrightarrow{CD}$ as X.

$\overline{CX}$ is the altitude to $\overline{AB}$ by definition.

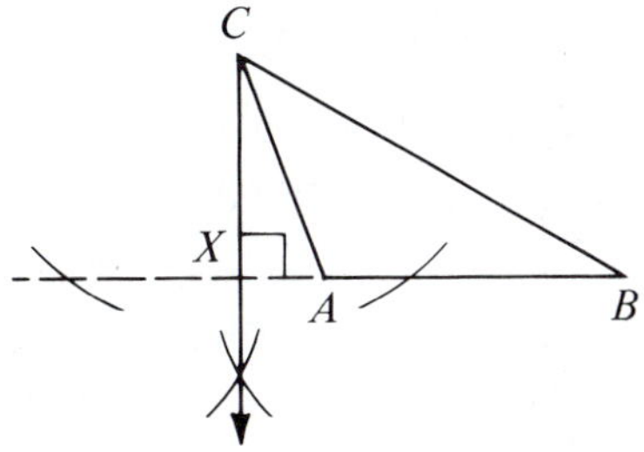

In order to construct a line perpendicular to a given segment, you may first have to extend the given segment. For example, to construct the altitude from C to $\overline{BA}$ in the obtuse triangle at the right, you first extend $\overline{BA}$ so that the arcs drawn with center C will intersect $\overleftrightarrow{BA}$. Thus, $\overline{CX}$ is the altitude to the line that contains side $\overline{BA}$.

ORAL EXERCISES

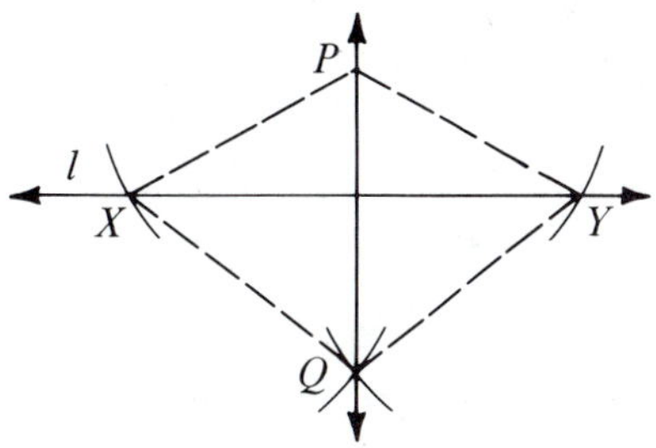

1. $\overleftrightarrow{PQ}$ has been constructed as in Construction 7. Supply the reason for each statement in the following justification of the construction.
   a. $PX = PY$; $QX = QY$
   b. P and Q lie on the perpendicular bisector of $\overline{XY}$.
   c. $\overleftrightarrow{PQ}$ is the perpendicular bisector of $\overline{XY}$.
   d. $\overleftrightarrow{PQ} \perp l$

Suppose you are given the figure shown in the diagram. Describe how you would perform the required construction.

2. Line k through point P parallel to line l (State the theorem that guarantees that line k will be parallel to line l.)
3. A line segment whose endpoints lie on the lines that contain the bases of trapezoid $ABCD$ and that is perpendicular to the bases (Such a line segment is called an *altitude* of the trapezoid.)

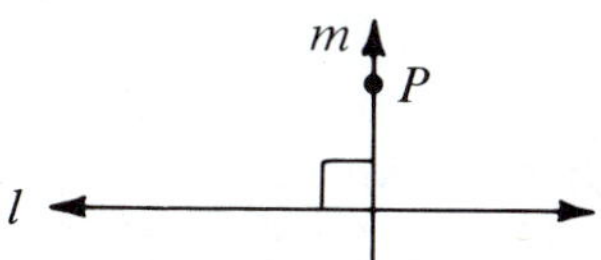

Ex. 2

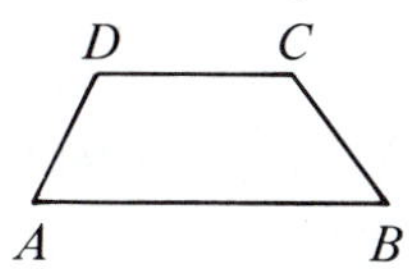

Ex. 3

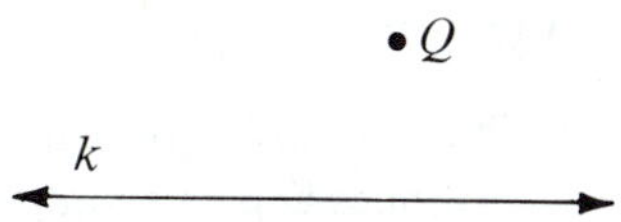

Exs. 4, 5

4. A square with one vertex at Q and one side contained in k
5. A right $\triangle QRS$ whose right angle has vertex Q and whose hypotenuse, $\overline{RS}$, lies on line k

WRITTEN EXERCISES

Draw a larger version of the given diagram and use it in your constructions.

A 1. Construct a line through P that is perpendicular to line l.

2. Construct a line through P that is perpendicular to line m.

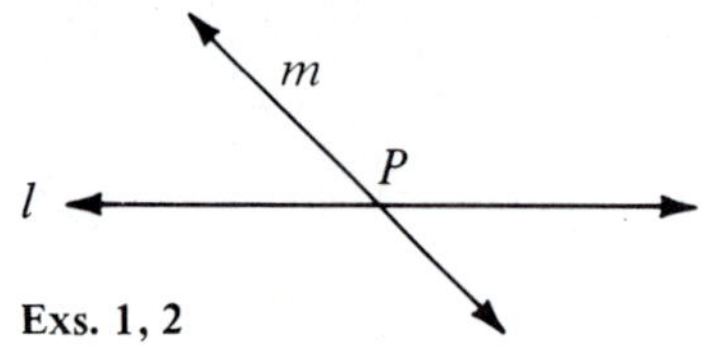

Exs. 1, 2

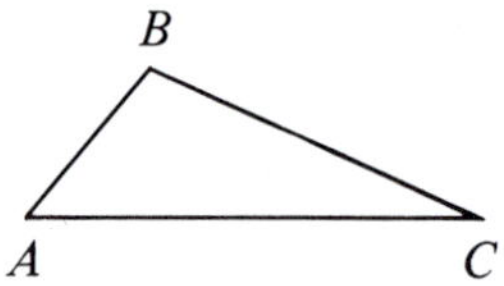

Exs. 3, 4

3. Construct the altitude of $\triangle ABC$ from B to $\overline{AC}$.

4. Construct the altitude of $\triangle ABC$ from A to the line containing $\overline{BC}$.

5. Construct a trapezoid one of whose legs is $\overline{XY}$ and whose other leg lies on line l.

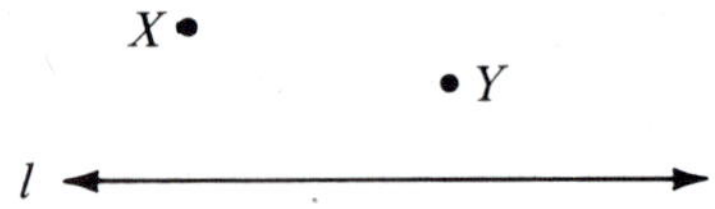

In Exercises 6-12, draw line segments and angles like those below. Use them in your constructions.

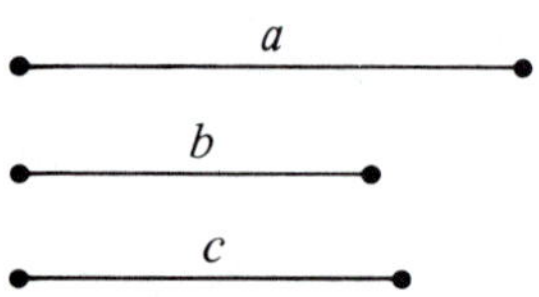

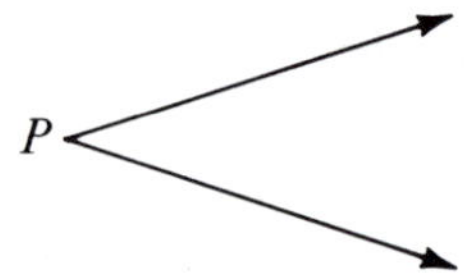

6. An angle complementary to $\angle P$

7. An angle of measure $(90 + m\angle P)$

8. A right triangle whose legs have lengths a and b

9. A right triangle one of whose acute angles has measure equal to $m\angle P$ and whose hypotenuse has length a

10. A rectangle whose adjacent sides have lengths c and $\frac{1}{2}a$

11. $\triangle XYZ$ in which the altitude from vertex X has length a and divides $\overline{YZ}$ into segments of lengths b and c

12. A trapezoid with height c and bases of lengths a and b

B 13. a. Draw an acute triangle and construct the altitudes to the three sides. Do the altitudes appear to intersect in a point?

b. Repeat part (a) for a right triangle.

c. Repeat part (a) for an obtuse triangle. Do the lines that contain the altitudes appear to intersect in a point?

d. For each of the three triangles you constructed, state whether the point of intersection of the altitudes lies inside, outside, or on the triangle.

14. Construct a regular octagon by constructing a circle, constructing two perpendicular diameters, and then bisecting the four angles formed. Justify your construction.

15. Draw a large acute triangle. By construction locate:
   a. the point of intersection, Y, of the perpendicular bisectors.
   b. the point of intersection, X, of the medians.
   c. the point of intersection, Z, of the altitudes.
   d. What appears to be true about the three points of intersection you found in parts (a), (b), and (c)?

C 16. Draw a segment and let its length be 1.
   a. Construct a segment of length $\frac{\sqrt{5}}{2}$. (*Hint*: Construct a right triangle whose hypotenuse has length $\frac{\sqrt{5}}{2}$.)
   b. Use the results of part (a) to construct a segment of length $\frac{\sqrt{5}-1}{2}$.
   c. Recall that a golden rectangle is a rectangle whose sides have lengths in the ratio $\frac{\sqrt{5}-1}{2}:1$. Use the results of part (b) to construct a golden rectangle.

17. Draw a segment like the one shown at the right. Construct an equilateral triangle with altitude of length h.

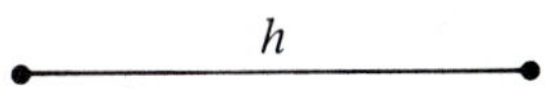

SELF-TEST 1

In Exercises 1-3, 7, and 8, draw segments and angles like those shown. Then construct the required figure.

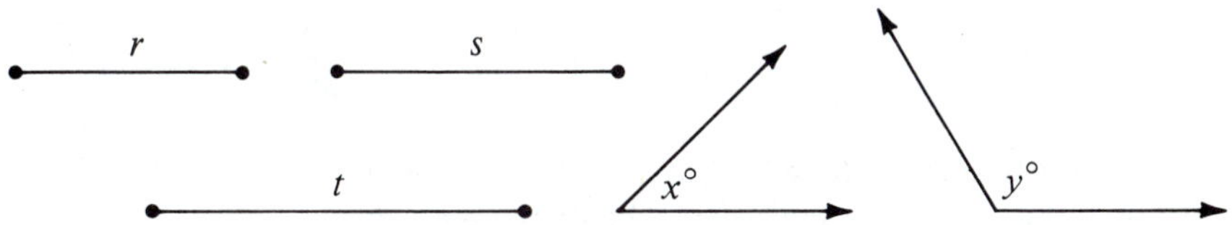

1. A segment of length $(2t - r)$ 2. An angle of measure $(x + y)$ Section 10-1

3. $\triangle ABC$ such that $AB = r$, $BC = s$, and $m\angle B = y$

4. Construct a $60°$ angle.

5. Draw a $\triangle JKL$. Section 10-2
   a. Bisect $\angle J$.
   b. Construct the perpendicular bisector of $\overline{JL}$.

6. Construct a $135°$ angle.

7. Construct $\triangle DEF$ such that $DE = r$, $DF = s$, and $EF = t$. Then construct the altitude from D to $\overline{EF}$. Section 10-3

8. Construct a right triangle with one leg of length r and hypotenuse of length t.

Section 10-4 CONSTRUCTING PARALLEL LINES

In Chapter 4 you learned that through a given point not on a given line, there is exactly one line parallel to the given line. In this section you will learn how to:

- construct a line through a given point parallel to a given line
- divide a line segment into a given number of congruent segments

The construction of a line parallel to a given line depends on the fact that if two lines are cut by a transversal so that corresponding angles are congruent, the lines are parallel.

Construction 8 Given a point not on a given line, construct a line through the given point parallel to the given line.

GIVEN: Line l; point P not on l
CONSTRUCT: A line through P parallel to l

Procedure:

1. Draw any line through P that intersects l at some point Q.

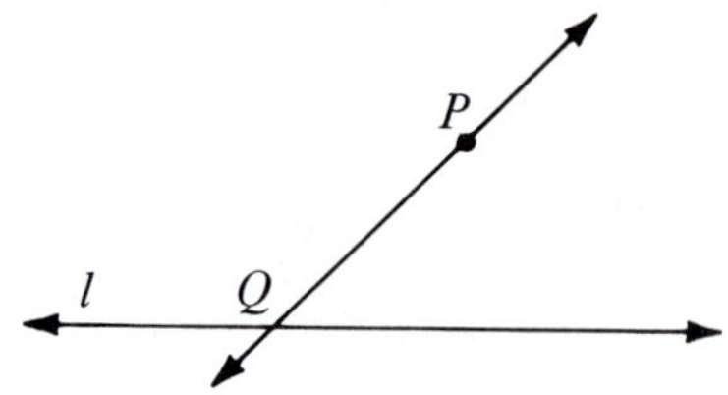

2. Use Construction 2 to construct $\angle 2$ with vertex P so that $\angle 1$ and $\angle 2$ are corresponding congruent angles. Label the line containing the ray you just constructed as m.

$m \parallel l$

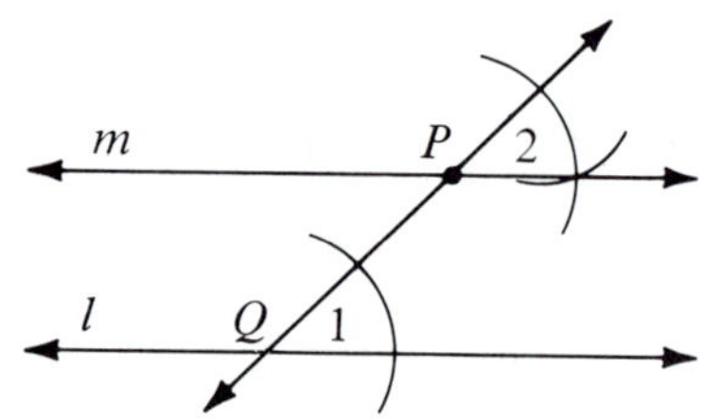

Justification: $\angle 2 \cong \angle 1$ by construction. Therefore, $m \parallel l$. (If two lines are cut by a transversal so that corresponding angles are congruent, the lines are parallel.)

Recall that the corollary to Theorem 5-14 states that if three parallel lines cut off congruent segments on one transversal, they cut off congruent segments on any transversal. It follows that if three *or more* parallel lines cut off congruent segments on one transversal, such as line j, then they cut off congruent segments on any transversal, such as line k. Our next construction is based on this idea.

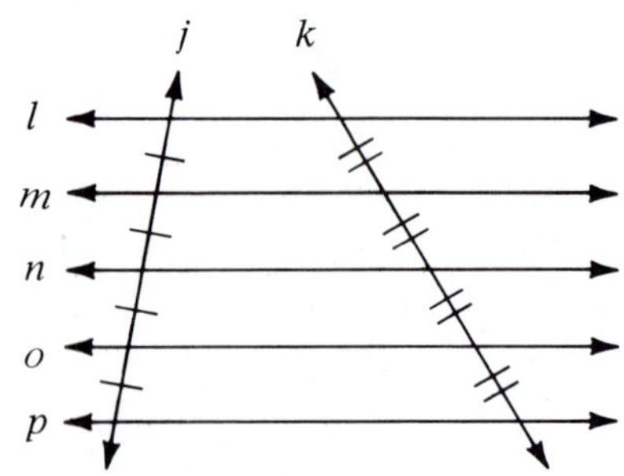

Construction 9 Divide a given segment into a given number of congruent segments. (3 shown here)

GIVEN: $\overline{AB}$

CONSTRUCT: Points C and D on $\overline{AB}$ such that $\overline{AC} \cong \overline{CD} \cong \overline{DB}$

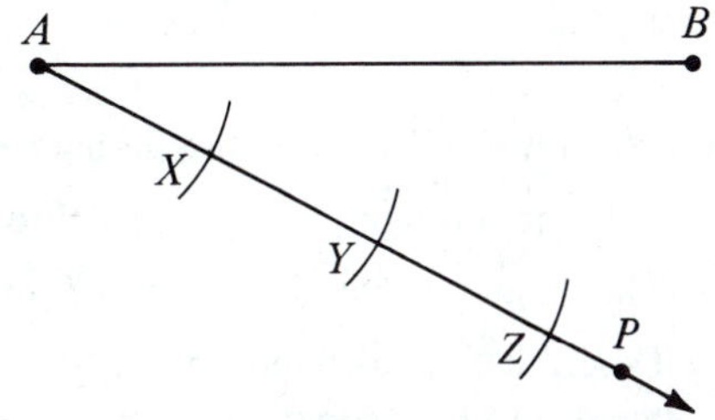

Procedure:

1. Draw any ray $\overrightarrow{AP}$ such that P, A, and B are noncollinear. Using any convenient radius, start with A as center and mark off X, Y, and Z so that $\overline{AX} \cong \overline{XY} \cong \overline{YZ}$.
2. Draw $\overleftrightarrow{BZ}$.
3. Use Construction 8 to construct lines parallel to $\overleftrightarrow{BZ}$ through X and Y. Label the points of intersection of these lines and $\overline{AB}$ as C and D.

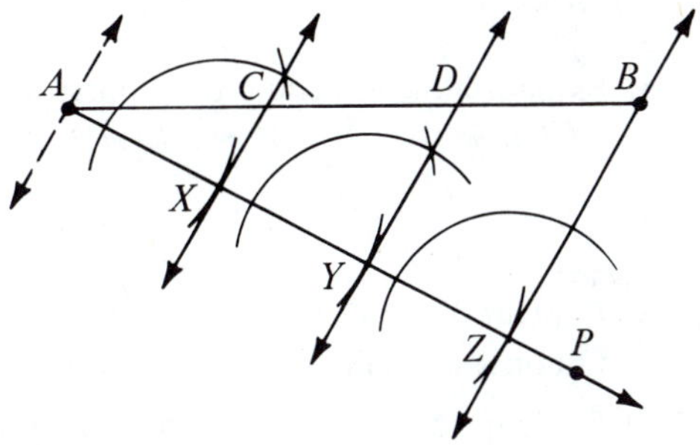

$\overline{AC} \cong \overline{CD} \cong \overline{DB}$

Justification: Consider the line through A parallel to $\overleftrightarrow{BZ}$. Since the four parallel lines cut off congruent segments on one transversal $\overleftrightarrow{AP}$, they cut off congruent segments on the other transversal $\overleftrightarrow{AB}$. Thus, $\overline{AC} \cong \overline{CD} \cong \overline{DB}$.

A simplified version of Construction 9 can be used to divide a given segment into two segments whose lengths are in a given ratio of positive integers.

EXAMPLE

GIVEN: $\overline{AB}$

CONSTRUCT: Point C on $\overline{AB}$ such that

$$AC:CB = 3:2$$

Procedure:

1. Draw $\overrightarrow{AP}$ as in Construction 9. Using any convenient radius, mark off five congruent segments on $\overrightarrow{AP}$. (The given ratio is 3:2, and $3 + 2 = 5$.)
2. Draw $\overleftrightarrow{BL}$.
3. Notice that J divides $\overline{AL}$ in the ratio 3:2. Therefore, use Construction 8 to construct a line through J parallel to $\overleftrightarrow{BL}$. Label the point of intersection of this line and $\overline{AB}$ as C.

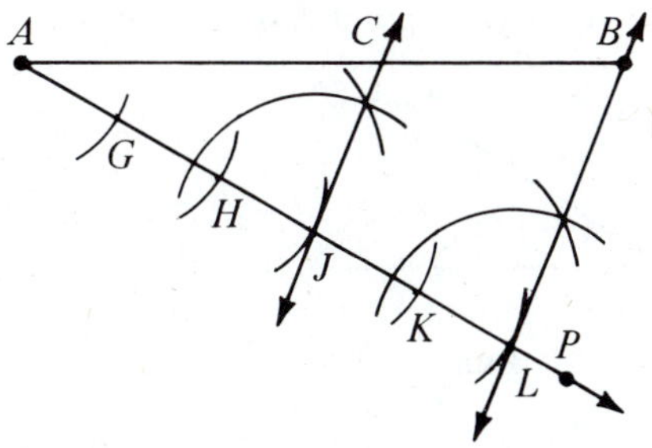

$AC:CB = 3:2$ by Theorem 8-2.

ORAL EXERCISES

1. How many congruent segments should be marked off on $\overrightarrow{AQ}$ to divide $\overline{AB}$ into two segments whose lengths are in the given ratio?
   a. 3:1 b. 5:2

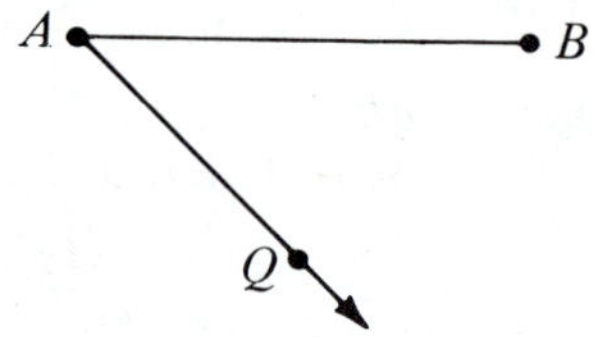

Describe a procedure for constructing each figure.

2. The line through point Q parallel to $\overline{OP}$
3. The diameter that is parallel to $\overline{PQ}$
4. Describe two different procedures for dividing a segment into four congruent segments.

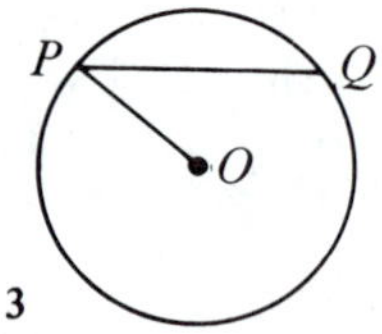

Exs. 2, 3

5. Explain how you could use Construction 8 to construct $\triangle HKL$ so that $\triangle HKL \sim \triangle HGJ$.

Ex. 5

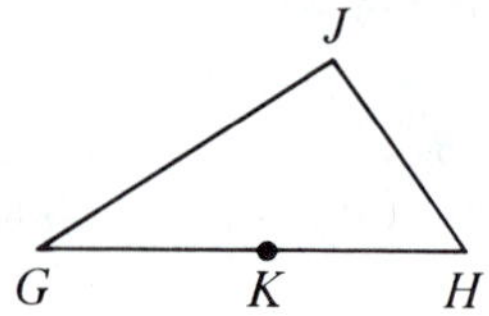

6. Explain how you could apply Theorem 4-1 (page 154) to construct a line through P parallel to l.
7. Explain how you could use Constructions 6 and 7 to construct a line through P parallel to l.

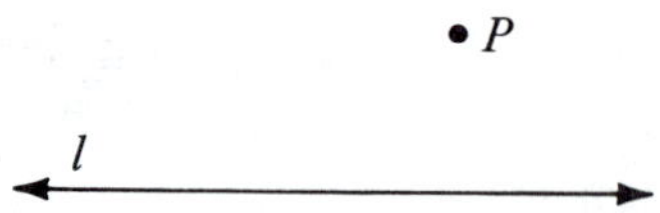

Exs. 6, 7

8. In the diagram at the right, $PQ = QR = RS = PS$. Describe the steps you would perform to construct a line through P parallel to l as shown in the diagram.

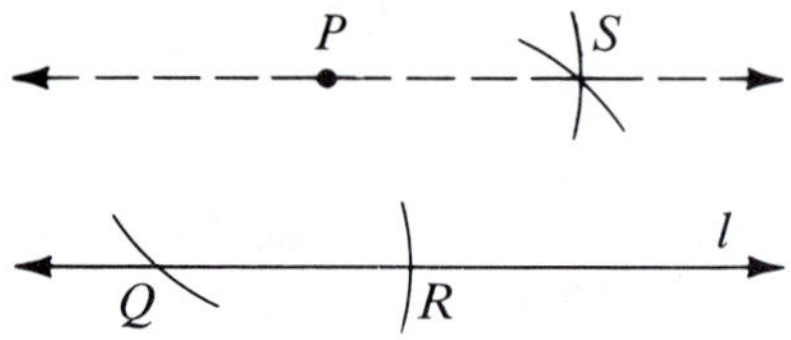

WRITTEN EXERCISES

Draw a larger version of the given diagram and use it in your construction.

A 1. Construct a line through R that is parallel to $\overline{ST}$.

2. Construct point E on AC so that $AE:EC = AD:DB$.

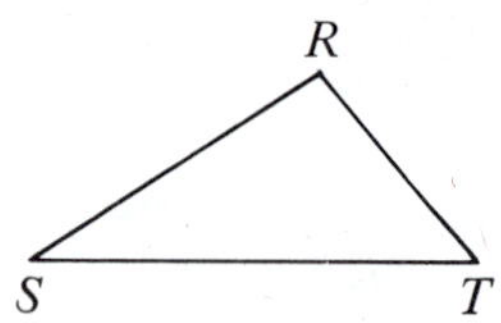

Ex.1

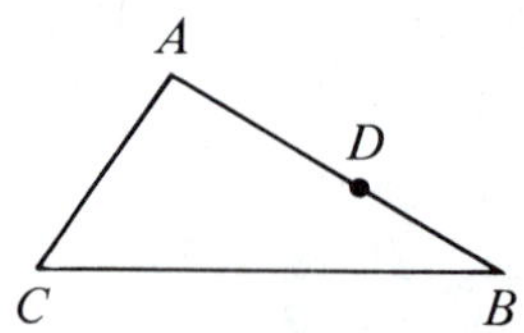

Ex.2

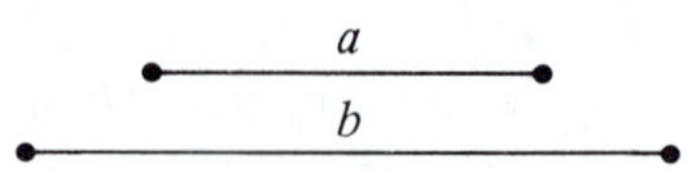

Exs. 3, 4

3. Construct a trapezoid whose bases have lengths a and b.
4. Construct a trapezoid whose legs have lengths a and b.

In Exercises 5-7 begin by drawing a line segment. Then use Construction 9 to divide the segment into the given number of congruent segments.

5. 4 **6.** 5 **7.** 6

In Exercises 8-10 begin by drawing a segment. Then use Construction 9 to divide the segment into two segments whose lengths have the given ratio.

8. 3:4 **9.** 2:1 **10.** 5:2

B **11.** Draw a larger version of the diagram at the right and use it in your construction. Construct a line through P that is parallel to l and a line through P that is parallel to m. What type of quadrilateral have you constructed?

Ex. 11

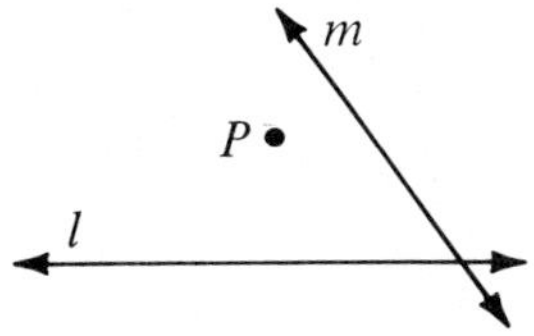

12. Draw a segment whose length is greater than half the width of the page on which you are working. Construct an equilateral triangle whose *perimeter* is equal to the length of the segment you have drawn.

13. Draw a segment as in Exercise 12. Construct a rhombus whose perimeter equals the length of the segment.

14. Draw line segments approximately equal in length to p, q, and r, and divide the segment of length r into two segments whose lengths are in the ratio $p:q$.

15. Draw a large obtuse $\triangle ABC$ with obtuse $\angle C$. Use the procedure in the example on page 401 to divide $\overline{AB}$ into two segments, $\overline{AD}$ and $\overline{DB}$, such that $AD:DB = AC:BC$. Can you think of a simpler way to divide $\overline{AB}$ in this ratio?

C **16.** In this exercise you will not be obeying the strict rules of straightedge-and-compass construction because you will use the markings on a ruler.

a. With a ruler marked in centimeters, draw a line and locate four points A, B, C, and D on the line so that $AB = BC = CD = 1$ cm. Construct parallel lines through A, B, C, and D. Then construct a line perpendicular to one (and hence to all) of the parallel lines. Explain why the parallel lines cut off congruent segments on the perpendicular.

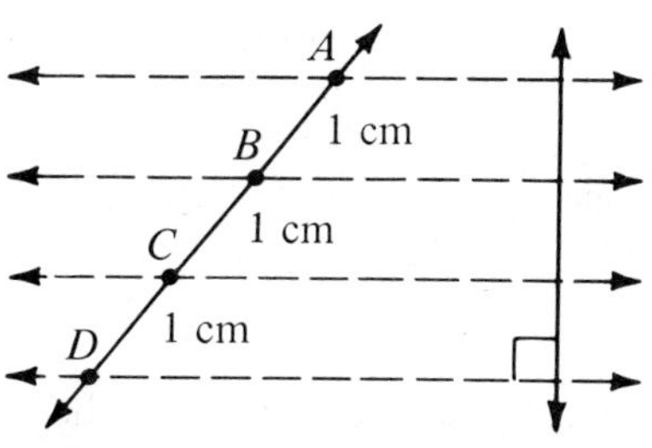

b. Carpenters use a tool called a try square to draw lines that are perpendicular to any straight edge of a board. Explain how a carpenter might use the idea presented in part (a) to divide the end of a board of any width into a particular number of congruent segments. (For your explanation, assume that the number of congruent segments is 7.)

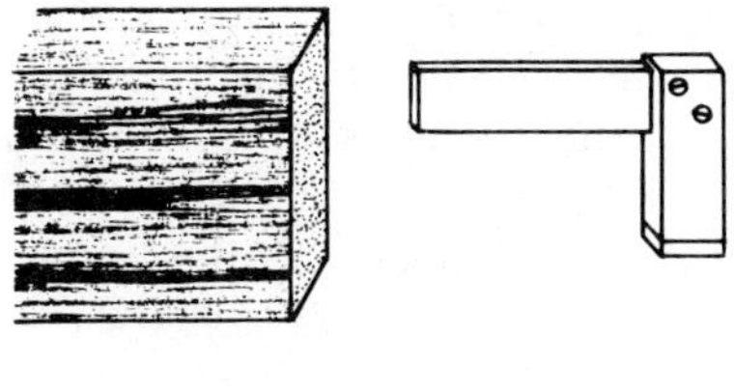

Section 10-5 CONSTRUCTIONS AND PROPORTIONS

In Section 10-4 you learned how to divide a given segment into two segments whose lengths are in a given ratio. Now you will learn how to construct a segment whose length:

- forms a proportion with the lengths of three given segments
- is the geometric mean between the lengths of two given segments

If you are given three segments, you can use a construction similar to the one in the example on page 401 to construct a fourth segment such that the lengths of all four are proportional.

Construction 10 Given three segments, construct a segment such that the lengths of the four segments form a proportion.

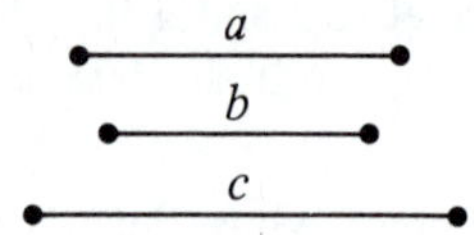

GIVEN: Segments of lengths a, b, and c

CONSTRUCT: A segment of length x so that $\frac{a}{b} = \frac{c}{x}$

Procedure:

1. Draw $\angle MPN$.
2. On $\overrightarrow{PM}$ mark off $PQ = a$ and $QR = b$.
3. On $\overrightarrow{PN}$ mark off $PS = c$.
4. Draw $\overleftrightarrow{QS}$.

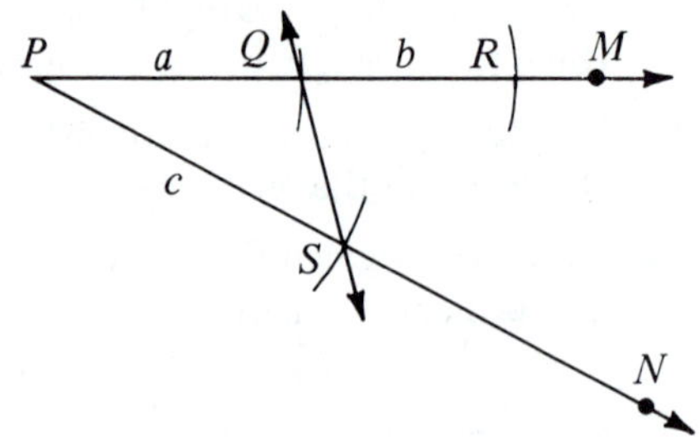

5. Use Construction 8 to construct the line through R parallel to $\overleftrightarrow{QS}$ and intersecting $\overrightarrow{PN}$ at point T.

The length of $\overline{ST}$ is x where $\frac{a}{b} = \frac{c}{x}$.

Justification: Since $\overleftrightarrow{RT} \parallel \overleftrightarrow{QS}$, $\frac{a}{b} = \frac{c}{x}$ by the Triangle Proportionality Theorem.

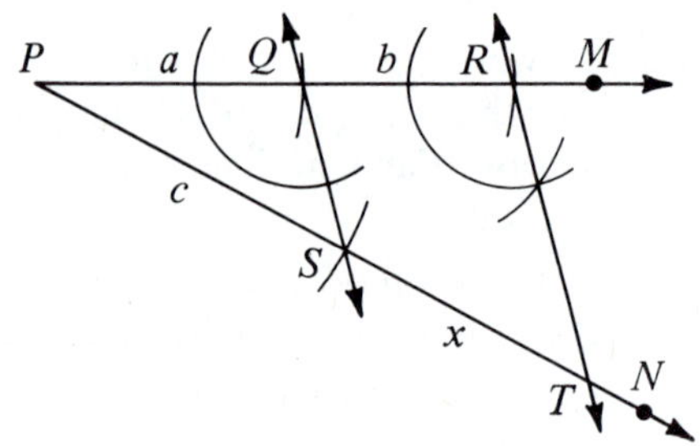

Recall that if x is the geometric mean between two positive real numbers a and d, then

$$\frac{a}{x} = \frac{x}{d} \quad \text{and hence} \quad x = \sqrt{ad}\,.$$

Our next construction shows how to construct a segment whose length is the geometric mean between the lengths of two given segments. This construction uses a theorem that you found in Book 1: An angle inscribed in a semicircle is a right angle. In the diagram at the right $\angle ABC$ is inscribed in a semicircle. Thus, $m\angle ABC = 90$.

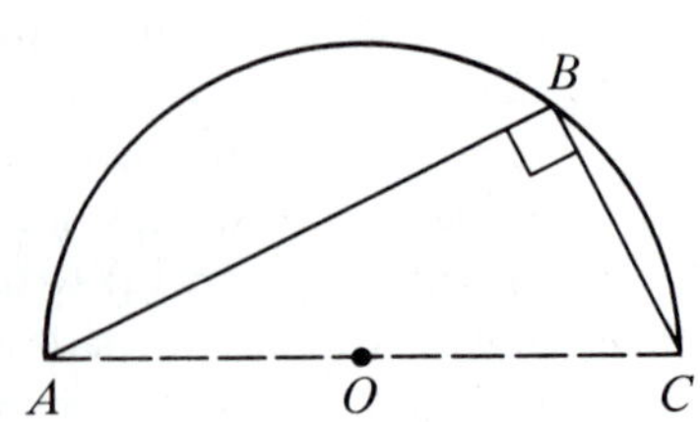

Construction 11 Given two segments, construct a third segment whose length is the geometric mean between the lengths of the given segments.

GIVEN: Segments of lengths a and d
CONSTRUCT: A segment of length $\sqrt{ad}$

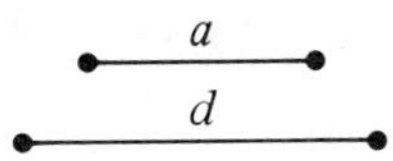

Procedure:

1. Draw a ray with endpoint Q and mark off $QR = a$ and $RS = d$.

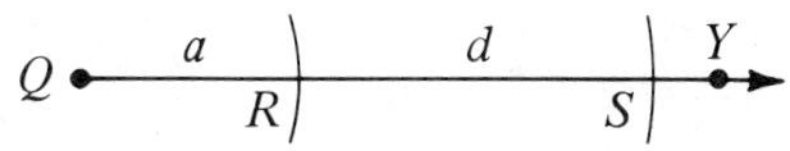

2. Use Construction 6 to construct $\overrightarrow{RT}$ perpendicular to $\overline{QS}$ at R.

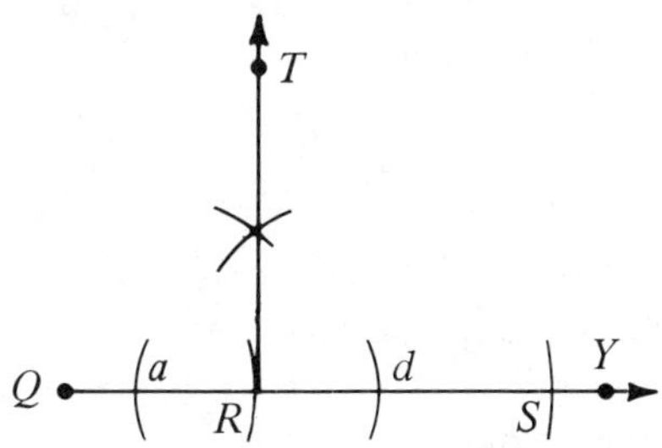

3. Use Construction 4 to locate M, the midpoint of $\overline{QS}$.
4. Draw a semicircle with center M and radius $\overline{MQ}$ (or $\overline{MS}$) that intersects $\overrightarrow{RT}$ at point P.

$PR = \sqrt{ad}$

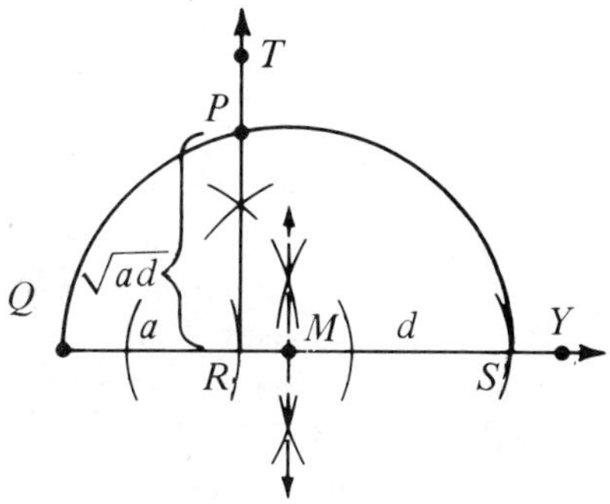

Justification: $\angle QPS$ is a right angle since it is inscribed in a semicircle. $\overline{PR} \perp \overline{QS}$ by construction and hence $\overline{PR}$ is the altitude to the hypotenuse of right $\triangle PQS$.

Corollary 1 to Theorem 9-1 states that the length of the altitude to the hypotenuse of a right triangle is the geometric mean between the lengths of the two segments of the hypotenuse. Therefore,

$$\frac{a}{PR} = \frac{PR}{d}, \text{ or } PR = \sqrt{ad}.$$

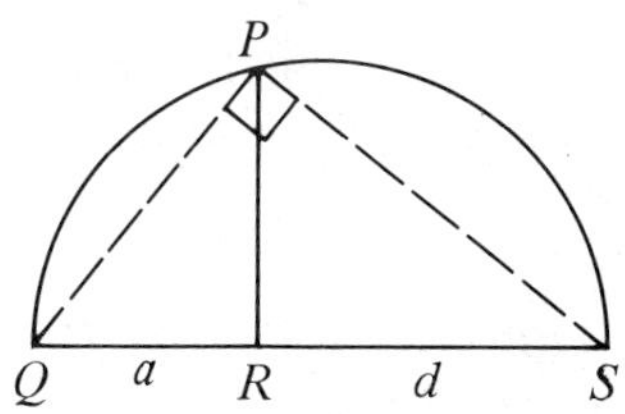

Given a segment defined to be of length 1, you can use Construction 11 to construct a segment whose length is the square root of any positive integer. The example on the following page illustrates the procedure.

EXAMPLE

GIVEN: $AB = 1$
CONSTRUCT: A segment of length $\sqrt{6}$

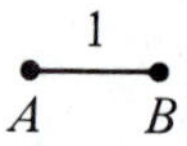

Procedure:

1. Choose two integers whose product is 6, say 2 and 3.

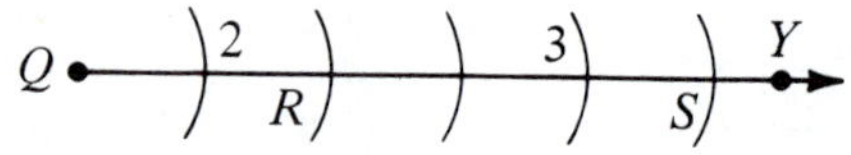

2. Draw $\overrightarrow{QY}$ and mark off $QR = 2$ and $RS = 3$. (Thus $QS = 2 + 3$, or 5.)
3. Use Construction 11 to construct the geometric mean of 2 and 3.

$PR = \sqrt{2 \cdot 3}$, or $\sqrt{6}$.

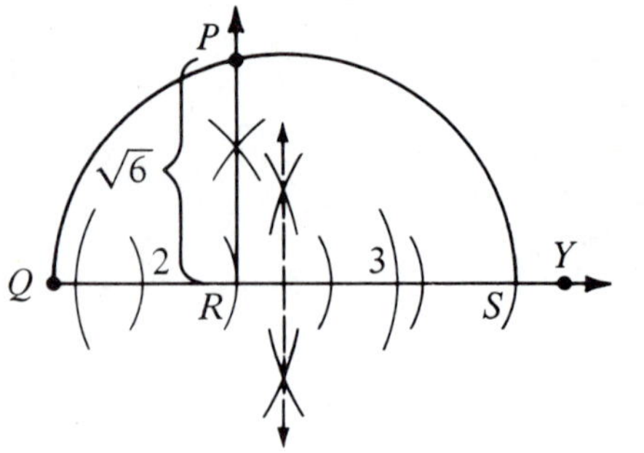

Note that you could have chosen as lengths for $\overline{QR}$ and $\overline{RS}$ any integers whose product is 6. If you chose $QR = 1$ and $RS = 6$, then the construction would require more space because QS would equal $1 + 6$, or 7, instead of $2 + 3$, or 5, as above.

ORAL EXERCISES

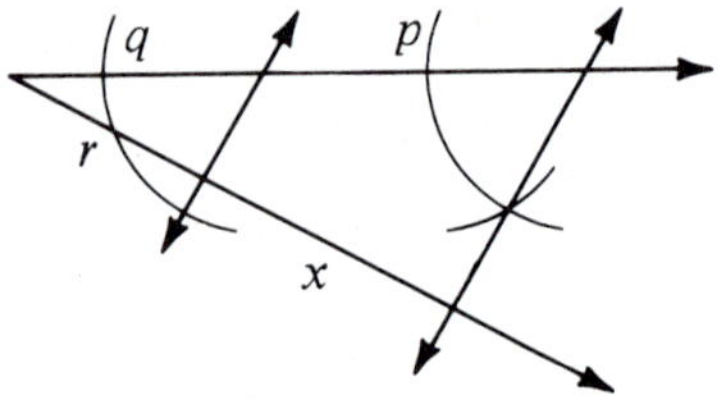

1. State as many proportions as you can for which the diagram at the right would represent a valid construction for a segment of length x given segments of lengths p, q, and r. (Do not include proportions that have compound expressions such as $p + q$ in a numerator or a denominator.)
2. Given segments of lengths a, b, and c, explain how you would construct a segment of length x such that:

   a. $\frac{c}{a} = \frac{b}{x}$ b. $x = \frac{ac}{b}$ c. $x = \frac{a^2}{c}$

3. In the diagram at the right Construction 11 was used to construct $\overline{PR}$ given segments $\overline{QR}$ and $\overline{RS}$. Find PR for each pair of values of QR and RS.

   a. $QR = 3$; $RS = 5$ b. $QR = 7$; $RS = 2$
   c. $QR = x$; $RS = y$ d. $QR = 2x$; $RS = 3y$

4. Suppose you wish to use Construction 11 to construct $\overline{PR}$ in the diagram so that $PR = \sqrt{30}$.

   a. State all possible pairs of integral values of QR and RS that you could use in your construction.
   b. Which pair of values that you listed in part (a) would you prefer to use? Explain.

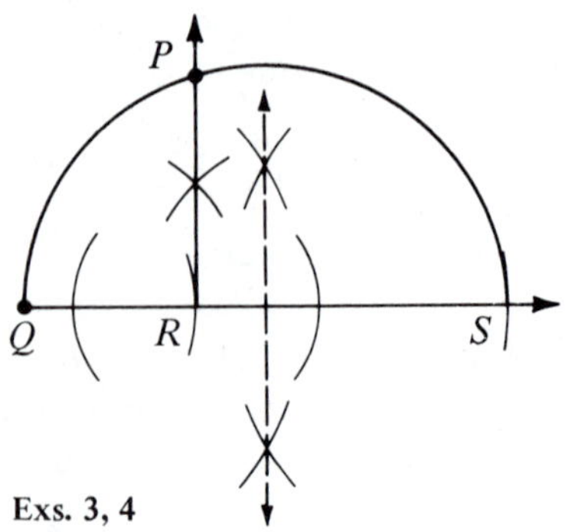

Exs. 3, 4

WRITTEN EXERCISES

In Exercises 1-6 draw segments like those shown and use them in your constructions.

Construct a segment whose length, x, satisfies the given proportion.

A 1. $\frac{a}{c} = \frac{b}{x}$ 2. $\frac{b}{c} = \frac{a}{x}$ 3. $\frac{b}{c} = \frac{x}{a}$

4. $\frac{x}{b} = \frac{b}{c}$ 5. $\frac{a}{x} = \frac{x}{c}$ 6. $\frac{x}{a} = \frac{b}{x}$

In Exercises 7-10 draw a segment like the one shown and let its length be 1. Use the segment to construct a segment of the given length.

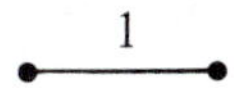

7. $\sqrt{3}$ 8. $\sqrt{5}$ 9. $\sqrt{8}$ 10. $\sqrt{10}$

Draw segments like those shown and use them to construct a segment of the given length.

EXAMPLE xz

SOLUTION Let $a = xz$. Since $a \cdot 1 = xz$, you can write the proportion $\frac{a}{x} = \frac{z}{1}$. You can then use Construction 10 to construct a segment of length a, or xz.

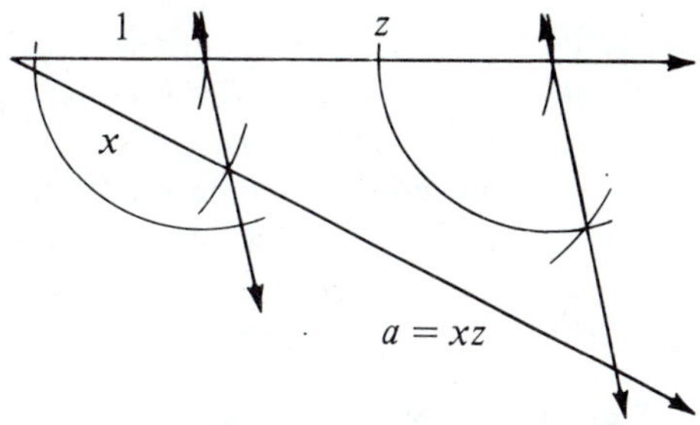

B 11. xy 12. x^2 13. yz 14. $\frac{xz}{3}$

15. $\frac{1}{x}$ 16. $\frac{y}{z}$ 17. $\sqrt{xy}$ 18. $\sqrt{yz}$

19. $\sqrt{y}$ 20. $\sqrt{2y}$ 21. $\sqrt{4x}$ 22. $\sqrt{3xy}$

C 23. $\sqrt{\frac{xy}{3}}$

24. First construct a rectangle that is not a square. Then construct a square that has the same area as the rectangle you constructed.

25. Draw two segments like the ones shown. Construct a right triangle in which the hypotenuse has length a and the altitude to the hypotenuse has length h.

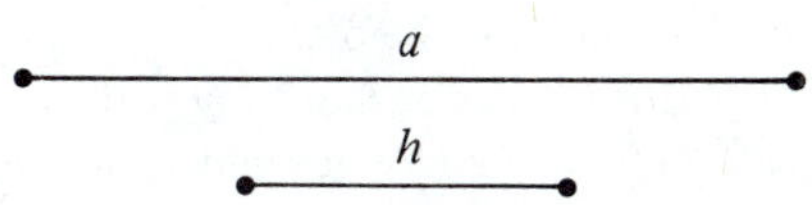

SELF-TEST 2

1. Draw any acute $\angle RST$. Construct a line through T that is parallel to $\overleftrightarrow{RS}$. — Section 10-4
2. Draw a segment 8 cm long. By construction, divide the segment into two segments whose lengths have the ratio 4 : 1.
3. Draw a segment 6 cm long. By construction, divide the segment into three congruent segments.

Draw segments like those shown. Then construct a segment whose length x satisfies the given proportion.

4. $\frac{c}{b} = \frac{x}{a}$
5. $\frac{x}{c} = \frac{b}{a}$
6. $\frac{b}{x} = \frac{x}{c}$
7. $\frac{x}{ac} = \frac{1}{x}$ — Section 10-5

Section 10-6 THE MEANING OF LOCUS

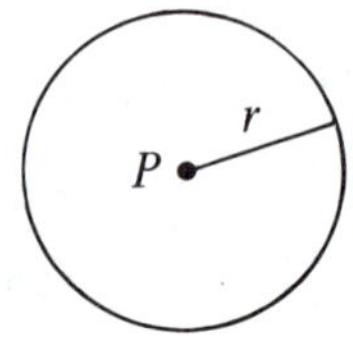

In Section 10-1 a circle was defined as the set of points a given distance (r in the diagram) from a given point (P). The requirement that each point on the circle be a distance r from point P is a condition that separates those points that lie on the circle from those that do not. In general, the set of all points that satisfy a given condition is called a *locus*. A circle could be defined as *the locus of points that are a given distance from a given point.*

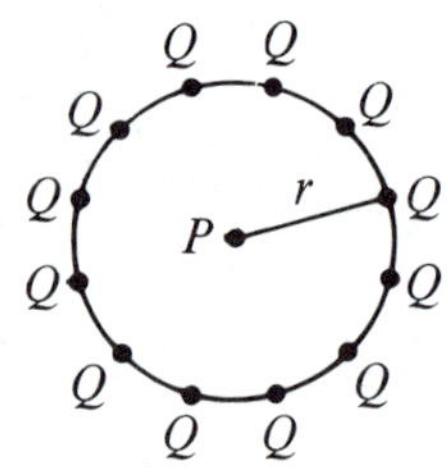

Since it is generally possible to draw a diagram of a locus, a locus is sometimes thought of as the *figure* or *path* traced by a general point that satisfies a given condition. For example, the circle with center P and radius r might be referred to as *the locus of the point Q such that Q is always a distance r from point P.*

In this section you will learn to describe and sketch various loci (plural of locus). Unless otherwise stated, we will assume that all loci mentioned in this book are in a plane.

We first need to define some terms. By Theorem 4-7, there is *exactly one* line perpendicular to a given line through a given point not on the given line. We define the *distance from a point to a line* as the length of the segment that is perpendicular to the given line and whose endpoints are the given point and a point on the given line. In the diagram, d is the distance from point P to line l. The distance d is unique because $\overline{PQ}$ is the only segment that is perpendicular to l and whose endpoints are P and a point on l.

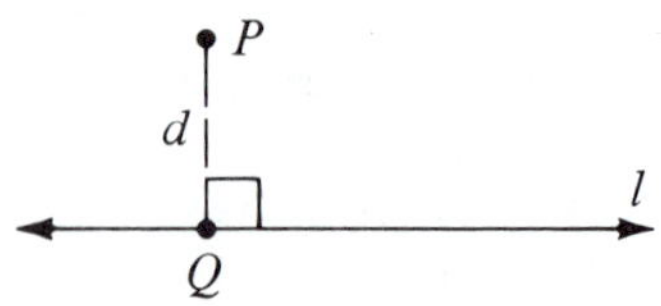

The *distance between two parallel lines* is the distance to one of the lines from any point on the other line. Thus, in the diagram, d is the distance between parallel lines m and n. This distance is the same no matter where along the parallel lines it is measured. (See Oral Exercises 11 and 12 on page 190.) Thus, parallel lines are said to be *everywhere equidistant.*

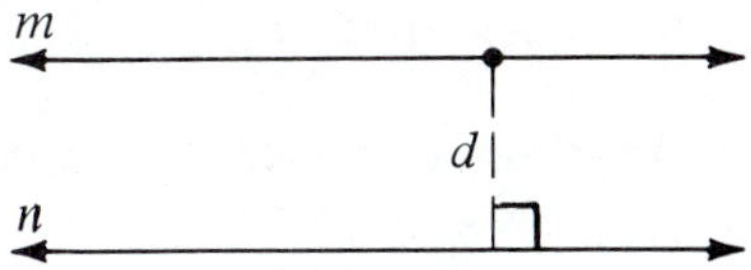

Suppose you are asked to describe the locus of points that are 2 cm from a given line l.

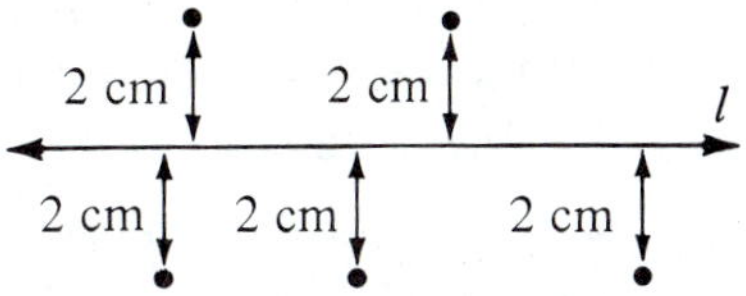

First, make a diagram and on it locate some points in the locus. A point is in the locus if its distance from l is 2 cm.

As you locate more points, the locus becomes clearer. As you can see, the locus is two lines parallel to line l, each 2 cm from line l.

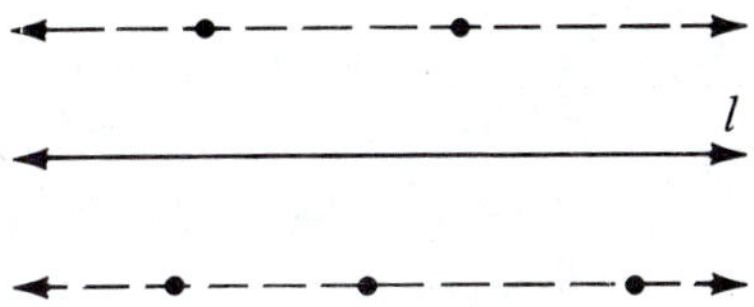

EXAMPLE Describe the locus of points that are equidistant from the sides of $\angle QPR$ and are in the interior of or on $\angle QPR$.

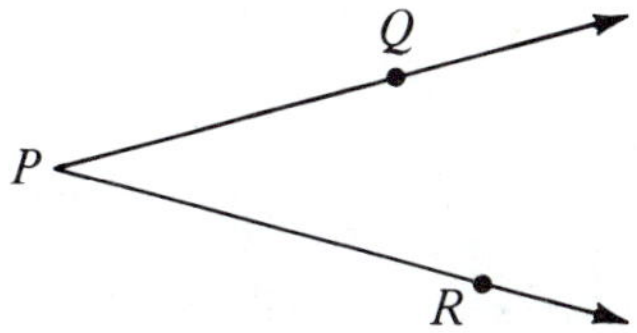

SOLUTION Locate several points that are equidistant from $\overrightarrow{PQ}$ and $\overrightarrow{PR}$. You can see that the locus is the bisector of $\angle QPR$.

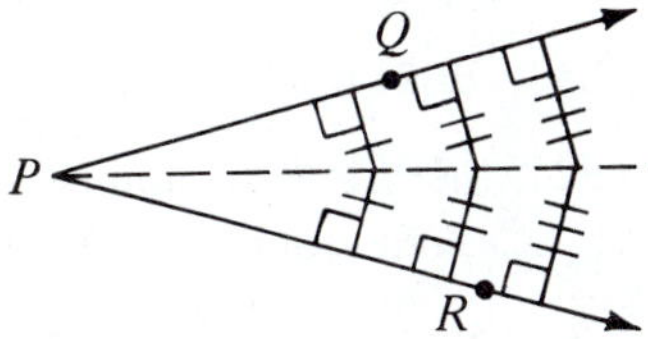

In Exercise 21 you will be asked to prove the following generalization of the previous example: The locus of points that are equidistant from the sides of an angle and are in the interior of or on the angle is the bisector of the angle.

The following example shows how you can sometimes apply theorems to find a locus of points.

EXAMPLE Describe the locus of points that are equidistant from two given points A and B.

SOLUTION Theorem 10-1 states that a point lies on the perpendicular bisector of a segment if and only if it is equidistant from the endpoints of the segment. Thus, the points (such as P and X in the diagram) that are equidistant from A and B are the points that lie on the perpendicular bisector of $\overline{AB}$. You can conclude that: The locus of points equidistant from two given points A and B is the perpendicular bisector of $\overline{AB}$.

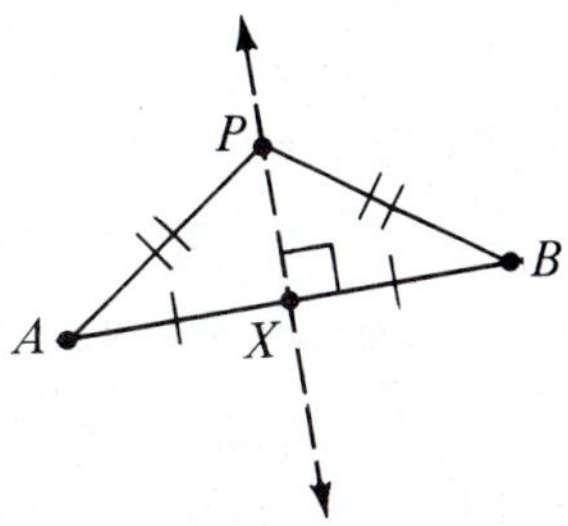

ORAL EXERCISES

1. Draw two parallel lines l and m about 2 cm apart.
   a. Locate 8 points that are equidistant from l and m.
   b. Describe the locus of all points that are equidistant from l and m.

Choose from the following list the set(s) of points that describes the given locus: l, m, $\overleftrightarrow{AE}$, $\overleftrightarrow{BD}$, $\overrightarrow{OC}$, circle O.

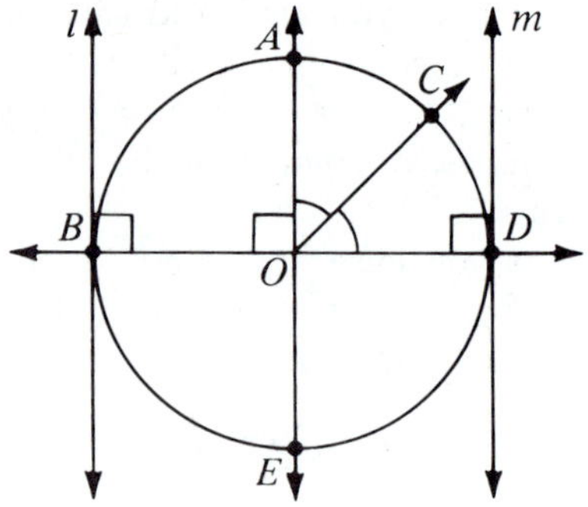

2. The locus of points 3 units from O
3. The locus of points equidistant from B and D
4. The locus of points equidistant from $\overrightarrow{OA}$ and $\overrightarrow{OD}$ and in the interior of or on $\angle AOD$
5. The locus of points equidistant from A and E
6. The locus of points equidistant from lines l and m
7. The locus of points 3 units from $\overleftrightarrow{AE}$

WRITTEN EXERCISES

Draw a diagram of the given locus. Then describe the locus.

A 1. The locus of points equidistant from two given points R and S

2. The locus of points 2 cm from a given point Q
3. The locus of points 1 cm from a given line n
4. The locus of points equidistant from two parallel lines j and k
5. The locus of points equidistant from the sides of right $\angle XYZ$ and in the interior of or on $\angle XYZ$

Exercises 6–15 refer to right $\triangle ABC$. Describe and sketch each locus.

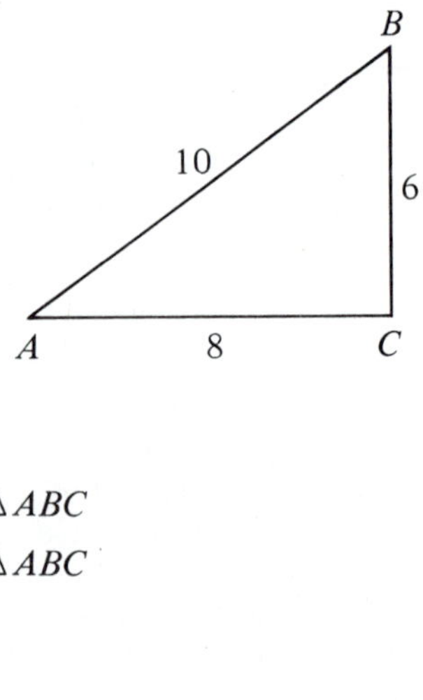

6. The locus of points equidistant from A and B
7. The locus of points 6 units from B
8. The locus of points 5 units from A
9. The locus of points equidistant from A and C
10. The locus of points 6 units from $\overleftrightarrow{AC}$
11. The locus of points 2 units from $\overleftrightarrow{BC}$
12. The locus of points 4 units from the midpoint of $\overline{AC}$

B 13. The locus of points equidistant from $\overrightarrow{CA}$ and $\overrightarrow{CB}$ and on or inside $\triangle ABC$

14. The locus of points equidistant from $\overrightarrow{BA}$ and $\overrightarrow{BC}$ and on or inside $\triangle ABC$
15. The locus of points 5 units from the perpendicular bisector of $\overline{AB}$.

Describe each locus and illustrate it with a diagram.

16. The locus of the midpoints of the radii of a circle of diameter 12 cm

17. The locus of the center of a wheel of a railroad car rolling along a piece of straight track 3 km long

18. The locus of points equidistant from $\overleftrightarrow{AB}$ and $\overleftrightarrow{CD}$ where $\overleftrightarrow{AB}$ and $\overleftrightarrow{CD}$ intersect at X

19. Given $\overline{XY}$ and a constant $k > 0$: Describe the locus of all points P such that area of $\triangle PXY = k$.

20. A penny and a quarter lie on a plane surface. The penny moves around the stationary quarter so that their edges are always touching. Describe the locus of the center of the penny.

21. Parts (a) and (b) prove the following theorem: The locus of points in the interior of or on an angle that are equidistant from the sides of the angle is the bisector of the angle.

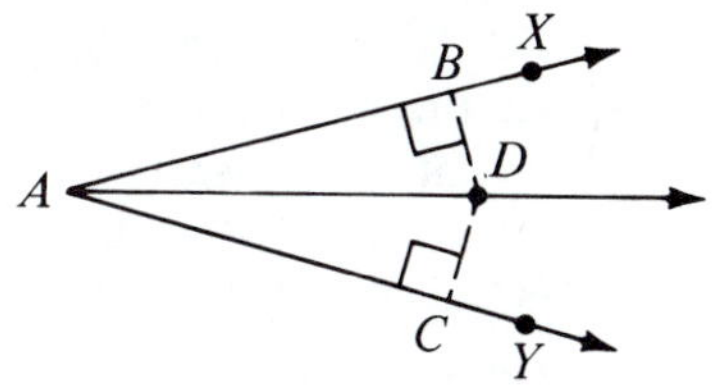

a. Given: D is equidistant from $\overrightarrow{AX}$ and $\overrightarrow{AY}$.
Prove: $\overrightarrow{AD}$ bisects $\angle XAY$.
(*Hint:* Draw $\overline{BD} \perp \overline{AX}$ and $\overline{DC} \perp \overline{AY}$.)

b. Given: $\overrightarrow{AD}$ bisects $\angle XAY$.
Prove: D is equidistant from $\overrightarrow{AX}$ and $\overrightarrow{AY}$.

C 22. Given line segment $\overline{AB}$, describe the locus of the points of intersection of the diagonals of all possible rectangles $ABCD$. Justify your conclusion.

23. Given $\overline{RS}$, describe the locus of all points P such that $\triangle PRS$ is a right triangle with right angle at P. Justify your conclusion.

Section 10-7 INTERSECTION OF LOCI

In this section you will use the basic loci introduced in Section 10-6 to describe loci defined by more than one condition.

Suppose you were asked to draw the locus of points that are 2 cm from a given line l *and* 3 cm from a given point P on l. Note that the desired locus is the intersection of two loci:

1. The locus of points 2 cm from line l; and
2. The locus of points 3 cm from point P.

To determine the locus, draw, on the same diagram, the locus that satisfies each separate condition, as described on the following page.

First draw locus (1). It consists of two lines parallel to l and 2 cm from l.

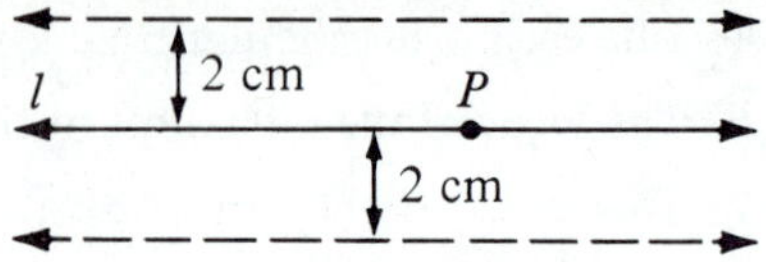

Then draw locus (2) on the same diagram. Locus (2) is a circle with center P and radius 3 cm.

As you can see, the desired locus consists of the four points A, B, C, and D, which are the points of intersection of locus (1) and locus (2).

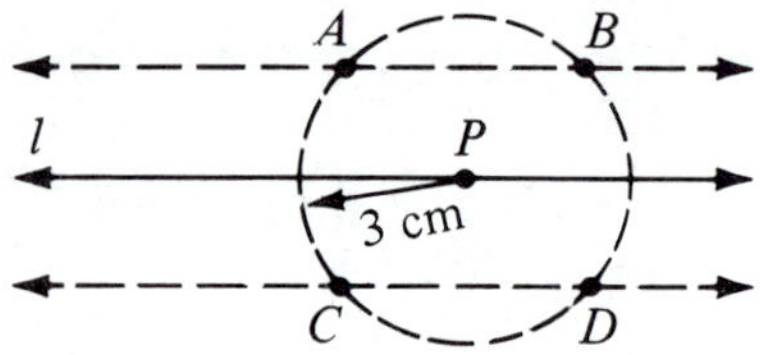

EXAMPLE Describe the locus of points equidistant from three points P, Q, and R.

SOLUTION You must consider two cases:

Case 1: Points P, Q, and R are *noncollinear*. The locus of points equidistant from P and Q is the perpendicular bisector of $\overline{PQ}$. The locus of points equidistant from Q and R is the perpendicular bisector of $\overline{QR}$. Thus, if A is the point of intersection of the two perpendicular bisectors, then $AP = AQ$ and $AQ = AR$, or $AP = AQ = AR$. Conversely, if $AP = AQ = AR$, then point A lies on the perpendicular bisectors of $\overline{PQ}$ and $\overline{QR}$.

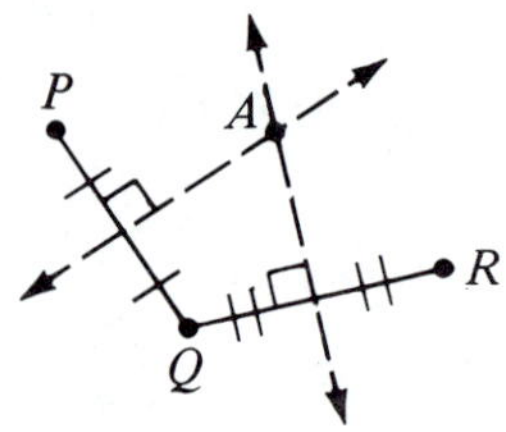

Therefore, the locus of points equidistant from all three points P, Q, and R is point A, the point of intersection of the perpendicular bisectors of $\overline{PQ}$ and $\overline{QR}$.

Case 2: Points P, Q, and R are *collinear*. Since the perpendicular bisectors of $\overline{PQ}$ and $\overline{QR}$ are both perpendicular to the same line $\overleftrightarrow{PR}$, they are parallel. Therefore, they do not intersect, and there is no point that is equidistant from P, Q, and R. The locus contains no points.

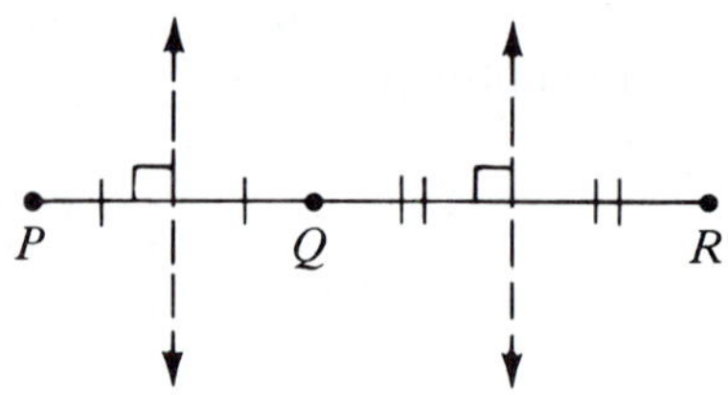

ORAL EXERCISES

1. Suppose you drew a circle with center P. In how many points would the circle intersect line l if the radius of the circle was:
   a. 1 cm b. 2 cm c. 3 cm

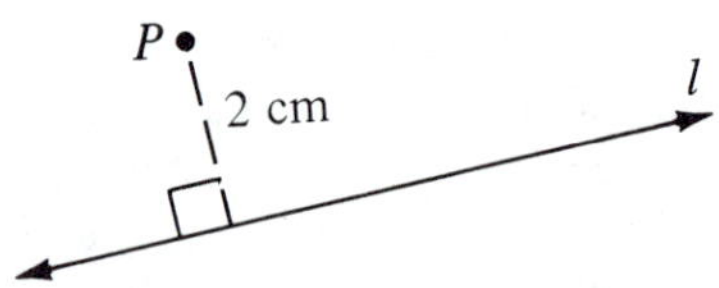

Using the diagram at the top of the next page, choose the point(s) or set(s) of points that best describes the given locus from the following list:

$A, B, C, D, E, F, l, m, \overleftrightarrow{CD}, \overleftrightarrow{GE}$, no points.

2. The locus of points on line l and 8 units from point O
3. The locus of points 8 units from point O and equidistant from lines l and m

4. The locus of points 8 units from point O and equidistant from points C and D

5. The locus of points equidistant from points A and B and equidistant from lines l and m

6. The locus of points 3 units from $\overleftrightarrow{CD}$ and 8 units from point O

7. The locus of points equidistant from points A and B and equidistant from points C and D

8. The locus of points 4 units from point O and equidistant from lines l and m

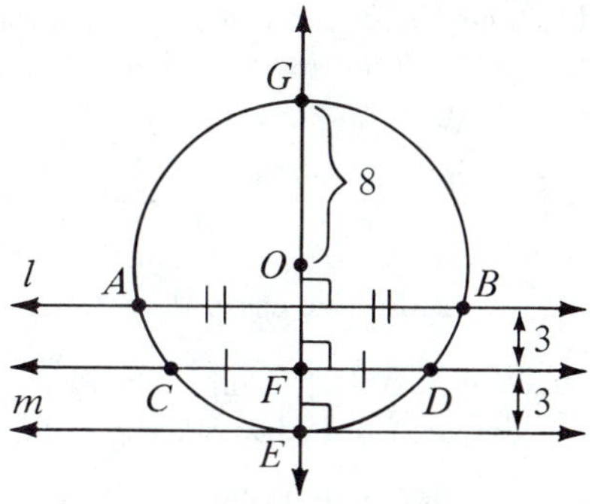

Exs. 2–8

WRITTEN EXERCISES

In Examples 1-8, draw a diagram of the given locus. Then describe the locus.

A 1. Parallel lines l and m are 2 cm apart and A is a point on line l. What is the locus of points equidistant from l and m and 3 cm from A?

2. Points P and Q are 4 cm apart. What is the locus of points equidistant from P and Q and:
   a. 3 cm from P b. 2 cm from P c. 1 cm from P

3. A and B are points on a given line l and P is a point not on l. What is the locus of points equidistant from A, B, and P?

4. Points R and S are 5 cm apart. What is the locus of points 2 cm from R and 1 cm from S?

5. Given obtuse $\angle Y$, what is the locus of points in the interior of $\angle Y$ that are equidistant from the sides of $\angle Y$ and 3 cm from Y?

6. Given acute $\triangle ABC$, what is the locus of points equidistant from $\overrightarrow{AB}$ and $\overrightarrow{AC}$ and also equidistant from A and B?

7. Given quadrilateral $WXYZ$, what is the locus of points equidistant from X, Y, and Z?

8. Given a line l and a point P that is 2 cm from l, what is the locus of points 4 cm from P and 1 cm from l?

In Exercises 9-12, describe each locus and illustrate it with a diagram.

B 9. The locus of points equidistant from the vertices of rectangle $JKLM$

10. The locus of points equidistant from the vertices of right $\triangle ABC$

11. The locus of points equidistant from the sides of rectangle $PQRS$ where $PQRS$ is not a square

12. The locus of points equidistant from the sides of rhombus $EFGH$

13. The locus of points equidistant from A and B and 5 units from A consists of two points, X and Y, such that $XY = 8$. Find the length AB.

14. The locus of points equidistant from two lines l and m and 13 units from a point P on line l consists of two points R and S such that $RS = 24$. How far apart are the lines l and m?

In Exercises 15-17, describe each locus and illustrate it with a diagram.

15. The locus of points 2 cm from a given point P and 1 cm from a given line k (Consider five cases.)

C 16. The locus of points equidistant from parallel lines m and n and also equidistant from parallel lines p and q (Consider three cases.)

17. The locus of points equidistant from the vertices W, X, Y, and Z of quadrilateral $WXYZ$ (Consider two cases.)

18. Line l intersects parallel lines m and n. Describe the locus of points equidistant from all three lines and illustrate the locus with a diagram.

19. State the number of points in each possible case of the following locus and illustrate each case with a diagram: The locus of points that are equidistant from intersecting lines l and m and at a given distance from a given point P.

Section 10-8 CONSTRUCTION OF LOCI

In this section you will use the loci and constructions discussed in previous sections to construct figures formed by points that satisfy one or more conditions.

Suppose you are asked to construct a $\triangle PQR$ where $QR = x$, the altitude to $\overline{QR}$ has length y, and the median to $\overline{QR}$ has length z. In order to determine a strategy for constructing the triangle, you would first make a sketch of the completed figure that contains all the given information.

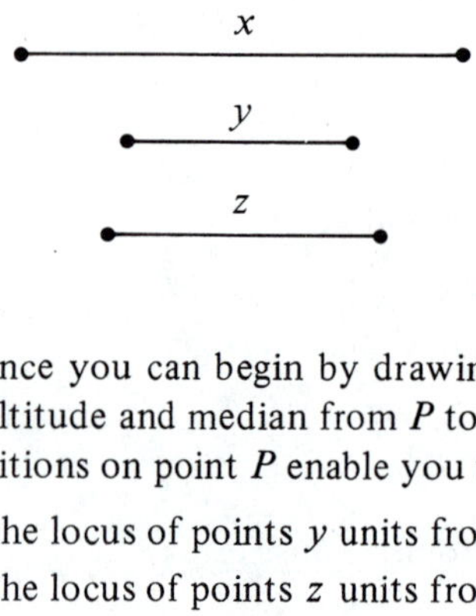

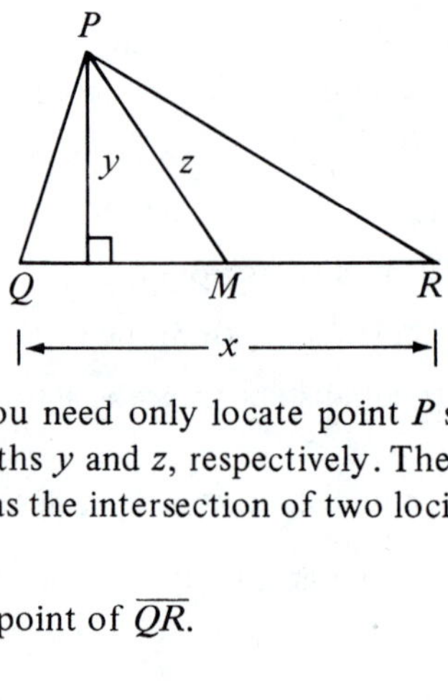

Since you can begin by drawing $QR = x$, you need only locate point P so that the altitude and median from P to $\overline{QR}$ have lengths y and z, respectively. These two conditions on point P enable you to think of P as the intersection of two loci:

(1) the locus of points y units from $\overleftrightarrow{QR}$; and

(2) the locus of points z units from M, the midpoint of $\overline{QR}$.

Locus (1) is two lines parallel to $\overleftrightarrow{QR}$ and y units from $\overleftrightarrow{QR}$ (you need only one of these lines for this construction). Locus (2) is a circle with center M and radius z.

Procedure:

1. Draw a ray and mark off $QR = x$.
2. Use Construction 6 to construct a line perpendicular to $\overrightarrow{QR}$ at any convenient point A.
3. Mark off $AB = y$.
4. Use Construction 6 to construct line l perpendicular to $\overleftrightarrow{AB}$ at B.
5. Use Construction 4 to locate the midpoint M of $\overline{QR}$.
6. With M as center and radius z, draw an arc that intersects line l at point P.
7. Draw $\overline{PQ}$ and $\overline{PR}$.

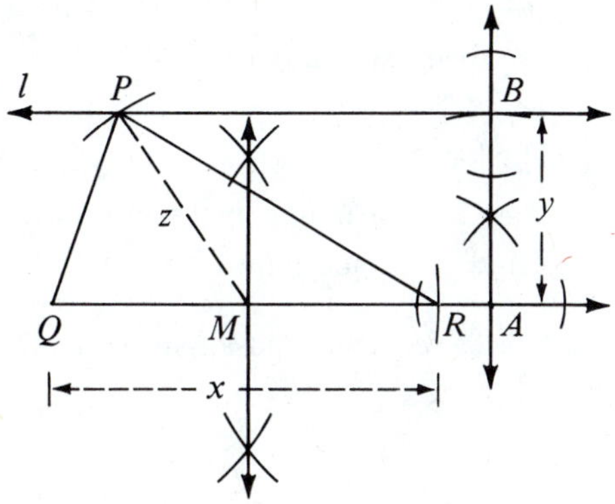

$\triangle PQR$ has $QR = x$, the altitude from P to $\overline{QR}$ of length y, and the median from P to $\overline{QR}$ of length z.

ORAL EXERCISE

The diagrams below represent the four steps in the construction of a $\triangle ABC$ in which $m\angle A = m\angle P$, the median to $\overline{AB}$ has length q, and the altitude to $\overline{AB}$ has length r. Explain each step of the construction.

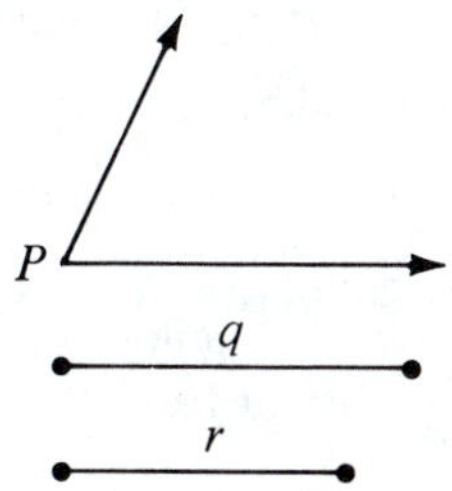

a.

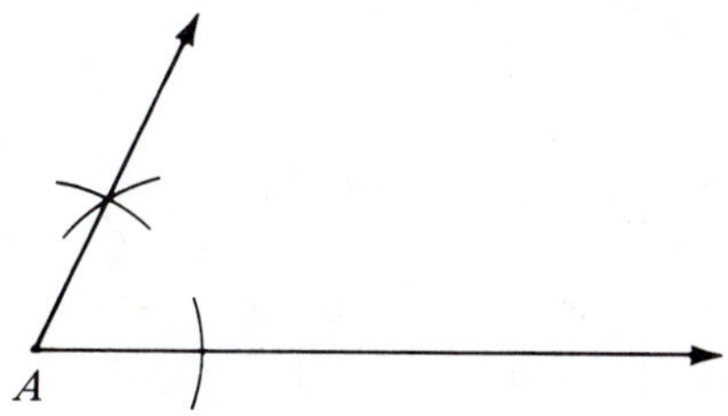

b.

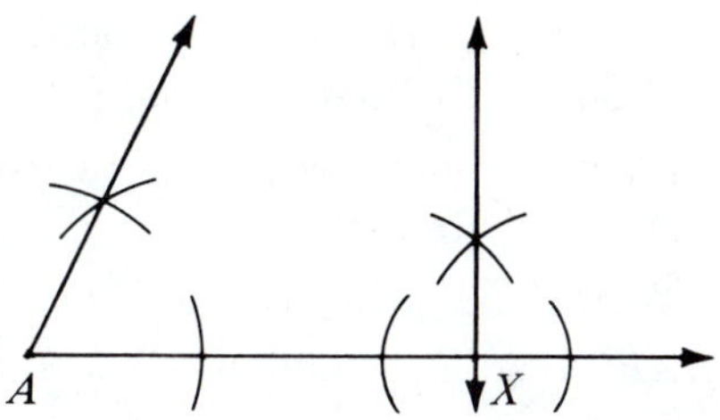

c.

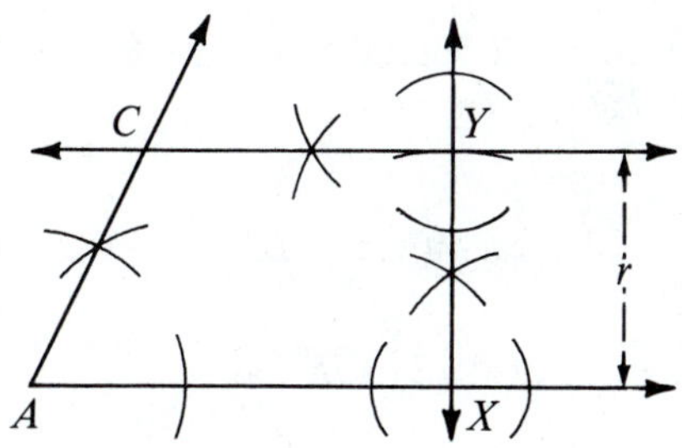

d.

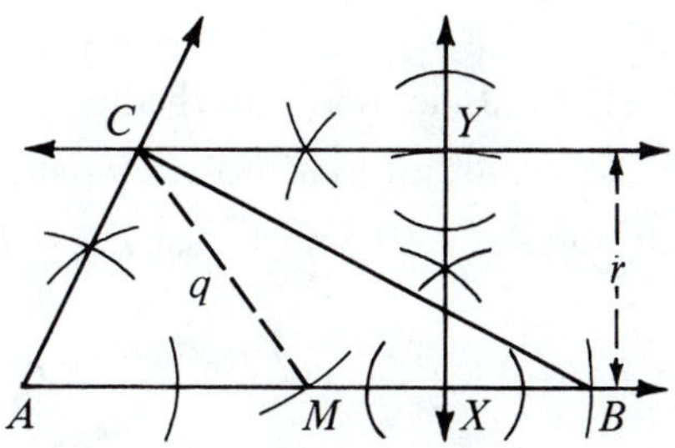

WRITTEN EXERCISES

In Exercises 1-4, draw an acute $\triangle XYZ$ and construct the given locus.

A 1. The locus of points equidistant from points Y and Z

2. The locus of points equidistant from $\overrightarrow{YX}$ and $\overrightarrow{YZ}$ and in the interior of or on $\angle XYZ$

3. The locus of points that lie on $\overline{YZ}$ and are equidistant from the sides of $\angle YXZ$

4. The locus of points in the interior of $\angle XZY$ that are equidistant from X and Z and also equidistant from $\overrightarrow{ZX}$ and $\overrightarrow{ZY}$

Draw three segments and an angle like those shown and use them in your constructions.

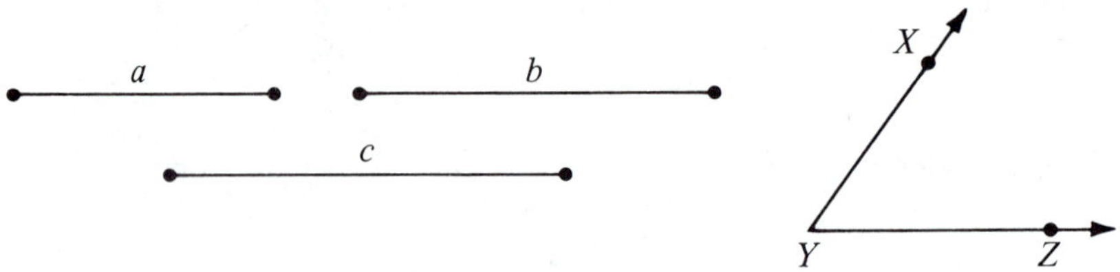

5. Draw two points V and W such that $VW = c$. Construct the locus of points that are a distance a from V and a distance b from W.

6. Draw a line k. Construct the locus of points a distance b from k.

7. Construct two parallel lines m and n a distance c apart. Construct the locus of points equidistant from m and n.

8. Draw two points P and Q a distance c apart. Construct the locus of points equidistant from P and Q and a distance b from P.

B 9. Draw two lines, k and l, and label their point of intersection as Q. Locate a point P on k such that $PQ = a$. Construct the locus of points that are equidistant from k and l and a distance c from P.

10. Draw a line l and a point P on l. Construct the locus of points a distance c from P and a distance a from l.

11. Construct an $\angle ABC$ congruent to $\angle Y$. Construct the locus of points in the interior of $\angle ABC$ that are equidistant from $\overrightarrow{BA}$ and $\overrightarrow{BC}$ and also a distance a from $\overleftrightarrow{BC}$.

12. Construct $\overline{PQ}$ so that $PQ = c$. Construct the locus of points R such that for every $\triangle PQR$, the altitude from R to the line containing $\overline{PQ}$ has length b.

13. Repeat Exercise 12, substituting median for altitude.

14. Construct an isosceles $\triangle DEF$ such that $DE = DF = b$ and the altitude to $\overline{DE}$ has length a.

15. Construct a $\triangle JKL$ such that $m\angle J = m\angle Y$, $JL = b$, and the median to $\overline{JK}$ has length c.

16. Draw a $\triangle XYZ$ that is not isosceles. Construct an isosceles $\triangle PQR$ that has the same area as $\triangle XYZ$.

C 17. Construct an equilateral $\triangle DEF$. Then construct an isosceles $\triangle ABC$ that is not equilateral but has the same area as $\triangle DEF$.

In Exercises 18 and 19, use the indicated line segment and angle shown on page 416.

18. Construct a right $\triangle PQR$ with right $\angle R$ such that $m\angle P = m\angle Y$ and the altitude to $\overline{PQ}$ has length a.

19. Construct a rectangle such that the shorter sides of the rectangle have length a and an angle formed by the intersection of the diagonals is congruent to $\angle Y$.

SELF-TEST 3

In the diagram for Exercises 1-5, $ABCD$ is a square. Choose the point(s) or set(s) of points that describes the given locus from the following list:
$A, B, N, Q, \overleftrightarrow{AB}, \overleftrightarrow{QN}, \overleftrightarrow{AC}, \overleftrightarrow{MP}, \overleftrightarrow{DC}, \overrightarrow{CA}, \overrightarrow{AC}$

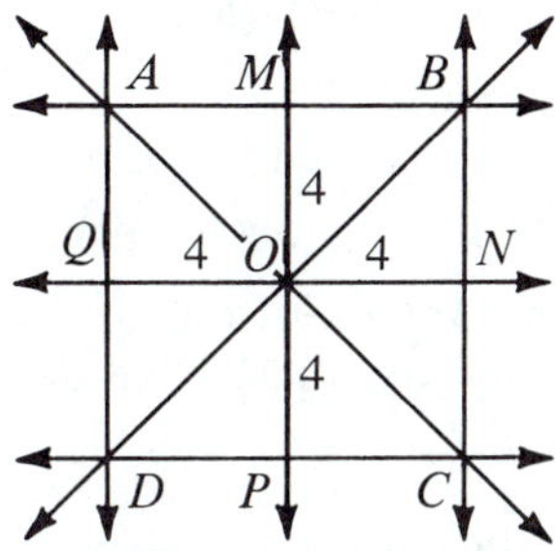

Section 10-6

1. The locus of points equidistant from $\overleftrightarrow{AD}$ and $\overleftrightarrow{BC}$
2. The locus of points equidistant from B and D
3. The locus of points 4 units from $\overleftrightarrow{QN}$
4. The locus of points equidistant from $\overrightarrow{AB}$ and $\overrightarrow{AD}$ and on or in the interior of $\angle BAD$
5. The locus of points equidistant from points A and D and 4 units from point O

Section 10-7

In Exercises 6–8, draw a diagram of the given locus. Then describe the locus.

6. Points R and S are 5 cm apart. What is the locus of points 3 cm from R and 4 cm from S?
7. Point X is 3 cm from line l. What is the locus of points 6 cm from X and 3 cm from l?
8. Parallel lines m and l are 2 cm apart. Points P and Q lie on line l and are 4 cm apart. What is the locus of points equidistant from P and Q and also equidistant from l and m?

In Exercises 9 and 10 draw segments like those shown.

Section 10-8

9. Draw a line l. Construct the locus of points a distance a from l.
10. Construct $\overline{PQ}$ of length c. Then construct the locus of points a distance a from P and a distance b from Q.
11. Construct an acute $\angle XYZ$ with $XY = b$. Then construct the locus of points that lie in the interior of $\angle XYZ$ and are equidistant from the sides of $\angle XYZ$ and also a distance a from point X.

CHAPTER REVIEW

Draw segments and angles like those shown. Use them as needed in your constructions.

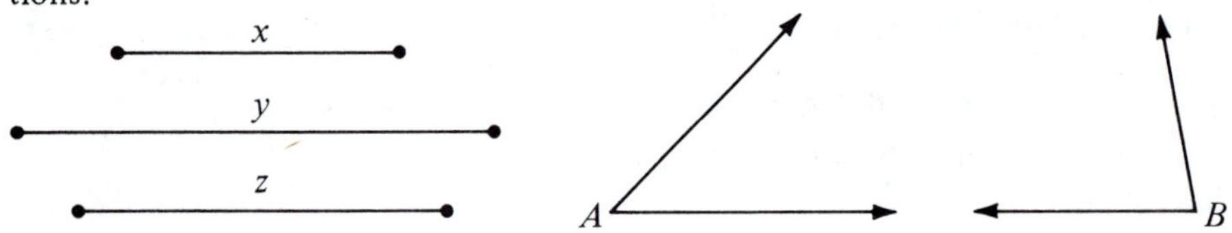

1. Construct $\triangle PQR$ such that $PR = x$, $QR = z$, and $\angle R \cong \angle B$. Section 10-1

2. Construct a $22\frac{1}{2}^\circ$ angle. Section 10-2

3. Draw an acute $\triangle DEF$.
   a. Construct the perpendicular bisector of $\overline{DE}$.
   b. Draw the median from F to $\overline{DE}$.

4. Draw an obtuse $\triangle PQR$. Construct the circumscribed circle of $\triangle PQR$.

5. Draw an acute $\triangle RST$. Section 10-3
   a. Construct the altitude from R to $\overline{ST}$.
   b. Construct a line through T perpendicular to $\overleftrightarrow{RT}$.

6. Draw a larger version of the given diagram and use it to construct a line through A parallel to m. Section 10-4

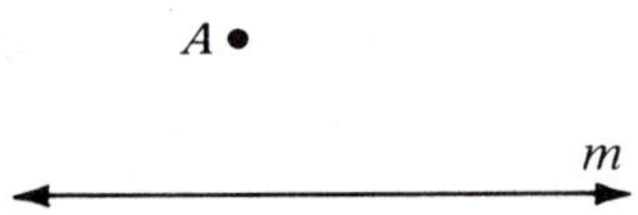

7. Construct a line segment of length $\frac{1}{3}y$.

8. Construct a segment of length t such that $\frac{y}{x} = \frac{z}{t}$. Section 10-5

9. Construct a segment whose length is the geometric mean between x and y.

10. Draw a line l. Describe the locus of points 3.5 cm from l and illustrate the locus with a diagram. Section 10-6

11. Parallel lines r and s are 4 cm apart. Point T is on line r. Draw a diagram showing the locus of points equidistant from r and s and 3 cm from T. How many points are in the locus? Section 10-7

12. Draw an acute $\angle XYZ$ such that $XY \neq YZ$. Construct the locus of points in the interior of $\angle XYZ$ that are equidistant from $\overrightarrow{YX}$ and $\overrightarrow{YZ}$ and are also equidistant from X and Y. Section 10-8

CHAPTER TEST

In Exercises 1-4 draw a larger version of the given diagram and use it in your construction.

1. Bisect $\angle P$.

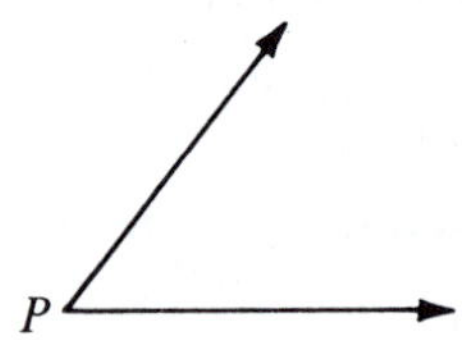

2. Construct $\angle R$ so that $\angle R \cong \angle A$.

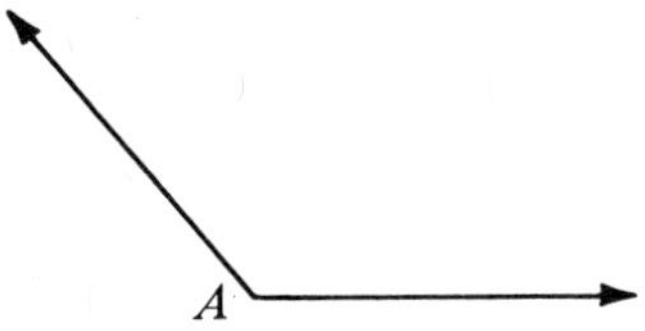

3. Construct $\triangle DEF$ so that $\triangle DEF \cong \triangle ABC$.

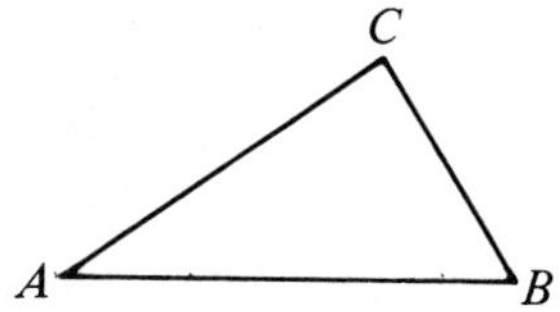

4. Construct the locus of points equidistant from points P and Q.

5. Construct a 30° angle.

6. Draw a larger version of the given diagram and use it to construct:
   a. a line through X perpendicular to l;
   b. a line through Y perpendicular to l.

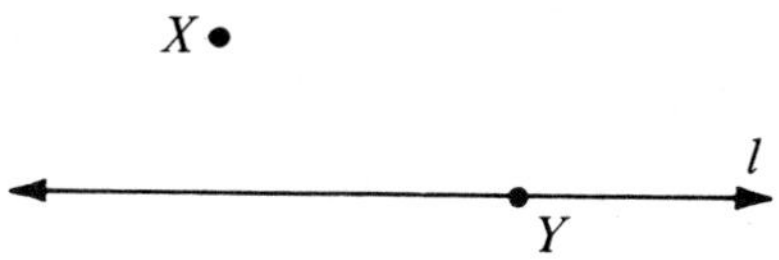

7. Draw an acute $\triangle RST$. Construct the circumscribed circle for $\triangle RST$.

In Exercises 8 and 9, choose from the following list the set(s) of points that describes the given locus: D, $\overrightarrow{BD}$, l, m, n

8. The locus of points 1 cm from line m

9. The locus of points in the interior of or on $\angle ABC$ and equidistant from $\overrightarrow{BA}$ and $\overrightarrow{BC}$

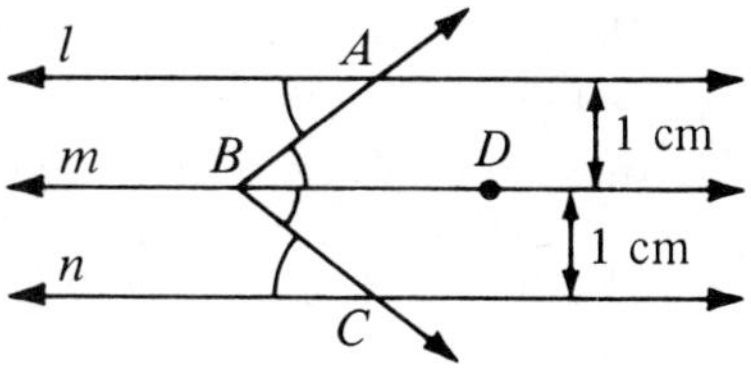

10. Draw a segment like the one shown at the right and let its length be 1. Construct a segment of length $\sqrt{3}$.

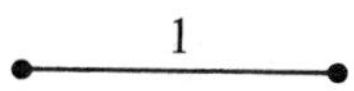

11. Draw a segment whose length is approximately one half the width of your paper. Divide the segment into two segments whose lengths are in the ratio 3 : 2.

12. Draw segments like those shown. Then construct a segment of length x such that $\frac{r}{p} = \frac{q}{x}$.

13. Given points F and G and parallel lines l and m as in the diagram at the right. Describe the locus of points equidistant from F and G and also equidistant from l and m and illustrate the locus with a diagram.

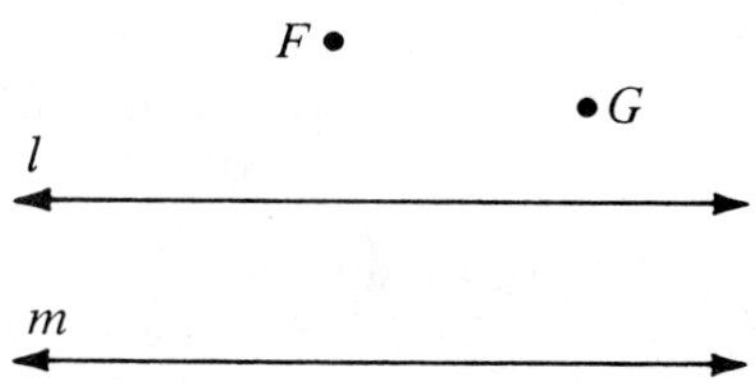

Indicate the best answer by writing the appropriate letter.

1. The locus of points equidistant from two points consists of:
   a. a point b. a line segment c. a line d. a pair of lines

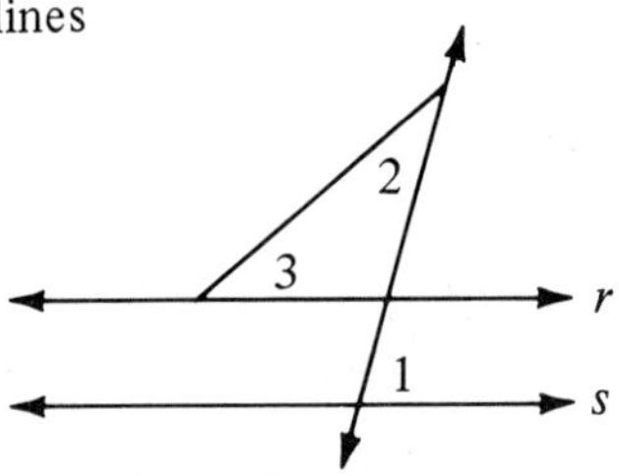

2. The measures of three angles of a triangle are in the ratio of 2:3:5. Find the measure of the *smallest* angle of the triangle.
   a. 18 b. 54 c. 40 d. 36

3. In the figure, $r \parallel s$, $m\angle 2 = x$, and $m\angle 3 = y$. Find $m\angle 1$.
   a. $x + y$ b. $x - y$
   c. $180 - (x + y)$ d. $180 - (x - y)$

4. A triangle has sides of length 8, 4, and 8. The shortest side of a similar triangle has length 6. Find the perimeter of the larger triangle.
   a. $13\frac{1}{3}$ b. 15 c. 30 d. 45

5. Find the value of sin 45°.
   a. $\sqrt{2}$ b. $\frac{\sqrt{2}}{2}$ c. 1 d. $2\sqrt{2}$

6. Two complementary angles have measures $5x + 5$ and $7x + 1$. Find the value of x.
   a. 2 b. 47 c. 14.5 d. 7

7. In $\triangle XYZ$, $XY = 6$, $YZ = 8$, and $XZ = 12$. M is the midpoint of $\overline{XY}$ and N is the midpoint of $\overline{XZ}$. Find MN.
   a. 3 b. 4 c. 5 d. 6

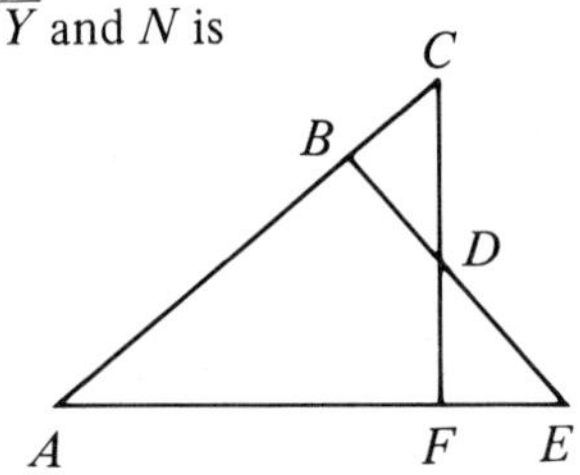

8. Which information would *not* be sufficient to prove two triangles congruent?
   a. $\overline{AC} \cong \overline{AE}$; $\angle C \cong \angle E$ b. $\overline{AC} \cong \overline{AE}$; $\overline{BC} \cong \overline{FE}$
   c. $\overline{DB} \cong \overline{DF}$; $\overline{DC} \cong \overline{DE}$ d. $\overline{AB} \cong \overline{AF}$; $\overline{BE} \cong \overline{CF}$

9. The altitude drawn to the hypotenuse of a right triangle divides the hypotenuse into two segments of lengths 4 and 25. What is the length of the altitude?
   a. 100 b. 50 c. 10 d. $6\frac{1}{4}$

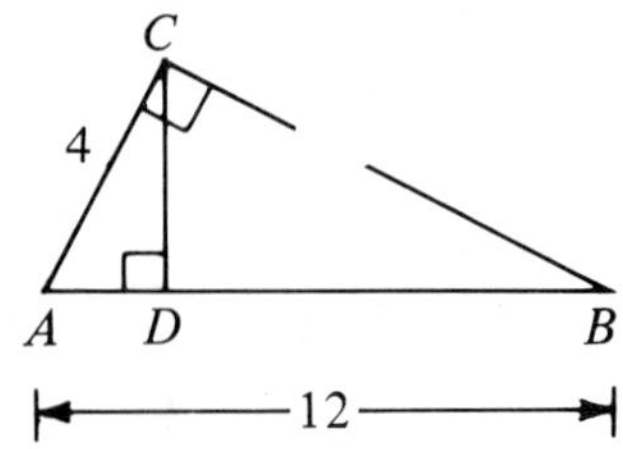

10. Use the information in the diagram to find AD.
    a. 2 b. $2\sqrt{3}$ c. $2\sqrt{2}$ d. $1\frac{1}{3}$

11. In $\triangle JKL$, $\overrightarrow{JM}$ bisects $\angle J$, and M lies on $\overline{KL}$. If $JK = 15$, $KM = 12$, and $ML = 8$, find JL.
    a. 12 b. 10 c. 25 d. 22.5

12. The length of the shorter diagonal $\overline{AC}$ of a rhombus $ABCD$ is 12, and $m\angle ABC = 60$. Find the length of a side of the rhombus.
    a. 24 b. $4\sqrt{3}$ c. $12\sqrt{3}$ d. 12

13. Which equation has $x = 2 \pm \sqrt{2}$ as its solution?
a. $x^2 - 4x + 2 = 0$ b. $x^2 - 9x + 16 = 0$
c. $x^2 + 9x + 16 = 0$ d. $x^2 + 4x + 2 = 0$

14. Given premises P_1: $p \to \sim q$ and P_2: q.
What can you conclude?
a. p b. $\sim p$ c. $\sim q$ d. no conclusion possible

15. If $\frac{w}{x} = \frac{y}{z}$, then which proportion is *not* necessarily true?
a. $\frac{w}{y} = \frac{x}{z}$ b. $\frac{w+x}{x} = \frac{y+z}{z}$ c. $\frac{w}{z} = \frac{y}{x}$ d. $\frac{w+y}{x+z} = \frac{w}{x}$

16. When $3a^2 + 5a + 1$ is divided by $a + 2$, what is the remainder?
a. 23 b. −21 c. 3 d. −1

17. What is the length of an altitude of an equilateral triangle whose sides each have length 6?
a. $3\sqrt{3}$ b. 6 c. 3 d. $6\sqrt{3}$

18. What is the negation of $\exists_x x < 3$?
a. $\exists_x x \geqslant 3$ b. $\forall_x x < 3$ c. $\forall_x x \geqslant 3$ d. $\exists_x x \leqslant 3$

19. Which statement is logically equivalent to $\sim(\sim p \vee q)$?
a. $p \vee \sim q$ b. $\sim p \wedge q$ c. $p \wedge \sim q$ d. $\sim p \vee \sim q$

20. Point P is 5 cm from line l. What is the total number of points 7 cm from P and 2 cm from l?
a. 4 b. 3 c. 2 d. 1

21. Which value of z will make the following sentence false?
$(2z - 7 > -3) \vee (2z - 7 \leqslant -13)$
a. 2 b. 3 c. −3 d. −4

22. Which of the following statements is true about the diagram at the right?
a. $\triangle ABC \sim \triangle DBE$ by the AA Similarity Theorem.
b. $\triangle ABC \sim \triangle EBD$ by the SAS Similarity Theorem.
c. $\triangle ABC \sim \triangle EBD$ by the AA Similarity Theorem.
d. The two triangles are not similar.

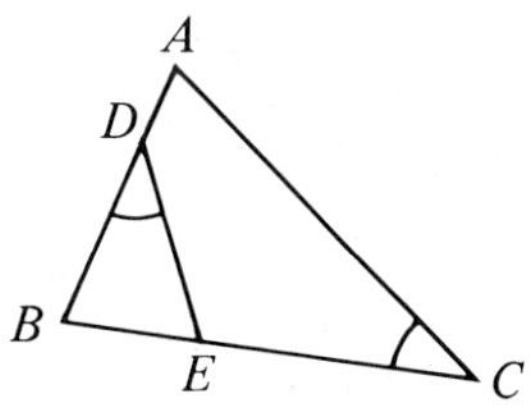

23. One isosceles right triangle has legs of length 3. Another isosceles right triangle has a hypotenuse of length 9. Find the ratio of their perimeters.
a. 1:3 b. $\sqrt{2}$:3 c. 1:9 d. 2:9

24. In the diagram, $\overline{DE} \parallel \overline{BC}$, $AB \neq AC$. Which equation is correct?
a. $\frac{AE + EC}{EC} = \frac{DE + BC}{BC}$ b. $\frac{AD}{AE} = \frac{DB}{AC}$
c. $\frac{AE}{AC} = \frac{DE}{BC}$ d. $AD \cdot AB = AE \cdot AC$

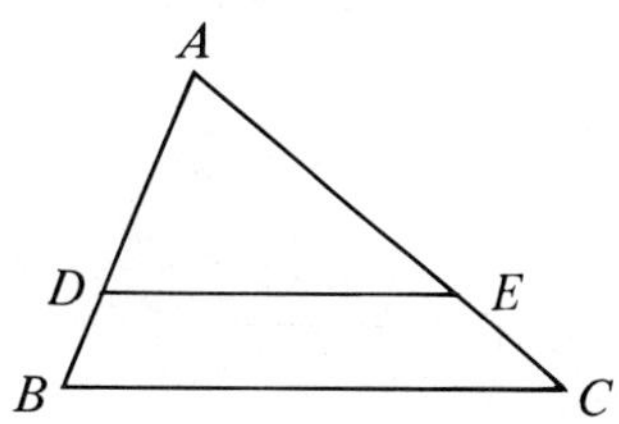

25. In $\triangle RST$, $m\angle R = 110$. Which statement is *false*?

a. $ST > RS$ b. $RS + RT \geqslant ST$
c. $m\angle R > m\angle S + m\angle T$ d. $m\angle S \geqslant 70$

26. Negate the sentence "No kindness goes unnoticed."

a. All kindnesses go unnoticed. b. Some kindnesses go unnoticed.
c. No kindness is noticed. d. Some kindnesses are noticed.

27. An isosceles trapezoid has bases of lengths 8 and 12 and an angle of measure 60. Find its area.

a. 28 b. 20 c. $20\sqrt{2}$ d. $20\sqrt{3}$

28. If $VWXYZ$ is a regular pentagon, find $m\angle WVX$.

a. 18 b. 36 c. 54 d. 72

29. The sum of two numbers is 4 and their product is -1. One of the numbers is:

a. $2 + \sqrt{5}$ b. $2 - \sqrt{3}$ c. $-2 - \sqrt{5}$ d. $-2 + \sqrt{3}$

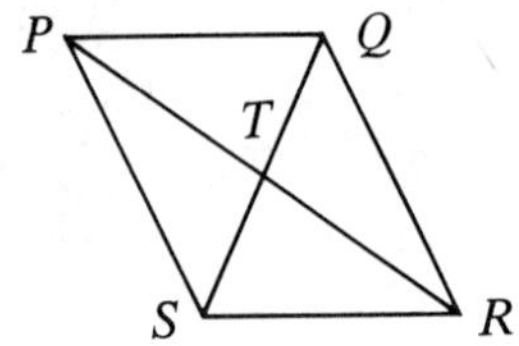

30. If T is the midpoint of $\overline{PR}$ and $\overline{QS}$, which statement need *not* be true?

a. $\angle QRS \cong \angle SPQ$ b. $\overline{QR} \parallel \overline{PS}$
c. $\overrightarrow{QS}$ bisects $\angle PQR$. d. $\overline{QP} \cong \overline{RS}$

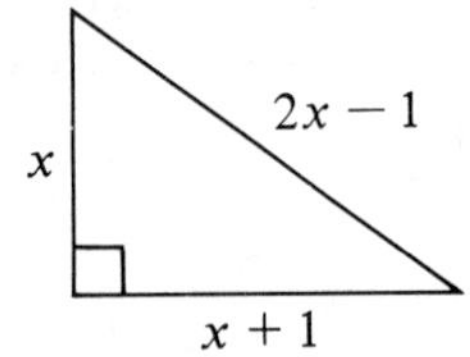

31. Find the area of the right triangle at the right.

a. 12 b. 3
c. 5 d. 6

32. In $\triangle PQR$, S and T are points on $\overline{QP}$ and $\overline{QR}$, respectively, so that $\overline{ST} \parallel \overline{PR}$. If $QS = 4$ and $SP = 5$, find the ratio of the area of $\triangle QST$ to the area of $\triangle QPR$.

a. 4:5 b. 16:25 c. 16:81 d. 4:9

Choose *one* of Exercises 33 and 34.

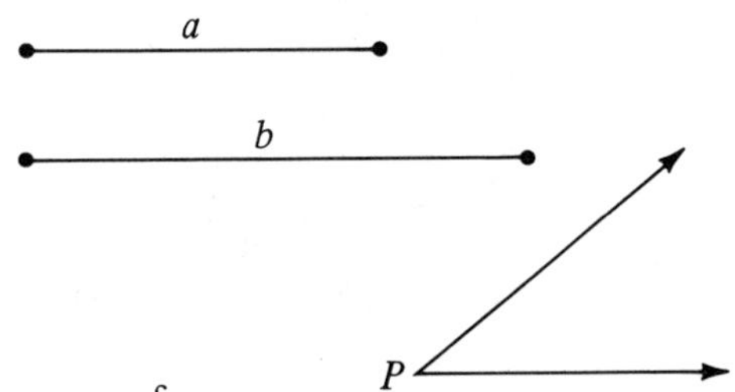

33. Draw line segments and an angle like those shown. Construct $\triangle XYZ$ with $XY = a$, $YZ = b$, and $m\angle Y = m\angle P$.

34. Draw any acute $\triangle ABC$. Construct the altitude from A to $\overline{BC}$.

Choose *one* of Exercises 35 and 36 and write a two-column proof.

35. Given: $\overline{AD} \perp \overline{BC}$;
$\angle B \cong \angle C$
Prove: $\overrightarrow{AD}$ bisects $\angle BAC$.

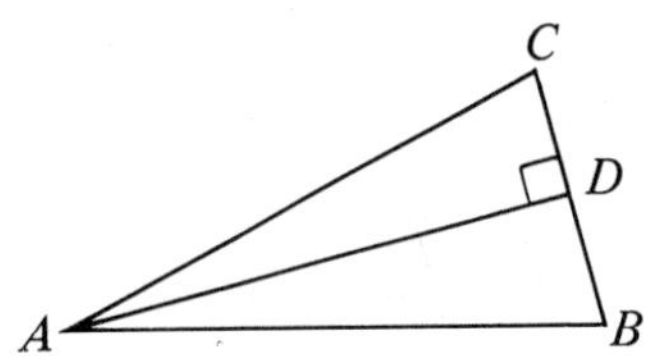

36. Given: $ABCD$ is a ▱;
$\overline{BE} \cong \overline{DF}$
Prove: $BEDF$ is a ▱.

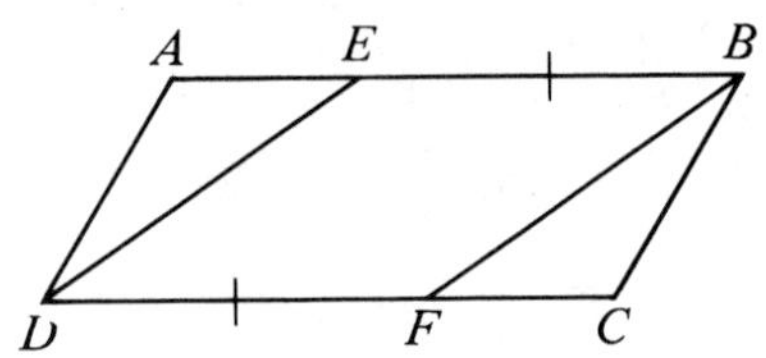

Careers — Oceanographer

The earth's oceans have long been a source of food and a means of transportation. In modern times the vastness of the oceans has encouraged their use as chemical and nuclear dumping grounds. The oceans have been overfished and liberally spewed with hydrocarbons from oil tankers and deep-sea rigs. Modern technology, however, has helped as well as hurt the oceans. Satellites, underwater craft, specialized measuring instruments, and computers are finally opening the secrets of the deep. This is the province of the oceanographer.

There are five primary specialties of oceanography. Biological oceanographers, or marine biologists, study the plant and animal life of the ocean. They gather statistics to determine the effects of pollution and commercial fishing on marine life. The physics of ocean water – the waves, tides, currents, and evaporation rates – fall in the domain of physical oceanographers.

Geological oceanographers study what is underneath the water: mountain ranges, rocks, sediments, and valuable minerals. The chemical composition of the water and ocean floor are left to the chemical oceanographers. Oceanographic engineers design and build the instruments for these types of research.

Oceanographers gather data and conduct experiments at sea. They also work in laboratories on land, examining specimens and analyzing theories such as continental drift. According to that theory, continents that were once joined together have drifted apart and caused the sea floor to spread.

Most oceanographers work for colleges and universities in research and teaching. The federal government is also a large employer, especially the Navy and the National Oceanic and Atmospheric Administration (NOAA).

A major in oceanography, biology, earth or physical sciences, mathematics, or engineering is good preparation for ocean studies. If chemical oceanography interests you, take courses in chemistry – and likewise for the other specialties. Statistics and computer science courses are important for building data collection and analysis skills. Most oceanographers obtain advanced degrees.

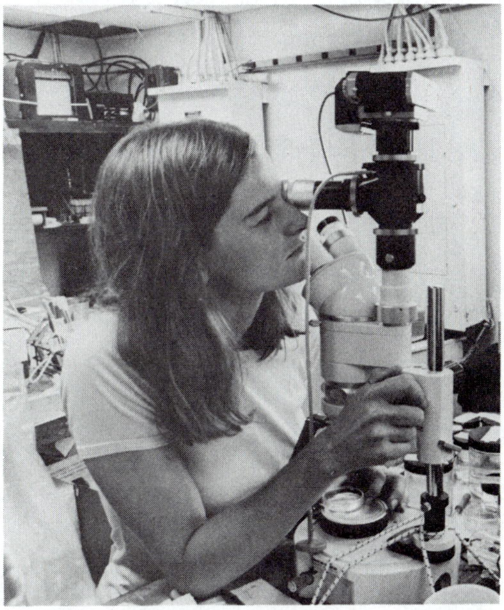

VOCABULARY REVIEW

Be sure that you understand the meaning of these terms:

circle, p. 386
diameter, p. 386
radius, p. 386
arc, p. 386
perpendicular bisector, p. 390
circumscribed circle, p. 392
locus, p. 408
distance from a point to a line, p. 408
distance between parallel lines, p. 409

MIXED REVIEW

1. Draw a line segment. Divide the line segment into two segments whose lengths are in the ratio 2:3.
2. In $\triangle ABC$, $AC = 14$ and $BC = 8$. Find all possible values of AB.

Exercises 3–5 refer to square $ABCD$. Describe and sketch each locus.

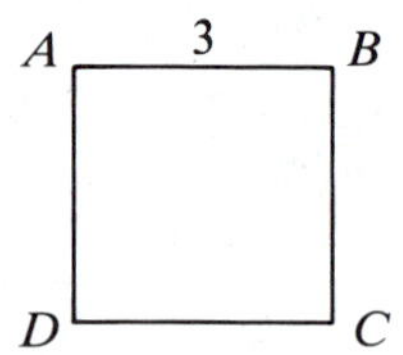

3. The locus of points 1 unit from A
4. The locus of points equidistant from A and C
5. The locus of points $1\frac{1}{2}$ units from $\overleftrightarrow{BC}$
6. The lengths of the sides of a right triangle are 8, 15, and 17. Find the measures of the acute angles to the nearest degree.

Write a quadratic equation in standard form with the given root(s).

7. $\{2 + \sqrt{2}, 2 - \sqrt{2}\}$
8. $\{-5\}$

Simplify.

9. $\dfrac{8y^3 - 4y^2 + 2y}{2y}$
10. $-\frac{4}{5}(35 - (-20)) + 8 \div 2 + 2$
11. Factor completely: $144a^2 - 16b^2$
12. Draw two non-perpendicular intersecting lines. Construct the bisectors of the two pairs of vertical angles formed.
13. Write a conclusion that follows from the pair of premises: $p \rightarrow \sim q$
    $s \rightarrow q$
14. A ladder 8 m long leans against a building so that its base is 3.4 m from the building. Find the measure to the nearest degree of the angle the ladder makes with the building.

Solve.

15. $3k^2 + 8k + 5 = 0$
16. $\dfrac{x}{x+4} = \dfrac{3}{x+2}$

In the diagram, M and P are the midpoints of $\overline{TR}$ and $\overline{TI}$, respectively.

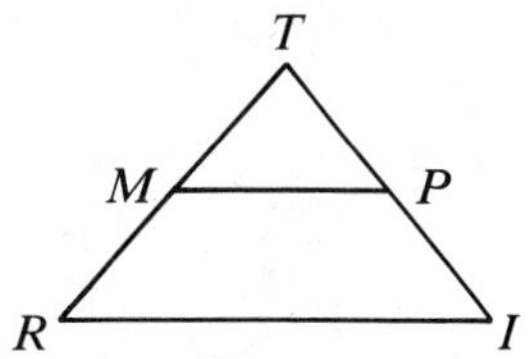

17. Name a pair of similar triangles. Justify your answer.
18. If $TR = 14$ and $MP = 9$, find RI.
19. What kind of quadrilateral is $RMPI$?

20. The ratio of the length of the corresponding sides of two similar triangles is 1:3. What is the ratio of their perimeters? their areas?

Draw segments and angles like those shown; then construct the required figures.

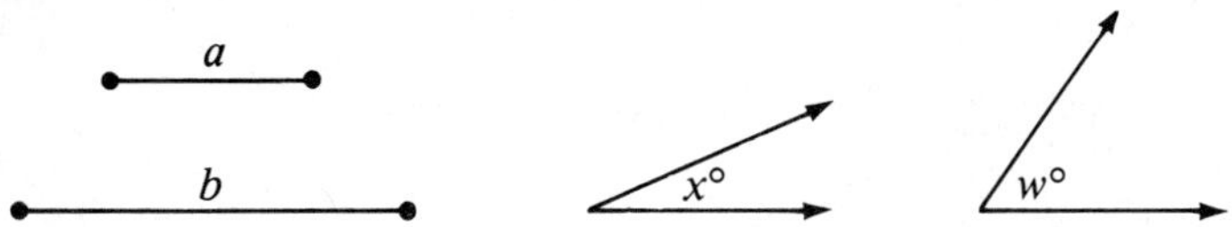

21. A segment of length $b - a$

22. An angle of measure $x + w$

In right $\triangle GHI$, $m\angle G = 30$.

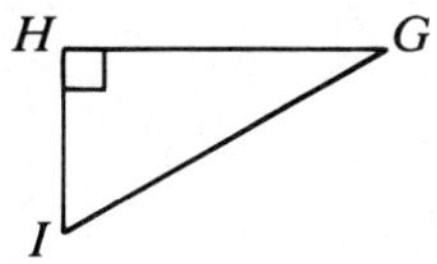

23. If $HI = 8$, find HG and GI.
24. If $HG = 7\sqrt{3}$, find HI and GI.
25. If $GI = 6$, find HI and HG.

26. A person 1.6 m tall casts a shadow 2.4 m long at the same time that a nearby pole casts a shadow 54 m long. Find the height of the pole.

Write two-column proofs.

27. Given: $\overline{TR} \parallel \overline{SA}$; $\angle S \cong \angle TPA \cong \angle A$
Prove: $TRAP$ is an isosceles trapezoid

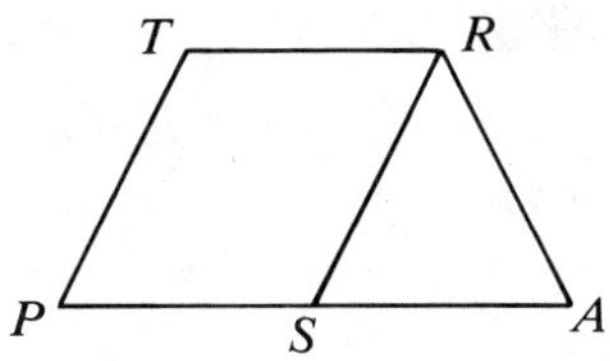

28. Given: $DC = 3$; $CE = 4$; $BC = 6$; $AC = 8$
Prove: $\triangle ABC \sim \triangle EDC$

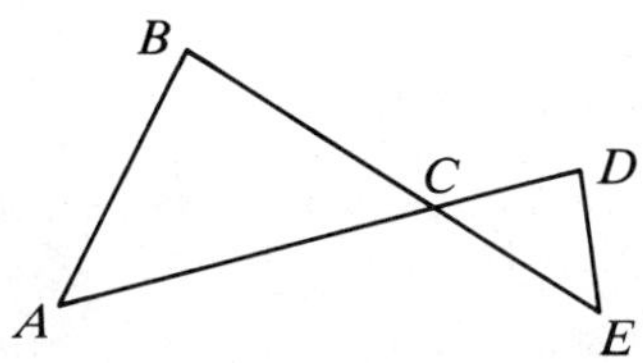

29. Given: $UWXY$ is a parallelogram; $UVWZ$ is a rectangle.
Prove: $UWXY$ is a rhombus.

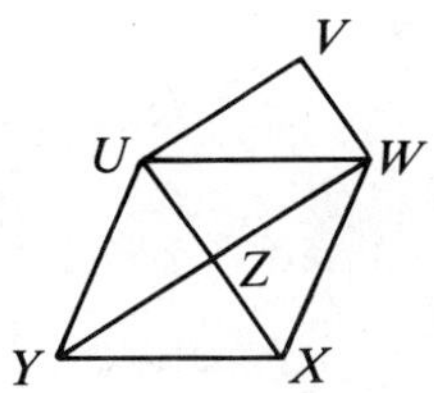

This circus performer cautiously traverses the tightrope from one end to the other, covering a distance that must seem much longer on the tightrope than it would on the ground. In this chapter, you will learn a mathematical formula for finding the distance between any two points.

Coordinate Geometry 11

The basis of coordinate, or analytic, geometry is the representation of positions of points by numerical coordinates. Thus, as you will see, lines and curves can be represented by equations, and algebraic techniques can be used to study geometric figures.

Coordinate systems have been used in surveying and map-making, for example, since the time of the ancient Egyptians and Greeks, but historians generally credit René Descartes and Pierre Fermat, two French mathematicians of the seventeenth century, with the invention of coordinate geometry.

As you will see in this chapter, the use of numbers to represent points frequently reduces complicated proofs in Euclidean geometry to straightforward calculations.

The development of coordinate geometry made possible the geometric representation of equations and thus enabled mathematicians to gain new insights into the analysis of equations. In fact, coordinate geometry made possible the development of differential and integral calculus. It is even possible to trace the development of the theory of relativity back to the beginnings of coordinate geometry, one of the most practical, widely used methods of analysis.

Section 11-1 POINTS IN A PLANE

In this section you will review graphing points in a coordinate plane. You will also learn about absolute value.

In Section 1-1 you learned how to graph a number as a point on a number line. However, you graph an *ordered pair* of numbers, say (2, 3), as a point in a *coordinate plane.* A coordinate plane contains two perpendicular number lines, the *x-axis* and the *y-axis*, that intersect at their origins. The point of intersection O of the *coordinate axes* is called the *origin* of the coordinate plane. The x-axis is referred to as horizontal and the y-axis is referred to as vertical. The positive direction on each axis is indicated by an arrowhead.

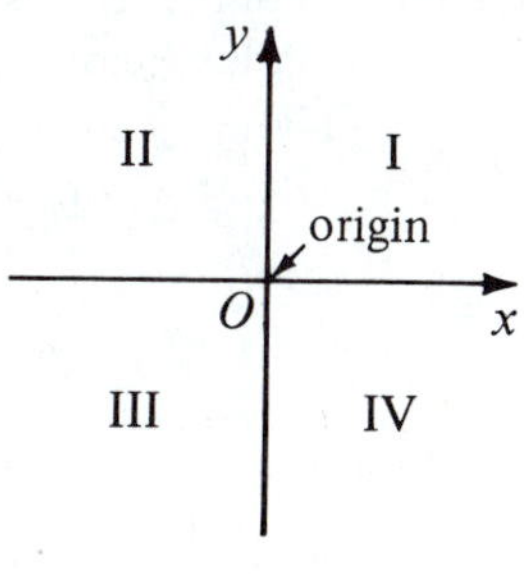

The x-axis and y-axis divide the points of the plane that are not on the axes into four regions, called *quadrants*, that are numbered as shown in the diagram.

Consider point P in the coordinate plane at the right. There exists a unique line through P parallel to the y-axis. The real number that represents the intersection of that line and the x-axis is called the *x-coordinate* (or *abscissa*) of P, namely 2. There is also a unique line through P parallel to the x-axis. The real number that represents the intersection of that line and the y-axis is called the *y-coordinate* (or *ordinate*) of P, namely 3. The abscissa and ordinate of P are called the *coordinates* of P and are written as an *ordered pair* with the abscissa first. Thus, the coordinates of P are (2, 3), and the *graph* of (2, 3) is point P. It is often convenient to name a point and its coordinates simultaneously. We do this by writing $P(2, 3)$, for example.

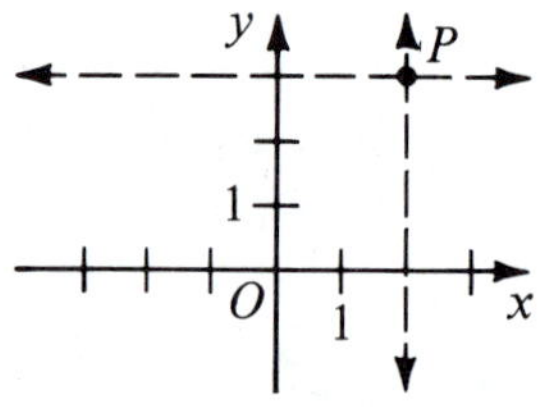

In graphing an ordered pair of real numbers in a coordinate plane, you use the following facts:

> Each ordered pair of real numbers corresponds to exactly one point of the coordinate plane.
> Each point in the coordinate plane corresponds to exactly one ordered pair of real numbers.

A shorter way of stating this is to say that there is a *one-to-one correspondence* between points in a plane and ordered pairs of real numbers.

EXAMPLE State the coordinates of points A, B, C, D, and E.

SOLUTION The coordinates of A are (1, 4).
The coordinates of B are (4, 1).
The coordinates of C are (−4, −3).
The coordinates of D are (−3, 0).
The coordinates of E are (0, −2).

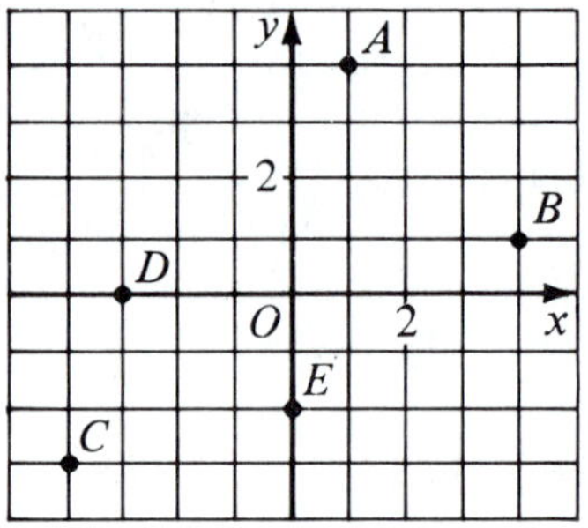

EXAMPLE Name the points that correspond to (1, 0), (0, 3), (−2, 4), and (3, −2).

SOLUTION Point T corresponds to (1, 0).
Point R corresponds to (0, 3).
Point Q corresponds to (−2, 4).
Point V corresponds to (3, −2).

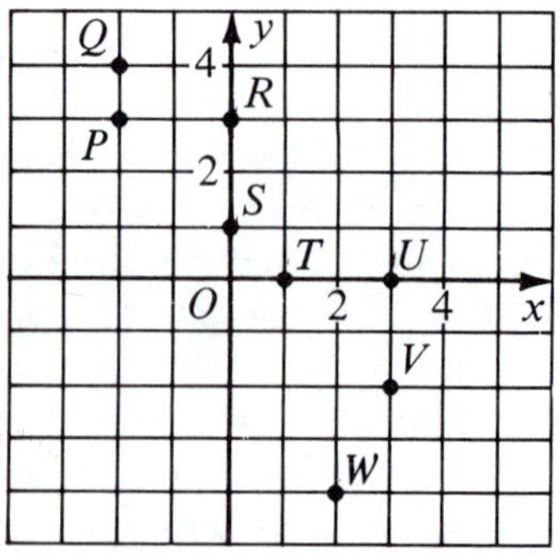

In the next section, you will learn how to find the distance between any two points in a coordinate plane. As you will see in the Written Exercises, to find the distance between two points, in a coordinate plane, that lie on the same vertical or horizontal line, you can think of the points as lying on a number line.

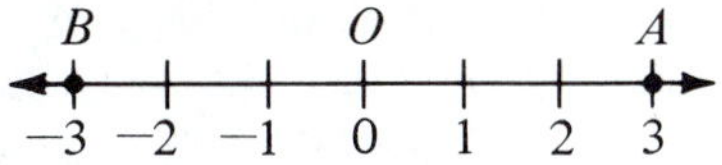

What is the distance from the origin to A? from the origin to B? Do you agree that these distances are equal? Although the coordinate of A is 3 and the coordinate of B is -3, OA and OB each contain 3 units. The *absolute value* of a number, x, written $|x|$, is the distance from the origin to the graph of x. Thus, $|3| = OA = 3$ and $|-3| = OB = 3$. More formally, the *absolute value* of x is defined as follows:

$$|x| = x \text{ if } x \geqslant 0$$
$$|x| = -x \text{ if } x < 0$$

In other words, the absolute value of any real number is the greater of x and its opposite, $-x$. Do you see that the absolute value of a number is *always* nonnegative?

EXAMPLE Simplify each of the following.

1. $|-7|$ 2. $|12|$ 3. $|4-9|$

SOLUTION
1. $|-7| = 7$
2. $|12| = 12$
3. $|4-9| = |-5| = 5$

ORAL EXERCISES

Exercises 1-3 refer to the diagram at the right.

1. Name the abscissa of each point.
2. Name the ordinate of each point.
3. Name the coordinates of each point.

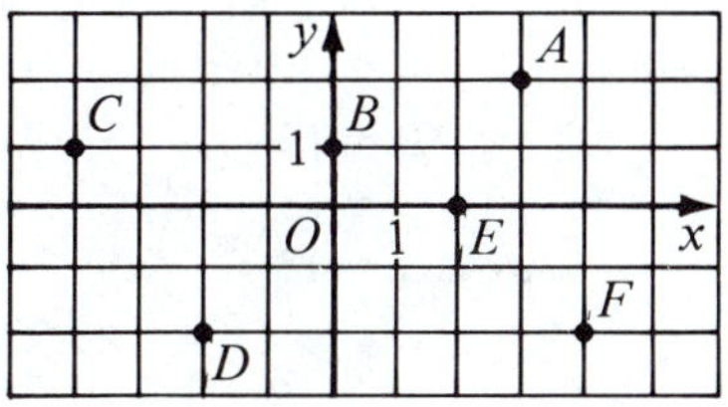

Exs. 1-3

4. Name the point that corresponds to the ordered pair.
   a. (0, 3) b. (2, 1) c. (−4, 2)
   d. (1, 2) e. (−4, −2) f. (2, −4)

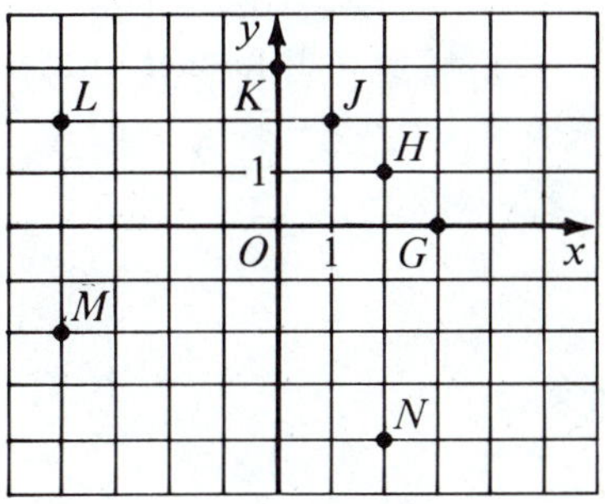

Ex. 4

Simplify each of the following.

5. $|3|$ 6. $|-28|$ 7. $|0|$ 8. $|15|$ 9. $|-5|$

WRITTEN EXERCISES

Graph each point on the same coordinate plane.

A 1. $A(1, 3)$ 2. $B(-3, -1)$ 3. $C(3, 1)$

4. $D(4, 0)$ 5. $E(5, -1)$ 6. $F\left(\frac{7}{2}, -\frac{5}{2}\right)$

7. $G(-1, 2)$ 8. $H(-2, 1)$ 9. $I(0, -4)$

10. $J\left(-\frac{5}{4}, -\frac{11}{4}\right)$ 11. $K\left(-3, \frac{1}{5}\right)$ 12. $L(1, -5)$

Name the quadrant in which the point (a, b) is located for the given values of a and b.

13. $a < 0,\ b < 0$ 14. $a < 0,\ b > 0$

15. $a > 0,\ b < 0$ 16. $a > 0,\ b > 0$

Simplify each of the following.

17. $|3 + 2|$ 18. $|3 - 2|$ 19. $|2 - 3|$

20. $|2| - |-3|$ 21. $|3(-2)|$ 22. $|2| \cdot |-3|$

Find the distance between points A and B.

EXAMPLES 1. $A(5, 6)$, $B(-8, 6)$

2. $A(1, -3)$, $B(1, 4)$

SOLUTION 1. A and B are on the same horizontal line. The distance from A to B equals the absolute value of the difference of their x-coordinates.

$$\begin{aligned} AB &= |5 - (-8)| \\ &= |5 + 8| \\ AB &= 13 \end{aligned}$$

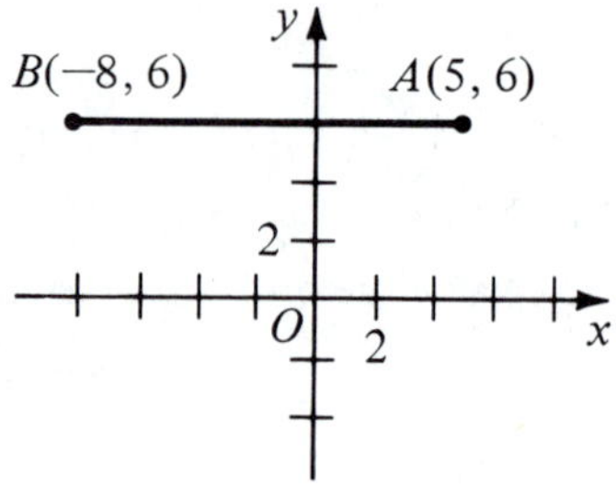

2. A and B are on the same vertical line. The distance from A to B equals the absolute value of the difference of their y-coordinates.

$$\begin{aligned} AB &= |-3 - 4| \\ &= |-7| \\ AB &= 7 \end{aligned}$$

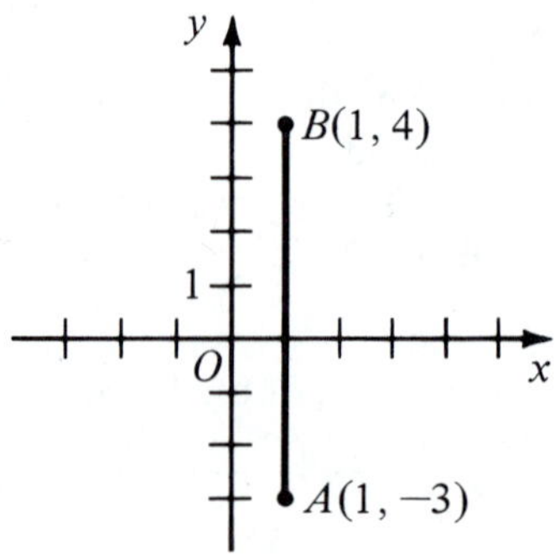

23. $A(6, 3)$, $B(2, 3)$ 24. $A(8, -2)$, $B(-2, -2)$ 25. $A(2, -6)$, $B(2, 6)$

26. $A(-3, 10)$, $B(-3, 15)$ 27. $A(-5, 4)$, $B(-6, 4)$ 28. $A(-1, -7)$, $B(-1, -1)$

Find the area of the figure formed by joining the given vertices in order.

EXAMPLE $A(-1, -2)$, $B(3, -2)$, $C(5, 2)$, and $D(1, 2)$

SOLUTION Plot the points. $ABCD$ is a parallelogram because $\overline{AB}$ and $\overline{DC}$ are horizontal and $AB = CD$. If you let $\overline{AB}$ be the base, then you can draw an altitude to $\overline{AB}$, say $\overline{DE}$.

$AB = 4$, $DE = 4$

$$\text{area } \square ABCD = (AB)(DE) = 4 \cdot 4 = 16$$

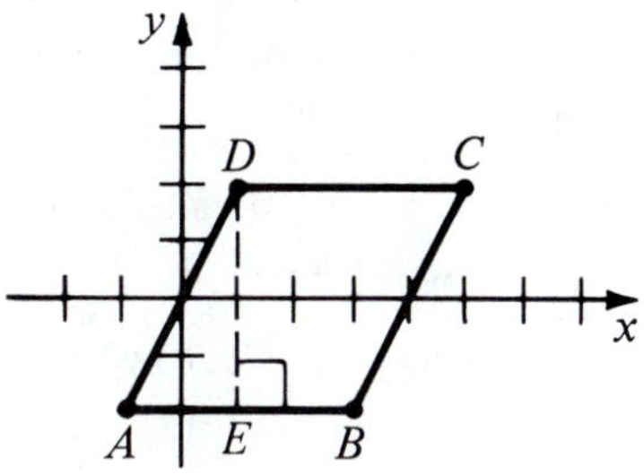

B 29. $Q(-3, -2)$, $R(1, -1)$, $S(1, 5)$, $T(-3, 4)$

30. $E(-2, 1)$, $F(3, 1)$, $G(1, 3)$, $H(-4, 3)$

31. $A(1, -3)$, $B(4, -3)$, $C(3, 3)$

32. $W(-1, -1)$, $X(3, -2)$, $Y(3, 4)$, $Z(-1, 2)$

EXAMPLE $A(3, 4)$, $B(1, -3)$, $C(-3, -1)$

SOLUTION Plot the points and join them to form $\triangle ABC$. Then sketch horizontal and vertical segments to enclose $\triangle ABC$ in a rectangle, rectangle $XAYZ$. Note that the vertices of the rectangle are $X(-3, 4)$, $A(3, 4)$, $Y(3, -3)$, and $Z(-3, -3)$.

$$\text{Area } \triangle ABC = \text{Area of rectangle } XAYZ - \text{Area } \triangle CXA - \text{Area } \triangle AYB - \text{Area } \triangle BZC$$

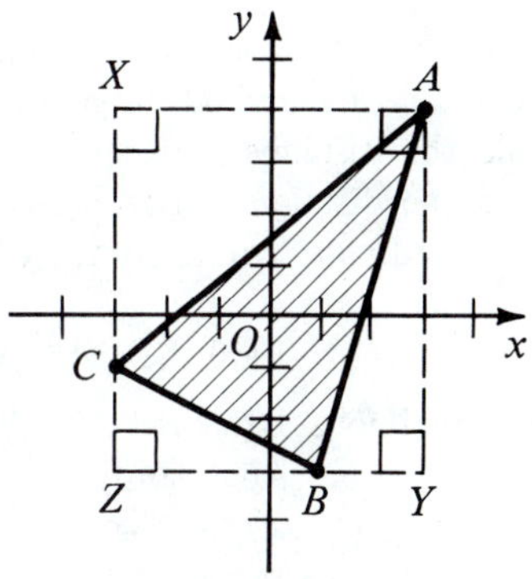

Recall that the area of a rectangle is the product of the length and the width; the area of a right triangle is half the product of the lengths of the legs.

$$\text{Area } \triangle ABC = (ZX)(XA) - \frac{1}{2}(CX)(XA) - \frac{1}{2}(AY)(YB) - \frac{1}{2}(BZ)(ZC)$$

$$= (7)(6) - \frac{1}{2}(5)(6) - \frac{1}{2}(7)(2) - \frac{1}{2}(4)(2)$$

$$= 16$$

33. $D(0, 0)$, $E(2, 5)$, $F(7, 2)$

34. $Q(-4, 2)$, $R(0, 0)$, $S(-2, -3)$

35. $A(3, 3)$, $B(1, -1)$, $C(-2, 2)$

36. $Q(0, 0)$, $R(0, 3)$, $S(5, 2)$, $T(4, 5)$

37. $A(-4, -2)$, $B(-4, 3)$, $C(-2, 5)$, $D(0, 0)$

38. $E(7, 5)$, $F(5, 0)$, $G(0, 0)$, $H(-2, 3)$

Quadrilateral $ABCD$ is a parallelogram. Find the value of a.

39. $A(0, 0)$, $B(0, 4)$, $C(5, 4)$, $D(a, 0)$

40. $A(0, 1)$, $B(a, 1)$, $C(-5, -4)$, $D(-2, -4)$

41. $A(0, a)$, $B(0, -4)$, $C(2, -1)$, $D(2, 2)$

42. $A(-2, -1)$, $B(-2, 4)$, $C(-4, a)$, $D(-4, -4)$

43. Quadrilateral $ABCD$ is an isosceles trapezoid with $\overline{AD} \parallel \overline{BC}$ and vertices $A(3, 0)$, $B(2, 3)$, and $C(-2, 3)$. Find the coordinates of D.

44. The coordinates of three vertices of a parallelogram are $(0, -2)$, $(-1, 0)$, and $(-2, -2)$. Find the coordinates of the fourth vertex. (*Hint:* There are three possible answers.)

C 45. Name the coordinates of the center of the circle that passes through the points $(3, 8)$, $(3, 4)$, and $(1, 6)$.

Graph all the points (a, b) whose coordinates satisfy the given condition.

46. $a = 3$ 47. $a = -4$ 48. $b = 2$ 49. $b = -5$

50. a is an integer such that $2 < a < 7$, and b is an integer such that $-4 < b < 0$.

Section 11-2 THE DISTANCE FORMULA AND THE MIDPOINT FORMULA

The distance formula is an excellent example of the interrelation of coordinate geometry and Euclidean geometry. In the preceding exercises, you learned how to find the distance between any two points with the same abscissa or ordinate. To find the distance between two points that do not have the same abscissa or the same ordinate, you apply the Pythagorean Theorem.

EXAMPLE Find the distance between $A(-2, -1)$ and $B(2, 2)$.

SOLUTION Draw right $\triangle ABC$ with hypotenuse $\overline{AB}$.
The coordinates of C are $(2, -1)$.
$AC = |-2 - 2| = 4$
$BC = |2 - (-1)| = 3$
By the Pythagorean Theorem,

$$(AB)^2 = (AC)^2 + (BC)^2$$
$$(AB)^2 = 4^2 + 3^2 = 25$$
$$AB = \sqrt{25} = 5$$

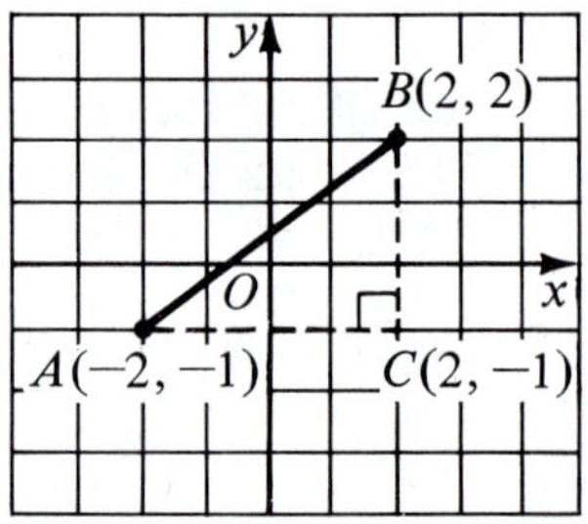

You can derive a general formula for the distance between any two points as follows.

Let $A(x_1, y_1)$ and $B(x_2, y_2)$ be the endpoints of a segment. Applying the method we used in the preceding example, we see that the coordinates of C are (x_2, y_1). Then,

$$AC = |x_2 - x_1|,\ (AC)^2 = |x_2 - x_1|^2 = (x_2 - x_1)^2$$
$$BC = |y_2 - y_1|,\ (BC)^2 = |y_2 - y_1|^2 = (y_2 - y_1)^2$$
$$(AB)^2 = (AC)^2 + (BC)^2$$
$$(AB)^2 = (x_2 - x_1)^2 + (y_2 - y_1)^2$$
$$AB = \sqrt{(x_2 - x_1)^2 + (y_2 - y_1)^2}$$

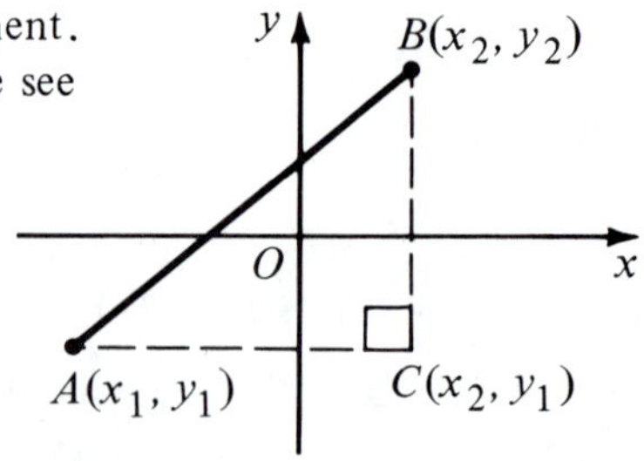

> **Distance Formula**
>
> The distance between $A(x_1, y_1)$ and $B(x_2, y_2)$ is
>
> $$\sqrt{(x_2 - x_1)^2 + (y_2 - y_1)^2}.$$

EXAMPLE Find the length of the line segment joining $J(-4, 3)$ and $K(2, -1)$.

SOLUTION If you let $(x_1, y_1) = (-4, 3)$ and $(x_2, y_2) = (2, -1)$, you have:

$$\begin{aligned} JK &= \sqrt{(2-(-4))^2 + (-1-3)^2} \\ &= \sqrt{6^2 + (-4)^2} = \sqrt{36 + 16} \\ &= \sqrt{52} = 2\sqrt{13} \end{aligned}$$

Notice that if you let $(x_1, y_1) = (2, -1)$ and $(x_2, y_2) = (-4, 3)$, you will get the same answer.

$$\begin{aligned} JK &= \sqrt{(-4-2)^2 + (3-(-1))^2} \\ &= \sqrt{(-6)^2 + 4^2} = \sqrt{36 + 16} \\ &= \sqrt{52} = 2\sqrt{13} \end{aligned}$$

Now we will see how the coordinates of the midpoint of a segment are related to the coordinates of the endpoints.

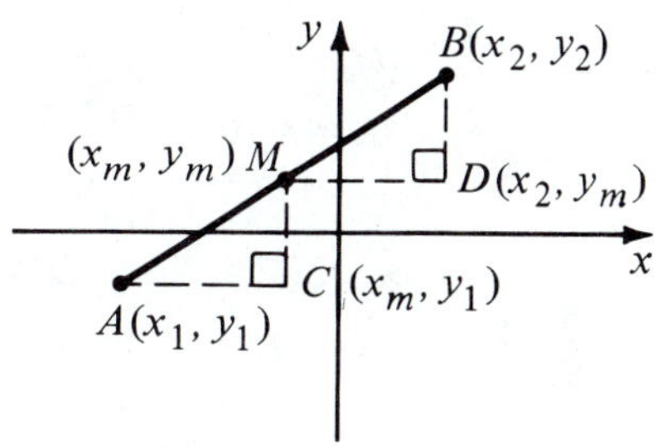

Let $A(x_1, y_1)$ and $B(x_2, y_2)$ be the endpoints of a segment. Let $M(x_m, y_m)$ be the midpoint of the segment. Draw lines parallel to the y-axis through B and M and draw lines parallel to the x-axis through A and M to form two right triangles, $\triangle ACM$ and $\triangle MDB$. $\triangle ACM \cong \triangle MDB$ (Why?) Therefore, $AC = MD$ and $CM = DB$.

$$\begin{aligned} AC = x_m - x_1 \text{ and } MD &= x_2 - x_m \\ x_m - x_1 &= x_2 - x_m \\ 2x_m &= x_1 + x_2 \\ x_m &= \frac{x_1 + x_2}{2} \end{aligned}$$

$$\begin{aligned} CM = y_m - y_1 \text{ and } DB &= y_2 - y_m \\ y_m - y_1 &= y_2 - y_m \\ 2y_m &= y_1 + y_2 \\ y_m &= \frac{y_1 + y_2}{2} \end{aligned}$$

Therefore, the coordinates of M, the midpoint, are

$$\left(\frac{x_1 + x_2}{2}, \frac{y_1 + y_2}{2}\right).$$

Midpoint Formula

Given points $A(x_1, y_1)$ and $B(x_2, y_2)$, the coordinates of the midpoint of $\overline{AB}$ are

$$\left(\frac{x_1 + x_2}{2}, \frac{y_1 + y_2}{2}\right).$$

EXAMPLE Given $T(-2, 5)$, $U(1, 1)$, and $V(2, 5)$, find the coordinates of the midpoints of $\overline{TU}$, $\overline{UV}$, and $\overline{TV}$.

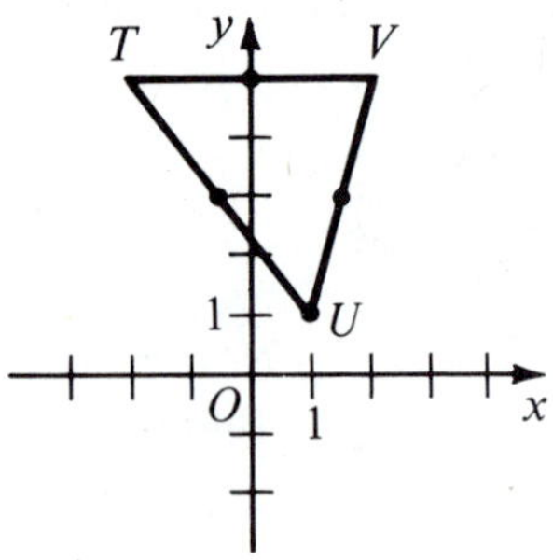

SOLUTION The coordinates of the midpoint of $\overline{TU}$ are

$$\left(\frac{-2+1}{2}, \frac{5+1}{2}\right), \text{ or } \left(-\frac{1}{2}, 3\right).$$

The coordinates of the midpoint of $\overline{UV}$ are

$$\left(\frac{1+2}{2}, \frac{1+5}{2}\right), \text{ or } \left(\frac{3}{2}, 3\right).$$

The coordinates of the midpoint of $\overline{TV}$ are

$$\left(\frac{-2+2}{2}, \frac{5+5}{2}\right), \text{ or } (0, 5).$$

You can use a coordinate system in proofs as well as in computational problems.

EXAMPLE Show that the triangle with vertices $H(2, 8)$, $J(0, 3)$, and $K(7, 6)$ is isosceles.

SOLUTION Use the distance formula to show that two sides of the triangle have equal length.

$$HJ = \sqrt{(2-0)^2 + (8-3)^2} = \sqrt{4+25} = \sqrt{29}$$
$$JK = \sqrt{(0-7)^2 + (3-6)^2} = \sqrt{49+9} = \sqrt{58}$$
$$HK = \sqrt{(7-2)^2 + (6-8)^2} = \sqrt{25+4} = \sqrt{29}$$

Since $HJ = HK$, $\triangle HJK$ is isosceles.

You may want to check your work by graphing H, J, and K, and drawing $\triangle HJK$.

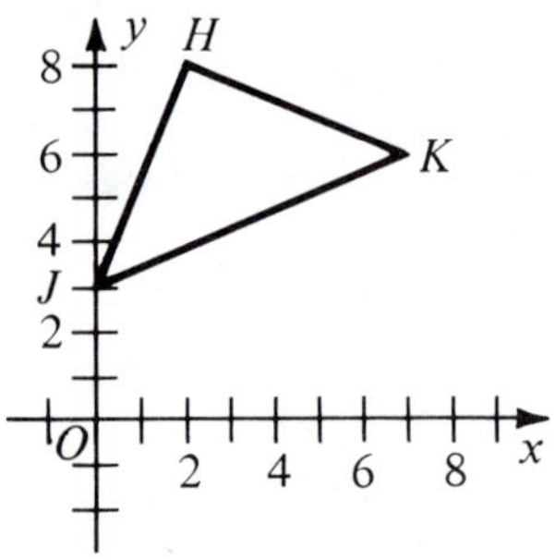

ORAL EXERCISES

Simplify.

1. $\sqrt{3^2 + 4^2}$
2. $\sqrt{5^2 + 12^2}$
3. $\sqrt{50}$
4. $\sqrt{24}$
5. $\sqrt{27}$
6. $\sqrt{63}$

Find the distance between the points. Express your answer in simplified radical form.

7. $(0, 0)$, $(-4, 3)$
8. $(8, 4)$, $(5, 4)$
9. $(-1, 1)$, $(3, 3)$

Find the coordinates of the midpoint of the line segment that joins the two points.

10. $(2, 5)$, $(2, 7)$
11. $(3, 7)$, $(9, 7)$
12. $(-3, 1)$, $(7, 9)$
13. (a, b), $(-a, b)$

WRITTEN EXERCISES

Find the coordinates of the midpoint of the segment that joins the given points.

A 1. $(-5, 2), (4, -2)$ 2. $(3, 0), (-5, 5)$ 3. $(1.3, 2.4), (2.5, 1.6)$

4. $\left(\frac{3}{2}, \frac{1}{2}\right), \left(-\frac{1}{4}, \frac{1}{4}\right)$ 5. $(a, b), (c, d)$ 6. $(a + 1, b), (a - 1, 2b)$

Find the distance between the points. Express your answer in simplified radical form.

7. $(0, 4), (-3, -2)$ 8. $(-1, 7), (2, 5)$ 9. $(-5, -3), (4, 6)$

10. $(-7, 5), (3, 0)$ 11. $(-2, 1), (-6, 5)$ 12. $(8, -3), (9, -10)$

In Exercises 13–16, M is the midpoint of $\overline{AB}$. Find the coordinates of B.

13. $M(-3, 0)$, $A(4, 6)$ 14. $M(2, 5)$, $A(-1, -2)$

15. $M(4, -8)$, $A(5, -3)$ 16. $M(3.2, -1.8)$, $A(1.4, -1.7)$

17. Quadrilateral $ABCD$ has vertices $A(1, 8)$, $B(10, 10)$, $C(9, 6)$, and $D(0, 4)$. Find:
 a. AB b. CD c. AD d. BC
 e. What kind of quadrilateral is $ABCD$? State a theorem that supports your answer.

18. The vertices of $\triangle QRS$ are $Q(8, 7)$, $R(-1, 1)$, and $S(-3, 4)$.
 a. Find QR, RS, and QS.
 b. Show that $\triangle QRS$ is a right triangle.
 c. Find the area of $\triangle QRS$.

19. Quadrilateral $FGHJ$ has vertices $F(-2, 5)$, $G(-4, 1)$, $H(-2, -3)$, and $J(0, 1)$. Show that $FGHJ$ is a rhombus.

20. $\triangle TUV$ has vertices $T(0, 8)$, $U(-2, 0)$, and $V(10, 8)$.
 a. Find the coordinates of the midpoint, M, of $\overline{UV}$.
 b. Find the length of the median $\overline{TM}$. Express your answer in simplified radical form.
 c. Find the area of $\triangle TUV$.

Find the value of t.

B 21. The midpoint of the segment joining $(t, 2)$ and $(3t, -4)$ is $(-6, -1)$.

22. The distance from $(3, 4)$ to $(t, 8)$ is $4\sqrt{2}$.

23. The distance from $(t, 0)$ to $(20, 5)$ is 13.

24. The distance from $(2t, t)$ to $(5, 0)$ is 5.

25. $\triangle RST$ has vertices $R(-5, 6)$, $S(-1, 2)$, and $T(5, 8)$.
 a. Use coordinate geometry to show that $\triangle RST$ is a right triangle.
 b. Use coordinate geometry to show that the length of the median to the hypotenuse is one half the length of the hypotenuse.

26. $\triangle DEF$ has vertices $D(4, 1)$, $E(2, -4)$, and $F(-1, -1)$.
   a. Use coordinate geometry to show that $\triangle DEF$ is isosceles.
   b. Find the coordinates of M, the midpoint of the base.
   c. Use coordinate geometry to show that $\overline{DM} \perp \overline{EF}$.
   d. Find the area of $\triangle DEF$.

27. Quadrilateral $ABCD$ has vertices $A(-3, 5)$, $B(2, 7)$, $C(4, 3)$, and $D(1, 1)$. P, Q, R, and S are the midpoints of $\overline{AB}$, $\overline{BC}$, $\overline{CD}$, and $\overline{AD}$, respectively. Use coordinate geometry to show that quadrilateral $PQRS$ is a parallelogram.

28. $\triangle ABC$ has vertices $A(-2, -2)$, $B(5, -1)$, and $C(-1, 5)$.
   a. Use coordinate geometry to show that $\triangle ABC$ is isosceles.
   b. Use coordinate geometry to show that the altitude to the base of $\triangle ABC$ divides $\triangle ABC$ into two congruent triangles.

Find the distance between each pair of points.

EXAMPLE (a, b), (b, a)

SOLUTION $d = \sqrt{(b-a)^2 + (a-b)^2}$

But, $(b-a)^2 = (a-b)^2$, so $d = \sqrt{(b-a)^2 + (b-a)^2}$

$d = \sqrt{2(b-a)^2}$

d cannot be negative, but we don't know which is greater, b or a. So, to simplify $\sqrt{2(b-a)^2}$ and guarantee that d is nonnegative, we use *absolute value.*

$d = |b-a|\sqrt{2}$

C 29. (c, d), $(0, 0)$

30. $(a + b, b)$, $(a, 0)$

31. $(a, a + c)$, $(a, a - c)$

Find the area of quadrilateral $ABCD$ with the given vertices.

32. $A(3, 0)$, $B(7, 4)$, $C(3, 10)$, $D(0, 8)$

33. $A(-2, -3)$, $B(7, -8)$, $C(10, -3)$, $D(-2, -1)$

34. $\triangle RST$ has vertices $R(0, 0)$, $S(2a, 0)$, and $T(2b, 2c)$.
   a. Find the coordinates of the midpoint of each side of $\triangle RST$.
   b. Use coordinate geometry to show that the perimeter of the triangle formed by joining the midpoints of the sides of $\triangle RST$ is one half the perimeter of $\triangle RST$.

Section 11-3 SLOPE OF A LINE

In this section you will study slope and the relationships of the slopes of:

- nonvertical parallel lines
- nonvertical perpendicular lines

Slope is a measure of the steepness of a line segment. It is the ratio of the change in y (the vertical change) to the change in x (the horizontal change). In each of the diagrams below, you can think of getting from point A to point B by moving vertically and then horizontally. Informally, the slope is $\frac{\text{rise}}{\text{run}}$.

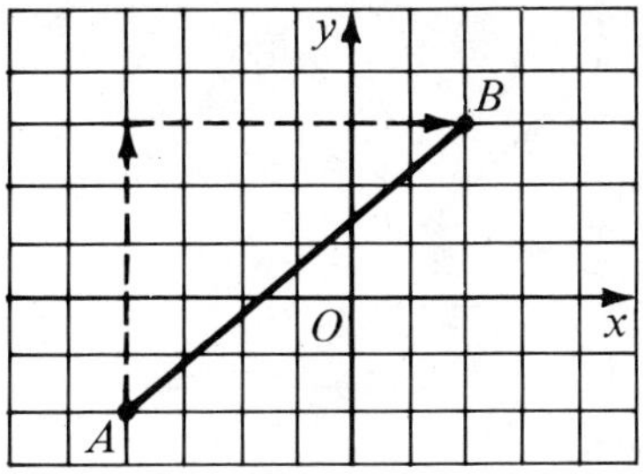

$$\text{slope of } \overline{AB} = \frac{\text{rise}}{\text{run}} = \frac{5}{6}$$

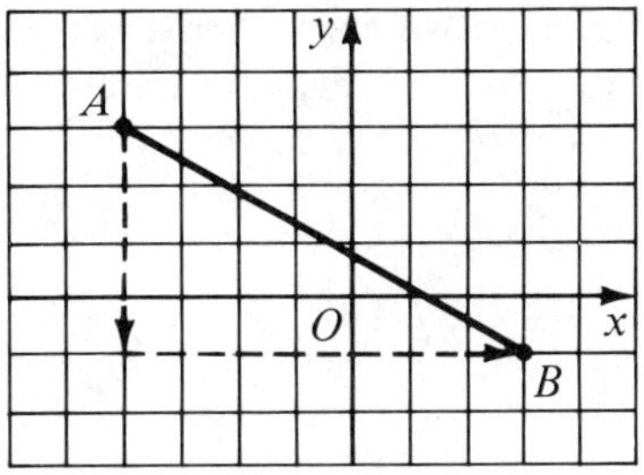

$$\text{slope of } \overline{AB} = \frac{\text{rise}}{\text{run}} = \frac{-4}{7} = -\frac{4}{7}$$

In a coordinate plane, the x-axis and all lines parallel to it are called horizontal lines. The y-axis and all lines parallel to it are called vertical lines.

If $A(x_1, y_1)$ and $B(x_2, y_2)$ are the endpoints of a nonvertical line segment, then the *slope*, denoted by m, of $\overline{AB}$ is $\frac{y_2 - y_1}{x_2 - x_1}$. Note the following facts about slope:

1. The slope of a segment does not depend on the order in which the points are named because $m = \frac{y_2 - y_1}{x_2 - x_1} = \frac{y_1 - y_2}{x_1 - x_2}$.

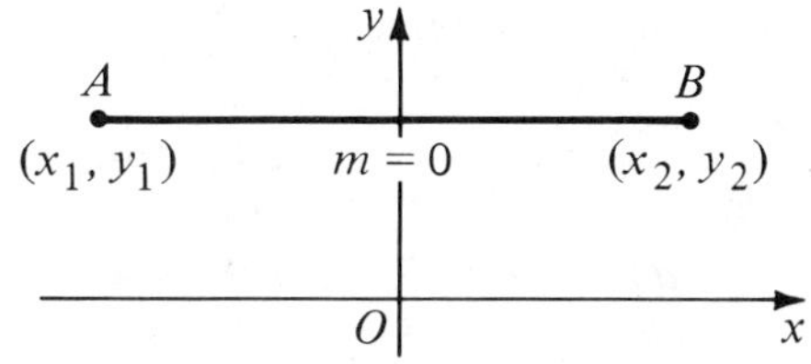

2. The slope of a horizontal segment is 0. Since $y_2 = y_1$, $m = \frac{y_2 - y_1}{x_2 - x_1} = \frac{0}{x_2 - x_1} = 0$.

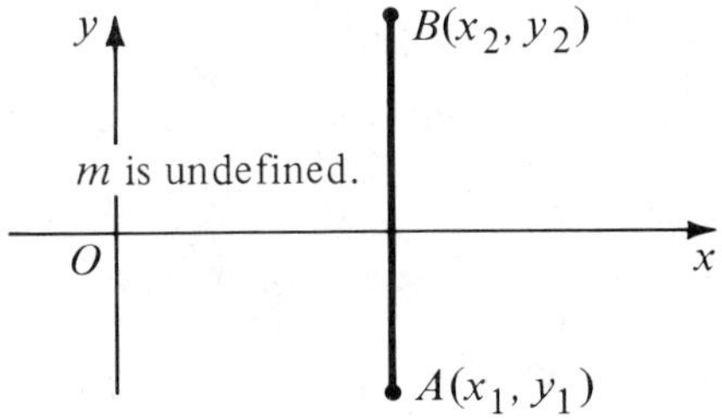

3. The slope of a vertical segment is undefined. Since $x_2 = x_1$, the denominator of $\frac{y_2 - y_1}{x_2 - x_1}$ is 0.

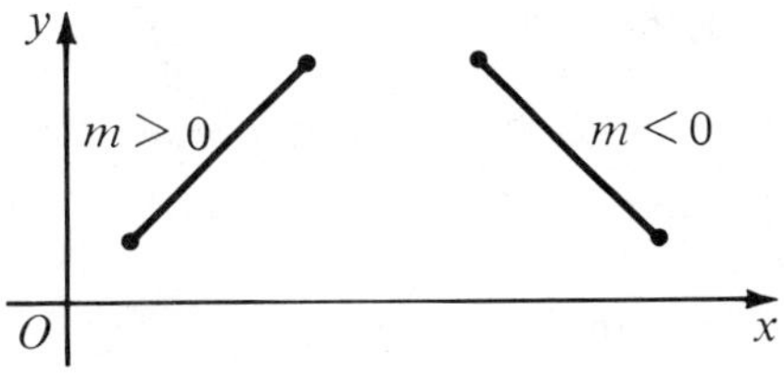

4. The slope of a segment that rises from left to right is positive. The slope of a segment that descends from left to right is negative.

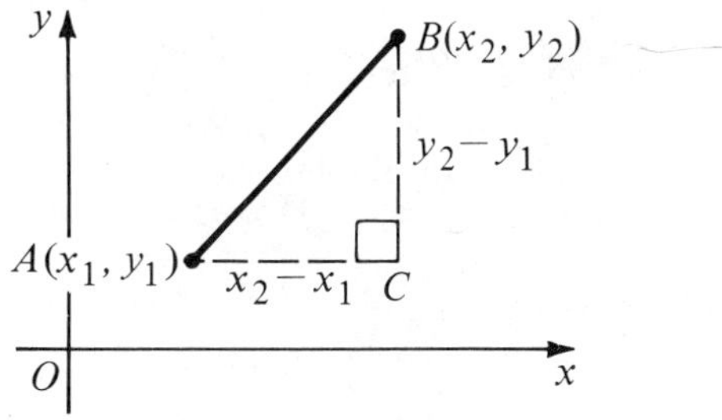

If a segment has positive slope, then you can think of the slope as the ratio of the lengths of the legs of a right triangle as shown at the right.

$$\text{slope of } \overline{AB} = \frac{y_2 - y_1}{x_2 - x_1} = \frac{BC}{AC}$$

If a segment has negative slope, then its slope is the *negative* of the ratio of the lengths of the legs of a right triangle as illustrated.

$$\text{slope of } \overline{AB} = \frac{y_2 - y_1}{x_2 - x_1} = \frac{y_2 - y_1}{-(x_1 - x_2)} = \frac{BC}{-AC} = -\frac{BC}{AC}$$

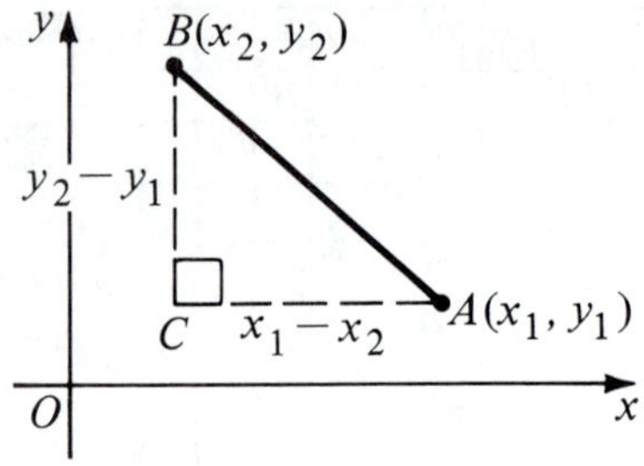

Theorem 11-1 All segments on a nonvertical line have the same slope.

GIVEN: $\overleftrightarrow{AD}$

PROVE: slope of $\overline{AB}$ = slope of $\overline{CD}$

Proof: Draw the lines parallel to the y-axis through B and D. Draw the lines parallel to the x-axis through A and C. Two right triangles are formed.

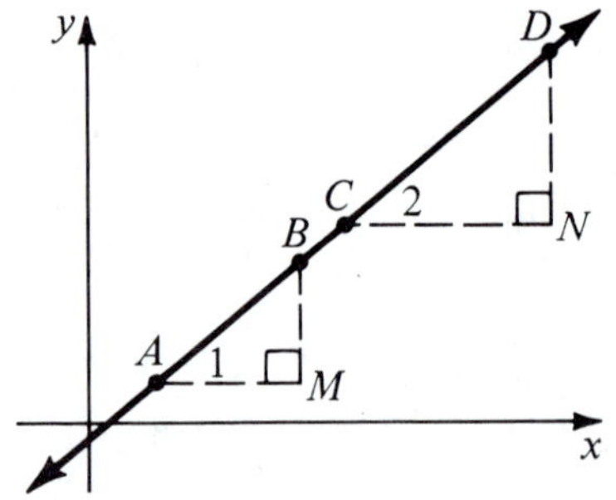

$\angle 1 \cong \angle 2$ (If 2 lines are ∥, corr. ∠s are ≅.)

$\angle M \cong \angle N$ (All rt. ∠s are ≅.)

$\triangle AMB \sim \triangle CND$ (AA Similarity Theorem)

$\frac{BM}{DN} = \frac{AM}{CN}$, or $\frac{BM}{AM} = \frac{DN}{CN}$ (Def. of ∼ △s)

But $\frac{BM}{AM}$ = slope of $\overline{AB}$, and $\frac{DN}{CN}$ = slope of $\overline{CD}$.

Therefore, $\overline{AB}$ and $\overline{CD}$ have equal slopes.

This theorem allows us to speak of *the* slope of a line. The *slope of a line* is the slope of any line segment contained in the line.

Theorem 11-2 If two nonvertical lines are parallel, then their slopes are equal.

The proof of Theorem 11-2 is left as Exercise 27.

Theorem 11-3 If the slopes of two lines are equal, then the lines are parallel.

The proof of Theorem 11-3 is left as Exercise 28.

Note that as a result of Theorems 11-2 and 11-3 you can conclude that *two distinct nonvertical lines intersect if and only if they have different slopes.*

EXAMPLE Use slope to determine whether the points $J(-5, -2)$, $K(-2, 2)$, and $L(4, 10)$ are collinear.

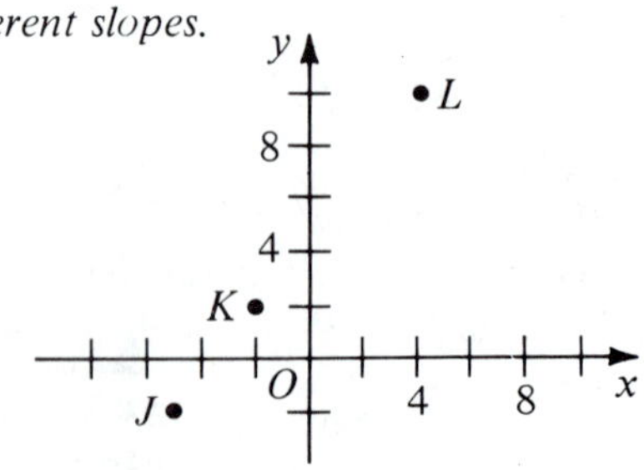

SOLUTION The slope of $\overleftrightarrow{JK}$ is $\frac{2-(-2)}{-2-(-5)} = \frac{4}{3}$.

The slope of $\overleftrightarrow{KL}$ is $\frac{10-2}{4-(-2)} = \frac{8}{6} = \frac{4}{3}$.

Clearly $\overleftrightarrow{JK}$ and $\overleftrightarrow{KL}$ are not parallel since they have at least point K in common. Therefore $\overleftrightarrow{JK}$ and $\overleftrightarrow{KL}$ are the same line and J, K, and L are collinear.

EXAMPLE Prove that the quadrilateral with vertices at $A(-4, -1)$, $B(3, 1)$, $C(4, 4)$, and $D(-3, 2)$ is a parallelogram.

SOLUTION Show that each side is parallel to the side opposite it.

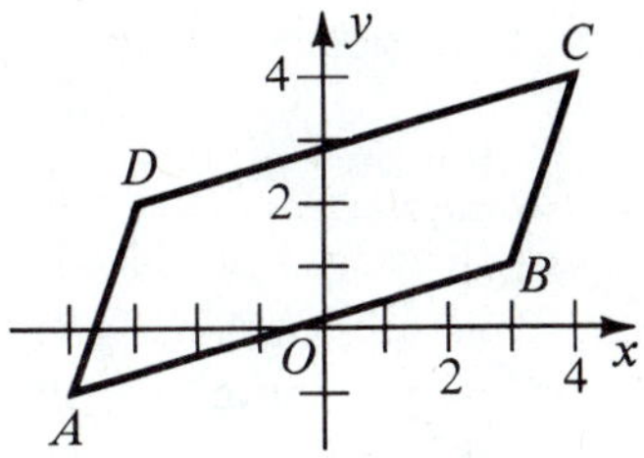

slope of $\overleftrightarrow{AB} = \frac{1-(-1)}{3-(-4)} = \frac{1+1}{3+4} = \frac{2}{7}$

slope of $\overleftrightarrow{BC} = \frac{4-1}{4-3} = \frac{3}{1} = 3$

slope of $\overleftrightarrow{CD} = \frac{2-4}{-3-4} = \frac{-2}{-7} = \frac{2}{7}$

slope of $\overleftrightarrow{AD} = \frac{2-(-1)}{-3-(-4)} = \frac{2+1}{-3+4} = \frac{3}{1} = 3$

Since $\overleftrightarrow{AB}$ and $\overleftrightarrow{CD}$ have equal slopes, $\overleftrightarrow{AB} \parallel \overleftrightarrow{CD}$.

Similarly, $\overleftrightarrow{BC}$ and $\overleftrightarrow{AD}$ have equal slopes so $\overleftrightarrow{BC} \parallel \overleftrightarrow{AD}$.

Therefore, $ABCD$ is a parallelogram. (Def. of a ▱)

Theorem 11-4 If two nonvertical lines are perpendicular, then the product of their slopes is -1.

GIVEN: $\overleftrightarrow{AB} \perp \overleftrightarrow{BC}$;
$\overleftrightarrow{AB}$ has slope m_1;
$\overleftrightarrow{BC}$ has slope m_2.

PROVE: $m_1 \cdot m_2 = -1$

Proof: See Oral Exercise 1.

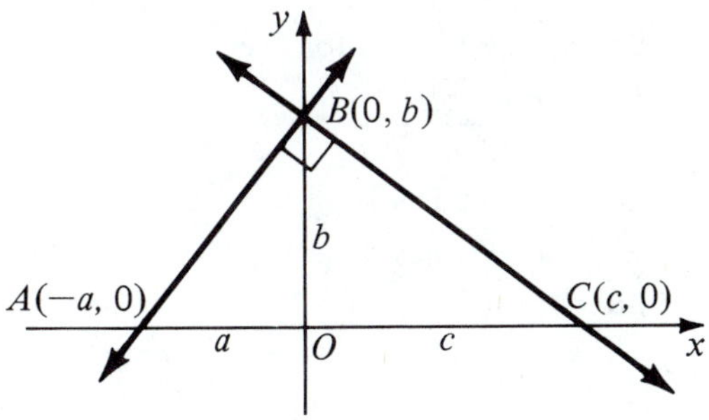

Theorem 11-5 If the product of the slopes of two lines is -1, then the lines are perpendicular.

GIVEN: $\overleftrightarrow{AB}$ has slope m_1;
$\overleftrightarrow{BC}$ has slope m_2;
$m_1 \cdot m_2 = -1$

PROVE: $\overleftrightarrow{AB} \perp \overleftrightarrow{CD}$

(The analysis and proof are on the next page.)

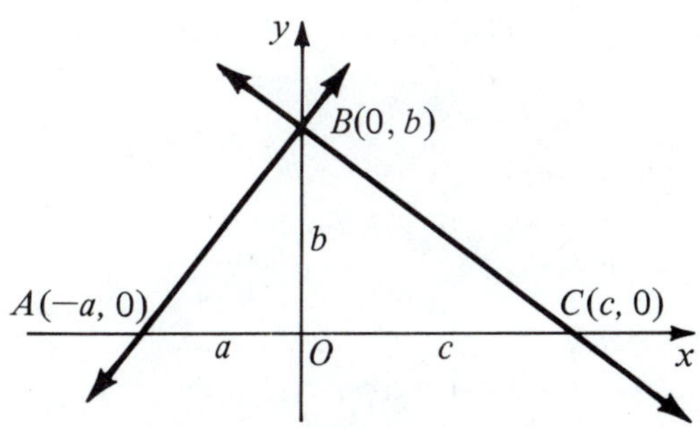

Analysis: Show that $\triangle ABC$ is a right triangle with right $\angle B$. By the converse of the Pythagorean Theorem, this will be true if $(AB)^2 + (BC)^2 = (AC)^2$.

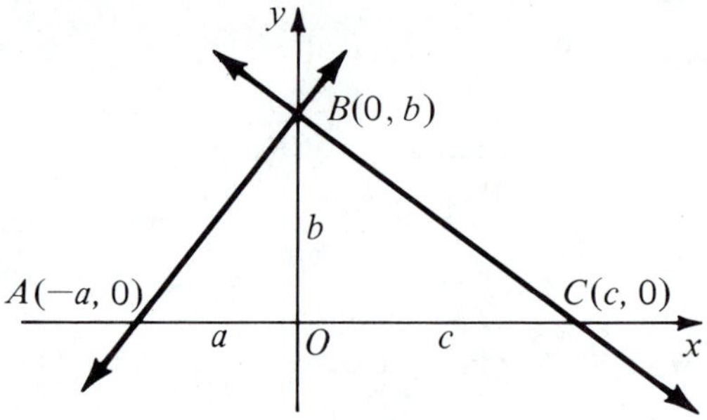

Proof: $AB = \sqrt{a^2 + b^2}$

$BC = \sqrt{c^2 + b^2}$

$AC = c + a$

$(AB)^2 + (BC)^2 = a^2 + 2b^2 + c^2$

$(AC)^2 = (c + a)^2 = a^2 + 2ac + c^2$

The slope of $\overleftrightarrow{AB} = m_1 = \frac{b}{a}$. The slope of $\overleftrightarrow{BC} = m_2 = -\frac{b}{c}$.

But $m_1 \cdot m_2 = -1$. Therefore, $\frac{b}{a}\left(\frac{-b}{c}\right) = -1$, or $\frac{b^2}{ac} = 1$.

Hence $b^2 = ac$.

By substitution, $(AB)^2 + (BC)^2 = a^2 + 2ac + c^2 = (AC)^2$.

By the converse of the Pythagorean Theorem, $\triangle ABC$ is a right triangle with right $\angle B$.

Therefore, $\overleftrightarrow{AB} \perp \overleftrightarrow{BC}$.

In other words, two nonvertical lines are perpendicular if and only if their slopes are *negative reciprocals* of each other.

ORAL EXERCISES

1. Supply the reasons for the following proof of Theorem 11-4.

   GIVEN: $\overleftrightarrow{AB} \perp \overleftrightarrow{BC}$;
   $\overleftrightarrow{AB}$ has slope m_1;
   $\overleftrightarrow{BC}$ has slope m_2.

   PROVE: $m_1 \cdot m_2 = -1$

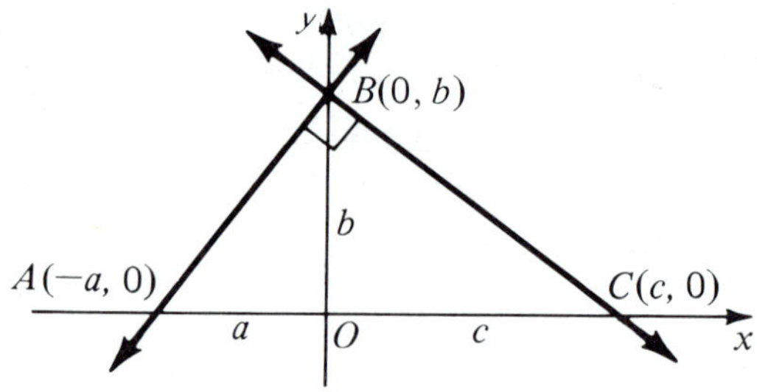

Statements	Reasons
1. $\overleftrightarrow{AB}$ has slope m_1; $\overleftrightarrow{BC}$ has slope m_2.	1. __?__
2. $m_1 = \frac{b}{a}$; $m_2 = \frac{-b}{c}$	2. __?__
3. $m_1 \cdot m_2 = -\frac{b^2}{ac}$	3. __?__
4. $\overleftrightarrow{AB} \perp \overleftrightarrow{BC}$	4. __?__
5. $\frac{a}{b} = \frac{b}{c}$	5. __?__
6. $b^2 = ac$	6. __?__
7. $m_1 \cdot m_2 = -1$	7. __?__

Find the slope of each line. If the slope is undefined, so state.

2.

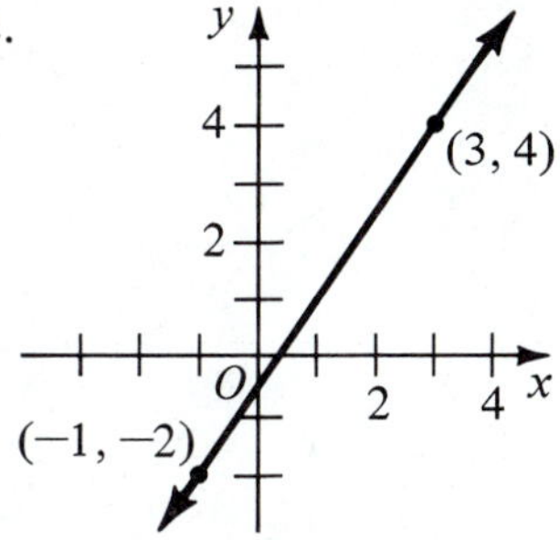

3.

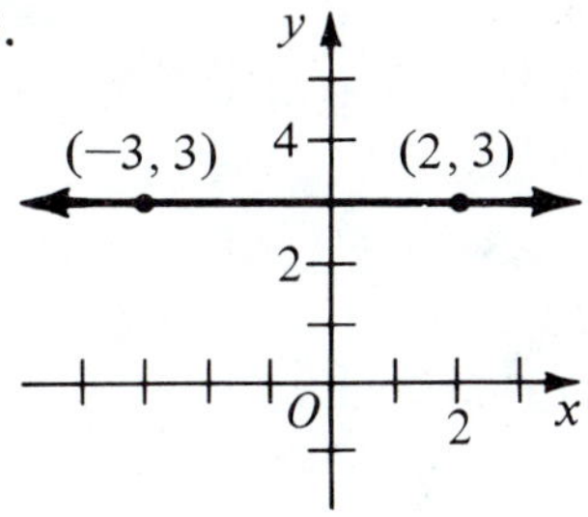

4.

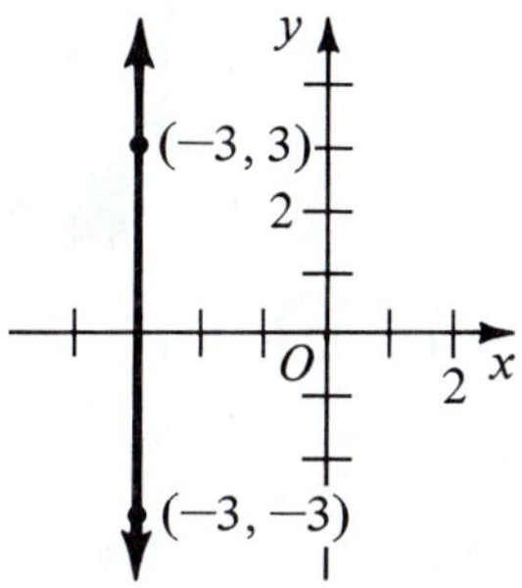

5.

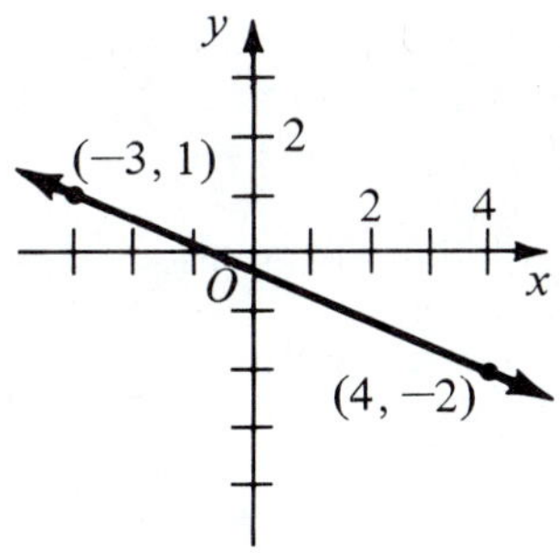

6. What can you say about the slope of a line that is parallel to a vertical line?

7. What can you say about the slope of a line that is perpendicular to a horizontal line?

Given points $A(1, 7)$, $B(6, 9)$, $C(8, 4)$, and $D(-2, 0)$, determine whether the statement is true or false.

8. $\overleftrightarrow{AB} \perp \overleftrightarrow{AD}$ 9. $\overleftrightarrow{AB} \perp \overleftrightarrow{BC}$ 10. $\overleftrightarrow{AB} \parallel \overleftrightarrow{CD}$ 11. $\overleftrightarrow{AD} \parallel \overleftrightarrow{BC}$

WRITTEN EXERCISES

Find the slope of the line through each pair of points. If the slope is undefined, so state.

A 1. $(-5, -4), (-3, -2)$ 2. $(0, 0.6), (-0.2, 0)$ 3. $(3, 8), (-1, 8)$

4. $(6, -1), (6, 4)$ 5. $\left(\frac{1}{2}, \frac{1}{3}\right), \left(\frac{5}{2}, \frac{10}{3}\right)$ 6. $\left(\frac{1}{3}, 3\right), \left(-\frac{1}{4}, 4\right)$

Find the value of k if $\overleftrightarrow{AB}$ has the given slope, m.

7. $A(k, 3), B(1, 9);\ m = 3$
8. $A(8, k), B(-3, 0);\ m = 1$
9. $A(-2, -5), B(k, 7);\ m = -2$
10. $A(-4, 1), B(-5, k);\ m = 0$
11. $A(k, 4), B(-10, 11);\ m$ is undefined.
12. $A(k, 1), B(4, k);\ m = -4$

13. Given points $A(1, -1)$, $B(4, 6)$, $C(0, -2)$, and $D(3, k)$.
   a. Find the slope of $\overleftrightarrow{AB}$.
   b. Express the slope of $\overleftrightarrow{CD}$ in terms of k.
   c. If $\overleftrightarrow{AB} \parallel \overleftrightarrow{CD}$, find the value of k.

14. Given points $A(-2, 3)$, $B(3, 1)$, $C(-1, 0)$, and $D(k, 10)$.
   a. Find the slope of $\overleftrightarrow{AB}$.
   b. Express the slope of $\overleftrightarrow{CD}$ in terms of k.
   c. If $\overleftrightarrow{AB} \perp \overleftrightarrow{CD}$, find the value of k.

15. Given trapezoid $ABCD$ with bases $\overline{AB}$ and $\overline{CD}$. The vertices are $A(5, -1)$, $B(-1, k)$, $C(6, 3k)$, and $D(3, 4)$.
   a. Express the slope of $\overline{AB}$ in terms of k.
   b. Express the slope of $\overline{CD}$ in terms of k.
   c. Find the value of k.

Use slopes to determine whether the points are collinear.

16. $(0, -3)$, $(2, -1)$, $(8, 5)$

17. $(3, 8)$, $(1, 4)$, $(-1, 1)$

18. $(1, 1)$, $(6, -9)$, $(-3, 10)$

19. $(-1, 5)$, $(0, 3)$, $(-7, 17)$

Find the value of x if the points are collinear.

20. $(2, 2)$, $(6, 4)$, $(x, 10)$

21. $(2, 1)$, $(3, 4)$, $(x, -8)$

22. $(3, 17)$, $(-1, -3)$, $(x, 12)$

23. $(-x, 4)$, $(3, -2)$, $(1, x)$

24. Quadrilateral $ABCD$ has vertices $A(1, -1)$, $B(11, 4)$, $C(22, 6)$, and $D(12, 1)$.
   a. Show that $\overline{AB} \cong \overline{CD}$.
   b. Show that $\overline{AB} \parallel \overline{CD}$.
   c. What kind of quadrilateral is $ABCD$? Explain.

25. Quadrilateral $ABCD$ has vertices $A(5, 2)$, $B(6, 7)$, $C(14, 15)$, and $D(19, 16)$.
   a. Show that $\overline{AB} \cong \overline{CD}$.
   b. Show that $\overline{BC} \parallel \overline{AD}$.
   c. Show that $\overline{AB}$ is not parallel to $\overline{CD}$.
   d. What kind of quadrilateral is $ABCD$? Explain.

26. $\triangle ABC$ has vertices $A(4, 5)$, $B(8, 13)$, and $C(-4, 9)$. Use slopes to show that $\triangle ABC$ is a right triangle.

B 27. Prove Theorem 11-2.
   Given: $\overleftrightarrow{AB} \parallel \overleftrightarrow{CD}$
   Prove: Slope of $\overleftrightarrow{AB}$ = slope of $\overleftrightarrow{CD}$
   (*Hint:* Draw lines through B and D perpendicular to the x-axis at M and N respectively, and prove that $\triangle ABM \sim \triangle CDN$.)

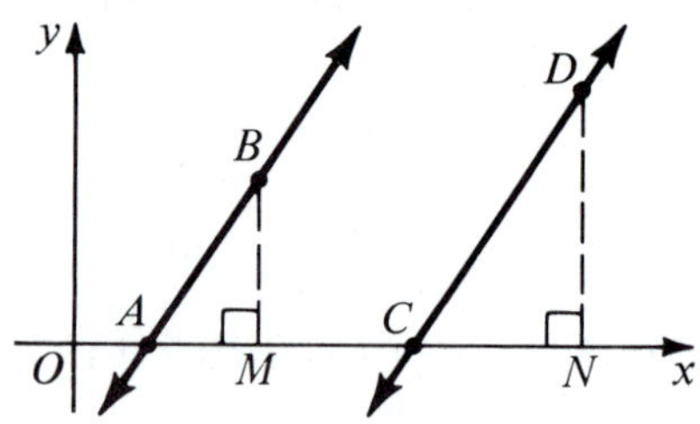

28. Use the diagram for Exercise 27 to prove Theorem 11-3.
   Given: Slope of $\overleftrightarrow{AB}$ = slope of $\overleftrightarrow{CD}$
   Prove: $\overleftrightarrow{AB} \parallel \overleftrightarrow{CD}$

29. $\triangle ABC$ has vertices $A(8, 6)$, $B(-1, 13)$, and $C(5, -5)$.
   a. Show that $\triangle ABC$ is isosceles.
   b. If M is the midpoint of $\overline{BC}$, show that $\overline{AM} \perp \overline{BC}$.

30. Quadrilateral $ABCD$ has vertices $A(0, -1)$, $B(6, 1)$, $C(8, 7)$, and $D(2, 5)$.
   a. Use slopes to show that $ABCD$ is a parallelogram.
   b. Show that $ABCD$ is a rhombus by showing that $AB = BC$.
   c. Show that the diagonals of $ABCD$ are perpendicular.

31. A and B have coordinates $A(2, 7)$ and $B(-7, 10)$. Find r if the line joining $(r, -3)$ and $(1, 3)$ is:
   a. parallel to $\overleftrightarrow{AB}$.
   b. perpendicular to $\overleftrightarrow{AB}$.

32. Show that if the line joining $(2, -1)$ and $(7, 4)$ is parallel to the line joining $(4, 9)$ and (a, b), then b must exceed a by 5.

33. Show that if the line joining $(1, 9)$ and $(8, 16)$ is to be perpendicular to the line joining $(1, 1)$ and (a, b), then the sum of a and b must be 2.

C 34. Find the values of c so that the slope of the line joining $(5, c)$ and $(6, c^2)$ is 12.

35. Use slopes to show that if the midpoints of the sides of the quadrilateral whose vertices are $P(-1, 6)$, $Q(3, 8)$, $R(5, 12)$, and $S(11, 2)$ are joined in order, then the figure formed is a parallelogram.

36. Find the value of t in terms of a and r if the line joining (a^2, a) and (r^2, r) is perpendicular to the line joining (t, r^2) and (r, t^2). (Assume that $r \neq \pm a$.)

37. Point P is 10 units from the origin. The slope of the line passing through P and the origin is 3. If P is in the first quadrant, find its coordinates in simplified radical form.

Section 11-4 PROOFS USING A COORDINATE SYSTEM

You are now ready to apply the distance formula, the midpoint formula, and the principles of parallelism and perpendicularity to prove theorems. A proof that uses the methods of coordinate geometry is called an *analytic proof.* The secret of a good analytic proof lies in the placement of the coordinate axes and the choice of the labels used in the diagrams. You must be careful to use a diagram that accurately describes the conditions of the problem without assuming any special conditions. You will acquire skill in this area as you do the proofs.

Suppose you want to prove that the segment joining the midpoints of two sides of a triangle is parallel to the third side. Several ways in which you could place the coordinate axes in relation to the triangle are discussed on the next page.

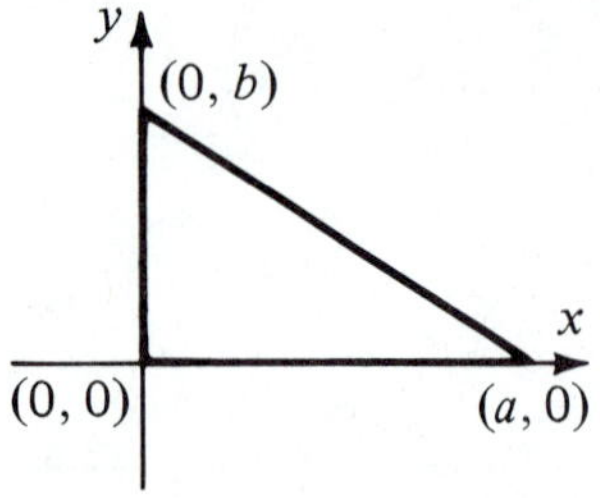

Figure 1

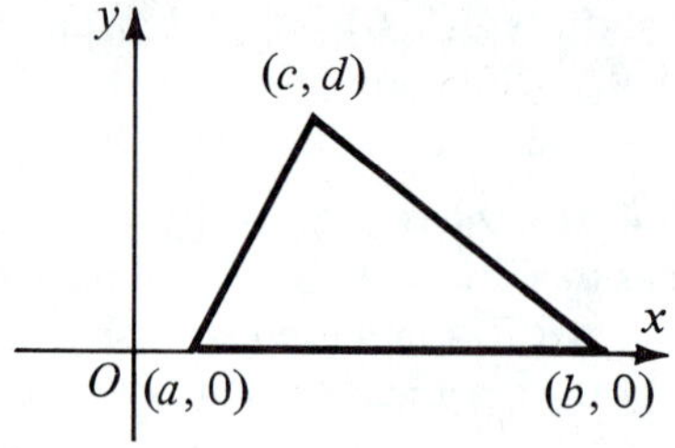

Figure 2

The diagram in Figure 1 is not general enough; it denotes a *right* triangle. The diagram in Figure 2 is general enough, but the placement of the axes will not simplify needed algebraic calculations. Compare Figure 2 to Figures 3 and 4.

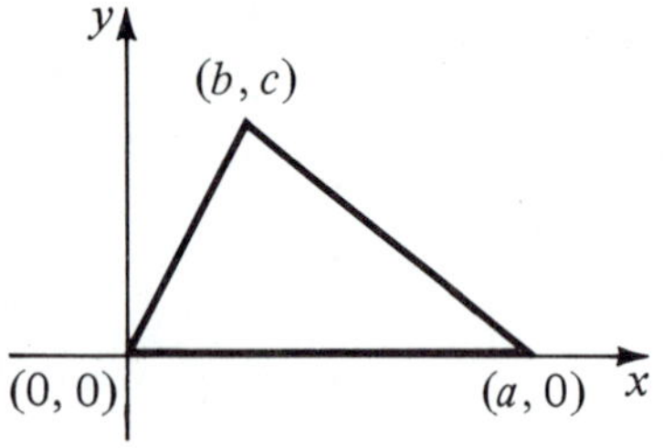

Figure 3

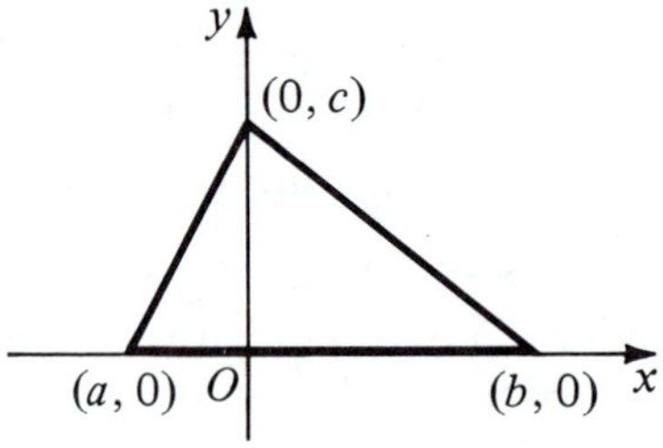

Figure 4

In Figure 3, one of the vertices has coordinates (0, 0). This choice of a vertex will make arithmetic calculations less complicated. The placement of the coordinate axes in Figure 4 also simplifies calculations since the x-coordinate or the y-coordinate of each vertex is equal to zero.

EXAMPLE Show that the segment joining the midpoints of two sides of a triangle is parallel to the third side.

SOLUTION Place the coordinate axes so that the origin is at A and the x-axis lies on $\overleftrightarrow{AB}$. Since we will be finding the coordinates of the midpoints of two sides, it will simplify our calculations if we use the coordinates in the diagram at the right.

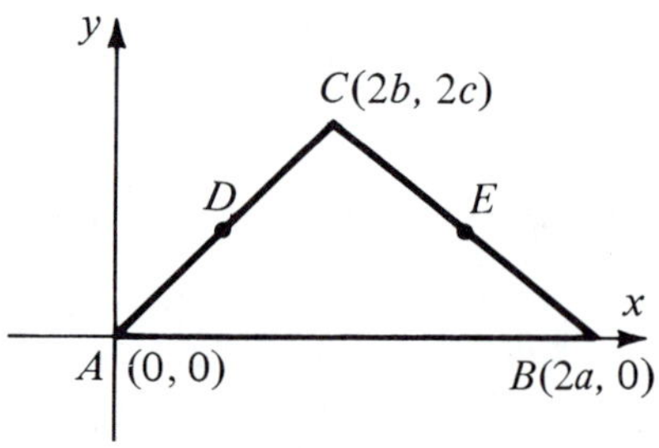

Let D be the midpoint of $\overline{AC}$. By the Midpoint Formula, the coordinates of D are $\left(\frac{2b+0}{2}, \frac{2c+0}{2}\right)$, or (b, c).

Let E be the midpoint of $\overline{BC}$. The coordinates of E are $\left(\frac{2b+2a}{2}, \frac{2c+0}{2}\right)$, or $(a+b, c)$.

The slope of $\overline{DE}$ is $\frac{c-c}{(a+b)-b} = 0$. The slope of $\overline{AB}$ is $\frac{0-0}{2a-0} = 0$.

Therefore, $\overline{DE} \parallel \overline{AB}$.

EXAMPLE Show that the diagonals of a square are congruent and perpendicular.

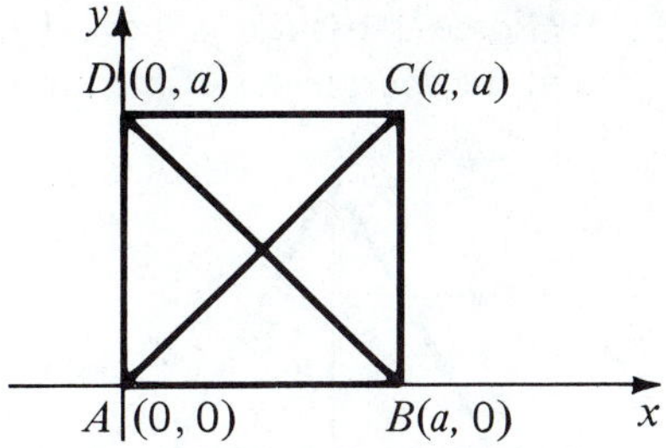

SOLUTION $ABCD$ is a square. Place the coordinate axes so that the origin coincides with one vertex of the square and the x-axis and y-axis contain two sides of the square.

$AC = \sqrt{(a-0)^2 + (a-0)^2} = \sqrt{a^2 + a^2}$
$= \sqrt{2a^2} = a\sqrt{2}$

$BD = \sqrt{(a-0)^2 + (0-a)^2} = \sqrt{a^2 + (-a)^2} = \sqrt{2a^2} = a\sqrt{2}$

Since $AC = BD$, $\overline{AC} \cong \overline{BD}$.

The slope of $\overline{AC} = \frac{a-0}{a-0} = \frac{a}{a} = 1 = m_1$.

The slope of $\overline{BD} = \frac{a-0}{0-a} = \frac{a}{-a} = -1 = m_2$.

Since $m_1 \cdot m_2 = -1$, $\overline{AC} \perp \overline{BD}$.

ORAL EXERCISES

Give the coordinates of the fourth vertex of the square.

1. 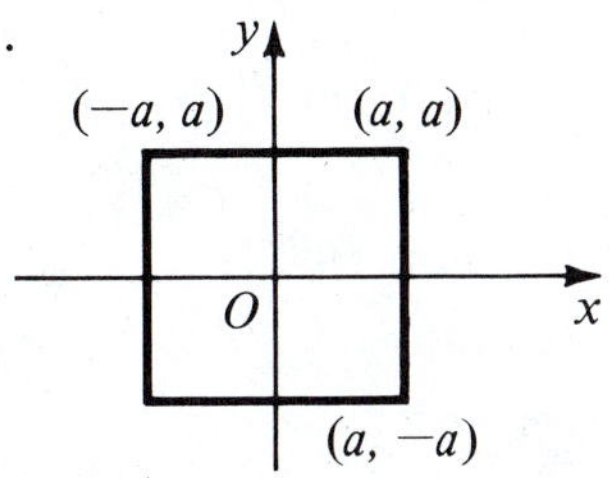

2. 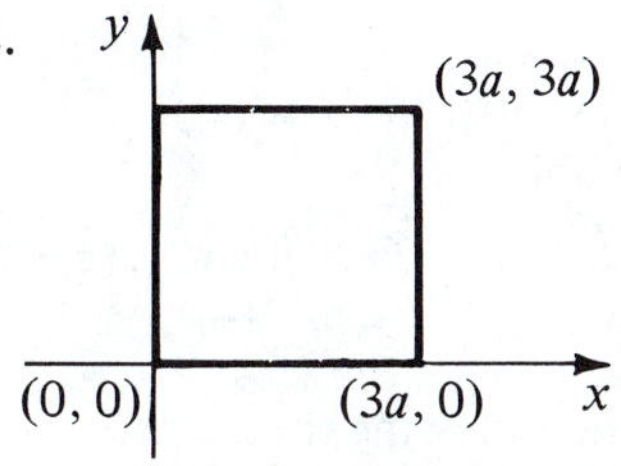

3. 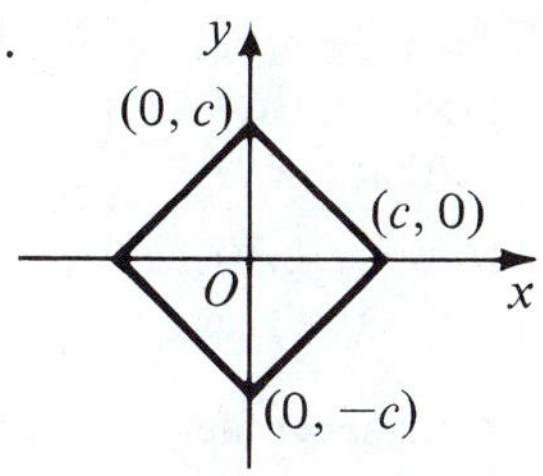

Supply the missing coordinates of the vertices of the rectangle.

4. 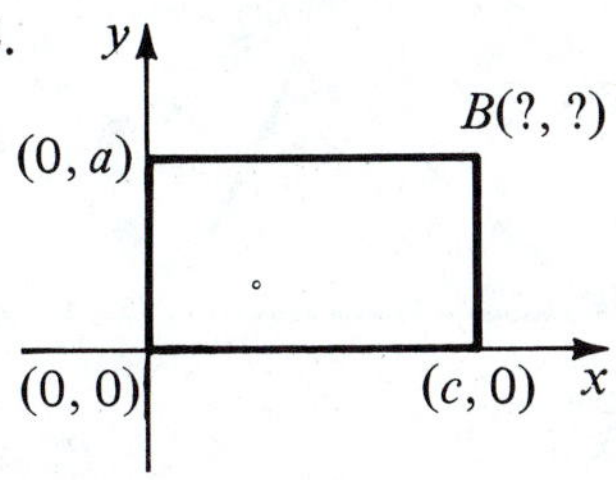

5. 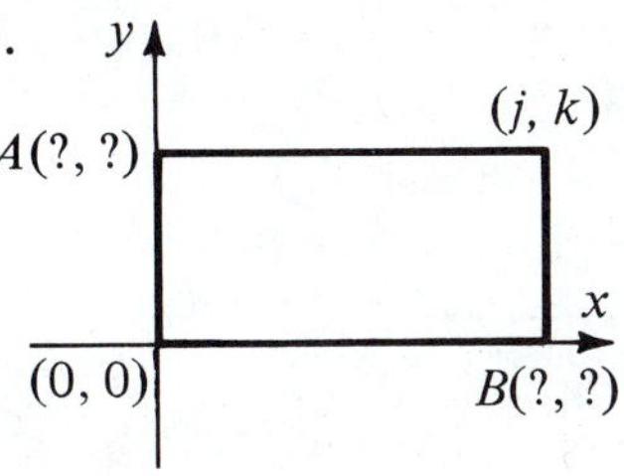

6. 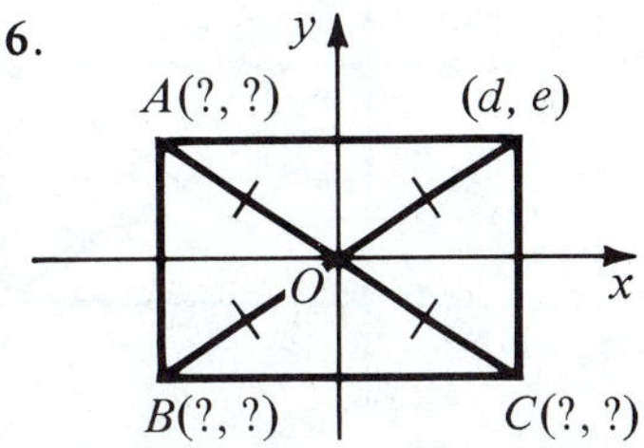

Give the coordinates of the fourth vertex of the parallelogram.

7. 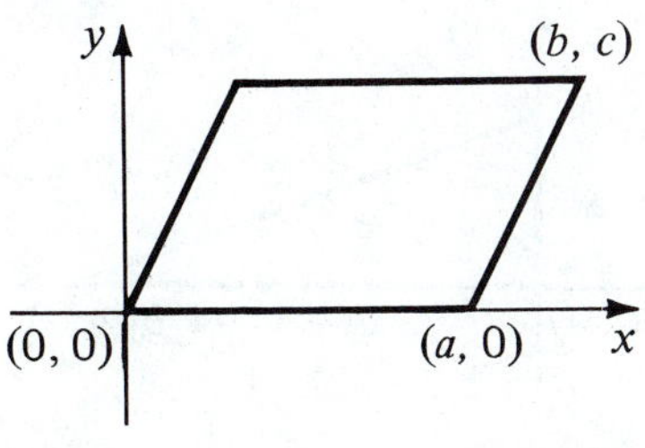

8. 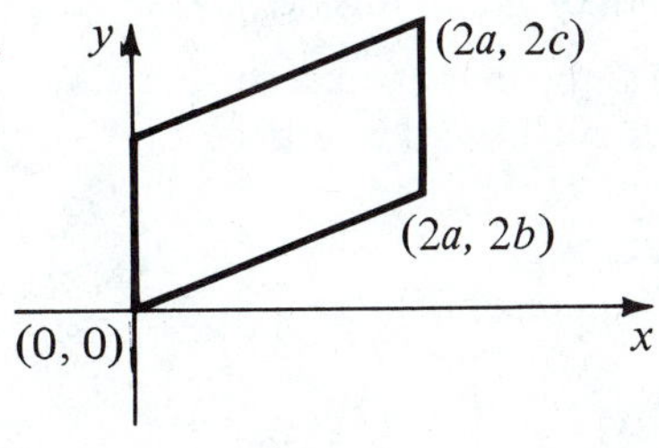

9. An isosceles triangle is shown below. The length of the altitude to the base is t. Find the coordinates of the third vertex.

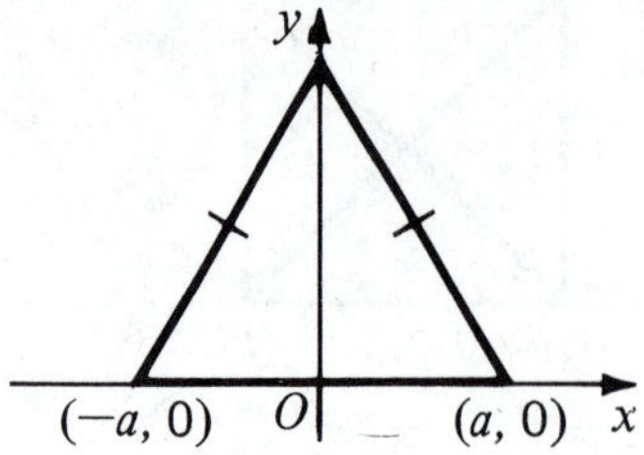

Ex. 9

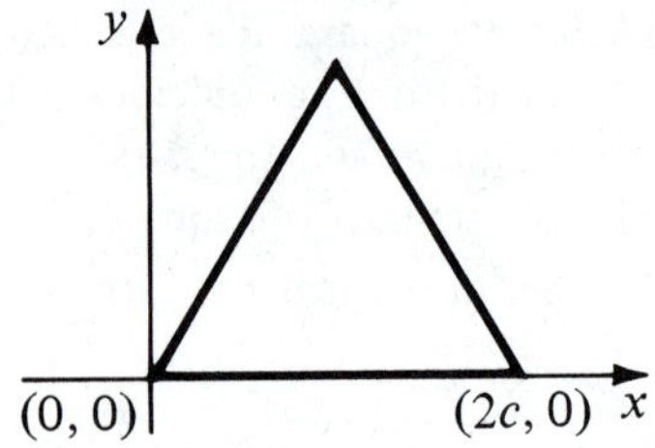

Ex. 10

10. An equilateral triangle is shown above. Find the coordinates of the third vertex.

WRITTEN EXERCISES

A 1. $\triangle ABC$ has vertices $A(-a, 0)$, $B(a, 0)$, and $C(0, b)$. What kind of triangle is $\triangle ABC$?

2. $\triangle DEF$ has vertices $D(0, 0)$, $E(a, 0)$, and $F(0, b)$. What kind of triangle is $\triangle DEF$?

3. ▱$ABCD$ has vertices $A(0, 0)$, $B(a, 0)$, and $D(b, c)$. What are the coordinates of vertex C?

4. Quadrilateral $QRST$ has vertices $Q(0, 0)$, $R(d, e)$, $S(d, e + f)$, and $T(0, f)$. Show that $QRST$ is a parallelogram.

5. Rhombus $DEFG$ has vertices $D(0, 0)$, $E(a, 0)$, and $F(a + b, c)$ where a, b, and c are positive real numbers. Show that $a^2 = b^2 + c^2$.

Use coordinate geometry to prove the following theorems.

6. The diagonals of a rectangle are congruent.

Ex. 6

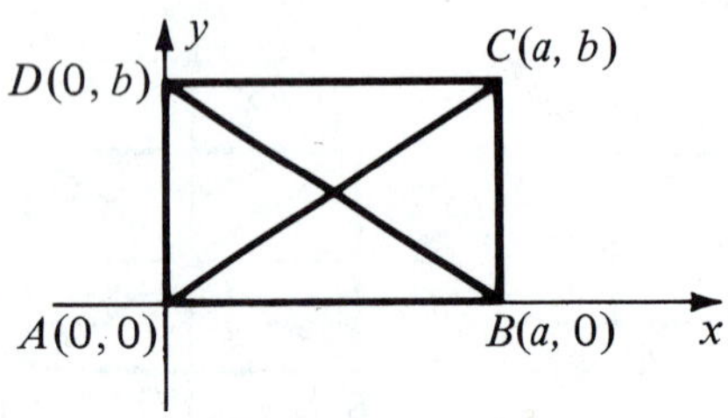

Ex. 7

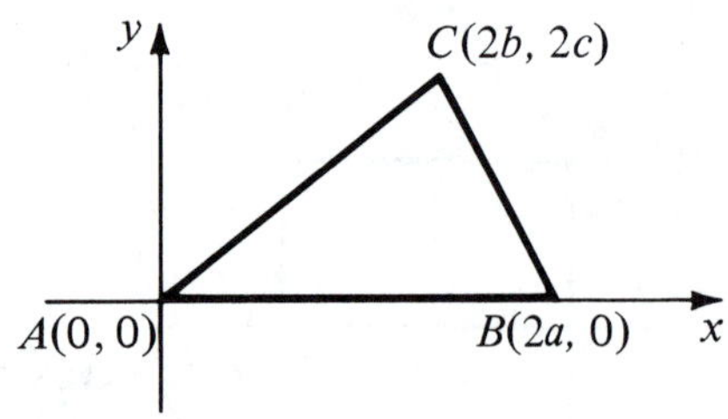

7. The length of the segment joining the midpoints of two sides of a triangle is one half the length of the third side.

8. The midpoint of the hypotenuse of a right triangle is equidistant from the vertices.

Ex. 8

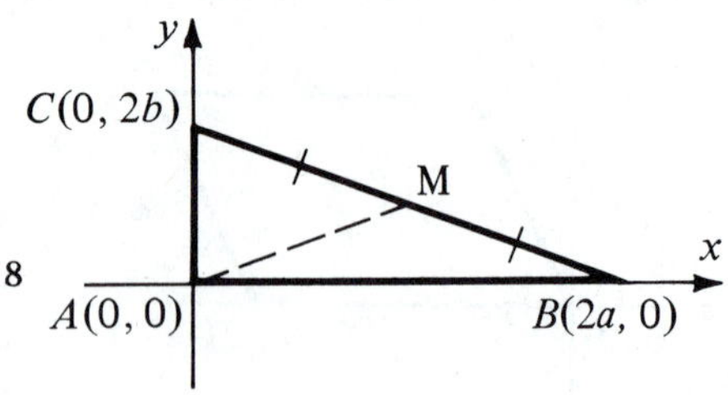

9. In a plane, two lines perpendicular to the same line are parallel. (*Hint:* Let the slope of $\overleftrightarrow{EF}$ be k and express the slopes of $\overleftrightarrow{AB}$ and $\overleftrightarrow{CD}$ in terms of k.)

Ex. 9

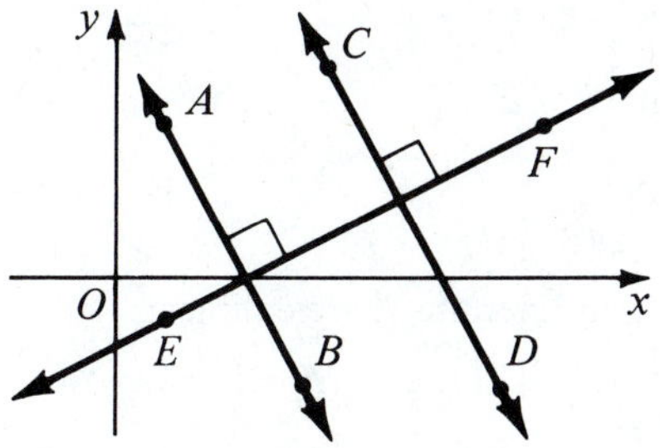

Ex. 10

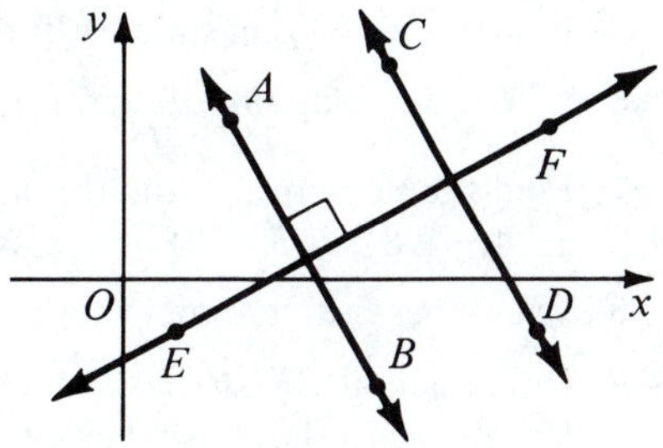

10. In a plane, a line perpendicular to one of two parallel lines is perpendicular to the other one.

B 11. The diagonals of a parallelogram bisect each other. (*Hint:* Find the coordinates of the midpoint of each diagonal.)

Ex. 11

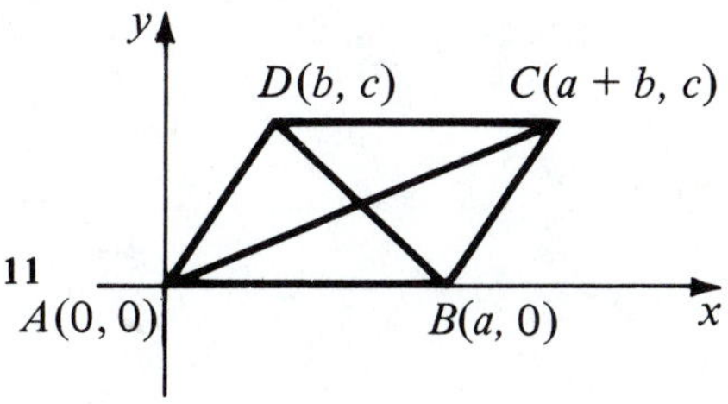

Ex. 12

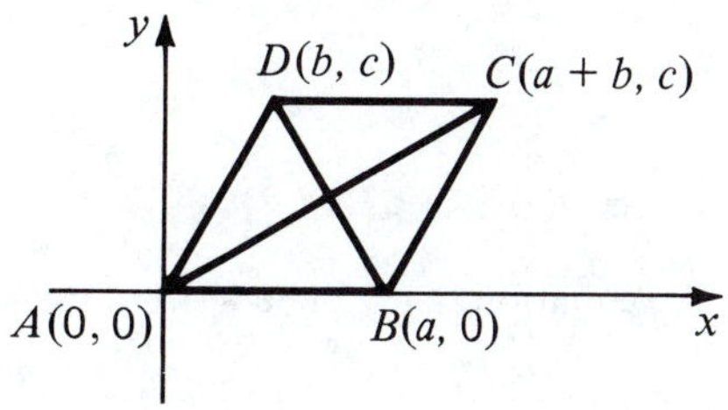

12. The diagonals of a rhombus are perpendicular. (*Hint:* See Exercise 5.)

13. If the diagonals of a parallelogram are congruent, then the parallelogram is a rectangle. (*Hint:* Show that $b = 0$.)

C 14. The length of the median of a trapezoid is one half the sum of the lengths of the bases.

Ex. 13

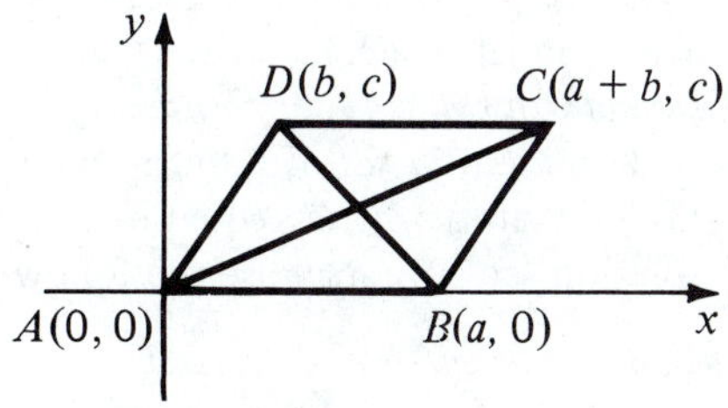

15. If the diagonals of a rectangle are perpendicular, then the rectangle is a square.

16. The segments joining the midpoints of the sides of any quadrilateral form a parallelogram.

SELF-TEST 1

1. Graph each point on the same coordinate plane. Section 11-1
   a. $J(4, -3)$ b. $K(0, -2)$ c. $L(-3, 1)$
2. Find the area of the triangle with vertices $A(-3, -2)$, $B(2, -2)$, and $C(1, 3)$.
3. Find the area of quadrilateral $QRST$ with vertices $Q(-2, -2)$, $R(-1, 3)$, $S(2, 3)$, and $T(0, 4)$.

(Self-Test continued on page 448.)

Exercises 4–6 refer to points $X(-4, 0)$, $Y(-2, -2)$, and $Z(4, 6)$.

4. Find the coordinates of the midpoint of $\overline{YZ}$. Section 11-2

5. Find XY and YZ in simplest form.

6. If X is the midpoint of $\overline{WY}$, find the coordinates of W.

7. Find the slope of the line through each pair of points. Section 11-3
   a. $A(-1, 8)$, $B(3, -2)$ b. $C(4, -7)$, $D(8, -7)$

8. Use slopes to determine whether points A, B, and C, in Exercise 7, are collinear.

9. Quadrilateral $EFGH$ has vertices $E(5, 0)$, $F(2, 9)$, $G(-4, 7)$, and $H(-1, -2)$. Use slopes to show that $EFGH$ is a rectangle.

10. $A(0, 0)$, $B(a, b)$, and $D(-a, b)$ are three vertices of rhombus $ABCD$. What are the coordinates of vertex C? Section 11-4

11. Use coordinate geometry to prove that if $A(0, 0)$, $B(a, b)$, and $C(c, d)$ are collinear (where a and c do not equal zero), then $bc = ad$.

Section 11-5 GRAPHING LINEAR EQUATIONS

An equation that can be written in the form

$$Ax + By = C$$

where A, B, and C are real numbers with A and B not both zero is called a *linear equation in two variables.* Note that in a linear equation each term is a constant or a monomial of degree 1. If A, B, and C are integers, the equation is said to be in *standard form.*

The *solution set* of a linear equation in two variables is the set of ordered pairs (x, y) that satisfy the equation. The graph of a linear equation is the graph of its solution set. You may use the following facts when you graph a linear equation:

> The graph of every linear equation in two variables is a line in a coordinate plane.
>
> Each line in a coordinate plane is the graph of a linear equation in two variables.

Since two points determine a line, you could graph a linear equation in two variables by graphing two points whose coordinates satisfy the equation and then drawing the unique line that contains the two points. In practice, it is wise to graph a third, or even a fourth, point to guard against errors. Convenient points to graph are the points where the line crosses the x-axis and the y-axis. To find the coordinate of the point where the line crosses the x-axis, you let $y = 0$ and solve for x. To find the coordinate of the point where the line crosses the y-axis, you let $x = 0$ and solve for y.

EXAMPLE Graph $2x + 3y = 6$.

SOLUTION Let $y = 0$; $2x + 3(0) = 6$
$2x = 6$
$x = 3$; Solution $(3, 0)$

Let $x = 0$; $2(0) + 3y = 6$
$3y = 6$
$y = 2$; Solution $(0, 2)$

Let $x = -3$; $2(-3) + 3y = 6$
$-6 + 3y = 6$
$3y = 12$
$y = 4$; Solution $(-3, 4)$

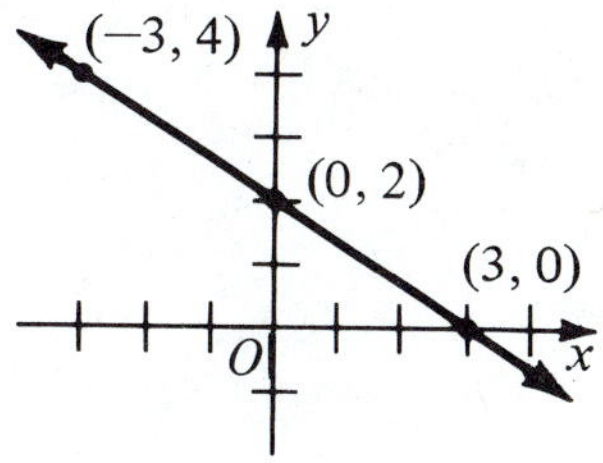

If a line crosses the x-axis at $(a, 0)$ and the y-axis at $(0, b)$, then a is called the *x-intercept* and b is called the *y-intercept*. In the preceding example, the x-intercept is 3 and the y-intercept is 2.

EXAMPLE Graph $x = 4$.

SOLUTION All points with x-coordinate 4 satisfy the equation. The graph is a vertical line with x-intercept 4.

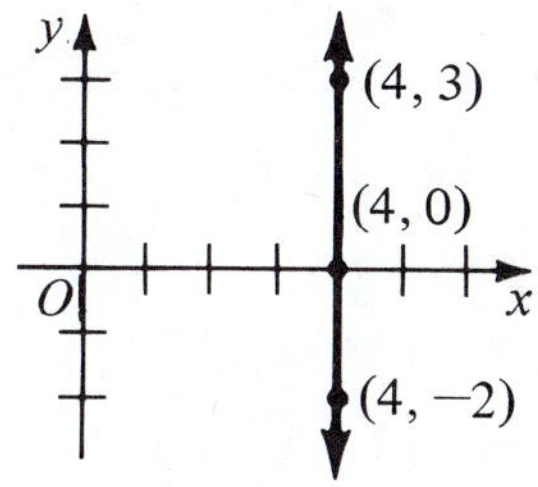

EXAMPLE Graph $y = -2$.

SOLUTION All points with y-coordinate -2 satisfy the equation. The graph is a horizontal line with y-intercept -2.

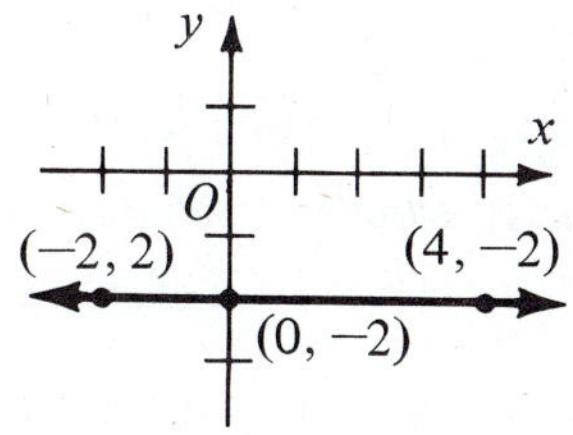

EXAMPLE Graph the line with y-intercept 1 and slope $\frac{2}{3}$.

SOLUTION The line intersects the y-axis at $(0, 1)$. Because the slope is $\frac{2}{3}$, a vertical change of 2 corresponds to a horizontal change of 3. Thus, the point $(3, 3)$ also lies on the line.

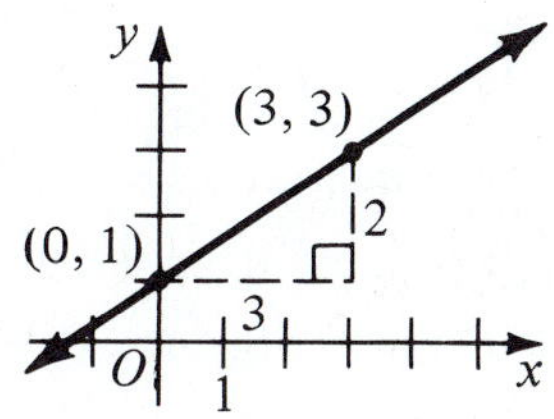

ORAL EXERCISES

State the x-intercept and y-intercept of each line whose equation is given.

1. $x + y = 7$
2. $x + 2y = 12$
3. $2x + y = 8$
4. $2x - 3y = 6$
5. $3x - y = 9$
6. $2x - 7y = 28$

Does the line whose equation is given contain the point $(2, -3)$?

7. $x - y = 5$
8. $2x + y = 1$
9. $x = -3$

Name three points that lie on the line with the given equation.

10. $x - y = 0$
11. $x - 2y = 6$
12. $2x + y = 10$

WRITTEN EXERCISES

Find the x-intercept and y-intercept of the line whose equation is given.

A

1. $x + 3y = 6$
2. $3x - 2y = 6$
3. $2x + y = 8$
4. $x - y = 10$
5. $2x - 3y = 15$
6. $5x + 4y = 18$

Draw the graphs of the following equations.

7. $x = 5$
8. $2x = 5$
9. $2y = -3$
10. $y = -3$
11. $x = y$
12. $x = -y$
13. $y = 2x$
14. $y = -3x$
15. $x + y = 6$
16. $x - y = 6$
17. $5x - y = 10$
18. $3x + y = 12$
19. $x - 5y = 10$
20. $x + 4y = 16$
21. $3x + 2y = 6$
22. $4x + 3y = 12$
23. $2x - 3y = 18$
24. $5x - 2y = 10$

Graph the line that satisfies the given conditions.

25. Its slope is 4 and its y-intercept is -8.
26. Its slope is -2 and its y-intercept is 3.
27. Its slope is $\frac{1}{2}$ and its x-intercept is -2.
28. Its slope is $-\frac{3}{4}$ and its x-intercept is 1.
29. Its x-intercept is -3 and its y-intercept is 1.
30. It contains the point $(4, 1)$ and is perpendicular to the x-axis.
31. It contains the point $(-1, 2)$ and has slope -2.
32. It contains the point $(-3, -2)$ and has slope $\frac{1}{3}$.

In Exercises 33–36, **(a)** find the coordinates of two points that lie on the line whose equation is given and **(b)** use them to determine the slope of the line.

B

33. $3x + y = 12$
34. $y = 4x + 5$
35. $y = -\frac{1}{2}x + 2$
36. $y = -2 - 3x$

Find the value of k so that the point with the given coordinates lies on the line with the given equation.

EXAMPLE $(3, k);\ 2x + 3y = 9$

SOLUTION Substitute 3 for x and k for y in the given equation and solve for k.

$$2 \cdot 3 + 3k = 9$$
$$3k = 3$$
$$k = 1$$

37. $(-2, k);\ x + 2y = 6$
38. $(4, k);\ 3x - y = 10$
39. $(k, -1);\ y = 2x - 5$
40. $(k, k);\ 2x - 4y = 5$

Find the value of k so that the line whose equation is given will contain the given point.

41. $kx + 3y = 6;\ (1, 7)$
42. $x + ky = 6;\ (1, 3)$
43. $2x - ky = 8;\ (-2, 2)$
44. $kx - 5y = 21;\ (-2, -1)$

Graph the set of points satisfying the given condition.

C 45. $|x| = 2$
46. $|y| = 3$
47. $y = |x|$
48. $y = |x - 1|$
49. $y = |x + 1|$
50. $y = |x| + 1$
51. $x = |y|$
52. $|x| = |y|$
53. $|y| = -|x| + 4$

Section 11-6 WRITING EQUATIONS OF LINES

In Section 11-5 you were given an equation of a line and asked to graph the equation. Now consider the reverse problem. How do you write an equation of a line given facts about its graph? You will study two forms that an equation of a line may take:

- the point-slope form
- the slope-intercept form

Consider a line with slope -2 that passes through the point $(-1, 3)$. Let $P(x, y)$ be any other point on the line. By the definition of slope,

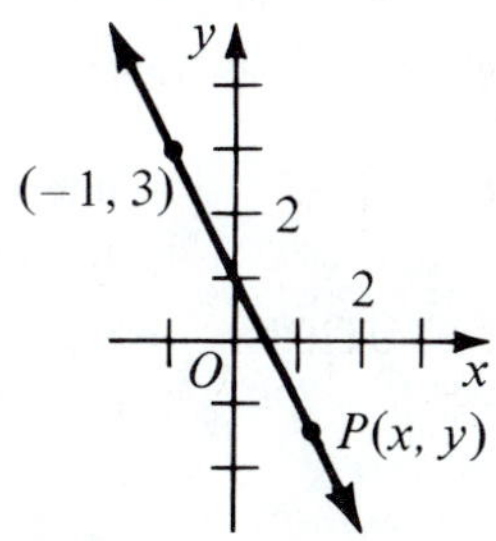

$$\frac{y - 3}{x - (-1)} = -2$$
$$\frac{y - 3}{x + 1} = -2$$
$$y - 3 = -2(x + 1)$$

This result can be generalized to give the *point-slope form* of an equation of a line.

> If a line has slope m and passes through a point with coordinates (x_1, y_1), then an equation of the line is:
>
> $$y - y_1 = m(x - x_1)$$

EXAMPLE Write an equation in standard form of the line that contains (1, −2) and (−1, 1).

SOLUTION Slope $= m = \dfrac{1-(-2)}{-1-1} = \dfrac{1+2}{-2}$

$$m = -\frac{3}{2}$$

Choose either point to be (x_1, y_1) and use the point-slope form. We will choose (1, −2) to be (x_1, y_1).

$$y - y_1 = m(x - x_1)$$
$$y - (-2) = -\frac{3}{2}(x - 1)$$
$$y + 2 = -\frac{3}{2}x + \frac{3}{2}$$
$$2y + 4 = -3x + 3$$
$$3x + 2y = -1$$

You should verify that (1, −2) and (−1, 1) satisfy this equation.

Suppose a line has slope m and y-intercept b. Then the line intersects the y-axis at $(0, b)$. If we substitute $(0, b)$ for (x_1, y_1) in the point-slope equation, we will have

$$y - b = m(x - 0)$$
$$y - b = mx$$
$$y = mx + b$$

> If a line has slope m and y-intercept b, then an equation of the line is:
>
> $$y = mx + b$$

This is called the *slope-intercept form* of an equation of a line. It is often the most convenient form to use.

EXAMPLE A line with slope 2 has a y-intercept of −4. Write an equation in slope-intercept form of the line.

SOLUTION In this case, $m = 2$ and $b = -4$. Substituting in the equation $y = mx + b$, you have $y = 2x + (-4)$, or

$$y = 2x - 4$$

Once an equation of a line is written in slope-intercept form, you can read off the line's slope and y-intercept.

EXAMPLE Graph $y = \frac{1}{2}x + 2$.

SOLUTION The y-intercept is 2, so the line contains (0, 2). The slope is $\frac{1}{2}$, so a vertical change of 1 corresponds to a horizontal change of 2. The line also contains (2, 3).

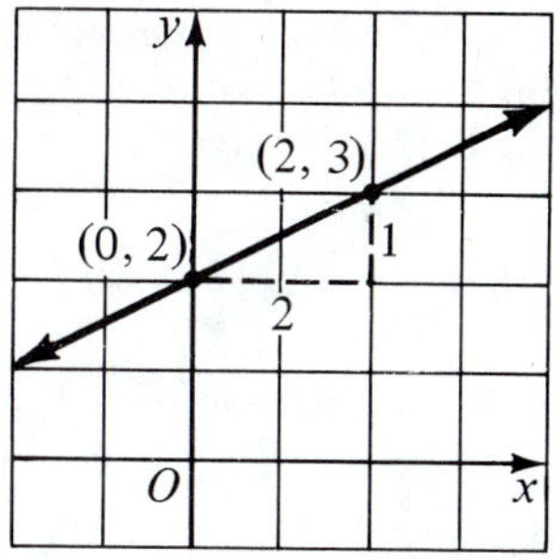

You can check your work by substituting 2 for x and 3 for y in the original equation.

$$y = \frac{1}{2}x + 2$$

$$3 \; ? \; \frac{1}{2}(2) + 2$$

$$3 = 1 + 2$$

EXAMPLE A line contains the points (−1, −2) and (3, 2). Write an equation of the line in **(a)** the point-slope form and **(b)** the slope-intercept form.

SOLUTION Find the slope of the line. $m = \frac{2 - (-2)}{3 - (-1)} = \frac{4}{4} = 1$

(a) There are two choices of equations depending on which point you use, (−1, −2) or (3, 2).

$$y + 2 = x + 1 \quad \text{and} \quad y - 2 = x - 3$$

are both correct equations in point-slope form.

(b) Solve either of the equations in part (a) for y and you will have:

$$y = x - 1$$

EXAMPLE Write an equation in slope-intercept form of the line with slope 5 that contains the point (2, −4).

SOLUTION Write the point-slope form of the equation.

$$y + 4 = 5(x - 2)$$

Solve for y.

$$y + 4 = 5x - 10$$

$$y = 5x - 14$$

ORAL EXERCISES

The following equations are written in point-slope form. For each equation, **(a)** find the slope of the line, and **(b)** find the coordinates of a point on the line.

1. $y - 3 = 4(x - 1)$ **2.** $y - 4 = 3(x - 2)$ **3.** $y - 6 = \frac{1}{2}(x + 2)$

4. $y - 1 = -\frac{2}{3}(x + 12)$ **5.** $y + 2 = 2(x + 5)$ **6.** $y + 5 = \frac{3}{4}x$

The following equations are written in slope-intercept form. For each equation, **(a)** find the slope of the line, and **(b)** find the y-intercept.

7. $y = 6x - 8$
8. $y = -3x + 2$
9. $y = x + 4$
10. $y = -x - 13$
11. $y = \frac{1}{2}x$
12. $y = -7$

State an equation of the line that passes through the given point and is **(a)** parallel to the x-axis; **(b)** parallel to the y-axis.

13. $(1, 3)$
14. $(2, 7)$
15. $(-1, 6)$
16. $(2, -3)$

WRITTEN EXERCISES

Find an equivalent equation in standard form.

A

1. $y = 3x + 2$
2. $y = -\frac{1}{2}x - 5$
3. $y + 1 = 2(x - 4)$

Find an equivalent equation in slope-intercept form and graph the equation.

4. $y - 3 = 2(x + 4)$
5. $3x + 4y = 8$
6. $y + 5 = -\frac{2}{3}(x - 1)$

For each pair of points, **(a)** find the slope of the line that contains them, and **(b)** find an equation in point-slope form of the line that contains them.

7. $(1, 2), (-2, -1)$
8. $(8, 6), (10, 8)$
9. $(3, -2), (4, 7)$
10. $(-3, -4), (5, 0)$
11. $(2, 7), (5, 8)$
12. $(3, 8), (9, 6)$

Find an equation in slope-intercept form of the line that contains the given point P and has slope m.

13. $P(0, -2), m = 3$
14. $P(2, 7), m = -7$
15. $P(-1, -1), m = 2$
16. $P(-1, -1), m = 0$
17. $P(-6, 0), m = \frac{1}{2}$
18. $P(2, 1), m = \frac{1}{3}$

Determine the slope and the y-intercept of the line with the given equation. Then graph the equation.

EXAMPLE $3x + 2y = 14$

SOLUTION Find an equivalent equation in slope-intercept form.

$$3x + 2y = 6$$
$$2y = -3x + 6$$
$$y = -\frac{3}{2}x + 3$$

The slope of the line is $-\frac{3}{2}$ and the y-intercept is 3.

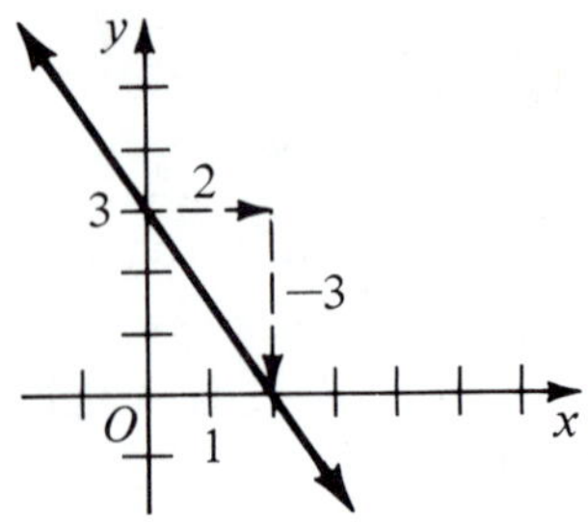

19. $2x - y = 2$ 20. $3x - 2y = 4$ 21. $x - 2y = 2$

22. $x + 2y = 2$ 23. $3x + 3y = 1$ 24. $2x + 3y = 1$

Find the slope of a line that is **(a)** parallel to the line whose equation is given; **(b)** perpendicular to the line whose equation is given. If the slope is undefined, so state.

25. $y = x + 2$ 26. $y = \frac{1}{2}x - 1$ 27. $y = -3x + 2$

28. $2x + y = -1$ 29. $2x - 3y = 12$ 30. $y = 6$

Find an equation in standard form of the line that passes through the point $(2, 3)$ and is parallel to the line whose equation is given.

B 31. $y = 3x + 1$ 32. $y = 2x - 1$ 33. $2x + y = 2$

34. $x + y = 7$ 35. $y = 2$ 36. $x = 4$

Find an equation in slope-intercept form of the line that passes through the given point and is perpendicular to the line whose equation is given.

EXAMPLE $(6, 7)$, $y = 3x + 1$

SOLUTION The given line has slope 3, so a line perpendicular to it has slope $-\frac{1}{3}$.

$$y = -\frac{1}{3}x + b$$

$$7 = -\frac{1}{3}(6) + b$$

$$7 = -2 + b$$

$$b = 9$$

$$y = -\frac{1}{3}x + 9$$

37. $(2, 1)$, $y = x + 2$ 38. $(-3, 2)$, $y = 2x + 1$ 39. $(0, 0)$, $2x + y = 1$

40. $(8, 1)$, $x - 2y = 1$ 41. $(2, -1)$, $x = 3$ 42. $(6, 2)$, $y = -1$

Write an equation in standard form of the line with the given x-intercept and y-intercept.

43. x-intercept 4, y-intercept 6 44. x-intercept -6, y-intercept 6

45. x-intercept 2, y-intercept -1 46. x-intercept -3, y-intercept -5

Write an equation in standard form of the line that is the perpendicular bisector of the line segment with the given endpoints.

47. $(8, 2)$, $(-6, 4)$ 48. $(-2, 1)$, $(8, 3)$

49. $(7, 3)$, $(3, 7)$ 50. $(2, 8)$, $(5, -5)$

C 51. Find an equation in standard form of the line that passes through (a, b) and $(0, c)$.

52. Find an equation in standard form of the perpendicular bisector of the segment joining $(2a, 2b)$ and $(0, 2c)$.

53. a. Find an equation of the perpendicular bisector of the segment with endpoints $A(5, 8)$ and $B(1, 2)$.

    b. Find the coordinates of the point P where the perpendicular bisector of $\overline{AB}$ intersects the y-axis, and show that $PA = PB$.

54. a. Write an equation of the perpendicular bisector of the segment with endpoints $A(-8, 2)$ and $B(8, -10)$.

    b. Find the coordinates of the point Q where the perpendicular bisector of $\overline{AB}$ intersects the x-axis, and show that $QA = QB$.

55. Show that if a line has x-intercept a and y-intercept b then its equation can be written as:

$$\frac{x}{a} + \frac{y}{b} = 1$$

Section 11-7 LOCI IN A COORDINATE PLANE

In Section 10-6 you learned how to write a description of a locus of points. Using a coordinate plane you can name a locus with an equation.

EXAMPLE Graph and find equations for the locus of points 4 units from the line with equation $x = 5$.

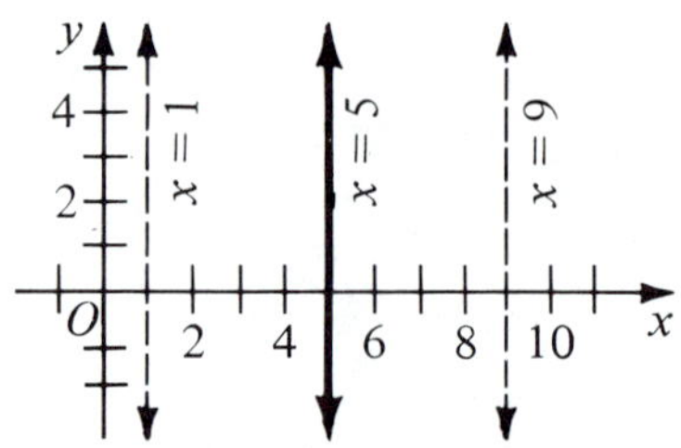

SOLUTION The locus of points 4 units from a given line is a pair of parallel lines, each parallel to the given line and 4 units from it. Since $5 - 4 = 1$ and $5 + 4 = 9$, the equations of these lines are $x = 1$ and $x = 9$.

The locus of points equidistant from two given parallel lines is a line parallel to each of the given lines and halfway between them.

EXAMPLE Graph and write an equation for the locus of points equidistant from the lines with equations $y = 2$ and $y = -4$.

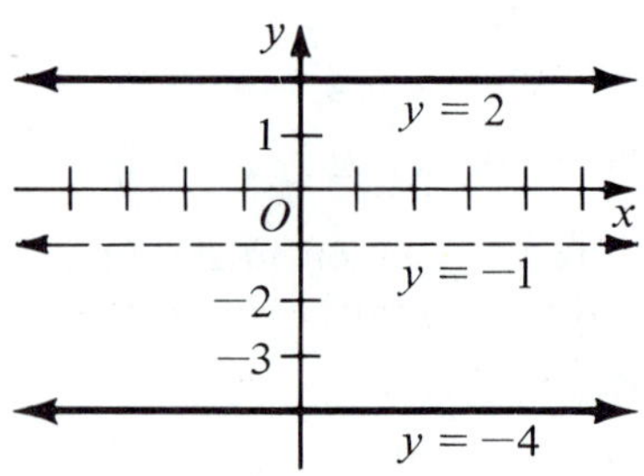

SOLUTION The distance from the line with equation $y = 2$ to the line with equation $y = -4$ is $2 - (-4) = 6$. The line equidistant from the two given lines must be 3 units from each. Such a line has equation $y = -1$.

The locus of points equidistant from the endpoints of a segment is the perpendicular bisector of the segment.

EXAMPLE Write an equation for the locus of points equidistant from $Q(2, 5)$ and $R(7, 3)$.

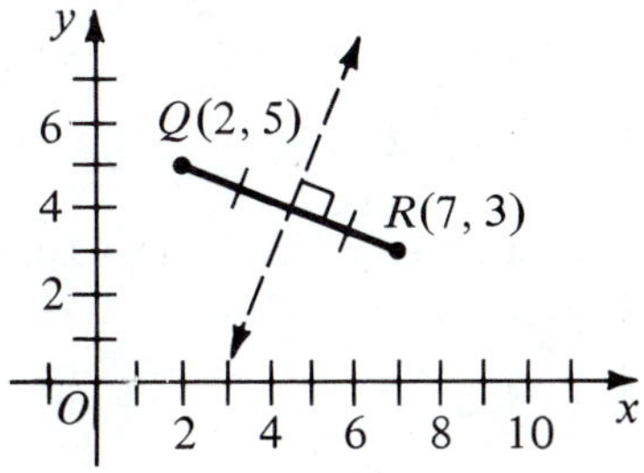

SOLUTION Write an equation of the line passing through M (the midpoint of $\overline{QR}$) and perpendicular to $\overline{QR}$.

The coordinates of M are

$$\left(\frac{2+7}{2}, \frac{5+3}{2}\right), \quad \text{or} \quad \left(\frac{9}{2}, 4\right).$$

The slope of $\overline{QR}$ is $\frac{3-5}{7-2} = -\frac{2}{5}$, so the slope of its perpendicular bisector is $\frac{5}{2}$.

Therefore, an equation is $y - 4 = \frac{5}{2}\left(x - \frac{9}{2}\right)$, or $10x - 4y = 29$.

ORAL EXERCISES

Give an equation for the locus of points:

1. 3 units to the right of the y-axis
2. 5 units to the left of the y-axis
3. 7 units below the x-axis
4. 20 units above the x-axis
5. equidistant from the points $(-6, 0)$ and $(6, 0)$
6. equidistant from the points $(0, 3)$ and $(0, -3)$

Describe each of the following as a locus of points in relation to the x-axis or y-axis.

7. the line with equation $x = -6$
8. the line with equation $y = 5$
9. the line with equation $y = -2$
10. the line with equation $x = 8$

WRITTEN EXERCISES

Write an equation (or equations) for the locus of points:

A

1. 1 unit below the line with equation $y = 3$
2. 2 units to the right of the line with equation $x = -5$
3. 3 units from the line with equation $x = 6$
4. 2 units from the line with equation $y = -1$
5. $1\frac{1}{2}$ units from the line with equation $y = \frac{1}{2}$
6. $5\frac{1}{2}$ units from the line with equation $x = -2\frac{1}{2}$

Write an equation (or equations) for the locus of points:

7. equidistant from the lines with equations $y = -6$ and $y = 4$
8. equidistant from the lines with equations $x = -5$ and $x = -2$
9. equidistant from the lines with equations $x = -2$ and $x = 1$
10. equidistant from the lines with equations $y = -3$ and $y = -8$

Write an equation for the locus of points that are equidistant from the given points.

11. (1, 0), (5, 0)
12. (3, −1), (3, 7)
13. (−4, 1), (−4, 4)
14. (−5, 3), (0, 3)
15. (2, 1), (4, 3)
16. (1, 1), (7, −5)

Draw on graph paper the locus of points that satisfy both conditions and name the locus.

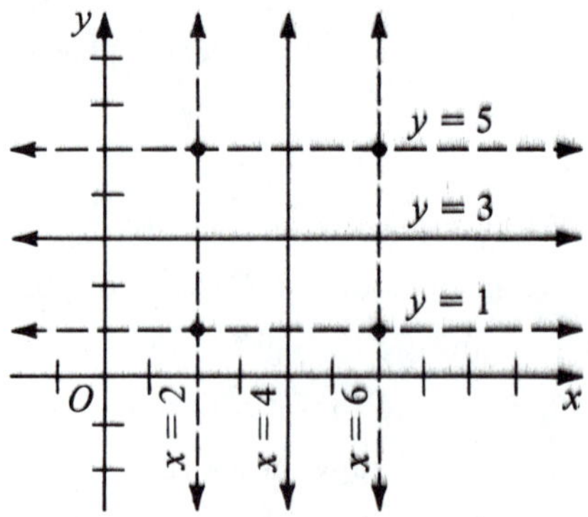

EXAMPLE 2 units from the line with equation $x = 4$ and 2 units from the line with equation $y = 3$

SOLUTION Graph the loci and find the coordinates of the points that satisfy both conditions. The locus consists of the points (2, 1), (6, 1), (2, 5), and (6, 5).

17. 3 units from the line with equation $x = 4$ and 2 units from the line with equation $y = 5$
18. 3 units from the line with equation $x = -5$ and 3 units from the line with equation $y = 2$
19. 1 unit from the line with equation $x = -1$ and 4 units from the line with equation $y = -3$
20. 3 units from the line with equation $x = 1$ and 1 unit from the line with equation $y = 2\frac{1}{2}$

Determine the total number of points in the given locus.

EXAMPLE 5 units from the origin and 3 units from the y-axis

SOLUTION The first locus is a circle with center at the origin and radius 5. The second locus is two lines parallel to the y-axis and 3 units from it. The total number of points satisfying *both* conditions is 4.

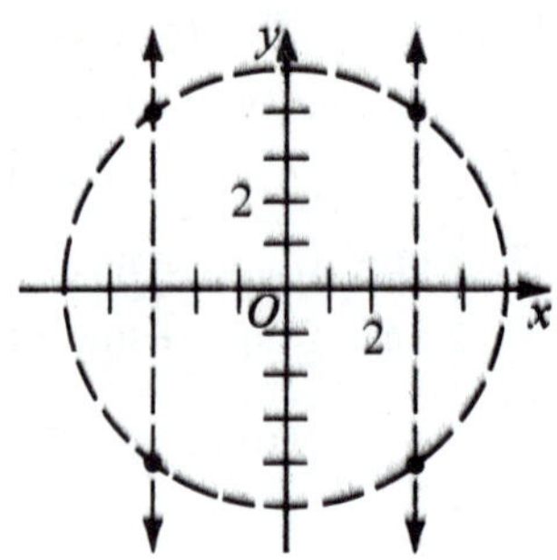

B **21.** 5 units from the origin and 3 units from the x-axis

22. 13 units from the origin and 5 units from the y-axis

23. 2 units from the origin and 4 units from the x-axis

24. 1 unit from the origin and 1 unit from the y-axis

25. 2 units from the origin and 1 unit from the line with equation $x = 2$

26. 3 units from the origin and 2 units from the line with equation $x = -5$

27. Equidistant from the x-axis and the y-axis and 4 units from the origin

28. Equidistant from the lines with equations $y = x$ and $y = -x$ and 2 units from the origin

29. Equidistant from the x-axis and the y-axis and 3 units from the x-axis

30. Equidistant from the x-axis and the y-axis and 1 unit to the right of the y-axis

C **31.** Write an equation for the locus of points $Q(x, y)$ such that the slope of the line through Q and the origin is twice the slope of the line through Q and $(0, 1)$.

32. Use the distance formula to find an equation for the locus of points equidistant from $(1, -3)$ and $(7, 9)$.

33. Use the distance formula to show that the locus of points equidistant from (x_1, y_1) and (y_1, x_1) is the line with equation $y = x$.

SELF-TEST 2

Draw the graph of each equation. State the x-intercept and the y-intercept of the line.

1. $3x = -2y$ **2.** $4x - 3y = 24$ Section 11-5

3. Graph the line that contains the point $(2, 4)$ and has slope $-\frac{2}{3}$.

4. Find an equation in point-slope form of the line that contains points $(-1, 4)$ and $(7, 0)$. Section 11-6

5. Find an equation in slope-intercept form of the line with slope 3 that contains the point $(-1, 2)$.

6. Find the slope and the y-intercept of the line with equation $4x + 5y = -20$.

Write an equation (or equations) for the locus of points:

7. 3 units from the line with equation $y = 0$ Section 11-7

8. equidistant from the lines with equations $x = -2$ and $x = 8$

9. equidistant from the points $(3, -7)$ and $(3, -4)$

Section 11-8 GRAPHING SYSTEMS OF LINEAR EQUATIONS

When two linear equations are graphed in a coordinate plane, the resulting lines will do exactly one of the following:

- intersect in exactly one point
- coincide and have infinitely many points in common
- be parallel and have no points in common

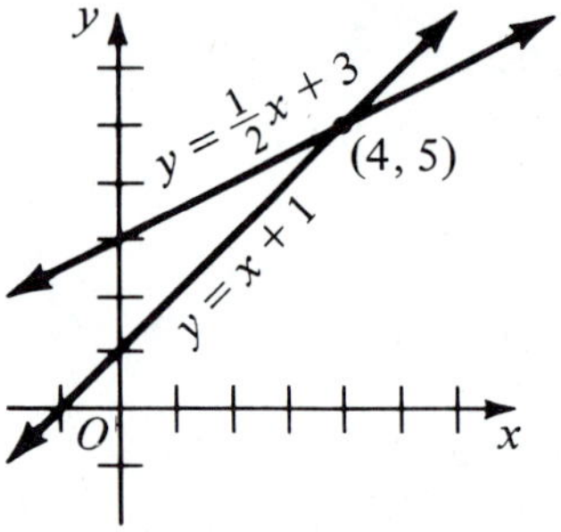

$y = x + 1$
$y = \frac{1}{2}x + 3$

Figure 1

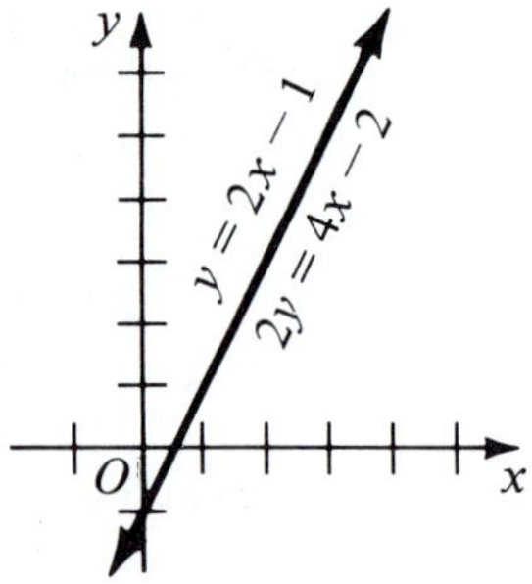

$y = 2x - 1$
$2y = 4x - 2$

Figure 2

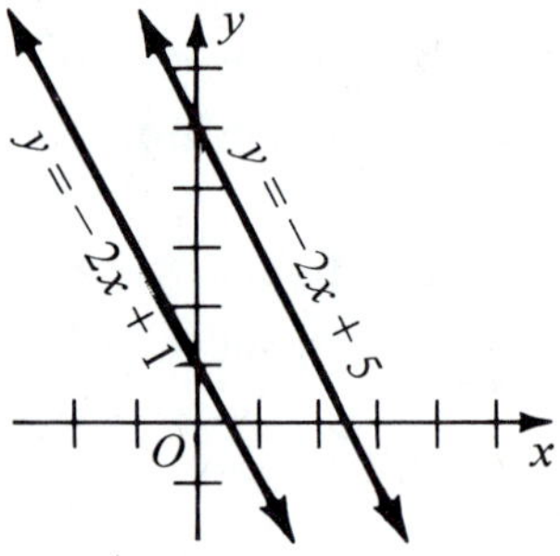

$y = -2x + 1$
$y = -2x + 5$

Figure 3

Two or more equations in two variables form a *system of equations in two variables*. The *solution set* of a system of linear equations in two variables is the set of ordered pairs that satisfy all of the equations in the system. In this chapter, we will consider only systems of *linear* equations in two variables.

A system of linear equations is *consistent*, Figures 1 and 2, if it has at least one solution. A consistent system of two equations is *independent*, Figure 1, if it has exactly one solution. A consistent system of two equations is *dependent*, Figure 2, if it has infinitely many solutions. The solution set of the system in Figure 2 is $\{(x, y)\colon y = 2x - 1\}$. A system is *inconsistent*, Figure 3, if its solution set is empty.

EXAMPLE Solve the system graphically and determine whether it is consistent.

$$3x - y = 1$$
$$y = x + 1$$
$$x + 2y = 5$$

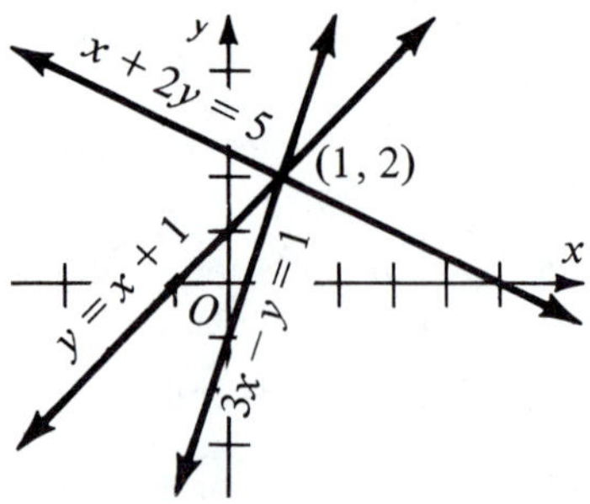

SOLUTION Graph the lines and check that their apparent point of intersection, (1, 2) satisfies each equation.

$3x - y = 1$	$y = x + 1$	$x + 2y = 5$
$3 \cdot 1 - 2 \ ? \ 1$	$2 = 1 + 1$	$1 + 2 \cdot 2 \ ? \ 5$
$3 - 2 = 1$		$1 + 4 = 5$

Since (1, 2) is a solution of the system, the system is consistent.

ORAL EXERCISES

Which ordered pair is a solution of the system?

1. $x + y = 3$ $\quad$ (0, 3), (1, 0), (1, 2), (2, 1)
   $x - y = 1$
2. $x - y = 6$ $\quad$ (6, 0), (3, −1), (4, −2), (7, 1)
   $x + y = 2$
3. $2x - y = 1$ $\quad$ (0, −1), (3, 2), (3, 5), (2, 3)
   $x + y = 5$
4. $3x + 2y = 8$ $\quad$ (3, −1), (2, 1), (7, 3), (−4, 10)
   $-2x + 5y = 1$

Supply the missing abscissa and/or ordinate so that the ordered pair satisfies both equations.

5. $x + y = 10$ $\quad$ (?, 3)
   $x - y = 4$
6. $x - y = 7$ $\quad$ (4, ?)
   $x + y = 1$
7. $2x + y = 3$ $\quad$ (−4, ?)
   $x - y = -15$
8. $x - 2y = -1$ $\quad$ (?, −2)
   $x + y = -7$
9. $x + y = 6$ $\quad$ (?, ?)
   $x - y = 0$
10. $x + y = 11$ $\quad$ (?, ?)
    $x - y = -10$

WRITTEN EXERCISES

Solve each system graphically. Check your solutions.

A

1. $y = x$
   $y = 2x - 2$
2. $y = x - 4$
   $y = -x$
3. $y = -x + 5$
   $y = x + 3$
4. $y = x$
   $y = 2x + 1$
5. $y = x - 6$
   $x + y = 4$
6. $y = x + 3$
   $x + y = 3$
7. $y = \frac{1}{2}x + 2$
   $x + y = -1$
8. $x + y = 2$
   $y = x - 4$
9. $x + 2y = 4$
   $y = -\frac{1}{2}x - 2$
10. $y - x = 3$
    $3y = x + 3$
11. $y = 2x - 4$
    $2x - y = 4$
12. $x + 5y = 10$
    $y = -\frac{1}{5}x + 2$
13. $y - 2x = 7$
    $x - y = -6$
14. $3x - 2y = 6$
    $y = \frac{3}{2}x - 3$
15. $y - 6 = x$
    $y - 2x = 8$

16–20. Classify each system in Exercises 8–12 as inconsistent, consistent and dependent, or consistent and independent.

Find the area bounded by the lines whose equations are given.

EXAMPLE $y = -\frac{3}{2}x - 6$

$y = \frac{3}{4}x - \frac{3}{2}$

$y = 0$

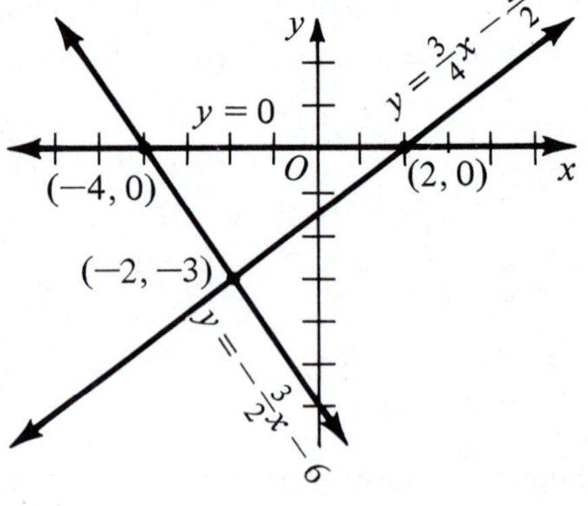

SOLUTION Draw the graphs of the lines and find the coordinates of the points of intersection. A triangular region is formed with vertices at (−4, 0), (2, 0), and (−2, −3). The area of the triangle is $\frac{1}{2}(6)(3) = 9$.

21. $y = 2x - 3$
$y = -x + 6$
$x = 0$

22. $y = x + 2$
$y = -2x + 8$
$y = 0$

23. $y = \frac{1}{2}x + 8$
$y = -\frac{3}{2}x$
$y = 0$

24. $y - x = 10$
$2y + x = -10$
$x = 0$

Solve each system graphically. Check your solutions.

B **25.** $3x - 2y = -2$
$x + 2y = -6$

26. $2x - y = 5$
$3x + 2y = 4$

27. $2x - 3y = 18$
$x + 3y = 0$

28. $3x - 2y = 2$
$2y = 5x + 2$

29. $4x + 3y = 12$
$2x + y = 6$
$x + 2y = 3$

30. $4x + y = 1$
$2y = x + 11$
$3x + 2y = 7$

Determine *without* graphing whether each system is inconsistent, consistent and dependent, or consistent and independent.

31. $y = \frac{1}{2}x + 3$
$x - 2y = 4$

32. $3y = 4x + 24$
$x + y = 5$

33. $x + 3y = 4$
$x = -3y$

34. $-\frac{3}{2}x = 4 + y$
$2y + 8 = -3x$

35. $4y - 8 = 10x$
$4 - 2y = -5x$

36. $2x = 4y + 4$
$y = \frac{1}{2}x + 6$

Find the area of the region bounded by the lines whose equations are given.

C **37.** $3y = 5x + 24$
$x + y = 0$
$3x + 4y = 32$
$x = 4y$

38. $2y + x = 0$
$2x + 3y = -12$
$3y = 2x$
$2y + 8 = 3x$

39. $2x + 3y = -22$
$7y - 6x = 66$
$2y + 3x = 0$
$4x = 5y$

Section 11-9 SOLVING SYSTEMS OF LINEAR EQUATIONS ALGEBRAICALLY

As you saw in the preceding section, a system of linear equations may be solved by graphing, but graphing is not always a practical method to use, especially if the solution is non-integral or very large. Now you will review two algebraic methods of solving a system of linear equations in two variables, namely:

- the substitution method
- the addition-or-subtraction method

To solve a system of linear equations by *substitution*, you solve one equation for one variable and then substitute the expression obtained in the other equation.

EXAMPLE Solve the system by substitution.

$$5x - 2y = 1 \quad (1)$$
$$x + 3y = 7 \quad (2)$$

SOLUTION Solve equation (2) for x.

$$x + 3y = 7$$
$$x = 7 - 3y$$

Substitute for x in equation (1).

$$5x - 2y = 1$$
$$5(7 - 3y) - 2y = 1$$
$$35 - 15y - 2y = 1$$
$$-17y = -34$$
$$y = 2$$

Since

$$x = 7 - 3y$$
$$x = 7 - 3(2)$$
$$x = 7 - 6$$
$$x = 1$$

Check Check the solution, (1, 2), in *both* original equations.

$$5x - 2y = 1 \qquad\qquad x + 3y = 7$$
$$5(1) - 2(2) \ ?\ 1 \qquad\qquad 1 + 3(2) \ ?\ 7$$
$$5 - 4 = 1 \qquad\qquad 1 + 6 = 7$$

Therefore, the solution is (1, 2).

The substitution method is easiest to use in a system of equations in which one of the variables has a coefficient of 1 or −1. A more versatile method is the *addition-or-subtraction method*, which we illustrate in the next example.

EXAMPLE Solve the system by addition or subtraction.

$$2x + 3y = 5 \quad (1)$$
$$5x - 2y = -16 \quad (2)$$

(The solution is given on the next page.)

SOLUTION Choose a variable to eliminate, say y.

Multiply both sides of equation (1) by 2 and both sides of equation (2) by 3.

$$\begin{aligned} (1)\times 2:&\quad 4x + 6y = 10 \\ (2)\times 3:&\quad 15x - 6y = -48 \\ \text{Add:}&\quad 19x = -38 \\ &\quad x = -2 \end{aligned}$$

Substitute -2 for x in one of the original equations, say equation (1), and solve for y.

$$\begin{aligned} 2(-2) + 3y &= 5 \\ -4 + 3y &= 5 \\ 3y &= 9 \\ y &= 3 \end{aligned}$$

Check Check the solution, $(-2, 3)$, in *both* original equations.

$$\begin{aligned} 2x + 3y &= 5 & \qquad 5x - 2y &= -16 \\ 2(-2) + 3(3) &\ ?\ 5 & \qquad 5(-2) - 2(3) &\ ?\ -16 \\ -4 + 9 &= 5 & \qquad -10 - 6 &= -16 \end{aligned}$$

This example illustrates the fact that to solve a system of linear equations by the addition-or-subtraction method, you multiply one or both equations by constants so that one variable is eliminated when the sides of the equations are added or subtracted.

ORAL EXERCISES

Solve for x in terms of y.

1. $x - y = 10$
2. $x + y = 5$
3. $2x - y = 0$
4. $3x = y + 6$

Solve for y in terms of x.

5. $2x + y = 4$
6. $2x - y = 2$
7. $6 - y = -3x$
8. $x - 4y = 8$

Describe what you would do to eliminate: **(a)** the variable x;
(b) the variable y.

9. $2x + 3y = 1$
$x - y = 3$

10. $21x - 2y = 2$
$3x + 4y = 1$

11. $5x + 2y = 10$
$-x + 6y = 14$

12. $5x + 7y = 1$
$2x + 3y = 3$

13. $7x - 2y = 2$
$-12x + 5y = -5$

14. $6x - 3y = 2$
$4x - 5y = 8$

WRITTEN EXERCISES

Solve each system by the substitution method.

A

1. $2g = 4$
$g + h = -6$

2. $3y = -15$
$2x - y = 1$

3. $a = 2c$
$a + 2c = 12$

4. $w = -3z$
$w + 4z = -2$

5. $x - y = 0$
$x + y = -4$

6. $c + d = 3$
$c - d = 1$

Solve each system by addition or subtraction.

7. $x - y = -9$
$x + y = -1$

8. $p + q = -2$
$p - q = 8$

9. $c + 2d = 3$
$4c + 5d = 6$

10. $2h + 3k = 0$
$5h + k = 26$

11. $4g + 7h = 1$
$8g + 2h = -34$

12. $5x - 4y = 2$
$3x + 2y = -12$

Solve each system.

13. $2m - n = 10$
$-m + n = 10$

14. $x - y = -1$
$2x - 4y = 2$

15. $w - 2z = 2$
$2w - 5z = 2$

16. $s - 2t = 10$
$2s + t = 2$

17. $c - d = 1$
$3c + 4d = 0$

18. $f - 3g = 1$
$4f - 10g = 1$

19. $3c - d = 6$
$2c = d + 2$

20. $5g - h = 12$
$4g + 3h = 14$

21. $6p = 5q + 8$
$3p - 10 = 2q$

22. $3y - x = 2$
$7x + 14 = 2y$

23. $y = -4x$
$4x + y = -9$

24. $3s - 2 = t$
$5t + 10 = 15s$

B 25. $5x + 2y = -1$
$4x + 3y = 9$

26. $3x - 2y = 5$
$2x - 3y = 5$

27. $3d - 7e = 9$
$2d - 5e = 7$

28. $3j + 4k = 24$
$10j + 9k = 15$

29. $10g + 3h = 2$
$6g - 7h = -12$

30. $16a + 22b = 57$
$18a + 5b = -5$

Find the area of the region bounded by the lines whose equations are given.

31. $y = 2,\ y = 8,\ y = 3x + 2,\ 3x - y = 28$

32. $x = 5,\ x = -3,\ x - y = -2,\ x - 4 = y$

33. $4y - 3x = 24,\ x + y = 13,\ y = 4$

34. $x = y - 9,\ 5x + 4y = 0,\ x = 8$

Solve each system.

C 35. $2x - 5y = 10$
$9x + y - 14.5 = 0$

36. $0.2c + 0.3d = 5$
$0.6c + 0.7d = -2$

37. $\frac{2}{3}g - \frac{1}{2}h = 11$
$\frac{5}{6}g + h = 4$

38. $\frac{j}{2} - \frac{2k}{3} = 3$
$\frac{j}{7} - \frac{k}{5} = 1$

39. $\frac{w + 2}{2} - \frac{z - 3}{5} = 3$
$\frac{w - 2}{2} - \frac{z - 1}{3} = 7$

40. $\frac{p - 2}{3} + \frac{q - 4}{6} = 1$
$\frac{p + 1}{4} - \frac{q - 1}{3} = 6$

Find the values of A and B so that the system of equations has the indicated solution.

41. $Ax - By = -3$
$Ax + By = 21$
Solution $(15, -12)$

42. $Ax + By = 0$
$2Ax - By = -12$
Solution $(-8, 12)$

43. $Ax - 2By = 3$
$Ax + By = 6$
Solution $\left(\frac{5}{2}, \frac{1}{3}\right)$

Exercises 44–47 refer to $\triangle ABC$ with vertices $A(-2, 0)$, $B(6, 0)$, and $C(0, 8)$.

44. a. Find equations for the three medians of the triangle.
b. Find the coordinates of the point of intersection, P, of two of the medians.
c. Verify that P is on the third median.

45. a. Find equations for the perpendicular bisectors of the sides of $\triangle ABC$.
b. Find the coordinates of the point of intersection, Q, of two of the perpendicular bisectors.
c. Verify that Q lies on the third perpendicular bisector.

46. a. Find equations for the lines containing the altitudes from A, B, and C.
b. Find the coordinates of the point of intersection, R, of the altitudes from A and B.
c. Verify that R lies on the altitude from C.

47. Show that points P, Q, and R (whose coordinates you found in Exercises 44–46) are collinear.

SELF-TEST 3

Solve the system graphically. Then classify it as inconsistent, consistent and dependent, or consistent and independent.

Section 11-8

1. $x + y = 4$
$y = -x - 4$

2. $y = \frac{1}{3}x - 2$
$x - 3y = 6$

3. $y = -\frac{1}{2}x$
$4y - x = 12$

Section 11-9

4. Solve the system by substitution.

$$7x + y = -3$$
$$3x + 2y = 5$$

Solve each system by addition or subtraction.

5. $3x + 2y = 2$
$2x - y = -15$

6. $2x - 3y = -3$
$9x - 8y = 36$

Computer Exercises

The midpoint formula enables you to find the coordinates of a point that divides a given segment into two congruent segments. A more general formula enables you to find the coordinates of a point that divides a given segment into two segments whose lengths are in any given ratio $r:s$, where r and s are positive real numbers.

Let $A(x_1, y_1)$ and $B(x_2, y_2)$ be the endpoints of a given segment and let $P(x, y)$ be the point on $\overline{AB}$ such that $\frac{AP}{PB} = \frac{r}{s}$. In the diagram at the right, $\overline{AE}$ and $\overline{PD}$ are parallel to the x-axis, and $\overline{BE}$ and $\overline{PC}$ are parallel to the y-axis. By the Triangle Proportionality Theorem,

$$\frac{AP}{PB} = \frac{AC}{CE} \quad \text{and} \quad \frac{AP}{PB} = \frac{ED}{DB}$$

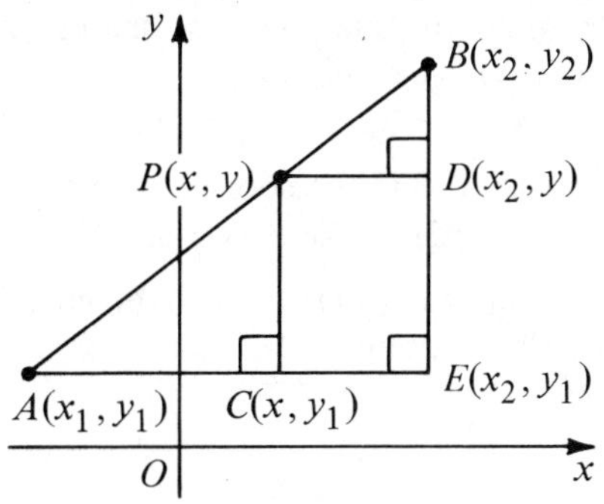

Thus,

$$\begin{aligned} \frac{AC}{CE} &= \frac{x - x_1}{x_2 - x} = \frac{r}{s} \\ s(x - x_1) &= r(x_2 - x) \\ sx - sx_1 &= rx_2 - rx \\ rx + sx &= rx_2 + sx_1 \\ x(r + s) &= rx_2 + sx_1, \\ x &= \frac{rx_2 + sx_1}{r + s} \end{aligned} \qquad \text{and} \qquad \begin{aligned} \frac{ED}{DB} &= \frac{y - y_1}{y_2 - y} = \frac{r}{s} \\ s(y - y_1) &= r(y_2 - y) \\ sy - sy_1 &= ry_2 - ry \\ ry + sy &= ry_2 + sy_1 \\ y(r + s) &= ry_2 + sy_1 \\ y &= \frac{ry_2 + sy_1}{r + s} \end{aligned}$$

Therefore, the coordinates of point P are $\left(\frac{rx_2 + sx_1}{r + s}, \frac{ry_2 + sy_1}{r + s}\right)$.

1. Copy and RUN the program for the given points A and B and the given ratio $r:s$. The computer will print the coordinates of point P.

```
10  REM***POINT A HAS COORDINATES (X1,Y1).
20  REM***POINT B HAS COORDINATES (X2,Y2).
30  PRINT "INPUT X1,Y1,X2,Y2 IN THAT ORDER";
40  INPUT X1,Y1,X2,Y2
50  PRINT "INPUT R,S";
60  INPUT R,S
70  LET X=(R*X2+S*X1)/(R+S)
80  LET Y=(R*Y2+S*Y1)/(R+S)
90  PRINT
100  PRINT "THE COORDINATES OF THE POINT P ON AB"
110  PRINT "SUCH THAT AP:PB=";R;":";S;"   ARE   ";
120  PRINT "(";X;",";Y;")"
130  END
```

a. $A(3, 4)$, $B(6, 10)$, $r:s = 2:1$ **b.** $A(6, 10)$, $B(3, 4)$, $r:s = 2:1$

c. $A(3, 4)$, $B(6, 10)$, $r:s = 1:2$ **d.** $A(-2, -5)$, $B(12, 2)$, $r:s = 4:3$

2. a. Use the distance formula and your answer to Exercise 1(d) to find AP and BP where A and B are the points in Exercise 1(d).

b. Use your answer to part (a) to verify that $\frac{AP}{BP} = \frac{4}{3}$.

3. RUN the program to find the coordinates of the midpoint of the segment whose endpoints are given below.
   a. $(3, 4)$, $(-5, 4)$ b. $(-3, 8)$, $(-6, 4)$
4. RUN the program to find the coordinates of the points that trisect the segment with endpoints $(2, -3)$ and $(8, 9)$. (*Hint:* RUN the program twice.)
5. RUN the program four times to find the coordinates of the points that divide the segment with endpoints $(-8, 7)$ and $(12, -3)$ into five congruent segments.

CHAPTER REVIEW

1. Graph the following points on the same coordinate plane: Section 11-1

$$A(-5, -2),\ B(3, -1),\ C(4, 0)$$

2. Find the area of the triangle with vertices $D(-1, -1)$, $E(3, -1)$, and $F(3, 4)$.
3. Find the area of quadrilateral $QRST$ with vertices $Q(4, -2)$, $R(5, 2)$, $S(0, 4)$, and $T(0, -1)$.
4. Find the coordinates of the midpoint of the segment with endpoints $(a, -2)$ and $(6 - a, 2)$. Section 11-2

Exercises 5-8 refer to points $L(2, 3)$, $M(4, -1)$, and $N(2, -2)$.

5. Use the distance formula to show that $\triangle LMN$ is a right triangle.
6. Find the slopes of $\overline{LM}$, $\overline{MN}$, and $\overline{LN}$. Use slopes to show that $\triangle LMN$ is a right triangle. Section 11-3
7. Find the slope of any line parallel to $\overleftrightarrow{LN}$.
8. Points M, N, and $P(k, 4)$ are collinear. Find the value of k.
9. In the figure, $TS = RS$. Find the coordinates of R.
10. Use coordinate geometry to show that the medians to the legs of isosceles $\triangle TRS$ are congruent. Section 11-4

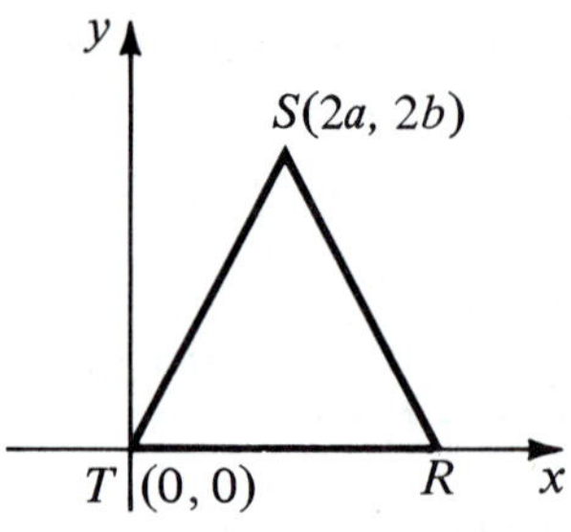

Exs. 9,10

11. Graph $3x - 2y = 12$. Section 11-5
12. Find the x-intercept and y-intercept of the line with equation $4x - 5y = 40$.
13. Graph the line with slope -1 and x-intercept 0.

Find an equation in the specified form for the line described.

14. Point-slope form; the line with slope 2 that contains point $(-7, 8)$ Section 11-6
15. Slope-intercept form; the line through points $(6, -2)$ and $(0, -5)$

16. Standard form; the line through point (5, 0) and parallel to the line with equation $y = -\frac{1}{3}x + 4$

Write an equation for the locus of points:

17. equidistant from the lines with equations $y = 1$ and $y = 9$ Section 11-7

18. equidistant from points $(-5, 3)$ and $(-2, 3)$

19. Draw on graph paper the locus of points that are 1 unit from the line with equation $x = 2$ and 3 units from the line with equation $y = -3$. Name the locus.

20. If two linear equations in two variables form an inconsistent system, what is true of their graphs? Section 11-8

Solve graphically.

21. $y = -x - 3$
$2x + 2y = -6$

22. $y = 3x + 7$
$y = -\frac{1}{2}x$

Solve and check your solutions.

23. $x + 3y = 2$
$2x + 5y = 2$

24. $2x - 3y = -1$
$5x - 4y = 15$ Section 11-9

CHAPTER TEST

Exercises 1-7 refer to points $A(6, -6)$, $B(2, 1)$, $C(-6, 2)$, and $D(-2, -5)$.

1. Find the coordinates of the midpoints of $\overline{AC}$ and $\overline{BD}$. What can you conclude about the diagonals of quadrilateral $ABCD$?
2. Find AB, BC, CD, and AD. What type of quadrilateral is quadrilateral $ABCD$? Be as specific as possible.
3. Graph points A, B, C, and D on the same coordinate plane.
4. Use slopes to show that $\overline{AC} \perp \overline{BD}$.
5. Use slopes to show that points A, C, and $E(3, -4)$ are collinear.
6. Find an equation in standard form for $\overleftrightarrow{AB}$.
7. Find the coordinates of the midpoints X and Y of $\overline{AB}$ and $\overline{BC}$. Use coordinate geometry to prove that $\overline{XY} \parallel \overline{AC}$.
8. $\triangle FGH$ has vertices $F(1, 4)$, $G(7, 7)$, and $H(1, 8)$. Find the area of $\triangle FGH$.

Graph the line:

9. With equation $4x - y = 8$

10. With slope $\frac{3}{4}$ and x-intercept -3

Write an equation in the specified form for the line described.

11. Standard form; the line with slope 0 and y-intercept $\frac{3}{2}$

12. Point-slope form; the line containing points $(-8, 5)$ and $(10, -1)$

13. Slope-intercept form; the line through $(0, -4)$ and perpendicular to the line with equation $y = x + 2$

Write an equation for the locus of points:

14. 5 units above the line with equation $y = -9$

15. equidistant from the lines with equations $x = -2$ and $x = -6$

16. How many points are in the locus of points 3 units from the origin and 3 units from the y-axis?

Solve graphically.

17. $y = -3x + 2$
$x + y = 4$

18. $3x + 2y = -6$
$6x + 4y = 6$

19. Solve by the substitution method: $5x + 3y = 9$
$2x - y = -3$

20. Solve by addition or subtraction: $3x - 7y = 16$
$4x - 2y = 3$

21. $\triangle ABC$ is an equilateral triangle. Find the coordinates of C.

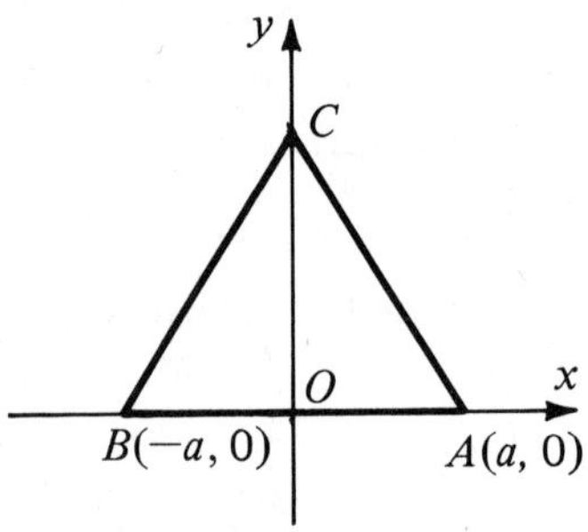

Biographical Note

René Descartes

Perhaps you have heard coordinates referred to as Cartesian coordinates or the coordinate plane referred to as the Cartesian plane, in honor of the French philosopher and mathematician René Descartes (1596–1650).

René Descartes was born at La Haye, France, in 1596. He received his early education at the Royal College at La Flèche, a small town north of Touraine. He remained there about eight years studying Latin, Greek, and French, literature, grammar and composition, logic, physics, astronomy, music, and mathematics. He especially enjoyed studying mathematics "because of the certainty of its proofs and the evidence of its reasonings."

After leaving the Royal College, Descartes studied at the University of Poitiers, graduating with a degree in law in 1616.

While in military service in the Netherlands, Descartes became acquainted with Isaac Beeckman, a Dutch mathematician who recognized Descartes's ability and further stimulated his interest in mathematics. In 1619 Descartes reported to Beeckman that he had discovered what would become analytic geometry, a "new science" that Descartes believed would be capable of solving philosophical as well as mathematical problems.

Descartes wrote several major works. In *La géométrie*, published in 1637, Descartes used exponents as we use them today. He also used the last letters of the alphabet to represent unknown quantities and the first letters of the alphabet to represent constants, giving his work a modern appearance. Probably his greatest contribution to the field of mathematics, however, was setting up a one-to-one correspondence between points in a plane and ordered pairs of numbers, thus making it possible to represent lines and other curves by equations.

The basic concepts of coordinate geometry, although revolutionary in the seventeenth century, are now part of standard high school mathematics courses. Graphs can give you greater insight into algebraic problems, and using coordinates can help you solve geometry problems.

VOCABULARY REVIEW

Be sure that you know the meaning of these terms:

coordinate plane, p. 427
coordinate axes, p. 427
origin, p. 427
quadrants, p. 427
x-coordinate (abscissa), p. 428
y-coordinate (ordinate), p. 428
absolute value, p. 429
slope, p. 396
linear equation, p. 448
x-intercept, p. 449
y-intercept, p. 449
consistent system, p. 460
independent system, p. 460
dependent system, p. 460
inconsistent system, p. 460

MIXED REVIEW

1. Find the x-intercept and the y-intercept of the line with equation $\frac{1}{2}x - 4y = 8$.

In $\triangle JKL$, $\angle J \cong \angle K$, $LJ = 14x - 5$, $LK = 3x + 28$, and $JK = 2x + 1$.

2. Find the length of each side of $\triangle JKL$.
3. Name the largest angle(s) of $\triangle JKL$.
4. Square $ABCD$ has sides of length 6 cm. Describe the locus of points equidistant from vertices A and C and 2 cm from $\overleftrightarrow{AB}$.

Graph in the coordinate plane.

5. $3y - x = 6$
6. $4 - 3y = 2$
7. $2x + 3y = 6$
8. Refer to the figure. Find the value of x.

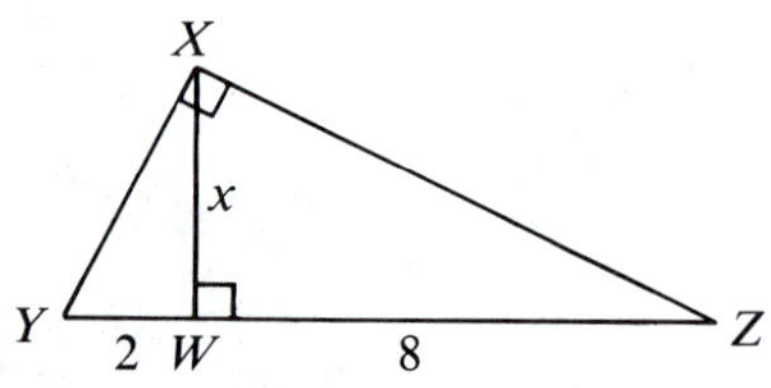

9. Draw a segment 2 cm long. Use it to construct a segment $\sqrt{10}$ cm long.
10. Three consecutive multiples of 8 have a sum of 120. Find the numbers.
11. The vertices of $\triangle ABC$ are $A(2, 7)$, $B(2, 1)$, and $C(7, 4)$. Show that $\triangle ABC$ is isosceles.
12. Find the area of the figure formed by joining $H(5, 2)$, $I(2, 5)$, and $J(-2, 1)$ in order.
13. A rhombus has a 120° angle and sides of length 4. Find the length of each diagonal.

14. During a storm, a tree 80 m high cracked. The base was still perpendicular to the ground and the treetop touched the ground at a point 60 m from the base. Find the height of the part left standing.

15. Find the distance between $C(-5, -4)$ and $D(-2, -3)$.

Solve.

16. $5x + 6y - 14 = 0$
$3y - x + 7 = 0$

17. $4x - 3y = 8$
$3x - 2y = 7$

Exercises 18 and 19 refer to $\triangle PQR$ with $m\angle Q = 90$. Use the table on page 373.

18. If $m\angle P = 47$ and $PR = 51$, find PQ to the nearest integer.

19. If $m\angle P = 35$ and $QR = 85$, find PR to the nearest integer.

20. Find the coordinates of the midpoint of the segment with endpoints $A(-3, 4)$ and $B(7, -3)$.

Write an equation in standard form for the line described.

21. The line through points $(2, 1)$ and $(-1, 4)$

22. The line with x-intercept -1 and slope 9

23. Factor completely: $k^2 - 25k + 150$

24. Determine the number of points in the locus of points 6 units from the origin and 3 units from the line with equation $y = 3$.

25. The length of a rectangle is 6 m more than its width. Find the length if the area is 216 m^2.

Write two-column proofs.

26. Given: $\triangle ADE \cong \triangle BDC \cong \triangle DAB$
Prove: $ABCE$ is an isosceles trapezoid.

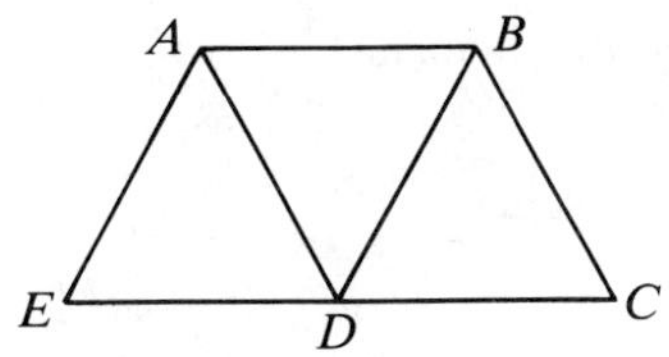

27. Given: $\overline{UV} \parallel \overline{ZY}$; $\overline{VW} \parallel \overline{XZ}$
Prove: $\triangle XUV \sim \triangle VWY$

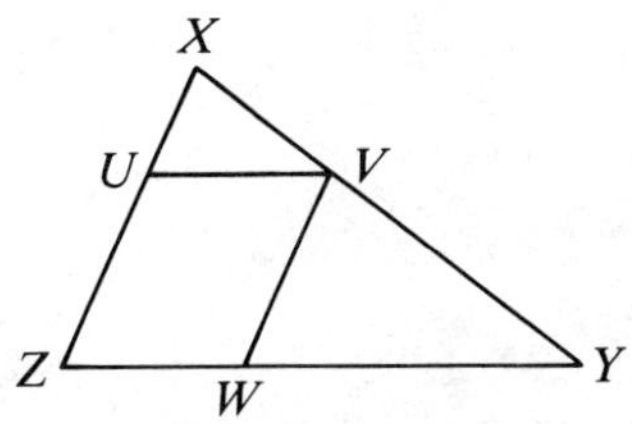

The reflecting surface of the radio telescope shown above is a parabolic dish. Incoming radio waves are reflected from the dish to the focal point. A parabolic dish approximates the type of geometric surface created by revolving a parabola about its axis of symmetry.

Quadratic Equations in Two Variables 12

The general *quadratic equation in two variables* is an equation of the form

$$Ax^2 + By^2 + Cxy + Dx + Ey + F = 0$$

where A, B, and C are not all zero. In other words, a quadratic equation in two variables contains at least one term of degree 2.

The graph of a quadratic equation in two variables is generally a parabola, a circle, an ellipse, or a hyperbola. Such graphs are called *conic sections* because they are formed by the intersection of a plane and a cone (or two cones placed vertex-to-vertex).

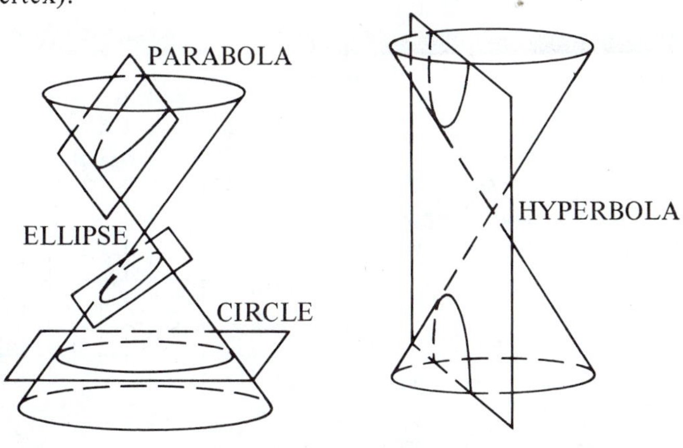

In the third century B.C., the Greek mathematician Apollonius studied conic sections extensively from a purely geometric point of view. But not until Descartes's publication of the coordinate method in the seventeenth century did mathematicians begin to study conic sections algebraically.

Today the conic sections have applications in such areas as optics (lenses), acoustics (auditorium design), astronomy (orbits of planets and comets), physics (paths of projectiles), and engineering (bridge design).

Section 12-1 THE GRAPH OF $y = a(x - h)^2 + k$

In Chapter 11 you learned that the graph of a linear equation in two variables is a line. Not all equations in two variables are linear, however. There are other types of equations whose graphs have interesting properties. In this section you will learn the meanings of the terms:

- parabola
- axis of symmetry
- vertex
- maximum point
- minimum point

Consider a simple example of a graph of a quadratic equation in two variables.

EXAMPLE Graph $y = x^2$.

SOLUTION Compute some coordinates and graph some ordered pairs.

x	x^2	y
-3	$(-3)^2$	9
-2	$(-2)^2$	4
-1	$(-1)^2$	1
0	0^2	0
1	1^2	1
2	2^2	4
3	3^2	9

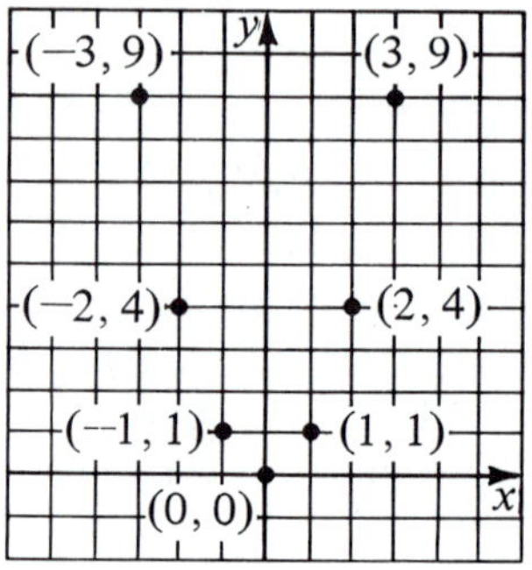

Join the points with a smooth curve to obtain the graph at the right. The arrows indicate that the graph continues upward on both sides of the y-axis.

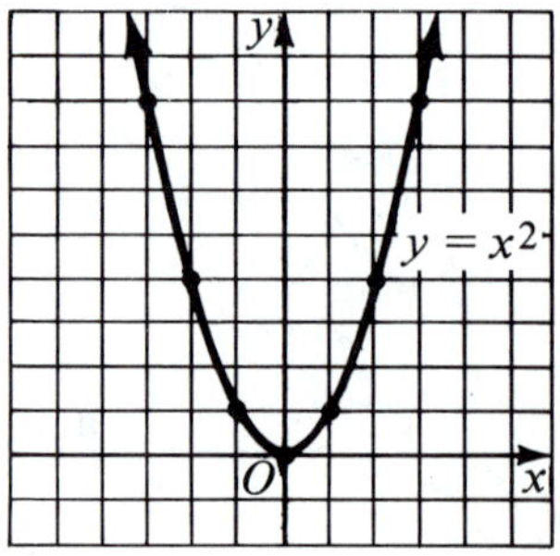

The graph of $y = x^2$ is called a *parabola.* For this graph, the y-axis acts as a "mirror" that reflects the points of one side of the parabola onto points of the other. Thus, the points $(-2, 4)$ and $(2, 4)$ are "mirror images" of each other. Every parabola has a "mirror line," which is called the *axis of symmetry* of the parabola. The point where a parabola intersects its axis of symmetry is called the *vertex,* or *turning point,* of the parabola. The vertex of the graph of $y = x^2$ is $(0, 0)$.

As you will see, any curve whose equation can be written $y = a(x - h)^2 + k$ is a parabola. First, we will consider the parabola $y = ax^2$. (In this book the phrase "the parabola $y = x^2$" will mean "the parabola whose equation is $y = x^2$.") By studying the graphs at the right, you can see that in general:

(1) If $a > 0$, the graph opens upward and the vertex is the *minimum* (lowest) point of the graph. If $a < 0$, the graph opens downward and the vertex is the *maximum* (highest) point of the graph.
(2) The smaller the absolute value of a, the "wider" the parabola.
(3) The graphs of $y = ax^2$ and $y = -ax^2$ are congruent.

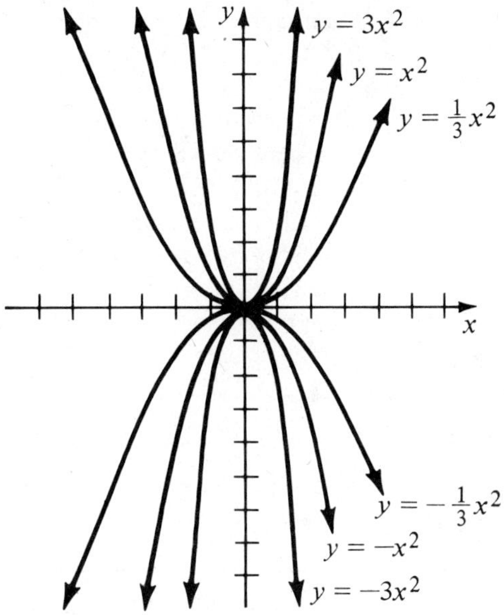

Next consider the parabola $y = x^2 + k$. Use the diagram below to compare the graphs of $y = x^2$, $y = x^2 - 3$, and $y = x^2 + 2$. The graphs are congruent. By changing the value of k, the vertex of the parabola is shifted up or down $|k|$ units along the axis of symmetry. In general, the graph of $y = ax^2 + k$ is obtained from the graph of $y = ax^2$ by shifting the graph of $y = ax^2$ vertically:

(1) up $|k|$ units if $k > 0$;
(2) down $|k|$ units if $k < 0$.

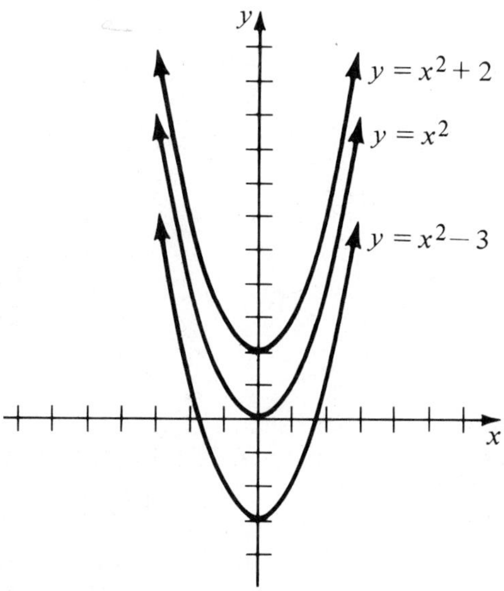

Finally, consider the parabola $y = (x - h)^2$. Compare the graphs of $y = x^2$, $y = (x - 4)^2$, and $y = (x + 3)^2$. The graphs are congruent. By changing the value of h, the parabola is shifted $|h|$ units to the right or left. In general, the graph of $y = a(x - h)^2$ is obtained from the graph of $y = ax^2$ by shifting the graph of $y = ax^2$ horizontally:

(1) to the right $|h|$ units if $h > 0$;
(2) to the left $|h|$ units if $h < 0$.

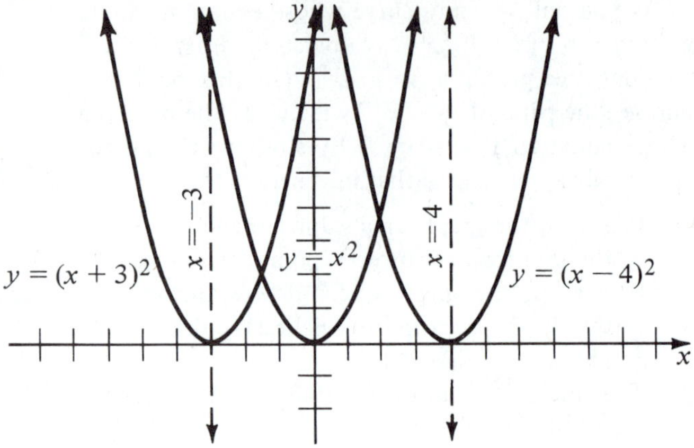

Combining all the preceding information, you have the following:

> If a parabola has equation $y = a(x - h)^2 + k$, then the parabola:
>
> (1) is congruent to the graph of $y = ax^2$.
>
> (2) has vertex (h, k).
>
> (3) opens upward and has a minimum point at (h, k) if $a > 0$, or opens downward and has a maximum point at (h, k) if $a < 0$.
>
> (4) has axis of symmetry with equation $x = h$.

EXAMPLE Find the vertex and the equation of the axis of symmetry of the parabola whose equation is $y = 2(x - 1)^2 - 3$. State whether the vertex is a maximum point or a minimum point. Graph the equation.

SOLUTION The equation $y = 2(x - 1)^2 - 3$ is in the form $y = a(x - h)^2 + k$ where $a = 2$, $h = 1$, and $k = -3$.

The vertex is $(1, -3)$, and the equation of the axis of symmetry is $x = 1$.

Since $a > 0$, the vertex is a minimum point.

Plot a few points and join them with a smooth curve.

x	$2(x-1)^2 - 3$	y
-1	$2(-2)^2 - 3$	5
0	$2(-1)^2 - 3$	-1
2	$2(1)^2 - 3$	-1
3	$2(2)^2 - 3$	5

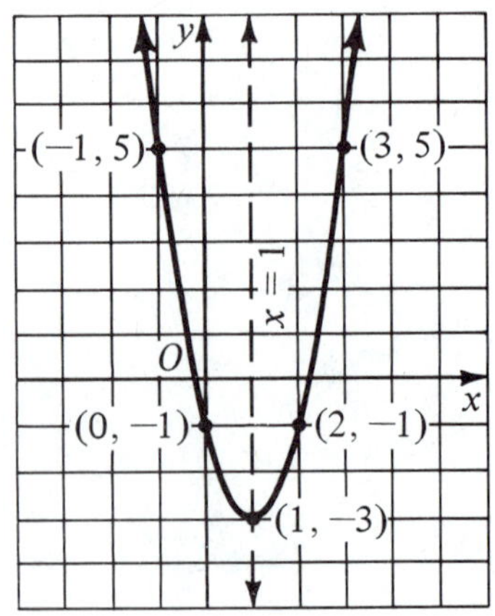

ORAL EXERCISES

Complete each table for the given values of x.

1. $y = (x + 2)^2 - 3$

	x	$(x + 2)^2 - 3$	y
	-1	$1^2 - 3$	-2
a.	3	$5^2 - 3$	?
b.	0	?	?
c.	1	?	?
d.	-4	?	?

2. $y = -(x - 3)^2 + 2$

	x	$-(x - 3)^2 + 2$	y
a.	4	$-(1)^2 + 2$	?
b.	2	?	?
c.	0	?	?
d.	-1	?	?
e.	-3	?	?

The figure shows the graphs of the equations in Exercises 3-6. Match each equation with its graph.

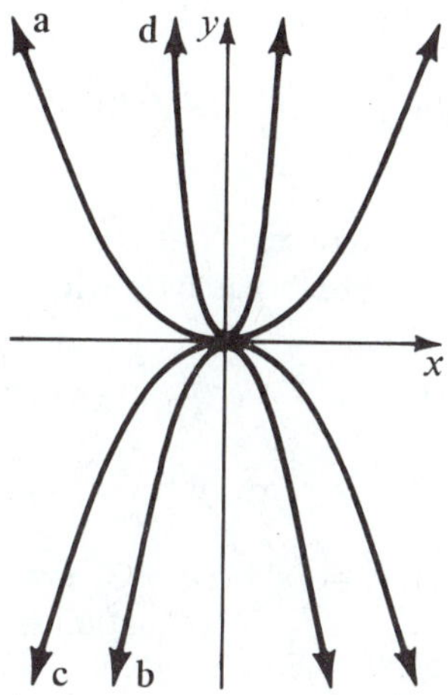

3. $y = -\frac{1}{3}x^2$
4. $y = 4x^2$
5. $y = -x^2$
6. $y = \frac{1}{4}x^2$

For each parabola whose equation is given below, state:
a. the values of a, h, and k
b. the coordinates of the vertex
c. the equation of the axis of symmetry
d. whether the parabola opens upward or downward
e. whether the vertex is a maximum point or a minimum point

7. $y = 5x^2 + 2$
8. $y = -2x^2 - 5$
9. $y = -4(x - 1)^2$
10. $y = \frac{1}{3}(x + 2)^2$
11. $y = (x - 7)^2 - 6$
12. $y = -\frac{3}{2}(x + 5)^2 + 1$

WRITTEN EXERCISES

In Exercises 1-12:
a. Make a table of values of y for integral values of x in the given interval.
b. Use the results of part (a) to graph the equation.
c. Indicate the axis of symmetry with a dashed line and state its equation.
d. Label the vertex as V and state its coordinates.
e. State whether the vertex is a maximum or a minimum point.

A 1. $y = x^2 + 1$; $-3 \leqslant x \leqslant 3$

2. $y = x^2 - 4$; $-3 \leqslant x \leqslant 3$

3. $y = \frac{1}{2}x^2 - 2$; $-3 \leqslant x \leqslant 3$

4. $y = -\frac{1}{3}x^2 + 1$; $-3 \leqslant x \leqslant 3$

Follow the directions given on the preceding page for Exercises 1-12.

5. $y = -(x-1)^2;\ -2 \leqslant x \leqslant 4$

6. $y = (x-3)^2;\ 0 \leqslant x \leqslant 6$

7. $y = -(x+1)^2;\ -4 \leqslant x \leqslant 2$

8. $y = (x+4)^2;\ -7 \leqslant x \leqslant -1$

9. $y = (x+3)^2 + 2;\ -6 \leqslant x \leqslant 0$

10. $y = (x-3)^2 + 4;\ 0 \leqslant x \leqslant 6$

11. $y = -(x+2)^2 + 5;\ -5 \leqslant x \leqslant 1$

12. $y = -(x-2)^2 - 2;\ -1 \leqslant x \leqslant 5$

Find an equivalent equation of the form $y = a(x-h)^2 + k$ for the parabola that is congruent to the given parabola and has the given point as vertex.

EXAMPLE $y = 4x^2;\ (3, -2)$

SOLUTIONS $y = 4(x-3)^2 - 2$ or $y = -4(x-3)^2 - 2$

13. $y = \frac{1}{2}x^2;\ (1, 5)$

14. $y = x^2;\ (1, 4)$

15. $y = -3x^2;\ (4, -7)$

16. $y = 2x^2;\ (-2, 0)$

17. $y = -\frac{1}{3}x^2;\ (6, 6)$

18. $y = 3x^2;\ (-1, -1)$

In Exercises 19-23, graph each group of equations on the same coordinate plane. Label each parabola with its equation. Compare the graphs in as many ways as you can.

B 19. $y = x^2 + 3,\ y = x^2 - 4,\ y = x^2 + 1$

20. $y = \frac{1}{4}x^2,\ y = \frac{3}{4}x^2,\ y = \frac{5}{4}x^2$

21. $y = (x+2)^2 + 3,\ y = -(x+2)^2 + 3$

22. $y = \frac{1}{2}(x-1)^2,\ y = \frac{1}{2}(x-4)^2,\ y = \frac{1}{2}(x+3)^2$

23. $y = -(x-2)^2 + 1,\ y = -2(x-2)^2 + 1,\ y = -\frac{1}{2}(x-2)^2 + 1$

24. Graph $y = x^2 + 2x + 1$ by computing coordinates and plotting ordered pairs for integral values of x from -4 to 2. How is this graph related to the graph of $y = (x+1)^2$?

C 25. a. Draw the graphs of $x = y^2$, $x = y^2 + 3$, and $x = y^2 - 2$ on the same coordinate plane.
b. From your graphs in part (a), how is the graph of $x = y^2 + k$ related to the graph of $x = y^2$?

26. a. Draw the graphs of $x = \frac{1}{2}y^2$ and $x = -\frac{1}{2}y^2$ on the same coordinate plane.
b. From your graphs in part (a), how is the graph of $x = ay^2$ related to the graph of $x = -ay^2$?

27. Draw the graphs of $x = y^2$, $y = \sqrt{x}$, and $y = -\sqrt{x}$ on separate coordinate planes. How are the three graphs related?

Section 12-2 THE GRAPH OF $y = ax^2 + bx + c$

To graph a parabola whose equation is of the form $y = ax^2 + bx + c$ $(a \neq 0)$, you use the method of completing the square to rewrite the equation in the form

$$y = a(x - h)^2 + k \quad (a \neq 0).$$

EXAMPLE Find the coordinates of the vertex and the equation of the axis of symmetry of the parabola $y = x^2 + 2x - 3$. Graph the equation and state whether the vertex is a maximum or a minimum point.

SOLUTION

$$y = x^2 + 2x - 3$$

1. Add 3 to both sides of the equation. $\quad y + 3 = x^2 + 2x$
2. Add 1 to both sides to complete the square on the right side. $\quad y + 4 = x^2 + 2x + 1$; $\quad y + 4 = (x + 1)^2$
3. Transform the equation to one of the form $y = a(x - h)^2 + k$. $\quad y = (x + 1)^2 - 4$

From the transformed equation, $y = (x + 1)^2 - 4$, you can see that the vertex is $(-1, -4)$ and that the equation of the axis of symmetry is $x = -1$. Plot the vertex and sketch the axis of symmetry. Substituting 0 and 1 for x in the original equation, you will find that $(0, -3)$ and $(1, 0)$ also satisfy the equation. Plot these points and their mirror images with respect to the axis of symmetry. Then sketch the parabola. Since $a > 0$, the vertex is a minimum point.

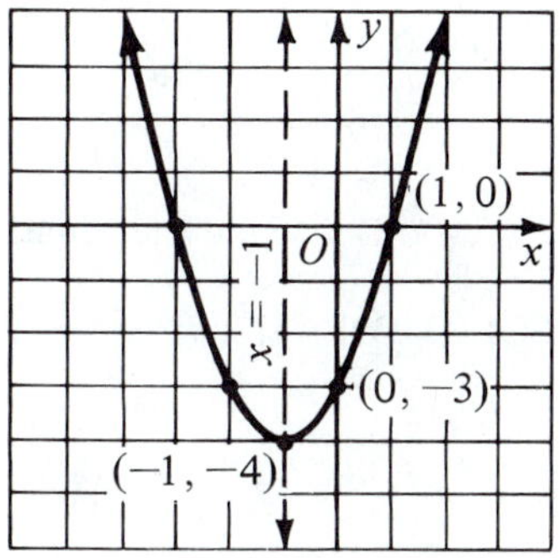

You find the y-intercept of a parabola the same way you find the y-intercept of a line: let $x = 0$, and solve for y. In the preceding example, the y-intercept is -3. The y-intercept is often convenient to use when graphing an equation of the form $y = ax^2 + bx + c$, since the y-intercept of the graph is c.

EXAMPLE Find the coordinates of the vertex, the equation of the axis of symmetry, and the y-intercept of the parabola $y = 2x^2 - 8x + 1$. Graph the equation.

SOLUTION

$$y = 2x^2 - 8x + 1$$

1. Subtract 1 from both sides of the equation. $\quad y - 1 = 2x^2 - 8x$
2. Divide both sides by 2. $\quad \dfrac{y - 1}{2} = x^2 - 4x$
3. Add 4 to both sides to complete the square on the right side. $\quad \dfrac{y - 1}{2} + 4 = x^2 - 4x + 4$; $\quad \dfrac{y - 1}{2} + 4 = (x - 2)^2$
4. Transform the equation into one of the form $y = a(x - h)^2 + k$. $\quad y - 1 + 2 \cdot 4 = 2(x - 2)^2$; $\quad y = 2(x - 2)^2 - 7$

(Solution continued on the next page)

From the transformed equation, $y = 2(x - 2)^2 - 7$, you can see that the vertex is $(2, -7)$, and the equation of the axis of symmetry is $x = 2$. From the original equation, the y-intercept is 1. Substituting 3 for x in the original equation, you will find that $(3, -5)$ and its mirror image $(1, -5)$ are on the graph, which is shown at the right.

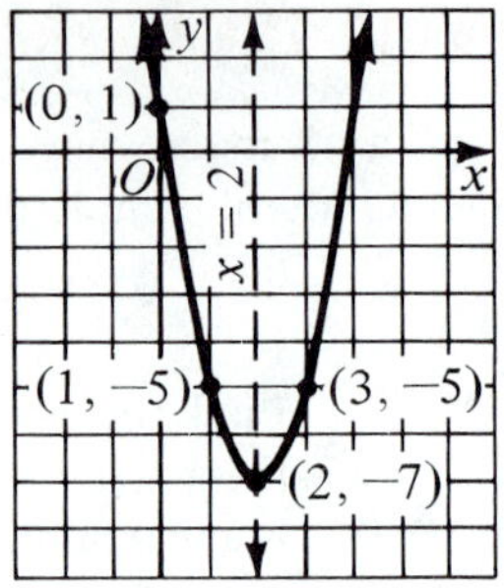

You can find the equation of the axis of symmetry of a parabola of the form $y = ax^2 + bx + c$ $(a \neq 0)$ without transforming the equation. (See Exercise 41.)

The equation of the axis of symmetry of the parabola $y = ax^2 + bx + c$ $(a \neq 0)$ is $x = -\frac{b}{2a}$.

EXAMPLE Find the equation of the axis of symmetry of the parabola $y = 3x^2 - 9x + 5$.

SOLUTION If $y = 3x^2 - 9x + 5$, then $a = 3$ and $b = -9$. Use the formula $x = -\frac{b}{2a}$ to find the equation of the axis of symmetry:

$$x = -\frac{-9}{2 \cdot 3} = \frac{9}{6}$$

$$x = \frac{3}{2}$$

The equation of the axis of symmetry of the parabola $y = 3x^2 - 9x + 5$ is $x = \frac{3}{2}$.

Since the vertex of a parabola is on the axis of symmetry, the x-coordinate of the vertex of the parabola $y = ax^2 + bx + c$ must be $-\frac{b}{2a}$. To find the y-coordinate of the vertex, substitute $-\frac{b}{2a}$ for x in the equation $y = ax^2 + bx + c$. Thus, you can find the coordinates of the vertex without transforming the equation.

EXAMPLE Find the coordinates of the vertex of the parabola $y = -4x^2 + 8x + 10$. State whether the vertex is a maximum or a minimum point.

SOLUTION If $y = -4x^2 + 8x + 10$, then $a = -4$ and $b = 8$.

$$x = -\frac{b}{2a}$$

$$x = -\frac{8}{2(-4)} = 1$$

$$\begin{aligned} y &= -4x^2 + 8x + 10 \\ &= -4(1)^2 + 8(1) + 10 \\ &= -4 + 8 + 10 \\ &= 14 \end{aligned}$$

The coordinates of the vertex are (1, 14).
Since $a < 0$, the vertex is a maximum point.

ORAL EXERCISES

For each parabola, points A, B, and C are the mirror images, with respect to the axis of symmetry, of the points whose coordinates are given. Find the coordinates of points A, B, and C.

1.
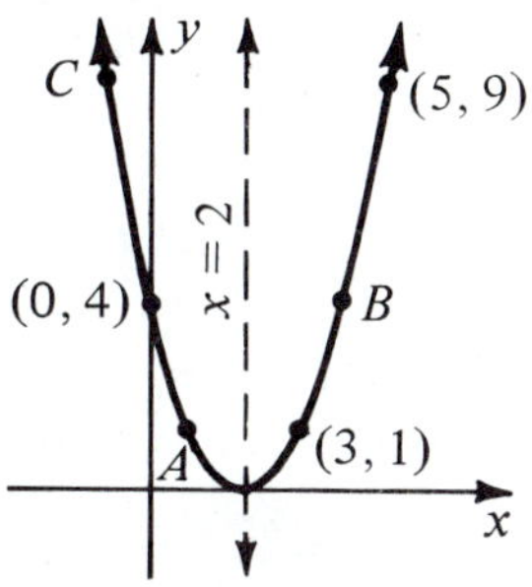

2.
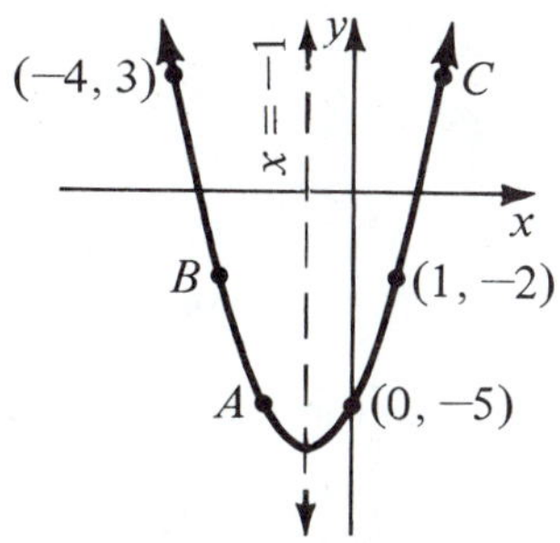

3.
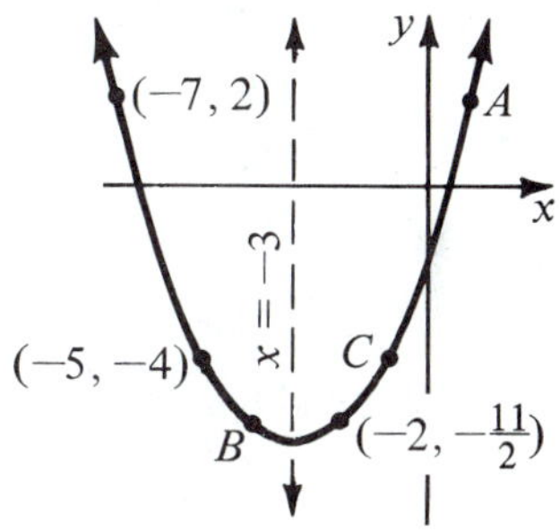

State the letter of the graph that corresponds to each equation.

4. $y = x^2 - 2x - 2$

5. $y = x^2 + 2x - 2$

6. $y = x^2 - 2x + 2$

7. $y = x^2 + 2x + 2$

a.
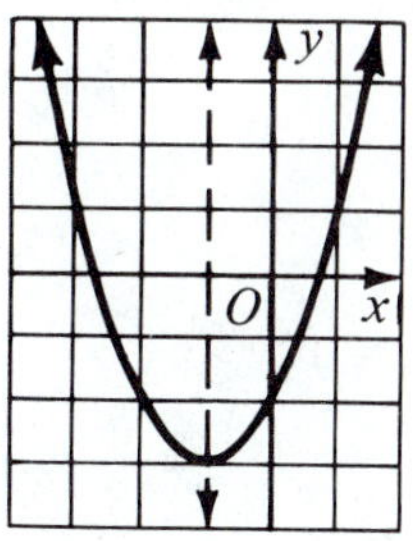

b.
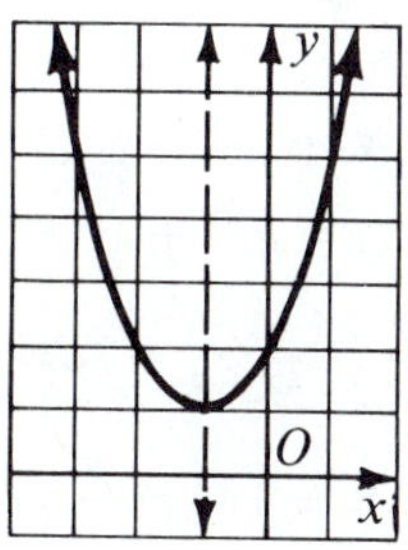

c.
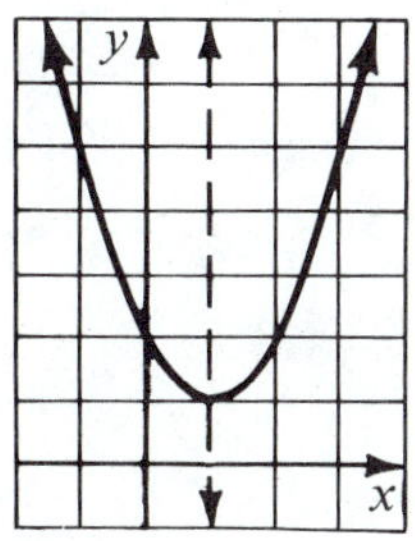

d.
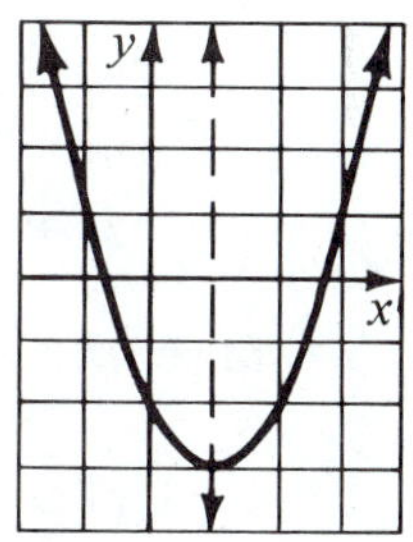

In Exercises 8–13, you are given equations of the form $y = ax^2 + bx + c$.

a. State the values of a and b.

b. Find the equation of the axis of symmetry of the parabola.

c. State whether the vertex is a maximum or a minimum point.

8. $y = x^2 - 4x + 1$

9. $y = -2x^2 + 3x + 5$

10. $y = 4x^2 - 2$

11. $y = x^2 - 2x$

12. $y = -3x^2 + 5$

13. $y = -2x^2 - 7x + 1$

WRITTEN EXERCISES

Find an equivalent equation of the form $y = a(x-h)^2 + k$ and state the coordinates of the vertex of the corresponding parabola.

A 1. $y = x^2 - 4x + 6$ 2. $y = x^2 + 6x + 8$ 3. $y = x^2 - 2x$

4. $y = x^2 - 8x$ 5. $y - 30 = x^2 + 10x$ 6. $y + 4x = 6 + x^2$

7. $y = x^2 - 3x + 1$ 8. $y = x^2 + 5x - 8$ 9. $y = x^2 + 9x - 3$

For each equation and corresponding interval:

a. Make a table of values of y for integral values of x in the given interval.
b. Use the results of part (a) to graph the equation.
c. Indicate the axis of symmetry with a dashed line and state its equation.
d. Label the vertex as V and state its coordinates.
e. State whether the vertex is a maximum or a minimum point.

10. $y = x^2 - 2x - 4;\ -2 \leqslant x \leqslant 4$ 11. $y = x^2 + 4x + 2;\ -5 \leqslant x \leqslant 1$

12. $y = x^2 - 6x;\ 0 \leqslant x \leqslant 6$ 13. $y = x^2 + 2x;\ -4 \leqslant x \leqslant 2$

14. $y = -x^2 + 10x - 20;\ 2 \leqslant x \leqslant 8$ 15. $y = -x^2 - 6x - 8;\ -6 \leqslant x \leqslant 0$

16. $y = 2x^2 + 4x - 1;\ -4 \leqslant x \leqslant 2$ 17. $y = -3x^2 + 6x - 2;\ -2 \leqslant x \leqslant 4$

18. $y = x^2 - 3x - 4;\ -1 \leqslant x \leqslant 4$ 19. $y = x^2 + 5x + 3;\ -5 \leqslant x \leqslant 0$

Without graphing the given equation, find the equation of the axis of symmetry.

20. $y = -3x^2 + 4x - 1$ 21. $y = \frac{1}{2}x^2 + x$ 22. $y = 1 - 2x - x^2$

Without graphing the given equation, find the coordinates of the vertex and state whether the vertex is a maximum or a minimum point.

23. $y = x^2 + 12x + 4$ 24. $y = x^2 - 4x + 5$ 25. $y = -x^2 + 2x - 7$

26. $y = 3x - x^2$ 27. $y = 4x^2 + 4x + 1$ 28. $y = -3x^2 + x + 5$

For the given parabola with the given axis of symmetry, find the value of d.

B 29. $y = 3x^2 + dx + 1;\ x = 2$ 30. $y = dx^2 + 4x - 3;\ x = 4$

31. $y = dx^2 - 6x - 4;\ x = -3$ 32. $y = \frac{1}{2}x^2 + 2dx - 2;\ x = 12$

33. $y = 2 - \frac{1}{2}x - dx^2;\ x = -2$ 34. $y = 5 - dx - 4x^2;\ x = -3$

Find an equivalent equation of the form $y = a(x-h)^2 + k$ and state the coordinates of the vertex of the corresponding parabola.

35. $y = 2x^2 - 4x + 2$ 36. $y = \frac{1}{2}x^2 + 3x - 1$ 37. $y = -x^2 + 5x + 6$

38. $y = 5x^2 - x$ 39. $y = \frac{1}{3}x^2 + 2x - 3$ 40. $y = 1 - 3x - 3x^2$

41. a. Prove that the equation of the axis of symmetry of the parabola $y = ax^2 + bx + c$ $(a \neq 0)$ is $x = -\frac{b}{2a}$. (*Hint:* Complete the square.)

b. Find the coordinates of the vertex of the parabola in part (a).

C 42. Prove that if the points $P(x_1, y_1)$ and $Q(x_2, y_2)$ are mirror images of each other with respect to the axis of symmetry of the parabola $y = ax^2 + bx + c$, then the midpoint M of $\overline{PQ}$ lies on the axis of symmetry, $x = -\frac{b}{2a}$. (*Hint:* $y_1 = ax_1{}^2 + bx_1 + c$, $y_2 = ax_2{}^2 + bx_2 + c$, and $y_1 = y_2$.)

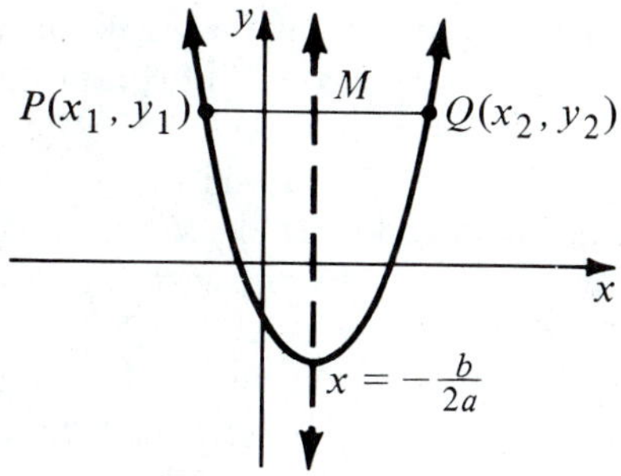

Section 12-3 LOOKING AT QUADRATIC EQUATIONS GRAPHICALLY

If a parabola $y = ax^2 + bx + c$ $(a \neq 0)$ intersects the x-axis, then $y = 0$ at the point or points of intersection. Therefore, $ax^2 + bx + c = 0$ at these points. Thus the real roots of the quadratic equation $ax^2 + bx + c = 0$ $(a \neq 0)$ are the x-intercepts of the parabola.

For example, from the figure at the right you can see that (4, 0) and (−2, 0) satisfy both of the following equations:

$$y = x^2 - 2x - 8$$
$$y = 0$$

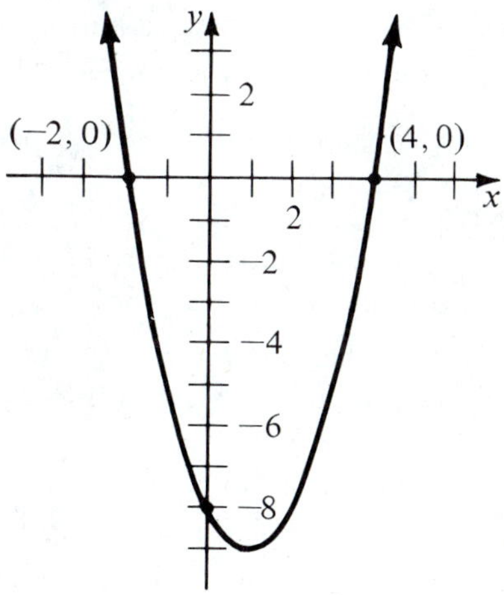

Therefore, the roots of $x^2 - 2x - 8 = 0$ are 4 and −2. You can check these by substituting in the equation. When $x = 4$, you have:

$$\begin{aligned} 4^2 - 2(4) - 8 &= 16 - 8 - 8 \\ &= 0 \end{aligned}$$

When $x = -2$, you have:

$$\begin{aligned} (-2)^2 - 2(-2) - 8 &= 4 - (-4) - 8 \\ &= 8 - 8 = 0 \end{aligned}$$

The following figures illustrate that a quadratic equation may have (i) two different real roots, (ii) one real root, or (iii) no real roots.

(i)

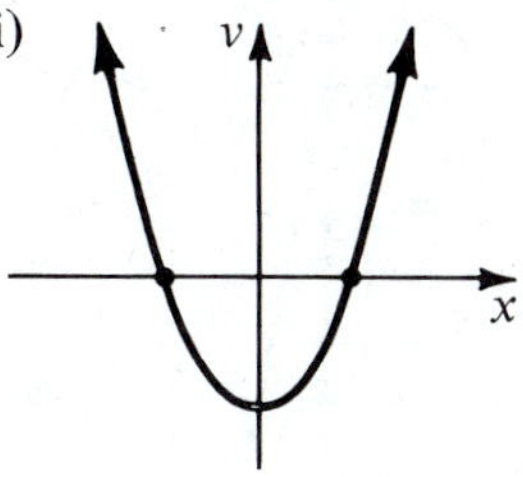

(ii)

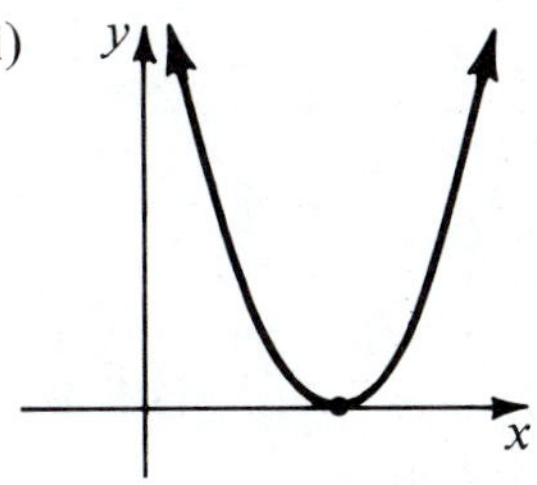

(iii)

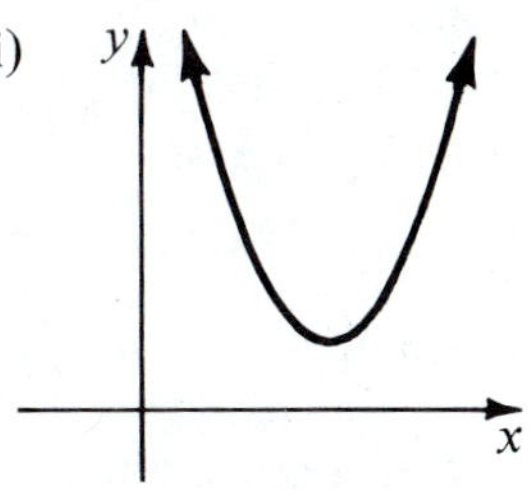

The diagram at the right shows the graph of $y = x^2 + 6x + 5$ and the graphs of $y = -6$, $y = -3$, $y = 0$, $y = 5$, and $y = 12$. From the graph, you can see that the roots of $x^2 + 6x + 5 = 0$ are -5 and -1. How would you find the roots of $x^2 + 6x + 5 = 12$? The coordinates of points A and B satisfy two equations: $y = 12$ and $y = x^2 + 6x + 5$. Therefore, the x-coordinates of A and B are the roots of $x^2 + 6x + 5 = 12$. From the graph you can see that these roots are -7 and 1. (You can check these roots by substitution.) How would you find the roots of $x^2 + 6x + 5 = -3$? of $x^2 + 6x + 5 = 5$?

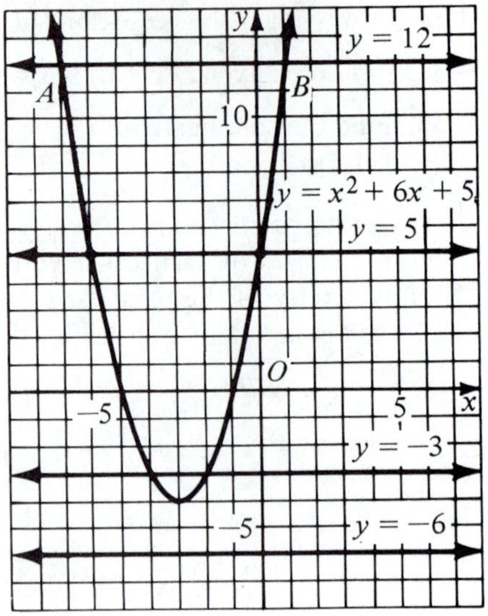

ORAL EXERCISES

Use the diagram at the right to determine the number of real roots of the given equation.

1. $x^2 + 3x - 4 = 0$
2. $x^2 - 8x + 17 = 0$
3. $-x^2 - 2x - 1 = 0$
4. Find the solution set of $x^2 + 3x - 4 = 0$. Check your solutions by substitution.
5. Find the solution set of $-x^2 - 2x - 1 = 0$. Check your solutions by substitution.

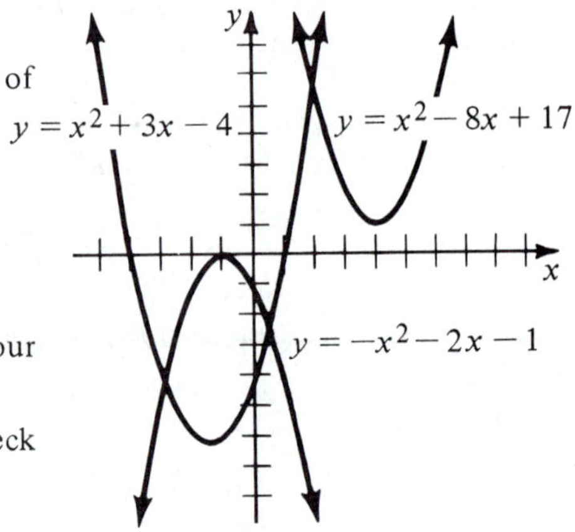

Exs. 1-5

Use the diagram at the right to determine the number of real roots of the given equation.

6. $-x^2 - 4x = -5$
7. $-x^2 - 4x = 0$
8. $-x^2 - 4x = 3$
9. $-x^2 - 4x = 4$
10. $-x^2 - 4x = 6$

11-15. Use the diagram to find the solution sets of the equations in Exercises 6-10. Check your solutions by substitution where possible.

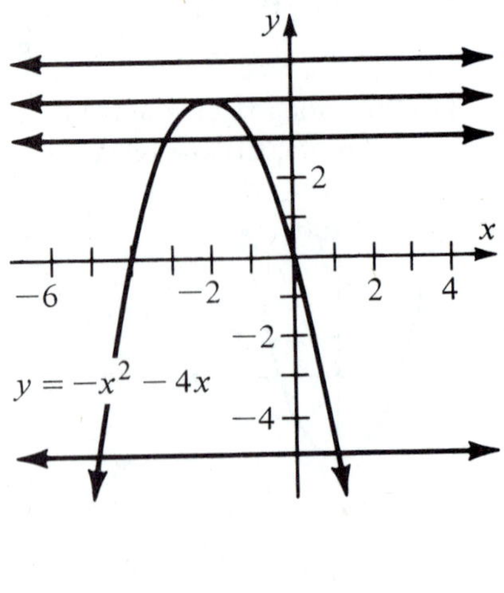

WRITTEN EXERCISES

Solve each equation and illustrate with a graph. Check your solutions.

A 1. $x^2 - 4x - 5 = 0$ 2. $x^2 + 2x - 8 = 0$ 3. $x^2 - 9 = 0$

4. $x^2 - 4 = 0$ 5. $x^2 + x - 12 = 0$ 6. $12 + 4x - x^2 = 0$

7. $4x^2 + 12x + 9 = 0$ 8. $x^2 + 3 = 0$ 9. $4x^2 = 0$

Graph the parabola $y = x^2 + 4x + 2$ and use your graph to solve each equation.

10. $x^2 + 4x + 2 = 2$ 11. $x^2 + 4x + 2 = -2$ 12. $x^2 + 4x + 2 = -1$

Solve each equation and illustrate with a graph. Check your solutions.

B 13. $x^2 + 3x + 4 = 0$ 14. $x^2 + 2x + 5 = 0$ 15. $x - x^2 = 0$

16. $x^2 - 6x + 2 = -3$ 17. $-2x^2 = 8$ 18. $6 + 4x - x^2 = 0$

19. Graph $y = x^2 - 2x + 2$. For what values of k will the equation $x^2 - 2x + 2 = k$ have exactly one real root? two real roots? no real roots?

The x-intercepts of the parabola $y = ax^2 + bx + c$ $(a \neq 0)$ are the roots of the equation $ax^2 + bx + c = 0$. Find the x-intercepts of each parabola *without* graphing its equation.

20. $y = x^2 - 8x - 9$ 21. $y = x^2 + 12x + 27$

22. $y = 4x^2 + 3x + 1$ 23. $y + 6 = 6x^2 + 5x$

24. Graph $y = x^2 - 1$. For what values of x is it true that $3 \leqslant y \leqslant 8$?

In Exercises 25-27:

a. Rewrite the inequality as one in which 0 is the right-hand member.
b. Solve the inequality in part (a) graphically.

C 25. $x^2 + x \geqslant 6$ 26. $x^2 + 3x + 1 < -2x - 5$ 27. $7x - 3 > 2x^2$

SELF-TEST 1

In Exercises 1 and 2:

a. Make a table of values of y for integral values of x in the given interval.
b. Use the results of part (a) to graph the equation.
c. Draw the axis of symmetry and state its equation.
d. Label the vertex as V and state its coordinates.
e. State whether the vertex is a maximum or a minimum point.

1. $y = -\frac{3}{4}x^2$; $-3 \leqslant x \leqslant 3$ Section 12-1

2. $y = (x + 1)^2 - 1$; $-4 \leqslant x \leqslant 2$

(Self-Test continued on next page)

Find an equivalent equation in the form $y = a(x - h)^2 + k$. State the coordinates of the vertex of the corresponding parabola.

3. $y = x^2 + 2x + 3$ 4. $y = x^2 - 5x + 6$ Section 12-2

5. a. Graph $y = x^2 - 4x + 3$ including all integral values of x from -1 to 5.
   b. Find an equation for the axis of symmetry of the parabola.

6. An equation of the axis of symmetry of the parabola $y = ax^2 + 6x - 1$ is $x = 1$. Find the value of a.

7. Solve $x^2 - 2x - 3 = 0$ and illustrate with a graph.   Section 12-3

8. a. Graph the equation $y = x^2 - 6x + 7$.
   b. What are the roots of $x^2 - 6x + 7 = 2$?

Section 12-4 CIRCLES

As you may recall from Section 10-6, a *circle* is the locus of all points in a plane at a given distance, the *radius* of the circle, from a given point, the *center* of the circle. Like equations of lines, you can find and graph equations of circles.

EXAMPLE Find an equation of the circle with center $C(2, 4)$ and radius 2.

SOLUTION If $P(x, y)$ is any point on the circle, then PC is the radius and $PC = 2$.

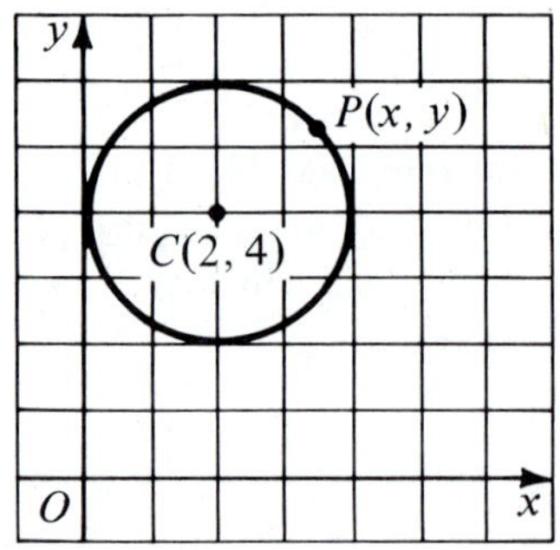

Apply the distance formula.

$$PC = \sqrt{(x-2)^2 + (y-4)^2}$$
$$2 = \sqrt{(x-2)^2 + (y-4)^2} \qquad (1)$$
$$4 = (x-2)^2 + (y-4)^2 \qquad (2)$$

Thus, any point on the circle satisfies equation (2). Conversely, if a point satisfies (2), it satisfies (1), and is 2 units from C and so, by definition, is a point on the circle. Therefore, an equation of the circle with center (2, 4) and radius 2 is $(x-2)^2 + (y-4)^2 = 4$.

You can use the method illustrated in the example to derive a general equation of a circle.

> An equation of the circle with radius r ($r > 0$) and center (h, k) is
> $$(x-h)^2 + (y-k)^2 = r^2.$$

This equation is sometimes called the *center-radius* form of an equation of a circle. In the special case where the center of the circle is (0, 0), an equation of the circle is $x^2 + y^2 = r^2$.

EXAMPLE State the coordinates of the center and the radius of the circle with equation $3(x+1)^2 + 3y^2 = 12$.

SOLUTION Rewrite the equation in center-radius form:

$$(x+1)^2 + y^2 = 4,$$

which is equivalent to $(x-(-1))^2 + (y-0)^2 = 2^2$.
The coordinates of the center are $(-1, 0)$, and the radius is 2.

EXAMPLE Write an equation of the circle with center $(4, -1)$ that passes through the point $(5, 2)$.

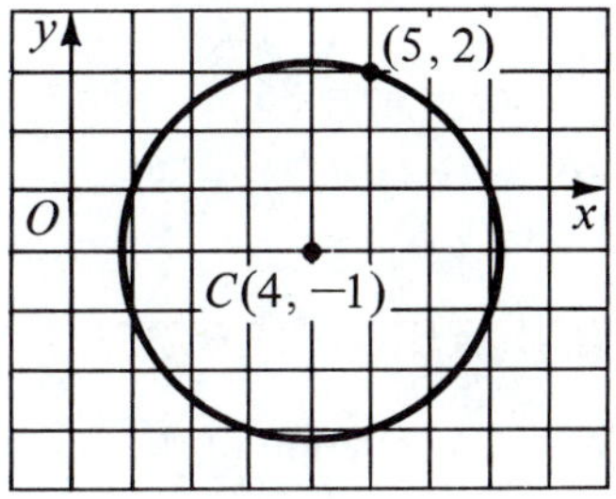

SOLUTION Since the center is $(4, -1)$, the equation has the form:

$$(x-4)^2 + (y+1)^2 = r^2$$

The point $(5, 2)$ satisfies the equation, so:

$$r^2 = (5-4)^2 + (2+1)^2 = 1 + 9 = 10$$

Then $(x-4)^2 + (y+1)^2 = 10$ is the required equation.

Notice that you could expand the equation $(x-4)^2 + (y+1)^2 = 10$ to get the equivalent equation

$$x^2 - 8x + 16 + y^2 + 2y + 1 = 10,$$

or

$$x^2 + y^2 - 8x + 2y + 7 = 0.$$

This illustrates the fact that an equation of the form $(x-h)^2 + (y-k)^2 = 0$ is equivalent to an equation of the form

$$x^2 + y^2 + ax + by + c = 0, \quad a, b, c \in \mathcal{R}$$

which is sometimes called the *expanded form* of an equation of a circle. To transform an equation of a circle written in expanded form into an equivalent equation written in center-radius form, complete the square twice, once for x and once for y.

EXAMPLE Graph $x^2 + y^2 + 6x - 2y + 6 = 0$.

SOLUTION Group the x-terms and the y-terms.

$$(x^2 + 6x \quad\quad) + (y^2 - 2y \quad\quad) = -6$$

Complete the square for x and for y.

$$(x^2 + 6x + 9) + (y^2 - 2y + 1) = -6 + 9 + 1$$
$$(x+3)^2 + (y-1)^2 = 4$$

From the last equation, you can see that the circle has center $(-3, 1)$ and radius $\sqrt{4}$, or 2. If you don't have a compass, you can sketch a circle by plotting the center and the four points that are 2 units horizontally and vertically from the center. Join the four points with a smooth curve to obtain the graph shown.

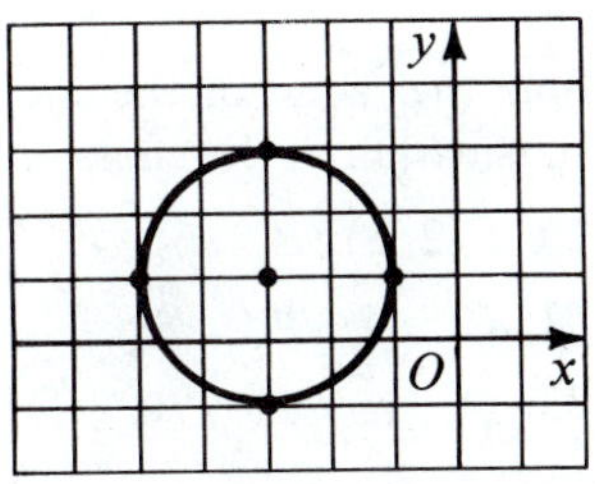

ORAL EXERCISES

The graph of each of the following equations is a circle whose center is at the origin. State the radius of each circle in simplest form.

1. $x^2 + y^2 = 100$
2. $x^2 + y^2 = 1$
3. $x^2 + y^2 = 5$
4. $x^2 + y^2 = 18$
5. $9x^2 + 9y^2 = 4$
6. $4x^2 + 4y^2 = 3$

State the radius and the coordinates of the center of each circle whose equation is given.

7. $x^2 + y^2 = 36$
8. $x^2 + y^2 = 4$
9. $(x - 6)^2 + y^2 = 25$
10. $x^2 + (y + 2)^2 = 49$
11. $2(x - 5)^2 + 2(y - 3)^2 = 2$
12. $(x + 13)^2 + (y + 5)^2 = 81$

WRITTEN EXERCISES

Write an equation of the circle with center at the origin and with the given radius r.

A

1. $r = 3$
2. $r = 6$
3. $r = 5$
4. $r = \frac{1}{2}$
5. $r = \sqrt{7}$
6. $r = 5\sqrt{2}$

Write an equation in center-radius form of the circle with center C and radius r.

7. $C(0, -4);\ r = 2$
8. $C(1, 2);\ r = 4$
9. $C(5, -4);\ r = 1$
10. $C(-1, 1);\ r = \frac{2}{3}$
11. $C(2, 0);\ r = \sqrt{7}$
12. $C(-3, -3);\ r = 2\sqrt{5}$

Find the radius and the coordinates of the center of each circle.

13. $x^2 + y^2 + 4x + 2y + 1 = 0$
14. $x^2 + y^2 - 2x + 6y - 15 = 0$
15. $x^2 + y^2 - 10x - 2y + 17 = 0$
16. $x^2 + y^2 + 6x - 8y + 24 = 0$

Find the radius and the coordinates of the center of each circle. Then sketch a graph of the equation.

17. $x^2 + y^2 = 25$
18. $4x^2 + 4y^2 = 9$
19. $(x - 2)^2 + (y + 1)^2 = 16$
20. $(x + 3)^2 + (y - 4)^2 = 4$
21. $x^2 + y^2 + 4x = 0$
22. $x^2 + y^2 - 6y = 0$
23. $x^2 + y^2 - 10x - 6y + 33 = 0$
24. $x^2 + y^2 + 4x - 12y + 31 = 0$

B

25. $9x^2 + 9y^2 + 36x - 12y - 185 = 0$
26. $4x^2 + 4y^2 - 12x - 16y + 21 = 0$
27. $16x^2 + 16y^2 - 16x + 24y - 51 = 0$
28. $2x^2 + 2y^2 + 10x + 6y - 1 = 0$

Find the radius of the circle with the given center C and containing P. Then write an equation of the circle in center-radius form.

29. $C(2, 3);\ P(2, 6)$
30. $C(0, 5);\ P(7, 5)$
31. $C(3, 4);\ P(0, 0)$
32. $C(1, -2);\ P(5, 6)$
33. $C(-2, -2);\ P(0, -1)$
34. $C(-1, -4);\ P(4, 8)$

Write an equation in expanded form of the circle whose diameter has the given endpoints.

EXAMPLE $(-3, 2), (5, 4)$

SOLUTION The midpoint of the diameter is the center of the circle.

The midpoint of the diameter is $\left(\frac{-3+5}{2}, \frac{2+4}{2}\right)$, or $(1, 3)$.

An equation of the circle will have the form

$$(x-1)^2 + (y-3)^2 = r^2.$$

The point $(5, 4)$ lies on the circle, so

$$r^2 = (5-1)^2 + (4-3)^2 = 16 + 1 = 17.$$

Therefore, an equation of the circle is

$$(x-1)^2 + (y-3)^2 = 17,$$

which is equivalent to

$$x^2 + y^2 - 2x - 6y - 7 = 0.$$

35. $(3, 4), (-3, -4)$

36. $(4, 2), (0, 0)$

37. $(6, 2), (-4, -8)$

38. $(0, 5), (10, -19)$

39. Show that the line with equation $x + 3y = -2$ passes through the center of the circle with equation $x^2 + y^2 + 10x - 2y - 3 = 0$.

40. Draw a segment that is 5 cm long. Draw several different right triangles that have this segment as the hypotenuse. Place several triangles on each side of the segment. Connect the vertices of the right angles with a smooth curve. What kind of curve does it appear to be?

C **41.** A circle has diameter $\overline{AB}$. $P(x, y)$ is any point on the circle except A or B. Use coordinate geometry to show that $\overline{PA} \perp \overline{PB}$. (*Hint:* Locate the circle on a coordinate plane so that its center is the origin, and A and B lie on the x-axis.)

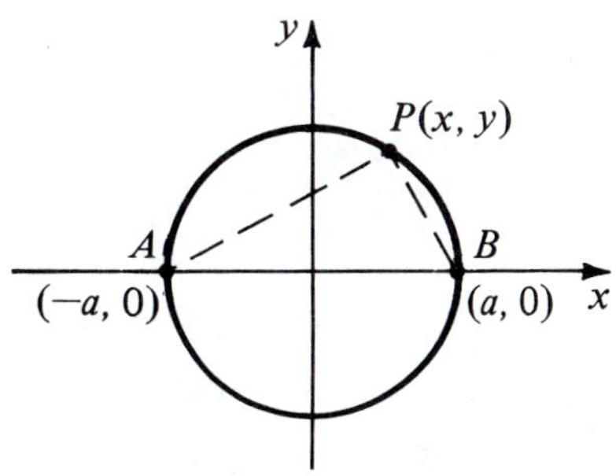

42. The center of a circle with radius 13 lies on the line with equation $y = 2$. If the circle passes through the point $(9, 7)$, find the coordinates of the center of the circle. (*Hint:* There are two solutions.)

43. A circle with equation $x^2 + y^2 - 2x + y - 8 = 0$ contains points $A(-2, 0)$ and $B(0.5, 2.5)$. Use coordinate geometry to prove that the diameter perpendicular to $\overline{AB}$ passes through the midpoint of $\overline{AB}$.

44. Find an equation of the circle that contains the points $(6, 5)$, $(-3, 2)$, and $(-2, -1)$.

45. Find an equation of the circle that contains the points $(0, 0)$, $(4, 0)$, and $(2, 2\sqrt{3})$.

Section 12-5 THE PARABOLA AS A LOCUS

In this section you will work with the parabola as a locus of points. To do this, you must learn the meanings of the terms:

- directrix
- focus

Figure 1 below shows the graphs of some points in the locus of points equidistant from point $F(0, 5)$ and the line with equation $y = -1$. Suppose you graphed more points in this locus and joined the points with a smooth curve. What kind of graph do you think you would have?

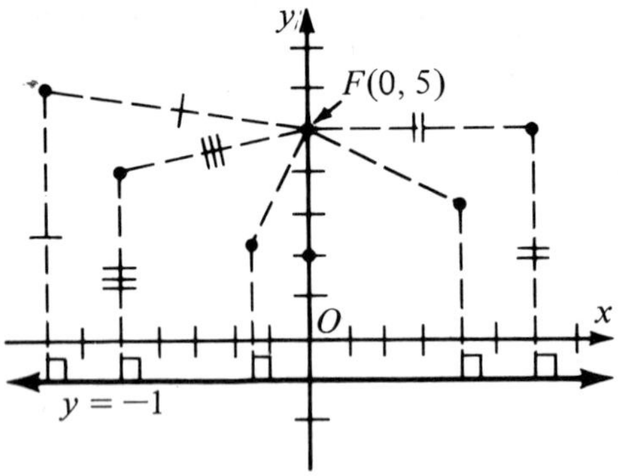

Figure 1

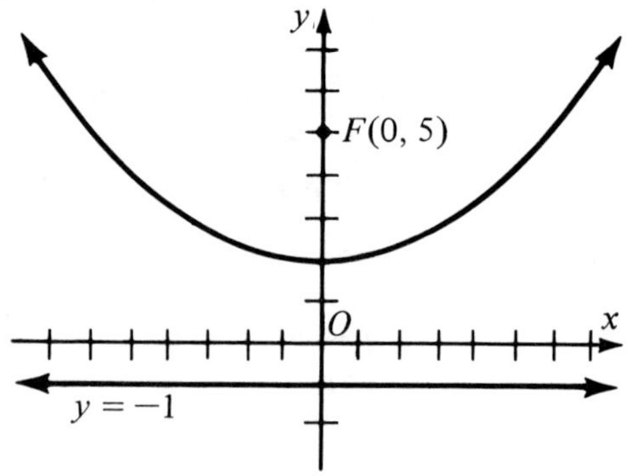

Figure 2

In Figure 2, we have joined with a smooth curve the points graphed in Figure 1. Perhaps you guessed correctly that the graph is a parabola. The locus of points $P(x, y)$ equidistant from a given line, called the *directrix*, and a given point not on the line, called the *focus*, is a *parabola*.

You can find an equation of the parabola with focus $F(0, 5)$ and directrix $y = -1$ by using the *focus-directrix* definition of a parabola. Let $P(x, y)$ be a point on the parabola. Then $D(x, -1)$ is the point such that PD is the distance from P to the directrix. By the focus-directrix definition of a parabola:

$$PF = PD$$

Since $\quad PF = \sqrt{(x-0)^2 + (y-5)^2}$

and $\quad PD = |(y-(-1)|,$

it follows that $\quad \sqrt{x^2 + (y-5)^2} = |y+1|.$

Square both sides. $\quad x^2 + (y-5)^2 = (y+1)^2$

Simplify. $\quad x^2 + y^2 - 10y + 25 = y^2 + 2y + 1$

$$12y = x^2 + 24$$

Solve for y. $\quad y = \frac{1}{12}x^2 + 2$

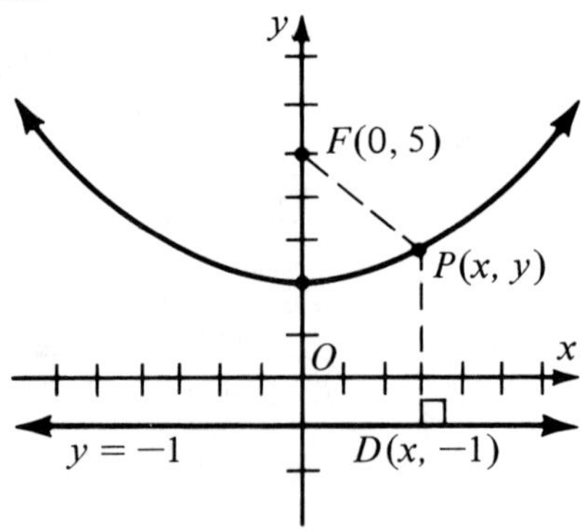

Some important facts about the parabola $y = \frac{1}{12}x^2 + 2$ are listed at the top of the next page.

(1) The directrix $y = -1$ is perpendicular to the axis of symmetry $x = 0$.
(2) The focus $F(0, 5)$ lies on the axis of symmetry.
(3) The vertex $V(0, 2)$ is the midpoint of the segment of the axis of symmetry that lies between the focus and the directrix.

EXAMPLE Write an equation of the form $y = ax^2 + bx + c$ of the parabola with focus $F(2, 1)$ and directrix $y = 3$.

SOLUTION Plot the focus and sketch the directrix. By studying the graph, you can see that the axis of symmetry of the parabola must be the line with equation $x = 2$. The vertex of the parabola is the midpoint of the segment of the axis of symmetry that lies between the focus and the directrix. Therefore, the coordinates of the vertex must be (2, 2). Sketch the axis of symmetry and plot the vertex. Since a parabola never intersects its directrix, you can tell that this parabola must open downward, meaning that $a < 0$. You can use these facts later to check your work.

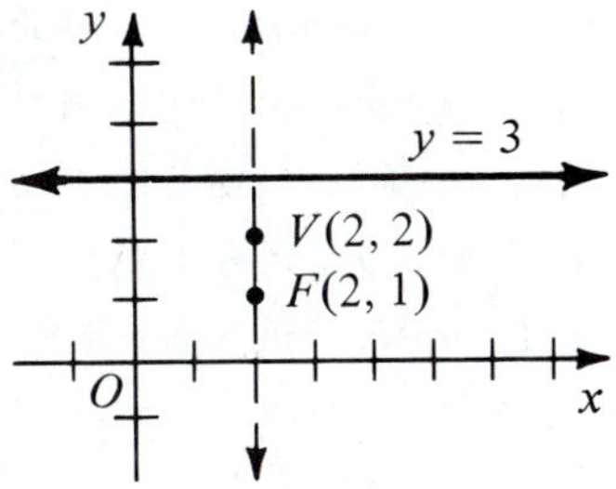

Let $P(x, y)$ be a point on the parabola. Then $D(x, 3)$ is the point such that PD is the distance from P to the directrix. By the focus-directrix definition of a parabola:

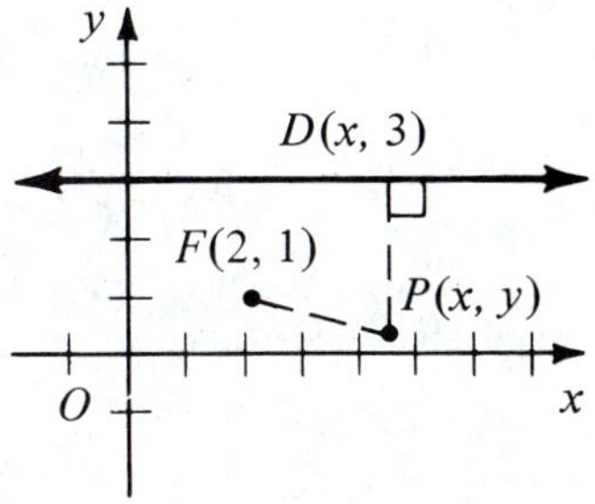

$$\begin{aligned} PF &= PD, \quad \text{or} \\ \sqrt{(x-2)^2 + (y-1)^2} &= |y - 3|. \\ (x-2)^2 + (y-1)^2 &= (y-3)^2 \\ x^2 - 4x + 4 + y^2 - 2y + 1 &= y^2 - 6y + 9 \\ x^2 - 4x + 5 - 2y &= -6y + 9 \\ 4y &= -x^2 + 4x + 4 \\ y &= -\frac{1}{4}x^2 + x + 1 \end{aligned}$$

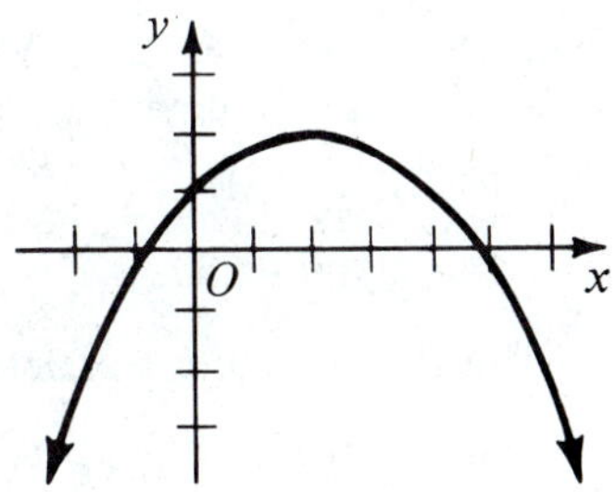

Check Does (2, 2) satisfy the equation $y = -\frac{1}{4}x^2 + x + 1$?

Yes, $2 = -\frac{1}{4}(2^2) + 2 + 1$.

Is $a < 0$? Yes, $a = -\frac{1}{4}$.

Now consider the locus of all points equidistant from point $F(3, 1)$ and the line with equation $x = -1$. By definition, this locus must be a parabola. Its equation is one of the form $x = ay^2 + by + c$, and its sketch is shown at the right. In this text, we have chosen to limit our study of parabolas to those of the form $y = ax^2 + bx + c$.

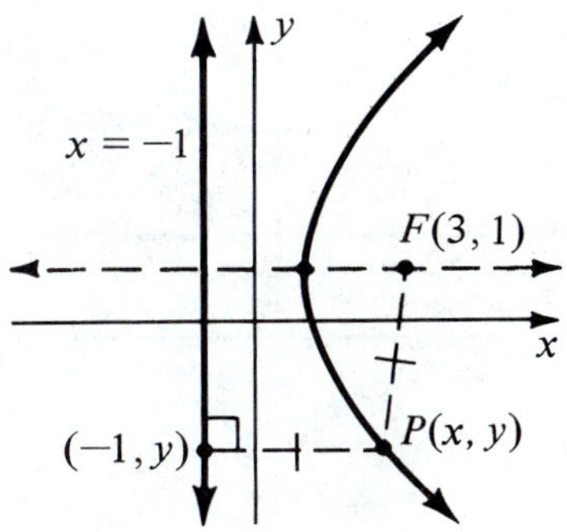

ORAL EXERCISES

1. Use the focus-directrix definition of a parabola to explain why the vertex of a parabola is the midpoint of the segment of the axis of symmetry joining the focus and the directrix of the parabola.

Exercises 2-5 refer to a parabola with focus $F(0, -2)$ and directrix $y = 3$, as shown at the right. $P(x, y)$ represents a point on the parabola.

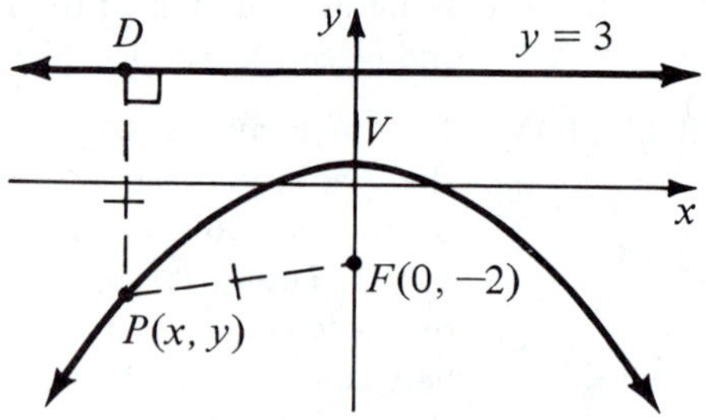

2. State the coordinates of the vertex, V.
3. Express the coordinates of D in terms of x.
4. Express PD in terms of y.
5. Express PF in terms of x and y.

Complete.

6. If the focus of a parabola lies below the directrix, then the parabola opens __?__.
7. If the focus of a parabola lies above the directrix, then the parabola opens __?__.

WRITTEN EXERCISES

In Exercises 1-6, you are given the focus F and the equation of the directrix of a parabola. State the coordinates of the vertex of the parabola.

A 1. $F(0, 2)$; $y = -2$ 2. $F(0, -3)$; $y = 3$ 3. $F(0, 0)$; $y = 4$

4. $F(0, 0)$; $y = -6$ 5. $F(1, 6)$; $y = 0$ 6. $F(-2, -5)$; $y = 0$

In Exercises 7-12, you are given the focus F and the vertex V of a parabola. State the equation of the directrix.

7. $F(0, 3)$; $V(0, 0)$ 8. $F(0, 0)$; $V(0, 2)$ 9. $F(0, 0)$; $V(0, -1)$

10. $F(0, -1)$; $V(0, 1)$ 11. $F(-2, -1)$; $V(-2, 0)$ 12. $F(3, 5)$; $V(3, 0)$

13. Use the focus-directrix definition to explain why a parabola with focus $F(0, k)$ and directrix $y = h$ has vertex $V\left(0, \frac{k+h}{2}\right)$.

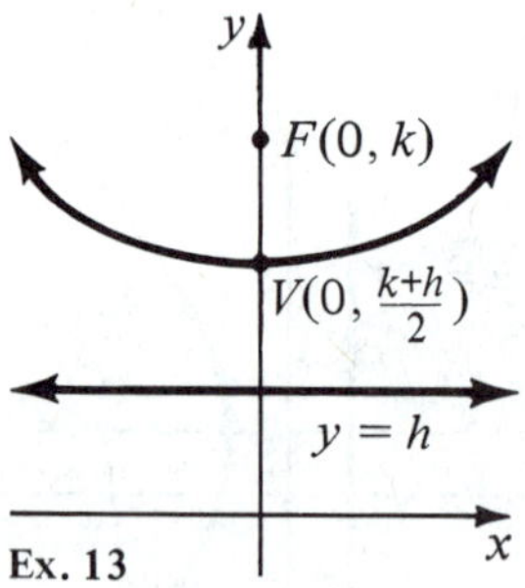

Ex. 13

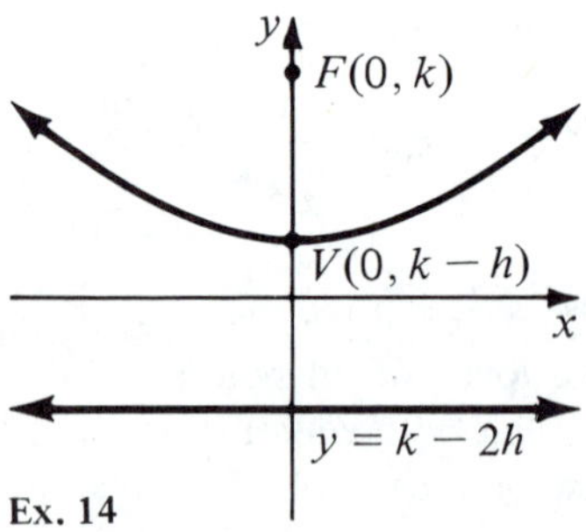

Ex. 14

14. Use the focus-directrix definition to explain why a parabola with focus $F(0, k)$ and vertex $V(0, k - h)$ has as its directrix $y = k - 2h$.

In Exercises 15-20, you are given the vertex V and the equation of the directrix of a parabola. State the coordinates of the focus.

15. $V(0, 0)$; $y = -2$
16. $V(0, 0)$; $y = 5$
17. $V(0, -1)$; $y = 2$
18. $V(0, 3)$; $y = -1$
19. $V(-1, 2)$; $y = \frac{1}{2}$
20. $V\left(1, \frac{1}{4}\right)$; $y = \frac{1}{2}$

Find an equation in the form $y = ax^2 + bx + c$ of the parabola with focus F and directrix having the given equation. Sketch the parabola.

B 21. $F(0, 5)$; $y = 3$
22. $F(0, 1)$; $y = -2$
23. $F(0, -3)$; $y = 1$
24. $F(0, -7)$; $y = -3$
25. $F(4, 2)$; $y = 6$
26. $F(1, 3)$; $y = -5$

Find an equation in the form $y = ax^2 + bx + c$ of the parabola with focus F and vertex V. Sketch the parabola.

27. $F(0, 4)$; $V(0, 0)$
28. $F(0, 6)$; $V(0, 8)$
29. $F(3, -6)$; $V(3, -2)$
30. $F(-5, 0)$; $V(-5, -2)$
31. $F(-2, 5)$; $V(-2, 3)$
32. $F\left(1, -\frac{5}{2}\right)$; $V(1, -2)$

C 33. A parabola has its vertex at the origin and its focus at $(0, r)$.
a. Show that the directrix has equation $y = -r$.
b. Show that an equation of the parabola is $x^2 = 4ry$.

Use the results of Exercise 33 to find the focus and directrix of each parabola.

34. $8y = x^2$
35. $y = -8x^2$
36. $3x^2 - 12y = 0$

37. A parabola has vertex (h, k) and focus $(h, k + r)$.
a. Show that the directrix has equation $y = k - r$.
b. Show that an equation of the parabola is $y = \frac{1}{4r}(x - h)^2 + k$.

Section 12-6 SOLVING LINEAR-QUADRATIC SYSTEMS

In Chapter 11, you learned how to use substitution to solve systems of linear equations in two variables. Now you will learn how to use substitution to solve a system of equations in two variables in which one equation is linear and the other equation is quadratic. Such a system is called a *linear-quadratic system of equations.*

To solve a linear-quadratic system, you solve the linear equation for one of the variables and then substitute in the quadratic equation. As you will see, a linear-quadratic system may have 0, 1, or 2 solutions.

EXAMPLE Solve the following system and illustrate with a graph.

$$x^2 + y = 5$$
$$y - x = 3$$

SOLUTION Solve the linear equation for one of the variables, say y.

$$y - x = 3$$
$$y = x + 3$$

Substitute in the quadratic equation.

$$x^2 + y = 5$$
$$x^2 + (x + 3) = 5$$
$$x^2 + x - 2 = 0$$
$$(x + 2)(x - 1) = 0$$
$$x = -2 \quad \text{or} \quad x = 1$$

Since $y = x + 3$, you find that $y = 1$ when $x = -2$, and $y = 4$ when $x = 1$. Therefore, there are two solutions, $(-2, 1)$ and $(1, 4)$.

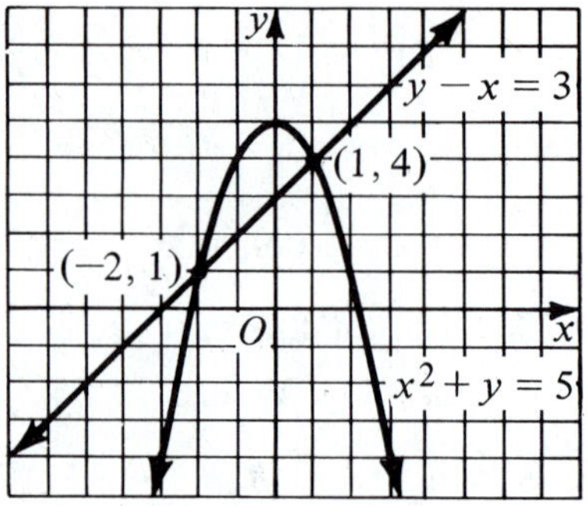

A graphic check is shown above. To check your answer algebraically, substitute in both original equations. This is left to you.

EXAMPLE Solve the following system and illustrate with a graph.

$$x^2 + y^2 = 25$$
$$x - y = 1$$

SOLUTION Solve the linear equation for one of the variables, say x.

$$x - y = 1$$
$$x = 1 + y$$

Substitute in the quadratic equation.

$$x^2 + y^2 = 25$$
$$(1 + y)^2 + y^2 = 25$$
$$1 + 2y + y^2 + y^2 = 25$$
$$2y^2 + 2y - 24 = 0$$
$$y^2 + y - 12 = 0$$
$$(y + 4)(y - 3) = 0$$
$$y = -4 \quad \text{or} \quad y = 3$$

Since $x = 1 + y$, you find that $x = -3$ when $y = -4$, and $x = 4$ when $y = 3$. Therefore, there are two solutions, $(-3, -4)$ and $(4, 3)$.

A graphic check is shown at the right. The algebraic check is left to you.

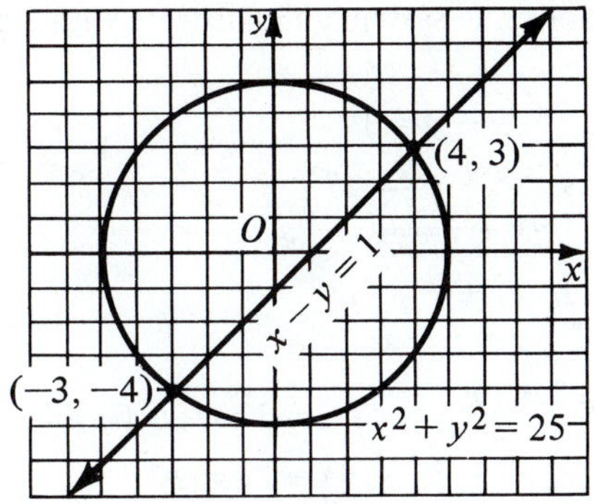

ORAL EXERCISES

State whether each ordered pair is a solution of the given system.

1. $y = x^2 + 2$ $\quad$ $(-1, 3), (3, 11), (4, -2), (0, 2)$
   $y = -x + 2$
2. $y = -x^2 + 1$ $\quad$ $(0, 1), (-4, -15), (3, 6), (2, -3)$
   $3x - y = 3$
3. $x^2 + y^2 = 13$ $\quad$ $(3, 2), (-3, -2), (2, 3), (-2, -3)$
   $y - x = 1$
4. $x^2 + y^2 = 10$ $\quad$ $(3, 1), (1, 3), (-1, -3), (-3, -1)$
   $y = x - 2$
5. $y = 3x^2 + 1$ $\quad$ $(0, 1), (-4, 5), (-1, 4), (2, -7)$
   $2x + y = -3$
6. The graph of the system

$$x^2 + (y - 3)^2 = 4$$
$$y = \frac{1}{2}x - 2$$

is shown at the right. State its solution set.

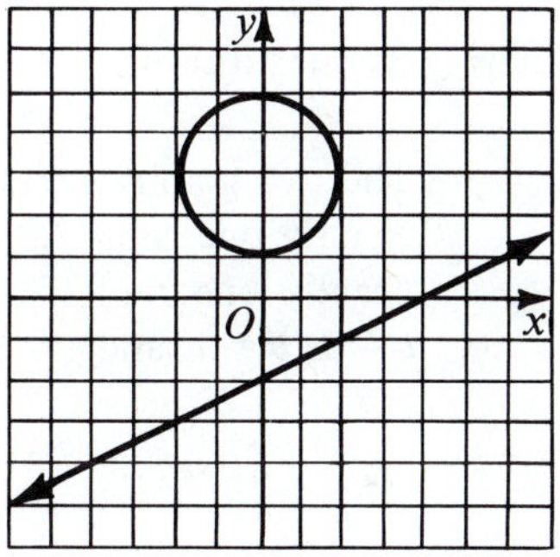

WRITTEN EXERCISES

Solve the system and illustrate with a graph.

A

1. $y = x^2 + 2x + 1$
   $y = 2x + 5$
2. $y = x^2 - 3x + 2$
   $y = x - 1$
3. $x^2 + y^2 = 8$
   $x = y$
4. $x^2 + y^2 = 32$
   $x + y = 0$
5. $y = x^2 + 2x - 1$
   $y = x + 1$
6. $y = x^2 + 1$
   $2x - y = -4$

Solve the system and illustrate with a graph.

7. $y = x^2 - 3x - 10$
   $3x - y = 19$

8. $2y = x^2 - 8x + 7$
   $x - y = 7$

9. $x^2 + y^2 = 25$
   $x + y = 8$

10. $x^2 + y^2 = 8$
    $y + 4 = x$

11. $x^2 + y^2 = 18$
    $y = 6 - x$

12. $y = -x^2 + x - 5$
    $2y - x = 4$

Solve each system and check the solutions if they exist.

B 13. $x^2 - 4x + y^2 - 2y = 20$
   $y = 2x - 8$

14. $x^2 - 2x + y^2 + 2y = 34$
    $x + y = 6$

15. $(x + 2)^2 + (y - 1)^2 = 9$
    $2y - x = 7$

16. $(x - 4)^2 + y^2 = 25$
    $x + 2y = -7$

17. $y = 2(x + 1)^2 - 3$
    $4x = 2y + 3$

18. $y - 5 = -(x - 2)^2$
    $4y = 12x + 5$

Find the value(s) of k that make the graphs of the given equations intersect in exactly one point.

EXAMPLE $y = x^2 + 4x - 1$
$y = 3x + k$

SOLUTION By substitution,

$$3x + k = x^2 + 4x - 1$$
$$x^2 + x - 1 - k = 0$$

The two graphs intersect in exactly one point if and only if $x^2 + x - 1 - k = 0$ has exactly one root. As you may recall, a quadratic equation, $ax^2 + bx + c = 0$, has exactly one root if and only if its discriminant, $b^2 - 4ac$, equals 0. In this case, $a = 1$, $b = 1$, and $c = 1 - k$, so we have

$$1^2 - 4(1)(-1 - k) = 0$$
$$1 + 4 + 4k = 0$$
$$k = -\frac{5}{4}$$

19. $y = x^2 - 3x + 2$
    $y = x + k$

20. $y = (x - 2)^2 + 1$
    $x + y = k$

21. $x^2 + y^2 = 9$
    $y = 2x + k$

22. $x^2 + y^2 = 8$
    $y = -3x + k$

C 23. A circle with center at the origin has radius 5. A line passes through the points (2, 9) and (−5, 2). Find the coordinates of each point of intersection between the circle and the line.

24. The endpoints of a diameter of a circle are the points of intersection of the graphs of $y = (3 - x)^2 + 2$ and $x + y = 5$. Find an equation of the circle.

25. The vertex of a parabola is the point where the circle with equation $(x + 4)^2 + (y + 5)^2 = 25$ intersects the x-axis. The focus of the parabola is the center of the circle. Find an equation in the form $y = ax^2 + bx + c$ of the parabola.

SELF-TEST 2

Write an equation in center-radius form of the circle with the given center C and radius r.

1. $C(0, 0)$; $r = 3$ 2. $C(2, -3)$; $r = \sqrt{2}$ Section 12-4

Find the radius and the coordinates of the center of the circle whose equation is given. Then graph the equation.

3. $(x-1)^2 + (y-2)^2 = 9$ 4. $x^2 + y^2 + 2y - 24 = 0$

5. A parabola has focus $F(0, 4)$ and directrix $y = 10$. State the coordinates of its vertex.   Section 12-5

6. A parabola has focus $F(0, -3)$ and vertex $V(0, -6)$. State an equation of its directrix.

7. Find an equation of the form $y = ax^2 + bx + c$ for the parabola with focus $F(0, 3)$ and directrix $y = 1$. Sketch the parabola.

8. a. Graph $y = x^2 + 4x - 5$.   Section 12-6
   b. On the same set of axes used in part (a), graph $x - y = 1$.
   c. Determine the coordinates of the points of intersection of the graphs in parts (a) and (b).

Solve the system and check the solutions (if they exist).

9. $x = y^2 - 9$
   $2y - x = 6$

10. $x^2 + y^2 = 5$
    $y = 3x - 5$

Section 12-7 THE ELLIPSE (Optional)

In this section you will learn how to write and graph an equation of an ellipse. You will also learn the meanings of several terms associated with an ellipse, including:

- foci
- focal radii
- intercepts
- major axis
- minor axis

If you tie the ends of a piece of string to two nails and place the nails at points F_1 and F_2, then the point P of a moving pencil that holds the string taut will trace a curve called an ellipse. Because the length of the string is $PF_1 + PF_2$, this sum is constant for all positions of P. This leads to the following definition.

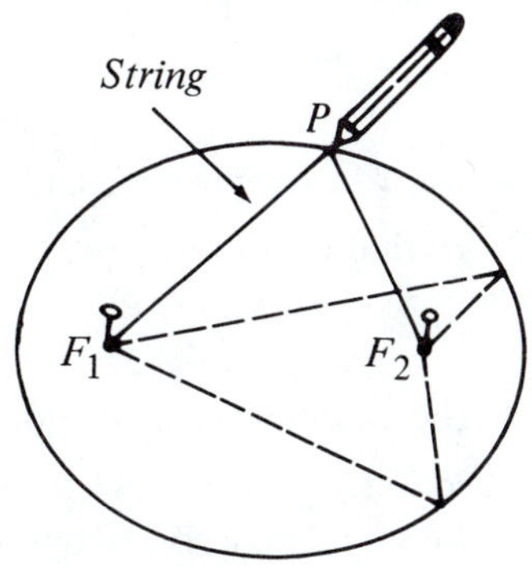

An *ellipse* is the locus of points such that the sum of the distances from any point in the locus to two fixed points is a constant. The fixed points are called the *foci* (plural of *focus*) of the ellipse and the lengths PF_1 and PF_2 are called the *focal radii* of the ellipse. (Note that the sum of the focal radii must be greater than the distance between the foci of an ellipse.)

To find an equation of an ellipse, apply the distance formula.

EXAMPLE Write an equation of the ellipse with foci at $F_1(-3, 0)$ and $F_2(3, 0)$ if the sum of the focal radii is 10.

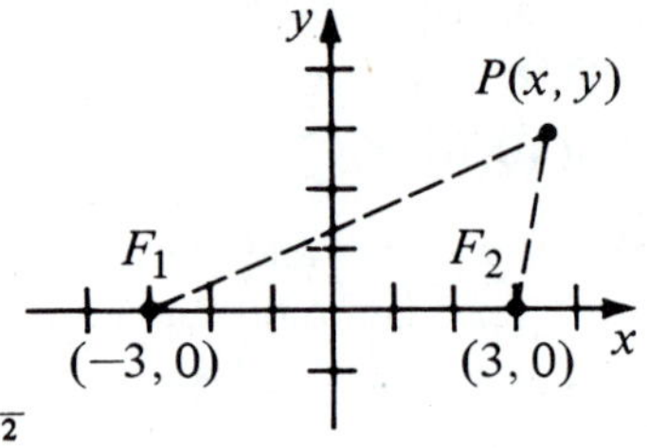

SOLUTION Let $P(x, y)$ be a point on the ellipse.
Use the distance formula to express PF_1 and PF_2 in terms of x and y.

$$PF_1 = \sqrt{(x-(-3))^2 + (y-0)^2} = \sqrt{(x+3)^2 + y^2}$$

$$PF_2 = \sqrt{(x-3)^2 + (y-0)^2} = \sqrt{(x-3)^2 + y^2}$$

$$PF_1 + PF_2 = 10$$

$$\sqrt{(x+3)^2 + y^2} + \sqrt{(x-3)^2 + y^2} = 10$$

$$\sqrt{(x-3)^2 + y^2} = 10 - \sqrt{(x+3)^2 + y^2}$$

Square each side and simplify.

$$(x-3)^2 + y^2 = 100 - 20\sqrt{(x+3)^2 + y^2} + (x+3)^2 + y^2$$

$$20\sqrt{(x+3)^2 + y^2} = 100 + 12x$$

Divide both sides by 4.

$$5\sqrt{(x+3)^2 + y^2} = 25 + 3x$$

Square each side again.

$$25[(x+3)^2 + y^2] = 625 + 150x + 9x^2$$

$$25x^2 + 150x + 225 + 25y^2 = 625 + 150x + 9x^2$$

$$16x^2 + 25y^2 = 400$$

In the long run it will be convenient to have the right-hand side of the equation equal to 1, so we divide both sides by 400.

$$\frac{x^2}{25} + \frac{y^2}{16} = 1$$

Now let's examine the graph of $\frac{x^2}{25} + \frac{y^2}{16} = 1$.

1. To find the x-intercepts, let $y = 0$ and solve for x.

$$\frac{x^2}{25} + \frac{0^2}{16} = 1; \quad x^2 = 25; \quad x = \pm 5$$

The x-intercepts are 5 and -5.

2. To find the y-intercepts, let $x = 0$ and solve for y.

$$\frac{0^2}{25} + \frac{y^2}{16} = 1; \quad y^2 = 16; \quad y = \pm 4$$

The y-intercepts are 4 and -4.

3. Solve for y in terms of x, as shown on the next page, and use a calculator to find the values of y corresponding to each of several values of x.

$$\frac{y^2}{16} = 1 - \frac{x^2}{25} = \frac{25 - x^2}{25}$$

$$y^2 = \frac{16(25 - x^2)}{25}$$

$$y = \pm\frac{4}{5}\sqrt{25 - x^2}$$

In the following table, the values of y have been rounded to the nearest tenth. The graph of $\frac{x^2}{25} + \frac{y^2}{16} = 1$ is shown at the right of the table.

x	y
±1	±3.9
±2	±3.7
±3	±3.2
±4	±2.4

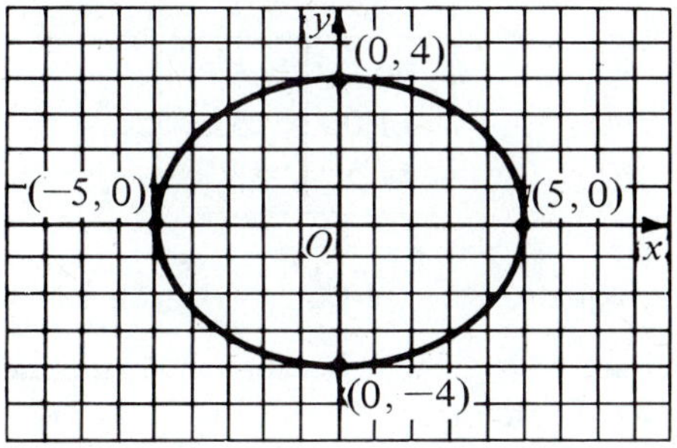

The segments that join the x-intercepts and the segments that join the y-intercepts are called the axes of the ellipse. The longer axis is referred to as the *major axis* and the shorter axis is referred to as the *minor axis* of the ellipse. The foci of an ellipse always lie on its major axis. In the ellipse illustrated, the length of the major axis is 10 and the length of the minor axis is 8.

You can use the method illustrated in the example to derive the general equation of an ellipse whose foci are on the x-axis. (See Exercise 35, page 504.)

An equation of the ellipse with foci $F_1(-c, 0)$ and $F_2(c, 0)$ and sum of the focal radii $2a$ is

$$\frac{x^2}{a^2} + \frac{y^2}{b^2} = 1,$$

where $b^2 = a^2 - c^2$.

If the foci of an ellipse lie on the y-axis, you can verify (Exercise 36, page 504) the following:

An equation of the ellipse with foci $F_1(0, -c)$ and $F_2(0, c)$ and sum of the focal radii $2a$ is

$$\frac{x^2}{b^2} + \frac{y^2}{a^2} = 1,$$

where $b^2 = a^2 - c^2$.

EXAMPLE Graph the ellipse $9x^2 + y^2 = 36$ and state the coordinates of its foci.

SOLUTION Divide each side of the equation by 36.

$$\frac{x^2}{4} + \frac{y^2}{36} = 1$$

1. The x-intercepts are 2 and -2; the y-intercepts are 6 and -6.
2. Solve for y in terms of x. $y = \pm 3\sqrt{4 - x^2}$
   Prepare a table of ordered pairs that satisfy the equation, rounding the values of y to the nearest tenth.

x	y
± 0.5	± 5.8
± 1.0	± 5.2
± 1.5	± 4.0

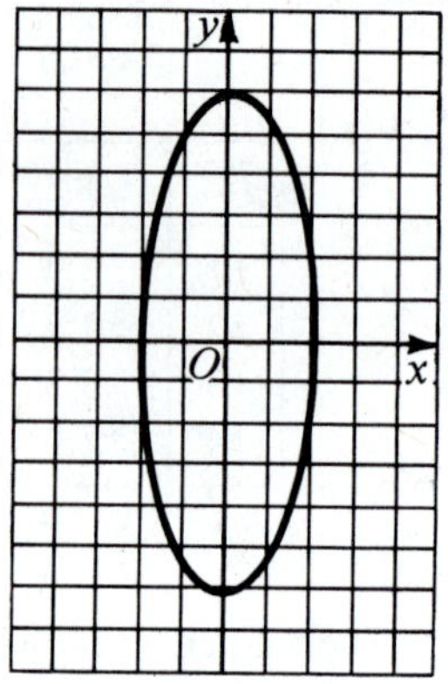

3. For the equation $\frac{x^2}{4} + \frac{y^2}{36} = 1$, you can see that the foci are on the y-axis because the major axis is on the y-axis. Therefore,

$$b^2 = 4$$
$$a^2 = 36$$

and thus,

$$4 = 36 - c^2$$
$$c^2 = 32$$
$$c = \pm\sqrt{32}$$
$$c = \pm 4\sqrt{2}$$

The coordinates of the foci are $(0, 4\sqrt{2})$ and $(0, -4\sqrt{2})$.

Of course, you don't have to draw an ellipse in order to find the coordinates of its foci.

EXAMPLE Find the coordinates of the foci of the ellipse

$$\frac{x^2}{64} + \frac{y^2}{25} = 1.$$

SOLUTION Since $b^2 = a^2 - c^2$, it follows that

$$c^2 = a^2 - b^2.$$

Because $c > 0$, $a^2 - b^2 > 0$, or $a^2 > b^2$.
Therefore, for the equation

$$\frac{x^2}{64} + \frac{y^2}{25} = 1$$

$a^2 = 64$ and $b^2 = 25$. The foci are on the x-axis.

$$c^2 = 64 - 25 = 39$$
$$c = \pm\sqrt{39}$$

The coordinates of the foci are $(\sqrt{39}, 0)$ and $(-\sqrt{39}, 0)$.

ORAL EXERCISES

For each ellipse whose equation is given, state the x-intercepts and the y-intercepts, and name the axis on which the foci lie.

1. $\frac{x^2}{100} + \frac{y^2}{64} = 1$
2. $\frac{x^2}{9} + \frac{y^2}{16} = 1$
3. $\frac{x^2}{25} + \frac{y^2}{36} = 1$
4. $\frac{x^2}{49} + \frac{y^2}{25} = 1$
5. $\frac{x^2}{10} + \frac{y^2}{5} = 1$
6. $\frac{x^2}{16} + \frac{y^2}{18} = 1$

Classify each of the following as an equation of an ellipse or an equation of a circle.

7. $\frac{x^2}{36} + \frac{y^2}{25} = 1$
8. $x^2 + y^2 = 25$
9. $x^2 + 9y^2 = 25$
10. $3x^2 + 3y^2 = 12$
11. $\frac{x^2}{12} + \frac{y^2}{12} = 1$
12. $16x^2 + 9y^2 = 64$

WRITTEN EXERCISES

In Exercises 1-12, **(a)** graph the equation and **(b)** give the coordinates of the foci.

A

1. $\frac{x^2}{25} + \frac{y^2}{9} = 1$
2. $\frac{x^2}{16} + \frac{y^2}{25} = 1$
3. $\frac{x^2}{16} + \frac{y^2}{9} = 1$
4. $\frac{x^2}{64} + \frac{y^2}{100} = 1$
5. $\frac{x^2}{49} + \frac{y^2}{36} = 1$
6. $\frac{x^2}{16} + \frac{y^2}{4} = 1$
7. $\frac{x^2}{81} + \frac{y^2}{121} = 1$
8. $\frac{x^2}{4} + \frac{y^2}{1} = 1$
9. $9x^2 + 36y^2 = 144$
10. $100x^2 + 25y^2 = 400$
11. $4x^2 + 25y^2 = 100$
12. $36x^2 + 25y^2 = 900$

Without graphing, find the coordinates of the foci of each ellipse.

13. $\frac{x^2}{225} + \frac{y^2}{289} = 1$
14. $\frac{x^2}{157} + \frac{y^2}{121} = 1$
15. $11x^2 + 7y^2 = 77$

Find an equation of the ellipse with the following properties.

16. Foci at $F_1(4, 0)$ and $F_2(-4, 0)$, and sum of the focal radii 10
17. Foci at $F_1(0, 8)$ and $F_2(0, -8)$, and sum of the focal radii 20
18. Foci at $F_1(12, 0)$ and $F_2(-12, 0)$, and sum of the focal radii 26

Find **(a)** the length of the major axis, and **(b)** the length of the minor axis of each ellipse.

19. $\frac{x^2}{16} + \frac{y^2}{49} = 1$
20. $\frac{x^2}{144} + \frac{y^2}{64} = 1$
21. $\frac{x^2}{20} + \frac{y^2}{10} = 1$
22. $\frac{x^2}{3} + \frac{y^2}{6} = 1$

In Exercises 23–26, **(a)** graph the equation and **(b)** give the coordinates of the foci.

B 23. $9x^2 + 4y^2 = 9$

24. $4x^2 + 25y^2 = 16$

25. $\frac{9x^2}{64} + \frac{y^2}{16} = 1$

26. $\frac{16x^2}{9} + \frac{y^2}{4} = 36$

In Exercises 27–34, an ellipse with center at the origin is described. Find an equation of the ellipse with the given properties.

27. The length of the major axis is 12, the length of the minor axis is 10, and the major axis lies on the y-axis.

28. The length of the major axis is 16, the length of the minor axis is 14, and the major axis lies on the x-axis.

29. The length of the major axis is 6 and a y-intercept is 2.

30. The length of the minor axis is 8 and a y-intercept is -7.

C 31. The foci are $F_1(0, 3)$ and $F_2(0, -3)$, and the y-intercepts are ± 6.

32. The foci are $F_1(2, 0)$ and $F_2(-2, 0)$, and the y-intercepts are ± 2.

33. The foci are $F_1(0, 4)$ and $F_2(0, -4)$, and the ellipse contains the point $\left(\frac{9}{5}, 4\right)$.

34. The foci are $F_1(8, 0)$ and $F_2(-8, 0)$, and the ellipse contains the point $\left(6, \frac{24}{5}\right)$.

35. Use the method illustrated on page 500 to show that an equation of the ellipse with foci $(c, 0)$ and $(-c, 0)$ and sum of focal radii $2a$ is

$$\frac{x^2}{a^2} + \frac{y^2}{a^2 - c^2} = 1.$$

36. Verify that an equation of the ellipse with foci $(0, c)$ and $(0, -c)$ and sum of focal radii $2a$ is $\frac{x^2}{a^2 - c^2} + \frac{y^2}{a^2} = 1$.

Section 12-8 THE HYPERBOLA (Optional)

In this section you will learn how to write and graph an equation of a hyperbola. You will also learn the meanings of several terms associated with a hyperbola, including:

- foci
- branches
- focal radii
- asymptotes
- intercepts

If you replace the word "sum" in the definition of an ellipse with the words "absolute value of the difference," you have the definition of a hyperbola. A *hyperbola* is the locus of points such that the absolute value of the difference of the distances from any point in the locus to two fixed points is a constant. The fixed points are called the *foci* of the hyperbola. Notice that the graph has two parts, or *branches*.

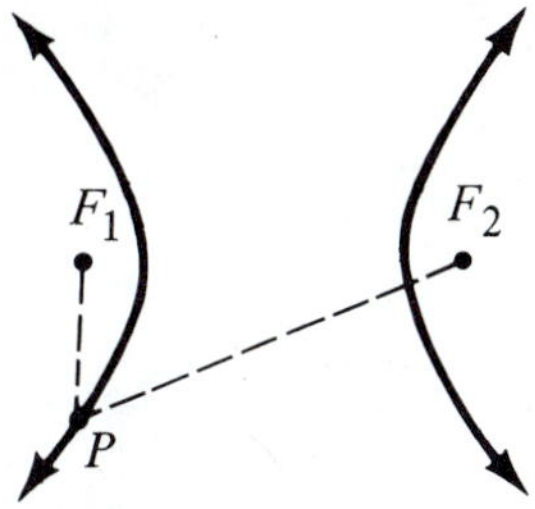

The diagram at the right shows a hyperbola with foci F_1 and F_2. By definition of a hyperbola,

$$|PF_2 - PF_1| = k$$

where k is a constant greater than 0. PF_1 and PF_2 are the *focal radii* of the hyperbola.

EXAMPLE Find an equation of the hyperbola with foci at $F_1(-5, 0)$ and $F_2(5, 0)$ if the difference of the focal radii is 8.

SOLUTION Let $P(x, y)$ be a point on the right-hand branch of the hyperbola. Use the distance formula to express PF_1 and PF_2 in terms of x and y.

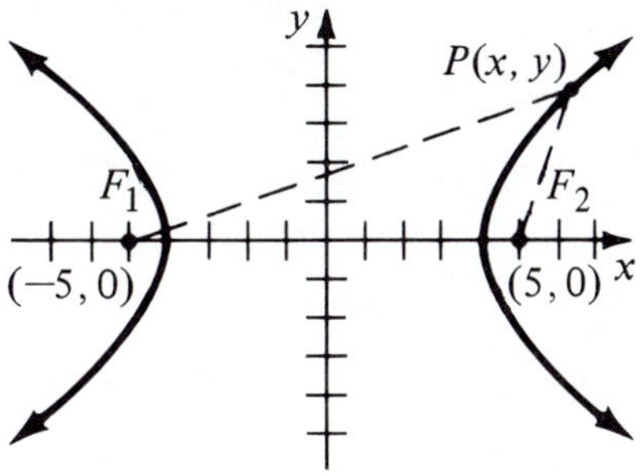

$$PF_1 = \sqrt{(x-(-5))^2 + (y-0)^2} = \sqrt{(x+5)^2 + y^2}$$
$$PF_2 = \sqrt{(x-5)^2 + (y-0)^2} = \sqrt{(x-5)^2 + y^2}$$

$$PF_1 \quad - \quad PF_2 \quad = 8$$
$$\sqrt{(x+5)^2 + y^2} - \sqrt{(x-5)^2 + y^2} = 8$$
$$\sqrt{(x+5)^2 + y^2} = \sqrt{(x-5)^2 + y^2} + 8$$

Square each side and simplify.

$$(x+5)^2 + y^2 = (x-5)^2 + y^2 + 16\sqrt{(x-5)^2 + y^2} + 64$$
$$20x - 64 = 16\sqrt{(x-5)^2 + y^2}$$

Divide each side by 4.

$$5x - 16 = 4\sqrt{(x-5)^2 + y^2}$$

Square each side again.

$$25x^2 - 160x + 256 = 16[(x-5)^2 + y^2]$$
$$25x^2 - 160x + 256 = 16x^2 - 160x + 400 + 16y^2$$
$$9x^2 - 16y^2 = 144$$

Divide both sides by 144.

$$\frac{x^2}{16} - \frac{y^2}{9} = 1$$

If we had let $P(x, y)$ be a point on the left-hand branch, we would have obtained the same result.

By examining the equation of a hyperbola, you can sketch its graph quickly. The example on the next page illustrates the method.

EXAMPLE Graph $\frac{x^2}{16} - \frac{y^2}{9} = 1$.

SOLUTION

1. To find the x-intercepts, let $y = 0$ and solve for x.

$$\frac{x^2}{16} - \frac{0^2}{9} = 1; \quad x^2 = 16; \quad x = \pm 4$$

2. To find the y-intercepts, let $x = 0$ and solve for y.

$$\frac{0^2}{16} - \frac{y^2}{9} = 1; \quad y^2 = -9$$

The hyperbola has no y-intercepts.

3. Solve the equation for y in terms of x.

$$\frac{x^2}{16} - \frac{y^2}{9} = 1$$

$$\frac{-y^2}{9} = 1 - \frac{x^2}{16}$$

$$y^2 = \frac{9(x^2 - 16)}{16}$$

$$y = \pm\frac{3}{4}\sqrt{x^2 - 16}$$

Therefore, y is a real number if and only if $x^2 - 16 \geqslant 0$. That is, y is real if and only if $x \geqslant 4$ or $x \leqslant -4$. Thus, there are no real values of y for $-4 < x < 4$, and consequently no part of the hyperbola lies between the lines with equations $x = -4$ and $x = 4$.

4. For $y = \pm\frac{3}{4}\sqrt{x^2 - 16}$, observe what happens as $|x|$ increases. As $|x|$ increases, -16 becomes negligible in relation to the value of $x^2 - 16$. Thus, the graph of the equation $y = \pm\frac{3}{4}\sqrt{x^2 - 16}$ approaches the graph of $y = \pm\frac{3}{4}\sqrt{x^2}$, or $y = \pm\frac{3}{4}x$. The lines whose equations are

$$y = \frac{3}{4}x \quad \text{and} \quad y = -\frac{3}{4}x$$

are called the *asymptotes* of the hyperbola.

5. Make a brief table of coordinates, rounding y-coordinates to the nearest tenth, and use the preceding facts to graph the hyperbola, as shown below. The dashed lines represent the asymptotes of the hyperbola.

x	y
± 5	± 2.3
± 6	± 3.4
± 7	± 4.3

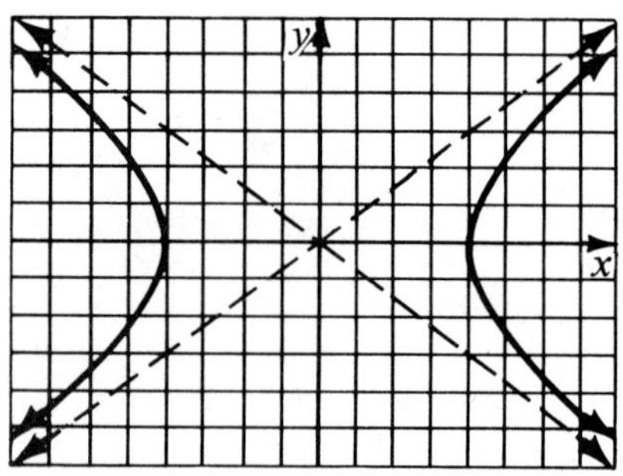

You can use the method illustrated in the first example to derive a general equation of a hyperbola. (See Exercise 31, page 509.)

An equation of the hyperbola with foci $F_1(-c, 0)$ and $F_2(c, 0)$ and difference of focal radii $2a$ is

$$\frac{x^2}{a^2} - \frac{y^2}{b^2} = 1, \text{ where } b^2 = c^2 - a^2.$$

The equations of the asymptotes are $y = \pm\frac{b}{a}x$.

If the foci of the hyperbola lie on the y-axis, you can verify (Exercise 32, page 510) the following.

An equation of the hyperbola with foci $F_1(0, -c)$ and $F_2(0, c)$ and difference of focal radii $2a$ is

$$\frac{y^2}{a^2} - \frac{x^2}{b^2} = 1, \text{ where } b^2 = c^2 - a^2.$$

The equations of the asymptotes are $y = \pm\frac{a}{b}x$.

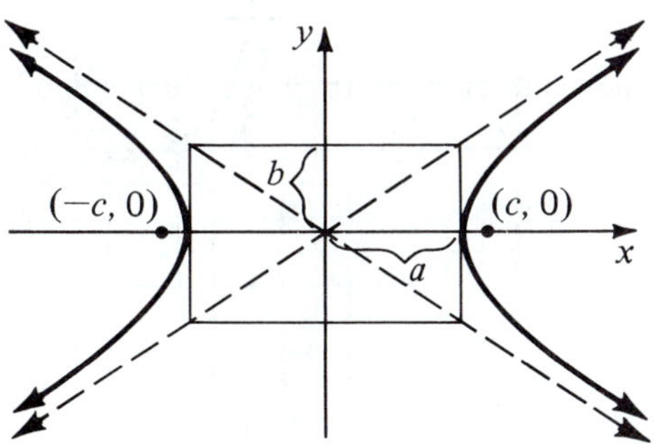

$$\frac{x^2}{a^2} - \frac{y^2}{b^2} = 1$$

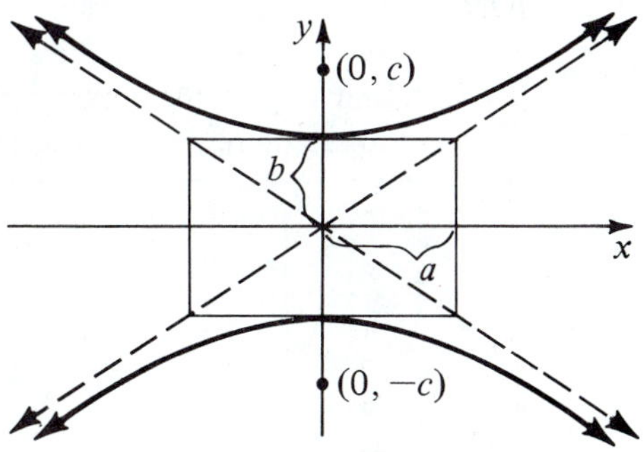

$$\frac{y^2}{a^2} - \frac{x^2}{b^2} = 1$$

EXAMPLE Graph the hyperbola $\frac{y^2}{4} - \frac{x^2}{16} = 1$, showing its asymptotes.

SOLUTION 1. There are no x-intercepts.

2. The y-intercepts are 2 and -2.

(Solution continued on the next page.)

3. $\frac{y^2}{4} = 1 + \frac{x^2}{16}$; $y^2 = \frac{x^2 + 16}{4}$; $y = \pm\frac{1}{2}\sqrt{x^2 + 16}$

$\frac{x^2}{16} = \frac{y^2}{4} - 1$; $x^2 = 4y^2 - 16$; $x = \pm 2\sqrt{y^2 - 4}$

Therefore, x is a real number if and only if $y \geqslant 2$ or $y \leqslant -2$.

4. $a = \pm 2$, $b = \pm 4$; the asymptotes are $y = \pm\frac{1}{2}x$.

5.

x	y
0	±2.0
±3	±2.5
±5	±3.2
±7	±4.0

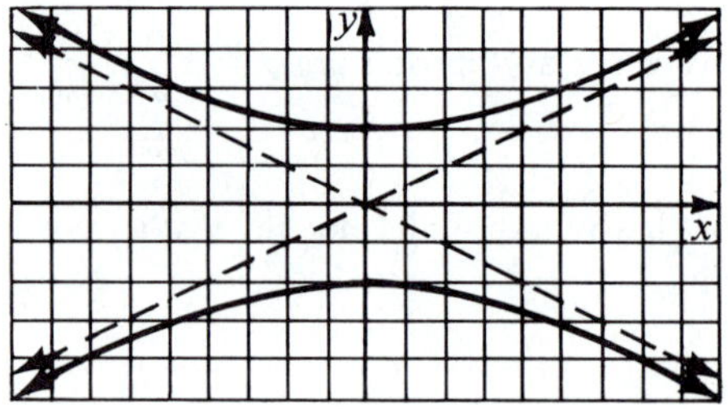

Another kind of quadratic equation that often arises is of the form $xy = k$, where k is a nonzero constant. The graph of this equation also turns out to be a hyperbola. If $k > 0$, the branches of the hyperbola lie in the first and third quadrants. If $k < 0$, the branches lie in the second and fourth quadrants. In both cases, the x-axis and the y-axis are the asymptotes of the hyperbola.

EXAMPLE Graph $xy = 4$.

SOLUTION The easiest way to graph a hyperbola of the form $xy = k$ is to make a table of values for x and y, plot the points, and draw the graph through these points as shown.

x	y
0.5	8
1	4
2	2
8	0.5

x	y
−0.5	−8
−1	−4
−2	−2
−8	−0.5

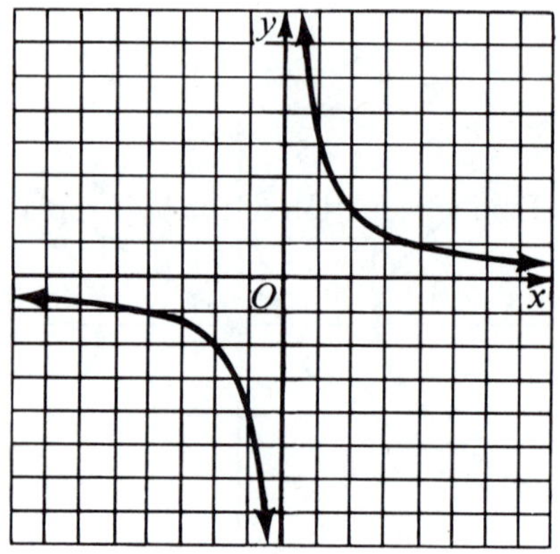

ORAL EXERCISES

Classify each equation as an equation of a parabola, circle, ellipse, or hyperbola.

1. $x^2 + y^2 = 9$
2. $x^2 - y^2 = 12$
3. $x^2 + y = 9$
4. $x^2 + 3y^2 = 27$
5. $\frac{x^2}{25} + \frac{y^2}{16} = 1$
6. $\frac{x^2}{3} + \frac{y^2}{3} = 3$
7. $y = x^2 + 7$
8. $x^2 + y^2 = 7$
9. $\frac{x^2}{9} - \frac{y^2}{4} = 1$
10. $xy = 12$
11. $9x^2 - 2y = 18$
12. $2x^2 + 3y^2 = 8$

WRITTEN EXERCISES

For each hyperbola, find **(a)** the intercepts, **(b)** the coordinates of the foci, **(c)** the equations of the asymptotes. Then graph the equation and indicate the asymptotes with dashed lines.

A 1. $\frac{y^2}{64} - \frac{x^2}{36} = 1$ 2. $\frac{y^2}{16} - \frac{x^2}{9} = 1$ 3. $\frac{x^2}{9} - \frac{y^2}{4} = 1$

4. $\frac{x^2}{4} - \frac{y^2}{4} = 1$ 5. $\frac{x^2}{4} - \frac{y^2}{16} = 1$ 6. $\frac{y^2}{16} - \frac{x^2}{25} = 1$

7. $\frac{y^2}{9} - \frac{x^2}{9} = 1$ 8. $\frac{x^2}{4} - \frac{y^2}{1} = 1$ 9. $\frac{y^2}{4} - \frac{x^2}{5} = 1$

10. $\frac{x^2}{9} - \frac{y^2}{7} = 1$ 11. $\frac{y^2}{12} - \frac{x^2}{4} = 1$ 12. $\frac{x^2}{24} - \frac{y^2}{12} = 1$

Find an equation of the hyperbola with the following properties.

13. Foci at $F_1(0, 5)$ and $F_2(0, -5)$, and difference of focal radii 6

14. Foci at $F_1(13, 0)$ and $F_2(-13, 0)$, and difference of focal radii 10

Plot a few points and sketch the graph of each equation.

15. $xy = 5$ 16. $xy = 8$ 17. $xy = -6$ 18. $xy = -10$

Graph the equation and indicate the asymptotes with dashed lines.

B 19. $9x^2 = 4y^2 + 36$ 20. $25y^2 - 225 = 9x^2$

21. $64 = -4x^2 + 64y^2$ 22. $25x^2 + 275 = 11y^2$

Find an equation of the hyperbola with the following properties.

23. y-intercepts of ± 2, and equations of asymptotes $y = \pm\frac{2}{5}x$

24. x-intercepts of ± 3, and equations of asymptotes $y = \pm\frac{5}{3}x$

25. x-intercepts of ± 5, and foci at $F_1(6, 0)$ and $F_2(-6, 0)$

26. y-intercepts of $\pm\sqrt{3}$, and foci at $F_1(0, 3)$ and $F_2(0, -3)$

27. $a = 11$, $b = 8$, and foci on the y-axis

28. $a = 4$, $b = 2\sqrt{2}$, and foci on the x-axis

C 29. Foci at $F_1(0, 2\sqrt{2})$ and $F_2(0, -2\sqrt{2})$, and equations of asymptotes $y = \pm x$

30. Foci at $F_1(5\sqrt{2}, 0)$ and $F_2(-5\sqrt{2}, 0)$, and equations of asymptotes $y = \pm\frac{4}{3}x$

31. Use the methods of the example in the text on page 505 to show that an equation of the hyperbola with foci $(c, 0)$ and $(-c, 0)$ and absolute value of the difference of the focal radii $2a > 0$, where $a < c$, is $\frac{x^2}{a^2} - \frac{y^2}{c^2 - a^2} = 1$.

32. Verify that an equation of the hyperbola with foci $(0, c)$ and $(0, -c)$ and absolute value of the difference of the focal radii $2a > 0$, where $a < c$, is
$\frac{y^2}{a^2} - \frac{x^2}{c^2 - a^2} = 1.$

SELF-TEST 3 (Optional)

In Exercises 1 and 2, **(a)** graph the equation and **(b)** give the coordinates of the foci.

1. $\frac{x^2}{100} + \frac{y^2}{36} = 1$ 2. $16x^2 + 9y^2 = 144$ Section 12-7

3. An ellipse has foci (0, 15) and (0, −15). The sum of its focal radii is 34. Find an equation of the ellipse.

For each hyperbola, find **(a)** the intercepts, **(b)** the coordinates of the foci, **(c)** the equations of the asymptotes. Then graph the equation.

4. $\frac{x^2}{36} - \frac{y^2}{9} = 1$ 5. $\frac{x^2}{9} - \frac{y^2}{16} = 1$ Section 12-8

6. A hyperbola has foci (6, 0) and (−6, 0). The difference of its focal radii is 8. Find an equation of the hyperbola.

7. Plot a few points and sketch the graph of $xy = 6$.

Computer Exercises

1. It is easy to write a program that will determine the center and radius of a circle. The circle with equation $(x - h)^2 + (y - k)^2 = r^2$ has center (h, k) and radius r. The center and radius of the circle with equation $x^2 + y^2 + dx + ey + f = 0$ can be found by completing the square for both the x and y terms.

   a. Verify that the coordinates of the center of the circle $x^2 + y^2 + dx + ey + f = 0$ are $\left(-\frac{d}{2}, -\frac{e}{2}\right)$ and that the radius is $\frac{1}{2}\sqrt{d^2 + e^2 - 4f}$.

   Copy and RUN the program for the circles with the given equations.

```
10   PRINT "ENTER D, E, AND F FROM"
20   PRINT "X↑2 + Y↑2 + DX + EY + F = 0";
30   INPUT D,E,F
40   PRINT "THE COORDINATES OF THE CENTER ARE   ";
50   PRINT "(";-D/2;",";-E/2;") AND THE RADIUS IS";
60   PRINT .5*SQR(D*D+E*E-4*F);"."
70   END
```

   b. $x^2 + y^2 - 6x + 4y + 9 = 0$
   c. $x^2 + y^2 + 10x + 2y + 1 = 0$
   d. $x^2 + y^2 + x - 6y + 0.25 = 0$
   e. $x^2 + y^2 + 3x + 2y + 1 = 0$
   f. $x^2 + y^2 - 8x - 9 = 0$
   g. $x^2 + y^2 + 10x - 14y + 25 = 0$

2. Since the radius of a circle must be a *positive real* number,

$$x^2 + y^2 + dx + ey + f = 0$$

is an equation of a circle if and only if $d^2 + e^2 - 4f > 0$. If $d^2 + e^2 - 4f < 0$, the computer will print an error message when it reaches line 60 of the program in Exercise 1 since the computer cannot take the square root of a negative number. (If $d^2 + e^2 - 4f = 0$, the computer will not print an error message even though the equation does not represent a circle.)

Modify the program in Exercise 1 to determine whether $d^2 + e^2 - 4f \leqslant 0$ and, if so, to print the message "NOT A CIRCLE." Test your program with the following equations.

a. $x^2 + y^2 - 4x + 10y + 36 = 0$
b. $x^2 + y^2 - 6y + 49 = 0$
c. $x^2 + y^2 - 4x - 4y - 8 = 0$
d. $x^2 + y^2 + 0.1x - 0.1y + 0.005 = 0$

3. The program that follows is designed to find the coordinates of the vertex and the focus of a parabola. You may recall from Section 12-2 that the x-coordinate of the *vertex* of the parabola $y = ax^2 + bx + c$ $(a \neq 0)$ is $-\frac{b}{2a}$. By substituting $-\frac{b}{2a}$ for x in $y = ax^2 + bx + c$, you can find the y-coordinate of the *vertex*.

The x-coordinate of the *focus* of $y = ax^2 + bx + c$ is the same as the x-coordinate of the vertex. In Exercise 37 of Section 12-5 you were asked to show that a parabola with vertex (h, k) and focus $(h, k + r)$ has an equation of the form $y = \frac{1}{4r}(x - h)^2 + k$. Since an equation of the form $y = ax^2 + bx + c$ is equivalent to an equation of the form $y = a(x - h)^2 + k$, you can see that $a = \frac{1}{4r}$, and hence $r = \frac{1}{4a}$. Thus the y-coordinate of the *focus* of a parabola is $k + \frac{1}{4a}$.

Copy and RUN the program for the parabolas with the given equations.

```
10   PRINT "ENTER A, B, AND C FROM"
20   PRINT "Y = AX↑2 + BX + C";
30   INPUT A,B,C
40   LET H=-B/(2*A)
50   LET K=A*H↑2+B*H+C
60   PRINT "THE COORDINATES OF THE VERTEX ARE  ";
70   PRINT "(";H;",";K;")."
80   LET F=K+1/(4*A)
90   PRINT "THE COORDINATES OF THE FOCUS ARE  ";
100  PRINT "(";H;",";F;")."
110  END
```

a. $y = x^2 + 16x + 62$
b. $y = 4x^2 + 8x + 1$
c. $y = -x^2 + 2x + 6$
d. $y = x^2 + x + 0.25$
e. $y = 0.25x^2 - x$
f. $y = -2x^2 + 6x - 3$

CHAPTER REVIEW

In Exercises 1 and 2:

a. Make a table of values of y for integral values of x in the given interval.
b. Use the results of part (a) to graph the equation.
c. Draw the axis of symmetry and state its equation.
d. Label the vertex as V and state its coordinates.
e. State whether the vertex is a maximum or a minimum point.

1. $y = -2x^2 - 1;\ -2 \leqslant x \leqslant 2$ 2. $y = (x-2)^2 + 3;\ -1 \leqslant x \leqslant 5$ Section 12-1

3. Write an equivalent equation in the form $y = a(x-h)^2 + k$ for $y = x^2 - 6x + 3$. State the coordinates of the vertex of the corresponding parabola. Section 12-2

4. a. Graph $y = x^2 + 2x + 2$.
   b. Find an equation for the axis of symmetry of the parabola in part (a).

5. An equation of the axis of symmetry of the parabola $y = 2x^2 + bx + 7$ is $x = -2$. Find the value of b.

6. Solve $x^2 + 4x + 4 = 0$ and illustrate with a graph. Section 12-3

7. Use your graph from Exercise 6 to solve the equation $x^2 + 4x + 4 = 1$.

8. Write an equation of the circle with the given center C and the given radius r. Section 12-4
   a. $C(0, 2),\ r = 3\sqrt{2}$ b. $C(2, -3),\ r = 5$

Find the radius and the coordinates of the center of the circle whose equation is given.

9. $(x-4)^2 + (y+2)^2 = 4$ 10. $x^2 + y^2 + 6x + 8y - 11 = 0$

11. Find the radius of the circle with center $(-3, 0)$ and containing the point $(3, -8)$. Then write an equation of the circle in center-radius form.

12. A parabola has focus $F(0, -1)$ and directrix $y = 7$. State the coordinates of its vertex. Section 12-5

13. A parabola has vertex $V(0, 8)$ and directrix $y = 12$. State the coordinates of its focus.

14. Find an equation of the form $y = ax^2 + bx + c$ for the parabola with focus $(-1, 4)$ and directrix $y = -2$. Sketch the parabola.

15. a. Graph $x^2 + y^2 = 25$. Section 12-6
    b. On the same set of axes used in part (a), graph $y = 3x - 15$.
    c. Determine the coordinates of the points of intersection of the graphs in parts (a) and (b).

16. Solve and check the solutions.

$$y = x^2 - 4x + 2$$
$$x - y = 2$$

For each ellipse find **(a)** the intercepts, **(b)** the coordinates of the foci, and **(c)** the length of the major axis. Then graph the equation.

17. $\frac{x^2}{9} + \frac{y^2}{25} = 1$

18. $\frac{x^2}{9} + \frac{y^2}{4} = 1$ Section 12-7 (Optional)

19. An ellipse has foci at $(0, 3)$ and $(0, -3)$. The sum of its focal radii is 10. Find an equation of the ellipse.

For the hyperbolas in Exercises 20 and 21, find **(a)** the intercepts, **(b)** the coordinates of the foci, and **(c)** the equations of the asymptotes. Then graph the equation.

20. $\frac{x^2}{9} - \frac{y^2}{27} = 1$

21. $y^2 - x^2 = 1$ Section 12-8 (Optional)

22. Graph $xy = -12$.

CHAPTER TEST

1. **a.** Graph $y = x^2 - 6x + 4$ using all integral values of x from 0 to 6 inclusive.
   b. On the same set of axes used in part (a), graph $y = -4$.
   c. Determine the coordinates of the points of intersection of the graphs in parts (a) and (b).
2. Write an equivalent equation in the form $y = a(x - h)^2 + k$ for $y = x^2 + 2x - 6$. State the equation of the axis of symmetry of the corresponding parabola.
3. Find the coordinates of the vertex of the parabola $y = -2x^2 + 12x + 4$.
4. An equation of the axis of symmetry of the parabola $y = ax^2 + 30x + 27$ is $x = -5$. Find the value of a.
5. Find an equation of the parabola shown at the right. The vertex of the parabola is $(0, 0)$.

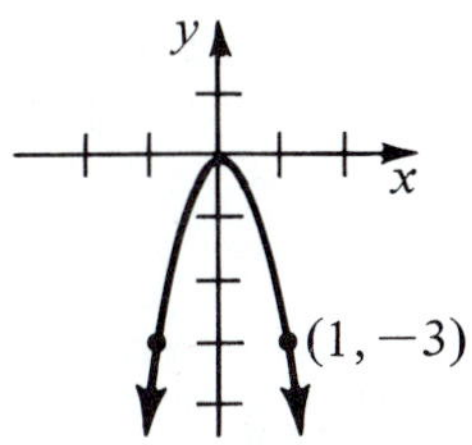

6. Find the radius and the coordinates of the center of the circle whose equation is $(x + 1)^2 + y^2 = 100$.
7. Find an equation of the circle with center at the origin and radius $\sqrt{3}$.
8. Find the radius and the coordinates of the center of the circle whose equation is $x^2 + y^2 - 10x - 4y - 52 = 0$.
9. **a.** Find the radius of the circle with center $(-3, 1)$ and containing the point $(-6, 1)$.
   b. Find an equation in center-radius form of the circle in part (a).

Solve each system and illustrate with a graph.

10. $x^2 + y^2 = 13$
    $y - x = 1$

11. $y = -x^2 + 5x - 8$
    $x - y = 4$

12. a. State the coordinates of the vertex of the parabola with focus $F(2, 0)$ and directrix $y = 4$.
    b. Find an equation of the form $y = ax^2 + bx + c$ of the parabola in part (a).

13. (Optional) Find **(a)** the intercepts, **(b)** the lengths of the major and minor axes, and **(c)** the coordinates of the foci of the ellipse with equation $\frac{x^2}{4} + \frac{y^2}{25} = 1$.

14. (Optional) An ellipse has foci at $(8, 0)$ and $(-8, 0)$. The sum of its focal radii is 20. Find an equation of the ellipse.

15. (Optional) A hyperbola has equation $\frac{x^2}{9} - y^2 = 1$. Find **(a)** the intercepts, **(b)** the coordinates of the foci, **(c)** the difference of the focal radii, and **(d)** the equations of the asymptotes. Then graph the hyperbola.

Application — Bridge Construction

Bridges are subject to both internal and external stresses. If they are not properly designed their weight alone can pull them apart, but bridges must also be able to withstand the forces of weather and heavy traffic. Engineers have used a number of well-known geometric shapes, such as triangles, parabolas, and trapezoids, to give their bridges strength and stability.

Triangles are called rigid figures because you cannot change the shape of a triangle without changing the length of at least one of its sides. As a result, triangular constructions are strong and stable. You can see triangles in the designs of many bridges. Notice the triangles in the design of the bridge shown below.

Engineers use the shape of a parabola in designing certain suspension bridges and concrete bridges that are supported from below. Notice that the vertical lines you can see in these bridge designs form parallel sides of figures that approximate trapezoids.

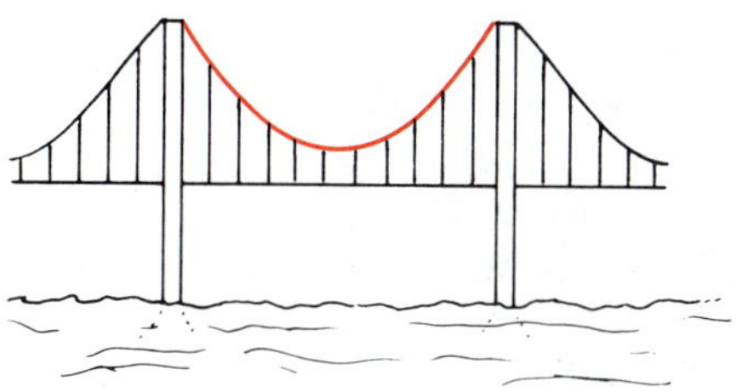

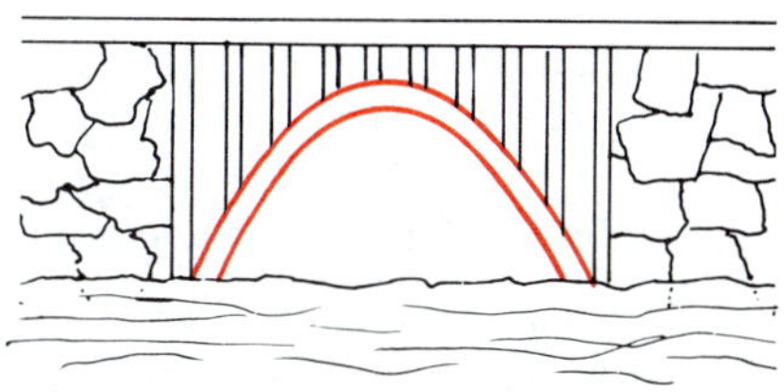

Trestles, which consist of a horizontal beam held up by supports that slant outward at the base, also resemble trapezoids, and are used to support bridges.

Look for pictures of bridges in books or magazines, and draw or take pictures of nearby bridges. Identify the geometric shapes used in their construction.

VOCABULARY REVIEW

Be sure that you understand the meaning of these terms:

parabola, p. 476
axis of symmetry, p. 476
vertex, p. 476
directrix, p. 492
focus of a parabola, p. 492

The following terms appear in Sections 12-7 and 12-8 (optional):

ellipse, p. 499
foci of an ellipse, p. 499
focal radii of an ellipse, p. 499
axes of an ellipse, p. 501
hyperbola, p. 504
foci of a hyperbola, p. 505
focal radii of a hyperbola, p. 505
asymptotes, p. 505

MIXED REVIEW

1. Find the radius and center of the circle with equation $(x-2)^2 + (y+3)^2 = 16$.
2. In the figure, $\overline{QT} \parallel \overline{RS}$. Find the value of x.

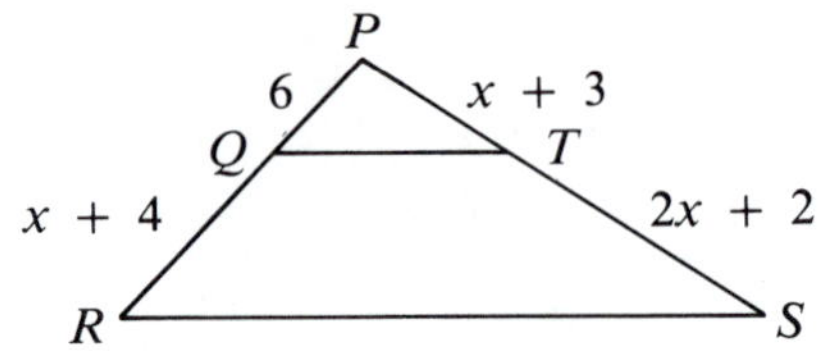

3. Quadrilateral $ABCD$ has vertices at $A(0, 0)$, $B(6, 1)$, $C(5, 4)$ and $D(-1, 3)$.
   a. Show that $\overline{BC} \cong \overline{AD}$.
   b. Show that $\overline{BC} \parallel \overline{AD}$.
   c. What kind of quadrilateral is $ABCD$? Explain.
4. Find the measure of each interior angle of a regular polygon with 30 sides.

Solve.

5. $y = x^2 - 2x - 3$
   $2x + y - 1 = 0$
6. $5y - 8x = -2$
   $12y - 6x + 18 = 0$
7. $\dfrac{2}{x+1} + \dfrac{1}{2} = \dfrac{2x+1}{10}$

8. Graph the line that contains the point $(3, -2)$ and has slope -1.
9. Is the following argument valid? If so, identify the rule of inference used. If not, identify the fallacy committed.

   P_1: If toves are slithy, then it is brillig.
   P_2: It is not brillig.
   C: Toves are not slithy.

10. Draw a line segment. Divide the line segment into 3 congruent segments.
11. The diagonal of a square is $\frac{4}{3}$ units long. Find the length of a side.

12. Determine without graphing if the system is inconsistent, consistent and dependent, or consistent and independent: $\frac{1}{2}x + 3y = 1$
$2x + 9y = 7$

13. Write an equation in center-radius form of the circle with center $\left(0, -\frac{1}{2}\right)$ and radius $3\sqrt{2}$.

14. Draw an acute triangle, $\triangle ABC$. Construct the locus of points equidistant from the sides of $\angle A$ and from the sides of $\angle B$.

15. Find the value of x.

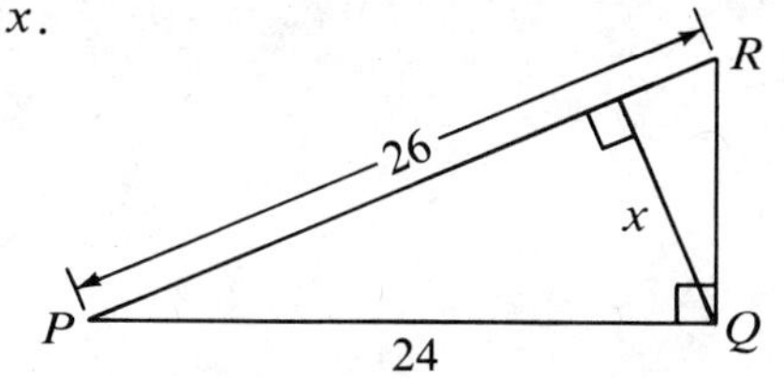

16. The areas of two similar octagons are in the ratio 9:25. The longest side of the larger octagon is 45 cm long. Find the length of the longest side of the smaller octagon.

17. In $\triangle DEF$, $m\angle E = 90$, $DE = 56$, and $EF = 74$. Find $m\angle F$ to the nearest degree.

18. Simplify: $\sqrt{20a^2b^3c^5}$

19. Graph: $6x - 5y = 15$

20. In quadrilateral $DEFG$, the diagonals intersect at X. If X is the midpoint of $\overline{DF}$ and $\overline{EG}$, is $DEFG$ a parallelogram? Explain.

21. Find the x-intercepts of the parabola with equation $y = 2x^2 + 5x - 3$.

22. True or false? If $\overline{AC} \perp \overline{BD}$ in quadrilateral $ABCD$, $ABCD$ is a rhombus.

23. Find an equation in the form $y = ax^2 + bx + c$ of the parabola with focus $F(1, 3)$ and vertex $V(1, -2)$.

24. Construct a $37\frac{1}{2}^\circ$ angle.

25. (Optional) Graph the equation $\frac{x^2}{9} + \frac{y^2}{100} = 1$ and give the coordinates of the foci.

26. (Optional) For the hyperbola with equation $\frac{x^2}{16} - \frac{y^2}{4} = 1$:

a. Find the intercepts.

b. Find the coordinates of the foci.

c. Find the equations of the asymptotes.

d. Graph the equation and show the asymptotes with dashed lines.

Probability theory is often used in analyzing situations involving waiting in line, such as cars waiting at a toll booth. An analysis of this type helps those operating the toll system make effective decisions regarding the number of toll booths to keep open.

Probability, Permutations, and Combinations 13

From worrying about the weather to participating in games of chance, people have long made intuitive use of the notion of probability. A simple rule in the related subject of combinations was first stated by Boethius around the year 510.

In 1494 Pacioli proposed the now famous "Problem of the Points," a question involving the division of stakes between two participants in an interrupted game of chance. Cardano (Cardan), Tartaglia, and Galileo each considered this or related problems. Around 1654 the Chevalier de Méré asked Pascal to solve a version of the Problem of the Points. Pascal was sufficiently intrigued to discuss the problem with Fermat, and in solving the problem the two developed some of the first general laws of probability.

During the next two hundred years great advances were made in both the theory and the applications of probability. The first texts on the theory were written by Jakob (Jacques) Bernoulli in 1713 and by De Moivre in 1718. Perhaps the most important probability text was written nearly one hundred years later by Laplace.

Today the theory of probability is applied in many fields besides recreational games, insurance, and horticulture. For example, water, gas, electric, and telephone companies need to calculate probable usage levels in order to meet varying service demands. It is hard to find areas of modern life not affected by our applications of the mathematical theory of probability.

Section 13-1 SAMPLE SPACES

In this section you will analyze some simple random experiments such as tossing coins and throwing dice. You will learn the meaning of these terms:

- random experiment
- outcome
- sample space
- event

A *random experiment* is an experiment in which you do not necessarily get the same results when you repeat the experiment under essentially the same conditions. The set of all possible *outcomes* is called the *sample space* of the experiment.

EXAMPLE Experiment: A single die is thrown and the number of spots showing is recorded.

Sample space: { 1, 2, 3, 4, 5, 6 }

An *event* is a subset of the sample space for a random experiment. When an event corresponds to a single outcome of the experiment, it is often called a *simple event.*

EXAMPLE Specify the following events for the experiment of throwing a single die.

a. An even number is rolled.
b. A number less than 3 is rolled.

SOLUTION a. { 2, 4, 6 } b. { 1, 2 }

You must be careful to list all the outcomes in the sample space, even for relatively simple experiments. Consider the experiment of tossing two coins at once and observing whether they show heads or tails. You can analyze this experiment by stating that one of these three events will occur: both heads, both tails, or one head and one tail. To construct the sample space, however, you must realize that the third event corresponds to two outcomes of the experiment, as the following example shows.

EXAMPLE A penny and a nickel are tossed and whether they show heads or tails is observed. Specify the following:

a. The sample space for the experiment
b. The event that one coin shows heads and the other shows tails
c. The event that at least one coin shows heads

SOLUTION a. The outcomes are shown in the table at the right. We will write the event that the penny shows heads and the nickel shows tails as (H, T). The sample space is {(H, H), (H, T), (T, H), (T, T)}.

b. { (H, T), (T, H) }
c. { (H, H), (H, T), (T, H) }

Penny	*Nickel*
H	H
H	T
T	H
T	T

Replacing the nickel with a second penny makes the two-coin toss more deceptive, but it doesn't change the sample space. In fact, the sample space and events listed in the example apply to any two-coin toss. Just think of the first coin as being a penny and the second coin as being a nickel.

The next example considers the experiment of throwing two dice. When two dice are thrown, we are usually interested in the total number of spots showing. For example, when one die shows 3 spots and the other shows 2 spots, we say that a 5 has been rolled.

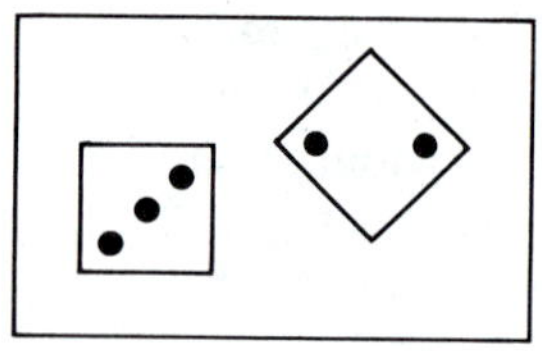

EXAMPLE Two dice are thrown. Specify the following:

a. The sample space
b. The event that an 11 is rolled
c. The event that a 4 is rolled

SOLUTION **a.** To construct the sample space for this experiment, you can think of one die as being red and the other as being green. Then, for example, there are two ways to roll a 2 and a 5—a red 2 and a green 5, or the other way around.

We can write these outcomes as (2, 5) and (5, 2), respectively. Using this ordered-pair notation, we can show the 36 outcomes that make up the sample space:

(1, 1)	(1, 2)	(1, 3)	(1, 4)	(1, 5)	(1, 6)
(2, 1)	(2, 2)	(2, 3)	(2, 4)	(2, 5)	(2, 6)
(3, 1)	(3, 2)	(3, 3)	(3, 4)	(3, 5)	(3, 6)
(4, 1)	(4, 2)	(4, 3)	(4, 4)	(4, 5)	(4, 6)
(5, 1)	(5, 2)	(5, 3)	(5, 4)	(5, 5)	(5, 6)
(6, 1)	(6, 2)	(6, 3)	(6, 4)	(6, 5)	(6, 6)

The sample space remains the same, of course, even for two dice of the same color.

b. $\{(5, 6), (6, 5)\}$ **c.** $\{(1, 3), (2, 2), (3, 1)\}$

ORAL EXERCISES

1. Each of the letters E, F, G, H, and I is written on a separate card. The cards are shuffled, and a single card is drawn. For this experiment specify each of the following:
   a. The sample space
   b. The event that a consonant is drawn
   c. The event that the seventh letter of the alphabet is drawn

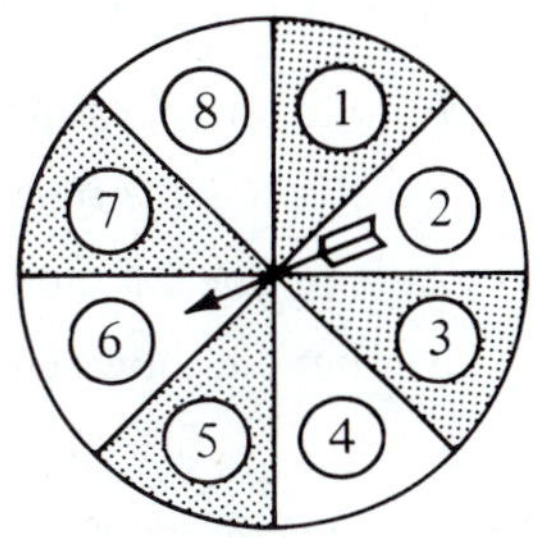

2. The pointer on the spinner shown at the right is spun in such a way that it does not stop on a line. For this experiment specify each of the following:
   a. The sample space
   b. The event that the spinner stops on a number divisible by 3
   c. The event that the spinner stops in one of the shaded regions

WRITTEN EXERCISES

A 1. Each of the letters M, N, O, P, Q, R, S, T, U, and V is written on a separate card. The cards are shuffled, and a single card is drawn. For this experiment specify each of the following:
   a. The sample space
   b. The event that a vowel is drawn
   c. The event that the twentieth letter of the alphabet is drawn
   d. The event that the letter drawn does not appear in the word QUART

2. Each of the integers from 10 to 20, inclusive, is written on a separate card. The cards are shuffled, and a single card is drawn. For this experiment specify each of the following:
   a. The sample space
   b. The event that an even number is drawn
   c. The event that a multiple of 5 is drawn
   d. The event that a multiple of 3 is drawn

In Exercises 3 and 4, refer to the two-coin toss discussed on page 520. Specify each of the events described.

3. At most one coin shows heads.

4. At least one coin shows tails.

In Exercises 5-7, refer to the following experiment: Keiko has five coins in his pocket: a penny, a nickel, a dime, a quarter, and a half dollar. Simultaneously, he takes two of the coins out of his pocket. Specify each of the following, for this experiment:

5. The sample space

6. The event that the total value of the coins is less than 30¢

7. The event that the total value of the coins exceeds 25¢ but is less than 60¢

In Exercises 8-17, refer to the experiment of throwing two dice discussed on pages 520 and 521. Specify each of the events described.

8. A 5 is rolled. (Recall that this event means that the sum of the spots showing on both dice is 5.)

9. A 6 is rolled.

10. A 7 is rolled.

11. An 8 is rolled.

12. Doubles are rolled; that is, both dice show the same number of spots.

13. A number greater than 10 is rolled.

B 14. An odd number is rolled.

15. Exactly one die shows 3 spots.

16. At least one die shows 3 spots.

17. Neither die shows an even number of spots.

Three coins are tossed, and the number of coins showing heads or tails is observed. For this experiment specify each of the following:

18. The sample space

19. The event that no coin shows heads

20. The event that all coins show tails

21. The event that exactly two coins show heads

22. The event that at least two coins show tails

Section 13-2 MEANING OF PROBABILITY

In the preceding section you learned about outcomes, sample spaces, and events associated with random experiments. In this section you will see how numbers called *probabilities* can be assigned to events. You will learn these terms:

- equally likely outcomes
- probability of an event

When you flip a fair coin, heads and tails are the *equally likely outcomes.* When you throw a die, the integers from 1 to 6 are the equally likely outcomes. In the rest of this chapter, you will study random experiments whose sample spaces consist of equally likely outcomes.

For a random experiment with n equally likely outcomes ($n > 0$):

1. The probability of any single outcome is $\frac{1}{n}$.
2. The probability of an event E consisting of k outcomes, which we denote by $P(E)$, is given by the formula:

$$P(E) = \frac{k}{n}$$

For any event E consisting of k outcomes, $0 \leqslant k \leqslant n$. Thus, an immediate consequence of the definition is that the probability of any event E is a number between 0 and 1, inclusive:

$$0 \leqslant P(E) \leqslant 1$$

Let us apply the definition to the simple experiment of tossing a coin. Since heads is one of the two equally likely outcomes, the probability of heads is $\frac{1}{2}$. This does not mean that the coin will show heads on alternate tosses, nor does it mean that the coin cannot show heads on three successive tosses. Rather, it means that the number of heads divided by the number of tosses will be approximately $\frac{1}{2}$ after the coin has been tossed many times. This quotient, the number of occurrences of the event divided by the number of trials, is called the *relative frequency* of the event. You may wish to interpret the probabilities found in the following material as predictions of relative frequencies.

EXAMPLE Two coins are tossed as described on page 520. Find the following probabilities.

a. P(exactly one head)

b. P(at least one head)

SOLUTION The sample space consists of 4 outcomes:

$$\{(H, H), (H, T), (T, H), (T, T)\}$$

a. Since $\{(H, T), (T, H)\}$ consists of 2 outcomes,

$$P(\text{exactly one head}) = \frac{2}{4} = \frac{1}{2}$$

b. Since $\{(H, H), (H, T), (T, H)\}$ consists of 3 outcomes,

$$P(\text{at least one head}) = \frac{3}{4}$$

EXAMPLE Two dice are thrown. Find the following probabilities:

a. $P(11)$ b. $P(4)$ c. $P(15)$

d. P(a number between 2 and 12, inclusive, is rolled)

SOLUTION Recall that the sample space (page 521) consists of 36 outcomes.

a. Since $\{(5, 6), (6, 5)\}$ consists of 2 outcomes,

$$P(11) = \frac{2}{36} = \frac{1}{18}$$

b. Since $\{(1, 3), (2, 2), (3, 1)\}$ consists of 3 outcomes,

$$P(4) = \frac{3}{36} = \frac{1}{12}$$

c. There is no outcome that corresponds to a total of 15. Thus:

$$P(15) = \frac{0}{36} = 0$$

d. This event includes all 36 possible outcomes. Thus,

$$P(\text{a number between 2 and 12, inclusive}) = \frac{36}{36} = 1$$

When, as in part (d) of the example above, the probability of an event is 1, the event is called a *certain event.* When, as in part (c) of the example above, the probability of an event is 0, the event is called an *impossible event.*

The study of probability often makes use of problems and examples that refer to a standard bridge deck of 52 cards. The deck consists of four suits: spades and clubs, which are black, and hearts and diamonds, which are red. Each suit consists of 13 cards: an ace, 2, 3, 4, 5, 6, 7, 8, 9, 10, jack, queen, and king. A jack, queen, or king can be referred to as a face card. When experiments involve drawing cards from a deck, it is understood that any card in the deck is equally likely to be drawn.

EXAMPLE A card is drawn at random from a standard bridge deck. Find the following probabilities.

a. P(heart) b. P(ace) c. P(face card) d. P(diamond or king)

SOLUTION a. Since there are 13 hearts,

$$P(\text{heart}) = \frac{13}{52} = \frac{1}{4}$$

b. Since there are 4 aces,

$$P(\text{ace}) = \frac{4}{52} = \frac{1}{13}$$

c. Since there are 4×3, or 12, face cards,

$$P(\text{face card}) = \frac{12}{52} = \frac{3}{13}$$

d. This event consists of the 13 diamonds and the kings of spades, hearts, and clubs. (Notice that we must not count the king of diamonds twice.) Thus,

$$P(\text{diamond or king}) = \frac{13 + 3}{52} = \frac{16}{52} = \frac{4}{13}$$

ORAL EXERCISES

Each of the letters D, E, F, G, H, I, J, and K is written on a card. The cards are shuffled, and a card is drawn at random. Find the following probabilities.

1. $P(D)$
2. $P(E)$
3. $P(K)$
4. $P(\text{a vowel})$
5. $P(\text{a consonant})$
6. $P(F \text{ or } D)$
7. $P(E, F, \text{ or } G)$
8. $P(M)$
9. $P(\text{letter of the alphabet})$

WRITTEN EXERCISES

A single die is thrown. Find the following probabilities.

A
1. $P(5)$
2. $P(6)$
3. $P(\text{an odd number})$
4. $P(\text{a number less than } 5)$
5. $P(7)$
6. $P(\text{a number less than } 7)$

Find the following probabilities for the two-coin toss described on page 520.

7. $P(\text{exactly one tail})$
8. $P(\text{at least one tail})$
9. $P(\text{exactly two tails})$
10. $P(\text{no tails})$

Two dice are thrown. Find the following probabilities.

11. $P(2)$
12. $P(12)$
13. $P(8)$
14. $P(9)$
15. $P(10)$
16. $P(5)$

The digits 2, 3, 5, and 6 are each written on separate slips of paper, which are then placed in a container. Simultaneously, two slips are chosen at random from the container. Find the following probabilities.

17. $P(\text{both digits are odd})$
18. $P(\text{both digits are even})$
19. $P(\text{the sum of the digits is less than } 8)$
20. $P(\text{both digits are odd or the sum is less than } 8)$
21. $P(\text{product of the digits is an even multiple of } 3)$

A card is drawn from a standard bridge deck. Find the following probabilities.

22. $P(\text{two of spades})$
23. $P(\text{four of hearts})$
24. $P(\text{three})$
25. $P(\text{club})$

B
26. $P(\text{ace or king})$
27. $P(\text{nine or ten})$
28. $P(\text{ace or face card})$
29. $P(\text{heart or queen})$
30. $P(\text{spade or heart})$
31. $P(\text{diamond or face card})$

Two dice are thrown. Find the following probabilities.

32. $P(7 \text{ or } 11)$
33. $P(\text{odd number})$
34. $P(\text{even number})$
35. $P(10, 11, \text{ or } 12)$
36. $P(\text{doubles})$

Three coins are tossed, and whether they show heads or tails is observed. Determine the sample space, and find the following probabilities.

37. P(two heads and a tail)
38. P(no heads)
39. P(at least two heads)
40. P(exactly one tail)
41. P(three heads or three tails)

C 42. P(exactly two coins show the same face)

43. P(at least two coins show the same face)

SELF-TEST 1

The name of each day of the week is written on a separate card. The cards are shuffled, and one card is randomly drawn.

Specify:

1. The sample space — Section 13-1
2. The event that a day whose name begins with the letter T is drawn
3. The event that the name of the day drawn contains no more than 7 letters

Find each probability.

4. P(Tuesday or Wednesday) — Section 13-2
5. P(a day whose name begins with a consonant)
6. P(a day whose name contains at least 7 letters)

Section 13-3 PROBABILITY FORMULAS

In this section you will learn about the relationships among the probabilities of different events within the sample space for an experiment. Specifically, you will learn rules for calculating probabilities for pairs of:

- mutually exclusive events
- intersecting events
- complementary events

If you take any two events within a given sample space, either they have some outcomes in common or they do not. For a pair of dice, the events of rolling a 6 and of rolling a double have the outcome (3, 3) in common. On the other hand, the events of rolling a 9 and of rolling a double have no outcomes in common.

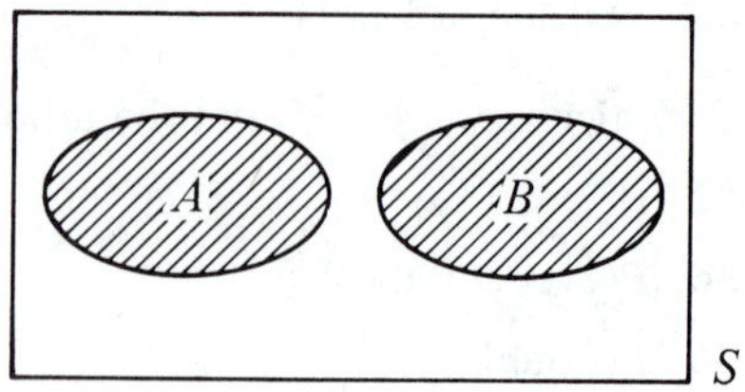

Two events that have no outcomes in common, such as those shown in the diagram at the right, are called *mutually exclusive.* For any two mutually exclusive events, the probability that an outcome will be in one event or the other is the sum of their individual probabilities.

If A and B are mutually exclusive events:

$$P(A \text{ or } B) = P(A) + P(B)$$

EXAMPLE For a pair of dice, if A is the event of rolling a 7 and B is the event of rolling a double, then find $P(A \text{ or } B)$.

SOLUTION These events are mutually exclusive. Since $P(A) = \frac{6}{36} = \frac{1}{6}$ and $P(B) = \frac{6}{36} = \frac{1}{6}$,

$$P(A \text{ or } B) = P(A) + P(B) = \frac{1}{6} + \frac{1}{6} = \frac{1}{3}.$$

Note that the combined event "A or B" consists of the two diagonals of the sample space for a pair of dice; "A or B" consists of 12 outcomes. Thus, the probability just calculated is correct since $\frac{12}{36} = \frac{1}{3}$.

EXAMPLE In drawing a card from a standard deck,

$$\begin{aligned} P(\text{face card or } 2) &= P(\text{face card}) + P(2) \\ &= \frac{12}{52} + \frac{4}{52} \\ &= \frac{16}{52} \\ &= \frac{4}{13} \end{aligned}$$

You can check this result by counting the number of cards in the event "face card or 2."

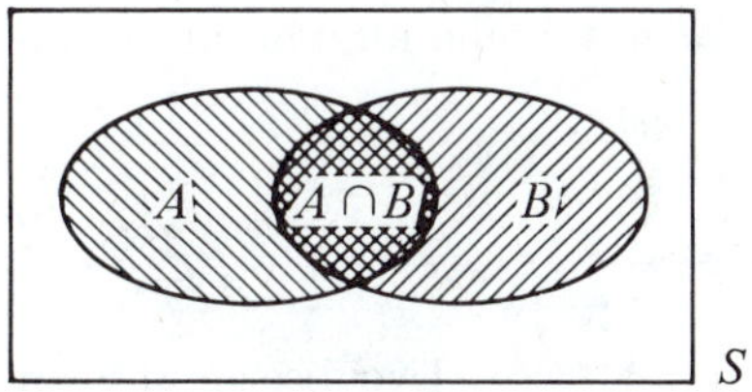

We will now consider events that are not mutually exclusive. When two events have some outcomes in common, as suggested in the diagram at the right, they are called *intersecting events.* An event is in $A \cap B$ if and only if both A and B have occurred. Thus, the event $A \cap B$ is the same event as "A and B." The following formula applies to intersecting events:

$$P(A \text{ or } B) = P(A) + P(B) - P(A \text{ and } B)$$

We subtract $P(A \text{ and } B)$ because the outcomes in "A and B" have been counted once as a part of A and once as a part of B, which is once too often. When two events are mutually exclusive, $P(A \text{ and } B) = 0$. Therefore, we can write the formula as shown below:

For any two events A and B:

$$P(A \text{ or } B) = P(A) + P(B) - P(A \text{ and } B)$$

EXAMPLE A card is drawn from a standard deck. Find the probability of drawing a diamond or a king.

SOLUTION $P(\text{diamond}) = \frac{13}{52}$ $\qquad P(\text{king}) = \frac{4}{52}$

$P(\text{king and diamond}) = P(\text{king of diamonds}) = \frac{1}{52}$

By the formula: $P(\text{diamond or king}) = P(\text{diamond}) + P(\text{king}) - P(\text{king of diamonds})$

$$= \frac{13}{52} + \frac{4}{52} - \frac{1}{52} = \frac{16}{52} = \frac{4}{13}$$

Note that this result agrees with the direct calculation from part (d) of the second example on page 524.

If A is an event within the sample space S of an experiment, the *complement* of A, denoted by A', consists of all outcomes in S that are *not* in A.

EXAMPLES For each experiment, an event and its complement are given.

1. Tossing a coin
   A: The coin shows heads.
   A': The coin shows tails.
2. Drawing a card from a standard deck
   B: Drawing a black card
   B': Drawing a red card

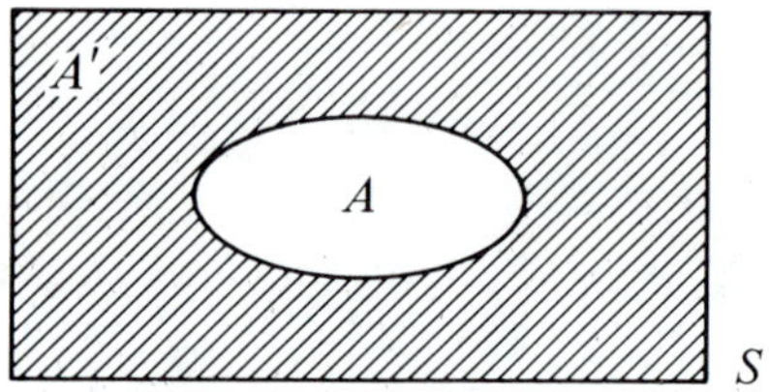

The Venn diagram at the right shows the relationships between A and A':

$$A \cap A' = \emptyset$$
$$A \cup A' = S$$

The second of these equations combined with the fact that $P(S) = 1$ enables us to conclude that

$$P(A) + P(A') = 1.$$

The most useful form of this relation is given below:

$$P(A') = 1 - P(A)$$

EXAMPLE Two dice are thrown. Find the probability of each of these events.

a. Not rolling doubles
b. Rolling 10 or less

SOLUTION a. Since $P(\text{doubles}) = \frac{1}{6}$,

$$P(\text{not rolling doubles}) = 1 - \frac{1}{6} = \frac{5}{6}$$

b. The complementary event to "rolling 10 or less" is "rolling 11 or 12." Thus:

$$P(10 \text{ or less}) = 1 - P(11 \text{ or } 12)$$
$$= 1 - [P(11) + P(12)]$$
$$= 1 - \left(\frac{2}{36} + \frac{1}{36}\right) = \frac{33}{36} = \frac{11}{12}$$

The *odds* in favor of an event A are the ratio of $P(A)$ to $P(A')$. In part (b) of the previous example, the odds in favor of rolling 10 or less are 11 to 1, found as follows:

$$P(A):P(A') = \frac{11}{12}:\left(1 - \frac{11}{12}\right) = \frac{11}{12}:\frac{1}{12} = 11:1$$

ORAL EXERCISES

In Exercises 1-7, refer to the following events in the sample space for throwing two dice.

A: A number greater than 8 is rolled.
B: An even number is rolled.
C: A number less than 5 is rolled.
D: An odd number is rolled.
E: A multiple of 5 is rolled.
F: A multiple of 3 is rolled.

1. Are events A and B mutually exclusive?
2. Which event is the complement of D?
3. Are events C and E mutually exclusive?
4. Are events B and E intersecting events?
5. State the complement of E.
6. List the common outcomes of events A and F.
7. Are events E and F mutually exclusive?

WRITTEN EXERCISES

A card is drawn from a standard deck. Find the following probabilities.

A 1. P(4 or 7)
2. P(jack or queen)
3. P(10 of diamonds or 6)
4. P(ace or red king)
5. P(face card or club)
6. P(heart or 5, 6, 8, 10)

Eight pieces of blue paper are numbered from 1 to 8 and seven pieces of white paper are numbered from 1 to 7. The slips of paper are placed in a container and shuffled. One slip of paper is drawn from the container. In Exercises 7-12, refer to the following events for this experiment:

A: An even number is chosen.
B: A number greater than or equal to 5 is chosen.
C: A blue slip is chosen.
D: An 8 is chosen.
E: A multiple of 4 is chosen.

Find each probability.

7. $P(A)$
8. $P(B)$
9. $P(C')$
10. $P(B \text{ or } C)$
11. $P(E \text{ or } D)$
12. $P(A' \text{ or } D)$

Two dice are thrown. Find the odds in favor of rolling the following numbers.

13. 11
14. 7
15. 5
16. 3

In Exercises 17-19, refer to the following experiment: A study of 150 commuters has found that 85 read the morning newspaper, 15 read both the morning and evening newspapers, while 25 read neither newspaper. If one of the commuters is chosen at random, what is the probability of each event?

17. The commuter reads only the morning newspaper.

18. The commuter reads only the evening newspaper.

19. The commuter reads either the morning or the evening newspaper, but not both.

In Exercises 20-28, refer to the following experiment: The calendar page at the right is placed face up on the ground. A dart is dropped at random from above in such a way that it lands in exactly one of the squares. If all the squares are the same size, find the following probabilities:

S	M	T	W	T	F	S
	1	2	3	4	5	6
7	8	9	10	11	12	13
14	15	16	17	18	19	20
21	22	23	24	25	26	27
28	29	30	31			

20. $P(\text{odd})$ **21.** $P(\text{blank})$

22. $P(\text{letter or even})$ **23.** $P(\text{a Wednesday})$

24. $P(\text{a Monday or even})$

25. $P(\text{a Thursday and odd})$

26. $P(\text{odd or blank})$ **27.** $P(\text{odd or a multiple of 5})$ **28.** $P(\text{Monday and blank})$

In Exercises 29-37, refer to events A, B, C, D, and E within the sample space S for a particular experiment. Use the facts given below to find the requested probabilities.

$$P(A) = \frac{2}{5} \qquad P(B) = \frac{1}{3} \qquad P(C) = \frac{1}{5}$$

$$P(A \text{ and } B) = \frac{1}{5} \qquad P(A \text{ or } C) = \frac{3}{5} \qquad P(D \text{ or } B) = \frac{2}{3}$$

$$P(D \text{ and } B) = \frac{1}{15} \qquad P(B \text{ or } C) = \frac{2}{5} \qquad P(E') = \frac{2}{3}$$

29. $P(A \text{ or } B)$ **30.** $P(A \text{ and } C)$ **31.** $P(B')$ **32.** $P([A \text{ or } C]')$

33. $P(D)$ **34.** $P(B \text{ and } C)$ **35.** $P(E)$ **36.** $P([B \cup B']')$

37. If B and E are mutually exclusive, find $P(B \text{ or } E)$.

In Exercises 38-43, refer to the following experiment: The integers from 1 to 100 are written on separate pieces of paper and placed in a hat. One piece of paper is drawn at random.
Find the probability that the number:

B **38.** is divisible by 3. **39.** is divisible by 5.

40. is divisible by 3 and 5. **41.** is not divisible by 7.

42. is not divisible by 6. **43.** is a prime less than 50.

44. There is an 80% probability of rain on Saturday and a 50% probability of rain on Sunday. If there is a 90% probability of rain on Saturday or Sunday, find the probability of rain on Saturday, but not Sunday.

Of the 263 students in the senior class of Fairmont High, 83 take art and 88 take music; 119 students do not take either of the two subjects. If a student is selected at random from the senior class, what is the probability that:

45. the student takes both art and music?

46. the student takes art or music?

47. the student takes art and not music?

48. the student takes music and not art?

C 49. Events A and B are in the sample space for a particular experiment; $P(A) = \frac{2}{3}$ and $P(B) = \frac{7}{12}$.

a. How do you know that A and B cannot be mutually exclusive?

b. If C consists of the outcomes common to A and B, show that $\frac{1}{4} \leqslant P(C) \leqslant \frac{7}{12}$.

50. A card is drawn from a standard deck, and events A, B, and C are defined as follows:

A: A heart is chosen.
B: A face card is chosen.
C: A red card higher than 7 (includes aces) is chosen.

a. Calculate $P(A \text{ or } B \text{ or } C)$ by counting the number of cards in the event "A or B or C." You may wish to diagram the entire deck as suggested below:

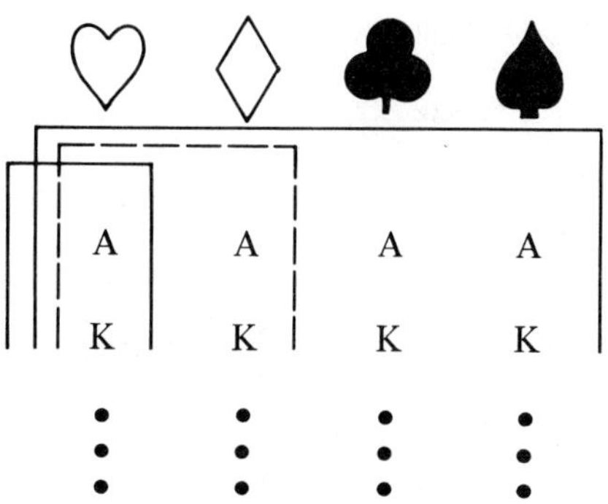

b. Calculate $P(A \text{ or } B \text{ or } C)$ using the following formula:

$$P(A \text{ or } B \text{ or } C) = P(A) + P(B) + P(C) - [P(A \text{ and } B) + P(A \text{ and } C) + P(B \text{ and } C)] + P(A \text{ and } B \text{ and } C)$$

c. Calculate $P(A \text{ or } B \text{ or } C)$ using the following formula:

$$P(A \text{ or } B \text{ or } C) = 1 - P(A' \text{ and } B' \text{ and } C')$$

Section 13-4 MULTISTAGE EVENTS

In this section you will learn how to analyze experiments that can be regarded as consisting of several stages. You will learn the following terms:

- multistage experiment
- tree diagram

EXAMPLE Construct the sample space for the experiment of tossing a coin and then rolling a die.

SOLUTION

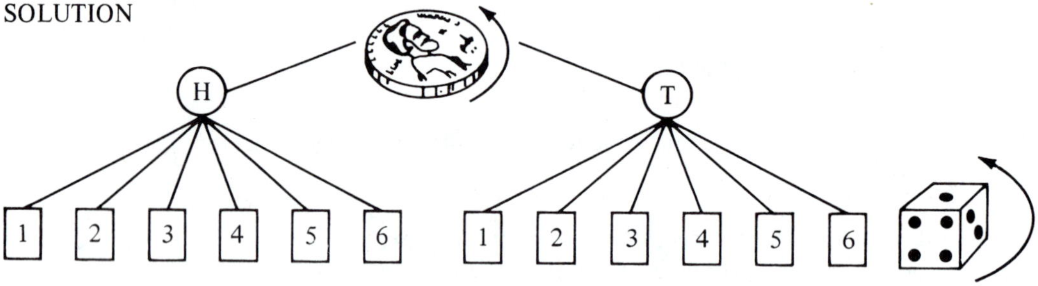

The sample space for this experiment is

$$\{(H, 1), (H, 2), (H, 3), (H, 4), (H, 5), (H, 6), (T, 1), (T, 2), (T, 3), (T, 4), (T, 5), (T, 6)\}.$$

Notice that the sample space for the experiment just presented consists of 2×6, or 12, outcomes. This result suggests the following general *counting principle:*

> If the first stage of a two-stage experiment can result in any of k outcomes and if for each of these outcomes the second stage can result in any of n outcomes, the sample space for the two-stage experiment will consist of kn outcomes.

An experiment that can be regarded as having two or more stages is referred to as a *multistage experiment.* This counting principle can be extended to multistage experiments with any number of stages.

To obtain the probability of an event within a multistage experiment, you can apply the definition of probability. For example, the probability of obtaining heads and an even number is $\frac{3}{12}$, or $\frac{1}{4}$. Another method is to multiply probabilities from each stage of the experiment:

$$P(\text{heads and even}) = P(\text{heads at first stage}) \cdot P(\text{even at second stage})$$
$$= \frac{1}{2} \cdot \frac{3}{6} = \frac{1}{4}$$

In many multistage experiments, the outcome of one stage directly influences what can occur at the next stage. Such experiments can be analyzed using a *tree diagram*, as shown in the next example.

EXAMPLE There are two identical urns. One urn contains 2 white balls and 1 red ball. The other contains 2 red balls. An urn is selected at random and a single ball is drawn from it.

a. What is the probability that the ball drawn is white?

b. What is the probability that the ball drawn is red?

SOLUTION

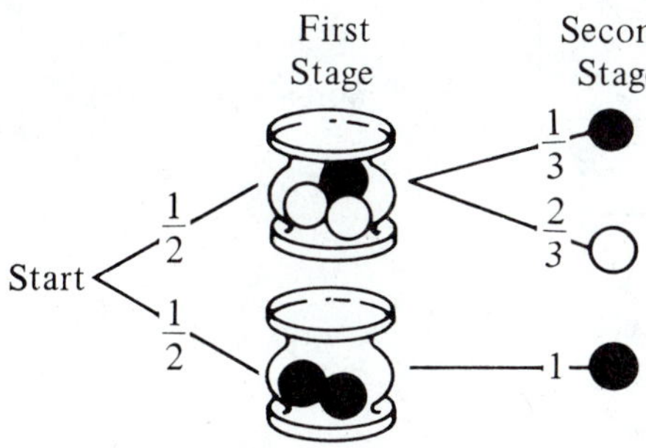

$$P(\text{first urn, red}) = \frac{1}{2} \cdot \frac{1}{3} = \frac{1}{6}$$

$$P(\text{first urn, white}) = \frac{1}{2} \cdot \frac{2}{3} = \frac{1}{3}$$

$$P(\text{second urn, red}) = \frac{1}{2} \cdot 1 = \frac{1}{2}$$

$$P(\text{second urn, white}) = 0$$

a. $P(\text{white}) = \frac{1}{3} + 0 = \frac{1}{3}$

b. $P(\text{red}) = \frac{1}{6} + \frac{1}{2} = \frac{2}{3}$

The principle used in the previous solution is this:

> The probability of any outcome in the sample space of a multistage experiment is the product of all probabilities along the path that represents that outcome on the tree diagram.

In the next example note that the first card is replaced in the deck before the second card is drawn. For this reason such an experiment can be described as *drawing with replacement.*

EXAMPLE A single card is drawn from a deck; its rank (ace, two, or three, for example) is noted. The first card is returned to the deck, and the deck is shuffled. A second card is drawn and its rank is noted. Analyze this experiment to find the probabilities of getting 0, 1, or 2 aces.

SOLUTION On either draw: $P(\text{ace}) = \frac{4}{52} = \frac{1}{13}$ and $P(\text{non-ace}) = \frac{48}{52} = \frac{12}{13}$.

The tree diagram and the analysis are shown on the next page.

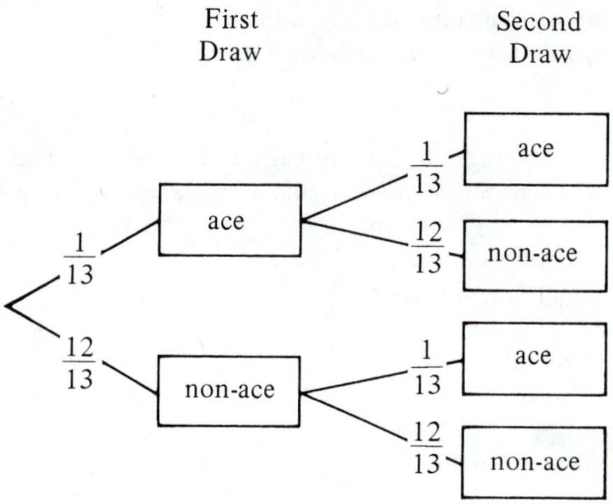

$$P(\text{ace, ace}) = \frac{1}{13} \cdot \frac{1}{13} = \frac{1}{169}$$

$$P(\text{ace, non-ace}) = \frac{1}{13} \cdot \frac{12}{13} = \frac{12}{169}$$

$$P(\text{non-ace, ace}) = \frac{12}{13} \cdot \frac{1}{13} = \frac{12}{169}$$

$$P(\text{non-ace, non-ace}) = \frac{12}{13} \cdot \frac{12}{13} = \frac{144}{169}$$

$$P(0 \text{ aces}) = \frac{144}{169}$$

$$P(1 \text{ ace}) = \frac{12}{169} + \frac{12}{169} = \frac{24}{169}$$

$$P(2 \text{ aces}) = \frac{1}{169}$$

In the next example this experiment is altered to require *drawing without replacement.* Note how this change affects the probabilities. *In this text you may assume that drawing is without replacement unless otherwise stated.*

EXAMPLE Two cards are drawn from a deck of cards (without replacement). Analyze this experiment to find the probabilities of getting 0, 1, or 2 aces.

SOLUTION

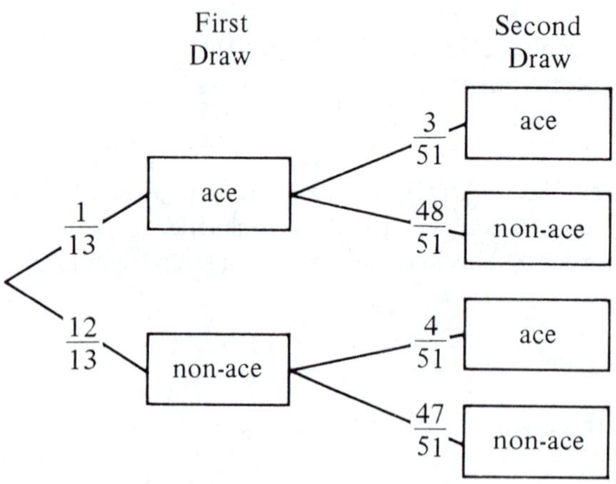

$$P(\text{ace, ace}) = \frac{1}{13} \cdot \frac{3}{51} = \frac{1}{221}$$

$$P(\text{ace, non-ace}) = \frac{1}{13} \cdot \frac{48}{51} = \frac{16}{221}$$

$$P(\text{non-ace, ace}) = \frac{12}{13} \cdot \frac{4}{51} = \frac{16}{221}$$

$$P(\text{non-ace, non-ace}) = \frac{12}{13} \cdot \frac{47}{51} = \frac{188}{221}$$

$$P(0 \text{ aces}) = \frac{188}{221}$$

$$P(1 \text{ ace}) = \frac{16}{221} + \frac{16}{221} = \frac{32}{221}$$

$$P(2 \text{ aces}) = \frac{1}{221}$$

ORAL EXERCISES

The four jacks are removed from a deck of cards. Exercises 1-3 refer to the following two-stage experiment: A die is thrown and one of the jacks is chosen.

1. How many outcomes are there in the sample space?
2. If the order of the stages were reversed, would the number of outcomes in the sample space change?
3. Find the following probabilities.
   a. P(even number)
   b. P(jack of clubs)
   c. P(4 and jack of hearts)
   d. P(even number and a red card)
4. When this experiment is altered by choosing the card from *all* the face cards in the deck, what is the number of outcomes in the sample space?

WRITTEN EXERCISES

A

1. A building superintendent decides to identify storage lockers with codes consisting of a letter followed by a numeral, such as J-5.
   a. How many such codes are possible if any letter and the digits 1 through 9 may be used?
   b. How many such codes are possible if any letter other than I or O and the digits 1 through 9 may be used?
2. There are 4 trails that go from Camp A to Camp B and 3 trails that go from Camp B to Camp C.
   a. How many different routes are there from Camp A to Camp C via Camp B?
   b. How many different routes are there from Camp A to Camp C and back that pass through Camp B both ways?

A card is chosen from a deck of cards and then a die is rolled. Find the probability of each of these events.

3. The card is a 4, and the die shows 4.
4. The card is not a face card, and the die shows an odd number.
5. The card is a club, and the die shows a number greater than 3.
6. The card is a red face card, and the die shows a prime number.

Two cards are drawn from a standard deck. Find the following probabilities if:
a. the first card is replaced before the second card is drawn.
b. the first card is not replaced.

7. P(a 9 and a 10 in that order)

8. P(a 10 and a face card in that order)

9. P(a 4 and an 8)

10. P(two face cards)

11. P(two hearts)

12. P(no hearts)

13. P(at least one heart)

14. P(one red and one black card)

An urn contains 3 green marbles and 1 blue marble. A second urn contains 2 green marbles and 3 blue marbles. An urn is chosen at random, and a single marble is drawn from it. Find the probabilities of the following events.

15. A blue marble is selected.

16. A green marble is selected.

B **17.** Show that the odds in favor of an event A are equal to the ratio of the number of outcomes in A to the number of outcomes in A'.

18. If the odds in favor of an event are $3:4$, what is the probability of the event?

19. If the odds in favor of an event are $a:b$, find an expression for the probability of the event in terms of a and b.

An urn contains 3 white marbles and 4 red marbles. A second urn contains 2 white marbles and 3 red marbles. An urn is chosen at random, and 2 marbles are drawn from the urn. Find the probabilities of the following events.

20. Both marbles are red.

21. Both marbles are white.

22. Both marbles are the same color.

23. One marble is red and the other is white.

C **24.** An urn contains one red and two white marbles. One marble is drawn and replaced with a marble of the other color, then a second marble is drawn. What is the probability that the second marble is white?

25. Three red marbles and three white marbles are to be placed in two urns, with at least one marble in each urn. An urn is to be picked at random and a marble drawn from it. How should the red marbles and the white ones be distributed in the two urns so that the probability of drawing a red marble is the greatest?

SELF-TEST 2

A card is randomly drawn from a standard deck. Find $P(A \text{ or } B)$.

1. A: drawing an ace; B: drawing a red card — Section 13-3

2. A: drawing a club; B: drawing a red face card

3. A: drawing a jack; B: drawing a card that *isn't* a face card

A bag contains 3 yellow gloves and 2 brown gloves. Find the required probabilities when 2 gloves are drawn **(a)** with replacement; **(b)** without replacement.

4. P(2 yellow gloves)

5. P(1 yellow glove and 1 brown glove) — Section 13-4

6. P(same color for both gloves)

7. P(at least one brown glove)

Section 13-5 PERMUTATIONS

In this section you will learn more about experiments in which each of the outcomes corresponds to a group of letters, numbers, or other objects arranged in a particular order. You will learn these terms:

- factorial
- permutation

Let us try to determine the number of arrangements of the letters A, B, C, D, and E using all five letters. We can regard this problem as determining the number of outcomes of a five-stage experiment. The first stage is to select the first letter at random from the given letters, the second stage is to select the second letter from the remaining four letters, and so on. We apply the counting principle on page 532 as follows:

$$\begin{aligned}
&5 \text{ outcomes for the first stage}\\
5 \times 4, \text{ or } &20, \text{ outcomes for the first two stages}\\
5 \times 4 \times 3, \text{ or } &60, \text{ outcomes for the first three stages}\\
5 \times 4 \times 3 \times 2, \text{ or } &120, \text{ outcomes for the first four stages}\\
5 \times 4 \times 3 \times 2 \times 1, \text{ or } &120, \text{ outcomes for the five-stage experiment}
\end{aligned}$$

Thus, there are 120 arrangements of the letters.

The last number in that list, $5 \times 4 \times 3 \times 2 \times 1$, is called five *factorial*, for which the symbol is 5!. Here are some other factorials:

$$\begin{aligned}
2! &= 2 \times 1 = 2\\
6! &= 6 \times 5 \times 4 \times 3 \times 2 \times 1 = 720\\
8! &= 8 \times 7 \times 6 \times 5 \times 4 \times 3 \times 2 \times 1 = 40{,}320\\
9! &= 9 \times 8 \times 7 \times 6 \times 5 \times 4 \times 3 \times 2 \times 1\\
&= 9 \times (8!) = 9 \times 40{,}320 = 362{,}880
\end{aligned}$$

For each natural number n, where $n > 1$:

$$n! = n(n-1)(n-2)(n-3)\cdots(3)(2)(1)$$

Also, $1! = 1$ and $0! = 1$ by definition.

Any collection of 5 different objects can be lined up in 5! different ways, since there are 5 ways to pick the first object, 4 ways to pick the second object, and so on. Each such arrangement is called a *permutation* of those objects. In general, a collection of n different objects has $n!$ permutations.

EXAMPLES

1. There are 3!, or 6, different numbers that can be formed using the digits 1, 2, and 3. They are, from least to greatest:

   123, 132, 213, 231, 312, 321

2. There are 4!, or 24, ways 4 cars can be parked in 4 adjacent spaces.
3. There are 5!, or 120, permutations of the letters in the word RISEN.
4. There are 6!, or 720, ways 6 different books can be arranged side by side on a shelf.

Note that you can make other arrangements with the letters of RISEN by using fewer than five letters. There are:

$5 \times 4 \times 3 \times 2$, or 120, 4-letter permutations;
$5 \times 4 \times 3$, or 60, 3-letter permutations;
5×4, or 20, 2-letter permutations;
5 1-letter permutations.

These are known, respectively, as the number of permutations of five things taken four at a time, written ${}_5P_4$, the number of permutations of five things taken three at a time, and so on. These can be expressed in terms of factorials as follows:

$$ {}_5P_4 = \frac{5 \times 4 \times 3 \times 2 \times 1}{1} = \frac{5!}{1!} = 5 \times 4 \times 3 \times 2 = 120 $$

$$ {}_5P_3 = \frac{5 \times 4 \times 3 \times 2 \times 1}{2 \times 1} = \frac{5!}{2!} = 5 \times 4 \times 3 = 60 $$

$$ {}_5P_2 = \frac{5 \times 4 \times 3 \times 2 \times 1}{3 \times 2 \times 1} = \frac{5!}{3!} = 5 \times 4 = 20 $$

$$ {}_5P_1 = \frac{5 \times 4 \times 3 \times 2 \times 1}{4 \times 3 \times 2 \times 1} = \frac{5!}{4!} = 5 $$

The preceding examples suggest the following general rule:

> The number of permutations of n things taken r at a time is given by the formula:
>
> $$ {}_nP_r = \frac{n!}{(n-r)!} $$

Another kind of permutation problem has to do with the number of arrangements of a collection of objects that are not all different.

EXAMPLE How many permutations are there of the letters ABAA?

SOLUTION 1 You can place the B in any one of four positions; the remaining positions will all be A's. Thus, there are 4 permutations.

SOLUTION 2 Turn the problem into one you already know by putting subscripts on the A's so you can tell them apart: $A_1BA_2A_3$. There are 4! permutations of these four distinct symbols. Notice, however, that these permutations fall into groups of 3! permutations such as the one shown at the right. The permutations within each group differ only in the arrangement of subscripts. The number of these groups is the total number of permutations divided by the number in each group, which is $\frac{4!}{3!}$, or 4.

$A_1BA_2A_3$
$A_1BA_3A_2$
$A_2BA_1A_3$
$A_2BA_3A_1$
$A_3BA_1A_2$
$A_3BA_2A_1$

Solution 2 to the preceding example suggests the general formula for permutations in which certain objects are indistinguishable.

In general, if a collection of n objects has a objects of one kind, b objects of a second kind, c of a third kind, and so on, then the number of distinct permutations of those n objects is given by the expression:

$$\frac{n!}{a!b!c!\ldots}$$

Permutations can be useful in answering probability questions in which order is significant.

EXAMPLE Three cards are drawn from a deck of cards without replacement. Find the probability that the cards are two 7's and a 6 in that order.

SOLUTION The sample space consists of ${}_{52}P_3$ outcomes. The event of drawing two 7's and a 6 in that order can be regarded as a two-stage event with ${}_4P_2$ outcomes for the first stage (drawing two 7's) and ${}_4P_1$ outcomes for the second stage (drawing a 6).

$$P(7, 7, 6 \text{ in that order}) = \frac{{}_4P_2 \cdot {}_4P_1}{{}_{52}P_3} = \frac{(4 \cdot 3)(4)}{52 \cdot 51 \cdot 50} = \frac{2}{5525}$$

When n objects are arranged in a circle, there are $(n - 1)!$ possible arrangements, called *circular permutations*, if no position is considered as fixed. For example, 6 place cards can be arranged around a circular table in $(6 - 1)!$, or 120, ways. It must be understood that we are observing only the relative positions of the cards; for example, we do not observe which card is closest to some fixed point in the room containing the table.

ORAL EXERCISES

Give each answer in the form ${}_nP_r$.

1. The number of 2-letter arrangements you can create from 7 different letters
2. The number of 5-letter arrangements you can create from 5 different letters
3. The number of ways 4 books from a collection of 12 different books can be arranged on a bookshelf
4. The number of ways 8 books from a collection of 12 different books can be arranged on a bookshelf
5. The number of 5-digit license plates that can be created using the digits 1 through 9 without repetition

Evaluate each of the following.

6. ${}_4P_2$ 7. ${}_8P_1$ 8. ${}_6P_3$ 9. ${}_{10}P_2$ 10. ${}_3P_3$

WRITTEN EXERCISES

Evaluate.

A 1. $_6P_2$ 2. $_3P_1$ 3. $_8P_5$ 4. $_7P_3$

5. $_9P_8$ 6. $_{10}P_4$ 7. $_{25}P_1$ 8. $_5P_5$

9. How many 3-letter permutations can you form from the letters of the word ABOUT?

10. How many 2-letter permutations can you form from the letters of the word MINOR?

11. How many 6-letter permutations can you form from the letters of the word HISTORY?

12. How many permutations can you form from the letters of the word COMPUTER if all the letters are used?

13. How many 4-letter permutations can you form from the letters of the word DYNAMIC?

14. How many 7-letter permutations can you form from the letters of the word SOFTWARE?

15. How many permutations can you form from the letters of the word VERTICAL if all the letters are used?

16. How many ways can 5 different colored flags be displayed on a flagpole if all the flags are used?

17. How many ways can 4 flags from a group of 5 different-colored flags be displayed on a flagpole?

18. In how many ways can a class elect its president, vice-president, treasurer, and secretary if there are 9 candidates, no one of whom may hold more than one office?

19. In how many ways can 5 students from a group of 10 students line up at a movie box office?

20. How many batting orders are possible for 9 starting players?

21. In how many ways can prizes be awarded to the first 3 finishers in a race if there are 12 runners and no ties?

How many different permutations are there of the letters in each of the following words?

22. ALGEBRA 23. GIGGLE 24. HORIZONTAL

25. ELEMENT 26. MINIMUM 27. STATISTICS

28. MISSISSIPPI 29. BOOKKEEPER 30. TROUBADOUR

A license plate consists of 2 letters followed by 3 digits. How many different license plates are possible under each of the following conditions?

31. No repetition of letters or digits is permitted.

32. Repetition of letters, but no repetition of digits, is permitted.

33. Letters and digits can be repeated.

B 34. A collection of books contains 5 history books, 5 math books, and 5 literature books, and 3 books are chosen from each group. If the 3 books of each subject must be next to each other, how many final displays are possible?

35. An actor's make-up kit contains 5 wigs, 3 moustaches, and 2 beards. How many different faces are possible if the actor uses:
   a. one from each category?
   b. a wig and either a moustache or a beard, but not both?
   c. items from any 2 of the 3 categories?

36. In how many ways can 8 people join hands for a circle dance?

37. In how many ways can 7 people be seated at a round table for 7?

38. In how many ways can 10 people occupy 10 seats of a Ferris wheel if each seat is occupied by 1 person?

39. In how many ways can 6 differently colored beads be strung on a necklace? (*Hint:* Consider what happens when you flip the necklace over.)

Two cards are chosen from a deck of cards without replacement. Find the probability of each event.

40. The first card is a jack, queen, or king, and the second card is an ace.

41. The first card is red and the second card is black.

42. They are both black.

C 43. Determine the number of ways 6 people from a group of 8 people may be seated at a round table for 6 people.

44. Consider the set of 3-digit numbers (those greater than 99). How many have no duplication of digits? How many have exactly two of the same digits?

45. How many 5-letter permutations can be formed from the letters in the word CALENDAR? (*Hint:* Consider separately those containing 0, 1, or 2 A's.)

Section 13-6 COMBINATIONS

In the previous section you studied selection problems in which order was an important factor. In this section you will study selection problems in which order is not considered. You will learn the term *combination*.

A *combination* is a set of objects. RISEN and SIREN are permutations of a single combination of letters: {R, I, S, E, N}.

EXAMPLE How many combinations of 3 letters can be chosen from the 5 letters in {R, I, S, E, N}?

SOLUTION There are ${}_5P_3$, or 60, permutations consisting of 3 letters each. For any given combination of 3 letters, such as {R, I, S}, there are 3!, or 6, permutations:

RIS RSI IRS ISR SRI SIR

When the 60 permutations are put in groups of 6 this way, there are $\frac{60}{6}$, or 10, groups, each representing a distinct combination of 3 letters. Thus, you can choose 10 combinations of 3 letters from the letters in {R, I, S, E, N}.

Let us denote the combinations of n things taken r at a time as ${}_nC_r$. Using this notation we can interpret the results of the previous example as follows:

$${}_5C_3 = \frac{{}_5P_3}{3!}$$

This approach suggests the following general result:

> The number of combinations of n things taken r at a time is given by the formula:
>
> $${}_nC_r = \frac{{}_nP_r}{r!} = \frac{n!}{r!(n-r)!}$$

For the preceding example, where $n = 5$ and $r = 3$,

$${}_5C_3 = \frac{5!}{3!(5-3)!} = \frac{5!}{3!2!}$$

Notice that each choice of a 3-letter combination, say {R, I, S}, determines a corresponding 2-letter combination, in this case {E, N}. Therefore the number of 2-letter combinations must be the same as the number of 3-letter combinations, and the formula confirms this fact:

$${}_5C_2 = \frac{5!}{2!(5-2)!} = \frac{5!}{2!3!}$$

This leads to the general symmetry formula for combinations:

$${}_nC_r = \frac{n!}{r!(n-r)!} = {}_nC_{n-r}$$

EXAMPLE The number of combinations of 3 digits that can be chosen from the digits 1 through 9 can be found as follows:

$${}_9C_3 = {}_9C_6 = \frac{9!}{6!3!} = \frac{9 \times 8 \times 7 \times (6!)}{6!3!}$$

$$= \frac{9 \times 8 \times 7}{3 \times 2 \times 1} = 84$$

The formula for ${}_nC_r$ can be restated in the following convenient form:

$$ {}_nC_r = \frac{n(n-1)\cdots[n-(r-1)]}{r!} $$

EXAMPLE The number of 5-card hands that can be dealt from a standard deck can be found as follows:

$$ {}_{52}C_5 = \frac{52 \times 51 \times 50 \times 49 \times 48}{5 \times 4 \times 3 \times 2 \times 1} = 52 \times 51 \times 5 \times 49 \times 4 $$
$$ = 2{,}598{,}960 $$

The sample spaces for many probability problems are made up of outcomes that can be regarded as combinations. In many such problems finding the answer in terms of ${}_nC_r$ is the most important step since it is the step in which mistakes are most likely to occur. For this reason we will ask that answers be given in terms of ${}_nC_r$, which we will refer to as being in *symbolic form*; we will also request that answers be given as decimals to a certain number of *significant digits.* We consider the first nonzero digit of the decimal answer to be the first significant digit, the next digit of the answer to be the second significant digit, and so on. We usually round the last significant digit in a division according to the value of the remainder.

EXAMPLE Find the probability that a 5-card hand dealt from a standard deck contains 3 hearts and 2 spades. Give the answer: **(a)** in symbolic form; **(b)** as a decimal with 3 significant digits.

SOLUTION **a.** To solve this problem, we need to know the number of 5-card hands that consist of 3 hearts and 2 spades. We can regard the selection of such a hand as a two-stage process. We can select the hearts in ${}_{13}C_3$ ways; we can select the spades in ${}_{13}C_2$ ways. Therefore, there are ${}_{13}C_3 \cdot {}_{13}C_2$ such hands.

$$ P(\text{3 hearts and 2 spades}) = \frac{{}_{13}C_3 \cdot {}_{13}C_2}{{}_{52}C_5} $$

b. We will now estimate the value of the answer to 3 significant digits. Note that the value of ${}_{52}C_5$ was found in the previous example.

$$ \frac{{}_{13}C_3 \cdot {}_{13}C_2}{{}_{52}C_5} = \frac{\frac{13 \cdot 12 \cdot 11}{3 \cdot 2 \cdot 1} \cdot \frac{13 \cdot 12}{2 \cdot 1}}{2{,}598{,}960} = \frac{286 \cdot 78}{2{,}598{,}960} \approx 0.00858 $$

Although ${}_nC_r$ is a useful notation, the more compact notation $\binom{n}{r}$ is often used to denote the combinations of n things taken r at a time. Frequently, $\binom{n}{r}$ is read as "n choose r."

EXAMPLE There are 4 blue marbles, 3 red marbles, and 5 white marbles in an urn. Two marbles are selected at random. Find the probability that both marbles are of the same color: **(a)** in symbolic form; **(b)** as a decimal with 3 significant digits.

(The solution is on the following page.)

SOLUTION a. There are $\binom{4}{2}$ ways to choose 2 blue marbles, $\binom{3}{2}$ ways to choose 2 red marbles, and $\binom{5}{2}$ ways to choose 2 white marbles. These combinations represent the three separate groups of outcomes that correspond to both marbles being the same color.

$$P(\text{both marbles are the same color}) = \frac{\binom{4}{2} + \binom{3}{2} + \binom{5}{2}}{\binom{12}{2}}$$

b. $$\frac{\binom{4}{2} + \binom{3}{2} + \binom{5}{2}}{\binom{12}{2}} = \frac{\frac{4 \cdot 3}{2 \cdot 1} + \frac{3 \cdot 2}{2 \cdot 1} + \frac{5 \cdot 4}{2 \cdot 1}}{\frac{12 \cdot 11}{2 \cdot 1}}$$

$$= \frac{6 + 3 + 10}{66} = \frac{19}{66} \approx 0.288$$

ORAL EXERCISES

Evaluate.

1. ${}_3C_2$
2. ${}_7C_5$
3. ${}_6C_3$
4. ${}_8C_8$
5. $\binom{5}{2}$
6. $\binom{8}{6}$
7. $\binom{8}{2}$
8. $\binom{12}{9}$

There are 12 students on a committee; 8 are juniors, and 4 are seniors. Determine the number of ways of forming the following subcommittees. Give answers in ${}_nC_r$ form.

9. A subcommittee of 5 juniors
10. A subcommittee of 3 seniors
11. A subcommittee of 6 students
12. A subcommittee of 3 juniors and 2 seniors

WRITTEN EXERCISES

Evaluate.

A

1. ${}_{12}C_2$
2. ${}_9C_5$
3. ${}_4C_3$
4. ${}_{13}C_7$
5. ${}_8C_3$
6. ${}_{52}C_1$
7. ${}_{12}C_0$
8. ${}_{15}C_{10}$
9. $\binom{39}{39}$
10. $\binom{9}{4}$
11. $\binom{7}{4}$
12. $\binom{18}{8}$
13. How many committees of 4 students can be formed from a class of 15 students?
14. There are 45 records on the shelf. Chris wants to take 5 of them to his friend's house. In how many ways can he make the selection?

15. Michelle must answer 3 of the 5 questions in Part A and 2 of the 4 questions in Part B of an essay test. In how many ways can she select the questions to be answered?

16. A pianist is selecting a program by choosing 3 of 12 preludes, 2 of 4 polonaises, and 1 of 3 sonatas. How many different programs are possible, disregarding the order in which the pieces are played?

17. There are 4 blue pencils, 3 black pencils, and 4 red pencils in a drawer. In how many ways can 3 pencils of the same color be selected?

18. A drawer contains 3 blue socks, 4 black socks, and 6 green socks. In how many ways can a pair of matching socks be selected?

Exercises 19–21 refer to an urn that contains 4 red marbles, 3 green marbles, and 3 white marbles. Give each answer in symbolic form and as a decimal with 3 significant digits.

19. If a marble is drawn at random, find the probability that it is:
    a. red   b. green   c. white

20. If two marbles are drawn at random (without replacement), find the probability that:
    a. One is red and one is white.
    b. One is red and one is green.
    c. One is green and one is white.
    d. Both are red.

21. Find the probability that both marbles are the same color in Exercise 20.

Three cards are drawn at random from a deck of cards. Find the probabilities of the following events. Give each answer in symbolic form and as a decimal with 3 significant digits.

22. All three cards are hearts.

23. All three cards are black.

24. No spade is drawn.

25. Two cards are aces; one card is not an ace.

26. Two kings and a queen are drawn.

In Exercises 27–29, refer to the diagram at the right. It represents 7 distinct points on the circumference of a circle.

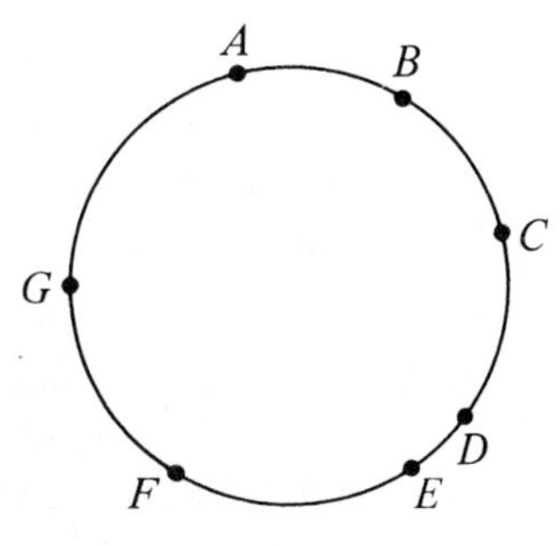

B 27. How many chords can be constructed whose endpoints are chosen from the 7 points?

28. How many triangles can be constructed whose vertices are chosen from the 7 points?

29. How many quadrilaterals can be constructed whose vertices are chosen from the 7 points?

30. In how many ways can a team of 9 players be selected from a group of 15 players if the team captain must be on the team?

Exercises 31–33 refer to a book collection that contains 6 different music books, 4 different art books, and 3 different literature books. A sample is selected from the collection.

31. How many 5-book samples are possible?

32. How many 5-book samples are there that have 3 music books and 2 literature books?

33. How many 5-book samples are there that have at least 2 literature books and at least 1 art book?

34. In how many ways can a set of 12 problems be divided into 2 equal problem sets? into 3 equal problem sets? into 4 equal problem sets?

A 5-card hand is dealt from a deck of cards. Determine the probabilities of being dealt a hand that contains each of the following combinations. Give each answer in symbolic form and as a decimal with 3 significant digits.

35. Four jacks

36. Four jacks or four queens

37. Four jacks or four queens or four kings

38. Four of a kind

39. Five hearts

40. Four hearts and a club

41. Three hearts and two black face cards

42. Exactly two hearts

In Exercises 43–53, give each answer in symbolic form and as a decimal with 3 significant digits.

43. An urn contains 4 white, 5 black, and 7 blue marbles. If 5 marbles are drawn at random, what is the probability that 3 are black and 2 are blue?

44. A cab driver keeps a container with change on the dashboard of the cab. There are 2 quarters, 3 dimes, 4 nickels, and 6 pennies in the container. What is the probability that two randomly selected coins will be sufficient to pay a 35¢ toll?

45. The president and vice president of a garden club are selected from a group of 10 people consisting of 7 members of the flower division and 3 members of the kitchen-garden division. What is the probability that both officers are from the same division?

46. A tennis player's athletic clothes are all jumbled together in a duffle bag. The player has 16 tennis socks of which 6 have red stripes. If the player draws a pair at random from the duffle bag, what is the probability that neither sock has red stripes?

C 47. A 13-card bridge hand is dealt from an ordinary deck of 52 cards. Find the probability that the hand contains:

a. Five hearts, four spades, three diamonds, and one club

b. Four aces or four kings

c. Four aces and four kings

48. Five cards are dealt from a deck of 52 cards. What is the probability that the hand contains at least 3 tens?

49. A committee of 3 persons is chosen at random from a group of 5 dentists, 4 doctors, and 6 plumbers. Find the probability that the group contains a majority of one profession.

50. Urn I contains 3 red, 2 blue, and 4 white marbles. Urn II contains 4 red, 3 blue, and 2 white marbles. An urn is chosen at random and then 3 marbles are drawn simultaneously. Find the probability that the three marbles are:
a. All white b. All red c. All the same color

A five-card hand is dealt from an ordinary deck of cards. Find the probability that the hand contains:

51. Five of one suit

52. Three of one suit and two of another suit

53. Four of a kind

MIXED EXERCISES

When you solve each of the following exercises, think carefully about whether to use combinations or permutations.

1. The dinner menu at Chez Annette offers 2 different soups, 5 main courses, 3 types of potato dishes, 4 choices of vegetables, and 5 different fruits. How many different dinner combinations are possible if an item is selected from each category?

A page of a postage stamp album has positions for 6 stamps.

2. In how many ways can a group of 6 different stamps be attached to the page?

3. In how many ways can 6 stamps chosen from a group of 11 different stamps be attached to the page?

4. At the back of the album is an envelope containing 20 different stamps. In how many ways can 7 stamps be selected from the envelope?

Each of the numbers 1, 2, 5, and 7 is written on a separate card. The cards are then arranged to form a 4-digit number.

5. How many even numbers are possible?

6. How many odd numbers are possible?

7. How many numbers ending in 1 are possible?

8. How many even numbers, starting with 5, are possible?

9. How many numbers less than 2700 are possible?

10. A collection of flags contains 4 red, 3 white, and 6 yellow flags. In how many different ways can the 13 flags be arranged on a vertical flagpole if all the flags of one color are identical?

Each of the letters of the word SECONDARY is written on a separate card. The cards are shuffled, and 4 cards are removed and placed next to one another to create a 4-letter arrangement. Find each of the following probabilities. Give the answer in symbolic form and as a decimal with 3 significant digits.

11. P(word ROSE)

12. P(word containing the letter Y)

13. P(word containing the letters E and O)

14. P(word that has A as the first letter)

15. P(word that ends with the letter D)

In a box of 12 batteries it is known that 5 are dead. Four batteries are selected simultaneously at random from the box to power an electronic game. What is the probability of each event? Give the answer in symbolic form and as a decimal with 3 significant digits.

16. Exactly one dead battery is selected.

17. All 4 of the selected batteries are dead.

18. At most 2 of the selected batteries are dead.

A deck of cards is shuffled and then 4 cards are removed one at a time, without replacement, and placed face up on a table. Find each probability. Give the answer in symbolic form and as a decimal with 3 significant digits.

19. P(4, 5, 6, 7 in that order)

20. P(all red)

21. P(club, club, club, heart in that order)

22. P(2 aces and 2 kings)

23. P(no diamonds)

SELF-TEST 3

1. Evaluate: **(a)** ${}_7P_2$; and **(b)** ${}_{100}P_3$. Section 13-5

2. How many 4-digit integers can be formed using the digits 1, 3, 5, 7, and 9? (Repetition of digits is not allowed.)

3. How many different permutations are there of the letters in the word CARAVAN?

4. Evaluate: **(a)** ${}_{50}C_{46}$; and **(b)** ${}_8C_4$. Section 13-6

5. In how many ways can a biologist choose a random sample of 3 plants from a group of 20 plants?

6. Two cards are randomly drawn from a deck of cards. Find the probability of each event. Give the answer in symbolic form and as a decimal with 3 significant digits.
   a. Both cards are face cards.
   b. At least one card is a club.

CHAPTER REVIEW

Exercises 1-6 refer to the following experiment: Each of the numbers 1, 2, 3, 4, 5, 6, 7, and 8 is written on a separate piece of paper. The slips with odd numbers on them are placed in a green hat; the slips with even numbers on them are placed in a red hat. A slip is selected at random from each hat, and the slip from the green hat is placed to the right of the slip from the red hat to form an odd number. Specify:

1. The sample space — Section 13-1
2. The event that the number is less than 42
3. The event that the number is divisible by 3

Find each probability.

4. P(the number is between 26 and 62) — Section 13-2
5. P(the number is divisible by 5)
6. P(the sum of the digits is 7)

Exercises 7-9 refer to the following experiment: The integers from 1 to 20 are written on separate cards and placed in a bag. One card is randomly drawn. Find each probability.

7. P(multiple of 7 or multiple of 9) — Section 13-3
8. P(odd integer and a perfect square)
9. P(*not* a factor of 12)

Exercises 10-12 refer to the following experiment: A jar contains 3 red marbles, 2 white marbles, and 5 blue marbles. Two marbles are randomly drawn without replacement.

10. Find the probability of each event. — Section 13-4
    a. P(both red)
    b. P(at least one white)
11. Give the odds in favor of the event in Exercise 10.
12. Use permutations to find the probability of drawing a red marble and then a blue marble. Give the answer in symbolic form and as a fraction. — Section 13-5
13. In how many ways can a baby be given a first name and a different middle name from a list of 1000 names?
14. Find an expression equal to the number of different permutations that can be formed from the letters of the word SASSAFRAS. (You need not simplify.)
15. Evaluate: **(a)** ${}_{12}C_{12}$; and **(b)** $\binom{42}{2}$. — Section 13-6

Exercises 16 and 17 refer to the following experiment: A club leader is choosing a committee of four from the 6 women and 4 men in the club.

16. How many different committees can be formed?
17. Find the probability that such a committee would include exactly 2 men.

Exercises 1–8 refer to the following experiment: A red die and a green die are thrown. The number of dots showing on each die is noted.

1. How many outcomes are there in the sample space?
2. Specify the event that both numbers are even.
3. Specify the event that the product of the numbers is 6.
4. Find the probability that the red die shows a greater number than the green die.
5. Find the probability that the sum of the numbers is no more than 4.
6. Find the probability that the product of the numbers is 28.
7. Find the probability that either one die shows a 4 or the sum of the numbers is 6.
8. Find the odds in favor of the event that at least one number is a multiple of 3.

Exercises 9 and 10 refer to the following experiment: A card is randomly chosen from a standard deck and a coin is tossed.

9. How many outcomes are there in the sample space?
10. Find the probability that:
    a. The card is not a heart.
    b. The card is a club or a 7, and the coin shows tails.
    c. The card is an ace or the coin shows heads.

Use permutations or combinations to answer each question.

11. In how many ways can 6 children be seated side-by-side on a bench?
12. How many different five-digit permutations are there of the digits in the number 12345? in the number 11234?
13. How many different amounts of money can be formed with a penny, a nickel, and a dime?
14. What is the probability that a four-card hand drawn randomly from a standard deck contains 4 face cards? Give the answer in symbolic form and as a decimal with 3 significant digits.
15. In how many different ways can 3 out of 20 applicants be hired as camp counselors for cabins 1, 2, and 3?

Biographical Note

Hypatia

Hypatia may not have been the first woman mathematician, but because we do not have any significant information about women mathematicians who may have lived before her, she is generally considered to be the first.

Hypatia was born about A.D. 370 in Alexandria, Egypt, a learning center of the world at that time. Daughter and pupil of the noted mathematician Theon, Hypatia studied mathematics, science, art, literature, philosophy, and public speaking.

Hypatia's keen intellect and her outstanding speaking ability attracted a great deal of attention. Hypatia became head of the Neo-Platonic school in Alexandria, and scholars from Europe, Asia, and Africa traveled there to hear her lectures on philosophy, astronomy, algebra, and geometry.

Besides teaching, Hypatia wrote on the most advanced mathematics of her day. It is believed that she helped her father write a commentary on the *Almagest*, a monumental work on astronomy compiled in A.D. 150 by the Greek mathematician and astronomer Ptolemy. She is thought to have written at least two other important mathematical commentaries. One was a commentary on the *Arithmetica*, a work consisting of thirteen books on number theory, which was written by Diophantus, a Greek algebraist of the third century. The other was a commentary on *Conic Sections*, a work in eight books written by Apollonius of Perga, a mathematician of the third century B.C. Unfortunately, only fragments of Hypatia's commentaries have been preserved.

Hypatia was also an inventor. She is credited with developing an apparatus for distilling water, one for measuring the level of water, and another for finding the specific gravity of liquids.

Unfortunately, this accomplished woman lived in Alexandria during turbulent times. She died tragically in A.D. 415 at the hands of an angry mob. When Hypatia wrote *On the Conics of Apollonius*, she must have realized how significant they would be to the study of mathematics.

VOCABULARY REVIEW

Be sure that you understand the meaning of these terms:

sample space, p. 519
event, p. 520
equally likely outcomes, p. 523
probability of an event, p. 523
relative frequency, p. 523
mutually exclusive events, p. 526
multistage experiment, p. 532
factorial, p. 537
permutation, p. 537
circular permutation, p. 539
combination, p. 541
significant digits, p. 543

MIXED REVIEW

1. How many different permutations are there of the letters in the word CIRCUIT?
2. Divide $6b^3 + 7b^2 - 25b + 8$ by $3b - 4$.

Two cards are drawn from a standard deck. Find the following probabilities if:

a. the first card is replaced before the second card is drawn.

b. the first card is not replaced.

3. P(a 6 and a queen in that order)
4. P(2 spades)
5. In $\triangle RSU$, $\overrightarrow{RT}$ bisects $\angle R$. If $SU = 18$, $RS = 10$, and $RU = 14$, find ST.

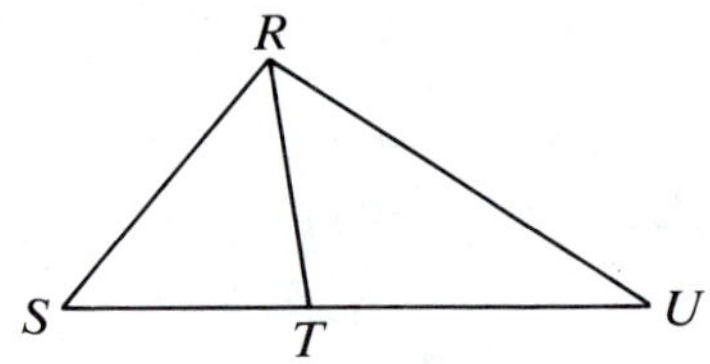

6. Find an equation in the form $y = ax^2 + bx + c$ of the parabola with focus $F(-1, -3)$ and directrix $y = 1$.
7. Graph the line with slope $-\frac{3}{2}$ and y-intercept -2.
8. Is the triangle with sides of length 1, 2, and $\sqrt{5}$ a right triangle?
9. The difference of a number and its reciprocal is $\frac{7}{12}$. Find the number.
10. From a point 18 m from the base of a flagpole, the angle of elevation of the top of the flagpole is 72°. Find the height of the flagpole to the nearest meter.
11. Show that the conclusion follows from the premises:

$$\begin{array}{ll} P_1: & \sim p \rightarrow r \\ P_2: & s \rightarrow \sim r \\ P_3: & s \\ \hline C: & p \end{array}$$

The name of each month of the year is written on a separate card. The cards are shuffled and one card is drawn at random.

12. Specify the sample space.

13. Specify the event that a month whose name contains exactly 4 letters is drawn.

14. Find P(a month whose name contains at least 6 letters).

15. In $\triangle ABC$, $m\angle C = 90$, $AB = 82$, and $AC = 58$. Find $m\angle A$ to the nearest degree.

16. Describe the locus of points equidistant from the sides of a square.

17. The vertices of $\triangle PQR$ are $P(-2, 1)$, $Q(2, 5)$, and $R(5, 2)$. Find the coordinates of the midpoints of $\overline{PQ}$ and $\overline{PR}$. Show that the length of the segment joining the two midpoints is half the length of the third side.

18. Solve: $3x - 4y = 3$
$2x - 3y = 3$

19. Graph: $3x - y = 4$

20. If you use three 4's and five 6's, how many different 8-digit numbers can you form?

21. Write an equation in standard form of the line through $(4, -3)$ and $(6, 2)$.

22. How many groups of 3 children can be selected from a group of 10 children?

23. Find the radius and center of the circle with equation $x^2 + y^2 + 10x - 2y = -6$.

24. Evaluate: **a.** ${}_5P_4$ **b.** ${}_7C_3$

25. (Optional) Graph the equation $4x^2 + y^2 = 100$ and give the coordinates of the foci.

Write two-column proofs.

26. Given: $\overline{BD}$ is the perpendicular bisector of $\overline{AC}$; $\overline{AB} \parallel \overline{DC}$
Prove: $ABCD$ is a rhombus.

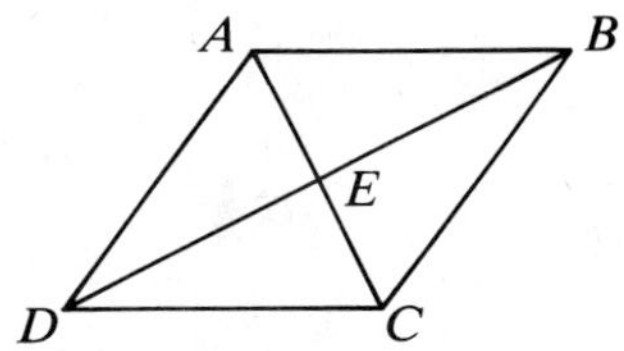

27. Given: $UVXY$ is a parallelogram; $\overline{VW} \cong \overline{ZY}$
Prove: $UWXZ$ is a parallelogram.

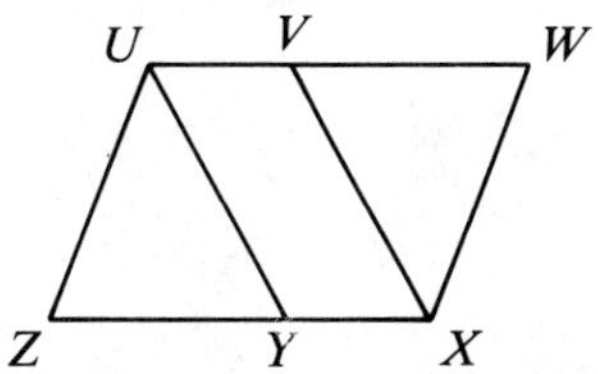

28. Write the negation and give its truth value: $\forall_x \left(x < 1 \rightarrow \frac{1}{x} > x\right)$

29. Rewrite the equation $y = -x^2 - 4x + 4$ in the form $y = a(x - h)^2 + k$ and state the coordinates of the vertex.

If classes begin at 8 o'clock and the school day lasts for 7 hours, at what time is school over? An arithmetic solution of this problem involves "clock-12 addition." (See page 560.) In this chapter you will study clock arithmetic and, more generally, modular systems.

Algebraic Systems

14

The ancient Greeks were able to solve certain simple cubic, or third-degree, equations. Nevertheless, as late as 1494, Pacioli asserted that no general solution existed for the cubic equation:

$$ax^3 + bx^2 + cx + d = 0 \quad (a \neq 0)$$

In the sixteenth century, contributions by the Italian mathematicians del Ferro, Tartaglia, and Cardano led to a general solution of the cubic equation. Ferrari, who was Cardano's pupil, devised the first general solution of the *quartic*, or fourth-degree, equation:

$$ax^4 + bx^3 + cx^2 + dx + e = 0 \quad (a \neq 0)$$

He solved the quartic equation by expressing it as a cubic equation, which could then be solved.

General solutions to higher-degree equations eluded mathematicians for almost two more centuries. Then, in the early eighteenth century, Abel proved that there is no real, general solution of the *quintic*, or fifth-degree, equation:

$$ax^5 + bx^4 + cx^3 + dx^2 + ex + f = 0 \quad (a \neq 0)$$

Finally, in 1832, Évariste Galois developed a theory that applies to equations of every degree. Using the theory of groups, which you will study in this chapter, he proved that a general solution does not exist for equations of degree greater than four.

Challenged to a duel, Galois, in anticipation of his death, wrote down his results the night before the duel. He died the next day, three days before his twenty-first birthday. His brilliant work, known as Galois Theory, was discovered by others after his death.

Section 14-1 FAMILIAR SYSTEMS

In this section we will review familiar concepts concerning integers, rational numbers, and real numbers. There will be a shift in our viewpoint, however. We will regard these sets of numbers coupled with certain arithmetic operations as *algebraic systems*. You will learn the following terms:

- binary operation
- closure
- associative
- commutative
- algebraic system

A *binary operation* on a set combines two elements of the set. For example, addition and multiplication are binary operations on the set of integers. Each operation combines two integers to produce a third (not necessarily different) integer. If a binary operation $*$ is defined on a set S, we can determine which, if any, of the following properties it possesses:

1. **Closure Property** For all a and b in S, $a * b$ is a unique element of S.
2. **Commutative Property** For all a and b in S, $a * b = b * a$.
3. **Associative Property** For all a, b, and c in S, $a * (b * c) = (a * b) * c$.

Recall that the set of real numbers is closed under the operations of addition and multiplication. These operations are both commutative and associative. As the following examples show, not all binary operations are commutative and associative.

EXAMPLE The operation # is defined on the set of real numbers as follows:

$$x \# y = x$$

Determine whether:

a. the set of real numbers is closed under #.

b. # is commutative. c. # is associative.

SOLUTION a. This operation can be described simply as taking the first of the two numbers; you can easily see that the set of real numbers is closed under #.

b. To prove that an operation is *not* commutative, you need find only one example showing that the operation is not commutative. Consider the following statements: $1 \# 2 = 1$; $2 \# 1 = 2$.
Thus, $1 \# 2 \neq 2 \# 1$, so # is not commutative.

c. The following equations show that # is associative:

$$a \# (b \# c) = a \# b = a$$
$$(a \# b) \# c = a \# c = a$$

Therefore, $$a \# (b \# c) = (a \# b) \# c$$

EXAMPLE The operation $*$ is defined on the set of real numbers as follows:

$$x * y = \frac{1}{2}(x + y)$$

Determine whether:

a. the set of real numbers is closed under $*$.

b. $*$ is commutative. c. $*$ is associative.

SOLUTION a. Notice that $*$ involves only the addition and multiplication of real numbers. Since the set of real numbers is closed under addition and multiplication, it is closed under $*$.

b. $x * y = \frac{1}{2}(x + y)$ $\qquad y * x = \frac{1}{2}(y + x)$

Because addition of real numbers is commutative, $\frac{1}{2}(x + y) = \frac{1}{2}(y + x)$,

so $x * y = y * x$.
Therefore, $*$ is commutative.

c. To prove that an operation is not associative, you need find only one example showing that the operation is not associative. Consider the following statements:

$$2 * (4 * 6) = 2 * \left[\frac{1}{2}(4+6)\right] = 2 * 5 = \frac{1}{2}(2+5) = 3\frac{1}{2}$$

$$(2 * 4) * 6 = \left[\frac{1}{2}(2+4)\right] * 6 = 3 * 6 = \frac{1}{2}(3+6) = 4\frac{1}{2}$$

Thus, $2 * (4 * 6) \neq (2 * 4) * 6$, so $*$ is not associative.

In the rest of this chapter we will frequently refer to the familiar subsets of the real numbers listed below.

Name	*Symbol*	*Definition*
Natural Numbers	N	$\{1, 2, 3, \ldots\}$
Whole Numbers	W	$\{0, 1, 2, 3, \ldots\}$
Integers	Z	$\{\ldots, -3, -2, -1, 0, 1, 2, 3, \ldots\}$
Rational Numbers	Q	$\{x: x = \frac{p}{q}$ where $p \in Z$ and $q \in N\}$
Real Numbers	$\mathcal{R}$	$\{x: x \in Q$ or x is irrational$\}$

We will also use Q^+ to indicate the positive rational numbers and $\mathcal{R}^+$ to indicate the positive real numbers. The symbols just presented give us a convenient notation. For example, $(W, +)$ denotes the whole numbers under addition.

An *algebraic system* consists of a set and one or more binary operations that are closed on the set.

EXAMPLES

1. The set of positive even integers under multiplication forms an algebraic system. The product of two positive even integers is known to be positive and must be even according to the result of Exercise 1 on page 85.
2. The system $(Q^+, \div)$, the positive rational numbers under division, forms an algebraic system. We will show that Q^+ is closed under division.
   Any two elements of Q^+ can be represented as:
   $$\frac{p}{q} \text{ and } \frac{r}{s} \text{ where } p, q, r, \text{ and } s \in N.$$
   Their quotient is given by the equation:
   $$\frac{p}{q} \div \frac{r}{s} = \frac{p}{q} \cdot \frac{s}{r} = \frac{ps}{qr}$$
   Since the product of two positive integers is always a positive integer, both ps and qr are elements of N. Therefore, $\frac{ps}{qr}$ is in Q^+.
3. The natural numbers under division do not form an algebraic system because the set N is not closed under division. For example, the quotient $1 \div 2$ is *not* a natural number.

ORAL EXERCISES

1. Addition and multiplication are two binary operations on the real numbers. Name two other familiar binary operations on the real numbers.
2. The examples on page 556 state rules for binary operations # and $*$ on the real numbers. Invent and state your own rule for a binary operation ★ on the real numbers. Then find 1 ★ 2.
3. Is addition of integers associative?
4. Is subtraction of integers commutative?
5. If you subtract one integer from another, is the result always an integer? Is the set of integers closed under subtraction?

The system of integers under multiplication can be written as $(Z, \cdot)$. In Exercises 6–8 describe in words each mathematical system that is expressed in symbols.

6. $(Q, -)$
7. $(W, +)$
8. $(\mathcal{R}^+, \div)$
9. $(N, \cdot)$

WRITTEN EXERCISES

In Exercises 1–10 a binary operation $*$ is defined for real numbers.

a. Is $*$ commutative?
b. Is $*$ associative?

A

1. $a * b = b$
2. $a * b = a - b$
3. $a * b = a \div 3$
4. $a * b = 2ab$
5. $a * b = ab + 1$
6. $a * b = \dfrac{a+b}{3}$
7. $a * b = a + 2b$
8. $a * b = ab^2$
9. $a * b = \sqrt{a^2 b^2}$
10. $a * b = |a - b|$

Exercises 11–14 refer to the following subsets of the real numbers:

$$N,\ W,\ Z,\ Q^+,\ Q,\ \mathcal{R}^+,\ \mathcal{R}$$

11. Which subsets are closed under addition?
12. Which subsets are closed under multiplication?
13. Which subsets are closed under subtraction?
14. Which subsets are closed under division?

Exercises 15–22 refer to the following subsets of the real numbers:

E	the even integers
D	the odd integers
Z^-	the negative integers
M	the integral multiples of 3

Is the given set closed under the given operation?

B

15. $(E, +)$
16. $(D, -)$
17. $(Z^-, -)$
18. $(Z^-, +)$
19. $(M, \div)$
20. $(E, \div)$
21. $(D, \cdot)$
22. $(M, \cdot)$

Not all operations are binary operations. Some operations are applied to a single element of a set to obtain a second (not necessarily different) element of the set. Such operations are called unary operations. For example, the operation SQR defined for the positive real numbers as

$$\text{SQR}(a) = \sqrt{a}$$

is a unary operation. Several other unary operations are defined as follows:

$$\text{ABS}(a) = |a| \qquad \text{RECIP}(a) = \frac{1}{a} \qquad \text{OPP}(a) = -a$$

A set S is said to be closed under a unary operation U if for all a in S, U(a) is a unique element of S.

Exercises 23–26 refer to the following subsets of the real numbers:

$$N,\ W,\ Z,\ Q^+,\ Q,\ \mathcal{R}^+,\ \mathcal{R}$$

23. Which subsets are closed under SQR?

24. Which subsets are closed under ABS?

25. Which subsets are closed under RECIP?

26. Which subsets are closed under OPP?

Section 14-2 IDENTITY ELEMENTS AND INVERSES

In Section 14–1 you learned about algebraic systems and operations within them. In this section you will study the behavior of individual elements within an algebraic system. You will learn these terms:

- identity element
- inverse

Recall that 0 is called the identity element for addition in ($\mathcal{R}$, +) because

$$x + 0 = 0 + x = x$$

for any real number x. Recall also that in the system ($\mathcal{R}$, +), $-x$ is the additive inverse of x. For example, -2 is the inverse of 2 since

$$2 + (-2) = -2 + 2 = 0.$$

These familiar facts should help you to understand the following definitions of an identity element and inverses within an algebraic system (S, $*$).

If there is an element e in S such that

$$a * e = e * a = a \text{ for every } a \text{ in } S,$$

then e is called an *identity element* for $*$.

If e is an identity element in (S, $*$) and if for an element x there is an element x^{-1} in S such that

$$x * x^{-1} = x^{-1} * x = e,$$

then x^{-1} is called an *inverse* of x under $*$. We usually read x^{-1} as "x inverse."

We will now develop a mathematical system based on the numbers shown on an ordinary clock. We will then look for an identity element and inverses within the system. The following example examines the process of adding numbers that represent time in hours.

EXAMPLE If classes begin at 8 o'clock and the school day lasts for 7 hours, at which time is school over? The answer, of course, is 3 o'clock, but how do you get that answer?

SOLUTION 1 One way is to start at 8 on a clock face and move 7 positions clockwise, landing on 3. Thus, school is over at 3 o'clock. Compare this geometric solution with the one below.

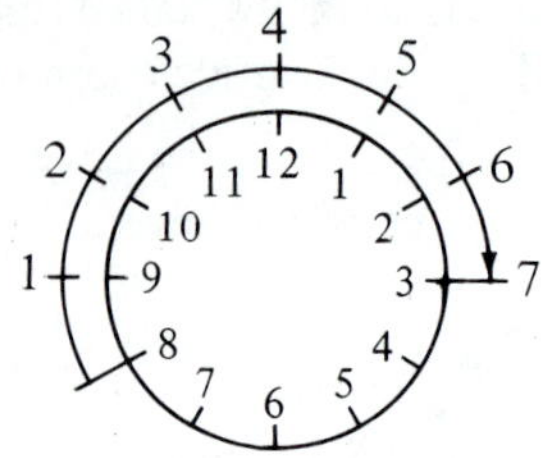

SOLUTION 2 The key to doing clock addition is that whenever we get to 12, we must start over again with 1 (instead of going on to 13). We will denote clock-12 addition with the symbol $\oplus$.

$$\begin{aligned} 8 \oplus 7 &= 8 + (4 + 3) \\ &= (8 + 4) + 3 \\ &= 12 + 3 \\ &= 3 \text{ (o'clock)} \end{aligned}$$

Using the second solution as a guide, we can make up the following addition table for clock-12 numbers, that is, the integers from 1 to 12 inclusive. We will continue to use $\oplus$ to indicate clock-12 additions.

← **column headings**

$\oplus$	1	2	3	4	5	6	7	8	9	10	11	12
1	2	3	4	5	6	7	8	9	10	11	12	1
2	3	4	5	6	7	8	9	10	11	12	1	2
3	4	5	6	7	8	9	10	11	12	1	2	3
4	5	6	7	8	9	10	11	12	1	2	3	4
5	6	7	8	9	10	11	12	1	2	3	4	5
6	7	8	9	10	11	12	1	2	3	4	5	6
7	8	9	10	11	12	1	2	3	4	5	6	7
8	9	10	11	12	1	2	3	4	5	6	7	8
9	10	11	12	1	2	3	4	5	6	7	8	9
10	11	12	1	2	3	4	5	6	7	8	9	10
11	12	1	2	3	4	5	6	7	8	9	10	11
12	1	2	3	4	5	6	7	8	9	10	11	12

↑ **row headings** ↖ **main diagonal**

Table 14–1. Addition of Clock-12 Numbers

EXAMPLE Find **(a)** $3 \oplus 4$ and **(b)** $6 \oplus 9$.

SOLUTION

a. To find $3 \oplus 4$, locate the entry in the row with heading 3 and the column with heading 4. The entry is 7, so $3 \oplus 4 = 7$.

b. The entry in the row with heading 6 and the column with heading 9 is 3, so $6 \oplus 9 = 3$.

Like ordinary addition of whole numbers, clock-12 addition is both commutative and associative. The identity for clock-12 addition, however, is 12, not 0 as in $(W, +)$. For example,

$$5 \oplus 12 = 12 \oplus 5 = 5.$$

Further, every number in the clock-12 addition system has an inverse. For example, 7 and 5 are inverses of each other since

$$7 \oplus 5 = 5 \oplus 7 = 12.$$

If an algebraic system $(S, *)$ can be defined by a table such as the one for clock-12 addition, we can analyze the system by examining the table. Before examining the table, we must make sure that the elements in the row and column headings are arranged in the same order top to bottom and left to right. Note that the headings in the clock-12 addition table are arranged this way.

1. The system is commutative if the table is symmetric about the main diagonal. (See Table 14–1.) Thus, clock-12 addition is commutative since the corresponding table is symmetric about its main diagonal.
2. The system has an *identity element* if the row with heading e matches the column headings of the table, and the column with heading e matches the row headings of the table. The identity element of the system is then e. Thus, the identity element for clock-12 addition is 12.
3. If e is an identity element of $(S, *)$, the *inverse* of a (if it exists) can be found by examining the row with heading a. Look for e in that row. Then note the column heading, say b, for the column in which that e appears.

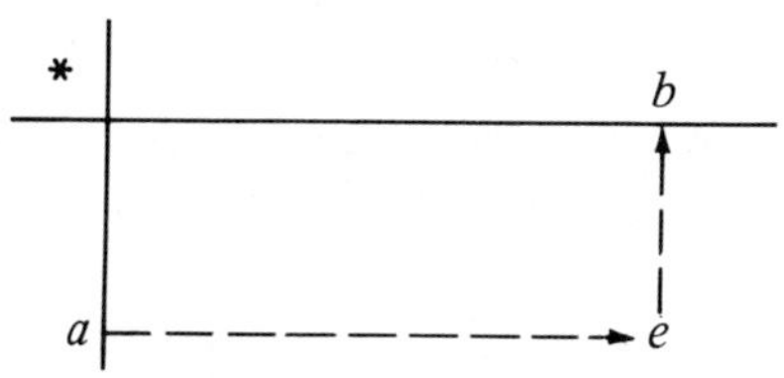

You now know that $a * b = e$. If you find that $b * a = e$ as well, then a and b are inverses of each other. As an example of this process consider finding the inverse of 3 in clock-12 addition.

$\oplus$	1	2	3	4	5	6	7	8	(9)	10	11	12
1	2	3	4	5	6	7	8	9	10	11	12	1
2	3	4	5	6	7	8	9	10	11	12	1	2
(3)	4	5	6	7	8	9	10	11	12	1	2	3

We note that the identity element 12 appears in the column under the 9. Therefore,

$$3 \oplus 9 = 12$$

Since clock-12 addition is commutative,

$$9 \oplus 3 = 12,$$

which enables us to conclude that 3 and 9 are inverses of each other.

It is not easy to tell whether or not a system is associative simply by examining its defining table, as you will see when we analyze the following systems.

EXAMPLE Let $S = \{0, 2, 4, 6, 8\}$. The binary operation $\odot$ is defined on S as taking the last digit of the usual product of the two numbers. The following table can be made for $\odot$.

$\odot$	2	4	6	8
2	4	8	2	6
4	8	6	4	2
6	2	4	6	8
8	6	2	8	4

Since each entry of the table is in S, S is closed under $\odot$, and therefore $(S, \odot)$ is an algebraic system. We can analyze the table in the way just described.

1. Since the table is symmetric about the main diagonal, $\odot$ is commutative.
2. The column beneath the heading 6 matches the row headings, and the row beside the heading 6 matches the column headings. Therefore, 6 is an identity element.
3. Examining the row by the heading 2, we discover that

$$2 \odot 8 = 6.$$

Since $\odot$ is commutative,

$$8 \odot 2 = 6.$$

Thus, 2 and 8 are inverses of each other. The table also shows that $6 \odot 6 = 6$ and $4 \odot 4 = 6$. Thus,

$$2^{-1} = 8 \qquad 4^{-1} = 4 \qquad 6^{-1} = 6 \qquad 8^{-1} = 2$$

It can be shown that $\odot$ is associative because of its relationship to ordinary multiplication. It is also possible to show that $\odot$ is associative by the time-consuming process of verifying 256 equations such as

$$2 \odot (4 \odot 6) = (2 \odot 4) \odot 6.$$

EXAMPLE The operation $*$ is defined as the set $\{x, y, z\}$ by the table at the right.

$*$	x	y	z
x	y	x	z
y	z	x	y
z	x	y	y

Since the table entries for $*$ are all members of the given set, $(\{x, y, z\}, *)$ is an algebraic system.

1. The table for $*$ is not symmetric about the main diagonal since, for example, $x * y = x$ while $y * x = z$. Therefore, $*$ is not commutative.
2. Since no row of table entries matches the column heading, there is no identity element for $*$.
3. Since $*$ has no identity element, it makes no sense to look for inverses.

The following two equations show that $*$ is not associative:

$$x * (y * z) = x * y = x$$
$$(x * y) * z = x * z = z$$

ORAL EXERCISES

1. Name the identity element in $(\mathcal{R}, \cdot)$. Justify your answer by using the definition of identity element on page 559.
2. If x is a nonzero real number, name the inverse of x in $(\mathcal{R}, \cdot)$. Justify your answer by using the definition of inverse on page 559.

Suppose that ordinary clocks used only *ten* numbers, as indicated on the clock shown here. Then you could do "clock-10" addition, which will be symbolized by $\bigtriangleup$. Use this "ten-clock" for Exercises 3 and 4.

3. If classes begin at 8 o'clock and the school day lasts for 7 hours, at which clock-10 time is school over?
4. What is the value of $3 \bigtriangleup 4$? of $6 \bigtriangleup 9$?
5. In the second example on page 562, you saw that $*$ is not commutative in the system $(\{x, y, z\}, *)$ because $x * y = x$ while $y * x = z$.
   a. Name two other elements in the system that do not commute under $*$. Justify your answer by stating the appropriate equations.
   b. Name two elements in the system that *do* commute under $*$. Justify your answer by stating the appropriate equations.

WRITTEN EXERCISES

In Exercises 1–5, use Table 1 on page 560. In Exercises 1–4, find the value of each expression.

A 1. $3 \oplus (7 \oplus 5)$ 2. $(4 \oplus 8) \oplus 2$ 3. $(9 \oplus 9) \oplus (5 \oplus 5)$ 4. $(10 \oplus 6) \oplus (11 \oplus 12)$

5. You saw that 7 is the inverse of 5 and 5 is the inverse of 7 under $\oplus$ because $5 \oplus 7 = 7 \oplus 5 = 12$. Name the inverse of each of the other elements in the clock-12 system. Justify your answers by using the definition of inverse.
6. Refer to the system $(\{0, 2, 4, 6, 8\}, \odot)$ defined in the example at the top of page 562. Demonstrate the Associative Property for the permutations: 4, 6, 8; 8, 6, 6; and 4, 4, 4.

In Exercises 7–12, refer to the system $(\{m, n, r\}, \star)$. A partial table for the commutative operation $\star$ is shown here.

$\star$	m	n	r
m	r	m	
n		n	
r	n	r	r

7. Copy the partial table. Then use the fact that $\star$ is commutative to complete the table.
8. Is there an identity element for $\star$? If so, name it.
9. Does m have an inverse under $\star$? If so, name it. Justify your answer.
10. Does n^{-1} exist? If so, name it. Justify your answer.
11. Does r^{-1} exist? If so, name it. Justify your answer.
12. Find $(m \star m) \star r$ and $m \star (m \star r)$. What can you conclude about the associativity of $\star$?

In Exercises 13–17, you will develop and analyze a "clock-7" mathematical system. It is similar to the clock-12 system except that when you do clock-7 addition you have to start over with 1 whenever you get to 7, not 12. Use the symbol ⊞ for clock-7 addition.

EXAMPLE Find 5 ⊞ 6.

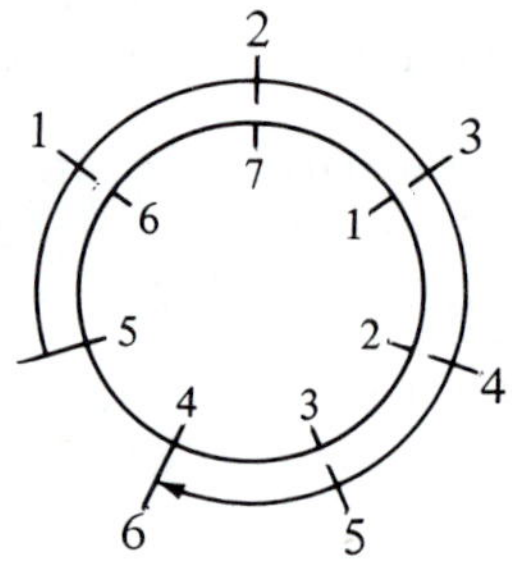

SOLUTION Note that
$$\begin{aligned} 5 + 6 &= 5 + (2 + 4) \\ &= (5 + 2) + 4 \\ &= 7 + 4. \end{aligned}$$

Thus, 5 ⊞ 6 = 4.

A partial addition table for clock-7 numbers is shown here.

⊞	1	2	3	4	5	6	7
1		3		5		7	
2	3		5		7		2
3		5		7		2	
4	5		7		2		4
5		7		2		4	
6	7		2		4		6
7		2		4		6	

13. Follow the method of the solution for the given example to find **(a)** 4 ⊞ 6 and **(b)** 3 ⊞ 5.
14. Copy and complete the addition table for clock-7 numbers.
15. Is ⊞ commutative? How can you tell?
16. Is there an identity element? How can you tell? Name the identity if it exists.
17. Name the inverse of each clock-7 number that has an inverse. Justify your answers by using the definition of inverse.

In Exercises 18–22, refer to the system $(\{a, b, c, d\}, \odot)$. A partial table for the commutative operation $\odot$ is shown at the right.

$\odot$	a	b	c	d
a	d	c	a	
b		b	b	
c			c	
d	a	b	d	d

B 18. Copy and complete the table.
19. Name an identity element, if one exists.
20. Name the inverse for each element that has an inverse.
21. Find $(a \odot b) \odot [(c \odot d) \odot a]$.
22. Prove that $\odot$ is not associative.

In Exercises 23–26, o represents any odd integer, e represents any even integer, p represents any positive real number, n represents any negative real number, and $+$, $\cdot$, and $\div$ are ordinary addition, multiplication, and division. We consider that $e + o = o$, for example, because the sum of any even number and any odd number is an odd number. Copy and complete the table for the given system. Then:

a. State whether or not the operation is commutative.
b. State whether or not the operation is associative.
c. Name an identity element, if one exists.
d. Name the inverse of each element that has an inverse.

23. $(\{e, o\}, +)$

+	e	o
e		
o		

24. $(\{e, o\}, \cdot)$

·	e	o
e		
o		

25. $(\{p, n\}, \div)$

÷	p	n
p		
n		

26. $(\{o, 1\}, \cdot)$

·	o	1
o		
1		

Section 14-3 MODULAR SYSTEMS

In the previous section you worked exercises involving clock-12 addition and clock-7 addition. In this section and the next you will study a type of mathematical system that includes clock arithmetic. You will learn these terms:

- integers mod N
- proper divisor of zero
- modular system

The integers in the set $\{0, 1, 2, \ldots, N-1\}$, with $N \geqslant 2$ are called the *integers mod N*; this set is denoted as Z_N. (The word *mod* is shorthand for *modulo*.)

EXAMPLE If $N = 12$, then $Z_{12} = \{0, 1, 2, 3, 4, 5, 6, 7, 8, 9, 10, 11\}$. Notice that 0 is included and 12 is not.

We can define the operation of addition mod N, denoted $\oplus$, as follows:

$a \oplus b$ is the remainder when the usual sum $a + b$ is divided by N.

We can define the operation of multiplication mod N, denoted $\otimes$, as follows:

$a \otimes b$ is the remainder when the usual product $a \times b$ is divided by N.

EXAMPLE Find the following sums and products in Z_5.

a. $1 \oplus 3 \oplus 3$ b. $3 \otimes 3$

SOLUTION
a. $1 + 3 + 3 = 7$
When 7 is divided by 5, the remainder is 2.
$1 \oplus 3 \oplus 3 = 2$

b. $3 \times 3 = 9$
When 9 is divided by 5, the remainder is 4.
$3 \otimes 3 = 4$

EXAMPLE Compute the following in Z_{12}.

a. $10 \oplus 11$ b. $10 \otimes 9$

SOLUTION
a. $10 + 11 = 21$
When 21 is divided by 12, the remainder is 9.
$10 \oplus 11 = 9$

b. $10 \times 9 = 90$
When 90 is divided by 12, the remainder is 6.
$10 \otimes 9 = 6$

Notice that part (a) of the preceding example gives the same sum as the table for clock-12 addition on page 560. In fact, when you call the identity element 0 instead of 12, the clock-12 integers become the integers mod 12. You can alter the clock-12 addition table on page 560 to make a table for mod 12 addition. (See Exercise 3, page 569.) Below is a table for mod 12 multiplication:

⊗	0	1	2	3	4	5	6	7	8	9	10	11
0	0	0	0	0	0	0	0	0	0	0	0	0
1	0	1	2	3	4	5	6	7	8	9	10	11
2	0	2	4	6	8	10	0	2	4	6	8	10
3	0	3	6	9	0	3	6	9	0	3	6	9
4	0	4	8	0	4	8	0	4	8	0	4	8
5	0	5	10	3	8	1	6	11	4	9	2	7
6	0	6	0	6	0	6	0	6	0	6	0	6
7	0	7	2	9	4	11	6	1	8	3	10	5
8	0	8	4	0	8	4	0	8	4	0	8	4
9	0	9	6	3	0	9	6	3	0	9	6	3
10	0	10	8	6	4	2	0	10	8	6	4	2
11	0	11	10	9	8	7	6	5	4	3	2	1

Table 14–2. Mod 12 Multiplication

The set of integers mod N is closed under the operations ⊕ and ⊗ because the remainders of the divisions must be whole numbers less than N. The commutativity of these operations follows from the commutativity of whole-number addition and multiplication. It can be shown that these operations are associative. It can also be shown that 0 is the identity element for addition, and 1 is the identity element for multiplication in any modulo system (Exercises 44 and 45 on page 570).

In the following discussion of the mod 5 system, we will use + and × to indicate addition and multiplication, respectively. In the rest of this chapter we will use direction lines or other introductory statements to indicate the system that we are using. Study the tables below for Z_5.

+	0	1	2	3	4
0	0	1	2	3	4
1	1	2	3	4	0
2	2	3	4	0	1
3	3	4	0	1	2
4	4	0	1	2	3

Table 14–3. Mod 5 Addition

×	0	1	2	3	4
0	0	0	0	0	0
1	0	1	2	3	4
2	0	2	4	1	3
3	0	3	1	4	2
4	0	4	3	2	1

Table 14–4. Mod 5 Multiplication

The additive inverse of an element a is usually denoted as $-a$. It is easy to see that each element in Z_5 has a unique additive inverse. For example,

$$-3 = 2 \quad \text{because} \quad 3 + 2 = 2 + 3 = 0.$$

Each nonzero element has a multiplicative inverse, denoted a^{-1}. For example,

$$2^{-1} = 3 \quad \text{because} \quad 3 \times 2 = 2 \times 3 = 1.$$

It follows that linear equations in Z_5 have unique solutions.

EXAMPLE Solve $2k = 3$ in Z_5. (Notice that we have written $2k$ for $2 \times k$.)

SOLUTION Multiply both sides of the equation by 2^{-1}, or 3.

$$3 \times 2k = 3 \times 3$$
$$k = 4$$

Therefore, the solution set is $\{4\}$.

Check $2 \times 4 = 3$ (from Table 14-4)

EXAMPLE Solve $3k + 4 = 0$ in Z_5.

SOLUTION Add -4, or 1, to each side of the equation.

$$3k + 4 + 1 = 0 + 1$$
$$3k + 0 = 1$$
$$3k = 1$$

Multiply both sides of the equation by 3^{-1}, or 2.

$$2 \times 3k = 2 \times 1$$
$$k = 2$$

Therefore, the solution set is $\{2\}$.

Check $3 \times 2 + 4 = 1 + 4 = 0$

If each element within a modular system has a unique additive inverse and each nonzero element has a unique multiplicative inverse, then each linear equation will have a unique solution. The first condition, a unique additive inverse for each element, is true of any modular system; the second condition, a unique inverse for each nonzero element, is not.

It is apparent from Table 14-2 that certain elements in the mod 12 system do not have multiplicative inverses (see Exercise 6 on page 569). Notice what happens when we solve simple linear equations in this system.

EXAMPLE Solve $3k = 9$ in the mod 12 system.

SOLUTION In Table 14-2 look at the row beside the 3. Notice that 9 occurs in the columns under 3, 7, and 11. Therefore, the solution set is $\{3, 7, 11\}$.

EXAMPLE Solve $4k + 5 = 0$ in Z_{12}.

SOLUTION Since 7 is the additive inverse of 5,

$$4k + 5 = 0$$
$$4k + 5 + 7 = 0 + 7$$
$$4k = 7$$

Table 14-2 shows that no element of Z_{12} satisfies the last equation. Therefore, the solution set is empty.

A pronounced difference between the real number system and mod 12 is that the Zero-Product Property (page 260) does *not* hold in mod 12:

$$2 \otimes 6 = 0 \qquad 3 \otimes 8 = 0 \qquad 4 \otimes 6 = 0 \qquad 6 \otimes 10 = 0$$

In each of the preceding equations two nonzero elements have zero as their product. Such elements, for example 2, 3, 4, 6, 8, and 10 in Z_{12}, are called *proper divisors of zero*. If you study Table 14-4 carefully, you will see that there are no proper divisors of zero in Z_5. In fact, whenever a modular system is based on a prime number, there are no proper divisors of zero:

> If a and b are in Z_P where P is a prime number,
>
> $$a \times b = 0$$
>
> if and only if $a = 0$ or $b = 0$.

EXAMPLE Solve $(y + 2)(y + 4) = 0$ in Z_5.

SOLUTION

$$y + 2 = 0 \quad \text{or} \quad y + 4 = 0$$
$$y = 3 \quad \text{or} \quad y = 1$$

Therefore, the solution set is $\{1, 3\}$.

Check

$$(3 + 2)(3 + 4) = 0 \times 1 = 0$$
$$(1 + 2)(1 + 4) = 3 \times 0 = 0$$

ORAL EXERCISES

Name the members of each of the following sets.

1. The integers mod 4, or Z_4
2. The integers mod 7, or Z_7
3. Z_8
4. Z_3
5. Z_2
6. Z_{11}
7. How many integers are in the set of integers mod N, or Z_N?

Use Tables 14-3 and 14-4 to compute the following in Z_5.

8. $4 + 3 + 2$
9. $4 \times 4 \times 2$
10. $3 + 4 \times 2$
11. $3 \times 4 + 1 \times 2$

WRITTEN EXERCISES

A

1. Use the definitions of addition and multiplication mod 4 to find the following sums and products in Z_4.
   a. $3 + 2$ b. $1 + 3$ c. 3×2
   d. 2×2 e. $(3 + 3) + 1$ f. $(3 \times 3) \times 2$
2. Use the definitions of addition and multiplication mod 11 to find the following sums and products in Z_{11}.
   a. 5×7 b. 8×3 c. $10 + 7$
   d. $6 + 6$ e. $(3 \times 4) \times 5$ f. $(7 + 8) + 4$

Exercises 9–12 refer to Z_3.

Section 14-3

9. Compute: a. $2 + 1$ b. 2×2

10. Name the additive inverses of 1 and 2.

11. Name the multiplicative inverses of 1 and 2.

12. Solve: a. $2 + x = 1$ b. $2x + 2 = 0$

Section 14-4 GROUPS

In Section 14-3 you learned that the solution set of the equation $3x = 9$ in mod 12 is $\{3, 7, 11\}$. This equation is an example of a linear equation without a unique solution. In this section you will study systems in which any equation of the form $a * x = b$ must have a unique solution.

An algebraic system $(G, *)$, consisting of a set G together with a binary operation $*$, is a *group* provided that the system possesses the following properties:

1. **Associative Property** For all a, b, and c in G, $(a * b) * c = a * (b * c)$.
2. **Existence of an Identity Element** There exists an element e in G such that $a * e = e * a = a$ for all a in G.
3. **Existence of Inverses** For each a in G, there exists an element a^{-1} in G such that $a^{-1} * a = a * a^{-1} = e$.

Notice that a group is closed under its operation since a group is an algebraic system. The identity element e must commute with each element, and each element must commute with its inverse, but the group operation need not be commutative for all pairs of elements. A group in which each pair of elements commutes under the operation is called a *commutative*, or *abelian, group.*

EXAMPLES Each of the following systems is a group.

1. $(Z, +)$, the integers under addition

   The system is closed since the sum of any two integers is an integer. Addition within this set is associative since real-number addition is associative. By inspection, 0 is the identity element. The inverse of any integer n is $-n$, which is also an integer. Thus, every element has an inverse.

2. $(Z_5 \backslash \{0\}, \times)$, the integers mod 5, excluding 0, under multiplication

   Notice that we write $Z_N \backslash \{0\}$ to denote the set Z_N with the element 0 eliminated.

   To obtain the operation table for this system, as shown at the right, we can delete from Table 14-4 the row and column that give products with 0. You can see from the table that this system is closed. Therefore, it is an algebraic system. Associativity in $(Z_5 \backslash \{0\}, \times)$ follows from associativity in $(Z_5, \times)$. By inspection, 1 is the identity element. Inverses, which were found in Exercise 5 on page 569, exist for all elements.

×	1	2	3	4
1	1	2	3	4
2	2	4	1	3
3	3	1	4	2
4	4	3	2	1

Each of the following examples shows a system that is *not* a group.

EXAMPLES

1. $(Z, \times)$, the integers under multiplication

   This system is not a group since no integer except 1 has a multiplicative inverse that is an integer.

2. $(Q^+, +)$, the positive rational numbers under addition

   This system is not a group since Q^+ contains no identity element.

Before we present some theorems about groups, we will state certain properties of equality. These properties of equality are true, regardless of the mathematical system being considered.

If a, b, and c are any elements of a mathematical system, the following properties of equality hold:

Reflexive Property $a = a$

Symmetric Property If $a = b$, then $b = a$.

Transitive Property If $a = b$ and $b = c$, then $a = c$.

Another important property that is true in any mathematical system is the Substitution Property, which was presented in Chapter 1.

Our first theorem about groups states that any group has exactly one identity element.

Theorem 14-1 In a group $(G, *)$, the identity element is unique.

GIVEN: $(G, *)$ is a group with identity elements e and f.
PROVE: $e = f$

Statements	Reasons
1. e and f are identity elements.	1. Given
2. $e * f = f$	2. Def. of identity element e
3. $e * f = e$	3. Def. of identity element f
4. $e = f$	4. Substitution Property

Theorem 14-2 In a group $(G, *)$, the inverse of each element is unique.

Corollary For any element a of a group $(G, *)$, $(a^{-1})^{-1} = a$.

You will prove Theorem 14-2 and its Corollary in Exercises 36 and 37.

When you see an equation in the real number system such as $5x = 5y$, you can immediately conclude that $x = y$. This type of cancellation can be done in any group. Note, however, that separate left and right cancellation laws must be proved since group operations are not necessarily commutative.

Theorem 14-3L In a group $(G, *)$, if $a * x = a * y$, then $x = y$.

GIVEN: $a * x = a * y$
PROVE: $x = y$

Statements	Reasons
1. The inverse of a, a^{-1}, exists.	1. Existence of inverses (Def. of a group)
2. $a * x = a * y$	2. Given
3. $a^{-1} * (a * x) = a^{-1} * (a * x)$	3. Reflexive Property
4. $a^{-1} * (a * x) = a^{-1} * (a * y)$	4. Substitution Prop. (Steps 1 and 3)
5. $(a^{-1} * a) * x = (a^{-1} * a) * y$	5. Associative Prop. (Def. of a group)
6. $e * x = e * y$	6. Def. of inverse
7. $x = y$	7. Def. of identity element e

A similar right cancellation law can be proved (Exercise 35 on page 574):

Theorem 14-3R In a group $(G, *)$, if $x * a = y * a$, then $x = y$.

Consider solving the equation $a * x = b$ within a group. Since a must have a unique inverse a^{-1}, we may proceed as follows:

$$a * x = b$$
$$a^{-1} * (a * x) = a^{-1} * (a * x)$$
$$a^{-1} * (a * x) = a^{-1} * b$$
$$(a^{-1} * a) * x = a^{-1} * b$$
$$e * x = a^{-1} * b$$
$$x = a^{-1} * b$$

This result is part (a) of the following theorem:

Theorem 14-4 For all a and b in a group $(G, *)$,

a. $a * x = b$ has as its unique solution: $x = a^{-1} * b$
b. $x * a = b$ has as its unique solution: $x = b * a^{-1}$

Exercise 38 on page 575 calls for a proof of this theorem.

ORAL EXERCISES

Use the definition of a group to explain why the following systems are groups. For each group state the identity and identify the inverse of the element 2.

1. $(\mathcal{R}, +)$
2. $(Z_{12}, +)$
3. $(Z_7 \setminus \{0\}, \times)$
4. Which of the systems in Exercises 1-3 are commutative groups?
5. Why is it necessary to exclude 0 in order for the remaining elements in Z_7 to form a group under multiplication?
6. Which property is not necessary for a system to be a group?
   a. Associative Property
   b. Existence of Inverses
   c. Closure Property
   d. Commutative Property

WRITTEN EXERCISES

Use the properties of real numbers and integers mod N to decide whether or not each of the following systems is a group. If the system is a group, state the identity element and identify the inverses. If the system is not a group, explain why not.

A

1. $(Q, +)$
2. $(Q, \times)$
3. $(Q^+, +)$
4. $(Q^+, \times)$
5. $(Q\backslash\{0\}, \times)$
6. $(Z, \times)$
7. $(Z\backslash\{0\}, \times)$
8. $(W, +)$
9. $(\mathcal{R}, \times)$
10. $(\mathcal{R}\backslash\{0\}, \times)$
11. $(\mathcal{R}^+, \times)$
12. $(Z_7, +)$
13. $(Z_3, +)$
14. $(Z_4, +)$

If the system is a group, state the identity element and identify the inverses. If the system is not a group, explain why not.

15.

#	x	y	z
x	x	y	z
y	y	z	x
z	z	x	y

16.

*	a	b	c
a	a	b	c
b	b	a	a
c	c	c	b

B

17.

·	□	○	△	▱
□	▱	△	□	○
○	△	▱	○	□
△	□	○	△	▱
▱	○	□	▱	△

18.

·	N	P	Q	R	S	T
N	N	P	Q	R	S	T
P	P	Q	N	S	T	R
Q	Q	N	P	T	R	S
R	R	T	S	N	Q	P
S	S	R	T	P	N	Q
T	T	S	R	Q	P	N

Use the properties of real numbers and integers mod N to decide whether or not each of the following systems is a group. If the system is a group, state the identity element and identify the inverses. If the system is not a group, explain why not.

19. $(Z_{12}\backslash\{0\}, \times)$
20. $(Z_3, \times)$
21. $(Z_3\backslash\{0\}, \times)$
22. $(Z_4\backslash\{0\}, \times)$
23. $(Z_4, \times)$
24. $(Z_9\backslash\{0\}, \times)$
25. $(Z_N, \times)$, where N is an integer $\geqslant 2$
26. $(Z_N, +)$, where N is an integer $\geqslant 2$
27. $(W, *)$, where $a * b = |a - b|$
28. $(Z, *)$, where $a * b = 2a + 2b$
29. $(\mathcal{R}^+, *)$, where $a * b = 2ab$
30. $(\mathcal{R}, *)$, where $a * b = \sqrt{a^2 + b^2}$

In Exercises 31-34, O is the set of odd integers and E is the set of even integers. Decide whether or not each of the following systems is a group. If the system is a group, state the identity element and identify the inverse of an element k in the group. If the system is not a group, explain why not.

31. $(O, +)$
32. $(E, +)$
33. $(O, *)$, where $a * b = a + b + 5$
34. $(E, *)$, where $a * b = a + b + 10$
35. Prove Theorem 14-3R on page 521.

36. Supply the reasons for the steps of the following proof of Theorem 14-2.

Given: a' and a'' are both inverses of a in $(G, *)$.
Prove: $a' = a''$

Statements	Reasons
1. a' and a'' are both inverses of a in $(G, *)$.	1. ?
2. $(a' * a) * a'' = (a' * a) * a''$	2. ?
3. $a' * (a * a'') = (a' * a) * a''$	3. ?
4. $a' * e = e * a''$	4. ?
5. $a' = a''$	5. ?

37. Write a two-column proof of the corollary to Theorem 14-2.

Given: a is an element of the group $(G, *)$.
Prove: $(a^{-1})^{-1} = a$

38. **a.** Supply the reasons for the steps of the following proof of Theorem 14-4(a) on page 573. We will prove this theorem in two steps: (1) a proof that the solution derived on page 573 is a solution of the equation; (2) a proof that the solution to the equation is unique.

(1) Given: a and b are elements of the group $(G, *)$.
Prove: $a * (a^{-1} * b) = b$

Statements	Reasons
1. $a * (a^{-1} * b) = (a * a^{-1}) * b$	1. ?
2. $(a * a^{-1}) * b = e * b$	2. ?
3. $e * b = b$	3. ?
4. $a * (a^{-1} * b) = b$	4. ?

(2) Given: $a * y = b$ and $a * z = b$ in $(G, *)$
Prove: $y = z$

Statements	Reasons
1. $a * y = b$ and $a * z = b$ in $(G, *)$.	1. ?
2. $a * y = a * z$	2. ?
3. $y = z$	3. ?

b. Follow the outline of the proof of Theorem 14-4(a) in part (a) to prove Theorem 14-4(b).

Section 14-5 SPECIAL GROUPS

In the previous section you learned the definition of a group. In this section you will study some particularly interesting groups. You will learn these terms:

- permutation group
- subgroup
- rigid motions of the rectangle

Consider cups numbered 1, 2, and 3 that contain, respectively, a red marble, a green marble, and a yellow marble, as shown at the top of the diagram. Suppose that you rearranged the marbles according to the following instructions: move the red marble to cup 2; move the green marble to cup 1; keep the yellow marble in cup 3. The resulting arrangement of marbles, which is a green marble in cup 1, a red marble in cup 2, and a yellow marble in cup 3, is shown at the bottom of the diagram.

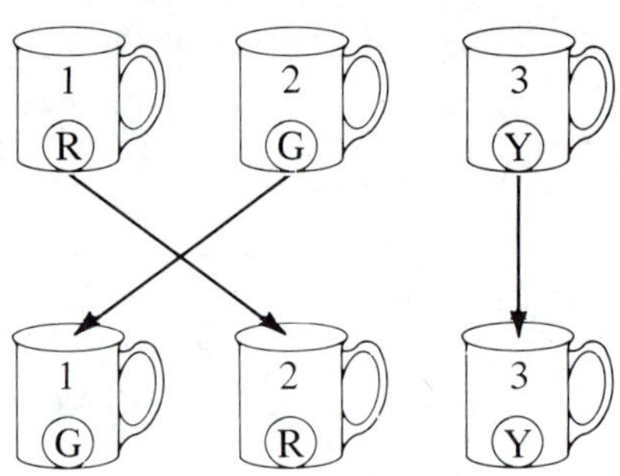

We can write the set of instructions just discussed as follows:

$$\begin{matrix} 1 & 2 & 3 \\ \downarrow & \downarrow & \downarrow \\ 2 & 1 & 3 \end{matrix} \quad \text{or} \quad \begin{pmatrix} 1 & 2 & 3 \\ 2 & 1 & 3 \end{pmatrix}$$

The set of instructions above can be referred to as a permutation since it determines a rearrangement of objects. In Chapter 13 you learned that there are six such permutations of the digits 1, 2, and 3. We will use these permutations to define the elements of a group:

$$A: \begin{pmatrix} 1 & 2 & 3 \\ 1 & 2 & 3 \end{pmatrix} \qquad B: \begin{pmatrix} 1 & 2 & 3 \\ 1 & 3 & 2 \end{pmatrix} \qquad C: \begin{pmatrix} 1 & 2 & 3 \\ 2 & 1 & 3 \end{pmatrix}$$

$$D: \begin{pmatrix} 1 & 2 & 3 \\ 2 & 3 & 1 \end{pmatrix} \qquad E: \begin{pmatrix} 1 & 2 & 3 \\ 3 & 1 & 2 \end{pmatrix} \qquad F: \begin{pmatrix} 1 & 2 & 3 \\ 3 & 2 & 1 \end{pmatrix}$$

Notice that C is the permutation discussed previously. Each permutation is determined by its columns, which may be rearranged. For example, we can write B as:

$$B: \begin{pmatrix} 3 & 1 & 2 \\ 2 & 1 & 3 \end{pmatrix}$$

We will define the operation $*$ on a set of permutations to mean "follow the first permutation by the second permutation." For example, $E * B$ means "follow E by B."

EXAMPLE

Determine the effect of $E * B$ on the original arrangement of marbles, which was a red marble in cup 1, a green marble in cup 2, and a yellow marble in cup 3.

SOLUTION

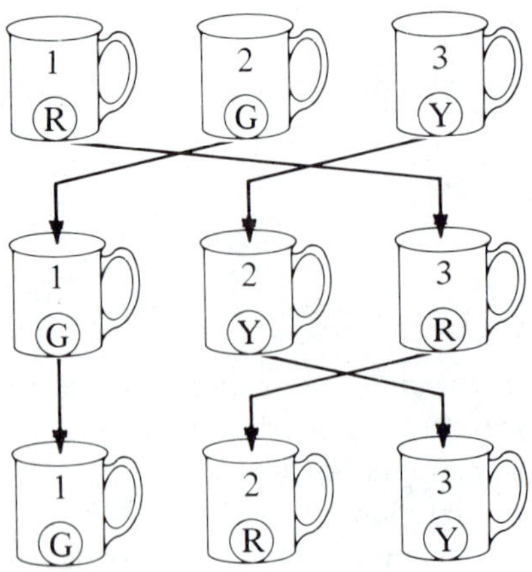

By studying the diagram, you should be able to see that: (1) applying E to the first row of the diagram results in the second row; (2) applying B to the second row of the diagram results in the third row.

Notice that the final arrangement of marbles is a green marble in cup 1, a red marble in cup 2, and a yellow marble in cup 3. Thus, applying $E * B$ gives the same results as applying C. For this reason, we can state that

$$E * B = C$$

We can also obtain $E * B$ by the symbolic method shown in part (a) below. We arrange the columns of B under E so that the first row of B matches the second row of E. The resulting element is found by omitting the two middle rows.

EXAMPLE Calculate: a. $E * B$ b. $F * B$ c. $B * F$

SOLUTION

a.
$$\begin{array}{cl} \begin{pmatrix} 1 & 2 & 3 \\ 3 & 1 & 2 \end{pmatrix} & E \\ * \begin{pmatrix} 3 & 1 & 2 \\ 2 & 1 & 3 \end{pmatrix} & B \\ \hline \begin{pmatrix} 1 & 2 & 3 \\ 2 & 1 & 3 \end{pmatrix} & C \end{array}$$
$E * B = C$

b.
$$\begin{array}{cl} \begin{pmatrix} 1 & 2 & 3 \\ 3 & 2 & 1 \end{pmatrix} & F \\ * \begin{pmatrix} 3 & 2 & 1 \\ 2 & 3 & 1 \end{pmatrix} & B \\ \hline \begin{pmatrix} 1 & 2 & 3 \\ 2 & 3 & 1 \end{pmatrix} & D \end{array}$$
$F * B = D$

c.
$$\begin{array}{cl} \begin{pmatrix} 1 & 2 & 3 \\ 1 & 3 & 2 \end{pmatrix} & B \\ * \begin{pmatrix} 1 & 3 & 2 \\ 3 & 1 & 2 \end{pmatrix} & F \\ \hline \begin{pmatrix} 1 & 2 & 3 \\ 3 & 1 & 2 \end{pmatrix} & E \end{array}$$
$B * F = E$

This example shows that $F * B = D$ and $B * F = E$. Thus, $*$ is not commutative. Nevertheless, it can be shown that $(\{A, B, C, D, E, F\}, *)$ is a group. This group is called a *permutation group*, and it is the first group you have seen that is *not commutative*. It is interesting to note that any group with fewer than 6 elements must be commutative.

When you complete the operation table for $*$ (Exercise 1 on page 578), you will see that $\{A, B, C, D, E, F\}$ is closed under $*$. If you think about how element A behaves in situations like the first example, you will see that A is the identity element. It can be shown that $*$ is associative. The operation table reveals that each element has an inverse (Exercise 2 on page 578).

For many groups $(G, *)$, it is possible to find one or more subsets of G that are also groups under the same operation $*$. If H is a *subset* of G and *if the system $(H, *)$ is itself a group*, then $(H, *)$ is called a *subgroup* of $(G, *)$. Note that $(G, *)$ and $(e, *)$, where e is the identity element of $(G, *)$, are always subgroups of the group $(G, *)$.

$*$	e
e	e

EXAMPLE Does $\{1, 2, 4\}$ form a subgroup of $(Z_7\backslash\{0\}, \times)$?

SOLUTION $\{1, 2, 4\}$ is a subgroup of $(Z_7\backslash\{0\}, \times)$. The operation table at the right shows that this set is closed under $\times$. The identity element is 1: $1^{-1} = 1$; $2^{-1} = 4$; $4^{-1} = 2$.

$\times$	1	2	4
1	1	2	4
2	2	4	1
4	4	1	2

ORAL EXERCISES

In Exercises 1-6 you are given the six permutations of the digits 1, 2, and 3. The columns have been rearranged. Match each permutation with the correct element of the permutation group $(\{A, B, C, D, E, F\}, *)$.

1. $\begin{pmatrix} 1 & 3 & 2 \\ 3 & 1 & 2 \end{pmatrix}$
2. $\begin{pmatrix} 3 & 1 & 2 \\ 3 & 2 & 1 \end{pmatrix}$
3. $\begin{pmatrix} 2 & 1 & 3 \\ 2 & 1 & 3 \end{pmatrix}$
4. $\begin{pmatrix} 3 & 2 & 1 \\ 1 & 3 & 2 \end{pmatrix}$
5. $\begin{pmatrix} 1 & 3 & 2 \\ 3 & 2 & 1 \end{pmatrix}$
6. $\begin{pmatrix} 2 & 3 & 1 \\ 3 & 2 & 1 \end{pmatrix}$

In Exercises 7 and 8 state the equation in $(\{A, B, C, D, E, F\}, *)$ that corresponds to the permutations described symbolically.

7. $$\begin{array}{c} \begin{pmatrix} 1 & 2 & 3 \\ 3 & 2 & 1 \end{pmatrix} \\ * \begin{pmatrix} 3 & 2 & 1 \\ 1 & 3 & 2 \end{pmatrix} \\ \hline \begin{pmatrix} 1 & 2 & 3 \\ 1 & 3 & 2 \end{pmatrix} \end{array}$$

8. $$\begin{array}{c} \begin{pmatrix} 1 & 2 & 3 \\ 3 & 1 & 2 \end{pmatrix} \\ * \begin{pmatrix} 3 & 1 & 2 \\ 3 & 2 & 1 \end{pmatrix} \\ \hline \begin{pmatrix} 1 & 2 & 3 \\ 3 & 2 & 1 \end{pmatrix} \end{array}$$

WRITTEN EXERCISES

A **1.** Complete the operation table for the permutation group $(\{A, B, C, D, E, F\}, *)$.

$*$	A	B	C	D	E	F
A		B		D		F
B	B					E
C						
D	D					
E		C				
F	F	D		B		

2. Examine the operation table in Exercise 1 to find the inverse of each element in the group $(\{A, B, C, D, E, F\}, *)$.

3. Use the table in Exercise 1 to show that for the permutation group $(\{A, B, C, D, E, F\}, *)$ the associative property holds in the given instances.

a. $C * (F * B) = (C * F) * B$ **b.** $E * (C * E) = (E * C) * E$

State whether or not the given subset forms a subgroup of the given group. If it does, make an operation table and identify the identity element and the inverses of the elements. If the given subset does not form a subgroup, state why.

4. $(Z_7 \backslash \{0\}, \times)$; $\{1, 6\}$ **5.** $(Z_6, +)$; $\{0, 2, 4\}$ **6.** $(Z_7, +)$; $\{1, 3, 5\}$

7. $(Z_7 \backslash \{0\}, \times)$; $\{1, 3, 5\}$ **8.** $(Z, +)$; $\{0\}$ **9.** $(Q \backslash \{0\}, \times)$; $\{1\}$

10. $(Z_{11} \backslash \{0\}, \times)$; $\{1, 10\}$ **11.** $(Z_5, +)$; $\{0, 2, 4\}$ **12.** $(Z_{13} \backslash \{0\}, \times)$; $\{1, 3, 9\}$

In Exercises 13–16: **a.** Make an operation table for the given group.
b. List all subsets that form subgroups of the given group.

B **13.** $(Z_4, +)$ **14.** $(Z_5 \backslash \{0\}, \times)$ **15.** $(Z_5, +)$ **16.** $(Z_6, +)$

Assume that rectangle $ABCD$ is positioned as shown with the axes of a rectangular coordinate system as the perpendicular bisectors of its sides. Consider the following four *rigid motions of the rectangle.* In each diagram the letters inside the rectangle identify the locations of the vertices after the rotation is completed.

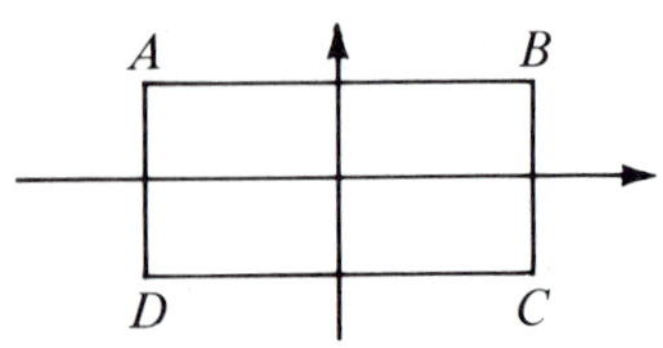

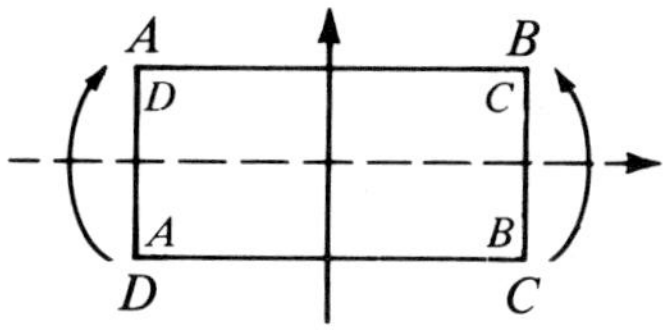

H: The reflection of rectangle $ABCD$ across the horizontal axis

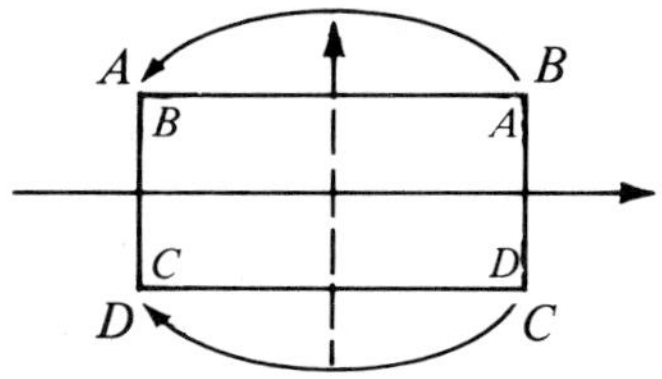

V: The reflection of rectangle $ABCD$ across the vertical axis

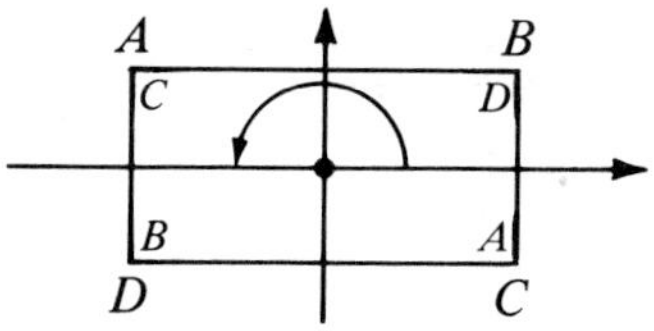

R: A 180° rotation of rectangle $ABCD$ about the origin

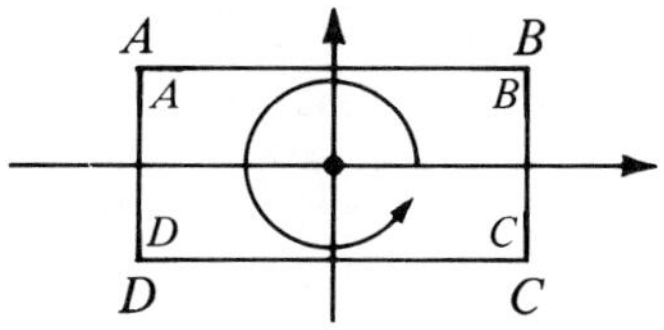

S: A 360° rotation of rectangle $ABCD$ about the origin

If we define the operation $*$ such that $H * V$ means H followed by V, then $(\{H, V, R, S\}, *)$ is a group.

17. Complete the operation table for the group $(\{H, V, R, S\}, *)$.

18. Is the group commutative?

19. State the identity element of the group.

20. State the inverse of each element of the group.

$*$	H	V	R	S
H	S	R		
V				
R			S	
S	H			

State whether the given subset is a subgroup of the given group. If it is, make an operation table and state the identity element. If it is not, state why.

21. $(Z_{31}\backslash\{0\}, \times)$; $\{1, 2, 4, 8, 16\}$

22. $(Z_{127}\backslash\{0\}, \times)$; $\{1, 2, 4, 8, 16, 32, 64\}$

23. **a.** Show that $(\{1, 2, 4, 8\}, \times)$ forms a group when $\times$ is defined to be multiplication mod 15.

b. Why is it not proper to call $(\{1, 2, 4, 8\}, \times)$ a subgroup of $(Z_{15}\backslash\{0\}, \times)$?

24. The subsets in Exercises 21–23 and in the example on page 577 fall into the pattern $\left\{1, 2, 4, \ldots, \frac{N+1}{2}\right\}$. State two facts about N that appear to be requirements for $\left\{1, 2, 4, \ldots, \frac{N+1}{2}\right\}$ to form a subgroup of $(Z_N\backslash\{0\}, \times)$.

C 25. Exercises 4 and 10 are examples of the following theorem:

If p is prime, then $\{1, p-1\}$ forms a subgroup of the group $(Z_p\backslash\{0\}, \times)$.

Show that the theorem is true.

SELF-TEST 2

1. State the three properties that an algebraic system $(S, *)$ must have to be a group. Section 14-4

In Exercises 2-6, decide whether or not each of the systems is a group. If the system is a group, state the identity element and identify the inverse of 4. If the system is not a group, explain why not.

2. $(Q^+, \cdot)$
3. $(\mathcal{R}^+, +)$
4. $(Z_6, +)$
5. $(Z_6 \setminus \{0\}, \times)$

6. $(\{p, q, r, s\}, *)$ where $*$ is defined by the operation table at the right.

7. If k were an element of the table in Exercise 6, which part of the definition of a group would not be satisfied?

$*$	p	q	r	s
p	p	p	p	p
q	p	p	q	p
r	p	q	r	s
s	p	p	s	p

8. Refer to the group of permutations defined on page 576 and the corresponding operation table in Exercise 1 on page 578. Verify the associative property in the following example: $D * (D * D) = (D * D) * D$ Section 14-5

9. State whether $\{0, 2, 4, 6\}$ forms a subgroup of $(Z_8, +)$. If it does, make an operation table and identify the identity element and the inverses of the elements. If it does not, explain why.

Section 14-6 FIELDS

In this section and the next you will study a type of mathematical system that consists of a set of elements and *two* operations. This type of system possesses many of the properties of the real numbers under addition and multiplication. This type of system is called a field.

Recall that the Distributive Property (page 9) relates the two operations real-number addition and multiplication as follows: $a(b + c) = ab + ac$. Informally, we say that multiplication distributes over addition.

EXAMPLE 1 $4(3 + 6) = 4 \cdot 3 + 4 \cdot 6.$

EXAMPLE 2 $4(3 \div 6) = 2 \neq \frac{1}{2} = (4 \cdot 3) \div (4 \cdot 6)$

The second example illustrates that multiplication does not distribute over division.

Before defining a field formally, we will note that a field is a mathematical system closed under two commutative operations that are related by a distributive property. As you will see, the two operations have other properties that are similar to those possessed by real-number addition and multiplication. For this reason, one field operation will be referred to as addition (+) and the other as multiplication ($\cdot$). It is also convenient to refer to the identity element for addition as 0 and the identity element for multiplication as 1.

An algebraic system $(F, +, \cdot)$, consisting of a set and two commutative operations, that possesses the following properties is called a *field*:

1. The elements of F under $+$ form a commutative group.
2. The elements of F excluding the identity element for $+$, which will be denoted as 0, form a commutative group under $\cdot$.
3. **Distributive Property** For all elements a, b, and c in F,

$$a(b+c) = ab + ac \quad \text{and} \quad (b+c)a = ba + ca.$$

EXAMPLE $(\mathcal{R}, +, \cdot)$ is a field since:

1. $(\mathcal{R}, +)$ is a commutative group.
2. $(\mathcal{R}\setminus\{0\}, \cdot)$ is a commutative group.
3. $a(b+c) = ab + ac$ and $(b+c)a = ba + ca$ for all $a, b, c \in \mathcal{R}$.

EXAMPLE $(Z_7, +, \times)$ is a field since:

1. In any modular system, $+$ and $\times$ are commutative operations.
2. $(Z_7, +)$ is a commutative group (Exercise 12 on page 574).
3. In Exercise 20 on page 569, you showed that multiplicative inverses exist for the elements of $Z_7\setminus\{0\}$. Therefore, $(Z_7\setminus\{0\}, \times)$ is a commutative group.
4. The Distributive Property follows from ordinary integer addition although we will not prove it. For example, $4(5+3) = 4 \cdot 1 = 4$
   $4 \cdot 5 + 4 \cdot 3 = 6 + 5 = 4$

EXAMPLE $(Z_{12}, +, \times)$ is not a field because $(Z_{12}\setminus\{0\}, \times)$ is not a group (Exercise 19 on page 574).

The theorems proved for groups can be applied to fields. The following theorem is an important result of the Distributive Property.

Theorem 14–5 If a is an element of the field $(F, +, \cdot)$, $a \cdot 0 = 0$ and $0 \cdot a = 0$. (Multiplicative Property of 0)

GIVEN: a is an element of the field $(F, +, \cdot)$, and 0 is the identity for $+$.
PROVE: $a \cdot 0 = 0$ and $0 \cdot a = 0$

Statements	Reasons
1. $0 + 0 = 0$	1. Def. of identity element
2. $a(0+0) = a \cdot 0$	2. Substitution Property
3. $a(0+0) = a \cdot 0 + a \cdot 0$	3. Distributive Property
4. $a \cdot 0 + a \cdot 0 = a \cdot 0$	4. Substitution Property (Steps 2 and 3)
5.* $\quad = a \cdot 0 + 0$	5. Def. of identity element
6. $a \cdot 0 + a \cdot 0 = a \cdot 0 + 0$	6. Transitive Prop. of = (Steps 4 and 5)
7. $a \cdot 0 = 0$	7. Cancellation (Theorem 14-3L)
8. $0 \cdot a = 0$	8. Commutative Property

*When the left side of an equation in a step is the same expression as the right side of the equation in the previous step, we will not write the expression again.

The following theorem can also be proved (Exercise 22 on page 583). Be aware that -1 stands for the additive inverse of the multiplicative identity element.

Theorem 14-6 If a is any element of the field $(F, +, \cdot)$, $a(-1) = -a$ and $(-1)a = -a$. (Multiplicative Property of -1)

EXAMPLE In $(Z_7, +, \times)$:

$$6(-1) = -6 = 1$$
$$(-1)5 = -5 = 2$$

The following corollaries to Theorem 14-6 can be proved (Exercises 23-25 on page 584).

If a and b are any elements of the field $(F, +, \cdot)$:

Corollary 1 $(-a)b = -(ab)$

Corollary 2 $a(-b) = -(ab)$

Corollary 3 $(-a)(-b) = ab$

EXAMPLE In $(Z_5, +, \times)$:

$$(-2)(3) = -(2 \times 3) = -1 = 4$$
$$4(-4) = -(4 \times 4) = -1 = 4$$
$$(-3)(-1) = (3 \times 1) = 3$$

ORAL EXERCISES

In Exercises 1-4, you are given a partial statement of the distributive property for two operations on the real numbers.

a. Complete the statement of the distributive property for the given operation.
b. State whether or not the distributive property is true for the two operations.

1. multiplication over addition
   $a \cdot (b + c) = \underline{\ ?\ }$
2. addition over multiplication
   $a + (b \cdot c) = \underline{\ ?\ }$
3. exponentiation over addition
   $(a + b)^n = \underline{\ ?\ }$
4. exponentiation over multiplication
   $(a \cdot b)^n = \underline{\ ?\ }$
5. The system $(\{a, b, c\}, +, \cdot)$ is a field as indicated by its operation tables below.
   a. Identify the identity elements for addition and multiplication.
   b. Identify the inverses for each element under addition and multiplication.
   c. Explain how the tables show the Multiplicative Property of 0.

+	*a*	*b*	*c*
a	*a*	*b*	*c*
b	*b*	*c*	*a*
c	*c*	*a*	*b*

·	*a*	*b*	*c*
a	*a*	*a*	*a*
b	*a*	*c*	*b*
c	*a*	*b*	*c*

WRITTEN EXERCISES

State whether or not the given system is a field. If the system is a field, state the identity element for each operation and identify the inverses of the elements. If the system is not a field, state at least one of the three field properties listed on page 581 that it fails to satisfy.

A 1. $(Q, +, \cdot)$ 2. $(Z, +, \cdot)$ 3. $(Q^+, +, \cdot)$

4. $(Z_5, +, \times)$ 5. $(Z_6, +, \times)$ 6. $(Z_7, +, \times)$

7. $(Z_9, +, \times)$ 8. $(Z_7\backslash\{0\}, +, \times)$ 9. $(\{-1, 0, 1\}, +, \cdot)$

10. $(Q_0^+, \#, \cdot)$, where $a \# b = |a - b|$ and Q_0^+ is the set of positive rational numbers together with zero

Use the multiplicative property of -1 and the corollaries on page 582 to evaluate the given expression in $(Z_7, +, \times)$.

11. $3(-1)$ 12. $(-1)2$ 13. $(-3)2$

14. $(-4)(-5)$ 15. $(-2)6$ 16. $4(-1)$

17. $(-2)(-5)$ 18. $3(-6)$ 19. $(-3)3$

20. Let $E_{10} = \{0, 2, 4, 6, 8\}$ and let $+$ and $\times$ be addition and multiplication mod 10. (It can be shown that $(E_{10}, +, \times)$ is a field.)
    a. Make operation tables for $+$ and $\times$ to verify that E_{10} is closed under both operations.
    b. State the identity elements for $+$ and $\times$.
    c. List the inverses for $+$ and $\times$.

B 21. Repeat Exercise 20 for the field $(O_{10}, \#, \times)$ where $O_{10} = \{1, 3, 5, 7, 9\}$, $a \# b = a + b + 5$ in mod 10, and $\times$ is multiplication mod 10.

22. Supply the reasons for the steps of the following proof of Theorem 14-6.

Given: a is any element of the field $(F, +, \cdot)$
Prove: $a(-1) = -a$ and $(-1)a = -a$

Statements	Reasons
1. a is an element of $(F, +, \cdot)$.	1. ?
2. $a \cdot 0 = 0$	2. ?
3. $-1 + 1 = 0$	3. ?
4. $a(-1 + 1) = 0$	4. ?
5. $a(-1 + 1) = a(-1) + a \cdot 1$	5. ?
6. $a(-1) + a \cdot 1 = 0$	6. ?
7. $a(-1) + a = 0$	7. ?
8. $a(-1) = -a$	8. ?
9. $(-1)a = -a$	9. ?

Exercises 23-25 prove Corollaries 1-3 of Theorem 14-6. Write two-column proofs.

23. Given: a and b are elements of the field $(F, +, \cdot)$.
Prove: $(-a)b = -ab$

24. Given: a and b are elements of the field $(F, +, \cdot)$.
Prove: $a(-b) = -(ab)$

25. Given: a and b are elements of the field $(F, +, \cdot)$.
Prove: $(-a)(-b) = ab$

26. Let $*$ be ordinary multiplication and let # be an operation such that $a \# b = \frac{a+b}{2}$, where a and b are any real numbers. Prove that $*$ is distributive over #.

C 27. Let $P = \{a + b\sqrt{2}\}$ where a and b are any rational numbers. State the values of a and b for the given element of P.

a. $3 + 5\sqrt{2}$ **b.** $-1 + \sqrt{2}$ **c.** 6 **d.** $\frac{1}{3}\sqrt{2}$

28. Prove the following facts about the system $(P, +, \cdot)$ defined in Exercise 27. (The system *is*, in fact, a field.)

a. P is closed under addition. (*Hint*: Show that $(w + x\sqrt{2}) + (y + z\sqrt{2})$ can be written in the form $(a + b\sqrt{2}.)$)

b. P is closed under multiplication.

c. P contains identity elements for addition and multiplication.

d. The system satisfies the inverse requirements of a field. (*Hint*: Begin by choosing a general element $a + b\sqrt{2}$; use the fact that $a^2 - 2b^2 \neq 0$ whenever a and b are rational numbers that are not both 0.)

Section 14-7 SOLVING EQUATIONS IN FIELDS

In this section you will learn more about multiplication in a field. In fact, you will learn that a zero-product property can be proved for any field. You will then learn to solve certain equations in a field.

You will prove the next theorem in Exercise 27 on page 587.

Theorem 14-7 For any elements a and b of the field $(F, +, \cdot)$:

$$ab = 0$$

if and only if $a = 0$ or $b = 0$. (Zero-Product Property)

We often find it convenient to write the multiplicative inverse of an element a as $\frac{1}{a}$ instead of a^{-1}. We also can refer to this inverse as the reciprocal of a. The following two theorems (Exercises 28 and 29 on pages 587 and 588) can be proved.

Theorem 14–8 For any two elements a and b of the field $(F, +, \cdot)$ such that $a \neq 0$ and $b \neq 0$:

$$\frac{1}{ab} = \frac{1}{a} \cdot \frac{1}{b}$$

Theorem 14–9 For any element a of the field $(F, +, \cdot)$ such that $a \neq 0$:

$$\frac{1}{-a} = -\frac{1}{a}$$

Suppose that a and b are elements of a field and that $a \neq 0$. Below we give the main steps in the derivation of a solution for the general linear equation:

$$\begin{aligned} ax + b &= 0 \\ ax + b + (-b) &= 0 + (-b) \\ ax + 0 &= -b \\ ax &= -b \\ x &= \frac{1}{a} \cdot (-b) \\ x &= -b \cdot \frac{1}{a} \end{aligned}$$

This result is proved as the next theorem. The proof (Exercise 30 on page 588) proves that the prospective solution is actually a solution to the equation and that the solution is unique.

Theorem 14–10 If a and b are elements of the field $(F, +, \cdot)$ and $a \neq 0$, then the equation

$$ax + b = 0$$

has $x = -b \cdot \frac{1}{a}$ as its unique solution.

EXAMPLE Solve $4x + 2 = 0$ in $(Z_5, +, \times)$.

SOLUTION Since there are so few elements in Z_5, the most efficient way to solve this equation is to test the elements of Z_5, one by one, to see which is the solution. We find that

$$\begin{aligned} 4 \times 0 + 2 &= 0 + 2 = 2 \\ 4 \times 1 + 2 &= 4 + 2 = 1 \\ 4 \times 2 + 2 &= 3 + 2 = 0 \end{aligned}$$

Therefore, the solution set is $\{2\}$.

This result agrees with that of the theorem.

We can apply the theorem as follows:

$$-2 = 3 \qquad 4^{-1} = 4$$
$$x = -2 \times 4^{-1} = 3 \times 4 = 2$$

Therefore, the solution set is $\{2\}$.

The following example shows how the Zero-Product Property can be used to solve a factorable quadratic equation in a field.

EXAMPLE Solve $x^2 + 2x + 6 = 0$ in $(Z_7, +, \times)$.

SOLUTION *Plan*: Factor the left side of the equation and then apply the Zero-Product Property.

First we will list the factors of 6 in mod 7 multiplication. (You may find it helpful to refer to the addition and multiplication tables you made for mod 7.)

6	
1	6
2	3
4	5

Test the corresponding factors of the left side of the equation:

$(x + 1)(x + 6) = x^2 + (6 + 1)x + 6 = x^2 + 0x + 6 = x^2 + 6$

$(x + 3)(x + 2) = x^2 + (2 + 3)x + 6 = x^2 + 5x + 6$

$(x + 4)(x + 5) = x^2 + (4 + 5)x + 6 = x^2 + 2x + 6$, the desired polynomial

Having factored the left side, we proceed as follows:

$$(x + 4)(x + 5) = 0$$

$x + 4 = 0$	$x + 5 = 0$
$x = -4$	$x = -5$
$x = 3$	$x = 2$

Therefore, the solution set is $\{3, 2\}$.

Check $3^2 + 2 \times 3 + 6 = 2 + 6 + 6 = 1 + 6 = 0$

$2^2 + 2 \times 2 + 6 = 4 + 4 + 6 = 1 + 6 = 0$

ORAL EXERCISES

State the additive inverse and the multiplicative inverse of the given element of the given field. If a given element has no multiplicative inverse, so state.

1. 2; $(Z_3, +, \times)$
2. 1; $(Z_{11}, +, \times)$
3. 6; $(Z_7, +, \times)$
4. 0; $(Z_7, +, \times)$
5. 3; $(Z_5, +, \times)$
6. 4; $(Z_5, +, \times)$
7. 2; $(Z_7, +, \times)$
8. -9; $(Q, +, \cdot)$
9. $\sqrt{3}$; $(\mathcal{R}, +, \cdot)$

WRITTEN EXERCISES

Solve the given equation in the field $(Z_p, +, \times)$ for the given set Z_p.

A

1. $x + 3 = 0$; Z_5
2. $2x + 1 = 0$; Z_7
3. $2x + 2 = 0$; Z_3
4. $4x + 2 = 0$; Z_5
5. $8x + 1 = 0$; Z_{11}
6. $5x + 3 = 0$; Z_7
7. $6x + 5 = 0$; Z_{13}
8. $6x + 5 = 0$; Z_7
9. $7x + 4 = 0$; Z_{11}
10. $12x + 5 = 0$; Z_{13}
11. $4x + 13 = 0$; Z_{17}
12. $9x + 10 = 0$; Z_{11}
13. $x^2 + 4x + 3 = 0$; Z_5
14. $x^2 + 4x + 3 = 0$; Z_7
15. $x^2 + 2x + 1 = 0$; Z_{13}
16. $x^2 + 4x + 4 = 0$; Z_5
17. $x^2 + 5x + 6 = 0$; Z_{11}
18. $x^2 + 17x + 72 = 0$; Z_{73}

B

19. $2x^2 + 3x + 1 = 0$; Z_5
20. $4x^2 + 5x + 1 = 0$; Z_7
21. $2x^2 + 5x + 2 = 0$; Z_7
22. $6x^2 + 7x + 1 = 0$; Z_{11}
23. $3x^2 + 2x + 2 = 0$; Z_5
24. $3x^2 + x + 2 = 0$; Z_5
25. $x^2 + 6x + 1 = 0$; Z_7
26. $x^2 + 10x + 10 = 0$; Z_{11}

27. The "if" part of Theorem 14-7 (Zero-Product Property) follows immediately from the Multiplicative Property of Zero. Supply the reasons for the following proof of the "only if" part of the Zero-Product Property. (We are justified in assuming in the "given" that $a \neq 0$ since if $a = 0$, the proof is complete.)

Given: a and b are elements of $(F, +, \cdot)$; $ab = 0$, $a \neq 0$
Prove: $b = 0$

Statements	Reasons
1. $ab = 0,\ a \neq 0$	1. ?
2. $\frac{1}{a}$ exists.	2. ?
3. $b = 0 \cdot \frac{1}{a}$	3. ?
4. $\quad = 0$	4. ?
5. $b = 0$	5. ?

28. Supply the reasons for the steps of the following proof of Theorem 14-8.

Given: a and b are elements of $(F, +, \cdot)$; $a \neq 0$ and $b \neq 0$

Prove: $\frac{1}{ab} = \frac{1}{a} \cdot \frac{1}{b}$

Statements	Reasons
1. a, b are elements of F; $a \neq 0,\ b \neq 0$	1. ?
2. $ab \neq 0$	2. ?
3. $(ab)\left(\frac{1}{ab}\right) = 1$	3. ?
4. $(ab)\left(\frac{1}{a} \cdot \frac{1}{b}\right) = (ab)\left(\frac{1}{b} \cdot \frac{1}{a}\right)$	4. ?
5. $\quad = a\left[b\left(\frac{1}{b} \cdot \frac{1}{a}\right)\right]$	5. ?
6. $\quad = a\left[\left(b \cdot \frac{1}{b}\right)\frac{1}{a}\right]$	6. ?
7. $\quad = a\left[1 \cdot \frac{1}{a}\right]$	7. ?
8. $\quad = a \cdot \frac{1}{a}$	8. ?
9. $\quad = 1$	9. ?
10. $(ab)\left(\frac{1}{a} \cdot \frac{1}{b}\right) = 1$	10. ?
11. $(ab)\left(\frac{1}{ab}\right) = (ab)\left(\frac{1}{a} \cdot \frac{1}{b}\right)$	11. ?
12. $\frac{1}{ab} = \frac{1}{a} \cdot \frac{1}{b}$	12. ?

29. Supply the reasons for the steps of the following proof of Theorem 14-9.

Given: a is an element of the field $(F, +, \cdot)$; $a \neq 0$

Prove: $\frac{1}{-a} = -\frac{1}{a}$

Statements	Reasons
1. a is an element of F; $a \neq 0$	1. ?
2. $(-1)a \neq 0$	2. ?
3. $(-1)a = -a$	3. ?
4. $-a \neq 0$	4. ?
5. $(-a)\left(\frac{1}{-a}\right) = 1$	5. ?
6. $(-a)\left(-\frac{1}{a}\right) = a \cdot \frac{1}{a}$	6. ?
7. $\quad = 1$	7. ?
8. $(-a)\left(-\frac{1}{a}\right) = 1$	8. ?
9. $(-a)\left(\frac{1}{-a}\right) = (-a)\left(-\frac{1}{a}\right)$	9. ?
10. $\frac{1}{-a} = -\frac{1}{a}$	10. ?

30. Supply the reasons for the following proof of Theorem 14–10.

(1) Given: a and b are elements of $(F, +, \cdot)$; $a \neq 0$.

Prove: $a\left(-b \cdot \frac{1}{a}\right) + b = 0$

Statements	Reasons
1. a and b are elements of F, $a \neq 0$	1. ?
2. $\frac{1}{a}$ exists.	2. ?
3. $a\left(-b \cdot \frac{1}{a}\right) = a\left[\frac{1}{a}(-b)\right]$	3. ?
4. $\quad = \left(a \cdot \frac{1}{a}\right)(-b)$	4. ?
5. $\quad = 1(-b)$	5. ?
6. $\quad = -b$	6. ?
7. $a\left(-b \cdot \frac{1}{a}\right) = -b$	7. ?
8. $a\left(-b \cdot \frac{1}{a}\right) + b = -b + b$	8. ?
9. $\quad = 0$	9. ?
10. $a\left(-b \cdot \frac{1}{a}\right) + b = 0$	10. ?

(2) Given: y and z are solutions of $ax + b = 0$ in $(F, +, \cdot)$, $a \neq 0$.
Prove: $y = z$

Statements	Reasons
1. $ay + b = 0$	1. ?
2. $az + b = 0$	2. ?
3. $ay + b = az + b$	3. ?
4. $ay = az$	4. ?
5. $y = z$	5. ?

C **31.** Prove the following theorem: For any two elements a and b of the field $(F, +, \cdot)$ such that $a \neq 0$ and $b \neq 0$, if $a \cdot \frac{1}{x} = b$, then $x = a \cdot \frac{1}{b}$.

SELF-TEST 3

1. Evaluate in Z_5: **a.** $-2(-4)$ **b.** $2(-4)$ Section 14-6

State whether or not the given system is a field. If the system is a field, state the identity element for each operation and identify the inverses of the elements. If the system is not a field, state at least one of the three field properties listed on page 581 that it fails to satisfy.

2. $(Z_3, +, \times)$ **3.** $(Z_8, +, \times)$ **4.** $(\mathcal{R}, -, \cdot)$

Solve the given equation in the field $(Z_N, +, \times)$ for the given set Z_N.

5. $x + 5 = 0$; Z_7 **6.** $4x + 1 = 0$; Z_5 Section 14-7

7. $x^2 + 3x + 2 = 0$; Z_5 **8.** $x^2 + x + 1 = 0$; Z_3

Computer Exercises

In the real numbers, $10 \cdot 9 = 90$, but $10 \times 9 = 6$ in multiplication mod 12 because $10 \cdot 9$, or 90, leaves a remainder of 6 when divided by 12. Thus we say that 90 is *equivalent* to 6 mod 12, or $90 = 6$ mod 12. Likewise, $27 = 2$ mod 5 because $27 \div 5$ leaves a remainder of 2. In general, X is *equivalent* to R mod N, or $X = R$ mod N, if X leaves a remainder of R when divided by N.

Suppose you want to use the computer to find a number, say R, in Z_{12} that is equivalent to 90 mod 12. (As we noted previously, $R = 6$.) What would happen if you gave the computer the following instruction?

```
LET Q = 90/12
```

The computer would compute Q as 7.5, *not* as 7 with a remainder of 6. You can use the following method to have the computer find the remainder.

Instruction	*Result*
1. LET Q=90/12	1. Q = 7.5
2. LET A=INT(Q)	2. A = 7
3. LET B=12∗A	3. B = 84
4. LET R=90−A	4. R = 6

These instructions can be combined into the single instruction:

```
LET R=90−12∗INT(90/12)
```

A more general example of this instruction is contained in line 40 of Exercise 1 below.

1. The following program computes the value of R in Z_N that is equivalent mod N to a given positive integer X. Copy and RUN the program for the given values of X and N.

```
10   PRINT "THIS PROGRAM COMPUTES R SUCH THAT X=R MOD N."
20   PRINT "INPUT THE VALUES OF X AND N";
30   INPUT X,N
40   LET R=X−N∗INT(X/N)
50   PRINT X;"=";R;"MOD";N
60   END
```

a. $X = 90$; $N = 12$ b. $X = 38$; $N = 5$ c. $X = 147$; $N = 17$
d. $X = 5$; $N = 8$ e. $X = 3626$; $N = 14$ f. $X = 52681$; $N = 619$

2. This program prints an operation table for multiplication mod N. Copy and RUN the program for the given values of N listed below. (If your computer uses a video screen that holds only 16 lines of print, insert the following instruction to prevent the computer from erasing the top of the output display: 175 IF A=N−1 THEN 175.)

```
10   PRINT "INPUT THE VALUE OF N";
20   INPUT N
30   PRINT "X";TAB(4);"!";
40   FOR B=0 TO N−1
50   PRINT TAB(4∗B+6);B;
60   NEXT B
70   PRINT
80   FOR L=1 TO 4∗N+5
90   PRINT "−";
100  NEXT L
110  PRINT
120  FOR A=0 TO N−1
130  PRINT A;TAB(4);"!";
140  FOR B=0 TO N−1
150  LET R=A∗B−N∗INT(A∗B/N)
160  PRINT TAB(4∗B+6);R;
170  NEXT B
180  PRINT
190  NEXT A
200  END
```

a. $N = 13$ b. $N = 14$

3. Using the operation table printed in Exercise 2(a), find the inverse of each element in Z_{13} under multiplication mod 13.

CHAPTER REVIEW

A binary operation $*$ is defined for all real numbers as $a * b = ab^2$.

1. Is $*$ commutative? **2.** Is $*$ associative? Section 14-1

3. Evaluate $(3 * 2) * 3$.

Exercises 4–7 refer to the following subsets of the real numbers:

Z, Z^- (negative integers), Q, Q^- (negative rational numbers)

4. Which sets are closed under addition?

5. Which sets are closed under subtraction?

6. Which sets are closed under multiplication?

7. Which sets are closed under division?

8. Use Table 14-2 on page 566 to find the value of each expression in the clock-7 system. Section 14-2

a. $(2 + 4) + 6$ **b.** $(1 + 6) + (3 + 5)$

Exercises 9–11 refer to the system $(\{-1, 0, 1\}, \cdot)$ where $a \cdot b$ is the usual product.

9. Make an operation table for the system.

10. Is the system commutative? associative?

11. Name the identity element and the inverse of each nonzero element.

Exercises 12–14 refer to the system $(Z_6, +, \times)$. (If you wish, you may prepare and use operation tables.)

12. Evaluate: **a.** $5 + 4$ **b.** $5(4)$ Section 14-3

13. Name the additive inverse and the multiplicative inverse of 5.

14. Name any proper divisors of 0.

15. You may use the tables on page 566 to solve the following equations in $(Z_5, +, \times)$:

a. $4x = 2$ **b.** $3x + 3 = 0$

Is the given system a group? If the system is a group, state the identity element and identify the inverses of the elements. If the system is not a group, explain why. Section 14-4

16. $(Z_2, +)$

17. $(\{a, b, c\}, \#)$ where # is defined by the table below

#	*a*	*b*	*c*
a	*c*	*b*	*a*
b	*b*	*a*	*c*
c	*a*	*c*	*b*

18. One of the six permutations defined on page 576 is listed below with its columns rearranged. Give the letter of the corresponding permutation. Section 14-5

$$\begin{pmatrix} 3 & 2 & 1 \\ 1 & 2 & 3 \end{pmatrix}$$

19. State whether $\{1, 12\}$ forms a subgroup of $(Z_{13} \setminus \{0\}, \times)$. If it does, make an operation table and identify the identity element and the inverses of the elements. If it does not, explain why.

20. Evaluate in $(Z_{13}, +, \times)$. Section 14-6

a. $(-1)8$ b. $(-8)6$ c. $(-8)(-4)$

Is the given system a field? If not, state at least one of the field properties listed on page 581 that it fails to satisfy.

21. $(\mathcal{R}, +, \div)$ 22. $(Z_{13}, +, \times)$

Solve the given equation in the field $(Z_N, +, \times)$ for the given set Z_N.

23. $9 + x = 0;\ Z_{11}$ 24. $4x + 3 = 0;\ Z_7$ Section 14-7

25. $x^2 + 7x + 10 = 0;\ Z_{13}$

CHAPTER TEST

1. A binary operation $*$ is defined for all real numbers as $a * b = \dfrac{ab}{2}$.

a. Is $*$ commutative? b. Is $*$ associative?

2. State whether the given set is closed under the given operation.

a. $(\mathcal{R}^+, \div)$ b. $(Z_4, \times)$

c. $(\{-1, -2, -3, \ldots\}, +)$ d. $(\{\text{irrational numbers}\}, \cdot)$

Exercises 3–5 refer to the commutative system $(\{a, b, c, d\}, *)$ defined by the partial operation table shown.

$*$	a	b	c	d
a	d		b	c
b	a	b		d
c		c	d	a
d		d		b

3. Copy and complete the table.

4. Name the identity element and the inverse of each element.

5. Assume that $*$ is associative. Is the system a group? Explain why or why not.

6. Is the system $(\{1, 3, 5, 7, \ldots\}, \cdot)$ a group? Explain.

7. Which of the following sets form subgroups of $(Z_5 \setminus \{0\}, \times)$?

$$\{1\}, \quad \{1, 2, 3\}, \quad \{1, 4\}$$

8. a. Simplify: $\begin{pmatrix} 1 & 2 & 3 \\ 2 & 3 & 1 \end{pmatrix} * \begin{pmatrix} 1 & 2 & 3 \\ 1 & 3 & 2 \end{pmatrix}$

b. What is the identity element of the six-element permutation group to which $\begin{pmatrix} 1 & 2 & 3 \\ 2 & 3 & 1 \end{pmatrix}$ belongs?

c. What is the inverse of the element given in part (b)?

Exercises 9–12 refer to the field $(Z_5, +, \times)$.

9. Name the identity elements for addition and multiplication.
10. Name the additive inverse and the multiplicative inverse of 4.
11. Name the proper divisors of 0, if any.
12. Compute: a. $3(-1)$ b. $(-3)2$ c. $(-2)(-3)$

13. Is $(Z_{10}, +, \times)$ a field? If not, state at least one field property that it fails to satisfy.

Solve the given equation in the field $(Z_N, +, \times)$ corresponding to the set Z_N.

14. $3 + x = 1;\ Z_{17}$
15. $2x + 4 = 0;\ Z_5$
16. $x^2 + 4x + 3 = 0;\ Z_5$

CUMULATIVE REVIEW: CHAPTERS 1–14

Indicate the best answer by writing the appropriate letter.

1. How many 7-letter permutations can be formed from the letters of the word INITIAL?
   a. $7!$ b. $(7-3)!$ c. $\frac{7!}{3!}$ d. $\binom{7}{3}$

2. A parallelogram has vertices $(-1, 2)$, $(-2, -1)$, and $(3, 2)$. The x-coordinate of the remaining vertex may *not* be:
   a. -1 b. 4 c. 2 d. -6

3. If $a * b$ is a binary operation defined as $a^2 - b$, evaluate $3 * 1$.
   a. 3 b. 2 c. 5 d. 8

4. The measures of the angles of a triangle are in the ratio 2:3:5. Find the measure of the *smallest* angle of the triangle.
   a. 18 b. 10 c. 20 d. 36

5. The sides of a triangle have lengths 8, 10, and 12. Find the longest side of a similar triangle whose perimeter is 45.
   a. 18 b. 12 c. 15 d. 20

6. If $q \rightarrow \sim p$ and $\sim r \rightarrow p$, then:
   a. $q \rightarrow \sim r$ b. $\sim q \rightarrow r$ c. $\sim q \rightarrow \sim r$ d. $q \rightarrow r$

7. The measure of the vertex angle of an isosceles triangle is half that of a base angle. Find the measure of an exterior angle of the triangle.
   a. 144 b. 135 c. 120 d. 90

8. An equation of the axis of symmetry of the parabola $y = ax^2 - 8x + 3$ is $x = -2$. Find the value of a.
   a. 4 b. 2 c. -2 d. -4

9. The hypotenuse of a right triangle has length 13, and one leg has length $2\sqrt{13}$. Find the length of the altitude to the hypotenuse.
   a. 6 b. $3\sqrt{13}$ c. 4 d. 9

10. Refer to the figure at the right. Find DB.

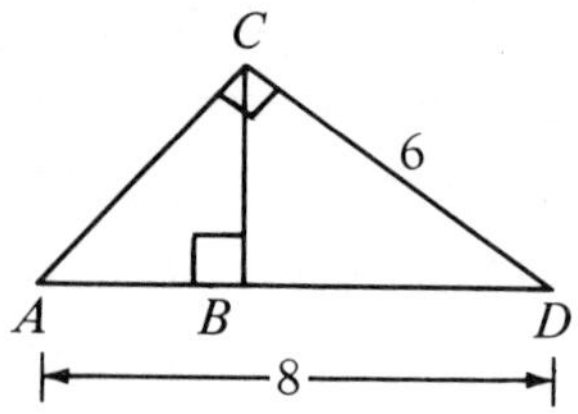

a. $2\sqrt{7}$ b. $4\frac{1}{2}$

c. 6 d. 2

11. Point M is the midpoint of $\overline{AB}$. The coordinates of B are $(-2, 3)$ and the coordinates of M are $(4, -1)$. Find the coordinates of A.

a. (1, 1) b. (−3, 2) c. (10, −5) d. (2, 2)

12. Use the table at the right to find the inverse element of a.

★	a	b	c
a	c	b	a
b	b	b	b
c	a	b	c

a. a b. b

c. c d. a does not have an inverse element.

13. Find the solution set of $2x = 4$ in the mod 6 (clock-6) system of arithmetic, where $x \in \{0, 1, 2, 3, 4, 5\}$.

a. $\{2, 5\}$ b. $\{2\}$ c. $\{5\}$ d. $\emptyset$

14. Point P lies on line l. How many points are in the locus of points 4 cm from P and 3 cm from l?

a. 4 b. 2 c. 1 d. 0

15. In how many ways can a committee of 4 be chosen from a group of 7 people?

a. 35 b. 840 c. 210 d. 28

16. Two marbles are drawn randomly from an urn containing 4 white marbles and 2 black marbles. Find P(same color).

a. $\frac{5}{9}$ b. $\frac{2}{5}$ c. $\frac{3}{5}$ d. $\frac{7}{15}$

17. Negate the sentence $\forall_x\ x^2 > -4$.

a. $\exists_x\ x^2 < 4$ b. $\exists_x\ x^2 \leqslant -4$ c. $\exists_x\ x^2 > 4$ d. $\forall_x\ x^2 \leqslant -4$

18. In the diagram at the right, if $\angle 1 \cong \angle 2$, find the value of a.

a. 7.1 b. 7.4

c. 2.4 d. 3.3

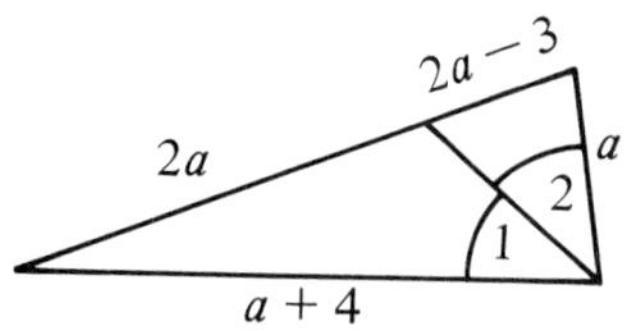

19. A picture is 22 cm long and 18 cm wide. A frame of uniform width increases the total area by 384 cm^2. How wide is the frame?

a. 2.5 cm b. 3 cm c. 3.5 cm d. 4 cm

20. Give segments of length 4, 5, and 11, which of the following can *not* be constructed?

a. a segment of length $2\sqrt{5}$ b. a segment of length 18

c. a rectangle of area 22 d. a triangle with sides of the given lengths

21. Find a root of $4x^2 + 25 = -20x$.

a. $-\frac{2}{5}$ b. $\frac{5}{2}$ c. $\frac{2}{5}$ d. $-\frac{5}{2}$

22. Find an equation of the circle with center $(-3, 4)$ and radius 2.
a. $(x-3)^2+(y+4)^2=2$ b. $(x+3)^2+(y-4)^2=2$
c. $(x-3)^2+(y+4)^2=4$ d. $(x+3)^2+(y-4)^2=4$

23. If $\left(\frac{5}{2}z+72\right)$ and $\left(\frac{3}{2}z+20\right)$ represent the measures of two angles, then the angles can *not* be:
a. corresponding angles formed by two parallel lines and a transversal
b. interior angles on the same side of the transversal of two parallel lines
c. complementary
d. an obtuse angle and an acute angle

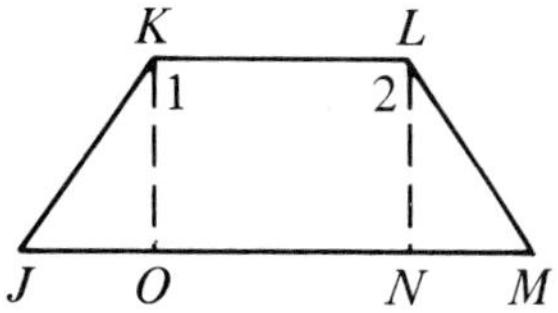

24. *JKLM* is an isosceles trapezoid. What additional information would allow you to prove the two triangles congruent?
a. $\overline{KO} \parallel \overline{LN}$ b. $\angle 1$ and $\angle 2$ are supplementary.
c. $\overline{KO} \cong \overline{LN}$ d. $\overline{KO} \perp \overline{JM}$, $\overline{LN} \perp \overline{JM}$

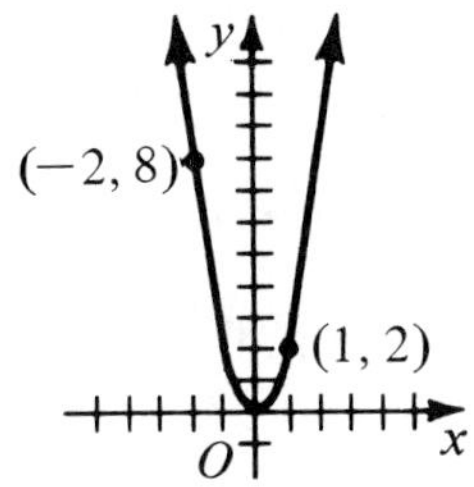

25. Which is an equation of the parabola shown at the right?
a. $y=2x^2$ b. $y=\frac{1}{2}x^2$
c. $y=-2x^2$ d. $y=x^2$

26. Find an equation of the line that is perpendicular to $2y=3x-5$ and passes through $(1, 2)$.
a. $2y=3x-1$ b. $3y=-2x+8$ c. $3y=2x+4$ d. $2y=-3x+7$

27. Under which operation is the set $\{1, 3, 5, 7, 9, \ldots\}$ closed?
a. addition b. subtraction c. multiplication d. division

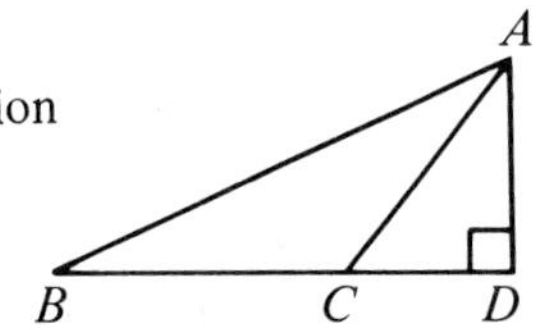

28. Which of the following is not necessarily true?
a. $m\angle B < m\angle ACD$ b. $AB > AC$
c. $m\angle B > m\angle CAD$ d. $\angle BAC$ is acute.

29. If you choose a card at random from a standard bridge deck, what is the probability that it will be a red card or a ten?
a. $\frac{1}{2}+\frac{1}{26}$ b. $\frac{1}{2}+\frac{1}{13}$ c. $\frac{1}{2}\left(\frac{1}{26}\right)$ d. $\frac{1}{2}\left(\frac{1}{13}\right)$

30. In $\triangle ABC$, M is on $\overline{AB}$, N is on $\overline{AC}$, and $\overline{MN} \parallel \overline{BC}$. If $AM=x+1$, $MB=3x$, $MN=x+2$, and $BC=5x+2$, find the value of x.
a. 0 b. 2 c. $\frac{1}{3}$ d. No real value of x exists.

31. Given the set $\{a, b, c, d\}$. Under the operation $*$, the identity element is c. Which statement is proved by the following sentence?

$$a * d = c \text{ and } d * a = c$$

a. The set is closed under $*$. b. The operation $*$ is commutative.
c. d is the inverse of a. d. a is the inverse of c.

Choose one of Exercises 32–34. Begin by drawing a larger version of the given diagram.

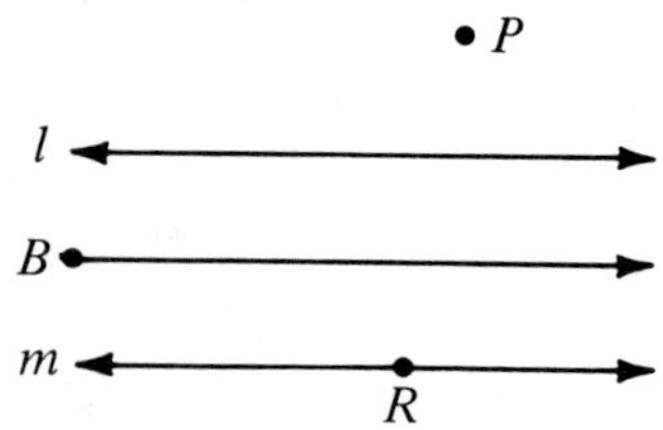

32. Construct a line through P parallel to l.

33. Construct an equilateral triangle with one vertex at B.

34. Construct a line through R perpendicular to m.

Choose one of Exercises 35–37.

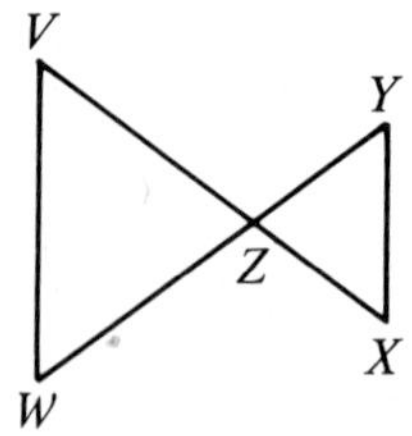

35. Write a two-column proof.

Given: $\frac{VZ}{XZ} = \frac{WZ}{YZ}$

Prove: $\angle W \cong \angle Y$

36. The vertices of $\triangle ABC$ are $A(-3, -1)$, $B(1, 7)$, and $C(5, 5)$.
   a. Show that $\triangle ABC$ is a right triangle.
   b. Find the area of $\triangle ABC$.

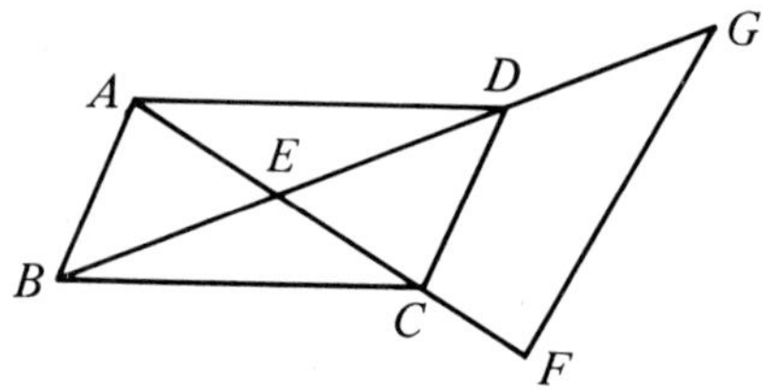

37. Write a two-column proof.

Given: $ABCD$ is a parallelogram; $\overline{CD} \parallel \overline{FG}$

Prove: $EA \cdot FG = EF \cdot AB$

Choose one of Exercises 38–40.

$*$	a	b	c	d
a	a	b	c	d
b	b	d	a	c
c	c	a	d	b
d	d	c	b	a

38. The table at the right represents the operation $*$ for the set $\{a, b, c, d\}$.
   a. What is the identity element of this system?
   b. What is the inverse of c?
   c. Find the value of $(b * d) * d$.
   d. Solve for x: $(d * c) * x = a$

39. a. Graph $y = -x^2 + 4x - 5$.
   b. Write an equation of the axis of symmetry.
   c. Write an equation of the line that passes through the vertex and the y-intercept of the parabola in part (a).

40. Given the following statements:

I cannot join the glee club or I cannot join the drama club.
If I don't join the drama club, then I don't attend rehearsals after school.
If I don't attend rehearsals after school, then I can keep my part-time job.
I cannot keep my part-time job.

Let g represent "join the glee club."
Let d represent "join the drama club."
Let r represent "attend rehearsals."
Let j represent "keep my part-time job."

   a. Using g, d, r, and j, and proper connectives, express each statement in symbolic form.
   b. Using laws of inference, show that I do not join the glee club.

Careers — School Food Planner/Buyer

School food planning and buying is one of a complex of careers in the hotel, restaurant, and institutional field. As one dietician said, "People, especially students, will always need good food. These jobs will always be there." Another advantage of a career in this field is that there is no one pathway of preparation. You may choose to learn by experience, for example, by working in your school's food service department or in a restaurant. You may also get a college degree in hotel and restaurant management or in dietetics.

The size of the operation you would be dealing with also varies tremendously. At one end of the scale is a small kindergarten for 100 children. These food programs offer little variety and are usually run on a small budget. At the other end of the scale is the large university or school system. These operations can be large-scale businesses, and planners within this kind of system need more management, statistical, and computer skills.

In the larger services, menus are planned long in advance, often with the help of dietitians who plan menus for their nutrient value as well as variety. Computers then work out the quantities of each item needed. Specifications are drawn up for both food and nonfood items. For example, a planner may request a certain size of disposable plate, or a certain type and size of meatball. These specifications are then submitted to vendors, who make bids on the contracts. The contracts may be for varying lengths of time and delivery schedules. These contracts are reviewed and updated on a regular basis.

If you want a less common path into this career, you might try taking courses in food services offered at schools that are associated with travel agencies. The military services also offer training for cooks and food planners. Dietitians have college degrees and usually a 1- or 2-year internship in the field of clinical or administrative nutrition.

VOCABULARY REVIEW

Be sure that you understand the meaning of these terms:

binary operation, p. 556
algebraic system, p. 557
identity element, p. 559
inverse, p. 559
integers mod N, p. 565
proper divisors of zero, p. 568
group, p. 571
commutative group, p. 571
subgroup, p. 577
field, p. 581

MIXED REVIEW

1. A binary operation $*$ is defined for real numbers by $a * b = a + b - ab$.
   a. Is $*$ commutative?   b. Is $*$ associative?
2. Draw an acute triangle, $\triangle ABC$.
   a. Construct the perpendicular bisector of $\overline{AB}$.
   b. Construct the bisector of $\angle C$.

Two cards are drawn from a standard deck. Find the following probabilities if:

a. the first card is replaced before the second card is drawn.

b. the first card is not replaced.

3. P(two red cards)
4. P(a king and a jack in that order)

In Exercises 5-7, use the definitions of addition mod 6 and multiplication mod 6.

5. Make a table for addition mod 6.
6. Make a table for multiplication mod 6.
7. Compute the following in Z_6.
   a. $4 + (3 + 5)$ b. $(3 \times 4) \times 4$ c. $(2 + 4) \times (3 + 3)$
8. In how many ways can you arrange 6 candles, each a different color, in a circle around a birthday cake?
9. Simplify: a. $\frac{4a^2 - 8a}{16a}$ b. $\frac{b}{x - b} + \frac{x}{b - x}$
10. Factor completely: $3d^2 - 33d + 54$
11. Solve: $-3x^2 - 5x = -4$
12. Find an equation in standard form of the line through $(-4, 2)$ and $(3, 5)$.
13. In $\triangle ABC$, $m\angle C = 90$, $AC = 4\sqrt{3}$, and $AB = 8$. Find CB.
14. Solve the system and illustrate with a graph: $y = x^2 - 2x + 4$
    $x - y = -8$
15. The sides of a rhombus are 82 m long and the measure of one angle is 64°. Find the length of the shorter diagonal to the nearest meter.
16. An executive has 8 suits and 3 coats. In how many ways can the executive select 5 suits and 1 coat to take on a business trip?

17. Determine the slope and y-intercept of the line with equation $5y - 6x = 12$.

18. Solve the system: $2x + 7y = 4$
$3x + 11y = 7$

19. Find an equation in the form $y = ax^2 + bx + c$ of the parabola with focus $F(1, 3)$ and directrix $y = 1$. Sketch the parabola.

20. Is the system a group? If so, state the identity element and identify the inverses. If the system is not a group, explain why not.

$\odot$	p	q	r
p	p	q	r
q	q	r	p
r	r	p	q

21. A park has 5 entrances and 3 exits. In how many ways can someone enter and exit the park?

22. Let p represent "flowers are blooming," q represent "the sun is shining," and r represent "it is snowing." Translate the following into English: $(\sim r \vee q) \rightarrow p$

23. Solve in Z_6: $4x + 2 = 4$

Determine the nature of the roots.

24. $4y^2 = 20y - 25$

25. $3a^2 = 12a$

26. A person 2 m tall casts a shadow 3.2 m long at the same time that a second person standing nearby casts a shadow 2.4 m long. How tall is the second person?

27. A circle has a diameter with endpoints $(8, -2)$ and $(4, -6)$. Write an equation of the circle.

Write two-column proofs.

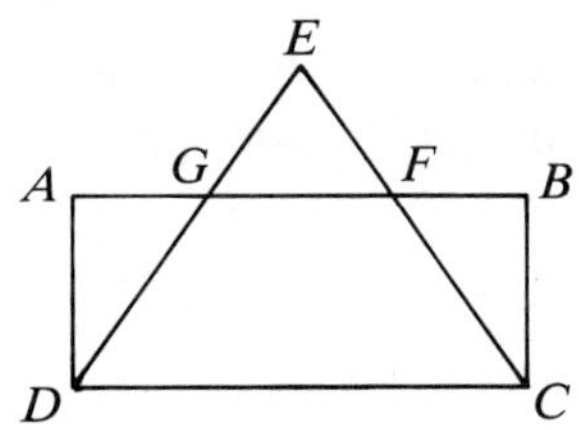

28. Given: $ABCD$ is a rectangle; $\overline{EF} \cong \overline{EG}$
Prove: $\overline{GD} \cong \overline{FC}$

29. Given: $\overline{ED} \cong \overline{EC}$; $\overline{AB} \cong \overline{DC}$;
$\overline{DA} \perp \overline{AB}$; $\triangle EGF \sim \triangle EDC$
Prove: $ABCD$ is a rectangle.

30. (Optional) Find an equation of the hyperbola with y-intercepts $\pm\sqrt{3}$ and foci $F_1(0, 3\sqrt{2})$ and $F_2(0, -3\sqrt{2})$.

Extra Practice

Chapter 1

A 1. Graph: {the positive multiples of 3}

Use the indicated property to make a true statement.

2. $3(x + 5) = \underline{\ ?\ }$ (Commutative Property of Multiplication)
3. $4x + 0 = \underline{\ ?\ }$ (Identity Property of Addition)

Calculate.

4. $-50 - (-42)$
5. $13 + [-(-9 + 4)]$
6. $3 - 4(3 - 4)$

Simplify.

7. $4x - 12x$
8. $5t - (-3t)$
9. $-4(2y - 5y)$

Solve and check.

10. $3x - 4 = 10 + x$
11. $12z - 31 = 5z + 25$
12. $y - 8 \leqslant 4(3 - y)$
13. $6 < \frac{1}{2}(3b + 2) < 9$

14. The sum of two numbers is 47. When twice the larger number is added to 3 times the smaller, the sum is 106. Find the numbers.

Let A = {the positive multiples of 2}, B = {the positive multiples of 3}, and C = {the counting numbers less than 11}. List the elements in each of the following sets.

B 15. $A \cap (B \cup C)$
16. $(A \cup B) \cap (B \cup C)$
17. $(A \cap C) \cup (B \cap C)$

18. Calculate: $2^3 - 6 \div 3 + 3(9 - 4)$

Solve.

19. $\frac{3a + 1}{4} - \frac{4a - 2}{12} = 0$
20. $7t - 3(3t + 2) = 4(t - 5) - 1$
21. $4(2.5 + 0.6b) \geqslant 0.4b$
22. $2y - 3 < 6 + y$ and $5y - 9 > 2y - 3$

23. Of three numbers, the first is 4 times the second and the third is 11 more than the first. If their sum is 200, what are the numbers?

24. A metallurgist has two alloys. One is 25% silver and the other is 40% silver. How much of each alloy should be used to make 120 kg of an alloy that is 30% silver?

Chapter 2

Rewrite each sentence as a related true statement beginning with *All*, *Some*, or *No*.

A 1. A triangle can be isosceles.

2. The diameters of a circle are always congruent.

Write the negation.

3. March is not windy or December is cold.

4. If I clean my room, then I can go to the movies.

5. $\forall_x (x < 1 \rightarrow \frac{1}{x} > x)$

6. Write the **(a)** converse, **(b)** inverse, and **(c)** contrapositive and compare their truth values: if x is divisible by 2, then x^2 is divisible by 4.

7. Write the conclusion that follows logically from the given premises.
   P_1: If a number is divisible by 2, then it's even.
   P_2: If a number is even, then it's not odd.

8. Prove that the sum of an odd number and an even number is odd.

B 9. Are the sentences "The counting number x is even" and "The counting number x is prime" contradictory? If not, explain how both can have the same truth value.

10. Let p represent "5 is odd" and q represent "5 is a multiple of 3."
    a. Translate "5 is odd or 5 is not a multiple of 3" into symbolic notation and give its truth value.
    b. Write in words the negation of the statement in part (a).

11. Use a variable and a quantifier to write in symbolic form: There is an integer whose square is its multiplicative inverse. Is the statement true or false?

12. Complete the argument: P_1: $(\sim p \wedge q) \vee (r \rightarrow s)$
    P_2: $p \vee \sim q$
    ______

13. Is the argument valid? If so, identify the rule of inference being used. If not, identify the fallacy.
    P_1: Either Nancy or David will cook dinner.
    P_2: Nancy will not cook dinner.
    ______
    C: David will cook dinner.

Chapter 3

Exercises 1-3 refer to the figure at the right.

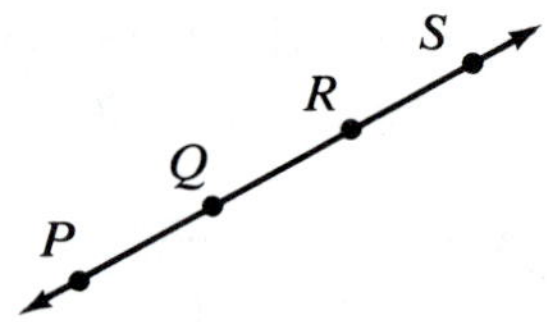

A 1. Name the set of points between P and Q together with all points on $\overrightarrow{QR}$ as a line, a segment, or a ray.

2. If $PS = 19.5$ and $PR = 13.75$, find RS.

3. If Q is the midpoint of $\overline{PR}$, $PQ = 3x + 9$, and $PR = 12x - 6$, find the value of x.

Exercises 4 and 5 refer to the figure at the right.

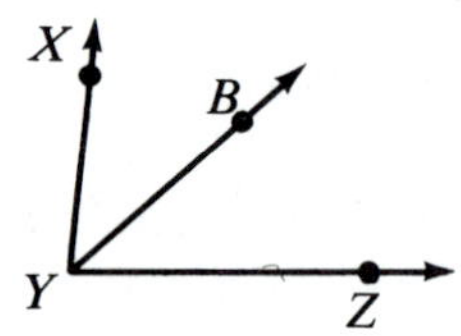

4. If $m\angle XYB = x + 10$, $m\angle BYZ = \frac{3}{2}x + 1$, and $m\angle XYZ = 3x + 9$, find the value of x.

5. If $\overrightarrow{YB}$ bisects $\angle XYZ$ and $m\angle BYZ = 42$, find $m\angle XYB$ and $m\angle XYZ$.

6. $\angle A$ and $\angle B$ are supplementary angles. If the measure of $\angle B$ is 15 less than twice the measure of $\angle A$, find the measure of each angle.

7. In $\triangle ABC$, $AC = 74$ and $BC = 55$. Find all possible values of AB.

Write two-column proofs.

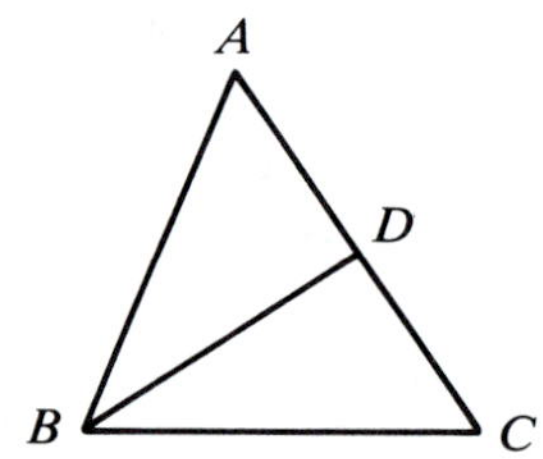

8. Given: $\overrightarrow{BD}$ bisects $\angle ABC$;
   $\overline{BD} \perp \overline{AC}$
   Prove: $\triangle ABD \cong \triangle CBD$

9. Given: $\overline{BD} \perp \overline{AC}$;
   $\overline{BA} \cong \overline{BC}$
   Prove: $\triangle ABD \cong \triangle CBD$

10. If $m\angle A = m\angle C$, $BA = \frac{1}{3}x + 5$, and $BC = x - 9$, find the value of x.

Write two-column proofs.

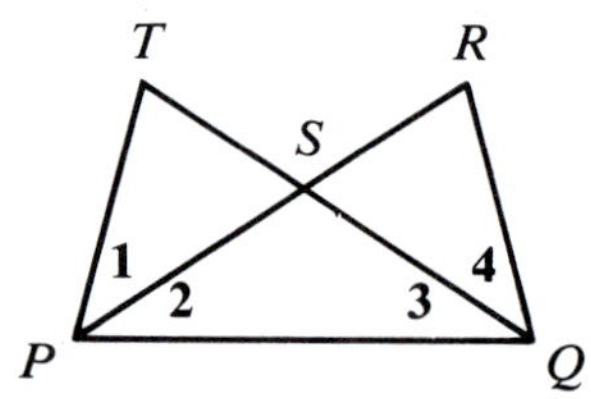

B 11. Given: $\angle 1 \cong \angle 4$;
   $\angle 2 \cong \angle 3$
   Prove: $\triangle TPQ \cong \triangle RQP$

12. Given: $\overline{TP} \cong \overline{RQ}$;
   $\angle TPQ \cong \angle RQP$
   Prove: $\triangle TSP \cong \triangle RSQ$

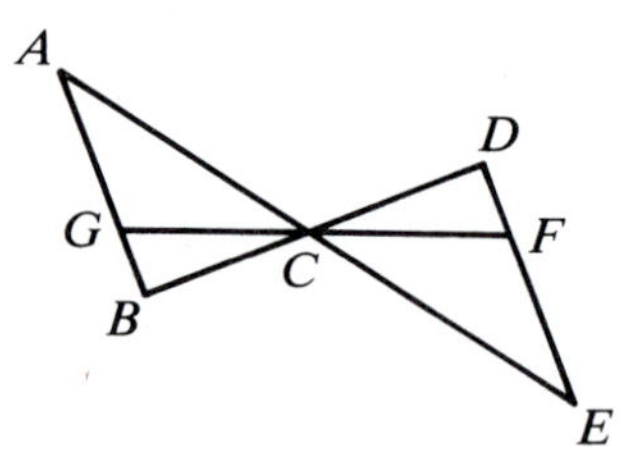

13. Given: $\overline{BC} \cong \overline{CD}$; $\overline{AC} \cong \overline{CE}$
   Prove: $\overline{GC} \cong \overline{FC}$

Chapter 4

Exercises 1-5 refer to the diagram.

Write two-column proofs.

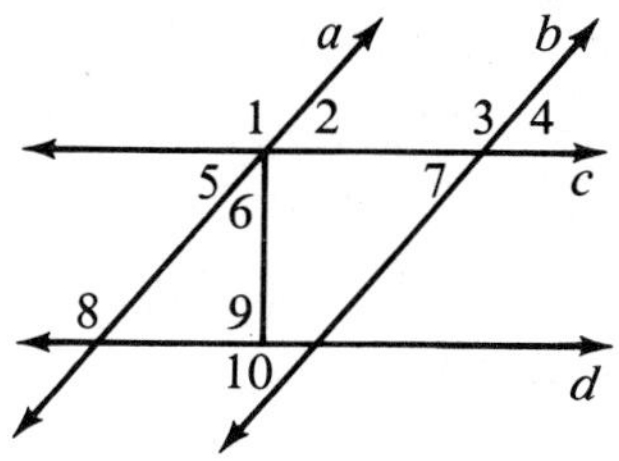

Exs. 1-5

A 1. Given: $\angle 5 \cong \angle 4$
Prove: $a \parallel b$

2. Given: $\angle 1$ and $\angle 10$ are supplementary; $a \parallel b$
Prove: $c \parallel d$

3. If $a \parallel b$ and $m\angle 2 = 50$, find $m\angle 4$.

4. If $c \parallel d$, $m\angle 5 = 3x + 4$, and $m\angle 8 = 11x - 52$, find the value of x.

5. If $m\angle 8 = 130$ and $m\angle 9 = 90$, find $m\angle 6$.

6. In $\triangle RST$, $m\angle R = 42$ and $m\angle S = 56$. Name the longest and shortest sides of $\triangle RST$.

7. In $\triangle JKL$, $m\angle J = 2x$, $m\angle K = 5x + 20$, and $m\angle L = 3x - 10$. Find the value of x and the measure of each angle.

8. Find the measure of each interior angle of a regular polygon with 15 sides.

B 9. Find the number of sides of a polygon if the sum of the measures of the interior angles is 4140.

Write two-column proofs.

10. Given: $\overrightarrow{AJ}$ bisects $\angle CAB$;
$\overrightarrow{BK}$ bisects $\angle ABD$; $\overline{AJ} \parallel \overline{BK}$
Prove: $\overleftrightarrow{CA} \parallel \overleftrightarrow{BD}$

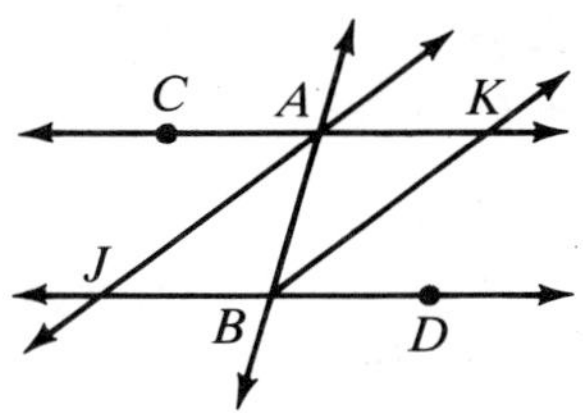

11. Given: $\overline{PS} \perp \overline{PQ}$; $\overline{QR} \perp \overline{PQ}$;
$\angle PSQ \cong \angle QRP$
Prove: $\overline{SQ} \cong \overline{RP}$

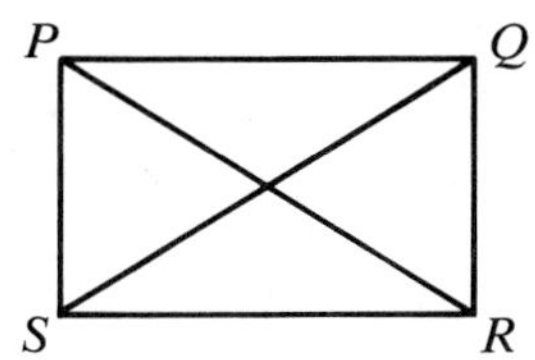

12. Given: $\overline{AB} \cong \overline{CB}$; $\overline{AD} \cong \overline{CD}$
Prove: $\overrightarrow{BD}$ bisects $\angle ABC$.

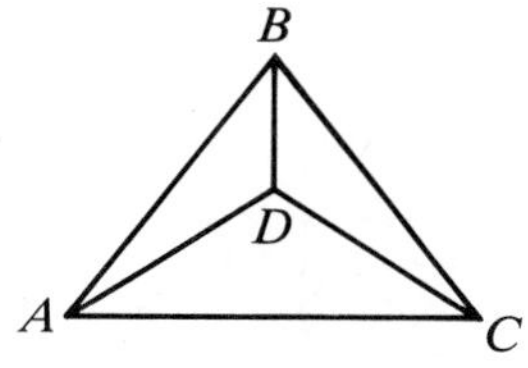

13. Given: $\overline{AC} \cong \overline{AE}$; $\overline{BD} \perp \overline{AC}$;
$\overline{DF} \perp \overline{AE}$
Prove: $\angle BDC \cong \angle FDE$

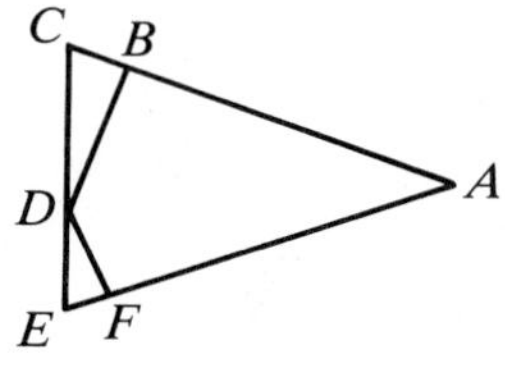

Chapter 5

Exercises 1–4 refer to the diagram.

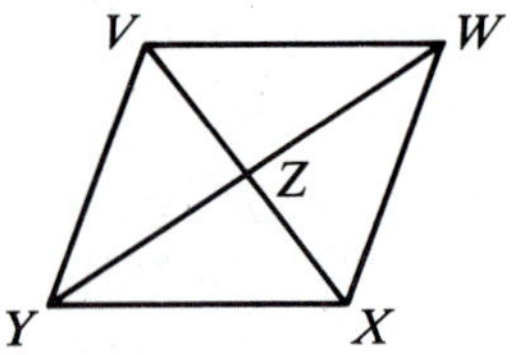

A 1. If $VWXY$ is a parallelogram, $m\angle VWX = 8x + 4$, and $m\angle VYX = 11x - 20$, find the value of x.

2. If $VWXY$ is a parallelogram, $YZ = 15 - \frac{2}{3}x$, and $YW = 8x + 2$, find the value of x.

3. If $m\angle VYX = 65$, $m\angle YXW = 12x + 7$, and $m\angle VWX = 7y - 5$, for which values of x and y is $VWXY$ a parallelogram?

4. Given that $\triangle ZWV \cong \triangle ZYX$, write a two-column proof that $VWXY$ is a parallelogram.

Exercises 5 and 6 refer to quadrilateral $QUAD$.

5. If $QUAD$ is a rectangle, $QA = 17 - 3x$ and $DU = 4x - 3$, find the value of x.

6. If $QUAD$ is a rhombus and $m\angle AQD = 44$, find the measure of each angle of $QUAD$.

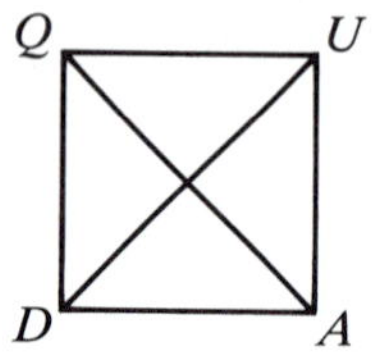

Exs. 5, 6

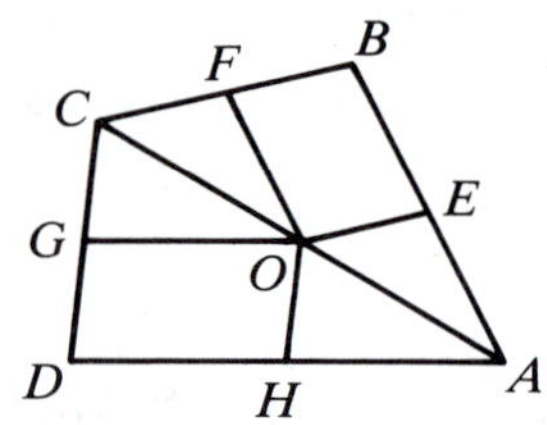

Exs. 7, 8

Exercises 7 and 8 refer to quadrilateral $ABCD$.

7. E, F, G, H, and O are the midpoints of $\overline{AB}$, $\overline{BC}$, $\overline{CD}$, $\overline{AD}$, and $\overline{AC}$, respectively. Find the perimeter of $ABCD$ if $OE = 1.5$, $OF = 2$, $GO = 2.5$, and $HO = 1.5$.

8. Find the length of the median of trapezoid $AOGD$.

Write two-column proofs.

B 9. Given: $\square MPQS$; N and R are the midpoints of $\overline{PM}$ and $\overline{QS}$, respectively.
Prove: $NPRS$ is a parallelogram.

10. Given: $\square NPRS$; $\triangle NMS \cong \triangle RQP$
Prove: $MPQS$ is a parallelogram.

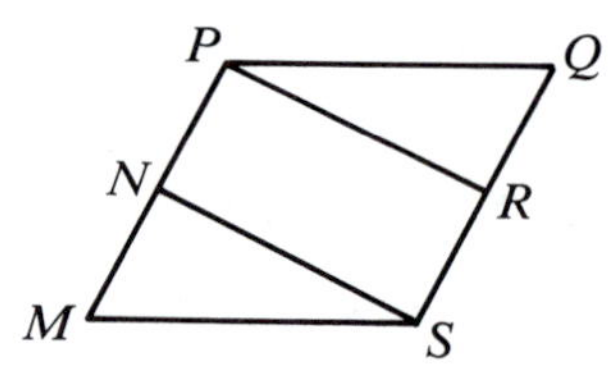

Exs. 9, 10

11. Given: $BCDE$ is a rectangle; $\overline{BA} \cong \overline{BD}$
Prove: $ABCE$ is a parallelogram.

12. Given: $\square ABCE$; $\triangle BAE \cong \triangle BDE$
Prove: $BCDE$ is a rectangle.

13. Given: $\square ABCE$; $\angle A \cong \angle ABD$
Prove: $ABFE$ is an isosceles trapezoid.

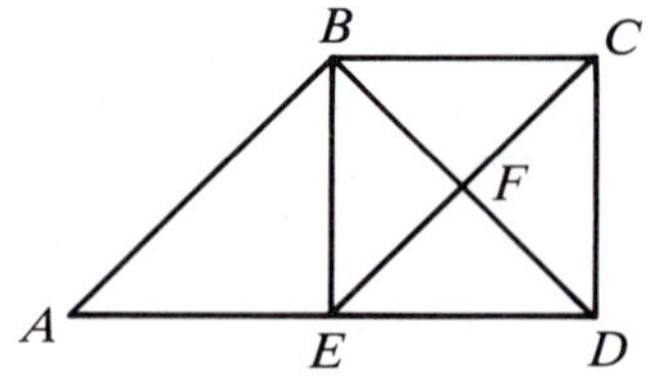

Exs. 11–13

Chapter 6

Simplify.

A 1. $5x^2y + 8x - 3x^2y - 5x$

2. $(7y^4 - y^3 + 8) - (3 - 6y + 3y^2 - 10y^4)$

3. $(-5a^2 + 3b + 2ab) + (18a^2 - 3b - 3ab)$

4. $(w^2 - 3xy + y^2) + (w^2 + 3xy + y^2) - (2w^2 + 2y^2)$

5. $(x - 11)(x + 4)$ 6. $(t + 4s)^2$ 7. $(a + 3b)(2a - 5b)$

Factor completely.

8. $n^2 - n - 20$ 9. $s^2 + 3st - 18t^2$ 10. $8c^3 + 1$

11. Divide $r^3 + 5r^2 + 9r + 5$ by $r + 2$.

Simplify.

12. $\dfrac{15y^3z^2}{-5zy}$ 13. $10xy \div 20x$ 14. $\dfrac{3x - 9y}{6y - 2x}$

15. $\dfrac{y^2 + 4y - 21}{3y - 9}$ 16. $\dfrac{cd - c}{2d + 6} \cdot \dfrac{3d + 9}{d^2 - c^2}$ 17. $\dfrac{2b}{3c} \cdot \dfrac{bd}{e} \div \dfrac{4bd^2}{6c^2e}$

18. $\dfrac{x}{x + 3} + \dfrac{3}{x + 3}$ 19. $\dfrac{1}{x} + \dfrac{2}{x^2} + \dfrac{3}{x^3}$

20. $(m + 4)(m - 3) - (m + 2)$ 21. $(-3r)^3(2r)^2$

Factor completely.

B 22. $10d^2 + 29d - 21$ 23. $-2b^2 - 8b - 8$ 24. $6y^2 + 21y + 15$

25. Divide $4x^3 + 3x^2 + 5x + 3$ by $x^2 + 1$.

Simplify.

26. $\dfrac{8t^3 - 27}{10t^2 + t - 24}$ 27. $\dfrac{4p^2 - 19p - 5}{p^2 - 3p - 10}$ 28. $\dfrac{16 - 9z^2}{9z^2 + 3z - 20}$

29. $\dfrac{a^2 - 49}{a - 5a - 14} \cdot \dfrac{2a^2 - 13a - 7}{a + 7}$ 30. $\dfrac{k^2 + 7k + 12}{k^2 + 6k + 8} \cdot \dfrac{k^2 - 3k - 10}{k^2 + 6k + 9}$

31. $\dfrac{2y^2 - 3y - 2}{3y^2 - 5y - 2} \div \dfrac{2y^2 - 15y - 8}{6y^2 - 7y - 3}$ 32. $\dfrac{7}{c + 2} + \dfrac{3}{2c + 2}$

33. $\dfrac{3y}{y^2 + 4y + 4} + \dfrac{4y}{y^2 - 4}$ 34. $\dfrac{x - y}{x^2 - y^2} - \dfrac{y}{x + y}$

Chapter 7

Solve.

A 1. $x^2 - 16 = 0$ 2. $b^2 - 5b + 6 = 0$ 3. $x^2 = 7x - 10$

4. $\frac{9}{y} - 2 = \frac{1}{4}$ 5. $\frac{5 - 3z}{12} + \frac{z}{4} = \frac{3}{2z}$ 6. $\frac{2x}{2x - 5} - 2 = \frac{15}{2x + 3}$

7. $z^2 - 4z = 12$ 8. $d^2 + 6d = 9$ 9. $t^2 + 4t = 5$

10. Simplify: $-\sqrt{121x^8}$

Use the discriminant to determine the nature of the roots of each quadratic equation. You do not need to solve the equation.

11. $9t^2 + 1 = 6t$ 12. $y^2 + 4y + 7 = 0$ 13. $5t^2 + 12t - 9 = 0$

Solve.

14. The product of two positive numbers is 20. If the second number is 3 less than twice the first, what are the numbers?

15. The sum of a number and its reciprocal is $\frac{41}{20}$. Find the number.

16. Two numbers have a sum of -3 and a product of -3. Find the numbers.

B 17. $8x^2 + 10x = 25$ 18. $9x^2 - 24x + 16 = 0$ 19. $12x^2 - 15 = -11x$

20. $\frac{y^2 - 5}{8 - y} + y = \frac{1}{8}$ 21. $\frac{2 - c}{2c + 12} = \frac{1}{2c} - \frac{9}{c^2 + 6c}$

22. $\frac{x}{2(x - 3)} - \frac{x - 2}{3(x - 3)} = \frac{3}{x^2 - 6x + 9}$ 23. $\left(x - \frac{2}{3}\right)^2 = \frac{1}{9}$

24. Write a quadratic equation that has solution set $\left\{-\frac{1}{2}, \frac{1}{3}\right\}$.

25. Simplify: $(2 - \sqrt{3})^2$

Solve.

26. $3b^2 + 8b - 2 = 0$ 27. $2(y + 1)^2 = 14$

28. $3a^2 + 7a + 3 = 0$ 29. $4z^2 + 8z + 1 = 0$

30. If twice the square of a negative number is added to the number, the sum is 120. What is the number?

31. The product of two numbers is 10 and the sum of their reciprocals is $\frac{7}{10}$. Find the numbers.

Chapter 8

For each relationship, find the indicated value.

A 1. $\frac{x}{5} = \frac{y}{7}$; $x:y$ 2. $3x - 8y = 0$; $x:y$ 3. $\frac{x-3}{16} = \frac{9}{24}$; x

4. The measures of the angles of a triangle are in the ratio 3:5:8. Find the three angle measures.

5. The sides of a triangle have lengths 7, 8, and 12. What is the length of the longest side of a similar triangle with perimeter 45?

Exercises 6–9 refer to the figure.

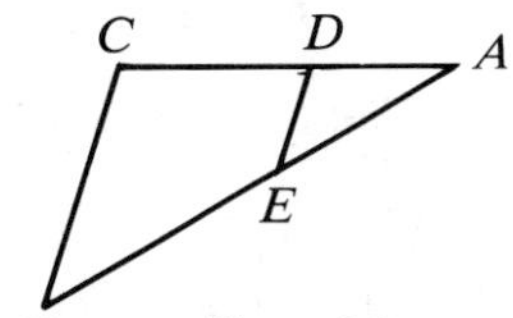

Exs. 6-9

6. If $\triangle ADE \sim \triangle ACB$, $AD = 5$, $DC = 8$, and $EB = 10$, find AE.

7. If $\overline{BC} \parallel \overline{ED}$, $AB = 21$, $BE = 6$, and $AD = 18$, find AC.

8. If $\overline{BC} \parallel \overline{ED}$, $AE = 4$, $BE = 2$, and $AC = 8$, find CD.

9. Given: D and E are the midpoints of $\overline{AC}$ and $\overline{AB}$, respectively.
   Prove: $\triangle AED \sim \triangle ABC$

10. The perimeters of two similar triangles are 20 and 35. If an altitude of the smaller triangle has length 8, what is the length of the corresponding altitude of the larger triangle?

Exercises 11 and 12 refer to the figure, in which $ABCDE \sim FGHIJ$.

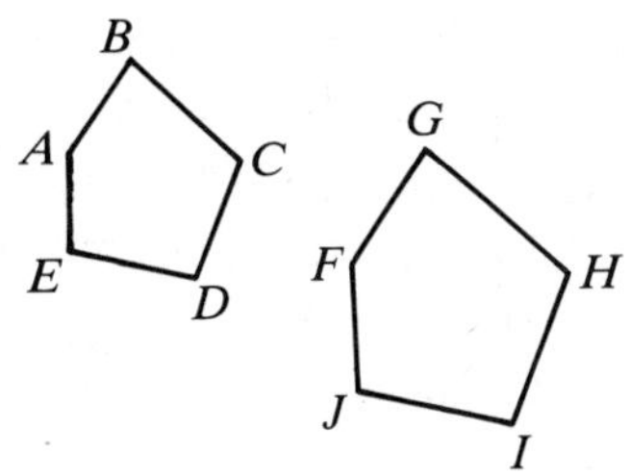

Exs. 11, 12

B 11. If $AB = 8$, $ED = x + 5$, $FG = 2x - 2$, and $JI = 2x + 4$, find all possible values of x.

12. If $\overline{AB}$ is 1 cm shorter than $\overline{FG}$, and the areas of $ABCDE$ and $FGHIJ$ are 27 cm^2 and 48 cm^2, respectively, find AB and FG.

Write two-column proofs.

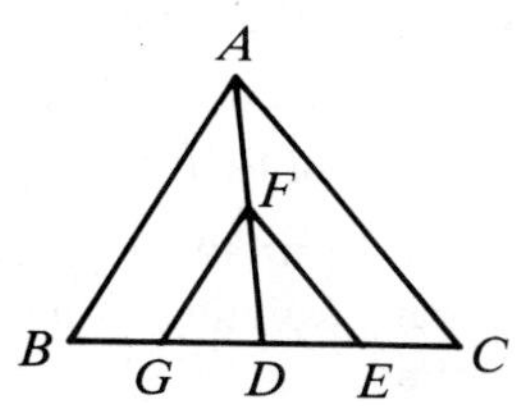

13. Given: $\overline{EF} \parallel \overline{AC}$; $\overline{FG} \parallel \overline{AB}$

    Prove: $\frac{DE}{EC} = \frac{DG}{GB}$

14. Given: $\overline{FG} \parallel \overline{AB}$; $\frac{FG}{AB} = \frac{GE}{BC}$

    Prove: $\triangle FGE \sim \triangle ABC$

15. An antenna casts a shadow 110 m when a vertical 10 m pole casts a shadow 8 m long. What is the height of the antenna?

Chapter 9

A 1. In right $\triangle XYZ$, $\overline{ZA}$ is the altitude to hypotenuse $\overline{XY}$. If $XA = 9$ and $AY = 4$, find ZA.

Right $\triangle ABC$ has legs of length a and b and hypotenuse of length c. Find the missing length. Simplify all radicals.

2. $a = 3$; $b = 3$; $c =$ ___?___
3. $b = 15$; $c = 17$; $a =$ ___?___
4. $b = 5$; $c = 20$; $a =$ ___?___
5. $a = \sqrt{7}$; $c = 7\sqrt{2}$; $b =$ ___?___

Is the triangle whose sides have the given lengths a right triangle?

6. 2, 2, 4
7. $4\sqrt{3}$, 6, $2\sqrt{3}$

8. Find the length of a diagonal of a rectangle with width 16 cm and length 30 cm.

Complete.

	9.	10.	11.	12.
XZ	3	?	?	$3\sqrt{3}$
YZ	?	?	5	?
XY	?	8	?	?

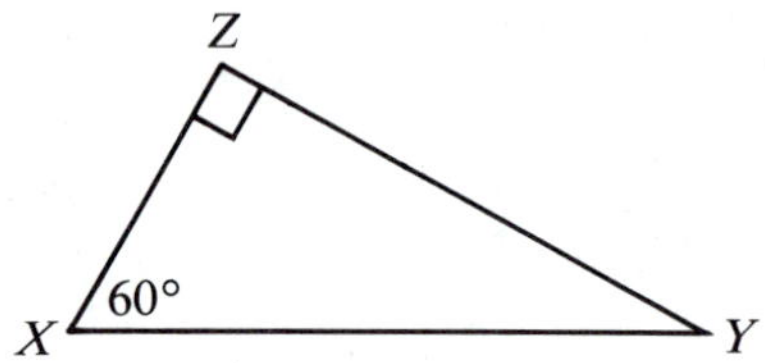

13. If $\triangle XYZ$ is such that $XZ = 9$, $XY = 41$, and $m\angle Z = 90$, find the sine, cosine, and tangent of $\angle X$ and $\angle Y$.

B 14. In right $\triangle ABC$, the length of the altitude $\overline{CD}$ to hypotenuse $\overline{AB}$ is 8. If the length of the hypotenuse is 60, find all possible values of AD.

15. In $\triangle TRI$, $\angle R$ is a right angle. If $TI = 3x + 1$, $TR = x + 1$, and $RI = 3x$, find all possible values of x.

16. Find the area of a regular hexagon with sides of length 4.

17. In $\triangle ABC$, $m\angle C = 90$ and $m\angle A = 35$. If $AC = 46$, find BC to the nearest integer. ($\sin 35° = 0.5736$; $\cos 35° = 0.8192$; $\tan 35° = 0.7002$)

18. A kite is attached to a string 35 m long. If the string makes an angle of 35° with the ground, find the height of the kite to the nearest meter.

19. From a window 18 m above ground, Bob determines the angle of elevation of the top of a bank building to be 68° and the angle of depression of a point on the ground directly below the top of the bank building to be 12°. Find the height of the bank building to the nearest meter.

Draw segments and angles like those shown and use them in Exercises 1-5, 11, 13, and 16.

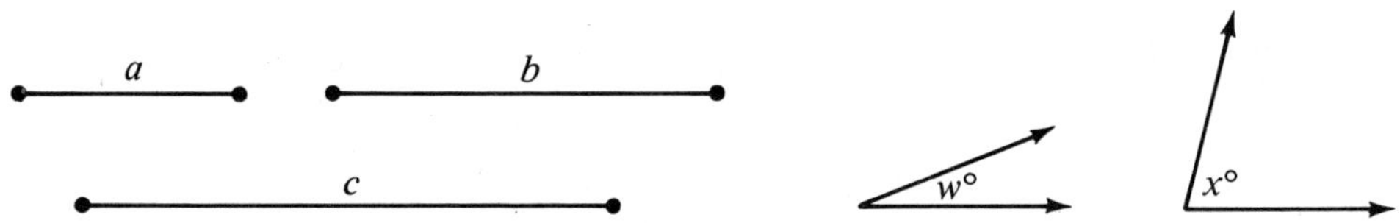

A 1. Construct a segment of length $a + 2b$.

2. Construct an angle of measure $x - w$.

3. Construct an equilateral triangle with sides of length b.

4. Construct a rectangle with sides of length a and b.

5. Construct a segment whose length, x, is such that $a:b = c:x$.

6. Draw any obtuse triangle and construct the three angle bisectors.

7. Draw any acute triangle and construct the three altitudes.

8. Draw any obtuse $\angle ABC$. Construct a line through A that is parallel to $\overleftrightarrow{BC}$. Construct a line through B that is perpendicular to $\overleftrightarrow{AB}$.

9. Draw a line segment and divide it into seven congruent segments.

10. Draw a line segment and divide it into two segments whose lengths are in the ratio 4:5.

11. Draw two points A and B such that $AB = c$. Construct the locus of points that are a distance a from A and a distance b from B.

B 12. Draw an acute triangle, $\triangle TRI$. Draw $\overrightarrow{TR}$ and $\overrightarrow{TI}$. Label a point on $\overrightarrow{TR}$ but not on $\overline{TR}$ as N and construct $\triangle TNG$ with G on $\overrightarrow{TI}$ so that $\triangle TRI \sim \triangle TNG$.

13. Use the figure at the top of the page.

a. Construct a segment of length $2c$ and divide it into two segments whose lengths are in the ratio $a:b$.

b. Construct a segment of length $\sqrt{ab}$.

Draw a diagram of each locus and then describe the locus.

14. The length of segment ST is 7 cm. What is the locus of points 5 cm from S and 5 cm from T?

15. Equilateral $\triangle ABC$ has sides of length 3 units. What is the locus of points equidistant from the three vertices?

16. Draw two points C and D a distance b apart. Construct the locus of points equidistant from C and D and a distance c from D.

Chapter 11

Find the distance between points C and D.

A 1. $C(3, 0)$, $D(-4, 0)$ 2. $C(2, 3)$, $D(7, 4)$ 3. $C(2, 4)$, $D(5, 3)$

Find the missing values if M is the midpoint of $\overline{AB}$.

4. $A(-1, -5)$; $B(6, -3)$; $M(\underline{\ ?\ }, \underline{\ ?\ })$
5. $A(4, 3)$; $B(\underline{\ ?\ }, \underline{\ ?\ })$; $M(6, 2)$

Find the slope of a line parallel to the line described and the slope of a line perpendicular to the line.

6. the line through $(8, 2)$ and $(-8, 2)$
7. the line through $(-3, -4)$ and $(1, 6)$
8. the line with equation $2x - y = 7$
9. the line with equation $4x + 3y = 5$
10. Use slopes to determine if $(1, 3)$, $(5, 6)$, and $(9, 9)$ are collinear.
11. Use coordinate geometry to prove that the diagonals of an isosceles trapezoid are congruent. (Use vertices $A(0, 0)$, $B(a, 0)$, $C(a - c, b)$, and $D(c, b)$.)
12. Graph the line with equation $3x - 2y = 6$.
13. Graph the line that contains the point $(3, -4)$ and has slope -4.

Find the area of the figure formed by joining the given vertices in order.

B 14. $P(1, 2)$, $Q(2, 5)$, $R(-1, 6)$
15. $A(0, -1)$, $B(4, -1)$, $C(6, 3)$, $D(2, 3)$
16. If the distance from $(t, 4)$ to $(8, 10)$ is $3\sqrt{5}$, find the value of t.
17. $\triangle XYZ$ has vertices $X(4, 8)$, $Y(-2, 0)$, and $Z(-3, 7)$.
    a. Use slopes to show that $\triangle XYZ$ is a right triangle.
    b. If M is the midpoint of $\overline{XY}$, show that $XM = MY = ZM$.

Write an equation in standard form of the line described.

18. the line with x-intercept -3 and y-intercept 5
19. the line that contains (4, 3) and is perpendicular to the line with equation $x - 2y = 4$
20. the perpendicular bisector of the segment with endpoints $(3, -3)$ and $(5, 7)$

Name and graph the locus of points that satisfy both conditions.

21. 2 units from the line with equation $3x + 6 = 0$ and 4 units from the line with equation $4y = 0$
22. 7 units from the line with equation $y - 3 = 0$ and 1 unit from the line with equation $2 - x = 0$
23. Solve each system: a. $4x + 5y = 6$, $5x + 2y = -1$ b. $7 = 3x - 2y$, $2x - 9 = 3y$ c. $3x = 2y + 4$, $3y = 5x - 5$

Chapter 12

A Graph each equation, give the coordinates of the vertex, and state the equation of the axis of symmetry.

1. $y = (x + 2)^2 + 2$
2. $y = -3(x - 4)^2$

Find an equivalent equation in form $y = a(x - h)^2 + k$.

3. $y = x^2 + 6x + 9$
4. $y = x^2 - 3x + 1$
5. $y = -3x^2 + 6x + 1$
6. $y = -4x^2 - 2$

Write an equation in center-radius form of the circle with center C and radius r.

7. $C(2, -1)$; $r = \frac{2}{3}$
8. $C(-6, 0)$; $r = 9$

9. Solve each system: **a.** $x^2 + y^2 = 9$, $x - y = 3$ **b.** $x^2 + 3y = 7$, $y = 2x - 3$

(Optional) Graph the equation and give the coordinates of the foci.

10. $\frac{x^2}{49} + \frac{y^2}{25} = 1$
11. $2x^2 + 3y^2 = 48$

(Optional) For each hyperbola, find **(a)** the intercepts, **(b)** the coordinates of the foci, **(c)** the equations of the asymptotes. Then graph the equation and show the asymptotes with dashed lines.

12. $\frac{x^2}{9} - \frac{y^2}{36} = 1$
13. $\frac{y^2}{5} - \frac{x^2}{4} = 1$

Find the x-intercepts of the parabola. Then graph the parabola.

B

14. $y = x^2 - x - 6$
15. $y = 2x^2 + 7x + 6$

Find the center and radius of each circle.

16. $4x^2 + 4y^2 + 4x - 4y - 14 = 0$
17. $-3(x + 4)^2 - 3y^2 + 9 = 0$
18. Write an equation of the circle with center $(4, 1)$ that contains the point $(8, 4)$.
19. Write an equation of the circle that has a diameter with endpoints at the x-intercept and the y-intercept of the line with equation $3x + 4y = 12$.

(Optional) Find an equation of the hyperbola with the given properties.

20. the hyperbola with y-intercepts $\pm\sqrt{3}$ and foci $(0, \pm 3)$
21. the hyperbola with $a = \sqrt{2}$, $b = \sqrt{3}$ and foci on the x-axis

Chapter 13

Exercises 1-7 (and Exercises 16 and 17) refer to the following experiment.

Each of the four prime numbers less than 10 is written on a separate card. The cards are shuffled and two cards are drawn at random. Specify each of the following.

A 1. The sample space

2. The event that 2 is not chosen

3. The event that 7 is chosen or the sum is 7

Find each probability.

4. P(5 or a sum of 5)

5. P(3 is chosen)

6. P(the product is even)

7. P(the sum is greater than 8)

In Exercises 8-13, refer to the following experiment.

The pointer on the spinner is spun in such a way that it does not stop on a line. Find each probability.

8. $P(1)$

9. $P(4)$

10. P(an even number)

11. P(1 or 4)

12. P(2 and not shaded)

13. P(2 or a number greater than 2)

Evaluate.

14. ${}_7P_4$

15. ${}_{13}C_9$

Refer to the experiment in Exercises 1-7.

B 16. Specify the event that at least one odd number is chosen.

17. Specify the event that exactly one odd number is chosen.

There are 315 girls and boys enrolled at a computer camp. Of these, 225 are also enrolled in the camp's sports program and 185 are enrolled in the crafts program. If a camper is selected at random, what is the probability that:

18. the camper is in both sports and crafts?

19. the camper is in sports or crafts?

20. the camper is in sports but not crafts?

21. the camper is in crafts but not sports?

22. An urn contains 5 red and 3 green marbles. A second urn contains 3 red and 4 green marbles. An urn is chosen at random and 2 marbles are drawn from the urn. Find the probability that both marbles are green.

23. A collection of 9 different books contains 4 tall books, 3 paperbacks, and 2 short books. In how many ways can the 9 books be arranged on a bookshelf if the books in each category must be next to each other?

Use Table 14–1 on page 560. Find the value of each expression.

A 1. $11 \oplus 11$ 2. $2 \oplus (7 \oplus 6)$

3. $(8 \oplus 10) \oplus 1$ 4. $(8 \oplus 2) \oplus (9 \oplus 4)$

In Exercises 5–7, refer to the system $(\{a, b, c, d\}, \#)$ shown at the right. A partial table for the commutative operation # is given.

#	a	b	c	d
a	b		d	
b	c	d		
c		a	b	
d	a	b	c	d

5. Copy and complete the table.
6. Name an identity element, if one exists.
7. Name the inverse for each element that has one.

Decide whether or not each of the following systems is a group.

8. $(\mathcal{R}, +)$ 9. $(Z_6 \setminus \{0\}, \times)$

10. $(W, *)$, where $a * b = |a + b|$ 11. $(N, +)$

State whether or not the following systems are fields. If the system is a field, state the identity element for each operation and identify the inverses of the elements.

12. $(Q^-, +, \cdot)$ 13. $(Z_6 \setminus \{0\}, +, \times)$

14. $(Z_3, +, \times)$ 15. $(Q, -, \cdot)$

Solve the equation in the field $(Z_p, +, \times)$ for the given set Z_p.

16. $2x + 1 = 0;\ Z_5$ 17. $x^2 + 2x + 1 = 0;\ Z_3$

Let D be the set of integral multiples of 2 and E be the set of odd natural numbers. State whether the given set is closed under the given operation.

B 18. $(E, +)$ 19. $(D, \cdot)$ 20. $(E, -)$

21. $(D, -)$ 22. $(D, \div)$ 23. $(E, \cdot)$

Define the following rigid motions of a square. Let P be the intersection of the diagonals, and let P_n represent a counterclockwise rotation of $n°$ about P. Define $*$ such that $P_n * P_k$ means P_n followed by P_k.

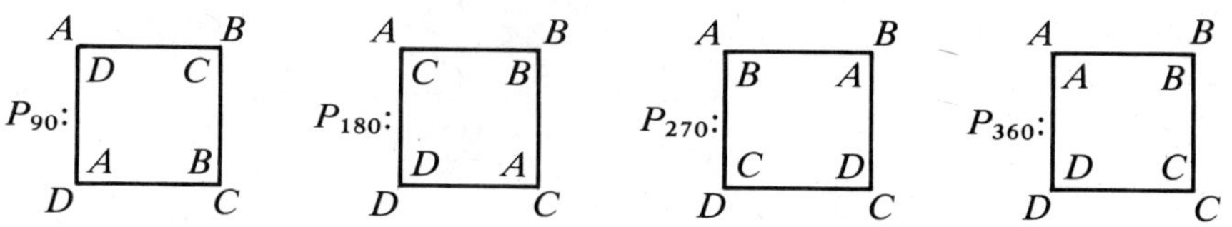

24. Make an operation table for $(\{P_{90}, P_{180}, P_{270}, P_{360}\}, *)$.
25. Is the system a group? If so, identify the identity element and the inverse of each element. If not, explain why not.

Algebra Review Exercises

Chapter 1

Simplify.

A 1. $2xy - 8x + 4xy + 8y$

2. $4x + 3x$

3. $8b(3 - (-2b))$

4. $9t - (-10t)$

5. $9(u - (-5u))$

6. $5(-3t)^2$

Solve and check.

7. $3t - 3 = 10$

8. $8 - 3w = 2(w + 3)$

9. $4r + 5 = 5r + 4$

10. $w + 2w + 3w + 4w = 123$

11. $4 - (x + 8) = 9$

12. $y - (5 - y) = 12y$

13. $2(x + 4) + 4(x + 2) = 1$

14. $12(x + 1) - 2(x - 4) = 5$

Solve.

15. $45 + 4t > 2$

16. $4 - 2w \leqslant 7 - 3w$

17. $3 + (3 - r) < r - 3$

18. $9(3 + p) > 27$

19. $7 \leqslant 3 + x < 10$

20. $22 \leqslant 4 - h \leqslant 33$

21. $5t < t + 4$ or $t < 9$

22. $3z - 5 \leqslant -11$ and $5 - z \leqslant 10$

Translate the problem into an equation or an inequality. Identify the variable you are using and state what it represents. Then solve the equation or inequality.

23. The value of the dimes in the box is \$5.60.

24. The sum of two consecutive odd numbers is 60.

25. The sum of two consecutive integers is at least 33.

Solve.

26. The sum of three consecutive odd integers is 45. Find the integers.

27. One alloy contains 8% iron and another contains 18% iron. How many grams of each should be mixed to produce 150 g of an alloy which is 11% iron?

B 28. $m - (3 - 2m) = 4m - (2m - 1)$

29. $5t - 13 = 2(3t + 4) - (3 - t)$

30. $\frac{y + 2}{6} - \frac{3y + 4}{2} = \frac{1}{3}$

31. $\frac{3z - 5}{5} + \frac{4z - 1}{2} = 5$

32. $-1.4x + 7.4 \leqslant 3 - 2.5x$

33. $8 - (5n + 2) \leqslant 4n - 3(n - 4)$

34. $7 - 3t \geqslant 2t + 5$ or $3t - 4 \geqslant 5 - 2t$

35. $2 \leqslant 5(z + 1)$ and $5(z + 1) < 9$

36. Two cars head in opposite directions on a straight road at rates of 50 km/h and 80 km/h. In how many hours will they be 455 km apart?

37. Andy left home at 12 noon and bicycled at a rate of 15 km/h. Anne then left at 1 P.M. to catch up with him and bicycled at a rate of 18 km/h. At what time did Anne catch up with Andy?

38. Harold and Claude began jogging at the same time and place. Harold maintained a steady pace of 20 km/h and Claude maintained a steady pace of 17 km/h. If Claude finished 1.8 min after Harold, find the distance each jogger ran.

Chapter 6

Simplify.

A 1. $(a + 4b - 6b^2) + (2a - 5b + 8b^2)$

2. $(8h^2 - 9h + 3) - (3h^2 - 9h + 8)$

3. $(x - 3y)(x + 6y)$

4. $(3a + 2b)(3a - 2b)$

5. $(y^2 - 2)(y + 4)$

6. $ab^2(2a^2 - b^2 + 5a - 3b)$

7. $\dfrac{2ab^2c^4}{10abc}$

8. $\dfrac{3x - 4y}{12}$

Factor completely.

9. $13x^2y^3 - 26x^3y^2$

10. $144a^2 + 441b^2$

11. $x^2 - 2x - 15$

12. $a^2 + 3a + 2$

13. $2p^2 + p - 3$

14. $16a^2 + 6ab - b^2$

15. $y^2 - 4xy - 21x^2$

16. $4x^2 - 4x + 1$

17. $4m^2 - 625$

18. $x^2 - x^2y^2$

19. $a^3 - 8$

20. $27x^3 + 1$

Simplify.

21. $\dfrac{3a - 12b}{9a + 15b}$

22. $\dfrac{3a - 12b}{5a - 20b}$

23. $\dfrac{2m + 8mn}{2m - 2mn}$

24. $\dfrac{3r^2 - 6rs}{12s - 6r}$

25. $\dfrac{x - 2}{x^2 - 3x + 2}$

26. $\dfrac{(x + y)(x - y)}{(y - x)^2}$

27. $\dfrac{m^2 - 4}{m^2 + m - 6}$

28. $\dfrac{9m^2 - 1}{9m^2 + 6m + 1}$

29. $\dfrac{1 - 4x^2}{1 - 4x + 4x^2}$

Simplify. Write the answer in lowest terms.

30. $\dfrac{2xy}{7ax} \cdot \dfrac{14a}{x^2}$

31. $\dfrac{a^2b}{b^2a} \cdot \dfrac{a^2b^2}{ab}$

32. $\dfrac{2m + 3}{5} \cdot \dfrac{5m + 10}{m + 2}$

33. $\dfrac{1}{2x} - \dfrac{2}{x}$

34. $\dfrac{x^2}{x + y} - \dfrac{y^2}{x + y}$

35. $1 - \dfrac{1}{t + 2}$

B 36. $\dfrac{(1 - m)^2}{(m - 1)^2}$

37. $\dfrac{a^2 - 36}{5a - 30} \div (5a + 30)$

38. $\dfrac{a - b}{(b - a)^2} \cdot \dfrac{(b - a)^3}{(a - b)^4}$

39. $\dfrac{(x + 8)^2}{2x + 8} \div \dfrac{x + 8}{4}$

40. $\dfrac{2}{x^2 - 4} + \dfrac{1}{2x - x^2}$

41. $\dfrac{4x}{3x^2 - 27} - \dfrac{1}{x^2 - 3x}$

Simplify.

42. $2t^2(3t+1)(3t-1)$

43. $(z-2y)(z+2y)-3(z+1)^2$

44. $\frac{1}{x}+\frac{2}{x+1}-\frac{1}{x-1}$

45. $\frac{m^2}{m^2-1}+\frac{m}{m+1}-\frac{m}{1-m}$

46. $\frac{x}{a}+\frac{x}{2a}+1$

47. $\frac{x+a}{x-a}-\frac{x-a}{x+a}+1$

48. Divide $4s^3+12s^2+13s-1$ by $2s+3$.

49. Divide $2z^4+3z^3+13z^2+15z+8$ by z^2+z+4.

50. One factor of x^3+3x^2+4x+4 is $x+2$. Find the other factor.

Chapter 7

Solve by factoring.

A

1. $x^2-7x+10=0$

2. $2a^2+9a+10=0$

3. $t^2+4t-21=0$

Solve by taking square roots.

4. $r^2=121$

5. $0.04m^2-1=0$

6. $n^2-169=0$

Solve by completing the square.

7. $t^2-2t=15$

8. $w^2+4w=-3$

9. $z^2+8z=25$

Solve by using the quadratic formula.

10. $4x^2-x-3=0$

11. $2m^2+3m-4=0$

12. $3z^2+5z-4=0$

Use the discriminant to determine the nature of the roots of each equation. You do not need to solve the equation.

13. $16r^2-8r+1=0$

14. $w^2+12w+3=0$

15. $11m^2+9m-2=0$

16. $x^2+5x+1=0$

17. $k^2-k-2=0$

18. $25v^2-30v+9=0$

Solve.

19. $w^2+7w=8$

20. $(x-4)^2=16$

21. $(2x-8)^2=1$

22. $m^2-3m=m+12$

23. $\frac{2y+1}{y}=\frac{1}{y}+\frac{y+1}{2}$

24. $\frac{2}{b}+\frac{b+2}{3}=\frac{5b}{3}$

25. The sum of a number and its reciprocal is $2\frac{1}{6}$. Find the number.

26. The sum of the squares of two consecutive odd integers is 394. Find the integers.

27. A positive number is 20 times as great as its reciprocal. Find the number.

B 28. It takes Susan 7 min longer to assemble a birdhouse than it does Karta. Working together, Susan and Karta can do the job in 12 min. How long does it take Karta if she works alone?

29. The sum of twice one number and three times a second is 50. Nine times the second number exceeds the first number by 10. Find the numbers.

30. The perimeter of a rectangular field is 96 m and the area is 320 m^2. Find the dimensions of the field.

31. Last year Jess bought some grain for \$28. This year she found that the price had increased by 25%, and that \$28 would buy forty fewer kilograms. Find the cost per kilogram of the grain now.

32. Andy and Barbara can refinish five tables in twelve hours when working together. To do one table, Andy needs two hours more than Barbara. How long does each need to refinish one table?

33. The width, length, and height of a rectangular prism are consecutive integers. The total surface area of the prism is 52 cm^2. Find the dimensions.

34. The sum of a number and its reciprocal is 3. Find the number.

Solve.

35. $(2x + 1)^2 + (x + 2)^2 = 26$

36. $x^2 + (3x + 2)^2 = 4$

37. $\left(1 + \frac{1}{x}\right)^2 = 4$

38. $\left(1 + \frac{3}{x}\right)^2 = 1$

39. $m + \frac{m + 1}{m + 2} = \frac{1}{m + 2}$

40. $\frac{c - 2}{c + 2} - \frac{c + 3}{c - 3} = 5$

Chapter 11

A 1. The vertices of $\triangle ABC$ are $A(6, 5)$, $B(9, 9)$, and $C(5, 12)$.
   a. Find AB, BC, and CA.
   b. Show that $\triangle ABC$ is a right triangle.
   c. Find the area of the triangle.

2. The vertices of $\triangle ABC$ are $A(-4, 4)$, $B(-4, -2)$, and $C(2, 4)$.
   a. Find AB, BC, and AC.
   b. Find the length of the median from B to $\overline{AC}$.
   c. Show that the median from A to $\overline{BC}$ is perpendicular to $\overline{BC}$.
   d. Find the area of $\triangle ABC$.

3. $A(0, 0)$, $B(a, b)$, C and $D(a, -b)$ are vertices of a rhombus. What are the coordinates of vertex C? (*Hint:* There are three possible answers.)

4. Find an equation in slope-intercept form of the line that passes through $(2, -5)$ and has slope -9.

5. Find an equation in standard form of the line that passes through the points $(4, -4)$ and $(-2, 2)$.

Write an equation (or equations) for the locus of points:

6. 2 units to the right of the line with equation $x = 4$
7. 9 units from the line with equation $y = 3$
8. equidistant from the lines with equations $y = 4$ and $y = -2$
9. equidistant from the points $(2, 7)$ and $(-4, 9)$

10. Find the coordinates of the intersection of the lines with equations $2x + 3y = 6$ and $y = x + 7$.

11. Use coordinate geometry to prove that the segments joining the midpoints of the opposite sides of a quadrilateral bisect each other.

12. Show that the lines with equations $x - 3y = 4$, $2x + y = 1$, and $3x - 2y = 5$ all pass through the same point.

B 13. Find the area of $\triangle ABC$ with vertices $A(-2, 4)$, $B(0, 0)$, and $C(4, 7)$.

14. The coordinates of three consecutive vertices of a parallelogram are $(0, -3)$, $(4, 1)$, and $(1, 5)$. Find the coordinates of the fourth vertex.

15. A and B have coordinates $A(0, 3)$ and $B(10, 8)$. Find the coordinates of the point P on the x-axis such that $PA = PB$.

16. A and B have coordinates $A(0, 3)$ and $B(10, 8)$. Find the coordinates of the points P on the x-axis such that $\overline{PA}$ is perpendicular to $\overline{PB}$.

17. Given the line with equation $y = 2x$ and the point $A(4, 3)$, find the coordinates of the point P such that the line is the perpendicular bisector of the segment $\overline{AP}$.

The trapezoid $ABCD$ has vertices $A(0, 0)$, $B(10, 0)$, $C(8, 4)$, and $D(2, 4)$.

18. Find the coordinates of M, the point of intersection of $\overleftrightarrow{AD}$ and $\overleftrightarrow{BC}$.
19. Find the coordinates of N, the point of intersection of $\overline{AC}$ and $\overline{BD}$.
20. Show that the line joining the points M and N bisects both $\overline{AB}$ and $\overline{CD}$.

Chapter 12

Find each x-intercept and y-intercept of the equation.

A 1. $y = 2x + 5$ 2. $x^2 = 16 - y^2$ 3. $y = (x + 5)^2 - 9$

4. $x^2 + y^2 = 12$ 5. $y = x^2 - 6x + 2$ 6. $y = (x + 3)^2 - 4$

7–9. Graph each of the equations in Exercises 1–3.

10. A parabola has focus at $(-2, 6)$ and directrix $y = 8$. State the coordinates of the vertex.

Graph the parabola $y = x^2 - 4x + 1$ and use your graph to determine whether each equation has no real solutions, one real solution, or two real solutions.

11. $x^2 - 4x + 1 = 8$

12. $x^2 - 4x + 1 = -8$

13. $x^2 - 4x + 1 = -5$

14. $x^2 - 4x + 1 = -1$

Write an equation for each of the following.

15. the circle with center at $(0, -2)$ and radius 5

16. the circle with center at $(4, 3)$ and radius 5

17. (Optional) the ellipse with foci at $(0, 4)$ and $(0, -4)$ if the sum of the focal radii is 10

18. Find the radius and the coordinates of the center of the circle with equation $x^2 + y^2 + 4y = 21$.

19. Solve the system and illustrate with a graph: $y = x^2 + 2x + 5$
$y = 2x + 5$

20. (Optional) Find the equations of the asymptotes of the hyperbola with equation $4x^2 - y^2 = 16$.

21. Express $y = 2x^2 + 8x + 1$ in the form $y = a(x - h)^2 + k$. Then state the coordinates of the vertex of the parabola.

22. Express the equation $x^2 + y^2 - 2x + 6y = 6$ in center-radius form.

B

23. Find an equation of the parabola with x-intercepts 7 and -7 and y-intercept 7.

24. Find the value of m if the line $y = x + m$ intersects the circle $x^2 + y^2 = 18$ in exactly one point.

25. Write an equation in expanded form of the circle that has a diameter with end points $(2, 1)$ and $(6, 5)$.

26. Write an equation in the form $y = ax^2 + bx + c$ of the parabola with focus $F(1, 0)$ and directrix with equation $y = 4$.

27. (Optional) Find an equation of an ellipse with center at the origin, x-intercepts ± 10, and y-intercepts ± 6.

28. (Optional) Find an equation of a hyperbola with x-intercepts ± 4 and asymptotes with equations $y = \pm\frac{3}{4}x$.

29. Solve the system and illustrate with a graph: $4x^2 - y^2 = 16$
$y = 2x + 8$

30. (Optional) Given the points $A(-2, 0)$ and $B(2, 0)$, find an equation of the locus of points $P(x, y)$ such that the product of the slopes of $\overleftrightarrow{PA}$ and $\overleftrightarrow{PB}$ is 2. Identify the locus.

Programming in BASIC

1. Introduction to Computers

A computer contains a *central processing unit* (CPU), which does the work, and a *memory*, in which information is stored. It must also have *input* and *output* devices.

COMPUTER

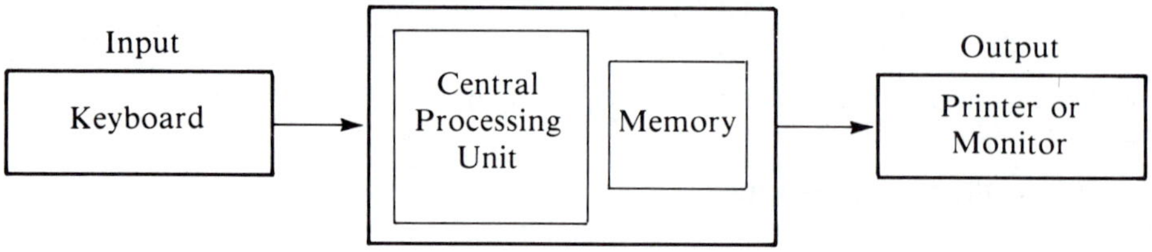

Computers are important because they can work with large amounts of data very fast. They work so rapidly that the directions for each project must be written out and stored ahead of time. Such a list of directions is called a *program*. Data and programs used with microcomputers (personal computers) are generally input by means of a *keyboard*. Everything that is typed in appears on the *monitor*, which is like a television screen. Sometimes a *printer* is also attached.

The actual instructions to the computer are given in *machine language*. Such a language may be written down by using 0's and 1's, where 0 and 1 may represent "off" and "on" switches respectively. Numbers written with 0's and 1's are said to be written in the *binary system* in contrast to the decimal system, which we ordinarily use. In the decimal system, 10 means ten. In the binary system, 10 means two, 11 means three, 100 means four, and so on. Other systems that are sometimes used are the octal (base eight) and the hexadecimal (base sixteen).

Most programmers use one of the so-called "higher level" languages, such as BASIC, COBOL, FORTRAN, or Pascal. The computer contains a program, called a *compiler*, or *interpreter*, that translates such a program into machine language. Since some form of BASIC is available on each of the microcomputers, as well as on larger computers, that language has been chosen for use in this book.

2. BASIC Symbols, FOR-NEXT, PRINT, END

BASIC uses many of the familiar signs and symbols of mathematics:

+ for addition	( ) grouping symbols
− for subtraction	> is greater than
= equals	< is less than

Other symbols are slightly different:

* for multiplication	>= for ≥
/ for division	<= for ≤
X↑3 or X^3 for x^3	<> for ≠

A computer program may be thought of as being made up of blocks like this:

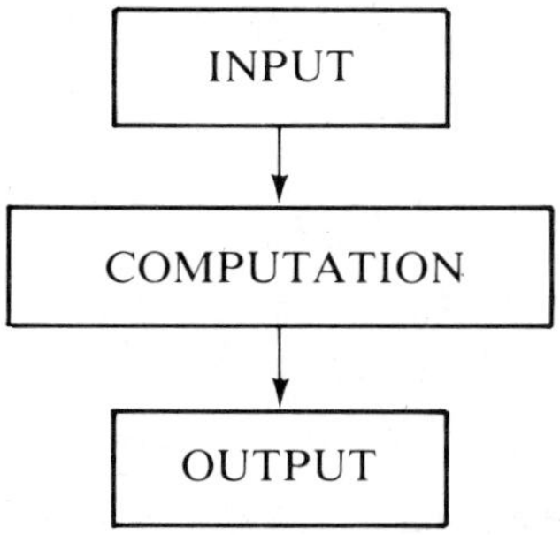

However, sometimes the blocks may overlap and sometimes one block may contain several processes.

A BASIC program consists of a list of numbered lines, or *statements*. It is customary to number the statements 10, 20, . . . , as shown below so that additional statements may be inserted when needed.

BASIC handles *variables* much as you do in mathematics. There are several ways of giving values to variables. When a fixed number of sequential values are to be given to a variable, a FOR-NEXT *loop* may be convenient, as shown below in lines 10 to 30.

```
10  FOR N=1 TO 10
20  PRINT N, N*N, N^3
30  NEXT N
40  END
```

(Some versions of BASIC do not require an END statement – line 40.) In this program, the value of the variable N is printed as 1, 2, . . . , 10. Thus, the input (line 10) and the computation and the output (line 20) are all contained in one block – the loop. Line 20 is a PRINT statement. It computes the values of N, N*N, and N^3 and then prints them. The *commas* in line 20 will cause the values to be printed on the same line but spaced apart. Type in this program and then type the *command* RUN. (A command has no line number because it is not part of a program.) The output will consist of ten lines and three columns.

You can use a PRINT statement with *quotation marks* and commas to print headings for the columns, as in line 5 below. Line 7 will "print" a blank line. Insert these lines:

```
5  PRINT "N", "N^2", "N^3"
7  PRINT
```

Type the command LIST to see the revised program, and then RUN it. The output should look something like this:

N	N^2	N^3
1	1	1
2	4	8
3	9	27
4	16	64
5	25	125
6	36	216
7	49	343
8	64	512
9	81	729
10	100	1000

3. IF-THEN(-ELSE), REM, GOTO, INPUT, DIM

A very powerful tool of computer programming is the IF-THEN(-ELSE) construction. This is illustrated in the program shown below, which illustrates the Example on page 21. Lines 30–100 form a loop within which are two *branches* resulting from the test in line 40. If $x + 5 > 7$, then "*" is printed (line 90). Otherwise (else) "-" is printed (line 60). The "else" and "then" branches are emphasized in this program by inserting REM (remark) statements as in lines 50 and 80. REM statements appear in a listing of a program but do not affect its execution. The GOTO statement in line 70 is necessary to skip over the other branch. The *semicolons* at the ends of lines 60 and 90 cause the symbols to be printed close together. Line 110 cancels this effect.

The loop in lines 120–190 will print "0" when I has the value zero. Then the program ends.

```
10   PRINT "TO GRAPH AN OPEN SENTENCE,"
20   PRINT "IN ONE VARIABLE:"
30   FOR X=-10 TO 10
40   IF X+5>7 THEN 90
50   REM: ELSE
60   PRINT "-";
70   GOTO 100
80   REM: THEN
90   PRINT "*";
100  NEXT X
110  PRINT
120  FOR I=-10 TO 10
130  IF I<>0 THEN 180
140  REM: ELSE
150  PRINT "0"
160  GOTO 200
170  REM: THEN
180  PRINT " ";        ← Single space between quotation marks
190  NEXT I
200  END
```

(Lines 30–100: LOOP; lines 120–190: LOOP)

RUN this program and compare the result with that on page 21. Notice that this program tests only *integers* from −10 to 10.

In this program, the inequality to be graphed appears in line 40. Therefore, this line must be changed for each exercise. Make this change:

```
40  IF 2 * X + 6 <= X + 3 THEN 90
```

RUN the program again and compare the result with that on page 22.

Some of the preceding BASIC statements are included in the summary on page 39. There we also see another way of giving a value to a variable – by using an INPUT statement. It is illustrated in the program on pages 39–40.

Lists of values are conveniently handled by using *subscripted variables*, which are described and used on pages 39–40.

When a subscripted variable is to have more than 10 (or, in some cases, 11) values, a DIM (dimension) statement must be included. See line 50 in the second program on page 251.

4. LET, SQR, INT, ABS

Still another way of giving a value to a variable is by using a LET statement. We can say simply, for example,

LET X = 59

or we can use a formula such as:

LET A = L * W

This latter usage appears in the program on page 180.

For H to be used as a *counter*, we start by giving it the value zero in:

```
70  LET H = 0
```

(This is read "Give the value 0 to the variable H.")
Then

```
110  LET H = H + 1
```

means: Take the current value of H, add 1 to it, and put the new value into H. See the program on page 251.

BASIC includes several functions. One of these is the *square root function*: SQR(X) = $\sqrt{X}$. This is used in the program on page 510.

BASIC also has an *absolute value function*: ABS(X) = $|X|$. Both the square root and absolute value functions are used in the program on page 252.

Another function is the *greatest integer function*, INT(X), which means "the greatest integer less than or equal to X." Thus:

INT(4) = 4 and INT(4.9) = 4

This function is used in factoring programs. For example, if

$$A/F = INT(A/F)$$

then F is a factor of A. In the program on page 251, we

$$LET\ Q = A/F$$

and then both F and Q are factors of A if it divides evenly.

5. Solving Quadratic Equations, GCF, STEP

The program to solve a quadratic equation on page 296 illustrates a three-way branch. For D, we have:

```
50  IF D < 0 THEN 230
70  IF D = 0 THEN 210
80  REM *** D IS POSITIVE.
```

There are two possibilities for $D > 0$:

(1) If the square root is a whole number (line 100)

(2) If the square root is not a whole number (line 110)

Thus, there are, in effect, four branches. copy or print out the program, and mark off the four branches, beginning with line 110.

A program for finding the GCF of two numbers by the Euclidean algorithm is embedded in the program on page 345. Here is an independent version:

```
10   PRINT "TO FIND THE GCF"
20   PRINT "BY THE EUCLID. ALG.:"
30   PRINT "INPUT A,B";
40   INPUT A,B
50   LET Q=INT(A/B)
60   LET R=A-Q*B
70   IF R=0 THEN 110
80   LET A=B
90   LET B=R
100  GOTO 50
110  PRINT "GCF = ";B
120  END
```

Insert the line

```
65  PRINT A, B, R
```

and RUN the program for A = 42 and B = 15. Compare the print-out with the table at the top of page 345. Take A = 15 and B = 42 and see what happens.

The GCF of two numbers can also be found by using factoring as in the following program. This program uses the STEP part of the FOR-NEXT arrangement (compare page 39). Here we use STEP −1, since A will be greater than 1.

```
10  PRINT "TO FIND GCF"
20  PRINT "BY FACTORING:"
30  PRINT "INPUT A, B";
40  INPUT A,B
50  FOR X=A TO 1 STEP -1
60  LET Q=A/X
70  IF Q <> INT(Q) THEN 100
80  LET Q1=B/X
90  IF Q1=INT(Q1) THEN 110
100 NEXT X
110 PRINT "GCF = ";X
120 END
```

Insert the lines

```
75 PRINT A; "/";X;" = ";Q
85 PRINT B; "/";X;" = ";Q1
```

and RUN the program for $A = 42$, $B = 15$, and also for $A = 15$, $B = 42$.

In general, in the statement

FOR I=M TO N STEP S

when $M < N$, so that I is increasing, $S > 0$. But when $M < N$, so that I is decreasing, $S < 0$.

Notice that in the program on page 379, STEP 5 is used in line 20. Problem 3 asks you to use STEP 3.

6. Rounding Numbers; TAB

The greatest integer function can be used to round numbers. For example, to round the trigonometric values found on page 379 to four decimal places, insert lines 32, 34, and 36, and change line 40:

```
32 LET S=INT(10000*SIN(R)+.5)/10000
34 LET C=INT(10000*COS(R)+.5)/10000
36 LET T=INT(10000*TAN(R)+.5)/10000
40 PRINT D;TAB(10);S;TAB(20);C;TAB(30);T
```

RUN the revised program and compare your results with the table on page 373.

We can also round the values to hundredths and then use TAB to print out a more compact table. Copy and RUN the following revised program. Note how the TAB function varies the output.

```
10    PRINT "DEG.";TAB(6);"SINE";TAB(12);
15    PRINT "COS";TAB(18);"TAN"
20    FOR D=5 TO 45 STEP 5
30    LET R=D*.017453
32    LET S=INT(100*SIN(R)+.5)/100
34    LET C=INT(100*COS(R)+.5)/100
36    LET T=INT(100*TAN(R)+.5)/100
40    PRINT D;TAB(6);S;TAB(12);
45    PRINT C;TAB(18);T
50    NEXT D
60    END
```

TAB is also used in printing the table in the program on page 590.

7. Graphing Parabolas

By using ideas from the program (page 622) for graphing open sentences and the program (page 511) that finds the vertex of a parabola, we can make a program to graph a parabola:

```
10    PRINT "TO GRAPH"
20    PRINT "Y = AX^2 + BX + C:"
30    PRINT "INPUT A, B, C";
40    INPUT A,B,C
50    LET H=-B/(2*A)
60    LET K=A*H*H+B*H+C
70    FOR Y=K+10 TO K-10 STEP -1
80    FOR X=H-10 TO H+10
90    IF Y=A*X*X+B*X+C THEN 180
100   IF X=0 THEN 160
110   IF Y=0 THEN 140
120   PRINT " "; ←Single space between quotation marks
130   GOTO 190
140   PRINT "-";
150   GOTO 190
160   PRINT "!";
170   GOTO 190
180   PRINT "*";
190   NEXT X
200   PRINT
210   NEXT Y
220   END
```

RUN this program for $y = x^2 + 2x - 3$ and compare the result with that shown on page 481. RUN it also for $y = x^2 - 2x$, $y = x^2 - 2x - 2$, and $y = x^2 - 2x + 2$.

Notice that in the program the loop for y is outside the loop for x. Such loops are called *nested loops*. Notice also that the values for y decrease while the values for x increase.

8. String Variables

A *string* is any set of characters enclosed in quotation marks. We have been using them simply as headings. However, they can also be used as values of *string variables*. A string variable may be denoted by a letter followed by $.

The following fragment of a program shows how strings may be used in chatty kinds of programs.

```
20   PRINT "WHAT IS YOUR NAME";
30   INPUT N$
40   PRINT "HELLO ";N$; "!"
50   PRINT "I AM GLAD TO MEET YOU."
60   PRINT "DO YOU ENJOY WORKING WITH ";
70   PRINT "ME (YES/NO)";
80   INPUT A$
90   IF A$ = "YES" THEN 170
100  REM: ELSE
110  PRINT "I AM VERY SORRY, ";N$; "."
120  PRINT "PERHAPS IF YOUR TEACHER ";
130  PRINT "HELPED YOU . . ."
140  PRINT "PLEASE TRY ME AGAIN LATER."
150  GOTO 200
160  REM: THEN
170  PRINT "I AM VERY GLAD, ";N$; "."
180  PRINT "WE CAN LEARN LOTS ";
190  PRINT "OF THINGS TOGETHER."
200  END
```

Type in and RUN this program twice – once with the answer YES and again with the answer NO. (You do not need to type quotation marks around the answer here.) Notice that the program will go the ELSE branch if you type in any string other than YES.

9. The Random Function, RND; Simulations

The properties of the *random function*, RND, vary from one version of BASIC to another. Type in and RUN the following program:

```
10  FOR I = 1 TO 20
20  PRINT RND(1),
30  NEXT I
40  END
```

This program may print out 20 different numbers N such that N is between 0 and 1. Several RUNs should give different lists. If your computer fails to give either of these results, consult your manual and find out what adjustments must be made in the program. Make the corresponding adjustments in the remaining programs given here.

Change line 20 to

```
20 PRINT INT(2 * RND(1) + 1),
```

and write down the list of twenty numbers. Repeat for:

```
20 PRINT INT(6 * RND(1) + 1),
```

Write out in words what each of these versions of line 20 does.

The RND function is often used in simulations. The following program uses it to simulate the tossing of a coin, followed by a roll of a die (compare page 532). RUN it several times for T = 48.

```
10  PRINT "TOSS A COIN, AND"
20  PRINT "THEN ROLL A DIE—"
30  PRINT "HOW MANY TIMES ";
40  INPUT T
50  DIM A(12)
60  FOR I = 1 TO 12
70  LET A(I) = 0
80  NEXT I
90  PRINT
100 FOR I = 1 TO T
110 LET C = INT (2 * RND (1) + 1)
120 LET D = INT (6 * RND (1) + 1)
130 IF C = 2 THEN 200
140 REM ****C=1
150 PRINT "H";
160 PRINT D; ",";
170 LET A(D) = A(D) + 1
180 GOTO 230
190 REM ****C=2
200 PRINT "T";
210 PRINT D; ",";
220 LET A(D + 6) = A(D + 6) + 1
230 NEXT I
240 PRINT
250 REM ****SUMMARY
255 REM ****P=THEORETICAL NO.
               OF OCCURRENCES
260 LET P = T / 12
265 PRINT
270 PRINT "P = ";P
275 PRINT TAB( 10);"DIFF. FROM P"
280 FOR I = 1 TO 6
290 PRINT "H";I;": ";A(I);
300 PRINT "            ";A(I) − P
310 NEXT I
320 FOR I = 7 TO 12
330 PRINT "T";I − 6;": ";A(I);
340 PRINT "            ";A(I) − P
350 NEXT I
360 END
```

Before using a larger value for T, delete lines 150, 160, 200, 210, and change:

```
130 IF C = 2 THEN 220
```

RUN the program several times, for T = 480 and observe the results.

Preparing for Regents Examinations

The Regents examination for Sequential Math–Course II is a comprehensive achievement test of the course objectives outlined in the New York State Syllabus. It is prepared by a committee of teachers and State Education Department specialists. The test is scheduled for January, June, and August each year.

The test is divided into three parts. Part I consists of 35 short-answer and multiple-choice questions. The student must answer 30 questions, each worth 2 points. Part II consists of 5 or 6 analytical problems related to major topic strands of the course. The student must answer 3 of these problems, each worth 10 points. Part III consists of 2 or 3 questions that require an analytic, geometric, or logical proof. The student must choose 1 form of proof, worth 10 points. A construction may also appear on Part I.

In addition to its comprehensiveness, the test is designed to measure a student's ability to apply what he or she has learned to a variety of different problem settings. The underlying course philosophy of topic integration and concept unification is clearly displayed in questions requiring insights and skills from several different areas. Most noteworthy is the integration of algebra and geometry, especially in the coordinate plane.

To prepare effectively for this examination, you should organize a list of the fundamental concepts, principles, and theorems. Then you should review and drill the various skills and procedures related to these topics. Throughout, you should strive for both accuracy and the ability to apply these skills and strategies to solve problems.

To assist you in this preparation, go back to the Cumulative Reviews on pages 145-148, 298-302, 420-422, 593-596. They provide ample practice for the short-answer and multiple-choice sections of the test and help you sharpen test-taking skills such as eliminating choices that are clearly in error.

The following pages contain checklists for each topic as well as many sample questions. These questions are similar in form and style to those found on the Regents exam but are longer in order to provide comprehensive review of the concepts involved. While a typical Regents question is shorter and more selective in its scope, these problems will help you to consider all aspects of topic applications. They are designed to help you further organize course objectives in a manner consistent with Regents testing philosophy.

Logic–Checklist

Statements
Equivalent statements
Quantifiers (all, some, no)
Negations
- quantified statements
- DeMorgan's Laws
- conditionals

Universal quantifier ($\forall$)
Existential quantifier ($\exists$)
Related conditionals

Arguments
- Law of the Contrapositive
- Law of Detachment
- Law of Contrapositive Inference
- Law of Disjunctive Inference
- Law of Conjunctive Simplification
- Law of the Syllogism

Logic—Sample Problems

1. Write a statement that is logically equivalent to the given statement.
   a. All students love math.
   b. Some numbers are prime numbers.
   c. No person under 18 will be admitted.
   d. $\sim(p \vee \sim q)$
   e. $\sim(p \rightarrow \sim q)$
   f. If x is a factor of 9, then x is not even.

2. Write the negation of each statement.
   a. All birds fly.
   b. Some eggs did not hatch.
   c. No goals were scored in the third period.
   d. $\sim p \vee \sim q$
   e. $\sim p \wedge \sim q$
   f. $p \rightarrow q$
   g. $\forall_x\ x^2 > 0$
   h. $\exists_x\ x^3 = x + 6$

3. Which is not always true?
   a. $p \wedge \sim p$
   b. $p \vee \sim p$
   c. $p \vee p$
   d. $p \rightarrow p$

4. If x is a positive odd integer, then which of the following is true?
   a. $\forall_x\ x^2 = -x$
   b. $\exists_x\ x^2 = -x$
   c. $\forall_x\ x^2 = x$
   d. $\exists_x\ x^2 = x$

5. Given the following statements:

   If Jeff has a B average, then he will try out for the team.
   If Jeff tries out for the team, then he'll make the team.
   Jeff has a B average or his parents will be disappointed.
   His parents are not disappointed.

   Let b represent "Jeff has a B average."
   Let t represent "He will try out for the team."
   Let m represent "He will make the team."
   Let d represent "His parents will be disappointed."

   a. Using b, t, m, d, and the proper connectives, express each sentence in symbolic form.
   b. Prove that Jeff will make the team.

6. Given the following statements:

   If it snows today, the skiing will be great.
   If it rains today, the skiing will be terrible.
   Either it snows or we will stay home.
   We will not stay home.

   Let s represent "It snows today."
   Let g represent "The skiing will be terrible."
   Let t represent "It rains today."
   Let h represent "We will stay home."

   a. Using s, g, t, h, and the proper connectives, express each sentence in symbolic form.
   b. Using the Laws of Inference, show that it will not rain today.

7. What conclusion can you draw from each of the following arguments?
   a. $(p \rightarrow \sim q) \wedge q$ b. $(p \rightarrow q) \wedge p$ c. $(p \rightarrow q) \wedge (q \rightarrow r)$
   d. $(p \rightarrow q) \wedge (r \rightarrow \sim q)$ e. $(p \vee q) \wedge (\sim q)$
   f. P_1: If Sally goes to the beach, she gets a sunburn.
      P_2: Sally does not have a sunburn.
   g. P_1: All squares are rectangles.
      P_2: All rectangles are parallelograms.

Euclidean Geometry–Checklist

Points
Lines, rays, segments
Midpoints
Angles and angle measure
- acute, right, obtuse, straight
- adjacent
- complementary
- supplementary
- vertical

Congruence of triangles
- methods for proving
- corresponding parts

Special triangles
- isosceles
- equilateral
- right

Medians, altitudes, and angle bisectors
Triangle inequalities
- sides
- angles

Parallel lines
- angles
- Parallel Postulate
- triangle midpoint theorems

Perpendicular lines
Angles of a triangle
- sum
- interior
- exterior

Quadrilaterals
- special types
- angle-sum
- analysis by diagonals

Constructions
- copy segments, angles, and triangles
- bisect angles and segments
- inscribe and circumscribe circles
- perpendicular and parallel lines
- proportional segments

Similarity and similar polygons
- ratios and proportions
- mean proportionals (geometric means)
- ratios of proportionality
  - lengths and perimeters
  - areas

Triangle Proportionality Theorem
Proving triangles similar
- methods
- corresponding sides
- corresponding angles

Right triangles
- altitude to hypotenuse
- Pythagorean Theorem
  - Pythagorean triples
  - special right triangles

Locus of points
- equidistant from:
  - a point
  - two points
  - a line
  - two parallel lines
  - the sides of an angle
  - three points (collinear or noncollinear)
- intersection of:
  - a point and a line
  - a point and two parallel lines

Euclidean Geometry—Sample Problems

1. Given: $\angle CFD$ is a right angle

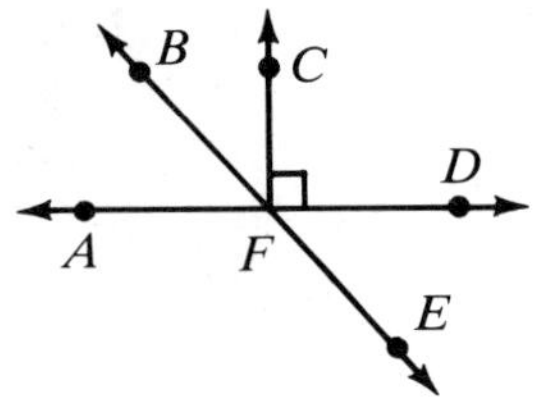

   a. If $m\angle AFB = 2x + 18$ and $m\angle BFC = 4x - 18$, find the value of x.
   b. If $m\angle AFB = 5z + 2$ and $m\angle AFE = 7z + 10$, find the value of z.
   c. If $m\angle DFE = 3y + 3$ and $m\angle BFC = 3y - 3$, find the value of y.

2. Given: $\triangle ABC$

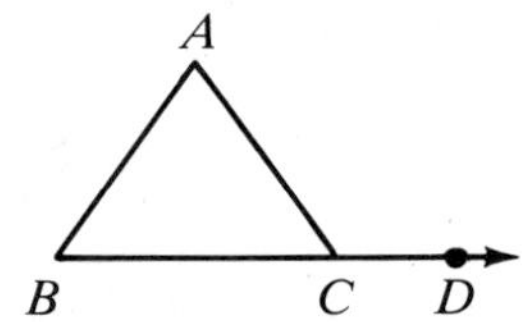

   a. If the ratio of the measures of the angles A, B, and C, respectively, is 1:4:7, find the measure of the largest angle of $\triangle ABC$ and name the longest side of $\triangle ABC$.
   b. If $AB = 5$ cm and $BC = 7$ cm, then $\overline{AC}$ measures less than __?__ cm.
   c. If $m\angle ACD = 2(m\angle BAC)$ and $m\angle ACB = 48$, find $m\angle ABC$.
   d. If $\triangle ABC$ is isosceles and vertex angle A has measure 42, find $m\angle ACD$. Is AB less than, greater than, or equal to BC?
   e. If $\angle A \cong \angle B$ and if $m\angle ACD = 110$, find $m\angle A$.

3. Given: $\overline{AB} \cong \overline{AC}$; $\overline{AF}$ is an altitude; D and E are the midpoints of $\overline{AB}$ and $\overline{AC}$, respectively.

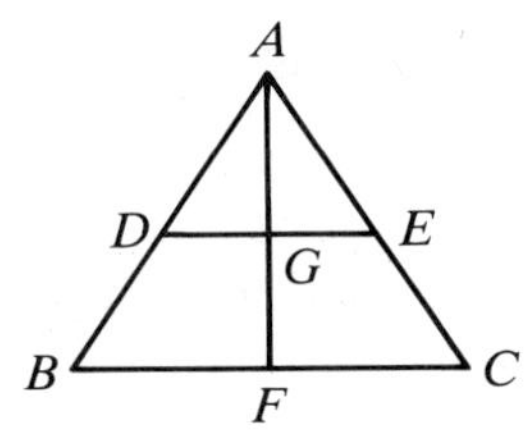

   a. Name an angle that is complementary to $\angle ABF$.
   b. If $AE = 7$, find BD.
   c. If $BC = 12$, find DE.
   d. If $BC = 12$ and $GF = 4$, find AE.
   e. If $m\angle BAF = 20$, find $m\angle AEG$.
   f. If $m\angle GEC = 100$, find $m\angle FAC$.
   g. If $DE = 6$ and $AF = 10$, find the area of $\triangle ABC$.

4. Given: $\overleftrightarrow{BC} \parallel \overleftrightarrow{DE} \parallel \overleftrightarrow{FG}$; $\overline{AB} \cong \overline{BD} \cong \overline{DF}$

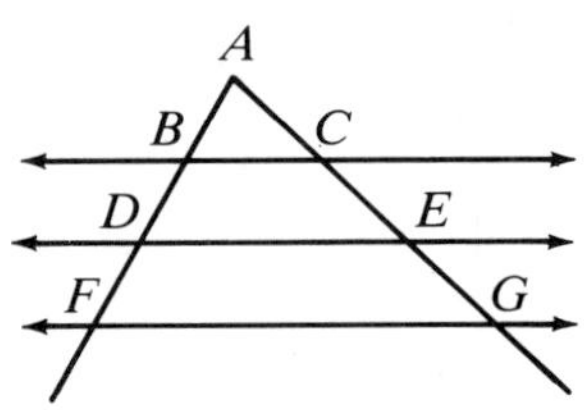

   a. If $\overline{AB} \cong \overline{AC}$ and if $AB = 6$, find AG.
   b. If $m\angle ABC = 3x + 3$ and $m\angle AFG = 4x - 4$, find $m\angle FDE$.
   c. If $m\angle FGE = 4x + 4$ and $m\angle BCE = 7x + 11$, find $m\angle AED$.
   d. If $\triangle ADE$ is equilateral and the perimeter of $\triangle ADE$ is 24, find BC.
   e. If $BC:FG = 1:3$, and if $FG = 12$, find DE.

5. Given: $\overleftrightarrow{AB} \parallel \overleftrightarrow{DE}$; $\overleftrightarrow{AE} \perp \overleftrightarrow{BD}$

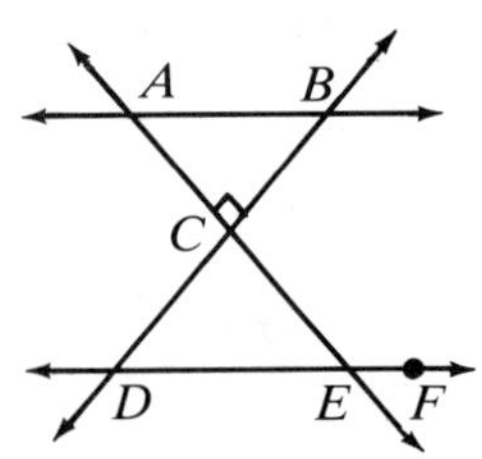

   a. If $m\angle CED = 40$, find $m\angle ABC$.
   b. If $m\angle CDE = x$ and $m\angle BAC = 2x$, find $m\angle CEF$.
   c. If $m\angle CED = 40$, name the shortest side of $\triangle ABC$.
   d. If $\overline{AC} \cong \overline{BC}$, find $m\angle CEF$.
   e. How many lines are there that contain point C and are parallel to $\overline{AB}$?

6. Given: $\square ABCD$

a. If $\overline{AC} \cong \overline{BD}$, find $m\angle ADC$.

b. If $\overline{AC}$ is perpendicular to $\overline{BD}$ and $AB = 10$, find BC.

c. If $\overline{AC}$ is perpendicular to $\overline{BD}$, $AD = 4x - 2$, and $CD = 3x + 3$, find the value of x.

d. If $m\angle A : m\angle B = 1:5$, find $m\angle C$.

e. If $ABCD$ is a rhombus, $AC = 8$, and $BD = 6$, find the perimeter of $ABCD$.

f. If $\overline{AC}$ is perpendicular to $\overline{BD}$, $AC = 6$, and $BD = 12$, find the area of $ABCD$.

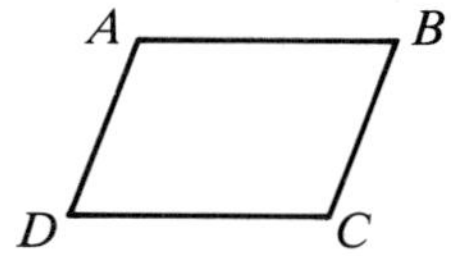

7. Given: Trapezoid $ABCD$; E, F, G, and H are the midpoints of $\overline{AB}$, $\overline{BC}$, $\overline{DC}$, and $\overline{AD}$, respectively.

a. If $\overline{AC} \cong \overline{BD}$ and $m\angle ADC = 60$, find $m\angle ABC$.

b. If $\overline{AD} \cong \overline{BC}$, what special type of quadrilateral is $EFGH$?

c. If $\overline{AD} \cong \overline{BC}$ and $BD = 20$, find the perimeter of $EFGH$.

d. If $\overline{AD} \cong \overline{BC}$ and $m\angle EFG = 80$, find $m\angle HGD$.

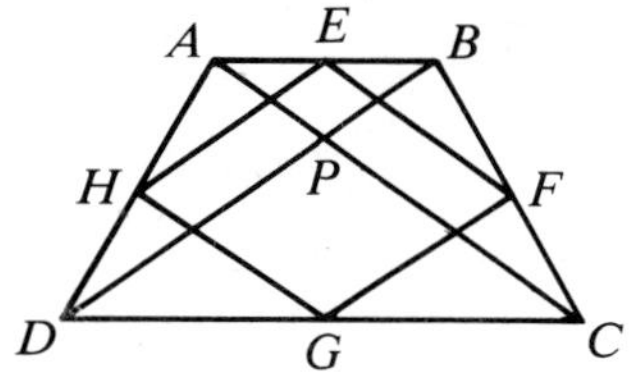

8. Given: D, E, and F are the midpoints of $\overline{AB}$, $\overline{AC}$, and $\overline{BC}$, respectively.

Prove: a. $\angle DAE \cong \angle FEC$

b. $\angle DEF \cong \angle DBF$

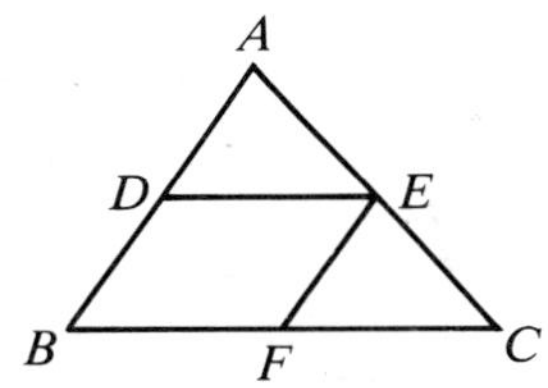

9. Given: $\overline{BC} \cong \overline{CD}$, $\overline{AC} \perp \overline{BD}$

Prove: $\triangle ABD$ is isosceles.

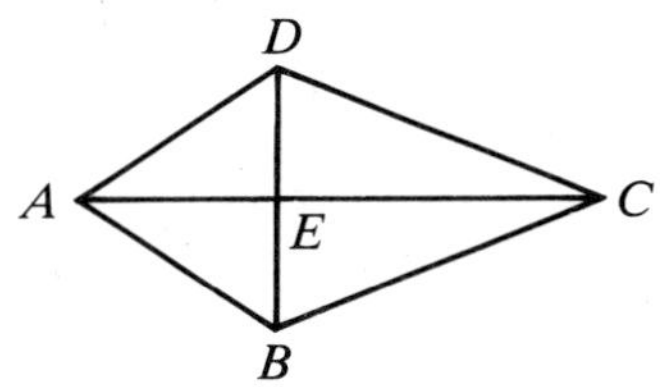

10. Given: $\overline{DE} \parallel \overline{BC}$

a. If $AD = 8$, $DB = 4$, and $DE = 6$, find BC.

b. If $AB = 8$, $AD = 6$, and $AC = 20$, find EC.

c. If $AD = 6$, $DB = 4$, and $AC = 15$, find EC.

d. If $AD = 4$, perimeter $\triangle ADE = 22$, and perimeter $\triangle ABC = 33$, find BD.

e. If $AD = 5$, $AB = 10$, and area $\triangle ADE = 6$, find the area of $\triangle ABC$.

f. If the ratio of the areas of $\triangle ADE$ and $\triangle ABC$ is 4:9, find the ratio of their perimeters.

g. If $AD = 2$, $AE = 5$, $DE = 6$, and $BC = 18$, find the perimeter of $\triangle ABC$.

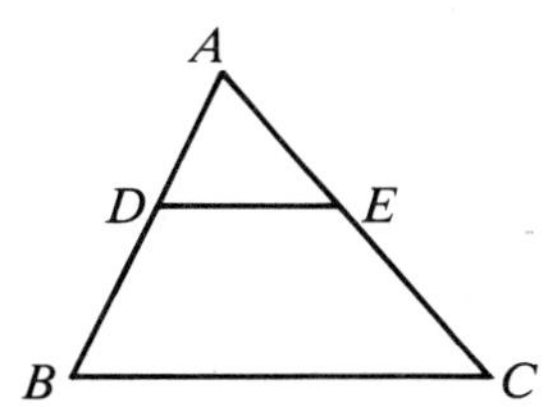

11. Given: Right triangle ABC with altitude $\overline{AD}$ drawn to hypotenuse $\overline{BC}$

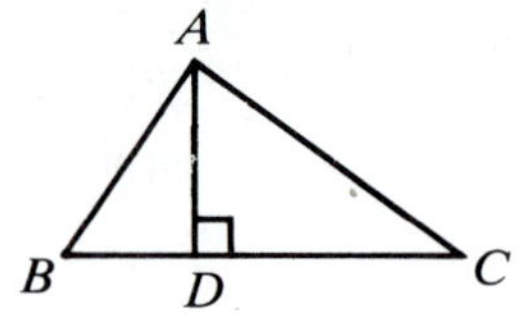

a. If $AD = 6$ and $BD = 9$, find DC.
b. If $BD = 3$ and $DC = 12$, find AD.
c. If $BD = 2$ and $DC = 9$, find AD.
d. If $\triangle ABC$ is isosceles with $AB = 6$, find AD.
e. If $AD = 3$, $BD = x + 3$, and $DC = x - 5$, find BC.
f. $AB:BC = BD:$ _?_
g. $AD \cdot BC = AB \cdot$ _?_
h. If $AD = 5$ and $DC = 12$, find AC.
i. If $AB = 15$ and $BD = 9$, find AD.
j. If $BC = 8$ and $AC = 6$, find AB.
k. If $AB = x$, $AC = 2x - 1$, and $BC = 2x + 1$, find the area of $\triangle ABC$.
l. If $BC = 18$ and $AB = 6$, find BD.

12. Given: $\square ABCD$

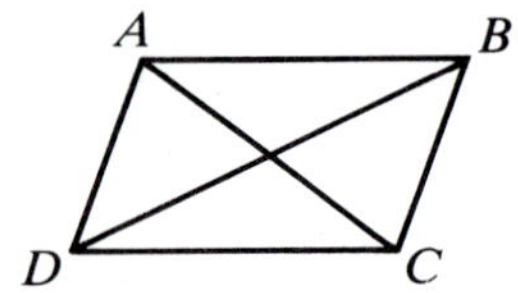

a. How many points of $ABCD$ are equidistant from:
   (1) C and D? (2) $\overline{AB}$ and $\overline{CD}$? (3) the sides of $\angle ACD$?
b. If points A and C are each equidistant from B and D, what special type of quadrilateral is $ABCD$?
c. If $AD = 5$ and $AB = 12$, how many points are both:
   (1) 5 units from A and 8 units from B?
   (2) 3 units from B and 2 units from D?
   (3) 7 units from C and 3 units from D?
   (4) 3 units from A and 2 units from $\overline{AB}$?
   (5) 4 units from C and equidistant from $\overline{AC}$ and $\overline{BD}$?
   (6) 4 units from C and equidistant from the sides of $\angle ACD$?

Algebraic Skills–Checklist

Operations with polynomials
Laws of exponents
Factoring polynomials
  greatest common monomial factors
  product of binomials
  complete factoring
  difference of squares
  sum and difference of cubes
Simplifying rational expressions
Operations with rational expressions
Solving quadratic equations
  by factoring
  by taking square root
  by completing the square
  by the quadratic formula
Analysis of discriminant
Solving fractional equations

Algebraic Skills–Sample Problems

1. Write a quadratic equation in the form $Ax^2 + Bx + C = 0$ with the given roots.
   a. $x = 2$ and $x = -3$
   b. $x = 0$ and $x = -4$
   c. $x = 5$ and $x = -5$
   d. $x = \frac{1}{2}$ and $x = \frac{3}{4}$
   e. $x = -3 \pm \sqrt{6}$
   f. $x = \frac{-1 \pm \sqrt{5}}{2}$

2. One root of each given quadratic equation is 4. Find the other root.
   a. $x^2 - x - 12 = 0$ b. $x^2 = 4x$ c. $4x^2 - 17x + 4 = 0$

3. If -4 is a root of $x^2 - x + a = 0$, find the value of a and the other root.

4. Solve.
   a. $x^2 - 5x - 3 = 0$
   b. $x^2 - 4x - 3 = 0$
   c. $3x^2 - 7x + 1 = 0$
   d. $3x^2 + 9x - 2 = 0$
   e. $x^3 - 4x^2 + 16x = 0$
   f. $\frac{7}{x} + \frac{3}{x^2} = \frac{5}{2}$
   g. $\frac{1}{x+2} + \frac{x}{x+1} = \frac{1}{2}$
   h. $x^2 - 7x = 81 - 7x$
   i. $\frac{2x}{x+3} = \frac{x-3}{x+6}$
   j. $49x^2 - 18 = 0$

5. Use the discriminant to determine the nature of the roots of each quadratic equation. Do not solve.
   a. $x^2 - 2x - 8 = 0$ b. $x^2 - 6x + 9 = 0$ c. $x^2 + 2x + 5 = 0$
   d. $2x^2 - 3x - 7 = 0$ e. $6x^2 - 7x + 2 = 0$

6. The perimeter of a rectangle is 28 cm. If a diagonal of the rectangle is 10 cm long, find the dimensions of the rectangle.

7. The sum of the squares of three consecutive positive odd integers is 83. Find the three integers.

8. The lengths of the base and the altitude of a triangle are in the ratio 5:8. The area of the triangle is 180 cm^2. Find the length of the base.

Coordinate/Analytic Geometry–Checklist

- Plotting points
- Distance
  - midpoint
  - perimeter and area
  - Pythagorean Theorem
  - distances in a circle
  - analytic proofs
- Slope
  - parallel and perpendicular lines
  - analytic proofs
- Linear equations
  - graphing
  - slope-intercept analysis
  - point-slope analysis
- Loci in coordinate plane
- Systems of equations
  - graphic solutions
  - algebraic solutions
    - substitution
    - addition and subtraction
- Conics
  - parabolas
    - vertex
    - axis of symmetry
    - directrix and focus
  - circles
  - intersection of conics with lines
    - graphic solutions
    - algebraic solutions

Coordinate/Analytic Geometry—Sample Problems

1. Given: M is the midpoint of $\overline{AB}$.
   a. If A has coordinates $(-3, 2)$ and B has coordinates $(5, -4)$, find the coordinates of M.
   b. If A has coordinates $(2, -4)$ and M has coordinates $(-1, -8)$, find the coordinates of B.
   c. If A has coordinates $(-3, -1)$ and M has coordinates $(0, 3)$, find AB.
   d. If A has coordinates $(4, 2)$ and B has coordinates $(1, 1)$, find the slope of $\overleftrightarrow{AB}$.
   e. If A has coordinates $(1, 3)$ and B has coordinates $(2, 7)$, find an equation for $\overleftrightarrow{AB}$.
   f. If M has coordinates $(2, 4)$ and the slope of $\overleftrightarrow{AB}$ is $\frac{1}{2}$, find an equation for $\overleftrightarrow{AB}$.
   g. If M has coordinates $(1, 4)$ and the slope of $\overleftrightarrow{AB}$ is 2, find an equation of the line perpendicular to $\overleftrightarrow{AB}$ at M.
   h. If A has coordinates $(0, -2)$ and B has coordinates $(1, 5)$, find an equation of the line parallel to $\overleftrightarrow{AB}$ that passes through the origin.
   i. If the slope of $\overleftrightarrow{AB}$ is 3, the coordinates of A are $(2, 4)$, and the coordinates of B are $(6, k)$, find the coordinates of M.
2. Given: Quadrilateral $ABCD$ with vertices $A(1, 1)$, $B(6, 1)$, $C(10, -2)$ and $D(5, -2)$
   a. Find the slope of $\overline{AB}$.
   b. Find the slope of $\overline{CD}$.
   c. Find the slope of $\overline{AD}$.
   d. Find the slope of $\overline{BC}$.
   e. Based on your answers to parts (a)-(d), what special type of quadrilateral is $ABCD$?
   f. Find the slope of $\overline{AC}$.
   g. Find the slope of $\overline{BD}$.
   h. Based on your answers to parts (e)-(g), what special type of quadrilateral is $ABCD$?
   i. Find the perimeter of quadrilateral $ABCD$.
   j. Find the area of quadrilateral $ABCD$.
   k. Find the coordinates of the point of intersection of $\overline{AC}$ and $\overline{BD}$.
3. Given: $\triangle ABC$ with vertices $A(-1, 2)$, $B(4, 2)$, and $C(4, 0)$
   a. Find the slope of $\overline{AB}$.
   b. Find the slope of $\overline{BC}$.
   c. Find $m\angle B$.
   d. Find the perimeter of $\triangle ABC$.
   e. Find the area of $\triangle ABC$.
4. Given: $\odot O$ with center at the origin, containing point $P(4, -3)$
   a. Find the length of the radius of $\odot O$.
   b. Find an equation of $\odot O$.
   c. Find the slope of $\overline{OP}$.
   d. Find the slope of the tangent line to $\odot O$ at P.
   e. Find an equation of the tangent line to $\odot O$ at P.

f. Find equations of the tangent lines to $\odot O$ that are parallel to the x-axis.
g. Find equations of the tangent lines to $\odot O$ that are parallel to the y-axis.
h. Find the coordinates of all points of intersection of the lines described in parts (f) and (g).
i. Find the coordinates of the points of intersection of $\odot O$ and the line with equation $y = x + 1$.

5. Given: The parabola with equation $y = x^2 - 8x + 12$ and the line with equation $y = 2x - 4$
   a. Find the equation of the axis of symmetry of the parabola.
   b. Find the coordinates of the vertex (turning point) of the parabola.
   c. Find the coordinates of the points where the parabola crosses the x-axis.
   d. Find the coordinates of the points of intersection of the parabola and the line.

6. Given: Points $P(2, 4)$ and $Q(-1, 0)$
   a. Find an equation of the locus of points 3 units from P.
   b. Find an equation of the locus of points equidistant from P and Q.
   c. How many points are both 4 units from P and 2 units from Q?

7. Given lines l, m, and n with equations as follows: l: $x = 4$, m: $2x - y = 1$, n: $x + 2y = 8$
   a. Graph l, m, and n on the same coordinate axes.
   b. Determine the coordinates of point A, the intersection of lines l and m.
   c. Determine the coordinates of point B, the intersection of lines l and n.
   d. Determine the coordinates of point C, the intersection of lines m and n.
   e. Prove: $\triangle ABC$ is a right triangle.
   f. Find the area of $\triangle ABC$.
   g. Find the perimeter of $\triangle ABC$.
   h. Determine the coordinates of point D, the y-intercept of line m.
   i. Determine the coordinates of point E, the y-intercept of line n.
   j. Prove: $\triangle ABC \cong \triangle DEC$

Probability, Permutations, and Combinations–Checklist

Sample space
Probabilities
- equally likely outcomes
- certain and impossible events
- mutually exclusive events
- intersecting events
- complementary events
- multi-stage events
  - with replacement
  - without replacement

Odds
Permutations
- distinct items
- items not all distinct
- probabilities

Combinations
- formula
- symmetry formula
- probabilities

Probability, Permutations, and Combinations—Sample Problems

1. Two dice are tossed. If the sum of the numbers showing is noted, find:
   a. $P(7)$ b. $P(\text{prime})$ c. $P(\text{at least } 9)$
   d. $P(\text{perfect square})$ e. $P(1)$
   If each number is noted and 2-digit numbers are formed, find:
   f. $P(\text{even number})$ g. $P(\text{prime number})$ h. $P(\text{perfect square})$

2. An urn contains 4 white marbles, 3 red marbles, and 1 blue marble. Three marbles are chosen at random without replacement.
   a. In how many ways can the marbles be chosen?
   b. In how many ways can 3 marbles of the same color be chosen?
   c. In how many ways can at least 2 red marbles be chosen?

3. One marble is drawn at random from the urn in Exercise 2. It is replaced and a second marble is drawn at random.
   a. Find the probability of choosing a white marble and a red marble (in that order).
   b. Find the probability of choosing two marbles of the same color.

4. a. How many distinct 6-letter permutations can be created using the letters of the word BANANA?
   b. How many combinations of 4 letters can be chosen from the 6 letters in the word ORANGE?

5. Six boys and 4 girls are members of the student government. Four of them are to be selected as officers (president, vice-president, secretary, and treasurer).
   a. In how many ways can the 10 students be arranged in a line?
   b. In how many ways can the 10 students be called during a vote if all 4 girls are called before the 6 boys?
   c. How many 4-member slates of officers can be formed from the 10 students?
   d. How many of these slates consist of exactly 2 boys and 2 girls?
   e. What is the probability that one of the slates of officers consists of exactly 2 boys and 2 girls?
   f. If Dave is a member of the student government, what is the probability that Dave will be:
      (1) an officer? (2) the president? (3) *not* chosen as an officer?

Algebraic Systems—Checklist

Binary operations
Properties
  closure
  associative
  commutative
  identity elements
  inverse elements

Clock or modular systems
  modular arithmetic
  solving equations
Groups
  commutative (abelian)
  noncommutative

1. Let E represent the set of even integers, O represent the set of odd integers, and M represent the set of integral multiples of 5.
   a. Which sets are closed under addition?
   b. Which sets are closed under subtraction?
   c. Which sets are closed under multiplication?
   d. Which sets are closed under division?
   e. Name the identity in each system if one exists:
   (1) $(E, +)$ (2) $(O, \cdot)$ (3) $(M, -)$
   f. In the system $(E, +)$, name the inverse of the number 2.
   g. Verify that $(E, +)$ is a group.
   h. Explain why $(M, -)$ is *not* a group.
2. Given: $Z_5 = \{0, 1, 2, 3, 4\}$
   a. Construct an addition table for the system $(Z_5, +)$.
   b. Using your table for $(Z_5, +)$, complete:
   (1) $3 + 4 + 2 =$ __?__. (2) $2 + 4 = 3 +$ __?__.
   (3) The identity element is __?__. (4) The inverse of 4 is __?__.
   (5) If $x + 4 = 1 + 2$, then $x =$ __?__. (6) Is $(Z_5, +)$ commutative?
   c. Construct a multiplication table for the system $(Z_5, \cdot)$.
   d. Using your table for $(Z_5, \cdot)$, complete:
   (1) $3 \cdot 4 \cdot 2 =$ __?__. (2) $(3)^2 =$ __?__.
   (3) $2 \cdot 4 = 3 \cdot$ __?__. (4) The identity element is __?__.
   (5) The inverse of 3 is __?__. (6) If $3x = 2$, then $x =$ __?__.
   e. Using your tables for both $(Z_5, +)$ and $(Z_5, \cdot)$, solve for x.
   (1) $3 \cdot x + 2 = 4$ (2) $x^2 = 4$
   (3) $3x^2 + 2 = 0$ (4) $(x + 1)(x + 4) = 0$
   (5) $2(x + 4) = 3$
3. Given the conditional $p \rightarrow q$, let S be the set of related conditions $\{O, C, I, P\}$.

   O: original conditional statement $(p \rightarrow q)$
   C: converse statement $(q \rightarrow p)$
   I: inverse statement $(\sim p \rightarrow \sim q)$
   P: contrapositive statement $(\sim q \rightarrow \sim p)$

   Define $*$ by $x * y =$ the x of the y. For example, $P * C =$ the contrapositive of the converse $= \sim p \rightarrow \sim q = I$.
   a. Construct the operation table for the system $(S, *)$.
   b. Is $(S, *)$ closed?
   c. Is $(S, *)$ associative?
   d. Is $(S, *)$ commutative?
   e. Does $(S, *)$ have an identity element? If so, what is it?
   f. Does every element of $(S, *)$ have an inverse?
   g. Is $(S, *)$ a group?
   h. Is $(S, *)$ an abelian group?

4. Given the set $T = \{1, 3, 5, 7\}$, define $*$ by $a * b = a + b + 1 \bmod 8$.
   a. Construct the operation table for the system $(T, *)$.
   b. Is $(T, *)$ closed?
   c. Is $(T, *)$ associative?
   d. Is $(T, *)$ commutative?
   e. Does $(T, *)$ have an identity element? If so, what is it?
   f. Does every element of $(T, *)$ have an inverse?
   g. Is $(T, *)$ a group?
   h. Is $(T, *)$ an abelian group?

APPENDIX 1

INVERSE VARIATION

One electric company bills its customers at a rate of \$.09 per kilowatt hour. This price structure is an example of *direct variation,* because as the amount (x) of electricity consumed increases (decreases), so does the total cost (y) of the electricity. This relationship can be written

$$y = 0.09x\,.$$

Any two variables that vary directly are related by an equation of the form

$$y = kx$$

for some constant k, called the *constant of variation*.

In another type of variation, as x increases, y decreases. For example, consider the length and width of a rectangular pen that will have an area of 48 square meters (m^2). A few of the choices for the width (x) and the length (y) are listed in the table at the right. Notice that as x increases, y decreases, and that the variables are related by the equation

$$xy = 48\,.$$

x Width (in m)	y Length (in m)
1	48
2	24
3	16
4	12
6	8
8	6
12	4
16	3
24	2
48	1

Two variables are said to *vary inversely* if they satisfy an equation of the form

$$xy = k$$

for some nonzero constant k. This equation is often written in the form

$$y = \frac{k}{x},$$

to emphasize the inverse relationship.

In the relation $xy = k$, the product of the x- and y-values of any point of the graph of the relation must be k. Thus, if (x_1, y_1) and (x_2, y_2) are two points of $xy = k$, then

$$x_1 y_1 = x_2 y_2\,,$$

because each side of the equation is equal to k.

The graph of an inverse relationship is shown at the right. The graph of the ordered pairs listed in the table above is shown by the solid curve. Corresponding negative values can also be graphed as shown by the broken curve. The complete graph is called a *hyperbola*. Notice that, even for negative values of x, as x increases, y decreases.

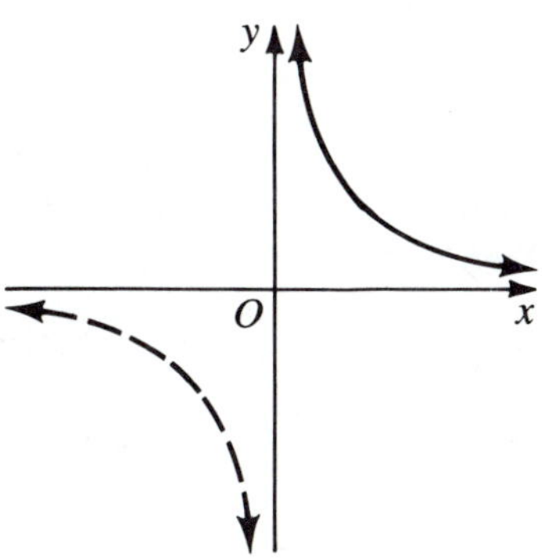

EXAMPLE At a constant temperature the volume of an ideal gas varies inversely with the pressure. At a pressure of 240 kilopascals (kPa), the volume of a certain amount of neon is 17.5 L. What will be the volume of this gas at 150 kPa? Express the relationship between pressure x and volume y in an equation of the form $xy = k$.

SOLUTION Let y be the volume in liters at a pressure of 150 kPa. Then

$$150y = (240)(17.5)$$
$$y = 28$$

To find the constant k in the equation $xy = k$, multiply either pressure and the corresponding volume.

$$k = (240)(17.5) = 4200 \quad \text{or} \quad k = (150)(28) = 4200$$

Therefore the equation is $xy = 4200$.

WRITTEN EXERCISES

1. If y varies inversely as x, and $y = 10$ when $x = 15$, find x when $y = 20$.
2. If y varies inversely as x, and $y = 360$ when $x = 8$, find y when $x = 12$.
3. The current in an electrical circuit varies inversely with the resistance in the circuit if the voltage is held constant.
   a. If a resistance of 150 Ω (ohms) produces a current of 0.8 A (amperes), how much current would a resistance of 250 Ω produce?
   b. In the same circuit, what resistance would produce a current of 1.2 A?
4. The frequency of a musical tone varies inversely with its wavelength. A has a frequency of 440 hertz (Hz) and a wavelength of 2.5 ft. What is the wavelength of high F#, which has a frequency of 550 Hz? State the relationship between the frequency y and the wavelength x in the form $xy = k$.
5. The slant height of a cone will vary inversely with the radius, if the surface area is kept constant. A given cone has a slant height of 18 and a radius of 25.
   a. What will be the slant height of a cone of radius 30 if the surface area is held constant?
   b. What is the radius of a cone with slant height 12.5 if the surface area is held constant?
6. For a camera lens of fixed focal length, the relative aperture of the lens (the "f-stop") varies inversely as its effective diameter. Suppose an f-stop of f/4 corresponds to an effective diameter of 12.5 mm.
   a. What is the effective diameter of this lens at f/16?
   b. What f-stop corresponds to an effective diameter of 20 mm?

APPENDIX 2

TRANSFORMATIONS

A. REFLECTIONS IN A LINE

A *reflection in a line* is a transformation that produces a mirror image of a figure on the opposite side of a given line. For example, the image of trapezoid $ABCD$ shown at the right is trapezoid $A'B'C'D'$.

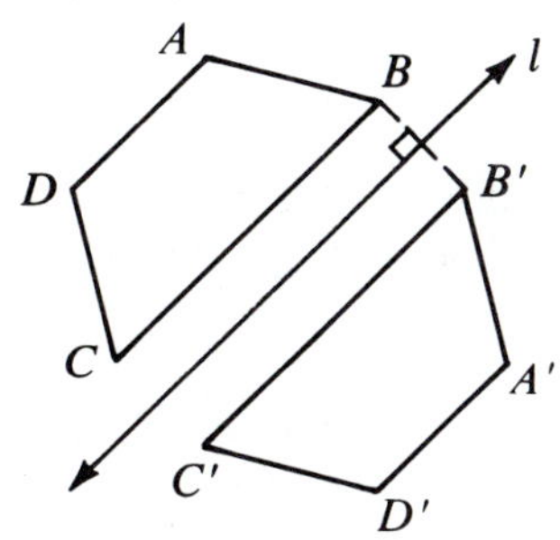

The image of a single point P is determined by these two rules:

1. If P is not on l, then the image P' is the point such that l is the perpendicular bisector of $\overline{PP'}$.
2. If P is on l, P is its own image.

Points in the coordinate plane can be reflected in either the x- or the y-axis. Consider the reflection of $P(3, 2)$ in the y-axis. The coordinates of its image P' are $(-3, 2)$, since the y-axis must be the perpendicular bisector of $\overline{PP'}$.

Reflection in the y-axis

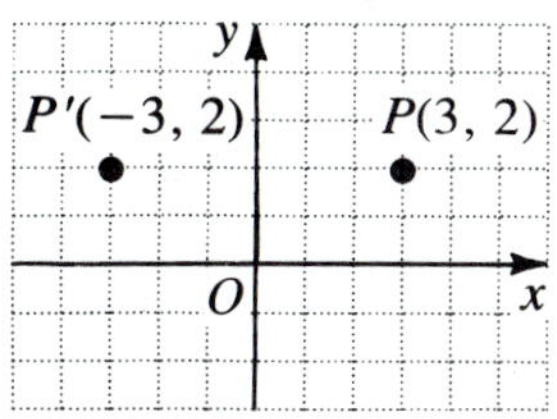

Reflection in the x-axis

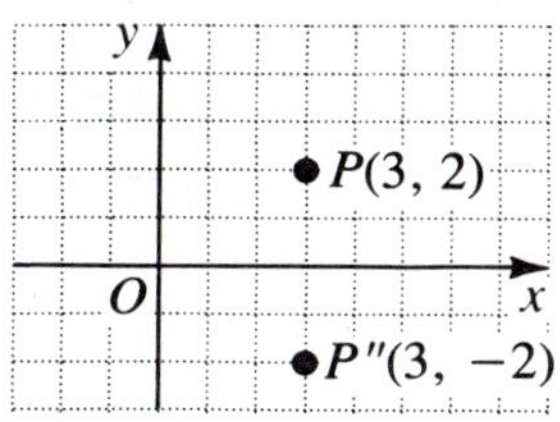

By a similar argument, if $P(3, 2)$ is reflected in the x-axis, its reflection is $P''(3, -2)$. If $P(a, b)$ is any point in a plane, then its reflection in the y-axis is $(-a, b)$, and its reflection in the x-axis is $(a, -b)$. Note that this statement is true even if either coordinate is a negative number or zero.

EXAMPLE Graph the reflection of the line $y = 2x + 3$ and write its equation when the line is reflected about

a. the x-axis;
b. the y-axis.

SOLUTION First construct a table of values for the equation $y = 2x + 3$.

x	y
−2	−1
−1	1
0	3
1	5
2	7

(Solution continued on next page)

a. A reflection about the x-axis replaces each y-value with its negative. The table of values for the reflected line and its graph are shown below.

x	y
-2	1
-1	-1
0	-3
1	-5
2	-7

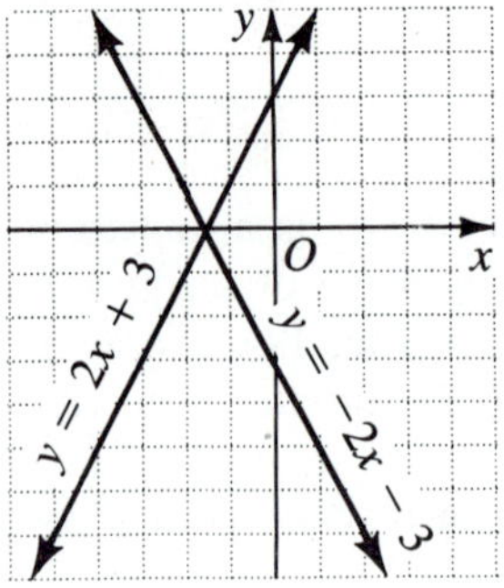

To find the equation of the image line, about the x-axis, replace each y-value with its negative in the original equation, to get

$$(-y) = 2x + 3\,,$$

or

$$y = -2x - 3\,.$$

b. A reflection about the y-axis replaces each x-value with its negative. The table of values for the reflected line and its graph are shown below.

x	y
2	-1
1	1
0	3
-1	5
-2	7

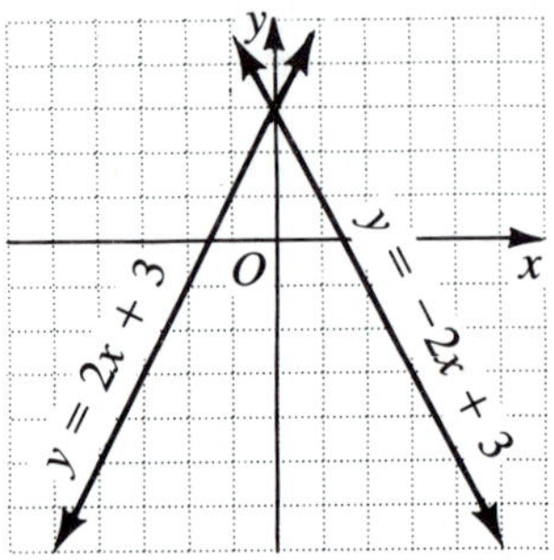

To find the equation of the image line about the y-axis, replace each x-value with its negative in the original equation, to get

$$y = 2(-x) + 3\,,$$

or

$$y = -2x + 3\,.$$

Lines other than the axes can be used as lines of reflection, as the following example shows.

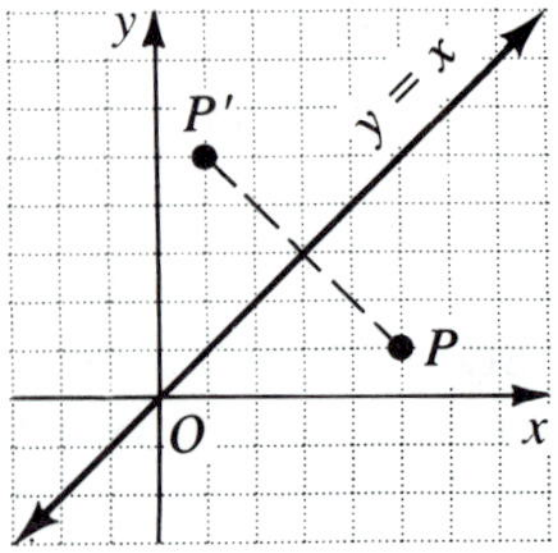

EXAMPLE Find the reflection of $P(5, 1)$ in the line $y = x$.

SOLUTION The image of P appears to be $P'(1, 5)$. To verify this, $\overline{PP'}$ must be perpendicular to the line $y = x$, and the midpoint of $\overline{PP'}$ must lie on the line.

Slope of $\overline{PP'} = \dfrac{1 - 5}{5 - 1} = -1$, so $\overline{PP'}$ is perpend...

... $= x$. The midpoint M of $\overline{PP'}$ has coordina...

so M does lie on the ... $= 3$ and $y = \dfrac{1 + 5}{2} = 3$,

In general the reflection of any point ... the line $y = x$ is the point $P'(b, a)$.

WRITTEN EXERCISES

1. Give the coordinates of each point under a reflection in th... ...xis.
   a. $P(-4, 6)$ **b.** $Q(0, -5)$ ... $R(7, 0)$
   d. $S(3, -4)$ **e.** $T(3, 3)$ ... $(-4, 4)$

2. For each point in Exercise 1, give the coordinates of the image under a ...ec-tion in the y-axis.

3. For each point in Exercise 1, give the coordinates of the image under a reflection in the line $y = x$.

4. For each point and its image listed in the table, write the equation for the line of reflection.

	Point	*Reflected point*
a.	$A(5, -1)$	$A'(5, 1)$
b.	$B(6, -3)$	$B'(-3, 6)$
c.	$C(0, 8)$	$C'(8, 0)$
d.	$D(-3, 4)$	$D'(3, 4)$
e.	$E(6, 1)$	$E'(-6, 1)$

5. Given $\triangle ABC$ with vertices $A(3, 1)$, $B(-2, 4)$, $C(5, 8)$. Sketch $\triangle A'B'C'$, the image of $\triangle ABC$ under a reflection in the x-axis. State the coordinates of the vertices of the image triangle.

6. Repeat Exercise 5, using the y-axis as the line of reflection.

7. Repeat Exercise 5, using the line $y = x$ as the line of reflection.

8. Graph the reflection of the line $2y = -x + 2$, using each of the following lines as a line of reflection:
   a. the x-axis **b.** the y-axis **c.** the line $y = x$

9. Using graph paper, plot and name the image of each point under each reflection:
   a. $P(-3, 5)$ in the line whose equation is $x = 2$
   b. $P(-3, 5)$ in the line whose equation is $y = 3$
   c. $P(-4, -1)$ in the line whose equation is $y = -1$
   d. $P(5, -2)$ in the line whose equation is $x = -1$

10. Graph the parabola whose equation is $y = x^2 - 2x + 4$, using values of x in the interval $-1 \le x \le 3$. Determine the line of reflection such that each point of the parabola will have an image point on the opposite side of the line and on the parabola. This line is called the *axis of symmetry*.

REFLECTIONS IN A POINT

In the figure at the left below, $\triangle ABC$ has be… 80° about point O. Because O is halfway between each point and …, the image is said to be a *reflection in point O*, and O is called the …ne reflection.

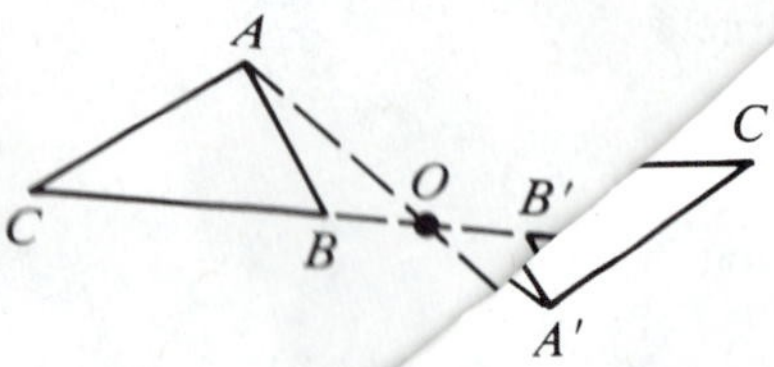

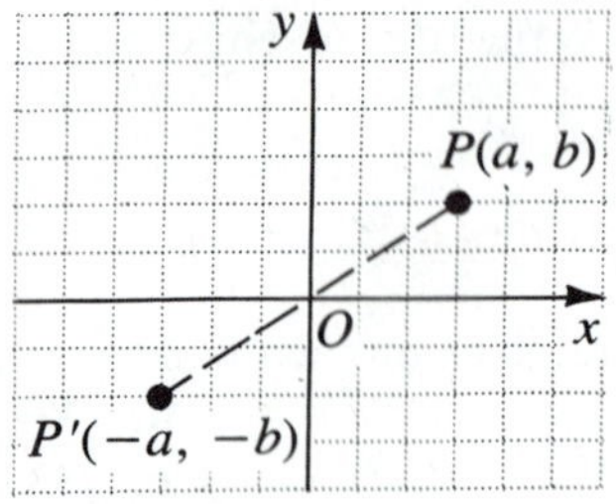

If $P(a, b)$ … any point of the plane, then its reflection in the origin is $P'(-a, -b)$ since (0, 0) is the midpoint of $\overline{PP'}$. If P is reflected in a point other than the origin, then the midpoint formulas are helpful in finding the image P'. (See the figure at the right above.)

EXAMPLE Given $P(4, 1)$, find the coordinates of its image P' under a reflection in $O(3, 5)$.

SOLUTION You know that $O(3, 5)$ must be the point halfway between $P(4, 1)$ and $P'(x, y)$. Using the midpoint formulas:

$$\frac{4 + x}{2} = 3 \quad \text{and} \quad \frac{1 + y}{2} = 5$$

$$4 + x = 6 \qquad 1 + y = 10$$

$$x = 2 \qquad y = 9$$

Therefore the coordinates of P' are (2, 9).

WRITTEN EXERCISES

1. P' is the image of P under a reflection in O. Copy and complete the table.

	Point P	Center O	Image P'
a.	(4, 1)	(0, 0)	?
b.	(5, 8)	(1, 2)	?
c.	(−3, 5)	(−1, 3)	?
d.	?	(0, 0)	(−3, 2)
e.	?	(−2, 2)	(−4, −6)
f.	(−6, 5)	?	(−2, 3)
g.	(0, 6)	?	(4, −6)
h.	(0, 0)	?	(10, −8)

2. The coordinates of one point P and those of its image P' under a reflection in O are given below. Find the image B' of $B(-3,5)$ under a reflection in O.
   a. $P(4,6)$ and $P'(-4,-6)$
   b. $P(4,-2)$ and $P'(6,-8)$
   c. $P(10,6)$ and $P'(2,8)$
3. Given $A(5,1)$, $B(4,7)$ and $C(-1,5)$. If $\triangle ABC$ is reflected in the origin, state the coordinates of the vertices of $\triangle A'B'C'$, the image of $\triangle ABC$.
4. Square $ABCD$ has vertices $A(2,0)$, $B(5,2)$, $C(3,5)$ and $D(0,3)$, and is reflected in $O(8,5)$. Find the coordinates of the vertices of the image, $A'B'C'D'$.

C. TRANSLATIONS

A *translation* is a transformation that slides a figure a given distance and direction in the plane. For example, the triangle below was translated 5 units in the direction of the arrow. Another way of saying this is that the triangle was moved 4 units in the positive x direction and 3 units in the positive y direction.

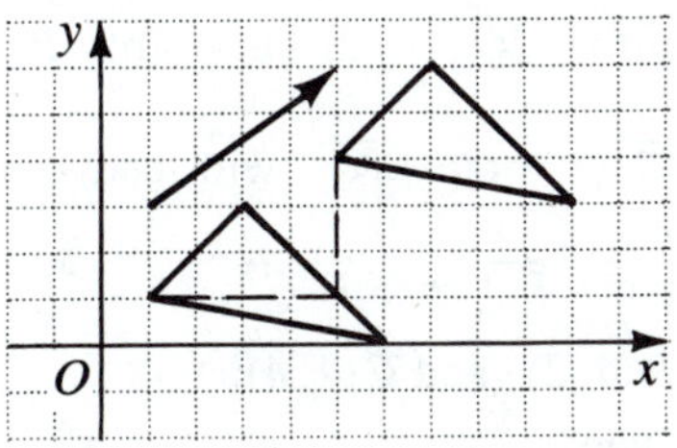

If a translation moves one point of the plane a units horizontally and b units vertically, then every other point of the plane is also moved the same distance horizontally and vertically. The point $P(x,y)$ is *mapped onto* $P'(x+a, y+b)$. This is written $P(x,y) \rightarrow P'(x+a, y+b)$.

EXAMPLE If $P(3,2) \rightarrow P'(1,6)$ under a translation, find the image B' of $B(-2,-1)$ under the same translation.

SOLUTION The translation moves P 2 units left and 4 units up. Therefore the coordinates of the image of $B(-2,-1)$ are $(-2-2, -1+4)$ or $B'(-4,3)$.

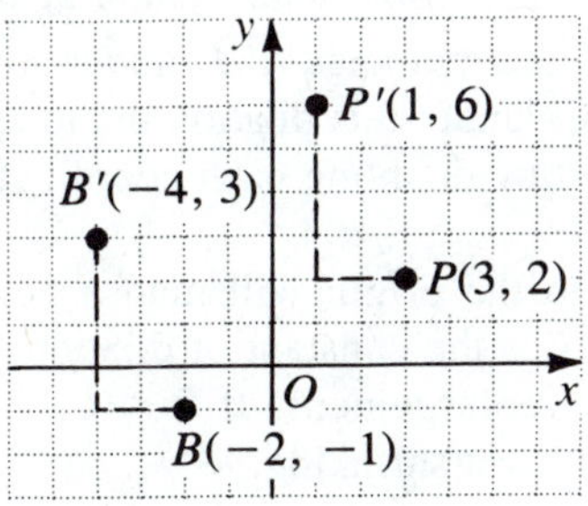

ITTEN EXERCISES

1. The coordinates of the vertices of $\triangle ABC$ are $A(2,4)$, $B(6,-1)$, and $C(5,6)$. Write the coordinates of the vertices of the image triangle $A'B'C'$ under a translation:
   a. 5 units left and 2 units down
   b. 3 units right
   c. 4 units up
   d. 2 units right and 1 unit down
   e. 4 units left and 3 units up
2. If a translation maps $P(-3,4)$ onto $P'(2,1)$, find the coordinates of Q', the image of $Q(0,6)$ under the same translation.
3. If a translation maps $P(0,6)$ onto $P'(6,0)$, and $Q'(4,2)$ is the image of Q, find the coordinates of Q.
4. A translation maps $M(3,-4)$ onto $M'(0,2)$. Under the same translation, $\triangle ABC$ maps onto $\triangle A'B'C'$ with coordinates $A'(3,-6)$, $B'(8,-1)$, and $C'(4,4)$. Find the coordinates of the vertices of the original triangle ABC.
5. A line segment has endpoints $P(3,4)$ and $Q(-1,2)$.
   a. Describe a translation of $\overline{PQ}$ onto $\overline{P'Q'}$ with endpoints $P'(1,2)$ and $Q'(-3,0)$.
   b. Describe a translation that maps $\overline{P'Q'}$ onto $\overline{P''Q''}$ with endpoints $P''(0,5)$ and $Q''(-4,3)$.
   c. Describe the transformation that maps $\overline{PQ}$ onto $\overline{P''Q''}$.
6. Given $\triangle ABC$ with vertices $A(1,2)$, $B(4,2)$, and $C(4,4)$:
   a. Draw $\triangle ABC$ on a piece of graph paper.
   b. Draw the reflection of $\triangle ABC$ in $O_1(3,1)$. Call the image $\triangle A'B'C'$.
   c. Draw the reflection of $\triangle A'B'C'$ in $O_2(2,2)$. Call this triangle $A''B''C''$.
   d. Describe a single transformation that would map $\triangle ABC$ onto $\triangle A''B''C''$.
7. A certain transformation maps $A(2,6)$ onto $A'(6,2)$. Find the image of $B(4,7)$ if
   a. the transformation is a translation.
   b. the transformation is a reflection in a point.
   c. the transformation is a reflection in a line.

D. DILATIONS

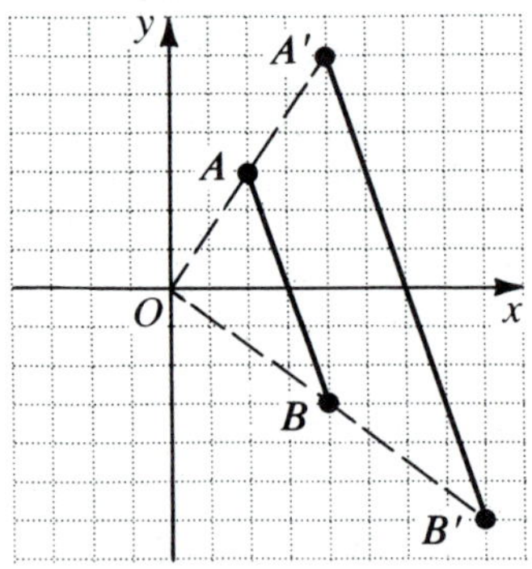

A *dilation* is a transformation which changes the size of a figure by stretching or shrinking. The coordinate system at the right shows a line segment $\overline{AB}$ and its image $\overline{A'B'}$ under a dilation with respect to the origin. Under this dilation the image P' of any point $P(x,y)$ is found by doubling each coordinate, that is, $P(x,y) \to P'(2x,2y)$.

Any dilation with respect to the origin will map a point $P(x,y)$ onto $P'(kx,ky)$, where k is the constant of dilation. If $k > 1$, then the dilation is an enlargement. If $0 < k < 1$, then the dilation is a reduction, or a shrinking.

EXAMPLE Given $\triangle ABC$ with vertices $A(-2, 2)$, $B(2, 4)$, and $C(-4, 3)$. Graph the image triangle $A'B'C'$ under a dilation with respect to the origin if the dilation maps A onto $A'(-6, 6)$. What is the constant of dilation?

SOLUTION Since $A(-2, 2) \rightarrow A'(-6, 6)$, the constant of dilation is 3. The images of B and C are $B'(6, 12)$ and $C'(-12, 9)$, respectively. This triangle is graphed below.

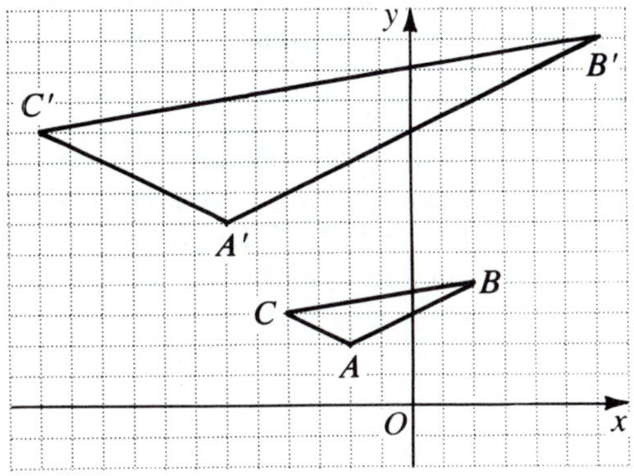

Notice that $\triangle A'B'C'$ is similar to $\triangle ABC$ and that the ratio of any two corresponding sides is equal to 3, the constant of dilation.

WRITTEN EXERCISES

1. $\triangle ABC$ has vertices $A(4, 2)$, $B(-2, 5)$, and $C(0, 6)$. Under a dilation with respect to the origin, the coordinates of the vertices of $\triangle A'B'C'$ are as follows. Find the constant of dilation.

	A'	B'	C'
a.	$(8, 4)$	$(-4, 10)$	$(0, 12)$
b.	$(2, 1)$	$(-1, 2.5)$	$(0, 3)$
c.	$(6, 3)$	$(3, 7.5)$	$(0, 9)$

2. A dilation with respect to the origin maps $P(1, 1)$ onto $P'(4, 4)$. Find the coordinates of the image of $\triangle ABC$ with vertices $A(-1, 2)$, $B(3, -1)$, and $C(1, 3)$.

3. A dilation with respect to the origin maps $A(2, 4)$ onto $A'(3, 6)$.
 a. What is the constant of dilation?
 b. Find the image B' of $B(2, -6)$ under this dilation.
 c. Find the lengths of $\overline{AB}$ and $\overline{A'B'}$. What is $\dfrac{A'B'}{AB}$?

4. A dilation with respect to the origin maps $P(8, 6)$ onto $P'(4, 3)$. The vertices of square $ABCD$ are $A(2, 0)$, $B(6, 6)$, $C(0, 10)$, and $D(-4, 4)$. Find the vertices of $A'B'C'D'$, the image square of $ABCD$, under the same dilation.

5. Use the distance formula to find the length of a side of each square in Exercise 4. What is the ratio of the length of a side of square $A'B'C'D'$ to the length of a side of square $ABCD$?

6. Given $\triangle ABC$ with vertices $A(0, 4)$, $B(2, 0)$, and $C(-2, 2)$. Its image under a dilation with respect to the origin is $\triangle A'B'C'$. A dilation was defined in the text for positive values of the constant of dilation. We shall now explore what would happen if we allow k to be negative.
   a. Find the coordinates of A', B', and C' if the constant of dilation, k, is -2. Graph $\triangle ABC$ and $\triangle A'B'C'$ on one coordinate system.
   b. Find the coordinates of A', B', and C' if the constant of dilation, k, is $-\frac{1}{2}$. Graph $\triangle ABC$ and $\triangle A'B'C'$ on a second coordinate system.
   c. What effect does a negative value for k seem to have on a figure?
   d. Find the coordinates of A', B', and C' if the constant of dilation is -1. Graph $\triangle ABC$ and $\triangle A'B'C'$ on a third coordinate system. What is another name for a dilation that has a constant of dilation equal to -1?

APPENDIX 3

THE TRIANGLE CONGRUENCE POSTULATES REVISITED

Although for the sake of simplicity we have stated as postulates the facts that two triangles are congruent by SAS, ASA, SSS, and HL, it is possible to prove ASA, SSS, and HL as theorems if we assume the SAS Postulate. In fact, *all four* statements can be proved as theorems if we assume the following postulate.

> **Postulate A** If two sides and the included angle of one triangle are congruent to two sides and the included angle of another triangle, then the two remaining angles of the first triangle are congruent to the corresponding angles of the second triangle.

The proofs of the SAS, ASA, SSS, and HL Theorems from this postulate are difficult, but outlines of them will be given here and in the exercises.

SAS Theorem

GIVEN: $\triangle ABC$ and $\triangle DEF$;
$\overline{AB} \cong \overline{DE}$;
$\overline{AC} \cong \overline{DF}$;
$\angle A \cong \angle D$

PROVE: $\triangle ABC \cong \triangle DEF$

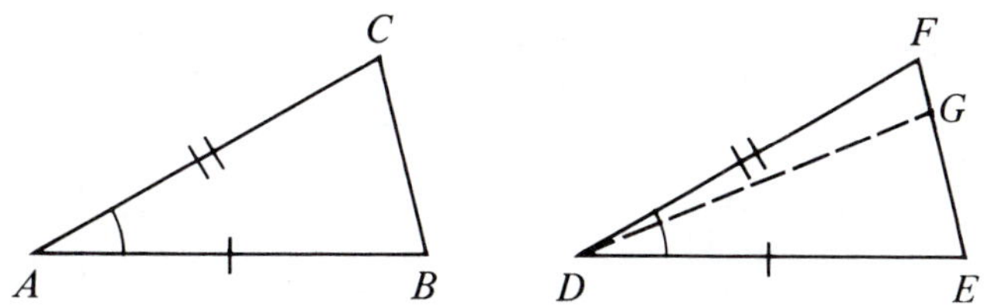

Proof: In order to prove $\triangle ABC \cong \triangle DEF$, we must show that $\angle B \cong \angle E$, $\angle C \cong \angle F$, and $\overline{BC} \cong \overline{EF}$. The first two of these congruences follow from Postulate A. We will give an indirect proof of the last congruence.

Temporarily assume that $BC \neq EF$. Then either $BC > EF$ or $EF > BC$. Since the proof is identical in either case, you may assume that $EF > BC$. If $EF > BC$, you can find a point G between E and F such that $BC = EG$. Since $\angle B \cong \angle E$, $\overline{AB} \cong \overline{DE}$, and $\overline{BC} \cong \overline{EG}$, you can apply Postulate A to $\triangle ABC$ and $\triangle DEG$ to conclude that $\angle A \cong \angle EDG$. Since $\angle A \cong \angle EDF$, you have $\angle EDF \cong \angle EDG$, or $m\angle EDF = m\angle EDG$. This fact contradicts Postulate 4 since $\overrightarrow{DF}$ and $\overrightarrow{DG}$ are different rays, with points F and G in the same half-plane determined by $\overleftrightarrow{DE}$. Therefore the temporary assumption that $BC \neq EF$ is false. Thus $BC = EF$, or $\overline{BC} \cong \overline{EF}$. Then, by the definition of congruent triangles, $\triangle ABC \cong \triangle DEF$.

ASA Theorem

GIVEN: $\triangle ABC$ and $\triangle DEF$;
$\angle A \cong \angle D$;
$\overline{AB} \cong \overline{DE}$;
$\angle B \cong \angle E$

PROVE: $\triangle ABC \cong \triangle DEF$

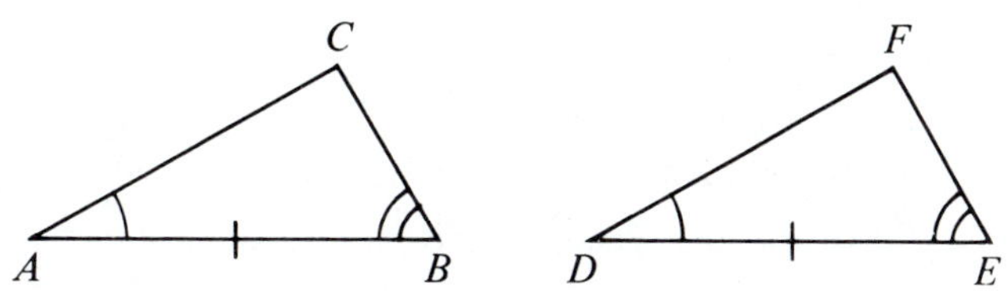

Analysis: If you can show that $\overline{AC} \cong \overline{DF}$, then you will have $\triangle ABC \cong \triangle DEF$ by the SAS Theorem. The proof of the ASA Theorem is left as Exercise 1.

In proving the SSS Theorem we will make use of Theorem 3-13, which states that the base angles of an isosceles triangle are congruent. Since Theorem 13-3 was proved using the SAS Postulate, it depends only on facts we have already developed in this section and thus can be used here.

SSS Theorem

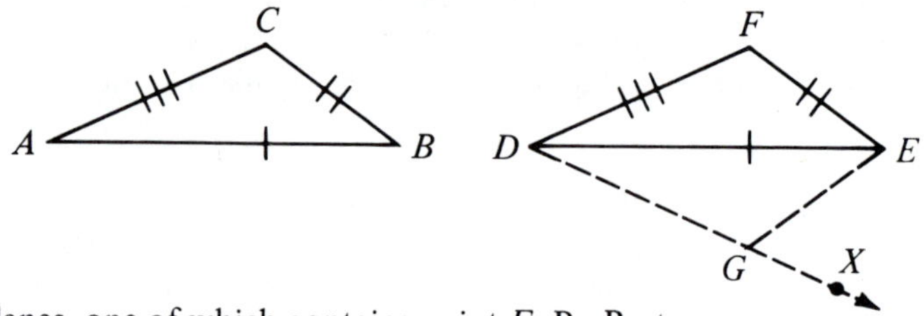

GIVEN: $\triangle ABC$ and $\triangle DEF$;
$\overline{AB} \cong \overline{DE}$;
$\overline{BC} \cong \overline{EF}$;
$\overline{AC} \cong \overline{DF}$

PROVE: $\triangle ABC \cong \triangle DEF$

Analysis: $\overleftrightarrow{DE}$ determines two half-planes, one of which contains point F. By Postulate 4, there is exactly one ray, $\overrightarrow{DX}$, such that X is in the half-plane that does not contain F and $\angle EDX \cong \angle A$. Let G be the point on $\overrightarrow{DX}$ such that $DG = AC$. Draw $\overline{GE}$. If you can prove that $\triangle ABC \cong \triangle DEG$ and that $\triangle DEG \cong \triangle DEF$, you will have $\triangle ABC \cong \triangle DEF$. The proof of the SSS Theorem is left as Exercise 2.

Using only the facts available, you can prove the HL Theorem.

HL Theorem

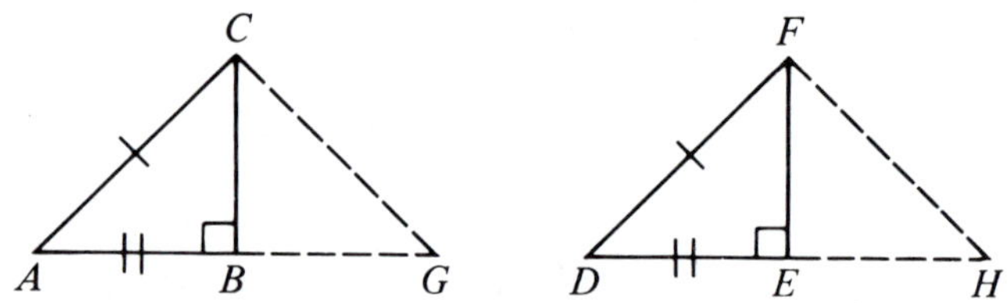

GIVEN: $\triangle ABC$ and $\triangle DEF$;
$\angle B$ and $\angle E$ are rt. $\angle$s;
$\overline{AC} \cong \overline{DF}$;
$\overline{AB} \cong \overline{DE}$

PROVE: $\triangle ABC \cong \triangle DEF$

Analysis: On the ray opposite to $\overrightarrow{BA}$ choose G so that $\overline{AB} \cong \overline{BG}$. On the ray opposite to $\overrightarrow{ED}$ choose H so that $\overline{DE} \cong \overline{EH}$. Draw $\overline{CG}$ and $\overline{FH}$. In Exercise 3 you will prove that $\triangle AGC \cong \triangle DHF$ and then that $\triangle ABC \cong \triangle DEF$.

WRITTEN EXERCISES

In Exercises 1-3, use the diagram in the text that accompanies the applicable theorem.

1. Prove the ASA Theorem by showing that $\overline{AC} \cong \overline{DF}$. Begin by assuming that $DF > AC$. Then choose G on $\overline{DF}$ such that $DG = AC$. Show that this leads to a contradiction. (*Hint:* Draw $\overline{GE}$.)
2. Prove the SSS Theorem. You may assume that $\triangle DEG$ has already been constructed. Use Theorem 3-13 to show that $\angle DFE \cong \angle DGE$. Then use the SAS Theorem and the Transitive Property of Congruence.

3. Prove the HL Theorem as suggested below. You may assume that point G has been determined on $\overrightarrow{AB}$ such that $BG = AB$ and that point H has been determined on $\overrightarrow{DE}$ such that $EH = DE$. Assume also that $\overline{CG}$ and $\overline{FH}$ have been drawn.
   a. Prove that $\triangle ABC \cong \triangle GBC$ and $\triangle DEF \cong \triangle HEF$.
   b. Prove that $\triangle AGC \cong \triangle DHF$.
   c. Prove that $\triangle ABC \cong \triangle DEF$.

Suppose that instead of Postulate A you start with Postulate B.

Postulate B If two angles and the included side of one triangle are congruent to two angles and the included side of another triangle, then the other two sides of the first triangle are congruent to the corresponding sides of the second triangle.

4. Use Postulate B to prove the ASA Theorem.

   Given: $\triangle PQR$ and $\triangle STU$;
   $\angle P \cong \angle S$;
   $\overline{PQ} \cong \overline{ST}$;
   $\angle Q \cong \angle T$

   Prove: $\triangle PQR \cong \triangle STU$

   (*Hint:* In view of Postulate B, you need only prove that $\angle R \cong \angle U$. Use an indirect proof and start by temporarily assuming that $m\angle U > m\angle R$. Then draw $\overline{UV}$ so that $\angle SUV \cong \angle R$.

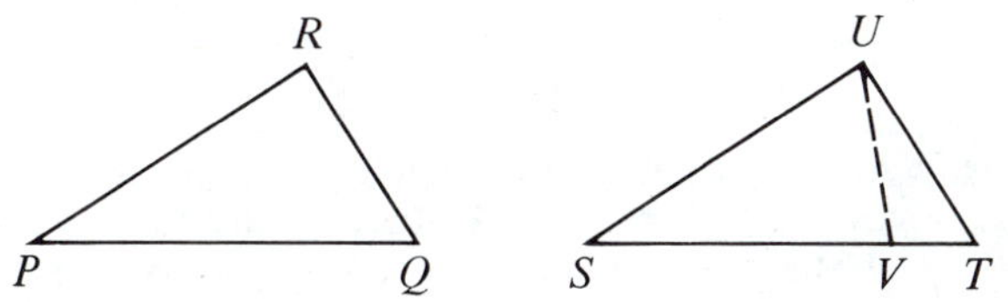

Exs. 4, 5

5. Prove the SAS Theorem *from the ASA Theorem.*

   Given: $\triangle PQR$ and $\triangle STU$;
   $\overline{PR} \cong \overline{SU}$;
   $\angle P \cong \angle S$;
   $\overline{PQ} \cong \overline{ST}$

   Prove: $\triangle PQR \cong \triangle STU$

   (*Hint:* You need only prove that $\angle R \cong \angle U$, and then the triangles will be congruent by the ASA Theorem. Use an indirect proof and start by temporarily assuming that $m\angle U > m\angle R$. Then draw $\overline{UV}$ so that $\angle SUV \cong \angle R$.)

ANSWERS TO SELF-TESTS

CHAPTER 1

SELF-TEST 1, page 16

1. $\subset$ 2. $\in$ 3. $\cap$

4. a. 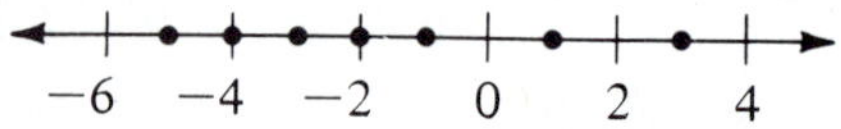 b. 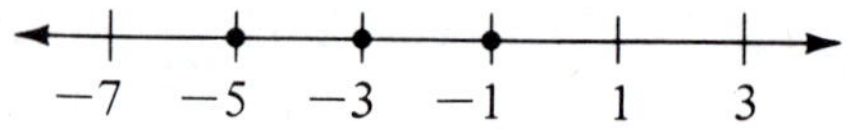

5. a. 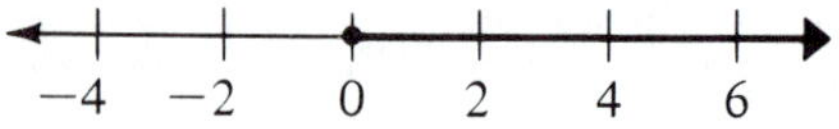 b. 

6. $(r + s)2$ 7. $2(s + r)$ 8. $2r + 2s$ 9. $-12t$ 10. $4j - 12$
11. $10y + z - 8yz$ 12. 82 13. -400 14. $-11v$ 15. $-12x + 25$
16. 48 17. -15

SELF-TEST 2, page 28

1. $x = -8$ 2. $a = 13$ 3. $y = 4$ 4. $d = 20$

5. $\{z: z \leqslant 5\}$ 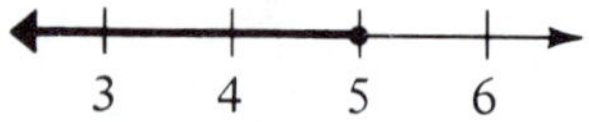

6. $\{x: x < -1\}$ 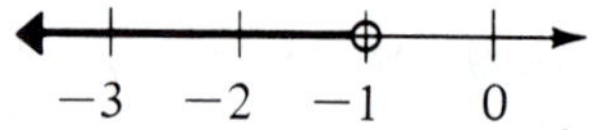

7. $\{t: t < 6\}$ 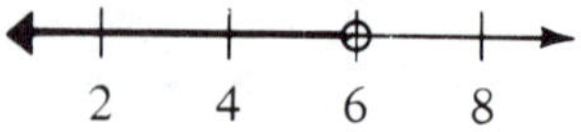

8. $\{n: n \geqslant 3\}$ 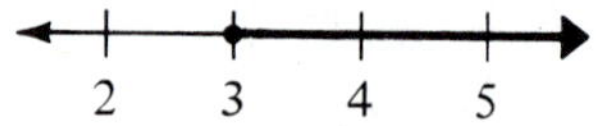

9. $\{b: -1 < b \leqslant 2\}$

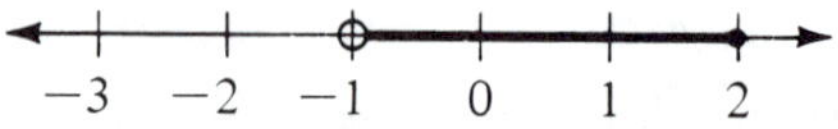

10. $\{y: y \leqslant -1 \text{ or } y \geqslant 3\}$

11. $\{x: 0 < x \leqslant 5\}$

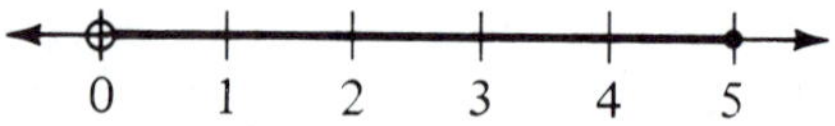

12. $\{x: 2 < x < 3\}$

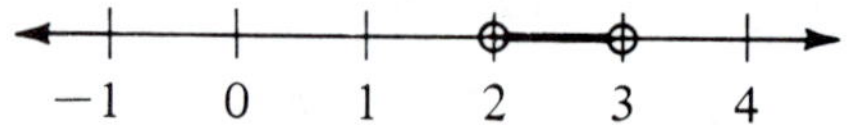

SELF-TEST 3, page 38

1. Let x = the first integer; then $x + 2$ = the next consecutive odd integer; $x + (x + 2) \geqslant 64$
2. Let x = Meena's present age in years; $x + 5 = 7$ 3. Let x = the number of half dollars; $50x = 1250$ or $0.50x = 12.50$ 4. Let x = the length, in centimeters, of one side. Then $2x - 7$ = the length of the adjacent side. 5. 11, 12, 13 6. The area is less than 2.25m^2.
7. 37.5 km 8. 20 mL of the 40% solution; 80 mL of the 25% solution

CHAPTER 2

SELF-TEST 1, page 61

1. No integer is irrational. 2. Some rhombuses are squares. 3. All squares are rhombuses.
4. Some prime numbers are not even (or some prime numbers are odd). 5. No integer is positive. 6. Some rectangles are triangles. 7. a. $p \vee \sim q$ b. True 8. a. $\sim q \wedge p$ b. True 9. a. $\sim p \wedge \sim q$ b. False 10. $\sim r \vee \sim s$ 11. $\sim r \wedge s$ 12. $s \vee r$
13. If a cat is a canary, then it has wings. 14. If $x > 5$, then x is positive. 15. $x = 3$ and $3x - 2 \neq 5$. 16. The number is divisible by three if and only if the sum of its digits is divisible by three.

SELF-TEST 2, page 75

1. $\exists_x\ x \geqslant x^2$; True 2. $\forall_x\ \frac{1}{x} \neq -x$; False 3. $\exists_x\ x = 3x - 2$ 4. $\forall_x\ x < 3x$

5. Converse: If it is raining, then the ground is wet. (T) Inverse: If the ground is not wet, then it is not raining. (T) Contrapositive: If it is not raining, then the ground is not wet. (F) 6. Rob doesn't eat properly. 7. It snows in Florida.

SELF-TEST 3, page 91

1. $p \rightarrow \sim r$ (or $r \rightarrow \sim p$) 2. If $3 = 6$, the moon is made of green cheese. 3. a. Given b. Subtraction Prop. of Equality c. $x = 3$ d. Div. Prop. of Equality 4. Given: $x > 0$ Prove $x^3 > 0$ (1) $x > 0$ (Given); (2) $x^2 > x \cdot 0$ (Mult. Prop. of Inequality); (3) $x^2 > 0$ (Mult. Prop. of 0 and Subst. Prop.); (4) $x \cdot x^2 > x \cdot 0$ (Mult. Prop. of Inequality); (5) $x^3 > 0$ (Mult. Prop. of 0 and Subst. Prop.) 5. Temporarily assume that $0 \cdot a = a$. (1) $a \neq 0$ and $0 \cdot a = a$ (Hypothesis and assumption); (2) $(0 \cdot a)\frac{1}{a} = a \cdot \frac{1}{a}$ (Mult. Prop. of Equality); (3) $0\left(a \cdot \frac{1}{a}\right) = a \cdot \frac{1}{a}$ (Assoc. Prop. of Mult.); (4) $0 \cdot 1 = 1$ (Inverse Prop. of Mult.); (5) $0 = 1$ (Identity Prop. of Mult. and Subst. Prop.); (6) Contradiction ($0 \neq 1$). Therefore, the temporary assumption that $0 \cdot a = 0$ must be false, and it follows that $0 \cdot a \neq a$.

CHAPTER 3

SELF-TEST 1, page 112

1. ray 2. line 3. 3, 5 4. 5 5. 9 6. A midpoint divides a segment into two segments, each half the length of the given segment. 7. Through any two points there is exactly one line (or two points determine a line). 8. There is exactly one point on a ray at a given distance from the endpoint of the ray. 9. $z = 14$; $m\angle DOE = 76$ 10. (1) $\overrightarrow{BX}$ bisects $\angle ABC$. (Given); (2) $\frac{1}{2} m\angle ABC$ (A bisector of an angle divides the angle into 2 $\angle$s, each of which has measure half that of the given angle.); (3) $m\angle ABC = 2m\angle 1$ (Mult. Prop. of Equality)

SELF-TEST 2, page 128

1. $m\angle 1 = 133$; $m\angle 2 = 47$ 2. $m\angle 1 = m\angle 2 = 4$ 3. (1) $\overrightarrow{OC} \perp \overleftrightarrow{AE}$ (Given); (2) $\angle AOB$ and $\angle BOC$ are comp. $\angle$s; $\angle DOE$ and COD are comp. $\angle$s (If ext. sides of 2 adj. acute $\angle$s are $\perp$, the $\angle$s are comp.); (3) $m\angle AOB = m\angle DOE$ (Given); (4) $m\angle BOC = m\angle COD$ (Two $\angle$s comp. to $\angle$s of = measure have = measures.); (5) $\overrightarrow{OC}$ bisects $\angle BOD$. (Def. of $\angle$ bisector) 4. no 5. yes; SSS Postulate 6. (1) $\overline{RL}$ bisects $\overline{TS}$. (Given); (2) $\overline{SM} \cong \overline{TM}$ (Def. of bisect); (3) $\angle S \cong \angle T$ (Given); (4) $\angle RMS \cong \angle LMT$ (Vert. $\angle$s are $\cong$.); (5) $\triangle RMS \cong \triangle LMT$ (ASA Post.) 7. HL Postulate 8. SAS Postulate 9. (1) $\angle 3 \cong \angle 4$ (Given); (2) $\angle ZXW$ and $\angle 3$ are supp. $\angle$s; $\angle ZXY$ and $\angle 4$ are supp. $\angle$s. (If the ext. sides of 2 adj. $\angle$s are opp. rays, the $\angle$s are supp.); (3) $\angle ZXW \cong \angle ZXY$ (Two $\angle$s supp. to $\cong$ $\angle$s are $\cong$.); (4) $\overline{WX} \cong \overline{YX}$ (Given); (5) $\overline{XZ} \cong \overline{XZ}$ (Refl. Prop. of $\cong$); (6) $\triangle XWZ \cong \triangle XYZ$ (SAS Post.)

SELF-TEST 3, page 142

1. (1) $\angle X$ and $\angle Z$ are right $\angle$s. (Given); (2) $\triangle WXY$ and $\triangle WZY$ are right $\triangle$s. (Def. of right $\triangle$); (3) $\overline{WX} \cong \overline{WZ}$ (Given); (4) $\overline{WY} \cong \overline{WY}$ (Refl. Prop. of $\cong$); (5) $\triangle WXY \cong \triangle WZY$ (HL Post.); (6) $\angle 1 = \angle 2$ (Corr. parts of $\cong$ $\triangle$s are $\cong$.) 2. 5 3. 24 4. (1) $\overline{AB} \cong \overline{CB}$; $\overline{AD} \cong \overline{CD}$ (Given); (2) $\overline{BD} \cong \overline{BD}$ (Refl. Prop. of Equality), *or* (2′) $\angle A \cong \angle C$ (Base $\angle$s of an isosceles $\triangle$ are $\cong$.); (3) $\triangle ABD \cong \triangle CBD$ (SSS Post.), *or* (3′) $\triangle ABD \cong \triangle CBD$ (SAS Post.); (4) $\angle ADB \cong \angle CDB$ (Corr. parts of $\cong$ $\triangle$s are $\cong$.); (5) $\overline{BD} \perp \overline{AC}$ (If 2 lines form $\cong$ adj. $\angle$s, the lines are $\perp$.) 5. $m\angle G > m\angle 1$ 6. $1 < EG < 17$ 7. The measure of an exterior $\angle$ of a triangle is greater than the measure of either remote interior $\angle$, so $m\angle 3 > m\angle 1$. But $\angle 1$ and $\angle 2$ are vertical $\angle$s, so $m\angle 1 = m\angle 2$. Then, by the Subst. Prop., $m\angle 3 > m\angle 2$.

CHAPTER 4

SELF-TEST 1, page 165

1. If two lines are cut by a transversal so that alternate interior ∠s are ≅, then the lines are ∥. 2. In a plane, 2 lines ⊥ to the same line are ∥. 3. If 2 lines are cut by a transversal so that corr. ∠s are ≅, the lines are ∥. 4. Answers may vary. Example: (1) ∠1 and ∠7 are supp. (Given); (2) ∠3 and ∠7 are supp. (If the ext. sides of 2 adj. ∠s form opp. rays, the ∠s are supp.); (3) ∠1 ≅ ∠3 (∠s supp. to the same ∠ are ≅.); (4) $a \parallel b$ (If 2 lines are cut by a trans. so that corr. ∠s are ≅, the lines are ∥.) 5. $m\angle 3 = m\angle 6 = m\angle 8 = x$; $m\angle 2 = m\angle 4 = m\angle 5 = m\angle 7 = 180 - x$ 6. 34 7. Answers may vary. Example: (1) $a \parallel b$ (Given); (2) ∠6 ≅ ∠8 (If 2 lines are ∥, corr. ∠s are ≅.); (3) ∠1 ≅ ∠6 (Vert. ∠s are ≅.); (4) ∠1 ≅ ∠8 (Trans. Prop. of ≅) 8. Exactly one. Through a given pt. not on a given line, there is exactly one line ⊥ to the given line.

SELF-TEST 2, page 179

1. a. 66 b. $\overline{BC}$ 2. a. 47 b. $\overline{AC}$ 3. a. 56 b. $\overline{AB}$ 4. $m\angle 6 + m\angle 4 = m\angle 7$ 5. 16 6. (1) ∠1 ≅ ∠2 (Given); (2) $\overline{ST} \cong \overline{SU}$ (Given); (3) ∠5 ≅ ∠6 (Vert. ∠s are ≅.); (4) $\triangle QST \cong \triangle RSU$ (AAS Thm.); (5) $\overline{QT} \cong \overline{RU}$ (Corr. parts of ≅ △s are ≅.) 7. 540 8. 360 9. measure of each interior angle = 135; measure of each exterior angle = 45

CHAPTER 5

SELF-TEST 1, page 201

1. 12 2. 13 3. 30 4. (1) *PRSU* is a ▱. (Given); (2) $\overline{RS} \cong \overline{PU}$ (Opp. sides of a ▱ are ≅.); (3) $m\angle S = m\angle P$ (Opp. ∠s of a ▱ are ≅.); (4) $m\angle RST = m\angle PUQ$ (Given); (5) $\triangle RST \cong \triangle UPQ$ (ASA Post.); (6) $\overline{RT} \cong \overline{UQ}$ (Corr. parts of ≅ △s are ≅.) 5. False 6. False 7. True 8. False 9. True 10. (1) *MNOP* is a rectangle. (Given); (2) $\overline{PN} \cong \overline{MO}$, or $PN = MO$ (Diagonals of a rectangle are ≅.); (3) $\overline{NP}$ and $\overline{MO}$ bisect each other. (Diagonals of a ▱ bisect each other.); (4) X is the midpt. of $\overline{NP}$; X is the midpt. of $\overline{MO}$. (Def. of a bisector); (5) $NX = \frac{1}{2}NP$, $MX = \frac{1}{2}MO$ (A midpt. divides a segment into 2 segs., each half the length of the given seg. (Theorem 3-2)); (6) $MX = NX$ (Subst. Prop.); (7) $\triangle MXN$ is isosceles. (Def. of an isosceles △) 11. (1) *ABCD* is a rhombus. (Given); (2) ∠A ≅ ∠C (Opp. ∠s of a ▱ are ≅.); (3) $\overline{AD} \cong \overline{CD}$; $\overline{AB} \cong \overline{CB}$, or $AB = CB$ (Def. of a rhombus); (4) $AB = AM + MB$; $CB = CN + NB$ (Property of Betweenness); (5) $AM + MB = CN + NB$ (Substitution Property); (6) $\overline{MB} \cong \overline{NB}$, or $MB = NB$ (Given); (7) $AM = CN$ (Subtraction Property of Equality); (8) $\triangle ADM \cong \triangle CDN$ (SAS Property); (9) $\overline{MD} \cong \overline{ND}$ (Corr. parts of ≅ △s are ≅.) 12. True 13. True 14. False 15. True

SELF-TEST 2, page 209

1. Because the segment joining the midpoints of 2 sides of a △ is ∥ to the third side, $\overline{XY} \parallel \overline{BC}$. Since ∠1 and ∠2 are corr. ∠s, they are ≅. 2. 5, 23 3. If 3 ∥ lines cut off ≅ segments on one transversal, they cut off ≅ segments on any transversal. 4. (1) $\overline{EF} \cong \overline{FG}$; $\overline{FH} \parallel \overline{EJ}$ (Given); (2) F is the midpoint of $\overline{EG}$. (Def. of midpoint); (3) $\overline{FH}$ bisects $\overline{JG}$. (A line that contains the midpoint of one side of a △ and is ∥ to another side bisects the third side.); (4) H is the midpoint of $\overline{JG}$. (Def. of midpoint); (5) $\overline{EH}$ is a median of $\triangle EGJ$. (Def. of a median of a △.) 5. Since the median, $\overline{UR}$, of trapezoid *PQST* is parallel to base $\overline{PQ}$, and ∠1 and ∠2 are interior angles on the same side of the transversal $\overline{PT}$, ∠1 and ∠2 are supplementary. 6. 12 7. 6 8. (1) *WXYZ* is an isosceles trapezoid, with $\overline{WZ} \cong \overline{XY}$. (Given); (2) $\overline{WY} \cong \overline{XZ}$ (The diagonals of an isosceles trapezoid are ≅.), *or* (2′) $\angle WZY \cong \angle XYZ$ (Base angles of an isosceles trapezoid are ≅.); (3) $\overline{ZY} \cong \overline{YZ}$ (Reflexive Property of Congruence); (4) $\triangle WZY \cong \triangle XYZ$ (SSS Postulate), *or* (4′) $\triangle WZY \cong \triangle XYZ$ (SAS Postulate)

CHAPTER 6

SELF-TEST 1, page 227

1. $12ab - 2a - 3b - 2b^2$; 2 2. $10x^2 - 2xy + y^2$; 2 3. $4d^3 - d^2 - 2$; 3
4. $4d^3 + d^2 + 14d - 14$; 3 5. $-27x^6y^3$ 6. $-12k^4 + 8k^3 - 4k^2$
7. $6g^2 + 11gh - 35h^2$ 8. $3m^3 - 8m^2 - 5m + 6$

SELF-TEST 2, page 240

1. $7a^3bc^2(3bc^3 - 8a^2c - 5ab^2)$ 2. $(t - 9)(t + 7)$ 3. $(7b - 2)(b - 1)$
4. $s(3s + 5)(2s - 1)$ 5. $(1 + 5n)(1 - 5n + 25n^2)$ 6. $(4d - 11)^2$
7. $6(3w - x)(3w + x)$ 8. $(j - 2k)(j^2 + 2jk + 4k^2)$ 9. $\frac{-27m^3}{125n^6}$ 10. $\frac{4x - 4y}{3x}$
11. $3z^2 - \frac{1}{4} + \frac{3}{2z^2}$ 12. $\frac{2y^3 - 3y^2 - 8y - 3}{y - 3} = 2y^2 + 3y + 1$ 13. $\frac{10x^3 + 21x^2 - 4}{5x - 2} =$
$2x^2 + 5x + 2$

SELF-TEST 3, page 251

1. $\frac{b}{5b - 2}$ 2. $-\frac{1}{2r + 1}$ 3. $\frac{s - 4}{s}$ 4. $\frac{y(x - y)}{x(x + y)}$ 5. $\frac{ab^2d}{3c^2}$ 6. -1
7. $\frac{r}{3}$ 8. -2 9. xy 10. $\frac{a^2 - 3ab + 18b}{6a^2b^2}$ 11. $\frac{m + 11}{3(m + 5)(m + 1)}$ 12. -2

CHAPTER 7

SELF-TEST 1, page 279

1. $\{-12, 2\}$ 2. $\{5, 9\}$ 3. $\{-10, 3\}$ 4. $\{\pm \frac{1}{2}\}$ 5. $\{3\}$ 6. $\{-4\}$
7. $\{-\frac{2}{3}, 2\}$ 8. $\{-\frac{6}{7}, 2\}$ 9. $-4x^2y\sqrt{3}$ 10. $\frac{4\sqrt{5}}{5}$ 11. $\{\pm \frac{10}{3}\}$
12. $\{-6, 8\}$ 13. $\{\pm 3\sqrt{7}\}$ 14. $\{\pm 2\sqrt{3}\}$ 15. $-6, -5$; 5, 6 16. width = 7 cm; length = 10 cm 17. The larger takes 2h; the smaller takes 3 h.

SELF-TEST 2, page 295

1. $\{-8, -2\}$ 2. $\{1 + \sqrt{3}, 1 - \sqrt{3}\}$ 3. $\{-\frac{1}{2}\}$ 4. $\{\frac{2 + \sqrt{2}}{3}, \frac{2 - \sqrt{2}}{3}\}$
5. $\{6\}$ 6. $\{\frac{-1 - \sqrt{29}}{2}, \frac{-1 + \sqrt{29}}{2}\}$ 7. $\{\frac{-2 + \sqrt{2}}{2}, \frac{-2 - \sqrt{2}}{2}\}$ 8. $\{-\frac{3}{4}, 2\}$
9. 625; 2 rational roots; $\{0, 25\}$ 10. 0; a rational double root; $\{\frac{1}{2}\}$ 11. 180;
2 irrational roots; $\{\pm \frac{3\sqrt{5}}{2}\}$ 12. 121; 2 rational roots; $\{-9, 2\}$ 13. $\frac{4}{5}$ or $\frac{5}{2}$
14. $3 + \sqrt{2}$ and $3 - \sqrt{2}$ 15. 20 km/h

CHAPTER 8

SELF-TEST 1, page 325

1. 1:2 2. $\frac{9}{s}$ 3. $\frac{8}{5}$, or 1.6 4. $\frac{4}{3}$ $\left(\text{or } \frac{3}{4}\right)$ 5. 9 6. 22.5 7. $RV = 5$, $VU = 10$ 8. (1) $\frac{RT}{ST} = \frac{RU}{VU}$ (Given); (2) $RS + ST = RT$; $RV + VU = RU$ (Prop. of Betweenness); (3) $\frac{RS + ST}{ST} = \frac{RV + VU}{VU}$ (Subst. Prop.); (4) $\frac{RS}{ST} = \frac{RV}{VU}$ (Prop. 4 of Proportions); (5) $\overline{SV} \parallel \overline{TU}$ (Thm. 8-4); (6) $\angle RVS \cong \angle U$ (If 2 lines are $\parallel$, corr. $\angle$s are $\cong$.) 9. 13.5

SELF-TEST 2, page 335

1. a. 2:3 b. 2:1 2. 12 cm and 20 cm 3. (1) $\angle 1 \cong \angle 2$ (Given); (2) $\angle 1 \cong \angle JKN$ (Vert. ∠s are ≅.); (3) $\angle 2 \cong \angle JNK$ (Subst. Prop.); (4) $\angle J \cong \angle J$ (Refl. Prop. of ≅); (5) $\triangle JKN \sim \triangle JLM$ (AA ~ Thm.) 4. no 5. yes; SSS ~ Thm. 6. yes; AA ~ Thm. 7. yes; SAS ~ Thm.

SELF-TEST 3, page 344

1. 16:9 2. 6:7 3. 3.5 cm 4. 27 m 5. 1.2 m, or 120 cm

CHAPTER 9

SELF-TEST 1, page 367

1. $\triangle WXY \sim \triangle WZX \sim \triangle XZY$ 2. $x = 9$, $y = 3\sqrt{13}$ 3. $z = \frac{48}{5}$, or 9.6

4. no; $20^2 + 21^2 \neq 28^2$. Thus by the contrapositive of the Pythagorean Theorem, the △ is not a right △. 5. 10 cm, $2\sqrt{7}$ cm 6. 34 m 7. $4\sqrt{2}$ and $8\sqrt{2}$ 8. a. 45–45–90 △; The diagonals of a rhombus bisect its ∠s. A square is a rhombus whose ∠s are each 90°. Thus, $m\angle MAB = m\angle MBA = 45$. By the Triangle-Angle-Sum Thm., $m\angle AMB = 90$. b. 32; 128

SELF-TEST 2, page 378

1. $\frac{\sqrt{5}}{3}$ 2. $\frac{2}{3}$ 3. $\frac{2}{3}$ 4. $\frac{2\sqrt{5}}{5}$ 5. 48 6. a. 0.6018 b. 0.6018 7. 266 m

CHAPTER 10

SELF-TEST 1, page 399

1. 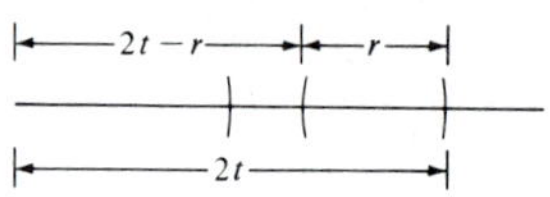

2. 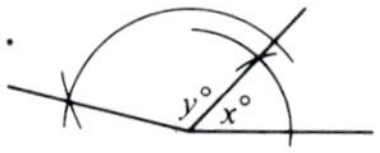

3. 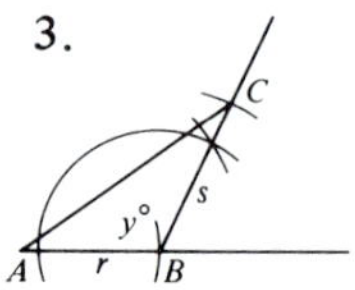

4. 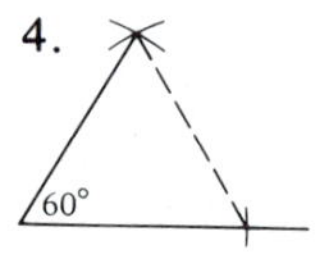

5. 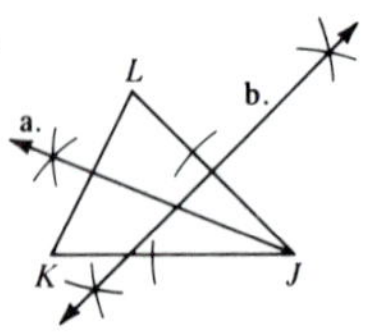

6. 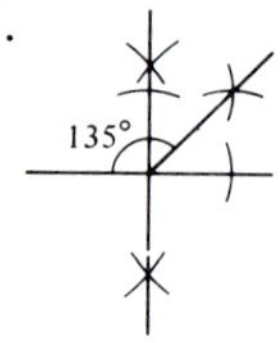

7. 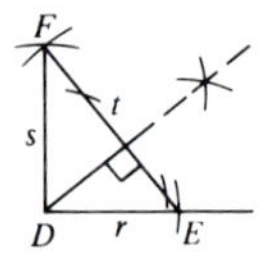

8. 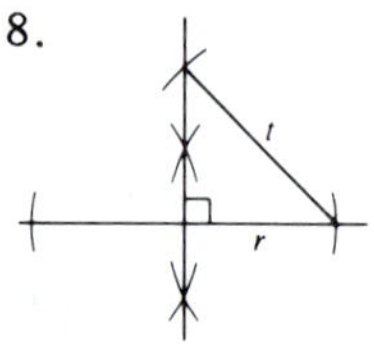

SELF-TEST 2, page 408

1. 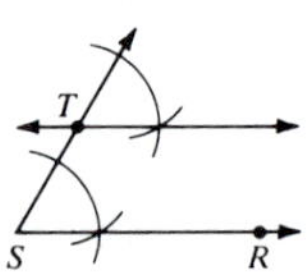

2. 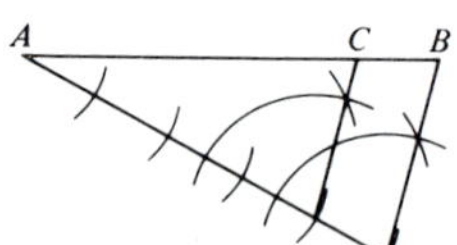

AC:CB = 4:1

3. 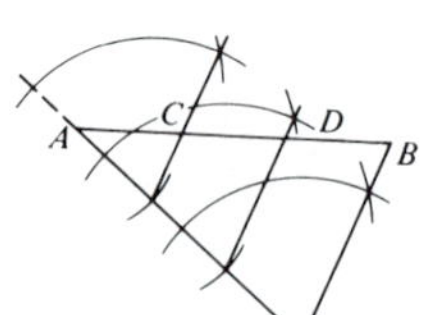

AC = CD = DB

4. Use Const. 10. Let $PQ = b$, $QR = c$ and $PS = a$. 5. Use Const. 10. Let $PQ = a$, $QR = b$, and $PS = c$. 6. Use Const. 11. Let $QR = b$ and $RS = C$. 7. Use Const. 11. Let $QR = a$ and $RS = c$.

SELF-TEST 3, page 417

1. $\overleftrightarrow{MP}$ 2. $\overleftrightarrow{AC}$ 3. $\overleftrightarrow{AB}$ and $\overleftrightarrow{DC}$ 4. $\overrightarrow{AC}$ 5. Q and N

6. The locus consists of two points that are the intersection of the circle with center R and radius 3 cm and the circle with center S and radius 4 cm.

7.. The locus consists of three points that are the intersection of the circle with center X and radius 6 cm and the two lines that are parallel to l and 3 cm from l.

8. The locus consists of one point that is on the perpendicular bisector of $\overline{PQ}$ and 1 cm from lines l and m.

9. 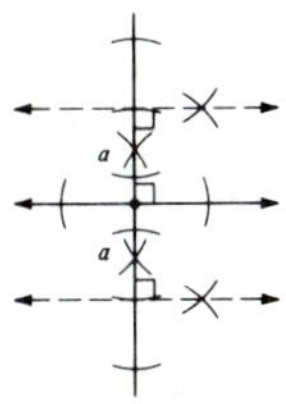

10. 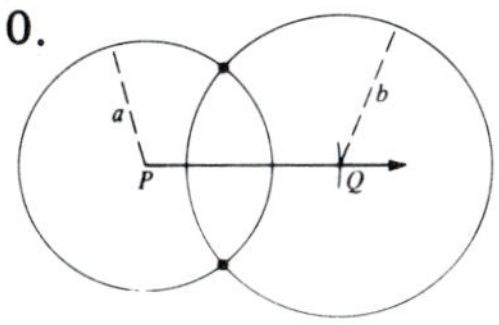

11. 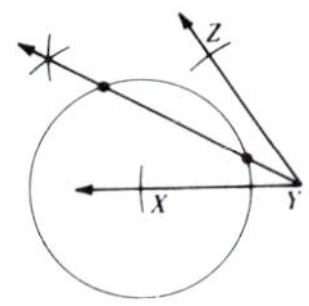

CHAPTER 11

SELF-TEST 1, page 447

1. 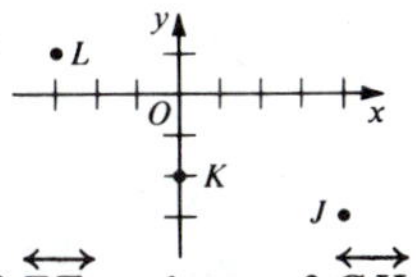

2. 12.5 3. 16.5 4. (1, 2) 5. $XY = 2\sqrt{2},\ YZ = 10$ 6. (−6, 2) 7. a. $-\frac{5}{2}$ b. 0 8. Slope of $\overleftrightarrow{AB} = -\frac{5}{2}$, slope of $\overline{BC} = -5$; A, B, and C are not collinear. 9. Slope of $\overleftrightarrow{EF}$ = slope of $\overleftrightarrow{GH} = -3$; slope of $\overleftrightarrow{FG}$ = slope of $\overleftrightarrow{EH} = \frac{1}{3}$. Since their slopes are =, $\overleftrightarrow{EF} \parallel \overleftrightarrow{GH}$ and $\overleftrightarrow{FG} \parallel \overleftrightarrow{EH}$, so $EFGH$ is a ▱. Since their slopes are negative reciprocals, $\overline{EF} \perp \overline{FG}$, for example, making $\angle F$ a rt. $\angle$. Therefore, $EFGH$ is a rectangle. 10. (0, 2b) 11. If A, B, and C are collinear, then they lie on a vertical segment or the slope of $\overline{AB}$ = the slope of $\overline{AC}$. Since the slope of $\overline{AB} = \frac{b}{a}$, $a \neq 0$, $\overline{AB}$ is *not* vertical. Thus, slope of $\overline{AB}$ = slope of $\overline{AC}$, or $\frac{b}{a} = \frac{d}{c}$. Cross-multiplying gives $bc = ad$.

SELF-TEST 2, page 459

1. 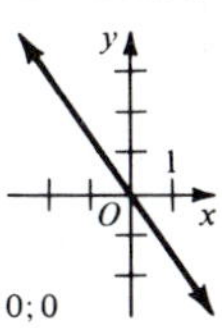

0; 0

2. 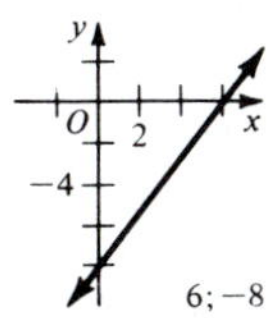

6; −8

3. 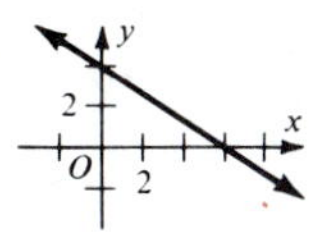

4. $y - 4 = -\frac{1}{2}(x + 1)$ or $y - 0 = -\frac{1}{2}(x - 7)$ 5. $y = 3x + 5$ 6. $m = -\frac{4}{5}$; $b = -4$

7. $y = 3$ or $y = -3$ 8. $x = 3$ 9. $x = -5.5$

SELF-TEST 3, page 466

1. 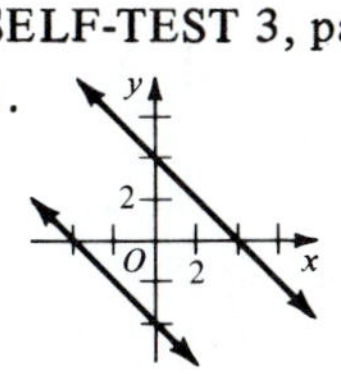

inconsistent

2. 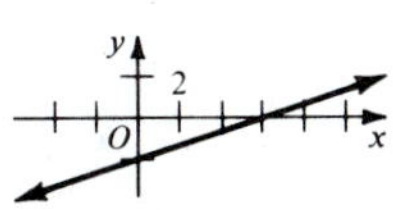

consistent and dependent

3. 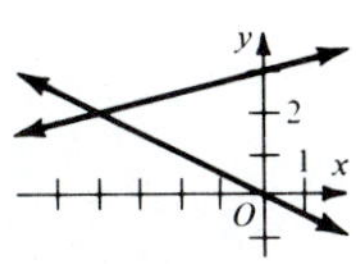

consistent and independent

4. (−1, 4)
5. (−4, 7)
6. (12, 9)

CHAPTER 12

SELF-TEST 1, page 487

1. a. $(-3, -6.75)$, $(-2, -3)$ $(-1, -0.75)$, $(0, 0)$, $(1, -0.75)$, $(2, -3)$, $(3, 6.75)$
   b–d. 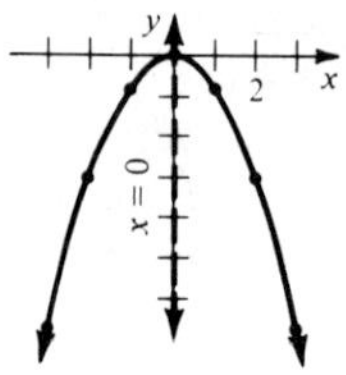
   e. maximum

2. a. $(-4, 8)$, $(-3, 3)$ $(-2, 0)$, $(-1, -1)$, $(0, 0)$, $(1, 3)$, $(2, 8)$
   b–d. 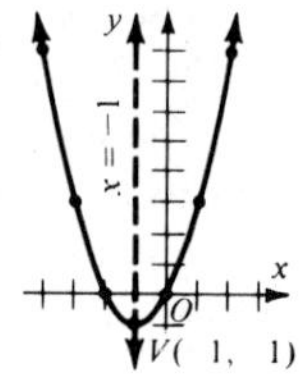
   e. minimum

3. $y = (x + 1)^2 + 2$; $(-1, 2)$

4. $y = \left(x - \frac{5}{2}\right)^2 - \frac{1}{4}$; $\left(\frac{5}{2}, -\frac{1}{4}\right)$

5. a. 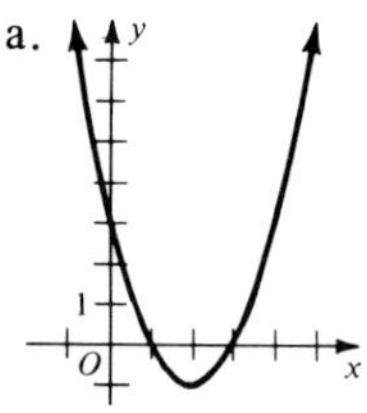
   b. $x = 2$

6. -3

7. $\{-1, 3\}$

8. a. 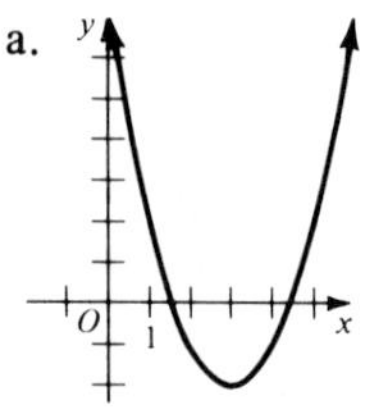
   b. 1 and 5

SELF-TEST 2, page 499

1. $x^2 + y^2 = 9$
2. $(x - 2)^2 + (y + 3)^2 = 2$
3. $r = 3$; $C(1, 2)$
4. $r = 5$; $C(0, -1)$
5. $(0, 7)$
6. $y = -9$
7. $y = \frac{1}{4}x^2 + 2$ 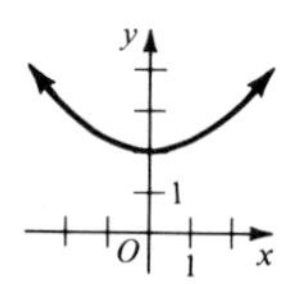
8. a, b. 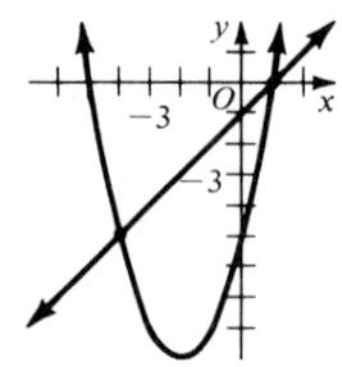
8. c. $(1, 0)$, $(-4, -5)$
9. $\{(0, 3), (-8, -1)\}$
10. $\{(1, -2), (2, 1)\}$

SELF-TEST 3, page 510

1. a. 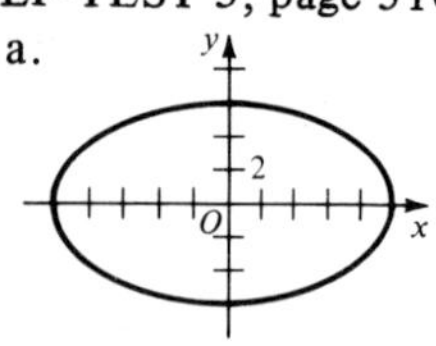
   b. $(8, 0)$, $(-8, 0)$

2. a. 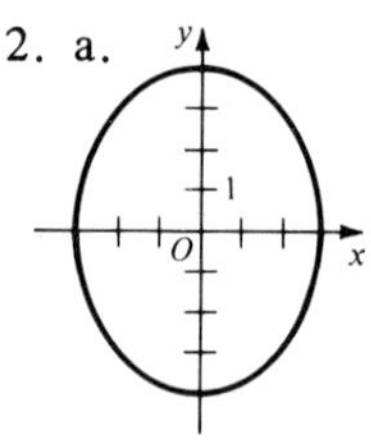
   b. $(0, \sqrt{7})$, $(0, -\sqrt{7})$

3. $\frac{x^2}{64} + \frac{y^2}{289} = 1$

4. a. x-intercepts: 6 and -6; no y-intercepts
   b. $(3\sqrt{5}, 0)$, $(-3\sqrt{5}, 0)$ c. $y = \pm\frac{1}{2}x$

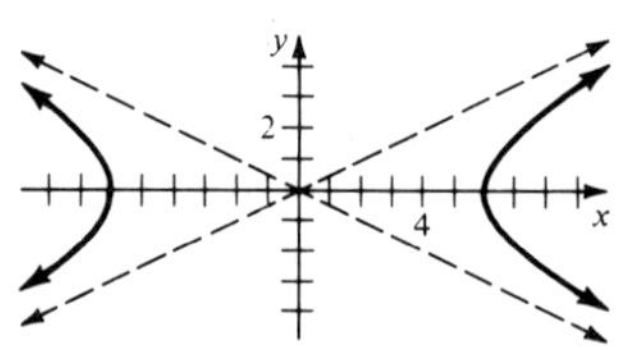

5. a. x-intercepts: 3 and -3; no y-intercepts
   b. $(5, 0)$, $(-5, 0)$ c. $y = \pm\frac{4}{3}x$

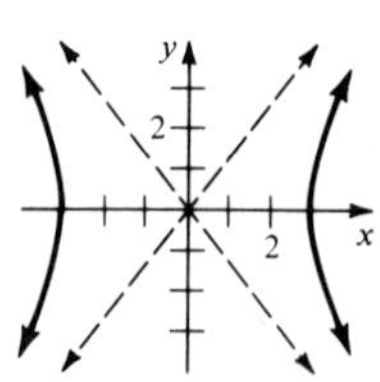

6. $\frac{x^2}{16} - \frac{y^2}{20} = 1$

7. 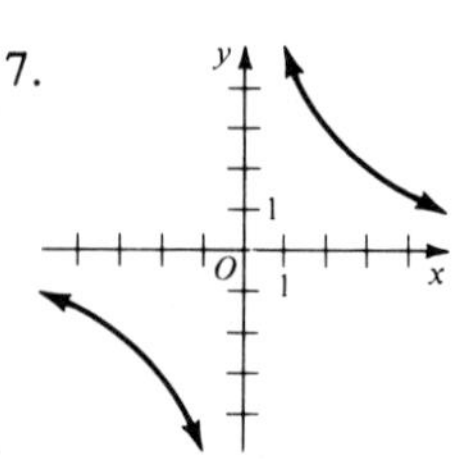

CHAPTER 13

SELF-TEST 1, page 526

1. {Sun., Mon., Tues., Wed., Thurs., Fri., Sat.}   2. {Tues., Thurs.}   3. {Sun., Mon., Tues., Fri.}   4. $\frac{2}{7}$ 5. 1   6. $\frac{4}{7}$

SELF-TEST 2, page 536

1. $\frac{7}{13}$ 2. $\frac{19}{52}$ 3. $\frac{11}{13}$ 4. a. $\frac{9}{25}$ b. $\frac{3}{10}$ 5. a. $\frac{12}{25}$ b. $\frac{3}{5}$ 6. a. $\frac{13}{25}$ b. $\frac{2}{5}$ 7. a. $\frac{16}{25}$ b. $\frac{7}{10}$

SELF-TEST 3, page 548

1. a. 42   b. 970,200   2. 120   3. 840   4. a. 230,300   b. 70   5. 1140

6. a. $\frac{\binom{12}{2}}{\binom{52}{2}} \approx 0.05$ b. $\frac{\binom{13}{1}\cdot\binom{39}{1}+\binom{13}{2}}{\binom{52}{2}}$ or $1-\frac{\binom{39}{2}}{\binom{52}{2}} \approx 0.441$

CHAPTER 14

SELF-TEST 1, page 570

1. a. yes   b. no   2. a. no   b. no   3. yes   4. no   5. yes   6. no

7. a. 10   b. 6

8. a.

#	*a*	*b*	*c*
a	*a*	*b*	*c*
b	*b*	*a*	*b*
c	*c*	*b*	*a*

b. yes   c. b d. yes; a e. $a^{-1}=a,\ b^{-1}=b,\ c^{-1}=c$

f. $(b \# b) \# c = c;\quad b \# (b \# c) = a$

9. a. 0   b. 1   10. $-1=2,\ -2=1$ 11. $1^{-1}=1,\ 2^{-1}=2$ 12. a. 2   b. 2

SELF-TEST 2, page 580

1. (1) Associative Property   (2) Existence of an identity element   (3) Existence of inverses

2. Yes. Identity element is 1; inverse of 4 is $\frac{1}{4}$.   3. No. No identity element or inverses.

4. Yes. Identity element is 0; inverse of 4 is 2.   5. No. The system is not closed since $2 \times 3 = 0$ and $3 \times 4 = 0$.   6. No; p, q, and s do not have inverses.   7. $\{p, q, r, s\}$ would not be closed under $*$; $\therefore (\{p, q, r, s\}, *)$ would not be an algebraic system.

8. $D * (D * D) = D * E = A$; $(D * D) * D = E * D = A$

9. Yes.
Identity element is 0;
$-0=0,\ -2=6,\ -4=4,\ -6=2$

+	0	2	4	6
0	0	2	4	6
2	2	4	6	0
4	4	6	0	2
6	6	0	2	4

SELF-TEST 3, page 589

1. a. 3   b. 2   2. Yes. Identity for + is 0; identity for · is 1. Inverses under + are $-0=0,\ -1=2,\ -2=1$; inverses under · are $1^{-1}=1;\ 2^{-1}=2$ 3. No $(Z_8\backslash\{0\}, \times)$ is not a commutative group because it is not closed. Also, 2, 4, and 6 do not have inverses under $\times$.

4. No. $(\mathcal{R}, -)$ is not a commutative group because $-$ is not commutative or associative.

5. {2}   6. {1}   7. {3, 4}   8. {1}

INDEX

Pages 5-7 1. a. $\{-2, -1, 0, 1, 3\}$ b. $\{-1, 1\}$ 3. a. $\{-5, -4, -3, 2, 3\}$ b. $\emptyset$ 5. a. {the rational numbers} b. {the integers}

7. 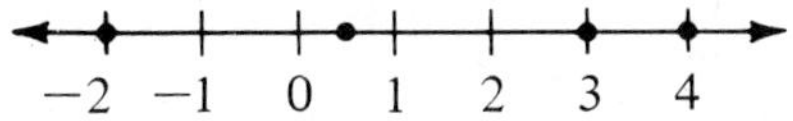

9. 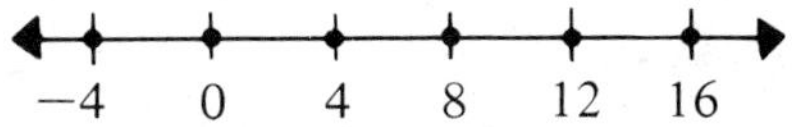

11. 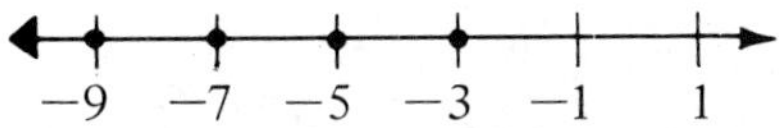

13. a. 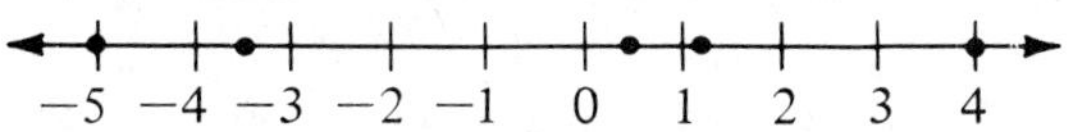

b. 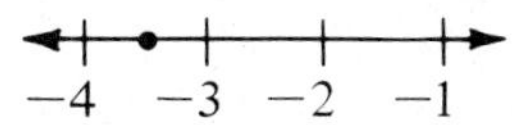

15. $\mathcal{R}, Q$ 17. $\mathcal{R}$, {the irrational numbers} 19. $\mathcal{R}, Z, W, Q$ 21. $\mathcal{R}, Z, W, Q$
23. $\emptyset, \{-2\}, \{-1\}, \{-2, -1\}$ 25. {the squares of the positive integers}
27. {the positive powers of 3} 29–33. Answers may vary. Examples are given.
29. $A = \{1, 2\}, B = \{2, 3\}, C = \{1, 2, 4\}$ 31. $A = \{1, 2\}, B = \{1, 2, 3\}, C = \{1, 2\}$
33. $A = \{1, 2, 3\}, B = \{1, 2\}, x = 3$ 35. $\emptyset$ 37. $\{1, 3\}$ 39. $\{1, 2, 3, 4, 5\}$

41. 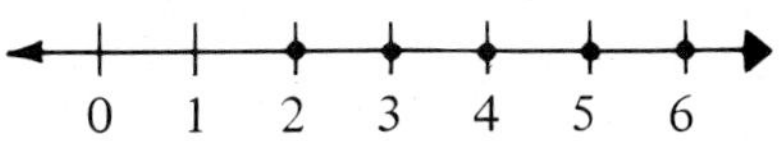

43. 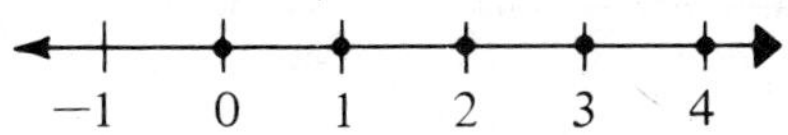

45. $\emptyset$

Pages 10-11 1. $6m + 7m$ 3. $(8 \cdot 17)x$ 5. $0 + 3$ 7. $1 \cdot y$
9. $((-x) + x) + 27$ 11. $13m$ 13. $4y - 4x$ 15. $3x - 21$ 17. $\frac{b}{2} + 1$
19. $0.3m - 1.5$ 21. $ab + 4a$ 23. $4y + 2y^2$ 25. $7b$ 27. $15m + 15n$
29. $q + 9r$ 31. $ab + 7a + 5b + 35$ 33. $x^2 + 7x + 12$ 35. $99 + 9w + 11x + xw$
37. $x = 0$ 39. $x = 8$ 41. $x = 1$

Page 15 1. 91 3. −147 5. −103 7. 0 9. 11 11. −24 13. 60
15. −10 17. 9 19. $-2a$ 21. $-18w$ 23. x 25. $10t$ 27. $-12x$
29. $-8m + 4n$ 31. −17 33. 50 35. 65 37. $7y + 15$ 39. $6y - w$
41. $25m - 8n$ 43. $2x + 2$ 45. $2a + 2b$ 47. $343y + x$ 49. $2a - 7b$
51. $8x^2 - 14xy + 3y^2$ 53. $-30w^2 - wv + v^2$ 55. $58x - 38$

Pages 20-21 1. 11 3. $-\frac{1}{2}$ 5. 5 7. −3 9. 2 11. 5 13. 7
15. −3 17. 3 19. $\mathcal{R}$ 21. 5 23. −12 25. 15 27. 9 29. 5
31. $\emptyset$ 33. $\frac{7}{2}$ 35. $\frac{5}{2}$ 37. −75 39. −6 41. $-\frac{37}{5}$ 43. 0 45. 4
47. −20 49. $\frac{F}{m}$ $(m \neq 0)$ 51. $\frac{C}{2\pi}$ 53. $\frac{P - 2w}{2}$ 55. $\frac{2A}{h} - c$ $(h \neq 0)$

Pages 23-24
1. $\{x: x < 15\}$

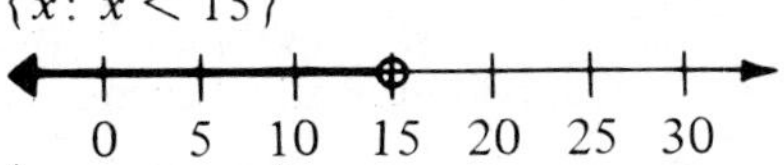

3. $\{y: y \geq 3\}$

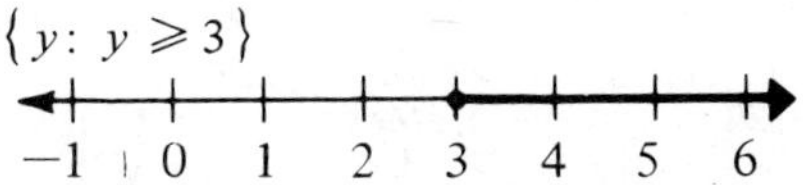

5. $\{y: y \geq -6\}$

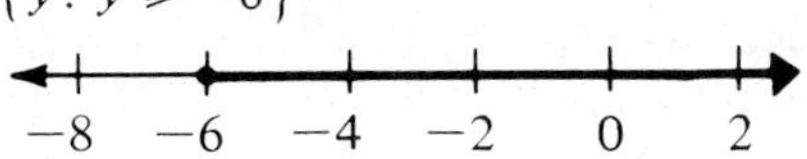

7. $\{t: t \leq 2\}$

9. $\{y: y > 4\}$

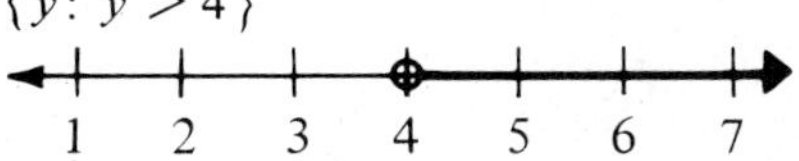

11. $\left\{x: x \leq \frac{5}{2}\right\}$ 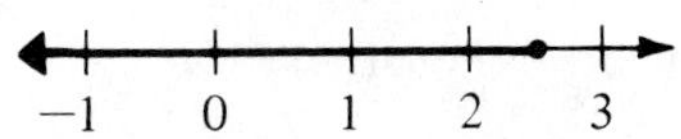

13. $\{z: z \leqslant 9\}$ 15. $\{z: z > 2\}$ 17. $\left\{n: n \geqslant \frac{5}{2}\right\}$ 19. $\{y: y < -9\}$

21. $\{w: w < 11\}$ 23. $\left\{y: y \geqslant \frac{5}{4}\right\}$ 25. $\{m: m < -3\}$ 27. $\left\{x: x \geqslant -\frac{4}{3}\right\}$

29. $\{x: x < 0.5\}$ 31. $\{j: j > -7\}$ 33. $\left\{x: x < \frac{9}{7}\right\}$ 35. $\left\{y: y \geqslant \frac{13}{4}\right\}$

37. $\{x: x \leqslant 15\}$

Pages 27–28 1. d 3. c 5. b

7. $\{x: -5 \leqslant x < -2\}$

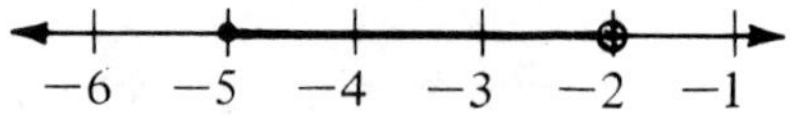

9. $\{x: -3 < x \leqslant 2\}$

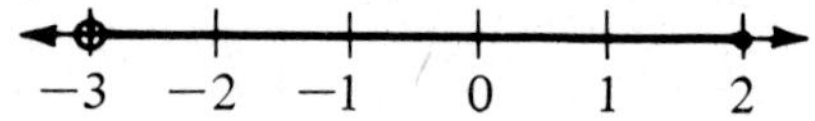

11. $\{x: -4 < x < 0\}$

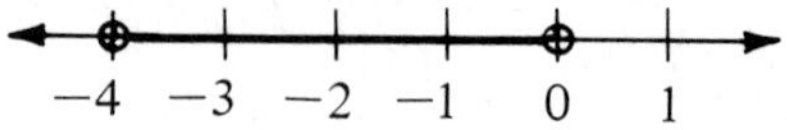

13. $\{y: y < 6 \text{ or } y > 6\}$, or $\{y: y \neq 6\}$

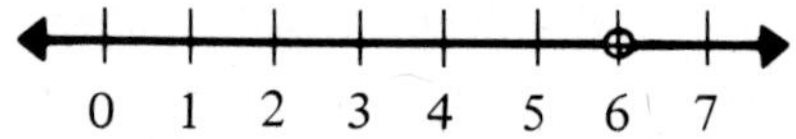

15. $\left\{y: y \leqslant -1 \text{ or } y \geqslant \frac{5}{2}\right\}$

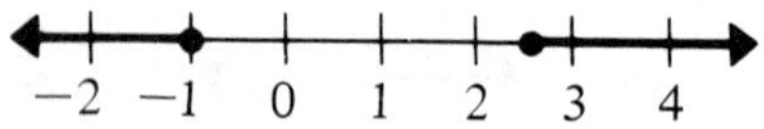

17. $\emptyset$

19. $\{y: y \leqslant 6 \text{ or } y > 10\}$

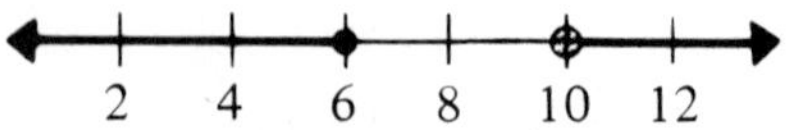

21. $\{x: x > 5\}$

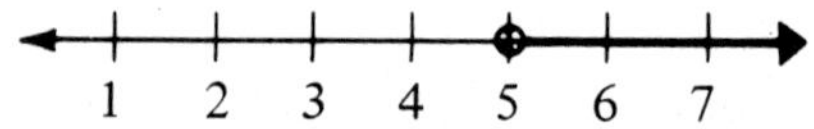

23. $\left\{a: a \leqslant -\frac{1}{3} \text{ or } a > \frac{1}{2}\right\}$ 25. $\{x: -7 < x < 1\}$ 27. $\{x: -2 < x \leqslant 5\}$

29. $\{y: y \leqslant 3 \text{ or } y \geqslant 7\}$ 31. $\mathcal{R}$ 33. $\emptyset$ 35. $\left\{y: -\frac{31}{2} \leqslant y < -\frac{1}{2}\right\}$

37. $\{x: 3 \leqslant x < 15\}$ 39. $\{x: -7 < x \leqslant -2\}$

Pages 31–33 1. Let x = the number; $x + 7 = 21$ 3. Let x = the number; $6x = 72$ 5. Let x = the first integer; $x + (x + 2) = 0$ 7. Let x = Miguel's age now; $x + 4 = 19$ 9. Let q = the number of quarters in the bank; $0.25q = 20.75$ 11. Let x = the number; $3(x + 5) > 12$ 13. Let x = the smaller integer; $x + (x + 1) - 1 \geqslant 2x$ 15. Let x = the number; $8x - 5 \leqslant 32$ 17. Let x = the cost per liter in dollars; $45x \geqslant 15.60$ 19. Let p = the number of pennies; then $3p$ = the number of nickels; $1p + 5(3p) \geqslant 192$. 21. Let b = Bill's age; then $2b$ = Nancy's age; $2b - b \geqslant 32$. 23. Let x = the length of the shorter piece in centimeters; then $2x + 4$ = the length of the longer piece in centimeters; $x + (2x + 4) = 421$. 25. Let x = the number of minutes it takes Sidney to walk to school; then $x + 10$ = the number of minutes it takes Sidney to walk home; $x + (x + 10) < 45$. 27. Let w = the width of each figure; then $w + 6$ = the length of the rectangle; $4w + (2w + 2(w + 6)) = 76$. 29. Let f = Tim's father's age; then $\frac{1}{3}f$ = Tim's age; $f - \frac{1}{3}f = 24$. 31. Let x = the first integer; then $x + 2$ = the second consecutive even integer, and $x + 4$ = the third such integer; $x + (x + 2) + (x + 4) \geqslant 156$. 33. Let x = the first integer; then $x + 2$ = the second consecutive even integer, $x + 4$ = the third such integer, and $x + 6$ = the fourth such integer; $x + (x + 2) + (x + 4) = 2(x + 6) + 12$.

35. Let d = the number of dimes Kay has; then $\frac{1}{2}d$ = the number of nickels, and $\frac{1}{2}d - 15$ = the number of quarters; $10d + 5\left(\frac{1}{2}d\right) + 25\left(\frac{1}{2}d - 15\right) = 11$.

Pages 36–38 1. rate = 56 3. rate = $\frac{x}{5}$ 5. time = $5x$ 7. −37, −35, and −33 9. 150 cm 11. at least 4001 13. 11 15. 36 17. The width is at least 4.5 cm. 19. −10, 0, 10, and 20 21. 2.5 L of 10% alcohol solution and 7.5 L of 30% alcohol solution 23. The amount of gold in the resulting alloy is less than 18 g. 25. 2.5 h 27. 50 km/h 29. 18 31. at least 68.75 L and at most 125 L

Pages 40–41 • Chapter Review 1. false; $Z \cup W = \{\ldots, -3, -2, -1, 0, 1, 2, \ldots\}$
2. true 3. false; the irrational numbers (which are elements of $\mathcal{R}$) are not elements of Q.

4. a. 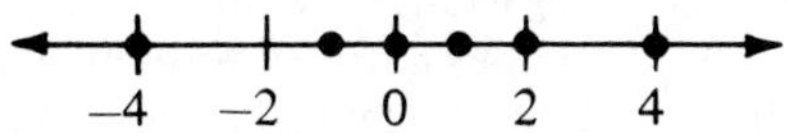 b. 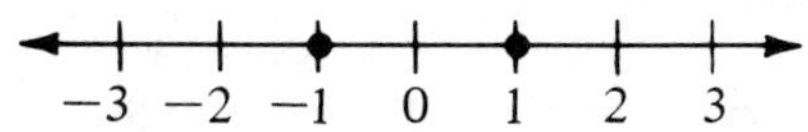

5. a. 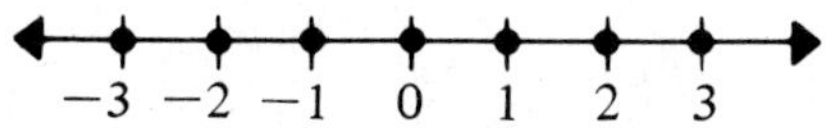 b. 

6. $t \cdot 1$ 7. $\left(t \cdot \frac{1}{x}\right)x$ 8. $-z$ 9. $2r - 2$ 10. $3br + 2bw + b$ 11. 3
12. −3 13. 25 14. $-4x$ 15. $12j - k$ 16. −3 17. 4 18. 1
19. 5 20. $\{a: a \geqslant -12\}$ 21. $\{t: t > 4\}$ 22. $\{m: m \leqslant -2\}$ 23. $\{n: n < 4\}$
24. $\{t: -4 < t < -1\}$ 25. $\{b: b \leqslant 2 \text{ or } b > 5\}$

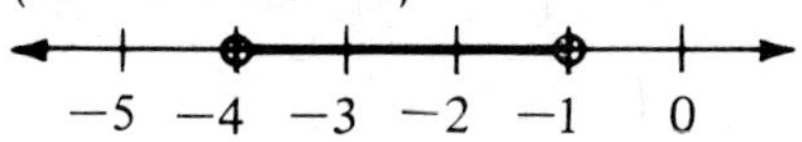 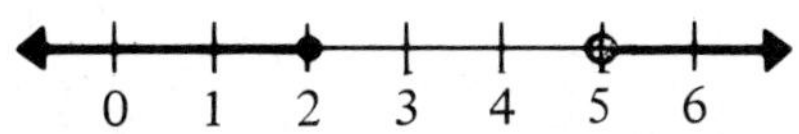

26. $\left\{x: \frac{3}{2} \leqslant x \leqslant 4\right\}$ 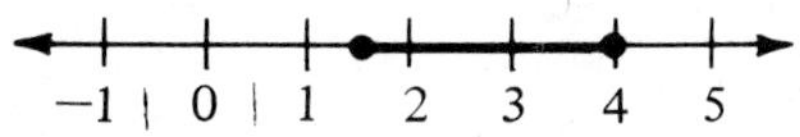

27. Let x = the number; $\frac{x}{3} - 4 = 1$ 28. Let x = Donald's age now; $x + 14 \geqslant 3x$
29. Let d = the number of dimes; then $26 - d$ = the number of quarters; $10d + 25(26 - d) \leqslant 395$. 30. 32, 34, and 36 31. 45 g of 18% copper alloy and 105 g of 8% copper alloy 32. 300 m/min

Pages 44–45 • Mixed Review
1. {3} 3. $\left\{-\frac{11}{4}\right\}$ 5. {−2} 7. $\left\{x: -\frac{15}{2} \leqslant x < 2\right\}$
9. a. 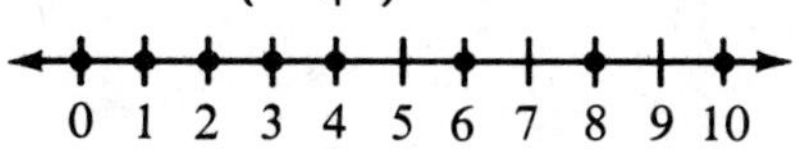 b. 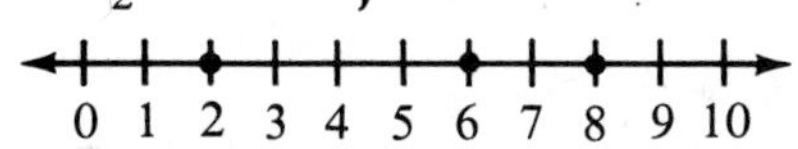

11. −4 13. $4x - 3y$ 15. Assoc. Prop. of Add. 17. Answers will vary.
19. Let x = the larger number; then $47 - x$ = the smaller number; $2x + 3(47 - x) = 106$; 12, 35
21. Let x = the second number; then $3x$ = the first number and $3x - 4$ = the third number; $3x + x + (3x - 4) = 108$; 48, 16, 44 23. {2, 4, 6, 8, 10}

Pages 51–52 1. All squares are rhombuses. 3. No regular pyramid has a square base. 5. All trapezoids have two opposite sides that are not parallel. 7. Some prime numbers are odd. 9. All negative numbers have negative reciprocals. 11. Some numbers have positive squares. 13–17. Forms of statements may vary. Examples are given.
13. All politicians are not dishonest; *or* All politicians are honest. 15. No student will not pass this course; *or* No student will fail this course. 17. All rhombuses do not fail to have perpendicular diagonals; *or* All rhombuses have perpendicular diagonals. 19–25. Forms of negations may vary. Examples are given. 19. 5 is a multiple of 3. 21. Some tall men are fat. 23. Some of my cousins are not cheerful. 25. Some physicists are philosophers.

27–31. Forms of negations may vary. Examples are given. 27. Some integers are irrational. The given statement is true; the negation is false. 29. Some irrational numbers have repeating decimals. The given statement is true; the negation is false. 31. No positive integer has an odd number of divisors. The given statement is true; the negation is false. 33. not contradictory; both are true when n is 6. 35. not contradictory; both are true when $p = 5$; both are false when p is not prime and not divisible by 5. 37. not contradictory; both are false when $\angle A = 90^\circ$. 39. a. even b. odd

Pages 55–56 1. d 3. e 5. f 7–11. Forms of negations may vary. Examples are given. 7. Harry is not hairy or Leroy is short. 9. Leo does not play football and Mimi does not play football. 11. Snow is not white and the Empire State Building is not tall. 13. a. $t \wedge \sim s$ b. F 15. a. $\sim t \wedge r$ b. T 17. a. $\sim r \vee \sim s$ b. T 19–25. Translations may vary. 19. It is March and the wind is blowing. 21. The wind is blowing or the sky is clear. 23. It is not true that the wind is blowing and the sky is clear; *or* The wind is not blowing or the sky is not clear. 25. It is not true that it is March or the wind is not blowing; *or* It is not March and the wind is blowing. 27. a. $p \wedge q$ b. F 29. a. $p \vee q$ b. T 31. a. $\sim(p \vee q)$ b. F 33. a. $p \vee \sim q$ b. F 35–41. Negations may vary. Examples are given. 35. Six is not prime or six is not even. 37. Six is not prime and six is not even. 39. Six is prime or even. 41. Six is not prime and six is even.

43.

p	q	$p \wedge q$	$\sim(p \wedge q)$	$\sim p$	$\sim q$	$\sim p \vee \sim q$
T	T	T	F	F	F	F
T	F	F	T	F	T	T
F	T	F	T	T	F	T
F	F	F	T	T	T	T

p	q	$p \vee q$	$\sim(p \vee q)$	$\sim p$	$\sim q$	$\sim p \wedge \sim q$
T	T	T	F	F	F	F
T	F	T	F	F	T	F
F	T	T	F	T	F	F
F	F	F	T	T	T	T

45. F 47. T 49. T

Pages 60–61 1. If two triangles are congruent, then they have the same area. 3. If x is an integer, then x is rational. 5. If it snows, then it is cold. 7. If Fido wags his tail, then he is happy. 9. If two lines intersect, then they are coplanar. 11. You turn down the thermostat and you don't use less heating oil. 13. I do my homework and I can't go to the movies. 15. Fish can fly and birds don't swim. 17–23. Translations may vary. Examples are given. 17. If it is snowing today, then it is cold today. 19. It is snowing today if and only if it is going to rain tomorrow. 21. If it is not cold today, then it is not snowing today. 23. If it is cold and not snowing today, then it is going to rain tomorrow. 25. a. $y > 3$ if and only if $3 < y$. b. T 27. a. Two rectangles have equal areas if and only if they have equal perimeters. b. F 29. a. x is even if and only if $2x$ is even. b. F 31. a. Line l is parallel to line m if and only if line m is parallel to line l. b. T 33. 4 is not divisible by 2 or 4 is divisible by 6. 35. $\sqrt{2}$ is rational or $(\sqrt{2})^2$ is irrational. 37. 4 is divisible by 2 and 4 is not divisible by 6. 39. $\sqrt{2}$ is irrational and $(\sqrt{2})^2$ is rational. 41. F 43. T 45. F

Pages 65-66 1-9. Forms of statements may vary. Examples are given. 1. For all nonzero x, $\frac{1}{x} > 0$; F 3. For all x, $x \leqslant 0$; F 5. There exists an x such that $x \leqslant 0$; T 7. For all x, $x \leqslant 0$ or $x > 0$; T 9. There exists a nonzero x such that $x \leqslant 0$ and $\frac{1}{x} > 0$; F 11. $\forall_x\ 2x \neq x + 1$; F 13. $\exists_x\ x + 0 \neq x$; F 15. $\exists_y\ y^2 = y$; T 17. $\exists_y\ 3y + 2 \neq 7$; T 19. a. $\exists_x\ x^2 < x$ b. T 21. a. $\forall_x\ x > 3x$ b. F 23. a. $\forall_x\ x > -x$ b. F 25. $\forall_x\ x^2 \geqslant x$ 27. $\exists_x\ x \leqslant 3x$ 29. $\exists_x\ (x \in Z \wedge 2x < 0)$ 31-33. Forms of statements may vary. Examples are given. 31. a. $\forall_t\ (P(t) \vee Q(t))$ b. F 33. a. $\exists_t\ (\sim P(t) \wedge \sim Q(t))$ b. T 35-39. Translations may vary. Examples are given. 35. There exists an x such that x is a multiple of 4 and x is not a multiple of 2; F 37. For every x, if x is a multiple of 2, then x is a multiple of 4; F 39. There exists an x such that if x is a multiple of 4, then x is a multiple of 2; T 41-43. Forms of negations may vary. Examples are given. 41. All cats have long hair or all cats do not; the original statement is true. 43. No integer is prime or some triangle is acute; the negation is true.

Page 68 a. $q \to p$ b. $\sim p \to \sim q$ c. $\sim q \to \sim p$ 3. a. $\sim p \to q$ b. $\sim q \to p$ c. $p \to \sim q$ 5. a. $\sim p \to \sim q$ b. $q \to p$ c. $p \to q$ 7-29. Forms of statements may vary. Examples are given. 7. a. If $ab = 0$, then $a = 0$. (F) b. If $a \neq 0$, then $ab \neq 0$. (F) c. If $ab \neq 0$, then $a \neq 0$. (T) 9. a. y is rational if y is an integer. (T) b. y is not an integer if y is irrational. (T) c. y is irrational if y is not an integer. (F) 11. a. If $x + 3 \neq 5$, then $x \neq 2$. (T) b. If $x = 2$, then $x + 3 = 5$. (T) c. If $x + 3 = 5$, then $x = 2$. (T) 13. a. $a + x = a + y$ implies $x = y$. (T) b. $x \neq y$ implies $a + x \neq a + y$. (T) c. $a + x \neq a + y$ implies $x \neq y$. (T) 15. a. $x > 3$ only if $x \neq 3$. (T) b. $x = 3$ only if $x \leqslant 3$. (T) c. $x \leqslant 3$ only if $x = 3$. (F) 17. For n to be even, it is sufficient for n^2 to be even. (T) b. For n^2 to be odd, it is sufficient for n to be odd. (T) c. For n to be odd, it is sufficient for n^2 to be odd. (T) 19. a. Numbers that are divisible by 2 are even. (T) b. Odd numbers are not divisible by 2. (T) c. Numbers that are not divisible by 2 are odd. (T) 21. a. Being a citizen of the United States is a sufficient condition for being born in New York. (F) b. Not being born in New York is a sufficient condition for not being a citizen of the United States. (F) c. Not being a citizen of the United States is a sufficient condition for not being born in New York. (F) 23. a. If $ab = 0$, then $a = 0$ or $b = 0$. (T) b. If $a \neq 0$ and $b \neq 0$, then $ab \neq 0$. (T) c. If $ab \neq 0$, then $a \neq 0$ and $b \neq 0$. (T) 25. a. If at least one of two numbers is divisible by 6, then their product is divisible by 6. (T) b. If the product of two numbers is not divisible by 6, then neither number is divisible by 6. (T) c. If neither of two numbers is divisible by 6, then their product is not divisible by 6. (F) 27. The given statement is equivalent to "All irrational numbers have nonterminating decimals." a. All numbers that have nonterminating decimals are irrational, or no numbers that have nonterminating decimals are rational. (F) b. All rational numbers have terminating decimals, or no rational numbers have nonterminating decimals. (F) c. All numbers that have terminating decimals are rational, or no numbers that have terminating decimals are irrational. (T) 29. The given statement is equivalent either to "If Paris is not the capital of England, then London is the capital of France" or to "If London is not the capital of France, then Paris is the capital of England." In either case, the converse, inverse, and contrapositive can be written as follows: a, b. Paris is not the capital of England or London is not the capital of France. (T) c. Paris is the capital of England or London is the capital of France. (F). 31. If $ab = 0$ and $b = 0$, then $a \neq 0$ (F); if $b = 0$ and $a \neq 0$, then $ab = 0$ (T).

Pages 73-75 1. Principle of Substitution 3. Fallacy of the Converse 4. Fallacy of Disjunction 5. Law of Contrapositive Inference 7. a. Ex. 5 b. Ex. 4 c. No; logic is concerned primarily with whether or not an argument is valid. If the premises of a *valid* argument are known to be true, then the conclusion must be true. 9. $\sqrt{2}$ is not a terminating or repeating decimal, or $\sqrt{2}$ is not a terminating decimal and $\sqrt{2}$ is not a repeating decimal; Law of Contrapositive Inference 11. Hamburgers are delicious; Law of Detachment 13. None; the argument has the form of the Fallacy of the Converse. 15. None; the argument has the form of the Fallacy of Disjunction. 17. x^2 is even; Law of Contrapositive Inference

19.

p	q	$p \vee q$	$p \to (p \vee q)$
T	T	T	T
T	F	T	T
F	T	T	T
F	F	F	T

21.

p	q	$p \to q$	$\sim q$	$(p \to q) \wedge \sim q$	$\sim p$	$[(p \to q) \wedge \sim q] \to \sim p$
T	T	T	F	F	F	T
T	F	F	T	F	F	T
F	T	T	F	F	T	T
F	F	T	T	T	T	T

23. C: $\sim(p \wedge q)$ 25. C: $r \wedge s$ 27. C: r 29. C: $p \vee q$; Law of Disjunctive Inference 31. $\sim r \to \sim q \vee \sim r$ or $\sim r \to \sim(q \wedge r)$ (Law of Addition and DeMorgan's Law); Conclusion: p (Law of Disjunctive Inference) 33. Conclusion 1: $\sim p$ (Law of Disjunctive Inference); Conclusion 2: q (Law of Disjunctive Inference)

Pages 79–81 1–15. Forms of answers may vary. Examples are given. 1. $p \to r$ 3. $r \to \sim t$ 5. If I pass this course, I will get sick. 7. Everyone who likes logic admires Julius Caesar. 9. If you don't let me carry your books, then I'm quite sure to fail the quiz. 11. If it's windy in Walla Walla, then it doesn't rain in Spain. 13. No lobsters expect impossibilities. 15. All bankers shun hyenas. 17–23. Reasoning chains may vary. Examples are given. 17. (1) $p \to q$ (Given); (2) $q \to r$ (Given); (3) $p \to r$ (Law of the Syllogism); (4) p (Given); (5) r (Law of Detachment) 19. (1) $p \vee \sim q$ (Given); (2) $\sim p \to \sim q$ ($p \vee \sim q$ is equivalent to $\sim p \to \sim q$.); (3) $q \to p$ (Law of the Contrapositive); (4) $r \to \sim p$ (Given); (5) $p \to \sim r$ (Law of the Contrapositive); (6) $q \to \sim r$ (Law of the Syllogism) 21. a. P_1: $\sim p \to \sim g$, P_2: $f \to \sim r$, P_3: $\sim r \to \sim p$ b. (1) $f \to \sim r$ (Given); (2) $\sim r \to \sim p$ (Given); (3) $f \to \sim p$ (Law of the Syllogism); (4) $\sim p \to \sim g$ (Given); (5) $f \to \sim g$ (Law of the Syllogism) 23. a. P_1: $r \to f$, P_2: r, P_3: $f \to \sim w$, P_4: $w \vee d$ b. (1) $r \to f$ (Given); (2) r (Given); (3) f (Law of Detachment); (4) $f \to \sim w$ (Given); (5) $\sim w$ (Law of Detachment); (6) $w \vee d$ (Given); (7) d (Law of Disjunctive Inference) 25. p': There exist two positive integers m and n such that $n = 2m^2$; p: There exist two positive integers m and n such that $n = m\sqrt{2}$. 26. b. Answers may vary. c. All chords joining six points of a circle divide it into 32 regions. d. 30 or 31 regions, depending on the locations of the points. e. The final conclusion should be that one can*not* conclude "and so on *ad infinitum*," since the fifth statement in the sequence is false.

Pages 85–87 1. a. Given b. Def. of even c. Mult. Prop. of Equality d. Closure for Z under Mult. e. xy is even. f. Def. of Even 3. a. xy is odd. b. Given c. $x = 2k + 1$ for some $k \in Z$; $y = 2m + 1$ for some $m \in Z$. d. Def. of Odd e. Mult. Prop. of Equality f. Dist. Prop. and Subst. Prop. g. Dist. Prop. h. $2km + k + m$ is an integer. i. Closure for Z under Mult., Closure for Z under Add. j. xy is odd. k. Def. of Odd 5. a. x is odd and

y is even. b. xy is even. c. x is odd and y is even. d. Given e. $x = 2k + 1$ for some $k \in Z$. f. Def. of Odd g. $y = 2m$ for some $m \in Z$. h. Def. of Even i. Mult. Prop. of Equality j. Comm. Prop. of Mult. k. $m(2k + 1)$ is an integer. l. xy is even. m. Def. of Even 6. a. $m\angle 1 = m\angle 3$ b. $\angle 1$ and $\angle 2$ are supp.; $\angle 3$ and $\angle 2$ are supp. c. Given d. $m\angle 1 + m\angle 2 = 180$; $m\angle 3 + m\angle 2 = 180$ e. Def. of supp. $\angle$s f. $m\angle 1 + m\angle 2 = m\angle 3 + m\angle 2$ g. Subst. Prop. h. $m\angle 1 = m\angle 3$ i. Subtr. Prop. of Equality
7. Prove the contrapositive: If x is rational, then x^2 is rational. GIVEN: x is rational. PROVE: x^2 is rational. (1) x is rational. (Given); (2) $x \cdot x$ is rational. (The product of two rational numbers is rational.); (3) x^2 is rational. (Subst. Prop.)

9. GIVEN: $\angle 1$ and $\angle 2$ are supp.;
$\angle 3$ and $\angle 4$ are supp.;
$m\angle 1 = m\angle 3$
PROVE: $m\angle 2 = m\angle 4$

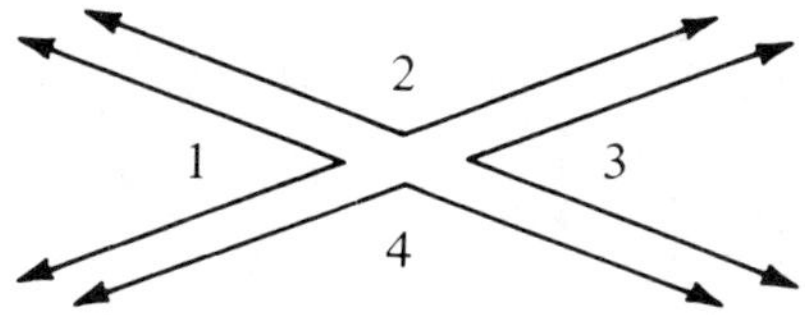

(1) $\angle 1$ and $\angle 2$ are supp.; $\angle 3$ and $\angle 4$ are supp. (Given); (2) $m\angle 1 + m\angle 2 = 180$; $m\angle 3 + m\angle 4 = 180$ (Def. of supp. $\angle$s); (3) $m\angle 1 + m\angle 2 = m\angle 3 + m\angle 4$ (Subst. Prop.); (4) $m\angle 1 = m\angle 3$ (Given); (5) $m\angle 2 = m\angle 4$ (Subtr. Prop. of Equality)
11. GIVEN: x is odd. PROVE: x^2 is odd. (1) x is odd. (Given); (2) $x = 2k + 1$ for some $k \in Z$. (Def. of Odd); (3) $x^2 = (2k + 1)(2k + 1)$ (Mult. Prop. of Equality); (4) $x^2 = 4k^2 + 4k + 1$ (Dist. Prop. and Subst. Prop.); (5) $x^2 = 2(2k^2 + 2k) + 1$ (Dist. Prop.); (6) $(2k^2 + 2k)$ is an integer. (Closure for Z under Add. and Mult.); (7) x^2 is odd. (Def. of Odd)
13. a. If the sum of two numbers is even, then both numbers are odd. b. Answers may vary. For example, $6 = 2 + 4$ 15. Prove the contrapositive: If an integer is odd, then the square of the integer is odd. GIVEN: n is odd. PROVE: n^2 is odd. (1) n is odd. (Given); (2) n^2 is odd. (Theorem in Exercise 11) 17. If $3x^2 = 6x$, then $x = 2$. Yes; the Division Prop. of Equality is used in the second step to justify dividing each side of the original equation by x. However, the property only applies when $x \neq 0$, and 0 is one of the two solutions of the original equation. 19. The premise of each argument is an equation (in part (b), two equations) whose left and right sides are equivalent quantities expressed in *different* units of measure. Before the properties of equality can be applied to such an equation, the quantities must be expressed in the *same* unit of measure. Each argument is invalid because the properties of equality are improperly applied.

Page 90 1. a. Mult. Prop. of Equality b. Subst. Prop. and Mult. Prop. of 0 c. $-1 \neq 0$
2. a. Hypothesis and assumption b. Mult. Prop. of Ineq. c. Subst. Prop. and Mult. Prop. of 0 d. $-1 \not> 0$ 3. a. Hypothesis and assumption b. Mult. Prop. of Ineq. c. Subst. Prop. and Mult. Prop. of 0 d. $-1 \not> 0$
5. GIVEN: $x^2 > 0$ PROVE: $x \neq 0$ Temporarily assume that $x = 0$. (1) $x^2 > 0$ and $x = 0$. (Hypothesis and assumption); (2) $0^2 > 0$ (Subst. Prop.); (3) $0 > 0$ (Mult. Prop. of 0); (4) Contradiction ($0 \not> 0$) Therefore, the temporary assumption that $x = 0$ is false. It follows that $x \neq 0$.
7. Recall that the counting numbers are the positive integers. Temporarily assume that there is a largest counting number C. Let $N = C + 1$. N is an integer since the set of integers is closed under addition, and $N > 0$ since $C + 1 > C > 0$. Thus, N is a counting number that is greater than C. But this contradicts the assumption that C is the largest counting number. Therefore, the temporary assumption must be false. It follows that there is no largest counting number.
9. The conditional $(p \wedge q) \rightarrow r$ is equivalent to the disjunction $\sim(p \wedge q) \vee r$, which is equivalent to the disjunction $\sim p \vee \sim q \vee r$ (DeMorgan's Laws). The conditional $(p \wedge \sim r) \rightarrow \sim q$ is equivalent to the disjunction $\sim(p \wedge \sim r) \vee \sim q$, which is equivalent to the disjunction $\sim p \vee r \vee \sim q$ (DeMorgan's Laws and the Law of Double Negation). Since $[(p \wedge q) \rightarrow r]$ and

$[(p \land \sim r) \rightarrow \sim q]$ are each equivalent to the disjunction of the statements $\sim p$, $\sim q$, and r, it follows that $[(p \land q) \rightarrow r]$ is equivalent to $[(p \land \sim r) \rightarrow \sim q]$.
10. Prove the partial contrapositive: If n is even and m is odd, then $m + n$ is odd. GIVEN: n is even and m is odd. PROVE: $m + n$ is odd. (1) n is even and m is odd. (Given); (2) $n = 2k$ for some $k \in Z$. (Def. of Even); (3) $m = 2x + 1$ for some $x \in Z$. (Def. of Odd); (4) $m + n = (2x + 1) + 2k$ (Add. Prop. of Equality); (5) $m + n = 2(x + k) + 1$ (Comm. and Associative Props. of Add., Dist. Prop.); (6) $x + k$ is an integer. (Closure for Z under Add.); (7) $m + n$ is odd. (Def. of Odd)

Pages 91–92 · Chapter Review 1–6. Forms of sentences may vary. Examples are given. 1. No perfect square is a prime number. 2. All hexagons do not have five sides. 3. Some clouds do not have silver linings. 4. All people should throw stones. 5. Fair is not foul or foul is not fair. 6. You don't help pay expenses and you can go. 7. a. $\sim p \lor \sim q$ b. T 8. a. $p \land \sim q$ b. F 9. If $x = 4$, then $3x = 12$; $x \neq 4$ or $3x = 12$. 10. If $x < 0$, then $x^2 > 0$; $x \geqslant 0$ or $x^2 > 0$. 11. Three hundred fifty is divisible by ten and the units' digit is not zero. 12. Quadrilateral $ABCD$ is a parallelogram if and only if it has congruent diagonals; F 13. $\exists_x\, x = x + 2$; F 14. $\forall_y\, \frac{1}{y} \neq 0$; T 15. $\exists_x\, x^3 < 0$ 16. $\forall_x\, -1 \cdot x = -x$ 17. $\sim q \rightarrow p$, $\sim p \rightarrow q$, $q \rightarrow \sim p$ 18. $p \rightarrow \sim q$, $q \rightarrow \sim p$, $\sim p \rightarrow q$ 19. Forms of sentences may vary. One example of each is given. The little dog laughs only if the cow jumps over the moon; the cow does not jump over the moon only if the little dog does not laugh; the little dog does not laugh only if the cow does not jump over the moon. 20. No conclusion; the argument has the form of the Fallacy of the Inverse. 21. That dog has fleas; Law of Disjunctive Inference 22. Charlotte is not a cat. 23. If the swallows are in Capistrano, then there is no snow on the ground; or if there is snow on the ground, then the swallows are not in Capistrano. 24. a. $x = y$ b. Given c. Add. Prop. of Equality d. $x - y = 0$ e. Inverse Prop. of Add. 25. GIVEN: $x = y + 1$ PROVE: $x \neq y$ Temporarily assume that $x = y$. (1) $x = y + 1$ and $x = y$. (Hypothesis and assumption); (2) $y = y + 1$ (Subst. Prop.); (3) $0 = 1$ (Subtr. Prop. of Equality); (4) Contradiction $(0 \neq 1)$ Therefore, the temporary assumption that $x = y$ is false. It follows that $x \neq y$.

Pages 96–97 · Mixed Review
1. It is Thursday or the book is interesting. 3. A circle is not round. 5. $\{y: y \leqslant 11\}$ 7. $\{x: 5 < x \leqslant 8\}$ 9. a. $\{-4, -2, 0, 1, 2, 3, 4\}$ b. $\{0, 2\}$ 11. $-t$ 13. No trapezoid is a parallelogram. 15.

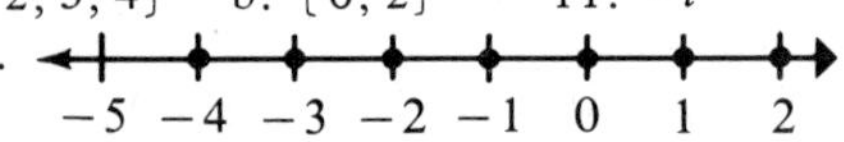

17. Some figs are big. 19. Let x = one number; then $2x - 5$ = the other number; $x + (2x - 5) = 44$ 21. If I work hard, then I will become an A student. 23. a. If x^2 is even, then x is even. (T) b. If x is not even, then x^2 is not even. (T) c. If x^2 is not even, then x is not even. (T) 25.

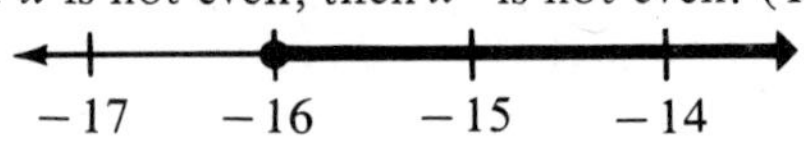

27. Money can buy happiness. 29. Donna's rate of speed is 80 km/h; Vera's is 85 km/h. 31. $-10a - 9b + 6$

Note regarding proofs: In many of the problems requiring a two-column proof, only the key concepts in the development of the proof are given here. It is the responsibility of the student to furnish a complete two-column proof.

Pages 102–103 1. $\overline{CD}$ 3. $\overleftrightarrow{CE}$ (or $\overleftrightarrow{CD}$ and so on) 5. $\overline{CE}$ 7. $\overrightarrow{CX}$ 9. 18

11. 4 13. 5 15. 7 17. 7 19. (1) Prop. of Betweenness (2) Given (3) Subst. Prop. (4) Given (5) Subtr. Prop. of Equality 21. R is between P and Q. 23. P is between Q and R. 25. R is between P and Q.

Pages 106–107 1. 30 3. 18 5. $R + 3$ 7. $QR = RT$, so R is the midpoint of $\overline{QT}$. 9. $PQ = QR$, so Q is the midpoint of $\overline{PR}$. 11. 21 13. 70 15. (1) Given (2) Def. of bisector (3) Def. of midpoint (4) Prop. of Betweenness (5) Subst. Prop. (6) Given (7) Def. of bisector (8) Def. of midpoint (9) Prop. of Betweenness (10) Subst. Prop. (11) Subst. Prop. 17. (1) $XZ = YZ$; $KZ = LZ$ (Given); (2) $XZ - KZ = YZ - LZ$ (Subtr. Prop. of Equality); (3) $XZ = XK + KZ$; $YZ = YL + LZ$ (Prop. of Betweenness); (4) $XZ - KZ = XK$; $YZ - LZ = YL$ (Subtr. Prop. of Equality); (5) $XK = YL$ (Subst. Prop.) 19. (1) K is the midpoint of $\overline{XZ}$; L is the midpoint of $\overline{YZ}$ (Given); (2) $KZ = \frac{1}{2}XZ$; $LZ = \frac{1}{2}YZ$ (Thm. 3–2); (3) $XZ = YZ$ (Given); (4) $KZ = LZ$ (Subst. Prop.)

Pages 110–112 1. $m\angle AOC = 60$ 3. $m\angle BOC = 17$ 5. $m\angle BOC = 36$; $m\angle AOC = 72$ 7. $m\angle AOB = m\angle BOC = 27$ 9. 18 11. 28 13. 28; 40 15. 6; 8 17. (1) Given (2) Def. of $\angle$ bisector (3) Given (4) Def. of $\angle$ bisector (5) Subst. Prop. (6) Add. Prop. of Equality (7) Angle Add. Post. (8) Angle Add. Post. (9) Subst. Prop. 19. 24 21. (1) $m\angle PXQ = m\angle RXS$ (Given); (2) $m\angle PXQ + m\angle QXR = m\angle RXS + m\angle QXR$ (Add. Prop. of Equality); (3) $m\angle PXQ + m\angle QXR = m\angle PXR$; $m\angle RXS + m\angle QXR = m\angle QXS$ (Angle Add. Post.); (4) $m\angle PXR = m\angle QXS$ (Subst. Prop.) 23. Use the definition of an angle bisector and the Angle Addition Postulate to show that $m\angle AOC = 2m\angle AOB$. 25. Show that $m\angle DOC + m\angle EOD = m\angle COB + m\angle BOA$ and that, consequently, $m\angle EOC = m\angle COA$ by the Angle Addition Postulate.

Pages 116–118 1. 45; 49 and 41 3. 16; 25 and 65 5. 18; 72 and 108 7. 12; 63 and 117 9. $m\angle A = 75$; $m\angle B = 15$ 11. $m\angle A = 116$; $m\angle B = 64$ 13. (1) Given (2) If the ext. sides of 2 adj. ∠s are ⊥, the ∠s are comp. (3) 2 ∠s comp. to the same $\angle$ or to ∠s of = meas. have = meas. 15. Use the definition of ⊥ lines and the definition of right ∠s. 17. Use Postulate 5 and Theorem 3–6. 19. Show that $\angle 1$ and $\angle 2$ are complementary, and that $\angle 3$ and $\angle 4$ are complementary. Then, $m\angle 2 = m\angle 4$ by Theorem 3-6. 21. Show that $m\angle 4 = m\angle 2$ and use the Substitution Property. 23. $m\angle JEK = 90$ and $m\angle 1 + 90 + m\angle 2 = 180$. By the Subtraction Property of Equality, $m\angle 1 + m\angle 2 = 90$. 25. Since $m\angle AOB + m\angle BOC = 180$, then by the Mult. Prop. of Equality, $\frac{1}{2}m\angle AOB + \frac{1}{2}m\angle BOC = 90$. By Theorem 3–4, $m\angle XOB = \frac{1}{2}m\angle AOB$ and $m\angle BOY = \frac{1}{2}m\angle BOC$.

Pages 122–123 1. $\triangle CAX$, $\triangle DAX$; SSS 3. $\triangle CAB$, $\triangle DAB$; ASA 5. $\triangle CAX$, $\triangle DAX$; ASA 7. Use the ASA Postulate. 9. Use the SSS Postulate. 11. Show that $\angle 6 \cong \angle 8$ by Theorem 3–7. Then use the ASA Postulate. 13. Use the SSS Postulate. 15. Use the ASA Postulate. 17. Use the SSS Postulate. 19. Use the ASA Postulate. 21. $\triangle ACX \cong \triangle BCX$; use the SSS Postulate. 23. $\triangle ACX \cong \triangle BCX$; use the ASA Postulate.

Pages 127–128 1. $\triangle ADX \cong \triangle BCX$; HL 3. $\triangle ADC \cong \triangle BCD$; SAS 5. $\triangle ACD \cong \triangle BDC$; SAS 7. Use the SAS Postulate. 9. Use the HL Postulate. 11. Use the ASA Postulate. 13. Use the HL Postulate. 15. Use the SSS Postulate. 17. Use the SAS Postulate. 19. Use the ASA Postulate. 21. Use the ASA Postulate. 23. Use the ASA Postulate. 25. Use the HL Postulate. 27. Show that $\angle 2 \cong \angle 3$ by Theorem 3–6. Then use the ASA Postulate.

Pages 130–132 1. $\triangle ART$, $\triangle ORT$ 3. $\triangle ATY$, $\triangle OTY$; $\triangle ARY$, $\triangle ORY$ 5. $\triangle ART$, $\triangle ORT$; $\triangle ARY$, $\triangle ORY$ 7. Use the ASA Postulate to show that $\triangle APB \cong \triangle DPC$.

9. Use the SSS Postulate to show that $\triangle AXY \cong \triangle BXY$. 11. Use the HL Postulate to show that $\triangle AXY \cong \triangle BXY$. 13. Use the SAS Postulate to show that $\triangle JHG \cong \triangle KHG$. 15. Use the SAS Postulate to show that $\triangle JHG \cong \triangle KHG$. 17. Use the SAS Postulate. 19. Use the SAS Postulate. 21. Use the SAS Postulate. 23. Use the ASA Postulate. 25. Use the SAS Postulate. 27. Show that $\triangle GKY \cong \triangle HKY$ by the ASA Postulate. Then use the SAS Postulate. 29. Show that $\triangle MRS \cong \triangle PRS$ by the SSS Postulate. Then use the SAS Postulate to show that $\triangle MRX \cong \triangle PRX$. 31. Show that $\triangle MSX \cong \triangle PSX$ by the HL Postulate. Then use the SAS Postulate to show that $\triangle MRS \cong \triangle PRS$.

Pages 135–137 1. 40 3. 14 5. 14 7. 9 9. Use Theorem 3–13 to show that $\angle 2 \cong \angle 3$. Then use Theorem 3–12. 11. Use Theorem 3–7 to show that $\angle 2 \cong \angle 3$. 13. Use the ASA Postulate. 15. Use the ASA Postulate to show that $\triangle PTQ \cong \triangle RVQ$. 17. Show that $\overline{LJ} \cong \overline{LK}$ by Theorem 3–14. Then use the ASA Postulate to show that $\triangle LJM \cong \triangle LKN$. 19. Use the ASA Postulate to show that $\triangle ABD \cong \triangle CBD$. 21. Use the HL Postulate to show that $\triangle ADB \cong \triangle CDB$. 23. Use the SAS Postulate. 25. Use the SAS Postulate to show that $\triangle AXC \cong \triangle BXD$. 27. Show that $\angle RSZ \cong \angle WTZ$ by Theorem 3–7. Then use the SAS Postulate to show that $\triangle RZS \cong \triangle WZT$. 29. Use Theorem 3–13 to show that $\angle A \cong \angle B$. Show $\angle DGH \cong \angle DHG$ by Theorem 3–13, and consequently $\angle AGE \cong \angle BHF$. Then use the ASA Postulate to show that $\triangle AGE \cong \triangle BHF$. 31. Use the ASA Postulate to show that $\triangle RPQ \cong \triangle SPQ$. Then show that $\overline{PR} \cong \overline{PS}$. 33. Use the ASA Postulate to show that $\triangle JMN \cong \triangle KNM$. Then show that $\triangle JMZ \cong \triangle KNZ$ by the ASA Postulate.

Pages 140–142 1. $\overline{EG}$ 3. $\angle K$ 5. $6 < AB < 24$ 7. $\frac{13}{4} < AB < \frac{17}{4}$ 9. $3 < AB < 2m + 1$ 11. (1) $m\angle 1 > m\angle A$ (Given); (2) $m\angle 2 = m\angle 1$ (Vert. ∠s are ≅.); (3) $m\angle 2 > m\angle A$ (Subst. Prop.); (4) $AC > BC$ (In a △, the longer side is opp. the greater ∠.) 13. (1) $m\angle C > m\angle A$ (Given); (2) $AB > CB$ (In a △, the longer side is opp. the greater ∠.); (3) $AC > AB$ (Given); (4) $AC > CB$ (Trans. Prop. of Ineq.); (5) $m\angle 2 > m\angle A$ (In a △, the greater ∠ is opp. the longer side.) 15. (1) $m\angle 4 > m\angle M$ (The meas. of an ext. ∠ of a △ is greater than that of either remote int. ∠.); (2) $KL > ML$ (Given); (3) $m\angle M > m\angle 5$ (In a △, the greater ∠ is opp. the longer side.); (4) $m\angle 4 > m\angle 5$ (Trans. Prop. of Ineq.) 17. $EF > EH > FH > GH > FG$ 19. By Theorem 3–15, $m\angle 1 > m\angle X$, and by Theorem 3–13, $m\angle 1 = m\angle Y$. Use this to show that $m\angle Y > m\angle X$. 21. By Theorem 3–13, $m\angle 3 = m\angle 2$, and by Theorem 3–15, $m\angle 3 > m\angle 1$. Use this to show that $m\angle 2 > m\angle 1$. 23. Use Theorem 3–15 and the Transitive Property of Inequality to show that $m\angle Q > m\angle P$. 25. Use Theorem 3–16 to show that $m\angle Q > m\angle P$ and use Theorem 3–15 to show that $m\angle 1 > m\angle Q$. Then use the Transitive Property of Inequality to show that $m\angle 1 > m\angle P$. 27. Use Theorem 3–15 to show that $m\angle 1 > m\angle Y$ and $m\angle 2 > m\angle X$. Then use the Transitive Property of Inequality and the given information to show that $m\angle 1 > m\angle 2$ and, consequently, $m\angle 1 > m\angle X$. 29. a. Use the SAS Postulate to show that $\triangle BXA \cong \triangle CXE$. Then $\angle B \cong \angle 1$ and $m\angle BCD = m\angle 1 + m\angle 2$; so $m\angle BCD > m\angle 1$. b. Show that $\triangle AXB \cong \triangle CXF$ and that $\angle FCX \cong \angle A$. Then $m\angle ACE > m\angle A$; but $\angle ACE$ and $\angle BCD$ are vertical angles.

Pages 143–144 · Chapter Review 1. a. 2 b. none c. 1 2. a. 13 b. 3 3. $3y$ 4. exactly one 5. (1) $RT = RS + ST$ (Prop. of Betweenness); (2) S is the midpoint of $\overline{RT}$. (Given); (3) $RS = ST$ (Def. of midpoint); (4) $ST = SQ$ (Given); (5) $RS = SQ$ (Subst. Prop.); (6) $RT = SQ + ST$ (Subst. Prop.) 6. a. A bisector of an ∠ divides the ∠ into 2 ∠s, each of which has meas. half that of the given ∠. b. Angle Add. Post. 7. 84 8. Vert. ∠s are ≅. 9. If 2 lines form ≅ adj. ∠s, the lines are ⊥. 10. If the ext. sides of 2 adj. ∠s are opp. rays, then the ∠s are supp. 11. 9 12. (1) $\angle 1$ and $\angle 3$ are comp. ∠s; $\angle 4$ and $\angle 2$ are comp. ∠s. (Given); (2) $m\angle 1 + m\angle 3 = 90$;

$m\angle 4 + m\angle 2 = 90$ (Def. of comp. $\angle$s); (3) $m\angle 1 + m\angle 3 = m\angle 4 + m\angle 2$ (Subst. Prop.); (4) $m\angle 2 = m\angle 3$, *or* $\angle 2 \cong \angle 3$ (Vert. $\angle$s are $\cong$.); (5) $m\angle 1 = m\angle 4$, *or* $\angle 1 \cong \angle 4$ (Subtr. Prop. of Equality); (6) $\overline{AC} \cong \overline{DC}$ (Given); (7) $\triangle ABC \cong \triangle DEC$ (ASA Post.) 13. $\overline{GH} \cong \overline{GJ}$ 14. $\angle GIH \cong \angle GIJ$ 15. (1) $\angle KLM$ and $\angle NML$ are rt. $\angle$s. (Given); (2) $\triangle KLM$ and $\triangle NML$ are rt. $\triangle$s. (Def. of rt. $\triangle$); (3) $\overline{KM} \cong \overline{NL}$ (Given); (4) $\overline{LM} \cong \overline{LM}$ (Refl. Prop. of $\cong$); (5) $\triangle KLM \cong \triangle NML$ (HL Post.) 16. Corr. parts of $\cong$ $\triangle$s are $\cong$. 17. $\angle XBC$, $\angle XDC$; $\angle ABX$, $\angle EDX$ 18. $\overline{XA}$, $\overline{XE}$ 19. (1) $\triangle XCB \cong \triangle XCD$ (Given); (2) $\overline{XB} \cong \overline{XD}$; $\angle XBC \cong \angle XDC$ (Corr. parts of $\cong$ $\triangle$s are $\cong$.); (3) $\angle ABX \cong \angle EDX$ (2 $\angle$s supp. to the same $\angle$ or to $\cong$ $\angle$s are $\cong$.); (4) $\overline{AB} \cong \overline{ED}$ (Given); (5) $\triangle XAB \cong \triangle XED$ (SAS Post.) 20. The measure of an ext. $\angle$ of a $\triangle$ is greater than that of either remote int. $\angle$. 21. XDA; A 22. $2 < RS < 12$

Pages 145–148 • Cumulative Review: Chapters 1–3 1. a 2. a 3. a 4. b 5. d 6. d 7. c 8. c 9. a 10. a 11. b 12. a 13. c 14. b 15. b 16. d 17. a 18. d 19. a 20. d 21. c 22. b 23. d 24. d 25. c 26. c 27. b 28. a 29. a 30. d 31. a 32. d 33. b 34. c 35. d 36. b 37. a. P_1: $v \rightarrow \sim t$; P_2: $b \rightarrow t$; P_3: v b. $v \rightarrow \sim t$ and v, so by the Law of Detachment, $\sim t$. $b \rightarrow t$ and $\sim t$, so by the Law of Contrapositive Inference, $\sim b$. 38. (1) $\overline{AC}$ bisects $\overline{BD}$. (Given); (2) $BX = DX$, *or* $\overline{BX} \cong \overline{DX}$ (Def. of bisector); (3) $\angle ADB \cong \angle CBD$ (Given); (4) $\angle AXD \cong \angle CXB$ (Vert. $\angle$s are $\cong$.); (5) $\triangle AXD \cong \triangle CXB$ (ASA Post.); (6) $\overline{AD} \cong \overline{CB}$ (Corr. parts of $\cong$ $\triangle$s are $\cong$.) 39. (1) $\overline{AD} \cong \overline{BC}$; $\overline{AC} \cong \overline{BD}$ (Given); (2) $\overline{DC} \cong \overline{DC}$ (Refl. Prop. of $\cong$); (3) $\triangle ADC \cong \triangle BCD$ (SSS Post.); (4) $\angle ADC \cong \angle BCD$ (Corr. parts of $\cong$ $\triangle$s are $\cong$.) 40. (1) $\triangle JKL \cong \triangle MNO$ (Given); (2) $\overline{JK} \cong \overline{MN}$, $\angle K \cong \angle N$; $\overline{KL} \cong \overline{NO}$, *or* $KL = NO$ (Corr. parts of $\cong$ $\triangle$s are $\cong$.); (3) $\overline{JY}$ and $\overline{MZ}$ are medians. (Given); (4) $KY = YL$; $NZ = ZO$ (Def. of median); (5) $KY + YL = KL$; $NZ + ZO = NO$ (Prop. of Betweenness); (6) $2KY = KL$; $2NZ = NO$; $2KY = 2NZ$ (Subst. Prop.); (7) $KY = NZ$, *or* $\overline{KY} \cong \overline{NZ}$ (Div. Prop. of Equality); (8) $\triangle JKY \cong \triangle MNZ$ (SAS Post.); (9) $\overline{JY} \cong \overline{MZ}$ (Corr. parts of $\cong$ $\triangle$s are $\cong$.)

Pages 150–151 • Mixed Review
1. $x = 4$; $m\angle BOC = 116$ 3. $\{6\}$ 5. $\{y: y < -7\}$ 7. 4.5 9. $\frac{10}{3}$ 11. A home does not have aluminum siding or it doesn't need painting. 13. (1) $\overline{AC} \cong \overline{BD}$; $\angle A \cong \angle B$; $\angle EDA \cong \angle FCB$ (Given); (2) $\overline{CD} \cong \overline{CD}$ (Refl. Prop. of $\cong$); (3) $AC + CD = BD + CD$ (Add. Prop. of Equality); (4) $AD = AC + CD$; $BC = BD + CD$ (Prop. of Betweenness); (5) $AD = BC$ (Subst. Prop.); (6) $\triangle ADE \cong \triangle BCF$ (ASA) 15. (1) $m\angle 1 = m\angle 2$ (Given); (2) $\angle 1$ and $\angle 3$ are supp. (If the ext. sides of 2 adj. $\angle$s are opp. rays, the $\angle$s are supp.); (3) $m\angle 1 + m\angle 3 = 180$ (Def. of Supp. $\angle$s); (4) $m\angle 2 + m\angle 3 = 180$ (Subst. Prop.); (5) $\angle 2$ and $\angle 3$ are supplementary (Def. of supp. $\angle$s) 17. $\forall_y\ y \neq -y$; F 19. $5\frac{1}{11}$ h

Pages 157–158 1. yes; $\overleftrightarrow{AC}$, $\overleftrightarrow{BD}$; if 2 lines are cut by a trans. so that alt. int. $\angle$s are $\cong$, the lines are $\parallel$. 3. no 5. yes; $\overleftrightarrow{AB}$, $\overleftrightarrow{CD}$; if 2 lines are cut by a trans. so that the alt. int. $\angle$s are $\cong$, the lines are $\parallel$. 7. yes; $\overleftrightarrow{AB}$, $\overleftrightarrow{CD}$; vert. $\angle$s are $\cong$; if 2 lines are cut by a trans. so that corr. $\angle$s are $\cong$, the lines are $\parallel$. 9. yes; $\overleftrightarrow{AB}$, $\overleftrightarrow{CD}$; if 2 lines are cut by a trans. so that int. $\angle$s on the same side of the trans. are supp., then the lines are $\parallel$. 11. (1) $\angle 2 \cong \angle 4$ (Vert. $\angle$s are $\cong$.); (2) $\angle 1 \cong \angle 4$ (Given); (3) $\angle 1 \cong \angle 2$ (Trans. Prop. of $\cong$); (4) $l \parallel m$ (If 2 lines are cut by a trans. so that corr. $\angle$s are $\cong$, the lines are $\parallel$.) 13. (1) $\overrightarrow{QT}$ bisects $\angle PQS$. (Given); (2) $\angle 1 \cong \angle 2$ (Def. of $\angle$ bisector); (3) $\angle 1 \cong \angle 4$ (Given); (4) $\angle 2 \cong \angle 4$ (Trans. Prop. of $\cong$); (5) $\overleftrightarrow{QT} \parallel \overleftrightarrow{RS}$ (If 2 lines are cut by a trans. so that alt. int. $\angle$s are $\cong$, the lines are $\parallel$.)

15. Show that $\angle 1 \cong \angle 3$. 17. Show that $\angle 5 \cong \angle 7$. 19. Show that $\angle 1 \cong \angle 2$. 21. Show that $\angle 2 \cong \angle 3$. 23. Show that $\angle 1 \cong \angle 4$. 25. Use the SAS Postulate to show that $\triangle BEA \cong \triangle CED$. Then show that $\angle B \cong \angle C$ (or $\angle A \cong \angle D$). 27. Show that $\overrightarrow{KR} \perp \overrightarrow{PZ}$ and $\overrightarrow{LX} \perp \overrightarrow{PZ}$. Then use Corollary 1 to Theorem 4–1. 29. Show that $\angle 2 \cong \angle 3$. 31. Use the SSS Postulate to show that $\triangle ACB \cong \triangle DBC$. Then show that $\angle 1 \cong \angle 4$.

Pages 162–164 1. 57 3. x 5. 62 7. $s \perp t$; in a plane, a line $\perp$ to one of two $\parallel$ lines is also $\perp$ to the other line. 9. 9 11. 46 13. 120 15. (1) $l \parallel m$ (Given); (2) $\angle 2 \cong \angle 3$ (If 2 lines are $\parallel$, corr. ∠s are $\cong$.); (3) $\angle 1 \cong \angle 2$ (Given); (4) $\angle 1 \cong \angle 3$ (Trans. Prop. of $\cong$) 17. (1) $l \parallel m$ (Given); (2) $\angle 1 \cong \angle 5$ (If 2 lines are $\parallel$, alt. int. ∠s are $\cong$.); (3) $\angle 4 \cong \angle 5$ (Given); (4) $\angle 1 \cong \angle 4$ (Trans. Prop. of $\cong$); (5) $\angle 3 \cong \angle 4$ (Vert. ∠s are $\cong$.); (6) $\angle 1 \cong \angle 3$ (Trans. Prop. of $\cong$) 19. Show that $l \parallel m$. 21. Show that $\overline{XY} \parallel \overline{ST}$. 23. Show that $\overleftrightarrow{KL} \perp \overrightarrow{LM}$. 25. Show that $\angle ABC \cong \angle 3$. 27. GIVEN: $l \parallel n$ PROVE: $\angle 1 \cong \angle 4$ *Proof:* (1) $l \parallel n$ (Given); (2) $\angle 4 \cong \angle 3$ (If 2 lines are $\parallel$, alt. int. ∠s are $\cong$.); (3) $\angle 1 \cong \angle 3$ (Vert. ∠s are $\cong$.); (4) $\angle 1 \cong \angle 4$ (Trans. Prop. of $\cong$) 29. Use the Corollary to Theorem 4–2 to show that $\overline{WZ} \parallel \overline{XY}$. Then use the Corollary to Theorem 4–4. 31. Use the ASA Postulate to show that $\triangle ADB \cong \triangle CBD$. 33. Use the SAS Postulate to show that $\triangle ADB \cong \triangle CBD$. Then use Theorem 4–1. 35. Use the SAS Postulate to show that $\triangle ROY \cong \triangle QOX$. Then use Theorem 4–1 to show that $\overline{PQ} \parallel \overline{RS}$. 37. Use the SAS Postulate to show that $\triangle GPF \cong \triangle HPE$. 39. Use the SAS Postulate to show that $\triangle AXY \cong \triangle BXY$. Then show that $\angle XYA \cong \angle XYB$. 41. (1) P is not on line l (Given); (2) Draw j, the line thru P that is $\parallel$ to l. (Thru a given point not on a given line, there is exactly one line $\parallel$ to the given line.); (3) Draw a line, k, thru P that is in the same plane as l and j and is $\perp$ to j. (In a plane containing a given line and a given point on the line, there is exactly one line $\perp$ to the line at that point.); (4) $k \perp l$ (In a plane, a line $\perp$ to one of two $\parallel$ lines is also $\parallel$ to the other line.)

Pages 167–169 1. 49 3. 56 5. 55 7. a. $\overline{PR}$ b. $\overline{QR}$ 9. a. $\overline{QR}$ b. $\overline{QP}$ 11. 38; $m\angle A = 41$; $m\angle B = 68$; $m\angle C = 71$ 13. 15; $m\angle A = 25$; $m\angle B = 5$; $m\angle C = 150$ 15. 38; $m\angle A = 36$; $m\angle B = 30$; $m\angle C = 114$ 17. $m\angle A = 69$; $m\angle B = 21$ 19. (1) $\angle 2 \cong \angle A$ (Given); (2) $\angle B \cong \angle B$ (Refl. Prop. of $\cong$); (3) $\angle 1 \cong \angle C$ (If 2 ∠s of one $\triangle$ are $\cong$ to 2 ∠s of another $\triangle$, the third ∠s are $\cong$.) 21. (1) $m\angle R = m\angle X$; $m\angle S = m\angle Y$ (Given); (2) $m\angle R + m\angle S = m\angle X + m\angle Y$ (Add. Prop. of Equality); (3) $m\angle R + m\angle S + m\angle T = 180$; $m\angle X + m\angle Y + m\angle Z = 180$ (Triangle Angle-Sum Thm.); (4) $m\angle R + m\angle S + m\angle T = m\angle X + m\angle Y + m\angle Z$ (Subst. Prop.); (5) $m\angle T = m\angle Z$ (Subtr. Prop. of Equality) 23. (1) $\overleftrightarrow{XY} \parallel \overleftrightarrow{VW}$ (Given); (2) $\angle 2 \cong \angle 6$ (If 2 lines are $\parallel$, alt. int. ∠s are $\cong$.); (3) $\angle 3 \cong \angle 5$ (Given); (4) $\angle 1 \cong \angle 4$ (If 2 ∠s of 1 $\triangle$ are $\cong$ to 2 ∠s of another $\triangle$, the third ∠s are $\cong$.); (5) $\overline{XV} \parallel \overline{YZ}$ (If 2 lines are cut by a trans. so that alt. int. ∠s are $\cong$, the lines are $\parallel$.) 25. Show that $\angle 3 \cong \angle 2$ and $\angle S \cong \angle 5$, then use Corollary 1 of Theorem 4–8. 27. 64, 64, 52 29. 39, 51 31. Use Theorem 4–8 in △s ADC and ABC. 33. $x = 15$, $y = 20$ 35. Since $m\angle 1 + m\angle 2 + m\angle 3 + m\angle 4 = 180$, $m\angle 2 + m\angle 3 = 90$. Now use Theorem 4–8 and the Subtr. Prop. of Equality.

Pages 172–175 1. AAS Theorem 3. ASA Post. 5. none 7. ASA Post. or AAS Theorem 9. 105 11. 85 13. $m\angle 6 = 47$; $m\angle 7 = 12$; $m\angle 8 = 35$ 15. $m\angle 6 = 46$; $m\angle 7 = 27$; $m\angle 8 = 19$ 17. (1) $\angle P \cong \angle R$; $\overline{PQ} \cong \overline{RS}$ (Given); (2) $\angle 1 \cong \angle 2$ (Vert. ∠s are $\cong$.); (3) $\triangle PQX \cong \triangle RSX$ (AAS Thm.); (4) $\overline{PX} \cong \overline{RX}$ (Corr. parts of $\cong$ △s are $\cong$.) 19. Use the AAS Thm. to show that $\triangle ABE \cong \triangle CBD$. 21. Use the Ext. Angle Thm. to show that $m\angle 1 + m\angle 2 = m\angle 3 + m\angle 4 + m\angle 5 + m\angle 6$. 23. shortest: $\overline{KL}$; longest: $\overline{KM}$ 25. 110 27. Use AAS Theorem to show that $\triangle ADT \cong \triangle BCT$. 29. Use the SAS Post. to show that $\triangle JAB \cong \triangle KAB$. 31. Use the AAS Theorem to show that $\triangle OLQ \cong \triangle PMQ$. 33. $m\angle 1 = 155$; $m\angle 2 = 65$; $m\angle 3 = 50$ 35. $x = 72$; $y = 36$; $z = 108$

Pages 178–179 1. 1260 3. 1440 5. 2160 7–10. 360 11. 108 13. 144 15. 162 17. 30 19. 18 21. 8 23. 13 25. 110 27. 40 29. 18

Page 181 · Chapter Review 1. $\overleftrightarrow{WZ} \parallel \overleftrightarrow{XY}$; in a plane, 2 lines $\parallel$ to the same line are $\parallel$. 2. $\overleftrightarrow{WX} \parallel \overleftrightarrow{ZY}$; if 2 lines are cut by a trans. so that corr. $\angle$s are $\cong$, the lines are $\parallel$. 3. $\overleftrightarrow{WZ} \parallel \overleftrightarrow{XY}$; if 2 lines are cut by a trans. so that int. $\angle$s on the same side of the trans. are supp. then the lines are $\parallel$. 4. Show that $\angle 8 \cong \angle 4$. 5. 11 6. 14 7. Show that $\angle 7 \cong \angle 4$ and $\angle 8 \cong \angle 4$. Then use Trans. Prop. of $\cong$. 8. 90; right 9. 49; acute 10. 29; 61 11. Use the AAS Theorem to show $\triangle QRS \cong \triangle TSR$. 12. $\overline{AC}$ 13. Show that $\angle 2 \cong \angle 5$. Then use Theorem 4–2. 14. a. 2520; b. 360 15. 20 16. 17

Pages 184–185 · Mixed Review
1. $m\angle C = 60$; acute

3.

r	s	$s \rightarrow r$	$r \rightarrow (s \rightarrow r)$
T	T	T	T
T	F	T	T
F	T	F	T
F	F	T	T

5. 115 7. $m\angle 9 = 65$, $m\angle 8 = 115$

9. 24 11. (1) $\overline{AB} \parallel \overline{DE}$; $\overline{AB} \cong \overline{DE}$ (Given); (2) $\angle B \cong \angle D$; $\angle A \cong \angle E$ (Theorem 4–4); (3) $\triangle ACB \cong \triangle ECD$ (ASA); (4) $\overline{AC} \cong \overline{EC}$; $\overline{BC} \cong \overline{DC}$ (Corr. parts of $\cong$ $\triangle$s are $\cong$.) 13. 39 15. $\angle C$ 17. $\{y\colon -4 \leq y \leq -2\}$ 19. $\{x\colon x < -2\}$ 21. 9000 23. Ed will visit us on Sunday; Law of Disjunctive Inference 25. (1) Given (2) Prop. of Betweenness (3) Subst. Prop. (4) Subtr. Prop. of Equality (5) Given (6) Theorem 3–14 (7) SAS (8) Corr. parts of $\cong$ $\triangle$s are $\cong$. (9) Theorem 3–13

Pages 190–192 1. $\overline{ML}$, $\overline{JK}$; $\overline{MJ}$, $\overline{LK}$; $\overline{MN}$, $\overline{NK}$; $\overline{JN}$, $\overline{NL}$ 3. 14 5. 116 7. 3 9. 24 11. 6 13. (1) $\overleftrightarrow{AB} \parallel \overleftrightarrow{DC}$; $\overleftrightarrow{AD} \parallel \overleftrightarrow{BC}$ (Given); (2) $ABCD$ is a ▱. (Def of a ▱); (3) $\angle 1 \cong \angle 2$ (The opp. $\angle$s of a ▱ are $\cong$.) 15. Show that $\overline{QR} \cong \overline{QT}$. 17. Draw $\overline{BD}$ and then show that $\triangle ABD \cong \triangle CDB$ by the SSS Postulate. Use corr. parts of $\cong$ $\triangle$s to show that $\angle ABD \cong \angle CDB$ and $\angle ADB \cong \angle CBD$. Then use the Add. Prop. of Equality to show that $\angle B \cong \angle D$. 19. Use the ASA Postulate to show that $\triangle JOP \cong \triangle LOQ$. 21. Use the SAS Postulate to show that $\triangle AFD \cong \triangle CEB$. 23. Use the AAS Theorem to show that $\triangle QJX \cong \triangle SKX$. 25. Use the AAS Theorem to show that $\triangle DRE \cong \triangle FSG$. 27. Use the SAS Postulate to show that $\triangle XAZ \cong \triangle CXY$. Then show that $\angle 1$ and $\angle 3$ are $\cong$ corr. $\angle$s. 29. Use the Trans. Prop. of $\cong$ and Theorem 5–1.

Pages 194–196 1. $x = 9$, $y = 7$ 3. $x = 3$, $y = 8$ 5. Use Theorem 5–5 to show that $PQRS$ is a ▱. 7. Show that $\overline{AK} \parallel \overline{BJ}$ and that $\overline{AK} \cong \overline{BJ}$. 9. Show that $\overleftrightarrow{AD} \parallel \overline{BC}$ and that $\overline{DE} \cong \overline{BC}$. 11. Use Theorem 5–7 to show that $ABCD$ is a ▱. Then use $\overline{DC} \cong \overline{AB}$ and the given information to show that $\overline{CD} \cong \overline{BE}$. 13. Use alt. int. $\angle$s to show that $\overline{RX} \parallel \overline{TZ}$. Use corr. parts of $\cong$ $\triangle$s to show that $\overline{RY} \cong \overline{TZ}$. Then use this fact, the given information, and the Trans. Prop. of $\cong$ to show that $\overline{RX} \cong \overline{TZ}$. 15. a. Draw $\overline{DB}$ and then use the SAS Postulate to show that $\triangle ABD \cong \triangle CDB$. Then use alt. int. $\angle$s to show that $\overline{AD} \parallel \overline{BC}$. b. Use corr. parts of $\cong$ $\triangle$s to show that $\overline{AD} \cong \overline{BC}$. 17. Use the AAS Theorem to show that $\triangle AXB \cong \triangle CXD$. Then use corr. parts of $\cong$ $\triangle$s and Theorem 5–5. 18. Draw $\overline{AC}$. Then use the AAS Theorem to show that $\triangle BAC \cong \triangle DCA$. Then use corr. parts of $\cong$ $\triangle$s and Theorem 5–5. 19. Use Theorem 4–11 ($n = 4$) to show that $m\angle A + m\angle B + m\angle C + m\angle D = 360$. Then use substitution with this equation and the given information to show that $2m\angle A + 2m\angle B = 360$ and $2m\angle A + 2m\angle D = 360$. Then show that $\angle A$ and $\angle B$ are supp. and that $\angle A$ and $\angle D$ are supp. Then use the corollary to Theorem 4-2 to show that $\overline{AD} \parallel \overline{BC}$ and $\overline{AB} \parallel \overline{DC}$. 20. Draw $\overline{DE} \perp \overleftrightarrow{AB}$ and $\overline{BF} \perp \overleftrightarrow{DC}$. Use the AAS Theorem to show that $\triangle AED \cong \triangle CFB$. Draw $\overline{DB}$. Then use the HL Postulate to show that $\triangle BFD \cong \triangle DEB$. Use the AAS Theorem to show that $\triangle DAB \cong \triangle BCD$. Then use alt. int. $\angle$s to show that $\overline{DC} \parallel \overline{AB}$ and $\overline{AD} \parallel \overline{BC}$. 21. Draw $\overleftrightarrow{AG} \perp \overleftrightarrow{DC}$ and $\overleftrightarrow{CH} \perp \overleftrightarrow{AB}$. Then draw $\overline{DG}$, $\overline{BH}$, and $\overline{AC}$. Use the AAS Theorem to show that $\triangle AGD \cong \triangle CHB$. Use the HL Postulate to show that $\triangle AGC \cong \triangle CHA$. Then use the AAS Theorem to show that $\triangle DAC \cong \triangle BCA$. Use corr. parts of $\cong$ $\triangle$s to show that $\overline{DC} \cong \overline{BA}$.

Pages 198–201 1. 90 3. a. 17 b. 63 5. a. 4 b. 13 7. 90 9. a. 8 b. 35 11. a. 5 b. 9 13. a. 8 b. rhomb.; since $TUVW$ is a ▱, $\overline{TU} \cong \overline{WV}$ and $\overline{TW} \cong \overline{UV}$. If $UV = z + 2 = 10$, $\overline{UV} \cong \overline{TW} \cong \overline{TU} \cong \overline{WV}$ and $TUVW$ is a rhomb. 15. Since $ABCD$ is a ▱, $\angle A$ and $\angle D$ are supp.; $2z + 20 + 3z - 15 = 180$; $5z = 175$; $z = 35$; $m\angle A = 90$. By Theorem 5-8, $ABCD$ is a rect. 17. Show that $\overline{BC} \cong \overline{CE}$. 19. Use the SAS Postulate to show that $\triangle XYM \cong \triangle SRM$. 21. Use Theorem 3-13 to show that $\angle 1 \cong \angle 3$ and $\angle 2 \cong \angle 4$. Then use alt. int. $\angle$s to show that $\angle 1 \cong \angle 4$ and $\angle 2 \cong \angle 3$. 22. See figure for Theorem 5-11 on p. 197. GIVEN: $PQRS$ is a ▱; $\overline{PR}$ bisects $\angle SPQ$. PROVE: $PQRS$ is a rhomb. Use alt. int. $\angle$s to show that $\angle 1 \cong \angle 4$ and use the given information to show that $\angle 1 \cong \angle 2$. Then $\angle 2 \cong \angle 4$, and by Theorem 3-14 $\overline{RQ} \cong \overline{PQ}$. 23. Use the SAS Postulate to show that $\triangle XEH \cong \triangle XEF$. Then use Theorem 3-11 to show that $\overline{EG} \perp \overline{HF}$. 24. See figure for Theorem 5-12 on p. 197. GIVEN: $EFGH$ is a ▱; $\overline{EG} \perp \overline{HF}$ PROVE: $EFGH$ is a rhomb. Use the SAS Postulate to show that $\triangle EXH \cong \triangle EXF$. Then use corr. parts of $\cong$ $\triangle$s to show that $\overline{EH} \cong \overline{EF}$. 25. Use the SAS Postulate to show that $\triangle DAB \cong \triangle CBA$. 26. See figure for Theorem 5-10 on p. 197. GIVEN: $ABCD$ is a ▱; $\overline{AC} \cong \overline{BD}$ PROVE: $ABCD$ is a rect. Use the SSS Postulate to show that $\triangle DAB \cong \triangle CBA$. Then $\angle A \cong \angle B$. Since $\angle A$ and $\angle B$ are supp., $m\angle A = 90$. 27. Show that $PQRS$ is a ▱. Then use the results of Ex. 22, p. 199. 29. Use the results of Ex. 22, p. 199 to show that $ABCD$ is a rhomb. 31. yes; if $x = 9$, then $RS = 60$, $ST = 60$, $TW = 60$, and $RW = 60$, so $RSTW$ is a rhomb. 33. Show that the area of $PQRS = \frac{1}{2}(PX)(QX) + \frac{1}{2}(RX)(QX) + \frac{1}{2}(PX)(XS) + \frac{1}{2}(RX)(XS)$. Then the area of $PQRS = \frac{1}{2}[(PX + RX)(QX) + (PX + RX)(XS)]$, or $\frac{1}{2}[PR(QX) + PR(XS)]$. Therefore, the area of $PQRS = \frac{1}{2}[PR(QX + XS)] = \frac{1}{2}(PR)(SQ)$. 35. Use the HL Postulate to show that $\triangle WVY \cong \triangle XYV$. Then use Theorem 5-5 to show that $VWXY$ is a ▱.

Pages 204–205 1. 7 3. $x - 2$ 5. 26, 39 7. 5 9. 4 11. 10 13. 5 15. 4 17. $XZ = 8$ cm, $XY = 16$ cm, $YZ = 14$ cm 19. 30 cm 21. 12 cm 23. Use Theorem 5-14 to show that Q is the midpt. of $\overline{OB}$. Then use Theorem 5-13. 25. a. Draw $\overline{AC}$. Then use Theorem 5-13 to show that $\overline{WX} \parallel \overline{AC}$ and $\overline{ZY} \parallel \overline{AC}$. Use Theorem 5-13 to show that $WX = \frac{1}{2}AC$ and $ZY = \frac{1}{2}AC$. Then use Theorem 5-5. b. The figure formed by joining the midpts. of each pair of adjacent sides of a quad. is a ▱. 29. Use Theorem 5-14 to show that $\overline{AH} \cong \overline{HF}$ and $\overline{EG} \cong \overline{GF}$. Then use Theorem 5-13.

Pages 208–209 1. 19 3. 10 5. 11 7. 9 9. $AB = 22$, $CD = 12$ 11. Use Theorem 5-15 to show that $\angle A \cong \angle B$. Then use Theorem 3-14. 13. Use the given information and Theorem 5-14. 15. Use the SAS Postulate to show that $\triangle DMA \cong \triangle CMB$. 17. Draw $\overline{HK}$ through $H \parallel$ to $\overline{GF}$ and intersecting $\overleftrightarrow{EF}$ at K. $HGFK$ is a ▱ with $\overline{HK} \cong \overline{GF}$. Show that $\overline{HE} \cong \overline{HK}$ and then use Theorem 3-13 to show that $\angle E \cong \angle HKE$. Use corr. $\angle$s to show that $\angle HKE \cong \angle F$. Use Corollary 1 of Theorem 4-5 to show that $\angle G$ and $\angle F$ are supp.

and that $\angle EHG$ and $\angle E$ are supp. Then use Theorem 3–7. 19. a. Draw $\overline{CE} \parallel \overline{DA}$. Then use corr. $\angle$s to show that $\angle 1 \cong \angle A$. Since $\angle A \cong \angle B$, $\angle 1 \cong \angle B$. Therefore, by Theorem 3–14, $\overline{CE} \cong \overline{CB}$. Show that $AECD$ is a ▱. Then show that $\overline{DA} \cong \overline{CB}$. b. Use Corollary 1 of Theorem 4–5 to show that $\angle A$ and $\angle D$ are supp. and that $\angle B$ and $\angle DCB$ are supp. Apply Theorem 3–7 to these facts and the given information to show that $\angle A \cong \angle B$. Then use part (a).

Pages 210–211 · Chapter Review 1. 4 2. 2 3. 9 4. Use the ASA Postulate. 5. yes; if the diags. of a quad. bisect each other, the quad. is a ▱. 6. no 7. yes; if two opp. $\angle$s of a quad. are $\cong$ and two opp. sides are $\parallel$, then the quad. is a ▱ (Ex. 18, p. 195). 8. Show that $WXYZ$ is a ▱. 9. $\cong$ 10. $\perp$ 11. square 12. Use Theorem 3–14 to show that $\overline{EH} \cong \overline{EF}$. 13. 7 14. The segment joining the midpts. of two sides of a $\triangle$ is $\parallel$ to the third side. If two lines are $\parallel$, alt. int. $\angle$s are $\cong$. 15. Use Corollary 1 of Theorem 4–1 to show that $\overline{SV} \parallel \overline{TU}$. Next use Theorem 5–14 to show that $\overline{SV}$ bisects $\overline{RU}$. Then use Theorem 5–13. 16. 108 17. no; the diagonals of an isos. trap. are $\cong$ and there is no real number y such that $3y - 5 = 3y + 5$. 18. $KL = 12$, $ON = 22$ 19. Use the SAS Postulate to show that $\triangle EAD \cong \triangle EBC$. Then show that $\triangle DEC$ is isos.

Pages 214–215 · Mixed Review
1. (1) $\overline{ED} \perp \overline{DC}$; $\overline{AC} \perp \overline{DC}$; B is the midpoint of $\overline{DC}$ (Given); (2) $\angle D$ and $\angle C$ are rt. $\angle$s. (Def. of $\perp$ lines); (3) $\angle D \cong \angle C$ (All rt. $\angle$s are $\cong$.) (4) $\overline{DB} \cong \overline{BC}$ (Def. of midpoint); (5) $\angle DBE \cong \angle CBA$ (Vertical $\angle$s are $\cong$.); (6) $\triangle EBD \cong \triangle ABC$ (ASA) 3. 55; 125; 55 5. 1 7. 135 9. (1) $\overline{MN} \cong \overline{MR}$; $\angle P \cong \angle Q$ (Given); (2) $\overline{MP} \cong \overline{MQ}$ (Theorem 3–14); (3) $\angle M \cong \angle M$ (Refl. Prop. of $\cong$); (4) $\triangle PMR \cong \triangle QMN$ (SAS); (5) $\overline{NQ} \cong \overline{RP}$ (Corr. parts of $\cong$ $\triangle$s are $\cong$.) 11. (1) $ABCD$ is a trapezoid with median $\overline{EF}$. (Given); (2) $\overline{EF} \parallel \overline{AB}$ (Theorem 5–17); (3) E is the midpoint of $\overline{AD}$. (Def. of the median of a trap.); (4) $\overline{EF}$ bisects diagonal $\overline{BD}$ (Theorem 5–14); Use similar reasoning to show that $\overline{EF}$ also bisects diagonal $\overline{AC}$. 13. -1 15. 16; obtuse 17. yes; Opposite sides are $\parallel$. 19. $r \wedge \sim s$ 21. $\{t: -14 \leqslant t < 4\}$ 23. a. $\exists_x\, x^2 > 0$ b. true c. No number has a positive square; $\forall_x\, x^2 \leqslant 0$ 24. 5

Pages 221–222 1. $x - 2y$ 3. $10m$ 5. $-4y^2 + 5y + 8$ 7. $9x^2 - 13x$ 9. $b + 6$ 11. $20x - 2$ 13. $6a - 4b$ 15. $11a - 3c$ 17. $5x^2 + 7y^2 - 3y - 5z$ 19. $13x^2 - 15xy + 21y^2$ 21. $-x^2 - 11y^3$ 23. $9ab - 6a + 6c$ 25. $14a - 19b - 16c - 12d$ 27. $-17x - 19y + 41z$ 29. $-4x^2 - 13xy + 18y^2$ 31. 3 33. r; 6 35. 5; 2 37. $58c^3 + 9c^2 - 11c - 37$ 39. $-7x^3 - 9x^2 - 7x - 35$ 41. $75a^3 - 73a^2 - 19a - 311$ 43. $69p^3 + 54p^2q + 49$ 45. $-x - 3$ 47. $4a + 4b$ 49. $15x - y$ 51. $6x - 7$ 53. $-3x^2$

Pages 225–227 1. $3x^5$ 3. $16z^8$ 5. $-x^6$ 7. y^7 9. $y^2 - 7y - 8$ 11. $q^2 + 11q + 18$ 13. $r^2 - r - 30$ 15. $v^2 - 81$ 17. $5x^2 - 27x + 10$ 19. $49p^2 + 28p + 4$ 21. $y^2 - 16$ 23. $4z^2 - 5z - 6$ 25. $y^2 + 4yz + 4z^2$ 27. $5a^2 - 8ab + 3b^2$ 29. $12h^2 + 19h + 5$ 31. $y^2 - 6yz - 7z^2$ 33. $y^3 + y^2 - 2y - 2$ 35. $2x^3 - 6x^2 + 8x$ 37. $28a^5 + 36a^4 - 4a^3$ 39. $6a^4bc - 18a^3b^2c + 24a^2b^3c$ 41. $-2a^2 - 33a + 16$ 43. $(x + 5)(2x + 5)$ 45. $3x^2 + x - 10$ 47. $3m^3 + 2m^2 - 20m + 3$ 49. $a^4 - 2a^3 + 3a^2 - 4a + 2$ 51. $2x^3 + 5x^2y - xy^2 - y^3$ 53. $10z^4 - 6z^3 + z^2 - 3z - 2$ 55. $5y^4 + 13y^3 + 3y^2 + 10y + 4$ 57. $8x^3y^6$ 59. $16a^2b^4$ 61. $54y^5$ 63. $288m^3n^2$ 65. $-54a^9b^4$ 67. $x^2 - 3x - 6$ 69. $3x^2 + y^2$ 71. $-2y^2 + 16y - 30$ 73. $-48t^4 + 3t^2$ 75. $-12x^4 + 6x^3 + 43x^2 - 14x - 35$ 77. $12x^2 + 16x + 3$ 79. $12x^2 + 27x - 35$ 81. $2n^2 + 10n - 52$ 83. $4a^2 + 5a + 1$ 85. $-r^2 - 38r - 85$ 87. $x^4 - 7x^3 + 12x^2 + 4x$

Pages 232–233 1. $2xy(11xy^2 + 7)$ 3. $7rs(2r - 3t)$ 5. $mn(m - n^2 + m^2n^3)$ 7. $3a(4a^2b^2 - 2b + 3a^2)$ 9. $-6y^3(6x^2 - 2y^3z^4 + 3x^3)$, or $6y^3(-6x^2 + 2y^3z^4 - 3x^3)$ 11. $(y + 18)(y + 2)$ 13. $(w + 3)(w + 8)$ 15. not factorable 17. $(x - 14)(x + 2)$ 19. $(x - 3)(x + 2)$ 21. $(3s + 8)(s - 2)$ 23. $(2q - 15)(q + 1)$ 25. not factorable 27. $(m - 4n)(m - 12n)$ 29. $(q - 12r)(q + 6r)$ 31. $(3x + y)(7x - y)$ 33. $(g - 9h)(g - 2h)$ 35. not factorable 37. $(7x - 8)(x + 2)$ 39. $(3r + 7)(2r - 1)$ 41. $(6b + 5)(2b + 3)$ 43. not factorable 45. $(4x + 1)(x + 6)$ 47. $(5a - 4b)(a - 3b)$ 49. $(8q + 3)(q - 4)$ 51. $(5r - 3s)(3r - 8s)$ 53. $(a + 5)(b - 1)$ 55. $6(x - 1)(x + 4)$ 57. $y(y + z)(y + 9)$ 59. $r^2(2q - 3)(q + 2)$ 61. $4(3a + b)(2a + b)$ 63. $2x(5x + 2)(3x - 1)$ 65. $4(z - 1)(z + 1)$ 67. $(x + 3)(y + 2)$ 69. $(u - 1)(v - 4)$ 71. $(x + 1)(3y - 1)$ 73. $(x^2 + 5)(x + 1)$ 75. $(w - 2)(v^2 + 3)$

Page 235 1. $(5x + 1)(5x - 1)$ 3. $(4b + 11)(4b - 11)$ 5. $(r - 5)^2$ 7. $(8 + x)^2$ 9. $(s - 1)(s^2 + s + 1)$ 11. $(5k + 3)(25k^2 - 15k + 9)$ 13. not factorable 15. $(xy + a)(xy - a)$ 17. $2(7h - 1)^2$ 19. $25(2y + z)(2y - z)$ 21. $(13x^2 + 3)^2$ 23. $3a(4 + 5z)(16 - 20z + 25z^2)$ 25. $4r(s - 13)^2$ 27. $(c + d - 1)(c - d - 1)$ 29. $(p + 2q + 3)(p - 2q - 3)$ 31. $(m^2 + 1)(m + 1)(m - 1)$ 33. $2(y^2 + 9)(y + 3)(y - 3)$ 35. $(9m^2 + 1)^2$ 37. $(x + 4)(x - 4)(x + 2)$ 39. $(m - 1)(m^2 + m + 1)(m + 2)$ 41. $(2 + a)(1 - b)(1 + b + b^2)$ 43. $-7(a - 6)^2(2a + 3)$ 45 $(t^n + 6v^{3n})(t^{2n} + 6t^nv^{3n} + 36v^{6n})$

Pages 239–240 1. $\frac{2}{x}$ 3. $-\frac{9ab^2}{2}$ 5. $-\frac{5x^3y^2}{3z}$ 7. $-\frac{7}{3(x + y)}$ 9. $\frac{3t^3}{4r^2}$ 11. $x + 3y$ 13. $xy + 4x^2$ 15. $\frac{b}{3a} - \frac{b^3}{2a^2}$ 17. $3x^3 - 5x^2y + 6xy^2$ 19. $\frac{x^2 + 12x - 28}{x - 2} = x + 14$ 21. $\frac{x^3 - 6x^2 + 13x - 12}{x - 3} = x^2 - 3x + 4$ 23. $\frac{2m^3 - 3m^2 + 5m - 2}{2m - 1} = m^2 - m + 2$ 25. $\frac{5t^3 + t^2 + t - 4}{5t - 4} = t^2 + t + 1$ 27. $\frac{x^3 - 9x^2 + 10x + 1}{x - 1} = x^2 - 8x + 2 + \frac{3}{x - 1}$ 29. $\frac{2y^3 - 11y^2 + 14y - 8}{y - 4} = 2y^2 - 3y + 2$ 31. $\frac{4z^3 + 4z^2 - z + 1}{2z + 1} = 2z^2 + z - 1 + \frac{2}{2z + 1}$ 33. $\frac{2a^3 + a^2 - 14a + 13}{a + 3} = 2a^2 - 5a + 1 + \frac{10}{a + 3}$ 35. $\frac{y^3 - 8y^2 + 15y - 18}{y^2 - 2y + 3} = y - 6$ 37. $\frac{w^3 + w^2 + w + 21}{w^2 - 2w + 7} = w + 3$ 39. $\frac{5m^3 + 2m^2 - 15m + 6}{m^2 - 3} = 5m + 2 + \frac{12}{m^2 - 3}$ 41. $\frac{6r^3 - 3r + 3}{3r + 3} = 2r^2 - 2r + 1$ 43. $\frac{6y^3 - 17y^2 + 8y + 8}{2y - 3} = 3y^2 - 4y - 2 + \frac{2}{2y - 3}$ 45. $\frac{3z^3 - 4z^2 - z + 2}{z^2 - 2z + 1} = 3z + 2$ 47. $\frac{y^4 + 3y^3 - 6y^2 - 6y + 8}{y^2 + 3y - 4} = y^2 - 2$ 49. $\frac{x^4 + x^3y - 2x^2y^2 + 3xy^3 + y^4}{x - y} = x^3 + 2x^2y + 3y^3 + \frac{4y^4}{x - y}$ 51. $\frac{x^4 - y^4}{x + y} = x^3 - x^2y + xy^2 - y^3$ 53. $\frac{2x^4 - 3x^3}{x^2 - 3} = 2x^2 - 3x + 6 - \frac{9x - 18}{x^2 - 3}$ 55. -19

Page 243 1. $\frac{1}{3x}$ 3. $\frac{4c^2}{7a^2}$ 5. $-\frac{1}{2}$ 7. $y + 1$ 9. $\frac{2}{3m}$ 11. $1 - 4x$

13. $-\frac{1}{n+2}$ 15. $\frac{p+4}{8}$ 17. $\frac{q-5}{q}$ 19. $\frac{h+3}{h+1}$ 21. $\frac{m^2-2m+4}{m-2}$
23. $\frac{x-1}{2}$ 25. $\frac{c+d}{c-d}$ 27. $\frac{y+2}{y-3}$ 29. $\frac{b-5}{2b-1}$ 31. $\frac{3(z+4)}{z+2}$
33. $-\frac{z+1}{5+2z}$ 35. $\frac{9k^2+6k+4}{5k+6}$ 37. $\frac{2x+3}{2(x+2)}$ 39. $m+1$ 41. $\frac{x^n+y^n}{x^n-y^n}$

Page 246 1. $\frac{20}{9}$ 3. $\frac{4x^2}{45}$ 5. $\frac{6c}{5}$ 7. $-36f^2$ 9. $\frac{4h^2}{3k}$ 11. $\frac{2}{x}$
13. $\frac{4x}{3y}$ 15. $\frac{2(t-2)}{3}$ 17. $\frac{y(x-2)}{2x(x+1)}$ 19. $\frac{2(x+2)}{3(x-y)}$ 21. $\frac{a-1}{3a}$ 23. $-\frac{9}{2}$
25. $\frac{r^2+s^2}{(r+s)^2}$ 27. $\frac{x+5}{x-4}$ 29. $\frac{x+3}{x-2}$ 31. $\frac{2u+3}{3u+2}$ 33. $\frac{t+3}{3t+1}$
35. $\frac{k^2+2k+4}{k-2}$ 37. $\frac{4(x+3)}{3(x+2)}$ 39. $\frac{x-3}{6x}$ 41. $\frac{1}{y(x-y)}$

Pages 249–250 1. $\frac{3x-4}{12x^2}$ 3. $\frac{3xy-40z}{8xyz}$ 5. $\frac{17x+18y}{6x^2}$ 7. $\frac{2a^2+4a+1}{20}$
9. $\frac{3x^2-4x+1}{6x^3}$ 11. $\frac{3a-11}{9b}$ 13. 1 15. $\frac{p^2-5}{p-5}$ 17. $\frac{b+1}{3}$ 19. $\frac{4x+3}{4}$
21. $\frac{3x+2y}{3y}$ 23. $\frac{2a+2b+3}{4}$ 25. $\frac{y}{3}$ 27. $\frac{7z+9}{12}$ 29. $\frac{2a^2+3a+4}{a^3}$
31. $\frac{9x+10y}{12xy}$ 33. $\frac{b-2a+3c}{abc}$ 35. $\frac{x^2-2x-1}{(x+2)(x-2)}$ 37. $\frac{4x+1}{6(x-1)}$
39. $\frac{2a^2+8}{(a+2)(a-2)}$ 41. $\frac{27-5a}{(a-4)(a-3)}$ 43. $\frac{11m-33}{(m+10)(m-10)(m+1)}$
45. $\frac{b^2-b}{(l+3)(l-3)(b+7)}$ 47. $\frac{3x^3+2x+12}{24x^2(x-3)}$ 49. $\frac{4x-2}{x(x+2)(x-2)}$
51. $\frac{-2x+12}{(2x+1)(2x-1)}$ 53. $\frac{3}{(a-6)(a+3)}$ 55. $\frac{7}{(m+1)(m-4)}$
57. $\frac{(r-5)(r^2-18r-7)}{(r-7)(r-18)}$

Pages 252–253 • Chapter Review 1. 3 2. 7 3. $-3a^2-2b^2$
4. x^3-4x^2+3x-7 5. $4c^3d^6$ 6. $12p^2-17p+6$ 7. $15w^2y^2-3wy^3$
8. $7a^3+20a^2-17a+2$ 9. $26ac^2$ 10. $5de(12d^2-5e)$ 11. $(f+14)(f-2)$
12. $(4g-h)(g-6h)$ 13. $(jk+3)^2$ 14. $(10-m)(100+10m+m^2)$
15. $3(2r+5s)(2r-5s)$ 16. $-\frac{5(c+d)^4}{3cd^2}$ 17. $2y^4-\frac{y}{2}+\frac{10}{y^2}$
18. $\frac{9a^3-3a^2-8a-1}{3a-2}=3a^2+a-2-\frac{5}{3a-2}$ 19. $\frac{4}{8-x}$ 20. $\frac{q^2}{q-1}$
21. $\frac{k+1}{k-5}$ 22. $\frac{2b}{a}$ 23. $\frac{s}{2r^4}$ 24. $-5y$ 25. $\frac{4}{z^2-2z+4}$ 26. $\frac{5w}{4}$
27. $\frac{3cd^2+2}{cd}$ 28. $-\frac{1}{t+2}$ 29. $-\frac{3}{2}$

Pages 256–257 • Mixed Review
1. $\{t: -1 \leqslant t < 1\}$ 3. $a(1+a)(1-a)$ 5. $(4a+7)(4a-1)$ 7. (1) $\overline{QW} \cong \overline{TW}$; $\angle 3 \cong \angle 8$ or $m\angle 3 = m\angle 8$ (Given); (2) $\angle 1 \cong \angle 7$ (Theorem 3–13); (3) $\angle RWS \cong \angle RWS$ or $m\angle RWS = m\angle RWS$ (Refl. Prop. of $\cong$); (4) $m\angle 3 + m\angle RWS = m\angle 8 + m\angle RWS$ (Add. Prop.

of Equality); (5) $m\angle 3 + m\angle RWS = m\angle QWS$; $m\angle 8 + m\angle RWS = m\angle TWR$ (Angle Add. Post.); (6) $m\angle QWS = m\angle TWR$ or $\angle QWS \cong \angle TWR$ (Subst. Prop.); (7) $\triangle WQS \cong \triangle WTR$ (ASA)
9. yes; Answers may vary. 11. (1) $QRST$ is a ▱. (Given); (2) $\overline{TO} \cong \overline{OR}$; $\overline{QO} \cong \overline{OS}$ (The diag. of a ▱ bisect each other.); (3) $\angle QOT \cong \angle SOR$ (Vert. ∠s are ≅.); (4) $\triangle QOT \cong \triangle SOR$ (SAS) 13. $b^2 + 9 - \frac{1}{b^2 - 6}$ 15. 8 17. 4
19. $-x^4 - x^3 + 2x^2 + x - 1$ 21. $\frac{2j^2 + 5j - 3}{2j + 1}$ 23. $\sim r \wedge s$ 25. $\frac{1}{b + 6}$
27. $2 < AB < 12$ 29. 20 31. 4 33. 6, 7, 8 35. $m\angle A = 54$; $m\angle B = 126$

Pages 262–263 1. $\{-2, 1\}$ 3. $\{5, -1\}$ 5. $\{0, -5\}$ 7. $\{-3, 2\}$
9. $\{-4, -3\}$ 11. $\{4\}$ 13. $\{10, -9\}$ 15. $\{-6, 2\}$ 17. $\{0, 3\}$ 19. $\{5, 7\}$
21. $\{2, -2\}$ 23. $\{5\}$ 25. $\{-4\}$ 27. $\left\{\frac{4}{3}, -2\right\}$ 29. $\{6, -1\}$ 31. 2
33. $-\frac{3}{5}$ 35. $\left\{\frac{2}{5}, -3\right\}$ 37. $\left\{\frac{1}{3}, -\frac{1}{2}\right\}$ 39. $\left\{\frac{1}{3}, -\frac{2}{3}\right\}$ 41. $\left\{\frac{3}{4}, \frac{2}{3}\right\}$
43. $\left\{0, -\frac{17}{3}\right\}$ 45. $\left\{\frac{3}{4}, 2\right\}$ 47. $\{0, -7, -8\}$ 49. $\left\{0, \frac{3}{2}, -2\right\}$ 51. $\left\{0, \frac{2}{7}, -\frac{5}{2}\right\}$
53, 55. Answers may vary. Examples are given. 53. $x^2 + x - 20 = 0$
55. $3x^2 + 5x - 2 = 0$ 57. $\{1, -1, -3\}$ 59. $\{8\}$ 61. $\left\{3, -3, \frac{3}{2}\right\}$ 63. -6
65. 18

Pages 266–267 1. $\{3\}$ 3. $\{-4\}$ 5. $\{-1\}$ 7. $\{5, 4\}$ 9. $\emptyset$ 11. $\{9, -1\}$
13. $\left\{\frac{9}{2}, -1\right\}$ 15. $\{6, -1\}$ 17. $\left\{\frac{4}{3}, \frac{3}{4}\right\}$ 19. $\left\{\frac{3}{2}, -\frac{2}{3}\right\}$ 21. $\{4, 20\}$
23. $\{-2\}$ 25. $\{1\}$ 27. $\left\{\frac{4}{3}, 2\right\}$ 29. $\{-4\}$ 31. $\left\{-\frac{1}{2}, -1\right\}$ 33. $\{1\}$
35. $\{2\}$ 37. $\emptyset$ 39. $\emptyset$ 41. $\emptyset$

Pages 271–272 1. 22 3. -16 5. $\frac{9}{20}$ 7. $6x$ 9. $-5x^2\sqrt{2}$
11. $-4xy^4\sqrt{5}$ 13. $-\frac{\sqrt{13}}{5}$ 15. $\frac{2a^2\sqrt{15}}{21}$ 17. $\frac{2}{5}$ 19. $\{\pm 11\}$ 21. $\{\pm 1.7\}$
23. $\left\{\pm\frac{7}{2}\right\}$ 25. $\{12, -2\}$ 27. $\{19, -41\}$ 29. $\{6, -18\}$ 31. $\{\pm 12\}$
33. $\emptyset$ 35. $\{\pm 6\sqrt{3}\}$ 37. $\left\{\pm\frac{4\sqrt{5}}{3}\right\}$ 39. $\left\{\pm\frac{5\sqrt{3}}{9}\right\}$ 41. $\left\{\pm\frac{1}{6}\right\}$
43. $7xy^2\sqrt{2x}$ 45. $-6x^3\sqrt{x}$ 47. $22 - 8\sqrt{6}$ 49. 22 51. $\{\pm 2\sqrt{14}\}$
53. $\{\pm\sqrt{86}\}$ 55. $\{\pm 7\}$ 57. $\{\pm 2\sqrt{2}\}$ 59. $\left\{\pm\frac{\sqrt{23}}{2}\right\}$ 61. $\left\{\pm\frac{2\sqrt{15}}{5}\right\}$
63. $\left\{\pm\frac{\sqrt{5}}{5}\right\}$ 65. $\left\{\pm\frac{4\sqrt{5}}{5}\right\}$ 67. $\emptyset$ 69. $\{4 \pm 4\sqrt{6}\}$ 71. $\{-3 \pm 4\sqrt{2}\}$
73. $\{-11 \pm 10\sqrt{5}\}$ 75. $\{1, 0\}$ 77. $\left\{0, -\frac{4}{3}\right\}$ 79. $\left\{\frac{\sqrt{7}}{7}, -\frac{13\sqrt{7}}{7}\right\}$
81. $\left\{\frac{9\sqrt{13}}{13}, -\frac{\sqrt{13}}{13}\right\}$ 83. $\{-5\sqrt{2} \pm 2\}$

Pages 276–279 1. 6 m 3. 3 cm 5. width: 5 cm; length: 8 cm 7. -11 or 9
9. 5 11. -8 or 5 13. 1 m 14. a. $3x^2 + 12x$ 15. 4 m 17. 12 m by 12 m

19. 7 21. −2 23. Wanda: 2 h; Jim: 8 h 25. 60 min 27. length: $9\sqrt{2}$; width: $6\sqrt{2}$ 29. 12 and 3 31. −13 and −9 or 9 and 13 33. 7 cm 35. −11, −9, and −7 or 7, 9, and 11 37. 5 h

Pages 281–282 1. 64 3. $\frac{1}{9}$ 5. $\{0, 2\}$ 7. $\{0, -7\}$ 9. $\{-8, -10\}$ 11. $\{-6, -20\}$ 13. $\{-3, 2\}$ 15. $\{7, -4\}$ 17. $\{-2 \pm 2\sqrt{7}\}$ 19. $\{4 \pm 2\sqrt{3}\}$ 21. $\left\{-\frac{1}{3}, -1\right\}$ 23. $\left\{\frac{1}{2}, -3\right\}$ 25. $\left\{\frac{-4 \pm \sqrt{10}}{2}\right\}$ 27. $\left\{\frac{-3 \pm 2\sqrt{6}}{3}\right\}$ 29. $\left\{\frac{1}{5}, -1\right\}$ 31. $\left\{\frac{1 \pm \sqrt{33}}{8}\right\}$ 33. $\{2 \pm \sqrt{10}\}$ 35. $\{-6 \pm 3\sqrt{3}\}$ 37. $\left\{\frac{-15 \pm \sqrt{385}}{8}\right\}$ 39. $\{-2\sqrt{3} \pm 3\sqrt{2}\}$ 41. $\left\{\frac{-3 \pm \sqrt{7}}{2}\right\}$ 43. $x = \frac{-b \pm \sqrt{b^2 - 4c}}{2}$ 45. $x = \frac{-a \pm \sqrt{a^2 + 8c}}{4}$

Pages 284–285 1. $\{-3, -4\}$ 3. $\{5, -2\}$ 5. $\{7, 1\}$ 7. $\{3\}$ 9. $\{-9\}$ 11. $\left\{\frac{3 \pm \sqrt{13}}{2}\right\}$ 13. $\left\{\frac{-5 \pm \sqrt{41}}{2}\right\}$ 15. $\{-4 \pm \sqrt{11}\}$ 17. $\left\{\frac{3}{2}, -2\right\}$ 19. $\left\{-3, -\frac{1}{3}\right\}$ 21. $\left\{-1, \frac{3}{5}\right\}$ 23. $\left\{\frac{3 \pm \sqrt{17}}{4}\right\}$ 25. $\left\{\frac{-2 \pm \sqrt{10}}{2}\right\}$ 27. $\left\{\frac{2 \pm \sqrt{22}}{3}\right\}$ 29. $\left\{-2, \frac{4}{3}\right\}$ 31. $\left\{\frac{4 \pm \sqrt{46}}{5}\right\}$ 33. $\left\{\frac{5 \pm \sqrt{13}}{4}\right\}$ 35. $\left\{\frac{-2 \pm \sqrt{2}}{2}\right\}$ 37. $\left\{\frac{1 \pm \sqrt{33}}{4}\right\}$ 39. $\left\{\frac{1 \pm \sqrt{5}}{2}\right\}$ 41. $\left\{\frac{3 \pm \sqrt{21}}{6}\right\}$ 43. $\left\{\frac{\sqrt{6} \pm \sqrt{2}}{2}\right\}$ 45. $\left\{\frac{-\sqrt{19} \pm \sqrt{35}}{2}\right\}$ 47. $\{16\}$ 49. $\{4, 9\}$ 51. $\{81\}$ 55. a. $\frac{7}{5}$ b. $-\frac{1}{5}$

Pages 289–290 1. $\{6\}$ 3. $\{0, 6\}$ 5. $\{8, 2\}$ 7. $\{\pm 2\sqrt{5}\}$ 9. $\left\{\pm \frac{5}{2}\right\}$ 11. $\left\{0, \frac{4}{9}\right\}$ 13. $\left\{-\frac{3}{2}\right\}$ 15. $\{4 \pm 4\sqrt{2}\}$ 17. $\left\{\frac{2}{5}\right\}$ 19. $\{1, -3\}$ 21. $\left\{\frac{3}{2}, -\frac{5}{2}\right\}$ 23. $\left\{\frac{-3 \pm \sqrt{89}}{4}\right\}$ 25. two rational roots 27. two irrational roots 29. two rational roots 31. two irrational roots 33. two irrational roots 35. two irrational roots 37. $\{-8, 2\}$ 39. $\left\{\frac{3}{2}\right\}$ 41. $\left\{\frac{1 \pm \sqrt{129}}{4}\right\}$ 43. $\{-10, -7\}$ 45. $\left\{\frac{3}{8}\right\}$ 47. $\frac{1}{2}$ 49. −3 51. $-\frac{2}{3}$ 53. two rational roots 55. two irrational roots 57. one rational double root 59–63. Answers may vary. Examples are given. 59. $x^2 + 2x\sqrt{3} + 3 = 0$ 61. $x^2 - 6x + 7 = 0$ 63. $2x^2 - 4x - 1 = 0$

Pages 293–295 1. $\frac{-8 + \sqrt{105}}{2}$ 3. 5 5. $\frac{5}{2}$ or $\frac{2}{5}$ 7. $-\frac{5}{3}$ or $\frac{2}{3}$ 9. −2 or $-\frac{1}{2}$ 11. $\frac{5}{2}$ 13. $\frac{-10 + \sqrt{149}}{2}$ m 15. $m\angle A = 162$, $m\angle B = 18$ 17. $3 + 2\sqrt{2}$ and $3 - 2\sqrt{2}$ 19. 13 21. 6 and 2 23. 2 km/h 25. 60 km/h 27. $(-2 + 2\sqrt{10})$ mm and $(2 + 2\sqrt{10})$ mm 29. 2 s and 3 s

Pages 296–297 · Chapter Review 1. $\{13\}$ 2. $\{0, -7\}$ 3. $\left\{-\frac{1}{3}, -2\right\}$ 4. $\{3\}$ 5. $\{6, -1\}$ 6. $\emptyset$ 7. a. $5xy^3\sqrt{6}$ b. $-\frac{7\sqrt{3}}{3}$ 8. $\{\pm 4\sqrt{10}\}$ 9. $\{5, -13\}$ 10. $\left\{\pm\frac{10\sqrt{2}}{3}\right\}$ 11. $\frac{1}{2}$ or $-\frac{1}{2}$ 12. 37.5 h 13. width: 6 cm; length: 8 cm 14. $\{-2, -14\}$ 15. $\{2 \pm \sqrt{11}\}$ 16. $\{-10, 3\}$ 17. $\{12, -3\}$ 18. $\left\{\frac{-5 \pm \sqrt{201}}{8}\right\}$ 19. $\left\{2, \frac{2}{3}\right\}$ 20. $\left\{\frac{1 \pm \sqrt{31}}{5}\right\}$ 21. a. Since the discriminant has value 16, there are two rational roots. b. $\left\{0, \frac{1}{5}\right\}$ 22. a. Since the discriminant has value 0, there is one rational double root. b. $\left\{\frac{3}{2}\right\}$ 23. a. Since the discriminant has value 5, there are two irrational roots. b. $\left\{\frac{-1 \pm \sqrt{5}}{2}\right\}$ 24. -3 and 1 or 2 and 6 25. 1.5 m

Pages 298–302 · Cumulative Review: Chapters 1–7 1. b 2. a 3. d 4. c 5. c 6. d 7. c 8. b 9. a 10. c 11. a 12. b 13. c 14. a 15. d 16. a 17. c 18. d 19. a 20. d 21. b 22. b 23. c 24. b 25. c 26. b 27. a 28. b 29. c 30. c 31. b 32. a 33. a 34. b 35. d 36. b 37. d 38. b 39. b 40. c 41. a 42. d 43. a 44. b 45. c 46. d 47. a. $\sim p \rightarrow \sim i$; $b \vee r$; $p \rightarrow \sim b$; $\sim r$ b. (1) $b \vee r$ (Given); (2) $\sim r$ (Given); (3) b (Law of Disjunctive Inference); (4) $\sim(\sim b)$ (Law of Double Negation); (5) $p \rightarrow \sim b$ (Given); (6) $\sim p$ (Law of Contrapositive Inference); (7) $\sim p \rightarrow \sim i$ (Given); (8) $\sim i$ (Law of Detachment) 48. (1) X is the midpoint of $\overline{RS}$ and $\overline{XY} \parallel \overline{ST}$. (Given); (2) $\overline{XY}$ bisects $\overline{RT}$. (Theorem 5–14); (3) Y is the midpoint of $\overline{RT}$. (Def. of a bisector of a segment); (4) $\overline{YZ} \parallel \overline{RS}$ (Given); (5) $\overline{YZ}$ bisects $\overline{ST}$. (Theorem 5–14); (6) Z is the midpoint of $\overline{ST}$. (Def. of a bisector of a segment); (7) $SZ = \frac{1}{2}ST$ (Theorem 3–2) 49. (1) $\triangle ADC \cong \triangle ABC$ (Given); (2) $\overline{DC} \cong \overline{BC}$; $\angle DCA \cong \angle BCA$ (Corr. parts of $\cong$ ⚠ are $\cong$.); (3) $\overline{CE} \cong \overline{CE}$ (Refl. Prop. of $\cong$); (4) $\triangle DCE \cong \triangle BCE$ (SAS Postulate); (5) $\angle DEC \cong \angle BEC$ (Corr. parts of $\cong$ ⚠ are $\cong$.); (6) $\overline{AC} \perp \overline{BD}$ (Theorem 3–11) 50. (1) $RSWY$ is a ▱; $\angle 1 \cong \angle 2$ (Given); (2) $\angle R \cong \angle W$ (Theorem 5–2); (3) $\overline{RY} \cong \overline{WS}$ (Theorem 5–1); (4) $\triangle XRY \cong \triangle ZWS$ (AAS Theorem); (5) $\overline{XY} \cong \overline{SZ}$ (Corr. parts of $\cong$ ⚠ are $\cong$.)

Pages 304–305 · Mixed Review

1. $6ab\sqrt{2ac}$ 3. -10 5. 7 and 8 7. $(\sim p \wedge q) \rightarrow r$ 9. $\frac{1}{y(x-y)}$

11. (1) $\overline{AC} \cong \overline{AD}$ or $AC = AD$; B and E are the midpoints of $\overline{AC}$ and $\overline{AD}$, respectively. (Given); (2) $BC = \frac{1}{2}AC$; $ED = \frac{1}{2}AD$ (Def. of midpoint); (3) $ED = \frac{1}{2}AC$ (Subst. Prop.); (4) $BC = ED$ (Subst. Prop.); (5) $\overline{BE} \parallel \overline{CD}$ (Theorem 5–13); (6) $BCDE$ is an isosceles trapezoid. (Def. of an isos. trap.) 13. $\{x: x < 3\}$ 15. $\{15\}$

17. a. 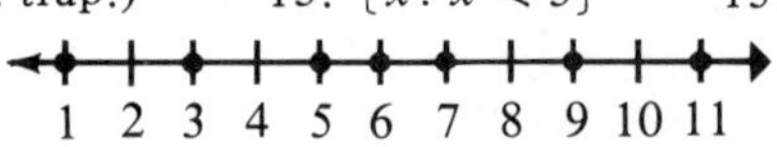
b. 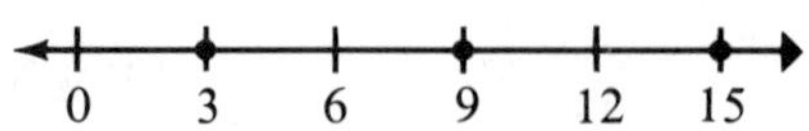

19. a. If $\frac{1}{x} > x$, then $0 < x < 1$. b. If $x \geq 1$ or $x \leq 0$, then $\frac{1}{x} \leq x$. c. If $\frac{1}{x} \leq x$, then $x \geq 1$

or $x \leqslant 0$.   d. F, F, T   21. $6b(2a+1)(2a-1)$ 23. $x^2 - x - 12 = 0$

25. two irrational roots   27. $5\frac{1}{2}$ 29. $\frac{a^2 + 2ab + b^2}{(a-b)^2}$ 31. 45

33. $\exists_x\ 2(x-1) \neq 2x - 2$

Pages 311–313   1. $\frac{5}{4}$ 3. 1:1   5. $\frac{2}{1}$ 7. $\frac{2}{5}$ 9. $6m$ 11. $\frac{4}{b}$ 13. $\frac{3}{2}$

15. $\frac{3}{2}$ 17. 10   19. 59   21. $x = 8$; $y = 10$ 23. $x = 35$; $y = 4$ 25. $\frac{10}{13}$

27. $\frac{63}{29}$ 29. 25, 65   31. 6, 10   33. $AC = 15$, $BC = 50$ 35. 36, 54, 90

37. (1) $\frac{a}{b} = \frac{c}{d}$ (Given); (2) $\frac{a}{b}(bd) = \frac{c}{d}(bd)$ (Mult. Prop. of Equality); (3) $ad = bc$ (Properties of Mult.)   39. a. (1) $\frac{a}{b} = \frac{c}{d}$ (Given); (2) $ad = bc$ (Prop. 1 of Proportions); (3) $bc = ad$ (Symm. Prop. of Equality); (4) $\frac{bc}{ac} = \frac{ad}{ac}$ (Div. Prop. of Equality); (5) $\frac{b}{a} = \frac{d}{c}$ (Properties of Mult.)   b. (1) $\frac{b}{a} = \frac{d}{c}$ (Given); (2) $bc = ad$ (Prop. 1 of Proportions); (3) $ad = bc$ (Symm. Prop. of Equality); (4) $\frac{ad}{bd} = \frac{bc}{bd}$ (Div. Prop. of Equality); (5) $\frac{a}{b} = \frac{c}{d}$ (Properties of Mult.)   41. Use the Add. Prop. of Equality with the given information to show that $\frac{a}{b} + 1 = \frac{c}{d} + 1$. Then show that $\frac{a}{b} + \frac{b}{b} = \frac{c}{d} + \frac{d}{d}$.   43. $\{-9, 3\}$ 45. $\{5, 4\}$

51. Use Prop. 3 of Proportions with the given information to show that $\frac{a}{x} = \frac{b}{y}$ and that $\frac{c}{x} = \frac{d}{y}$. Then use the Subtr. Prop. of Equality and the Subst. Prop. to show that $\frac{a-c}{x} = \frac{b-d}{y}$. Then use Prop. 3 of Proportions.

Pages 317–319   1. a. $\frac{1}{2}$ b. $x = 6$, $y = 14$ 3. a. $\frac{3}{2}$ b. $x = 4\frac{1}{2}$, $y = 13\frac{1}{2}$

	DE	EF	DF	Per. of $\triangle DEF$	RS	ST	RT	Per. of $\triangle RST$
5.	12	15	9	36	8	10	6	24
7.	3	7.5	9	19.5	8	20	24	52
9.	9	15	12	36	12	20	16	48

11. $4\frac{1}{2}$ 13. $DE = 7\frac{1}{2}$, $EF = 10\frac{1}{2}$, $FD = 15$ 15. 17   17. $2\frac{2}{3}$ 19. 3, 5

21. Suppose the two polygons each have n sides, one has sides of length x, and one has sides of length y. All the ∠s have measure $\frac{(n-2)180}{n}$ and are ≅. The ratio of any pair of corr. sides is $\frac{x}{y}$. Thus the polygons are ~.   23. (1) $\triangle ABC \cong \triangle DEF$ (Given); (2) $\angle A \cong \angle D$; $\angle B \cong \angle E$; $\angle C \cong \angle F$; $\overline{AB} \cong \overline{DE}$; $\overline{BC} \cong \overline{EF}$; $\overline{AC} \cong \overline{DF}$ (Corr. parts of ≅ △s are ≅.); (3) $\triangle DEF \sim \triangle JKL$ (Given); (4) $\angle D \cong \angle J$; $\angle E \cong \angle K$; $\angle F \cong \angle L$; $\frac{DE}{JK} = \frac{EF}{KL} = \frac{DF}{JL}$ (Def. of ~ poly.); (5) $\angle A \cong \angle J$; $\angle B \cong \angle K$; $\angle C \cong \angle L$ (Trans. Prop. of ≅); (6) $\frac{AB}{JK} = \frac{BC}{KL} = \frac{AC}{JL}$ (Subst. Prop.);

(7) $\triangle ABC \sim \triangle JKL$ (Def. of $\sim$ poly.) 25. Use Thm. 5–13 to show that $\overline{DE} \parallel \overline{BA}$, $\overline{EF} \parallel \overline{CB}$, and $\overline{FD} \parallel \overline{AC}$. Then show that $EDBF$, $FAED$, and $EFDC$ are ▱s. Use Thm. 5–2 to show that $\angle DEF \cong \angle B$, $\angle EDF \cong \angle A$, and $\angle EFD \cong \angle C$. Use Thm. 5–13 to show that $DE = \frac{1}{2} \cdot BA$, $EF = \frac{1}{2} \cdot CB$, and $FD = \frac{1}{2} \cdot AC$. Then by the Div. Prop. of Equality and the Trans. Prop. of Equality $\frac{EF}{BC} = \frac{DE}{BA} = \frac{FD}{AC}$.

Pages 322–324 1. 12 3. $5\frac{1}{3}$ 5. 15 7. 20 9. 33 11. 4 13. 7 15. yes 17. yes 19. 8 21. 12 23. 8 25. a. (1) $\overline{DE} \parallel \overline{BC}$ (Given); (2) $\frac{AD}{DB} = \frac{AE}{EC}$ (Triangle Prop. Thm.); (3) $\frac{AD + DB}{DB} = \frac{AE + EC}{EC}$ (Prop. 4 of Proportions); (4) $\frac{AB}{DB} = \frac{AC}{EC}$ (Prop. of Betweenness and Subst. Prop.); (5) $\frac{DB}{AB} = \frac{EC}{AC}$ (Prop. 3 of Proportions) b. Use Props. 3 + 4 of Proportions. 27. Use the Subst. Prop. with the $\triangle$ Prop. Thm. and the given information. 29. 4 31. 6 33. 18 35. 3 37. Use Thm. 8–3 to show that $\frac{BE}{CE} = \frac{AB}{AC}$. Use the $\triangle$ Prop. Thm. to show that $\frac{BE}{CE} = \frac{BF}{FD}$. Then use the Subst. Prop. 39. Draw $\overline{BC}$ and let G be the pt. where $\overline{BC}$ intersects $\overline{EF}$. Use the $\triangle$ Prop. Thm. to show that $\frac{BE}{EA} = \frac{BG}{GC}$ and $\frac{BG}{GC} = \frac{DF}{FC}$. Then use the Subst. Prop. 41. 6 43. Draw $\overleftrightarrow{FJ}$ through $F \parallel$ to $\overleftrightarrow{HE}$. Use alt. int. ∠s to show that $\angle 4 \cong \angle 1$ and use corr. ∠s to show that $\angle 3 \cong \angle 2$. Then use the given information and the Subst. Prop. to show that $\angle 3 \cong \angle 4$. So $\overline{FE} \cong \overline{JE}$ and by the $\triangle$ Prop. Thm. $\frac{DH}{FH} = \frac{DE}{JE}$.

Pages 329–330 1. QR 3. PQ 5. SQ, YT 7. 6 9. 16 11. 15 13. 18 15. 70 17. (1) $\angle 1 \cong \angle 2$ (Given); (2) $\angle L \cong \angle L$ (Refl. Prop. of $\cong$); (3) $\triangle MLN \sim \triangle JLK$ (AA $\sim$ Thm.) 19. Use the Trans. Prop. of $\cong$ with the given information and $\cong$ vert. ∠s. 21. Show that $\triangle CAB \sim \triangle BAD$. 23. Use the AA $\sim$ Thm. to show that $\triangle FXY \sim \triangle FDG$ and that $\triangle FYZ \sim \triangle FGE$. Then show that $\frac{XY}{DG} = \frac{FY}{FG}$ and that $\frac{FY}{FG} = \frac{YZ}{GE}$. 25. Use Thm. 3–9 and the given information. 27. Use the AA $\sim$ Thm. to show that $\triangle SRV \sim \triangle QPV$ and that $\triangle VRT \sim \triangle VPS$. Then show that $\frac{QV}{SV} = \frac{PV}{RV}$ and that $\frac{PV}{RV} = \frac{SV}{TV}$. 29. Use the AA $\sim$ Thm. to show that $\triangle CPQ \sim \triangle CAB$. Then use $\frac{CP}{CA} = \frac{PQ}{AB}$ with the given information to show that $\frac{CP}{CA} = \frac{PQ}{BD}$. Use the AA $\sim$ Thm. to show that $\triangle PQR \sim \triangle DBR$. Then $\frac{PQ}{BD} = \frac{PR}{RD}$.

Pages 333–335 1. $\angle 1 \cong \angle 2$; $\frac{DF}{EF} = \frac{CF}{BF}$ 3. $\angle ACB \cong \angle GCD$; $\frac{CA}{CG} = \frac{CB}{CD}$ 5. $\frac{AB}{DF} = \frac{CA}{CD} = \frac{CB}{CF}$ 7. $\frac{AB}{GD} = \frac{CA}{CG} = \frac{CB}{CD}$ 9. yes; $\angle CXD \cong \angle BXA$, $\angle C \cong \angle A$ (or $\angle D \cong \angle B$); AA $\sim$ Thm. 11. yes; $\angle Z \cong \angle Y$, $\angle W \cong \angle X$ (or $\angle ZVW \cong \angle YVX$);

AA $\sim$ Thm. 13. Use the SAS $\sim$ Thm. to show that $\triangle AXB \sim \triangle CXD$. 15. Show that $\frac{OM}{OP} = \frac{RQ}{RS}$ and then use Prop. 3 of Proportions to show that $\frac{OM}{RQ} = \frac{OP}{RS}$. Then use the SAS $\sim$ Thm. with this and the given information to show that $\triangle OMP \sim \triangle RQS$. 17. Use Thm. 3–2 to show that $AM = \frac{1}{2} \cdot AB$ and $DN = \frac{1}{2} \cdot DE$. Then $\frac{AM}{AB} = \frac{1}{2} = \frac{DN}{DE}$ or $\frac{AM}{AB} = \frac{DN}{DE}$. By Prop. 3 of Proportions, $\frac{AM}{DN} = \frac{AB}{DE}$. Use the SAS $\sim$ Thm. with this and the fact that $\angle A \cong \angle D$ to show that $\triangle CAM \sim \triangle FDN$. 19. Use the SAS $\sim$ Thm. to show that $\triangle APB \sim \triangle DPC$. Then use the Subst. Prop. with $\frac{CP}{BP} = \frac{CD}{BA}$ and the given information to show that $\frac{CX}{BY} = \frac{CD}{BA}$. 20. Draw $\overline{AC}$ and $\overline{PR}$. Use the SAS $\sim$ Thm. to show that $\triangle ABC \sim \triangle PQR$. 21. Use the SSS $\sim$ Thm. to show that $\triangle ABC \sim \triangle PQR$, $\triangle BCD \sim \triangle QRS$, $\triangle CDE \sim \triangle RST$, $\triangle DEA \sim \triangle STP$, and $\triangle EAB \sim \triangle TPQ$. Then use the Def. of $\sim$ poly.

Pages 339–340 1. 25:36 3. 25:64 5. 1:8 7. a. $\frac{3}{4}$ b. $\frac{9}{16}$ c. 96 9. 144 11. 25:36 13. 64 15. a. $\frac{9}{49}$ b. $\frac{16}{9}$ 17. Use the results of Ex. 20, Section 8–5, to show that $\frac{AB}{FG} = \frac{BC}{GH} = \frac{AC}{FH}$ and that $\frac{AC}{FH} = \frac{CD}{HJ} = \frac{AD}{FJ}$. Then use the SSS $\sim$ Thm. 19. 2.5 cm; 1.5 cm 21. $\frac{AB}{AD} = \frac{x+y}{x}$; Use Thm. 8–9 and Prop. 4 of Proportions.

Pages 342–343 1. 36 m 3. approximately 1,300,000 km 5. 1.44 m 7. 26 m 9. Show that $\triangle PBX \cong \triangle PB'X$ and then use SAS to prove that $\triangle QBX \cong \triangle QB'X$. $AB' < AQ + QB'$ (Triangle Ineq.), thus $AP + PB' < AQ + QB'$.

Page 346 · Chapter Review 1. a. $3y$ b. $\frac{x}{y}$ c. $\frac{y-4}{4}$ 2. $4\sqrt{2}$ 3. $-\frac{3}{2}$ 4. 9 cm, 18 cm, 36 cm 5. $\angle C \cong \angle T$; Corr. ∠s of $\sim$ polygons are $\cong$. 6. 2:3 7. 15, 36 8. 3 9. 9 10. 6 11. (1) $\overline{WZ} \parallel \overline{XY}$ (Given); (2) $\angle VWZ \cong \angle VXY$ (If 2 lines are ∥, corr. ∠s are $\cong$); (3) $\angle V \cong \angle V$ (Ref. Prop. of $\cong$); (4) $\triangle VWZ \sim \triangle VXY$ (AA $\sim$ Thm.); (5) $\frac{WZ}{XY} = \frac{VW}{VX}$ (The lengths of corr. sides of $\sim$ △s are prop.)

9. (1) $\frac{VW}{WX} = \frac{VZ}{ZY}$ (Given); (2) $\overline{WZ} \parallel \overline{XY}$ (Thm. 8–4); (3) $\angle 3 \cong \angle 2$ (If 2 lines are ∥, alt. int. ∠s are $\cong$.); (4) $\angle 1 \cong \angle 2$ (Given); (5) $\angle 1 \cong \angle 3$ (Subst. Prop.); (6) $WX = WZ$ (Thm. 3–14) 13. 168 cm 14. yes; SAS 15. no 16. $CD = 16$ 17. 16:49 18. 36 19. 7.2 m

Pages 350–351 · Mixed Review

1. $\{4\}$ 3. $\left\{\frac{8}{5}\right\}$ 5. $\forall_x\ 3x + 1 \leqslant -5$ 7. $\frac{m}{4n}$ 9. $3t^3 - 4t^2 + t + 1 - \frac{10}{2t+3}$ 11. (1) $\triangle QRS$; X, Y, and Z are midpoints of $\overline{RS}$, $\overline{QS}$, and $\overline{QR}$, respectively. (Given); (2) $XZ = \frac{1}{2}QS$; $YZ = \frac{1}{2}RS$; $XY = \frac{1}{2}QR$ (Theorem 5–13); (3) $\frac{XZ}{QS} = \frac{1}{2}$; $\frac{YZ}{RS} = \frac{1}{2}$; $\frac{XY}{QR} = \frac{1}{2}$ (Div. Prop. of Equality); (4) $\frac{XZ}{QS} = \frac{YZ}{RS} = \frac{XY}{QR}$ (Subst. Prop.); (5) $\triangle XYZ \sim \triangle QRS$

(SSS $\sim$ Thm.) 13. (1) $\overline{AC}$ bisects $\angle BAD$; $\angle B \cong \angle D$ (Given); (2) $\angle BAC \cong \angle DAC$; (Def. of $\angle$ bisector); (3) $\overline{AC} \cong \overline{AC}$ (Refl. Prop. of $\cong$); (4) $\triangle BAC \cong \triangle DAC$ (AAS); (5) $\overline{BC} \cong \overline{CD}$ (Corr. parts of $\cong$ ⩟ are $\cong$.); (6) C is the midpoint of $\overline{BD}$. (Def. of midpoint)

15. You do not rush in. 17. $\frac{3}{2}\left(\text{or } \frac{2}{3}\right)$ 19. 27 21. $\left\{\frac{1 \pm 2\sqrt{2}}{3}\right\}$

23. 9 cm, 5 cm 25. $-(3x + 1)(x - 36)$ 27. 50 m^2 29. (1) $\angle 1 \cong \angle 10$ (Given); (2) $\angle 10 \cong \angle 7$ (Vert. ⦣ are $\cong$.); (3) $\angle 1 \cong \angle 7$ (Trans. Prop. of $\cong$); (4) $p \parallel q$ (Theorem 4-2)

Pages 356-357 1. 9 3. 30 5. $5\sqrt{3}$ 7. ab 9. 10 11. 6 13. 25 15. 12 17. 3 19. 15 21. 4 23. 50; 8 25. 3; 27; 30

27. Use the AA $\sim$ Thm. 29. In $\triangle DFE$, use Cor. 1 of Thm. 9-1 to show that $\frac{DG}{FG} = \frac{FG}{GE}$. In $\triangle DGF$, use Cor. 2 of Thm. 9-1 to show that $\frac{DF}{FG} = \frac{FG}{HF}$. Then use Prop. 1 of Proportions to rewrite each proportion.

Pages 360-362 1. 20 3. 24 5. 17 7. $4\sqrt{3}$ 9. $\sqrt{13}$ 11. a. yes b. yes c. yes d. yes 13. 13 15. 40 17. $3\sqrt{2}$ 19. 12; 60 21. 25 km 23. Use Thm. 9-3 to show that $\triangle AXB$ is a rt. $\triangle$. Then use Ex. 24, Section 5-3.

25. 3 27. 12 29. $l = 60$; $w = 11$ 31. a. 30 b. $\frac{60}{13}$ c. $\frac{144}{13}$ d. $\frac{4320}{169}$ 33. $DB = 32$; $AD = 18$; $AB = 50$; $AC = 30$ 35. a. Use the SSS $\sim$ Thm. b. Use corr. ⦣ of $\sim$ poly. are $\cong$. 37. a. area of $BDGE = cy$, area of $BHJC = a^2$, area of $ADGF = cx$, and area of $ACKL = b^2$. By Cor. 2 of Thm. 9-1, $\frac{c}{a} = \frac{a}{y}$ and $\frac{c}{b} = \frac{b}{x}$. Then $a^2 = cy$ and $b^2 = cx$. b. area of $ABEF$ = area of $BDGE$ + area of $ADGF$ = area of $BHJC$ + area of $ACKL$; that is, $c^2 = a^2 + b^2$. 39. $5\sqrt{6}$

Pages 365-367 1. $6\sqrt{2}$ 3. 10 5. 10 7. $6\sqrt{2}$ 9. $ST = 5\sqrt{3}$; $RT = 10$ 11. $RS = 7$; $ST = 7\sqrt{3}$ 13. $RS = 4\sqrt{3}$; $RT = 8\sqrt{3}$ 15. $ST = \sqrt{6}$; $RT = 2\sqrt{2}$

17. 2 19. $3\sqrt{3}$ 21. $9\sqrt{3}$ 23. 2; $2\sqrt{3}$ 25. $\frac{\sqrt{3}}{3}$; $\frac{\sqrt{3}}{3}$; $\frac{4\sqrt{3}}{3} + 2$

27. $40\sqrt{3}$ 29. a. $\frac{s\sqrt{3}}{2}$ b. Area $= \frac{s^2\sqrt{3}}{4}$ c. 4 31. $150\sqrt{3}$

33. a. $12 + 4\sqrt{3}$; $4\sqrt{6}$ b. $12 + 12\sqrt{3} + 4\sqrt{6}$ c. $24 + 24\sqrt{3}$ 35. $6\sqrt{2}$; $18 + 6\sqrt{3}$ 37. $3\sqrt{6}$ 39. $\frac{\sqrt{3}}{2}$

Pages 371-372 1. $\frac{\sqrt{7}}{4}$ 3. $\frac{\sqrt{7}}{4}$ 5. $\frac{\sqrt{7}}{3}$ 7. $\frac{\sqrt{2}}{2}$ 9. $\frac{1}{2}$

11. P: $\frac{3}{5}, \frac{4}{5}, \frac{3}{4}$; Q: $\frac{4}{5}, \frac{3}{5}, \frac{4}{3}$ 13. P: $\frac{5}{13}, \frac{12}{13}, \frac{5}{12}$; Q: $\frac{12}{13}, \frac{5}{13}, \frac{12}{5}$

15. P: $\frac{7}{25}, \frac{24}{25}, \frac{7}{24}$; Q: $\frac{24}{25}, \frac{7}{25}, \frac{24}{7}$ 17. 9 19. 59 21. 46 23. 1

27. 1 31. a. $\frac{ST}{SR}$ b. $\sin R$ c. $\frac{\sin R}{\cos R}$

Pages 376-378 1. 20 3. 27 5. 33 7. 247 9. 238 11. 68 13. 53 15. 26 17. 100 m 19. 45 m 21. 22° 23. $w = 4$, $l = 11$ 25. 29, 29, 122

Pages 379–380 • Chapter Review 1. y; z 2. $y+z$; z 3. $y+z$; v 4. $z=144$; $v=60$; $x=156$ 5. yes; $1^2+(\sqrt{3})^2=1+3=4=2^2$. By Thm. 9–3, the $\triangle$ is a rt. $\triangle$. 6. 126 7. 17 cm 8. 62.5 9. $XA=6\sqrt{3}$; $XM=2\sqrt{3}$; $MA=4\sqrt{3}$ 10. $\frac{3\sqrt{13}}{13}$ 11. $\frac{3\sqrt{13}}{13}$ 12. $\frac{3}{2}$ 13. $\frac{2}{3}$ 14. 0 15. 0.9945; 0.1045 16. 16 17. 2 km

Pages 382–383 • Mixed Review

1. no 3. 10 5. $\frac{\sqrt{10}}{10}$ 7. $\frac{1}{3}$ 9. 3 h 11. $\{x: 1 \leq x < 3\}$ 13. $\{-1, 4\}$ 15. 32 cm 17. $\frac{-6n+8}{(n-1)(n-2)}$ 19. 97 21. 98 m 23. $\sim q \vee \sim p$; T 25. 66 m 27. 26 cm 29. (1) $\angle 1 \cong \angle 4$; $\angle 2 \cong \angle 3$ (Given); (2) $\overline{AB} \parallel \overline{DC}$ (Theorem 4–1); (3) $\overline{DB} \cong \overline{DB}$ (Refl. Prop. of $\cong$); (4) $\triangle DAB \cong \triangle BCD$ (AAS); (5) $\overline{AB} \cong \overline{CD}$ (Corr. parts of $\cong$ $\triangle$s are $\cong$.); (6) $ABCD$ is a parallelogram. (Theorem 5–5)

Pages 388–389 1. Use Construction 1. 3. Use Const. 1 twice. 7. Use Const. 2. 9. Use Const. 2 to const. $\angle ABC \cong \angle P$. Use Const. 2 to const. $\angle ABD \cong \angle Q$ ($\overrightarrow{BD}$ in the exterior of $\angle ABC$). $m\angle DBC = m\angle P + m\angle Q$ 11. Use the method of the example on p. 386; SSS Post. 13. Use Const. 1 to const. $\overline{XY}$, $XY = c$. Then use Const. 2 to const. $\angle P \cong \angle D$ and $\angle Q \cong \angle E$; AA $\sim$ Thm. 17. Use the method of the example on p. 386; SSS $\sim$ Thm.; 1:4. 19. *Hint*: $120 = 180 - 60$.

Pages 393–395 1. Use Const. 3. 3. Use Const. 4. 5. Use the method of the example on p. 391. 7. Draw an obtuse $\angle ABC$. Use Const. 3 to construct the bisector $\overrightarrow{BX}$ of $\angle ABC$. Use Const. 3 to bisect $\angle ABX$ and $\angle CBX$. 9. Use the method of the example on p. 391. 13. Use Const. 4 to construct a rt. $\angle$. Use Const. 3 to bisect one rt. $\angle$, creating two $45°$ $\angle$s. Use Const. 3 again to bisect one of the $45°$ $\angle$s. 15. *Hint*: $30 = \frac{1}{2}(60)$ 17. *Hint*: $150 = 180 - 30$; $75 = \frac{1}{2}(150)$ 21. Use the SAS Post. to show that $\triangle PAC \cong \triangle PBC$. 25. Use the SAS Post. to show that $\triangle OFC \cong \triangle OFA$. 27. a. Construct $\angle W \cong \angle C$ and use Const. 3 to construct the bis. of $\angle W$. On the bisector, construct a segment $\overline{WY}$ of length d. Use Const. 4 to construct the $\perp$ bis. of $\overline{WY}$, intersecting the sides of $\angle W$ at X and Z. Draw $\overline{XY}$ and $\overline{YZ}$. $WXYZ$ is a rhombus with longer diagonal of length d. b. Construct an $\angle \cong$ to $\angle C$ and extend one side of $\angle C$ in the opp. direction to construct $\angle Q \cong$ to the supp. of $\angle C$. Proceed as in part (a). $QRST$ is a rhombus with shorter diagonal of length d.

Pages 398–399

1. Use Const. 6. 3. Use the method of the example on page 397. 5. Other answers are possible.

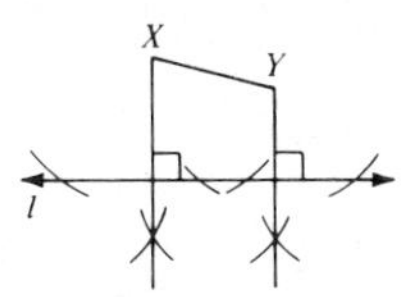

7. Use Const. 2, then Const. 6. 15. d. They appear to be collinear. 17.

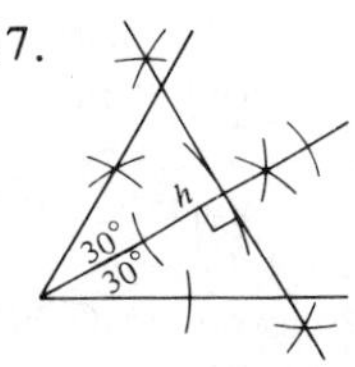

Pages 402–403 1. Use Const. 8. 3. Use Const. 1 and Const. 8. 11. a parallelogram 13. Draw a segment and use Const. 9 to divide it into 4 $\cong$ segments. Label one of the segments $\overline{PQ}$. Then construct a rhombus, each of whose sides has length PQ, using the construction suggested in the diagram that accompanies Oral Exercise 8. 15. Yes; use Const. 3 to construct $\overrightarrow{CD}$, the bisector of $\angle C$ (Thm. 8–3).

Page 407 1, 3. Use Construction 10. 5. Use Construction 11. 7. Follow the example on page 406, but let $QR = 1$ and $RS = 3$. 9. Follow the example on page 406, but let $QR = 2$ and $RS = 4$. 11. Use Construction 10, but let $PQ = 1$, $QR = x$, and $PS = y$ (or $QR = y$ and $PS = x$). Then $ST = xy$. 13. Use Construction 10, but let $PQ = 1$, $QR = y$, and $PS = z$ (or $QR = z$ and $PS = y$). Then $ST = yz$. 15. Use Construction 10, but let $PQ = x$, $QR = 1$, and $PS = 1$. Then $ST = \frac{1}{x}$. 17. Use Construction 11, but let $QR = x$ and $RS = y$ (or $QR = y$ and $RS = x$). Then $PR = \sqrt{xy}$. 19. Use Construction 11, but let $QR = 1$ and $RS = y$. Then $PR = \sqrt{y}$. 21. Use Construction 11, but let $QR = 4$ and $RS = x$. Then $PR = \sqrt{4x}$. (Or let QR and RS be x and 4, or 2 and $2x$, or $2x$ and 2.)

23.

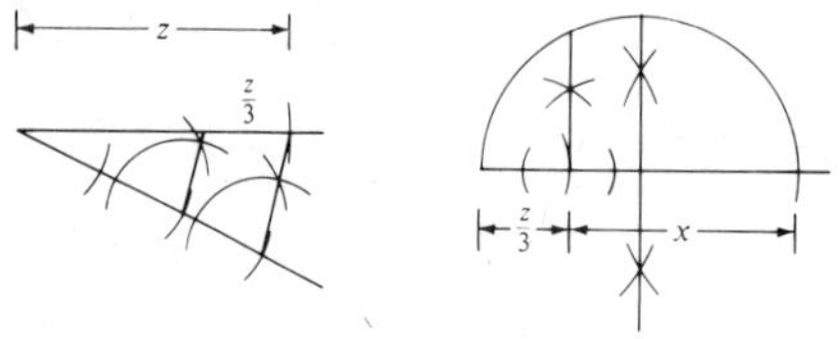

25. More than one answer is possible. An example follows.

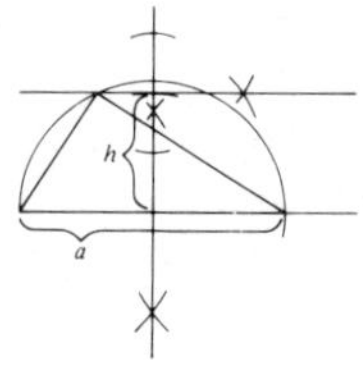

Pages 410–411 1. the ⊥ bis. of $\overline{RS}$ 3. 2 lines, both ∥ to n and 1 cm from n 5. the bis. of $\angle XYZ$ 7. the circle with center B, radius 6 units 9. the ⊥ bis. of $\overline{AC}$ 11. two lines, both ∥ to $\overleftrightarrow{BC}$ and 2 units from $\overleftrightarrow{BC}$ 13. Let $\overrightarrow{CD}$ be the bis. of $\angle ACB$, intersecting $\overline{AB}$ at E. The desired locus is $\overline{CE}$. 15. 2 lines, both ⊥ to $\overline{AB}$, one line containing A, the other containing B 17. a segment of length 3 km ∥ to the track at a distance from the track equal to the radius of the wheel 19. two lines, both ∥ to $\overline{XY}$ and at a distance $\frac{2k}{XY}$ from $\overline{XY}$ 21. a. Use the HL Theorem to show that $\triangle BAD \cong \triangle CAD$. Then show that $\angle BAD \cong \angle CAD$. b. Use the AAS ≅ Theorem to show that $\triangle BAD \cong \triangle CAD$. Then show that $\overline{BD} \cong \overline{CD}$. 23. Since an ∠ inscribed in a semicircle is a rt. ∠, the locus is the circle (excluding points R and S) that has $\overline{RS}$ as a diameter.

Pages 413–414

1. 2 points, the intersection of the circle with center A and radius 3 cm and the line ∥ to, and 1 cm from, both m and l
3. 1 point, the intersection of the ⊥ bis. of the side of $\triangle ABP$
5. 1 point, the intersection of the bis. of $\angle Y$ and the circle with center Y and radius 3 cm
7. 1 point, the intersection of the ⊥ bis. of $\overline{XY}$ and $\overline{YZ}$
9. 1 point, the intersection of the ⊥ bis. of $\overline{JK}$ (and $\overline{ML}$) and $\overline{JM}$ (and $\overline{KL}$) 11. no points
13. 6
15. Let x = distance from P to k. The locus is the intersection of the circle with center P and radius 2 cm and 2 lines, both ∥ to and 1 cm from l. Case 1 ($x < 1$): 4 points; case 2 ($x = 1$): 3 points; case 3 ($1 < x < 3$): 2 points; case 4 ($x = 3$): 1 point; case 5 ($x > 3$): no points
17. The locus is the intersection of the ⊥ bis. of the sides of $WXYZ$. Case 1, points W, X, Y, and Z lie on a circle: 1 point; case 2, points W, X, Y, and Z do not lie on a circle: no points
19. The bisectors of the 4 ∠s formed by the intersection of lines m and n form two ⊥ lines, r and s. A point is on the locus if it is on either r or s and also on the circle with center P and radius the given distance. Case 1, the circle intersects neither r nor s: no points; case 2, the circle is tangent to one of r or s but does not intersect the other: 1 point; case 3, the circle is tangent to both r and s or intersects one of them twice and the other not at all: 2 points; case 4, the circle is tangent to one of r or s and intersects the other twice: 3 points; case 5, the circle intersects both r and s twice: 4 points.

Pages 416–417 1. Use Constr. 4 to construct the $\perp$ bis. of $\overline{YZ}$. 3. Use Constr. 3 to construct the bis. of $\angle YXZ$, intersecting $\overline{YZ}$ at V. The locus is V. 5. Use Constr. 1 to construct $\overline{VW}$ such that $VW = c$. Draw 2 circles, one with center V and radius a, the other with center W and radius b. The locus consists of the two points where the circles intersect. 7. Draw line m. Choose any point X on m and use Constr. 6 to construct line p through X and $\perp$ to m. Mark off $XY = c$ and use Constr. 6 to construct line n through $Y \perp$ to p. Use Constr. 4 to construct line q, the $\perp$ bis. of $\overline{XY}$. The locus is q. 9. With center Q and radius a, draw an arc intersecting k at P. Use Constr. 3 to construct the $\angle$ bis. of both pairs of vert. $\angle$s formed by k and l. Draw the circle with center P and radius c. The locus consists of the 4 points where the circle intersects the $\angle$ bisectors. 11. Use Constr. 3 to construct $\overrightarrow{BD}$, the bis. of $\angle ABC$. Use Constr. 6 to construct line $m \perp \overleftrightarrow{BC}$ through any point X on $\overleftrightarrow{BC}$, and mark off $XE = a$ on m. Use Constr. 6 to construct line n through $E \perp$ to m. The locus is the point where n intersects $\overrightarrow{BD}$.
13. Use Constr. 1 to construct $\overline{PQ}$ such that $PQ = c$. Use Constr. 4 to construct the $\perp$ bis. of $\overline{PQ}$ and locate M, the midpt. of $\overline{PQ}$. Draw the circle with center M and radius b. The locus is the circle, excluding the 2 points where the circle intersects $\overleftrightarrow{PQ}$. 15. Use Constr. 2 to construct $\angle J \cong \angle Y$. Mark off $JL = b$ on one side of $\angle J$. Draw an arc with center L and radius c intersecting the other side of $\angle J$ at M. Mark off $MK = JM$. Draw $\overline{KL}$.
17. Answers may vary. Construct an equilateral $\triangle DEF$. Use Constr. 7 to construct $\overline{DG}$, the alt. to $\overline{EF}$. Use Constr. 1 to construct $\overline{BC} \cong \overline{DG}$. Use Constr. 4 to construct the $\perp$ bis. of $\overline{BC}$ and locate H, the midpt. of $\overline{BC}$. Mark off $HA = EF$ on the $\perp$ bis. $\triangle DEF$ and $\triangle ABC$ both have area $\frac{1}{2}(EF \cdot DG)$. 19. Use Constr. 2 to construct $\angle ABC \cong \angle Y$. Extend $\overrightarrow{CB}$ through B to form $\angle ABD$, the supp. of $\angle ABC$. Use Constr. 3 to construct $\overrightarrow{BE}$, the bis. of $\angle ABD$. Use Constr. 1 to construct $\overline{PQ}$ such that $PQ = a$. Use Constr. 2 to construct $\angle$s $\cong$ to $\angle Y$ at P and Q, and form isos. $\triangle RPQ$. Extend $\overrightarrow{PR}$ and $\overrightarrow{QR}$. Mark off $RS = PR$ on $\overrightarrow{PR}$ and $RT = PR$ on $\overrightarrow{QR}$. Draw $\overline{PT}, \overline{TS}, \overline{SQ}$.

Page 418 • Chapter Review

1.

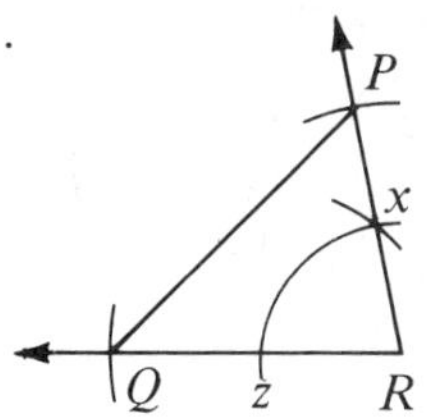

2. See Ex. 13, page 393. 3. a. Use Constr. 4. b. See the example, page 391. 4. Use Constr. 5. 5. a. See the example, page 397. b. Use Constr. 6. 6. Use Constr. 8. 7. Use Constr. 1 to construct $\overline{AB}$ such that $AB = y$. Then use Constr. 9. 8. Use Constr. 10. 9. Use Constr. 11. 10. 2 lines both $\parallel$ to l and 3.5 cm from l.

11.

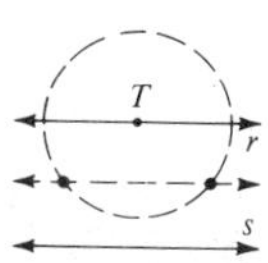

There are 2 points in the locus.

12.

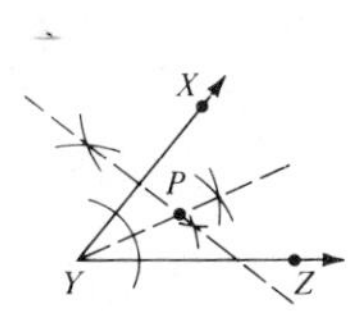

The locus is point P.

Pages 420–422 • Cumulative Review: Chapters 1–10 1. c 2. d 3. a 4. c 5. b 6. d 7. b 8. d 9. c 10. d 11. b 12. d 13. a 14. b 15. c 16. c 17. a 18. c 19. c 20. b 21. a 22. c 23. b 24. c 25. d 26. b 27. d 28. b 29. a 30. c 31. d 32. c

33. 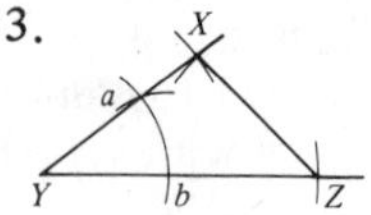
34. 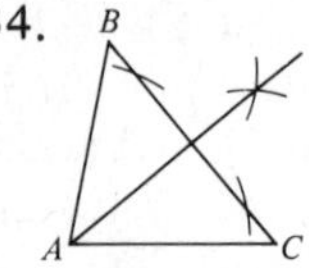
35. (1) $\overline{AD} \perp \overline{BC}$ (Given); (2) $\angle ADB \cong \angle ADC$ (Thm. 3–10); (3) $\angle B \cong \angle C$ (Given); (4) $\overline{AD} \cong \overline{AD}$ (Reflexive Prop. of $\cong$); (5) $\triangle ABD \cong \triangle ACD$ (AAS Thm.); (6) $\angle BAD \cong \angle CAD$ (Corr. parts of $\cong$ ▲s are $\cong$.); (7) $\overrightarrow{AD}$ bisects $\angle BAC$. (Def. of an $\angle$ bisector) 36. (1) $ABCD$ is a ▱. (Given); (2) $\overline{AB} \parallel \overline{DC}$ (Def. of a ▱); (3) $\overline{BE} \cong \overline{DF}$ (Given); (4) $BEDF$ is a ▱. (If 2 sides of a quad. are both $\cong$ and $\parallel$, then the quad. is a ▱.) 40. a. P_1: $h \to p$; P_2: $\sim h \to \sim s$; P_3: $t \to s$; P_4: $\sim p$ b. (1) $h \to p$ (Given); (2) $\sim p$ (Given); (3) $\sim h$ (Law of Contrapositive Inference); (4) $\sim h \to \sim s$ (Given); (5) $\sim s$ (Law of Detachment); (6) $t \to s$ (Given); (7) $\sim t$ (Law of Contrapositive Inference)

Pages 424–425 · Mixed Review

1. Use Construction 9. 3. the circle with center A, radius 1 unit 5. 2 lines, both $\parallel$ to $\overleftrightarrow{BC}$ and $1\frac{1}{2}$ units from $\overleftrightarrow{BC}$ 7. $x^2 - 4x + 2 = 0$ 9. $4y^2 - 2y + 1$ 11. $16(3a + b)(3a - b)$ 13. $p \to \sim s$ 15. $\left\{-\frac{5}{3}, -1\right\}$ 17. $\triangle TMP \sim \triangle TRI$; $\overline{MP} \parallel \overline{RI}$ (Theorem 5–13); $\angle TMP \cong \angle TRI$; $\angle TPM \cong \angle TIR$ (Theorem 4–5); $\therefore \triangle TMP \sim \triangle TRI$ (AA $\sim$ Thm.) 19. trapezoid 21. Use Construction 1 twice. 23. $HG = 8\sqrt{3}$; $GI = 16$ 25. $HI = 3$; $HG = 3\sqrt{3}$ 27. (1) $\overline{TR} \parallel \overline{SA}$ or $\overline{TR} \parallel \overline{PA}$; $\angle S \cong \angle TPA \cong \angle A$ (Given); (2) $TRAP$ is a trapezoid. (Def. of a trap.); (3) $TRAP$ is an isosceles trapezoid. (If 2 base $\angle$s of a trap. are $\cong$, then the trap. is isos.; see Ex. 19 on page 209.) 29. (1) $UWXY$ is a parallelogram; $UVWZ$ is a rectangle. (Given); (2) $\angle UZW$ is a rt. $\angle$ (Def. of a rect.); (3) $\overline{UX} \perp \overline{YW}$ (Def. of $\perp$ lines); (4) $UWXY$ is a rhombus. (If the diags. of a ▱ are $\perp$, then the ▱ is a rhombus; see Ex. 24 on page 200.)

Pages 430–432

1–11. 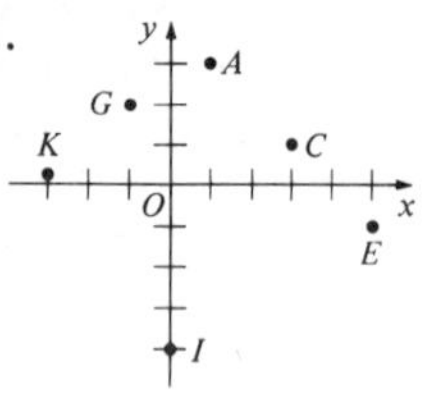
13. III 15. IV 17. 5 19. 1 21. 6 23. 4 25. 12 27. 1 29. 24 31. 9 33. $\frac{31}{2}$ 35. 9 37. 17 39. 5 41. -1 43. $(-3, 0)$ 45. $(3, 6)$
47. 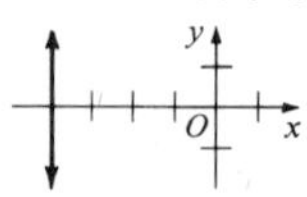
49. 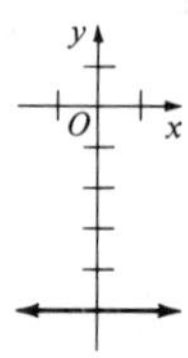

Pages 435–436 1. $\left(-\frac{1}{2}, 0\right)$ 3. $(1.9, 2)$ 5. $\left(\frac{a+c}{2}, \frac{b+d}{2}\right)$ 7. $3\sqrt{5}$ 9. $9\sqrt{2}$ 11. $4\sqrt{2}$ 13. $(-10, -6)$ 15. $(3, -13)$ 17. a. $\sqrt{85}$ b. $\sqrt{85}$ c. $\sqrt{17}$ d. $\sqrt{17}$ e. a ▱; Theorem 5–4 19. Use the distance formula to show that $AB = BC = CD = DA$. 21. -3 23. 8 or 32 25. a. Use the distance formula to show that $(RS)^2 + (ST)^2 = (TR)^2$. b. By the midpoint formula, hypotenuse $\overline{TR}$ has midpoint $M(0, 7)$. By the distance formula, $SM = \sqrt{26} = \frac{1}{2}(TR)$. 27. Use the distance formula to show that $PQ = RS$ and $QR = PS$. 29. $\sqrt{c^2 + d^2}$ 31. $2|c|$ 33. 42

Pages 441–443 1. 1 3. 0 5. $\frac{3}{2}$ 7. -1 9. -8 11. -10

13. a. $\frac{7}{3}$ b. $\frac{k+2}{3}$ c. 5 15. a. $-\frac{k+1}{6}$ b. $\frac{3k-4}{3}$ c. 1 17. noncollinear 19. collinear 21. −1 23. −6 or 1 25. a. By the distance formula, $AB = CD = \sqrt{26}$. b. Slope of $\overline{BC}$ = slope of $\overline{AD}$ c. Slope of $\overline{AB} = 5$; slope of $\overline{CD} = \frac{1}{5}$ d. $ABCD$ is an isosceles trapezoid. 27. Prove that $\triangle ABM \sim \triangle CDN$. Then use the properties of ~ ▲s and the properties of proportions to show that $\frac{BM}{AM} = \frac{DN}{CN}$. 29. a. Use the distance formula to show that two sides are ≅. b. Use the midpoint formula and Theorem 11–5. 31. a. 19 b. −1 33. Use the def. of slope and Theorem 11–5. 37. $(\sqrt{10}, 3\sqrt{10})$

Pages 446–447 1. isosceles 3. $(a + b, c)$ 5. Use the distance formula and the def. of a rhombus. 7. By the midpoint formula, the midpoints of $\overline{AC}$ and $\overline{BC}$ are $M(b, c)$ and $N(a + b, c)$, respectively. $AB = 2a$, and $MN = (a + b) - b = a$. Therefore, $MN = \frac{1}{2}AB$. 9. Use Theorem 11–4 and Theorem 11–3. 11. Show that both diagonals have the same midpoint. 13. Show that the coordinates of D must be $(0, c)$. 15. Let the vertices be $A(0, 0)$, $B(a, 0)$, $C(a, b)$, and $D(0, b)$, where $b > 0$. Use Thm. 11–5.

Pages 450–451 1. x-intercept, 6; y-intercept, 2 3. x-intercept, 4; y-intercept, 8 5. x-intercept, $\frac{15}{2}$; y-intercept, −5

11. 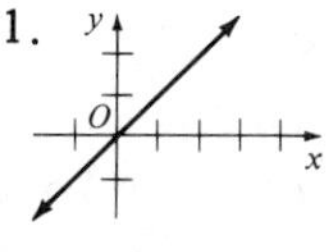

13. 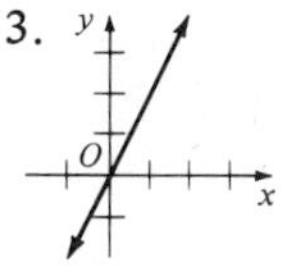

21. 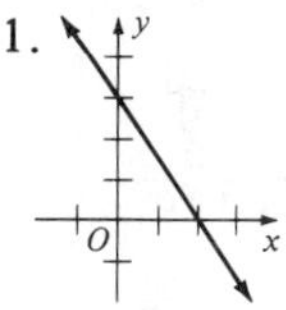

27. 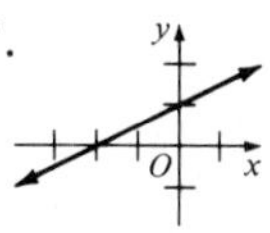

33. b. −3 35. b. $-\frac{1}{2}$ 37. 4 39. 2 41. −15 43. −6

45. 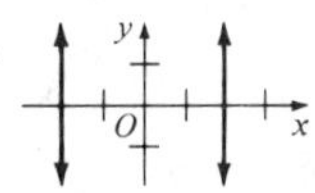

47. 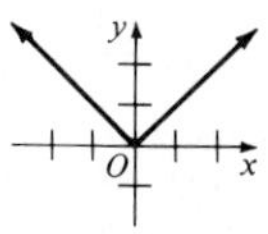

49. 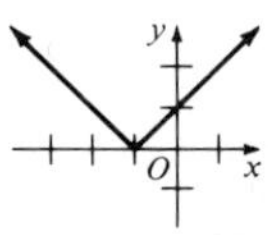

51. 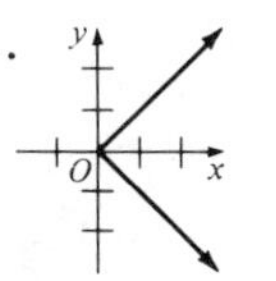

53. 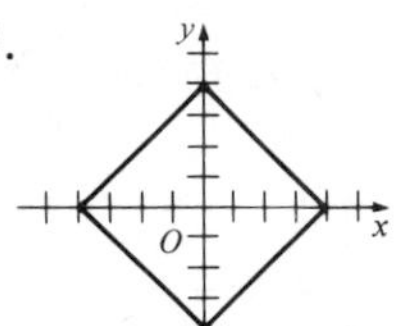

Pages 454–456 1, 3. Form of equation may vary. An example is given. 1. $3x - y = -2$ 3. $2x - y = 9$ 5. $y = -\frac{3}{4}x + 2$ 7. a. 1 b. $y - 2 = x - 1$ or $y + 1 = x + 2$ 9. a. 9 b. $y + 2 = 9(x - 3)$ or $y - 7 = 9(x - 4)$ 11. a. $\frac{1}{3}$ b. $y - 7 = \frac{1}{3}(x - 2)$ or $y - 8 = \frac{1}{3}(x - 5)$ 13. $y = 3x - 2$ 15. $y = 2x + 1$ 17. $y = \frac{1}{2}x + 3$ 19. slope, 2; y-intercept, −2 21. slope, $\frac{1}{2}$; y-intercept, −1 23. slope, −1; y-intercept, $\frac{1}{3}$ 25. a. 1 b. −1 27. a. −3 b. $\frac{1}{3}$ 29. a. $\frac{2}{3}$ b. $-\frac{3}{2}$ 31, 33. Form of equation may vary. An example is given. 31. $3x - y = 3$ 33. $2x + y = 7$ 35. $y = 3$ 37. $y = -x + 3$ 39. $y = \frac{1}{2}x$ 41. $y = -1$

43–53. Form of equation may vary. An example is given. 43. $3x + 2y = 12$ 45. $x - 2y = 2$ 47. $7x - y = 4$ 49. $x - y = 0$ 51. $(b - c)x - ay = -ac$ 53. a. $2x + 3y = 21$ b. $P(0, 7)$; use the distance formula to show that $PA = PB$. 55. Use the fact that the line contains the points $(a, 0)$ and $(0, b)$.

Pages 457–459 1. $y = 2$ 3. $x = 3$ and $x = 9$ 5. $y = -1$ and $y = 2$ 7. $y = -1$ 9. $x = -\frac{1}{2}$ 11. $x = 3$ 13. $y = \frac{5}{2}$ 15. $x + y = 5$ 17. (1, 3), (7, 3), (1, 7), (7, 7) 19. (−2, −7), (0, −7), (−2, 1), (0, 1) 21. 4 23. none 25. 2 27. 4 29. 4 31. $y = 2$, except for the point (0, 2) 33. Let $P(x, y)$ be any point on the locus. Use the distance formula to express the fact that P is equidistant from (x_1, y_1) and (y_1, x_1). Then simplify the resulting equation.

Pages 461–462 1. (2, 2) 3. (1, 4) 5. (5, −1) 7. (−2, 1) 9. $\emptyset$ 11. $\{(x, y)\colon y = 2x - 4\}$ 13. (−1, 5) 15. (−2, 4) 17. inconsistent 19. consistent and independent 21. 13.5 23. 48 25. (−2, −2) 27. (6, −2) 29. (3, 0) 31. inconsistent 33. inconsistent 35. consistent and dependent 37. 44 39. 55

Pages 464–466 1–39. Note that variables should be listed alphabetically within ordered pairs. 1. (2, −8) 3. (6, 3) 5. (−2, −2) 7. (−5, 4) 9. (−1, 2) 11. (−5, 3) 13. (20, 30) 15. (6, 2) 17. $\left(\frac{4}{7}, -\frac{3}{7}\right)$ 19. (4, 6) 21. $\left(\frac{34}{3}, 12\right)$ 23. $\emptyset$ 25. (−3, 7) 27. (−4, −3) 29. $\left(-\frac{1}{4}, \frac{3}{2}\right)$ 31. 60 33. $\frac{175}{6}$ 35. $\left(\frac{165}{94}, -\frac{61}{47}\right)$ 37. (12, −6) 39. (−16, −47) 41. $A = \frac{3}{5}$, $B = -1$ 43. $A = 2$, $B = 3$ 45. a. The ⊥ bisectors of $\overline{AB}$, $\overline{BC}$, and $\overline{AC}$ have equations $x = 2$, $3x - 4y = -7$, and $x + 4y = 15$, respectively. b. $Q\left(2, \frac{13}{4}\right)$ c. Verify that the coordinates of Q satisfy the equation of the third ⊥ bis. 47. Show that the slope of $\overline{PQ}$ = the slope of $\overline{QR}$.

Pages 468–469 • Chapter Review

1. 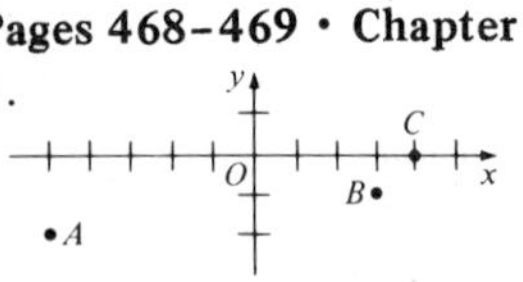

2. 10 3. 21 4. (3, 0) 5. By the distance formula, $LM = 2\sqrt{5}$, $MN = \sqrt{5}$, and $LN = 5$. Then $(LM)^2 + (MN)^2 = 20 + 5 = 25 = (LN)^2$. Thus $\triangle LMN$ is a rt. △ by Theorem 9–3. 6. Slope of $\overline{LM} = -2$; slope of $\overline{MN} = \frac{1}{2}$; slope of $\overline{LN}$ is undefined. Since $-2 \cdot \frac{1}{2} = -1$, $\overline{LM} \perp \overline{MN}$ by Theorem 11–5. Thus $\angle LMN$ is a rt. ∠ and $\triangle LMN$ is a rt. △ by definition. 7. undefined 8. 14 9. $(4a, 0)$ 10. By the midpoint formula, the midpoints of $\overline{TS}$ and $\overline{RS}$ have coordinates $M_1(a, b)$ and $M_2(3a, b)$, respectively. The medians to the legs of isosceles $\triangle TRS$ are $\overline{RM_1}$ and $\overline{TM_2}$. $RM_1 = \sqrt{9a^2 + b^2}$ and $TM_2 = \sqrt{9a^2 + b^2}$. Thus, $\overline{RM_1} \cong \overline{TM_2}$.

11. 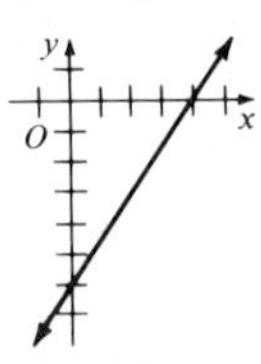

12. x-intercept, 10; y-intercept, −8

13. 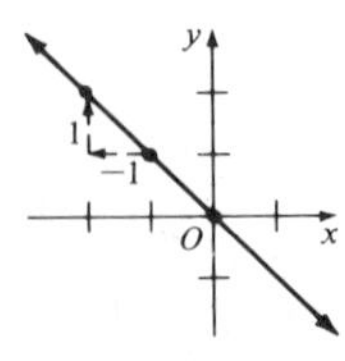

14. $y - 8 = 2(x + 7)$ 15. $y = \frac{1}{2}x - 5$ 16. Form of equation may vary. For example, $x + 3y = 5$. 17. $y = 5$ 18. $x = -\frac{7}{2}$ 19. $\{(1, -6), (3, -6), (1, 0), (3, 0)\}$
20. The graphs are parallel lines. 21. $\{(x, y)\colon x + y = -3\}$ 22. $(-2, 1)$
23. $(-4, 2)$ 24. $(7, 5)$

Pages 472–473 • Mixed Review

1. x-intercept, 16; y-intercept, -2 3. $\angle J$, $\angle K$

5.

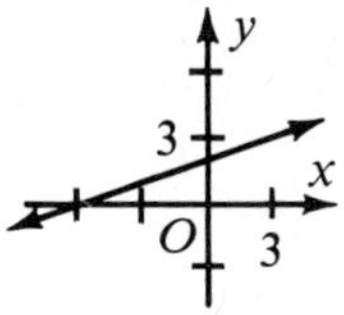

7.

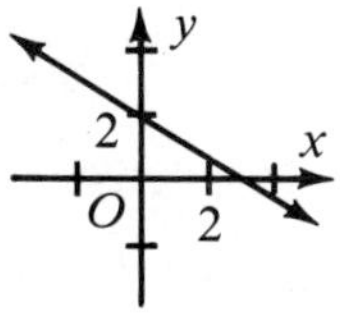

9. Use Constructions 4 and 11 and the procedure outlined in the Example on page 406.

11. $AC = \sqrt{(7-2)^2 + (4-7)^2} = \sqrt{5^2 + (-3)^2} = \sqrt{25 + 9} = \sqrt{34}$;
$BC = \sqrt{(7-2)^2 + (4-1)^2} = \sqrt{5^2 + 3^2} = \sqrt{25 + 9} = \sqrt{34}$ 13. $4, 4\sqrt{3}$ 15. $\sqrt{10}$
17. $(5, 4)$ 19. 148 21. $x + y = 3$ 23. $(k - 10)(k - 15)$ 25. 18 m
27. (1) $\overline{UV} \parallel \overline{ZY}$; $\overline{VW} \parallel \overline{XZ}$ (Given); (2) $\angle XUV \cong \angle Z$; $\angle Z \cong \angle VWY$; $\angle XVU \cong \angle VYW$ (Theorem 4–5); (3) $\angle XUV \cong \angle VWY$ (Trans. Prop. of $\cong$); (4) $\triangle XUV \sim \triangle VWY$ (AA $\sim$ Thm.)

Pages 479–480

1. b–d.

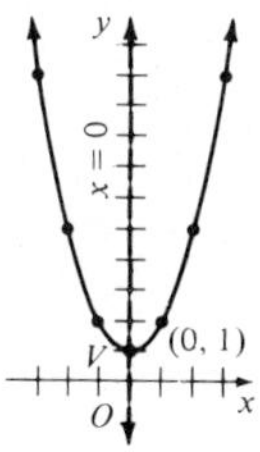

e. minimum point

3. b–d.

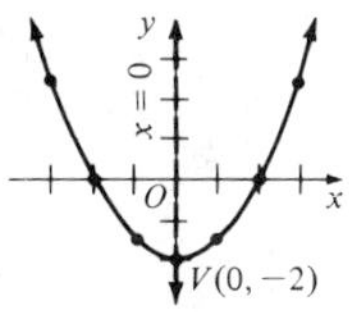

e. minimum point

5. b–d.

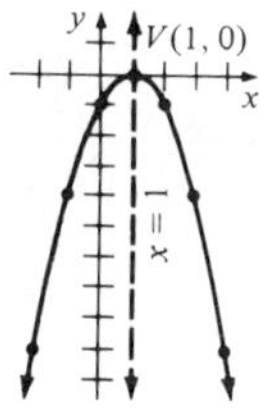

e. maximum point

11. b–d.

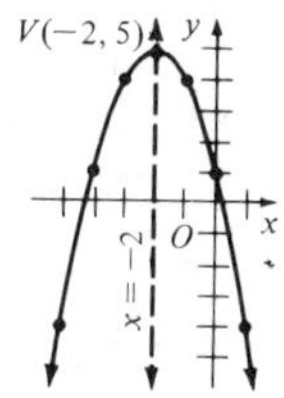

e. maximum point

13. $y = \frac{1}{2}(x - 1)^2 + 5$ or $y = -\frac{1}{2}(x - 1)^2 + 5$ 15. $y = -3(x - 4)^2 - 7$ or $y = 3(x - 4)^2 - 7$

17. $y = -\frac{1}{3}(x - 6)^2 + 6$ or $y = \frac{1}{3}(x - 6)^2 + 6$ 19. All three parabolas are congruent to the graph of $y = x^2$, open upward, and have axis of symmetry $x = 0$. Their vertices are all minimum points and lie on the vertical line $x = 0$. The graphs of $y = x^2 + 3$, $y = x^2 - 4$, and $y = x^2 + 1$ are obtained by shifting the graph of $y = x^2$ up 3 units, down 4 units, and up 1 unit, respectively.
21. Both parabolas are congruent to the graph of $y = x^2$, have axis of symmetry $x = -2$, and have vertex $(-2, 3)$. The parabola $y = (x + 2)^2 + 3$ opens upward (vertex is a minimum point); the parabola $y = -(x + 2)^2 + 3$ opens downward (vertex is a maximum point).
23. All three parabolas open downward, have axis of symmetry $x = 2$, and have vertex and maximum point $(2, 1)$. The parabola $y = -\frac{1}{2}(x - 2)^2 + 1$ is "wider" than the parabola $y = -(x - 2)^2 + 1$, which is "wider" than the parabola $y = -2(x - 2)^2 + 1$.

25. a.

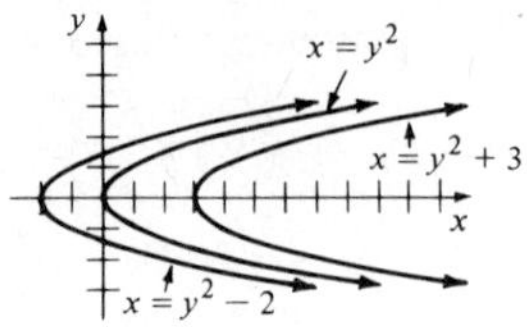

b. The graph of $x = y^2 + k$ is congruent to the graph of $x = y^2$, has the same axis of symmetry, and is obtained by shifting the graph of $x = y^2$ horizontally by $|k|$ units: to the right if $k > 0$ and to the left if $k < 0$.

27.

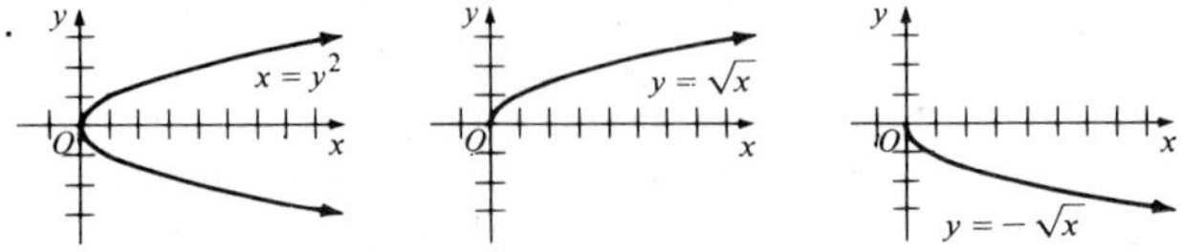

The graph of $y = \sqrt{x}$ is the upper half of the parabola $x = y^2$; the graph of $y = -\sqrt{x}$ is the lower half.

Pages 484–485 1. $y = (x - 2)^2 + 2$; $(2, 2)$ 3. $y = (x - 1)^2 - 1$; $(1, -1)$

5. $y = (x + 5)^2 + 5$; $(-5, 5)$ 7. $y = \left(x - \frac{3}{2}\right)^2 - \frac{5}{4}$; $\left(\frac{3}{2}, -\frac{5}{4}\right)$

9. $y = \left(x + \frac{9}{2}\right)^2 - \frac{93}{4}$; $\left(-\frac{9}{2}, -\frac{93}{4}\right)$

11. b–d.

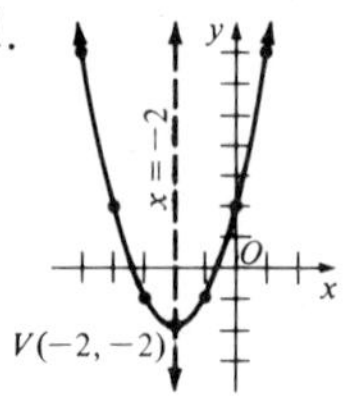

e. minimum point

15. b–d.

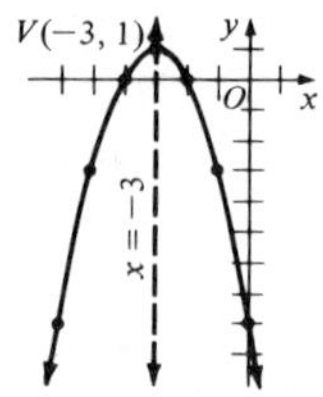

e. maximum point

21. $x = -1$ 23. $(-6, -32)$; minimum point 25. $(1, -6)$; maximum point

27. $\left(-\frac{1}{2}, 0\right)$; minimum point 29. -12 31. -1 33. $\frac{1}{8}$ 35. $y = 2(x - 1)^2$; $(1, 0)$ 37. $y = -\left(x - \frac{5}{2}\right)^2 + \frac{49}{4}$; $\left(\frac{5}{2}, \frac{49}{4}\right)$ 39. $y = \frac{1}{3}(x + 3)^2 - 6$; $(-3, -6)$

41. b. $\left(-\frac{b}{2a}, \frac{4ac - b^2}{4a}\right)$.

Page 487 $\{-1, 5\}$ 3. $\{-3, 3\}$ 5. $\{-4, 3\}$ 7. $\left\{-\frac{3}{2}\right\}$ 9. $\{0\}$

11. $\{-2\}$ 13. $\emptyset$ 15. $\{0, 1\}$ 17. $\emptyset$ 19. $k = 1$; $k > 1$; $k < 1$ 21. $-9, -3$

23. $-\frac{3}{2}, \frac{2}{3}$ 25. $\{x: x \leq -3 \text{ or } x \geq 2\}$ 27. a. $\left\{x: \frac{1}{2} < x < 3\right\}$

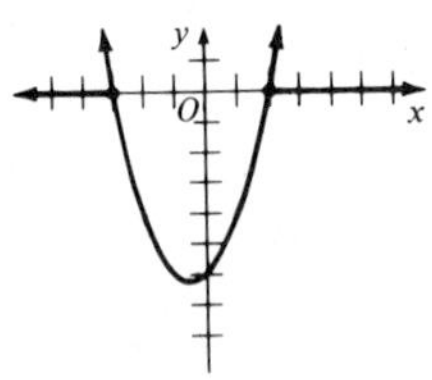

Pages 490–491 1. $x^2 + y^2 = 9$ 3. $x^2 + y^2 = 25$ 5. $x^2 + y^2 = 7$

7. $x^2 + (y + 4)^2 = 4$ 9. $(x - 5)^2 + (y + 4)^2 = 1$ 11. $(x - 2)^2 + y^2 = 7$

13. $r = 2$; $C(-2, -1)$ 15. $r = 3$; $C(5, 1)$ 17. $r = 5$; $C(0, 0)$ 19. $r = 4$; $C(2, -1)$

21. $r = 2$; $C(-2, 0)$ 23. $r = 1$; $C(5, 3)$ 25. $r = 5$; $C\left(-2, \frac{2}{3}\right)$
27. $r = 2$; $C\left(\frac{1}{2}, -\frac{3}{4}\right)$ 29. $r = 3$; $(x-2)^2 + (y-3)^2 = 9$ 31. $r = 5$; $(x-3)^2 + (y-4)^2 = 25$ 33. $r = \sqrt{5}$; $(x+2)^2 + (y+2)^2 = 5$ 35. $x^2 + y^2 - 25 = 0$
37. $x^2 + y^2 - 2x + 6y - 40 = 0$ 41. Since the circle has center $(0, 0)$ and radius a, its equation is $x^2 + y^2 = a^2$. If $P(x, y)$ is any point on the circle except A or B, then the slope of $\overline{PA}$ is $\frac{y-0}{x-(-a)}$, or $\frac{y}{x+a}$, and the slope of $\overline{PB}$ is $\frac{y-0}{x-a}$, or $\frac{y}{x-a}$. The product of the slopes of $\overline{PA}$ and $\overline{PB}$ is $\left(\frac{y}{x+a}\right)\left(\frac{y}{x-a}\right) = \frac{y^2}{x^2-a^2} = \frac{y^2}{x^2-(x^2+y^2)} = \frac{y^2}{-y^2} = -1$, so that $\overline{PA} \perp \overline{PB}$.
45. $(x-2)^2 + \left(y - \frac{2\sqrt{3}}{3}\right)^2 = \frac{16}{3}$

Pages 494–495 1. $(0, 0)$ 3. $(0, 2)$ 5. $(1, 3)$ 7. $y = -3$ 9. $y = -2$
11. $y = 1$ 13. The axis of symmetry intersects the directrix at the point $D(0, h)$. The focus is the point $F(0, k)$. The vertex is the midpoint of $\overline{FD}$ and thus has coordinates $\left(\frac{0+0}{2}, \frac{k+h}{2}\right)$, or $\left(0, \frac{k+h}{2}\right)$. 15. $(0, 2)$ 17. $(0, -4)$ 19. $\left(-1, \frac{7}{2}\right)$ 21. $y = \frac{1}{4}x^2 + 4$
23. $y = -\frac{1}{8}x^2 - 1$ 25. $y = -\frac{1}{8}x^2 + x + 2$ 27. $y = \frac{1}{16}x^2$
29. $y = -\frac{1}{16}x^2 + \frac{3}{8}x - \frac{41}{16}$ 31. $y = \frac{1}{8}x^2 + \frac{1}{2}x + \frac{7}{2}$ 35. $F\left(0, -\frac{1}{32}\right)$; directrix: $y = \frac{1}{32}$

Pages 497–498
1. $\{(-2, 1), (2, 9)\}$

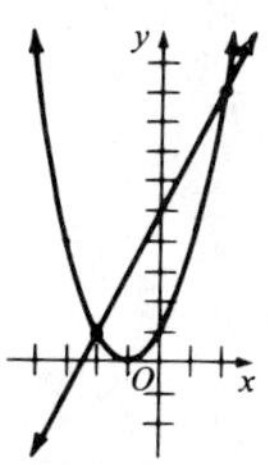

3. $\{(-2, -2), (2, 2)\}$ 5. $\{(-2, -1), (1, 2)\}$ 7. $\{(3, -10)\}$
9. $\emptyset$ 11. $\{(3, 3)\}$ 13. $\{(2, -4), (6, 4)\}$
15. $\left\{(-5, 1), \left(-\frac{1}{5}, \frac{17}{5}\right)\right\}$ 17. $\left\{\left(-\frac{1}{2}, -\frac{5}{2}\right)\right\}$ 19. -2
21. $\pm 3\sqrt{5}$ 23. $(-4, 3), (-3, 4)$ 25. $y = -\frac{1}{20}x^2 - \frac{2}{5}x - \frac{4}{5}$

Pages 503–504
1. a.

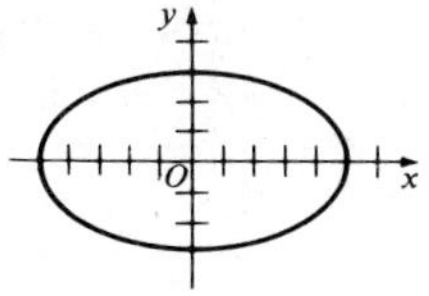

b. $(4, 0), (-4, 0)$

5. a.

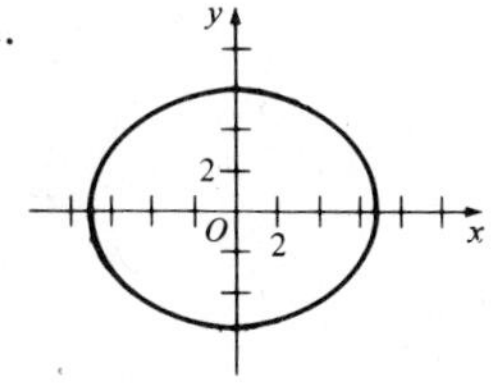

b. $(\sqrt{13}, 0), (-\sqrt{13}, 0)$

9. a.

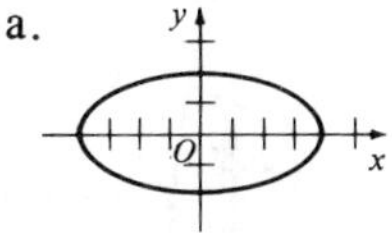

b. $(2\sqrt{3}, 0), (-2\sqrt{3}, 0)$

13. $(0, 8), (0, -8)$ 15. $(0, 2), (0, -2)$ 17. $\frac{x^2}{36} + \frac{y^2}{100} = 1$
19. a. 14 b. 8 21. a. $4\sqrt{5}$ b. $2\sqrt{10}$

25. a. 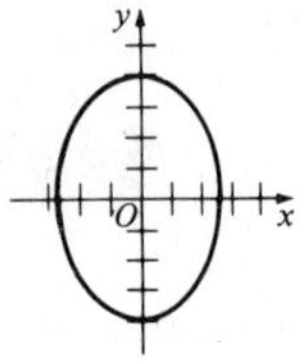

b. $\left(0, \frac{4\sqrt{5}}{3}\right), \left(0, -\frac{4\sqrt{5}}{3}\right)$

27. $\frac{x^2}{25} + \frac{y^2}{36} = 1$ 29. $\frac{x^2}{9} + \frac{y^2}{4} = 1$ 31. $\frac{x^2}{27} + \frac{y^2}{36} = 1$

33. $\frac{x^2}{9} + \frac{y^2}{25} = 1$

Pages 509–510

1. a. no x-intercepts; y-intercepts, ± 8
   b. $(0, 10)$, $(0, -10)$
   c. $y = \pm\frac{4}{3}x$

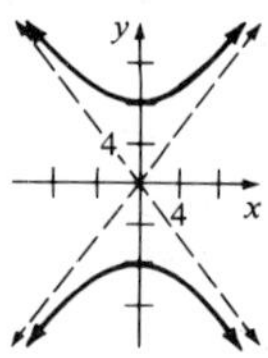

3. a. x-intercepts, ± 3; no y-intercepts
   b. $(\sqrt{13}, 0)$, $(-\sqrt{13}, 0)$
   c. $y = \pm\frac{2}{3}x$

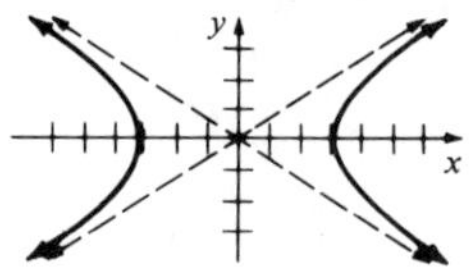

5. a. x-intercepts, ± 2; no y-intercepts
   b. $(2\sqrt{5}, 0)$, $(-2\sqrt{5}, 0)$
   c. $y = \pm 2x$

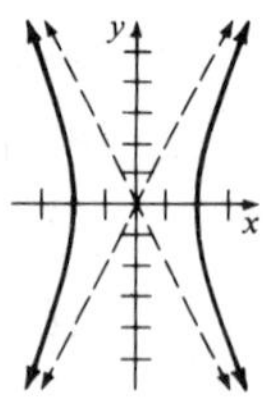

9. a. no x-intercepts; y-intercepts, ± 2
   b. $(0, 3)$, $(0, -3)$
   c. $y = \pm\frac{2\sqrt{5}}{5}x$

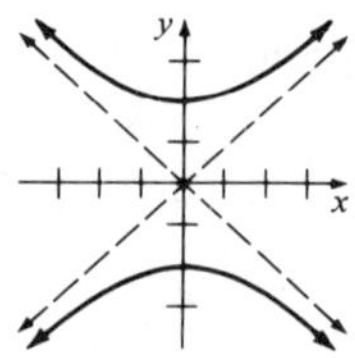

13. $\frac{y^2}{9} - \frac{x^2}{16} = 1$

17. 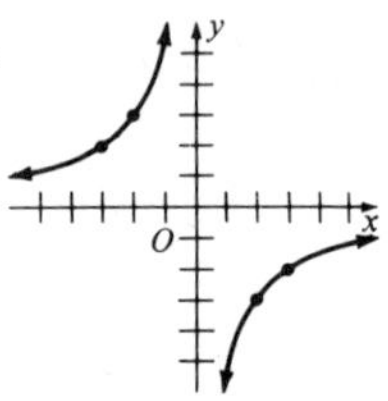

19. 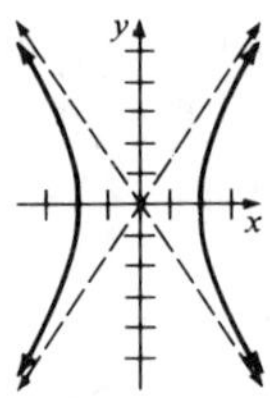

23. $\frac{y^2}{4} - \frac{x^2}{25} = 1$ 25. $\frac{x^2}{25} - \frac{y^2}{11} = 1$

27. $\frac{y^2}{121} - \frac{x^2}{64} = 1$ 29. $\frac{y^2}{4} - \frac{x^2}{4} = 1$

Pages 512–513 · Chapter Review

1. b–d. 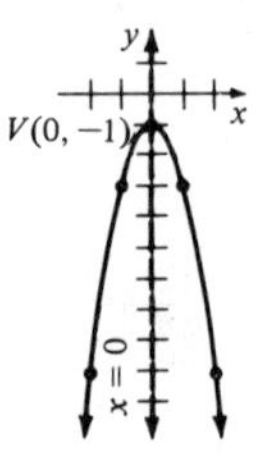

e. maximum point

2. b–d. 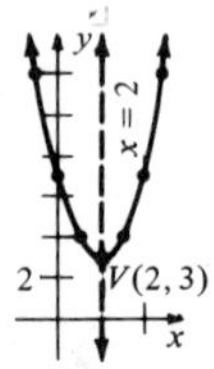

e. minimum point

3. $y = (x - 3)^2 - 6$; $(3, -6)$

4. a, b. 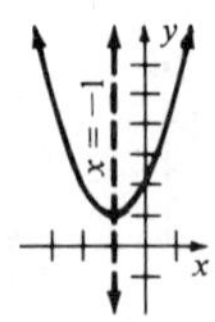

5. 8

6. $\{-2\}$ 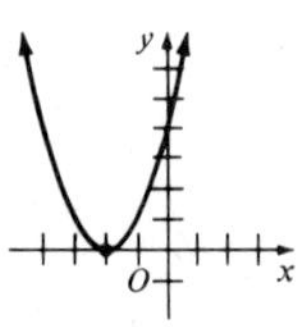

7. $\{-3, -1\}$ 8. a. $x^2 + (y - 2)^2 = 18$
b. $(x - 2)^2 + (y + 3)^2 = 25$
9. $r = 2$; $C(4, -2)$ 10. $r = 6$; $C(-3, -4)$
11. $r = 10$; $(x + 3)^2 + y^2 = 100$
12. $(0, 3)$ 13. $(0, 4)$

14. $y = \frac{1}{12}x^2 + \frac{1}{6}x + \frac{13}{12}$

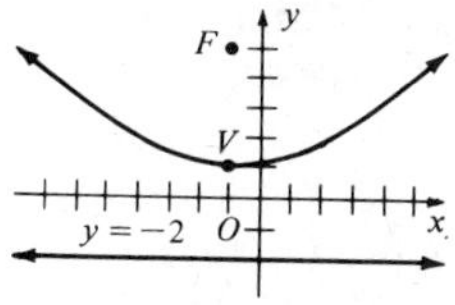

15. a, b.

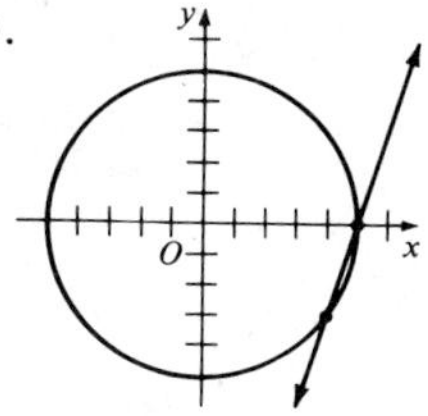

c. $(4, -3), (5, 0)$

16. $\{(1, -1), (4, 2)\}$

17. a. x-intercepts, ± 3; y-intercepts, ± 5
b. $(0, 4), (0, -4)$
c. 10

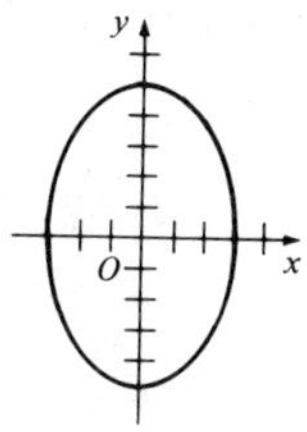

18. a. x-intercepts, ± 3; y-intercepts, ± 2
b. $(\sqrt{5}, 0), (-\sqrt{5}, 0)$
c. 6

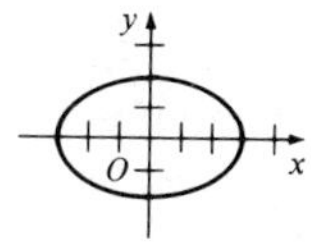

19. $\frac{x^2}{16} + \frac{y^2}{25} = 1$

20. a. x-intercepts, ± 3; no y-intercepts
b. $(6, 0), (-6, 0)$
c. $y = \pm\sqrt{3}\,x$

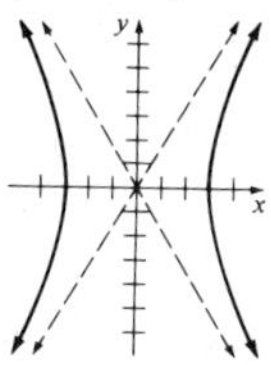

21. a. no x-intercepts; y-intercepts, ± 1
b. $(0, \sqrt{2}), (0, -\sqrt{2})$
c. $y = \pm x$

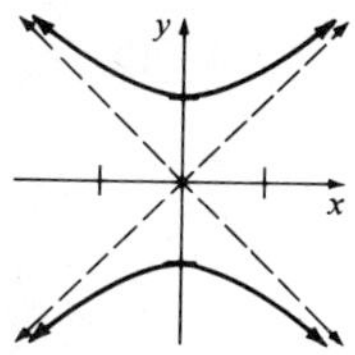

22.

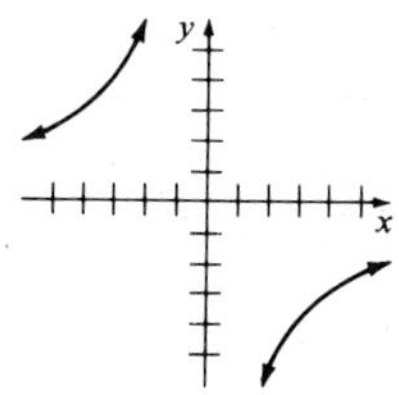

Pages 516–517 • Mixed Review

1. $r = 4$, $C(2, -3)$ 3. a. $BC = \sqrt{(6-5)^2 + (1-4)^2} = \sqrt{1+9} = \sqrt{10}$; $AD = \sqrt{(-1-0)^2 + (3-0)^2} = \sqrt{1+9} = \sqrt{10}$ b. slope of $\overline{BC} = \frac{1-4}{6-5} = \frac{-3}{1} = -3$; slope of $\overline{AD} = \frac{3-0}{-1-0} = \frac{3}{-1} = -3$ c. a parallelogram; one pair of opposite sides are $\parallel$ and $\cong$. 5. $\{(2, -3), (-2, 5)\}$ 7. $\{-3, 4\}$ 9. yes; Law of Contrapositive Inference 11. $\frac{2\sqrt{2}}{3}$ units 13. $x^2 + \left(y + \frac{1}{2}\right)^2 = 18$ 15. $9\frac{3}{13}$ 17. $37°$

19.

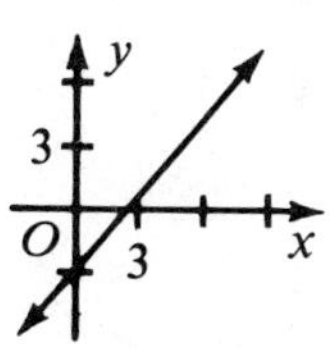

21. $-3, \frac{1}{2}$ 23. $y = \frac{1}{20}x^2 - \frac{1}{10}x - \frac{39}{20}$

25.

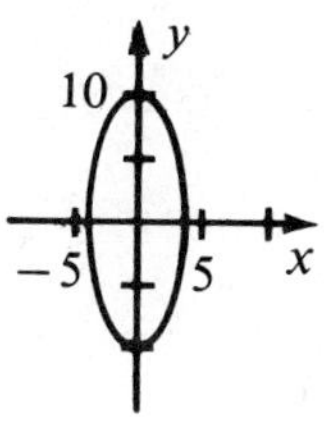

$(0, \sqrt{91}), (0, -\sqrt{91})$

Pages 521–522 1. a. {M, N, O, P, Q, R, S, T, U, V} b. {O, U} c. {T} d. {M, N, O, P, S, V} 3. {(H, T), (T, H), (T, T)} 5, 7. Let the penny, nickel, dime, quarter, and half dollar be represented by P, N, D, Q, and H. 5. {P and N, P and D, P and Q, P and H, N and D, N and Q, N and H, D and Q, D and H, Q and H} 7. {P and Q, P and H, N and Q, N and H, D and Q} 9. {(1, 5), (2, 4), (3, 3), (4, 2), (5, 1)} 11. {(2, 6), (3, 5), (4, 4), (5, 3), (6, 2)} 13. {(5, 6), (6, 5), (6, 6)} 15. {(1, 3), (2, 3), (4, 3), (5, 3), (6, 3), (3, 1), (3, 2), (3, 4), (3, 5), (3, 6)} 17. {(1, 1), (1, 3), (1, 5), (3, 1), (3, 3), (3, 5), (5, 1), (5, 3), (5, 5)} 18. {(H, H, H), (H, H, T), (H, T, H), (T, H, H), (H, T, T), (T, H, T), (T, T, H), (T, T, T)} 19. {(T, T, T)} 21. {(H, H, T), (H, T, H), (T, H, H)}

Pages 525–526 1. $\frac{1}{6}$ 3. $\frac{1}{2}$ 5. 0 7. $\frac{1}{2}$ 9. $\frac{1}{4}$ 11. $\frac{1}{36}$ 13. $\frac{5}{36}$ 15. $\frac{1}{12}$ 17. $\frac{1}{6}$ 19. $\frac{1}{3}$ 21. $\frac{2}{3}$ 23. $\frac{1}{52}$ 25. $\frac{1}{4}$ 27. $\frac{2}{13}$ 29. $\frac{4}{13}$ 31. $\frac{11}{26}$ 33. $\frac{1}{2}$ 35. $\frac{1}{6}$ 37. $\frac{3}{8}$ 39. $\frac{1}{2}$ 41. $\frac{1}{4}$ 43. 1

Pages 529–531 1. $\frac{2}{13}$ 3. $\frac{5}{52}$ 5. $\frac{11}{26}$ 7. $\frac{7}{15}$ 9. $\frac{7}{15}$ 11. $\frac{1}{5}$ 13. 1 to 17 15. 1 to 8 17. $\frac{7}{15}$ 18. $\frac{4}{15}$ 19. $\frac{11}{15}$ 21. $\frac{2}{21}$ 23. $\frac{5}{42}$ 25. $\frac{1}{21}$ 27. $\frac{19}{42}$ 29. $\frac{8}{15}$ 31. $\frac{2}{3}$ 33. $\frac{2}{5}$ 35. $\frac{1}{3}$ 37. $\frac{2}{3}$ 39. $\frac{1}{5}$ 41. $\frac{43}{50}$ 43. $\frac{3}{20}$ 45. $\frac{27}{263}$ 46. $\frac{144}{263}$ 47. $\frac{56}{263}$

Pages 535–536 1. a. 234 b. 216 3. $\frac{1}{78}$ 5. $\frac{1}{8}$ 7. a. $\frac{1}{169}$ b. $\frac{4}{663}$ 9. a. $\frac{2}{169}$ b. $\frac{8}{663}$ 11. a. $\frac{1}{16}$ b. $\frac{1}{17}$ 13. a. $\frac{7}{16}$ b. $\frac{15}{34}$ 15. $\frac{17}{40}$ 19. $\frac{a}{a+b}$ 21. $\frac{17}{140}$ 23. $\frac{41}{70}$ 25. Put one red marble in one urn and the remaining marbles (two red and three white) in the other.

Pages 540–541 1. 30 3. 6720 5. 362,880 7. 25 9. 60 11. 5040 13. 840 15. 40,320 17. 120 19. 30,240 21. 1320 23. 120 25. 840 27. 50,400 29. 151,200 31. 468,000 33. 676,000 35. a. 30 b. 25 c. 31 37. 720 39. 60 41. $\frac{13}{51}$ 43. 3360 45. 3720

Pages 544–547 1. 66 3. 4 5. 56 7. 1 9. 1 11. 43,758 13. 1365 15. 60 17. 9 19. a. $\frac{{}_4C_1}{{}_{10}C_1}$, or 0.4 b. $\frac{{}_3C_1}{{}_{10}C_1}$, or 0.3 c. $\frac{{}_3C_1}{{}_{10}C_1}$, or 0.3 21. $\frac{{}_4C_2 + {}_3C_2 + {}_3C_2}{{}_{10}C_2}$, or 0.267 23. $\frac{{}_{26}C_3}{{}_{52}C_3}$, or 0.118 25. $\frac{{}_4C_2 \cdot {}_{48}C_1}{{}_{52}C_3}$, or 0.0130 27. 21 29. 35 31. 1287 33. 330 35. $\frac{{}_4C_4 \cdot {}_{48}C_1}{{}_{52}C_5}$, or 0.0000185 37. $\frac{{}_4C_4 \cdot {}_{48}C_1 + {}_4C_4 \cdot {}_{48}C_1 + {}_4C_4 \cdot {}_{48}C_1}{{}_{52}C_5}$, or 0.0000554 39. $\frac{{}_{13}C_5}{{}_{52}C_5}$, or 0.000495 41. $\frac{{}_{13}C_3 \cdot {}_6C_2}{{}_{52}C_5}$, or 0.00165 43. $\frac{{}_5C_3 \cdot {}_7C_2}{{}_{16}C_5}$, or 0.0481 45. $\frac{{}_7C_2 + {}_3C_2}{{}_{10}C_2}$, or 0.533 47. a. $\frac{{}_{13}C_5 \cdot {}_{13}C_4 \cdot {}_{13}C_3 \cdot {}_{13}C_1}{{}_{52}C_{13}}$, or 0.00539

b. $\dfrac{{}_4C_4 \cdot {}_{48}C_9 + {}_4C_4 \cdot {}_{48}C_9 - {}_4C_4 \cdot {}_4C_4 \cdot {}_{44}C_5}{{}_{52}C_{13}}$, or 0.00528 c. $\dfrac{{}_4C_4 \cdot {}_4C_4 \cdot {}_{44}C_5}{{}_{52}C_{13}}$, or 0.00000171

49. $1 - \dfrac{{}_5C_1 \cdot {}_4C_1 \cdot {}_6C_1}{{}_{15}C_3}$, or $\dfrac{{}_5C_2 \cdot {}_{10}C_1 + {}_4C_2 \cdot {}_{11}C_1 + {}_6C_2 \cdot {}_9C_1 + {}_5C_3 + {}_4C_3 + {}_6C_3}{{}_{15}C_3}$, or 0.736

51. $\dfrac{{}_{13}C_5 + {}_{13}C_5 + {}_{13}C_5 + {}_{13}C_5}{{}_{52}C_5}$, or 0.00198 53. $13\left(\dfrac{{}_4C_4 \cdot {}_{48}C_1}{{}_{52}C_5}\right)$, or 0.000240

Pages 547–548 1. 600 3. 332,640 5. 6 7. 6 9. 10

11. $\dfrac{{}_1P_1 \cdot {}_1P_1 \cdot {}_1P_1 \cdot {}_1P_1}{{}_9P_4}$, or 0.000331 13. $\dfrac{{}_1C_1 \cdot {}_1C_1 \cdot {}_7C_2}{{}_9C_4}$, or 0.167

15. $\dfrac{{}_8P_3 \cdot {}_1P_1}{{}_9P_4}$, or 0.111 17. $\dfrac{{}_5C_4}{{}_{12}C_4}$, or 0.0101 19. $\dfrac{{}_4P_1 \cdot {}_4P_1 \cdot {}_4P_1 \cdot {}_4P_1}{{}_{52}P_4}$, or 0.0000394

21. $\dfrac{{}_{13}P_3 \cdot {}_{13}P_1}{{}_{52}P_4}$, or 0.00343 23. $\dfrac{{}_{39}C_4}{{}_{52}C_4}$, or 0.304

Page 549 · Chapter Review 1. {(2, 1), (2, 3), (2, 5), (2, 7), (4, 1), (4, 3), (4, 5), (4, 7), (6, 1), (6, 3), (6, 5), (6, 7), (8, 1), (8, 3), (8, 5), (8, 7)} 2. {(2, 1), (2, 3), (2, 5), (2, 7), (4, 1)} 3. {(2, 1), (2, 7), (4, 5), (6, 3), (8, 1), (8, 7)} 4. $\frac{3}{8}$ 5. $\frac{1}{4}$ 6. $\frac{3}{16}$

7. $\frac{1}{5}$ 8. $\frac{1}{10}$ 9. $\frac{7}{10}$ 10. a. $\frac{1}{15}$ b. $\frac{17}{45}$ 11. a. 1:14 b. 17:28

12. $\dfrac{{}_3P_1 \cdot {}_5P_1}{{}_{10}P_2}$, or $\frac{1}{6}$ 13. 999,000 14. $\dfrac{9!}{4!\,3!}$, or 2520 15. a. 1 b. 861

16. 210 17. $\frac{3}{7}$

Pages 552–553 · Mixed Review

1. 1260 3. a. $\frac{1}{169}$ b. $\frac{4}{663}$ 5. $7\frac{1}{2}$ 7.

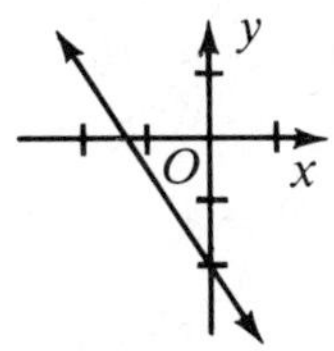

9. $\frac{4}{3}$ or $-\frac{3}{4}$

11. (1) s (Given); (2) $s \rightarrow \sim r$ (Given); (3) $\sim r$ (Law of Detachment); (4) $\sim p \rightarrow r$ (Given); (5) p (Law of Contrapositive Inference) 13. {June, July} 15. 45°

17. midpt. of $\overline{PQ}$: (0, 3), midpt. of $\overline{PR}$: $\left(\frac{3}{2}, \frac{3}{2}\right)$; length of the segment joining the midpts. of $\overline{PQ}$ and $\overline{PR}$ = $\sqrt{\left(0 - \frac{3}{2}\right)^2 + \left(3 - \frac{3}{2}\right)^2} = \sqrt{\left(-\frac{3}{2}\right)^2 + \left(\frac{3}{2}\right)^2} = \sqrt{\frac{9}{4} + \frac{9}{4}} = \sqrt{\frac{18}{4}} = \frac{3\sqrt{2}}{2}$;

$QR = \sqrt{(5-2)^2 + (2-5)^2} = \sqrt{3^2 + (-3)^2} = \sqrt{9+9} = \sqrt{18} = 3\sqrt{2}$; $\frac{1}{2} \cdot 3\sqrt{2} = \frac{3\sqrt{2}}{2}$

19.

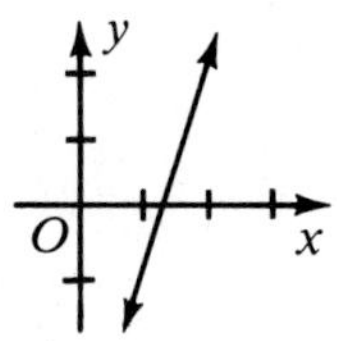

21. $5x - 2y = 26$ 23. $r = 2\sqrt{5}$; $C(-5, 1)$

25.

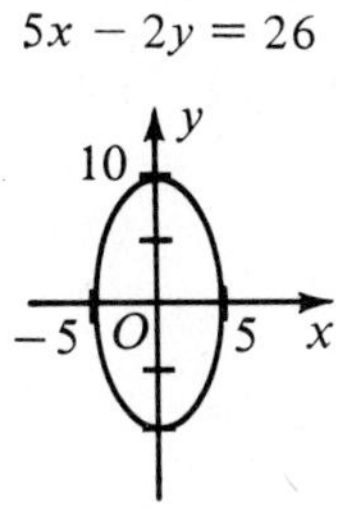

$(0, 5\sqrt{3})$, $(0, -5\sqrt{3})$

27. (1) $UVXY$ is a parallelogram; $\overline{VW} \cong \overline{ZY}$ or $VW = YZ$ (Given); (2) $\overline{UV} \parallel \overline{YX}$ or $\overline{UW} \parallel \overline{XZ}$ (Def. of a ▱); (3) $\overline{UV} \cong \overline{XY}$ or $UV = XY$ (Theorem 5-1); (4) $UV + VW = XY + YZ$ (Add. Prop. of Equality); (5) $UV + VW = UW$; $XY + YZ = XZ$ (Prop. of Betweenness); (6) $UW = XZ$ or $\overline{UW} \cong \overline{XZ}$ (Subst. Prop.); (7) $UWXZ$ is a parallelogram. (Theorem 5-5)
29. $y = -(x + 2)^2 + 8$; $(-2, 8)$

Pages 558–559 1. a. no b. yes 3. a. no b. no 5. a. yes b. no 7. a. no b. no 9. a. yes b. yes 11. $N, W, Z, Q^+, Q, \mathcal{R}^+, \mathcal{R}$ 13. $Z, Q, \mathcal{R}$ 15. yes 17. no 19. no 21. yes 23. $\mathcal{R}^+$ 25. $Q^+, \mathcal{R}^+$

Pages 563–565 1. 3 3. 4 5. $1^{-1} = 11$, $2^{-1} = 10$, $3^{-1} = 9$, $4^{-1} = 8$, $6^{-1} = 6$, $8^{-1} = 4$, $9^{-1} = 3$, $10^{-1} = 2$, $11^{-1} = 1$, $12^{-1} = 12$

7.

★	m	n	r
m	r	m	n
n	m	n	r
r	n	r	r

8. yes; n 9. yes; $m^{-1} = r$ because $m \star r = r \star m = n$.
11. yes; $r^{-1} = m$ because $r \star m = m \star r = n$.
13. a. $4 \boxplus 6 = 3$ b. $3 \boxplus 5 = 1$

14.

⊞	1	2	3	4	5	6	7
1	2	3	4	5	6	7	1
2	3	4	5	6	7	1	2
3	4	5	6	7	1	2	3
4	5	6	7	1	2	3	4
5	6	7	1	2	3	4	5
6	7	1	2	3	4	5	6
7	1	2	3	4	5	6	7

15. yes; the table is symmetric about the main diagonal.
16. The identity element is 7. 17. $1^{-1} = 6$, $2^{-1} = 5$, $3^{-1} = 4$, $4^{-1} = 3$, $5^{-1} = 2$, $6^{-1} = 1$

18.

⊙	a	b	c	d
a	d	c	a	a
b	c	b	b	b
c	a	b	c	d
d	a	b	d	d

19. c 21. a

23.

+	e	o
e	e	o
o	o	e

a. commutative
b. associative
c. e
d. $e^{-1} = e$; $o^{-1} = o$

24.

·	e	o
e	e	e
o	e	o

a. commutative
b. associative
c. o
d. $o^{-1} = o$

25.

÷	p	n
p	p	n
n	n	p

a. commutative
b. associative
c. p
d. $p^{-1} = p$; $n^{-1} = n$

26.

·	o	1
o	o	o
1	o	1

a. commutative
b. associative
c. 1
d. $1^{-1} = 1$

Pages 568–570 1. a. 1 b. 0 c. 2 d. 0 e. 3 f. 2

3.

+	0	1	2	3	4	5	6	7	8	9	10	11
0	0	1	2	3	4	5	6	7	8	9	10	11
1	1	2	3	4	5	6	7	8	9	10	11	0
2	2	3	4	5	6	7	8	9	10	11	0	1
3	3	4	5	6	7	8	9	10	11	0	1	2
.	.	.	.	.	.	.	.	.	.	.	.	.
.	.	.	.	.	.	.	.	.	.	.	.	.
.	.	.	.	.	.	.	.	.	.	.	.	.

5. $1^{-1} = 1$, $2^{-1} = 3$, $3^{-1} = 2$, $4^{-1} = 4$ 7. $1^{-1} = 1$, $5^{-1} = 5$, $7^{-1} = 7$, $11^{-1} = 1$ 9. 4 11. 2 13. 2 15. 3

17.

+	0	1	2	3	4	5	6
0	0	1	2	3	4	5	6
1	1	2	3	4	5	6	0
2	2	3	4	5	6	0	1
3	3	4	5	6	0	1	2
4	4	5	6	0	1	2	3
5	5	6	0	1	2	3	4
6	6	0	1	2	3	4	5

18.

X	0	1	2	3	4	5	6
0	0	0	0	0	0	0	0
1	0	1	2	3	4	5	6
2	0	2	4	6	1	3	5
3	0	3	6	2	5	1	4
4	0	4	1	5	2	6	3
5	0	5	3	1	6	4	2
6	0	6	5	4	3	2	1

19. $-0 = 0$, $-1 = 6$, $-2 = 5$, $-3 = 4$, $-4 = 3$, $-5 = 2$, $-6 = 1$ 20. $1^{-1} = 1$, $2^{-1} = 4$, $3^{-1} = 5$, $4^{-1} = 2$, $5^{-1} = 3$, $6^{-1} = 6$ (The element 0 has no multiplicative inverse.) 21. no
23. $\{1\}$ 25. $\{6\}$ 27. $\{2\}$ 29. $\{3\}$ 31. $\{3, 4\}$ 33. $\{8\}$ 35. $\{8\}$
37. $\emptyset$ 39. $\{0, 2, 4, 6, 8, 10\}$ 41. $\{2\}$ 43. a. $\{1\}$ b. $\{2\}$ c. $\emptyset$

Pages 574–575 1. group; 0 is the identity element; $-q$ is the additive inverse of q for every q in Q. 3. not a group; there is no identity element. 5. group; 1 is the identity element; $\frac{1}{q}$ is the multiplicative inverse of q for every q in $Q\backslash\{0\}$. 7. not a group; no element of $Z\backslash\{0\}$ except 1 has a multiplicative inverse in $Z\backslash\{0\}$. 9. not a group; the element 0 has no multiplicative inverse. 11. group; 1 is the identity element; $\frac{1}{r}$ is the multiplicative inverse of r for every r in $\mathcal{R}^+$. 13. group; 0 is the identity element; $-0 = 0$, $-1 = 2$, and $-2 = 1$. 15. group; x is the identity element; $x^{-1} = x$, $y^{-1} = z$, and $z^{-1} = y$. 17. group, Δ is the identity element; $\square^{-1} = \bigcirc$, $\bigcirc^{-1} = \square$, $\Delta^{-1} = \Delta$, and $\diagup\!\!\square^{-1} = \diagup\!\!\square$. 19. not a group; only elements 1, 5, 7, and 11 have multiplicative inverses. 21. group; 1 is the identity element; $1^{-1} = 1$ and $2^{-1} = 2$. 23. not a group; elements 0 and 2 do not have multiplicative inverses. 25. not a group; the element 0 does not have a multiplicative inverse (and for certain values of N, elements other than 0 do not have multiplicative inverses) 27. not a group; $*$ is not associative. 29. group; $\frac{1}{2}$ is the identity element; $\frac{1}{4r}$ is the inverse of r for every r in $\mathcal{R}^+$. 31. not a group; the system is not closed under + since the sum of any two odd integers is an even integer. 33. group; -5 is the identity element; $-(k + 10)$ is the inverse of k for every k in O. 37. Show that a is the inverse of a^{-1}.

Pages 578–579

1.

*	A	B	C	D	E	F
A	A	B	C	D	E	F
B	B	A	D	C	F	E
C	C	E	A	F	B	D
D	D	F	B	E	A	C
E	E	C	F	A	D	B
F	F	D	E	B	C	A

3. a. $C * (F * B) = C * D = F$ and $(C * F) * B = D * B = F$
b. $E * (C * E) = E * B = C$ and $(E * C) * E = F * E = C$

4.

X	1	6
1	1	6
6	6	1

subgroup; 1 is the identity element; $1^{-1} = 1$ and $6^{-1} = 6$

5.

+	0	2	4
0	0	2	4
2	2	4	0
4	4	0	2

subgroup; 0 is the identity element; $0^{-1} = 0$, $2^{-1} = 4$, $4^{-1} = 2$

7. not a subgroup; the subset is not closed under the operation because $3 \times 3 = 2$ and $5 \times 5 = 4$.

9.

X	1
1	1

subgroup;
1 is the identity element;
$1^{-1} = 1$

10.

X	1	10
1	1	10
10	10	1

subgroup;
1 is the identity element;
$1^{-1} = 1$ and $10^{-1} = 10$

11. not a subgroup; the subset is not closed under the operation because $2 + 4 = 4 + 2 = 1$ and $4 + 4 = 3$.

13. a.

+	0	1	2	3
0	0	1	2	3
1	1	2	3	0
2	2	3	0	1
3	3	0	1	2

b. $\{0\}$, $\{0, 2\}$, $\{0, 1, 2, 3\}$

15. a.

+	0	1	2	3	4
0	0	1	2	3	4
1	1	2	3	4	0
2	2	3	4	0	1
3	3	4	0	1	2
4	4	0	1	2	3

b. $\{0\}$, $\{0, 1, 2, 3, 4\}$

17.

*	*H*	*V*	*R*	*S*
H	*S*	*R*	*V*	*H*
V	*R*	*S*	*H*	*V*
R	*V*	*H*	*S*	*R*
S	*H*	*V*	*R*	*S*

19. *S*

21.

X	1	2	4	8	16
1	1	2	4	8	16
2	2	4	8	16	1
4	4	8	16	1	2
8	8	16	1	2	4
16	16	1	2	4	8

subgroup; 1

23. a.

X	1	2	4	8
1	1	2	4	8
2	2	4	8	1
4	4	8	1	2
8	8	1	2	4

b. Because $(Z_{15}\backslash\{0\}, \times)$ is not closed, $(Z_{15}\backslash\{0\}, \times)$ is not an algebraic system and hence not a group.

25. Since $\{1, p - 1\}$ is a subset of $Z_p\backslash\{0\}$, it is necessary to show that if p is prime, then $(\{1, p - 1\}, \times)$ forms a group.

Pages 583–584 1. field; 0 and 1 are the identity elements for + and ·, respectively; $-q$ is the additive inverse for every q in Q and $\frac{1}{q}$ is the multiplicative inverse for every q in $Q\backslash\{0\}$.
3. not a field; $(Q^+, +)$ is not a group since there is no identity element. 5. not a field; $(Z_6\backslash\{0\}, \times)$ is not a group since 2, 3, and 4 do not have multiplicative inverses. 7. not a field; $(Z_9\backslash\{0\}, \times)$ is not a group since 3 and 6 do not have multiplicative inverses. 9. not a field; $(\{-1, 0, 1\}, +)$ is not a group since the system is not closed. For example, $-1 + (-1) = -2$ and $1 + 1 = 2$. 11. 4 13. 1 15. 2 17. 3 19. 5

21. a.

#	1	3	5	7	9
1	7	9	1	3	5
3	9	1	3	5	7
5	1	3	5	7	9
7	3	5	7	9	1
9	5	7	9	1	3

X	1	3	5	7	9
1	1	3	5	7	9
3	3	9	5	1	7
5	5	5	5	5	5
7	7	1	5	9	3
9	9	7	5	3	1

b. 5 is the identity element for #; 1 is the identity element for X. c. $-1 = 5$, $-3 = 3$, $-5 = 1$, $-7 = 9$, $-9 = 7$; $1^{-1} = 1$, $3^{-1} = 7$, $7^{-1} = 3$, $9^{-1} = 9$ 23. Use the Mult. Prop. of -1.

25. Use the results of Exs. 23 and 24. 27. a. $a = 3, b = 5$ b. $a = -1, b = 1$ c. $a = 6, b = 0$ d. $a = 0, b = \frac{1}{3}$

Pages 586–589 1. {2} 3. {2} 5. {4} 7. {10} 9. {1} 11. {1} 13. {2, 4} 15. {12} 17. {8, 9} 19. {2, 4} 21. {3, 5} 23. {3} 25. {3, 5} 27. 1. Given 2. Existence of Multiplicative Inverses 3. Thm. 14-4 4. Multiplicative Prop. of Zero 5. Transitive Prop. of = 29. 1. Given 2. Zero-Product Prop. 3. Multiplicative Prop. of -1 4. Substitution Prop. (Steps 2 and 3) 5. Def. of multiplicative inverse 6. Corollary 3 to Multiplicative Prop. of -1 7. Def. of Multiplicative Inverse 8. Transitive Prop. of = 9. Substitution Prop. (Steps 5 and 8) 10. Left Cancellation 31. (1) $\frac{1}{x} \cdot x = 1, \frac{1}{b} \cdot b = 1$ (Def. of mult. inverses); (2) $a \cdot 1 = a \cdot 1$ (Ref. Prop. of =); (3) $a\left(\frac{1}{x} \cdot x\right) = a\left(\frac{1}{b} \cdot b\right)$ (Substitution Prop.); (4) $\left(a \cdot \frac{1}{x}\right)x = \left(a \cdot \frac{1}{b}\right)b$ (Associative Prop.); (5) $a \cdot \frac{1}{x} = b, a \neq 0, b \neq 0$ (Given); (6) $bx = \left(a \cdot \frac{1}{b}\right)b$ (Substitution Prop.); (7) $xb = \left(a \cdot \frac{1}{b}\right)b$ (Commutative Prop.); (8) $x = a \cdot \frac{1}{b}$ (Right Cancellation)

Pages 591–592 • Chapter Review 1. no 2. no 3. 108 4. Z, Z^-, Q, Q^- 5. Z, Q 6. Z, Q 7. none 8. a. 5 b. 1 9.

•	−1	0	1
−1	1	0	−1
0	0	0	0
1	−1	0	1

10. yes 11. identity element, 1; $(-1)^{-1} = -1, 1^{-1} = 1$ 12. a. 3 b. 2 13. $-5 = 1, 5^{-1} = 5$ 14. 2, 3, and 4 15. a. {3} b. {4} 16. yes; 0 is the identity element; $-0 = 0, -1 = 1$ 17. no; there is no identity element. (The system is not assoc.) 18. F 19.

×	1	12
1	1	12
12	12	1

yes; 1 is the identity element; $1^{-1} = 1$, $12^{-1} = 12$

20. a. 5 b. 4 c. 6 21. no; $(\mathscr{R}\backslash\{0\}, \div)$ is not a *commutative* group. 22. yes 23. {2} 24. {1} 25. {11, 8}

Pages 593–596 • Cumulative Review: Chapters 1–14 1. c 2. a 3. d 4. d 5. a 6. d 7. a 8. c 9. a 10. b 11. c 12. a 13. a 14. a 15. a 16. d 17. b 18. c 19. d 20. d 21. d 22. d 23. a 24. d 25. a 26. b 27. c 28. c 29. a 30. b 31. c

32. 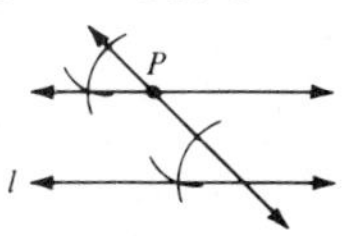

33. 

34. 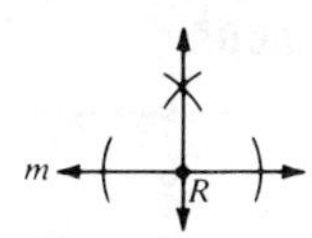

35. (1) $\frac{VZ}{XZ} = \frac{WZ}{YZ}$ (Given); (2) $\angle VZW \cong \angle XZY$ (Vert. ∠s are ≅.); $\triangle VZW \sim \triangle XZY$ (SAS ~ Thm.); (4) $\angle W \cong \angle Y$ (Corr. ∠s of ~ △s are ≅.) 36. a. Answers will vary: either

slope or the distance formula may be used. Example: The slope of $\overline{AB}$ is 2. The slope of $\overline{BC}$ is $-\frac{1}{2}$. By Thm. 11-5, $\overline{AB} \perp \overline{BC}$, so $\angle ABC$ is a rt. $\angle$. b. 20 37. (1) $ABCD$ is a ▱. (Given); (2) $\overline{AB} \parallel \overline{CD}$ (Opp. sides of a ▱ are ∥.); (3) $\overline{CD} \parallel \overline{FG}$ (Given); (4) $\overline{AB} \parallel \overline{FG}$ (Corollary 2 to Thm. 4-5); (5) $\angle EAB \cong \angle EFG$, $\angle EBA \cong \angle EGF$ (If 2 lines are ∥, alt. int. ∠s are ≅.); (6) $\triangle EAB \sim \triangle EFG$ (AA ~ Thm.); (7) $\frac{EA}{EF} = \frac{AB}{FG}$ (The lengths of corr. sides of ~ △s are in prop.); (8) $EA \cdot FG = EF \cdot AB$ (Prop. 1 of Proportions) 38. a. a b. b c. h d. c 39. a.

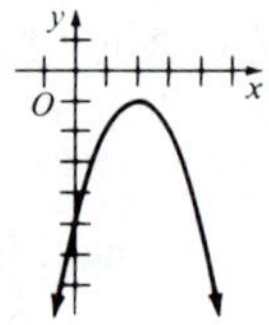

b. $x = 2$ c. $y = 2x - 5$

40. a. P_1: $\sim g \vee \sim d$; P_2: $\sim d \rightarrow \sim r$; P_3: $\sim r \rightarrow j$; P_4: $\sim j$ b. (1) $\sim d \rightarrow \sim r$ (Given); (2) $\sim r \rightarrow j$ (Given); (3) $\sim d \rightarrow j$ (Law of the Syllogism); (4) $\sim j \rightarrow d$ (Law of the Contrapositive); (5) $\sim j$ (Given); (6) d (Law of Detachment); (7) $\sim g \vee \sim d$ (Given); (8) $\sim g$ (Law of Disjunctive Inference)

Pages 598–599 · Mixed Review

1. a. yes b. yes 3. a. $\frac{1}{4}$ b. $\frac{25}{102}$ 5.

+	0	1	2	3	4	5
0	0	1	2	3	4	5
1	1	2	3	4	5	0
2	2	3	4	5	0	1
3	3	4	5	0	1	2
4	4	5	0	1	2	3
5	5	0	1	2	3	4

7. a. 0 b. 0 c. 0 9. a. $\frac{a-2}{4}$ b. -1 11. $\left\{\frac{-5 \pm \sqrt{73}}{6}\right\}$ 13. 4

15. 87 m 17. slope, $\frac{6}{5}$; y-intercept, $\frac{12}{5}$ 19. $y = \frac{1}{4}x^2 - \frac{1}{2}x + \frac{9}{4}$

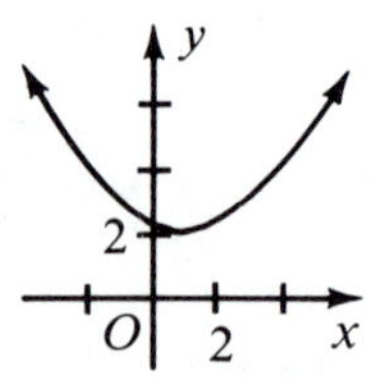

21. 15 23. $\{2, 5\}$ 25. two rational roots 27. $(x - 6)^2 + (y + 4)^2 = 8$ 29. (1) $\overline{ED} \cong \overline{EC}$; $\overline{AB} \cong \overline{DC}$; $\overline{DA} \perp \overline{AB}$; $\triangle EGF \sim \triangle EDC$ (Given); (2) $\angle EGF \cong \angle EDC$ (Corr. ∠s of ~ polygons are ≅.); (3) $\overline{AB} \parallel \overline{DC}$ (Theorem 4-2); (4) $ABCD$ is a ▱. (Theorem 5-5); (5) $\angle A$ is a rt. $\angle$. (Def. of ⊥ lines); (6) $ABCD$ is a rectangle. (Theorem 5-8).

EXTRA PRACTICE

Chapter 1, page 600

1.

3. $4x$ 5. 18 7. $-8x$ 9. $12y$

11. $\{8\}$ 13. $\left\{b: \frac{10}{3} < b < \frac{16}{3}\right\}$ 15. $\{2, 4, 6, 8, 10, 12, 18, 24, 30, 36, \ldots\}$

17. $\{2, 3, 4, 6, 8, 9, 10\}$ 19. $\{-1\}$ 21. $\{b: b \geqslant -5\}$ 23. 84, 21, 95

Chapter 2, page 601

1. Some triangles are isosceles. 3. March is windy and December is not cold. 5. $\exists_x \left(x < 1 \wedge \frac{1}{x} \leqslant x\right)$ 7. If a number is divisible by 2, then it's not odd. 9. no; If $x = 2$, then both statements are true since the number 2 is both an even number and a prime number. 11. $\exists_x\ x^2 = \frac{1}{x}$; T 13. yes; Law of Disjunctive Inference

Chapter 3, page 602

1. $\overrightarrow{PQ}$ 3. 4 5. $m\angle XYB = 42$, $m\angle XYZ = 84$ 7. $19 < AB < 129$ 9. (1) $\overline{BD} \perp \overline{AC}$; $\overline{BA} \cong \overline{BC}$ (Given); (2) $\angle ADB$ and $\angle CDB$ are rt. ∠s. (Def. of ⊥ lines); (3) $\overline{BD} \cong \overline{BD}$ (Refl. Prop. of ≅); (4) $\triangle ABD \cong \triangle CBD$ (HL Post.) 11. (1) $\angle 1 \cong \angle 4$; $\angle 2 \cong \angle 3$ (Given); (2) $m\angle 1 + m\angle 2 = m\angle 3 + m\angle 4$ (Add. Prop. of Equality); (3) $m\angle TPQ = m\angle 1 + m\angle 2$; $m\angle RQP = m\angle 3 + m\angle 4$ (Angle Add. Post.); (4) $m\angle TPQ = m\angle RQP$ (Subst. Prop.); (5) $\overline{PQ} \cong \overline{PQ}$ (Refl. Prop. of ≅); (6) $\triangle TPQ \cong \triangle RQP$ (ASA) 13. (1) $\overline{BC} \cong \overline{CD}$; $\overline{AC} \cong \overline{CE}$ (Given); (2) $\angle ACB \cong \angle ECD$ (Vert. ∠s are ≅.); (3) $\triangle ACB \cong \triangle ECD$ (SAS); (4) $\angle ABC \cong \angle EDC$ (Corr. parts of ≅ △s are ≅.); (5) $\angle GCB \cong \angle FCD$ (Vert. ∠s are ≅.); (6) $\triangle GCB \cong \triangle FCD$ (ASA); (7) $\overline{GC} \cong \overline{FC}$ (Corr. parts of ≅ △s are ≅.)

Chapter 4, page 603

1. (1) $\angle 5 \cong \angle 4$ (Given); (2) $\angle 5 \cong \angle 2$ (Vert. ∠s are ≅.); (3) $\angle 4 \cong \angle 2$ (Subst. Prop.); (4) $a \parallel b$ (Theorem 4-2) 3. 50 5. 40 7. $x = 17$, $m\angle J = 34$, $m\angle K = 105$, $m\angle L = 41$ 9. 25 11. (1) $\overline{PS} \perp \overline{PQ}$; $\overline{QR} \perp \overline{PQ}$; $\angle PSQ \cong \angle QRP$ (Given); (2) $\angle SPQ$ and $\angle RQP$ are rt. ∠s. (Def. of ⊥ lines); (3) $m\angle SPQ = m\angle RQP$ (All rt. ∠s are ≅.) (4) $\overline{PQ} \cong \overline{PQ}$ (Refl. Prop. of ≅); (5) $\triangle SPQ \cong \triangle RQP$ (AAS); (6) $\overline{SQ} \cong \overline{RP}$ (Corr. parts of ≅ △s are ≅.) 13. (1) $\overline{AC} \cong \overline{AE}$; $\overline{BD} \perp \overline{AC}$; $\overline{DF} \perp \overline{AE}$ (Given); (2) $\angle ACE \cong \angle AEC$ (Base ∠s of an isos. △ are ≅.); (3) $\angle CBD$ and $\angle EFD$ are rt. ∠s. (Def. of ⊥ lines); (4) $\angle CBD \cong \angle EFD$ (All rt. ∠s are ≅.); (5) $\angle BDC \cong \angle FDE$ (Corollary 1, Theorem 4-8)

Chapter 5, page 604

1. 8 3. $x = 9$, $y = 10$ 5. $\frac{20}{7}$ 7. 15 9. (1) ▱$MPQS$; N and R are midpoints of $\overline{PM}$ and $\overline{QS}$, respectively. (Given); (2) $\overline{PM} \parallel \overline{QS}$ or $\overline{PN} \parallel \overline{RS}$ (Def. of a ▱); (3) $PN = \frac{1}{2}PM$; $RS = \frac{1}{2}QS$ (Theorem 3-2); (4) $\overline{PM} \cong \overline{QS}$ (Theorem 5-1); (5) $\overline{PN} \cong \overline{RS}$ (Subst. Prop.); (6) $NPRS$ is a ▱. (Theorem 5-5) 11. (1) $BCDE$ is a rect.; $\overline{BA} \cong \overline{BD}$ (Given); (2) $\overline{BD} \cong \overline{CE}$ (The diag. of a rect. are ≅.); (3) $\overline{BA} \cong \overline{CE}$ (Subst. Prop.); (4) $\angle BDA \cong \angle BAD$ (Base ∠s of an isos. △ are ≅.); (5) $\overline{BD}$ and $\overline{CE}$ bisect each other. (The diag. of a ▱ bisect each other.); (6) $FD = \frac{1}{2}BD$; $FE = \frac{1}{2}CE$ (Def. of bisect); (7) $FD = FE$ (Subst. Prop.); (8) $\angle FED \cong \angle FDE$ or $\angle FED \cong \angle BDA$ (Base ∠s of an isos. △ are ≅.); (9) $\angle FED \cong \angle BAD$ (Subst. Prop.); (10) $\overline{BA} \parallel \overline{CE}$ (Theorem 4-2); (11) $ABCE$ is a ▱. (Theorem 5-5) 13. (1) ▱$ABCE$; $\angle A \cong \angle ABD$ (Given); (2) $\overline{AB} \parallel \overline{EC}$ or $\overline{AB} \parallel \overline{EF}$ (Def. of a ▱); (3) $ABFE$ is a trapezoid. (Def. of a trap.); (4) $ABFE$ is an isosceles trapezoid. (If 2 base ∠s of a trap. are ≅, then the trap. is isos.; see Ex. 19 on page 209.)

Chapter 6, page 605

1. $2x^2y + 3x$ 3. $13a^2 - ab$ 5. $x^2 - 7x - 44$ 7. $2a^2 + ab - 15b^2$ 9. $(s + 6t)(s - 3t)$ 11. $r^2 + 3r + 3 - \frac{1}{r+2}$ 13. $\frac{y}{2}$ 15. $\frac{y+7}{3}$ 17. $\frac{bc}{d}$ 19. $\frac{x^2 + 2x + 3}{x^3}$ 21. $-108r^5$ 23. $-2(b + 2)^2$ 25. $4x + 3 + \frac{x}{x^2+1}$ 27. $\frac{4p+1}{p+2}$ 29. $\frac{2a^2 - 13a - 7}{a+2}$ 31. $\frac{2y-3}{y-8}$ 33. $\frac{y(7y+2)}{(y+2)^2(y-2)}$

Chapter 7, page 606

1. $\{\pm 4\}$ 3. $\{2, 5\}$ 5. $\left\{\frac{18}{5}\right\}$ 7. $\{6, -2\}$ 9. $\{-5, 1\}$ 11. one rational double root 13. two rational roots 15. $\frac{4}{5}, \frac{5}{4}$ 17. $\left\{\frac{5}{4}, -\frac{5}{2}\right\}$ 19. $\left\{\frac{3}{4}, -\frac{5}{3}\right\}$ 21. $\{4, -3\}$ 23. $\left\{1, \frac{1}{3}\right\}$ 25. $7 - 4\sqrt{3}$ 27. $\{-1 \pm \sqrt{7}\}$ 29. $\left\{\frac{-2 \pm \sqrt{3}}{2}\right\}$ 31. 2, 5

Chapter 8, page 607

1. 5:7 3. 9 5. 20 7. $25\frac{1}{5}$ 9. (1) D and E are the midpoints of $\overline{AC}$ and $\overline{AB}$, respectively. (Given); (2) $\overline{DE} \parallel \overline{BC}$ (Theorem 5–13); (3) $\angle AED \cong \angle ABC$ (Theorem 4–5); (4) $\angle A \cong \angle A$ (Refl. Prop. of $\cong$); (5) $\triangle AED \sim \triangle ABC$ (AA $\sim$ Thm.) 11. 7 13. (1) $\overline{EF} \parallel \overline{AC}$; $\overline{FG} \parallel \overline{AB}$ (Given); (2) $\frac{DE}{EC} = \frac{DF}{FA}$; $\frac{DG}{GB} = \frac{DF}{FA}$ ($\triangle$ Prop. Thm.); (3) $\frac{DE}{EC} = \frac{DG}{GB}$ (Subst. Prop.) 15. 137.5 m

Chapter 9, page 608

1. 6 3. 8 5. $\sqrt{91}$ 7. yes

	9.	11.
XZ	3	$\frac{5\sqrt{3}}{3}$
YZ	$3\sqrt{3}$	5
XY	6	$\frac{10\sqrt{3}}{3}$

13. $\sin X = \frac{40}{41}$, $\cos X = \frac{9}{41}$, $\tan X = \frac{40}{9}$; $\sin Y = \frac{9}{41}$, $\cos Y = \frac{40}{41}$, $\tan Y = \frac{9}{40}$ 15. 4 17. 32 19. 228 m

Chapter 10, page 609

1. Use Construction 1 three times. 3. Use the method of the example on pages 386–387. 5. Use Construction 10. 7. Use the method of the example on page 397. 9. Use Construction 9. 11. Use Construction 1 to construct $\overline{AB}$ such that $AB = c$. Construct 2 circles, one with center A and radius a, the other with center B and radius b. The locus consists of the two points where the circles intersect. 13. a. Use Construction 1 twice to construct $\overline{AB}$ of length $2c$. Then use the method of the example on page 401 with $AJ = a$ and $JL = b$. Then $AC:CB = a:b$. b. Use Construction 11. 15. 1 point, the intersection of the $\perp$ bis. of the sides of $\triangle ABC$

Chapter 11, page 610

1. 7 3. $\sqrt{10}$ 5. 8; 1 7. $\frac{5}{2}$; $-\frac{2}{5}$ 9. $-\frac{4}{3}$; $\frac{3}{4}$ 11. The length of diagonal $\overline{AC}$ is $\sqrt{(a-c)^2+b^2} = \sqrt{a^2-2ac+c^2+b^2}$; the length of diagonal $\overline{BD}$ is $\sqrt{(c-a)^2+b^2} = \sqrt{c^2-2ac+a^2+b^2} = \sqrt{a^2-2ac+c^2+b^2}$; $\therefore \overline{AC} \cong \overline{BD}$

13. 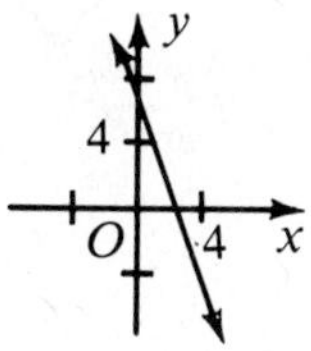

15. 16 17. slope of $\overline{YZ} = -7$; slope of $\overline{ZX} = \frac{1}{7}$.

Since $(-7)\left(\frac{1}{7}\right) = -1$, $\overline{YZ} \perp \overline{ZX}$ $\therefore \angle Z$ is a rt. $\angle$ and $\triangle XYZ$ is a rt. $\triangle$.

b. By the midpt. formula, the coordinates of point M are (1, 4). By the distance formula, $XM = MY = ZM = 5$.

19. $2x + y = 11$ 21. (0, 4), (0, −4), (−4, 4), (−4, −4) 23. a. (−1, 2) b. $\left(\frac{3}{5}, -\frac{13}{5}\right)$ c. (−2, −5)

Chapter 12, page 611

1. 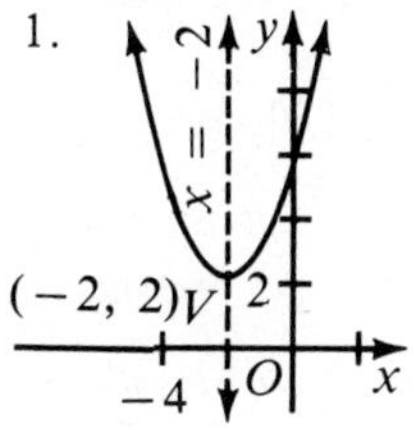

3. $y = (x+3)^2$ 5. $y = -3(x-1)^2 + 4$ 7. $(x-2)^2 + (y+1)^2 = \frac{4}{9}$

9. a. {(3, 0), (0, −3)} b. {(2, 1), (−8, −19)}

11. 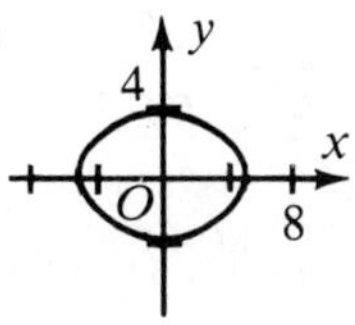

$(\pm 2\sqrt{2}, 0)$

13. a. no x-intercepts; y-intercepts, $\pm\sqrt{5}$
b. $(0, \pm 3)$
c. $y = \pm\frac{\sqrt{5}}{2}x$

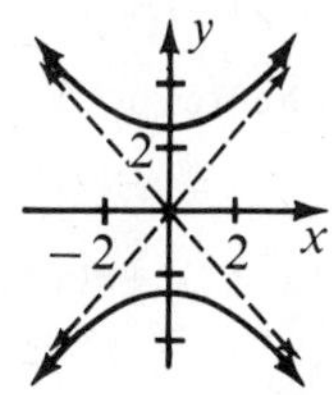

15. $\left(-\frac{3}{2}, 0\right)$, $(-2, 0)$

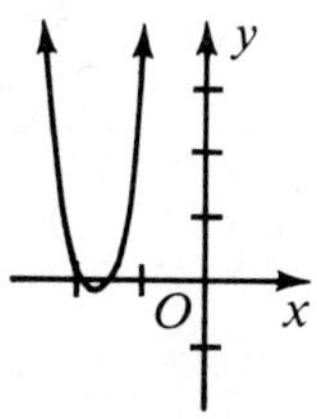

17. $C(-4, 0)$; $r = \sqrt{3}$ 19. $(x-2)^2 + \left(y - \frac{3}{2}\right)^2 = \frac{25}{4}$ or $x^2 + y^2 - 4x - 3y = 0$ 21. $\frac{x^2}{2} - \frac{y^2}{3} = 1$

Chapter 13, page 612

1. {(2, 3), (2, 5), (2, 7), (3, 5), (3, 7), (3, 2), (5, 7), (5, 2), (5, 3), (7, 2), (7, 3), (7, 5)}

3. {(2, 5), (2, 7), (3, 7), (5, 7), (5, 2), (7, 2), (7, 3), (7, 5)} 5. $\frac{1}{2}$ 7. $\frac{1}{2}$ 9. $\frac{1}{8}$

11. $\frac{3}{8}$ 13. $\frac{3}{4}$ 15. 715 17. {(2, 3), (2, 5), (2, 7), (3, 2), (5, 2), (7, 2)}

19. 1 21. $\frac{2}{7}$ 23. 1728

Chapter 14, page 613

1. 10 3. 7 5.

#	a	b	c	d
a	b	c	d	a
b	c	d	a	b
c	d	a	b	c
d	a	b	c	d

7. $a^{-1} = c$, $b^{-1} = b$, $c^{-1} = a$, $d^{-1} = d$

9. no 11. no 13. no 15. no 17. $\{2\}$ 19. yes 21. yes
23. yes 25. yes; The identity element is P_{360}; $P_{90}^{-1} = P_{270}$, $P_{180}^{-1} = P_{180}$, $P_{270}^{-1} = P_{90}$, $P_{360}^{-1} = P_{360}$

ALGEBRA REVIEW EXERCISES

Chapter 1, pages 614-615 1. $6xy - 8x + 8y$ 3. $16b^2 + 24b$ 5. $54u$ 7. $\frac{13}{3}$

9. 1 11. -13 13. $-\frac{5}{2}$ 15. $\left\{t: t > -\frac{43}{4}\right\}$ 17. $\left\{r: r > \frac{9}{2}\right\}$
19. $\{x: 4 \leqslant x < 7\}$ 21. $\{t: t < 9\}$ 23. Let d = the number of dimes; $.10d = 5.60$; 56 dimes 25. Let x = the smaller of the two consecutive integers; $x + (x + 1) \geqslant 33$; $\{16, 17, 18\}$ 27. 105 g of the 8% iron alloy, 45 g of the 18% iron alloy 29. $\{-9\}$
31. $\left\{\frac{5}{2}\right\}$ 33. $\{n: n \geqslant -1\}$ 35. $\left\{z: -\frac{3}{5} \leqslant z < \frac{4}{5}\right\}$ 37. 6 P.M.

Chapter 6, pages 615-616 1. $3a - b + 2b^2$ 3. $x^2 + 3xy - 18y^2$
5. $y^3 + 4y^2 - 2y - 8$ 7. $\frac{bc^3}{5}$ 9. $13x^2y^2(y - 2x)$ 11. $(x - 5)(x + 3)$
13. $(2p + 3)(p - 1)$ 15. $(y - 7x)(y + 3x)$ 17. $(2m - 25)(2m + 25)$
19. $(a - 2)(a^2 + 2a + 4)$ 21. $\frac{a - 4b}{3a + 5b}$ 23. $\frac{1 + 4n}{1 - n}$ 25. $\frac{1}{x - 1}$ 27. $\frac{m + 2}{m + 3}$
29. $\frac{1 + 2x}{1 - 2x}$ 31. a^2 33. $-\frac{3}{2x}$ 35. $\frac{t + 1}{t + 2}$ 37. $\frac{1}{25}$ 39. $\frac{2(x + 8)}{x + 4}$
41. $\frac{4x^2 - 3x - 9}{3x(x - 3)(x + 3)}$ 43. $-2z^2 - 4y^2 - 6z - 3$ 45. $\frac{3m^2}{(m + 1)(m - 1)}$
47. $\frac{x^2 + 4ax - a^2}{(x - a)(x + a)}$ 49. $2z^2 + z + 4 + \frac{7z - 8}{z^2 + z + 4}$

Chapter 7, pages 616-617 1. $\{2, 5\}$ 3. $\{-7, 3\}$ 5. $\{\pm 5\}$ 7. $\{-3, 5\}$
9. $\{-4 \pm \sqrt{41}\}$ 11. $\left\{\frac{-3 \pm \sqrt{41}}{4}\right\}$ 13. one rational double root 15. two rational roots 17. two rational roots 19. $\{-8, 1\}$ 21. $\left\{\frac{7}{2}, \frac{9}{2}\right\}$ 23. $\{3\}$ 25. $\frac{2}{3}$ or $\frac{3}{2}$
27. $2\sqrt{5}$ 29. $\frac{10}{3}$ and 20 31. \$.175/kg 33. 2 cm, 3 cm, 4 cm 35. $\left\{\frac{7}{5}, -3\right\}$
37. $\left\{1, -\frac{1}{3}\right\}$ 39. $\{0, -3\}$

Chapter 11, pages 617-618 1. a. $AB = 5$, $BC = 5$, $CA = 5\sqrt{2}$ b. Since $5^2 + 5^2 = 25 + 25 = 50$ and $(5\sqrt{2})^2 = 25 \cdot 2 = 50$, $\triangle ABC$ is a rt. $\triangle$ by the converse of the Pythagorean Theorem. c. 12.5 3. $(2a, 0)$, $(0, 2b)$, or $(0, -2b)$ 5. $x + y = 0$

7. $y = -6$ and $y = 12$ 9. $3x - y = -11$

11.

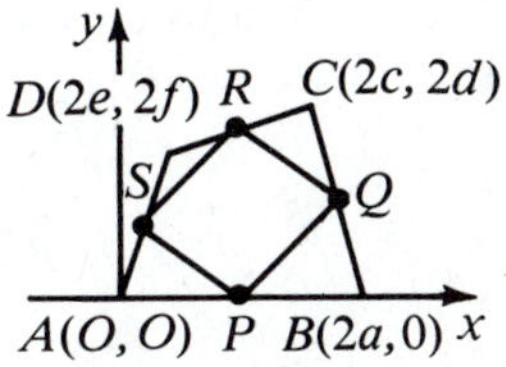

Show that $\overline{SQ}$ and $\overline{PR}$ have the same midpoint.

13. 15 15. $(7\frac{3}{4}, 0)$ 17. (0, 5) 19. $\left(5, \frac{5}{2}\right)$

Chapter 12, pages 618–619 1. x-intercept, $-\frac{5}{2}$; y-intercept, 5 3. x-intercepts, -8 and -2; y-intercept, 16 5. x-intercepts, $3 \pm \sqrt{7}$; y-intercept, 2

7.

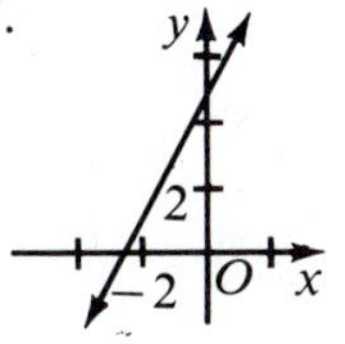

9.

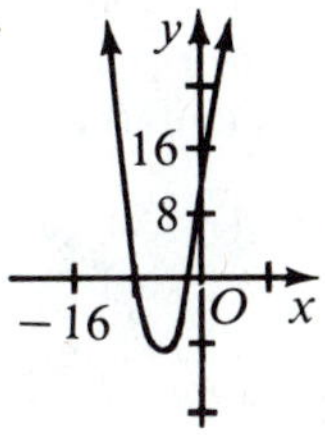

11. two real solutions 13. no real solutions 15. $x^2 + (y + 2)^2 = 25$ 17. $\frac{x^2}{9} + \frac{y^2}{25} = 1$

19.

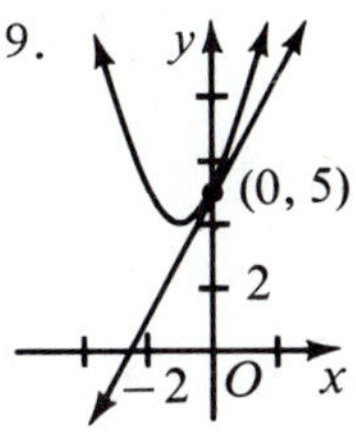

21. $y = 2(x + 2)^2 - 7$; $(-2, -7)$ 23. $y = -\frac{1}{7}x^2 + 7$ 25. $x^2 + y^2 - 8x - 6y + 17 = 0$ 27. $\frac{x^2}{100} + \frac{y^2}{36} = 1$

29.

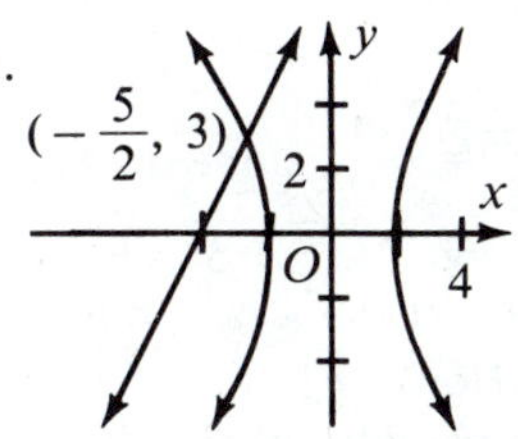

PREPARING FOR REGENTS EXAMINATIONS

Logic

1. Answers may vary. Examples are given. a. No student hates math. b. At least one number is a prime number. c. Any person under 18 will not be admitted. d. $\sim p \wedge q$ e. $p \wedge q$ f. If x is even, then x is not a factor of 9. 3. a and c 5. a. $b \rightarrow t$; $t \rightarrow m$; $b \vee d$; $\sim d$ b. (1) $b \vee d$; $\sim d$ (Given); (2) b (Law of Disjunctive Inference); (3) $b \rightarrow t$; $t \rightarrow m$ (Given); (4) $b \rightarrow m$ (Law of the Syllogism); (5) m (Law of Detachment) 7. a. $\sim p$ b. q c. $p \rightarrow r$ d. $p \rightarrow \sim r$ e. p f. Sally does not go to the beach. g. All squares are parallelograms.

Euclidean Geometry

1. a. 15 b. 14 c. 15 3. a. $\angle BAF$ b. 7 c. 6 d. 5 e. 70 f. 10 g. 60 square units 5. a. 50 b. 120 c. $\overline{BC}$ d. 135 e. one 7. a. 120 b. a rhombus c. 40 d. 40 9. (1) $\overline{BC} \cong \overline{CD}$, $\overline{AC} \perp \overline{BD}$ (Given); (2) $\angle BEC$ and $\angle DEC$ are rt. $\angle$s. (Def. of $\perp$ lines); (3) $\overline{CE} \cong \overline{CE}$ (Refl. Prop. of $\cong$); (4) $\triangle BEC \cong \triangle DEC$ (HL Post.);

(5) $\overline{BE} \cong \overline{ED}$ (Corr. parts of $\cong$ ▲ are $\cong$.); (6) $\overline{AE} \cong \overline{AE}$ (Refl. Prop. of $\cong$); (7) $\angle AED \cong \angle AEB$ (2 adj. ∠s formed by ⊥ lines are $\cong$.); (8) $\triangle AEB \cong \triangle AED$ (SAS Post.); (9) $\overline{AB} \cong \overline{AD}$ (Corr. parts of $\cong$ ▲ are $\cong$.); (10) $\triangle ABD$ is isosceles. (Def. of an isos. $\triangle$)
11. a. 4 b. 6 c. $3\sqrt{2}$ d. $3\sqrt{2}$ e. 10 f. AB g. AC h. 13 i. 12 j. $2\sqrt{7}$ k. 60 square units l. 2

Algebraic Skills

1. a. $x^2 + x - 6 = 0$ b. $x^2 + 4x = 0$ c. $x^2 - 25 = 0$ d. $8x^2 - 10x + 3 = 0$ e. $x^2 + 6x + 3 = 0$ f. $x^2 + x - 1 = 0$ 3. $a = -20$; 5 5. a. two rational roots b. one double rational root c. no real roots d. two irrational roots e. two rational roots 7. 3, 5, 7

Coordinate/Analytic Geometry

1. a. (1, −1) b. (−4, −12) c. 10 d. $\frac{1}{3}$ e. $y = 4x - 1$ f. $y = \frac{1}{2}x + 3$ g. $y = -\frac{1}{2}x + \frac{9}{2}$ h. $y = 7x$ i. (4, 10) 3. a. 0 b. undefined c. 90 d. $7 + \sqrt{29}$ e. 5 square units 5. a. $x = 4$ b. (4, −4) c. (2, 0), (6, 0) d. (2, 0), (8, 12)

7. a.

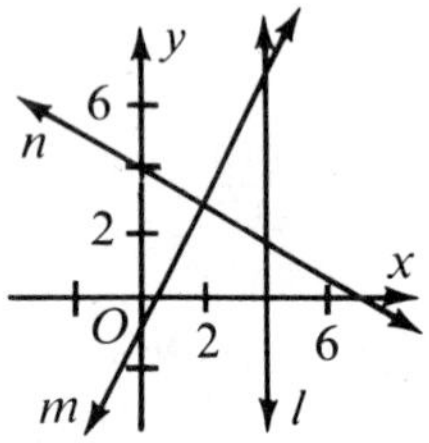

b. (4, 7) c. (4, 2) d. (2, 3)

e. Proofs may vary. Example: slope of $\overline{BC} = -\frac{1}{2}$; slope of $\overline{AC} = 2$; ∴ $\overline{AC} \perp \overline{BC}$, $\angle C$ is a rt. ∠, and $\triangle ABC$ is a rt. $\triangle$. f. 5 square units g. $3\sqrt{5} + 5$ h. (0, −1) i. (0, 4) j. Methods may vary. Students may prove that the ▲ are $\cong$ by finding the lengths of the sides of the ▲ and by using the SSS Post. However, since $\overleftrightarrow{AB} \parallel \overleftrightarrow{ED}$ and $\overline{AB}$ and $\overline{ED}$ can each be seen from the graph to have length 5, it would probably be less time-consuming to use the ASA Post.

Probability, Permutations, and Combinations

1. a. $\frac{1}{6}$ b. $\frac{5}{12}$ c. $\frac{5}{18}$ d. $\frac{7}{36}$ e. 0 f. $\frac{1}{2}$ g. $\frac{2}{9}$ h. $\frac{1}{9}$ 3. a. $\frac{3}{16}$ b. $\frac{13}{32}$
5. a. 3,628,800 b. 17,280 c. 5040 d. 2160 e. $\frac{3}{7}$ f. (1) $\frac{2}{5}$ (2) $\frac{1}{10}$ (3) $\frac{3}{5}$

Algebraic Systems

1. a. E and M b. E and M c. E, O, and M d. none e. (1) 0 (2) 1 (3) none f. −2 g. $(E, +)$ is a group because addition within set E is associative, 0 is the identity element, and each element of set E has an inverse. h. $(M, -)$ is not a group because subtraction is not associative. 3. a.

*	O	C	I	P
O	O	C	I	P
C	C	O	P	I
I	I	P	O	C
P	P	I	C	O

b. yes c. yes d. yes e. yes, O f. yes g. yes h. yes

Appendix 1 • Inverse Variation
Page 642 • Written Exercises
1. $x = 7.5$ 3. a. 0.48 A b. 100 Ω 5. a. 15 b. 36

Appendix 2 • Transformations
Page 645 • Reflections in a Line
1. a. $P'(-4, -6)$ b. $Q'(0, 5)$ c. $R'(7, 0)$ d. $S'(3, 4)$ e. $T'(3, -3)$ f. $U'(-4, -4)$
3. a. $P'(6, -4)$ b. $Q'(-5, 0)$ c. $R'(0, 7)$ d. $S'(-4, 3)$ e. $T'(3, 3)$ f. $U'(4, -4)$
5. $A'(3, -1)$; $B'(-2, -4)$; $C'(5, -8)$ 7. $A'(1, 3)$; $B'(4, -2)$; $C'(8, 5)$
9. a. $P'(7, 5)$ b. $P'(-3, 1)$ c. $P'(-4, -1)$ d. $P'(-7, -2)$

Pages 646–647 • Reflections in a Point
1. a. $(-4, -1)$ b. $(-3, -4)$ c. $(1, 1)$ d. $(3, -2)$ e. $(0, 10)$ f. $(-4, 4)$ g. $(2, 0)$
h. $(5, -4)$ 3. $A'(-5, -1)$; $B'(-4, -7)$; $C'(1, -5)$

Page 648 • Translations
1. a. $A'(-3, 2)$; $B'(1, -3)$; $C'(0, 4)$ b. $A'(5, 4)$; $B'(9, -1)$; $C'(8, 6)$ c. $A'(2, 8)$; $B'(6, 3)$; $C'(5, 10)$
d. $A'(4, 3)$; $B'(8, -2)$; $C'(7, 5)$ e. $A'(-2, 7)$; $B'(2, 2)$; $C'(1, 9)$ 3. $Q(-2, 8)$
5. a. 2 units left and 2 units down b. 1 unit left and 3 units up c. 3 units left and 1 unit up
7. a. $B'(8, 3)$ b. $B'(4, 1)$ c. $B'(7, 4)$

Page 649–650 • Dilations
1. a. 2 b. $\frac{1}{2}$ c. 1.5 3. a. 1.5 b. $B'(3, -9)$ c. $AB = 10$; $A'B' = 15$; $\frac{A'B'}{AB} = 1.5$
5. $AB = 2\sqrt{13}$; $A'B' = \sqrt{13}$; $\frac{1}{2}$

Appendix 3 • The Triangle Congruence Postulates Revisited
Pages 652–653 • Written Exercises
1. Begin by temporarily assuming that $DF > AC$. Then by Postulate 2 there is exactly one point G on $\overline{DF}$ such that $DG = AC$. Thus, since $\angle A \cong \angle D$ and $\overline{AB} \cong \overline{DE}$, $\triangle ABC \cong \triangle DEG$ by the SAS Theorem, and consequently $\angle B \cong \angle DEG$. But $\angle B \cong \angle DEF$, so $\angle DEG \cong \angle DEF$. This statement contradicts the assumption that $DF > AC$. The assumption that $DF < AC$ leads to a similar contradiction. Thus $DF = AC$, and $\triangle ABC \cong \triangle DEF$ by the SAS Theorem. 3. a. Since $\angle ABC$ and $\angle GBC$ are rt. ∠s, $\overline{AB} \cong \overline{BG}$, and $\overline{BC} \cong \overline{BC}$, $\triangle ABC \cong \triangle GBC$ by the SAS Theorem. By similar reasoning, $\triangle DEF \cong \triangle HEF$. b. Since $\triangle ABC \cong \triangle GBC$, $\overline{AC} \cong \overline{GC}$. Similarly, $\overline{DF} \cong \overline{HF}$. Thus $\overline{GC} \cong \overline{HF}$ by the Trans. Prop. of ≅. Since $BG = AB = DE = EH$, $AG = DH$ by the Add. Prop. of Equality, and $\triangle AGC \cong \triangle DHF$ by the SSS Theorem. c. Since $\triangle AGC \cong \triangle DHF$, $\angle A \cong \angle D$, and thus $\triangle ABC \cong \triangle DEF$ by the SAS Theorem. 5. By Postulate 4, there is exactly one ray $\overrightarrow{UW}$ in the half-plane determined by $\overleftrightarrow{SU}$ and point T such that $\angle SUW \cong \angle R$. Assume temporarily that $m\angle SUT > m\angle R$. Then $\overrightarrow{UW}$ intersects $\overline{ST}$ in some point V. Since $\overline{PR} \cong \overline{SU}$ and $\angle P \cong \angle S$, $\triangle PQR \cong \triangle SVU$ by the ASA Theorem. Thus $\overline{PQ} \cong \overline{SV}$ and consequently $SV = PQ = ST$. Thus V and T are the same distance, PQ, from point S, a fact that contradicts Postulate 2. The assumption that $m\angle SUT < m\angle R$ leads to a similar contradiction. Thus $\angle SUT \cong \angle R$, and $\triangle PQR \cong \triangle SUT$ by the ASA Theorem.

ANSWERS TO COMPUTER EXERCISES

PAGES 39–40 · COMPUTER EXERCISES

1. $\{1, 2, 3, 4, 5, 6\}$ 2. $\{1, 2, 3\}$ 3. $\{2, 5, -1, -2, 3\}$ 4. $\{1, 2, 3, 4, 5, 6\}$, $\{1, 2, 3\}$
5. Delete lines 130–150 and line 200. Make the following changes and additions:

```
120   PRINT "INTERSECTION OF A AND B:"
180   IF B[J]<>A[I] THEN 190
185   PRINT A[I];
187   GOTO 210
```

PAGE 180 · COMPUTER EXERCISES

1. a. 60, 120 b. 90, 90 c. 108, 72 d. 120, 60 e. 128.571, 51.4286 f. 135, 45 g. 140, 40 h. 144, 36 2. If N = 2, 1, or −1, the computer will print a meaningless answer. If N = 0, the computer will print an error message because it was asked to divide by 0. 3. a. increase b. decrease c. remains constant (180)

PAGES 251–252 · COMPUTER EXERCISES

1. 1, 576; 2, 288; 3, 192; 4, 144; 6, 96; 8, 72; 9, 64; 12, 48; 16, 36; 18, 32; 24, 24 2. a. $(1x + (2))(1x + (3))$ b. $(1x + (3))(1x + (-8))$ c. $(4x + (3))(5x + (4))$ d. $(2x + (3))(3x + (-4))$ e. $(2x + (-5))(3x + (-1))$ f. not factorable

PAGE 296 · COMPUTER EXERCISES

1. a. two rational roots, $\{2, -2\}$ b. two rational roots, $\{8, 0\}$ c. two rational roots, $\{-3, -5\}$ d. two irrational roots, {(−4 + SQR(20))/2, (−4 − SQR(20))/2} e. no real roots f. double root, 2.5
2. (25) two rational roots, $\{2, -10\}$; (26) two irrational roots, {(−6 + SQR(40))/2, (−6 − SQR(40))/2)}; (27) two irrational roots, {(2 + SQR(12))/4, (2 − SQR(12))/4}; (28) one double root, {−.333333}; (29) two rational roots, {10, 4}; (30) two rational roots, {0, −.333333}; (31) two irrational roots, {(0 + SQR(28))/2, (0 − SQR(28))/2}; (32) two irrational roots, {(0 + SQR(768))/12, (0 − SQR(768))/12}; (33) two irrational roots, {(−10 + SQR(12))/−4, (−10 − SQR(12))/−4}; (34) one double root, {.666667}; (35) two irrational roots, {(−7 + SQR(17))/2, (−7 − SQR(17))/2}; (36) two rational roots, {1.5, .25}

PAGES 344–345 · COMPUTER EXERCISES

1. 2:3, 2:3, 2:3; the △ are $\sim$. 2. 1:2, 1:2, 1:3; the △ are not $\sim$. 3. 10:7, 5:3, 5:3; the △ are not $\sim$. 4. 2:7, 2:7, 2:7; the △ are $\sim$.

PAGES 378–379 · COMPUTER EXERCISES

a. DEGREES	SINE	COSINE	TANGENT
5	8.71543E–02	.996195	8.74872E–02
10	.173645	.984808	.176324
15	.258815	.965927	.267944
20	.342015	.939695	.363964
25	.422612	.906311	.466299
30	.499992	.86603	.577339
35	.573568	.819158	.700192
40	.642779	.766052	.83908
45	.707098	.707116	.999974

b. 20 FOR D=5 TO 85 STEP 5

DEGREES	SINE	COSINE	TANGENT
5	8.71543E-02	.996195	8.74872E-02
10	.173645	.984808	.176324
15	.258815	.965927	.267944
20	.342015	.939695	.363964
25	.422612	.906311	.466299
30	.499992	.86603	.577339
35	.573568	.819158	.700192
40	.642779	.766052	.83908
45	.707098	.707116	.999974
50	.766035	.642799	1.19172
55	.819143	.57359	1.4281
60	.866017	.500015	1.73198
65	.9063	.422635	2.1444
70	.939686	.342039	2.7473
75	.96592	.25884	3.73172
80	.984804	.173671	5.67051
85	.996193	8.71804E-02	11.4268

Yes, the corresponding values are approximately the same.

c. 20 FOR D=1 TO 43 STEP 3

DEGREES	SINE	COSINE	TANGENT
1	1.74521E-02	.999848	1.74548E-02
4	6.97553E-02	.997564	6.99256E-02
7	.121867	.992546	.122782
10	.173645	.984808	.176324
13	.224947	.974371	.230864
16	.275633	.961263	.28674
19	.325563	.94552	.344321
22	.374601	.927186	.404019
25	.422612	.906311	.466299
28	.469464	.882951	.531699
31	.51503	.857172	.600848
34	.559185	.829043	.674494
37	.601806	.798642	.753537
40	.642779	.766052	.83908
43	.681989	.731362	.932492

PAGES 466–468 · COMPUTER EXERCISES

1. a. (5, 8) b. (4, 6) c. (4, 6) d. (6, −1) 2. a. $AP = 4\sqrt{5}$, $BP = 3\sqrt{5}$
b. $\frac{AP}{BP} = \frac{4\sqrt{5}}{3\sqrt{5}} = \frac{4}{3}$ 3. a. (−1, 4) b. (−4.5, 6) 4. (4, 1), (6, 5)
5. (−4, 5), (0, 3), (4, 1), (8, −1)

PAGES 510–511 · COMPUTER EXERCISES

1. a. $x^2 + y^2 + dx + ey + f = 0$; $x^2 + dx + y^2 + ey = -f$;
$x^2 + dx + \frac{d^2}{4} + y^2 + ey + \frac{e^2}{4} = \frac{d^2}{4} + \frac{e^2}{f} - f$; $\left(x + \frac{d}{2}\right)^2 + \left(y + \frac{e}{2}\right)^2 = \frac{d^2 + e^2 - 4f}{4}$;

$\therefore$ the center has coordinates $\left(-\frac{d}{2}, -\frac{e}{2}\right)$ and $r = \frac{1}{2}\sqrt{d^2 + e^2 - 4f}$

b. $C(3, -2)$; $r = 2$ c. $C(-5, -1)$; $r = 5$ d. $C(-0.5, 3)$; $r = 3$
e. $C(-1.5, -1)$; $r = 1.5$ f. $C(4, 0)$; $r = 5$ g. $C(-5, 7)$; $r = 7$

2. Insert the following instructions:

```
35   IF D*D+E*E-4*F<=0 THEN 65
63   STOP
65   PRINT "NOT A CIRCLE"
```

a. not a circle b. not a circle c. $C(2, 2)$; $r = 4$ d. not a circle

3. a. $V(-8, -2)$; $F(-8, -1.75)$ b. $V(-1, -3)$; $F(-1, -2.9375)$ c. $V(1, 7)$; $F(1, 6.75)$
d. $V(-0.5, 0)$; $F(-0.5, 0.25)$ e. $V(2, -1)$; $F(2, 0)$ f. $V(1.5, 1.5)$; $F(1.5, 1.375)$

PAGES 589–590 • COMPUTER EXERCISES

1. a. 90 = 6 MOD 12 b. 38 = 3 MOD 5 c. 147 = 11 MOD 17 d. 5 = 5 MOD 8
e. 3626 = 0 MOD 14 f. 52681 = 66 MOD 619

2. a.

X	!	0	1	2	3	4	5	6	7	8	9	10	11	12
0	!	0	0	0	0	0	0	0	0	0	0	0	0	0
1	!	0	1	2	3	4	5	6	7	8	9	10	11	12
2	!	0	2	4	6	8	10	12	1	3	5	7	9	11
3	!	0	3	6	9	12	2	5	8	11	1	4	7	10
4	!	0	4	8	12	3	7	11	2	6	10	1	5	9
5	!	0	5	10	2	7	12	4	9	1	6	11	3	8
6	!	0	6	12	5	11	4	10	3	9	2	8	1	7
7	!	0	7	1	8	2	9	3	10	4	11	5	12	6
8	!	0	8	3	11	6	1	9	4	12	7	2	10	5
9	!	0	9	5	1	10	6	2	11	7	3	12	8	4
10	!	0	10	7	4	1	11	8	5	2	12	9	6	3
11	!	0	11	9	7	5	3	1	12	10	8	6	4	2
12	!	0	12	11	10	9	8	7	6	5	4	3	2	1

b.

X	!	0	1	2	3	4	5	6	7	8	9	10	11	12	13
0	!	0	0	0	0	0	0	0	0	0	0	0	0	0	0
1	!	0	1	2	3	4	5	6	7	8	9	10	11	12	13
2	!	0	2	4	6	8	10	12	0	2	4	6	8	10	12
3	!	0	3	6	9	12	1	4	7	10	13	2	5	8	11
4	!	0	4	8	12	2	6	10	0	4	8	12	2	6	10
5	!	0	5	10	1	6	11	2	7	12	3	8	13	4	9
6	!	0	6	12	4	10	2	8	0	6	12	4	10	2	8
7	!	0	7	0	7	0	7	0	7	0	7	0	7	0	7
8	!	0	8	2	10	4	12	6	0	8	2	10	4	12	6
9	!	0	9	4	13	8	3	12	7	2	11	6	1	10	5
10	!	0	10	6	2	12	8	4	0	10	6	2	12	8	4
11	!	0	11	8	5	2	13	10	7	4	1	12	9	6	3
12	!	0	12	10	8	6	4	2	0	12	10	8	6	4	2
13	!	0	13	12	11	10	9	8	7	6	5	4	3	2	1

3. 0 has no inverse; $1^{-1} = 1$, $2^{-1} = 7$, $3^{-1} = 9$, $4^{-1} = 10$, $5^{-1} = 8$, $6^{-1} = 11$, $7^{-1} = 2$, $8^{-1} = 5$, $9^{-1} = 3$, $10^{-1} = 4$, $11^{-1} = 6$, $12^{-1} = 12$